Applied Mathematical Sciences
Volume 107

Applied Mathematical Sciences

1. *John:* Partial Differential Equations, 4th ed.
2. *Sirovich:* Techniques of Asymptotic Analysis.
3. *Hale:* Theory of Functional Differential Equations, 2nd ed.
4. *Percus:* Combinatorial Methods.
5. *von Mises/Friedrichs:* Fluid Dynamics.
6. *Freiberger/Grenander:* A Short Course in Computational Probability and Statistics.
7. *Pipkin:* Lectures on Viscoelasticity Theory.
8. *Giacoglia:* Perturbation Methods in Non-linear Systems.
9. *Friedrichs:* Spectral Theory of Operators in Hilbert Space.
10. *Stroud:* Numerical Quadrature and Solution of Ordinary Differential Equations.
11. *Wolovich:* Linear Multivariable Systems.
12. *Berkovitz:* Optimal Control Theory.
13. *Bluman/Cole:* Similarity Methods for Differential Equations.
14. *Yoshizawa:* Stability Theory and the Existence of Periodic Solution and Almost Periodic Solutions.
15. *Braun:* Differential Equations and Their Applications, 3rd ed.
16. *Lefschetz:* Applications of Algebraic Topology.
17. *Collatz/Wetterling:* Optimization Problems.
18. *Grenander:* Pattern Synthesis: Lectures in Pattern Theory, Vol. I.
19. *Marsden/McCracken:* Hopf Bifurcation and Its Applications.
20. *Driver:* Ordinary and Delay Differential Equations.
21. *Courant/Friedrichs:* Supersonic Flow and Shock Waves.
22. *Rouche/Habets/Laloy:* Stability Theory by Liapunov's Direct Method.
23. *Lamperti:* Stochastic Processes: A Survey of the Mathematical Theory.
24. *Grenander:* Pattern Analysis: Lectures in Pattern Theory, Vol. II.
25. *Davies:* Integral Transforms and Their Applications, 2nd ed.
26. *Kushner/Clark:* Stochastic Approximation Methods for Constrained and Unconstrained Systems.
27. *de Boor:* A Practical Guide to Splines.
28. *Keilson:* Markov Chain Models—Rarity and Exponentiality.
29. *de Veubeke:* A Course in Elasticity.
30. *Shiatycki:* Geometric Quantization and Quantum Mechanics.
31. *Reid:* Sturmian Theory for Ordinary Differential Equations.
32. *Meis/Markowitz:* Numerical Solution of Partial Differential Equations.
33. *Grenander:* Regular Structures: Lectures in Pattern Theory, Vol. III.
34. *Kevorkian/Cole:* Perturbation Methods in Applied Mathematics.
35. *Carr:* Applications of Centre Manifold Theory.
36. *Bengtsson/Ghil/Källén:* Dynamic Meteorology: Data Assimilation Methods.
37. *Saperstone:* Semidynamical Systems in Infinite Dimensional Spaces.
38. *Lichtenberg/Lieberman:* Regular and Chaotic Dynamics, 2nd ed.
39. *Piccini/Stampacchia/Vidossich:* Ordinary Differential Equations in $\mathbf{R}^n$.
40. *Naylor/Sell:* Linear Operator Theory in Engineering and Science.
41. *Sparrow:* The Lorenz Equations: Bifurcations, Chaos, and Strange Attractors.
42. *Guckenheimer/Holmes:* Nonlinear Oscillations, Dynamical Systems and Bifurcations of Vector Fields.
43. *Ockendon/Taylor:* Inviscid Fluid Flows.
44. *Pazy:* Semigroups of Linear Operators and Applications to Partial Differential Equations.
45. *Glashoff/Gustafson:* Linear Operations and Approximation: An Introduction to the Theoretical Analysis and Numerical Treatment of Semi-Infinite Programs.
46. *Wilcox:* Scattering Theory for Diffraction Gratings.
47. *Hale et al:* An Introduction to Infinite Dimensional Dynamical Systems—Geometric Theory.
48. *Murray:* Asymptotic Analysis.
49. *Ladyzhenskaya:* The Boundary—Value Problems of Mathematical Physics.
50. *Wilcox:* Sound Propagation in Stratified Fluids.
51. *Golubitsky/Schaeffer:* Bifurcation and Groups in Bifurcation Theory, Vol. I.

(continued following index)

Stuart S. Antman

Nonlinear Problems of Elasticity

With 105 Illustrations

Springer-Verlag
New York Berlin Heidelberg London Paris
Tokyo Hong Kong Barcelona Budapest

Stuart S. Antman
Department of Mathematics
and
Institute for Physical Science and Technology
University of Maryland
College Park, MD 20742
USA

Mathematics Subject Classification (1991): 73C50, 73Kxx, 49Rxx, 70Bxx

Library of Congress Cataloging-in-Publication Data
Antman, S.S. (Stuart S.)
Nonlinear problems of elasticity / Stuart S. Antman.
p. cm. – (Applied mathematical sciences ; v. 107)
Includes bibliographical references and index.
ISBN 0-387-94199-1
1. Elasticity. 2. Nonlinear theories. I. Series: Applied mathematical sciences (Springer-Verlag New York Inc.) ; v. 107.
QA1.A647 vol. 107
[QA931]
510 s–dc20
[624.1′71′01515355] 94-25684

Printed on acid-free paper.

Production managed by Hal Henglein; manufacturing supervised by Genieve Shaw.
Photocomposed copy prepared from the author's AMSTeX file.
Printed and bound by R.R. Donnelley & Sons, Harrisonburg, VA.
Printed in the United States of America.

9 8 7 6 5 4 3 2 1

ISBN 0-387-94199-1 Springer-Verlag New York Berlin Heidelberg
ISBN 3-540-94199-1 Springer-Verlag Berlin Heidelberg New York

To the Memory of My Parents,

Gertrude and Mitchell Antman

Preface

The scientists of the seventeenth and eighteenth centuries, led by Jas. Bernoulli and Euler, created a coherent theory of the mechanics of strings and rods undergoing planar deformations. They introduced the basic concepts of strain, both extensional and flexural, of contact force with its components of tension and shear force, and of contact couple. They extended Newton's Law of Motion for a mass point to a law valid for any deformable body. Euler formulated its independent and much subtler complement, the Angular Momentum Principle. (Euler also gave effective variational characterizations of the governing equations.) These scientists breathed life into the theory by proposing, formulating, and solving the problems of the suspension bridge, the catenary, the velaria, the elastica, and the small transverse vibrations of an elastic string. (The level of difficulty of some of these problems is such that even today their descriptions are seldom vouchsafed to undergraduates. The realization that such profound and beautiful results could be deduced by mathematical reasoning from fundamental physical principles furnished a significant contribution to the intellectual climate of the Age of Reason.) At first, those who solved these problems did not distinguish between linear and nonlinear equations, and so were not intimidated by the latter.

By the middle of the nineteenth century, Cauchy had constructed the basic framework of three-dimensional continuum mechanics on the foundations built by his eighteenth-century predecessors. The dominant influence on the direction of further work on elasticity (and on every other field of classical physics) up through the middle of the twentieth century was the development of effective practical tools for solving linear partial differential equations on suitably shaped domains. So thoroughly did the concept of linearity pervade scientific thought during this period that mathematical physics was virtually identified with the study of differential equations containing the Laplacian. In this environment, the respect of the scientists of the eighteenth century for a (typically nonlinear) model of a physical process based upon fundamental physical and geometrical principles was lost.

The return to a serious consideration of nonlinear problems (other than those admitting closed-form solutions in terms of elliptic functions) was led by Poincaré and Lyapunov in their development of qualitative methods for the study of ordinary differential equations (of discrete mechanics) at the end of the nineteenth century and at the beginning of the twentieth century. Methods for handling nonlinear boundary-value problems were slowly developed by a handful of mathematicians in the first half of the twentieth century. The greatest progress in this area was attained in the study of direct methods of the calculus of variations (which are very useful in nonlinear elasticity).

A rebirth of interest in nonlinear elasticity occurred in Italy in the 1930's under the leadership of Signorini. A major impetus was given to the subject in the years following the Second World War by the work of Rivlin. For special, precisely formulated problems he exhibited concrete and elegant solutions valid for arbitrary nonlinearly elastic materials. In the early 1950's, Truesdell began a critical examination of the foundations of continuum thermomechanics in which the roles of geometry, fundamental physical laws, and constitutive hypotheses were clarified and separated from the unsystematic approximation then and still prevalent in parts of the subject. In consequence of the work of Rivlin and Truesdell, and of work inspired by them, continuum mechanics now possesses a clean, logical, and simple formulation and a body of illuminating solutions.

The development after the Second World War of high-speed computers and of powerful numerical techniques to exploit them has liberated scientists from dependence on methods of linear analysis and has stimulated growing interest in the proper formulation of nonlinear theories of physics. During the same time, there has been an accelerating development of methods for studying nonlinear equations. While nonlinear analysis is not yet capable of a comprehensive treatment of nonlinear problems of continuum mechanics, it offers exciting prospects for certain specific areas. (The level of generality in the treatment of large classes of operators in nonlinear analysis exactly corresponds to that in the treatment of large classes of constitutive equations in nonlinear continuum mechanics.) Thus, after two hundred years we are finally in a position to resume the program of analyzing illuminating, well-formulated, specific nonlinear problems of continuum mechanics.

The objective of this book is to carry out such studies for problems of nonlinear elasticity. It is here that the theory is most thoroughly established, the engineering tradition of treating specific problems is most highly developed, and the mathematical tools are the sharpest. (Actually, more general classes of solids are treated in our studies of dynamical problems, and Chap. XV is devoted to a presentation of a general theory of nonlinear elasto-plasticity.) This book is directed toward scientists, engineers, and mathematicians who wish to see careful treatments of uncompromised problems. My aim is to retain the orientation toward fascinating problems that characterizes the best engineering texts on structural stability while retaining the precision of modern continuum mechanics and employing powerful, but accessible, methods of nonlinear analysis.

My approach is to lay down a general theory for each kind of elastic body, carefully formulate specific problems, introduce the pertinent mathematical methods (in as unobtrusive a way as possible), and then conduct rigorous analyses of the problems. This program is successively carried out for strings, rods, shells, and three-dimensional bodies. This ordering of topics essentially conforms to their historical development. (Indeed, we carefully study modern versions of problems treated by Huygens, Leibniz, and the Bernoullis in Chap. III, and by Euler and Kirchhoff in Chaps. IV, V, and VIII.) This ordering is also the most natural from the viewpoint of pedagogy: Chaps. II–VI, VIII–X constitute what might be considered a modern course in nonlinear structural mechanics. From these chapters the novice in solid mechanics can obtain the requisite background in the common heritage of applied mechanics, while the experienced mechanician can gain an appreciation of the simplicity of geometrically exact, nonlinear (re)formulations of familiar problems of structural mechanics and an appreciation of the power of nonlinear analysis to treat them. At the same time, the novice in nonlinear analysis can see the application of this theory in simple, concrete situations.

The remainder of the book is devoted to a thorough formulation of the three-dimensional continuum mechanics of solids, the formulation and analysis of three-dimensional problems of nonlinear elasticity, a general treatment of theories of rods and shells on the basis of the three-dimensional theory, an account of nonlinear plasticity, and a treatment of nonlinear wave propagation and related questions in solid mechanics. The book concludes with a few self-contained appendices on analytic tools that are used throughout the text. The exposition beginning with Chap. XI is logically independent of the preceding chapters. Most of the development of the mechanics is given a material formulation because it is physically more fundamental than the spatial formulation and because it leads to differential equations defined on fixed domains.

The theories of solid mechanics are each mathematical models of physical processes. Our basic theories, of rods, shells, and three-dimensional bodies, differ in the dimensionality of the bodies. These theories may not be constructed haphazardly: Each must respect the laws of mechanics and of geometry. Thus, the only freedom we have in formulating models is essentially in the description of material response. Even here we are constrained to constitutive equations compatible with invariance restrictions imposed by the underlying mechanics. Thus, both the mechanics and mathematics in this book are focused on the formulation of suitable constitutive hypotheses and the study of their effects on solutions. I tacitly adopt the philosophical view that the study of a physical problem consists of three distinct steps: formulation, analysis, and interpretation, and that the analysis consists solely in the application of mathematical processes exempt from ad hoc physical simplifications.

The notion of solving a nonlinear problem differs markedly from that for linear problems: Consider boundary-value problems for the linear ordinary differential equation

$$\frac{d^2}{ds^2}\theta(s) + \lambda\theta(s) = 0, \tag{1}$$

which arises in the elementary theory for the buckling of a uniform column. Here λ is a positive constant. Explicit solutions of the boundary-value problems are immediately found in terms of trigonometric functions. For a nonuniform column (of positive thickness), (1) is replaced with

$$\frac{d}{ds}\left[B(s)\frac{d\theta}{ds}(s)\right] + \lambda\theta(s) = 0 \tag{2}$$

where B is a given positive-valued function. In general, (2) cannot be solved in closed form. Nevertheless, the Sturm-Liouville theory gives us information about solutions of boundary-value problems for (2) so detailed that for many practical purposes it is as useful as the closed-form solutions obtained for (1). This theory in fact tells us what is essential about solutions. Moreover, this information is not obscured by complicated formulas involving special functions. We accordingly regard this qualitative information as characterizing a solution.

The elastica theory of the Bernoullis and Euler, which is a geometrically exact generalization of (1), is governed by the semilinear equation

$$\frac{d^2}{ds^2}\theta(s) + \lambda \sin\theta(s) = 0. \tag{3}$$

It happens that boundary-value problems for (3) can be solved explicitly in terms of elliptic functions, and we again obtain solutions in the traditional sense. On the other hand, for nonuniform columns, (3) must be replaced by

$$\frac{d}{ds}\left[B(s)\frac{d\theta}{ds}(s)\right] + \lambda\sin\theta(s) = 0, \tag{4}$$

for which no such solutions are available. In Chap. V we develop a nonlinear analog of the Sturm-Liouville theory that gives detailed qualitative information on solutions of boundary-value problems for (4). The theory has the virtues that it captures all the qualitative information about solutions of (3) available from the closed-form solutions and that it does so with far less labor than is required to obtain the closed-form solutions. We shall not be especially concerned with models like (4), but rather with its generalizations in the form

$$\frac{d}{ds}\left[\hat{M}\left(\frac{d\theta}{ds}(s), s\right)\right] + \lambda\sin\theta(s) = 0. \tag{5}$$

Here $\hat{M}$ is a given constitutive function that characterizes the ability of the column to resist flexure. When we carry out an analysis of equations like (5), we want to determine how the properties of $\hat{M}$ affect the properties of solutions. In many cases, we shall discover that different kinds of physically reasonable constitutive functions give rise to qualitatively different kinds of solutions and that the distinction between the kinds of solutions has great physical import. We regard such analyses as constituting solutions.

The prerequisites for reading this book, spelled out in Sec. I.2, are a sound understanding of Newtonian mechanics, advanced calculus, and linear algebra, and some elements of the theories of ordinary differential equations and linear partial differential equations. More advanced mathematical topics are introduced when needed. I do not subscribe to the doctrine that the mathematical theory must be fully developed before it is applied. Indeed, I feel that seeing an effective application of a theorem is often the best motivation for learning its proof. Thus, for example, the basic results of global bifurcation theory are explained in Chap. V and immediately applied there and in Chaps. VI and IX to a variety of buckling problems. A self-contained treatment of degree theory leading to global bifurcation theory is given in the Appendix (Chap. XIX).

A limited repertoire of mathematical tools is developed and broadly applied. These include methods of global bifurcation theory, continuation methods, and perturbation methods, the latter justified whenever possible by implicit function theorems. Direct methods of the calculus of variations are the object of only Chap. VII. The theory is developed here only insofar as it can easily lead to illuminating insights into concrete problems; no effort is made to push the subject to its modern limits. Special techniques for dynamical problems are mostly confined to Chap. XVI (although several dynamical problems are treated earlier).

This book encompasses a variety of recent research results, a number of unpublished results, and refinements of older material. I have chosen not to present any of the beautiful modern research on existence theories for three-dimensional problems, because the theory demands a high level of technical expertise in modern analysis, because very active contemporary research, much inspired by the theory of phase transformations, might very strongly alter our views on this subject, and because there are very attractive accounts of earlier work in the books of Ciarlet (1988), Dacorogna (1989), Hanyga (1985), Marsden & Hughes (1983), and Valent (1988). My treatment of specific problems of three-dimensional elasticity differs from the classical treatments of Green & Adkins (1970), Green & Zerna (1968), Ogden (1984), Truesdell & Noll (1965), and Wang & Truesdell (1973) in its emphasis on analytic questions associated with material response. In practice, many of the concrete problems treated in this book involve but one spatial variable, because it is these problems that lend themselves most naturally to detailed global analyses. The choice of topics naturally and strongly reflects my own research interests in the careful formulation of geometrically exact theories of rods, shells, and three-dimensional bodies, and in the global analysis of well-set problems.

There is a wealth of exercises, which I have tried to make interesting, challenging, and tractable. They are designed to cause the reader to (i) complete developments outlined in the text, (ii) carry out formulations of problems with complete precision (which is the indispensable skill required of workers in mechanics), (iii) investigate new areas not covered in the text, and, most importantly, (iv) solve concrete problems. Problems, on the other hand, represent what I believe are short, tractable research projects on generalizing the extant theory to treat minor, open questions. They afford a natural entrée to bona fide research problems.

This book had its genesis in a series of lectures I gave at Brown University in 1978–1979 while I was holding a Guggenheim Fellowship. It has been progressively refined in courses I have subsequently given at the University of Maryland and elsewhere. I am particularly indebted to many students and colleagues who have caught errors and made useful suggestions. Among those who have made special contributions have been Carlos Castillo-Chavez, P. M. Fitzpatrick, James M. Greenberg, Leon Greenberg, Timothy Healey, Massimo Lanza de Cristoforis, John Maddocks, Pablo Negrón, Robert Rogers, Felix Santos, Friedemann Schuricht, and Li-Sheng Wang. Over the years my research, some of which is presented here, has been supported by the National Science Foundation and the Air Force Office of Scientific Research. I am grateful to these organizations and to the taxpayers who support them.

Contents

CHAPTER I

Background

1. Notational and Terminological Conventions

Mathematical statements such as formulas, theorems, figures, and exercises are numbered consecutively in each section. Thus formula (III.4.11) and Theorem III.4.12 are the eleventh and twelfth numbered statements in Sec. 4 of Chap. III. Within Chap. III, these statements are designated simply by (4.11) and Theorem 4.12.

Ends of proofs are designated by $\square$. The symbols $\exists$ and $\forall$, which appear only in displayed mathematical expressions, respectively stand for *there exists* and *for all* (or *for any* or *for every*). In definitions of mathematical entities, I follow the convention that the expression *iff* designates the logically correct *if and only if*, which is usually abbreviated by *if*. In the statements of necessary and sufficient conditions, the phrase *if and only if* is always written out.

I follow the somewhat ambiguous mathematical usage of the adjective *formal*, which here means *systematic, but without rigorous justification*, as in *a formal calculation*. A common exception to this usage is *formal proof*, which is not employed in this book because it smacks of redundancy.

An elastic body is often described by an adjective referring to its shape, e.g., a *straight rod* or a *spherical shell*. In each such case, it is understood that the adjective refers to the reference configuration of the body and not to a typical deformed configuration. When a restricted class of deformations is studied, the restrictions are explicitly characterized by further adjectives, as in *axisymmetric deformations of a spherical shell*.

Passages in small type contain refinements of fundamental results, proofs that are not crucial for further developments, advanced mathematical arguments (typically written in a more condensed style), discussions of related problems, bibliographical notes, and historical remarks. Although none of this material is essential for a first reading, I regard much of it as interesting.

2. Prerequisites

The essential mathematical prerequisite for understanding this book is a sound knowledge of advanced calculus, linear algebra, and the elements of the theories of ordinary differential equations and partial differential equations, together with enough mathematical sophistication, gained by an exposure to upper-level undergraduates courses in pure or applied mathematics, to follow careful mathematical arguments. Some of the important topics from these fields that will be repeatedly used are the Implicit Function Theorem, the conditions for the minimization of a real-valued function, variants of the Divergence Theorem, the Stokes Theorem, standard results of vector calculus, eigenvalues of linear transformations, positive-definiteness of linear transformations, the basic theorems on existence, uniqueness, continuation, and continuous dependence on data of solutions to initial-value problems of ordinary differential equations, phase-plane methods, the classification of partial differential equations as to type, and orthogonal expansions of solutions to linear partial differential equations.

A number of more esoteric mathematical concepts, most of which deal with methods for treating nonlinear equations, will be given self-contained developments. For the sake of added generality or precision, certain presentations are couched in the language of modern real-variable theory. The reader having but a nodding familiarity with the intuitive interpretations of these concepts, presented in Sec. 7, can blithely ignore their technical aspects, which play no essential role in the exposition. Those few arguments that rely on real-variable theory in a crucial way are presented in small type; they can be skipped by the novice.

The prerequisites in physics or engineering are not so sharply delineated. In principle, all that is necessary is a thorough understanding of Newtonian mechanics. In practice, the requisite understanding is gained by exposure to serious undergraduate courses in mechanics.

The rest of this chapter explains the conventions used in this book. The next two sections contain important statements of notational philosophy.

3. Functions

Consider the following little exercise: Suppose that a function f is defined by the formula

$$f(x,y) = x^2 + y^2. \tag{3.1}$$

Let

$$x = r\cos\theta, \quad y = r\sin\theta. \tag{3.2}$$

What is $f(r,\theta)$? The answer that $f(r,\theta) = r^2$ is false (although traditional). The correct answer is that $f(r,\theta) = r^2 + \theta^2$: We do not change the form of the function f by changing the symbols for the independent variables. The transformation (3.2) is irrelevant; it was introduced expressly to be misleading. To make sense of the incorrect answer and to account for (3.2) we define a function g by $g(r,\theta) = f(r\cos\theta, r\sin\theta)$. Then we find that $g(r,\theta) = r^2$. Thus g and f are different functions. The definition of g shows how they are related. Now $f(x,y)$ could represent the value of some physical quantity, e.g., the temperature, at a point of the plane with Cartesian coordinates (x,y). If (3.2) is used to replace (x,y) with polar coordinates (r, θ), then the function g that delivers the same temperature at the same point now represented by polar coordinates is a function different from f, but "it has the same values".

In short, a function is a rule. We consistently distinguish between the function f and its value $f(x, y)$ and we consistently avoid using the same notation for different functions with the same values. We never refer to 'the function $f(x, y)$'. We can of course define a function f by specifying its values $f(x, y)$ as in (3.1).

Formally, a *function* ϕ *from set* $\mathcal{A}$ *to set* $\mathcal{B}$ is a rule that associates with each element a of $\mathcal{A}$ a unique element $\phi(a)$ of $\mathcal{B}$. $\phi(a)$ is called the *value* of ϕ at a. $\mathcal{A}$ is called the *domain* (*of definition*) of ϕ. $\mathcal{B}$ may be called the *target* of ϕ. If we wish to emphasize the domain and target of ϕ, we refer to it as the function $\phi : \mathcal{A} \to \mathcal{B}$. If we wish to emphasize the form of the function ϕ, we refer to it as the function $a \mapsto \phi(a)$. For example, we can denote the function f defined by (3.1) by $(x, y) \mapsto x^2 + y^2$. We give maximum information about ϕ by denoting it as $\mathcal{A} \ni a \mapsto \phi(a) \in \mathcal{B}$. Finally, in certain circumstances it is convenient to refer to a function ϕ by $\phi(\cdot)$. For example, suppose y is fixed at some arbitrary value. Then (3.1) defines a function of x (parametrized by y) that we denote by either $x \mapsto f(x, y)$ or $f(\cdot, y)$. If $\mathcal{D}$ is any subset of $\mathcal{A}$, we define the *range* or *image* of $\mathcal{D}$ under ϕ to be the set $\phi(\mathcal{D}) \equiv \{\phi(a) : a \in \mathcal{D}\}$ of all the values assumed by ϕ when its arguments range over $\mathcal{D}$. (The terminology is not completely standardized.)

To anyone exposed to the standard texts in elementary and applied mathematics, physics, or engineering, such a refined notational scheme might seem utterly pretentious. But I find that the use of the traditional simpler notation of such texts, though adequate for linear problems, typically produces undue confusion in the mind of the unsophisticated reader confronting nonlinear problems not only in continuum mechanics, but also in rigid-body mechanics, calculus of variations, and differential equations (because each of these fields requires the precise manipulation of different functions having the same values).

Consequently, the refined notations for functions described above (found in modern books on real variables) will be used consistently throughout this book. In particular, if ϕ is a function, then an equation of the form $\phi = 0$ means that ϕ is the zero function; there is no need to write $\phi(a) \equiv 0$. We reserve the symbol '$\equiv$' to mean 'is defined to equal'.

We often abbreviate the typical partial derivative $\frac{\partial}{\partial x} g(x, y)$ by either $g_x(x, y)$ or $\partial_x g(x, y)$, whichever leads to clearer formulas. (There is a notational scheme in which g_x and g_y are denoted by g_1 and g_2. This scheme does not easily handle arguments that are vector-valued.)

A function is said to be *affine* iff it differs from a linear function by a constant.

The *support* of a function is the closure of the set on which it is not zero. A function (defined on a finite-dimensional space) thus has *compact support* if the set on which it is not zero is bounded.

If two functions f and g have the same domain of definition $\mathcal{D}$, then we write $f \geq g$ iff $f(x) \geq g(x)$ for all x in $\mathcal{D}$. We write $f > g$ iff $f \geq g$ and$f \neq g$.

We generally avoid using f^{-1} to designate the inverse of a function f. In the rare cases in which it is used, $f^{-1}(y)$ denotes the value of the inverse at y, while $f(x)^{-1}$ denotes the reciprocal of the value of a real-valued function f at x.

We apply the adjective *smooth* informally to any function that is continuously differentiable and that has as many derivatives as are needed to make the mathematical processes valid in the classical sense. (We do not follow the convention in which a smooth function is defined to be infinitely differentiable.)

4. Vectors

There are three definitions of the concept of three-dimensional vectors corresponding to three levels of sophistication: (i) Vectors are directed line segments that obey the parallelogram rule of addition and that can be multiplied by scalars. (ii) Vectors are triples of real numbers that can be added and be multiplied by scalars in the standard componentwise fashion. (iii) Vectors are elements of a three-dimensional real vector space. We regard the most primitive definition (i) and the most abstract definition (iii) as being essentially equivalent, the latter giving a mathematically precise realization of the concepts of the former. The vectors we deal with are either geometric or physical objects. If we refer such vectors to a rectilinear coordinate system, then their coordinate triples satisfy definition (ii). We eschew this definition on the practical grounds that its use makes the formulas look more complicated and makes conversion to curvilinear coordinates somewhat less efficient and on the philosophical ground that its use suppresses the invariance of the equations of physics under the choice of coordinates: Even a boxer on awakening from a knockout punch knows that the impulse vector applied to his chin has a physical significance independent of any coordinate system used to describe it.

In light of these remarks, we define *Euclidean 3-space* $\mathbb{E}^3$ to be abstract three-dimensional real inner-product space. Its elements are called *vectors*. They are denoted by lower-case, boldface, italic symbols $\boldsymbol{u}$, $\boldsymbol{v}$, etc. The inner product of vectors $\boldsymbol{u}$ and $\boldsymbol{v}$ is denoted by the dot product $\boldsymbol{u} \cdot \boldsymbol{v}$. $\mathbb{E}^3$ is defined by assigning to this dot product the usual properties. Since we ignore relativistic effects, we take $\mathbb{E}^3$ as our the model for physical space. We define the length of vector $\boldsymbol{u}$ by $|\boldsymbol{u}| \equiv \sqrt{\boldsymbol{u} \cdot \boldsymbol{u}}$. On $\mathbb{E}^3$ we can define the cross-product $\boldsymbol{u} \times \boldsymbol{v}$ of $\boldsymbol{u}$ and $\boldsymbol{v}$.

A *basis* for $\mathbb{E}^3$ is a linearly independent set of vectors in $\mathbb{E}^3$. A necessary and sufficient condition for $\{\boldsymbol{u}, \boldsymbol{v}, \boldsymbol{w}\}$ to be a basis is that $(\boldsymbol{u} \times \boldsymbol{v}) \cdot \boldsymbol{w} \neq 0$. Such a basis is *right-handed* iff $(\boldsymbol{u} \times \boldsymbol{v}) \cdot \boldsymbol{w} > 0$. Throughout this book we denote a fixed right-handed orthonormal basis by

$$\{\boldsymbol{i}, \boldsymbol{j}, \boldsymbol{k}\} \equiv \{\boldsymbol{i}_1, \boldsymbol{i}_2, \boldsymbol{i}_3\} \equiv \{\boldsymbol{i}^1, \boldsymbol{i}^2, \boldsymbol{i}^3\}$$

using whichever notation is most convenient.

A linear transformation taking vectors into vectors (i.e., a linear transformation taking $\mathbb{E}^3$ into itself) is called a (*second-order*) *tensor*. Such tensors are denoted by upper-case boldface symbols $\boldsymbol{A}$, $\boldsymbol{B}$, etc. We denote the value of a tensor $\boldsymbol{A}$ at $\boldsymbol{u}$ by $\boldsymbol{A}\cdot\boldsymbol{u}$ (in place of the more customary $\boldsymbol{Au}$). We correspondingly denote the product of $\boldsymbol{A}$ and $\boldsymbol{B}$ by $\boldsymbol{A}\cdot\boldsymbol{B}$ (in place of the more customary $\boldsymbol{AB}$). The identity tensor is denoted $\boldsymbol{I}$ and the zero tensor is denoted $\boldsymbol{O}$. We set $\boldsymbol{u}\cdot\boldsymbol{A}\cdot\boldsymbol{v} \equiv \boldsymbol{u}\cdot(\boldsymbol{A}\cdot\boldsymbol{v})$. A tensor $\boldsymbol{A}$ is said to be *symmetric* iff $\boldsymbol{u}\cdot\boldsymbol{A}\cdot\boldsymbol{v} = \boldsymbol{v}\cdot\boldsymbol{A}\cdot\boldsymbol{u}$ for all $\boldsymbol{u}, \boldsymbol{v}$. A tensor $\boldsymbol{A}$ (symmetric or not) is said to be *positive-definite* iff the quadratic form $\boldsymbol{u}\cdot\boldsymbol{A}\cdot\boldsymbol{u} > 0$ for all $\boldsymbol{u} \neq \boldsymbol{0}$. The determinant $\det \boldsymbol{A}$ of $\boldsymbol{A}$ is defined by

$$(4.1)\qquad \det \boldsymbol{A} \equiv \frac{[(\boldsymbol{A}\cdot\boldsymbol{u})\times(\boldsymbol{A}\cdot\boldsymbol{v})]\cdot(\boldsymbol{A}\cdot\boldsymbol{w})}{[\boldsymbol{u}\times\boldsymbol{v}]\cdot\boldsymbol{w}}$$

for any basis $\{\boldsymbol{u}, \boldsymbol{v}, \boldsymbol{w}\}$. It is independent of the basis chosen. A detailed discussion of tensor algebra and calculus is postponed until it is needed.

Let $\mathbb{R}^n$ denote the set of n-tuples of real numbers. We denote typical elements of this set by lower-case boldface sans-serif symbols, like $\mathsf{a} \equiv (a_1, \ldots, a_n)$. As noted above, we distinguish $\mathbb{R}^3$ from the Euclidean 3-space $\mathbb{E}^3$. Nevertheless, if necessary, we can assign any one of several equivalent norms to $\mathbb{R}^n$. When it is physically meaningful we define $\mathsf{a}\cdot\mathsf{b} \equiv \sum_{k=1}^n a_k b_k$. In particular, in Chap. VIII we introduce a variable orthonormal basis $(s,t) \mapsto \{\boldsymbol{d}_k(s,t)\}$. We represent a vector-valued function $\boldsymbol{v}$ by $\boldsymbol{v} = v_1\boldsymbol{d}_1 + v_2\boldsymbol{d}_2 + v_3\boldsymbol{d}_3$ and we denote the triple (v_1, v_2, v_3) by v. Thus $\boldsymbol{v}\cdot\boldsymbol{v} = \mathsf{v}\cdot\mathsf{v}$. It is essential to note that the components $\partial_t v_k$ of v_t are generally not equal to the components $\boldsymbol{v}_t\cdot\boldsymbol{d}_k$ of $\boldsymbol{v}_t$.

If $\boldsymbol{u} \mapsto \boldsymbol{f}(\boldsymbol{u})$ is defined in a neighborhood of $\boldsymbol{v}$, then $\boldsymbol{f}$ is said to be (*Fréchet-*) *differentiable* at $\boldsymbol{v}$ iff there is a tensor $\boldsymbol{A}$ and a function $\boldsymbol{r}$ such that

$$(4.2)\quad \boldsymbol{f}(\boldsymbol{v}+\boldsymbol{h}) = \boldsymbol{f}(\boldsymbol{v}) + \boldsymbol{A}\cdot\boldsymbol{h} + \boldsymbol{r}(\boldsymbol{h},\boldsymbol{v}) \quad \text{with} \quad \frac{|\boldsymbol{r}(\boldsymbol{h},\boldsymbol{v})|}{|\boldsymbol{h}|} \to 0 \quad \text{as} \quad \boldsymbol{h} \to \boldsymbol{0}.$$

In this case, $\boldsymbol{A}$ is denoted $\frac{\partial \boldsymbol{f}}{\partial \boldsymbol{u}}(\boldsymbol{v})$ or $\boldsymbol{f}_{\boldsymbol{u}}(\boldsymbol{v})$ or $\partial \boldsymbol{f}(\boldsymbol{v})/\partial \boldsymbol{u}$ and is called the (*Fréchet*) *derivative of* $\boldsymbol{f}$ *at* $\boldsymbol{v}$. A much weaker notion of a derivative is that of a directional derivative: If for fixed $\boldsymbol{v}$ and $\boldsymbol{h}$ there is a number $\varepsilon > 0$ such that $\boldsymbol{f}(\boldsymbol{v}+t\boldsymbol{h})$ is defined for all $t \in [0,\varepsilon]$ and if $\frac{d}{dt}\boldsymbol{f}(\boldsymbol{v}+t\boldsymbol{h})\big|_{t=0}$ exists, then it is called the (*Gâteaux*) *differential of* $\boldsymbol{u} \mapsto \boldsymbol{f}(\boldsymbol{u})$ *at* $\boldsymbol{v}$ *in the direction* $\boldsymbol{h}$. When it is linear in $\boldsymbol{h}$, we denote this differential by the same notation we have used for the Fréchet differential: $\frac{\partial \boldsymbol{f}}{\partial \boldsymbol{u}}(\boldsymbol{v})\cdot\boldsymbol{h}$. In this case, we call $\frac{\partial \boldsymbol{f}}{\partial \boldsymbol{u}}(\boldsymbol{v})$ the *Gâteaux derivative of* $\boldsymbol{f}$ *at* $\boldsymbol{v}$. Fréchet derivatives are Gâteaux derivatives. We shall make the distinction between these derivatives explicit when we use them. Obvious analogs of these notations for spaces that are not Euclidian will also be used.

Let Ω be a domain in $\mathbb{R}^n$ and let $\Omega \ni \mathsf{u} \mapsto \mathsf{f}(\mathsf{u}) \in \mathbb{R}^m$ be continuously differentiable. Let points x and y of Ω be joined by the straight line segment $\{\alpha\mathsf{x} + (1-\alpha)\mathsf{y} : 0 \leq \alpha \leq 1\}$ lying entirely in Ω. Then the Fundamental

Theorem of Calculus implies that

$$
\begin{aligned}
\mathsf{f}(\mathsf{x})-\mathsf{f}(\mathsf{y}) &= \mathsf{f}(\alpha\mathsf{x}+(1-\alpha)\mathsf{y})\big|_{\alpha=0}^{\alpha=1} = \int_0^1 \frac{\partial}{\partial\alpha}\mathsf{f}(\alpha\mathsf{x}+(1-\alpha)\mathsf{y})\,d\alpha \\
&= \left[\int_0^1 \frac{\partial}{\partial\mathsf{u}}\mathsf{f}(\alpha\mathsf{x}+(1-\alpha)\mathsf{y})\,d\alpha\right]\cdot(\mathsf{x}-\mathsf{y}).
\end{aligned} \tag{4.3}
$$

The equality of the leftmost and rightmost terms of (4.3) constitutes the form of the Mean-Value Theorem we consistently use. Multiple applications of it yields Taylor's Formula with Remainder.

A *(parametrized) curve* in $\mathbb{E}^3$ is a continuous function $s \mapsto \boldsymbol{r}(s) \in \mathbb{E}^3$ defined on an interval of $\mathbb{R}$. We often distinguish this curve from its image, which is the geometrical figure consisting of all its values. The curve $\boldsymbol{r}$ is *continuously differentiable* iff it admits a parametrization in which $\boldsymbol{r}$ is continuously differentiable with respect to the parameter and its derivative with respect to the parameter never vanishes.

Let f and g be defined on an interval and let a belong to the closure of this interval. Then we write that $\mathsf{f}(t) = O\{\mathsf{g}(t)\}$ as $t \to a$ iff $|\mathsf{f}/\mathsf{g}|$ is bounded near a and we write that $\mathsf{f}(t) = o\{\mathsf{g}(t)\}$ as $t \to a$ iff $|\mathsf{f}(t)/\mathsf{g}(t)| \to 0$ as $t \to a$. O and o are the *Landau order symbols.*

5. Differential Equations

We denote ordinary derivatives by primes.

Let $\mathcal{D}$ be an open connected subset of $\mathbb{R}^3$ and let $\mathcal{I}$ be an open interval of real numbers. Let $\mathcal{D}\times\mathcal{I} \ni (x,y,z,s) \mapsto f(x,y,z,s)$ be a given continuous function. A *classical solution of the second-order ordinary differential equation*

$$f(u,u',u'',s)=0 \tag{5.1}$$

is a twice continuously differentiable function $\mathcal{I} \ni s \mapsto u(s)$ such that $\big(u(s),u'(s),u''(s)\big) \in \mathcal{D}$ for all s in $\mathcal{I}$ and such that

$$f(u(s),u'(s),u''(s),s)=0 \quad \forall\, s\in\mathcal{I}. \tag{5.2}$$

If $f(\cdot,\cdot,\cdot,s)$ is affine for all s in $\mathcal{I}$, then the differential equation (5.1) is said to be *linear.* If $f(x,y,\cdot,s)$ is affine for every (x,y,s) at which it is defined, then the differential equation (5.1) is said to be *quasilinear.* If $f(x,y,z,s)$ has the form $l(z,s)+g(x,y,s)$ where $l(\cdot,s)$ is linear for all s in $\mathcal{I}$, then the differential equation (5.1) is said to be *semilinear.* Analogous definitions apply to ordinary differential equations of any order, to systems of ordinary differential equations, to partial differential equations of any order, and to systems of partial differential equations. In each case, the highest-order derivatives assume the role of u'' here.

The fundamental partial differential equations of nonlinear solid mechanics are typically quasilinear. The fundamental ordinary differential equations of one-dimensional static problems, likewise typically quasilinear, can often be converted to semilinear systems. For example, consider the quasilinear second-order ordinary differential equation

$$\frac{d}{ds}\left[u'(s) - \frac{1}{u'(s)}\right] + g(u(s)) = 0 \tag{5.3}$$

defined for u' everywhere positive. Here g is a given function. (We seek a solution u so that (5.3) holds for all s in a given interval. We exhibit the independent variable s in (5.3) rather than convert (5.3) to the form (5.1), so that we can avoid complicating the simple given form by carrying out the differentiation d/ds by the chain rule.) By setting $v = u' - 1/u'$, we readily convert (5.3) to the system

$$u' = \frac{v + \sqrt{v^2+4}}{2}, \quad v' = -g(u), \tag{5.4}$$

which we term semilinear because the system is linear in the highest derivatives u' and v'.

Throughout this book, we often tacitly scale an independent spatial variable s so that it lies in an interval of a simple form, such as [0,1].

6. Notation for Sets

To describe parts of a physical body or collections of functions, we use the notation of set theory. A *set* is a collection of objects called its *elements.* We denote the membership of an element a in a set $\mathcal{A}$ by $a \in \mathcal{A}$. $\mathcal{A}$ is a *subset* of set $\mathcal{B}$, denoted $\mathcal{A} \subset \mathcal{B}$, iff every element of $\mathcal{A}$ is a member of $\mathcal{B}$. A set is clearly a subset of itself. A set may be defined by listing its elements (within braces) or by specifying its defining properties. The set of all elements a in set $\mathcal{B}$ enjoying property $P(a)$ is denoted $\{a \in \mathcal{B} : P(a)\}$. For example, the set of all positive numbers is $\{x \in \mathbb{R} : x > 0\}$. The *empty set* $\emptyset$ is the set with no elements.

The *union* of $\mathcal{A}$ and $\mathcal{B}$, denoted $\mathcal{A} \cup \mathcal{B}$, is the set of all elements that belong to either $\mathcal{A}$ or $\mathcal{B}$ (or both). The *intersection* of $\mathcal{A}$ and $\mathcal{B}$, denoted $\mathcal{A} \cap \mathcal{B}$, is the set of all elements that belong to both $\mathcal{A}$ and $\mathcal{B}$. The set of all elements in $\mathcal{A}$ and not in $\mathcal{B}$ is denoted $\mathcal{A} \setminus \mathcal{B}$. The complement $\setminus\mathcal{B}$ of $\mathcal{B}$ is the set of all elements (in the universe) not in $\mathcal{B}$. The set of all ordered pairs (a, b) with $a \in \mathcal{A}$ and $b \in \mathcal{B}$ is denoted $\mathcal{A} \times \mathcal{B}$; sets of ordered n-tuples are denoted analogously. We set $\mathcal{A} \times \mathcal{A} = \mathcal{A}^2$, etc. Thus the set of points (x_1, x_2) with $0 \le x_1 \le 1$ and $0 \le x_2 \le 1$ is denoted $[0,1]^2$.

The closure, interior, and boundary of a set $\mathcal{A}$ are denoted by $\operatorname{cl} \mathcal{A}$, $\operatorname{int} \mathcal{A}$, and $\partial\mathcal{A}$. A subset of $\mathbb{E}^n$ or $\mathbb{R}^n$ is said to be a *domain* iff it is open and connected. A subset of these spaces is said to be *compact* iff it is closed and bounded. (For definitions of standard topological notions used here, see elementary books on analysis.)

7. Real Analysis

Most of the fundamental laws of continuum mechanics are expressed as relations among integrals. In traditional approaches, the integrands are typically presumed continuous. Such a concession to mathematical convenience sacrifices the generality that enables the laws to encompass such diverse phenomena as shock waves, domain walls, and fracture. Accordingly, we shall require that the integrands in our fundamental integral laws merely be integrable in a general sense. We thereby separate the statement of fundamental principle from the regularity problem of deducing precisely where the integrands enjoy more smoothness.

To make these notions precise we must employ the modern theory of functions of a real variable. The purpose of this little section is not to give an indigestible capsulization of real analysis, but merely to introduce a couple of useful concepts.

The *Lebesgue measure*, or simply, the *measure* $|\mathcal{A}|$ of a subset $\mathcal{A}$ of $\mathbb{R}$ is a generalized length of $\mathcal{A}$, which reduces to the usual length when $\mathcal{A}$ is an interval. Likewise, the Lebesgue measure $|\mathcal{B}|$ of a subset $\mathcal{B}$ of $\mathbb{R}^2$ or of $\mathbb{E}^2$ is a generalized area and the Lebesgue measure $|\mathcal{C}|$ of a subset $\mathcal{C}$ of $\mathbb{R}^3$ or of $\mathbb{E}^3$ is a generalized volume, etc. Not all sets in $\mathbb{R}^n$ or $\mathbb{E}^n$ admit a Lebesgue measure. There are other kinds of measures, such as mass measures, useful in mechanics; see Sec. XII.6.

A set $\mathcal{C}$ of $\mathbb{R}^3$ has a Lebesgue measure, or is (*Lebesgue-*) *measurable* iff it can be suitably approximated by a *countable* number of rectangular blocks. For $\mathcal{C}$ to have the classical notion of volume, which is its *Jordan content*, it must be suitably approximated by a *finite* number of rectangular blocks. Consequently, the collection of measurable sets is much larger than the collection of sets with Jordan content.

A set $\mathcal{C}$ of $\mathbb{R}^3$ has (*Lebesgue*) *measure* 0 iff for each $\varepsilon > 0$, $\mathcal{C}$ can be covered by a countable collection of rectangular blocks (possibly overlapping) whose total volume is $\leq \varepsilon$. (This is the first bona fide definition given in this section.) Analogous definitions hold on $\mathbb{R}$ and $\mathbb{R}^2$. A property that holds everywhere on a set $\mathcal{C}$ except on a subset of measure 0 is said to hold *almost everywhere*, abbreviated *a.e.* (on $\mathcal{C}$). Thus it is easy to show that the set of rational numbers, though everywhere dense in $\mathbb{R}$, has measure 0 in $\mathbb{R}$.

The Lebesgue integral is defined in a way naturally compatible with the definition of measure. The Lebesgue measure and integral afford not only greater generality than the corresponding Jordan content and Riemann integral, but also support a variety of powerful theorems, such as the Lebesgue Dominated Convergence Theorem and the Fubini Theorem, which give easily verified conditions justifying the interchange of the orders of infinite processes.

If f is Lebesgue-integrable on an interval $\mathcal{I}$ of $\mathbb{R}$ containing the point a, then (its indefinite integral) F, defined by

$$F(x) = c + \int_a^x f(\xi)\, d\xi \quad \text{for } x \in \mathcal{I} \tag{7.1}$$

with c a constant, belongs to the very useful space $\mathrm{AC}(\mathcal{I})$ of absolutely continuous functions on $\mathcal{I}$. It can be shown that if F is absolutely continuous on $\mathcal{I}$, then it is continuous on $\mathcal{I}$, it has a well-defined derivative f a.e. on $\mathcal{I}$, f is Lebesgue-integrable on $\mathcal{I}$, and F is related to f by (7.1) with c replaced by $F(a)$. (A function F is *absolutely continuous* on $\mathcal{I}$ iff for arbitrary $\varepsilon > 0$ there is a $\delta > 0$ such that

$$\sum_{k=1}^{n} |f(y_k) - f(x_k)| < \varepsilon \tag{7.2}$$

for every finite collection $(x_1, y_1), \dots (x_n, y_n)$ of nonoverlapping intervals with $\sum_{k=1}^{n} |y_k - x_k| < \delta$.) The absolutely continuous functions play a fundamental role in the general treatment of ordinary differential equations.

8. Function Spaces

Many processes in analysis are systematized by the introduction of collections of functions having certain useful properties in common. A *function space* is such a collection having the defining property that if any two functions f and g belong to the collection, then so does every linear combination $\alpha f + \beta g$ where α and β are numbers. For example, let Ω be a connected region of $\mathbb{R}^n$ or of $\mathbb{E}^n$ and let m be a positive integer. Then the collection of all continuous functions from Ω to $\mathbb{R}^m$ is the function space denoted by $C^0(\Omega; \mathbb{R}^m)$. Since the range $\mathbb{R}^m$ is obvious in virtually all our work (because the notational scheme described in Sec. 4 tells when the range consists of scalars, vectors, tensors, or some other objects), we suppress the appearance of the range, and simply write $C^0(\Omega)$. By $C^0(\operatorname{cl} \Omega)$ we denote the functions continuous on the closure of Ω, which are the functions uniformly continuous on Ω. Likewise, for any positive integer k we denote by $C^k(\Omega)$ the space of all k-times continuously differentiable functions on Ω.

Of comparable utility for mechanics are the *real Lebesgue spaces* $L_p(\Omega)$, $p \geq 1$, consisting of (equivalence classes of) all real-valued functions u on Ω (differing on a set of measure 0) such that $|u|^p$ is Lebesgue-integrable. Thus if $u \in L_p(\Omega)$, then

$$\int_{\Omega} |u(\boldsymbol{z})|^p \, dv(\boldsymbol{z}) < \infty \tag{8.1}$$

where $dv(\boldsymbol{z})$ is the differential volume at $\boldsymbol{z}$ in Ω (i.e., v is the Lebesgue measure on Ω). If $u \in L_p(\Omega)$ and $v \in L_q(\Omega)$ where $\frac{1}{p} + \frac{1}{q} = 1$, then they satisfy the very useful *Hölder inequality:*

$$\int_{\Omega} |u(\boldsymbol{z}) v(\boldsymbol{z})| \, dv(\boldsymbol{z}) \leq \left[\int_{\Omega} |u(\boldsymbol{z})|^p \, dv(\boldsymbol{z}) \right]^{\frac{1}{p}} \left[\int_{\Omega} |v(\boldsymbol{z})|^q \, dv(\boldsymbol{z}) \right]^{\frac{1}{q}}. \tag{8.2}$$

(If $p = 1$ so that $q = \infty$ here, then the second integral on the right-hand side of (8.2) can be interpreted as the (essential) supremum of $|v|$ on Ω.) When $p = 2 = q$, (8.2) is called the *Cauchy-Bunyakovskiĭ-Schwarz inequality.*

The *Sobolev spaces* $W_p^1(\Omega)$, $p \geq 1$, consist of (equivalence classes of) all real-valued functions u on Ω (differing on a set of measure 0) such that $|u|^p$ and $|u_{\boldsymbol{z}}|^p$ are Lebesgue-integrable. Here $u_{\boldsymbol{z}}$ denotes the 'generalized derivative' of u, which is defined in Sec. XVII.1.

If Ω is an interval $\mathcal{I}$ on $\mathbb{R}$, and if $x, y \in \mathcal{I}$ with $x < y$, then the classical formula

$$u(y) - u(x) = \int_x^y u'(\xi) \, d\xi \tag{8.3}$$

can be shown to have meaning for $u \in W_p^1(\mathcal{I})$. If, furthermore, $p > 1$, then the Hölder inequality implies that

$$
\begin{aligned}
|u(y) - u(x)| &\leq \int_x^y 1|u'(\xi)|\,d\xi \\
&\leq \left[\int_x^y 1^q\,d\xi\right]^{\frac{1}{q}} \left[\int_x^y |u'(\xi)|^p\,d\xi\right]^{\frac{1}{p}} \leq C(y-x)^{\frac{1}{q}}
\end{aligned}
\tag{8.4}
$$

where the constant $C \equiv \left[\int_{\mathcal{I}} |u'(\xi)|^p\,d\xi\right]^{\frac{1}{p}} < \infty$ depends on u. This inequality says that if $u \in W_p^1(\mathcal{I})$ with $p > 1$, then u is continuous. It actually says more (about the modulus of continuity of u):

$$
\sup_{x,y\in\mathcal{I}:x\neq y} \frac{|u(x) - u(y)|}{|x-y|^a} \leq \text{const.}
\tag{8.5}
$$

Here $\alpha = \frac{1}{q} = 1 - \frac{1}{p}$. Thus u belongs to the space $C^{0,\alpha}(\mathcal{I})$ of *Hölder-continuous functions with exponent* α on $\mathcal{I}$. A function that is Hölder-continuous with exponent 1 is said to be *Lipschitz-continuous.*

If Ω is an interval such as $[a, b]$ or (a, b), then we abbreviate $C^0([a, b])$ and $C^0((a, b))$ by $C^0[a, b]$ and $C^0(a, b)$, etc.

The spaces $C^0(\operatorname{cl}\Omega)$, $C^{0,\alpha}(\operatorname{cl}\Omega)$, $L_p(\Omega)$, and $W_p^1(\Omega)$ are Banach spaces, which are discussed in Chap. XVII. Each function in such a space is endowed with a size, its norm, naturally related to the linear structure of the space, and the space has nice convergence properties expressed in terms of the norm.

CHAPTER II

The Equations of Motion for Extensible Strings

1. Introduction

The main purpose of this chapter is to give a derivation, which is mathematically precise, physically natural, and conceptually simple, of the quasilinear system of partial differential equations governing the large motion of nonlinearly elastic and viscoelastic strings. A part of this treatment is a careful study the Principle of Virtual Power and the equivalent Impulse-Momentum Law, which are physically and mathematically important generalizations of the governing equations. These formulations form the foundation of the treatment of concrete problems, which is begun in this chapter for very simple problems and is pursued in greater depth in Chaps. III and VI. These formulations also serve as models for those for more complicated bodies, such as rods, shells, and three-dimensional bodies, which we study later.

One of the central themes of this chapter is the pervasive role played by the Principle of Virtual Power, which is manifested in the treatment of jump conditions in Sec. 5, of variational formulations in Sec. 10, and of approximation methods in Sec. 11. A second such theme, which recurs continually in this book, is the role of constitutive restrictions. These restrictions, which must reflect physically natural material response, are consistently used to justify the nonlinear analysis of the equations. In this chapter, these restrictions support the study of equilibrium states, of perturbation methods, and of the classification of partial differential equations.

The exact equations for the large planar motion of a string were derived by Euler (1751) in 1744 and those for the large spatial motion by Lagrange (1762). By some unfortunate analog of Gresham's law, the simple and elegant derivation of Euler (1771), which is based on Euler's (1752) straightforward combination of geometry with mechanical principles, has been driven out of circulation and supplanted with baser derivations, relying on ad hoc geometrical and mechanical assumptions. (Evidence for this statement can be found in numerous introductory texts on partial differential equations and on mathematical physics. A distinguished exception to this unhappy tradition is the text of Weinberger (1965, Sec. 1).) A goal of this chapter is to show that it is easy to derive the equations correctly, much easier than following most modern expositions, which ask the reader to emulate the Red Queen by believing six impossible things before breakfast.

The correct derivation is simple because Euler made it so. Modern authors should be faulted not merely for doing poorly what Euler did well, but also for failing to copy from the master. A typical ad hoc assumption found in the textbook literature is that the motion of each material point is confined to the plane through its equilibrium position perpendicular to the line joining the ends of the string. In Sec. 7 we show that scarcely any elastic strings can execute such a motion. Most derivations suppress the role of material properties and even the extensibility of the string by assuming that the tension is approximately constant for all small motions. Were it exactly constant, then no segment of a uniform string could change its length, and if the ends of such a string were fixed, then the string could not move. (One author of a research monograph on one-dimensional wave propagation derived the wave equation governing the motion of an inextensible string. Realizing that an inextensible string with its ends separated by its natural length could not move, however pretty its governing equations, he assumed that one end of the string was joined to a fixed point by a spring.) One can make sense out of such assumptions as those of purely transverse motion and of the constancy of tension by deriving them as consequences of a systematic perturbation scheme applied to the exact equations, as we do in Sec. 8.

Parts of Secs. 1–4, 6, 8 of this chapter are adapted from Antman (1980b) with the kind permission of the Mathematical Association of America.

2. The Classical Equations of Motion

In this section we derive the classical form of the equations for the large motion of strings of various materials. A *classical* solution of these equations has the defining property that all its derivatives appearing in the equations are continuous on the interiors of their domains of definition. To effect our derivation, we accordingly impose corresponding regularity restrictions on the geometrical and mechanical variables. Since it is well known on both physical and mathematical grounds that solutions of these equations need not be classical, we undertake in Secs. 3 and 4 a more profound study of their derivation, which dispenses with simplified regularity assumptions.

Our derivation of the equations for strings, just as all our subsequent derivations of equations governing the behavior of more complicated bodies, is broken down into the description of (i) the kinematics of deformation, (ii) fundamental mechanical laws (such as the generalization of Newton's Second Law to continua), and (iii) material properties by means of constitutive equations. This scheme separates the treatment of geometry and mechanics in steps (i) and (ii), which are regarded as universally valid, from the treatment of constitutive equations, which vary with the material. Since this derivation serves as a model for all subsequent derivations, we examine each aspect of it with great care.

Let $\{\boldsymbol{i}, \boldsymbol{j}, \boldsymbol{k}\}$ be a fixed right-handed orthonormal basis for the Euclidean 3-space $\mathbb{E}^3$. A *configuration of a string* is defined to be a curve in $\mathbb{E}^3$. A *string* itself is defined to be a set of elements called *material points* (or *particles*) having the geometrical property that it can occupy curves in $\mathbb{E}^3$ and having the mechanical property that it is 'perfectly flexible'. The definition of perfect flexibility is given below.

We refrain from requiring that the configurations of a string be simple (nonintersecting) curves for several practical reasons: (i) Adjoining the global requirement that configurations be simple curves to the local requirement that configurations satisfy a system of differential equations can lead to severe analytical difficulties. (ii) If two different parts of a string come into contact, then the nature of the resulting mechanical interaction must be carefully specified. (iii) A configuration with self-intersections may serve as a particularly convenient model for a configuration in which distinct parts of a string are close, but fail to touch. (iv) It is possible to show that configurations corresponding to solutions of certain problems must be simple (see, e.g., Chap. III).

We distinguish the configuration $[0,1] \ni s \mapsto s\boldsymbol{k}$, in which the string lies along the unit interval in the $\boldsymbol{k}$-direction, as the *reference configuration.* We identify each material point in the string by its coordinate s in this reference configuration. For mathematical simplicity, we have, without loss of generality, scaled the length variable s to lie in the unit interval. Below we shall formally assume that this reference configuration is a *natural* configuration, i.e., a configuration in which the resultant force and torque on any part of the string is zero.

Suppose the string is undergoing some motion. Let $\boldsymbol{r}(s,t)$ denote the position of the material point (with coordinate) s at time t. For the purpose of studying initial-boundary-value problems, we take the domain of $\boldsymbol{r}$ to be $[0,1] \times [0,\infty)$. The function $\boldsymbol{r}(\cdot,t)$ defines the configuration of the string at time t. In this section we adopt the convention that every function of s and t, such as $\boldsymbol{r}$, whose values are exhibited here is ipso facto assumed to be continuous on the interior of its domain. (We critically examine this assumption in the next two sections.) The vector $\boldsymbol{r}_s(s,t)$ is tangent to the curve $\boldsymbol{r}(\cdot,t)$ at $\boldsymbol{r}(s,t)$. (By our convention, $\boldsymbol{r}_s$ is assumed to be continuous on $(0,1)\times(0,\infty)$.) Note that we do not parametrize the curve $\boldsymbol{r}(\cdot,t)$ with its arc length. The parameter s, which identifies material points, is far more convenient on mathematical and physical grounds.

The length of the material segment (s_1,s_2) in the configuration at time t is the integral $\int_{s_1}^{s_2} |\boldsymbol{r}_s(s,t)|\,ds$. The *stretch* $\nu(s,t)$ of the string at (s,t) is

$$\nu(s,t) \equiv |\boldsymbol{r}_s(s,t)|. \tag{2.1}$$

(It is the local ratio at s of the deformed to reference length, i.e., it is the limit of $\int_{s_1}^{s_2} |\boldsymbol{r}_s(s,t)|\,ds/(s_2-s_1)$ as the material segment (s_1,s_2) shrinks down to the material point s.) An attribute of a 'regular' motion is that this length ratio never be reduced to zero:

$$\nu(s,t) > 0 \quad \forall\ (s,t) \in [0,1]\times[0,\infty). \tag{2.2}$$

Where $\nu(s,t) > 1$, the string is *elongated*, and where $\nu(s,t) < 1$, the string is *compressed*. (The difficulty one encounters in compressing a real string is a consequence of an instability due to its great flexibility. Below we adopt as a defining property of a string the requirement that it be perfectly flexible.)

We assume that the ends $s=0$ and $s=1$ of the string are fixed at the points $\boldsymbol{0}$ and $L\boldsymbol{k}$ where L is a given positive number. In the optimistic spirit that led us to assume that $\boldsymbol{r}$ is continuous on $(0,1)\times(0,\infty)$, we further suppose that $\boldsymbol{r}(\cdot,t)$ is continuous on $[0,1]$ for all $t>0$. In this case, our prescription of $\boldsymbol{r}$ at $s=0$ and at $s=1$ leads to boundary conditions expressed by the following pointwise limits:

$$\lim_{s\to 0}\boldsymbol{r}(s,t)=\boldsymbol{0},\quad \lim_{s\to 1}\boldsymbol{r}(s,t)=L\boldsymbol{k}\quad\text{for}\quad t>0, \tag{2.3a}$$

which imply that $\boldsymbol{r}(\cdot,t)$ is continuous up to the ends of its interval of definition. These conditions are conventionally denoted by

$$\boldsymbol{r}(0,t)=\boldsymbol{0},\quad \boldsymbol{r}(1,t)=L\boldsymbol{k}. \tag{2.3b}$$

We assume that the string is released from configuration $s\mapsto\boldsymbol{u}(s)$ with velocity field $s\mapsto\boldsymbol{v}(s)$ at time $t=0$. If $\boldsymbol{r}_t(s,\cdot)$ is assumed to be continuous on $[0,\infty)$ for each $s\in(0,1)$, then these initial conditions have the pointwise interpretations

$$\lim_{t\to 0}\boldsymbol{r}(s,t)=\boldsymbol{u}(s),\quad \lim_{t\to 0}\boldsymbol{r}_t(s,t)=\boldsymbol{v}(s)\quad\text{for}\quad s\in(0,1), \tag{2.4a}$$

which are conventionally written as

$$\boldsymbol{r}(s,0)=\boldsymbol{u}(s),\quad \boldsymbol{r}_t(s,0)=\boldsymbol{v}(s). \tag{2.4b}$$

The requirement that the data given on the boundary of $[0,1]\times[0,\infty)$ by (2.3) and (2.4) be continuous, so that $\boldsymbol{r}_t$ could be continuous on its domain, is expressed by the *compatibility conditions*

$$\boldsymbol{u}(0)=\boldsymbol{0},\quad \boldsymbol{u}(1)=L\boldsymbol{k},\quad \boldsymbol{v}(0)=\boldsymbol{0},\quad \boldsymbol{v}(1)=\boldsymbol{0}. \tag{2.5}$$

These considerations complete our study of the kinematics of deformation of a string. We now study the mechanics.

Let $0<a<b<1$. We assume that the forces acting on (the material of) (a,b) in configuration $\boldsymbol{r}(\cdot,t)$ consist of a *contact force* $\boldsymbol{n}^+(b,t)$ exerted on (a,b) by $[b,1]$, a *contact force* $-\boldsymbol{n}^-(a,t)$ exerted on (a,b) by $[0,a]$, and a *body force* exerted on (a,b) by all other agents. We assume that the body force has the form $\int_a^b \boldsymbol{f}(s,t)\,ds$. The contact force $\boldsymbol{n}^+(b,t)$ has the defining property that it is the same as the force exerted on (c,b) by $[b,d]$ for each c and d satisfying $0<c<b<d<1$. Analogous remarks apply to $-\boldsymbol{n}^-$. Thus $\boldsymbol{n}^\pm(\cdot,t)$ are defined on an interval $(0,1)$ of real numbers, as indicated (and not on a collection of pairs of disjoint intervals). We shall see that the distinction between open and closed sets in the definitions of contact forces will evaporate. The minus sign before $\boldsymbol{n}^-(a,t)$ is introduced for mathematical convenience. (It corresponds to the sign convention of structural mechanics.)

Let $(\rho A)(s)$ denote the mass density per unit length at s in the reference configuration. This rather clumsy notation, using two symbols for one function, is employed because it is traditional and because it suggests that the density per unit reference length at s in a real three-dimensional string is the integral of the density per unit reference volume, traditionally denoted by ρ, over the cross section at s with area $A(s)$. It is important to note, however, that the notion of a cross-sectional area never arises in our idealized model of a string.

The integrand $\boldsymbol{f}(s,t)$ of the the body force is the *body force per unit reference length* at s,t. The most common example of the body force on a segment is the weight of this segment, in which case $\boldsymbol{f}(s,t) = -(\rho A)(s)g\boldsymbol{e}$ where g is the acceleration of gravity and $\boldsymbol{e}$ is the unit vector pointing in the vertical direction. $\boldsymbol{f}(s,t)$ could depend on $\boldsymbol{r}$ in quite complicated ways. For example, $\boldsymbol{f}$ could have the composite form

$$\boldsymbol{f}(s,t) = \boldsymbol{g}\big(\boldsymbol{r}(s,t), \boldsymbol{r}_t(s,t), s, t\big) \tag{2.6}$$

where $\boldsymbol{g}$ is a prescribed function, which describes the effects of the environment. The dependence of $\boldsymbol{g}$ on the velocity $\boldsymbol{r}_t$ could account for air resistance and its dependence upon the position $\boldsymbol{r}$ could account for variable gravitational attraction.

The requirement that at typical time t the resultant force on the typical material segment $(a,s) \subset (0,1)$ equal the time derivative of the *linear momentum* $\int_a^s (\rho A)(\xi)\boldsymbol{r}_t(\xi,t)\,d\xi$ of that segment yields the following integral form of the *equation of motion*

$$\begin{aligned} \boldsymbol{n}^+(s,t) - \boldsymbol{n}^-(a,t) + \int_a^s \boldsymbol{f}(\xi,t)\,d\xi \\ = \frac{d}{dt}\int_a^s (\rho A)(\xi)\boldsymbol{r}_t(\xi,t)\,d\xi = \int_a^s (\rho A)(\xi)\boldsymbol{r}_{tt}(\xi,t)\,d\xi. \end{aligned} \tag{2.7}$$

This equation is to hold for all $(a,s) \subset (0,1)$ and all $t > 0$.

The continuity of $\boldsymbol{n}^+$ implies that $\boldsymbol{n}^+(a,t) = \lim_{s\to a} \boldsymbol{n}^+(s,t)$. Since $\boldsymbol{f}$ and $\boldsymbol{r}_{tt}$ are continuous, we let $s \to a$ in (2.7) to obtain

$$\boldsymbol{n}^+(a,t) = \boldsymbol{n}^-(a,t) \quad \forall\, a \in (0,1). \tag{2.8}$$

Since the superscripts $\pm$ on $\boldsymbol{n}$ are thus superfluous, we drop them. We differentiate (2.7) with respect to s to obtain the *classical form of the equations of motion:*

$$\boldsymbol{n}_s(s,t) + \boldsymbol{f}(s,t) = (\rho A)(s)\,\boldsymbol{r}_{tt}(s,t) \quad \text{for} \quad s \in (0,1),\ t > 0. \tag{2.9}$$

These equations represent the culmination of the basic mechanical principles for strings.

We describe those material properties of a string that are relevant to mechanics by specifying how the contact force $\boldsymbol{n}$ is related to the change of shape suffered by the string in every motion $\boldsymbol{r}$. Such a specification, called a *constitutive relation*, must distinguish the material response of a rubber band, a steel band, a cotton thread, and a filament of chewing gum. The system consisting of (2.9) and the constitutive equation is formally determinate: It has as many equations as unknowns.

A defining property of a string is its perfect flexibility, which is expressed mathematically by the requirement that $\boldsymbol{n}(s,t)$ be tangent to the curve $\boldsymbol{r}(\cdot,t)$ at $\boldsymbol{r}(s,t)$ for each s,t:

$$\text{(2.10a)} \qquad \boldsymbol{r}_s(s,t) \times \boldsymbol{n}(s,t) = \boldsymbol{0} \quad \forall\, s,t$$

or, equivalently, that there exist a scalar-valued function N such that

$$\text{(2.10b)} \qquad \boldsymbol{n}(s,t) = N(s,t)\frac{\boldsymbol{r}_s(s,t)}{|\boldsymbol{r}_s(s,t)|}.$$

(Note that (2.2) ensures that $\boldsymbol{r}_s(s,t) \neq \boldsymbol{0}$ for each s,t.) Why (2.10) should express perfect flexibility is not obvious from the information at hand. The actual motivation for this tangency condition comes from outside our self-consistent theory of strings, namely, from the theory of rods, which is developed in Chaps. IV and VIII. The motion of a rod is governed by (2.9) and a companion equation expressing the equality of the resultant torque on any segment of the rod with the time derivative of the angular momentum for that segment. In the degenerate case that the rod offers no resistance to bending, this second equation reduces to (2.10a).

The force (component) $N(s,t)$ is the *tension* at (s,t). It may be of either sign. Where N is positive it is said to be *tensile* and the string is said to be *under tension*; where N is negative it is said to be *compressive* and the string is said to be *under compression*. (This terminology is typical of the inhospitability of the English language to algebraic concepts.)

From primitive experiments, we might conclude that the tension $N(s,t)$ at (s,t) in a rubber band depends only on the stretch $\nu(s,t)$ at (s,t) and on the material point s. Such experiments would not suggest that this tension depends on the rate at which the deformation is occurring, on the past history of the deformation, or on the temperature. Thus we might be led to assume that the string is *elastic*, i.e., that there is a constitutive function $(0,\infty) \times [0,1] \ni (\nu,s) \mapsto \hat{N}(\nu,s) \in \mathbb{R}$ such that

$$\text{(2.11)} \qquad N(s,t) = \hat{N}\big(\nu(s,t),s\big).$$

Note that (2.11) does not allow $N(s,t)$ to depend upon $\boldsymbol{r}(s,t)$ through $\hat{N}$. Were there such a dependence, then we could change the material properties of the string simply by translating it from one position to another. (In this case, it would be impossible to use springs to measure the acceleration of gravity at different places, as Hooke did, by measuring the elongation produced in a given spring by the suspension of a given mass.) Similarly, (2.11) does not allow $N(s,t)$ to depend upon all of $\boldsymbol{r}_s(s,t)$, but only on its magnitude, the stretch $\nu(s,t)$. A dependence on $\boldsymbol{r}_s(s,t)$ would mean that we could change the material response of the string by merely changing its orientation. Finally, (2.11) does not allow $N(s,t)$ to depend explicitly on absolute time t (i.e., $\hat{N}$ has no slot for the argument t alone). At first sight, this omission seems like an unwarranted restriction of generality, because a real rubber band becomes more brittle with the passage of time. But a careful consideration of this question suggests that the degradation of a rubber band depends on the time elapsed since its manufacture, rather than on the absolute time. Were the constitutive function to depend explicitly on t, then the outcome of an experiment performed today on a material manufactured yesterday would differ from the outcome of the same experiment performed tomorrow on the same material manufactured today. This dependence on time lapse can be generalized by allowing $N(s,t)$ to depend on the past history of the deformation at (s,t). We shall soon show how to account for this dependence. In using (2.11) one chooses to ignore such effects.

That the material response should be unaffected by rigid motions and by time translations is called the *Principle of Frame-Indifference* (or the *Principle of Objectivity*). In Chap. VIII and elsewhere, we show how its use leads to a systematic method for reducing a constitutive equation in a general form such as

$$N(s,t) = N_0\big(\boldsymbol{r}(s,t), \boldsymbol{r}_s(s,t), s, t\big) \tag{2.12}$$

to a very restricted form such as (2.11).

There is no physical principle preventing the constitutive function from depending in a frame-indifferent way on higher s-derivatives of $\boldsymbol{r}$. Such a dependence allows for surface-tension effects, which seem to be of limited physical importance except for problems of shock structure and phase changes (see Hagan & Slemrod (1983) and Carr, Gurtin, & Slemrod (1984)). We shall not devote much attention to these materials.

Anyone who rapidly deforms a rubber band feels an appreciable increase in temperature θ. One can also observe that the mechanical response of the band is influenced by its temperature. To account for these effects we may replace (2.11) with the mechanical constitutive equation for a *thermoelastic* string:

$$N(s,t) = N_{00}\big(\nu(s,t), \theta(s,t), s\big). \tag{2.13}$$

When this equation is used, the equation of motion must be supplemented with the energy equation, and the new variables entering the energy equation must be related by constitutive equations.

The motion of a rubber band fixed at its ends and subject to zero body force is seen to die down in a short time, even if the motion occurs in a vacuum. The chief source of this decay is internal friction, which is intimately associated with thermal effects. The simplest model for this friction, which ignores thermal effects, is obtained by assuming that the tension $N(s,t)$ depends on the stretch $\nu(s,t)$, the rate of stretch $\nu_t(s,t)$, and the material point s; that is, there is a function $(0,\infty) \times \mathbb{R} \times [0,1] \ni (\nu, \xi, s) \mapsto N_1(\nu, \xi, s) \in \mathbb{R}$ such that

$$N(s,t) = N_1\big(\nu(s,t), \nu_t(s,t), s\big). \tag{2.14}$$

(Note that in general $\nu_t \equiv |\boldsymbol{r}_s|_t$ is not equal to $|\boldsymbol{r}_{st}|$.) When (2.14) holds, the string may be called *viscoelastic of differential type with complexity 1.* (This terminology is not universally accepted.) It is clear that (2.14) ensures that the material response is unaffected by rigid motions and translations of time.

The form of (2.14) suggests the generalization in which $N(s,t)$ depends upon the first k t-derivatives of $\nu(s,t)$ and on s. (Such a string is termed *viscoelastic of differential type with complexity* k.) This generalization is but a special case of that in which $N(s,t)$ depends upon the past history of $\nu(s,\cdot)$ and upon s. To express the constitutive equation for such a material, we define the *history* $\nu^t(s,\cdot)$ *of* $\nu(s,\cdot)$ *up to time* t on $[0,\infty)$ by

(2.15) $$\nu^t(s,\tau) \equiv \nu(s,t-\tau) \quad \text{for} \quad \tau \geq 0.$$

Then the most general constitutive equation of the class we are considering has the form

(2.16) $$N(s,t) = N_\infty\big(\nu^t(s,\cdot),s\big).$$

The domain of $N_\infty(\cdot,s)$ is a class of positive-valued functions. A material described by (2.16) (that does not degenerate to (2.11)) and that is dissipative may be called *viscoelastic.* This term is rather imprecise; in modern continuum mechanics it is occasionally used as the negation of *elastic* and is thus synonymous with *inelastic.*

Note that (2.14) reduces to (2.11) where the string is in equilibrium. Similarly, if the string with constitutive equation (2.16) has been in equilibrium for all time before t (or, more generally, for all such times $t-\tau$ for which $\nu(s,t-\tau)$ influences N_∞), then (2.16) also reduces to (2.11). Thus "the equilibrium response of all strings (in a purely mechanical theory) is elastic." We shall pay scant attention to constitutive equations of the form (2.16). There is a fairly new and challenging mathematical theory for such materials with nonlinear constitutive equations; see Renardy, Hrusa, & Nohel (1987).

A string is said to be *uniform* if ρA is constant and if its constitutive function $\hat{N}$, $N_1, \ldots$ does not depend explicitly on s. A real (three-dimensional) string fails to be uniform because its material properties vary along its length or, more commonly, because its cross section varies along its length. If only the latter occurs, we can denote the cross-sectional area at s by $A(s)$. Then $(\rho A)(s)$ reduces to $\rho A(s)$ where ρ is the given constant mass density per reference volume. In this case, the constitutive function $\hat{N}$ might well have the form $\hat{N}(\nu,s) = A(s)\overline{N}(\nu)$, etc.

Not every choice of the constitutive functions $\hat{N}$, etc., is physically reasonable: We do not expect a string to shorten when we pull on it and we do not expect friction to speed up its motion. We can ensure that an increase in tension accompany an increase in stretch for an elastic string by assuming that:

(2.17) $$\nu \mapsto \hat{N}(\nu,s) \quad \text{is strictly increasing.}$$

Since $\nu \mapsto N_1(\nu,0,s)$ describes elastic response, we could require it to satisfy (2.17). A stronger, though reasonable, restriction on N_1 is that:

(2.18) $$\nu \mapsto N_1(\nu,\xi,s) \quad \text{is strictly increasing.}$$

Similar restrictions could be placed on other constitutive functions. Models satisfying (2.17) except for ν in a small interval have been used to describe instabilities associated with phase transitions (see Ericksen (1975, 1977b), James (1979, 1980), Magnus & Poston (1979), and Carr, Gurtin, & Slemrod (1984)).

The discussion of armchair experiments in the preceding paragraph is intentionally superficial. If we pull on a real string, we prescribe either its total length or the tensile forces at its ends. But in pulling the string we may produce local effects such as the stretch at each point (which may differ from point to point), the integral of which is the total actual length. In a typical experiment, we measure the tensile force at the ends if we prescribe the total length, and we measure the total length if we prescribe the tensile force at the ends. These experimental measurements of global quantities correspond to information coming from the solution of a boundary-value problem. It is in general a very difficult matter to determine the constitutive function, which has a local significance and which determines the governing equations, from a family of solutions.

For an elastic string the requirements that an infinite tensile force must accompany an infinite stretch and that an infinite compressive force must accompany a total compression to zero stretch are embodied in

$$(2.19\text{a,b}) \qquad \hat{N}(\nu, s) \to \infty \quad \text{as} \quad \nu \to \infty, \qquad \hat{N}(\nu, s) \to -\infty \quad \text{as} \quad \nu \to 0.$$

The reference configuration is *natural* if the tension vanishes in it. Thus for elastic strings this property is ensured by the constitutive restriction

$$(2.20) \qquad \hat{N}(1, s) = 0.$$

It is easy to express assumptions corresponding to those of this paragraph for other materials.

That (2.14) describes a material with a true internal friction, i.e., a material for which energy is dissipated in every motion, is ensured by the requirement that

$$(2.21) \qquad [N_1(\nu, \xi, s) - N_1(\nu, 0, s)]\xi > 0 \quad \text{for} \quad \xi \neq 0.$$

A proof that (2.21) ensures that (2.14) is 'dissipative' is given in Ex. 2.29. A stronger restriction, which ensures that the frictional force increases with the rate of stretch, is that

$$(2.22) \qquad \xi \mapsto N_1(\nu, \xi, s) \quad \text{is strictly increasing.}$$

Clearly, (2.22) implies (2.21). Condition (2.22) is mathematically far more tractable than (2.21).

There are a variety of mathematically useful consequences of the constitutive restrictions we have imposed. In particular, hypothesis (2.19) and the continuity of $\hat{N}$ enable us to deduce from the Intermediate Value Theorem that for each given $s \in [0, 1]$ and $N \in \mathbb{R}$ there is a ν satisfying $\hat{N}(\nu, s) = N$. Hypothesis (2.17) implies that this solution is unique. We denote it by $\hat{\nu}(N, s)$. Let us strengthen (2.17) by requiring that $\hat{N}_\nu$ be everywhere positive. Then the classical Local Implicit Function Theorem implies that $\hat{\nu}$ is continuously differentiable because $\hat{N}$ is. These results constitute a simple example of a *global implicit function theorem.* We shall employ a variety of generalizations of it throughout this book. Thus (2.11) is equivalent to

$$\nu(s,t) = \hat{\nu}\big(N(s,t), s\big). \tag{2.23}$$

One can impose hypotheses on $\hat{N}$ short of differentiability that ensure that $\hat{\nu}$ has properties somewhat better than mere continuity: Suppose that $\hat{N}$ is continuous and further that there is a function f on $[0,\infty)$ with $x \mapsto f(x)/x$ strictly increasing from 0 to ∞ such that

$$[\hat{N}(\nu_1, s) - \hat{N}(\nu_2, s)](\nu_1 - \nu_2) \geq f(|\nu_1 - \nu_2|).$$

This condition strengthens (2.17). Let g be the inverse of $x \mapsto f(x)/x$. Then we immediately find that

$$|\hat{\nu}(N_1, s) - \hat{\nu}(N_2, s)| \leq g\left(|N_1 - N_2|\right),$$

which implies that $\hat{\nu}$ is continuous and gives a modulus of continuity for it.

We substitute (2.11) or (2.14) into (2.10b) and then substitute the resulting expression into (2.9). We obtain a quasilinear system of partial differential equations for the components of $\boldsymbol{r}$. The full *initial-boundary-value problem for elastic strings* consists of (2.3), (2.4), (2.9), (2.10b), and (2.11). That for the viscoelastic string of differential type is obtained by replacing (2.11) with (2.14). If we use (2.16), then in place of a partial differential equation we obtain a partial functional-differential equation, for which we must supplement the initial conditions (2.4) by specifying the history of $\boldsymbol{r}$ up to time 0.

It proves mathematically convenient to recast these initial-boundary-value problems in an entirely different form, called the weak form of the equations by mathematicians and the Principle of Virtual Power (or the Principle of Virtual Work) by physicists and engineers. The traditional derivation of this formulation from (2.9) is particularly simple: We introduce the class of functions $\boldsymbol{y} \in C^1([0,1] \times [0,\infty))$ such that $\boldsymbol{y}(0,t) = \boldsymbol{0} = \boldsymbol{y}(1,t)$ (for all $t \geq 0$) and such that $\boldsymbol{y}(s,t) = \boldsymbol{0}$ for all t sufficiently large. These functions are termed *test functions* by mathematicians and *virtual velocities* (or *virtual displacements*) by physicists and engineers. We take the dot product of (2.9) with a test function $\boldsymbol{y}$ and integrate the resulting expression by parts over $[0,1] \times [0,\infty)$. Using (2.4) and the properties of $\boldsymbol{y}$ we obtain

$$\begin{aligned} &\int_0^\infty \int_0^1 [\boldsymbol{n}(s,t) \cdot \boldsymbol{y}_s(s,t) - \boldsymbol{f}(s,t) \cdot \boldsymbol{y}(s,t)]\, ds\, dt \\ &= \int_0^\infty \int_0^1 (\rho A)(s)[\boldsymbol{r}_t(s,t) - \boldsymbol{v}(s)] \cdot \boldsymbol{y}_t(s,t)\, ds\, dt \quad \text{for all test functions } \boldsymbol{y}. \end{aligned} \tag{2.24}$$

Equation (2.24) expresses a version of the *Principle of Virtual Power* for any material. We can substitute our constitutive equations into it to get a version of this principle suitable for specific materials.

Under the smoothness assumptions in force in this section, we have shown that (2.7) and (2.4) imply (2.24). An equally simple procedure (relying on the Fundamental Lemma of the Calculus of Variations) shows that the converse is true.

2.25. Exercise. Derive (2.24) from (2.9) and (2.4) and then derive (2.9) and (2.4) from (2.24). The Fundamental Lemma of the Calculus of Variations states that if f is integrable on a measurable set $\mathcal{E}$ of $\mathbb{R}^n$ and if $\int_{\mathcal{E}} fg\,dv = 0$ for all continuous g, then $f = 0$ (a.e.). Here dv is the differential volume of $\mathbb{R}^n$.

Equation (2.9) is immediately integrated to yield (2.7) with $\boldsymbol{n}^+ = \boldsymbol{n}^- = \boldsymbol{n}$. Then the integral form (2.7), the classical form (2.9), and the weak form (2.24) of the equations of motion are equivalent under our smoothness assumptions. In Sec. 4 we critically reexamine this equivalence in the absence of such smoothness.

2.26. Exercise. When undergoing a steady whirling motion about the $\boldsymbol{k}$-axis, a string lies in a plane rotating about $\boldsymbol{k}$ with constant angular velocity ω and does not move relative to the rotating plane. Let $\boldsymbol{f}(s,t) = g(s)\boldsymbol{k}$, where g is prescribed. Let (2.3) hold. Find a boundary-value problem for a system of ordinary differential equations, independent of t, governing the steady whirling motion of an elastic string under these conditions. Show that the steady whirling of a viscoelastic string described by (2.14) is governed by the same boundary-value problem. How is this result influenced by the frame-indifference of (2.14)? (Suppose that N were to depend on $\boldsymbol{r}_s$ and $\boldsymbol{r}_{st}$.)

2.27. Exercise. For an elastic string, let $W(\nu,s) \equiv \int_1^\nu \hat{N}(\bar{\nu},s)\,d\bar{\nu}$. Suppose that $\boldsymbol{f}$ has the form $\boldsymbol{f}(s,t) = \boldsymbol{g}(\boldsymbol{r}(s,t),s)$ where $\boldsymbol{g}(\cdot,s)$ is the Fréchet derivative (gradient) of the scalar-valued function $-\omega(\cdot,s)$, i.e., $\boldsymbol{g}(\boldsymbol{r},s) = -\omega_{\boldsymbol{r}}(\boldsymbol{r},s)$, where ω is prescribed. (Thus $\boldsymbol{f}$ is conservative.) W is the *stored-energy* or *strain-energy function* for the elastic string and ω is the *potential-energy density function* for the body force $\boldsymbol{f}$. Show that the integration by parts of the dot product of (2.9) with $\boldsymbol{r}_t$ over $[0,1] \times [0,\tau)$ and the use of (2.3) and (2.4) yield the *conservation of energy*:

$$(2.28)\qquad \int_0^1 \left[W(\nu(s,\tau),s) + \omega(\boldsymbol{r}(s,\tau),s) + \tfrac{1}{2}(\rho A)(s)|\boldsymbol{r}_t(s,\tau)|^2\right] ds$$
$$= \int_0^1 \left[W(|\boldsymbol{u}_s(s)|,s) + \omega(\boldsymbol{u}(s),s) + \tfrac{1}{2}(\rho A)(s)|\boldsymbol{v}(s)|^2\right] ds.$$

(This process parallels that by which (2.24) is obtained from (2.9) and (2.4).) Show that (2.28) can be obtained directly from (2.24) and (2.3) by choosing $\boldsymbol{y}(s,t)$ in (2.24) to equal $\boldsymbol{r}_t(s,t)\chi(t,\tau,\varepsilon)$ where

$$\chi(t,\tau,\varepsilon) \equiv \begin{cases} 1 & \text{for } \ 0 \le t \le \tau, \\ 1 + (\tau - t)/\varepsilon & \text{for } \ \tau \le t \le \tau + \varepsilon, \\ 0 & \text{for } \ \tau + \varepsilon \le t, \end{cases}$$

and then taking the limit of the resulting version of (2.24) as $\varepsilon \to 0$. See Sec. 10 for further material on energy.

2.29. Exercise. Let (2.14) hold and set $\hat{N}(\nu,s) = N_1(\nu,0,s)$. Define W as in Ex. 2.27. Let $\boldsymbol{f}$ have the conservative form shown in Ex. 2.26. Define the *total energy of the string at time* τ to be the left-hand side of (2.28). Form the dot product of (2.9) with $\boldsymbol{r}_t$, integrate the resulting expression with respect to s over $[0,1]$, and use (2.3) to obtain an expression for the time derivative of the total energy at time t. This formula gives a precise meaning to the remarks surrounding (2.21).

2.30. Exercise. Formulate the boundary conditions in which the end $s = 1$ is constrained to move along a frictionless continuously differentiable curve in space. Let this curve be given parametrically by $a \mapsto \bar{r}(a)$. (Locate the end at time t with the parameter $a(t)$.) A mechanical boundary condition is also needed.

2.31. Exercise. Formulate a suitable Principle of Virtual Power for the initial-boundary-value problem of this section modified by the replacement of the boundary condition at $s = 1$ with that of Ex. 2.30. The mechanical boundary condition at $s = 1$ should be incorporated into the principle.

The first effective steps toward correctly formulated equations for the vibrating string were made by Taylor (1714) and Joh. Bernoulli (1729). D'Alembert (1743) derived the first explicit partial differential equation for the small motion of a heavy string. The correct equations for the large vibrations of a string in a plane, equivalent to the planar version of (2.9), (2.10b), were derived by Euler (1751) in 1744 by taking the limit of the equations of motion for a finite collection of beads joined by massless elastic springs as the number of beads approaches infinity while their total mass remains fixed. The correct linear equation for the small planar transverse motion of an elastic string, which is just the wave equation, was obtained and beautifully analyzed by d'Alembert (1749). Euler (1752) stated 'Newton's equations of motion' and in his notebooks used them to derive the planar equations of motion for a string in a manner like the one just presented. A clear exposition of this derivation together with a proof that $\boldsymbol{n}^+ = \boldsymbol{n}^-$ was given by Euler (1771). Lagrange (1762) used the bead model to derive the spatial equations of motion for an elastic string. The Principle of Virtual Power in the form commonly used today was laid down by Lagrange (1788). A critical historical appraisal of these pioneering researches is given by Truesdell (1960), upon whose work this paragraph is based.

We note that the quasilinear system (2.9), (2.10b), (2.11) arising from the conceptually simple field of classical continuum mechanics is generally much harder to analyze than semilinear equations of the form $u_{tt} - u_{ss} = f(u, u_s)$, which arise in conceptually difficult fields of modern physics.

3. The Linear Impulse-Momentum Law

The partial differential equations for the longitudinal motion of an elastic string are the same as those for the longitudinal motion of a naturally straight elastic rod (for which compressive states are observed). It has long been known that solutions of these equations can exhibit shocks, i.e., discontinuities in $\boldsymbol{r}_s$ or $\boldsymbol{r}_t$. (See the discussion and references in Chap. XVI.) Shocks can also arise in strings with constitutive equations of the form (2.16) (see Renardy, Hrusa, & Nohel (1987)). On the other hand, Greenberg, MacCamy, & Mizel (1968), Kanel' (1968), Dafermos (1969), MacCamy (1970), and Antman & Seidman (1995), among others, have shown that the longitudinal motions of nonlinearly viscoelastic strings (or rods) for special cases of (2.14) satisfying a uniform version of (2.22) do not exhibit shocks. The burden of these remarks is that the smoothness assumptions made in Sec. 2 are completely unwarranted for nonlinearly elastic strings and for certain kinds of nonlinearly viscoelastic strings.

It is clear that the integral form (2.7) of the equations of motion makes sense under smoothness assumptions weaker than those used to derive (2.9). In this section we study natural generalizations of (2.3), (2.4), (2.7), and (2.8) under such weaker assumptions. In the next section we demonstrate the equivalence of these generalizations with a precisely formulated version of the Principle of Virtual Power.

We formally integrate (2.7) with respect to t over $[0, \tau]$ and take account of (2.4) to obtain the *Linear Impulse-Momentum Law*:

(3.1)
$$\int_0^\tau \left[\boldsymbol{n}^+(s,t) - \boldsymbol{n}^-(a,t)\right] dt + \int_0^\tau \int_a^s \boldsymbol{f}(\xi,t)\, d\xi\, dt = \int_a^s (\rho A)(\xi)[\boldsymbol{r}_t(\xi,\tau) - \boldsymbol{v}(\xi)]\, d\xi,$$

which is to hold for (almost) all a, s, τ. The left-hand side of (3.1) is the *linear impulse of the force system* $\{\boldsymbol{n}^\pm, \boldsymbol{f}\}$ and the right-hand side is the *change in linear momentum* for the material segment (a, s) over the time interval $(0, \tau)$. We regard (3.1) as the natural generalization of the equations of motion (2.7).

We now state virtually the weakest possible conditions on the functions entering (3.1) for its integrals to make sense as Lebesgue integrals and for our boundary and initial conditions to have consistent generalizations. These generalizations are the highlights of the ensuing development, the details of which can be omitted by the reader unfamiliar with real analysis.

We assume that there are numbers σ^- and σ^+ such that

(3.2)
$$0 < \sigma^- \le (\rho A)(s) \le \sigma^+ < \infty \quad \forall\, s \in [0,1].$$

We assume that $\boldsymbol{r}_s$ and $\boldsymbol{r}_t$ are locally integrable on $[0,1] \times [0,\infty)$, that $\boldsymbol{r}$ satisfies the boundary conditions (2.3) in the sense of *trace* (see Adams (1975), Nečas (1967)), i.e., that

(3.3)
$$\lim_{s\to 0} \int_{t_1}^{t_2} \boldsymbol{r}(s,t)\, dt = \boldsymbol{0}, \quad \lim_{s\to 1} \int_{t_1}^{t_2} [\boldsymbol{r}(s,t) - L\boldsymbol{k}]\, dt = \boldsymbol{0} \quad \forall\, (t_1,t_2) \subset [0,\infty),$$

that $\boldsymbol{u}$ is integrable on $[0,1]$, that the first initial condition of (2.4) is assumed in the sense of trace:

(3.4)
$$\lim_{t\to 0} \int_a^b (\rho A)(s)[\boldsymbol{r}(s,t) - \boldsymbol{u}(s)]\, ds = \boldsymbol{0} \quad \forall\, [a,b] \subset [0,1],$$

and that $\boldsymbol{v}$ is integrable on $[0,1]$. Conditions (3.3) and (3.4) are consistent with the local integrability of $\boldsymbol{r}_s$ and $\boldsymbol{r}_t$ (see Adams (1975), Nečas (1967)). We do not prescribe a generalization of the second initial condition of (2.4) because we shall show that it is inherent in (3.1), as the presence there of $\boldsymbol{v}$ suggests. We finally assume that $\boldsymbol{n}^\pm$ and $\boldsymbol{f}$ are locally integrable on $[0,1] \times [0,\infty)$.

Since we are merely assuming that our variables are integrable over compact subsets of $[0,1] \times [0,\infty)$, we must show that the single integrals in (3.1) make sense: By Fubini's Theorem the local integrability of $\boldsymbol{n}^+$ implies that for each $\tau \in (0,\infty)$ there is a set $\mathcal{A}^+(\tau) \subset [0,1]$ with Lebesgue measure $|\mathcal{A}^+(\tau)| = 1$ such that $\boldsymbol{n}^+(s,\cdot)$ is integrable over $[0,\tau]$ for $s \in \mathcal{A}^+(\tau)$. Moreover, the Lebesgue Differentiation Theorem implies that there is a subset $\mathcal{A}_0^+(\tau)$ of $\mathcal{A}^+(\tau)$ with $|\mathcal{A}_0^+(\tau)| = 1$ such that for $s \in \mathcal{A}_0^+(\tau)$, the integral $\int_0^\tau \boldsymbol{n}^+(s,t)\, dt$ has the 'right' value in the sense that it is the limit of its averages over intervals centered at s. The corresponding statements obtained by replacing the superscript '+' by '−' are likewise true. Let $\mathcal{A}(\tau) \equiv \mathcal{A}_0^+(\tau) \cap \mathcal{A}_0^-(\tau)$. (Thus $|\mathcal{A}(\tau)| = 1$ for each τ.) Let $\mathcal{B}$ be the set of $t \ge 0$ for which $(\rho A)(\cdot)\boldsymbol{r}_t(\cdot,t)$ is integrable over $[0,1]$ and for which $\int_0^1 (\rho A)(s)\boldsymbol{r}_t(s,t)\, ds$ has the 'right' value. (Fubini's Theorem and Lebesgue's Differentiation Theorem imply that $|\mathcal{B} \cap [0,T]| = T$ for all $T \ge 0$.) Thus each term in

(3.1) is well-defined for each $\tau \in \mathcal{B}$ and for each a and s in $\mathcal{A}(\tau)$ with $a \leq s$. Hence (3.1) holds a.e.

We now derive some important consequences from (3.1). Since Fubini's Theorem allows us to interchange the order of integration in the double integral, we can represent the left-hand side of (3.1) as an integral over (a, s) of an integrable function of $\bar{s}$ for $\tau \in \mathcal{B}$. Thus for each $\tau \in \mathcal{B}$, the function $s \mapsto \int_0^\tau \boldsymbol{n}^+(s,t)\,dt$ is absolutely continuous, not merely on $\mathcal{A}(\tau)$, but on all of $[0,1]$. Consequently,

$$\int_0^\tau \boldsymbol{n}^+(a,t)\,dt = \lim_{s\to a} \int_0^\tau \boldsymbol{n}^+(s,t)\,dt \quad \forall\, \tau \in \mathcal{B}. \tag{3.5}$$

Then (3.1) implies that

$$\int_0^\tau \boldsymbol{n}^+(a,t)\,dt = \int_0^\tau \boldsymbol{n}^-(a,t)\,dt \quad \forall\, \tau \in \mathcal{B}. \tag{3.6}$$

Thus the superscripts '+' and '−' are superfluous even in this more general setting and will accordingly be dropped.

The properties of the Lebesgue integral imply that if $a, s \in \mathcal{A}(T)$, then $a, s \in \mathcal{A}(\tau)$ for all $\tau \in [0,T]$. Let us fix $T > 0$. Let $a, s \in \mathcal{A}(T)$. Since the left-hand side of (3.1) is an integral over $(0,\tau)$ of an integrable function of t, the right-hand side of (3.1) defines an absolutely continuous function of τ for $a, s \in \mathcal{A}(T)$. Thus

$$\lim_{\tau\to 0} \int_a^s (\rho A)(\xi)[\boldsymbol{r}_t(\xi,\tau) - \boldsymbol{v}(\xi)]\,d\xi = \boldsymbol{0} \quad \forall\, a, s \in \mathcal{A}(T). \tag{3.7}$$

This generalization of the second initial condition of (2.4), which has the same form as (3.4), is thus implicit in (3.1).

It is important to note that the generalizations (3.3), (3.4), (3.7) of the boundary and initial conditions (2.3) and (3.3) represent averages of the classical pointwise conditions. As such, the limiting processes they embody correspond precisely to the way they could be tested experimentally.

Our basic smoothness assumption underlying the development of this section is the local integrability of $\boldsymbol{r}_s$, $\boldsymbol{r}_t$, and $\boldsymbol{n}$. Since $\boldsymbol{n}$ is to be given as a constitutive function of the stretch and possibly other kinematic variables, the local integrability of $\boldsymbol{n}$ imposes restrictions on the class of suitable constitutive functions.

In the modern study of shocks, physically realistic solutions $\boldsymbol{r}$ are sought in larger classes of functions, such as functions of bounded variation, which need not have locally integrable derivatives. Thus there is a need for mathematically sound and physically realistic generalizations of the development of this and the next section.

3.8. Exercise. Repeat Ex. 2.26, but now obtain the same equations for the steady whirling of the string directly from (3.1) and (3.6). This derivation can easily be performed with complete mathematical rigor.

4. The Equivalence of the Linear Impulse-Momentum Law with the Principle of Virtual Power

In this section we prove that the Linear Impulse-Momentum Law formulated in Sec. 3 is equivalent to a generalized version of the Principle of Virtual Power stated in (2.24). Our proof is completely rigorous and technically simple. Although we couch our presentation in the language of real analysis to ensure complete precision, all the steps have straightforward interpretations in terms of elementary calculus.

The demonstration of equivalence given at the end of Sec. 2, which is universally propounded by mathematicians and physicists alike, pivots on the classical form (2.9) of the equations of motion. But this form is devoid of meaning in the very instances when the Linear Impulse-Momentum Law and the Principle of Virtual Power are essential, i.e., when there need not be classical solutions. In our approach given below, Eq. (2.9) never appears.

Since (2.9) never appears, it is therefore never exposed to abuse. The most dangerous sort of abuse would consist in multiplying (2.9) by a positive-valued function depending on the unknowns appearing in (2.9), thereby converting (2.9) to an equivalent classical form. But its corresponding weak form, obtained by the procedure leading to (2.24), would not be equivalent to (2.24), because the integration by parts would produce additional terms caused by the presence of the multiplicative factor. Consequently, the corresponding jump conditions at discontinuities (see Sec. 5) would have forms we deem wrong because they are incompatible with the jump conditions coming from the generalization of (2.24). This generalization is deemed correct because we shall show that it is equivalent to the Linear Impulse-Momentum Law, which we regard as a fundamental postulate of mechanics.

Note that the Principle of Virtual Power as stated in (2.24) makes sense when the smoothness restrictions imposed on $\boldsymbol{r}$ and $\boldsymbol{n}$ in Sec. 2 are replaced by the much weaker conditions of Sec. 3. The resulting form of (2.24) can be further extended to apply to all test functions $\boldsymbol{y}$ that have essentially bounded generalized derivatives, that vanish for large t, and that vanish in the sense of trace on the boundaries $s = 0$ and $s = 1$. (These functions form a subspace of the Sobolev space $W^1_\infty([0,1]\times[0,\infty))$.) The smoothness assumptions on the variables entering these formulations are the weakest that allow all the integrals to make sense as Lebesgue integrals. We refer to the resulting version of (2.24) as the *generalized* Principle of Virtual Power.

We now derive this principle from the Linear Impulse-Momentum Law under the assumptions of Sec. 3. Let ϕ be a polygonal (piecewise affine) function of s with support in (a, b) and let ψ be a polygonal function of t with support in $[0, \tau)$. (The *support* of a function is the closure of the set on which it is not zero.) Note that the support of ψ is contained in a half-closed interval. Let $\boldsymbol{e}$ be a fixed but arbitrary constant unit vector. Then (3.1) implies that

$$\begin{aligned}(4.1)\quad \int_0^\tau \int_a^b \phi_s(s)\psi_t(t) \bigg\{ \int_0^t \boldsymbol{e}\cdot[\boldsymbol{n}(s,\bar{t}) - \boldsymbol{n}(a,\bar{t})]\, d\bar{t} \\ + \int_0^t \int_a^s \boldsymbol{e}\cdot \boldsymbol{f}(\bar{s},\bar{t})\, d\bar{s}\, d\bar{t} \bigg\}\, ds\, dt \\ = \int_0^\tau \int_a^b \phi_s(s)\psi_t(t) \int_a^s (\rho A)(\bar{s})\cdot[\boldsymbol{r}_t(\bar{s},t) - \boldsymbol{v}(\bar{s})]\, d\bar{s}\, ds\, dt.\end{aligned}$$

Since ψ and ϕ are absolutely continuous, we can integrate the triple integral on the left-hand side of (4.1) by parts with respect to t, we can integrate the quadruple integral on the left-hand side of (4.1) by parts with respect to t and s, and we can integrate the right-hand side of (4.1) by parts with respect to s. Since $\psi(\tau) = 0$, $\phi(a) = 0 = \phi(b)$, we thereby convert (4.1) to

$$\text{(4.2)} \quad \int_0^\tau \int_a^b \phi_s(s)\psi(t)\boldsymbol{e} \cdot \boldsymbol{n}(s,t)\, ds\, dt - \int_0^\tau \int_a^b \phi(s)\psi(t)\boldsymbol{e} \cdot \boldsymbol{f}(s,t)\, ds\, dt$$
$$= \int_0^\tau \int_a^b \phi(s)\psi_t(t)\boldsymbol{e} \cdot [\boldsymbol{r}_t(s,t) - \boldsymbol{v}(s)]\, ds\, dt.$$

Let us set

$$\text{(4.3)} \qquad \boldsymbol{y}(s,t) = \phi(s)\psi(t)\boldsymbol{e}.$$

Since this $\boldsymbol{y}$ has support in $(a,b) \times [0,\tau)$, we can write (4.2) in the form (2.24) for all $\boldsymbol{y}$'s of the form (4.3). More generally, (2.24) holds for all $\boldsymbol{y}$'s in the space that is the completion in the norm of $W^1_\infty([0,1] \times [0,\infty))$ of finite linear combinations of functions of the form (4.3). (Some properties of this space are discussed by Antman & Osborn (1979).)

Equation (2.24) for this large class of $\boldsymbol{y}$'s expresses the *generalized Principle of Virtual Power* or the *Weak Form* of (2.9), (2.3), and (2.4). We henceforth omit the adjective *generalized.* (If we allow $\boldsymbol{n}$ and $\boldsymbol{f}$ to be smoother, we can allow the $\boldsymbol{y}$'s to be rougher.) The *Weak Form of the Initial-Boundary Value Problem* for elastic strings is obtained by inserting (2.10b) and (2.11) into (2.24) and appending (3.3) and (3.4). Analogous definitions hold for other materials.

Without making unwarranted smoothness assumptions, we have thus shown that the Linear Impulse-Momentum Law implies the Principle of Virtual Power. Conversely, we can likewise recover (3.1) (without the superscripts '±') from (2.24) by taking ε to be a small positive number, taking $\boldsymbol{y}$ to have the form (4.3) (which reduces (2.24) to (2.42)), taking ϕ and ψ to have the forms

$$\text{(4.4a)} \qquad \phi(\bar{s}) = \begin{cases} 0 & \text{for } 0 \le \bar{s} \le a, \\ \frac{\bar{s}-a}{\varepsilon} & \text{for } a \le \bar{s} \le a+\varepsilon, \\ 1 & \text{for } a+\varepsilon \le \bar{s} \le s-\varepsilon, \\ \frac{s-\bar{s}}{\varepsilon} & \text{for } s-\varepsilon \le \bar{s} \le s, \\ 0 & \text{for } s \le \bar{s} \le 1, \end{cases}$$

$$\text{(4.4b)} \qquad \psi(t) = \begin{cases} 1 & \text{for } 0 \le t \le \tau, \\ 1 - \frac{t-\tau}{\varepsilon} & \text{for } \tau \le t \le \tau+\varepsilon, \\ 0 & \text{for } \tau+\varepsilon \le t, \end{cases}$$

and then letting $\varepsilon \to 0$. In this process, we must evaluate the typical expression

$$\text{(4.5)} \qquad \lim_{\varepsilon \to 0} \int_0^{\tau+\varepsilon} \frac{1}{\varepsilon} \int_a^{a+\varepsilon} \boldsymbol{n}(s,t) \cdot \boldsymbol{e}\psi(t)\, ds\, dt,$$

which Fubini's Theorem and (4.4b) allow us to rewrite as

$$\begin{aligned}(4.6)\quad &\lim_{\varepsilon\to 0}\frac{1}{\varepsilon}\int_a^{a+\varepsilon}\int_0^{\tau+\varepsilon}\boldsymbol{n}(s,t)\cdot\boldsymbol{e}\,dt\,ds\\ &\qquad -\lim_{\varepsilon\to 0}\left\{\varepsilon\left[\frac{1}{\varepsilon^2}\int_a^{a+\varepsilon}\int_\tau^{\tau+\varepsilon}\boldsymbol{n}(s,t)\cdot\boldsymbol{e}\frac{t-\tau}{\varepsilon}\,dt\,ds\right]\right\}.\end{aligned}$$

The Lebesgue Differentiation Theorem implies that the first term in (4.6) is

$$(4.7)\qquad \int_0^\tau \boldsymbol{n}(a,t)\cdot\boldsymbol{e}\,dt$$

for almost all a in $(0,1)$ and that the supremum of the absolute value of the bracketed expression in the second term of (4.6) is finite for almost all a in $(0,1)$ and τ in $(0,\infty)$. (Note that $|(t-\tau)/\varepsilon|\le 1$ for $t\in[\tau,\tau+\varepsilon]$.) Thus (4.5) equals (4.7). The other terms are treated similarly. The arbitrariness of $\boldsymbol{e}$ allows it to be cancelled in the final expression. Thus (2.24) implies (3.1) and these two principles are equivalent.

The Principle of Virtual Power can be used to exclude certain naive solutions of differential equations as unphysical. We illustrate this property with a differential equation simpler than that for a string. Consider the boundary-value problem

$$(4.8)\qquad u''(s)+\pi^2u(s)=0\quad\text{on}\quad(-1,1),\quad u(\pm1)=0.$$

The continuous function u^* defined by $u^*(s)=|\sin\pi s|$ is in $W_1^1(-1,1)$, satisfies the boundary conditions, and satisfies the differential equation everywhere except at 0. Other than its failure to be a classical solution of the boundary-value problem, there is nothing intrinsically wrong with u^* from a purely mathematical standpoint. Now suppose that (4.8) is regarded as a symbolic representation for the weak problem

$$(4.9)\qquad \int_{-1}^1[u'v'-\pi^2uv]\,ds=0\quad\forall v\in C^1[-1,1]\quad\text{with}\quad v(\pm1)=0,\quad u(\pm1)=0,$$

solutions of which are sought in $W_1^1(-1,1)$. We presume that (4.9) embodies a Principle of Virtual Power, representing a description of the underlying physics more fundamental than that given by (4.8). It is easy to show that u^* does not satisfy (4.9) and can therefore be excluded as unrealistic. Indeed, by using methods like those of (4.4)–(4.6) or of Sec. 6, we can show that every (weak) solution of (4.9) is a classical solution of (4.8). In the next section we show how the Principle of Virtual Power enables us to classify precisely those kinds of jumps that are compatible with it.

In much of modern mathematical literature, the classical form of an equation is regarded as merely an abbreviation for the weak form. Since weak formulations of equivalent classical formulations need not be equivalent, this convention should be used with care. The weak form is also sometimes termed the *variational form*, an expression we never employ because it connotes far more generality than the notion of variational structure introduced in Sec. 10.

If there are concentrated or impulsive forces applied to the string, then $\boldsymbol{f}$ would not be locally integrable, and the development of these last two sections would not be valid. Distribution theory, which was designed to handle linear equations with such forces, has recently been extended to handle nonlinear equations (see Colombeau (1990), Rossinger (1987)). But it is not evident how to obtain (3.6) in such a more general setting. In Sec. 6 we comment further on this question for a degenerately simple static problem.

5. Jump Conditions

We now show how the Principle of Virtual Power yields jump conditions that (certain kinds of) weak solutions must satisfy at their discontinuities. Let $\mathcal{C} \in [0,1] \times [0,\infty)$ be (the image of) a simple curve. We assume that $\mathcal{C}$ is so smooth that it possesses a unit normal (γ_1, γ_2) at almost every point. (It suffices for $\mathcal{C}$ to be uniformly Lipschitz-continuous, i.e., that there be a finite number of open sets covering $\mathcal{C}$ such that in each such set $\mathcal{E}$ there is a coordinate system with respect to which $\mathcal{C} \cap \mathcal{E}$ can be described as the graph of a Lipschitz continuous function.) Suppose that there are two disjoint, simply-connected open sets $\mathcal{G}_1$ and $\mathcal{G}_2$ such that $\emptyset \neq \partial\mathcal{G}_1 \cap \partial\mathcal{G}_2 \subset \mathcal{C}$ (see Fig. 5.1), that (2.9) holds in the classical sense in $\mathcal{G}_1$ and in $\mathcal{G}_2$, and that there are integrable functions $\boldsymbol{n}^1, \boldsymbol{n}^2, \boldsymbol{r}_t^1, \boldsymbol{r}_t^2$ on $\mathcal{C}$ such that

$$(5.2) \quad \boldsymbol{n} \to \boldsymbol{n}^\alpha, \ \boldsymbol{r}_t \to \boldsymbol{r}_t^\alpha \quad \text{(in the sense of trace)} \qquad \text{as} \quad \mathcal{G}_\alpha \ni (s,t) \to \mathcal{C}, \quad \alpha = 1,2.$$

Set

$$(5.3) \qquad [\![\boldsymbol{n}]\!] \equiv \boldsymbol{n}^2 - \boldsymbol{n}^1, \quad [\![\boldsymbol{r}_t]\!] \equiv \boldsymbol{r}_t^2 - \boldsymbol{r}_t^1 \quad \text{on } \mathcal{C}.$$

$[\![\boldsymbol{n}]\!]$ is called the *jump* in $\boldsymbol{n}$ across $\mathcal{C}$.

If $\boldsymbol{y}$ is taken to have support in $\mathcal{G}_1 \cup \mathcal{G}_2 \cup \mathcal{C}$, then (2.24) reduces to

$$(5.4) \qquad \int_{\mathcal{G}_1 \cup \mathcal{G}_2} [\boldsymbol{n} \cdot \boldsymbol{y}_s - \boldsymbol{f} \cdot \boldsymbol{y} - (\rho A)(\boldsymbol{r}_t - \boldsymbol{v}) \cdot \boldsymbol{y}_t] \, ds \, dt = 0$$

for all such $\boldsymbol{y}$'s. We separately integrate (5.4) by parts over $\mathcal{G}_1$ and $\mathcal{G}_2$ (by means of the divergence theorem), noting that (2.9) is satisfied in each region and that $\boldsymbol{y}$ vanishes on $\partial\mathcal{G}_1 \setminus \mathcal{C}$ and on $\partial\mathcal{G}_2 \setminus \mathcal{C}$. We obtain

$$(5.5) \qquad \int_{\mathcal{C}} \boldsymbol{y} \cdot \{[\![\boldsymbol{n}]\!]\gamma_1 - \rho A [\![\boldsymbol{r}_t]\!]\gamma_2\} \, d\sigma = 0$$

for all such $\boldsymbol{y}$'s. Here $d\sigma$ is the differential arc length along $\mathcal{C}$. Since $\boldsymbol{y}$ is arbitrary on $\mathcal{C}$, Eq. (5.5) implies that

$$(5.6) \qquad [\![\boldsymbol{n}]\!]\gamma_1 - \rho A [\![\boldsymbol{r}_t]\!]\gamma_2 = \boldsymbol{0} \quad \text{a.e. on } \mathcal{C}.$$

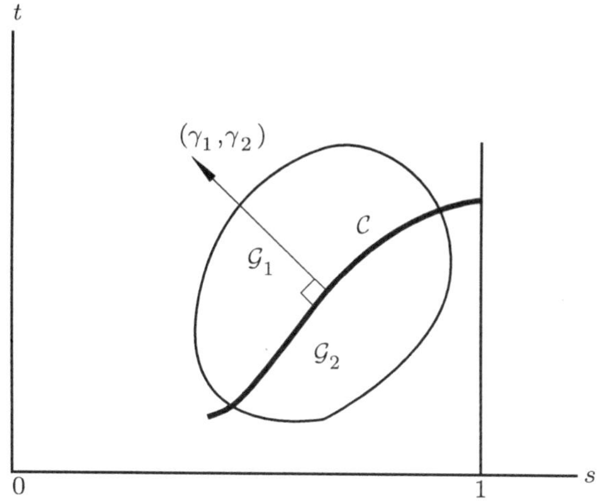

Figure 5.1. The neighborhood of a curve of discontinuity.

These are the *Rankine-Hugoniot jump conditions* for (2.24). A curve in the (s,t)-plane across which there are jumps in $\boldsymbol{n}$ or $\boldsymbol{r}_t$ is called a *shock (path)*. A solution suffering such a jump is said to have (or be) a *shock*. Suppose the shock path has the equation $s = \sigma(t)$. Then $\sigma'(t)$ is the *shock speed* at $(\sigma(t), t)$. Equation (5.6) thus has the form

$$[\![\boldsymbol{n}]\!] + \rho A \sigma' [\![\boldsymbol{r}_t]\!] = \boldsymbol{0}. \tag{5.7}$$

The foregoing analysis leading to (5.6) is formal to the extent that solutions are presumed classical except on isolated curves. For further information on jump conditions and shocks, see Chap. XVI and the references cited there.

6. The Existence of a Straight Equilibrium State

When none of the variables appearing in the Linear Impulse-Momentum Law depends on the time, it reduces to the static form of (2.7), (2.8):

$$\boldsymbol{n}(s) - \boldsymbol{n}(a) + \int_a^s \boldsymbol{f}(\xi)\, d\xi = \boldsymbol{0} \tag{6.1}$$

for (almost) all a and s in $[0,1]$. If $\boldsymbol{f}$ is Lebesgue-integrable, then (6.1) implies that $\boldsymbol{n}$ is absolutely continuous and has a derivative almost everywhere. Thus the classical equilibrium equation

$$\boldsymbol{n}'(s) + \boldsymbol{f}(s) = \boldsymbol{0} \tag{6.2}$$

holds a.e. When (2.3a) holds, the Principle of Virtual Power (2.24) reduces to

$$\int_0^1 [\boldsymbol{n}(s) \cdot \boldsymbol{y}'(s) - \boldsymbol{f}(s) \cdot \boldsymbol{y}(s)]\, ds = 0 \tag{6.3}$$

for all sufficiently smooth $\boldsymbol{y}$ that vanish at 0 and 1.

Note that in equilibrium the constitutive equation (2.14) reduces to that for an elastic string, namely (2.11). (Indeed, if the string has been in equilibrium for its entire past history, then the general constitutive equation (2.16) itself reduces to (2.11). This observation must be interpreted with care, because a string described by (2.16) can creep under the action of an equilibrated system of forces not varying with time.) We accordingly limit our attention to elastic strings, described by (2.10b) and (2.11):

$$\boldsymbol{n}(s) = \hat{N}\big(\nu(s), s\big) \frac{\boldsymbol{r}'(s)}{\nu(s)}. \tag{6.4}$$

We assume that $\hat{N}$ is continuously differentiable, although much can be done with $\hat{N}$'s that are merely continuous. (See the remarks following (2.23).)

For integrable $\boldsymbol{f}$ we now study the boundary-value problem of finding a function $\boldsymbol{r}$, whose (distributional) derivative $\boldsymbol{r}'$ is integrable, that satisfies the system (6.1), (6.4), (2.3), which we record as

$$\hat{N}\big(|\boldsymbol{r}'(\xi)|, \xi\big) \frac{\boldsymbol{r}'(\xi)}{|\boldsymbol{r}'(\xi)|} \bigg|_a^s + \int_a^s \boldsymbol{f}(\xi)\, d\xi = \boldsymbol{0}, \tag{6.5}$$

$$\boldsymbol{r}(0) = \boldsymbol{0}, \quad \boldsymbol{r}(1) = L\boldsymbol{k}. \tag{6.6}$$

(Conditions (6.6) interpreted as (2.3a) make sense because $\boldsymbol{r}$ is the indefinite integral of the integrable function $\boldsymbol{r}'$ and is accordingly absolutely continuous.)

In Chap. III we shall study a rich collection of problems for (6.5) and (6.6) in which $\boldsymbol{f}$ depends on $\boldsymbol{r}$. Here we content ourselves with the study of straight equilibrium configurations $s \mapsto \boldsymbol{r}(s) = z(s)\boldsymbol{k}$, in which z is an absolutely continuous, increasing function, when $\boldsymbol{f}$ has the special form

$$\boldsymbol{f}(s) = g(s)\boldsymbol{k}. \tag{6.7}$$

In keeping with the local integrability of $\boldsymbol{f}$ assumed in Secs. 3 and 4, we take g to be Lebesgue-integrable. The requirement that z be increasing ensures (2.2). Under these conditions, the problem (6.5), (6.6) reduces to finding z and a constant K such that

$$\hat{N}(z'(s), s) = G(s) + K \quad \forall\, s \in [0,1], \quad G(s) \equiv -\int_0^s g(\xi)\, d\xi, \tag{6.8}$$

$$z(0) = 0, \quad z(1) = L. \tag{6.9a,b}$$

In view of the equivalence of (2.11) with (2.23), Equation (6.8) is equivalent to

$$z'(s) = \hat{\nu}\big(G(s) + K, s\big). \tag{6.10}$$

We integrate (6.10) subject to (6.9a) to obtain

$$z(s) = \int_0^s \hat{\nu}\big(G(\xi) + K, \xi\big)\, d\xi. \tag{6.11}$$

The boundary-value problem (6.8), (6.9) has this z as a solution provided K can be chosen so that (6.11) satisfies (6.9b), i.e., so that

$$\Phi(K) \equiv \int_0^1 \hat{\nu}\big(G(\xi) + K, \xi\big)\, d\xi = L. \tag{6.12}$$

Note that since G is the indefinite integral of the integrable function g, it is absolutely continuous. It follows that the function $\Phi(\cdot)$, just like $\hat{\nu}(\cdot, s)$, strictly increases from 0 to ∞ as its argument increases from $-\infty$ to ∞. Since $L > 0$, we can reproduce the argument justifying the existence of $\hat{\nu}$ to deduce that (6.12) has a unique solution K (depending on G and L). The solution of (6.8), (6.9) is then obtained by substituting this K into (6.11).

Since G and $\hat{\nu}$ are continuous, (6.11) implies that the solution z is continuously differentiable and its derivative is given by (6.10). If g happens to be continuous, then G is continuously differentiable. Since $\hat{\nu}$ is continuously differentiable, (6.10) implies that the solution z is twice continuously differentiable. Now (6.10) is equivalent to (6.8). When g is continuous, we can accordingly differentiate (6.8) to show that z is a classical solution of the ordinary differential equation

$$\tfrac{d}{ds}\hat{N}\big(z'(s), s\big) + g(s) = 0. \tag{6.13}$$

(The regularity theory of this paragraph is called a *bootstrap* argument.) We summarize our results:

6.14. Theorem. *Let $\hat{N}$ be continuously differentiable on $(0,\infty)\times[0,1]$, $\hat{N}_\nu(\nu, s) > 0$ for all ν and s, $\hat{N}(\nu, s) \to \infty$ as $\nu \to \infty$, and $\hat{N}(\nu, s) \to -\infty$ as $\nu \to 0$. Let g be Lebesgue-integrable. Then (6.8), (6.9) has a unique solution z, which is continuously differentiable. If g is continuous, then z is twice continuously differentiable and satisfies (6.13).*

Note that (6.10) implies that the stretch z' is constant if the material is uniform, i.e., if $\hat{N}_s = 0$ and if G is constant, i.e., if $g = 0$.

Equation (6.8) and its equivalent, (6.10), make sense if G is merely integrable. In this case, g is defined as the distributional derivative of $-G$. Our analysis goes through with the solution z continuous by (6.11). The only trouble with such a solution lies in its mechanical interpretation: Our proof in Sec. 3 that $\boldsymbol{n}^+ = \boldsymbol{n}^-$ is no longer applicable. Much of the difficulty with this question evaporates if the distribution g were to equal an integrable function a.e. In particular, if G were a Heaviside (i.e., a step) function, then g would be a Dirac delta, and our problem, which would be solvable, would also make mechanical sense.

Note that the unique solution of (6.8), (6.9) may well represent a compressed straight state. This certainly occurs if $g = 0$ and $L < 1$. Such a solution should certainly be unstable under any reasonable physical criterion. (It is unique only among all straight equilibrium states.) This solution is nevertheless worthy of study because its equations are exactly those for the straight equilibrium state of a naturally straight rod, whose bending stiffness allows it to sustain a certain amount of compression without losing stability. A knowledge of the properties of the straight states of a straight rod is necessary for the study of its buckling from that state.

Note that (6.5) and (6.6) may admit straight folded solutions in which z is not increasing. These can have a very complicated structure (see Reeken (1984a) and the treatments of Chaps. III and VI). These solutions are not accounted for by Theorem 6.14.

7. Purely Transverse Motions

The ad hoc assumption that the motion of each material point is confined to a plane perpendicular to the line joining the ends of the string, frequently used in textbook derivations of the equations of motion of strings and discussed in Sec. 1, motivates our study in this section of conditions under which such special motions can occur.

J. B. Keller (1959) and B. Fleishman (1959) independently observed that if an elastic string has a constitutive equation of the form

$$\hat{N}(\nu, s) = (EA)(s)\nu \tag{7.1}$$

where EA is a given positive-valued function, then the equations of motion (2.9), (2.10b), and (2.11) reduce to the special form

$$[(EA)(s)\boldsymbol{r}_s(s,t)]_s + \boldsymbol{f}(s,t) = (\rho A)(s)\boldsymbol{r}_{tt}(s,t). \tag{7.2}$$

(The ungainly symbol EA is used because it roughly conforms to traditional engineering notation. See the discussion of the notation ρA in Sec. 2.) If $\boldsymbol{f}$ does not depend on $\boldsymbol{r}$ through a relation such as (2.6), then (7.2) is a system of three uncoupled nonhomogeneous wave equations. In particular, if f satisfies

$$\boldsymbol{k} \cdot \boldsymbol{f}(s,t) = -G'(s) \tag{7.3}$$

and if the initial data satisfy

$$\boldsymbol{k} \cdot \boldsymbol{u} = z, \quad \boldsymbol{k} \cdot \boldsymbol{v} = 0 \tag{7.4}$$

where z is the unique solution of (6.8) and (6.9), then the initial-boundary-value problem consisting of (7.2), (2.3), and (2.4) has a unique solution with $\boldsymbol{k} \cdot \boldsymbol{r} = z$. (The existence and uniqueness of such a solution, under mild conditions on the data, follows from the theory of partial differential equations.) This solution describes a purely transverse motion.

Of course, (7.1) satisfies neither (2.19b) nor (2.20). Keller noted, however, that (7.1) could closely approximate the behavior of certain rubber strings when ν is large. This observation does not imply that the motion of a string satisfying a constitutive equation close to (7.1) is close to the motion given by (7.2), because a small nonlinear coupling can shift energy from one mode to another, as is well known in rigid-body mechanics. In particular, even if (7.3) and (7.4) hold, a string with a constitutive equation close to (7.1) could undergo motions with a significant longitudinal component.

We now address the converse problem of determining what restrictions are imposed on the constitutive functions by the assumption that the string must execute a nontrivial purely transverse motion with

$$\boldsymbol{r}(s,t) \cdot \boldsymbol{k} = z(s), \quad (\boldsymbol{r} \cdot \boldsymbol{i})^2 + (\boldsymbol{r} \cdot \boldsymbol{j})^2 \neq 0 \tag{7.5a,b}$$

for every $\boldsymbol{f}$ satisfying (7.3) and for all initial conditions satisfying (7.4) when z satisfies (6.8) and (6.9) and when ν lies in a certain interval (ν^-, ν^+) in $(0, \infty)$ with $\nu^- \le \min z'$. The substitution of (7.5a) and (2.10b) into the $\boldsymbol{k}$-component of (2.9) yields

$$\left[\hat{N}(\nu(s,t),s)\frac{z'(s)}{\nu(s,t)}\right]_s - G'(s) = 0, \tag{7.6}$$

so that

$$\Omega(\nu(s,t),s) \equiv \frac{\hat{N}(\nu(s,t),s)z'(s)}{\nu(s,t)} - G(s) = \Omega(\nu(0,t),0). \tag{7.7}$$

The theory of initial-value problems for quasilinear partial differential equations, applied to the full system of governing equations, says that in a small neighborhood of the initial time, solutions depend continuously on smooth initial data. Thus smooth initial data satisfy (7.7) for $t = 0$. For any fixed s we can prescribe the initial data $\nu(s,0)$ and $\nu(0,0)$ arbitrarily in (ν^-, ν^+). Thus from (7.7) at $t = 0$ we conclude that Ω is a constant function. This constancy of Ω ensures that $\hat{N}(\cdot, s)$, restricted to (ν^-, ν^+), has the form (7.1). We summarize this argument, a modified version of that of J. B. Keller (1959):

7.8. Theorem. *Let* (7.3) *and* (7.4) *hold. If every solution of* (2.9), (2.10b), (2.11), (2.3), (2.4) *for which* $\nu^- < \nu < \nu^+$ *is purely transverse, i.e., satisfies* (7.5), *then* $\hat{N}(\cdot, s)$ *restricted to* (ν^-, ν^+) *has the form* (7.1).

7.9. Problem. Let (7.3) and (7.4) hold. Suppose that Ω is independent of s and that the initial-boundary-value problem admits *a* nontrivial purely transverse motion satisfying (7.5). What restrictions are thereby imposed on $\hat{N}$?

The following exercise, proposed by J. M. Greenberg, also indicates the role played by linear, or more generally, affine constitutive relations.

7.10. Exercise. Consider the free motion of a uniform, nonlinearly elastic string of doubly infinite length. Thus $\boldsymbol{f} = \boldsymbol{0}$, $\hat{N}_s = 0$, $s \in (-\infty, \infty)$. A solution $\boldsymbol{r}$ of the governing equations is called a *travelling wave* iff it has the form

$$\boldsymbol{r}(s,t) = \boldsymbol{p}(s - ct) \tag{7.11}$$

where c is a real number. Show that if there is no nonempty open interval of $(0, \infty)$ on which $\hat{N}$ is affine, then the travelling waves in a string have very special and uninteresting forms. Determine those forms. (In Sec. IX.3 we shall see that the equations for rods have a very rich collection of travelling waves.)

8. Perturbation Methods and the Linear Wave Equation

In Sec. 6 we proved the existence of a unique straight equilibrium configuration $z\boldsymbol{k}$ for an elastic string when $\boldsymbol{f} = g(s)\boldsymbol{k}$. In this section we study the motion of an elastic string near this equilibrium state by formal perturbation methods. We first outline their application to the initial-boundary-value problem (2.9), (2.10b), (2.11), (2.3), (2.4) and then give a detailed treatment of time-periodic solutions. We discuss the validity of the perturbation methods in the next section.

We begin by studying the initial-boundary-value problem when the data are close to those yielding the straight equilibrium state: Let ε represent a small real parameter and let the data have the form

$$\boldsymbol{u}(s) = z(s)\boldsymbol{k} + \varepsilon\boldsymbol{u}_1(s), \quad \boldsymbol{v}(s) = \varepsilon\boldsymbol{v}_1(s), \quad \boldsymbol{f}(s,t) = g(s)\boldsymbol{k} + \varepsilon\boldsymbol{f}_1(s,t). \tag{8.1}$$

We suppose that $\hat{N}(\cdot, s)$ is $(p+1)$-times continuously differentiable. We seek formal solutions of the initial-boundary-value problem whose dependence on the parameter ε is specified by a representation of the form

$$\boldsymbol{r}(s,t,\varepsilon) = z(s)\boldsymbol{k} + \sum_{k=1}^{p} \frac{\varepsilon^k}{k!}\boldsymbol{r}_k(s,t) + o(\varepsilon^p). \tag{8.2}$$

Since (8.2) implies that

$$\boldsymbol{r}_k(s,t) = \left.\frac{\partial^k \boldsymbol{r}(s,t,\varepsilon)}{\partial \varepsilon^k}\right|_{\varepsilon=0}, \tag{8.3}$$

we can find the problem formally satisfied by $\boldsymbol{r}_k$ by substituting $\boldsymbol{r}(s,t,\varepsilon)$ into the equations of the nonlinear problem, differentiating the resulting equations k times with respect to ε, and then setting $\varepsilon = 0$. We find that the equation for $\boldsymbol{r}_k$ is linear and involves $\boldsymbol{r}_1, \dots, \boldsymbol{r}_{k-1}$; thus the equations for $\boldsymbol{r}_1, \dots, \boldsymbol{r}_p$ can be solved successively.

To compute these equations vectorially, we define

$$\hat{\boldsymbol{n}}(\boldsymbol{q}, s) \equiv \hat{N}(|\boldsymbol{q}|, s)\boldsymbol{q}|\boldsymbol{q}|^{-1}. \tag{8.4}$$

Thus (2.9), (2.10b), (2.11) has the form

$$\hat{\boldsymbol{n}}(\boldsymbol{r}_s, s)_s + \boldsymbol{f} = \rho A \boldsymbol{r}_{tt}. \tag{8.5}$$

We use the definition of Gâteaux derivative following (I.4.2) to obtain

$$\hat{\boldsymbol{n}}_{\boldsymbol{q}}(\boldsymbol{q}, s)\cdot \boldsymbol{c} = \hat{N}_\nu(|\boldsymbol{q}|, s)\frac{\boldsymbol{q}\,\boldsymbol{q}\cdot\boldsymbol{c}}{|\boldsymbol{q}|^2} + \frac{\hat{N}(|\boldsymbol{q}|, s)}{|\boldsymbol{q}|}\left[\boldsymbol{c} - \frac{\boldsymbol{q}\,\boldsymbol{q}\cdot\boldsymbol{c}}{|\boldsymbol{q}|^2}\right]. \tag{8.6}$$

(To differentiate $|\boldsymbol{q}|$ with respect to $\boldsymbol{q}$, we write it as $\sqrt{\boldsymbol{q}\cdot\boldsymbol{q}}$ so that $\partial\sqrt{\boldsymbol{q}\cdot\boldsymbol{q}}/\partial\boldsymbol{q} = \boldsymbol{q}/\sqrt{\boldsymbol{q}\cdot\boldsymbol{q}}$.) Thus we find that

$$\partial_\varepsilon \hat{\boldsymbol{n}}(\boldsymbol{r}_s(s,t,\varepsilon), s)|_{\varepsilon=0} = \hat{\boldsymbol{n}}_{\boldsymbol{q}}(z'(s)\boldsymbol{k}, s)\cdot \partial_s \boldsymbol{r}_1(s,t). \tag{8.7}$$

Note that $\hat{\boldsymbol{n}}_{\boldsymbol{q}}$ is symmetric.

Differentiating (8.5) once with respect to ε and using (8.6) and (8.7) we reduce the equation for $\boldsymbol{r}_1$ to

$$(\boldsymbol{L}\cdot\boldsymbol{r}_1)(s,t) = \boldsymbol{f}_1(s,t) \tag{8.8a}$$

where the vector-valued partial differential operator $\boldsymbol{L}$ is defined by

$$\begin{aligned}(\boldsymbol{L}\cdot\boldsymbol{r})(s,t) &\equiv (\rho A)(s)\boldsymbol{r}_{tt}(s,t)\\ &- \left\{\frac{N^0(s)}{z'(s)}[\boldsymbol{r}_s(s,t)\cdot\boldsymbol{i}\,\boldsymbol{i} + \boldsymbol{r}_s(s,t)\cdot\boldsymbol{j}\,\boldsymbol{j}]_s + N^0_\nu(s)\boldsymbol{r}_s(s,t)\cdot\boldsymbol{k}\,\boldsymbol{k}\right\}_s\end{aligned} \tag{8.8b}$$

with

$$N^0(s) \equiv \hat{N}(z'(s), s), \quad N_\nu^0(s) \equiv \hat{N}_\nu(z'(s), s). \tag{8.9}$$

We use an analogous notation for higher derivatives. Note that the components of (8.8) in the $\boldsymbol{i}$-, $\boldsymbol{j}$-, and $\boldsymbol{k}$-directions uncouple into three scalar wave equations.

$\boldsymbol{r}_1$ must satisfy the boundary conditions

$$\boldsymbol{r}_1(0,t) = \boldsymbol{0}, \quad \boldsymbol{r}_1(1,t) = \boldsymbol{0} \tag{8.10}$$

and the initial conditions

$$\boldsymbol{r}_1(s,0) = \boldsymbol{u}_1(s), \quad \partial_t \boldsymbol{r}_1(s,0) = \boldsymbol{v}_1(s). \tag{8.11}$$

The component $\boldsymbol{r}_1 \cdot \boldsymbol{k}$ satisfies the following wave equation obtained by dotting (8.8) with $\boldsymbol{k}$:

$$(\rho A)(s) w_{tt}(s,t) - [N_\nu^0(s) w_s(s,t)]_s = \boldsymbol{f}_1(s,t) \cdot \boldsymbol{k}. \tag{8.12}$$

We can simplify the equations for the other two components of (8.8) by introducing the change of variable

$$\zeta = z(s) \quad \text{or, equivalently,} \quad s = \tilde{s}(\zeta) \tag{8.13}$$

where $\tilde{s}$ is the inverse of z, which exists by virtue of the positivity of z'. We set

$$\tilde{\boldsymbol{r}}_1(\zeta,t) \equiv \boldsymbol{r}_1(\tilde{s}(\zeta),t), \quad \overline{\rho A}(\zeta) \equiv \frac{(\rho A)(\tilde{s}(\zeta))}{z'(\tilde{s}(\zeta))}. \tag{8.14}$$

Then $\tilde{\boldsymbol{r}}_1 \cdot \boldsymbol{i}$ and $\tilde{\boldsymbol{r}}_1 \cdot \boldsymbol{j}$ satisfy

$$\overline{\rho A}(\zeta) u_{tt}(\zeta,t) - [N^0(\tilde{s}(\zeta)) u_\zeta(\zeta,t)]_\zeta = h(\zeta,t) \tag{8.15}$$

where $h(\zeta,t)$ respectively equals $\boldsymbol{f}_1(\tilde{s}(\zeta),t) \cdot \boldsymbol{i}$ and $\boldsymbol{f}_1(\tilde{s}(\zeta),t) \cdot \boldsymbol{j}$. Note that (6.8) and (8.9) imply that

$$N^0(\tilde{s}(\zeta)) = K + G(\tilde{s}(\zeta)) \tag{8.16}$$

where K satisfies (6.12). $\overline{\rho A}$ is the mass per unit length in the configuration $z\boldsymbol{k}$. The change of variables (8.13) and (8.14) is tantamount to taking the stretched equilibrium configuration $z\boldsymbol{k}$ as the reference configuration.

Equation (8.12) describes the small longitudinal motion of a string (or of a rod) about its straight stretched equilibrium state. The nonuniformity of the string and the presence of G cause the coefficients to depend on s. A slight strengthening of assumption (2.17) ensures that N_ν^0 is positive and that (8.12) is consequently hyperbolic.

Equation (8.15) describes the small transverse vibrations of the string. $N^0(\tilde{s}(\zeta))$ is the tension at $\tilde{s}(\zeta)$ in the configuration $z\boldsymbol{k}$. If $G = 0$, Eq. (8.16) implies that this tension is constant whether or not the string is uniform. Under the hypotheses (2.17) and (2.20), Eq. (6.13) implies that for $G = 0$ this constant tension is positive if and only if $L > 1$. Where this tension (constant or not) is positive, (8.15) is hyperbolic, and where the tension is negative, (8.15) is elliptic. In the latter case, expected physical instabilities are reflected by the ill-posedness of initial-boundary-value problems for (8.15). Analogous statements apply to the full nonlinear system. By endowing the string with resistance to bending and twisting (i.e., by replacing the string theory with a rod theory), we remove this ill-posedness at the cost of enlarging the system. See Chaps. IV and VIII.

8.17. Exercise. Find the linearized equations (satisfied by $\boldsymbol{r}_1$) for the motion of a viscoelastic string satisfying (2.14). Classify the equations as to type.

8.18. Exercise. Suppose that $L > 1$ and that $\boldsymbol{v} = \boldsymbol{0}$ and $\boldsymbol{f} = \boldsymbol{0}$. Find equations for $\boldsymbol{r}_1$, $\boldsymbol{r}_2$, $\boldsymbol{r}_3$ for the perturbation solution for (2.9), (2.10b), (2.11), (2.3), (2.4). If $\boldsymbol{u}_1 \cdot \boldsymbol{k} = 0$, how do these equations illuminate the role of purely transverse motions discussed in Sec. 7?

We now turn to the more interesting problem of determining the properties of free time-periodic motions of an elastic string near the straight equilibrium state. We seek motions satisfying (8.5) with $\boldsymbol{f}(s,t) = g(s)\boldsymbol{k}$, satisfying (2.3), and having an as yet undetermined period $2\pi/\sqrt{\lambda}$ with $\lambda > 0$ so that

$$\boldsymbol{r}(s, t + 2\pi/\sqrt{\lambda}) = \boldsymbol{r}(s,t). \tag{8.19}$$

Let us set $\bar{t} = \sqrt{\lambda}t$, $\bar{\boldsymbol{r}}(s,\bar{t}) = \boldsymbol{r}(s, \bar{t}/\sqrt{\lambda})$, introduce these variables into the governing equations, and then omit the superposed bars. In this case, (8.5) is modified by having λ precede ρA. Equation (8.19) reduces to

$$\boldsymbol{r}(s, t + 2\pi) = \boldsymbol{r}(s,t). \tag{8.20}$$

Note that a small parameter ε is not supplied in this problem. We may think of it as an amplitude characterizing the departure of time-periodic solutions from the trivial straight equilibrium state.

The following exercise shows that we cannot attack this problem by blindly following the approach used for the initial-boundary-value problem.

8.21. Exercise. Substitute (8.2) into the problem for time-periodic solutions. Find the frequencies λ for which the problem for $\boldsymbol{r}_1$ has solutions of period 2π in time. Show that if the corrections $\boldsymbol{r}_2$ and $\boldsymbol{r}_3$ have the same frequencies, then $\hat{N}$ is subjected to unduly severe restrictions.

We circumvent this difficulty by allowing λ also to depend on ε. The dependence of frequency on the amplitude thereby permitted is a typical physically important manifestation of nonlinearity. We accordingly supplement (8.2) with

$$\lambda(\varepsilon) = \omega^2 + \sum_{k=1}^{p} \frac{\varepsilon^k}{k!}\lambda_k + o(\varepsilon^{p+1}). \tag{8.22}$$

Equations (8.2) and (8.22) give a parametric representation (i.e., a curve) for the configuration and the frequency in a neighborhood of the trivial state. We obtain the equations satisfied by $\boldsymbol{r}_k$ and λ_{k-1} by substituting (8.2) and (8.22) into the governing equations, differentiating them k times with respect to ε, and then setting $\varepsilon = 0$. We find that $\boldsymbol{r}_k$ satisfies the boundary conditions and periodicity conditions

(8.23a,b) $$\boldsymbol{r}_k(0,t) = \boldsymbol{0} = \boldsymbol{r}_k(1,t), \quad \boldsymbol{r}_k(s,t+2\pi) = \boldsymbol{r}_k(s,t).$$

$\boldsymbol{r}_1$ satisfies

(8.24) $$\boldsymbol{L}(\omega^2)\cdot\boldsymbol{r}_1 = \boldsymbol{0}$$

where $\boldsymbol{L}(\lambda)$ is defined by (8.8b) with ρA replaced with $\lambda\rho A$. We assume that system (8.24) is hyperbolic, i.e., we assume that N^0 is everywhere positive. We can solve (8.23), (8.24) for $\boldsymbol{r}_1$ by separation of variables. We find that nontrivial solutions $\boldsymbol{r}_1$ have the form

(8.25a,b) $$\boldsymbol{r}_1(s,t) = \begin{cases} u_l(s)[\boldsymbol{a}_{lm}\cos mt + \boldsymbol{b}_{lm}\sin mt] & \text{when} \quad \omega^2 = \sigma_l^2/m^2, \\ w_l(s)[\alpha_{lm}\cos mt + \beta_{lm}\sin mt]\boldsymbol{k} & \text{when} \quad \omega^2 = \tau_l^2/m^2, \end{cases}$$
$$l = 0,1,2,\dots, \quad m = 1,2,\dots,$$

where the $\{\boldsymbol{a}_{lm}\}$ and the $\{\boldsymbol{b}_{lm}\}$ are arbitrary vectors in span$\{\boldsymbol{i},\boldsymbol{j}\}$, where the $\{\sigma_l^2\}$ are the eigenvalues and $\{u_l\}$ are the corresponding eigenfunctions of the Sturm-Liouville problem

(8.26) $$\frac{d}{ds}\left[\frac{N^0(s)u'}{z'(s)}\right] + \sigma^2(\rho A)(s)u = 0, \quad u(0) = 0 = u(1),$$

where the $\{\alpha_{lm}\}$ and the $\{\beta_{lm}\}$ are arbitrary real numbers, and where the $\{\tau_l^2\}$ are the eigenvalues and the $\{w_l\}$ are the corresponding eigenfunctions of the Sturm-Liouville problem

(8.27) $$\frac{d}{ds}\left[N_\nu^0(s)w'\right] + \tau^2(\rho A)(s)w = 0, \quad w(0) = 0 = w(1).$$

We normalize $\{u_l\}$ by the requirement that

(8.28) $$\int_0^1 (\rho A)(s)u_l(s)u_n(s)\,ds = \delta_{ln} \equiv \begin{cases} 1 & \text{if} \quad l = n, \\ 0 & \text{if} \quad l \neq n, \end{cases}, \quad u_l'(0) > 0$$

and adopt the same conditions for $\{w_l\}$. δ_{ln} is the *Kronecker delta*. The positivity everywhere of N^0 and z' ensures that $0 < \sigma_0^2 < \sigma_1^2 < \cdots$ and that $\sigma_l^2 \to \infty$ as $l \to \infty$ by the Sturm-Liouville theory (see Ince (1926) or Coddington & Levinson (1955), e.g.). The positivity everywhere of N_ν^0 ensures that $\{\tau_l^2\}$ has the same properties. The representations of (8.25a,b) respectively correspond to transverse and longitudinal motions. Since (8.23) with $k = 1$ and (8.24) are invariant under translations of time and under rotations about the $\boldsymbol{k}$-axis, we could without loss of generality fix an appropriate two of the four components of $\boldsymbol{a}_{lm}$ and $\boldsymbol{b}_{lm}$ appearing in (8.25a).

Note that for fixed $\omega^2 = \sigma_l^2/m^2$, there are as many different solutions of the form (8.25) as there are distinct pairs (j,p) of integers, $j = 1,2,\dots, p = 0,1,2,\dots$, satisfying

(8.29a,b) $$\frac{\sigma_l^2}{m^2} = \frac{\sigma_j^2}{p^2} \quad \text{or} \quad \frac{\sigma_l^2}{m^2} = \frac{\tau_j^2}{p^2}.$$

(Note that the special condition that $\hat{N}(\cdot,s)$ be linear, discussed in Sec. 7, ensures that $\sigma_l^2 = \tau_l^2$ for all l.) If N^0/z' and ρA are constant, then there are infinitely many pairs (j,p) satisfying (8.29a).

Suppose that $\omega^2 = \sigma_l^2/m^2$ and that there are no pairs of integers (j,p) such that (8.29b) holds. In this case we find that $\boldsymbol{r}_1\cdot\boldsymbol{k} = 0$. Then the perturbation procedure yields

(8.30) $$\boldsymbol{L}(\sigma_l^2/m^2)\cdot\boldsymbol{r}_2 = 2\lambda_1\rho A\partial_{tt}\boldsymbol{r}_1 + \partial_s\left\{\left[\frac{N^0 - z'N_\nu^0}{(z')^2}\right]\partial_s\boldsymbol{r}_1\cdot\partial_s\boldsymbol{r}_1\right\}\boldsymbol{k}.$$

Before blindly lurching toward a solution of (8.30), it is useful to take a preliminary step that can greatly simplify the analysis: We take the dot product of (8.30) with $\boldsymbol{r}_1$ of (8.25) and then integrate the resulting expression by parts twice over $[0,1]\times[0,2\pi]$. Since $\boldsymbol{r}_1$ satisfies the homogeneous equation and since $\boldsymbol{r}_1$ has no $\boldsymbol{k}$-component, the resulting equation reduces to

$$\lambda_1 = 0. \tag{8.31}$$

Thus the equations for the $\boldsymbol{i}$- and $\boldsymbol{j}$-components of $\boldsymbol{r}_2$ are exactly the same as those for these components of $\boldsymbol{r}_1$. It follows that the contribution of these terms of $\boldsymbol{r}_2$ to (8.2) has exactly the same form as the corresponding components of $\boldsymbol{r}_1$, but with a coefficient of $\varepsilon^2/2$ in place of ε. We accordingly absorb $\boldsymbol{r}_2$ into $\boldsymbol{r}_1$ by taking

$$\boldsymbol{r}_2 \cdot \boldsymbol{i} = 0 = \boldsymbol{r}_2 \cdot \boldsymbol{j}, \tag{8.32}$$

In view of (8.31) and (8.32), problem (8.30) reduces to a nonhomogeneous linear equation for $\boldsymbol{r}_2 \cdot \boldsymbol{k}$:

$$\begin{aligned} &\frac{\sigma_l^2}{m^2}\rho A \partial_{tt}\boldsymbol{r}_2 \cdot \boldsymbol{k} - \partial_s[N_\nu^0 \partial_s \boldsymbol{r}_2 \cdot \boldsymbol{k}] \\ &= \tfrac{1}{2}\partial_s\left[\frac{N^0 - z' N_\nu^0}{(z')^2}(u_l')^2\right]\left[|\boldsymbol{a}_{lm}|^2 + |\boldsymbol{b}_{lm}|^2 + (|\boldsymbol{a}_{lm}|^2 - |\boldsymbol{b}_{lm}|^2)\cos 2mt \right. \\ &\qquad\qquad \left. + \boldsymbol{a}_{lm}\cdot \boldsymbol{b}_{lm} \sin 2mt\right]. \end{aligned} \tag{8.33}$$

Since we know that the homogeneous problem for (8.33) has only the trivial solution, we can seek a solution in the form $f(s) + g(s)\cos 2mt + h(s)\sin 2mt$ and obtain boundary-value problems for f, g, and h like (8.27). The solutions of these boundary-value problems can be represented in terms of a Green function associated with the operator of (8.27) or alternatively by an expansion in terms of the eigenfunctions associated with (8.27). We can also represent the solution of (8.33) directly as an eigenfunction expansion with respect to the basis

$$\{(s,t) \mapsto \tfrac{1}{\pi} w_q(s)\cos nt,\ \tfrac{1}{\pi} w_q(s)\sin nt, \quad q = 0,1,\dots,\ n = 1,2,\dots\}. \tag{8.34}$$

We find the Fourier coefficients of $\boldsymbol{k}\cdot\boldsymbol{r}_2$ by multiplying (8.33) by a member of (8.34) and integrating the resulting equation by parts over $[0,1]\times[0,2\pi]$. (See Stakgold (1979), e.g.) We get

$$\begin{aligned} &\boldsymbol{k}\cdot\boldsymbol{r}_2(s,t) \\ &= \frac{m^2}{2\pi}\sum_{q=0}^{\infty} \mu_{lq} w_q(s)\left[(|\boldsymbol{a}_{lm}|^2 - |\boldsymbol{b}_{lm}|^2)\cos 2mt + (\boldsymbol{a}_{lm}\cdot\boldsymbol{b}_{lm})\sin 2mt\right], \end{aligned} \tag{8.35a}$$

$$\left(\tfrac{1}{4}\tau_q^2 - \sigma_l^2\right)\mu_{lq} \equiv \int_0^1 \partial_s\left\{\frac{N^0(s) - z'(s)N_\nu^0(s)}{z'(s)^2}u_l'(s)^2\right\} w_l(s)\,ds \tag{8.35b}$$

when (8.25a) holds and when $\tau_q^2 \neq 4\sigma_l^2$ for each q. The properties of $\{\tau_q^2\}$ developed in Sturm-Liouville theory ensure that (8.35a) converges. Equation (8.35) shows that the first correction to the purely transverse linear motion is a longitudinal motion.

Using (8.25), (8.31), and (8.32) we find that

$$\begin{aligned} &\boldsymbol{L}(\sigma_l^2/m^2)\cdot\boldsymbol{r}_3 \\ &= 3\lambda_2\rho A\partial_{tt}\boldsymbol{r}_1 + 3\partial_s\left\{(N^0 - z'N_\nu^0)\frac{[\partial_s\boldsymbol{r}_1\cdot\partial_s\boldsymbol{r}_1 + (\boldsymbol{k}\cdot\partial_s\boldsymbol{r}_2)z']\partial_s\boldsymbol{r}_1}{(z')^3}\right\}. \end{aligned} \tag{8.36}$$

We treat this equation just like (8.30): We dot it with $\boldsymbol{r}_1$ and integrate the resulting equation by parts over $[0,1]\times[0,2\pi]$ to get

$$\begin{aligned} &m^2\pi(|\boldsymbol{a}_{lm}|^2 + |\boldsymbol{b}_{lm}|^2)\lambda_2 \\ &= \int_0^1\int_0^{2\pi} \partial_s\left\{(N^0 - z'N_\nu^0)\frac{[\partial_s\boldsymbol{r}_1\cdot\partial_s\boldsymbol{r}_1 + (\boldsymbol{k}\cdot\partial_s\boldsymbol{r}_2)z']\partial_s\boldsymbol{r}_1}{(z')^3}\right\}\cdot\partial_s\boldsymbol{r}_1\,dt\,ds. \end{aligned} \tag{8.37}$$

In view of (8.22) the sign of this expression for λ_2 gives the important physical information of whether the frequency λ increases or decreases with the amplitude ε of the motion. Note how λ_2 depends crucially on the behavior of the constitutive function $\hat{N}$. The procedures we have used in this analysis are quite general.

8.38. Exercise. Obtain an explicit representation for λ_2 when the material is uniform, so that $\hat{N}_s = 0$ and z' and ρA are constant.

A computation analogous to that leading to (8.37) can be carried out for the purely longitudinal motion. But the results are purely formal because it can be shown that no purely longitudinal periodic motion is possible (see Keller & Ting (1966) and Lax (1964)). The solutions must exhibit shocks. For the transverse motions (which have a longitudinal component as we have seen), periodic solutions are possible. (The energy could be shifted about and avoid being concentrated. We give a transparent example of such a phenomenon in Sec. XIII.14.) This possibility of shocks makes it hard to justify the method and to interpret the results. The formal results clearly say something important about the nonlinear system, but it is difficult to give a mathematically precise and physically illuminating explanation of exactly what is being said. In other words, it is not clear what the linear wave equations say about solutions of the nonlinear equations. By introducing a strong dissipative mechanism, corresponding to (2.14) subject to a strengthened version of (2.22), it is likely that we could prevent our equations from having shocks. But this dissipation would prevent periodic solutions unless we introduced periodic forcing. The resulting perturbation scheme would be more complicated, but there is some hope that the approach could be justified. It is physically attractive but notoriously difficult to study the undamped system by taking the limit as the dissipation goes to zero. We comment on related questions at the end of Sec. 11.

Carrier (1945, 1949) used such perturbation methods to study periodic planar vibrations of an elastic string for which $\hat{N}(\cdot, s)$ is taken to be affine, although this restriction is inessential. The work of this section is largely based on Keller & Ting (1966) and J. B. Keller (1968). For other applications of this formalism, see Millman & Keller (1969) and Iooss & Joseph (1990).

9. The Justification of Perturbation Methods

In this section we give precise conditions justifying perturbation methods for static problems. The fundamental mathematical tool for our analysis is the Implicit Function Theorem in different manifestations. The difficulties that would attend the justification of perturbation methods for dynamical problems are touched on at the end of the last section. Our basic result is

9.1. Theorem. *Let z be as in Sec. 6. If $\hat{N}$ is continuous, if $\hat{N}_\nu(z'(s), s) > 0$ and $\hat{N}(z'(s), s) > 0$ for each s, if $\hat{N}(\cdot, s) \in C^{p+1}(0, \infty)$, if $g \in C^0[0,1]$, and if $\boldsymbol{f}_1 \in C^0[0,1]$, then there is a number $\eta > 0$ such that for $|\varepsilon| < \eta$ the boundary-value problem*

$$\text{(9.2)} \qquad \frac{d}{ds}\left[\hat{N}\left(|\boldsymbol{r}'(s)|, s\right)\frac{\boldsymbol{r}'(s)}{|\boldsymbol{r}'(s)|}\right] + g(s)\boldsymbol{k} + \varepsilon \boldsymbol{f}_1(s) = \boldsymbol{0},$$

$$\text{(9.3a,b)} \qquad \boldsymbol{r}(0) = \boldsymbol{0}, \quad \boldsymbol{r}(1) = L\boldsymbol{k}$$

has a unique solution $\boldsymbol{r}(\cdot, \varepsilon)$ with $\boldsymbol{r}(\cdot,\varepsilon) \in C^2[0,1]$ and $\boldsymbol{r}(s, \cdot) \in C^{p+1}(-\eta, \eta)$. (Thus $\boldsymbol{r}(s, \varepsilon)$ admits an expansion like (8.2).)

Proof. From (6.5) with $a = 0$ and from (8.4) we get

$$\text{(9.4)} \qquad \hat{\boldsymbol{n}}\left(\boldsymbol{r}'(s), s\right) - \hat{\boldsymbol{n}}\left(\boldsymbol{r}'(0), 0\right) - G(s)\boldsymbol{k} + \varepsilon \int_0^s \boldsymbol{f}_1(\xi)\, d\xi = \boldsymbol{0},$$

which can be obtained from the integration of (9.2). From (8.6) we obtain

$$\mathbf{c} \cdot \hat{\boldsymbol{n}}_{\boldsymbol{q}}(z'(s)\boldsymbol{k}, s) \cdot \mathbf{c} = N_\nu^0(s)(\boldsymbol{k} \cdot \mathbf{c})^2 + \frac{N^0(s)}{z'(s)} \left[\mathbf{c} \cdot \mathbf{c} - (\boldsymbol{k} \cdot \mathbf{c})^2\right]. \tag{9.5}$$

Thus $\hat{\boldsymbol{n}}_{\boldsymbol{q}}(z'(s)\boldsymbol{k}, s)$ is positive-definite and therefore nonsingular. The classical Implicit Function Theorem thus implies that for $\boldsymbol{q}$ near $z'(s)\boldsymbol{k}$, the function $\boldsymbol{q} \mapsto \hat{\boldsymbol{n}}(\boldsymbol{q}, s)$ has an inverse, which we denote by $\boldsymbol{n} \mapsto \boldsymbol{m}(\boldsymbol{n}, s)$. We use it to solve (9.4) for $\boldsymbol{r}'(s)$. We integrate the resulting equation from 0 to s subject to (9.3a) to obtain

$$\boldsymbol{r}(s) = \int_0^s \boldsymbol{m}\left(\hat{\boldsymbol{n}}(\boldsymbol{r}'(0), 0) + G(\xi)\boldsymbol{k} - \varepsilon \int_0^\xi \boldsymbol{f}_1(\sigma)\, d\sigma,\ \xi\right) d\xi. \tag{9.6}$$

The requirement that (9.6) satisfy (9.3b) yields

$$\begin{aligned} &\boldsymbol{l}(\boldsymbol{r}'(0), \varepsilon) \\ &\quad \equiv \int_0^1 \boldsymbol{m}\left(\hat{\boldsymbol{n}}(\boldsymbol{r}'(0), 0) + G(\xi)\boldsymbol{k} - \varepsilon \int_0^\xi \boldsymbol{f}_1(\sigma)\, d\sigma,\ \xi\right) d\xi = L\boldsymbol{k}. \end{aligned} \tag{9.7}$$

If there is a unique solution $\boldsymbol{r}'(0) = \boldsymbol{p}(\varepsilon)$ of this equation, then its substitution for $\boldsymbol{r}'(0)$ in (9.6) yields the solution $\boldsymbol{r}(\cdot, \varepsilon)$ of (9.2), (9.3). We now verify that (9.7) meets the hypotheses of the classical Implicit Function Theorem: First of all, we must show that

$$\boldsymbol{l}(z'(0)\boldsymbol{k}, 0) = L\boldsymbol{k}. \tag{9.8}$$

But this is equivalent to (6.12) because (8.4) and (6.8) imply that

$$\begin{aligned} \hat{\boldsymbol{n}}\left(z'(0)\boldsymbol{k}, 0\right) + G(\xi)\boldsymbol{k} &= [N(z'(0), 0) + G(\xi)]\boldsymbol{k} = [K + G(\xi)]\boldsymbol{k} \\ &= \hat{N}(z'(\xi), \xi)\boldsymbol{k} = \hat{\boldsymbol{n}}\left(z'(\xi)\boldsymbol{k}, \xi\right). \end{aligned} \tag{9.9a}$$

Next, (9.7) implies that

$$\boldsymbol{l}_{\boldsymbol{p}}(\boldsymbol{p}, 0) = \int_0^1 \boldsymbol{m}_{\boldsymbol{n}}\left(\hat{\boldsymbol{n}}(\boldsymbol{p}, 0) + G(\xi)\boldsymbol{k},\ \xi\right) \cdot \hat{\boldsymbol{n}}_{\boldsymbol{q}}(\boldsymbol{p}, 0)\, d\xi. \tag{9.9b}$$

From (9.9a) we find that

$$\boldsymbol{m}_{\boldsymbol{n}}\left(\hat{\boldsymbol{n}}(z'(0)\boldsymbol{k}, 0) + G(\xi)\boldsymbol{k},\ \xi\right) = \boldsymbol{m}_{\boldsymbol{n}}\left(\hat{N}(z'(\xi), \xi)\boldsymbol{k},\ \xi\right), \tag{9.10}$$

so that (9.10) is the inverse of the symmetric positive-definite tensor $\hat{\boldsymbol{n}}_{\boldsymbol{q}}(z'(\xi)\boldsymbol{k}, \xi)$. It follows that (9.10) is positive-definite. Since the product of two symmetric positive-definite tensors is positive-definite, we find that

$$\boldsymbol{l}_{\boldsymbol{p}}(z'(0)\boldsymbol{k}, 0) \quad \text{is nonsingular.} \tag{9.11}$$

Conditions (9.8) and (9.11) are the requisite hypotheses for the classical Implicit Function Theorem, which says that there is a number $\eta > 0$ such that (9.7) has a unique solution $\boldsymbol{p}(\varepsilon)$ for $|\varepsilon| < \eta$ and that $\boldsymbol{p}(\cdot) \in C^{p+1}(-\eta, \eta)$. It then follows from (9.5) that $\boldsymbol{r}(s, \cdot)$ itself is in this space. The regularity of $\boldsymbol{r}(\cdot, \varepsilon)$ can be read off from (9.6). (It is correspondingly enhanced for increased smoothness of g and $\boldsymbol{f}_1$.) □

Note that this theorem is purely local in the sense that it gives information about solutions of the nonlinear problem (9.2), (9.3) only in a neighborhood of a known solution. In contrast, the elementary analysis of Sec. 6 is global. In Chap. III we shall give global analyses of equilibrium states of strings under several more interesting force systems.

In this proof we have avoided the use of determinants. They are not suitable for proving (9.11) because $\boldsymbol{l_p}(z'(0)\boldsymbol{k}, 0)$ is an integral. If an integrand is a positive-definite tensor everywhere, then its integral is likewise, but if an integrand is merely nonsingular everywhere, then its integral need not be nonsingular.

9.12. Exercise. Prove the last assertion about nonsingular tensors.

The proof of Theorem 9.1 relied on the special nature of (9.2). If $\boldsymbol{f}_1$, say, were to depend upon $\boldsymbol{r}$, then (9.5) would be an integral equation for $\boldsymbol{r}$ and would require a subtler analysis. Procedures for such analyses have been systematized, the most comprehensive methods employing an abstract version of the local Implicit Function Theorem XVIII.1.27 in Banach Space, which is applied in several places in this book. Here we present a related concrete approach, the *Poincaré shooting method,* applicable to systems of ordinary differential equations (more complicated than (9.2)).

Proof of Theorem 9.1 by the Poincaré shooting method. We seek a vector $\boldsymbol{a}$ such that (9.2) subject to the *initial conditions*

$$\boldsymbol{r}(0) = \boldsymbol{0}, \quad \boldsymbol{r}'(0) = \boldsymbol{a} \tag{9.13}$$

has a solution satisfying (9.3b). To apply the basic theory of ordinary differential equations to this problem, it is convenient to write (9.2) as a first-order system in which the derivatives of the unknowns are expressed as functions of the unknowns. This reduction can be effected in two ways: For $\boldsymbol{r}'$ close enough to $z'\boldsymbol{k}$, we can use the tools developed in the above proof of Theorem 9.1 to write (9.2), (9.3) as

$$\boldsymbol{n}' = -g(s)\boldsymbol{k} - \varepsilon \boldsymbol{f}_1(s), \quad \boldsymbol{r}' = \boldsymbol{m}(\boldsymbol{n}, s), \quad \boldsymbol{r}(0) = \boldsymbol{0}, \quad \boldsymbol{n}(0) = \boldsymbol{b} \equiv \hat{\boldsymbol{n}}(\boldsymbol{a}, 0). \tag{9.14}$$

Alternatively, we could carry out the differentiation in (9.2). For $\boldsymbol{r}'$ close enough to $z'\boldsymbol{k}$, we can solve (9.2) for $\boldsymbol{r}''$, obtaining an equation of the form $\boldsymbol{r}'' = \boldsymbol{h}(\boldsymbol{r}', s, \varepsilon)$. We set $\boldsymbol{v} = \boldsymbol{r}'$ and thereby convert this second-order system to the equivalent first-order system

$$\boldsymbol{v}' = \boldsymbol{h}(\boldsymbol{v}, s, \varepsilon), \quad \boldsymbol{r}' = \boldsymbol{v}. \tag{9.15}$$

To be specific, we limit our attention to (9.14). Since the results of Sec. 6 imply that it has a unique solution $\boldsymbol{r} = z\boldsymbol{k}$, $\boldsymbol{n} = \hat{\boldsymbol{n}}(z'(\cdot)\boldsymbol{k}, \cdot))$ for $\varepsilon = 0$ and $\boldsymbol{b} = \hat{\boldsymbol{n}}(z'(0)\boldsymbol{k}, 0))$, the basic theory of ordinary differential equations (see Coddington & Levinson (1955, Chaps. 1,2) or Hale (1969, Chap. 1), e.g.) implies that (9.14) has a unique solution $\boldsymbol{r}(\cdot, \boldsymbol{b}, \varepsilon)$ defined on the whole interval $[0, 1]$ if ε and $\boldsymbol{b}$ are close enough to 0 and $\hat{\boldsymbol{n}}(z'(0)\boldsymbol{k}, 0))$. Moreover, $\boldsymbol{r}(s, \cdot, \cdot)$ is $(p+1)$-times continuously differentiable. $\boldsymbol{r}(\cdot, \boldsymbol{b}, \varepsilon)$ would correspond to a solution of (9.2), (9.3) for small nonzero ε if $\boldsymbol{b}$ can be chosen so that

$$\boldsymbol{r}(1, \boldsymbol{b}, \varepsilon) = L\boldsymbol{k}. \tag{9.16}$$

We know that this system for $\boldsymbol{b}$ has the solution $\boldsymbol{b}_0 \equiv \hat{\boldsymbol{n}}(z'(0)\boldsymbol{k}, 0))$ for $\varepsilon = 0$. The Implicit Function Theorem then implies that there is a number $\eta > 0$ such that (9.16) has a unique solution $(-\eta,\eta) \ni \varepsilon \mapsto \hat{\boldsymbol{b}}(\varepsilon)$ with $\hat{\boldsymbol{b}} \in C^{p+1}(-\eta,\eta)$ and with $\hat{\boldsymbol{b}}(0) = \boldsymbol{b}_0$ provided that

$$\det \boldsymbol{R}(1) \neq 0, \quad \boldsymbol{R}(s) \equiv \frac{\partial \boldsymbol{r}}{\partial \boldsymbol{b}}(s, \boldsymbol{b}_0, 0). \tag{9.17}$$

The theory of ordinary differential equations implies that the matrix $\boldsymbol{R}$ satisfies the initial-value problem obtained by formally differentiating (9.14) with respect to $\boldsymbol{b}$ and then setting $(\boldsymbol{b},\varepsilon) = (\boldsymbol{b}_0, 0)$. This process yields

$$\boldsymbol{n}_b' = \boldsymbol{O}, \quad \boldsymbol{r}_b' = \boldsymbol{m}_n\left(\hat{\boldsymbol{n}}(z'(s)\mathbf{k}, s)\right) \cdot \boldsymbol{n}_b, \quad , \boldsymbol{n}_b(0) = \boldsymbol{I}, \quad \boldsymbol{r}_b(0) = \boldsymbol{O}, \tag{9.18}$$

whence we obtain

$$\boldsymbol{R}' = \boldsymbol{m}_n\left(\hat{\boldsymbol{n}}(z'(s)\mathbf{k}, s)\right), \quad \boldsymbol{R}(0) = \boldsymbol{O}. \tag{9.19}$$

We obtain $\boldsymbol{R}(1)$ by integrating (9.19). It is positive-definite because (9.10) is. □

9.20. Problem. Investigate the validity of the perturbation process when N^0 is not everywhere positive.

10. Variational Characterization of the Equations for an Elastic String

If $f : \mathbb{R} \to \mathbb{R}$ is continuous, then the equation

$$f(x) = 0 \tag{10.1}$$

is equivalent to

$$\phi'(x) = 0 \tag{10.2}$$

where

$$\phi(x) = \int_0^x f(\xi)\, d\xi. \tag{10.3}$$

Thus we might be able to study the existence of solutions of (10.1) by showing that ϕ has an extremum (a maximum or a minimum) on $\mathbb{R}$. If ϕ is merely continuous, we can still study the minimization of ϕ, although the corresponding problem (10.1) for f need not be meaningful. (The present situation in three-dimensional nonlinear elastostatics has precisely this character: Under certain conditions the total energy is known to have a minimizer, but it is not known whether the equilibrium equations, which correspond to the vanishing of the Gâteaux derivative of the energy, have solutions.)

If $\mathsf{f} : \mathbb{R}^n \to \mathbb{R}^n$ is continuous, then the system

$$\mathsf{f}(\mathsf{x}) = \mathbf{0} \tag{10.4}$$

may not be equivalent to the vanishing of a gradient

$$\partial\phi(\mathsf{x})/\partial\mathsf{x} = \mathbf{0} \tag{10.5}$$

because there may not be a scalar-valued function ϕ such that $\partial\phi/\partial\mathbf{x} = \mathbf{f}$. If $\mathbf{f} \in C^1(\mathbb{R}^n)$, then a necessary and sufficient condition for the existence of such a ϕ is that

$$\partial\mathbf{f}/\partial\mathbf{x} \quad \text{is symmetric.} \tag{10.6}$$

In this case, ϕ is defined by the line integral

$$\phi(\mathbf{x}) \equiv \int_C \mathbf{f}(\mathbf{y}) \cdot d\mathbf{y} \tag{10.7}$$

where C is a sufficiently smooth curve joining a fixed point to $\mathbf{x}$. (For a proof, see Sec. XI.3.) We could then study (10.4) by studying extrema of ϕ.

In this section we show how the equations of motion for an elastic string can be characterized as the vanishing of the Gâteaux differential of a scalar-valued function of the configuration. For this purpose, we must first extend the notion of directional derivative of a real-valued function defined on some part of $\mathbb{R}^n$ to a real-valued function defined on a set of functions, which are to be candidates for solutions of the governing equations.

Let $\mathcal{E}_1$ and $\mathcal{E}_2$ be normed spaces (e.g., spaces of continuous functions) and let $\mathcal{A} \subset \mathcal{E}_1$. Let $f[\cdot] : \mathcal{E}_1 \to \mathcal{E}_2$. (When the argument of a function lies in a function space, we often enclose this argument with brackets instead of parentheses. Examples of such f's are forthcoming.) If

$$\tfrac{d}{d\varepsilon} f[u + \varepsilon y]|_{\varepsilon=0} \tag{10.8}$$

exists for $u \in \mathcal{A}$, $y \in \mathcal{E}_1$, and $\varepsilon \in \mathbb{R} \setminus \{0\}$, then it is called the *Gâteaux differential, directional derivative*, or *first variation of* of *f at u in the direction y*. If (10.8) exists for all y in $\mathcal{E}_1$ (for which it is necessary that u be an interior point of $\mathcal{A}$) and if it is a bounded linear operator acting on y (i.e., if (10.8) is linear in y and if the $\mathcal{E}_2$-norm of (10.8) is less than a constant times the $\mathcal{E}_1$-norm of y), then f is said to be *Gâteaux-differentiable* at u. In this case, (10.8) is denoted by $f_u[u] \cdot y$, and $f_u[u]$ is called the *Gâteaux derivative* of f at u. (This terminology is not completely standardized. See Vaĭnberg (1956) for a comprehensive treatment of the interrelationship of various kinds of differentiations.) If $\mathcal{E}_2 = \mathbb{R}$, then f is called a *functional.*

We ask whether the (weak form of) the governing equations for a string can be characterized by the vanishing of the Gâteaux differential of some suitable functional. We show that this can be done for elastic strings under conservative forces. We study the formulation of the equations for time-periodic motions of such strings; the formulation of initial-boundary-value problems by variational methods proves to be somewhat unnatural.

The *kinetic energy* of the string at time t is

$$K[\boldsymbol{r}](t) \equiv \tfrac{1}{2} \int_0^1 (\rho A)(s)|\boldsymbol{r}_t(s,t)|^2 \, ds. \tag{10.9}$$

The *stored-energy* (or *strain-energy*) (*density*) *function* for an elastic string is the function W defined by

$$W(\nu, s) \equiv \int_1^\nu \hat{N}(\beta, s)\, d\beta. \tag{10.10}$$

The (*total*) *stored energy* in the string at time t is

$$\Psi[\boldsymbol{r}](t) \equiv \int_0^1 W\big(\nu(s,t), s\big)\, ds. \tag{10.11}$$

Suppose that $\boldsymbol{f}$ has the form

$$\boldsymbol{f}(s,t) = \boldsymbol{g}(\boldsymbol{r}(s,t), s) \tag{10.12}$$

and that there is a scalar-valued function ω, called the *potential-energy density* of $\boldsymbol{g}$, such that

$$\boldsymbol{g}(\boldsymbol{r}, s) = -\partial\omega(\boldsymbol{r}, s)/\partial\boldsymbol{r} \tag{10.13}$$

(see the remarks following (10.5)). Thus $\boldsymbol{g}$ is conservative. The *potential energy of the body force* $\boldsymbol{g}$ *at time* t is

$$\Omega[\boldsymbol{r}](t) \equiv \int_0^1 \omega\big(\boldsymbol{r}(s,t), s\big)\, ds. \tag{10.14}$$

The *potential-energy functional for the string* is $\Psi + \Omega$.

Let $\mathcal{E}$ consist of all continuously differentiable vector-valued functions $[0,1] \times \mathbb{R} \ni (s,t) \mapsto \boldsymbol{y}(s,t)$ that satisfy

$$\boldsymbol{y}(0,t) = \boldsymbol{0} = \boldsymbol{y}(1,t), \quad \boldsymbol{y}(s,\cdot) \quad \text{has period } T. \tag{10.15}$$

The norm on $\mathcal{E}$ can be taken to be

$$\|\boldsymbol{y}\| \equiv \max\{|\boldsymbol{y}(s,t)| + |\boldsymbol{y}_s(s,t)| + |\boldsymbol{y}_t(s,t)| \colon (s,t) \in [0,1] \times [0,T]\}. \tag{10.16}$$

We introduce the *Lagrangian functional* Λ by

$$\Lambda[\boldsymbol{r}] \equiv \int_0^T \{K[\boldsymbol{r}](t) - \Psi[\boldsymbol{r}](t) - \Omega[\boldsymbol{r}](t)\}\, dt. \tag{10.17}$$

We study this functional on the class of admissible motions

$$\mathcal{A} \equiv \{\boldsymbol{r} \colon \boldsymbol{r}(s,t) = \boldsymbol{y}(s,t) + Ls\boldsymbol{k},\ \boldsymbol{y} \in \mathcal{E},\ |\boldsymbol{r}_s(s,t)| > 0\}. \tag{10.18}$$

(We can take $\mathcal{E}_1$ in the definition of Gâteaux derivative above to be $C^1([0,1] \times [0,T])$.)

For $\boldsymbol{r} \in \mathcal{A}$ and $\boldsymbol{y} \in \mathcal{E}$, we obtain from (10.9)–(10.14) that

$$\begin{aligned} \Lambda_{\boldsymbol{r}}[\boldsymbol{r}] \cdot \boldsymbol{y} = \frac{d}{d\varepsilon} \int_0^T \int_0^1 & \left[\tfrac{1}{2}(\rho A)(s)|\boldsymbol{r}_t(s,t) + \varepsilon \boldsymbol{y}_t(s,t)|^2 \right. \\ & - W\big(|\boldsymbol{r}_s(s,t) + \varepsilon \boldsymbol{y}_s(s,t)|, s\big) \\ & \left. - \omega\big(\boldsymbol{r}(s,t) + \varepsilon \boldsymbol{y}(s,t), s\big)\right] ds\, dt \bigg|_{\varepsilon=0} \\ = \int_0^T \int_0^1 & \Big[(\rho A)(s)\boldsymbol{r}_t(s,t) \cdot \boldsymbol{y}_t(s,t) \\ & - \hat{N}\big(|\boldsymbol{r}_s(s,t)|, s\big) \frac{\boldsymbol{r}_s(s,t) \cdot \boldsymbol{y}_s(s,t)}{|\boldsymbol{r}_s(s,t)|} \\ & + \boldsymbol{g}\big(\boldsymbol{r}(s,t), s\big) \cdot \boldsymbol{y}(s,t)\Big]\, ds\, dt. \end{aligned} \tag{10.19}$$

The mild difference between the vanishing of (10.19) and the Principle of Virtual Power (2.24), embodied in the presence of $\boldsymbol{v}$ in (2.24), reflects the fact that (2.24) accounts for initial conditions, whereas (10.19) accounts for periodicity conditions. *Hamilton's Principle* for elastic strings under conservative forces states that (the weak form of) the governing equations can be characterized by the vanishing of the Gâteaux differential of the Lagrangian functional Λ. Any system of equations that can be characterized by the vanishing of the Gâteaux differential of a functional is said to have a *variational structure*; the equations are called the *Euler-Lagrange equations* for that functional.

Hamilton's principle does not require that variables entering it be periodic in time. In fact, in the mechanics of discrete bodies, the configuration is typically required to satisfy boundary conditions at an initial and terminal time. Such conditions are artificial; they are devised so as to yield the governing equations as Euler-Lagrange equations. On the other hand, periodicity conditions define an important class of problems.

In continuum mechanics, Hamilton's principle is applicable only to frictionless systems acted on solely by conservative forces. A criterion telling whether a system of equations admits a natural variational structure is given by Vaĭnberg (1956) and is exploited by Tonti (1969). (Its derivation is just the generalization to function spaces of that for (10.6).) There is also a theory, akin to the theory of holonomicity in classical mechanics, that tells when a system can be transformed into one having a variational principle. The use of such a theory for quasilinear partial differential equations is very dangerous because the requisite transformations may change the weak form of the equations. For physical systems, the altered form may not be physically correct because it does not conform to the Principle of Virtual Power and accordingly does not deliver the correct jump conditions.

For Hamilton's Principle to be useful, it must deliver something more than an alternative derivation of the governing equations with theological overtones. One way for it to be useful would be for it to promote the proof of existence theorems for solutions characterized as extremizers of Λ. Serious technical difficulties have so far prevented this application to the equations of motion of nonlinear elasticity. Hamilton's Principle, however, has recently proved to be very effective in supporting the demonstration of the existence of multiple periodic solutions of systems of ordinary differential equations (see Ekeland (1990), Rabinowitz et al. (1987), e.g.). The specialization of Hamilton's Principle to static problems, called the *Principle of Minimum Potential Energy*, is very useful for existence theorems and for the interpretation of the stability of equilibrium states, as we shall see in Chap. VII. Moreover, Hamiltonian structure has been effectively exploited to derive stability theorems for certain elastic systems (see Simo, Posbergh, & Marsden (1991), e.g.).

11. Discretization

In this section we briefly survey some numerical methods for solving partial differential equations like those for the string. This text is not directly concerned with numerical methods; we examine these questions here because they are intimately related to the Principle of Virtual Power. (They can also be used to produce constructive existence theorems for certain problems.)

We describe a simple method that leads to the formal approximation of the partial differential equations for an elastic string by a system of ordinary differential equations. This procedure, associated with the names of Galerkin, Faedo, and Kantorovich, is sometimes called the method of lines. A special case of it is the semi-discrete finite-element method.

Let $\{s \mapsto \phi_k(s),\ k = 1, 2, \dots\}$ be a given set of functions in $W_p^1(0,1)$ with the properties that $\phi_k(0) = 0 = \phi_k(1)$ and that given an arbitrary function in $W_p^1(0,1)$ and an error, there exists a finite linear combination of the ϕ_k that approximate the given function to within the assigned error in the W_p^1-norm. (The set $\{s \mapsto \sin k\pi s\}$ has these properties. Another such set is defined in (11.9).) We seek to approximate solutions of the initial-boundary-value problem for elastic strings of Sec. 2 by functions $\boldsymbol{r}^K$ of the form

$$\boldsymbol{r}^K(s,t) = Ls\boldsymbol{k} + \sum_{k=1}^{K} \phi_k(s)\boldsymbol{r}_k(t) \tag{11.1}$$

where the functions $\boldsymbol{r}_k$ are to be determined. We approximate the given initial position $\boldsymbol{u}(s)$ and initial velocity $\boldsymbol{v}(s)$ by

$$\boldsymbol{u}^K(s) = Ls\boldsymbol{k} + \sum_{k=1}^{K} \phi_k(s)\boldsymbol{u}_k, \quad \boldsymbol{v}^K(s) = \sum_{k=1}^{K} \phi_k(s)\boldsymbol{v}_k \tag{11.2a,b}$$

where the vectors $\{\boldsymbol{u}_k, \boldsymbol{v}_k\}$ are given. In the Principle of Virtual Power (2.24), (2.10b), (2.11) for elastic strings let us replace $\boldsymbol{r}$ and $\boldsymbol{v}$ with $\boldsymbol{r}^K$ and $\boldsymbol{v}^K$ and let us choose

$$\boldsymbol{y}(s,t) = \phi_l(s)\boldsymbol{y}_l(t) \tag{11.3}$$

where $\boldsymbol{y}_l$ is an arbitrary absolutely continuous function that vanishes for large t. (There is no need for $\boldsymbol{y}_l$ to be indexed with l. No summation is intended on the right-hand side of (11.3).) Then this principle reduces to the following weak formulation of the system of ordinary differential equations for $\{\boldsymbol{r}_k\}$:

$$\begin{aligned} &\int_0^\infty \int_0^1 \hat{\boldsymbol{n}}\left(L\boldsymbol{k} + \sum_{k=1}^{K} \phi_k'(s)\boldsymbol{r}_k(t),\ s\right) \cdot \boldsymbol{y}_l(t)\phi_l'(s)\,ds\,dt - \int_0^\infty \boldsymbol{f}_l \cdot \boldsymbol{y}_l\,dt \\ &\qquad = \sum_{k=1}^{K} \langle \phi_k, \phi_l \rangle \int_0^\infty (\dot{\boldsymbol{r}}_k - \boldsymbol{v}_k) \cdot \dot{\boldsymbol{y}}_l\,dt \end{aligned} \tag{11.4}$$

for all absolutely continuous $\boldsymbol{y}_l$, $l = 1, \dots, K$. Here the superposed dot denotes differentiation with respect to t and

$$\boldsymbol{f}_l(t) \equiv \int_0^1 \boldsymbol{f}(s,t)\phi_l(s)\,ds, \quad \langle \phi_k, \phi_l \rangle \equiv \int_0^1 \rho A \phi_k \phi_l\,ds. \tag{11.5}$$

In consonance with (2.4) we require that $\{r_k\}$ satisfy the initial conditions

$$r_k(0) = u_k, \quad \dot{r}_k(0) = v_k, \tag{11.6a,b}$$

the second of which is incorporated into (11.4) as we shall see.

11.7. Exercise. Suppose that (11.4) has a continuously differentiable (or more generally an absolutely continuous) solution $\{r_1, \ldots, r_K\}$. Take $y_l(t) = \psi(t)e$ where ψ is defined in (4.4b) and and where e is an arbitrary constant vector. Use the method described at the end of Sec. 4 to prove that $\sum_{k=1}^K \langle \phi_k, \phi_l \rangle (\dot{r}_k - v_k)$ is continuously differentiable (whichever smoothness hypothesis is made about the solution) and accordingly satisfies (11.6b). Show that the solution of (11.4) is thus a classical solution of the system

$$\sum_{k=1}^K \langle \phi_k, \phi_l \rangle \ddot{r}_k + \int_0^1 \hat{n}\left(Lk + \sum_{k=1}^K \phi_l'(s) r_k(t),\ s\right) \phi_k'(s)\, ds - f_l = 0. \tag{11.8}$$

If we make the very reasonable assumption that the Gram matrix with components $\langle \phi_k, \phi_l \rangle$ is nonsingular, then (11.8) can be put into standard form. In particular, if $\phi_k(s) = \sin k\pi s$ and if ρA is constant, then this matrix as well as the corresponding matrix with components $\langle f_k', \phi_l' \rangle$ is diagonal. For practical computation, this virtue is counterbalanced by the high cost of the numerical evaluation of the integrals in (11.4) and (11.8).

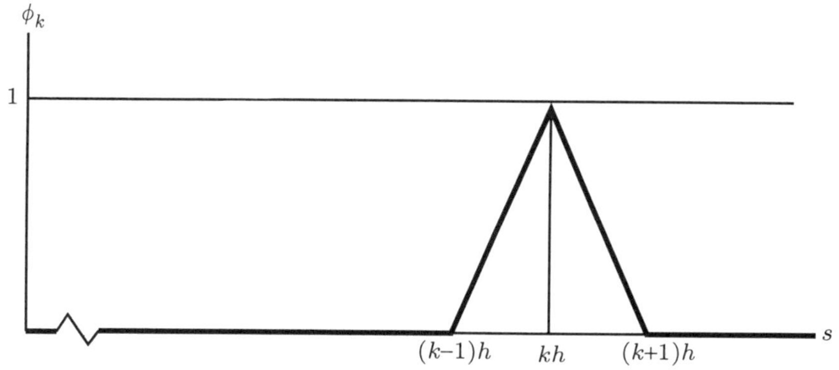

Figure 11.10. The function ϕ_k.

Let us set $h = 1/(K+1)$ and

$$\phi_k(s) = \begin{cases} h^{-1}[s-(k-1)h] & \text{for} \quad (k-1)h \le s \le kh, \\ 1 - h^{-1}(s-kh) & \text{for} \quad kh \le s \le (k+1)h, \\ 0 & \text{elsewhere}. \end{cases} \tag{11.9}$$

(This function is shown in Fig. 11.10.) When (11.9) is used, the matrices whose elements are $\langle \phi_k, \phi_l \rangle$ and $\langle \phi_k', \phi_l' \rangle$ are tridiagonal. The cost of the numerical evaluation of the integrals in (11.4) and (11.8) is low. (Matrices with components $\langle \phi_k', \phi_l' \rangle$ arise naturally in the linearization of (11.8) and are associated with the finite-difference approximation of r_{ss}.) The choice (11.9) gives the simplest (semi-discrete) finite-element approximation to our nonlinear initial-boundary-value problem. If ρA is constant, say $\rho A = 1$, then

$$\langle \phi_k, \phi_k \rangle = \frac{4h}{6}, \quad \langle \phi_k, \phi_{k+1} \rangle = \frac{h}{6}, \quad \langle \phi_k, \phi_l \rangle = 0 \quad \text{for} \quad l \neq k-1, k, k+1. \tag{11.11}$$

If, however, we were to approximately evaluate these integrals approximately by using the trapezoid rule, then we would find that $\langle \phi_k, \phi_l \rangle = \delta_{kl}$. In this case ,the left-hand side of (11.8) would uncouple and the resulting equations could be identified with the equations of motion of K beads joined by massless nonlinearly elastic springs.

11.12. Exercise. Replace $\boldsymbol{r}$ in (10.17) with $\boldsymbol{r}^K$ of (11.1). Show that the vanishing of the Gâteaux derivative of the resulting functional of $\{\boldsymbol{r}_1, \dots, \boldsymbol{r}_K\}$ is equivalent to (11.4).

11.13. Exercise. Using the principles of classical particle mechanics, find the equations of motion of K beads joined in sequence by massless nonlinearly elastic springs, with the first and the Kth bead joined to fixed points by such springs. Compare the resulting equations with (11.8). Formally obtain (2.9)–(2.11) by letting $K \to \infty$ while the total mass of the beads stays constant. (See the discussion at the end of Sec. 2.)

Even though the form of the governing equations for discrete models converges to the form of the governing equations for string models, it does not follow that the solutions of the former converge to solutions of the latter in any physically reasonable sense. Von Neumann (1944) (in a paper filled with valuable insights) advanced the view, now recognized as false, that the solutions for the positions of the beads in the equations of Ex. 11.13, together with their time derivatives and suitable difference quotients should converge respectively to the position, velocity, and strain fields for (2.9)–(2.11). This convergence is valid only where the partial differential equations have classical solutions. Where the velocity and strain suffer jump discontinuities (shocks), the solutions of the discrete problem develop high-frequency oscillations that persist in the limit as $K \to \infty$. Consequently it can be shown that the limiting stress is incorrect. For thorough discussions of this this phenomenon and related issues, see Greenberg (1989, 1992), Hou & Lax (1991), and the references cited therein.

Likewise, finite-element discretizations of dynamic problems of nonlinear elasticity may fail to give sharp numerical results because they are not well adapted to capture the shocks such systems may possess. (There is an effort to change this state of affairs.) There are a variety of effective numerical schemes, originally developed for gas dynamics, that can effectively handle shocks. One such scheme, that of Godunov (see Bell, Colella, & Trangenstein (1989), e.g.), may be regarded as a discretization of the Impulse-Momentum Law of Sec. 3 in a way that exploits the characteristics of the underlying hyperbolic system. There is an extensive literature on the finite-element method for equilibrium problems. Among the more mathematical works oriented toward solid mechanics are those of Brezzi & Fortin (1991), Ciarlet (1978), Ciarlet & Lions (1991), Johnson (1987), Hughes (1987), Oden & Carey (1981–1984), and Szabó & Babuška (1991).

Although we do not know in what sense the solution of (11.8) converges to the solution of the partial differential equations for elastic strings, we might be able to resolve this question for viscoelastic strings by using modern analytic techniques associated with the Faedo-Galerkin method. (Cf. Ladyzhenskaya (1973), Ladyzhenskaya, Solonnikov, & Ural'tseva (1967), and Lions (1969).) An analysis along these lines for a quasilinear engineering model of an elastic string was carried out by Dickey (1973). He proved that the solutions of a system like (11.8) converge to the classical solution of the partial differential equations until the advent of shocks. Antman & Seidman (1995) used the Faedo-Galerkin method to treat the longitudinal motion of a viscoelastic string with a constitutive equation of the form (2.14).

CHAPTER III

Elementary Problems for Elastic Strings

1. Introduction

The development of both continuum mechanics and mathematics during the eighteenth century was profoundly influenced by the successful treatment of conceptually simple, but technically difficult problems. (Indeed, one can argue that the dominant philosophical attitudes of the Age of Reason were founded on an awareness of these scientific triumphs, if not on their understanding.) Among the most notable of these classical problems are those of determining the equilibrium states of inextensible strings hung between two points and subjected to various systems of loads. (An *inextensible* string is one for which the stretch ν is constrained to equal 1, no matter what force system is applied to the string.) The problem of the *catenary* is to determine the equilibrium states of such a string when the applied force is the weight of the string. The problem of the *suspension bridge* is to determine these states when the applied force is a vertical load of constant intensity per horizontal distance. (The string does not correspond to the bridge, but to the wires from which it is suspended.) The problem of the *velaria* is to determine these states when the applied force is a normal pressure of constant intensity. In the related problem of the *lintearia*, the applied force is a normal pressure varying linearly with depth. (This problem describes the deformation of a cylindrical membrane holding a liquid, the string representing a typical section of the membrane.) In a fifth problem, the applied force is the attraction to a fixed point. In Sec. 8 we outline the progression from haphazard conjecture to elegant solution of these problems in the seventeenth and eighteenth centuries.

Catenary comes from the Latin *catena*, meaning *chain*. It is the curve assumed by a suspended chain. *Velaria* comes from the Latin *velarium*, meaning *awning* or *sail*. In the terminology of Jas. Bernoulli, it is the curve assumed by a cylindrical surface subjected to simple hydrostatic pressure. *Lintearia* comes from the Latin *lintea*, also meaning *sail*. (*Lintea* is the plural of *linteum* meaning *linen cloth*.) In the terminology of Jas. Bernoulli, the lintearia is the curve assumed by a horizontal cylindrical surface holding water.

Jas. Bernoulli formulated corresponding problems for elastic strings. Dickey (1969) observed that the catenary problem for nonlinearly elastic strings offers novel challenges on questions of existence, multiplicity, and qualitative behavior of solutions, which were contemplated neither by the savants of the seventeenth and eighteenth centuries nor by their successors. In this chapter we study these and related questions for generalizations of the problems mentioned above to nonuniform, nonlinearly elastic strings. One feature of these problems is that many detailed properties of the deformed shape are independent of the material response, which intervenes most significantly in the study of existence and multiplicity. The next section illustrates how the essential geometry is unaffected by constitutive equations for strings under vertical loads. Throughout this chapter we retain the notation of Chap. II.

2. Equilibrium of Strings under Vertical Loads

Let a string have a natural reference length 1. Let $\boldsymbol{f}(s)$ denote the force per unit reference length at s applied to the string. We assume that $\boldsymbol{f}$ is integrable. Then the equilibrium equations are

$$\text{(2.1a,b,c)} \qquad \boldsymbol{n}(s) - \boldsymbol{n}(c) + \int_c^s \boldsymbol{f}(\xi)\, d\xi = \boldsymbol{0}, \quad \boldsymbol{n} = N\boldsymbol{e}, \quad \boldsymbol{e} \equiv \frac{\boldsymbol{r}'}{|\boldsymbol{r}'|}$$

for almost all c and s. (See (II.6.1) and (II.2.10b). Note that (II.6.2) holds a.e.) Given $\boldsymbol{f}$, we seek absolutely continuous configurations $\boldsymbol{r}$ satisfying (2.1), satisfying

$$\text{(2.2)} \qquad \nu \equiv |\boldsymbol{r}'| > 0 \quad \text{a.e.,}$$

and satisfying the boundary conditions

$$\text{(2.3)} \qquad \boldsymbol{r}(0) = \boldsymbol{0}, \quad \boldsymbol{r}(1) = a\boldsymbol{i} + b\boldsymbol{j}, \quad a \geq 0, \quad b \geq 0.$$

(There is no loss in generality in taking position boundary conditions in this form.)

We assume that $\boldsymbol{f}$, which may be a composite function depending on $\boldsymbol{r}$ in a general way, has the form

$$\text{(2.4)} \qquad \boldsymbol{f}(s) = -F'(s)\boldsymbol{j}, \quad F'(s) > \alpha \quad \text{a.e.,} \quad F(0) = 0$$

where F is absolutely continuous and α is a given positive number. We give specific forms for F in subsequent sections. The unit vector $\boldsymbol{j}$ is interpreted as pointing upward.

We postpone the introduction of a constitutive equation for N. Our first results apply to the underdetermined system (2.1)–(2.4). As the comments following (II.6.3) explain, the equilibrium response of strings is elastic, so we shall employ the constitutive equation (II.6.4). But the next few results would apply equally well to an N depending on $|\boldsymbol{r}'|, |\boldsymbol{r}'|', \ldots$.

It may seem physically obvious that the string whose configuration is to be determined by (2.1)–(2.4) should lie in the $\{\boldsymbol{i}, \boldsymbol{j}\}$-plane. One's confidence in this intuition may be shaken by observing nonplanar equilibrium states of a real thread held at its ends under its own weight. One could attribute such states to the presence of small flexural and torsional stiffness that can dominate the small weight of the thread. If the planarity of the equilibrium state is deemed physically obvious, then it should be proved to follow from (2.1)–(2.4) in order to demonstrate the soundness of our model; if it is not deemed obvious, then it should be proved because it is an important and unexpected consequence of these equations. We now prove that the solutions of these equations are in fact planar (thereby supporting in an oblique way the explanation just adduced for the skewness of real threads).

2.5. Proposition. *Let $\boldsymbol{r}$ and $\boldsymbol{n}$ satisfy* (2.1)–(2.4) *with $\boldsymbol{r}$ absolutely continuous and with N continuous. Then $\boldsymbol{r}$ and $\boldsymbol{n}$ are planar: $\boldsymbol{r}\cdot\boldsymbol{k} = 0 = \boldsymbol{n}\cdot\boldsymbol{k}$.*

Proof. Condition (2.2) and the absolute continuity of $\boldsymbol{r}$ ensure that $\boldsymbol{e}$ is defined as a unit vector by (2.1c) a.e. Equation (2.1a) implies that $\boldsymbol{n}$ is absolutely continuous. Consequently (2.1a) holds for all a and s. Equations (2.1) and (2.4) imply that

(2.6a,b) $$\boldsymbol{n} = N\boldsymbol{e} = F\boldsymbol{j} + \boldsymbol{n}(0), \quad \text{whence} \quad |N| = |F\boldsymbol{j} + \boldsymbol{n}(0)|.$$

Thus $|N|$ is absolutely continuous. If N vanishes at a point s_0 in [0,1], then (2.6b) implies that $\boldsymbol{n}(0)$, which is not known beforehand, must equal $-F(s_0)\boldsymbol{j}$, whence (2.6) reduces to

(2.7) $$\boldsymbol{n}(s) \equiv N(s)\boldsymbol{e}(s) = [F(s) - F(s_0)]\boldsymbol{j}.$$

Thus $\boldsymbol{n}\cdot\boldsymbol{k} = 0$. Since $F' > \alpha$, Eq. (2.7) implies that N vanishes only at s_0. Thus (2.7) implies that

(2.8) $$\boldsymbol{e}(s) = \operatorname{sign}\{[F(s) - F(s_0)]N(s)^{-1}\}\boldsymbol{j} \quad \text{for} \quad s \neq s_0.$$

(Equation (2.8) is also valid without 'sign'.) Thus

(2.9a,b) $$\boldsymbol{r}(s)\cdot\boldsymbol{k} = \int_0^s \nu(\xi)\boldsymbol{e}(\xi)\cdot\boldsymbol{k}\,d\xi = 0.$$

If N vanishes nowhere, then (2.3) and (2.6) imply that

(2.10)
$$0 = (a\boldsymbol{i}+b\boldsymbol{j})\cdot\boldsymbol{k} = \int_0^1 \boldsymbol{r}'(s)\cdot\boldsymbol{k}\,ds = \int_0^1 \nu(s)\boldsymbol{e}(s)\cdot\boldsymbol{k}\,ds = \boldsymbol{n}(0)\cdot\boldsymbol{k}\int_0^1 \frac{\nu(s)}{N(s)}\,ds.$$

Since N is continuous, the integrand in the rightmost term of (2.10) has but one sign, so that $\boldsymbol{n}(0)\cdot\boldsymbol{k} = 0$. Thus $\boldsymbol{n}\cdot\boldsymbol{k} = 0$ and $\boldsymbol{e}\cdot\boldsymbol{k} = 0$ by (2.6). Equation (2.9a) then implies that $\boldsymbol{r}\cdot\boldsymbol{k} = 0$. □

That the vanishing of N at any point leads to a degenerate solution, as suggested by the first part of this proof, is confirmed by

2.11. Proposition. *Let $\boldsymbol{r}$ and $\boldsymbol{n}$ satisfy the hypotheses of Proposition 2.5. If there is an s_0 in $[0,1]$ such that $\boldsymbol{n}(s_0)\cdot\boldsymbol{i} = 0$ (i.e., if either $\boldsymbol{e}(s_0)\cdot\boldsymbol{i} = 0$, so that the tangent to the string at s_0 is vertical, or $N(s_0) = 0$), then $\boldsymbol{n}\cdot\boldsymbol{i} = 0$, $\boldsymbol{e}\cdot\boldsymbol{i} = 0$, $a = 0$, and the string lies along the vertical $\boldsymbol{j}$-axis.*

Otherwise, N never vanishes, so that the string is either everywhere in tension ($N > 0$) or everywhere in compression ($N < 0$) and the string is nowhere vertical.

Proof. Proposition 2.5 and Eq. (2.6) imply that if $\boldsymbol{n}(s_0) \cdot \boldsymbol{i} = 0$, then $\boldsymbol{n} \cdot \boldsymbol{i} = 0$. If $N(s_0) = 0$, then (2.8) holds, implying that $\boldsymbol{e} \cdot \boldsymbol{i} = 0$, so that

$$\boldsymbol{r}(s) \cdot \boldsymbol{i} = \int_0^s \nu(\xi)\boldsymbol{e}(\xi) \cdot \boldsymbol{i}\, d\xi = 0, \quad \boldsymbol{r}(1) \cdot \boldsymbol{i} \equiv a = 0. \tag{2.12}$$

If $\boldsymbol{n}(s_0) \cdot \boldsymbol{i} = 0$ but N vanishes nowhere, then the equation $\boldsymbol{n} \cdot \boldsymbol{i} = 0$ implies that $\boldsymbol{e} \cdot \boldsymbol{i} = 0$ with the same consequence (2.12). □

2.13. Problem. Generalize Propositions 2.5 and 2.11 when the requirement that F' have a positive lower bound a.e. is suspended.

In view of Proposition 2.5, we may introduce the representations

$$\boldsymbol{e}(s) = \cos\theta(s)\boldsymbol{i} + \sin\theta(s)\boldsymbol{j}, \tag{2.14}$$

$$\boldsymbol{n}(0) = \lambda\boldsymbol{i} + \mu\boldsymbol{j}. \tag{2.15}$$

Then by dotting (2.6) successively with $\boldsymbol{i}$, $\boldsymbol{j}$, and $\boldsymbol{e}$ we obtain

$$N\cos\theta = \lambda, \tag{2.16a}$$

$$N\sin\theta = \mu + F, \tag{2.16b}$$

$$N = \lambda\cos\theta + (\mu + F)\sin\theta. \tag{2.16c}$$

If $\lambda = \boldsymbol{n}(0) \cdot \boldsymbol{i} \neq 0$, then (2.16a,b) imply that

$$\tan\theta = \frac{\mu + F}{\lambda}, \tag{2.17}$$

(2.18)

$$\cos\theta = \pm\frac{\lambda}{\delta}, \quad \sin\theta = \pm\frac{\mu + F}{\delta}, \quad \delta \equiv \sqrt{\lambda^2 + (\mu + F)^2}, \quad N = \pm\delta.$$

Since F is absolutely continuous, (2.17) implies that $\tan\theta$ is also. We may accordingly take θ to be absolutely continuous and have range in $(-\pi/2, \pi/2)$. Since F is strictly increasing, Eq. (2.17) implies that θ is strictly increasing if $\lambda > 0$, and strictly decreasing if $\lambda < 0$. Thus

2.19. Proposition. *Let the hypotheses of Proposition 2.5 hold and let $a > 0$. The configuration $\boldsymbol{r}$ admits the usual Cartesian parametrization*

$$\boldsymbol{r} \cdot \boldsymbol{i} = x, \quad \boldsymbol{r} \cdot \boldsymbol{j} = y(x). \tag{2.20}$$

y is strictly convex for $\lambda > 0$ and strictly concave for $\lambda < 0$.

2.21. Exercise. For $\boldsymbol{n} \cdot \boldsymbol{i} \neq 0$, derive (2.17) by the following alternative process: Substitute (2.14) into (2.1) and use (2.4) to get

$$N' = F'\sin\theta, \quad N\theta' = F'\cos\theta, \quad \frac{N'}{N} = \theta'\tan\theta. \tag{2.22a,b,c}$$

Integrate (2.22c) subject to (2.15) to obtain (2.16a), substitute (2.16a) into (2.22b) to get

$$\theta'\sec^2\theta = F'/\lambda, \tag{2.23}$$

and integrate (2.23) to obtain (2.17).

3. The Catenary Problem

We now study the existence and multiplicity of absolutely continuous solutions of the boundary-value problem (2.1)–(2.3) for elastic strings, with constitutive equation (II.6.4), loaded by their own weights. Thus we specialize (2.4) to

$$(3.1) \qquad \boldsymbol{f}(s) = -(\rho A)(s)g\boldsymbol{j}, \quad F(s) = \int_0^s (\rho A)(\xi)g\,d\xi,$$

under the assumption that there is a positive number ω such that $(\rho A)(s) \geq \omega$. Here $(\rho A)(s)$ is the mass density of the string per unit unstretched length at s, and g is the acceleration of gravity. We assume that $\hat{N}$ is continuously differentiable, satisfies $\hat{N}_\nu > 0$ everywhere, and satisfies the growth conditions (II.2.19). By the argument preceding (II.2.23), $\hat{N}(\cdot, s)$ has an inverse $\hat{\nu}(\cdot, s)$ with $\hat{\nu}$ continuously differentiable. Thus (II.2.11) and (II.2.23) are equivalent.

Before studying the number of solutions of this boundary-value problem, we use a bootstrap argument to show that any solution must be classical:

3.2. Proposition. *Let F of (3.1) be absolutely continuous on $[0, 1]$ and let $\hat{N}$ have the properties just described. Let $a > 0$. Then every absolutely continuous solution $\boldsymbol{r}$ of the boundary-value problem (2.1)–(2.3), (II.6.4), (3.1) for which N is continuous has a derivative $\boldsymbol{r}'$ that is absolutely continuous. At any s at which F is continuously differentiable (i.e., at which ρA is continuous), $\boldsymbol{r}$ is continuously differentiable. If F is continuously differentiable, then $\boldsymbol{r}$ is a classical solution of the boundary-value problem.*

Proof. Proposition 2.11 implies that $\lambda \neq 0$ if and only if $a \neq 0$. Thus (2.17) implies that θ has the same regularity as F, and (2.16c) accordingly implies the same about N. (Products of absolutely continuous functions are absolutely continuous. A Lipschitz-continuous function of an absolutely continuous function is absolutely continuous. See Natanson (1961, Chap. IX), e.g.) Thus $\hat{\nu}(N(\cdot), \cdot)$ has the same regularity as F. The rest of the proposition follows from the representation $\boldsymbol{r}' = \nu\boldsymbol{e}$. □

3.3. Exercise. For $a = 0$ show that the boundary-value problem for the elastic catenary has an uncountable number of absolutely continuous solutions if the requirement that N be continuous is suspended. (Most of these solutions are pathological. Since they satisfy the integral form of the equilibrium equations or, equivalently, the Principle of Virtual Power, their physical unacceptability can only be attributed to the absence of bending stiffness. The requirement that N be continuous is an artificial admissibility condition in lieu of a characterization of boundary-value problems for strings as singular limits of those for rods. I expect that such a characterization would identify as physically reasonable only those solutions for which N is continuous.)

To show that the boundary-value problem has a solution, we merely have to show that λ and μ can be determined from the data a, b, F. There are exactly as many solutions of the boundary-value problem as there are distinct pairs (λ, μ). Indeed, if $a > 0$, then $\lambda \neq 0$, so that (2.17) yields θ, and (2.16c) yields N. The function $\boldsymbol{r}$ is then found by integrating $\boldsymbol{r}' = \hat{\nu}(N(\cdot), \cdot)\boldsymbol{e}$ subject to $\boldsymbol{r}(0) = \boldsymbol{0}$. (The same conclusion holds when $a = 0$.)

In the rest of this section, we assume that $a > 0$. We obtain a pair of equations for λ and μ from the boundary conditions (2.3) by using (2.1c), (2.14), and (2.18):

$$(3.4^{\pm})\qquad \begin{aligned} a\boldsymbol{i} + b\boldsymbol{j} = \boldsymbol{r}(1) - \boldsymbol{r}(0) &= \int_0^1 \nu(s)[\cos\theta(s)\boldsymbol{i} + \sin\theta(s)\boldsymbol{j}]\,ds \\ &= \int_0^1 \frac{\hat{\nu}\big(\pm\delta(s), s\big)}{\pm\delta(s)}\{\lambda\boldsymbol{i} + [\mu + F(s)]\boldsymbol{j}\}\,ds \\ &\equiv P^{\pm}(\lambda,\mu;F)\boldsymbol{i} + Q^{\pm}(\lambda,\mu;F)\boldsymbol{j}. \end{aligned}$$

Equation (3.4^+) describes tensile solutions and (3.4^-) describes compressive solutions.

Let us introduce the function W^* *conjugate to the stored-energy function* W of (II.10.10) by the Legendre transformation

$$(3.5)\qquad W^*(N,s) \equiv N\hat{\nu}(N,s) - W\big(\hat{\nu}(N,s), s\big),$$

so that

$$(3.6)\qquad \hat{\nu}(N,s) = W^*_N(N,s), \qquad W^*(N,s) = \int_0^N \hat{\nu}(\overline{N}, s)\,d\overline{N}.$$

We illustrate the form of $\hat{\nu}$ and W^* in Fig. 3.7. We set

$$(3.8)\qquad \Phi^{\pm}(\lambda,\mu;a,b,F) = \int_0^1 W^*\big(\pm\delta(s), s\big)\,ds - \lambda a - \mu b.$$

Equation $(3.4^{\pm})$ is equivalent to the vanishing of the gradient of $\Phi^{\pm}(\cdot,\cdot;a,b,F)$. (There is no problem with the differentiability of this function because $F(s) > 0$ for $s > 0$. Note that given $P^{\pm}$ and $Q^{\pm}$, we can construct $\Phi^{\pm}$ by (II.10.7).)

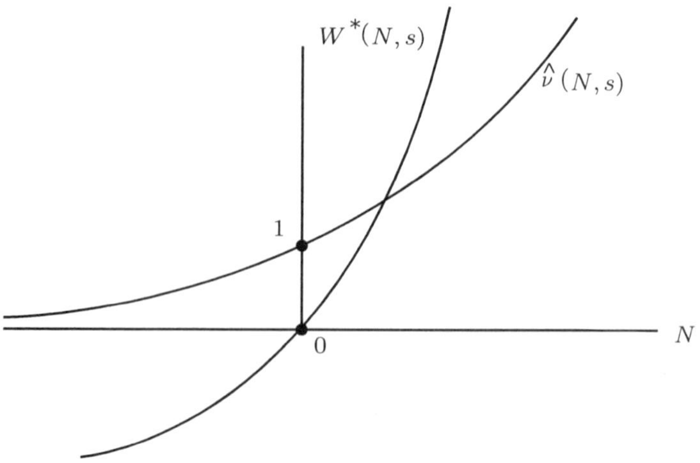

Fig. 3.7. Graphs of $\hat{\nu}$ and W^*.

We now analyze (3.4$^+$). We show that the gradient of Φ^+ vanishes at one point at least, because Φ^+ has a minimum: The properties of W^* inherited from $\hat{N}$ and illustrated in Fig. 3.7 imply that

$$\Phi^+(\lambda, \mu; a, b, F) \to \infty \quad \text{as} \quad \lambda^2 + \mu^2 \to \infty. \tag{3.9}$$

Thus the set $\mathcal{K} \equiv \{(\lambda, \mu) : \Phi^+(\lambda, \mu; a, b, F) \leq \Phi^+(0, 0; a, b, F)\}$ must be bounded. $\mathcal{K}$ is closed because Φ^+ is continuous. Therefore, $\Phi^+(\cdot, \cdot; a, b, F)$ restricted to $\mathcal{K}$ is a continuous function on the closed and bounded set $\mathcal{K}$ and therefore has a minimum at some point $(\bar{\lambda}, \bar{\mu})$ in $\mathcal{K}$. But this minimum is clearly a minimum for the unrestricted function. Thus for each a, b, F, the continuous function $\Phi^+(\cdot, \cdot; a, b, F)$ on $\mathbb{R}^2$ has an absolute minimum at a point $(\bar{\lambda}, \bar{\mu})$, at which the gradient of $\Phi^+(\cdot, \cdot; a, b, F)$ must vanish. Hence $(\bar{\lambda}, \bar{\mu})$ is a solution of (3.4$^+$). The positivity of $\hat{\nu}_N$, which is a consequence of that of $\hat{N}_\nu$, implies that Φ^+ is strictly convex. Thus its Hessian matrix

$$\begin{pmatrix} P_\lambda^+ & P_\mu^+ \\ Q_\lambda^+ & Q_\mu^+ \end{pmatrix} \quad \text{is positive-definite.} \tag{3.10}$$

This condition implies that (3.4$^+$) has at most one solution. Since $a > 0$, we can use (2.16a) to show that $\bar{\lambda} > 0$ and use (2.17) to show that θ is a strictly increasing function of s. Therefore, we have proved

3.11. Theorem. *Let $\hat{\nu}$ have the properties described above. There is exactly one absolutely continuous solution $\boldsymbol{r}$ of the boundary-value problem for the catenary for which N is everywhere positive. For this solution, $\bar{\lambda} \equiv \boldsymbol{n}(0) \cdot \boldsymbol{i} > 0$ and the function y of (2.20) is strictly convex.*

3.12. Exercise. Carry out the details of the proof of this theorem, paying particular attention to (3.9), (3.10), and the proof of uniqueness. Note that these critical results are direct consequences of our constitutive hypotheses.

3.13. Exercise. Show that

$$\Phi^-(\lambda, \mu; a, b, F) = -\lambda a - \mu b + o\left(\sqrt{\lambda^2 + \mu^2}\right) \quad \text{as} \quad \lambda^2 + \mu^2 \to \infty. \tag{3.14}$$

3.15. Exercise. For *inextensible* strings, i.e., strings for which $\hat{\nu} = 1$, show that the variational method used to prove Theorem 3.11 can be readily adapted to treat compressive solutions: Prove that if $0 < a$ and if $a^2 + b^2 < 1$, then there is exactly one tensile solution, for which y is convex, and there is exactly one compressive solution, for which y is concave. Specialize these results to a uniform string, for which $F = \gamma s$, and determine how the shape of the compressive configuration is related to that of the tensile configuration. Show that the solutions (λ, μ) of (3.4$^+$) parametrized by a, b have the property that $\lambda^2 + \mu^2 \nearrow \infty$ as $a^2 + b^2 \nearrow 1$. In this limit, the constraint that the supports are separated by the length of the inextensible string permits only the unique configuration in which $\boldsymbol{r}$ describes the straight line joining the supports. Thus the tension necessary to hold a heavy inextensible string taut is infinite. (This paradox has been known at least since the nineteenth century.) If $a^2 + b^2 > 1$, then there can be no solutions.

Exercise 3.13 shows that there is no straightforward way to exploit the variational structure of (3.4$^-$) (for elastic strings). A common modern response to a system like this is to abandon any hope of deducing information analytically and instead appeal to numerical processes. Each numerical study requires a choice of data $a, b, F, \hat{\nu}$. Consequently each numerical study, presumed reliable, yields information about just one set of data; such studies cannot distinguish between what is typical of all data and what is special. In particular, such studies cannot readily identify thresholds in data across which solution properties change drastically. We accordingly approach our problem by other analytic techniques. Such analyses of course form a very useful concomitant to any serious numerical study.

In the rest of this section, we drop the minus signs ornamenting P and Q in (3.4^-). We first observe that if (3.4^-) has a solution, then $\operatorname{sign}\lambda = -\operatorname{sign} a$ and $\mu < 0$. If (3.4^-) has a solution with $a^2 + b^2 > 1$, then there must be an s_0 such that $\nu(s_0) > 1$, so that $N(s_0) > 0$. Proposition 2.11 then implies that $N > 0$, which is incompatible with the fact that (3.4^-) corresponds to a negative N. Thus if (3.4^-) has a solution, then $a^2+b^2 \leq 1$. In fact, we have a sharper result: Since $\lambda \leq 0$ and since $\mu + F$ can vanish at most once in $[0, 1]$, it follows that $\delta(s) > 0$ and $\hat{\nu}\big(-\delta(s), s\big) < 1$ for all except possibly one s in $[0, 1]$. Hence the Cauchy-Bunyakovskiĭ-Schwarz inequality implies that $1 > P^2 + Q^2 = a^2 + b^2$. We summarize these results:

3.16. Proposition. *Let $\hat{\nu}$ have the properties described above. If (3.4^-) has a solution, then*

$$a^2 + b^2 < 1, \quad \operatorname{sign}\lambda = -\operatorname{sign} a, \quad \mu < 0. \tag{3.17}$$

Thus if $a > 0$, then $\lambda < 0$ and solutions correspond to strictly concave y's by Proposition 2.19. Proposition 3.16 says that these arciform compressive states exists only if the distance between the supports is less than the natural length of the string. For the degenerate case that $a = 0$, we shall easily show that there are exactly two solutions (with N continuous) when (3.17) holds: In one solution, the string is compressed between the two supports, and in the other, the string is balanced precariously in a folded vertical configuration above its supports.

It is evident that these compressive configurations should be unstable according to any physically reasonable criterion. But as both Hooke and Jas. Bernoulli observed, such configurations correspond to moment-free states of arches composed of materials that resist bending. They accordingly play an important role in optimal design. There is also evidence that such configurations have an asymptotic significance in rod theory. Finally, these configurations correspond to critical points in any dynamical problem for these strings, and as such influence the global evolution of solutions. (These observations escaped a reviewer who criticized Dickey (1969) for studying unstable solutions.)

The most direct approach to analyzing (3.4^-) consists in determining the nature of the graphs $(\lambda, \mu) \mapsto P(\lambda, \mu; F)$, $Q(\lambda, \mu; F)$, using this information to discover what the level curves $P(\lambda, \mu; F) = a$, $Q(\lambda, \mu; F) = b$ look like, and then finding simple conditions that ensure that these level curves

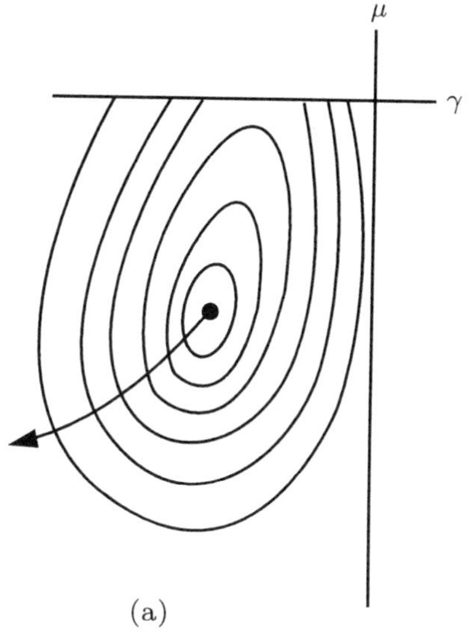

Fig. 3.19(a). The level curves

$$P(\lambda, \mu; \gamma\Gamma) = a.$$

The arrow indicates the direction of decreasing a.

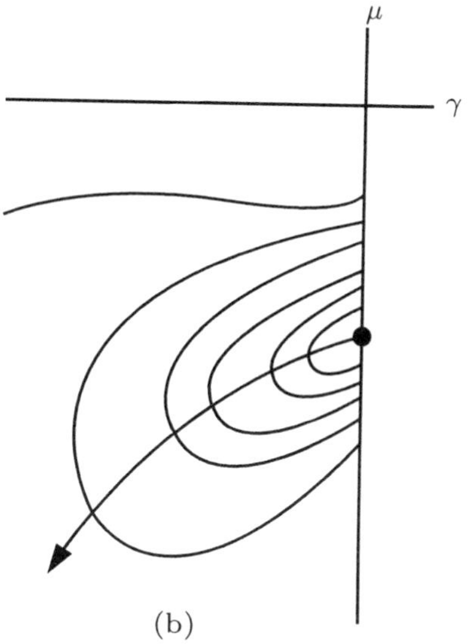

Fig. 3.19(b). The level curves

$$Q(\lambda, \mu; \gamma\Gamma) = b.$$

The arrow indicates the direction of decreasing b.

intersect (at solutions of (3.4^-)). For this purpose, it is useful to set

$$\gamma \equiv F(1) = \max F, \quad \Gamma(s) = \gamma^{-1}F(s) \tag{3.18}$$

and regard Γ, which gives the shape of F, as fixed.

For fixed $\gamma > 0$, we determine the properties of the graphs of $(\lambda, \mu) \mapsto P(\lambda, \mu; \gamma\Gamma), Q(\lambda, \mu; \gamma\Gamma)$ by showing how they are related to the graphs of $(\lambda, \mu) \mapsto P(\lambda, \mu; 0), Q(\lambda, \mu; 0)$. The presence of the radical δ in the integrands of P and Q prevents $P(\lambda, \mu; \gamma\Gamma)$ and $Q(\lambda, \mu; \gamma\Gamma)$ from converging uniformly to $P(\lambda, \mu; 0)$ and $Q(\lambda, \mu; 0)$ for $\lambda \leq 0$, $\mu \leq 0$ as $\gamma \searrow 0$. The same lack of uniformity occurs in other limit processes. Although we must accordingly treat various limit processes with care, we can nevertheless deduce detailed qualitative information about the graphs $(\lambda, \mu) \mapsto P(\lambda, \mu; \gamma\Gamma)$, $Q(\lambda, \mu; \gamma\Gamma)$. The details, omitted here, are given by Antman (1979b). The dependence of the level curves $P(\lambda, \mu; \gamma\Gamma) = a$, $Q(\lambda, \mu; \gamma\Gamma) = b$ on a and b is shown in Fig. 3.19. From it and related figures, we can deduce

3.20. Theorem. *Let $\hat{\nu}$ have the properties described above.* (i) *Let $\gamma > 0$ be given. If a and b are chosen sufficiently small with $a > 0$, then there are at least two distinct solutions of (3.4^-) and consequently at least*

two distinct solutions of the boundary-value problem. (ii) *Let a and b be given with $a > 0$ and with $a^2 + b^2 < 1$. If γ is chosen sufficiently small, then there are likewise at least two distinct solutions of the boundary-value problem. For these solutions, $\lambda \equiv \boldsymbol{n}(0) \cdot \boldsymbol{i} < 0$, and y, defined by (2.20), is strictly concave.*

We now approach this problem by an elementary topological method that does not require the detailed computations underlying the proof of Theorem 3.20. The basic idea is that solutions of (3.4⁻) correspond to points at which the continuous vector field

$$\mathbb{E}^2 \ni \lambda \boldsymbol{i} + \mu \boldsymbol{j} \mapsto [P(\lambda, \mu; F) - a]\boldsymbol{i} + [Q(\lambda, \mu; F) - b]\boldsymbol{j}$$

vanishes. Degree theory provides a way to reduce the study of this vanishing to the study of the degenerate problem in which $a = 0$. We now give an intuitive outline of degree theory in a more general setting, deferring a careful exposition to Chap. XIX.

Let Ω be a bounded domain in $\mathbb{R}^n$ and let I be an interval on $\mathbb{R}$. Let $\mathsf{f} : \mathrm{cl}\,\Omega \times I \to \mathbb{R}^n$ be continuous. We wish to study how solutions $\mathsf{x} \in \mathrm{cl}\,\Omega$ of the equation

$$\mathsf{f}(\mathsf{x}, a) = \mathbf{0} \tag{3.21}$$

depend on the parameter a. (We shall identify x with (λ, μ) and $\mathsf{f}(\mathsf{x}, a)$ with $\big([P(\lambda, \mu; F) - a], [Q(\lambda, \mu; F) - b]\big)$, holding b and F fixed.) If $\mathsf{f}(\cdot, a)$ does not vanish on $\partial\Omega$, then we can define an integer $\deg\big(\mathsf{f}(\cdot, a), \Omega\big)$, called the (*Brouwer*) *degree of* $\mathsf{f}(\cdot, a)$ *on* Ω, with the properties that it is easy to compute and that $|\deg\big(\mathsf{f}(\cdot, a), \Omega\big)|$ gives a lower bound for the number of solutions of (3.21). In essence, we assign the number ± 1 to each solution x of (3.21) according to an appropriate rule and define the degree to be the sum of these numbers. The method we employ is a generalization of the Intermediate Value Theorem for real-valued functions of a real variable.

If $\mathsf{f}(\cdot, a)$ is continuously differentiable on $\mathrm{cl}\,\Omega$, does not vanish on $\partial\Omega$, and has a Jacobian that does not vanish where $\mathsf{f}(\cdot, a)$ does, we define $\deg\big(\mathsf{f}(\cdot, a), \Omega\big)$ to be the sum of the signs of the Jacobians at the zeros of $\mathsf{f}(\cdot, a)$:

$$\deg\big(\mathsf{f}(\cdot, a), \Omega\big) \equiv \sum \operatorname{sign} \det \frac{\partial \mathsf{f}}{\partial \mathsf{x}}(\mathsf{x}) \tag{3.22}$$

where the sum is taken over all x in Ω for which $\mathsf{f}(\mathsf{x}, a) = \mathbf{0}$. This definition can be extended by approximation to continuously differentiable functions $\mathsf{f}(\cdot, a)$ for which the Jacobian can vanish at its zeros, and then to continuous functions. As the presence of the Jacobian suggests, the degree can be represented as an integral, from which it can be shown that the degree depends continuously on $\mathsf{f}(\cdot, a)$ in $C^0(\mathrm{cl}\,\Omega)$ provided that $\mathsf{f}(\cdot, a)$ does not vanish on $\partial\Omega$. We shall vary $\mathsf{f}(\cdot, a)$ by varying a. Since the degree is an integer, it stays fixed for continuous variations of $\mathsf{f}(\cdot, a)$ that do not vanish on $\partial\Omega$. It follows from this invariance that $\deg\big(\mathbf{f}(\cdot,a),\Omega\big)$ is determined by the restriction of $\mathbf{f}(\cdot,a)$ to $\partial\Omega$. This fact simplifies the computation of degree. In particular, if $n=2$, then it can be shown that $2\pi\deg\big(\mathbf{f}(\cdot,a),\Omega\big)$ is the angle through which $\mathbf{f}(\cdot,a)$ rotates as $\partial\Omega$ is traversed in a counterclockwise sense. (For continuous complex-valued functions of complex variables, degree theory reduces to the Argument Principle.) If $\mathbf{f}(\cdot,a)$ does not vanish on $\partial\Omega$, then the algebraic number of solutions of (3.21) in Ω is *even* or *odd* according as $\deg\big(\mathbf{f}(\cdot,a),\Omega\big)$ is even or odd.

To avoid some minor technical difficulties in the application of these ideas, we begin by assuming that the string is uniform so that $\hat{\nu}_s=0$ and $F(s)=\gamma s$. We first study the degenerate problem in which $a=0$ and $b=b_0$, where b_0 is a sufficiently small positive number (whose permissible range will be specified shortly). Our goal is to determine how solutions evolve from solutions of the degenerate problem as the parameters are varied. We shall determine the degree of $\lambda\boldsymbol{i}+\mu\boldsymbol{j}\mapsto[P(\lambda,\mu;F)]\boldsymbol{i}+[Q(\lambda,\mu;F)-b]\boldsymbol{j}$ on certain sets and then use the invariance of degree under changes in a to determine properties of the solutions of (3.4^-).

For (3.4^-) to have a solution with $a=0$, it must reduce to

(3.23a)
$$Q(0,\mu;F)\equiv-\int_0^1\frac{\hat{\nu}(-|\mu+\gamma s|)}{|\mu+\gamma s|}(\mu+\gamma s)\,ds\equiv-\frac{1}{\gamma}\int_\mu^{\mu+\gamma}\hat{\nu}(-|u|)\frac{u}{|u|}\,du=b_0.$$

Using (3.6), we find that

$$\text{(3.23b)}\quad \gamma Q(0,\mu;F)=\begin{cases}W^*(\mu+\gamma)-W^*(\mu) & \text{for}\quad \mu\le-\gamma,\\ W^*(-\mu-\gamma)-W^*(\mu) & \text{for}\quad -\gamma\le\mu\le0,\\ W^*(-\mu-\gamma)-W^*(-\mu) & \text{for}\quad 0\le\mu,\end{cases}$$

from which we obtain

(3.24)
$$Q(0,\mu;F)<0\quad\text{for}\quad 0\le\mu,\quad Q(0,\mu;F)\to0\quad\text{as}\quad\mu\to-\infty,$$

(3.25)
$$Q_\mu(0,\mu;F)=-\frac{\hat{\nu}(-|\mu+\gamma|)(\mu+\gamma)}{\gamma|\mu+\gamma|}-\frac{\hat{\nu}(\mu)}{\gamma}\quad\text{for}\quad\mu\le0,\ \mu\ne-\gamma,$$

(3.26)
$$Q_\mu(0,\mu;F)>0\quad\text{for}\quad\mu\le-\gamma,\quad Q_\mu(0,\mu;F)<0\quad\text{for}\quad-\gamma<\mu\le0.$$

$Q_\mu(0,\cdot;F)$ is discontinuous at $-\gamma$, where $Q(0,\cdot;F)$ assumes its maximum on $(-\infty,0]$. The form of $Q(0,\cdot;F)$ is shown in Fig. 3.27. This figure shows that if $0<b_0<Q(0,-\gamma;F)$, then (3.4^-) has exactly two solutions $\mu_1(b_0)$ and $\mu_2(b_0)$, with $\mu_1(b_0)<-\gamma$ and with $-\gamma<\mu_2(b_0)<0$. The configuration corresponding to $\mu_1(b_0)$ is a straight line joining the supports at $(0,0)$ and $(0,b_0)$. The configuration corresponding to $\mu_2(b_0)$ is a folded line in which the fold lies above the higher support $(0,b_0)$. In the former case, $N = -\delta$ is everywhere negative; in the latter case, N must vanish exactly once, at the fold. As b_0 is reduced to 0, $\mu_1(b_0) \to -\infty$ and the corresponding unfolded compressed state is ultimately reduced to a point, this configuration being maintained by the reaction $\mu_1(0) = -\infty$. The existence of these two degenerate solutions for $0 < b_0 < Q(0, -\gamma; F)$ suggests why there would be two compressive states for $a > 0$, as ensured by Theorem 3.20. Note that these solutions, just like those of Ex. 3.3, satisfy the integral form of the equilibrium equations or, equivalently, the Principle of Virtual Power.

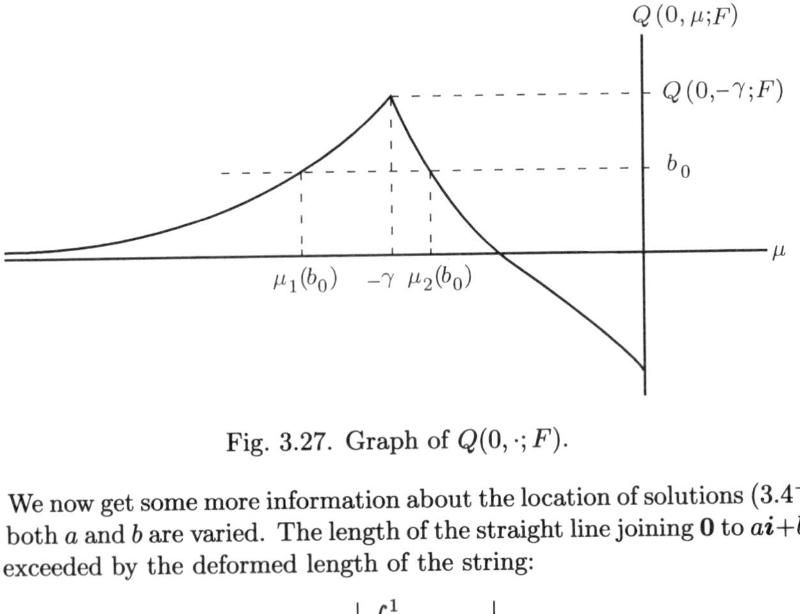

Fig. 3.27. Graph of $Q(0, \cdot; F)$.

We now get some more information about the location of solutions (3.4⁻) as both a and b are varied. The length of the straight line joining $\mathbf{0}$ to $a\boldsymbol{i}+b\boldsymbol{j}$ is exceeded by the deformed length of the string:

$$(3.28)\qquad \begin{aligned} \sqrt{a^2+b^2} &= |a\boldsymbol{i}+b\boldsymbol{j}| = \left|\int_0^1 \boldsymbol{r}'(s)\,ds\right| \\ &\le \int_0^1 |\boldsymbol{r}'(s)|\,ds = \int_0^1 \hat{\nu}\left(-\sqrt{\lambda^2+[\mu+F(s)]^2}, s\right)\,ds. \end{aligned}$$

(For generality, we treat nonuniform strings in (3.28).) If there is a number $\varepsilon > 0$ such that $a^2+b^2 \ge \varepsilon^2$, then (3.28) implies that λ and μ must each have lower bounds depending only on ε (and F and $\hat{\nu}$), but independent of a and b. (If not, the properties of $\hat{\nu}$ illustrated in Fig. 3.7 would ensure the violation of (3.28) for large negative λ or μ.) Thus for fixed ε (and for fixed F and $\hat{\nu}$), we can choose Λ and M so large that all solutions of (3.4⁻) lie in the rectangle bounded by the lines $\lambda = \pm\Lambda$, $\mu = 0$, and $\mu = -M$. In particular, we take $-M < \mu_1(b_0)$. We first examine problems in which b is held fixed at b_0, and a is increased from 0. In this case, we choose $\varepsilon = b_0$.

The vector field $\lambda\boldsymbol{i} + \mu\boldsymbol{j} \mapsto [P(\lambda,\mu;F) - a]\boldsymbol{i} + [Q(\lambda,\mu;F) - b]\boldsymbol{j}$ is continuous on $\mathbb{E}^2$. In Fig. 3.29 we use the definitions of P and Q in (3.4⁻)

to sketch this field for $a = 0$ and $b = b_0$ on the lines $\lambda = \pm\Lambda$, $\mu = 0$, $\mu = -\gamma$, and $\mu = -M$. As the rectangle with sides $\lambda = \pm\Lambda$, $\mu = 0$, and $\mu = -\gamma$ is traversed in the counterclockwise sense, the vector field $P(\cdot,\cdot;F)\boldsymbol{i} + [Q(\cdot,\cdot;F) - b_0]\boldsymbol{j}$ rotates through an angle of 2π. Thus its degree on the domain bounded by this rectangle is 1. Since the only singular point of this field on this domain is at $(0, \mu_2(b_0))$, the degree of this field on every small neighborhood containing $(0, \mu_2(b_0))$, called the (*Brouwer*) *index* of $(0, \mu_2(b_0))$, is also 1 (essentially as a consequence of definition (3.22)). Similarly, the index of $(0, \mu_1(b_0))$ is -1. The degree of $P(\cdot,\cdot;F)\boldsymbol{i} + [Q(\cdot,\cdot;F) - b_0]\boldsymbol{j}$ on the large rectangular domain bounded by the lines $\lambda = \pm\Lambda$, $\mu = 0$, and $\mu = -M$ is 0, the sum of its indices, as can be determined directly by computing its rotation.

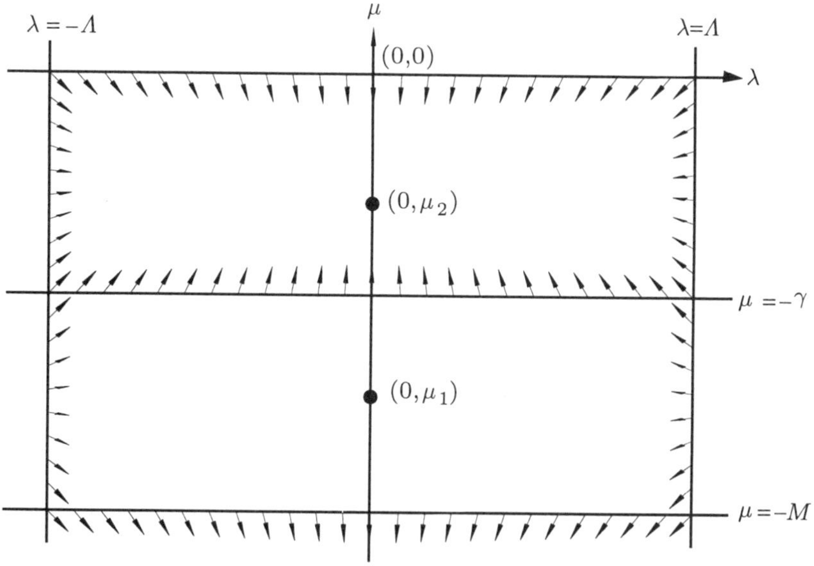

Fig. 3.29. The vector field $P(\cdot,\cdot;F)\boldsymbol{i} + [Q(\cdot,\cdot;F) - b_0]\boldsymbol{j}$ on the (λ, μ)-plane.

We regard the *state* variables (λ, μ) as constituting a single entity, distinct from the parameter a. We say that $\big((\lambda,\mu), a\big)$ is a *solution pair* of (3.4^-) if it satisfies this equation.

3.30. Theorem. *Let $b = b_0 \in \big(0, Q(0, -\gamma; F)\big)$, let $\hat{\nu}$ be independent of s and have the properties described above, and let $F(s) = \gamma s$. Then the set of solution pairs $\big((\lambda,\mu), a\big)$ of (3.4^-), which lie in $[-\Lambda, 0] \times [-M, 0] \times [0, 1]$, contains a connected set joining $\big((0, \mu_1(b_0)), 0\big)$ to $\big((0, \mu_2(b_0)), 0\big)$. (See Fig. 3.31.)*

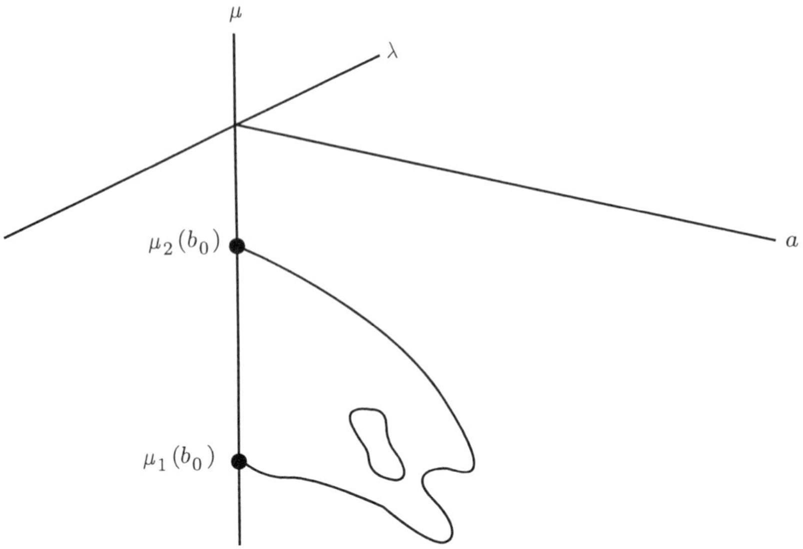

Fig. 3.31. Schematic diagram of solution pairs of (3.4⁻). A plane perpendicular to the a-axis through a small value of a intersects the set of solution pairs at an algebraically even number of points. It is conceivable that there are also disconnected sets of solutions.

Sketch of Proof. The set of solution pairs of (3.4^-), being the inverse image of $\mathbf{0}$ under the continuous function $((\lambda,\mu),a) \mapsto [P(\cdot,\cdot;F)-a]\boldsymbol{i} + [Q(\cdot,\cdot;F)-b_0]\boldsymbol{j}$, is closed. Conditions (3.17) and the consequences of (3.28) imply that this set is also bounded. Suppose for contradiction that $\big((0,\mu_1(b_0)),0\big)$ and $\big((0,\mu_2(b_0)),0\big)$ were not connected by a set of solution pairs. Let $\mathcal{C}$ be the maximal connected set of solution pairs containing $\big((0,\mu_1(b_0)),0\big)$. (We admit the possibility that $\mathcal{C}$ might consist only of the point $\big((0,\mu_1(b_0)),0\big)$.) Since the set of solution pairs is compact, so is $\mathcal{C}$. Since the distance between two disjoint compact sets is positive, the set $\mathcal{C}$ can be enclosed in a bounded open set $\mathcal{O}$ whose closure contains no solution pairs other than those of $\mathcal{C}$. Let $\mathcal{O}(\alpha)$ represent the intersection of $\mathcal{O}$ with the plane $a=\alpha$: $\mathcal{O}(\alpha) \equiv \{((\lambda,\mu),a) \in \mathcal{O}\colon a=\alpha\}$. We know that $\deg\big(P\boldsymbol{i}+(Q-b_0)\boldsymbol{j},\mathcal{O}(0)\big)=1$. The compactness of $\mathcal{C}$ and the construction of $\mathcal{O}$ ensure that there is a number $\bar{a}$ such that $\operatorname{cl}\mathcal{O}(\bar{a})$ contains no solution pairs. Thus, by a result that can be ultimately traced back to definition (3.22), it follows that $\deg\big((P-a)\boldsymbol{i}+(Q-b_0)\boldsymbol{j},\mathcal{O}(\bar{a})\big)=0$. But this result is incompatible with the invariance of degree, which implies that these two degrees should be equal. □

There are many strategies by which Theorem 3.30 can be extended to more general choices of a, b, F, and $\hat{\nu}$. Instead of varying a alone, we can vary all these quantities by replacing them with continuous functions $[0,\infty) \ni t \mapsto a^\star(t)$, $b^\star(t)$, $F^\star(\cdot,t)$, and $\nu^\star(\cdot,\cdot,t)$ with $a^\star(0)=0$, $b^\star(0)=b_0$, $F^\star(s,0)=\gamma s$, and $\nu^\star(N,s,0)=\hat{\nu}(N)$. The role of a as parameter in (3.4^-) is played by t. If $t \mapsto a^\star(t)^2+b^\star(t)^2$ has a positive lower bound and exceeds 1 for large t, then the conclusions of Theorem 3.30 are still valid.

We can also use degree theory to treat problems in which $F^\star(s,0)$ and $\nu^\star(N,s,0)$ can have arbitrary form. Let $P^\star(\lambda,\mu;t)$ and $Q^\star(\lambda,\mu;t)$ be the values of P and Q obtained by replacing F and $\hat{\nu}$ with $F^\star$ and $\nu^\star$. We find that $Q^\star(0,\cdot;0)$ increases on $(-\infty,-\gamma)$ but need not be decreasing on $(-\gamma,0]$ from its positive value at $-\gamma$ to its negative value at 0. Thus if $0<b_0<Q^\star(0,-\gamma;0)$, then the equation $Q^\star(0,\mu;0)=b_0$ has exactly one

solution $\mu_1(b_0)$ in $(-\infty, -\gamma)$ and an algebraically odd number of solutions in $(-\gamma, 0)$. If $Q^\star(0, \cdot; 0) - b_0$ changes sign at $\bar{\mu}$, then the index of $(\lambda, \mu) \mapsto \big(P^\star(\lambda, \mu; 0), Q^\star(\lambda, \mu; 0)\big)$ at $(0, \bar{\mu})$ is ± 1, as an examination of the rotation of this vector field shows. (Since this field is not differentiable, we cannot use definition (3.22) to compute the degree.) Following the proof of Theorem 3.30, we obtain

3.32. Theorem. *Let $0 < b_0 < Q^\star(0, -\gamma; 0)$. Let $[0, \infty) \ni t \mapsto a^\star(t)$, $b^\star(t)$, $F^\star(\cdot, t)$, and $\nu^\star(\cdot, \cdot, t)$ be continuous functions with $a^\star(0) = 0$, $b^\star(0) = b_0$, with $a^\star(\cdot)^2 + b^\star(\cdot)^2$ having a positive lower bound and ultimately exceeding 1, and with $\nu^\star(\cdot, \cdot, t)$ having the same properties as $\hat{\nu}$. Then the set of solution pairs $\big((\lambda, \mu), t\big)$ of*

$$P^\star(\lambda, \mu, t) = a^\star(t), \quad Q^\star(\lambda, \mu, t) = b^\star(t) \tag{3.33}$$

contains a connected set joining $\big((0, \mu_1(b_0)), 0\big)$ to one of the solutions $\bar{\mu}$ of $Q^\star(0, \mu; 0) = b_0$ in $(-\gamma, 0)$. If $\bar{\mu}$ is any point at which $Q^\star(0, \cdot; 0) - b_0$ changes sign, then the set of solution pairs of (3.33) contains a connected set joining $\big((0, \bar{\mu}), 0\big)$ to $\big((0, \bar{\bar{\mu}}), 0\big)$ where $\bar{\bar{\mu}}$ is another point at which $Q^\star(0, \cdot; 0) = b_0$.

3.34. Exercise. Let $\hat{\nu}(N, \frac{1}{2} - \xi) = \hat{\nu}(N, \frac{1}{2} + \xi)$ and $F'(\frac{1}{2} - \xi) = F'(\frac{1}{2} + \xi)$. For $b = 0$ prove that the equation $Q(\lambda, \mu; F) = 0$ has a unique solution for μ, which is $\mu = -\gamma/2$. (Thus each support bears half the weight of the symmetric string. Hint: Let $\omega = \mu + \frac{\gamma}{2}$, $G(\xi) = \Gamma(\frac{1}{2} + \xi) - \frac{\gamma}{2}$. Thus $G(\cdot)$ is odd. Use the modified version of $Q(\lambda, \mu; F) = 0$ to isolate ω.) Thus (3.4^-) is reduced to the problem of solving $P(\lambda, -\gamma/2; F) = a$ for λ. By studying the form of this P, show that there are at least two solutions λ if a is small enough. (This is essentially the problem treated by Dickey (1969).) Now use degree theory to study the multiplicity and connectivity of solutions as b varies from 0.

The connectivity results of Theorem 3.30 and Theorem 3.32 essentially rely on the variation of only one parameter. It is natural, however, to choose parameters, say (a, b, γ), that vary in a finite-dimensional space, or even parameters $(a, b, F, \hat{\nu})$ that vary in an infinite-dimensional space. In these cases, we might expect the dimension of the set of solution pairs to be that of the set of parameters. Topological tools more subtle than degree theory are required to handle such problems. See Alexander & Antman (1981, 1983), Alexander & Yorke (1976), Fitzpatrick, Massabò, & Pejsachowicz (1983), and Ize, Massabò, Pejsachowicz, & Vignoli (1985). The paper of Alexander & Antman (1983) actually addresses the catenary problem with an infinite-dimensional parameter.

It is instructive to obtain an ordinary differential equation for y of (2.20), which explicitly gives the shape of the elastic catenary. We assume that $a > 0$. Let $\tilde{s}(x)$ be the arc length in the reference configuration to the material point having abscissa x in the deformed configuration. It is well-defined by virtue of Proposition 2.11. Then (2.17) and (2.20) imply that

$$\lambda y'(x) = \mu + F\big(\tilde{s}(x)\big), \quad \tilde{s}'(x) = \frac{\sqrt{1 + y'(x)^2}}{\nu(\tilde{s}(x))}. \tag{3.35a,b}$$

Equation (2.16a) implies that

$$\nu(\tilde{s}(x)) = \hat{\nu}\left(\lambda\sqrt{1 + y'(x)^2}, \tilde{s}(x)\right). \tag{3.36}$$

Since F is invertible, (3.35a) is equivalent to

$$\tilde{s}(x) = F^{-1}\big(\lambda y'(x) - \mu\big). \tag{3.37}$$

By differentiating (3.35a) with respect to x and substituting (3.35b)–(3.37) into the resulting equation, we obtain the desired equation

$$\lambda y''(x) = \frac{F'\left(F^{-1}\big(\lambda y'(x) - \mu\big)\right)\sqrt{1 + y'(x)^2}}{\hat{\nu}\left(\lambda\sqrt{1 + y'(x)^2}, F^{-1}\big(\lambda y'(x) - \mu\big)\right)}. \tag{3.38}$$

This equation implies that $\lambda y''(x) > 0$ for all x, in confirmation of Proposition 2.19.

3.39. Exercise. Find a set of four boundary conditions for y and y' appropriate for the catenary problem that should be appended to (3.38) to produce a boundary-value problem from which λ, μ, and the two constants of integration can be found.

If the string is inextensible so that $\hat{\nu} = 1$ and if it is uniform so that $F' = \gamma$, then (3.38) reduces to the classical equation

$$\lambda y''(x) = \gamma\sqrt{1 + y'(x)^2} \tag{3.40}$$

for the inextensible uniform catenary found by Leibniz and Joh. Bernoulli and integrated by Leibniz to show that y could be expressed as a hyperbolic cosine. Since the simplicity of (3.40) does not extend to (3.38), we used the intrinsic formulation of the problem culminating in (3.4) in our analysis of the multiplicity of solutions. Our approach of using the intrinsic geometric variables θ and ν, which originated with Jas. Bernoulli, is more natural for the description of the underlying physics. The problem of the inextensible uniform catenary is treated in detail in books on the calculus of variations. Its equations are the Euler-Lagrange equations for the extremization of the potential energy functional for the string (See Sec. II.10). Note how easy the variational treatments of Theorem 3.11 and Ex. 3.15 are as compared with those in books on the calculus of variations. I know of no book on the calculus of variations treating even the nonuniform inextensible string.

4. The Suspension Bridge Problem

We now take the vertical load $\boldsymbol{f}$ to have a prescribed intensity per unit horizontal length in its deformed configuration. Thus we assume that there is given an absolutely continuous function $[0, a] \ni x \mapsto G(x)$ and a positive number α such that

$$\boldsymbol{f}(s) = -G'(x(s))x'(s)\boldsymbol{j}, \quad G(0) = 0, \quad G' > \alpha \quad \text{a.e.}, \tag{4.1}$$

where $x(\cdot)$ is defined by (2.20). Thus the function F of (2.4) is defined by

$$F(s) \equiv G(x(s)). \tag{4.2}$$

Let $\tilde{s}(x)$ be the arc length in the reference configuration to the material point whose abscissa in the deformed configuration is x. We restrict our attention to the nondegenerate case in which $a > 0$ so that $\lambda \neq 0$. Then (2.16)–(2.18), and (2.20) imply that

$$\lambda y'(x) = \mu + G(x), \tag{4.3}$$

$$N(\tilde{s}(x)) = \lambda\sqrt{1 + y'(x)^2} = \pm\sqrt{\lambda^2 + [\mu + G(x)]^2}, \tag{4.4}$$

$$\tilde{s}'(x) = \frac{\sqrt{1 + y'(x)^2}}{\hat{\nu}(N(\tilde{s}(x)), \tilde{s}(x))} = \frac{\sqrt{\lambda^2 + [\mu + G(x)]^2}}{|\lambda|\hat{\nu}\left(\pm\sqrt{\lambda^2 + [\mu + G(x)]^2}, \tilde{s}(x)\right)}. \tag{4.5}$$

We obtain y by integrating (4.3). If the load is uniform so that there is a positive number γ for which $G(x) = \gamma x$, then y describes a parabola, no matter what the constitutive function $\hat{\nu}$ is! More generally, from (4.3) and Proposition 2.11 we have

4.6. Proposition. *Let $a > 0$. The shape of every string satisfying (2.1), (2.3), and (4.1) for which $\boldsymbol{r}$ is absolutely continuous and N is continuous is given by*

$$\lambda y(x) = \mu x + \int_0^x G(\xi)\, d\xi \tag{4.7}$$

no matter what its constitutive function and what values are assigned to a and b.

Note that (4.7) implies that any such y has an absolutely continuous derivative.

For $a > 0$, the problem for the suspension bridge for an elastic string is reduced to Eqs. (4.5) and (4.7), subject to the boundary conditions

$$y(0) = 0, \quad y(a) = b, \quad \tilde{s}(0) = 0, \quad \tilde{s}(a) = 1, \tag{4.8a,b,c,d}$$

the first of which has already been accounted for by (4.7). We get a solution if and only if λ and μ can be chosen to satisfy these conditions. The substitution of (4.7) into (4.8b) yields a linear relation between λ and μ:

$$\mu a = \lambda b - \int_0^a G(\xi)\, d\xi. \tag{4.9}$$

We substitute (4.9) into (4.5) and require its integral to satisfy (4.8c,d):

$$R^{\pm}(\lambda) \equiv \pm \int_0^a \frac{\tilde{\delta}(x,\lambda)}{\lambda \hat{\nu}\big(\pm\tilde{\delta}(x,\lambda), \tilde{s}(x)\big)} = 1 \tag{4.10$^\pm$}$$

where

(4.11)

$$\tilde{\delta}(x,\lambda) \equiv \sqrt{\lambda^2 + [\lambda\beta + H(x)]^2}, \quad \beta \equiv \frac{b}{a}, \quad H(x) \equiv G(x) - \frac{1}{a}\int_0^a G(\xi)\, d\xi.$$

We obtain a solution to our boundary-value problem whenever (4.10) has a solution λ. We first study tensile solutions corresponding to λ's satisfying (4.10$^+$).

4.12. Theorem. *Let $\hat{\nu}$ have the properties described at the beginning of Sec. 3.* (i) *There is at least one tensile solution of the boundary-value problem for $a > 0$.* (ii) *If $\hat{\nu}$ is independent of s and if $b = 0$, then there is exactly one such tensile solution.* (iii) *For each tensile solution, $\lambda \equiv \boldsymbol{n}(0) \cdot \boldsymbol{i} > 0$ and y is strictly convex.*

Proof. By the Intermediate Value Theorem, statement (i) follows from the continuity of R^+ on $(0, \infty)$ and from the limits

$$R^+(\lambda) \to 0 \quad \text{as} \quad \lambda \to \infty, \qquad R^+(\lambda) \to \infty \quad \text{as} \quad \lambda \to 0. \tag{4.13}$$

Under the hypotheses of (ii), the integrand of $dR^+(\lambda)/d\lambda$ is everywhere negative, so that (ii) holds. Equation (4.10$^+$) implies that $\lambda > 0$. Equation (4.7) then yields (iii). □

4.14. Problem. The proof of Theorem 4.12(i) shows that the algebraic number of solutions of (4.10$^+$) is odd. Either prove a uniqueness statement without the conditions of statement (ii) or else show that there are circumstances under which (4.10$^+$) has more than one solution.

As in Sec. 3, the existence question for compressive states is more delicate:

4.15. Theorem. (i) *If there is a compressive solution to the boundary-value problem, then* $a < 1$. (ii) *The algebraic number of solutions of compressive solutions is even.* (iii) *If* a, b/a, *and* G *are sufficiently small, then there are at least two compressive solutions.* (iv) *For each compressive solution,* $\lambda \equiv \boldsymbol{n}(0) \cdot \boldsymbol{i} < 0$ *and* y *is strictly concave.*

Proof. Statement (i) follows from the estimate that $R^-(\lambda) > a$. Statement (ii) follows from the limits $R^-(\lambda) \to \infty$ as $\lambda \nearrow 0$ and as $\lambda \to -\infty$. Statement (iii) follows from this last result and from the observation that $R^-(\lambda)$ can be made < 1 by taking a, b/a, and G small enough. Statement (iv) is proved as is statement (iii) of Theorem 4.12. □

4.16. Exercise. Use the ideas leading to Theorems 3.30 and 3.32 to determine the connectivity of solution pairs $(\lambda, (a, b))$ satisfying (4.10⁻).

4.17. Exercise. Let $\{\boldsymbol{i}, \boldsymbol{j}\}$ be an orthonormal pair of vectors. Let $x(s)\boldsymbol{i} + y(s)\boldsymbol{j}$ be the position of the material point of a string (of any material) constrained to deform in the $\{\boldsymbol{i}, \boldsymbol{j}\}$-plane. Suppose that the ends $s = 0$ and $s = 1$ of the string are fixed:

$$x(0) = 0 = y(0), \quad x(1) = a > 0, \quad y(1) = b \geq 0.$$

Suppose that the force applied to an arbitrary material segment (c, d) is

$$\left[\int_{x(c)}^{x(d)} f(x)\, dx \right] \boldsymbol{j}.$$

For what class of functions f does every possible equilibrium configuration of the string satisfy an equation of the form $y(s) = \omega(x(s))$ where ω is a polynomial of fourth-order?

4.18. Exercise. State and prove an analog of Theorem 4.15 for inextensible strings.

5. Equilibrium of Strings under Normal Loads

Let a string of reference length 2α be confined to the $\{\boldsymbol{i}, \boldsymbol{j}\}$-plane, so that (2.14) holds. If the string is subject to a load distribution $\boldsymbol{f}$ that lies in the $\{\boldsymbol{i}, \boldsymbol{j}\}$-plane and acts normal to the string, then $\boldsymbol{f}$ has the form

$$\boldsymbol{f}(s) = -h(s)\boldsymbol{k} \times \mathbf{e}. \tag{5.1}$$

We assume that h, which could depend on the configuration $\boldsymbol{r}$, is (Lebesgue-) integrable. We may interpret the string as a section of a cylindrical membrane with generators parallel to $\boldsymbol{k}$. Forces of the form (5.1) are exerted by arbitrary fluids at rest and by inviscid fluids in motion.

Let $\boldsymbol{r}$ and $\boldsymbol{n}$ satisfy (2.1), (2.2), and (5.1) with $\boldsymbol{r}$ absolutely continuous and with N continuous. We take $s \in [-\alpha, \alpha]$. Then (2.1) and (5.1) imply that

$$\boldsymbol{n}(s) = N(s)\mathbf{e}(s) = \boldsymbol{n}(0) + \int_0^s h(\xi)\boldsymbol{k} \times \mathbf{e}(\xi)\, d\xi. \tag{5.2}$$

Thus $\boldsymbol{n}$ and $|N|$ are absolutely continuous. Since N is assumed to be continuous, it is positive on an open set and it is negative on an open set, each of which can be represented as countable disjoint unions of open intervals. Let $\mathcal{I}$ denote any such interval. On $\mathcal{I}$, N is absolutely continuous and nowhere vanishing. For s in $\mathcal{I}$, we can divide (5.2) by $N(s)$ and thus deduce that $\boldsymbol{e}$ is absolutely continuous on $\mathcal{I}$. We know that the equilibrium equation (II.6.2) holds a.e. We can actually compute $\boldsymbol{n}' = N'\boldsymbol{e} + N\boldsymbol{e}'$ a.e. on $\mathcal{I}$. Thus (II.6.2) and (2.14) yield

$$N'\boldsymbol{e} + N\boldsymbol{e}' - h\boldsymbol{k} \times \boldsymbol{e} = \boldsymbol{0} \quad \text{a.e. on} \quad \mathcal{I}, \tag{5.3}$$

whence

$$N = \text{const.}, \quad N\theta' = h \quad \text{a.e. on} \quad \mathcal{I}. \tag{5.4a,b}$$

Since N is required to be continuous, we obtain

5.5. Proposition. *Let* $\boldsymbol{r}$ *and* $\boldsymbol{n}$ *satisfy* (2.1), (2.2), *and* (5.1) *with* $\boldsymbol{r}$ *absolutely continuous and with* N *continuous. Then* $N = \text{const.}$ *on* $[-\alpha, \alpha]$. *If* $N \neq 0$, *then* θ *is absolutely continuous. If* $N = 0$, *then* $h = 0$. *If* h *is a.e. positive or negative, then* ($N \neq 0$ *and thus*) θ' *is a.e. positive or negative, so that the curve* $\boldsymbol{r}$ *is convex.*

We shall limit our attention to problems for which the continuity of N is compatible with the form of h.

Let $\sigma(s)$ denote the arc length of $\boldsymbol{r}$ from $\boldsymbol{r}(0)$ to $\boldsymbol{r}(s)$. Then $\sigma' = \nu$. The positivity of ν, ensured by (2.2), implies that σ has an inverse $\tilde{s}$. Let $\tilde{\theta}(\sigma) \equiv \theta(\tilde{s}(\sigma))$. In view of (5.4b), the curvature of $\boldsymbol{r}$ at $\tilde{s}(\sigma)$ is given by

$$\tilde{\theta}'(\sigma) = \frac{h(\tilde{s}(\sigma))}{N\nu(\tilde{s}(\sigma))}. \tag{5.6}$$

The force $\boldsymbol{f}$ of (5.1) is called a *simple hydrostatic load* iff

$$h(s) = p\nu(s) \tag{5.7}$$

where p is a constant called the *pressure*. To account for the pressure exerted by a gas, we may let the pressure depend on the shape $\boldsymbol{r}$ of the container, so that it is a given functional with value $p[\boldsymbol{r}]$. (The equation of state of a gas typically prescribes the pressure as a function of the temperature and specific volume. For a uniform gas at rest within a closed cylindrical membrane, the specific volume can be replaced by the cross-sectional area enclosed by the membrane, which is determined by $\boldsymbol{r}$.) From (5.7) we find that the curvature (5.6) is constant:

5.8. Proposition. *Let the hypotheses of Proposition* 5.5 *hold with* h *given by* (5.7). *Then* $\boldsymbol{r}$ *is a circular arc.*

There are other cases in which circular equilibrium states occur. From (5.6) we immediately obtain

5.9. Proposition. *Let the hypotheses of Proposition 5.5 hold and let h have the form $h(s) = pj(\nu(s))$ where j is a given function. Let the string be uniform and elastic, so that $\hat{\nu}$ is independent of s. Then $\boldsymbol{r}$ is a circular arc.*

The case that the function $j = 1$ is often treated in the engineering literature. It represents an approximation to (5.7). The approximation process results in a weakening of Proposition 5.8 by the inclusion of the additional hypothesis that the string be uniform.

There is an illuminating alternative proof of Proposition 5.8: for $p \neq 0$, the substitution of (5.7) into (5.2) yields

$$\begin{aligned} N(s)\boldsymbol{e}(s) &= \boldsymbol{n}(0) + p\boldsymbol{k} \times \int_0^s \nu(\xi)\boldsymbol{e}(\xi)\,d\xi \\ &= \boldsymbol{n}(0) + p\boldsymbol{k} \times [\boldsymbol{r}(s) - \boldsymbol{r}(0)] = p\boldsymbol{k} \times [\boldsymbol{r}(s) - \boldsymbol{c}] \end{aligned} \tag{5.10}$$

where $p\boldsymbol{k} \times \boldsymbol{c} \equiv -\boldsymbol{n}(0) + p\boldsymbol{k} \times \boldsymbol{r}(0)$. We now use Proposition 5.5 to obtain

$$N = p|\boldsymbol{r}(s) - \boldsymbol{c}|, \tag{5.11}$$

which is the equation of a circle.

We now focus our attention on elastic strings subject to simple hydrostatic loads (5.7). We shall determine the existence and multiplicity of circular equilibrium states. Without loss of generality, we fix the orientation of $\boldsymbol{r}$ by taking $\theta' > 0$ and accordingly require that N and p have the same sign. In this case, (5.4b) and (5.11) can be written as

$$\theta'(s) = \nu(pq, s)/q, \quad N = pq, \quad q \equiv |\boldsymbol{r} - \boldsymbol{c}|. \tag{5.12a,b,c}$$

We first take $\alpha = \pi$ and study the existence of simple, closed circular configurations with

$$\theta(\pi) = \theta(-\pi) + 2\pi. \tag{5.13}$$

Since (5.12a) implies that $\tilde{\theta}(\sigma) = \tilde{\theta}(0) + \sigma/q$, we find from (5.6), (5.7), and (5.13) that

$$\begin{aligned} \boldsymbol{r}(\pi) - \boldsymbol{r}(-\pi) &= \int_{-\pi}^{\pi} \hat{\nu}(pq, s)[\cos\theta(s)\boldsymbol{i} + \sin\theta(s)\boldsymbol{j}]\,ds \\ &= \int_{\sigma(-\pi)}^{\sigma(\pi)} [\cos\tilde{\theta}(\sigma)\boldsymbol{i} + \sin\tilde{\theta}(\sigma)\boldsymbol{j}]\,d\sigma \\ &= q\int_{\tilde{\theta}(\sigma(\pi))-2\pi}^{\tilde{\theta}(\sigma(\pi))} [\cos\tilde{\theta}(\sigma)\boldsymbol{i} + \sin\tilde{\theta}(\sigma)\boldsymbol{j}]\,d\tilde{\theta} = \boldsymbol{0}. \end{aligned} \tag{5.14}$$

Thus (5.13) ensures that $\boldsymbol{r}$ is closed and therefore a full circle. We use (5.12a) to reduce (5.13) to

$$2\pi q = \int_{-\pi}^{\pi} \hat{\nu}(pq, s)\,ds. \tag{5.15}$$

If p and q satisfy this equation, then (5.12a) generates a solution of the full problem. By examining the properties of $\hat{\nu}$ as sketched in Fig. 3.7, or by examining Fig. 5.23, corresponding to the more complicated equation (5.21) below, we deduce the following result from (5.15):

5.16. Theorem. *Let $\hat{\nu}$ have the properties described at the beginning of Sec. 3. Let p be a given nonpositive number. Then (5.15) has exactly one solution q, which corresponds to a unique compressive state. There is a positive number $P[\hat{\nu}]$ depending on the constitutive function $\hat{\nu}$ such that if $0 < p < P[\hat{\nu}]$, then (5.15) has at least one solution q, which corresponds to a tensile state. For given positive p, there are materials $\hat{\nu}$ for which (5.15) has any specified number of tensile solutions. If $\hat{\nu}(N, s)/N \to 0$ as $N \to \infty$, then for any given positive p, (5.15) has at least one solution q. If $\hat{\nu}(N, s)/N$ has a positive lower bound for $N \in [0, \infty)$, then there is a positive number M such that (5.15) has no solutions for $p > M$. If q is a given positive number, then (5.15) has exactly one solution p.*

This theorem stands in marked contrast with Theorems 3.11, 3.20, 3.30, 4.12, and 4.15 because its statements on existence and multiplicity depend crucially on a threshold in material response distinguishing what may be called soft and hard materials. The presence of such a threshold, characteristic of problems in nonlinear elasticity, is a consequence of the nature of the dependence of the load intensity on the configuration in (5.6). A load depending on a configuration is said to be *live*. If $\hat{\nu}$ is independent of s, then (5.15) is equivalent to the much simpler equation $pq = \hat{N}(q)$, which comes directly from (5.4a) and the constitutive function for N (see Antman (1972, Sec. 23)).

5.17. Exercise. Suppose that a given quantity of gas is introduced inside the membrane whose section is $\boldsymbol{r}$. Then neither p nor q is known a priori. Instead, an equation of state relates the pressure to the density. For example, in a polytropic gas, the pressure is taken proportional to ρ^γ where ρ is the density of the gas and γ is a given number, usually taken to lie in $[1, \frac{5}{3}]$. The density is the reciprocal of the specific volume. In our setting, the equation of state for a polytropic gas reduces to $p = \beta q^{-2\gamma}$ where β is a prescribed positive number. Prove an analog of Theorem 5.16 for the problem in which neither p nor q is prescribed, but for which this equation of state holds.

We now turn to the more difficult problem (of the velaria), analogous to those treated in Secs. 3 and 4, in which the ends of the string are fixed:

(5.18) $$\boldsymbol{r}(\pm\alpha) = \pm\boldsymbol{i}.$$

Equation (5.12a) implies that

(5.19) $$\theta(\alpha) - \theta(-\alpha) = \frac{1}{q}\int_{-\alpha}^{\alpha} \hat{\nu}(pq, s)\, ds.$$

Since $\theta' > 0$ and since $\boldsymbol{r}$ must be circular and satisfy (5.18), elementary geometry implies that

(5.20) $$\theta(\alpha) = -\theta(-\alpha) \in [0, \pi), \quad \sin\theta(\alpha) = 1/q.$$

Let S^- be the inverse of sin on $[0, \frac{\pi}{2}]$ and S^+ be the inverse of sin on $[\frac{\pi}{2}, \pi]$. Then (5.19) and (5.20) imply that

(5.21) $$qS^\pm\left(\frac{1}{q}\right) = \frac{1}{2}\int_{-\alpha}^{\alpha} \hat{\nu}(pq, s)\, ds \equiv \alpha\bar{\nu}(pq).$$

If (5.20) and (5.21) hold, then we follow (5.14) to find that

(5.22) $$\boldsymbol{r}(\alpha) - \boldsymbol{r}(-\alpha) = \int_{\sigma(-\alpha)}^{\sigma(\alpha)} [\cos\tilde{\theta}(\sigma)\boldsymbol{i} + \sin\tilde{\theta}(\sigma)\boldsymbol{j}]\, d\sigma = 2\boldsymbol{i}.$$

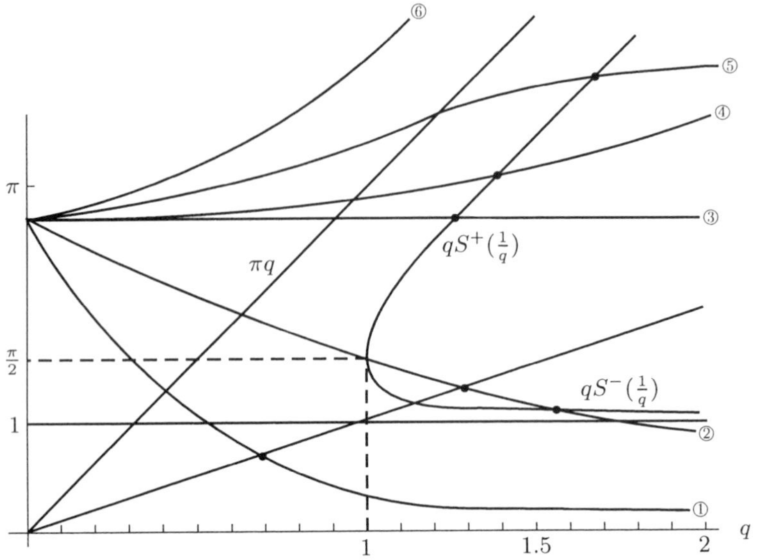

Fig. 5.23. Graphs of $q \mapsto qS^{\pm}(1/q)$, $q \mapsto q$, and the functions $q \mapsto \bar{\nu}(pq)$ for different p's and different rates of growth of $\hat{\nu}(\cdot, s)$. (Note that the vertical and horizontal scales are different.) The intersections of the curves $q \mapsto \bar{\nu}(pq)$ with the straight line of slope 1 give solutions of (5.15). The intersections of these curves with the graphs of $q \mapsto qS^{\pm}(1/q)$ give solutions of (5.21). The function $q \mapsto \bar{\nu}(pq)$ is represented by curve 1 for p large and negative, by curve 2 for p small and negative, by curve 3 for $p = 0$, by curve 4 for p small and positive, by curve 5 for a sufficiently large positive p when $\bar{\nu}$ is asymptotically sublinear, and by curve 6 for p positive when $\bar{\nu}$ is asymptotically superlinear. Curve 3 also corresponds to an inextensible string.

Thus (5.18) is automatically satisfied whenever (5.21) is satisfied. Each solution of the latter generates an equilibrium state, which is *tensile* if $\bar{\nu} > 1$ and *compressive* if $\bar{\nu} < 1$. In Fig. 5.23 we plot graphs of the functions $q \mapsto qS^{\pm}(1/q)$ and $q \mapsto \bar{\nu}(pq)$ when $\alpha > \frac{\pi}{2}$. From this figure and from its analogs for $1 \leq \alpha \leq \frac{\pi}{2}$ and for $\alpha < 1$, we deduce

5.24. Theorem. *Let $\hat{\nu}$ have the properties described at the beginning of Sec. 3.*

(i) *Let $\alpha > 1$ and let p be given. There is a positive number $P_1[\hat{\nu}]$ depending on the constitutive function $\hat{\nu}$ such that if $p < -P_1[\hat{\nu}]$, then (5.21) has no solution q. There is a positive number $P_2[\hat{\nu}]$ such that if $-P_2[\hat{\nu}] < p < 0$, then (5.21) has at least two solutions q; all solutions correspond to compressive equilibrium states. There is a positive number $P_3[\hat{\nu}]$ such that if $0 < p < P_3[\hat{\nu}]$, then (5.21) has at least one solution q; each solution corresponds to a tensile equilibrium state. For given positive p there are materials $\hat{\nu}$ for which (5.21) has any specified number of tensile solutions. If $\hat{\nu}(N, s)/N \to 0$ as $N \to \infty$, then for any given positive p, (5.21) has at least one solution q. If $\hat{\nu}(N, s)/N$ has a positive lower bound for $N \in [0, \infty)$, then there is a positive number M such that (5.21) has no solutions for $p > M$.*

(ii) *Let $\alpha > 1$ and let q be given. If $q < 1$, then (5.21) has no solutions p (as is geometrically obvious). If $q = 1$, then (5.21) has exactly one solution p, which is negative if $\alpha > \frac{\pi}{2}$, zero if $\alpha = \frac{\pi}{2}$, and positive if $\alpha < \frac{\pi}{2}$. Let $q^\star(\alpha)$ be the solution of $qS^-(1/q) = \alpha$ if $\alpha \leq \frac{\pi}{2}$ and be the solution of $qS^+(1/q) = \alpha$ if*

$\alpha \geq \frac{\pi}{2}$. If $q \in (1, q^\star)$ and if $\alpha > \frac{\pi}{2}$, then (5.21) has exactly two solutions p, each of which is negative. If $q \in (1, q^\star)$ and if $\alpha = \frac{\pi}{2}$, then (5.21) has exactly two solutions p, one positive and one negative. If $q \in (1, q^\star)$ and if $\alpha < \frac{\pi}{2}$, then (5.21) has exactly two solutions p, each of which is positive. If $q = q^\star$ and if $\alpha > \frac{\pi}{2}$, then $q^\star$ must equal 1, and (5.21) has exactly two solutions p, one negative and one zero. If $q = q^\star$ and if $\alpha = \frac{\pi}{2}$, then $q^\star$ must equal 1, and (5.21) has the unique solution $p = 0$. If $q = q^\star$ and if $\alpha < \frac{\pi}{2}$, then (5.21) has exactly two solutions p, one positive and one zero. If $q > q^\star$, then (5.21) has exactly two solutions p, one positive and one negative.

(iii) *Let $\alpha \leq 1$ and let p be given. Equation (5.21) has no solutions for $p \leq 0$. For $p > 0$, the results of* (i) *hold.*

(iv) *Let $\alpha \leq 1$ and let q be given. Equation (5.21) has no solutions p for $q < 1$, has exactly one solution p (which is positive and corresponds to a tensile state) if $q = 1$, and has exactly two solutions p (which are positive and correspond to tensile states) if $q > 1$.*

We now study the problem (of the lintearia) in which the pressure increases linearly with depth. In particular, we regard the string as a cross section of a cylindrical membrane supporting a heavy incompressible liquid. We take $\boldsymbol{j}$ to point in the vertical direction and define $x = \boldsymbol{r} \cdot \boldsymbol{i}$, $y = \boldsymbol{r} \cdot \boldsymbol{j}$. Let $s \in [0, 1]$ and let (2.3) hold. Let h be a given number, representing the elevation of the free surface of the liquid. We assume that the pressure force is

$$\boldsymbol{f}(s) = \begin{cases} -p\boldsymbol{k} \times \boldsymbol{r}'(s) & \text{if} \quad y(s) \geq h, \\ -\{p + \omega[h - y(s)]\}\boldsymbol{k} \times \boldsymbol{r}'(s) & \text{if} \quad y(s) \leq h. \end{cases} \tag{5.25}$$

Here p is a given nonnegative number, representing atmospheric (simple hydrostatic) pressure, and ω is a given positive number, representing the weight of the liquid per unit length of the $\boldsymbol{k}$-direction. The height h can be prescribed if there is an infinite reservoir of liquid available, as shown in Fig. 5.26. For simplicity, we assume that the entire string is in contact with the liquid: $y(s) \leq h$ for $0 \leq s \leq 1$. In particular, we assume that $b \leq h$.

Equations (5.4b) and (5.25) imply that

$$N\theta'(s) = \{p + \omega[h - y(s)]\}\hat{\nu}(N, s). \tag{5.27}$$

From (2.14) we obtain

$$x'(s) = \hat{\nu}(N, s)\cos\theta(s), \quad y'(s) = \hat{\nu}(N, s)\sin\theta(s). \tag{5.28a,b}$$

This system (5.27), (5.28) of ordinary differential equations is supplemented by the four boundary conditions (2.3), which correspond to the three constants of integration for the system and the unknown parameter N.

5.29. Problem. Carry out a full existence and multiplicity theory for this boundary-value problem along the lines of Sec. 3. The difficulty here is caused by the absence of integrals of the ordinary differential equations, which reduce their study to that of finite-dimensional equations. The Poincaré shooting method, described in Sec. II.9, can be used to reduce the boundary-value problem to a finite-dimensional system. This system

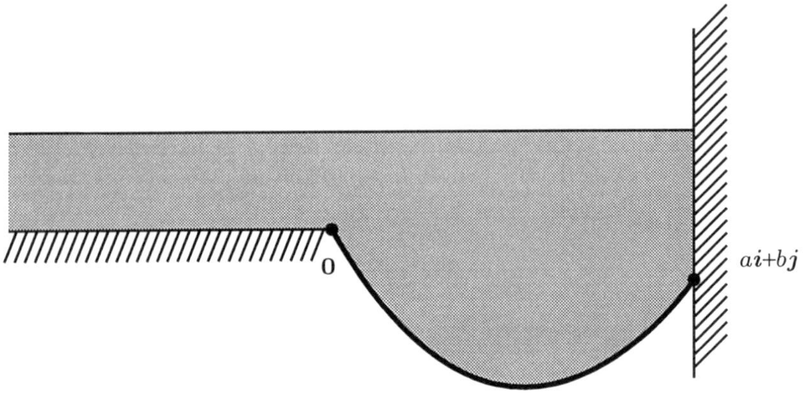

Fig. 5.26. A string (cylindrical membrane) supporting a heavy liquid with its free surface maintained at a prescribed height by an infinite reservoir of liquid.

can be studied by the degree-theoretic methods used in Sec. 3. (A similar approach is used by Antman & Wolfe (1983) for problems of Sec. 6.)

Using the notation of (5.6) and its obvious extensions, we can write (5.27) and (5.28) as

$$(5.30\text{a,b,c}) \quad N\tilde{\theta}'(\sigma) = \omega[h - \tilde{y}(\sigma)], \quad \tilde{x}'(\sigma) = \cos\tilde{\theta}(\sigma), \quad \tilde{y}'(\sigma) = \sin\tilde{\theta}(\sigma).$$

Since N is constant, we eliminate $\tilde{y}$ from (5.30a,b) to find that $\tilde{\theta}$ satisfies

$$(5.30\text{d}) \qquad \tilde{\theta}''(\sigma) + (\omega/N)\sin\tilde{\theta}(\sigma) = 0,$$

which is the equation of the pendulum and of the elastica (the latter discussed in Chap. IV and elsewhere). Boundary conditions for (5.30d) come from (2.3) and (5.30a,b):

$$(5.31) \qquad N\tilde{\theta}'(0) = \omega h, \quad N\tilde{\theta}'(\sigma(1)) = \omega(h - b).$$

5.32. Exercise. Suppose that (5.25) holds, but that h is not prescribed. Instead, the membrane is required to support a given amount of liquid. Thus the cross-sectional area α occupied by the liquid is prescribed. See Fig. 5.33. Show how the problem just described must be modified to accommodate this change. (Hint: Represent the prescribed area as a line integral over its boundary, consisting partly of $\boldsymbol{r}$, partly of the free surface $y = h$, where h is not known, and partly of rigid walls. Assume that these rigid walls are vertical as in Fig. 5.33.)

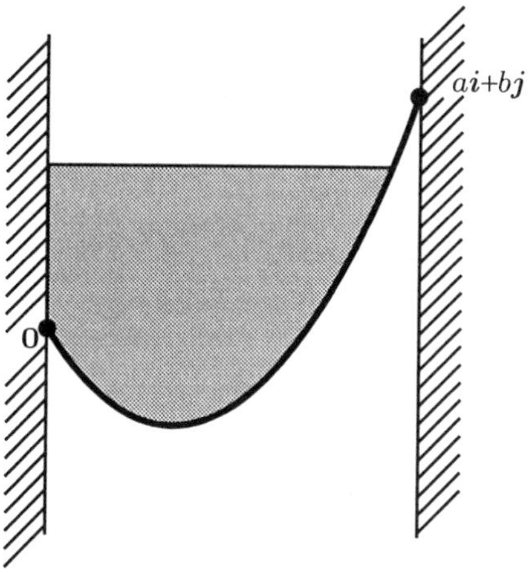

Fig. 5.33. Membrane supporting a fixed amount of liquid.

5.34. Exercise. Under certain standard simplifying assumptions, the force on a wire in equilibrium carrying an electric current I in an ambient magnetic field $\boldsymbol{b}$ has intensity $\boldsymbol{f} = I\boldsymbol{r}' \times \boldsymbol{b}$ per unit reference length. Suppose that $\boldsymbol{b} = B\boldsymbol{k}$ and that IB is a nonzero constant. Interpret the wire as an elastic string. From the analog of (5.2), prove that $\boldsymbol{n}$ and $|N|$ are absolutely continuous when $\boldsymbol{r}$ is absolutely continuous. Assuming that N itself is continuous, prove that N is constant. Show that if $N \neq 0$, then

$$\boldsymbol{r}'(s) \cdot \boldsymbol{k} = \nu(s)\boldsymbol{e}(s) \cdot \boldsymbol{k} = \frac{\hat{\nu}(N,s)}{N}\boldsymbol{n}(0) \cdot \boldsymbol{k}. \tag{5.35}$$

Let $\boldsymbol{P}$ be the projection of $\mathbb{E}^3$ onto the $\{\boldsymbol{i},\boldsymbol{j}\}$-plane. Thus $\boldsymbol{P} \cdot \boldsymbol{v} = -\boldsymbol{k} \times (\boldsymbol{k} \times \boldsymbol{v}) = \boldsymbol{v} - (\boldsymbol{v} \cdot \boldsymbol{k})\boldsymbol{k}$. Show that

$$\boldsymbol{P} \cdot [\boldsymbol{r} - \boldsymbol{r}(0)] - (IB)^{-1}\boldsymbol{k} \times \boldsymbol{n}(0) = -(IB)^{-1}\boldsymbol{k} \times \boldsymbol{n}. \tag{5.36}$$

(This equation is equivalent to (5.10) for strings in a plane under simple hydrostatic pressure.) Show that $\boldsymbol{n} \cdot \boldsymbol{k} = \boldsymbol{n}(0) \cdot \boldsymbol{k}$ and thus that

$$\begin{gathered} \boldsymbol{P} \cdot [\boldsymbol{n} - \boldsymbol{n}(0)] = \boldsymbol{n} - \boldsymbol{n}(0), \quad |\boldsymbol{P} \cdot \boldsymbol{n}| = \text{const.}, \\ |\boldsymbol{P} \cdot [\boldsymbol{r} - \boldsymbol{r}(0)] - (IB)^{-1}\boldsymbol{k} \times \boldsymbol{n}(0)| = |(IB)^{-1}\boldsymbol{P} \cdot \boldsymbol{n}| \equiv q \text{ (const)}. \end{gathered} \tag{5.37}$$

Thus $\boldsymbol{r}$ lies on a circular cylinder of radius q with generators parallel to the $\boldsymbol{k}$-axis. We can set (5.36) equal to $q\boldsymbol{a} \equiv q(\cos\theta\boldsymbol{i} + \sin\theta\boldsymbol{j})$. Use (5.36) to show that $q\theta'\boldsymbol{k} \times \boldsymbol{a} = \boldsymbol{P} \cdot \boldsymbol{r}' = \nu IBq\boldsymbol{k} \times \boldsymbol{a}/N$ so that

$$\theta' = \nu IB/N. \tag{5.38}$$

Introduce the variables of (5.6) into (5.35) and (5.38) and deduce that the image of $\boldsymbol{r}$ is a helix (no matter what is the constitutive equation of the string). Let $0 \le s \le 1$. Let the string satisfy the boundary conditions

$$\boldsymbol{r}(0) = \boldsymbol{0}, \quad \boldsymbol{r}(1) = c\boldsymbol{k} \tag{5.39}$$

where c is a given number. Prove that the resulting boundary-value problem has a solution corresponding to each solution N of the equations

$$2m\pi N = IB \int_0^1 \hat{\nu}(N,s)\, ds \tag{5.40}$$

where m is an integer. Show how the number of solutions of (5.40) is affected by the behavior of $\hat{\nu}(\cdot, s)$. (This problem is based on Wolfe (1983). If the wire is also rotating, the analysis becomes much richer. See Wolfe (1985) and Healey (1990).)

If the magnetic field is perpendicular to the line joining the ends of the wire, then the wire admits an equilibrium configuration lying in the plane perpendicular to the magnetic field. The governing equations are exactly those for a string under a simple hydrostatic pressure. To make physical sense of this hydrostatic pressure, we interpret the string as a section of a cylindrical membrane. In contrast, the physical interpretation of a string lying in a plane perpendicular to a magnetic field is intrinsically one-dimensional.

6. Equilibrium of Strings under Central Forces

We study general properties of solutions of the equilibrium equations for elastic strings under central forces. Our development, which is similar in spirit to that of Sec. 2, uses techniques from the theory of the motion of a particle in a central force field.

We assume that the force intensity $\boldsymbol{f}$ is central:

$$\boldsymbol{f}(s) = f(\boldsymbol{r}(s), s)\frac{\boldsymbol{r}(s)}{r(s)} \tag{6.1}$$

where $r \equiv |\boldsymbol{r}|$. We assume that $f(\cdot, s)$ is continuous on $\mathbb{E}^3 \setminus \{\boldsymbol{0}\}$ and that $f(\boldsymbol{p}, \cdot)$ is integrable for each $\boldsymbol{p} \in \mathbb{E}^3 \setminus \{\boldsymbol{0}\}$. $\boldsymbol{f}$ could account for a central gravitational, electrostatic, or magnetostatic force. In this case, f is given by the inverse-square law:

$$f(\boldsymbol{r}, s) = \mu(s) r^{-2}, \tag{6.2}$$

where μ is a given function of fixed sign. (Problems for strings in which (6.2) holds may offer good models for the tethering of space structures.) $\boldsymbol{f}$ could also account for a centrifugal force on a string constrained to lie in a plane due to the rotation of the string with constant angular velocity ω about an axis perpendicular to the plane. In this case,

$$f(\boldsymbol{r}, s) = \omega^2 (\rho A)(s) r. \tag{6.3}$$

We focus on force fields like (6.2) that are singular at the point $\boldsymbol{0}$, which we treat with circumspection. We assume that on any compact subset of $[\mathbb{E}^3 \setminus \{\boldsymbol{0}\}] \times [0, 1]$ the function f either has a positive lower bound or a negative upper bound. We seek an absolutely continuous position field $\boldsymbol{r}$, a contact force field $\boldsymbol{n}$ that is absolutely continuous where $r > 0$, and a tension field N that is continuous where $r > 0$ satisfying (2.1), (2.2), (6.1) wherever $r > 0$ and satisfying the boundary conditions

$$\boldsymbol{r}(0) = \boldsymbol{r}_0, \quad \boldsymbol{r}(1) = \boldsymbol{r}_1, \tag{6.4}$$

where $\boldsymbol{r}_0$ and $\boldsymbol{r}_1$ are given vectors in $\mathbb{E}^3$.

Let $\mathcal{R}$ be the set of material points s on which $r > 0$. $\mathcal{R}$ is open because r is assumed to be continuous. We show that N can vanish only at isolated points of $\mathcal{R}$ (although the zeros of N could accumulate at the boundary points of $\mathcal{R}$ where $r = 0$). Suppose for contradiction that there were a sequence $\{s_k\}$ of points of $\mathcal{R}$ converging to a point $s_\infty \in \mathcal{R}$ with $N(s_k) = 0$, $N(s_\infty) = 0$. Then (2.1) and (6.1) imply that

$$\int_{s_\infty}^{s_k} f(\boldsymbol{r}(s), s)\frac{\boldsymbol{r}(s)}{r(s)}\, ds = \boldsymbol{0}, \quad k = 1, 2, \ldots. \tag{6.5}$$

Since $\boldsymbol{r}$ is absolutely continuous and nonzero on $\mathcal{R}$, the function $s \mapsto \frac{\boldsymbol{r}(s_\infty)}{r(s_\infty)} \cdot \frac{\boldsymbol{r}(s)}{r(s)} \equiv R(s)$ is absolutely continuous on $\mathcal{R}$ and equals 1 at s_∞. Thus (6.5) implies that

$$\int_{s_\infty}^{s_k} f(\boldsymbol{r}(s), s) R(s)\, ds = 0, \quad k = 1, 2, \ldots. \tag{6.6}$$

But R is positive near s_∞ and f vanishes nowhere. Therefore the integrand in (6.6) has one sign for large enough k, so that (6.6) is impossible.

Let $\mathcal{N}$ be the subset (necessarily open) of $\mathcal{R}$ on which N does not vanish. As in the preceding sections, we find that N is absolutely continuous on $\mathcal{R}$, that $\boldsymbol{e}$ is absolutely continuous on $\mathcal{N}$, and that

$$\boldsymbol{n}'(s) + \frac{f(\boldsymbol{r}(s), s)\boldsymbol{r}}{r} = \boldsymbol{0} \tag{6.7}$$

a.e. on $\mathcal{N}$. In view of the isolation of the zeros of N on $\mathcal{R}$, Eq. (6.7) holds a.e. on $\mathcal{R}$. We operate on (6.7) with $\boldsymbol{r}\times$ to obtain

$$(\boldsymbol{r} \times \boldsymbol{n})' = \boldsymbol{0} \tag{6.8}$$

a.e. on $\mathcal{R}$ since $\boldsymbol{r}' \times \boldsymbol{n} = \boldsymbol{0}$ by (2.1b,c). By properties of absolutely continuous functions, (6.8) implies that on each component interval $\mathcal{P}$ of $\mathcal{R}$ there is a vector $\boldsymbol{c}$ (depending on $\mathcal{P}$) such that

$$\boldsymbol{r} \times \boldsymbol{n} = \boldsymbol{c}, \quad \boldsymbol{r} \cdot \boldsymbol{c} = 0, \quad \boldsymbol{n} \cdot \boldsymbol{c} = 0, \tag{6.9a,b,c}$$

so that the restrictions of $\boldsymbol{r}$ and $\boldsymbol{n}$ to $\mathcal{P}$ lie in a plane perpendicular to $\boldsymbol{c}$.

Now suppose that N vanishes somewhere on $\mathcal{P}$. Then (6.9a) implies that $\boldsymbol{c} = \boldsymbol{0}$. Since N vanishes only at isolated points of $\mathcal{P}$, Eqs. (2.1) and (6.9a) imply that $\boldsymbol{r} \times \boldsymbol{r}' = \boldsymbol{0}$ a.e. on $\mathcal{P}$. Since the positivity of r and ν on $\mathcal{P}$ prevent $\boldsymbol{r}$ and $\boldsymbol{r}'$ from vanishing except on a set of measure 0, it follows that $\boldsymbol{r}$ and $\boldsymbol{r}'$ must be parallel on $\mathcal{P}$. Thus there must be a locally integrable function γ on $\mathcal{P}$ that can vanish at most on a set of measure 0 such that

$$\boldsymbol{r}'(s) = \gamma(s)\boldsymbol{r}(s) \tag{6.10}$$

a.e. on $\mathcal{P}$, whence it follows that

$$\boldsymbol{r}(s) = \boldsymbol{r}(a) \exp \int_a^s \gamma(\xi)\, d\xi \tag{6.11}$$

on $\mathcal{P}$ where a is any point in $\mathcal{P}$. This equation says that $\boldsymbol{r}(\mathcal{P})$ lies on a ray. We get the same conclusion if $\boldsymbol{r} \times \boldsymbol{r}'$ should vanish at a point of $\mathcal{P}$. Thus the material of $\mathcal{P}$ must either lie on a ray or else be nowhere radial. The latter case occurs if and only if $\boldsymbol{c} \neq \boldsymbol{0}$. When this happens, N cannot vanish on $\mathcal{P}$ (by (6.9a)) so the configuration of $\mathcal{P}$ is either tensile or compressive and so that $\boldsymbol{e}$ is absolutely continuous on $\mathcal{P}$. We summarize these results:

6.12. Proposition. *Let $\boldsymbol{r}$ be absolutely continuous and let N be continuous where $r > 0$. Let $\boldsymbol{r}$ and $\boldsymbol{n}$ satisfy* (2.1), (2.2), *and* (6.1) *on $\mathcal{P}$, a component open interval of $\mathcal{R}$. Then N can vanish only at isolated points of $\mathcal{P}$, N is absolutely continuous on $\mathcal{P}$, and $\boldsymbol{e}$ is absolutely continuous on each open subinterval of $\mathcal{P}$ on which N does not vanish. If N vanishes on $\mathcal{P}$ or if the tangent to the string is radial at a point of $\mathcal{P}$, then the configuration of $\mathcal{P}$ is radial. Otherwise, this configuration is a plane curve that is nowhere radial and the material of $\mathcal{P}$ is either everywhere in tension or everywhere in compression.*

We readily strengthen the regularity theory of this proposition by introducing our standard constitutive hypotheses:

6.13. Proposition. *Let $\boldsymbol{r}$ and $\boldsymbol{n}$ satisfy the hypotheses of Proposition* 6.12 *and satisfy* (II.6.4). *Let $\hat{\nu}$ have the properties described at the beginning of Sec.* 3. *Then $\boldsymbol{r}'$ is absolutely continuous on $\mathcal{N} \cap \mathcal{P}$. If f is continuous on $(\mathbb{E}^3 \setminus \{\boldsymbol{0}\}) \times \mathcal{P}$, then $\boldsymbol{r}$ is twice continuously differentiable on $\mathcal{N} \cap \mathcal{P}$.*

We now study nonradial configurations on $\mathcal{N} \cap \mathcal{P}$, for which $\boldsymbol{c} \neq \boldsymbol{0}$, N vanishes nowhere, and the string is confined to a plane perpendicular to $\boldsymbol{c}$, which we take to be spanned by the orthonormal pair $\{\boldsymbol{i}, \boldsymbol{j}\}$. We can then locate $\boldsymbol{r}$ with polar coordinates r and ϕ:

$$\boldsymbol{r}(s) = r(s)[\cos\phi(s)\boldsymbol{i} + \sin\phi(s)\boldsymbol{j}]. \tag{6.14}$$

Thus

$$\nu = \sqrt{(r')^2 + (r\phi')^2}, \quad r' = \nu\cos(\theta - \phi), \quad r\phi' = \nu\sin(\theta - \phi). \tag{6.15a,b,c}$$

We substitute (6.14) and (2.1b,c) into (6.9a) and dot the resulting equation with $\boldsymbol{k}$ to obtain

$$r^2\phi' = \boldsymbol{c}\cdot\boldsymbol{k}\nu/N = c\nu/|N| \tag{6.16}$$

in which we ensure that $\phi' > 0$ by choosing $\{\boldsymbol{i}, \boldsymbol{j}, \boldsymbol{k}\}$ so that $\boldsymbol{c}\cdot\boldsymbol{k}$ has the same sign as N, and $c = |\boldsymbol{c}| = |\boldsymbol{c}\cdot\boldsymbol{k}|$. Under the hypotheses of Proposition 6.13, ϕ' is absolutely continuous on $\mathcal{P}$.

Now we substitute (2.1b,c) and (2.14) into (6.7), which holds everywhere on $\mathcal{P}$ in virtue of Proposition 6.13, and then dot the resulting equation with $\boldsymbol{k}\times\boldsymbol{e}$ to obtain

$$N\theta' = \frac{f\boldsymbol{k}\cdot(\boldsymbol{r}\times\boldsymbol{e})}{r} = f\sin(\theta - \phi). \tag{6.17}$$

From (6.15c) and (6.17) we obtain

$$N\theta' = rf\phi'/\nu. \tag{6.18}$$

Since ϕ' is positive on $\mathcal{P}$, both θ' and the curvature θ'/ν have the same sign as Nf. We say that the curve $\boldsymbol{r}$ is *bowed out* on $\mathcal{P}$ if $\theta' > 0$ on $\mathcal{P}$ and *bowed in* if $\theta' < 0$ on $\mathcal{P}$. In particular, if f is attractive, so that $f < 0$, then each bowed-out configuration is compressive and each bowed-in configuration is tensile.

Since ϕ' is positive on $\mathcal{P}$ (when $c > 0$), the function $\mathcal{P} \ni s \mapsto \phi(s)$ has an inverse $\tilde{s}$, which is continuously differentiable when the hypotheses of Proposition 6.13 hold. We set

$$u(\phi) \equiv 1/r(\tilde{s}(\phi)). \tag{6.19}$$

The substitution of (6.14), (6.16), and (6.19) into the radial component of (6.7) yields

$$u''(\phi) + u(\phi) = \frac{N(\tilde{s}(\phi))f(\boldsymbol{r}(\tilde{s}(\phi)), \tilde{s}(\phi))}{c^2\nu(\tilde{s}(\phi))u(\phi)^2} \tag{6.20}$$

a.e. on $\mathcal{P}$ and everywhere on $\mathcal{P}$ that f and ν are continuous. The substitution of (6.16) and (6.19) into (6.15) yields

$$N(\tilde{s}(\phi))^2 = c^2[u(\phi)^2 + u'(\phi)^2]. \tag{6.21}$$

We replace ν in (6.20) and (6.16) by its constitutive function and then substitute (6.21) into the resulting equations to obtain the following semilinear system of ordinary differential equations for u and $\tilde{s}$:

$$u'' + u = \pm\frac{\sqrt{u^2 + (u')^2}\, f((\cos\phi\boldsymbol{i} + \sin\phi\boldsymbol{j})/u, \tilde{s})}{c\hat{\nu}\left(\pm c\sqrt{u^2 + (u')^2}, \tilde{s}\right)u^2}, \tag{6.22$^\pm$}$$

$$\tilde{s}' = \frac{\sqrt{u^2 + (u')^2}}{\hat{\nu}\left(\pm c\sqrt{u^2 + (u')^2}, \tilde{s}\right)u^2}. \tag{6.23$^\pm$}$$

In the important case that $\mathcal{P} = [0,1]$, these equations are to be supplemented by boundary conditions equivalent to (6.4), namely,

(6.24a,b,c,d) $$u(0) = u_0, \quad u(\phi_1) = u_1, \quad \tilde{s}(0) = 0, \quad \tilde{s}(\phi_1) = 1.$$

Here we have taken $\phi(0) = 0$ and $\phi_1 = \phi(1)$ without loss of generality. $\phi(1) - \phi(0)$ is determined by (6.4). The four conditions of (6.24) correspond to the three unknown constants of integration for (6.22) and (6.23) and to the unknown parameter c.

If f depends only on r and s, then (6.22) and (6.23) reduces to an autonomous system for u and $\tilde{s}$. If furthermore the string and force field are uniform so that $\hat{\nu}$ and f are independent of s, in which case we can set $f(r,s) \equiv g(u)u^2$, then (6.22) and (6.23) uncouple, with the former reducing to the autonomous equation

(6.25±) $$u'' + u = \pm \frac{\sqrt{u^2 + (u')^2}\, g(u)}{c\hat{\nu}(\pm c\sqrt{u^2 + (u')^2})}.$$

In this case, we obtain a regular configuration, i.e., one with r everywhere positive, provided that c can be found so that u satisfies (6.25±), (6.24a,b), and the auxiliary condition

(6.26±) $$\int_0^{\phi_1} \frac{\sqrt{u^2 + (u')^2}}{\hat{\nu}(\pm c\sqrt{u^2 + (u')^2}, \tilde{s})u^2}\, d\phi = 1,$$

which comes from (6.23±) and (6.24c,d).

By multiplying (6.25) by u' and then rearranging and integrating the resulting equation we obtain the integral

(6.27) $$c^2[u^2 + (u')^2] = [h(E + G(u))]^2$$

where E is a constant, h is the inverse of $N \mapsto W^*(N) \equiv \int_0^N \hat{\nu}(n)\, dn$, and $G(u) \equiv \int_1^u g(v)\, dv$. Thus the phase portrait of (6.25) corresponding to parameters c and E consists of those points (u, u') at which the paraboloid $z = c^2[u^2 + (u')^2]$ intersects the cylindrical surface $z = [h(E + G(u))]^2$ in (u, u', z)-space. From the phase portrait we can determine detailed qualitative information about all regular solutions of the differential equations (6.25) and about possible regular solutions of the boundary-value problem (6.25), (6.24a,b), (6.26). This behavior depends upon the strength of the string, characterized by $\hat{\nu}$, the nature of the force field g, and the placement of $\boldsymbol{r}_0$ and $\boldsymbol{r}_1$.

6.28. Exercise. Sketch the phase portrait of (6.25) for various choices of g (and $\hat{\nu}$).

6.29. Exercise. A uniform force field f satisfies the inverse-square law (6.2) if and only if $g(u) = \mu$ (const.). In this case, (6.25) and (6.27) become simpler. If furthermore the string is inextensible (so that $\hat{\nu} = 1$), then (6.27) can be integrated in closed form. (This is the problem treated by Joh. Bernoulli.) Carry out the integration and express $\boldsymbol{r}$ in terms of the constants of the problem. Discuss the existence and multiplicity of solutions to the boundary-value problem (6.25), (6.24a,b), (6.26).

The treatment of singular solutions, given by Antman & Wolfe (1983) requires a careful extension of the laws of mechanics to handle infinite forces. These solutions have a very rich structure. These authors give a full existence and multiplicity theory by combining the Poincaré shooting method, described in Sec. II.6, with degree theory, described in Sec. 3.

7. Dynamical Problems

Radial vibrations. The classical equations of motion of a uniform elastic string confined to the $\{\boldsymbol{i}, \boldsymbol{j}\}$-plane and subjected to a simple hydrostatic pressure consist of (II.2.9), (II.2.10b), (II.2.11), (5.1), and (5.7):

(7.1) $$\left[\hat{N}(|\boldsymbol{r}_s|)\frac{\boldsymbol{r}_s}{|\boldsymbol{r}_s|}\right]_s - p\boldsymbol{k} \times \boldsymbol{r}_s = \rho A \boldsymbol{r}_{tt}.$$

(Our assumption of uniformity means that $\hat{N}$ and ρA are independent of s.) We assume that the configuration of the string is a simple closed loop in which s increases as the loop is traversed in the counterclockwise sense. In this case, a positive p tends to inflate the string.

We seek solutions of (7.1) in which each material point s executes a purely radial motion so that each configuration of the string is a circle of radius $r(t)$. Specifically, we suppose that the reference configuration is an unstretched circle of radius 1 and that the motion is restricted to have the form

$$\boldsymbol{r}(s,t) = r(t)[\cos s\boldsymbol{i} + \sin s\boldsymbol{j}] \equiv r(t)\boldsymbol{a}(s). \tag{7.2}$$

The substitution of (7.2) into (7.1) yields the following second-order autonomous ordinary differential equation for r:

$$\rho A r_{tt} + \hat{N}(r) - pr = 0. \tag{7.3a}$$

This equation has the integral

$$\rho A {r_t}^2 + 2W(r) - pr^2 = 2c \text{ (const.)}, \tag{7.3b}$$

where the stored-energy function W is defined in (II.10.10). These equations also govern the motion of some elastic rings.

7.4. Exercise. Investigate the way the phase portrait of (7.3) depends upon the sign of p and the nature of W. (This is an open-ended problem.) In particular, for $p > 0$ discuss how the behavior of the collection of solutions of (7.3) is influenced by conditions such as $W(r)/r^2 \to \infty$ as $r \to \infty$ and $W(r)/r^2 \to 0$ as $r \to \infty$. Discuss the different kinds of periodic solutions that (7.3) can have for different choices of W. Show how the phase portrait enables one to determine the restricted stability under radial perturbations of the equilibrium states discussed in Theorem 5.16.

7.5. Exercise. Do Ex. 7.4 under the assumption that the string is viscoelastic and has constitutive equation (II.2.14) satisfying (II.2.22).

Steady motions. Let $[0,l] \ni \sigma \mapsto \boldsymbol{p}(\sigma)$ denote a smooth, simple closed curve in $\mathbb{E}^3$. We take σ to be the arc-length parameter, so that $|\boldsymbol{p}'| = 1$. We study steady free motions of a loop of a uniform nonlinearly elastic string of natural reference length L occupying the curve $\boldsymbol{p}$ and having the form

$$\boldsymbol{r}(s,t) = \boldsymbol{p}(\lambda s - ct) \tag{7.6}$$

where λ is a positive number and c is a real number. Since l is the length of $\boldsymbol{p}$, we readily find that $\lambda = l/L$ is the constant stretch. Since $\boldsymbol{p}$ is closed, $\boldsymbol{p}'' \neq \boldsymbol{0}$. Thus the substitution of (7.6) into the equations of motion (II.2.9), (II.2.10b), (II.2.11) for uniform elastic strings under zero applied force leads to

$$\lambda \hat{N}(\lambda) = \rho A c^2. \tag{7.7}$$

Since $\hat{N}(\nu) < 0$ for $\nu < 1$, Eq. (7.7) has a unique solution for c^2 in terms of $\lambda = l/L$ if and only if $\lambda \geq 0$. Thus for any shape $\boldsymbol{p}$ there are exactly two such steady motions, having opposite directions, whenever the length of $\boldsymbol{p}$ is greater than or equal to the natural length of the string.

If $\boldsymbol{p}$ were interpreted as a frictionless tube, then a body force $\boldsymbol{f}$ acting in the direction $\boldsymbol{p}''(\sigma)$ at $\boldsymbol{p}(\sigma)$, which is normal to $\boldsymbol{p}'(\sigma)$, can always be found to enforce this constraint at any speed c. This problem represents a generalization by Healey & Papadopoulos (1990) of a problem for inextensible strings discussed by Routh (1905, Chap. 13). Healey & Papadopoulos discuss the stability of motions (7.6).

7.8. Problem. Investigate the cable-laying problem of Routh (1905, Art. 597) for extensible strings under various models of viscous drag.

Massless springs. We study the longitudinal motions along the $\boldsymbol{k}$-axis of a string whose end $s = 0$ is fixed at $\boldsymbol{0}$ and to whose end $s = 1$ is attached a material point of mass m that can move freely along this axis. Compressed states are clearly unstable for a string. However, the equations governing the longitudinal motion of a string are exactly the same as those governing the longitudinal motion of a straight rod, and rods have flexural stiffness, which increases the stability of longitudinal motions. Thus this problem, like that of radial vibrations just discussed, has physical meaning.

We set $\boldsymbol{r}(s,t) = z(s,t)\boldsymbol{k}$. Thus one boundary condition is

$$z(0,t) = 0. \tag{7.9}$$

If there are no body forces, then the appropriate version of the linear momentum law (cf. (II.2.7)) is

$$-N(s,t) = \int_s^1 (\rho A)(\xi) z_{tt}(\xi,t)\, d\xi + m z_{tt}(1,t), \tag{7.10}$$

from which we immediately obtain the governing partial differential equation

$$(\rho A)(s) z_{tt} = N_s \tag{7.11}$$

and the boundary condition

$$m z_{tt}(1,t) = -N(1,t) \tag{7.12}$$

at $s = 1$. We impose initial conditions

$$z(s,0) = z_0(s), \quad z_t(s,0) = z_1(s). \tag{7.13a,b}$$

We assume that z_0 and z_1 are continuously differentiable. To obtain a full initial-boundary-value problem for z, we need only choose a constitutive equation for N. In keeping with (II.2.2), we require that $z_s(s,t) > 0$. We shall study elastic and viscoelastic materials with constitutive equations coming from (II.2.11) and (II.2.14).

This one-dimensional motion of a string is a model for the motion of a spring. In elementary courses in mechanics, springs are assumed massless; their only role is to communicate forces to the bodies they join. The motions of such bodies are accordingly governed by ordinary differential equations. To study the status of the ordinary differential equations governing systems with massless springs, we may assume that the mass $\int_0^1 (\rho A)(s)\,ds$ of the string is small relative to the mass m by setting

$$(\rho A)(s) = \varepsilon\sigma(s) \tag{7.14}$$

where ε is a small positive parameter. We content ourselves here with studying the *reduced* initial-boundary-value problems for elastic and viscoelastic strings obtained by setting $\varepsilon = 0$.

We first study the reduced problem for elastic strings. Substituting (2.11) into (7.12), we obtain

$$m z_{tt}(1,t) = -N\big(z_s(1,t),1\big). \tag{7.15}$$

The presence of the s-derivative of z in (7.15) prevents it from being the expected ordinary differential equation for the motion $z(1,\cdot)$ of the tip mass. We seek conditions under which (7.15) yields a bona fide ordinary differential equation. For this purpose, we must exploit (7.11). Substituting (2.11) and (7.14) into (7.11) and then setting $\varepsilon = 0$, we obtain

$$\hat{N}(z_s(s,t),s) = \hat{N}(z_s(1,t),1). \tag{7.16a}$$

Assuming that $\hat{N}$ satisfies the restrictions stated at the beginning of Sec. 3, we can put (7.16a) into the equivalent form

$$z_s(s,t) = \hat{\nu}\big(\hat{N}(z_s(1,t),1),s\big). \tag{7.16b}$$

We integrate this equation over $[0,1]$ subject to the boundary condition (7.9) to obtain

$$z(1,t) = \int_0^1 \hat{\nu}\big(\hat{N}(z_s(1,t),1),s\big)\,ds \equiv J\big(z_s(1,t)\big). \tag{7.17}$$

Our constitutive hypotheses imply that J increases strictly from 0 to ∞ as its argument increases from 0 to ∞. Thus J has an inverse and (7.17) is equivalent to

$$z_s(1,t) = J^{-1}\big(z(1,t)\big). \tag{7.18}$$

The substitution of (7.18) into (7.15) yields a true ordinary differential equation for $z(1,\cdot)$. The solution of it subject to initial conditions coming from (7.13) can be substituted into (7.18), which in turn can be substituted into (7.16b), which yields $z(\cdot,\cdot)$. In general, the initial conditions (7.13) are incompatible with the form of z.

7.19. Exercise. For a uniform elastic string, find the form of the initial conditions (7.13) that are compatible with the form of z.

Let us now try to repeat this development for viscoelastic strings with constitutive equation (II.2.14). We assume that N_1 satisfies (II.2.18) and a uniform version of (II.2.22), namely, that there is a positive number α such that $\partial N_1(\nu, \xi, s)/\partial \xi \geq \alpha$ everywhere. Then $N_1(\nu, \cdot, s)$ has an inverse $N \mapsto P(\nu, N, s)$. Following the development for elastic strings, we find that the reduced problem for viscoelastic strings is governed by

$$N_1\big(z_s(s,t), z_{st}(s,t), s\big) = N_1\big(z_s(1,t), z_{st}(1,t), 1\big) \equiv q(t) \tag{7.20a}$$

or, equivalently,

$$z_{st}(s,t) = P\big(z_s(s,t), q(t), s\big), \tag{7.20b}$$

and the boundary condition

$$m z_{tt}(1,t) + q(t) = 0. \tag{7.21}$$

We get an expression for $z_t(1,t)$ by integrating (7.20b) with respect to s from 0 to 1. If we differentiate this expression with respect to t, use (7.20b) and (7.21), and solve the resulting equation for q_t, we obtain

$$q_t(t) = -\frac{q(t)/m + \int_0^1 P_\nu\big(z_s(s,t), q(t), s\big) P\big(z_s(s,t), q(t), s\big)\, ds}{\int_0^1 P_N\big(z_s(s,t), q(t), s\big)\, ds}. \tag{7.22}$$

Equations (7.20b) and (7.22) form a semilinear system of ordinary differential equations for $t \mapsto z_s(\cdot, t), q(\cdot)$. If we impose reasonable initial conditions on these variables, coming from (7.13), then a standard application of the Contraction Mapping Principle implies that there is a time interval on which the resulting initial-value problem has a unique solution. As before, the solution of this initial-value problem generates a solution to the reduced problem. We wish to determine whether the initial-value problem has a unique solution defined for all positive time and then whether the motion of the tip mass is governed by an ordinary differential equation.

For this purpose, let us assume that for each fixed ξ the function N_1 satisfies (II.2.19). Then it follows that

$$P(\nu, \xi, s) \to \begin{Bmatrix} \infty \\ -\infty \end{Bmatrix} \quad \text{as} \quad \nu \to \begin{Bmatrix} 0 \\ \infty \end{Bmatrix}. \tag{7.23}$$

Now let us suppose that q is a given continuous function. We regard (7.20) as an ordinary differential equation for $z_s(s, \cdot)$ parametrized by s. Its initial data come from (7.13a). By the continuous dependence of solutions of initial-value problems for ordinary differential equations on their data (see Coddington & Levinson (1955, Thm. 2.4.2)), we find that solutions z_s depend continuously upon s as long as they exist, and by the continuation theory (op. cit. Thm. 2.1.3), the solutions exist as long as $z_s(s, \cdot)$ remains in $(0, \infty)$. But (7.23) prevents $z_s(s, \cdot)$ from leaving $(0, \infty)$, as a sketch of the slope field $(t, \nu) \mapsto P(\nu, q(t), s)$ immediately shows. Thus, solutions of the full initial-value problem for (7.20) and (7.22) exist, are unique, and depend continuously on s as long as q stays bounded.

We now give conditions that ensure that q stays bounded for all time and thus that solutions of this initial-value problem have these properties for all time. We define

$$W(\nu, s) \equiv \int_0^\nu N_1(y, 0, s)\, dy \tag{7.24}$$

(see Ex. II.2.27). We require that

$$W(\nu, s) \to \infty \quad \text{as} \quad \nu \to 0, \infty. \tag{7.25}$$

Let us multiply (7.20a) by z_{st} and use (7.21) to obtain

$$\begin{aligned} W_\nu(z_s(s,t),s)z_{st}(s,t) + [N_1(z_s(s,t), z_{st}(s,t), s) - N_1(z_s(s,t),0,s)]z_{st}(s,t) \\ = q(t)z_{st}(s,t) = -mz_{tt}(1,t)z_{st}(s,t). \end{aligned} \tag{7.26}$$

Integrating (7.26) with respect to s over $[0,1]$, we obtain the energy equation

$$\begin{aligned} \frac{d}{dt}\left\{\tfrac{1}{2}mz_t(1,t)^2 + \int_0^1 W(z_s(s,t),s)\,ds\right\} \\ = -\int_0^1 [N_1(z_s(s,t), z_{st}(s,t), s) - N_1(z_s(s,t),0,s)]z_{st}(s,t)\,ds \le 0, \end{aligned} \tag{7.27}$$

whence there is a number C, independent of t, such that

$$\tfrac{1}{2}mz_t(1,t)^2 + \int_0^1 W(z_s(s,t),s)\,ds \le C. \tag{7.28}$$

We shall deduce the requisite bounds from this energy inequality. Note that the left-hand side of (7.28), the total energy of the massless string, consists of the kinetic energy of the tip mass and the potential energy of string; the string has no kinetic energy. Inequality (7.28) implies that there is a (different) number C such that

$$|z_t(1,t)| \le C, \quad |z(1,t)| \le C(1+t). \tag{7.29}$$

Next we observe that

$$z(1,t) > 0 \tag{7.30}$$

for t in any bounded subinterval of the the interval of existence, for if not, there would be a τ in the interval of existence such that

$$z(1,t) \equiv \int_0^1 z_s(s,t)\,ds \searrow 0 \quad \text{as} \quad t \nearrow \tau. \tag{7.31}$$

Thus $z_s(\cdot,t)$ would converge in measure to 0 as $t \nearrow \tau$, in contradiction to (7.25) and (7.28).

Now we show that $|q|$ cannot blow up in finite time. If $q(t) \to \infty$ as $t \to \tau$, then (7.20a) would imply that

$$z_s(s,t) \to \infty \quad \text{or} \quad z_{st}(s,t) \to \infty \quad \text{as} \quad t \to \tau \tag{7.32}$$

for each $s \in [0,1]$ as a consequence of the growth conditions on N_1. But (7.31) would then imply that

$$z(1,t) \to \infty \quad \text{as} \quad t \to \tau \quad \text{or} \quad z_t(1,t) \to \infty \quad \text{as} \quad t \to \tau, \tag{7.33}$$

in contradiction to (7.29). If $q(t) \to -\infty$ as $t \to \tau$, then analogously

$$z_s(s,t) \to 0 \quad \text{or} \quad z_{st}(s,t) \to -\infty \quad \text{as} \quad t \to \tau \tag{7.34a,b}$$

for each $s \in [0,1]$. Condition (7.34b) cannot hold on a set of positive measure because it would imply by (7.31) that $z_t(1,t) \to -\infty$, in contradiction to (7.29). Thus (7.34a) would have to hold for almost all s in $[0,1]$. Then (7.31) would imply that $z(1,t) \to 0$ as $t \to \tau$, in contradiction to (7.30). Therefore, when (7.25) and our other constitutive assumptions hold, solutions of initial-value problems for (7.20b) and (7.22) exist for all time and are unique. In particular, q is continuous.

7.35. Exercise. Let the restriction that $W(\nu, s) \to \infty$ as $\nu \to 0$ of (7.25) be suspended. Instead, suppose that there is a strictly decreasing, positive-valued, C^1 function $(0, \infty) \ni \nu \mapsto \beta(\nu)$ with $\beta(\nu) \to \infty$ as $\nu \to 0$ and there are numbers $\nu_* \in (0, 1]$ and $a \geq 0$ such that

$$N_1(\nu, \xi, s) - N_1(\nu, 0, s) \leq \beta'(\nu)\xi + a \quad \text{when} \quad \nu \leq \nu_*. \tag{7.36}$$

(This hypothesis, due to Antman (1988b), says that the internal frictional effects become infinitely large in a certain way where the spring suffers a total compression.) Prove that the analysis beginning with (7.24) is still valid. (Hint: It suffices to preclude (7.34a). Without loss of generality, assume that $\nu_* < \min z_s(s, 0)$. Suppose that there is a point (s, ω) for which $z_s(s, \omega) < \nu_*$. If not, there would be nothing to prove. Let τ be the largest value of t less than ω at which $z_s(s, \tau) = \nu_*$. Study the equation $N_1(z_s, z_{st}, s) = -mz_{tt}(1, t)$ on the interval (τ, ω).) Use the Gronwall inequality to extend this development to a weaker version of (7.36):

$$N_1(\nu, \xi, s) - N_1(\nu, 0, s) \leq \beta'(\nu)\xi + a\beta(\nu) \quad \text{when} \quad \nu \leq \nu_*. \tag{7.37}$$

Let us again treat (7.20b) as an ordinary differential equation for $z_s(s, \cdot)$ with q given. The solution of the initial-value problem for this equation depends continuously on q. Specifically, $z_s(s, t)$ depends continuously on the history of q up to time t (in the C^0-norm of q) on the interval $[0, t]$. In view of (7.21), $z_s(s, t)$ thus depends on the history of $z_{tt}(1, \cdot)$ or, equivalently, on the history of $z(1, \cdot)$ up to time t. We thus find that $z_s(1, t)$ and $z_{st}(1, t)$ depend on the history of $z(1, \cdot)$ up to time t. It thus appears that the substitution of representations for these functions into (7.21) yields ordinary-functional differential equations for the motion $z(1, \cdot)$ of the tip mass. It might happen that this dependence on the past history is trivial and that (7.21) is just an ordinary differential equation. But Antman (1988a), in a work on which this discussion is based, shows that if the material is not uniform or if the initial conditions are not special, then it is impossible to describe the motion of the end mass by the traditional ordinary differential equation. Moreover, it can be shown that the solution of the reduced problem is the leading term of the regular part of a rigorous asymptotic expansion of the solutions of the full initial-boundary-value problem with respect to the parameter ε (see Antman (1995)).

8. Comments and Historical Notes

Each of the specific problems of this chapter is treated by a different formulation embodying different variables to exploit the underlying geometry. Only for the suspension bridge problem did we use a Cartesian description of the configuration of the string. The special character of the loading, which made the inextensible problem tractable in the seventeenth century, made this Cartesian description effective here. On the other hand, the Cartesian formulation of the inextensible catenary problem, which Joh. Bernoulli and especially Leibniz exploited with great success, makes the treatment of elastic catenary problems excessively complicated, as the work of Jas. Bernoulli suggests and as (3.38) demonstrates.

The static problems we have treated have the common feature that the shape of equilibrium configurations is to a large extent independent of the material response. The ease with which the shape could be determined for certain problems no doubt hid the necessity of finding conditions to ensure that there actually are such solutions.

Compressive solutions of the static problems tend to come in pairs. In the singular limit of inextensible strings, these arciform solutions coalesce into a single compressive state. For strings under vertical loads, this compressive state is a reflection of the unique tensile state (see Ex. 3.15).

The problems of Secs. II.6–II.8 become degenerate if $\hat{N}(\cdot, s)$ is linear. The same remark applies to many of the problems of this section. The nineteenth-century authors Todhunter, Minchin, and Routh, who wrote on statics, advocated the use of such linear functions because they deemed such laws to be firmly established by experiment. The ubiquity of rubber bands in this century makes their view seem archaic.

The problem of determining the catenary was posed by Leonardo da Vinci, who made false assertions about its mathematical character. In 1638 Galileo incorrectly stated that the curve (for a uniform inextensible string) is a parabola. The problem of the suspension bridge was studied by Stevin in 1608. The correct form of the curve (for an inextensible string under constant horizontal load intensity), a parabola, was found by Beeckman in 1614–1615, Huygens in 1646, and Pardies in 1673, the successive analyses exhibiting increasing sophistication and accuracy. Both Huygens and Pardies also noted that the catenary could not be a parabola. Pardies incorrectly stated that the curve for the suspension bridge problem with an elastic string (under constant horizontal load intensity) is not a parabola. In 1646 Huygens incorrectly stated that the velaria is a parabola. He acknowledged his error in 1668. In 1691 Jas. Bernoulli correctly asserted the velaria to be a circle.

In 1675 Hooke observed that a (uniform inextensible) moment-free arch that supports its own weight is obtained by turning the catenary, whatever its form, upside-down. This observation may be regarded as a primitive multiplicity theorem. In 1690 Jas. Bernoulli challenged the scientific world to find the catenary. Within a year, Joh. Bernoulli, Leibniz, and Huygens did so, with Leibniz obtaining its graph in the explicit form of a hyperbolic cosine. (Full proofs of this result were finally given by Hermann in 1716 and Taylor in 1715–1717.) Between 1691 and 1704, Jas. Bernoulli intensively studied static problems for nonuniform elastic strings, the formulation of which forced him to penetrate far deeper toward the fundamental physical principles than any of his predecessors and contemporaries. Between 1713 and 1728, Joh. Bernoulli found the shape of an inextensible string gravitationally attracted to a fixed point. (This superficial historical sketch is based on Truesdell (1960), which should be consulted for full details.)

By the second half of the nineteenth century, problems for both inextensible and linearly elastic strings were routinely treated in the British books on statics by Todhunter (1853), Minchin (1887), and Routh (1891). These texts are virtually devoid of references to earlier work. In my (far from exhaustive) examination of French, German, Italian, and Russian books, I found none that had treatments of these questions comparable to those of Todhunter, Minchin, and Routh.

The first advance beyond these works was made by Dickey (1969), who saw the need for a full existence and multiplicity theory and met it in his treatment of the symmetric elastic catenary. (cf. Ex. 3.34). The development of Secs. 2–5 extends that of Antman (1979b, © Society for Industrial and Applied Mathematics). (The statement of Theorem 5.24 corrects his faulty statement.) The topological approach to the catenary developed at the end of Sec. 3 is a special version of the general theory of Alexander & Antman (1983). The treatment of Sec. 6 is adapted from the elementary part of that of Antman & Wolfe (1983) by kind permission of Academic Press. The treatment of massless springs in Sec. 7 is adapted from Antman (1988b, © Society for Industrial and Applied Mathematics).

CHAPTER IV

Planar Equilibrium Problems for Elastic Rods

A *theory of rods* is the characterization of the motion of slender solid bodies by a finite number of equations in which there is but one independent spatial variable. (The theory of strings, formulated in Chap. II, is thus an example of a theory of rods.) In this chapter we formulate and analyze equilibrium problems for the planar deformation of elastic rods. The intrinsically one-dimensional theory that we employ, which may be called the *special Cosserat theory of rods*, has several virtues: It is exact in the same sense as the theory of strings of Chap. II is exact, namely, it is not based upon ad hoc geometrical approximations or mechanical assumptions. It is much more general than the standard theories used in structural mechanics. Many important concrete problems for the theory admit detailed global analyses, some of which are presented below.

In Sec. 1 we give a concise development of the fundamental theory for planar equilibrium problems, deferring to Chaps. VIII and XIV detailed discussions of dynamical theories of rods deforming in space. There we present generalizations, refinements, and a thorough analyses of all the concepts introduced here in a simplified setting. (Chapter VIII can be read right now.) In Secs. 2–5 we solve a variety of elementary problems. (We continue this program in Chap. VI, where we treat a variety of global bifurcation problems for the planar deformation of rods, and in Chap. VII, where we use variational methods for some of these problems.)

1. Formulation of the Governing Equations

Let a thin two-dimensional body occupy a region $\mathcal{B}$ of the the plane spanned by $\{\boldsymbol{i}, \boldsymbol{j}\}$. We call this configuration the *reference configuration* of the body. See Fig. 1.1a. This configuration is often taken to be one that is free of applied forces. We may think of it as the intersection with the $\{\boldsymbol{i}, \boldsymbol{j}\}$-plane of the reference configuration of a slender three-dimensional body that is symmetric about this plane. In this case, we limit our studies here to deformations that preserve this symmetry. Alternatively, we could regard $\mathcal{B}$ as a typical cross section of a very long (or infinitely long) cylindrical body with generators parallel to $\boldsymbol{k}$. In this case, we limit our studies here to deformations that preserve the cylindricity, so that the deformation of $\mathcal{B}$ is typical of the deformation of every cross section.

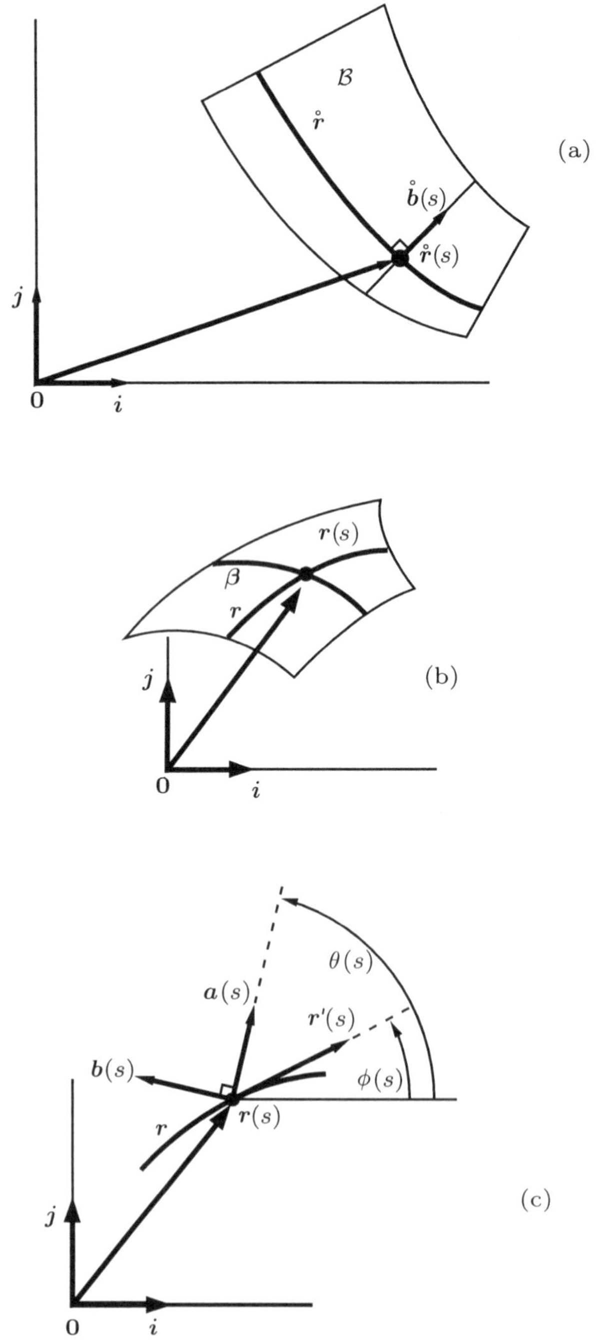

Fig. 1.1. (a) Reference configuration of a two-dimensional body. (b) Deformed configuration of a two-dimensional body. (c) Deformed planar configuration of a Cosserat rod.

Within $\mathcal{B}$ we identify a smooth plane *base curve* $[s_1, s_2] \ni s \mapsto \mathring{\boldsymbol{r}}(s) \in \mathcal{B}$. We take s to be the arc-length parameter of $\mathring{\boldsymbol{r}}$. This curve should be chosen so that $\mathcal{B}$ represents a 'thickening' of $\mathring{\boldsymbol{r}}$. In particular, the lines normal to $\mathring{\boldsymbol{r}}$ should not intersect within $\mathcal{B}$. The intersection with $\mathcal{B}$ of the normal line through $\mathring{\boldsymbol{r}}(s)$ is the *material* (*cross*) *section* (at) s. The vector $\mathring{\boldsymbol{b}}(s) \equiv \boldsymbol{k} \times \mathring{\boldsymbol{r}}'(s)$ is a unit vector normal to $\mathring{\boldsymbol{r}}$ at $\mathring{\boldsymbol{r}}(s)$; it lies along the section s. It is sometimes desirable to choose $\mathring{\boldsymbol{r}}(s)$ to bisect the cross sections (provided that such a 'curve of centroids' exists).

Under the action of forces and couples the body shown in Fig. 1.1a is deformed into the configuration shown in Fig. 1.1b. The material points forming $\mathring{\boldsymbol{r}}$ now occupy a plane curve $[s_1, s_2] \ni s \mapsto \boldsymbol{r}(s)$ for which s is in general not the arc-length parameter. The straight section at s is typically deformed into a plane curve $\boldsymbol{\beta}$. An exact description of such a deformation would require partial differential equations with two independent spatial variables (see Chap. XII). We now use the thinness of the body to motivate the construction of a simpler theory governed by ordinary differential equations in s. It is clear that if we know $\boldsymbol{r}$, we know the gross shape of the deformed body. In the model we employ, the planar version of the *special Cosserat theory of rods*, we find ordinary differential equations not only for $\boldsymbol{r}$ but also for the unit vector field $\boldsymbol{b}$, which characterizes some average orientation of the deformed cross section $\boldsymbol{\beta}$. See Fig. 1.1c. The theory we now formulate, though motivated by the considerations embodied in Fig. 1.1, stands on its own as a coherent independent mathematical model of the two-dimensional deformation of thin bodies.

Geometry of deformation. A *planar configuration of a special Cosserat rod* is defined by a pair of vector-valued functions

(1.2) $$[s_1, s_2] \ni s \mapsto \boldsymbol{r}(s),\ \boldsymbol{b}(s) \in \operatorname{span}\{\boldsymbol{i}, \boldsymbol{j}\}$$

where $\boldsymbol{b}(s)$ is a unit vector, called the *director* at s. We can accordingly represent $\boldsymbol{b}$ and the vector $\boldsymbol{a} \equiv -\boldsymbol{k} \times \boldsymbol{b}$ by

(1.3) $$\boldsymbol{a} = \cos\theta \boldsymbol{i} + \sin\theta \boldsymbol{j}, \quad \boldsymbol{b} = -\sin\theta \boldsymbol{i} + \cos\theta \boldsymbol{j}.$$

Thus a configuration can be alternatively defined by $\boldsymbol{r}$ and θ.

Throughout the remainder of this book up to Chap. XIV, we take the word *rod* to be synonymous with *special Cosserat rod*, unless there is a specific statement to the contrary. We have defined the configuration of a rod, but have not defined a rod itself. (See the comments following (VIII.3.5).)

Since the basis $\{\boldsymbol{a}, \boldsymbol{b}\}$ is natural for the intrinsic description of deformation, we decompose all vector-valued functions with respect to it. In particular, we set

(1.4) $$\boldsymbol{r}' = \nu \boldsymbol{a} + \eta \boldsymbol{b}.$$

The functions

$$\nu, \quad \eta, \quad \mu \equiv \theta' \tag{1.5}$$

are the *strain* variables corresponding to the configuration (1.2). We identify the values of all variables in the reference configuration by a superposed circle. Thus in this configuration,

$$\mathring{\boldsymbol{r}}' = \mathring{\boldsymbol{a}} = \cos \mathring{\theta} \boldsymbol{i} + \sin \mathring{\theta} \boldsymbol{j}, \quad \mathring{\nu} = 1, \quad \mathring{\eta} = 0. \tag{1.6}$$

To interpret the strain variables, we first observe that that stretch of the base curve $\mathring{\boldsymbol{r}}$, i.e., the local ratio of deformed to reference length, is $|\boldsymbol{r}'| = \sqrt{\nu^2 + \eta^2}$. The strain $\nu = \boldsymbol{r}' \cdot \boldsymbol{a} = \boldsymbol{k} \cdot (\boldsymbol{r}' \times \boldsymbol{b})$ is the local ratio of deformed to reference area of the parallelogram with sides $\boldsymbol{r}'$ and $\boldsymbol{b}$. In a pure elongation of the rod, $|\boldsymbol{r}'|$ reduces to ν. It is convenient, though imprecise, to think of ν as measuring stretch. The dot product

$$\boldsymbol{b} \cdot \frac{\boldsymbol{r}'}{|\boldsymbol{r}'|} = \frac{\eta}{\sqrt{\nu^2 + \eta^2}} \equiv \sin \beta \tag{1.7}$$

is the projection of $\boldsymbol{b}$ on the unit vector $\boldsymbol{r}'/|\boldsymbol{r}'|$ or vice versa; it measures the shear angle β, which is the reduction in the the angle between $\mathring{\boldsymbol{r}}'$ and $\mathring{\boldsymbol{b}}$. Let $\phi(s)$ be the angle between $\boldsymbol{i}$ and $\boldsymbol{r}'(s)$. Thus $\phi = \theta + \beta$. Its derivative with respect to the actual arc length of $\boldsymbol{r}$ is the curvature of $\boldsymbol{r}$. Thus this curvature is $\phi'/|\boldsymbol{r}'|$, which is in general not equal to $\mu = \theta'$. To appreciate the significance of μ, consider the pure inflation of an initially circular ring. The curvature changes, but μ remains fixed. On the other hand, if a straight rod is bent into a circle without elongation, then μ and the curvature both change. The virtue of μ is that it isolates the effect of pure bending from changes in curvature associated with stretching.

The requirement that the deformation not be so severe that the stretch of $\boldsymbol{r}$ is ever reduced to zero or that the shear is so large that $\boldsymbol{b}$ and $\boldsymbol{r}$ ever coincide is ensured by the inequality

$$\nu(s) > 0 \quad \forall\, s \in [s_1, s_2]. \tag{1.8}$$

The more restrictive requirement that distinct nearby cross sections of the body $\mathcal{B}$ of Fig. 1.1 cannot intersect within the deformed image of $\mathcal{B}$ is discussed in Sec. VIII.6. It leads to the following generalization of (1.8): There is a convex function $V(\cdot, s)$ on $\mathbb{R}$ with $V(0, s) = 0$ for all s such that

$$\nu(s) > V\big(\theta'(s), s\big). \tag{1.9}$$

V depends on the disposition of $\mathring{\boldsymbol{r}}$ in $\mathcal{B}$ (see Fig. 1.1). We restrict our attention to the important case that $V \geq 0$.

1.10. Exercise. Prove that ν, η, μ are unaffected by rigid displacements and that they determine $\boldsymbol{r}, \boldsymbol{b}$ to within a rigid displacement. Thus they account for change of shape.

We have not introduced conditions preventing $\boldsymbol{r}$ from intersecting itself. We can therefore model problems for thin rods in which different parts of a rod slide past each other without effect. See Fig. 2.15. On the other hand, if we regard a rod as a cross section of a cylindrical shell perpendicular to the generators, then such self-intersections have no physical significance. A characterization of the family of nonintersecting configurations useful for analytic purposes is given by Lanza de Cristoforis & Antman (1991).

Statics. Let $s_1 < c < s < s_2$. In a deformed configuration, the material of $(s, s_2]$ exerts a *resultant contact force* $\boldsymbol{n}^+(s)$ and a *resultant contact torque* $\boldsymbol{r}(s) \times \boldsymbol{n}^+(s) + \boldsymbol{m}^+(s)$ *about* $\boldsymbol{0}$ on the material of $[c, s]$. The vector $\boldsymbol{m}^+(s)$ is the corresponding *resultant contact couple*. Since we are restricting our attention to planar problems, we are assuming that $\boldsymbol{n}^+$ takes values in span$\{\boldsymbol{i}, \boldsymbol{j}\}$ and $\boldsymbol{m}^+$ takes values in span$\{\boldsymbol{k}\}$. Just as in Sec. II.2, $\boldsymbol{n}^+(\cdot)$ and $\boldsymbol{m}^+(\cdot)$ depend only on the section separating the two bodies, and are independent of any other property of the bodies. The resultant contact force and contact torque about $\boldsymbol{0}$ exerted on $[c, s]$ by $[s_1, c)$ are denoted by $-\boldsymbol{n}^-(c)$ and $-\boldsymbol{r}(c) \times \boldsymbol{n}^-(c) - \boldsymbol{m}^-(c)$. The resultant of all other forces acting on $[c, s]$ in this configuration is assumed to have the form

$$\int_c^s \boldsymbol{f}(\xi)\, d\xi \tag{1.11}$$

where $\boldsymbol{f}$ takes values in span$\{\boldsymbol{i}, \boldsymbol{j}\}$. The resultant of all other torques is assumed to have the form

$$\int_c^s [\boldsymbol{r}(\xi) \times \boldsymbol{f}(\xi) + \boldsymbol{l}(\xi)]\, d\xi \tag{1.12}$$

where $\boldsymbol{l} = l\boldsymbol{k}$. The vectors $\boldsymbol{f}(s)$ and $\boldsymbol{l}(s)$ are the *body force* and *body couple per unit reference length at s*.

The rod is in equilibrium if the resultant force and moment acting on each material segment $[c, s]$ are each zero:

$$\boldsymbol{n}^+(s) - \boldsymbol{n}^-(c) + \int_c^s \boldsymbol{f}(\xi)\, d\xi = \boldsymbol{0}, \tag{1.13}$$

$$\begin{aligned} \boldsymbol{m}^+(s) - \boldsymbol{m}^-(c) + \boldsymbol{r}(s) \times \boldsymbol{n}^+(s) - \boldsymbol{r}(c) \times \boldsymbol{n}^-(c) \\ + \int_c^s [\boldsymbol{r}(\xi) \times \boldsymbol{f}(\xi) + \boldsymbol{l}(\xi)]\, d\xi = \boldsymbol{0}. \end{aligned} \tag{1.14}$$

The extension of these laws to intervals for which $c = s_1$ or $s = s_2$ is straightforward: Prescribed forces and couples acting at the ends of the rod must be accounted for. (No such forces are treated in (II.3.1).)

By imitating the procedures of Secs. II.2 and II.3, we can obtain the obvious analogs of the results found there. In particular, we find that the superscripts '±' are superfluous. Assuming that all the functions appearing in (1.13) and (1.14) are sufficiently regular, we can differentiate these equations with respect to s to obtain

$$\boldsymbol{n}' + \boldsymbol{f} = \boldsymbol{0}, \tag{1.15}$$

$$\boldsymbol{m}' + (\boldsymbol{r} \times \boldsymbol{n})' + \boldsymbol{r} \times \boldsymbol{f} + \boldsymbol{l} = \boldsymbol{0}. \tag{1.16}$$

The substitution of (1.15) into (1.16) reduces the latter to

$$\boldsymbol{m}' + \boldsymbol{r}' \times \boldsymbol{n} + \boldsymbol{l} = \boldsymbol{0}. \tag{1.17}$$

Equations (1.15) and (1.17) are the classical forms of the *equations of equilibrium for (the special theory of Cosserat) rods.*

By integrating (1.17) over $[c, s]$, we obtain a reduced version of (1.14):

$$\boldsymbol{m}(s) - \boldsymbol{m}(c) + \int_c^s [\boldsymbol{r}'(\xi) \times \boldsymbol{n}(\xi) + \boldsymbol{l}(\xi)]\, d\xi = \boldsymbol{0}. \tag{1.18}$$

By imposing mild conditions on the functions entering (1.13) and (1.14) we could deduce (1.18) directly from (1.13) and (1.14) without resorting to the classical equations. Thus the Balance Laws (1.13) and (1.14) are equivalent to (1.13) and (1.18).

In accord with our assumption of planarity, we set

$$\boldsymbol{n} = N\boldsymbol{a} + H\boldsymbol{b}, \quad \boldsymbol{m} = M\boldsymbol{k}. \tag{1.19}$$

The substitution of (1.19) into (1.15) and (1.17) yields componential forms for these equations:

$$\begin{aligned} N' - \mu H + \boldsymbol{f} \cdot \boldsymbol{a} &= 0, \\ H' + \mu N + \boldsymbol{f} \cdot \boldsymbol{b} &= 0, \\ M' + \nu H - \eta N + l &= 0. \end{aligned} \tag{1.20}$$

As we shall see, the componential forms are often less convenient than their vectorial counterparts.

We may call H the *shear force* and M the *bending couple* (or *moment*). The *tension* is $\boldsymbol{n} \cdot \boldsymbol{r}'/|\boldsymbol{r}'|$. When $\eta \neq 0$, there is no standard name for N, and the definition of shear force differs from that used in structural mechanics, in which it is usually assumed that $\boldsymbol{a} = \boldsymbol{r}'/|\boldsymbol{r}'|$.

Under the very mild conditions prevailing in this chapter, the integral balance laws (1.13) and (1.18) are equivalent to (1.15) and (1.18) (see Sec. II.6).

Constitutive equations. The rod is called *elastic* if there are constitutive functions $\hat{N}, \hat{H}, \hat{M}$ such that

$$\begin{gathered} N(s) = \hat{N}\big(\nu(s), \eta(s), \mu(s), s\big), \quad H(s) = \hat{H}\big(\nu(s), \eta(s), \mu(s), s\big), \\ M(s) = \hat{M}\big(\nu(s), \eta(s), \mu(s), s\big). \end{gathered} \tag{1.21}$$

For each fixed s, the common domain of these constitutive functions consists of those ν, η, μ satisfying

$$\nu > V(\mu, s) \tag{1.22}$$

(cf. (1.9)). In Sec. VIII.7 we show that these constitutive equations are invariant under rigid displacements. The substitution of (1.21) into (1.19) and its insertion into (1.15) and (1.17) yields a sixth-order quasilinear system of ordinary differential equations for $\boldsymbol{r}$ and θ. The study of concrete problems for this system is the object of the next few chapters.

For simplicity of exposition, we assume that $\hat{N}, \hat{H}, \hat{M}$ are continuously differentiable. We ensure that an increase in the tension $\boldsymbol{n} \cdot \boldsymbol{r}'/|\boldsymbol{r}'|$ accompany an increase in the stretch $|\boldsymbol{r}'|$, that an increase in the shear force H accompany an increase in the shear strain η, and that an increase in the bending couple M accompany an increase in the bending strain μ by requiring that:

$$\text{the matrix } \frac{\partial(\hat{N}, \hat{H}, \hat{M})}{\partial(\nu, \eta, \mu)} \text{ is positive-definite.} \tag{1.23}$$

We term (1.23) the *monotonicity condition.* (This condition is equivalent to the positive-definiteness of the symmetric part of this matrix.) We assume that extreme values of the resultants accompany extreme values of the strains in a way that is compatible with (1.23):

$$\hat{N}(\nu, \eta, \mu, s) \to \begin{Bmatrix} +\infty \\ -\infty \end{Bmatrix} \quad \text{as} \quad \nu \to \begin{Bmatrix} +\infty \\ V(\mu, s) \end{Bmatrix}, \tag{1.24}$$

$$\hat{H}(\nu, \eta, \mu, s) \to \pm\infty \quad \text{as} \quad \eta \to \pm\infty, \tag{1.25}$$

$$\hat{M}(\nu, \eta, \mu, s) \to \pm\infty \quad \text{as} \quad \mu \to \begin{Bmatrix} \sup \\ \inf \end{Bmatrix} \{\mu : \nu > V(\mu, s)\}. \tag{1.26}$$

In (1.24) the variables η, μ, s are fixed; analogous remarks hold for (1.25) and (1.26).

We assume that it is no more difficult to shear the rod in one sense than in the opposite sense and that N and M are unaffected by the sense of shearing:

$$\hat{N}(\nu, \cdot, \mu, s) \text{ and } \hat{M}(\nu, \cdot, \mu, s) \text{ are even,} \quad \hat{H}(\nu, \cdot, \mu, s) \text{ is odd.} \tag{1.27}$$

We finally require that N, H, M vanish in the reference configuration:

$$\hat{N}(1, 0, \overset{\circ}{\mu}, s) = 0, \quad \hat{M}(1, 0, \overset{\circ}{\mu}, s) = 0. \tag{1.28}$$

In Chap. XIV, we undertake an intensive study of the constitutive conditions from the viewpoint of three-dimensional elasticity.

We assume that the mapping

$$(1.29a) \qquad (\nu, \eta, \mu) \mapsto \big(\hat{N}(\nu, \eta, \mu, s), \hat{H}(\nu, \eta, \mu, s), \hat{M}(\nu, \eta, \mu, s)\big)$$

has the inverse

$$(1.29b) \qquad (N, H, M) \mapsto \big(\hat{\nu}(N, H, M, s), \hat{\eta}(N, H, M, s), \hat{\mu}(N, H, M, s)\big),$$

so that (1.21) is equivalent to constitutive equations of the form

$$(1.30) \qquad \begin{gathered} \nu(s) = \hat{\nu}\big(N(s), H(s), M(s), s\big), \quad \eta(s) = \hat{\eta}\big(N(s), H(s), M(s), s\big), \\ \mu(s) = \hat{\mu}\big(N(s), H(s), M(s), s\big). \end{gathered}$$

When (1.30) is used, it is assumed that $\hat{\nu}$ and $\hat{\mu}$ satisfy (1.22), and that (1.29b) enjoys the same smoothness as (1.29a).

It is not hard to show that conditions (1.23)–(1.26) support a global implicit function theorem to the effect that (1.29a) has the inverse (1.29b) with the indicated properties; see Theorems XIX.2.30 and XVIII.1.27.

1.31. Exercise. Let N, H, M, s be given numbers. Prove that (1.23) ensures that the system

$$N = \hat{N}(\nu, \eta, \mu, s), \quad H = \hat{H}(\nu, \eta, \mu, s), \quad M = \hat{M}(\nu, \eta, \mu, s)$$

(cf. (1.21)) can have at most one solution for ν, η, μ.

If the rod is constrained so that $\eta = 0$ in all circumstances, then the rod is said to be *unshearable*. We construct such a theory from (1.30) by simply defining $\hat{\eta} = 0$. In this case, H (which is the Lagrange multiplier corresponding to this constraint) is not defined constitutively by any version of (1.21); it is just an unknown of any problem in which it appears. (For a full treatment of constraints, see Secs. VIII.15 and XII.12.) Likewise, if the rod is further constrained so that $\nu = 1$, then the rod is said to be *inextensible*, and N is (a Lagrange multiplier) not defined constitutively by any version of (1.21). These two constraints are commonly imposed in structural mechanics. The *elastica* theory is based on these two constraints and on the Bernoulli-Euler constitutive equation that says that M is linear in the change in μ (which is now the curvature):

$$(1.32) \qquad M(s) = (EI)(s)[\mu(s) - \overset{\circ}{\mu}(s)].$$

(This theory was created by Jas. Bernoulli (1694), Euler (1727, 1732), and D. Bernoulli (1732). See Truesdell (1960).)

If the matrix of (1.23) is symmetric, then there exists a scalar-valued function $(\nu, \eta, \mu, s) \mapsto W(\nu, \eta, \mu, s)$, called the *stored energy function*, such that

$$(1.33a) \qquad \hat{N} = W_\nu, \quad \hat{H} = W_\eta, \quad \hat{M} = W_\mu.$$

In this case, which is motivated by thermodynamical considerations (see Sec. XII.14), the rod is said to be *hyperelastic*. Condition (1.23) ensures that the Hessian matrix of second derivatives of W with respect to ν, η, μ is positive-definite, so that $W(\cdot,\cdot,\cdot,s)$ is strictly convex. The symmetry of the matrix of (1.23) implies the symmetry of the matrix of partial derivatives of $(\hat{\nu}, \hat{\eta}, \hat{\mu})$ with respect to (N, H, M). Thus there exists a scalar-valued function $(N, H, M, s) \mapsto W^*(N, H, M, s)$, called the *conjugate stored-energy function*, such that

$$\text{(1.33b)} \qquad \hat{\nu} = W^*_N, \quad \hat{\eta} = W^*_H, \quad \hat{\mu} = W^*_M.$$

$(W(\cdot,\cdot,\cdot,s)$ is the *Legendre transform* of $W(\cdot,\cdot,\cdot,s)$.)

1.34. Exercise. Prove that

$$\text{(1.35)} \qquad \begin{aligned} W^*(N,H,M,s) = {} & N\hat{\nu}(N,H,M,s) + H\hat{\eta}(N,H,M,s) + M\hat{\mu}(N,H,M,s) \\ & - W\big(\hat{\nu}(N,H,M,s), \hat{\eta}(N,H,M,s), \hat{\mu}(N,H,M,s), s\big). \end{aligned}$$

Boundary conditions. Let us now briefly describe some simple types of boundary conditions (deferring to Sec. VIII.12 a comprehensive treatment). We just describe conditions at end s_1; those at end s_2 are completely analogous.

We can prescribe the position of the end s_1 by requiring that there be a given vector $\boldsymbol{r}_1$ such that

$$\text{(1.36)} \qquad \boldsymbol{r}(s_1) = \boldsymbol{r}_1.$$

In this case, we do not prescribe the force $\boldsymbol{n}(s_1)$ acting at the end, so that it remains free to accommodate (1.36). We could alternatively prescribe the force acting at the end by requiring that there be a given vector $\boldsymbol{n}_1$ such that

$$\text{(1.37)} \qquad \boldsymbol{n}(s_1) = \boldsymbol{n}_1.$$

In this case, we do not prescribe $\boldsymbol{r}(s_1)$. (For equilibrium problems, one is not free to prescribe forces at both ends in an arbitrary way because it is necessary, but not sufficient, for equilibrium that the resultant force on the rod vanish.) We can restrict the end s_1 to move in a groove under the action of a tangential force by requiring that there be a C^1-curve $\sigma \mapsto \boldsymbol{r}_1(\sigma)$ and a scalar-valued function $\sigma \mapsto n_1(\sigma)$ such that

$$\text{(1.38)} \qquad \boldsymbol{r}(s_1) = \boldsymbol{r}_1(\sigma), \quad \boldsymbol{n}(s_1) \cdot \frac{\boldsymbol{r}_1'(\sigma)}{|\boldsymbol{r}_1'(\sigma)|} = n_1(\sigma).$$

Here σ is an unknown. In this case, which is intermediate to (1.36) and (1.37), we prescribe no further restrictions.

We can prescribe the orientation $\boldsymbol{b}(s_1)$ of the end section by requiring that there be a given number θ_1 such that

$$\text{(1.39)} \qquad \theta(s_1) = \theta_1.$$

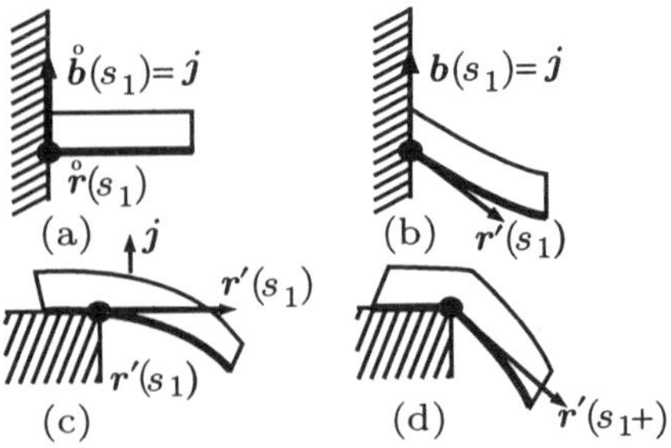

Fig. 1.40. Welding and clamping of an end. Here the base curve is taken to be the bottom edge of the rod. (a) The reference configuration. (b) A deformation satisfying the boundary condition $\theta(s_1) = 0$. (c) A deformation satisfying the boundary condition $\phi(s_1) = 0$. (d) A deformation with a support like that of (c) in which ϕ is discontinuous at s_1. It can be shown that the configuration in (d) is natural, whereas that in (c) is artificial: The configuration of (c) must be maintained by a feedback couple applied at $\boldsymbol{r}(s_1)$ (see Antman & Lanza de Cristoforis (1995)).

This condition can be effected by welding the end section to a rigid wall whose normal makes an angle θ_1 with $\boldsymbol{i}$. See Fig. 1.40b. In this case, we do not prescribe $M(s_1)$. (Notice that conditions like (1.39) when prescribed at both ends carry far more information than the specification of $\boldsymbol{b}(s_1)$ and $\boldsymbol{b}(s_2)$ because conditions for θ give a winding number for $\boldsymbol{b}$.)

We can prescribe the bending couple at s_1 by requiring that there be a given number M_1 such that

$$M(s_1) = M_1. \tag{1.41}$$

In this case, we may assume that the end s_1 can freely rotate about a hinge (under the action of the prescribed couple).

The various alternatives such as (1.36) or (1.37) seem physically reasonable. They can be put into a formal mathematical setting within the Principle of Virtual Power (see Sec. VIII.13). Their full mathematical justification is effected by suitable existence theorems. We do not pause to develop systematically such theorems, though many arise as by-products of our analyses of specific problems in Chaps. VI and VII.

A common alternative to boundary condition (1.39), which is used for less sophisticated models, is that of clamping, one version of which is illustrated in Fig. 1.40c). In it the tangent to $\boldsymbol{r}$ at s_1 is fixed by requiring that there be a given number ϕ_1 such that

$$\phi(s_1) = \phi_1. \tag{1.42a}$$

This condition is equivalent to

(1.42b,c)

$$\tan\theta(s_1) = \frac{\nu(s_1)\tan\phi_1 - \eta(s_1)}{\nu(s_1) - \eta(s_1)\tan\phi_1}, \quad \text{or, equivalently,} \quad \frac{\eta(s_1)}{\nu(s_1)} = \tan(\phi_1 - \theta(s_1)).$$

Note that for unshearable rods, (1.39) and (1.42) are equivalent. It can be shown that (1.42) in general must be maintained by a feedback couple, and is therefore unnatural. (This condition, however, is useful for certain problems of fluid-solid interaction.) For shearable rods, the boundary conditions (1.42) present many unpleasant analytical challenges and mechanical surprises, and should be avoided in the modelling of natural problems. See Antman & Lanza de Cristoforis (1994).

Boundary-value problems. Our boundary-value problems consist of the geometrical relations (1.4) and the identity in (1.5), the equilibrium equations (1.15) and (1.17) or, equivalently, (1.20), the constitutive equations in the form (1.21) or (1.30), and a suitable collection of boundary conditions, e.g., either (1.36) or (1.37) or (1.38), and either (1.39) or (1.41) at s_1, with analogous choices at s_2. In particular, if we substitute (1.30) into (1.20), we obtain a system involving just the unknowns N, H, M. The substitution of (1.30) into (1.4) and (1.5) gives three more equations for $\boldsymbol{r}$ and θ in terms of (N, H, M).

Equations of Motion. It is convenient to include the study of the ordinary differential equations of steady motions under the study of the ordinary differential equations of pure equilibrium problems. Furthermore, to study the stability of steady states, we need to examine associated equations of motion. For this purpose, we record the form of the equations of motion, deferring to Chap. VIII a full discussion of this subject. We obtain the equations of motion by replacing the $\mathbf{0}$'s on the right-hand sides of (1.15) and (1.16) with time derivatives of the linear and angular momentum, respectively. We might expect the linear momentum to be linear in $\boldsymbol{r}_t$ and $\boldsymbol{b}_t$ and the angular momentum to be linear in cross products of $\boldsymbol{r}, \boldsymbol{b}$ with $\boldsymbol{r}_t, \boldsymbol{b}_t$. Alternatively, we might be motivated to choose forms for these momenta from those for the two-dimensional body of Fig. 1.1. In any event, we show in Secs. VIII.2 and VIII.4 for a rod undergoing a planar deformation that a suitable form for the linear momentum per unit reference length at (s, t) is

$$(\rho A)(s)\boldsymbol{r}_t(s,t) + (\rho I)(s)\boldsymbol{b}_t(s,t) \tag{1.43}$$

and that a suitable form for the angular momentum per unit reference length at (s, t) is

$$\begin{aligned}(\rho A)(s)\boldsymbol{r}(s,t) \times \boldsymbol{r}_t(s,t) + (\rho I)(s)\boldsymbol{r}(s,t) \times \boldsymbol{b}_t(s,t)&\\ + (\rho I)(s)\boldsymbol{b}(s,t) \times \boldsymbol{r}_t(s,t) + (\rho J)(s)\boldsymbol{b}(s,t) \times \boldsymbol{b}_t(s,t)&.\end{aligned} \tag{1.44}$$

Here $(\rho A)(s)$, $(\rho I)(s)$, $(\rho J)(s)$ may be interpreted as the mass, first moment of mass, and second moment of mass per unit reference length about $\mathring{\boldsymbol{r}}(s)$ of the section s of the reference configuration. We take ρA and ρJ to be everywhere positive on (s_1, s_2). In the important case of a straight rod in which $\mathring{\boldsymbol{r}}$ is regarded as the curve of mass centers of the cross sections of the three-dimensional rod, we can take $\rho I = 0$. (For curved rods, we can make ρI vanish by a suitable choice of $\mathring{\boldsymbol{r}}$, but there are compensatory disadvantages, as we discuss in Chap. VIII.) The replacement of the $\mathbf{0}$'s on the right-hand sides of (1.15) and (1.16) with the time derivatives of (1.43) and (1.44) yields the equations of motion corresponding to (1.15) and (1.17):

$$\boldsymbol{n}_s + \boldsymbol{f} = \rho A \boldsymbol{r}_{tt} + \rho I \boldsymbol{b}_{tt}, \tag{1.45}$$

$$\boldsymbol{m}_s + \boldsymbol{r}_s \times \boldsymbol{n} + l\boldsymbol{k} = \rho I \boldsymbol{b} \times \boldsymbol{r}_{tt} + \rho J \boldsymbol{b} \times \boldsymbol{b}_{tt}. \tag{1.46}$$

Note that $\boldsymbol{b} \times \boldsymbol{b}_{tt} = \theta_{tt}\boldsymbol{k}$ if $\boldsymbol{b}$ has the form (1.3) with θ time-dependent, i.e., if it is confined to a fixed plane. We shall soon be considering problems in which $\boldsymbol{b}$ is confined to a rotating plane, for which we need the generality of (1.46).

2. Planar Equilibrium States of Straight Rods under Terminal Loads

Let us assume that $\boldsymbol{f} = \mathbf{0}$ and $l = 0$ and take $s_1 = 0$, $s_2 = 1$. We set

$$\boldsymbol{n}(1) = -\Lambda(\cos\alpha\boldsymbol{i} + \sin\alpha\boldsymbol{j}) \tag{2.1}$$

with $\Lambda \geq 0$. Then (1.3), (1.4), (1.13), (1.18), and (1.19) imply that if $\boldsymbol{r}$ and θ are absolutely continuous, then

$$\boldsymbol{n}(s) = -\Lambda(\cos\alpha\boldsymbol{i} + \sin\alpha\boldsymbol{j}), \tag{2.2a}$$

$$M' = -\Lambda[\nu\sin(\theta - \alpha) + \eta\cos(\theta - \alpha)]. \tag{2.2b}$$

We set

$$\gamma \equiv \theta - \alpha. \tag{2.3}$$

Equations (1.30) and (2.2a) imply that

$$\nu = \hat{\nu}(-\Lambda\cos\gamma, \Lambda\sin\gamma, M, s), \tag{2.4a}$$

$$\eta = \hat{\eta}(-\Lambda\cos\gamma, \Lambda\sin\gamma, M, s), \tag{2.4b}$$

$$\gamma' = \hat{\mu}(-\Lambda\cos\gamma, \Lambda\sin\gamma, M, s). \tag{2.4c}$$

Eq. (2.4c) and Eq. (2.2b) with ν and η replaced with (2.4a,b) form a second-order semilinear system for (θ, M).

Had we chosen in place of (1.30) constitutive equations of the form

$$\nu = \nu^\sharp(N, H, \mu, s), \quad \eta = \eta^\sharp(N, H, \mu, s), \quad M = M^\sharp(N, H, \mu, s), \tag{2.5}$$

then in place of this semilinear system we obtain the single second-order quasilinear equation

$$\frac{d}{ds}M^\sharp(-\Lambda\cos\gamma, \Lambda\sin\gamma, \gamma', s) \\ + \Lambda[\nu^\sharp(-\Lambda\cos\gamma, \Lambda\sin\gamma, \gamma', s)\sin\gamma\eta^\sharp(-\Lambda\cos\gamma, \Lambda\sin\gamma, \gamma', s)\cos\gamma] = 0. \tag{2.6}$$

It is somewhat more convenient to use the semilinear system.

If we specialize our equations to the elastica by constraining $\nu = 1$, $\eta = 0$, and by adopting (1.32), then we reduce (2.4c) and (2.2b) to

$$\gamma'(s) = \frac{M(s)}{(EI)(s)}, \quad M' = -\Lambda \sin\gamma, \tag{2.7a,b}$$

which is equivalent to the correspondingly degenerate form of (2.6):

$$\tfrac{d}{ds}[(EI(s)\gamma'(s)] + \Lambda \sin\gamma(s) = 0. \tag{2.8}$$

We assume that the reference configuration of the rod is straight and we adopt the symmetry restrictions:

$$\begin{aligned} &\hat{\nu} \ \text{ is even in } H \text{ and even in } M, \\ &\hat{\eta} \ \text{ is odd in } H \text{ and even in } M, \\ &\hat{\mu} \ \text{ is even in } H \text{ and odd in } M. \end{aligned} \tag{2.9}$$

These conditions say that it is no harder to shear or bend the rod in one sense than to do so the same amount in the opposite sense. These conditions are degenerate versions of the transverse isotropy conditions described in Sec. VIII.9. (In Sec. XIV.5 three-dimensional considerations are used to show that the symmetry with respect to H can generally be expected and that if each cross section of the rod is doubly symmetric about its centroid, then the symmetry with respect to M holds, but is otherwise unwarranted.) Let us assume that the material of the rod is uniform, so that the constitutive functions of (1.21) or (1.30) are independent of s. In this case, (2.4c) and (2.2b) reduce to the autonomous system

$$\gamma' = \hat{\mu}(-\Lambda\cos\gamma, \Lambda\sin\gamma, M), \tag{2.10a}$$

$$\begin{aligned} M' = -\,&\Lambda[\hat{\nu}(-\Lambda\cos\gamma, \Lambda\sin\gamma, M)\sin\gamma \\ &+ \hat{\eta}(-\Lambda\cos\gamma, \Lambda\sin\gamma, M)\cos\gamma]. \end{aligned} \tag{2.10b}$$

We study the qualitative properties of all solutions to (2.10) by examining its phase portrait in the (γ, M)-phase plane. Conditions (2.9) and the form of (2.10) imply that the phase portrait is symmetric about the γ-axis, is symmetric about the M-axis, and has period 2π in γ. Conditions (2.9) and (1.23) imply that the right-hand side of (2.10a) vanishes if and only if $M = 0$. Thus the singular points of (2.10) have the form $(\gamma, 0)$ where γ satisfies

$$\tan\gamma = -\frac{\hat{\eta}(-\Lambda\cos\gamma, \Lambda\sin\gamma, 0)}{\hat{\nu}(-\Lambda\cos\gamma, \Lambda\sin\gamma, 0)} \equiv h(\gamma, \Lambda). \tag{2.11}$$

Since $\hat{\nu}$ is everywhere positive, the denominator of (2.11) does not vanish. For an unshearable rod, the numerator is zero, in which case the singular points have the form $(k\pi, 0)$ where k is an integer. Condition (2.9) implies that (2.11) always has roots $k\pi$. We want to determine if it has others.

Since $\Lambda \geq 0$, conditions (2.9) and (1.23) imply that $h(\gamma, \Lambda)$ has the sign opposite to that of $\sin\gamma$ for $\Lambda > 0$. By sketching the graphs of tan and $h(\cdot, \Lambda)$ and by noting that $h(\gamma, 0) = 0 = h_\gamma(\gamma, 0)$, we find that there is a positive number Λ^+ (possibly infinite) such that the only solutions of (2.11) are $k\pi$ for $\Lambda < \Lambda^+$.

2.12. Exercise. Linearize (2.10) about its singular points to determine their type. In particular, let

$$q(\Lambda) \equiv \frac{\Lambda\hat{\nu}(-\Lambda\cos\gamma, \Lambda\sin\gamma, 0)}{\cos\gamma}[1 - h_\gamma(\gamma, \Lambda)\cos^2\gamma] \tag{2.13}$$

when γ satisfies (2.10). Prove that the singular point is a center when $q(\Lambda) > 0$ and a saddle point when $q(\Lambda) < 0$. Show that the points $(2k\pi, 0)$ are centers and that the points $((2k+1)\pi, 0)$ are saddle points when $h_\gamma((2k+1)\pi, \Lambda) < 1$ and centers when $h_\gamma((2k+1)\pi, \Lambda) > 1$. More generally, show that a solution γ of (2.11) in $(\frac{\pi}{2}, \frac{3\pi}{2})$ corresponds to a saddle point if the slope of $h(\cdot, \Lambda)$ is less than the slope of tan at γ and corresponds to a center if the slope of $h(\cdot, \Lambda)$ exceeds the slope of tan at γ.

In view of the results of Ex. 2.12, the phase portraits of (2.10) have the form shown in Fig. 2.14. (The symmetry conditions ensure that these portraits have the character of those for conservative problems in that there are no spiral points or nodes or their global analogs.) The phase portrait for the elastica is just like Fig. 2.14a.

We reconstruct the possible shapes of the deformed rod by integrating (1.4). Since γ' behaves like the curvature

$$\frac{\gamma'}{[\hat{\nu}^2 + \hat{\eta}^2]^{1/2}} + \frac{\hat{\nu}\hat{\eta}' - \hat{\eta}\hat{\nu}'}{[\hat{\nu}^2 + \hat{\eta}^2]^{3/2}},$$

at least for moderate shear, we can practically read off the shapes from the phase portraits. For example, a closed orbit of Fig. 2.14a encircling the origin corresponds to a segment of a curve like that shown in Fig. 2.15a, while an unbounded trajectory lying outside the separatrix corresponds to a segment of the noninflexional curve shown in Fig. 2.15b. We emphasize that these shapes are merely candidates for solutions of specific boundary-value problems. We have not demonstrated that it is possible to choose unprescribed constants and integration constants so that a trajectory uses up exactly one unit of the independent variable in joining termini corresponding to prescribed boundary conditions. We address important aspects of this question in the next chapter and elsewhere.

Figures 2.14b,c have additional closed orbits, in which γ is confined to a subinterval of $(\frac{\pi}{2}, \frac{3\pi}{2})$. Here the rod is everywhere in tension. In each case, there is a singular point with γ equal to a nonintegral multiple of π. This singular point describes a straight rod with $\eta \neq 0$ and corresponds to a shear instability that can arise in rods sufficiently weak in shear when they are in tension. Figure 2.16 show the configuration of a rod corresponding to a closed orbit encircling such a singular point.

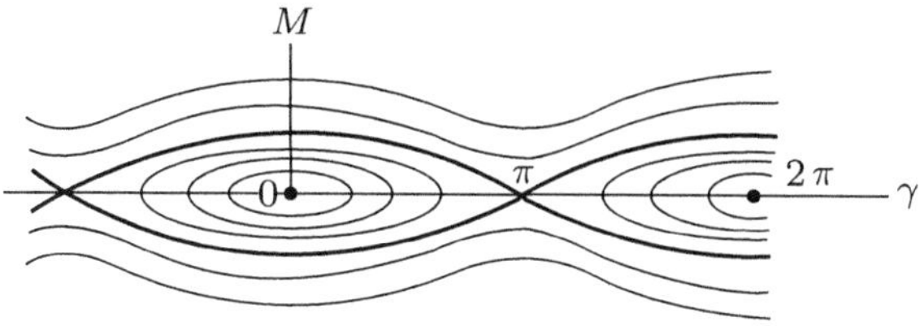

Fig. 2.14a. Typical phase portrait of (2.10) when $\Lambda < \Lambda^{+}$.

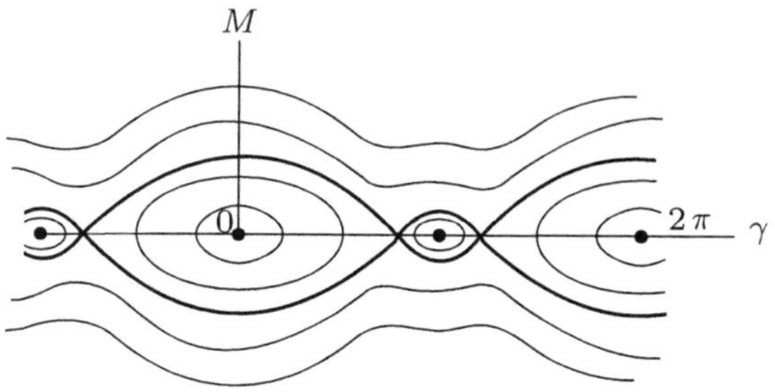

Fig. 2.14b. Typical phase portrait of (2.10) when $\Lambda > \Lambda^{+}$ and $h_{\gamma}(\pi, \Lambda) > 1$.

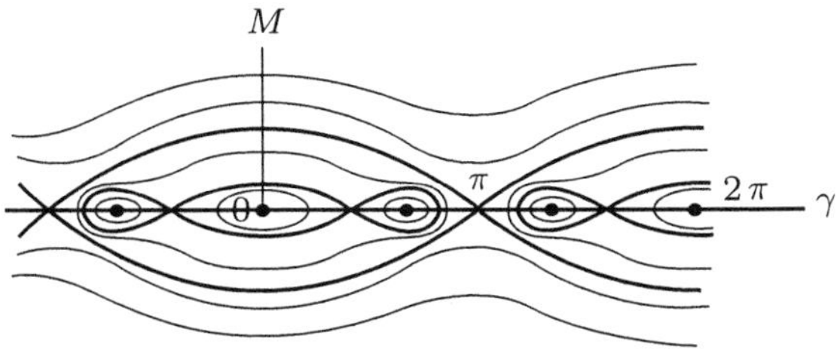

Fig. 2.14c. Typical phase portrait of (2.10) when $\Lambda > \Lambda^{+}$ and $h_{\gamma}(\pi, \Lambda) < 1$.

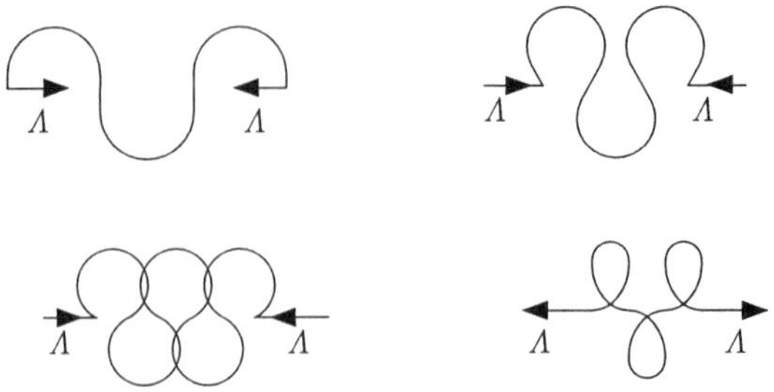

Fig. 2.15a. Inflexional configurations of $\boldsymbol{r}$ corresponding to closed orbits of Fig. 2.14a encircling the origin.

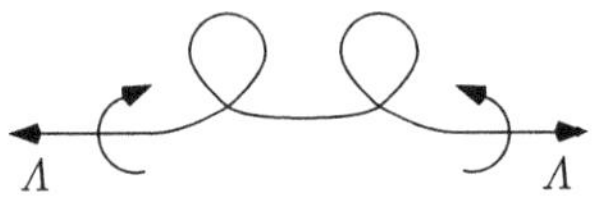

Fig. 2.15b. Noninflexional configuration of $\boldsymbol{r}$ corresponding to an unbounded trajectory of Fig. 2.14a.

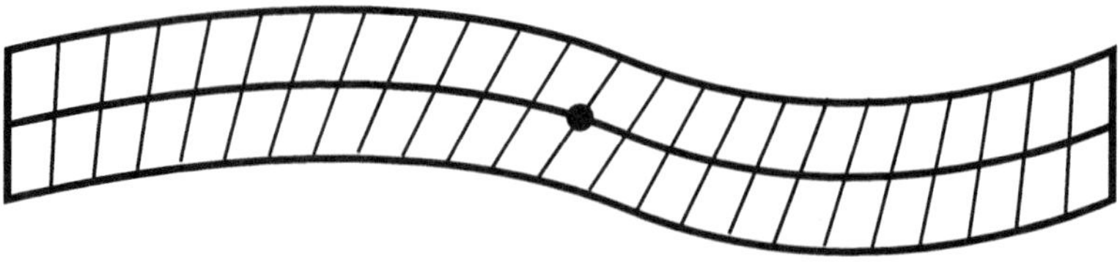

Fig. 2.16. Configuration of a rod corresponding to a closed orbit encircling a singular point of Fig. 2.14c with γ equal to a nonintegral multiple of π.

2.17. Exercise. Suppose that the rod is hyperelastic and uniform, so that the functions in (1.30) have the form (1.33b) with W^* independent of s. Show that the trajectories of the phase portrait of (2.10) have the form

$$W^*(-\Lambda\cos\gamma, \Lambda\sin\gamma, M) = \text{const.} \tag{2.18}$$

Alternatively, let the functions of (2.5) have the form $\nu^\sharp = W^\sharp_N$, $\eta^\sharp = W^\sharp_H$, $M^\sharp = -W^\sharp_\mu$ with $W^\sharp$ independent of s. Show that the trajectories of the phase portrait of (2.6) in (γ, γ')-space have the form

$$\gamma' W^\sharp_\mu(-\Lambda\cos\gamma, \Lambda\sin\gamma, \gamma') - W^\sharp(-\Lambda\cos\gamma, \Lambda\sin\gamma, \gamma') = \text{const.} \tag{2.19}$$

The symmetry condition (2.9) implies that $W^*(N, H, \cdot)$ is even. Thus, under mild regularity assumptions, W^* has the form $W^*(N,H,M) = W^\flat(N,H,K)$ with $K = M^2$. Prove that the Monotonicity Condition (1.23) implies that (2.18) can be solved uniquely for M^2 in terms of $-\Lambda\cos\gamma$, $\Lambda\sin\gamma$, and the constant, by showing that $W^\flat_K > 0$. Use a similar construction to show that (2.19) can be solved for $(\gamma')^2$ in terms of the other variables.

2.20. Exercise. The component of $\boldsymbol{n}$ perpendicular to $\boldsymbol{r}'$ is $Q \equiv \boldsymbol{n}\cdot(\boldsymbol{k}\times\boldsymbol{r}')/|\boldsymbol{r}'|$. (It is termed the *vertical shear* in the engineering literature.) Prove that (2.11) is equivalent to $Q = 0$.

2.21. Exercise. Show that (2.9) is equivalent to the corresponding statement in which the roles of (ν, η, μ) and (N, H, M) are switched.

2.22. Problem. Carry out the phase-plane analysis of the planar deformation of a rod whose natural state has a circular axis. Here the symmetry conditions (2.9) are weakened by dropping those involving M.

In his remarkable work of 1744, Euler catalogued all possible equilibrium states for the elastica and thereby established the qualitative theory of ordinary differential equations. See Love (1927, Secs. 262, 263) and Truesdell (1960) for accounts of this work. Among the many references showing how solutions of the problem for the elastica can be solved in terms of elliptic functions are Love (1927), Frisch-Fay (1962), and Reiss (1969). The treatment given here, which is in the spirit of Euler's, generalizes that of Antman (1974a), which in turn is based on that of Antman (1968).

3. Equilibrium of Rings under Hydrostatic Pressure

We now study planar equilibrium states of an initially circular elastic ring subjected only to simple hydrostatic pressure

$$\boldsymbol{f} = p\boldsymbol{k}\times\boldsymbol{r}'. \tag{3.1}$$

(See Sec. III.5. The ring may be regarded as a typical cross section of an infinitely long cylindrical shell.) Under our standard minimal regularity assumptions, solutions of (1.13), (1.18), and (1.30) are classical solutions of the differential equations

$$\boldsymbol{n}' + p\boldsymbol{k}\times\boldsymbol{r}' = \boldsymbol{0}, \quad M' + \boldsymbol{k}\cdot(\boldsymbol{r}'\times\boldsymbol{n}) = 0, \tag{3.2a,b}$$

and (1.30). We dot (3.2a) with $\boldsymbol{n}$ and substitute the resulting expression into (3.2b), which is the exact derivative of the integral

$$|\boldsymbol{n}|^2 - 2pM \equiv N^2 + H^2 - 2pM = a \text{ (const.)}. \tag{3.3}$$

We assume that the natural reference configuration is a circle of unit radius so that in this configuration $\mu = 1$. Thus we take $s_1 = 0$, $s_2 = 2\pi$. The requirement that $\boldsymbol{r}$ have a winding number of 1 leads to the side condition

$$2\pi = \theta(2\pi) - \theta(0) = \int_0^{2\pi} \mu(s)\, ds. \tag{3.4}$$

We require that the position field $\boldsymbol{r}$, the strains ν, η, μ, and the resultants N, H, M, when extended to the whole real line, each have period 2π in s. (The requirement that $\boldsymbol{r}'$ have period 2π, which is necessary for the smoothness of $\boldsymbol{r}$, is ensured by the periodicity of the strain.) The problem consisting of (1.30), (3.2), (3.4), and these periodicity conditions is called the *ring problem.*

Unshearable rings. If the ring is unshearable, then (1.20) or (3.2) reduces to

$$N' - H\theta' = 0, \quad H' + N\theta' + p\nu = 0, \quad M' + \nu H = 0. \tag{3.5a,b,c}$$

As we pointed out in the paragraph containing (1.32), H is a Lagrange multiplier, not determined constitutively, and the constitutive equations have the form (1.30) with $\hat{\eta} = 0$. We assume that $\hat{\nu}$ and $\hat{\mu}$ are independent of H, as is true for a hyperelastic rod. Methods for treating the problem when these functions depend on H are like those described below for shearable rings. We wish to eliminate the multiplier H from our formulation.

The substitution of (3.5c) into (3.3) yields

$$(M')^2 = \nu^2(a + 2pM - N^2). \tag{3.6}$$

From (3.5a,c) we get

$$\nu N' + \theta' M' = 0. \tag{3.7}$$

If the material is hyperelastic and uniform, so that the constitutive functions have one of the pairs of forms

$$\hat{\nu}(N, M) = \frac{\partial W^*}{\partial N}(N, M), \quad \hat{\mu}(N, M) = \frac{\partial W^*}{\partial M}(N, M), \tag{3.8a,b}$$

$$\hat{N}(\nu, \mu) = \frac{\partial W}{\partial \nu}(\nu, \mu), \qquad \hat{M}(\nu, \mu) = \frac{\partial W}{\partial \mu}(\nu, \mu), \tag{3.9a,b}$$

then the substitution of (3.8) or (3.9) into (3.7) yields the integral

$$\begin{gathered} W^*(N, M) = b\,(\text{const.}) \quad \text{or} \\ \nu W_\nu(\nu, \theta') + \theta' W_\mu(\nu, \theta') - W(\nu, \theta') = b. \end{gathered} \tag{3.10}$$

Let us substitute our specialization of the constitutive equations (1.30) into (3.7). The resulting equation says that the vector (M', N') is orthogonal to the vector $(\hat{\mu}(N,M), \hat{\nu}(N,M))$ at (M,N). The positivity of $\hat{\nu}$ everywhere, a consequence of (1.8), implies that $(\hat{\mu}(\cdot,\cdot), \hat{\nu}(\cdot,\cdot))$ is nowhere horizontal on the (M,N)-plane. Thus (M',N') is nowhere vertical. It follows that trajectories in the (M,N)-plane satisfying (3.7) and the specialization of the constitutive equations (1.30) lie on graphs of solutions of the ordinary differential equation

$$\frac{dN}{dM} = -\frac{\hat{\mu}(N,M)}{\hat{\nu}(N,M)} \equiv -\hat{\kappa}(N,M). \tag{3.11}$$

A specific planar version of the inequality (1.22), which is established in Sec. VIII.6, is just that

$$\nu > V(\mu) \equiv \begin{cases} h_2\mu & \text{if} \quad \mu \geq 0, \\ h_1\mu & \text{if} \quad \mu \leq 0 \end{cases} \tag{3.12}$$

where h_1 and h_2 are numbers characterizing the geometry of the cross-section. We take $h_1 \leq 0 \leq h_2$. We assume that the constitutive functions $\hat{\nu}$ and $\hat{\mu}$ satisfy (3.12). Then the absolute value of the right-hand side of (3.11) is bounded by $1/\min\{|h_1|, h_2\}$. (Note that the right-hand side of (3.11) is just the negative of the curvature of $\boldsymbol{r}$.) Therefore, the continuation theory for ordinary differential equations implies that solutions of initial-value problems for (3.11) are defined for all M. We denote the solution of (3.11) satisfying the initial condition $N = c$ when $M = 0$ by $M \mapsto \check{N}(M,c)$.

Under the monotonicity conditions that support (3.10), we find that the Legendre transform W^* of W is convex and can be taken to vanish at $(0,0)$. But it is not coercive like W. Consequently its level sets do not form closed orbits. (Such closed orbits would also be incompatible with the global existence theory for $\check{N}(\cdot,c)$.) Our existence theory for (3.11) is equivalent to showing the existence of an integrating factor for (3.7) and (1.30).

We now substitute $\check{N}(M,c)$ into (3.6) to obtain a two-parameter family of equations for phase-plane trajectories in the (M,M')-plane:

$$M' = \pm\hat{\nu}(\check{N}(M,c),M)\sqrt{a + 2pM - \check{N}(M,c)^2}. \tag{3.13}$$

In order that (3.13) correspond to an equilibrium state of a ring it must represent a closed orbit (touching no singular points). Let $a+2pM-\check{N}(M,c)^2$, which is a well-defined constitutive function depending on M and the parameters p,a,c, be positive on an interval of the form $(M^-(p,a,c), M^+(p,a,c))$ and have simple zeros at the end points of this interval. Standard phase-plane methods imply that there is a one-to-one correspondence between such intervals and closed orbits for (3.13). Moreover, each such closed orbit has the property that $M^-(p,a,c) \leq M \leq M^+(p,a,c)$. Let $s \mapsto M(s;p,a,c)$ denote a solution of (3.13) corresponding to a closed orbit. (The integration of (3.13) would produce another constant of integration that we can fix by taking $M(0;p,a,c) = M^-(p,a,c)$.) For $M(\cdot;p,a,c)$

to generate a solution of the equilibrium problem for the ring, it must (i) satisfy (3.4):

$$(3.14)\qquad \int_0^{2\pi} \hat{\mu}\big(\check{N}(M(s;p,a,c),c), M(s;p,a,c)\big)\,ds = 2\pi,$$

(ii) have period 2π, and (iii) generate an $\boldsymbol{r}$ with period 2π. Condition (ii) is equivalent to the statement that there is a positive integer k such that $M(\cdot;p,a,c)$ has least period $2\pi/k$. Thus s increases from 0 to π/k in the first transit of (M, M') on the upper half of the trajectory corresponding to (3.13), from which it follows that

$$(3.15)\qquad \frac{\pi}{k} = \int_{M^-(p,a,c)}^{M^+(p,a,c)} \frac{ds}{dM}\,dM = \int_{M^-(p,a,c)}^{M^+(p,a,c)} \frac{dM}{\hat{\nu}\big(\check{N}(M,c),M\big)\sqrt{a+2pM-\check{N}(M,c)^2}}.$$

By (1.3) and (1.4), condition (iii) is equivalent to

$$(3.16a)\qquad \mathbf{0} = \boldsymbol{r}(2\pi) - \boldsymbol{r}(0) = \int_0^{2\pi} \hat{\nu}\big(\check{N}(M(s;p,a,c),c), M(s;p,a,c)\big)[\cos\theta(s)\boldsymbol{i} + \sin\theta(s)\boldsymbol{j}]\,ds$$

where, without loss of generality, we take

$$(3.16b)\qquad \theta(s) = \int_0^s \hat{\mu}\big(\check{N}(M(\xi;p,a,c),c), M(\xi;p,a,c)\big)\,d\xi.$$

We do not consider here with whether there are equilibrium states of the ring, i.e., whether there are numbers a and c satisfying (3.14)–(3.16). We merely determine what equilibrium states look like. In this process, we show that in a certain sense (3.16a) is automatically satisfied, so we do not have to concern ourselves with it. Our first observation is a summary of the consequences of the phase-plane analysis of (3.13):

3.17. Lemma. *Let $a + 2pM - \check{N}(M,c)^2$ be positive on the interval $\big(M^-(p,a,c), M^+(p,a,c)\big)$ and have simple zeros at its end points. Let $s \mapsto M(s;p,a,c)$ denote a solution of (3.13) corresponding to a closed orbit. Then*

$$(3.18)\qquad M^-(p,a,c) \le M(s;p,a,c) \le M^+(p,a,c),$$

the graph of $M(\cdot;p,a,c)$ is symmetric about its points of tangency to the lines $M = M^\pm(p,a,c)$, the derivative $M'(\cdot;p,a,c)$ vanishes only on these lines, and $M(\cdot;p,a,c)$ is monotone between successive contact points with these lines.

From (3.11) and (3.7) we compute

$$(3.19)\qquad \tfrac{d}{ds}\hat{\kappa}\big(N(s),M(s)\big) = \frac{[\hat{\mu}^2\hat{\nu}_N - \hat{\mu}\hat{\nu}\hat{\nu}_M - \hat{\mu}\hat{\nu}\hat{\mu}_N + \hat{\nu}^2\hat{\mu}_M]}{\hat{\nu}^3} M'(s)$$

where the arguments of the constitutive functions on the right-hand side of (3.19) are $\big(N(s), M(s)\big)$. It follows from (1.22) and (1.23) that the quadratic form in brackets in (3.19) is everywhere positive. Therefore

$$\kappa(\cdot; p, a, c) \equiv \hat{\kappa}\big(\check{N}(M(\cdot; p, a, c), c), M(\cdot; p, a, c)\big)$$

and $M(\cdot; p, a, c)$ have extrema at exactly the same points. Thus $\kappa(\cdot; p, a, c)$ enjoys properties analogous to those of $M(\cdot; p, a, c)$ in Lemma 3.17.

Let us denote the actual arc length from 0 to s by

$$\sigma(s) \equiv \int_0^s \nu(\xi)\, d\xi \tag{3.20}$$

and define $2l = \sigma(2\pi)$. We denote the inverse of σ by $\tilde{s}$ and define

$$\tilde{\kappa}(\sigma; p, a, c) \equiv \kappa(\tilde{s}(\sigma); p, a, c). \tag{3.21}$$

3.22. Lemma. *Let $M(\cdot; p, a, c)$ have least period $2\pi/k$ (so that (3.15) is satisfied). Then $\tilde{\kappa}(\cdot; p, a, c)$ has least period $2l/k$. If s_0 is a zero of $M'(\cdot; p, a, c)$, then $\tilde{\kappa}(\cdot; p, a, c)$ assumes a maximum or minimum at $\sigma(s_0)$, is symmetric about this point, and enjoys properties analogous to those of $M(\cdot; p, a, c)$ in Lemma 3.17.*

3.23. Exercise. Prove Lemma 3.22.

We now obtain the following striking result:

3.24. Theorem. *Every simple (nonintersecting) configuration of the ring must have at least two axes of symmetry.*

Proof. The Four-Vertex Theorem (see Jackson (1944)) asserts that any simple twice continuously differentiable closed curve $\boldsymbol{r}$ must have at least four points at which the curvature has a local extremum. It follows from Lemma 3.22 that $\tilde{\kappa}(\cdot; p, a, c)$ must have least period $2l/k$ with $k \geq 2$. The symmetry of the shape follows from the symmetry of the curvature. □

The proof of the Four-Vertex Theorem of course relies on the use of (3.16a). We now prove that Theorem 3.24 implies that (3.16a) holds. This means that (3.16a) does not further restrict the solutions whose properties we have extracted from the phase portrait corresponding to (3.13). Setting $\tilde{\theta}(\sigma) = \theta(\tilde{s}(\sigma))$, we can write (3.16a) as

$$\int_0^{2l} [\cos\tilde{\theta}(\sigma)\boldsymbol{i} + \sin\tilde{\theta}(\sigma)\boldsymbol{j}]\, d\sigma = \boldsymbol{0} \quad \text{or} \quad \int_0^{2l} e^{i\tilde{\theta}(\sigma)}\, d\sigma = 0. \tag{3.25a,b}$$

We set

$$\tilde{\theta}(\sigma) \equiv \frac{\pi\sigma}{l} + \tilde{\psi}(\sigma). \tag{3.26}$$

Theorem 3.24 implies that $\tilde{\psi}$ has period $2l/k$ with $k \geq 2$. Thus the left-hand side of (3.25b) becomes

$$\int_0^{2l} \exp\left(i\frac{\pi\sigma}{l}\right) \exp(i\tilde{\psi}(\sigma))\, d\sigma, \tag{3.27}$$

which vanishes identically because the two exponentials in (3.27) are orthogonal by virtue of their periodicities.

Let us denote by M_k any function that generates a solution of the ring problem with least period $2\pi/k$. Corresponding to it is the bending strain μ_k, which we write as

$$\mu_k = 1 + \zeta_k. \tag{3.28}$$

Condition (3.14) implies that the integral of ζ_k is zero, so that if $\zeta_k \neq 0$, then

$$\min_s \zeta_k(s) < 0 < \max_s \zeta_k(s). \tag{3.29}$$

It follows from our preceding results that ζ_k has exactly $2k$ extrema (and $2k$ zeros) on any half-open interval of length 2π.

3.30. Theorem. *Let j and k be unequal positive integers. Let $\mathcal{C}$ denote any connected set of solution pairs (p, ζ) of the ring problem. Suppose $\mathcal{C}$ contains a solution pair (p_j, ζ_j) with ζ_j having least period $2\pi/j$ and another solution pair (p_k, ζ_k) with ζ_k having least period $2\pi/k$. Then $\mathcal{C}$ must contain a trivial solution pair (for which $\zeta = 0$) and the ζ's of $\mathcal{C}$ can change their least period only at a trivial solution pair. Indeed, the only solutions that, with their first derivatives, are uniformly near a given nonzero solution are solutions with the same least period as the given solution.*

We set $\|\zeta\| \equiv \max\{|\zeta(s)| : s \in [0, 2\pi]\}$. The proof consists in using elementary calculus to show that

$$\|\zeta_k\| \leq \frac{2j\|\zeta_k - \zeta_j\| + \pi\|\zeta_k' - \zeta_j'\|}{|k-j|}. \tag{3.31}$$

Proof of (3.31). Assume without loss of generality that $\zeta_k(s^\star) = \max_s \zeta_k(s) \geq \max_s \zeta_j(s)$. Then (3.29) implies that

$$|\max_s \zeta_k(s) - \max_s \zeta_j(s)| = \zeta_k(s^\star) - \max_s \zeta_j(s) \leq \zeta_k(s^\star) - \zeta_j(s^\star) \leq \|\zeta_k - \zeta_j\|. \tag{3.32a}$$

Similarly, we obtain

$$|\min_s \zeta_k(s) - \min_s \zeta_j(s)| \leq \|\zeta_k - \zeta_j\|. \tag{3.32b}$$

Since ζ_k has the same properties as $M(\cdot; p, a, c)$ in Lemma 3.17, it follows that

$$\int_0^{2\pi} |\zeta_k'(s)|\, ds = 2k[\max_s \zeta_k(s) - \min_s \zeta_k(s)]. \tag{3.33}$$

Thus

$$\begin{aligned}
\pi\|\zeta_k' - \zeta_j'\| &\geq \tfrac{1}{2} \int_0^{2\pi} |\zeta_k'(s) - \zeta_j'(s)|\, ds \\
&\geq \tfrac{1}{2} \left| \int_0^{2\pi} |\zeta_k'(s)|\, ds - \int_0^{2\pi} |\zeta_j'(s)|\, ds \right| \\
&= \left| k[\max_s \zeta_k(s) - \min_s \zeta_k] - j[\max_s \zeta_j(s) - \min_s \zeta_j] \right| \\
&= |(k-j)[\max_s \zeta_k(s) - \min_s \zeta_k(s)] + j[\max_s \zeta_k(s) - \max_s \zeta_j(s)] \\
&\quad - j[\min_s \zeta_k(s) - \min_s \zeta_j(s)]| \\
&\geq |k-j|[\max_s \zeta_k(s) - \min_s \zeta_k(s)] - 2j\|\zeta_k - \zeta_j\|,
\end{aligned} \tag{3.34}$$

the last step coming from (3.32). We now use (3.29) to get (3.31). □

It must be emphasized that this result depends critically on consequences of Lemma 3.22 and is not a property of periodic functions in general. Indeed, by appropriately perturbing a function with period P one can obtain functions with least period nP where n is any positive integer. Theorem 3.30 has the following important consequence:

3.35. Corollary. *On every connected set (or branch) of solution pairs of the ring problem not containing a trivial pair, the ζ's are globally characterized by their least period or, equivalently, by their number of zeros.*

Shearable rings. We now sketch the generalization of the preceding theory to shearable rings, for which (1.20) or (3.2) reduces to

(3.36a,b,c) $$N' = H\theta' + p\eta, \quad H' = -N\theta' - p\nu, \quad M' = \eta N - \nu H$$

(cf. (3.5)), from which we obtain

(3.37a,b) $$NN' + HH' - pM' = 0, \quad \nu N' + \eta H' + \mu M' = 0$$

(cf. (3.3) and (3.7)). The integral of (3.37a) is (3.3). If the material is hyperelastic and uniform, with constitutive equations generalizing those of (3.8), then (3.37b) has the integral $W^*(N, H, M) = \text{const}$. We adopt (1.30) with the constitutive functions independent of s and satisfying the symmetry conditions that $\hat{\nu}$ and $\hat{\mu}$ are even in H and $\hat{\eta}$ is odd in H. We can accordingly represent the constitutive functions of (1.30) by

(3.38) $$\hat{\nu}(N, H, M) = \bar{\nu}(N, \Omega, M), \quad \hat{\eta}(N, H, M) = \bar{\omega}(N, \Omega, M)H, \quad \hat{\mu}(N, H, M) = \bar{\mu}(N, \Omega, M)$$

where

(3.39) $$\Omega \equiv \tfrac{1}{2}H^2.$$

We substitute the constitutive functions (3.38) into (3.37) to obtain

(3.40a,b) $$NN' + \Omega' - pM' = 0, \quad \bar{\nu}(N, \Omega, M)N' + \bar{\omega}(N, \Omega, M)\Omega' + \bar{\mu}(N, \Omega, M)M' = 0.$$

Now (3.36c) implies that M' can vanish only at a point s_0 where either $H(s_0) = 0$ or

(3.41) $$N(s_0)\bar{\omega}(N(s_0), \Omega(s_0), M(s_0)) = \bar{\nu}(N(s_0), \Omega(s_0), M(s_0)).$$

Since the monotonicity condition implies that $\hat{\eta}(N, \cdot, M)$ is strictly increasing, it follows that $\bar{\omega}$ must be everywhere positive and thus (3.41) could hold only for certain positive N's. Condition (3.41) is associated with a shear instability of the sort discussed in the preceding section. For materials satisfying the constitutive restriction that

(3.42) $$\bar{\nu}(N, \Omega, M) > N\bar{\omega}(N, \Omega, M) \quad \text{everywhere},$$

we need not bother with (3.41).

Let $\mathcal{I}$ be any open interval on the s-axis on which M' nowhere vanishes. On $\mathcal{I}$ we can replace (3.40) with

(3.43) $$\frac{dN}{dM} = -\frac{p\bar{\omega} + \bar{\mu}}{\bar{\nu} - N\bar{\omega}}, \quad \frac{d\Omega}{dM} = \frac{p\bar{\nu} + N\bar{\mu}}{\bar{\nu} - N\bar{\omega}}$$

where $\bar{\nu}, \bar{\omega}, \bar{\mu}$ have the arguments shown in (3.40). Since this system possesses the integral (3.3), solutions of initial-value problems for it can be defined everywhere on

$\operatorname{cl}\mathcal{I}$ that $\Omega \geq 0$. We denote the two-parameter family of solutions of (3.43) by $M \mapsto \check{N}(M,a,c,p), \check{\Omega}(M,a,c,p)$ (where a is defined by (3.3)).

The substitution of these functions into (3.36c) yields the analog of (3.13):

$$(3.44) \quad M' = \pm\sqrt{2\check{\Omega}(M,a,c,p)}\,\big[\bar{\omega}\big(\check{N}(M,a,c,p),\check{\Omega}(M,a,c,p),M\big)\check{N}(M,a,c,p) \\ -\bar{\nu}\big(\check{N}(M,a,c,p),\check{\Omega}(M,a,c,p),M\big)\big]$$

where $\check{\Omega}(M,a,c,p) = a + 2pM - \check{N}(M,a,c,p)^2$ by (3.3).

Note that the route to (3.44) differs considerably from that to (3.13). The treatment of curvature is likewise more convoluted: Let ϕ be the angle between $\boldsymbol{i}$ and $\boldsymbol{r}'$ so that

$$(3.45) \qquad \boldsymbol{r}' = |\boldsymbol{r}'|(\cos\phi\boldsymbol{i} + \sin\phi\boldsymbol{j}).$$

We set

$$(3.46) \qquad \phi = \theta + \beta, \quad \text{so that} \quad \beta = \arctan\frac{\eta}{\nu}.$$

(β is the *shear angle*.) The curvature is $\phi'/|\boldsymbol{r}'|$.

3.47. Exercise. Suppose that there is no s_0 for which (3.41) holds. Then $M' = 0$ if and only if $H = 0$, which happens if and only if $\eta = 0$. By inserting the constitutive functions into the formula for the curvature and repeatedly using (3.36), show that the derivative of the curvature vanishes where M' vanishes. Show that the converse need not be true because the expression playing the role of the bracketed term in (3.19) typically involves second derivatives of the constitutive functions and products of the first derivatives. (Thus, in the absence of further constitutive assumptions, this expression can vanish, so that the number of relative extrema of the curvature may exceed that of M, and the Four-Vertex Theorem cannot be invoked as above.) Under suitable constitutive assumptions, prove an analog of Theorem 3.24 for shearable rings.

We now develop an analog of Theorem 3.24 by a method alternative to that suggested by Ex. 3.47. We observe that where (3.43) holds, M has properties just like those given in Lemma 3.17. Thus, without loss of generality, we may assume that M is an even function of s having a maximum at $s = 0$ with least period $2\pi/k$. We indicate by a subscript k any solution of the ring problem corresponding to such an M. By tracing through the effects of the constitutive functions and the functions $\check{\Omega}$ and $\check{N}$, we find that we can fix a rigid rotation such that the ϕ_k corresponding to M_k has the property that $s \mapsto \phi_k(s) - s$ is odd and has period $2\pi/k$. Thus $\phi_k(\pi/k) = \pi/k$. Since we now define σ and l by (3.20) with ν replaced by $|\boldsymbol{r}'| = \sqrt{\nu^2+\eta^2}$, condition (3.16a) now implies that (3.25) holds with $\tilde{\theta}(\sigma)$ replaced with $\tilde{\phi}_k(\sigma) \equiv \phi_k(\tilde{s}(\sigma))$. As before, we find that this condition holds identically if $k \geq 2$.

The requirement that the reference configuration be natural implies that

$$(3.48) \qquad \bar{\mu}(0,0,0) = 1.$$

Suppose that we provisionally adopt the far more stringent condition that

$$(3.49) \qquad \bar{\mu}(N,\Omega,0) = 1 \quad \forall\, N,\Omega.$$

(We discuss this condition in Sec. XIV.5.) Condition (3.4) says that μ_k must have mean value 1. The monotonicity of $\bar{\mu}(N,\Omega,\cdot)$ then implies that $\mu_k(s) - 1$ must have the same sign as $M_k(s)$. Since M_k has its maximum at $s = 0$ and has the properties given in Lemma 3.17, it follows that $M_k' < 0$ on $(0,\pi/k)$. Let us assume that

$$(3.50) \qquad \nu_k(s) - \omega_k(s)N_k(s) > 0 \quad \forall\, s$$

(cf. (3.42)). Since M' can then vanish only where H and therefore η vanish, we conclude that $\eta_k(0) = 0$, so that $\theta_k(0) = 0$, by (3.6) and by our fixing of a rigid rotation so that

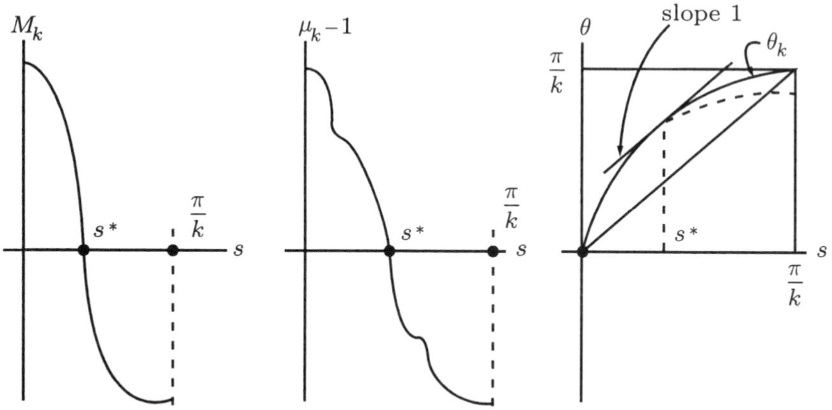

Fig. 3.52. Construction leading to the proof of (3.51). There is a unique $s^\star$ in $[0, \pi/2]$ such that $M_k > 0$ and $\mu_k = \theta_k' > 1$ for $0 \leq s < s^\star$ and such that $M_k < 0$ and $\mu_k = \theta_k' < 1$ for $s^\star < s \leq \pi/k$. Suppose that the graph of θ_k were to fall below the graph of the straight line $s \mapsto s$ in $(0, \pi/k)$, as shown by the dashed line in (c). This could only happen in the interval $(s^\star, \pi/k)$. Since $\theta_k' < 1$ here, it would be impossible for θ_k to satisfy the condition that $\theta_k(\pi/k) = \pi/k$.

$\phi_k(0) = 0$. It then follows from (3.4) and the oddness of θ_k that $\theta_k(\pi/k) = \pi/k$. Then these conditions imply that

$$\theta_k(s) \geq s \quad \text{for} \quad s \in [0, \pi/k]. \tag{3.51}$$

(See Fig. 3.52.) Since (3.50) implies that $H_k(s)$, $\eta_k(s)$, $\beta_k(s)$ each have the sign opposite to that of $M_k'(s)$, it follows from (3.46) that

$$\tilde{\phi}(\sigma) \geq \frac{\pi\sigma}{l} \quad \text{for} \quad \sigma \in [0, l/k]. \tag{3.53}$$

We use these observations to prove:

3.54. Theorem. *Let* (3.49) *hold. Any configuration of the ring problem for which* (3.50) *holds, for which* $\check{N}$ *and* $\check{\Omega}$ *are well defined, and for which*

$$\frac{\pi\sigma}{l} \leq \tilde{\phi}(\sigma) \leq 2\pi - \frac{\pi\sigma}{l} \quad \text{for} \quad \sigma \in [0, l/k] \tag{3.55}$$

must have at least two axes of symmetry.

Note that (3.55) does not force $\boldsymbol{r}$ to be convex.

Proof of Theorem 3.54. Were k to equal 1, then (3.55) would imply that $\cos\tilde{\phi}(\sigma) \leq \cos(\pi\sigma/l)$ for $\sigma \in [0, 2l]$. Thus the condition that $\int_0^{2l} \cos\tilde{\phi}(\sigma)\, d\sigma = 0$ of the modified form of (3.25) could not hold. □

We can easily generalize this result. All we need to do is to ensure that (3.53) holds so that (3.55) makes sense. We suspend (3.49), retain (3.48), and incorporate in the theorem the restriction that the conclusion holds only in the region of solution-parameter space for which $\mu - 1$ has a single zero s_0 on $[0, \pi]$ with $\mu - 1 > 0$ on $[0, s_0)$ and $\mu - 1 < 0$ on $(s_0, \pi]$.

Dickey & Roseman (1993) pointed out that symmetry theorems like 3.24 and 3.54 do not hold for rings subject to a central force (for which they discovered an interesting secondary bifurcation), and they adduced this fact as an explanation for experimental evidence indicating that observed large buckled states may have but one axis of symmetry. I believe that such an unnatural loading system is inappropriate for explaining the reduced symmetry observed in experiment. Rather, restrictions on size of solutions such as that embodied in (3.55), which would presumably be more stringent for more accurate rod theories, possibly augmented with the results of an analysis of imperfections, seem quite consistent with the available experimental data.

3.56. Exercise. Use the methods just developed for shearable rings to prove Theorems 3.24 and 3.30 for unshearable rings with constitutive equations of the form $\nu = \hat{\nu}(N, H, M)$, $\eta = 0$, $\mu = \hat{\mu}(N, H, M)$.

In our study of the buckling of shearable arches in Sec. VI.12 we shall use a more direct approach to obtain analogs of Theorems 3.24 and 3.30. We shall characterize qualitative properties of solution branches by the number of zeros of H or, equivalently, of η and for appropriate boundary-value problems we shall show that this number can change only where H has a double zero. If this happens, the system (3.36), (3.38) has a critical point $(N, H, M) = (N^0, 0, M^0)$, corresponding to a trivial circular equilibrium state, with $N^0 \bar{\mu}(N^0, 0, M^0) = -p\bar{\nu}(N^0, 0, M^0)$. For (3.4) to hold, it is necessary that $\bar{\mu}(N^0, 0, M^0) = 1$. We thus obtain a system of two equations for N^0 and M^0. The monotonicity of $(\bar{\nu}(\cdot, 0, \cdot), \bar{\mu}(\cdot, 0, \cdot))$ supports a global inverse function theorem to the effect that this system has a unique solution (N^0, M^0) if $p \geq 0$. A much easier approach to this conclusion follows from the observation that this system for (N^0, M^0) is equivalent to $N^0 = \hat{N}(\nu_0, 0, 1)$, $M^0 = \hat{M}(\nu_0, 0, 1)$ where ν_0 satisfies $\hat{N}(\nu_0, 0, 1) = -p\nu_0$, the analysis of which has been carried out in Sec. III.5 (see the comments following the statement of Theorem III.5.16). This second formulation is easy only because we have already established the invertibility of our constitutive functions by means of the very global inverse function theorems that would have enabled us to resolve the original formulation of the problem for (N^0, M^0).

The theory for rods under simple hydrostatic pressure was developed by Lévy (1884). The solution of the equilibrium equations for Bernoulli-Euler rings was expressed in terms of elliptic functions by Halphen (1884). Tadjbakhsh & Odeh (1967) initiated the mathematical analysis of the buckling of Bernoulli-Euler rings. The presentation of this section represents a simplification and generalization of the work of Antman (1970a, 1973b, 1974a) and of Antman & Dunn (1980).

4. Asymptotic Shape of Inflated Rings

We study the equilibrium of nonuniform, noncircular rings under an internal simple hydrostatic pressure. Our governing equations are (1.3), (1.4), (3.36) with p replaced with $-p$, and (1.21). We take $s \in [-L, L]$ and require that

$$\theta(L) - \theta(-L) = 2\pi, \quad \boldsymbol{r}(L) = \boldsymbol{r}(-L). \tag{4.1}$$

In the natural reference state

$$\nu = 1, \quad \eta = 0, \quad \theta = \overset{\circ}{\theta}. \tag{4.2}$$

We require that constitutive restrictions (1.23)–(1.28) hold. Let ε be a small positive parameter. We characterize large deformations of the ring by the requirement that the area enclosed by $\boldsymbol{r}$ (computed by Green's Theorem) be the large number π/ε^2:

$$\tfrac{1}{2}\int_{-L}^{L} \boldsymbol{k}\cdot(\boldsymbol{r}\times\boldsymbol{r}')\,ds = \frac{\pi}{\varepsilon^2}. \tag{4.3}$$

We wish to determine how solutions of this *ring problem* depend on the parameter ε. (We require the area, rather than the pressure, to be large, because there are materials having arbitrarily large circular equilibrium configurations for bounded internal pressures; see Sec. III.5.)

For (4.3) to hold, ν, $\boldsymbol{r}$, and the radius of curvature, which is the reciprocal of the curvature

$$\kappa = \frac{\theta'}{[\nu^2+\eta^2]^{1/2}} + \frac{\nu\eta'-\eta\nu'}{[\nu^2+\eta^2]^{3/2}} \tag{4.4}$$

(cf. (3.46)), must be correspondingly large. We consequently introduce new variables

$$\bar{\nu} = \varepsilon\nu, \quad \bar{\boldsymbol{r}} = \varepsilon\boldsymbol{r}, \quad \bar{\kappa} = \kappa/\varepsilon. \tag{4.5}$$

Motivated by a three-dimensional interpretation of constitutive equations for nonlinearly elastic materials under large extension (see Sec. XIV.5), we adopt the scaling

$$\overline{N} = \alpha(\varepsilon)N, \quad \varepsilon\overline{H} = \alpha(\varepsilon)H, \quad \overline{M} = \alpha(\varepsilon)M \tag{4.6}$$

where α is a strictly increasing function defined on an interval of the form $[0, E]$ with $\alpha(0) = 0$. (Other scalings are possible; they are treated similarly.) Looking ahead to the role played by p, we adopt the scaling

$$\varepsilon\bar{p} = \alpha(\varepsilon)p. \tag{4.7}$$

Thus our governing differential equations (1.3), (1.4), and (3.36) with p replaced by $-p$ assume the form

$$\bar{\boldsymbol{r}}' = \bar{\nu}\boldsymbol{a} + \varepsilon\eta\boldsymbol{b}, \tag{4.8}$$

$$\overline{N}' = \varepsilon\theta'\overline{H} - \varepsilon\bar{p}\eta, \tag{4.9}$$

$$\varepsilon\overline{H}' = -\theta'\overline{N} + \bar{p}\bar{\nu}, \tag{4.10}$$

$$\overline{M}' = \eta\overline{N} - \bar{\nu}\overline{H}. \tag{4.11}$$

They are subject to the side conditions (4.1) and (4.3):

$$\int_{-L}^{L} \theta'(s)\,ds = 2\pi, \tag{4.12}$$

$$\bar{\boldsymbol{r}}(L) = \bar{\boldsymbol{r}}(-L), \tag{4.13}$$

$$\int_{-L}^{L} \boldsymbol{k}\cdot[\bar{\boldsymbol{r}}\times(\bar{\nu}\boldsymbol{a}+\varepsilon\eta\boldsymbol{b})]\,ds = 2\pi. \tag{4.14}$$

Note that

$$\bar{\kappa} = \frac{\theta'}{(\bar{\nu}^2 + \varepsilon^2\eta^2)^{1/2}} + \frac{\varepsilon(\bar{\nu}\eta' - \eta\bar{\nu}')}{(\bar{\nu}^2 + \varepsilon^2\eta^2)^{3/2}}. \tag{4.15}$$

We say that a sequence of functions $\{\varepsilon \mapsto \alpha_k(\varepsilon), k = 0, \dots, K\}$, with K finite or infinite, is an *asymptotic sequence* iff (i) each α_k is continuous on a common interval of the form $[0, E]$, (ii) each α_k is positive on $(0, E]$, and (iii) $\alpha_k(\varepsilon) = o(\alpha_{k-1}(\varepsilon))$ as $\varepsilon \to 0$. Let

$$\mathbf{q} \equiv (\bar{\nu}, \eta, \mu), \quad \mathbf{Q} \equiv (\overline{N}, \overline{H}, \overline{M}). \tag{4.16}$$

We adopt constitutive equations of the form

$$\mathbf{Q} = \sum_{k=0}^{K} \alpha_k(\varepsilon)\mathbf{Q}^k(\mathbf{q}, s) + \mathbf{Q}^+(\alpha_K(\varepsilon), \mathbf{q}, s) \tag{4.17}$$

where $\{\alpha_k\}$ is an asymptotic sequence, $\alpha_0 = 1$, $\mathbf{Q}^k \equiv (N^k, H^k, M^k)$, $\mathbf{Q}^+(\cdot, \mathbf{q}, s)$ is continuously differentiable, and $\mathbf{Q}^+(\alpha_K(\varepsilon), \mathbf{q}, s) = o(\alpha_K(\varepsilon))$ uniformly for $(\mathbf{q}, s)$ in a compact subset of $(0, \infty) \times \mathbb{R}^2 \times [-L, L]$ (see (3.12)). There is no loss of generality in using the same α_k for each component of $\mathbf{Q}_k$ because $\mathbf{Q}^k$ may have zero components. We assume that distinct members of $\{\varepsilon \mapsto \alpha_k(\varepsilon), \varepsilon\alpha_k(\varepsilon), k = 0, \dots, K\}$ can be put into an asymptotic sequence and that any two members of this set that are asymptotically equivalent are actually equal. We furthermore require that the constitutive functions have as much smoothness in $\mathbf{q}$ as required in the analysis.

The behavior of the material for ε small corresponds to very large extensions by virtue of (4.5). We accordingly use a very general representation for the dependence of the constitutive functions on ε, in place of merely taking the $\alpha_k(\varepsilon)$ to be powers of ε, because we have no experimental evidence to guide us and we are not so sanguine as to believe that nature must accommodate herself to our mathematical convenience.

We assume that

$$N^k \text{ and } M^k \text{ are even in } \eta, \quad H^k \text{ is odd in } \eta, \tag{4.18}$$

$$N^0_\eta = 0 = M^0_\eta, \tag{4.19}$$

$$N^0(0, \mu, s) = 0, \tag{4.20}$$

$$N^0(\bar{\nu}, \bar{\nu}, s) \to \infty \quad \text{as} \quad \bar{\nu} \to \infty, \tag{4.21a}$$

$$\frac{\partial}{\partial \bar{\nu}} N^0(\bar{\nu}, \mu, s) > 0, \quad \frac{\partial}{\partial \bar{\nu}} N^0(\bar{\nu}, \bar{\nu}, s) > 0, \tag{4.21b,c}$$

$$\eta \mapsto \eta N^0(\bar{\nu}, \bar{\nu}, s) - \bar{\nu} H^0(\bar{\nu}, \eta, \bar{\nu}, s) \text{ is invertible } \forall\, \bar{\nu} > 0. \tag{4.22}$$

Condition (4.18) is analogous to the symmetry condition (3.38). Conditions (4.19) and (4.21) are inspired by three-dimensional considerations. Condition (4.20) is a natural concomitant of the asymptotic approach we employ. Condition (4.22) is a technical requirement like (3.42) that enables us to avoid dealing with shear instabilities.

We now seek solutions of the ring problem in the form

$$(4.23)\quad \mathbf{q}(s,\varepsilon) = \sum_{k=0}^{K} \beta_k(\varepsilon)\mathbf{q}_k(s) + o(\beta_K(\varepsilon)), \quad \bar{p}(\varepsilon) = \sum_{k=0}^{K} \beta_k(\varepsilon)p_k + o(\beta_K(\varepsilon))$$

with the β_k forming an asymptotic sequence with $\beta_0 = 1$. We define

$$(4.24)\qquad \theta_k' = \mu_k.$$

We must determine the β_k, $\mathbf{q}_k$, p_k.

We get the leading terms of (4.23) by substituting it into the the governing equations (4.8)–(4.14) and then letting $\varepsilon \to 0$:

$$(4.25)\qquad \boldsymbol{r}_0' = \nu_0 \boldsymbol{a}_0 \equiv \nu_0(\cos\theta_0 \boldsymbol{i} + \sin\theta_0 \boldsymbol{j}),$$

$$(4.26\text{a})\qquad (N^0)' = 0,$$

$$(4.26\text{b})\qquad -\theta_0' N^0 + p_0\nu_0 = 0,$$

$$(4.26\text{c})\qquad (M^0)' = \eta_0 N^0 - \nu_0 H^0,$$

$$(4.27\text{a})\qquad \int_{-L}^{L} \theta_0'\, ds = 2\pi,$$

$$(4.27\text{b})\qquad \int_{-L}^{L} \nu_0(\cos\theta_0 \boldsymbol{i} + \sin\theta_0 \boldsymbol{j})\, ds = \mathbf{0},$$

$$(4.27\text{c})\qquad \boldsymbol{k}\cdot\int_{-L}^{L} (\boldsymbol{r}_0 \times \boldsymbol{r}_0')\, ds = 2\pi.$$

Thus $N^0 =$ const. If p_0 were to equal 0, then (4.26b) would imply that either $N^0 = 0$ or $\theta_0' = 0$. But if $N^0 = 0$, then (4.20) and (4.21) would imply that $\nu_0 = 0$, in contradiction to (4.27c). If $\theta_0' = 0$, then (4.27a) would be violated. Thus $p_0 \neq 0$. It then follows from (4.26b) that $N^0 \neq 0$ and that ν_0 and θ_0' can vanish nowhere. Thus (4.15) yields the leading term for the curvature:

$$(4.28)\qquad \kappa_0 = \frac{\theta_0'}{\nu_0} = \frac{p_0}{N^0} = \text{const.}$$

Thus if there is a solution for the leading term (an issue we soon resolve), then $\boldsymbol{r}_0$ represents a circle. Indeed, (4.25) and (4.26b) imply that

$$(4.29)\qquad \boldsymbol{r}_0' = \frac{N^0}{p_0}\theta_0' \boldsymbol{a}_0 = -\frac{N^0}{p_0}\boldsymbol{b}_0'.$$

By fixing the center of the circle at the origin we obtain from (4.29) that

$$(4.30)\qquad \boldsymbol{r}_0 = -\frac{N^0}{p_0}\boldsymbol{b}_0.$$

We now substitute (4.29) and (4.30) into (4.27c) and use (4.27a) to obtain $(N^0/p_0)^2 = 1$. Fixing the orientation of $\boldsymbol{r}_0$ so that $\kappa_0 > 0$, we thus obtain from (4.26b) that

(4.31a,b,c,d) $$N^0 = p_0, \quad \theta_0' = \nu_0, \quad \boldsymbol{r}_0 = -\boldsymbol{b}_0, \quad \kappa_0 = 1.$$

We must now find $\nu_0 = \mu_0, \eta_0, p_0$ to show that there are solutions. Equation (4.31b) reduces (4.27a) to

(4.32) $$\int_{-L}^{L} \nu_0 \, ds = 2\pi$$

and reduces (4.31a) to

(4.33) $$N^0(\nu_0(s), \nu_0(s), s) = p_0.$$

Properties (4.21) ensure that if $p_0 > 0$, then (4.33) can be uniquely solved for $\nu_0(s)$:

(4.34) $$\nu_0(s) = \nu_0^\sharp(p_0, s).$$

$\nu_0^\sharp(\cdot, s)$ strictly increases from 0 to ∞ as its argument increases over this range. Thus the substitution of (4.34) into (4.32) yields an equation that can be uniquely solved for p_0. We then have ν_0 from (4.34) and $\mu_0 = \nu_0$ from (4.31b). Condition (4.22) ensures that η_0 can be uniquely determined from (4.26c).

We now study the first-order corrections. We again substitute (4.23) into (4.8)–(4.14) and use the results about the leading-order terms to obtain

(4.35) $$\tfrac{d}{ds}[\beta_1(\varepsilon)(\partial N^0/\partial \mathbf{q}) \cdot \mathbf{q}_1 + \alpha_1(\varepsilon) N^1] = \varepsilon \nu_0 H^0 - \varepsilon p_0 \eta_0 + \cdots,$$

(4.36) $$\varepsilon \tfrac{d}{ds} H^0 = \beta_1(\varepsilon)[p_0(\nu_1 - \mu_1) + \nu_0(p_1 - (\partial N^0/\partial \mathbf{q}) \cdot \mathbf{q}_1)] - \alpha_1(\varepsilon)\nu_0 N^1 + \cdots,$$

(4.37) $$\tfrac{d}{ds}[\beta_1(\varepsilon)(\partial M^0/\partial \mathbf{q}) \cdot \mathbf{q}_1 + \alpha_1(\varepsilon) M^1] = \beta_1(\varepsilon)[\eta_1 p_0 + \eta_0(\partial N^0/\partial \mathbf{q}) \cdot \mathbf{q}_1 - \nu_1 H^0 - \nu_0(\partial H^0/\partial \mathbf{q}) \cdot \mathbf{q}_1] + \alpha_1(\varepsilon)[\eta_0 N^1 nu_0 H^1] + \cdots,$$

(4.38) $$0 = \int_{-L}^{L} \mu_1(s) \, ds + \cdots,$$

(4.39) $$0 = \int_{-L}^{L} [\nu_0 \theta_1 \boldsymbol{b}_0 + \nu_1 \boldsymbol{a}_0] \, ds + \cdots,$$

(4.40) $$0 = \int_{-L}^{L} \nu_1(s) \, ds + \cdots,$$

(4.41) $\nu_1, \ \eta_1, \ \mu_1$ have period $2L$.

Here $(\partial N^0/\partial \mathbf{q}) \cdot \mathbf{q}_1 \equiv N^0_\nu \nu_1 + N^0_\eta \eta_1 + N^0_\mu \mu_1$, etc., the arguments of N^0_ν, N^1, etc., are ν_0, η_0, μ_0, s, and the ellipses represent terms of order higher than those explicitly exhibited. In obtaining (4.40), it is convenient to use integration by parts together with (4.31c).

There are two given functions of ε, namely, $\varepsilon \mapsto \varepsilon, \alpha_1(\varepsilon)$, in these equations. We show that there is essentially only one way to choose the unknown function β_1 so as to obtain ν_1, η_1, μ_1, p_1 as nontrivial solutions. Our results depend on the relative size of ε and $\alpha_1(\varepsilon)$.

Case 1. $\alpha_1(\varepsilon) = o(\varepsilon)$. We can eliminate the terms containing α_1 from (4.35) and (4.36). There are three possibilities for β_1:

(a) $\varepsilon = o(\beta_1(\varepsilon))$. We can drop terms containing ε from (4.35) and (4.36) and we can drop the terms containing α_1 from (4.37). We thus find (in the limit that $\varepsilon \to 0$) that

$$(4.42a) \qquad (\partial N^0/\partial \mathbf{q}) \cdot \mathbf{q}_1 = c \text{ (const.)},$$

$$(4.42b) \qquad p_0(\nu_1 - \mu_1) + \nu_0(p_1 - (\partial N^0/\partial \mathbf{q}) \cdot \mathbf{q}_1) = 0,$$

$$(4.42c) \qquad \tfrac{d}{ds}[(\partial M^0/\partial \mathbf{q}) \cdot \mathbf{q}_1] - [\eta_1 p_0 + \eta_0 (\partial N^0/\partial \mathbf{q}) \cdot \mathbf{q}_1] = -[\nu_1 H^0 + \nu_0 (\partial H^0/\partial \mathbf{q}) \cdot \mathbf{q}_1].$$

Now we integrate (4.42b) over $(-L, L)$ and use (4.38) and (4.40) to obtain $p_1 = c$. From (4.42b) we then obtain that $\nu_1 = \mu_1$. We now use (4.42a) and (4.19) to find that $\mu_1 = p_1/(N^0_\nu + N^0_\mu)$, which is well-defined by virtue of (4.21). We substitute this expression into (4.38) to find that $p_1 = 0$, whence $\nu_1 = \mu_1 = 0$. Equation (4.42c) thus reduces to $\eta_1 p_0 - \nu_0 H^0_\eta = 0$, so that $\eta_1 = 0$ by virtue of (4.22). Thus our assumption (a) yields only trivial solutions, in which case there is no justification for the presence of β_1 in (4.23). We accordingly regard assumption (a) as worthless. Now we try

(b) $\beta_1(\varepsilon) = o(\varepsilon)$. Then we can also drop the terms containing β_1 from (4.35) and (4.36) to obtain $\nu_0 H^0 = p_0 \eta_0$ and $H^0 = \text{const}$. These equations impose further restrictions on the variables ν_0, η_0, μ_0 already found. It is very unlikely that these restrictions and any that come from (4.37) would hold. If they do not, then assumption (b) is invalid. If these conditions happen to be met, then the visible terms in (4.35)–(4.41) need not constitute all the candidates for lower-order terms. When the missing candidates are restored, we can analyze the resulting equations by the same process as we are now carrying out. Our remaining alternative is

(c) $\beta_1(\varepsilon) = \varepsilon$. Then (4.35)–(4.37) reduce to

$$(4.43a) \qquad \tfrac{d}{ds}[(\partial N^0/\partial \mathbf{q}) \cdot \mathbf{q}_1] = \nu_0 H^0 - p_0 \eta_0,$$

$$(4.43b) \qquad \tfrac{d}{ds} H^0 = p_0(\nu_1 - \mu_1) + \nu_0[p_1 - (\partial N^0/\partial \mathbf{q}) \cdot \mathbf{q}_1],$$

together with (4.42c).

4.44. Exercise. Show how to solve (4.43) and (4.42c) subject to (4.38)–(4.41).

4.45. Exercise. Carry out the analysis for Case 2: $\alpha_1(\varepsilon) = \varepsilon$ and Case 3: $\varepsilon = o(\alpha_1(\varepsilon))$.

The treatment of higher-order corrections proceeds along the same lines, with a rapidly growing level of complexity. The approach is analogous to the ideas underlying the Newton Polygon (see Dieudonné (1949), Vaĭnberg & Trenogin (1969), or Chow & Hale (1982), e.g.). The development of this section is adapted from that of Antman & Calderer (1985a) with the kind permission of Cambridge University Press. They work out a concrete example and show that the first-order correction is useful for values of ε as large as $\frac{1}{4}$. They prove that in the absence of shear instabilities the ring problem has a unique solution of the form (4.23) under assumptions of the sort we have made above. Their proof (which requires some minor adjustments) uses the theory of ordinary differential equations with periodic coefficients to replace the ring problem with an operator equation that can be analyzed by an implicit function theorem.

5. Straight Configurations of a Whirling Rod

In this section we study a conceptually simple problem whose solutions have the rich complexity that comes from the interaction of a nonlinear constitutive equation with a 'live' load. The problem has the attraction that virtually every standard method of analysis is easily applied and yields distinctive information.

The end $s = 1$ of a naturally straight rod of length 1 is welded to the inside of a rigid ring of radius R so that the rod lies on the ray from the weld to the center of the ring. (It is mathematically more convenient to fix the length of the rod rather than the radius of the ring.) We take the ring to lie in the $\{\boldsymbol{i}, \boldsymbol{j}\}$-plane. See Fig. 5.1. The end $s = 0$ is free. We seek straight steady configurations of the rod in which the system is rotated at constant angular velocity ω about the $\boldsymbol{k}$-axis. (We discuss bifurcation from such states in Sec. VI.13.)

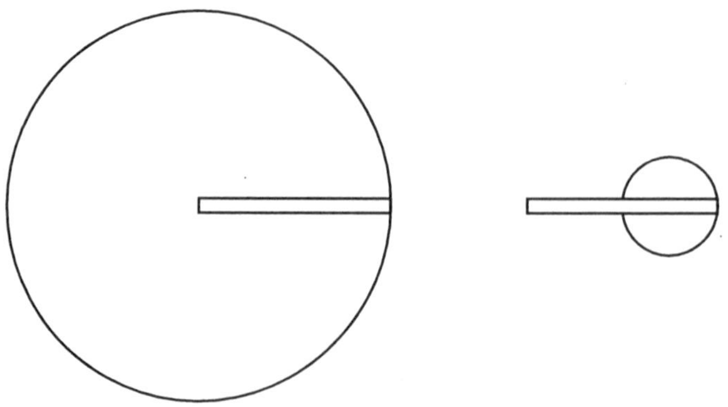

Fig. 5.1. Reference configurations of the rod attached to large and small rings.

We accordingly seek constrained motions of the form

$$\text{(5.2)} \quad \boldsymbol{r}(s,t) = r(s)\boldsymbol{a}(t), \quad \boldsymbol{a}(t) \equiv \cos\omega t\boldsymbol{i} + \sin\omega t\boldsymbol{j}, \quad \boldsymbol{b}(s,t) = \boldsymbol{k} \times \boldsymbol{a}(t)$$

satisfying the boundary condition

$$\text{(5.3)} \qquad r(1) = R.$$

We assume that the rod is nonlinearly elastic with constitutive functions satisfying (1.23)–(1.28) and the symmetry conditions (2.9). Consequently, the contact loads corresponding to (5.2) have the form $H = 0 = M$, so that

$$\text{(5.4)} \qquad \boldsymbol{n}(s,t) = N(s)\boldsymbol{a}(t).$$

We assume that there are no forces applied to the rod except the contact force at $s = 1$. In this case, the integral version of the equation of motion (1.45) (with $\rho I = 0$) reduces to

$$\text{(5.5)} \qquad N(s) = -\omega^2 \int_0^s (\rho A)(\xi) r(\xi)\, d\xi.$$

The integration of (1.4) subject to (5.3) yields

$$\text{(5.6)} \qquad r(s) = R - \int_s^1 \nu(\xi)\, d\xi.$$

For simplicity of exposition, we assume that the rod is uniform, define

$$\text{(5.7)} \qquad \lambda \equiv \omega^2 \rho A,$$

and write our nontrivial constitutive equation in one of the equivalent forms:

$$\text{(5.8)} \quad N(s) = \hat{N}\big(\nu(s)\big) \equiv \hat{N}\big(\nu(s), 0, 0\big), \quad \nu(s) = \hat{\nu}\big(N(s)\big) \equiv \hat{\nu}\big(N(s), 0, 0\big).$$

Let the constitutive functions $\hat{N}$ and $\hat{\nu}$ of (5.8) be continuously differentiable. Our *boundary-value problem* is (5.5)–(5.8). It is clear that if this problem has a solution with r' integrable, then r is twice continuously differentiable and N is thrice continuously differentiable. Thus this problem is equivalent to

$$\text{(5.9a,b,c,d)} \qquad r' = \hat{\nu}(N), \quad N' = -\lambda r, \quad r(1) = R, \quad N(0) = 0$$

or

$$\text{(5.10)} \qquad N'' + \lambda \hat{\nu}(N) = 0, \quad N(0) = 0, \quad N'(1) = -\lambda R.$$

We introduce the stored-energy function and its Legendre transform by

$$W(\nu) \equiv \int_1^\nu \hat{N}(\bar{\nu})\,d\bar{\nu}, \quad W^*(N) \equiv \int_0^N \hat{\nu}(\bar{N})\,d\bar{N}. \tag{5.11}$$

See Fig. III.3.7 for a sketch of the graph of W^*.

Before embarking on our mathematical analysis, let us first use physical intuition to anticipate the kind of solutions we obtain. Suppose first of all that $R \geq 1$ and that the angular velocity of the ring is slowly raised to a constant value. Then the centrifugal force on the rod would tend to produce compression everywhere and we would expect there to be a solution $s \mapsto r(s;\lambda)$ with every part of the rod in compression: $r'(s;\lambda) < 1$ for each s. Moreover, we expect $r'(s,\cdot)$ to be decreasing. Now suppose that $R < 1$ and that the angular velocity of the ring is slowly raised to a constant value. In this case, the centrifugal force would tend to produce compression on part of the rod (at least for angular velocities that are not too large) and produce tension on the remainder. It may happen that the material in tension is incapable of resisting the centrifugal force, whose intensity per unit natural length is proportional to the distance of the material point from the center of the ring. In this case, there would be no equilibrium solution. For small enough ω, we expect there to be an equilibrium state in this case, but for large ω, there could be no solution or any number of solutions, depending on the parameters and the material properties.

If $R < 1$, or even if $R < \frac{1}{2}$, it should be possible to produce a solution for which the rod is everywhere in compression by the artificial process of compressing it to the desired deformed length and then spinning the ring so rapidly that centrifugal force can maintain the rod in compression. Conversely, if $R > 1$, it might be possible to have configurations with deformed length exceeding $2R$ when the spin is sufficiently rapid and the material is sufficiently strong in resisting tension.

Before studying the existence of solutions, we first study their qualitative behavior because it suggest ways to handle existence questions effectively. System (5.9) has the integral

$$\lambda r^2 + 2W^*(N) = 2c \equiv \lambda r(0)^2 = \lambda R^2 + 2W^*(N(1)), \tag{5.12}$$

which describes the trajectories of the phase portrait of (5.9), sketched in Fig. 5.13. Only those trajectories intersecting the nonnegative N-axis are consistent with the requirement that $c \geq 0$, coming from (5.12). This figure implies that there could be three kinds of solutions: (i) those with N everywhere ≤ 0, (ii) those with N positive on an open interval with endpoint $s = 0$ and negative on an open interval with endpoint $s = 1$, these intervals having a common endpoint, and (iii) those with N everywhere ≥ 0. Note that Fig. 5.13 implies that N is symmetric about the point lying over the center of the ring.

We now prove a collection of theorems ensuring the existence of solutions of various types to our boundary-value problem for various ranges of the parameters λ and R.

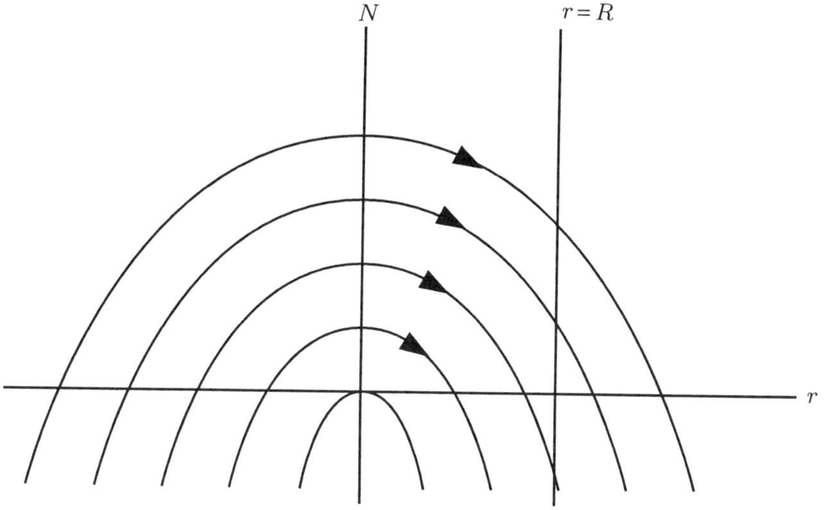

Fig. 5.13. Phase portrait of (5.9). Solutions correspond to trajectories that must start on the line $N = 0$, terminate on the line $r = R$, and use up exactly one unit of independent variable s in making the transition.

5.14. Theorem. *Let $R > 0$ be arbitrary. Then* (5.9) *has a solution if λ is sufficiently small* (*its size depending on R*).

Proof. We use the Poincaré shooting method, introduced in Sec. II.9. We consider the initial-value problem consisting of (5.9a,b,d) and the auxiliary initial condition

$$r(0) = R - 1 + \alpha. \tag{5.15}$$

For $\alpha = 0 = \lambda$, this problem has the unique solution defined by $N(s) = 0$ and $r(s) = s + R - 1$. Standard results from the theory of ordinary differential equations imply that this initial-value problem has a unique solution, which we denote by $\big(r(\cdot;\alpha,\lambda,R), N(\cdot;\alpha,\lambda,R)\big)$, if $|\alpha|$ and λ are small enough. This solution is a solution of the boundary-value problem (5.9) if α can be chosen so that

$$F(\alpha,\lambda,R) \equiv r(1;\alpha,\lambda,R) - R = 0. \tag{5.16}$$

We know that $F(0,0,R) = 0$ and that F is differentiable (by basic theory of ordinary differential equations). The Implicit Function Theorem implies that we can find an α satisfying (5.16) for small λ if $F_\alpha(0,0,R) \neq 0$. To compute $F_\alpha(0,0,R)$, we first determine $\big(r_\alpha(\cdot;0,0,R), N_\alpha(\cdot;0,0,R)\big)$, which satisfies

$$\begin{aligned} r'_\alpha(s;0,0,R) = \hat{\nu}_N(0)N_\alpha(s;0,0,R), \quad N'_\alpha(s;0,0,R) = 0, \\ N_\alpha(0;0,0,R) = 0, \quad r_\alpha(0;0,0,R) = 1, \end{aligned} \tag{5.17}$$

from which we find $F_\alpha(0,0,R) = r_\alpha(1;0,0,R) = 1$. □

5.18. Exercise. Let $\alpha^\star(\lambda, R)$ denote the (small) solution of (5.16) for small λ. Then the solution of (5.9) is $\big(r(\cdot;\alpha^\star(\lambda,R),\lambda,R), N(\cdot;\alpha^\star(\lambda,R),\lambda,R)\big)$. Formulate and solve the problem for

$$\frac{\partial}{\partial\lambda}\big(r(\cdot;\alpha^\star(\lambda,R),\lambda,R),\ N(\cdot;\alpha^\star(\lambda,R),\lambda,R)\big)\Big|_{\lambda=0}$$

and give a physical interpretation of the solution.

We now find conditions on R so that for all λ there is a completely compressed configuration corresponding to a trajectory of Fig. 5.13 that starts on the positive r-axis. We apply the shooting method, in a global way, replacing the initial-value problem (5.9a,b,d) and (5.15) with the equivalent integral equations:

$$N(s) = -\lambda\int_0^s r(\xi)\,d\xi, \quad r(s) = R - 1 + \alpha + \int_0^s \hat{\nu}(N(\xi))\,d\xi. \tag{5.19}$$

We denote solutions as before. The integral (5.12) ensures that the solution is defined on the entire interval $[0,1]$ for any choice of parameters. We seek an α so that (5.9c) holds, i.e., so that

$$\alpha + \int_0^1 \hat{\nu}\big(N(s;\alpha,\lambda,R)\big)\,ds = 1. \tag{5.20}$$

From (5.19) we obtain

$$\begin{aligned} R - 1 + \alpha \le r(s;\alpha,\lambda,R) \le R - 1 + \alpha + \int_0^1 \hat{\nu}(N(s))\,ds, \\ -\lambda\left[R - 1 + \alpha + \int_0^1 \hat{\nu}(N(s))\,ds\right] \le N(s;\alpha,\lambda,R) \le -\lambda(R-1+\alpha)s. \end{aligned} \tag{5.21}$$

Thus the monotonicity of $\hat{\nu}$ implies that

$$\begin{aligned} \alpha &\le \alpha + \int_0^1 \hat{\nu}\left(-\lambda\left[R - 1 + \alpha + \int_0^1 \hat{\nu}(N(\xi))\,d\xi\right]\right) ds \\ &\le \alpha + \int_0^1 \hat{\nu}\big(N(s;\alpha,\lambda,R)\big)\,ds \le \alpha + \int_0^1 \hat{\nu}\big(-\lambda(R-1+\alpha)s\big)\,ds. \end{aligned} \tag{5.22}$$

If $R \ge 1$ and if $\alpha = 0$, then the rightmost term of (5.22) is < 1. If $R \ge 1$ and if α approaches ∞, then the leftmost term of (5.22) approaches ∞. It follows from the Intermediate Value Theorem that (5.20) has a positive solution $\alpha^\star(\lambda, R)$.

5.23. Exercise. Show that the solution just described satisfies

$$-\lambda Rs < N(s;\alpha^\star(\lambda,R),\lambda,R) < \lambda(-Rs + s - s^2/2), \quad N'(s;\alpha^\star(\lambda,R),\lambda,R) < 0. \tag{5.24}$$

Thus the solution corresponds to a configuration that is everywhere in compression.

We have just proved

5.25. Theorem. *If $R \geq 1$, then for each $\lambda \geq 0$, problem* (5.9) *has a solution satisfying* (5.24).

Alternative proof of Theorem 5.25. Let

$$\mathcal{K} \equiv \{(r, N) \in C^0[0,1] : 0 \leq r(s) \leq R,\ -\lambda R s \leq N(s) \leq 0\}. \tag{5.26}$$

We represent (5.5), (5.6), and (5.8) in the form

$$(r, N) = \boldsymbol{f}[r, N, \lambda, R] \tag{5.27}$$

where $\boldsymbol{f}[\cdot, \cdot, \lambda, R]$ is a compact and continuous mapping taking the closed convex subset $\mathcal{K}$ of the Banach space $C^0[0,1]$ into itself. The Schauder Fixed-Point Theorem XIX.3.17 implies that (5.27) has a fixed point in $\mathcal{K}$, i.e., that (5.5), (5.6), (5.8) have a solution meeting the restrictions defining $\mathcal{K}$. □

Though moderately sophisticated, the Schauder Fixed-Point Theorem is easy to apply and can lead to sharp results. We now use it to show that a completely compressed solution exists for any λ provided that λ is large enough:

5.28. Theorem. *Let R and A be given with $0 < A < R$. Then for sufficiently large λ, problem* (5.10) *has a solution N satisfying*

$$-\lambda R s < N(s) < -\lambda A s \quad \text{and} \quad N'(s) < 0 \quad \text{for } s > 0. \tag{5.29}$$

5.30. Exercise. Let $Y = N/\lambda$. Prove that (5.10) is equivalent to the integral equation:

$$Y(s) = -Rs + \int_0^1 \min(s, \xi)\hat{\nu}(\lambda Y(\xi))\, d\xi \equiv T[Y, \lambda, R](s). \tag{5.31}$$

Proof of Theorem 5.28. Since

$$-Rs \leq T[Y, \lambda, R](s) \leq -Rs + s\int_0^1 \hat{\nu}(\lambda Y(\xi))\, d\xi, \tag{5.32}$$

we immediately see that we can take λ so large that $T[\cdot, \lambda, R]$ is a compact and continuous mapping taking the closed convex subset $\{Y \in C^0[0,1] : -Rs \leq Y(s) \leq -As\}$ of the Banach space C^0 into itself. The conclusion of the theorem follows immediately from the Schauder Fixed-Point Theorem. □

5.33. Exercise. Use a similar approach to treat states that are not necessarily compressive. Prove:

5.34. Theorem. *Suppose that either* (i) *λ is so small that the algebraic equation $2N = \lambda\hat{\nu}(N)$ has a (necessarily positive) solution N, or* (ii) *λ is so small and R is so large that the algebraic equation $N = \lambda[\hat{\nu}(N) - R]$ has a solution N. (If there are numbers $\alpha \leq 1$ and $K > 0$ such that $\hat{\nu}(N) \leq K(N^\alpha + 1)$, then each of these equations has a solution for all λ and R.) Let C denote the smallest solution of either of these equations. Then* (5.10) *has a solution satisfying*

$$-\lambda R s \leq N(s) \leq \lambda[-Rs + \hat{\nu}(C)(s - s^2/2)]. \tag{5.35}$$

Let us denote any compressive solution of (5.10) by $N(\cdot, \lambda, R)$. Thus $N(s, \lambda, R) < 0$ for $s > 0$. We now give a criterion ensuring that $N(s, \lambda, R)$ decreases as λ increases:

5.36. Theorem. *Let $\big(r(\cdot,\lambda,R), N(\cdot,\lambda,R)\big)$ be a solution of (5.5), (5.6), (5.8) for which there are a nonnegative number λ_0 and a continuous function $g:[0,1]\to\mathbb{R}$ with $g(0)=0$, $g(s)>0$ for $s>0$ such that*

$$-\lambda R < N(s,\lambda,R) < -\lambda g(s) \quad \text{and} \quad N_s(s,\lambda,R) > 0 \quad \text{for } s>0, \lambda \geq \lambda_0. \tag{5.37}$$

(This condition is ensured by the various existence theorems described above.) Let $\hat{\nu}$ satisfy

$$N\hat{\nu}_N(N) \to 0 \quad \text{as } N \to -\infty. \tag{5.38}$$

Then for λ sufficiently large, $N_\lambda(\cdot,\lambda,R)$ exists and satisfies

$$N_\lambda(s,\lambda,R) < 0 \quad \text{for } s \in (0,1]. \tag{5.39}$$

Proof. By hypothesis, problem (5.5)–(5.8) has a solution. The Implicit Function Theorem XVIII.1.27 (in Banach space) implies that $N_\lambda(\cdot,\lambda,R)$ exists if the only solution of the linearization

$$\begin{aligned}\overset{\vartriangle}{N}(s) &= -\lambda \int_0^s \overset{\vartriangle}{r}(\xi)\,d\xi,\\ \overset{\vartriangle}{r}(s) &= -\int_s^1 \hat{\nu}_N\big(N(\xi,\lambda,R)\big)\overset{\vartriangle}{N}(\xi,la,R)\,d\xi\end{aligned} \tag{5.40}$$

is the trivial solution $(\overset{\vartriangle}{r},\overset{\vartriangle}{N})=(0,0)$. Problem (5.40) is equivalent to the boundary-value problem

$$\overset{\vartriangle}{N}{}' = -\lambda\overset{\vartriangle}{r}, \quad \overset{\vartriangle}{r}{}' = \hat{\nu}_N\big(N(s,\lambda,R)\big)\overset{\vartriangle}{N}, \quad \overset{\vartriangle}{N}(0)=0, \quad \overset{\vartriangle}{r}(1)=0. \tag{5.41}$$

Let us make the Prüfer transformation

$$\overset{\vartriangle}{r} = \sigma\cos\psi, \quad \overset{\vartriangle}{N} = \lambda\sigma\sin\psi, \tag{5.42}$$

which converts (5.41) to

$$\begin{aligned}\sigma' &= \sigma[-1+\lambda\hat{\nu}_N\big(N(s,\lambda,R)\big)]\cos\psi\sin\psi,\\ \psi' &= -[\cos^2\psi + \lambda\hat{\nu}_N\big(N(s,\lambda,R)\big)\sin^2\psi],\\ &\psi(0)=0, \quad \psi(1) = -\tfrac{\pi}{2} \pmod{\pi}.\end{aligned} \tag{5.43}$$

A nontrivial solution of (5.43) (for which $\sigma \neq 0$) satisfies

$$\tfrac{\pi}{2} \pmod{\pi} = \int_0^1 [\cos^2\psi + \lambda\hat{\nu}_N(N)\sin^2\psi]\,ds \leq 1 + \int_0^1 \lambda\hat{\nu}_N(N)\sin^2\psi\,ds. \tag{5.44}$$

Conditions (5.37) and (5.38) ensure that (5.44) cannot hold for λ sufficiently large. Thus $Y \equiv N_\lambda(\cdot,\lambda,R)$ exists and satisfies

$$Y(s) = -Rs + \int_0^s\int_\xi^1 [\hat{\nu}(N(\sigma)) + \lambda\hat{\nu}_N\big(N(\sigma,\lambda,R)\big)Y(\sigma)]\,d\sigma\,d\xi. \tag{5.45}$$

Conditions (5.37) and (5.38) imply that for arbitrary $\varepsilon>0$ we can find a $\lambda_1>0$ and a $k>0$ such that

$$Y(s) \leq -ks + \varepsilon\int_0^s\int_\xi^1 Y(\sigma)\,d\sigma\,d\xi \tag{5.46}$$

for $\lambda \geq \lambda_1$. Thus

$$U(s) \equiv \int_0^s \int_\xi^1 Y(\sigma)\, d\sigma\, d\xi \tag{5.47}$$

satisfies the differential inequality

$$U'' + \varepsilon U \geq ks \geq 0, \quad U(0) = 0, \quad U'(1) = 0. \tag{5.48a,b,c}$$

It is not hard to show that if $2 > 2\delta > \varepsilon$, then (5.48) implies that $s \mapsto U(s)[1 - \delta s^2]^{-1}$ has its maximum at $s = 0$, which (5.47) implies is 0. It then follows from (5.46) that

$$Y(s) \equiv N_\lambda(s, \lambda, R) \leq -ks, \tag{5.49}$$

a result stronger than (5.39). □

5.50. Exercise. Prove that $U \leq 0$ by showing that (5.48a) prevents $s \mapsto U(s)[1 - \delta s^2]^{-1}$ from having an interior maximum on $(0, 1)$ and then showing that (5.48c) and (5.48a) prevent this function from having a maximum at $s = 1$. (This result is an ad hoc application of the Maximum Principle. For the general theory, see Protter & Weinberger (1967), especially Theorems 1.3–1.5.)

We can supplement the qualitative information given by Theorem 5.36 with an existence and uniqueness theorem:

5.51. Theorem. *For λ sufficiently large, there is exactly one solution of* (5.5), (5.6), (5.8) *satisfying the hypotheses of Theorem 5.36.*

5.52. Exercise. Prove Theorem 5.51 by showing that the Mean Value Theorem implies that the difference of two solutions satisfies a system like (5.41). Then follow the proof of Theorem 5.36.

Theorems 5.14 and 5.34 are the only results developed so far for the tricky case of solutions whose deformed length exceeds $2R$. Theorems VII.6.12 and VII.6.19, which rely on the calculus of variations, apply to this case.

The results of this section are adapted from the work of Antman & Nachman (1980, Sec. 7) (with kind permission from Elsevier Science Ltd., The Boulevard, Langford Lane, Kidlington OX5 1GB, UK). For further developments in this direction, see Burton (1986).

6. Bibliographical Notes

We shall discuss planar buckling problems for rods in Chap. VI and variational treatments of problems for rods in Chap. VII. There are several works devoted to the treatment of large displacements for the elastica, often by the use of elliptic functions. They include Born (1906), Frisch-Fay (1962), Funk (1970), Ilyukhin (1979), Love (1927), Nikolai (1955), Popov (1948), and Saalschütz (1880). These books contain extensive bibliographies.

Nonlinear constitutive equations seem to have been first considered in practical applications by Haringx (1942, 1948–1949).

Truesdell (1954) solved the design problem of determining the reference shape of an elastica in order that it have a prescribed shape under a specific loading. To outline his work in a slightly more general setting, suppose that the rod is inextensible and unshearable and that its constitutive equation is $M(s) = M_*\big(\theta'(s), \mathring{\theta}'(s)\big)$ where $\mathring{\theta}'(s)$ is the reference curvature. For given $\boldsymbol{f}$ (possibly configuration-dependent) and θ, the equilibrium equations reduce to a degenerately simple ordinary differential equation for $\mathring{\theta}$. Truesdell analyzed this equation when M_* is linear in $\theta'(s) - \mathring{\theta}'$. In Chap. XIV, we show how the constitutive equations depend on the reference configuration.

CHAPTER V

Introduction to Bifurcation Theory and its Applications to Elasticity

1. The Simplest Buckling Problem

If a naturally straight thin rod, such as a plastic or metal ruler, is subjected to a small compressive thrust applied to its ends, it remains straight. If the thrust is slowly increased beyond a certain critical value, called the *buckling load*, the rod assumes a configuration, called a *buckled state*, that is not straight. See Fig. 1.1. This process is called *buckling*. Depending on the precise mode of loading and the nature of the rod, the transition to a buckled state can be very rapid. If the thrust is further increased, the deflection of the rod from its straight state is likewise increased. If this entire process is repeated, the rod may well buckle into another configuration such as the reflection of the first through a plane of symmetry. The performance of a whole series of such experiments on different rods would lead to the observation that the buckling loads and the nature of buckled states depend upon the material and shape of the rod and upon the manner in which it is supported at its ends. It can also be observed that the results of experiments are highly sensitive to slight deviations of the rod from perfect straightness or of the thrust from perfect symmetry. The study of buckling for different bodies is one of the richest sources of important problems in nonlinear solid mechanics.

An important aspect of the example we have just described is that there are multiple equilibrium states, among which is an *unbuckled* or *trivial* state in which the rod remains straight. This state is not readily observed when the thrust exceeds the buckling load because it is unstable. Rather than inferring its existence from our intuitive notions of symmetry, we shall prove its existence for all thrusts in the mathematical models we adopt to describe buckling. Nevertheless, a simple experiment makes plausible the existence of an unbuckled state for all thrusts: We introduce restraints preventing the lateral deflection of the rod at a certain height (see Fig. 1.2) and observe that the rod does not buckle for a range of thrusts exceeding the buckling load. The lateral constraints apply no net force to the rod in its observed straight state, for if so, the contact couple exerted by one segment of the rod on an adjacent segment would not be zero and bending would result. (See Chap. IV.)

In many experiments, the rod can exhibit dynamic effects. There are challenging mathematical difficulties in accounting correctly for these ef-

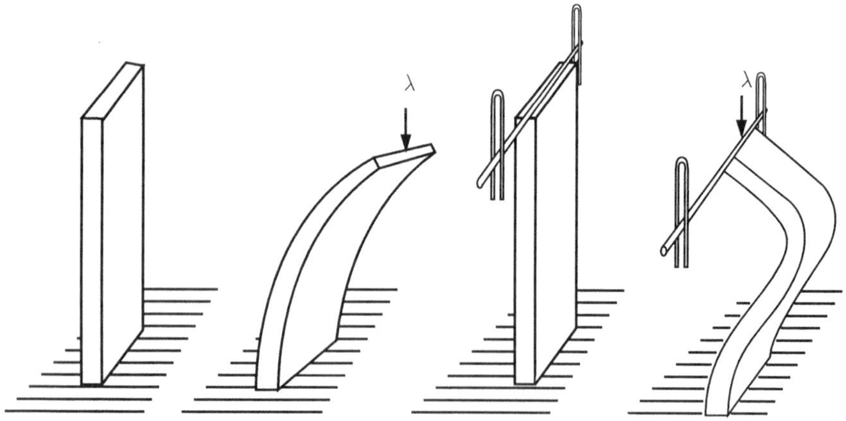

(a) (b) (c) (d)

Fig. 1.1. Buckling of a naturally straight rod. (a) The natural state of a rod whose lower end is welded to a rigid foundation and whose upper end is free. (b) A buckled state of this rod, which occurs when a sufficiently large thrust λ is applied to the upper end. In this case, the base applies a reactive force equal and opposite to the thrust applied at the upper end. (c), (d) Analogous states for a rod whose upper end is hinged in such a way that it is constrained to remain directly above the lower end.

fects. Fortunately, we can get good insights into many buckling processes by studying the multiplicity and stability of equilibrium states. From these, we can often infer useful information about the dynamical processes taking the structure from one equilibrium state to another. We comment briefly on this issue in Sec. 7.

The simplest model for describing the planar buckling of a rod is the elastica, described in Sec. IV.1. Suppose that its reference configuration is defined by

$$\text{(1.3)} \qquad \boldsymbol{r}(s) = s\boldsymbol{i}, \quad s \in [0,1], \quad \boldsymbol{a} = \boldsymbol{i} \quad \text{so that} \quad \theta = 0.$$

We assume that the end $s = 0$ is welded to a rigid wall perpendicular to the $\boldsymbol{i}$-axis at the origin $\mathbf{0}$ and that the end $s = 1$ is free of geometrical restraint and is subject to a compressive thrust of magnitude λ acting in the $-\boldsymbol{i}$-direction. (This is the situation illustrated in Fig. 1.1b.) Then the boundary conditions are

$$\text{(1.4a,b,c,d)} \qquad \boldsymbol{r}(0) = \mathbf{0}, \quad \theta(0) = 0, \quad \boldsymbol{n}(1) = -\lambda\boldsymbol{i}, \quad M(1) = 0.$$

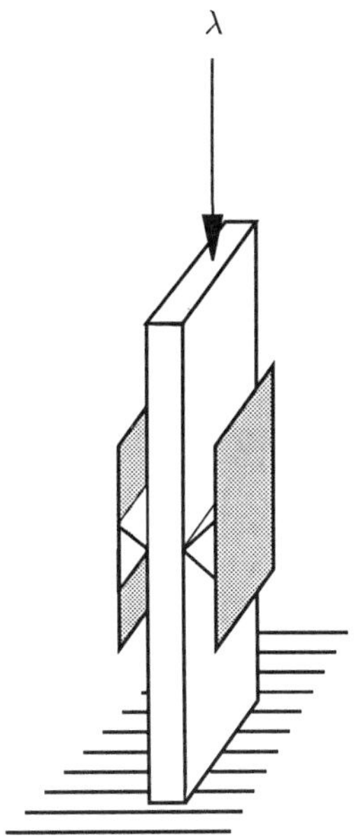

Fig. 1.2. A device that raises the effective buckling load.

We assume that no body force or couple is applied to the rod. Then the integral form of the equilibrium equations, corresponding to (IV.1.32) and to (IV.2.10b) with $\gamma = \theta$, reduce to

(1.5a,b)

$$\theta(s) = \int_0^s \frac{M(\xi)}{(EI)(\xi)}\, d\xi, \quad M(s) = \boldsymbol{k} \cdot \int_s^1 \boldsymbol{r}_s(\xi) \times \boldsymbol{n}(\xi)\, d\xi = \lambda \int_s^1 \sin\theta(\xi)\, d\xi.$$

We assume for simplicity that EI is continuous and everywhere positive. Equations (1.5) represent a pair of integral equations for (θ, M) depending on the parameter λ. To each solution of this system there corresponds an $\boldsymbol{r}$, which is the solution of the initial-value problem consisting of the differential equation $\boldsymbol{r}' = \cos\theta\boldsymbol{i} + \sin\theta\boldsymbol{j}$ (the specialization of (IV.1.4)) and the initial condition (1.4a). By taking limits in (1.5) and by differentiating this system, we can reduce it to the familiar boundary-value problem

$$\text{(1.6a,b,c)} \qquad \tfrac{d}{ds}[(EI(s))\theta'(s)] + \lambda \sin\theta(s) = 0, \quad \theta(0) = 0, \quad \theta'(1) = 0.$$

For EI = const., the solution of (1.6) can be found in terms of elliptic functions (see Reiss (1969), e.g.). Nevertheless, even when EI = const., we prefer to use (1.5), which directly accounts for the boundary conditions, directly captures the underlying mechanics, and allows us to seek solutions in a class of functions larger than those for which (1.6) is meaningful.

1.7. Exercise. Convert (1.6) to the alternative integral equation

$$\text{(1.8)} \qquad \theta(s) = \lambda \int_0^1 G(s,\xi) \sin\theta(\xi)\, d\xi$$

where

$$G(s,\xi) \equiv \begin{cases} \int_0^\xi \frac{d\sigma}{(EI)(\sigma)} & \text{for} \quad \xi \le s, \\ \int_0^s \frac{d\sigma}{(EI)(\sigma)} & \text{for} \quad s \le \xi \end{cases} \tag{1.9}$$

by substituting (1.5b) into (1.5a). G is the Green function for the differential operator $\theta \mapsto -\frac{d}{ds}[EI\theta']$ subject to (1.6b,c). This result can also be obtained from (1.6) by using distribution theory (see Stakgold (1979)). Suppose that (1.6c) is replaced with $\theta(1) = 0$. Find the corresponding Green function.

Let us represent problem (1.5) symbolically by

$$\mathsf{f}[\lambda, \mathsf{u}] = \mathbf{0} \tag{1.10}$$

where u stands for the pair (θ, M). We thus regard u as a pair of continuous functions and regard f as an operator taking such pairs into pairs of continuous functions. We could equivalently represent (1.6) and the integral equation of Ex. 1.7 in the abstract form (1.10) by making different identifications. Indeed, we regard (1.10) as the abstract form for all parameter-dependent problems.

We assume that the problem (1.10) has enough symmetry that a family of 'trivial' solutions can be readily identified for all values of λ and that the variable u is so chosen that all these trivial solutions can be characterized by the equation $\mathsf{u} = \mathbf{0}$. Thus the operator f must have the property that

$$\mathsf{f}[\lambda, \mathbf{0}] = \mathbf{0} \quad \forall\ \lambda. \tag{1.11}$$

Property (1.11) is certainly enjoyed by (1.5).

As we shall see, there are cases in which it is not immediately obvious how to formulate (1.10) so that it satisfies (1.11). For example, consider the planar buckling of an extensible rod. Then the trivial solution, which corresponds to an unbuckled state, is defined by an expression of the form $\boldsymbol{r}(s) = r(s,\lambda)\boldsymbol{i}$. If the rod is not uniform, then the existence of r must be demonstrated by an argument like that of Sec. II.6. For more complicated problems, the proof of the existence of a family of trivial solutions may require a far deeper analysis. In fact, there need not be a unique trivial solution for each value of the parameter λ. Suppose that the governing equations for the buckling of this extensible rod have the symbolic form

$$\boldsymbol{g}[\lambda, \boldsymbol{r}] = \mathbf{0}. \tag{1.12}$$

We are given that

$$\boldsymbol{g}[\lambda, r\boldsymbol{i}] = \mathbf{0}. \tag{1.13}$$

To put (1.12) into the form (1.10) satisfying (1.11), we set

$$\boldsymbol{u} \equiv \boldsymbol{r} - r(\cdot,\lambda)\boldsymbol{i}, \quad \boldsymbol{f}[\lambda, \boldsymbol{u}] \equiv \boldsymbol{g}[\lambda, r(\cdot,\lambda)\boldsymbol{i} + \boldsymbol{u}]. \tag{1.14}$$

Thus we need only study bifurcation problems for which (1.11) holds, realizing, however, that the reduction to the form (1.10) may be far easier in principle than in practice.

We can now represent solutions of (1.10) by the simple device of a *bifurcation diagram.* Let any (λ, u) satisfying (1.10) be called a *solution pair.*

Since $\mathbf{u}$ represents a function, it lies in an infinite-dimensional space. Therefore plotting the solution pairs in $(\lambda, \mathbf{u})$-space is impossible. Instead, we let φ be some convenient real-valued function of $\mathbf{u}$, e.g., an amplitude, and plot all points $(\lambda, \varphi(\mathbf{u}))$ in $\mathbb{R}^2$ corresponding to solution pairs $(\lambda, \mathbf{u})$. For our example, we could take $\varphi(\mathbf{u})$ to be $M(0)$ or $\theta(1)$. In Fig. 1.15 we sketch such a bifurcation diagram, obtained in Sec. 5, for our example problem. The diagram can be constructed explicitly if EI is constant; see Ex. 1.16. We follow the mathematical convention of taking the parameter axis to be the abscissa. (Engineers usually take it to be the ordinate.) We label the ordinate $\mathbf{u}$ instead of $\varphi(\mathbf{u})$ because we choose to interpret this figure as a schematic diagram of all solution pairs in the space of $(\lambda, \mathbf{u})$. The points on the λ-axis represent solution pairs by virtue of (1.11). This axis is called the *trivial branch of solution pairs.* A *branch* of solution pairs is any connected set of them. For our example, the other branches, termed *nontrivial*, correspond to *buckled states* of the rod. In Fig. 1.15 we see that there are three solution pairs when $\lambda_1^0 < \lambda < \lambda_2^0$, five solution pairs when $\lambda_2^0 < \lambda < \lambda_3^0$, etc. We shall see that the numbers $\lambda_1^0, \lambda_2^0, \ldots$ are the eigenvalues of the linearization of (1.5) about the trivial branch. For our simple model of an elastic structure, there are a countable infinity of such eigenvalues.

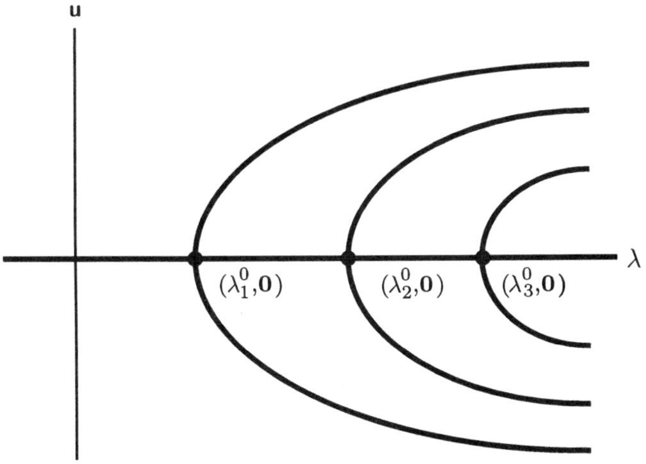

Fig. 1.15. Bifurcation diagram for (1.5) or (1.6) corresponding to the buckling process of Fig. 1.1b.

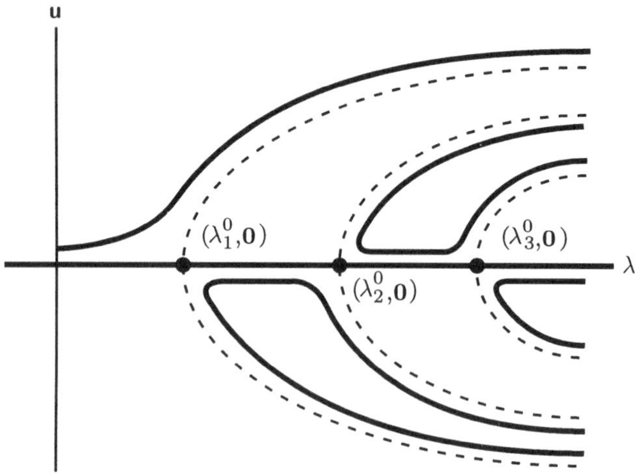

Fig. 1.17. One possible bifurcation diagram corresponding to an imperfect form of Fig. 1.15. In this case, some number slightly less than λ_0 may be regarded as an effective buckling load because the most pronounced deviation from straightness shows up when such a value of thrust is reached. See Fig. VI.9.33 for a different 'unfolding' of Fig. 1.15.

1.16. Exercise. Use elliptic functions to construct explicitly a bifurcation diagram for (1.6) when EI is constant. In particular, let $\alpha = \theta(1)$ and plot $\sin(\alpha/2)$ vs. λ. (A detailed treatment of (1.6) subject to different boundary conditions was carried out by Reiss (1969). Also see Love (1927, Secs. 262, 263) and Frisch-Fay (1962).)

The word *bifurcation* (from the Latin *furca* for *fork*) describes the local appearance of Fig. 1.15 near $(\lambda_k^0, \mathbf{0})$. We shall give a formal mathematical definition of bifurcation in Sec. 3. The interpretation of a bifurcation diagram such as Fig. 1.12 should actually be done with care. The diagram is just a two-dimensional projection of an infinite-dimensional figure, so that there may be fewer connections and more branches than it shows. We shall find that many of the bifurcation diagrams to be encountered below have a far richer structure than that of Fig. 1.15. This figure gives no information on which of several possible configurations would be occupied by the body of Fig. 1.b for a given λ, i.e., it says nothing about the stability of these multiple configurations. Neither does Fig. 1.15 say anything about the dynamical process by which the rod moves from one configuration to another. In the engineering literature there are certain accepted ways of interpreting the stability and dynamics of the rod from a bifurcation diagram, which we discuss below. Much, but not all, of this doctrine can be mathematically justified.

If the rod in Fig. 1.1 is not perfectly straight or if the thrust is applied eccentrically, then the bifurcation diagram of Fig. 1.15 does not apply. Suppose that we introduce a small parameter ε accounting for the lack of straightness. In particular, let us interpret ε as a magnitude of the curvature of the natural reference state. If ε is small and if there are no other such *imperfections*, then the bifurcation diagram might assume the form of Fig. 1.17. Such a result is somewhat unsatisfactory, because there are many conceivable kinds of imperfections that are physically reasonable. The mathematical study of the effects of many imperfection parameters (such as ε) would seem to be prohibitively complicated. On the other hand, the very imperfections that one might wish to ignore in order to avoid mathematical complications might give rise to dangerous instabilities not evident in such innocuous diagrams as Fig. 1.17. Fortunately, it is possible to use the recently created subject of singularity theory (see Golubitsky & Schaeffer (1985)) to determine precisely how many imperfection parameters are needed to describe 'nearly' all possible local responses to any system of imperfections. (In many respects, the development of singularity theory consists in the complicated unification of elementary concepts. The theory nevertheless is easy to apply.) We describe applications of it in Sec. VI.6. There we show that there can be imperfections producing bifurcation diagrams qualitatively different from Fig. 1.17.

2. Classical Buckling Problems of Elasticity

We now briefly describe some of the classical buckling problems of elasticity, many of which have features that are far from being completely understood. For generality, in this section we regard a rod as a slender three-dimensional body. For brevity of exposition we assume that the reference configuration of the rod has a curve of centroids. We denote a typical configuration of this curve by $\boldsymbol{r}$. We likewise regard a shell as a thin three-dimensional body. (See Chap. X for an introduction to the Cosserat theory of shells.)

The planar buckling problem described in Sec. 1 was given its first definitive treatment by Euler (1744). The terminal thrust causing the buckling can be either replaced by or supplemented by the rod's weight. Euler (1780a,b,c) posed and solved the problem of determining the critical height at which a uniform rod buckles under its own weight (see Truesdell (1960)).

The interesting version of Euler's buckling problem in which the end thrust is required to remain tangent to the deformed $\boldsymbol{r}$ was first studied by Beck (1952). For this problem, the only compressed equilibrium states are straight, but the rod buckles by going into motion.

Far richer variants of these problems arise when $\boldsymbol{r}$ is allowed to deform into a space curve. In this case, one can also study the effects of a terminal couple acting in concert with the terminal thrust and the weight. The nature of the equilibrium states depends crucially on the precise way the boundary conditions are maintained. The study of such problems was initiated by Greenhill (1883). If the rod is uniform, then this problem has three buckling parameters: the thrust, a suitable component of the applied

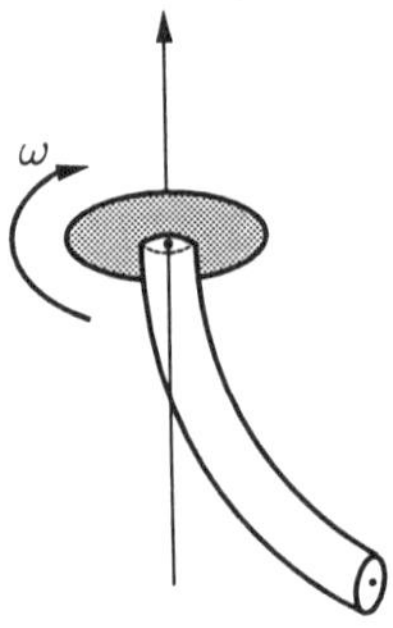

Fig. 2.1. Buckling of a whirling rod.

couple, and the constant mass density per unit natural length. This last parameter can be replaced by the length of the rod. The abstract version of this problem is obtained by replacing the real parameter λ of (1.10) by a triple $\boldsymbol{\lambda}$ of real parameters.

If the rod is not uniform, then the effect of the weight depends upon the mass distribution ρA along the rod's length. In this case the 'parameter' ρA is a function and therefore belongs to an infinite-dimensional space of functions (e.g., the space of continuous functions or the space of bounded measurable functions). In the optimal design of rods against buckling, the cross section is varied along the length to maximize the 'minimum buckling load'.

In a related problem, also first studied by Greenhill (1883), one end of a straight (heavy) rod is welded to a rigid horizontal plane rotating at constant angular velocity ω about the vertical axis that coincides with the trivial configuration of $\boldsymbol{r}$. See Fig. 2.1. For this problem the straight configuration is destabilized by the centrifugal force, and steadily rotating buckled states are observed. (This problem differs from that in which a naturally straight rod with a circular cross section, with one end welded to a *fixed* rigid horizontal plane, executes steady nontrivial rotations about the vertical; see Healey (1992).) A degenerate version of these problems for rotating rods, having a particularly beautiful analysis, is obtained by replacing the rod with an elastic string. (This is the first bifurcation problem to be analyzed in the next chapter.) In another class of buckling problems for whirling rods (which cannot be regarded as classical), one end of the rod is welded to the inside of a rigid ring rotating with constant angular velocity about the line perpendicular to the plane of the ring and passing through its center. See Fig. 2.2. The analyses carried out in Secs. IV.5 and VII.6 show that the set of trivial configurations is very rich, depending crucially on both the angular velocity and on the ratio of the natural length of the rod to the radius of the ring.

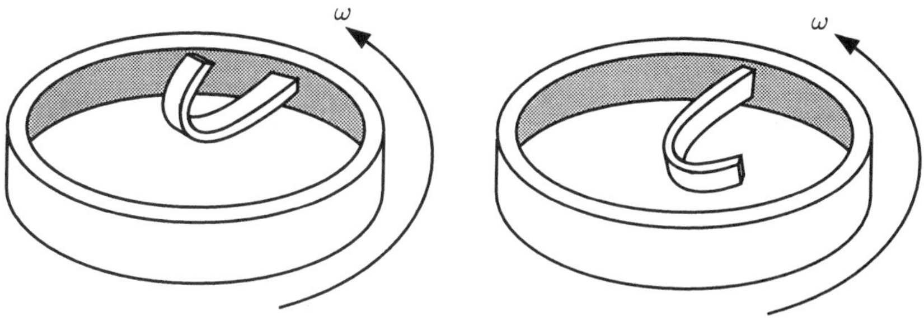

Fig. 2.2. Buckled states of a rod welded to a rotating rigid ring. In (a) the rod buckles out of the plane of the ring and in (b) it buckles in the plane of the ring.

Now consider a rod with a plane of symmetry in its reference configuration that is subjected to a system of forces having the same plane of symmetry and a system of couples acting about axes perpendicular to this plane. The reference configuration of the rod need not be straight. The rod may have a family of equilibrium configurations parametrized by these loads and having the same plane of symmetry. These equilibrium configurations are termed *trivial* even though their determination may be far from trivial. Buckled states corresponding to a *lateral instability* occur when the configuration of the rod loses its plane of symmetry. See Fig. 2.3. (For the special class of problems in which the only loads are equal and opposite bending couples applied to the ends of the rod, the technical difficulties are greatly reduced.)

A circular ring or arch with its material curve of centroids constrained to lie in a plane can buckle under hydrostatic pressure or related load systems by losing its circularity. (See Fig. 2.4.) The ring or arch can be regarded as a cross section of a complete or incomplete cylindrical tube. (A force system identical to that exerted by stationary fluid on a cylinder is exerted by a transverse magnetic field on a current-carrying arch.) As our preliminary analyses of Chap. IV indicate, the mathematical structure of these buckling problems has a character quite different from that for straight rods.

For certain ideal boundary conditions, a straight rod subjected to tensile forces applied to its ends has a family of trivial solutions in which the axis of the rod (the material line of centroids) remains straight and the cross sections remain perpendicular to it. When the applied force exceeds a critical value, a rod may suffer a shear or necking instability. In a shear instability (Fig. 2.5), cross sections originally perpendicular to the axis are

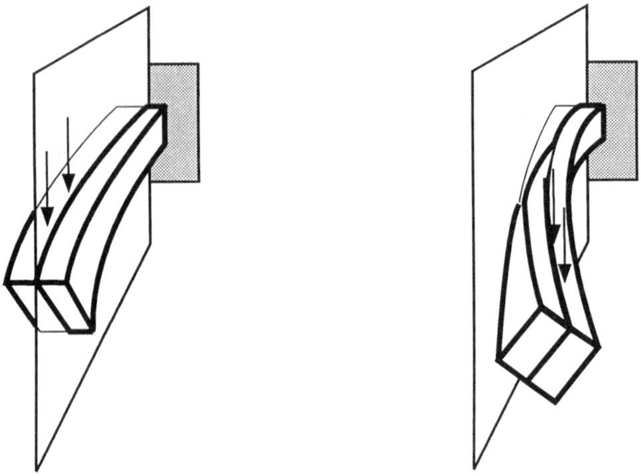

Fig. 2.3. Lateral buckling of a beam. The vertical plane is the plane of symmetry of the problem. (a) A trivial configuration. (b) A buckled state, in which $\boldsymbol{r}$ is not planar and the rod has suffered twisting. The rod could also buckle into the reflection of the configuration shown through the plane of symmetry.

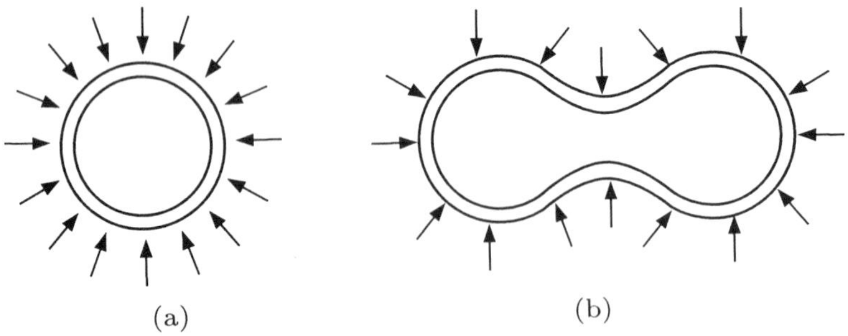

Fig. 2.4. (a) A uniformly compressed trivial state of a hydrostatically loaded ring. (b) A buckled state of the ring. Similar figures hold for arches.

sheared with respect to this line. In a necking instability (Fig. 2.6), the thickness suffers a large localized reduction. (See the photographs in Nadai (1950).) Although these instabilities are usually associated with plastic response, we shall find that they actually occur for certain nonlinearly elastic materials. Neither of these instabilities is termed *buckling*. Nevertheless,

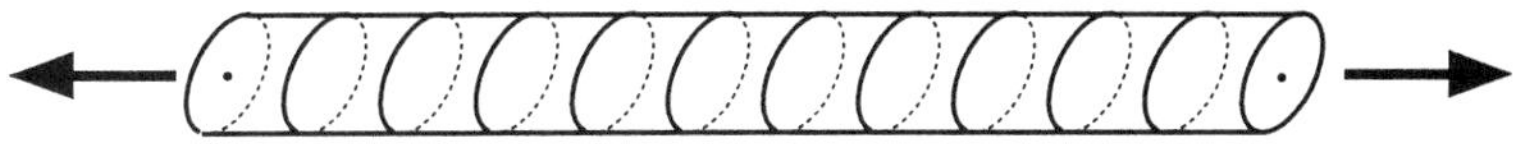

Fig. 2.5. Configuration of a doubly hinged, initially cylindrical rod undergoing a shear instability. The corresponding instability for a doubly welded rod is illustrated in Fig. IV.2.6.

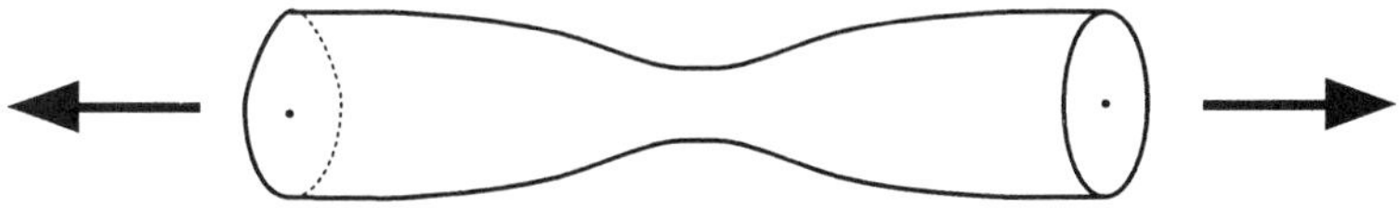

Fig. 2.6. Configuration of an initially cylindrical rod undergoing a necking instability.

their mathematical treatment within the context of bifurcation theory is the same as that for buckling problems.

For each of the examples just discussed, there are models that describe the equilibrium states by ordinary differential equations. The buckling problems for shells, on the other hand, lead to partial differential equations for the description of equilibrium. Under special conditions of symmetry, these problems possess a family of equilibrium states described by ordinary differential equations. But the physical interpretation of the stability of

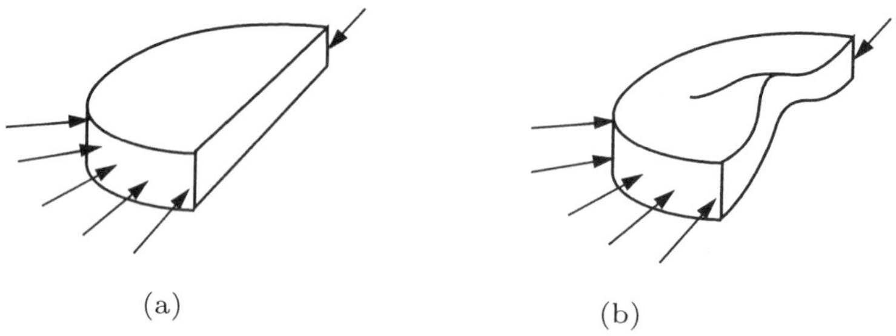

Fig. 2.7. Buckling of a plate under a planar system of loads applied to its edge. A trivial state is shown in (a) and a buckled state in (b). The most frequently studied cases are those of circular and rectangular plates subjected to normal pressures applied to their edges.

the special symmetric states requires the consideration of the full partial differential equations. The three classical problems of shell buckling are (i) the buckling of a plate under a planar system of forces applied to its edge (Fig. 2.7), (ii) the buckling of a (circular) cylindrical shell under a compressive system of forces acting parallel to its generators (Fig. 2.8), and (iii) the buckling of a spherical shell or cap under a hydrostatic pressure or similar load system (Fig. 2.9).

The buckling of the spherical cap illustrated in Fig. 2.9 exhibits a new effect that does not arise in the simplest models for rods. If the cap is shallow, then simple models for it lead to bifurcation diagrams of the sort illustrated in Fig. 2.10. Here λ represents the hydrostatic pressure and $\mathbf{u}$ represents some measure of the distortion from the uniformly compressed spherical configuration. Thus the λ-axis is the branch of trivial solutions.

Suppose that the pressure is very slowly raised from 0. (This is a so-called quasistatic process. There are serious difficulties, seldom confronted, in making the notion of such a process mathematically precise and in justifying the conclusions and interpretations made on the basis of such a notion; see Sec. 7.) Initially the shell remains spherical until the value λ_0 at the point A of Fig. 2.10 is reached. λ_0 is the lowest eigenvalue of the linearization of the governing equations about the trivial state. (This process is an idealization: In practice, the sphericity is lost due to imperfections before point A is reached.) As λ is increased past λ_0, it is observed that solution pairs switch at A from the trivial branch to the branch AB in a process that is the same as that for the buckled rod described by Fig. 1.15. The observed branch AB need not correspond to axisymmetric configu-

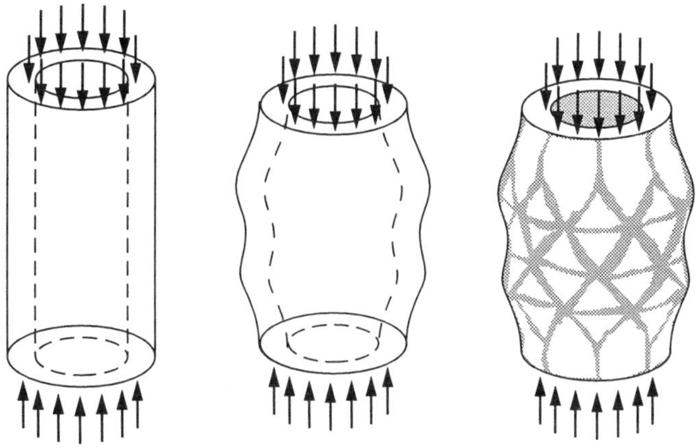

Fig. 2.8. Buckling of a cylindrical shell under forces applied to its ends that are parallel to its generators. A trivial state is shown in (a), an axisymmetric buckled state (having a *barrelling instability*) is shown in (b), and a state exhibiting the commonly observed pattern of lozenges is shown in (c). This pattern can also be found behind the knees of a pair of denim trousers. (The case in which the cylinder is circular has received by far the most attention.)

rations. We do not yet attempt to justify the standard terminology that branch A∞ is unstable and that branch AB is stable.

As λ is increased slowly beyond λ_0, the solution pairs move along branch AB until point B is reached. As λ is increased beyond its value at B, there are no longer any nearby equilibrium states on which solution pairs remain. The accepted doctrine, borne out by experience, is that the solution 'jumps' to another equilibrium state. We indicate this jumping process by the dotted line joining B to F. Point F is on the 'stable' branch DEFG∞. The jump BF represents the phenomenon of *snapping*, wherein the shell moves suddenly from a slight perturbation of Fig. 2.9a to a configuration like that of Fig. 2.9b. This motion is usually accompanied by a popping sound. As λ is slowly increased beyond its value at F, the solution pairs move out along branch FG∞. The progression along this branch represents the slow inflation of the configuration of Fig. 2.9b.

The snapping process, which is not well understood, clearly is the antithesis of a quasistatic process: It is exceedingly rapid and its observed motion dies out very rapidly. Moreover, our description of this buckling process gives no explanation for why the shell should jump from state B to branch DEFG, rather to some other branch not shown in Fig. 2.10. (The mathematical description of snapping requires a study of the equations of motion, which must account for a powerful internal damping mechanism to bring the shell rapidly to rest after snapping. The branch to which $\mathbf{u}$

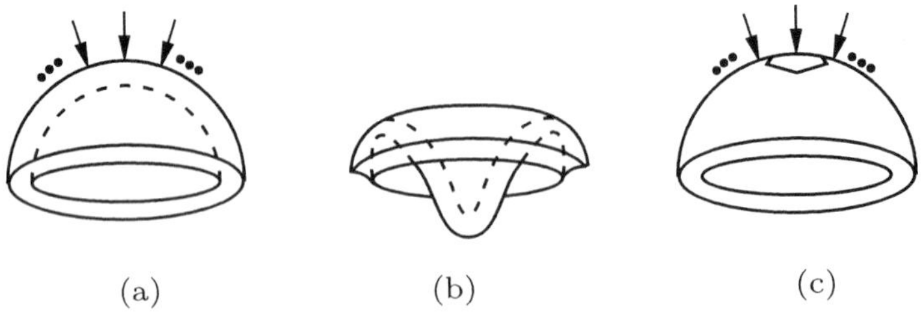

Fig. 2.9. Buckling of a spherical cap under hydrostatic pressure. A uniformly compressed trivial state is shown in (a), a large axisymmetric deformation is shown in (b), and a state exhibiting a pentagonal dimple is shown in (c).

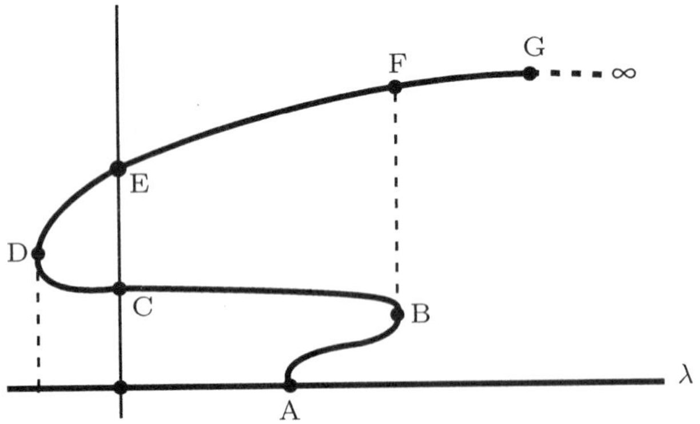

Fig. 2.10. Bifurcation diagram for the buckling of a shallow spherical cap under hydrostatic pressure. Only one branch of nontrivial solutions is shown.

is carried during a snapping is that containing the equilibrium point in whose region of attraction is found the initial data for $\mathbf{u}$ corresponding to state B. (An inkling of the complexity of this process is given by Reiss & Matkowsky (1971) in their study of a simplified model for the dynamics of a rod not undergoing snapping.)

Let us now look at the unloading process in which the shell begins in a state G on the stable branch DEFG and the pressure is gradually lowered. Solution pairs then move from G to F to E, which corresponds to a nontrivial equilibrium state under zero pressure. Such a state would look like that of Fig. 2.9b. Configurations of this form persist when λ assumes negative values, which correspond to pressures applied from below in Fig. 2.9b. When λ reaches the negative value corresponding to D, the configuration snaps back to a trivial state by a process just like that of BF. Note that if λ is slowly raised and lowered over an interval containing the values of λ corresponding to states D and B, then the system exhibits hysteresis, even though there are no memory terms in the equations. Branch BCD is unstable and is therefore not readily observed in experiment.

Two problems in which the trivial solution is time-dependent are that of a straight rod subjected to time-dependent thrusts applied to its ends and that of a circular ring subjected to a time-varying pressure. The former problem seems to be technically more difficult. There are obvious analogs of these problems for plates and shells. None of these problems for geometrically exact theories has been given a satisfactory nonlinear analysis.

There are numerous important examples of buckling caused by effects of nonmechanical origin, such as that of a thermoelastic rod whose ends are welded to rigid walls at a fixed distance apart. When the rod is heated, the walls thwart the natural propensity of the rod to elongate. The resulting compressive stresses cause buckling when the temperature field exceeds a critical level. Similar phenomena arise for plates and shells. Structures carrying electric currents and structures made of magnetoelastic materials can buckle in magnetic fields. In some of these problems, the deformed shape of the structure affects the ambient field. Such difficult interactions also occur in the buckling of structures in contact with moving fluids. This buckling can be either steady-state or dynamic.

There is a voluminous and interesting engineering literature dealing with buckling problems. Much of this is devoted to finding the eigenvalues of problems linearized about the trivial unbuckled state, which for degenerate theories is taken to be the reference configuration. The remainder typically treat small solutions of engineering theories of structures by perturbation methods. Recently some studies have taken up sophisticated local analyses based on catastrophe or singularity theory. Though this classical literature typically uses formulations of the equations and methods of analysis far more primitive than those used here, it nevertheless represents an excellent source of fascinating problems and of suggestive methods of analysis.

Among the many texts containing large collections of buckling problems are those of Biezeno & Grammel (1953), Bleich (1952), Bolotin (1956, 1961), Brush & Almroth (1975), Burgermeister, Steup, & Kretzschmar (1957, 1963), Dinnik (1935), Dowell (1975), Dowell & Il'gamov (1988), Dym (1974), Grigolyuk & Kabanov (1978), Hartmann (1937), Herrmann (1967b), Huseyin (1975), Kollbrunner & Meister (1961), Leipholz (1968, 1971, 1978a,b), Moon (1984), Nikolai (1955), Panovko & Gubanova (1979), Pflüger (1964), Prescott (1924), Rzhanitsyn (1955), Thompson & Hunt (1973, 1984), Timoshenko & Gere (1961), Vol'mir (1967, 1979), and Ziegler (1977). Among the most useful review articles are those of Babcock (1983), Budiansky (1974), Bushnell (1981), Eisley (1963), Herrmann (1967a), Hutchinson & Koiter (1970), Moon (1978), Nemat-Nasser (1970), and Pogorelov & Babenko (1992). Most of these works have extensive bibliographies. The article by Knops & Wilkes (1973), which approaches the subject of elastic stability from a very careful mathematical viewpoint, gives excellent coverage to the engineering literature. Surveys of geometrically exact problems for nonlinearly elastic structures are given by Antman (1977, 1980c, 1981, 1989).

3. Mathematical Concepts and Examples

We study operator equations of a form that generalizes (1.10) by allowing for an n-tuple $\boldsymbol{\lambda}$ of parameters in place of a single parameter:

$$\mathbf{f}[\boldsymbol{\lambda}, \mathbf{u}] = \mathbf{0}, \quad \text{where} \quad \mathbf{f}[\cdot, \mathbf{0}] = \mathbf{0}. \tag{3.1}$$

We assume that the domain of $\mathbf{f}$ is a subset of $\mathcal{X} \times \mathbb{R}^n$ where $\mathcal{X}$ is a real Banach space and that its target is a real Banach space. (Banach spaces provide a convenient and natural way to describe the size and convergence of functions. See Sec. XVII.1 for a discussion of these spaces. In many of our applications, $\mathcal{X}$ is just the Banach space of continuous functions on a closed and bounded interval.) If $\mathcal{X}$ is finite-dimensional, we replace $\mathbf{f}[\boldsymbol{\lambda}, \mathbf{u}]$ with $\mathbf{f}(\boldsymbol{\lambda}, \mathbf{u})$.

Our restriction of $\mathbf{u}$ and $\boldsymbol{\lambda}$ to real rather than complex spaces, which corresponds to the applications we study, would seem to simplify matters. The opposite is in fact true. The most elementary application of the principle that the complex case is simple and that the real case is complex arises in the study of the roots of a polynomial. The basic existence theorem for polynomial equations is the Fundamental Theorem of Algebra, which asserts that a polynomial of degree n with complex coefficients has n complex roots counted according to multiplicity. On the other hand, there are no convenient, universally applicable tests telling exactly how many real roots are possessed by an arbitrary polynomial with real coefficients.

A point $(\boldsymbol{\lambda}^0, \mathbf{0})$ on the trivial branch of (3.1) is called a *bifurcation point* on this branch iff in every neighborhood of this point there is a solution pair $(\boldsymbol{\lambda}, \mathbf{u})$ of (3.1) with $\mathbf{u} \neq \mathbf{0}$. (Some authors call this a *branching point*. The terminology is not completely standardized for this and other concepts.)

Bifurcation theory may be divided onto two main areas: the *local theory*, which treats the behavior of solutions near a bifurcation point, and the *global theory*, which treats the behavior of solutions wherever they occur. The local theory may be further subdivided into *perfect bifurcation*, which we simply call *bifurcation*, and *imperfect bifurcation*, which describes how the bifurcation diagram is altered by the presence of imperfections. The study of the detailed effects of imperfections within the global theory is in a primitive state. Our main interest is in the study of the applications of global bifurcation theory.

The simplest example of bifurcation occurs for linear equations of the form

$$\mathbf{A}(\boldsymbol{\lambda}) \cdot \mathbf{u} = \mathbf{0}. \tag{3.2}$$

Here $\mathbf{A}(\boldsymbol{\lambda})$ is a linear operator, depending on the n-dimensional parameter $\boldsymbol{\lambda}$, that acts on the unknown $\mathbf{u}$. (An outline of linear operator theory is given in Chap. XVII, which should be consulted for explanations of any unfamiliar concepts used in this section.) An n-tuple $\boldsymbol{\lambda}^0$ is called an *eigenvalue* of the operator-valued function $\boldsymbol{\lambda} \mapsto \mathbf{A}(\boldsymbol{\lambda})$ iff the equation $\mathbf{A}(\boldsymbol{\lambda}^0) \cdot \mathbf{v} = \mathbf{0}$ has a nonzero solution $\mathbf{w}$, called an *eigenvector*, corresponding to $\boldsymbol{\lambda}^0$. The set of all eigenvectors corresponding to $\boldsymbol{\lambda}^0$ form the *null space* or *kernel* $\ker \mathbf{A}(\boldsymbol{\lambda}^0)$ of the operator $\mathbf{A}(\boldsymbol{\lambda}^0)$ or, equivalently, form the *eigenspace* corresponding to $\boldsymbol{\lambda}^0$. The linearity of $\mathbf{A}(\boldsymbol{\lambda}^0)$ ensures that $\ker \mathbf{A}(\boldsymbol{\lambda}^0)$ is a vector space. (Thus, if $\mathbf{w}_1$ and $\mathbf{w}_2$ are eigenvectors corresponding to the eigenvalue $\boldsymbol{\lambda}^0$, then $\alpha_1\mathbf{w}_1 + \alpha_2\mathbf{w}_2$ for all real α_1 and α_2 is also a solution of (3.2) when $\boldsymbol{\lambda} = \boldsymbol{\lambda}^0$.) The *geometric multiplicity* of $\boldsymbol{\lambda}^0$ is the dimension of $\ker \mathbf{A}(\boldsymbol{\lambda}^0)$. The algebraic multiplicity, which also plays an important role in bifurcation theory, is discussed below. These considerations show that the bifurcation diagram for (3.2) has the form shown in Fig. 3.3.

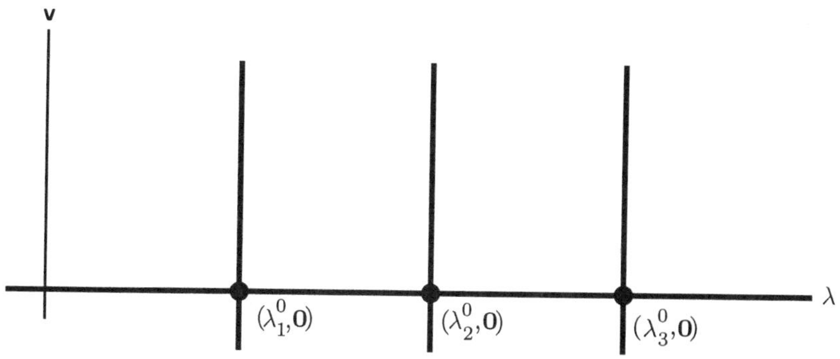

Fig. 3.3. Bifurcation diagram for (3.2) when $\boldsymbol{\lambda}$ is a scalar λ. The trivial branch is the λ-axis. The nontrivial solution pairs are represented by straight vertical lines through the points $\{(\lambda_k^0, \mathbf{0})\}$, where the λ_k^0 are the (real) eigenvalues of (3.2). This representation should not be construed as implying that the collection of solution pairs $(\lambda_k^0, \mathbf{w})$ is one-dimensional; it is in fact a plane in the $(\lambda, \mathbf{u})$-space whose dimension is the geometric multiplicity of λ_k^0.

Our definition of eigenvalue is more general than usual in mathematics. In the special case that λ is a real or complex parameter, it is standard to say that λ^0 is an eigenvalue of the operator $\mathbf{A}$ iff the the equation $\mathbf{A} \cdot \mathbf{v} = \lambda^0\mathbf{v}$ has a nontrivial solution, called the eigenvector corresponding to λ^0. In our terminology, this λ^0 is an eigenvalue of the operator-valued function $\lambda \mapsto \mathbf{A} - \lambda\mathbf{I}$. Some authors refer to the eigenvalues of $\lambda \mapsto \mathbf{I} - \lambda\mathbf{A}$ as *characteristic values*. We do not adhere to this policy. We employ our general definition because the operators that arise in the bifurcation problems of nonlinear elasticity are typically not affine functions of the parameter $\boldsymbol{\lambda}$.

We now begin our study of the relationship of the bifurcation diagram for a nonlinear problem such as (3.1) to that for its linearization about $\mathbf{u} = \mathbf{0}$. Recall from Sec. II.10 that if $\lim_{\varepsilon \to 0} \mathbf{f}[\boldsymbol{\lambda}, \varepsilon \mathbf{v}]/\varepsilon$ exists for all $\mathbf{v}$ in $\mathcal{X}$ and if this limit is a bounded linear operator acting on $\mathbf{v}$, then we denote it by $(\partial \mathbf{f}/\partial \mathbf{u})[\boldsymbol{\lambda}, \mathbf{0}] \cdot \mathbf{v}$. In this case $(\partial \mathbf{f}/\partial \mathbf{u})[\boldsymbol{\lambda}, \mathbf{0}]$ is called the *Gâteaux derivative* of $\mathbf{f}[\boldsymbol{\lambda}, \cdot]$ at $\mathbf{0}$. We then say that the equation

$$(3.4) \qquad (\partial \mathbf{f}/\partial \mathbf{u})[\boldsymbol{\lambda}, \mathbf{0}] \cdot \mathbf{v} = \mathbf{0},$$

which has the form (3.2), is the *linearization* of (3.1) about $\mathbf{u} = \mathbf{0}$. If, furthermore, $\mathbf{f}[\boldsymbol{\lambda}, \cdot]$ has the form

$$(3.5) \qquad \mathbf{f}[\boldsymbol{\lambda}, \mathbf{u}] = \mathbf{A}(\boldsymbol{\lambda}) \cdot \mathbf{u} + o(\|\mathbf{u}\|) \quad \text{as } \|\mathbf{u}\| \to 0$$

where $\mathbf{A}(\boldsymbol{\lambda})$ is a *bounded* linear operator and where $\|\mathbf{u}\|$ denotes the $\mathcal{X}$-norm of $\mathbf{u}$, then $\mathbf{f}[\boldsymbol{\lambda}, \cdot]$ is said to be (*Fréchet-*) *differentiable* at $\mathbf{0}$, and $\mathbf{A}(\boldsymbol{\lambda})$ is called the (*Fréchet*) *derivative* there. $\mathbf{A}(\boldsymbol{\lambda})$ is also denoted by $(\partial \mathbf{f}/\partial \mathbf{u})[\boldsymbol{\lambda}, \mathbf{0}]$. Indeed, if $\mathbf{f}[\boldsymbol{\lambda}, \cdot]$ is Fréchet-differentiable at $\mathbf{0}$, then it is Gâteaux-differentiable there and the derivatives are equal.

If we identify (θ, M) of (1.5) with $\mathbf{u}$ in $C^0[0,1] \times C^0[0,1] \equiv \mathcal{X}$, then its linearization is

$$(3.6) \qquad \theta_1(s) = \int_0^s \frac{M_1(\xi)}{(EI)(\xi)}\, d\xi, \quad M_1(s) = \lambda \int_s^1 \theta_1(\xi)\, d\xi,$$

which is readily seen to be equivalent to the problem

$$(3.7\text{a,b,c}) \qquad \tfrac{d}{ds}[(EI(s))\theta_1'(s)] + \lambda\theta_1(s) = 0, \quad \theta_1(0) = 0, \quad \theta_1'(1) = 0,$$

corresponding to (1.6).

3.8. Exercise. Prove that the operator

$$(3.9) \qquad C^0[0,1] \ni \theta \mapsto f(\theta, \lambda) \equiv \theta - \lambda \int_0^1 G(\cdot, \xi) \sin\theta(\xi)\, d\xi \in C^0[0,1],$$

where G is defined in (1.9), is Fréchet-differentiable at 0. Write the linearization of (1.8). Repeat this exercise when C^0 on the left and right of (3.9) is replaced with C^1.

It was long an unwritten law of physics, still widely believed, that the behavior of small solutions of (3.1), presumed to describe physical processes, is accurately accounted for by the behavior of solutions of its linearization (3.4). It is fortunate that this law is unwritten, because it is not true. One mission of bifurcation theory is to determine what accurate information about (3.1) is provided by (3.4).

Examples. We now present a selection of concrete, simple mathematical examples showing the variety of behavior that is possible for solutions of bifurcation problems and that must be accounted for by the basic theorems stated in Sec. 4. The simplest problems are those for which $\mathcal{X}$ is $\mathbb{R}$ or $\mathbb{R}^2$, for then we can represent precisely the entire set of solution pairs for one-parameter problems in the the two- or three-dimensional space of $(\lambda, \mathbf{u})$. Problems for which $\mathcal{X}$ is finite-dimensional are of major importance because virtually all the tractable infinite-dimensional bifurcation problems can be approximated effectively by finite-dimensional problems.

Virtually the simplest nonlinear bifurcation problem is

$$(3.10) \qquad x = (\lambda - x^2)x, \quad x \in \mathbb{R}.$$

Its linearization about the trivial branch $x = 0$ is the linear eigenvalue problem $y = \lambda y$ which has as its only eigenvalue $\lambda = 1$. The bifurcation diagram for (3.10), immediately found, consists of the trivial branch $x = 0$ and the closure of the nontrivial branches, which has the equation $\lambda = 1 + x^2$. Note that the nontrivial branches are curved, as is typical for nonlinear problems, and that the nontrivial branches bifurcate from $(\lambda, x) = (1, 0)$, where 1 is the eigenvalue of the linearized problem. In the following examples, when bifurcation occurs, it occurs at the eigenvalues of the linearization. We shall prove a precise version of this statement in Theorem 4.1.

It can happen that nonlinear problems have bifurcations in which nontrivial solution pairs lie on lines, just as for linear eigenvalue problems. For example, consider the system

$$(3.11) \qquad x^2 + y^2 = \lambda x^2, \quad 2xy = \lambda y^2, \quad (x, y) \in \mathbb{R}^2,$$

whose bifurcation diagram is readily found, or the boundary-value problem

$$(3.12) \qquad \frac{d}{ds}[g(u(s), u'(s), s, \lambda)] - h(u(s), u'(s), s, \lambda) = 0, \quad u(0) = 0 = u(1)$$

where $g(\cdot, \cdot, s, \lambda)$ and $h(\cdot, \cdot, s, \lambda)$ are homogeneous of degree $\alpha > 0$. The analysis of problems like (3.12) by Lyusternik (1937, 1938) furnished an early illustration of the application of the ideas of the powerful Lyusternik-Shnirel'man category theory to bifurcation problems. Problems lacking homogeneous nonlinearities also may admit solutions lying on lines: For $\lambda > 0$, consider the following problem proposed by Wolkowisky (1969):

$$(3.13) \qquad u'' + \lambda^2 u[1 + (\lambda \cos \lambda s)u - (\sin \lambda s)u'] = 0, \quad u(0) = 0 = u(1).$$

It is easy to see that (3.13) has nontrivial solution pairs of the form $(u, \lambda) = (u_k, k\pi)$ with $u_k(s) =$const. $\sin k\pi s$ for each integer $k > 0$. We shall be able to prove that these nontrivial branches are the only branches bifurcating from the trivial branch.

One of the central results justifying the need for a theory is that there need not be bifurcation from an eigenvalue of the linearized problem. Consider the following problem for $(x, y) \in \mathbb{R}^2$:

$$(3.14\text{a,b}) \qquad x + y(x^2 + y^2) = \lambda x, \quad y - x(x^2 + y^2) = \lambda y,$$

which has the trivial branch defined by $x = 0 = y$. The linearization of (3.14) about the trivial branch is

$$(3.15) \qquad x = \lambda x, \quad y = \lambda y,$$

which has $\lambda = 1$ as its only eigenvalue. This eigenvalue has geometric multiplicity 2, i.e., every nonzero pair (x, y) is an eigenvector of (3.15). Nevertheless, if we multiply (3.14a) by y, multiply (3.14b) by x, and subtract one of the resulting equations from the other, then we obtain $(x^2+y^2)^2 = 0$. It is significant to note that (3.14) cannot be characterized as the vanishing of the gradient of a scalar function of x and y when λ is fixed because

$$\frac{\partial}{\partial y}[x + y(x^2+y^2) - \lambda x] \neq \frac{\partial}{\partial x}[y - x(x^2+y^2) - \lambda y]. \tag{3.16}$$

Thus (3.14) is not variational. A similar effect occurs for differential equations:

3.17. Exercise. Combine the techniques used for (3.14) with integration by parts to prove that the boundary-value problem

$$\begin{aligned} -u'' = \lambda[u + v(u^2+v^2)], \quad -v'' = \lambda[v - u(u^2+v^2)], \\ u(0) = 0 = v(0), \quad u(1) = 0 = v(1) \end{aligned} \tag{3.18}$$

has the unique solution $u = 0 = v$ for each real λ. Find the eigenvalues and eigenfunctions of the linearization of (3.18) about the trivial branch.

To gain some insight into the uniqueness of the trivial solution of (3.14), we embed it into a one-parameter family of problems

$$(1+2\alpha)x + y(x^2+y^2) = \lambda x, \quad y - x(x^2+y^2) = \lambda y. \tag{3.19a,b}$$

When $\alpha = 0$, this system reduces to (3.14). For $\alpha \neq 0$, the linearization of (3.19) about the trivial branch has eigenvalues 1 and $1+2\alpha$ with corresponding normalized eigenvectors $(0,1)$ and $(1,0)$. Equation (3.19) implies that if $x = 0$, then $y = 0$ and vice versa. For nontrivial solutions, (3.19) implies that

$$[\lambda - (1+\alpha)]^2 + (x^2+y^2)^2 = \alpha^2, \tag{3.20}$$

which is the equation of a sphere-like surface in (λ, x, y)-space. We substitute (3.20) into (3.19b) to obtain

$$\sqrt{\lambda - 1}\, y = -\sqrt{1 + 2\alpha - \lambda}\, x \quad \text{for } 1 \leq \lambda \leq 1 + 2\alpha \tag{3.21}$$

for $\alpha \geq 0$. The bifurcating branches lie on the curve formed by the intersection of the two surfaces of (3.20) and (3.21), which is shown in Fig. 3.22. We see that as $\alpha \to 0$, this curve shrinks to the point on the λ-axis with $\lambda = 1$, which is the eigenvalue of geometric multiplicity 2 of the linearization of (3.14). This example suggests that the absence of nontrivial branches for (3.14) may be associated with the multiplicity of the eigenvalue. This example has the additional feature that the nontrivial branches are bounded and that they connect their bifurcation points.

We further examine the source of difficulties for (3.14) by studying a variational problem whose linearization has an eigenvalue with geometric multiplicity 2. The equations expressing the vanishing of the gradient of

$$(x, y) \mapsto \tfrac{1}{2}(1-\lambda)(x^2+y^2) + \tfrac{1}{4}(x^2+y^2)^2 + \tfrac{1}{4}\alpha x^4$$

are

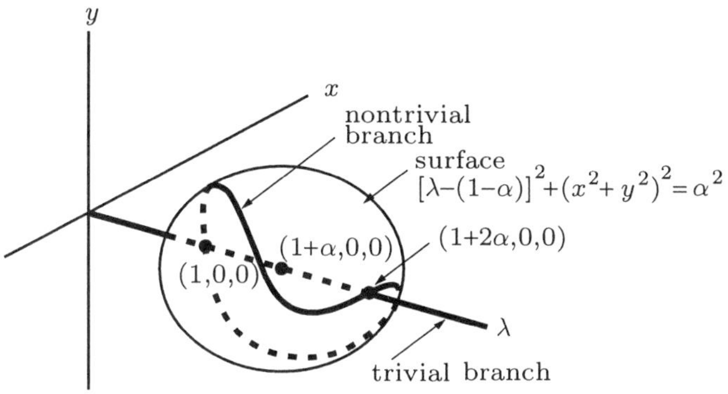

Fig. 3.22. Bifurcation diagram for (3.19) with $\alpha > 0$.

(3.23) $$x + (x^2 + y^2)x + \alpha x^3 = \lambda x, \quad y + (x^2 + y^2)y = \lambda y$$

(which for $\alpha = 0$ differs from (3.14) by the replacement of the coefficients $(y, -x)$ of $x^2 + y^2$ with (x, y)). Here α is a real parameter. The only eigenvalue of the linearization of (3.23) about the trivial branch is $\lambda = 1$, which has geometric multiplicity 2. Nevertheless, (3.23) has exactly two nontrivial branches:

(3.24) $$\begin{aligned} y = 0, &\quad \lambda = 1 + x^2 + \alpha x^2; \\ x = 0, &\quad \lambda = 1 + y^2, \end{aligned}$$

each of which bifurcates from $(\lambda, x, y) = (1, 0, 0)$. This example suggests that the absence of nontrivial branches for (3.14) may be associated jointly with its lack of variational structure and with the multiplicity of the eigenvalue of the linearized problem. Indeed, in the next section we state theorems to the effect that if $\mathbf{f}$ of (3.1) with a scalar eigenvalue parameter is sufficiently well behaved and if either (i) its linearization about the trivial branch has an eigenvalue λ^0 of odd algebraic multiplicity, or (ii) $\mathbf{f}[\cdot, \mathbf{u}]$ is affine, $\mathbf{f}[\lambda, \cdot]$ is the gradient of a scalar-valued function, and λ^0 is an eigenvalue of (3.4) of *any* multiplicity, then $(\lambda^0, \mathbf{0})$ is a bifurcation point.

Note that if $\alpha = 0$, then (3.23) has a two-dimensional paraboloidal surface of nontrivial solution pairs, bifurcating from $(1, 0, 0)$. This phenomenon is associated with the invariance of (3.23) under rotations about 0 in the (x, y)-plane.

Another mechanism by which a point $(\lambda^0, \mathbf{0})$ may fail to be a bifurcation point for (3.1) when λ^0 is an eigenvalue of (3.4) is evidenced by the scalar problem $x = (1 - \lambda^2)x - x^3$ (cf. (3.10)). The linearization of this equation about the trivial branch has $\lambda = 0$ as its only eigenvalue. It has algebraic multiplicity 2 (defined in the next section) and geometric multiplicity 1. That this problem cannot have nontrivial solutions is associated with the evenness of the algebraic multiplicity.

Now consider the following variational problem, due to Pimbley (1969),

$$\lambda x = x + \tfrac{3}{4}x^3 + \tfrac{3}{2}xy^2, \quad \alpha\lambda y = y + \tfrac{3}{4}y^3 + \tfrac{3}{2}yx^2 \tag{3.25}$$

where α is a real parameter. The bifurcation diagrams for this problem are shown in Fig. 3.26 for various values of α. There is *secondary bifurcation* for $\alpha \in (1, 2)$. $\lambda = 1$ is the only eigenvalue of the linearization of (3.25) about the trivial branch for $\alpha = 1$. It has geometric and algebraic multiplicity 2. There are, however, exactly four pairs of distinct continuously differentiable branches of nowhere trivial solution pairs bifurcating from (1,0,0). This result and the previous ones indicate that there are no simple rules relating the multiplicity of the eigenvalue to the number of bifurcating branches.

3.27. Exercise. Prove that the bifurcation diagrams for (3.25) indeed have the forms shown in Fig. 3.26 and find the diagrams for $\alpha < 1$.

Bauer, Keller, & Reiss (1975) have shown how some examples of secondary bifurcation, which arise in applications, have a mathematical structure like (3.25): At a critical value of an auxiliary parameter α, (3.4) has a multiple eigenvalue μ and there are multiple nontrivial branches of (3.1) bifurcating from $(\mu, \mathbf{0})$. As α passes through the critical value, nontrivial branches coalesce at $(\mu, \mathbf{0})$ and then break up into primary branches, which bifurcate from the trivia branch, and secondary branches, which bifurcate from the primary branches. Ķeener (1979) has analyzed phenomena of this kind and supplied an interesting collection of examples.

The solutions of the nonlinear Hammerstein integral equation

$$\lambda w(s) = \frac{2}{\pi}\int_0^\pi \left(\sin s \sin t + \frac{1}{\alpha}\sin 2s \sin 2t\right)[w(t) + w(t)^3]\,dt, \quad s \in [0, \pi] \tag{3.28}$$

(with a degenerate kernel) must have the form

$$w(s) = x \sin s + y \sin 2s. \tag{3.29}$$

The substitution of (3.29) into (3.28) yields (3.25).

A more spectacular example of multiple bifurcation is afforded by the problem

$$\lambda u''(s) + u(s)\int_0^1 u(t)^2\,dt = 0 \quad \text{for } s \in (0,1), \quad u(0) = 0 = u(1) \tag{3.30}$$

(proposed by Kirchgässner). The only eigenvalue of the linearization

$$\lambda v'' = 0 \quad \text{for } s \in (0,1), \quad v(0) = 0 = v(1) \tag{3.31}$$

of (3.30) about the trivial branch is $\lambda = 0$, which has infinite geometrical multiplicity. We multiply (3.30) by $u(s)$ and integrate the resulting expression by parts over $(0, 1)$ to obtain

$$-\lambda \int_0^1 u'(s)^2\,ds + \left[\int_0^1 u(s)^2\,ds\right]^2 = 0, \tag{3.32}$$

which implies that the only solution of (3.30) for $\lambda \le 0$ is trivial. For $\lambda > 0$, we regard (3.30) as a linear ordinary differential equation with constant

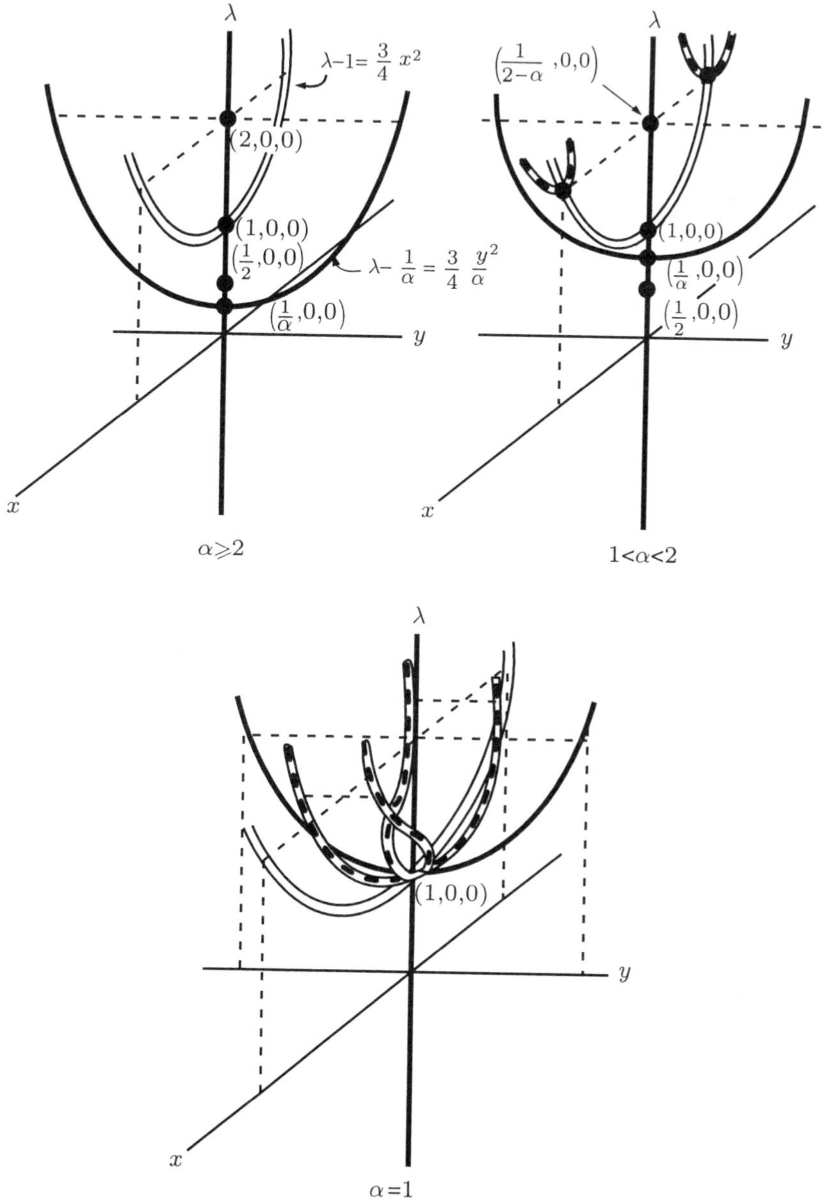

Fig. 3.26. Bifurcation diagrams for (3.25) for $\alpha \geq 1$. The diagrams for $\alpha < 1$ are similar.

coefficients that happen to depend on the unknown u. The nontrivial solution pairs must have the form $u_n(s) = c_n \sin n\pi s$ when $n^2\pi^2\lambda = \int_0^1 u(t)^2\, dt$, where c_n is a constant to be determined and n is a nonzero integer. The substitution of u_n into (3.30) yields a formula for c_n, from which we obtain

$$u_n(s) = \pm\sqrt{2\lambda}\sin n\pi s. \tag{3.33}$$

The bifurcation diagram Fig. 3.34 defined by (3.30) exhibits an infinity of nontrivial branches bifurcating from (0,0). If λ is replaced with $1/\lambda$, then the modified equation has an infinity of nontrivial branches, none of which bifurcate from the trivial branch. For more complicated equations, the demonstration that there are such nonbifurcating branches, sometimes called *isolas*, and their construction can be quite difficult (see Antman & Pierce (1990), H. B. Keller (1981), and Peitgen, Saupe, & Schmitt (1981).)

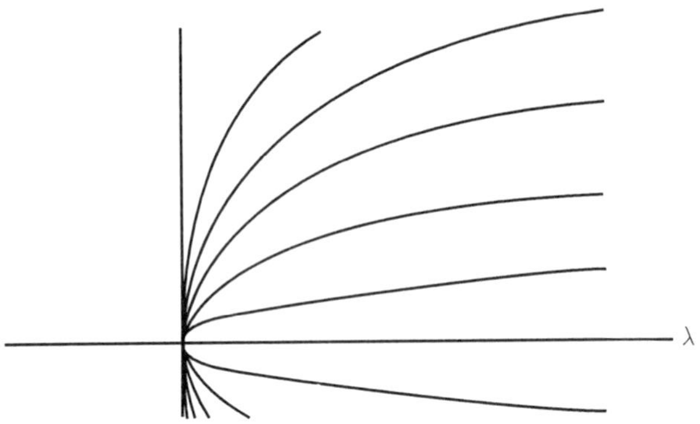

Fig. 3.34. Bifurcation diagram for (3.30) showing a plot of $\operatorname{sign} u_n'(0) \max |u_n| = \operatorname{sign} u_n'(0)\sqrt{2\lambda}n\pi$ as a function of λ, which comes from (3.33).

3.35. Exercise. Bifurcation diagrams like Fig. 3.34 occur in mechanics. Suppose that the reduction δ of the distance between the ends of an inextensible elastica is prescribed in place of the thrust λ. Then θ and the unknown thrust λ must satisfy (1.6) and the additional condition

$$\int_0^1 \cos\theta(s)\, ds = 1 - \delta. \tag{3.36}$$

Show that the bifurcation diagram for this problem has the form of Fig. 3.34 with δ replacing λ. How is this bifurcation diagram related to Fig. 1.15? (When the elastica is uniform, this problem can be solved explicitly by using the results of Ex. 1.16.) Discuss the role of inextensibility in producing a bifurcation diagram like Fig. 3.34.

The following artificial problem disabuses us of the optimistic hope that solution branches of a reasonably well-behaved bifurcation problems are curves. The nontrivial branch for

$$\lambda x(x-1)^2 = x(x-1)^2\left[2 + \sin\left(\frac{\pi}{x-1}\right)\right], \tag{3.37}$$

which is easily plotted, does not form a curve, although it is connected. (A set $\mathcal{E}$ is called *connected* iff however it is expressed as a union of two disjoint nonempty sets $\mathcal{A}$ and $\mathcal{B}$, at least one of these sets must contain a limit point of the other. Thus $\mathcal{E}$ is connected iff there do not exist nonempty sets $\mathcal{A}$ and $\mathcal{B}$ such that $\mathcal{E} = \mathcal{A} \cup \mathcal{B}$ with $\mathcal{A} \cap \mathrm{cl}\, \mathcal{B} = \emptyset = \mathcal{B} \cap \mathrm{cl}\, \mathcal{A}$; see Alexander (1981), Whyburn (1964), and the paragraph preceding Theorem 4.19.) The factor $(x-1)^2$ is put into (3.37) so that the right-hand side is continuously differentiable.

We can construct examples like this for boundary-value problems of the form

$$u'' + \lambda[1 + h(\lambda^2 u^2 + (u')^2, \lambda)]u = 0 \quad \text{for } s \in (0,1), \quad u(0) = 0 = u(1) \tag{3.38}$$

where $h(0,\lambda) = 0$. (This example, which generalizes (3.13), is due to Rabinowitz (1973a).) This problem has nontrivial branches bifurcating from $(k^2\pi^2, 0)$ of the form (λ, u) with $u(s) = c \sin k\pi s$ and

$$\lambda[1 + h(k^2\pi^2 c^2, \lambda)] = k^2\pi^2. \tag{3.39}$$

A bifurcation diagram can be determined from (3.39). By varying h we can construct a very rich variety of diagrams, including those with branches that are not connected. One of the difficult problems of bifurcation theory for differential equations is to determine when the branches are actually curves.

Our next objective is to show that there may be no nontrivial branches passing through a bifurcation point. Our example gives a further illustration of the difficulty in deducing statements about the multiplicity of bifurcating branches of (3.1) from the multiplicity of eigenvalues of (3.4). To construct such examples when $\mathbf{f}$ is very well behaved requires great care. We follow the approach of Böhme (1972).

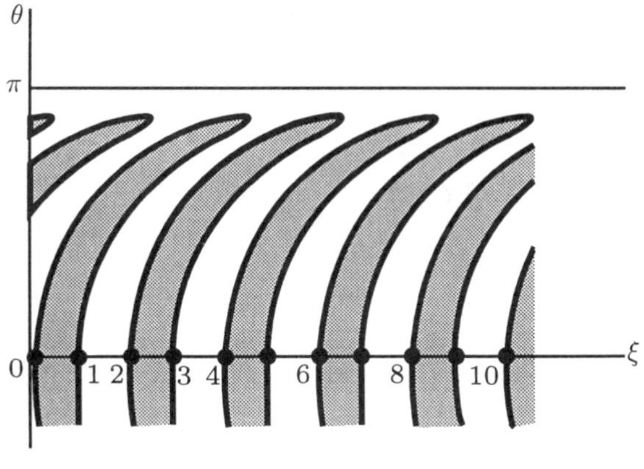

Fig. 3.40. Φ_θ is negative in the shaded regions.

Let $\mathbb{R}^2 \ni (\xi, \theta) \mapsto \Phi(\xi, \theta) \in \mathbb{R}$ be infinitely differentiable, let $\Phi(\cdot, \theta)$ have period 2, let $\Phi(\xi, \cdot)$ have period 2π, and let the locus of points where Φ_θ vanishes have the form shown in Fig. 3.40. We define

$$F(r\cos\theta, r\sin\theta) \equiv \Phi(1/r, \theta)\exp(-1/r^2). \tag{3.41}$$

Then F is infinitely differentiable and $F(0,0) = F_x(0,0) = F_y(0,0) = 0$. We study the variational problem

$$(\lambda - 1)x = F_x(x,y), \quad (\lambda - 1)y = F_y(x,y), \tag{3.42}$$

which is readily shown to be equivalent to

$$\lambda - 1 = \left[\frac{2\Phi(1/r,\theta)}{r^4} - \frac{\Phi_\xi(1/r,\theta)}{r^3}\right]\exp\left(-\frac{1}{r^2}\right), \tag{3.43}$$

$$\Phi_\theta(1/r,\theta) = 0. \tag{3.44}$$

Let us fix $r^2 \equiv x^2 + y^2 = \alpha^2$ (const.). Figure 3.40 implies that for each such $\alpha^2 > 0$, Eq. (3.44) has at least six solutions θ. For each such solution, Eq. (3.43) yields a corresponding value of λ. Thus there are at least six solution pairs on each cylinder $\{(\lambda, x, y) : x^2 + y^2 = \alpha^2\}$ in $\mathbb{R}^3$. A bifurcation diagram in (λ, r)-space for (3.42) is shown in Fig. 3.45. Thus problem (3.42), involving the vanishing of the gradient of an infinitely differentiable function, has no nontrivial branches passing through the bifurcation point $(\lambda, x, y) = (1, 0, 0)$ (i.e., there is no branch of nontrivial solution pairs whose closure contains this bifurcation point), has a complicated multiplicity structure that bears no obvious relation to the geometric multiplicity 2 of the eigenvalue $\lambda = 1$ of the linearization, and has a countable infinity of nonbifurcating branches. Theorem 4.19 will show that some of the pathologies of this example cannot occur if $\mathbf{f}$ is real-analytic.

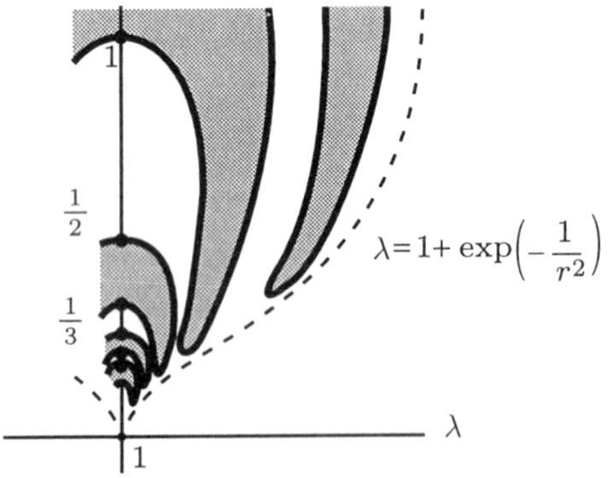

Fig. 3.45. A schematic bifurcation diagram for (3.42) when F has the properties described in the paragraph containing (3.41).

3.46. Problem. Choose an appropriate F with the properties described in the paragraph containing (3.41) and sketch the full bifurcation diagram for (3.42) in perspective in (λ, x, y)-space.

The following exercise describes related bifurcation phenomena.

3.47. Exercise. Sketch the full bifurcation diagrams for (3.42) in (λ, x, y)-space when F is defined by (3.41) with (i) $\Phi(\xi,\theta) = \sin \pi\xi$, (ii) $\Phi(\xi,\theta) = \theta^2 - \pi\theta \sin \pi\xi$, and (iii) $\Phi(\xi,\theta) = \theta^2 - 2\theta\xi$.

Another example showing the existence of a connected nontrivial branch that is not a curve is given by

3.48. Exercise. Sketch the full bifurcation diagrams for (3.42) in (λ, x, y)-space when

$$F(r\cos\theta, r\sin\theta) = [\theta^2 - 2\theta\log(1+\cos\pi r)]\exp[-1/(1+\cos\pi r)].$$

It is of course easy to construct examples of bifurcation problems with worse pathologies when $\mathbf{f}$ is very badly behaved. For example, for the problem $\lambda x = x\sin(1/x)$, which does not have a linearization at the trivial branch, each point of the segment $[-1,1]\times\{0\}$ is a bifurcation point.

A well-known multiparameter bifurcation problem from rigid-body mechanics is to find solutions of the ordinary differential equation

$$\frac{d^2\theta}{dt^2} + (\lambda_1 + \lambda_2\cos t)\sin\theta = 0 \tag{3.49}$$

having a prescribed period. This equation describes the motion of a pendulum whose support is given a vertical sinusoidal motion. Two artificial two-parameter problems are

$$u'' + \varphi(\lambda_1,\lambda_2)u + \psi(\lambda_1,\lambda_2)u^3 = 0, \qquad u(0) = 0 = u(1), \tag{3.50}$$

$$u'' + \varphi(\lambda_1,\lambda_2)u + \psi(\lambda_1,\lambda_2)u\int_0^1 u(t)^2\,dt = 0, \quad u(0) = 0 = u(1) \tag{3.51}$$

where φ and ψ are given functions. These problems have the same linearizations, the set of their eigenvalues consisting of all pairs (λ_1, λ_2) that satisfy

$$\varphi(\lambda_1,\lambda_2) = n^2\pi^2, \quad n = 1, 2, \ldots. \tag{3.52}$$

Problem (3.50) can be solved in closed form in terms of elliptic functions and problem (3.51) can be solved in closed form by elementary means. If φ and ψ are continuously differentiable and if (3.52) represents a family of smooth disjoint curves (as would happen if $\varphi(\lambda_1,\lambda_2) = \lambda_1 - (\lambda_2)^2$, e.g.), then the nontrivial solution pairs of (3.51) and (3.52) form smooth surfaces of dimension 2 in $\mathbb{R}^2 \times C^2[0,1]$. A properly stated generalization of this observation is the content of Theorem 4.14.

3.53. Exercise. Study the nature of solution pairs of (3.51) with $\varphi(\lambda_1,\lambda_2) = \lambda_1 - (\lambda_2)^2$ for various choices of ψ.

4. Basic Theorems of Bifurcation Theory

The examples of the last section expose the need for effective theorems describing the local and global behavior of bifurcating branches of solution pairs. In this section we provide such a collection of theorems. The proofs of all these theorems except the first are either omitted or deferred.

Our first theorem makes precise the observation made in the preceding section that well-behaved problems (3.1) have bifurcation points of the form $(\boldsymbol{\lambda}^0, \mathbf{0})$ where $\boldsymbol{\lambda}^0$ is an eigenvalue of the linearization (3.4).

4.1. Theorem. *Let $\mathcal{X}$ and $\mathcal{Y}$ be Banach spaces. Let $\mathcal{A}$ be a neighborhood of $\boldsymbol{\lambda}^0$ in $\mathbb{R}^n$ and let $\mathcal{U}$ be a neighborhood of $\mathbf{0}$ in $\mathcal{X}$. Let $\mathsf{f} : \mathcal{A}\times\mathcal{U} \to \mathcal{Y}$ and let $\mathsf{f}[\boldsymbol{\lambda}, \cdot]$ be Fréchet-differentiable at $\mathbf{0}$ for each $\boldsymbol{\lambda} \in \mathcal{A}$. (Thus f satisfies (3.5) where $\mathsf{A}(\boldsymbol{\lambda})$ is a bounded linear operator from $\mathcal{X}$ to $\mathcal{Y}$.) Let $\mathsf{A}(\cdot)$ be continuous (from $\mathcal{A}$ to the space $\mathcal{L}(\mathcal{X},\mathcal{Y})$ of bounded linear operators from $\mathcal{X}$ to $\mathcal{Y}$). If $(\boldsymbol{\lambda}^0, \mathbf{0})$ is a bifurcation point on the trivial branch of (3.1), then $\boldsymbol{\lambda}^0$ is in the spectrum of $\mathsf{A}(\cdot)$, i.e., $\mathsf{A}(\boldsymbol{\lambda}^0)$ does not have a bounded inverse.*

Proof. If $(\boldsymbol{\lambda}^0, \mathbf{0})$ is such a bifurcation point, then by its definition there is a sequence of solution pairs $\{(\boldsymbol{\lambda}_k, \mathsf{u}_k)\}$ of (3.1) with $\mathsf{u}_k \neq \mathbf{0}$ such that $(\boldsymbol{\lambda}_k, \mathsf{u}_k) \to (\boldsymbol{\lambda}^0, \mathbf{0})$ as $k \to \infty$. These solution pairs satisfy $\mathsf{f}[\boldsymbol{\lambda}_k, \mathsf{u}_k] = \mathbf{0}$, which is equivalent to

$$\mathsf{A}(\boldsymbol{\lambda}^0)\cdot\mathsf{u}_k = [\mathsf{A}(\boldsymbol{\lambda}^0) - \mathsf{A}(\boldsymbol{\lambda}_k)]\cdot\mathsf{u}_k + o(\|\mathsf{u}_k\|), \tag{4.2}$$

by (3.5). Were $\mathsf{A}(\boldsymbol{\lambda}^0)$ to have a bounded inverse $\mathsf{A}(\boldsymbol{\lambda}^0)^{-1}$, then (4.2) would imply that

$$\mathsf{u}_k = \mathsf{A}(\boldsymbol{\lambda}^0)^{-1}\{[\mathsf{A}(\boldsymbol{\lambda}^0) - \mathsf{A}(\boldsymbol{\lambda}_k)]\cdot\mathsf{u}_k + o(\|\mathsf{u}_k\|)\}. \tag{4.3}$$

We divide this equation by $\|\mathsf{u}_k\|$ and then equate the norms of the resulting equation to obtain

$$1 = \left\|\mathsf{A}(\boldsymbol{\lambda}^0)^{-1}\left\{[\mathsf{A}(\boldsymbol{\lambda}^0) - \mathsf{A}(\boldsymbol{\lambda}_k)]\cdot\frac{\mathsf{u}_k}{\|\mathsf{u}_k\|} + \frac{o(\|\mathsf{u}_k\|)}{\|\mathsf{u}_k\|}\right\}\right\|. \tag{4.4}$$

The continuity of A implies that the right-hand side of (4.4) would then approach 0 as $k \to \infty$, which is absurd. □

We now introduce the notation to be used in expressing further results. Let $\mathcal{X}$ be a real Banach space with the norm $\|\cdot\|$ and let $\mathcal{D}$ be an open subset of $\mathbb{R}^n\times\mathcal{X}$ that has a nonempty intersection $\mathcal{D}_0\times\{\mathbf{0}\} \equiv \mathcal{D}\cap[\mathbb{R}^n\times\{\mathbf{0}\}]$ with the parameter space. Let $\mathcal{D} \ni (\boldsymbol{\lambda}, \mathsf{u}) \mapsto \mathsf{f}[\boldsymbol{\lambda}, \mathsf{u}] \in \mathcal{X}$ have the form

$$\mathsf{f}[\boldsymbol{\lambda}, \mathsf{u}] = \mathsf{u} - \mathsf{L}(\boldsymbol{\lambda})\cdot\mathsf{u} - \mathsf{g}[\boldsymbol{\lambda}, \mathsf{u}] \tag{4.5}$$

where $\mathsf{L}(\boldsymbol{\lambda})$ is a linear operator from $\mathcal{X}$ to itself, $\mathsf{L}(\cdot)$ is continuous (i.e., $\mathsf{L}(\cdot) \in C^0(\mathcal{D}_0, \mathcal{L}(\mathcal{X},\mathcal{X}))$), $\mathsf{g} : \mathcal{D} \to \mathcal{X}$ is continuous, and $\mathsf{g}[\boldsymbol{\lambda}, \mathsf{u}] = o(\|\mathsf{u}\|)$ as $\mathsf{u} \to \mathbf{0}$ uniformly for $\boldsymbol{\lambda}$ in any bounded set. Thus $\mathsf{f}[\boldsymbol{\lambda}, \mathbf{0}] = \mathbf{0}$ and all members of $\mathcal{D}_0 \times \{\mathbf{0}\}$, called the *trivial family* of solution pairs, satisfy (3.1). The closure of the set of nontrivial solution pairs of (3.1) is denoted $\mathcal{S}$. The set of bifurcation points of (3.1) at the trivial family is $\mathcal{F} \equiv \mathcal{S}\cap[\mathcal{D}_0 \times \{\mathbf{0}\}]$. The linearization of (3.1) about the trivial family now has the form

$$\mathsf{v} = \mathsf{L}(\boldsymbol{\lambda})\mathsf{v}. \tag{4.6}$$

Let $\mathcal{E}$ denote the set of eigenvalues of (4.6).

We now prepare the groundwork for our next theorem, which gives some representative conditions ensuring that from $(\boldsymbol{\lambda}^0, \mathbf{0}) \in \mathcal{E}\times\{\mathbf{0}\}$ there bifurcates a connected set of nontrivial solution pairs forming a C^1-surface of dimension n near $(\boldsymbol{\lambda}^0, \mathbf{0})$ and which describes precisely the relationship between nontrivial solution pairs and the eigenpairs from which their branches bifurcate. To appreciate the technical content of the constructions, we consider the case in which $n = 2$. By the examples of the last section, we may expect $\mathcal{E}$ to contain eigencurves. We require that $(\boldsymbol{\lambda}^0, \mathbf{0})$ lie on a sufficiently smooth eigencurve. We express the requisite smoothness by requiring a neighborhood of $(\boldsymbol{\lambda}^0, \mathbf{0})$ to be the image under a smooth mapping of a rectangular region in which the inverse image of the eigencurve is a line. We can exploit the theory for one-parameter problems by restricting the eigenvalue parameter $\boldsymbol{\lambda}$ to a curve through $(\boldsymbol{\lambda}^0, \mathbf{0})$ transversal to the eigencurve. The reader should specialize the following development to the simple case of $n = 1$ and should try to visualize the geometrical ideas when $n \geq 3$.

Let $\boldsymbol{\lambda}^0 \in \mathcal{E}$ and let $\operatorname{cl} \mathcal{A} \in \mathcal{D}_0$ be the closure of a neighborhood of $\boldsymbol{\lambda}^0$ consisting of points $\boldsymbol{\lambda}$ of the form

$$\boldsymbol{\lambda} = \tilde{\boldsymbol{\lambda}}(\boldsymbol{\sigma}, \tau), \quad \boldsymbol{\sigma} \in [-1, 1]^{n-1}, \quad \tau \in [-1, 1] \tag{4.7}$$

where $\tilde{\boldsymbol{\lambda}}$ is an invertible, continuously differentiable mapping of $[-1, 1]^n$ onto $\operatorname{cl} \mathcal{A}$ for which $\mathcal{E} \cap \operatorname{cl} \mathcal{A}$ consists of those $\boldsymbol{\lambda}$'s of the form $\boldsymbol{\lambda} = \tilde{\boldsymbol{\lambda}}(\boldsymbol{\sigma}, 0)$ with $\boldsymbol{\sigma} \in [-1, 1]^{n-1}$ and for which $\boldsymbol{\lambda}^0 = \tilde{\boldsymbol{\lambda}}(\mathbf{0}, 0)$. See Fig. 4.8.

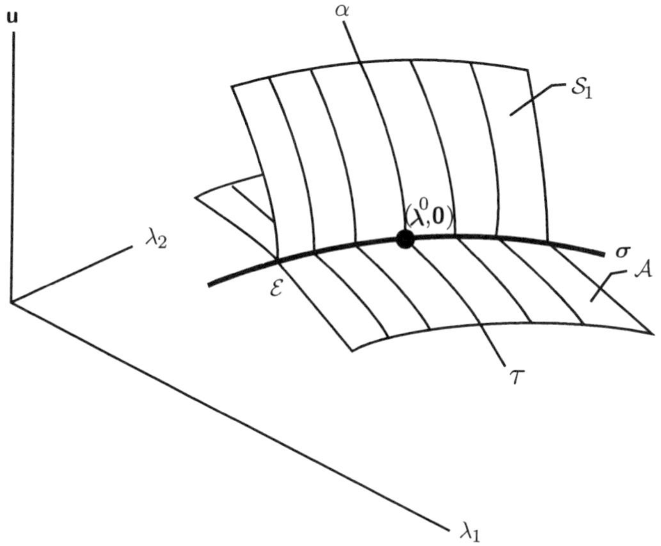

Fig. 4.8. Bifurcation diagram illustrating the variables associated with (4.7) and the nature of the connected set $\mathcal{S}_1$ of nontrivial solution pairs bifurcating from $(\boldsymbol{\lambda}^0, \mathbf{0})$ when the hypotheses of Theorem 4.14 are met. Near this point, $\mathcal{S}_1$ is an n-dimensional surface.

Assume that $\mathsf{L}(\cdot) \in C^1(\mathrm{cl}\,\mathcal{A})$. The restriction of $\mathsf{L}(\cdot)$ to a τ-coordinate curve through $\tilde{\boldsymbol{\lambda}}(\boldsymbol{\sigma}, 0)$ is

$$[-1,1] \ni \tau \mapsto \mathsf{L}(\tilde{\boldsymbol{\lambda}}(\boldsymbol{\sigma}, \tau)). \tag{4.9}$$

Following Magnus (1976), we say that the eigenvalue 0 of $\tau \mapsto \mathsf{I} - \mathsf{L}(\tilde{\boldsymbol{\lambda}}(\boldsymbol{\sigma}, \tau))$ is *simple* iff

$$\dim \ker [\mathsf{I} - \mathsf{L}(\tilde{\boldsymbol{\lambda}}(\boldsymbol{\sigma}, 0))] = 1, \quad \dim \ker [\mathsf{I} - \mathsf{M}(\boldsymbol{\sigma})] = 0 \tag{4.10}$$

where

$$\mathsf{M}(\boldsymbol{\sigma}) \equiv \mathsf{L}(\tilde{\boldsymbol{\lambda}}(\boldsymbol{\sigma}, 0)) + [(\partial \mathsf{L}/\partial \boldsymbol{\lambda})(\tilde{\boldsymbol{\lambda}}(\boldsymbol{\sigma}, 0)) \cdot \tilde{\boldsymbol{\lambda}}_\tau(\boldsymbol{\sigma}, 0)] \cdot \mathsf{P}(\boldsymbol{\sigma}) \tag{4.11}$$

with $\mathsf{P}(\boldsymbol{\sigma})$ being any projection of $\mathcal{X}$ onto $\ker [\mathsf{I} - \mathsf{L}(\tilde{\boldsymbol{\lambda}}(\boldsymbol{\sigma}, 0))]$.

For linear operators depending affinely on a scalar parameter λ, there are elegant and powerful formulations of spectral theory that contain a precise definitions of algebraic multiplicity (see Kato (1976) and Taylor & Lay (1980)). For finite-dimensional operators (matrices), there is an elementary theory with the same effect (of crucial importance in the theory of linear systems of ordinary differential equations). For linear operators that do not depend affinely on λ, the definition of algebraic multiplicity is more delicate. If $\mathsf{A}(\lambda)$ is a square matrix depending smoothly on λ, then λ^0 is an eigenvalue of $\mathsf{A}(\cdot)$ if and only if $\det \mathsf{A}(\lambda^0) = 0$. Its algebraic multiplicity is just the order of the first nonvanishing derivative of $\lambda \mapsto \det \mathsf{A}(\lambda)$ at λ^0. It is this notion that must be extended to general linear operators for which the notion of determinant is not natural. For a definitive discussion of algebraic multiplicity of eigenvalues of linear operators, see Fitzpatrick & Pejsachowicz (1991).

We require the following technical result.

4.12. Lemma. *Let $\mathsf{L}(\cdot) \in C^1(\mathrm{cl}\,\mathcal{A})$. For each $\boldsymbol{\sigma} \in [-1,1]^{n-1}$, let 0 be a simple eigenvalue of $\mathsf{I} - \mathsf{L}(\tilde{\boldsymbol{\lambda}}(\boldsymbol{\sigma}, \cdot))$ and let $\mathsf{v}(\boldsymbol{\sigma})$ denote a corresponding eigenvector. If*

$$\dim \ker [\mathsf{I} - \mathsf{L}(\tilde{\boldsymbol{\lambda}}(\boldsymbol{\sigma}, 0))]^2 = \dim \ker [\mathsf{I} - \mathsf{L}(\tilde{\boldsymbol{\lambda}}(\boldsymbol{\sigma}, 0))] \equiv \operatorname{span} \mathsf{v}(\boldsymbol{\sigma}) = 1, \tag{4.13}$$

then $\mathsf{v}(\cdot)$ and $\mathsf{P}(\cdot)$, which are defined above, can be chosen to be continuously differentiable on $[-1,1]^{n-1}$.

Our main result on the local structure of solutions to (3.1) merely requires that g be Lipschitz-continuous:

4.14. Local Bifurcation Theorem. *Let $\boldsymbol{\lambda}^0 \in \mathcal{E}$ and let $\mathrm{cl}\,\mathcal{A} \subset \mathcal{D}_0$ be a neighborhood of $\boldsymbol{\lambda}^0$ consisting of points $\boldsymbol{\lambda}$ of the form (4.7) where these variables have the properties described above. Let $\mathsf{L}(\cdot) \in C^1(\mathrm{cl}\,\mathcal{A})$. Let (4.5), (4.10), and (4.13) hold. Let there be a number $\varepsilon > 0$ and a continuous, increasing function $\varphi : [0,\infty) \to [0,\infty)$ with $\varphi(0) = 0$ such that*

$$\|\mathsf{g}[\boldsymbol{\alpha}, \mathsf{x}] - \mathsf{g}[\boldsymbol{\beta}, \mathsf{y}]\| \le \varphi(\|\mathsf{x}\| + \|\mathsf{y}\|)\,[\,\|\mathsf{x} - \mathsf{y}\| + (\|\mathsf{x}\| + \|\mathsf{y}\|)|\boldsymbol{\alpha} - \boldsymbol{\beta}|\,] \tag{4.15}$$

$$\forall\, (\boldsymbol{\alpha}, \mathsf{x}), (\boldsymbol{\beta}, \mathsf{y}) \in \mathrm{cl}\,\mathcal{Z} \equiv \{(\boldsymbol{\lambda}, \mathsf{u}) \in \mathcal{D} : \boldsymbol{\lambda} \in \mathrm{cl}\,\mathcal{A},\ \|\mathsf{u}\| \le \varepsilon\}.$$

Let $\mathbf{v}(\cdot)$ *and* $\mathbf{P}(\cdot)$ *be taken to be continuously differentiable in accord with Lemma 4.12. Then there exist a number* $\eta > 0$ *and continuous functions*

$$(4.16)\qquad [-1,1]^{n-1}\times[-\eta,\eta]\ni(\boldsymbol{\sigma},\rho)\mapsto\begin{cases}\kappa(\boldsymbol{\sigma},\rho)\in\mathbb{R},\\ \mathbf{w}(\boldsymbol{\sigma},\rho)\in[\mathbf{I}-\mathbf{P}(\boldsymbol{\sigma})]\mathcal{X}\end{cases}$$

with $\kappa(\cdot,\rho)$ *and* $\mathbf{w}(\cdot,\rho)$ *continuously differentiable, with* $\kappa(\cdot,0)=0$, *and with* $\mathbf{w}(\cdot,0)=\mathbf{0}$ *such that*

$$(4.17)\qquad \big(\tilde{\boldsymbol{\lambda}}(\boldsymbol{\sigma},0)+\kappa(\boldsymbol{\sigma},\rho)\tilde{\boldsymbol{\lambda}}_\tau(\boldsymbol{\sigma},0),\ \rho[\mathbf{v}(\boldsymbol{\sigma})+\mathbf{w}(\boldsymbol{\sigma},\rho)]\big)$$

is a solution pair of (3.1) *for* $\boldsymbol{\sigma}\in[-1,1]^{n-1}$ *and* $|\rho|\le\eta$. *Moreover, there is a subneighborhood* $\operatorname{cl}\mathcal{W}$ *of* $\operatorname{cl}\mathcal{Z}$ *containing* $(\operatorname{cl}\mathcal{A})\times\{\mathbf{0}\}$ *such that if* $(\boldsymbol{\lambda},\mathbf{u})$ *is a solution pair of* (3.1) *in* $\operatorname{cl}\mathcal{W}$, *then either* $\mathbf{u}=\mathbf{0}$ *or else there is a* $(\boldsymbol{\sigma},\rho)\in[-1,1]^{n-1}\times[-\eta,\eta]$ *such that* $(\boldsymbol{\lambda},\mathbf{u})$ *is given by* (4.17).

The local behavior ensured by Theorem 4.14 is illustrated in Fig. 4.8. The conditions of the following useful corollary are often met.

4.18. Corollary. *Let the hypotheses of Theorem 4.14 hold, and furthermore let* $\tilde{\boldsymbol{\lambda}}\in C^r(\operatorname{cl}\mathcal{A})$, $\mathbf{L}(\cdot)\in C^r(\operatorname{cl}\mathcal{A})$, $\mathbf{g}\in C^r(\operatorname{cl}\mathcal{Z})$, *where* r *is a positive integer or* ∞. *Then* κ *and* $\mathbf{w}$ are in $C^r\big([-1,1]^{n-1}\times[-\eta,\eta]\big)$. *Thus, the closure of the connected set of nontrivial solution pairs of* (3.1) *bifurcating from* $(\boldsymbol{\lambda}^0,\mathbf{0})$ *forms a* C^r*-surface of dimension* n *near* $(\boldsymbol{\lambda}^0,\mathbf{0})$.

The proofs of Lemma 4.12 and Theorem 4.14 are given by Alexander & Antman (1981). Corollary 4.18 is proved like Theorem XVIII.1.16. These results are generalizations to n-parameter problems of Theorem 1.19 of Rabinowitz (1973a), which is a refinement of the results of Crandall & Rabinowitz (1971). Their work generalizes earlier implicit function theorems by others going back to the work of Lyapunov and of Schmidt. For details, see Chow & Hale (1982), Ize (1976), and Vaĭnberg & Trenogin (1969). The basic mathematical tools for proving these results are described in Chap. XVIII.

Our next few theorems are central to the bifurcation problems treated in the following chapters. The traditional versions of the first theorem employ the notion of odd algebraic multiplicity of an eigenvalue. Rather than extending our definition (4.10) to that for an eigenvalue of arbitrary multiplicity (see Fitzpatrick & Pejsachowicz (1991)), we can use the natural generalization of the observation: If $\mathbf{A}(\lambda)$ is a square matrix depending continuously on λ and if λ^0 is an eigenvalue of $\mathbf{A}(\cdot)$ of odd algebraic multiplicity, then $\det\mathbf{A}(\cdot)$ changes sign at λ^0. By virtue of the elementary properties of the Brouwer degree, sketched in Sec. III.3, especially (3.22), and derived in Chap. XIX, we see that $\det\mathbf{A}(\cdot)$ changes sign if and only if the Brouwer index of $\mathbf{A}(\lambda)$ (at the zero solution) changes as λ crosses λ^0. The Brouwer degree for operators on finite-dimensional spaces can be extended to the Leray-Schauder degree for operators on infinite-dimensional spaces having the form $\mathbf{I}-\mathbf{h}(\cdot)$ where $\mathbf{h}$ is compact (defined below). Thus in place of the hypothesis that an eigenvalue has odd algebraic multiplicity, we use the hypothesis that the Leray-Schauder index of $\mathbf{I}-\mathbf{L}(\cdot)$ (at the zero solution) changes at an eigenvalue. Of course, to verify that the index changes, it might be necessary to compute the algebraic multiplicity of the eigenvalue. We shall not need to do so in any of the concrete problems we treat.

The two technical concepts used here are those of a compact operator and of connectivity. Formally, a *compact operator* from a metric space $\mathcal{X}$ to a metric space $\mathcal{Y}$ is one that takes every bounded sequence in $\mathcal{X}$ into a sequence in $\mathcal{Y}$ having a convergent subsequence. The convergence of the subsequence is the device that allows the construction of solutions of equations as subsequential limits. There are simple tests ensuring that a given operator is compact, which we describe and use below. The reader unfamiliar with this concept may be content for the time being in knowing that nice integral operators, such as those of (1.5), are compact. (A set $\mathcal{E}$ in metric space $\mathcal{Y}$ is *precompact* iff any sequence from $\mathcal{E}$ has a convergent subsequence. Thus, a compact operator from $\mathcal{X}$ to $\mathcal{Y}$ is equivalently defined as one that takes every bounded set into a precompact set.)

Let us now define connectivity. Let $\mathcal{X}$ be a metric space (for simplicity of exposition). A separation of $\mathcal{X}$ is a pair $\mathcal{U}, \mathcal{V}$ of its nonempty open subsets with $\mathcal{X} = \mathcal{U} \cup \mathcal{V}$ and $\mathcal{U} \cap \mathcal{V} = \emptyset$. $\mathcal{X}$ is *connected* iff it does not admit a separation. Two sets $\mathcal{A}$ and $\mathcal{B}$ in $\mathcal{X}$ are *connected to each other in* $\mathcal{X}$ iff there exists a connected set $\mathcal{Y}$ in $\mathcal{X}$ with $\mathcal{A} \cap \mathcal{Y} \neq \emptyset$, $\mathcal{B} \cap \mathcal{Y} \neq \emptyset$. Two nonempty sets $\mathcal{A}$ and $\mathcal{B}$ in $\mathcal{X}$ are *separated from each other in* $\mathcal{X}$ iff there exists a separation $\mathcal{U}, \mathcal{V}$ of $\mathcal{X}$ with $\mathcal{A} \subset \mathcal{U}$, $\mathcal{B} \subset \mathcal{V}$. (If $\mathcal{A}$ and $\mathcal{B}$ are separated from each other in $\mathcal{X}$, they are not connected in $\mathcal{X}$, but if they are not connected, they need not be separated.) The set of points in $\mathcal{X}$ connected to a given point $\mathbf{x}$ is called the (*connected*) *component of* $\mathbf{x}$. For a useful and accessible discussion of connectivity and its applications, see Alexander (1981).

4.19. One-Parameter Global Bifurcation Theorem. *Let* $\mathbf{f}$ *have the form* (4.5) *and satisfy the accompanying restrictions with* $\boldsymbol{\lambda}$ *replaced by a scalar* λ *and with* $\mathcal{D}_0$ *an interval. Let* $(\lambda, \mathbf{u}) \mapsto \mathbf{L}(\lambda)\mathbf{u}$, $\mathbf{g}[\lambda, \mathbf{u}]$ *be compact and continuous. Let* $\alpha, \beta \in \mathcal{D}_0 \setminus \mathcal{E}$ *and let the Leray-Schauder indices of* $\mathbf{I} - \mathbf{L}(\alpha)$ *and* $\mathbf{I} - \mathbf{L}(\beta)$ *at* $\mathbf{0}$ *be different. Then between* α *and* β *there is an eigenvalue* λ^0 *of* (4.6) *with the property that* $(\lambda^0, \mathbf{0})$ *belongs to a maximal connected subset* $\mathcal{S}(\lambda^0)$ *of the closure* $\mathcal{S}$ *of the set of nontrivial solution pairs. Moreover,* $\mathcal{S}(\lambda^0)$ *satisfies at least one of the following two alternatives:*

(i) $\mathcal{S}(\lambda^0)$ *does not lie in a closed and bounded subset of* $\mathcal{D}$. (*In particular, if* $\mathcal{D} = \mathbb{R} \times \mathcal{X}$, *then* $\mathcal{S}(\lambda^0)$ *is unbounded in* $\mathbb{R} \times \mathcal{X}$.)

(ii) *There is another eigenvalue* μ *of* $\lambda \mapsto \mathbf{I} - \mathbf{L}(\lambda)$ *such that* $\mathcal{S}(\lambda^0)$ *contains* $(\mu, \mathbf{0})$. *If, moreover, the eigenvalues of* $\lambda \mapsto \mathbf{I} - \mathbf{L}(\lambda)$ *are isolated* (*as would be the case if* $\mathbf{L}(\lambda)$ *has the form* $\lambda\mathbf{A}$), *then there is an integer* $k \geq 2$ *such that* $\operatorname{cl} \mathcal{S}(\lambda^0)$ *contains exactly* k *points of the form* $(\lambda_j^0, \mathbf{0})$, $j = 1, \dots, k$, *where* λ_j^0 *is an eigenvalue of* (4.6) *and* λ^0 *is counted as a member of* $\{\lambda_j^0\}$. *Moreover, the number of such points at which* $\mathbf{I} - \mathbf{L}(\cdot)$ *changes its Leray-Schauder index is even.*

If, furthermore, $\mathbf{f}$ *is real-analytic, then* $\mathcal{S}(\lambda^0)$ *is a locally finite union of analytic arcs and is arcwise-connected. It contains a curve* $\mathcal{K}(\mu)$ *that satisfies statements* (i) *and* (ii).

Clearly, if λ^0 is an eigenvalue of (4.6) at which the Leray-Schauder index of $\mathbf{I} - \mathbf{L}(\cdot)$ (at the zero solution) changes, then $\mathcal{S}(\lambda^0)$ has the properties indicated in this theorem. In Sec. XIX.4 we give the proof of those parts of this theorem we shall use. The proof has the character of the proof of Theorem III.3.30.

The essential form of this theorem was proved by Rabinowitz (1971a). It represents a major generalization of both the local theory of Krasnosel'skiĭ (1956) and the global results of Crandall & Rabinowitz (1970) for nonlinear Sturm-Liouville problems. Expositions of Rabinowitz's results together with numerous mathematical applications are given by Rabinowitz (1971b, 1973a, 1975), Nirenberg (1974), and Ize (1976). For a related development, see Turner (1970). For a historical analysis of the genesis of this theory see Antman (1983b). The results on real-analytic problems are due to Dancer (1973). Other refinements are due to Dancer (1974a), Ize (1976), and Magnus (1976). Further generalizations in which the hypotheses on the compactness of L and g are replaced by weaker assumptions have been obtained by Alexander & Fitzpatrick (1979, 1980, 1981), Lev (1978), Stuart (1973a), and Toland (1976, 1977). For generalizations in which our hypotheses on the form (4.5) are relaxed, see Fitzpatrick & Pejsachowicz (1993) and Fitzpatrick, Rabier, & Pejsachowicz (1994).

There are many important singular bifurcation problems that cannot be handled by Theorem 4.19. Some of these can be handled by theorems developed in the references of the last paragraph. Alternatively, we might have recourse to the following procedure. Suppose that a one-parameter bifurcation problem (3.1) fails to meet the hypotheses of Theorem 4.19. Suppose that we can replace (3.1) by a sequence of approximate problems

$$\mathsf{f}^m[\lambda, \mathsf{u}] = \mathbf{0}, m = 1, 2, \dots, \tag{4.20}$$

each of which does satisfy the hypotheses of Theorem 4.19. Then for each m, we obtain a family of connected solution branches. Do the individual branches converge to connected sets and are these limiting sets branches of solution pairs of (3.1)?

For example, suppose that (3.1) corresponds to an ordinary differential equation on $(0,1)$ containing a term $1/s$ whose presence destroys the compactness of an appropriate integral operator. (Here s denotes the independent variable lying in $(0,1)$, say.) Then we could construct a sequence of approximating problems by replacing $1/s$ by $1/\sqrt{s^2 + m^{-2}}$. Or suppose that (3.1) corresponds to an ordinary differential equation on the unbounded interval $(0,\infty)$. Then we could replace the original problem with a sequence of ordinary differential equations on $(0,m)$ subject to suitable boundary conditions at $s = m$. Analogous procedures are used to treat singular problems for linear ordinary differential equations (see Coddington & Levinson (1955, Chap. 9), e.g.).

Our basic result on approximation is the following theorem. For bifurcation problems the set $\mathcal{A}$ may be thought of as part of the trivial branch and $\mathcal{B}$ may be thought of as a neighborhood of ∞.

4.21. Theorem. *Let $\mathcal{X}$ and $\mathcal{Y}$ be Banach spaces and let $\mathcal{D} \subset \mathcal{X}$. Let $\mathsf{f} : \mathcal{D} \to \mathcal{Y}$, $\mathsf{f}^m : \mathcal{D} \to \mathcal{Y}$, $m = 1, 2, \dots$. Let*

$$\mathcal{S} \equiv \{\mathsf{x} : \mathsf{f}[\mathsf{x}] = \mathbf{0}\}, \quad \mathcal{S}^m \equiv \{\mathsf{x} : \mathsf{f}^m[\mathsf{x}] = \mathbf{0}\}$$

denote the zero sets of these mappings. Let $\mathcal{A}$ and $\mathcal{B}$ be two closed subsets of $\mathcal{D}$ for which $\mathcal{A} \cap \mathcal{S}^m$ and $\mathcal{B} \cap \mathcal{S}^m$ are (not empty and are) not separated in $\mathcal{S}^m$. Suppose that

(i) *$\mathcal{S}$, $\mathcal{S}^m$, $m = 1, 2, \dots$, are closed in $\mathcal{D}$.*
(ii) *If $\mathsf{x}^m \in \mathcal{S}^m$, $m = 1, 2, \dots$, and $\{\mathsf{x}^m\}$ converges, then $\lim \mathsf{x}^m \in \mathcal{S}$.*
(iii) *If $\mathsf{x}^m \in \mathcal{S}^m$, $m = 1, 2, \dots$, then $\{\mathsf{x}^m\}$ has a convergent subsequence.*

Then $\mathcal{A} \cap \mathcal{S}$ and $\mathcal{B} \cap \mathcal{S}$ are not separated in $\mathcal{S}$. Moreover, if each $\mathcal{S}^m$ is compact, then so is $\mathcal{S}^\infty \equiv \lim \mathcal{S}^m \subset \mathcal{S}$.

For a proof and applications of this theorem, see Alexander (1981). We describe applications to elasticity in Secs. VI.3 and VI.4.

To motivate the statement of the Multiparameter Global Bifurcation Theorem 4.23 below and to furnish a caricature of the ideas underlying its proof (which we do not provide), we describe an intuitive topological way of unifying the statements of the One-Parameter Global Bifurcation Theorem 4.19. Suppose for simplicity that the domain $\mathcal{D}$ of f of (4.5) is the whole of $\mathbb{R} \times \mathcal{X}$. In Fig. 4.22a we illustrate on the horizontal plane a typical bifurcation diagram for a one-parameter problem. We place a sphere on top of this plane with the south pole of the sphere lying on the origin of the plane. We then project the bifurcation diagram stereographically onto the sphere. The projection of the trivial branch is a great circle joining the south pole to the north pole. Every maximal bifurcating branch intersects this trivial great circle at at least two distinct points. Unbounded branches intersect it at the north pole. These observations unify the alternatives of Theorem 4.19.

In more technical language, in projecting $\mathcal{D}$ stereographically onto the unit sphere we are adjoining to the metric space $\mathcal{D}$ a point ∞ (which we identify with the north pole of the sphere), thereby producing a topological space $\mathcal{D}^+ \equiv \mathcal{D} \cup \{\infty\}$ in which a neighborhood basis of ∞ consists of complements of bounded sets. Below we apply this process to subsets of $\mathcal{D}$. There are technical issues in giving a full justification of Fig. 4.22, which we avoid. In Secs. VI.3 and VI.4 this construction is combined with Theorem 4.21 to handle some tricky problems of connectivity.

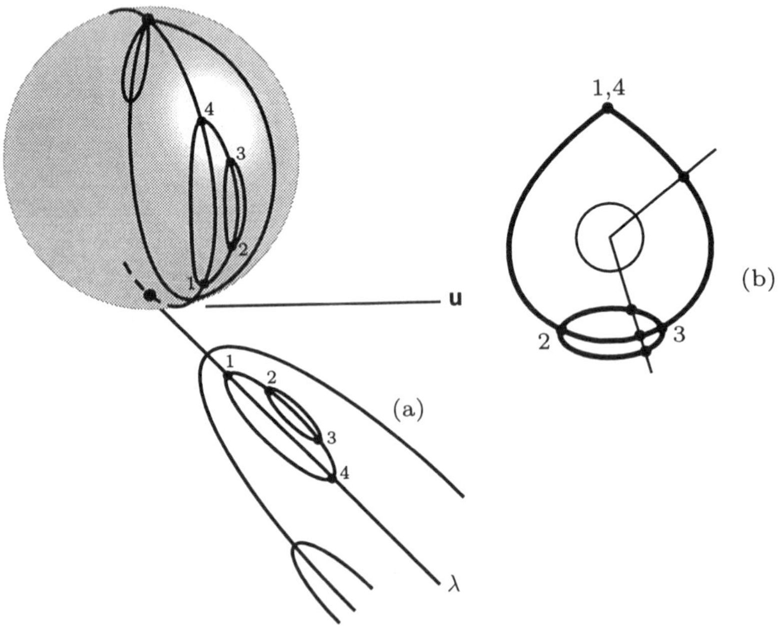

Fig. 4.22. Stereographic projection of the bifurcation diagram on the sphere and subsequent transformations.

Let us now continue our interpretation of Theorem 4.19 We take the closure of the image of a nontrivial branch, such as 1234 or 567∞, on the sphere and 'identify' its termini, where it meets the trivial great circle. We thereby obtain the closed curve-like set $\mathcal{G}$, the image of our branch, shown in Fig. 4.22b. We now intuitively develop a mathematically precise way to say that $\mathcal{G}$ has no gaps: We flatten out $\mathcal{G}$ so that it lies in a plane. Inside of it we place a circle, i.e., a one-dimensional sphere. We radially project the flattened $\mathcal{G}$ onto this circle. Since $\mathcal{G}$ has no gaps in it, we cannot continuously deform the projection into another mapping that takes $\mathcal{G}$ onto a single point on the circle.

How can this process be distinguished from one applied to a Figure $\mathcal{H}$ with gaps? Let us think of the flattened $\mathcal{G}$ as a circle of radius 2 and the flattened $\mathcal{H}$ as the arc of a spiral consisting of all points in the plane with Cartesian coordinates of the form $\big((2+\theta)\cos\theta, (2+\theta)\sin\theta\big)$, $0 \le \theta \le 3\pi$. The radial projection of this arc onto the unit circle takes each such point parametrized by θ to $(\cos\theta, \sin\theta)$. But since this arc has a gap, we can continuously deform this mapping through a family of continuous mappings into a constant mapping that takes the arc onto a single point on the circle of radius 1: We simply embed the projection into the one-parameter family of mappings that take the point parametrized by θ to $(\cos\alpha\theta, \sin\alpha\theta)$, where $0 \le \alpha \le 1$. For $\alpha = 1$, we have the projection, and for $\alpha = 0$ we have the constant mapping. By elementary topological arguments (based on degree theory, e.g.), it can be shown that the radial projection of a circle of radius 2 onto the concentric circle of radius 1 cannot be continuously deformed through a family of continuous mappings into a constant mapping. A continuous mapping that cannot be continuously deformed through a family of continuous mappings into a constant mapping is said to be *essential*. That $\mathcal{G}$ has no gaps is expressed by saying that there is an essential mapping of $\mathcal{G}$ onto the unit circle. In the next theorem we replace the unit circle on which $\mathcal{G}$ is projected with a unit sphere having the dimension equal to the number of parameters.

4.23. Multiparameter Global Bifurcation Theorem. *Let* f *have the form* (4.5) *and satisfy the accompanying restrictions.*

(i) *Let* $\mathcal{D} = \mathbb{R} \times \mathcal{X}$. *Let* $\mathcal{D} \ni (\boldsymbol{\lambda}, \mathsf{u}) \mapsto \mathsf{L}(\boldsymbol{\lambda})\mathsf{u}$, $\mathsf{g}[\boldsymbol{\lambda}, \mathsf{u}]$ *be compact and continuous. Let* $\boldsymbol{\alpha}, \boldsymbol{\beta} \in \mathcal{D}_0 \setminus \mathcal{E}$ *and let the Leray-Schauder indices of* $\mathsf{I} - \mathsf{L}(\boldsymbol{\alpha})$ *and* $\mathsf{I} - \mathsf{L}(\boldsymbol{\beta})$ *be different. Then each continuously differentiable arc* $\mathcal{C}$ *in* $\mathcal{D}_0$ *that contains* $\boldsymbol{\alpha}$ *and* $\boldsymbol{\beta}$ *contains an eigenvalue* $\boldsymbol{\lambda}^0$ *of* (4.6) *with the property that* $(\boldsymbol{\lambda}^0, \mathbf{0})$ *belongs to a maximal connected subset* $\mathcal{S}(\boldsymbol{\lambda}^0)$ *of the closure* $\mathcal{S}$ *of the set of nontrivial solution pairs of* (3.1). $\mathcal{S}(\boldsymbol{\lambda}^0)$ *contains a connected subset* $\mathcal{T}(\boldsymbol{\lambda}^0)$, *with* $\mathcal{T}(\boldsymbol{\lambda}^0) \cap \mathcal{F} = \mathcal{S}(\boldsymbol{\lambda}^0) \cap \mathcal{F}$, *that has Lebesgue dimension at least* n *at each of its points. Indeed, if* $\mathcal{T}(\boldsymbol{\lambda}^0)^+/\mathcal{F}^+$ *represents* $\mathcal{T}(\boldsymbol{\lambda}^0)^+$ *with its points on* $\mathcal{F}^+$ *identified, then there is an essential map that takes* $\mathcal{T}(\boldsymbol{\lambda}^0)^+/\mathcal{F}^+$ *onto the* n*-dimensional sphere.*

(ii) *Let* $\mathcal{D} = \cup\{\mathcal{D}(\gamma) \colon \gamma \in (0,1)\}$ *where the* $\mathcal{D}(\gamma)$ *are open sets with the property that* $\operatorname{cl}\mathcal{D}(\gamma_1) \subset \mathcal{D}(\gamma_2)$ *when* $0 < \gamma_1 < \gamma_2 < 1$. *Let* $\mathcal{D} \ni (\boldsymbol{\lambda}, \mathsf{u}) \mapsto \mathsf{L}(\boldsymbol{\lambda})\mathsf{u}$ *be compact and continuous. Let* $\mathcal{D} \ni (\boldsymbol{\lambda}, \mathsf{u}) \mapsto \mathsf{g}[\boldsymbol{\lambda}, \mathsf{u}]$ *be continuous and let* $\mathcal{D}(\gamma) \ni (\boldsymbol{\lambda}, \mathsf{u}) \mapsto \mathsf{g}[\boldsymbol{\lambda}, \mathsf{u}]$ *be compact for each* γ. *Let* $\boldsymbol{\alpha}, \boldsymbol{\beta} \in \mathcal{D}(\gamma)_0 \setminus \mathcal{E}$ *for each* γ *and let the Leray-Schauder indices of* $\mathsf{I} - \mathsf{L}(\boldsymbol{\alpha})$ *and* $\mathsf{I} - \mathsf{L}(\boldsymbol{\beta})$ *be different. Then each continuously differentiable arc* $\mathcal{C}$ *in* $\mathcal{D}(\gamma)_0$ *that contains* $\boldsymbol{\alpha}$ *and* $\boldsymbol{\beta}$ *contains an eigenvalue* $\boldsymbol{\lambda}^0$ *of* (4.6) *with the property that* $(\boldsymbol{\lambda}^0, \mathbf{0})$ *belongs to a maximal connected subset* $\mathcal{S}(\boldsymbol{\lambda}^0)$ *of the closure* $\mathcal{S}$ *of the set of nontrivial solution pairs. Then at least one of the following statements is true:*

(a) $\mathcal{S}(\boldsymbol{\lambda}^0)$ *is bounded and there is an* $\gamma^* \in (0,1)$ *such that* $\mathcal{S}(\boldsymbol{\lambda}^0) \subset \mathcal{D}(\gamma^*)$. $\mathcal{S}(\boldsymbol{\lambda}^0)$ *contains a connected subset* $\mathcal{T}(\boldsymbol{\lambda}^0)$ *with the properties stated in* (i).

(b) $\mathcal{S}(\boldsymbol{\lambda}^0) \setminus \mathcal{D}(\gamma) \ne \emptyset$ *for all* $\gamma \in (0,1)$ *or* $\mathcal{S}(\boldsymbol{\lambda}^0)$ *is unbounded. For each* $\gamma \in (0,1)$ *there is a modified equation* $\mathsf{u} = \varphi(\boldsymbol{\lambda}, \mathsf{u}, \gamma)\{\mathsf{L}(\boldsymbol{\lambda}) \cdot \mathsf{u} + \mathsf{g}[\boldsymbol{\lambda}, \mathsf{u}]\}$ *defined on all of* $\mathbb{R}^n \times \mathcal{X}$ *that agrees with* (3.1) *on* $\mathcal{D}(\gamma)$. *The function* φ *can be defined to satisfy*

$$\varphi(\boldsymbol{\lambda}, \mathsf{u}, \gamma) = \begin{cases} 1 & \text{for } (\boldsymbol{\lambda}, \mathsf{u}) \in \mathcal{D}(\gamma), \\ 0 & \text{for } (\boldsymbol{\lambda}, \mathsf{u}) \in \mathbb{R}^n \times \mathcal{X} \setminus \mathcal{D}(\tfrac{1}{2}(\gamma+1)). \end{cases}$$

Then the maximal connected subset $\mathcal{S}(\boldsymbol{\lambda}^0, \gamma)$ *of the closure of the set of nontrivial solution pairs of the modified equation that bifurcate from* $(\boldsymbol{\lambda}^0, \mathbf{0})$ *has the same properties as* $\mathcal{S}(\boldsymbol{\lambda}^0)$ *given in statement* (i).

(iii) *If, furthermore,* $\boldsymbol{\lambda}^0$ *is an eigenvalue of* (4.6), *if* $\mathcal{E}$ *near* $\boldsymbol{\lambda}^0$ *is an isolated* C^1*-surface of dimension* $n-1$, *and if the restriction of* $\mathsf{I} - \mathsf{L}(\cdot)$ *to a smooth curve transversal to* $\mathcal{E}$ *through* $\boldsymbol{\lambda}^0$ *changes its Leray-Schauder index at* $\boldsymbol{\lambda}^0$ (*in particular, if* $\boldsymbol{\lambda}^0$ *is an eigenvalue of this restriction of odd algebraic multiplicity*), *then there is a bifurcating set* $\mathcal{S}(\boldsymbol{\lambda}^0)$ *with all the properties described above.*

(iv) *The solutions of a one-parameter bifurcation problem corresponding to the restriction of* f *to* $\mathcal{C} \times \mathcal{X}$ *where* $\mathcal{C}$ *is a curve in* $\mathcal{D}_0$ *have the properties described in Theorem 4.19.*

Statement (i) of this theorem was proved by Alexander & Antman (1981) by using Čech cohomology. Statement (ii) incorporates refinements introduced by Lanza & Antman (1991, Thm. 5.1). The notion of Lebesgue (topological) dimension is explained by Hurewicz & Wallman (1948). This theorem is related to the Global Implicit Function Theorem of Alexander & Yorke (1976). Related multiparameter global bifurcation and continuation theorems are given by Fitzpatrick, Massabò, & Pejsachowicz (1983) and by Ize, Massabò, Pejsachowicz, & Vignoli (1985). Alexander & Antman (1983) treat bifurcation problems with infinite-dimensional parameters, which arise, e.g., when the equation itself is regarded as the parameter. (In elasticity, this means that the constitutive function or the shape of the body can be taken as the parameter.)

5. Applications of the Basic Theorems to the Simplest Buckling Problem

We now show that by supplementing the abstract theorems of Sec. 4 with some elementary methods we can obtain very detailed global information about the nontrivial bifurcating branches of (1.5) or (1.6), comparable to that given by Ex. 1.16, without using the explicit construction of this exercise (which is not available when EI is not constant).

We take u to be the pair (θ, M) and take the Banach space $\mathcal{X}$ to be $C^0[0,1] \times C^0[0,1]$. We put (1.5) into the mold (1.10) with f having the form (4.5) by simply decomposing the integrand $\sin\theta(\xi)$ in (1.5b) as $\theta(\xi) + [\sin\theta(\xi) - \theta(\xi)]$. Then we identify

$$\mathsf{g}[\lambda, \mathsf{u}](s) = \left(0,\ \lambda \int_s^1 [\sin\theta(\xi) - \theta(\xi)]\, d\xi\right). \tag{5.1}$$

$\mathsf{L}(\lambda)$, which is affine in λ, is defined similarly.

To use Theorem 4.19, we must establish that L and g are compact. The verification of the remaining hypotheses is immediate. In accord with the definition of a compact operator given in Sec. 4, to show that g is compact from $\mathbb{R} \times \mathcal{X}$ to $\mathcal{X}$, we must prove that if there is a positive number M and a sequence $\{\lambda_k, \theta_k, M_k\}$ such that

$$|\lambda_k| \le M, \quad \max_s |\theta_k(s)| \le M, \quad \max_s |M_k(s)| \le M, \tag{5.2}$$

then $\mathsf{g}[\lambda_k, \theta_k, M_k]$ has a subsequence that is uniformly convergent. For this purpose, we can easily apply the

5.3. Arzelà-Ascoli Theorem. *Let* $\{g_k\}$ *be a bounded, equicontinuous sequence of functions defined on a compact subset* $\mathcal{K}$ *of* $\mathbb{R}^n$, *i.e., let*

there be a positive number N independent of k, and for every $\varepsilon > 0$ let there be a $\delta(\varepsilon) > 0$ independent of k for which

$$\max\{|g_k(\mathbf{x})| : \mathbf{x} \in \mathcal{K}\} \le N, \\ |g_k(\mathbf{x}) - g_k(\mathbf{y})| < \varepsilon \quad \textit{if} \ |\mathbf{x} - \mathbf{y}| < \delta(\varepsilon), \quad \mathbf{x}, \mathbf{y} \in \mathcal{K} \tag{5.4}$$

for all k. Then $\{g_k\}$ has a uniformly convergent subsequence.

The proof of this theorem is given in a variety of books on advanced calculus, differential equations, and real analysis.

We now show that $\mathbf{g}$ is compact. We identify g_k with the second component of $\mathbf{g}[\lambda_k, \theta_k, M_k]$. A trivial computation shows that if (5.2) holds, then $|g_k| \le M(M+1) \equiv N$. Next we observe that if $0 \le x \le y \le 1$, then

$$|g_k(x) - g_k(y)| \le M \int_x^y |\sin\theta(\xi) - \theta(\xi)|\, d\xi \le N|x - y|. \tag{5.5}$$

Thus $\{g_k\}$ is equicontinuous, with $\delta(\varepsilon) = \varepsilon/N$. The compactness follows from Theorem 5.3. The proof of the compactness of $(\lambda, \mathbf{u}) \mapsto \mathbf{L}(\lambda) \cdot \mathbf{u}$ is proved the same way. Note that the same proof applies if $(EI)^{-1}$ is merely integrable and bounded above. In this case, however, (1.6) is not satisfied in the classical sense, although (1.5) is.

5.6. Exercise. Suppose EI is continuous and positive on $[0, 1)$ and $(EI)(1) = 0$. Find the necessary and sufficient condition on EI for $\mathbf{L}(\lambda)$ to be compact from $\mathcal{X}$ to itself. Suppose that each section of the rod is rectangular and that the width is constant. Let EI be the product of a constant E and a variable I defined according to (VIII.4.7). If the depth of the rod near $s = 1$ is a multiple of $(1 - s)^\alpha$, find the critical α separating compact from noncompact operators.

5.7. Exercise. Suppose that the linear constitutive equation corresponding to (1.5a) is replaced by the nonlinear equation $\theta'(s) = \hat{\mu}(M(s), s)$ where $\hat{\mu}$ is continuously differentiable and where $\hat{\mu}(\cdot, s)$ is odd and has a derivative with a positive lower bound. Prove that the resulting modification of (1.5) satisfies the compactness restrictions of Theorem 4.19.

5.8. Exercise. If (1.4d) is replaced with $\theta(1) = 0$, find an appropriate set of integral equations like (1.5). To do this, replace (1.5b) with

$$M(s) = M(0) - \lambda \int_0^s \sin\theta(\xi)\, d\xi, \tag{5.9}$$

substitute this expression into (1.5a) evaluated at $s = 1$, set the resulting expression equal to 0 in consonance with the new boundary condition, and solve the resulting equation for the unknown constant $M(0)$ as a functional of (θ, M). Now substitute this representation for $M(0)$ into (5.9). Show that (1.5a) and (5.9) satisfy the compactness restrictions of Theorem 4.19. Repeat this exercise using the constitutive equation of Ex. 5.7.

5.10. Exercise. Suppose that the linear constitutive equation (IV.1.32) with $\mathring{\mu} = 0$ is replaced by a nonlinear equation $M(s) = \hat{M}(\theta'(s), s)$ with $\hat{M}$ continuously differentiable and with $M(\cdot, s)$ odd. We now entertain the possibility that $\hat{M}(\mu, s)$ need not approach $\pm\infty$ as $\mu \to \pm\infty$, although we require that $\hat{M}_\mu > 0$ everywhere. Then the inverse $\hat{\mu}(\cdot, s)$ of $\hat{M}(\cdot, s)$ is defined only on a bounded open interval and blows up at its end points. Consequently, the system of integral equations used in Exs. 5.7 and 5.8

need not be well defined. Instead, one can directly attack the second-order quasilinear problem, which is the analog of (1.6). Prove that the analog of the problem of Ex. 5.7 can be converted to the following system for (θ, θ'):

$$\theta(s) = \int_0^1 G(s,\xi) \frac{\hat{M}_s(\theta'(\xi),\xi) + \lambda \sin\theta(\xi)}{\hat{M}_\mu(\theta'(\xi),\xi)}\, d\xi, \tag{5.11}$$

$$\theta'(s) = \int_s^1 \frac{\hat{M}_s(\theta'(\xi),\xi) + \lambda \sin\theta(\xi)}{\hat{M}_\mu(\theta'(\xi),\xi)}\, d\xi \tag{5.12}$$

where $G(s,\xi) = \xi$ for $\xi \le s$ and $G(s,\xi) = s$ for $s \le \xi$; cf. (1.9). (Were $\hat{M}(\mu, 1)$ to vanish as in Ex. 5.6, then we would introduce EI explicitly as advocated in (VIII.10.3) and replace G with that given by (1.9).) Show that the right-hand sides of (5.11) and (5.12) define a continuous and compact operator taking any (θ, θ') in $C^0[0,1]^2$ into $C^0[0,1]^2$. (That $\hat{M}_\mu$ might fail to have a positive lower bound causes no trouble.) We can replace the system (5.11) and (5.12) with a single integral equation by the following device. We introduce a new variable u by

$$u = \theta'', \quad \theta(0) = 0 = \theta'(1). \tag{5.13}$$

The integration of this system gives

$$\theta(s) = \int_0^1 G(s,\xi) u(\xi)\, d\xi \equiv A[u](s), \quad \theta'(s) = \int_s^1 u(\xi)\, d\xi \equiv B[u](s). \tag{5.14}$$

Show that the substitution of these representations into the second-order quasilinear boundary-value problem converts it into the integral equation

$$u(s) = -\frac{\hat{M}_s(B[u](s), s) + \lambda \sin\big(A[u](s)\big)}{\hat{M}_\mu(B[u](s), s)}. \tag{5.15}$$

Note that in all the nonlinear integral equations we had previously obtained, a linear integral operator acts on a nonlinear function of the unknown, whereas in (5.15), nonlinear functions act on the linear integral operators A and B. For $u \in C^0[0,1]$, show that (5.15) is equivalent to the original quasilinear boundary-value problem. Show that the right-hand side of (5.15) defines a compact and continuous operator from $C^0[0,1]$ to itself. Discuss the modifications necessary to handle the analog of Ex. 5.8. (Formulations like (5.15) are particularly useful for axisymmetric problems for plates and shells, for which the full range of invertibility results of Secs. IV.1 and VIII.8 are not available. See Chap. X.)

The linearization (4.5) of (1.5) is equivalent to the Sturm-Liouville problem

$$\tfrac{d}{ds}[(EI)(s)\psi'(s)] + \lambda\psi(s) = 0, \quad \psi(0) = 0 = \psi'(1). \tag{5.16}$$

The Sturmian theory (see Coddington & Levinson (1955) or Ince (1926), e.g.) asserts that the eigenvalues $\lambda_1^0, \lambda_2^0, \dots$ of this problem are simple and positive, that $\lambda_k^0 \to \infty$ as $k \to \infty$, and that the eigenfunction ψ_k corresponding to λ_k has exactly k zeros on $[0,1]$, each simple. (For EI constant, these results follow from an elementary computation.)

Thus, the One-Parameter Global Bifurcation Theorem 4.19 implies that bifurcating from each point $(\lambda_k^0, \mathbf{0})$ on the trivial branch of (1.5) is the branch $\mathcal{S}(\lambda_k^0)$, which is a maximal connected subset of the closure $\mathcal{S}$ of nontrivial solution pairs. $\mathcal{S}(\lambda_k^0)$ is unbounded in $\mathbb{R} \times \mathcal{X}$ or joins $(\lambda_k^0, \mathbf{0})$ to another such bifurcation point. We wish to determine what θ looks like on the branches $\mathcal{S}(\lambda_k^0)$ and where $\mathcal{S}(\lambda_k^0)$ are located.

The symmetry of our problem implies that $\mathcal{S}(\lambda_k^0)$ can be decomposed into the union of two branches $\mathcal{S}^{\pm}(\lambda_k^0)$, which are reflections of each other. We may define $\mathcal{S}^{+}(\lambda_k^0)$ to consist of all elements of $\mathcal{S}(\lambda_k^0)$ for which $\theta'(0) \geq 0$, and $\mathcal{S}^{-}(\lambda_k^0)$ to consist of all elements of $\mathcal{S}(\lambda_k^0)$ for which $\theta'(0) \leq 0$.

In the ensuing development, we focus on the function θ. If (θ, M) in $C^0[0,1] \times C^0[0,1]$ satisfies (1.5), then θ in $C^1[0,1]$ satisfies (1.6). We accordingly study branches with θ continuously differentiable.

A very simple piece of information that greatly illuminates the nature of θ on a nontrivial branch is the number of its zeros. A number s_0 in $[0,1]$ is a *simple zero* or a *node* of a continuously differentiable function θ iff $\theta(s_0) = 0$, $\theta'(s_0) \neq 0$. For EI everywhere positive, the zeros of a nontrivial solution θ must be simple, for if not, θ would satisfy the initial-value problem consisting of (1.6a) and a set of initial conditions of the form $\theta(s_0) = 0 = \theta'(s_0)$. But this initial-value problem, which must have a unique solution (when EI is everywhere positive), clearly has the unique solution 0. (If EI should vanish as in Ex. 5.6, then the standard uniqueness theory for ordinary differential equations fails if θ has a double zero where EI vanishes. A more sophisticated analysis is then required. We shall treat examples of such analyses in the following chapters.) In Fig. 5.17 we show the graph of a function θ satisfying (1.6b,c) and having exactly three zeros, each of which is simple. In Fig. 5.18, we illustrate corresponding configurations $\boldsymbol{r}$ satisfying the specialization $\boldsymbol{r}' = \cos\theta \boldsymbol{i} + \sin\theta \boldsymbol{j}$ of (IV.1.3) to the elastica.

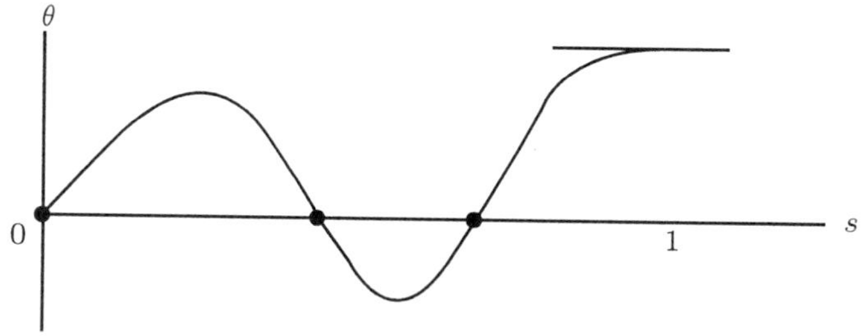

Fig. 5.17. Graph of a function θ satisfying (1.6b,c) and having exactly three zeros, each of which is simple. Knowing the number of its zeros and knowing something about the size of θ between its zeros, we have a very accurate qualitative picture of the behavior of θ.

If we knew that the closure of a nontrivial branch were a curve, then by the definition of a curve, θ would depend continuously on a parameter for the branch. Were θ to change the number of its zeros on such a branch, then, as Fig. 5.17 suggests, it would have to evolve through a state in which it would have a double zero. By the argument given above, such a state

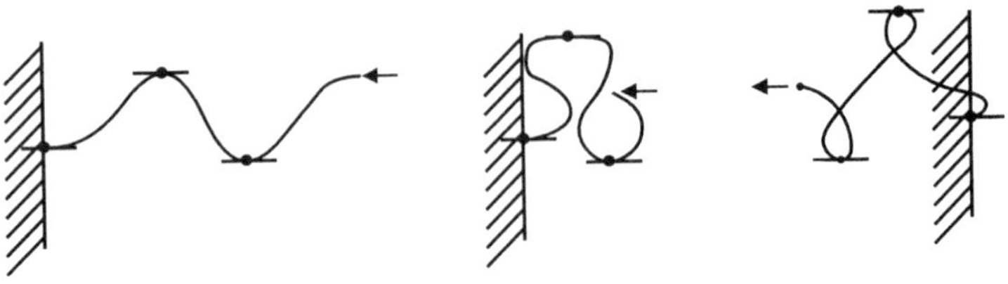

Fig. 5.18. Sketches of configurations of the elastica when θ has the behavior shown in Fig. 5.17. That in (a) corresponds to a θ with $\max_s |\theta(s)| < \pi/2$. That in (b) corresponds to a θ with the maxima of $|\theta|$ slightly exceeding $\pi/2$ on each of the three intervals bounded by the zeros of θ, and that in (c) corresponds to a θ with the maxima of $|\theta|$ slightly less than π.

is trivial. Thus along a curve of solution pairs, θ could change its nodal properties only at a trivial solution pair. Each curve of nontrivial solution pairs is therefore characterized by the number of zeros of θ. We now show that this same conclusion remains valid for any branch of solutions, which may merely be connected.

Let $\mathcal{Z}$ denote the set of all continuously differentiable functions on [0,1] that satisfy (1.6b,c). Let $\mathcal{Z}_j$ denote the subset of all functions in $\mathcal{Z}$ that have exactly j zeros, each of which is simple. $\mathcal{Z}_j$ is open in $\mathcal{Z}$ because each element of $\mathcal{Z}_j$ has a C^1-neighborhood of elements of $\mathcal{Z}$ that lies in $\mathcal{Z}_j$. (That is, if $\theta \in \mathcal{Z}_j$, then there is an $\varepsilon > 0$ (depending upon θ) such that if $\phi \in \mathcal{Z}$ and if $\max_s |\theta(s) - \phi(s)| + \max_s |\theta'(s) - \phi'(s)| \leq \varepsilon$, then $\phi \in \mathcal{Z}_j$. Note that $\mathcal{Z}_j$ is not open in $C^0[0,1]$.) The sets $\mathcal{Z}_j$ are clearly disjoint.

The complement $\mathcal{Z} \setminus \cup_{j=1}^{\infty} \mathcal{Z}_j$ of the union of the $\mathcal{Z}_j$ consists of all those functions in $\mathcal{Z}$ having a zero of multiplicity 2 or more. Since $\partial \mathcal{Z}_j$ is in this complement, an element of $\partial \mathcal{Z}_j$ must have a zero of multiplicity 2 or more. Now the last statement of Theorem 4.14 implies that if a nontrivial solution pair $(\lambda, (\theta, M))$ of (1.5) is in $\mathcal{S}(\lambda_k^0)$ and lies sufficiently close to $(\lambda_k^0, \mathbf{0})$, then $\theta \in \mathcal{Z}_k$. But this fact implies that if $(\lambda, (\theta, M))$ is any nontrivial solution pair $\mathcal{S}(\lambda_k^0)$, then $\theta \in \mathcal{Z}_k$: Suppose that there were a nonzero $(\tilde{\lambda}, (\tilde{\theta}, \tilde{M}))$ in $\mathcal{S}(\lambda_k^0)$ with $\tilde{\theta}$ not in $\mathcal{Z}_k$. We could represent the maximal connected set of nontrivial solution pairs containing $(\tilde{\lambda}, (\tilde{\theta}, \tilde{M}))$ as the union of those pairs with θ in $\mathcal{Z}_k$ and those pairs with θ not in $\mathcal{Z}_k$. Neither of these sets could be empty. Since $\mathcal{S}(\lambda_k^0)$ is connected (see the definition following (3.37)), the second set would have to contain a boundary point of the first set and therefore must contain a nontrivial solution pair with $\theta \in \partial \mathcal{Z}_k$. But this is impossible because this θ must have a double zero and therefore be the zero function. Thus, each $\mathcal{S}(\lambda_k^0)$ is globally characterized by the nodal property that each nontrivial θ has exactly k zeros, each simple. This property is inherited from the corresponding eigenfunction of linearized problem.

Moreover, $\mathcal{S}(\lambda_k^0)$ cannot contain a point $(\lambda_l^0, \mathbf{0})$, $l \neq k$, for, if so, our preceding argument would show that if $(\lambda, (\theta, M)) \in \mathcal{S}(\lambda_k^0)$, then $\theta \in \mathcal{Z}_k \cap \mathcal{Z}_l = \emptyset$. By the alternative of Theorem 4.19, each branch $\mathcal{S}(\lambda_k^0)$ is unbounded in $(\lambda, (\theta, M))$-space.

5.19. Exercise. Prove that if $(\lambda, (\theta, M)) \in \mathcal{S}(\lambda_k^0)$, then

$$\max_s |\theta(s)| < \pi. \tag{5.20}$$

Use this fact to show that each bifurcating branch is also characterized by the number of zeros of M.

We now obtain information about the disposition of the branches $\mathcal{S}(\lambda_k^0)$. Let us multiply (1.6a) by θ and use (1.6b,c) to integrate the resulting expression by parts. We obtain

$$\int_0^1 (EI)(s)[\theta'(s)]^2 \, ds = \lambda \int_0^1 \theta(s) \sin\theta(s) \, ds. \tag{5.21}$$

Inequality (5.20) implies that the integral on the right-hand side of (5.21) is positive for any nontrivial solution, so that a corresponding λ must be positive. It then follows that on a bifurcating branch the right-hand side of (5.21) is less than or equal to $\lambda \int_0^1 \theta(s)^2 \, ds$. The theory of Rayleigh quotients (see Courant & Hilbert (1953) or Stakgold (1979), e.g.) implies that if $\theta \neq 0$, then

$$\frac{\int_0^1 \theta'(s)^2 \, ds}{\int_0^1 \theta(s)^2 \, ds} \geq \frac{\pi^2}{4} \tag{5.22}$$

where $\pi^2/4$ is the smallest eigenvalue for $\psi'' + \lambda\psi = 0$ subject to (1.6b,c). Thus

$$\lambda \geq \frac{\pi^2}{4} \min_s (EI)(s) \tag{5.23}$$

on any bifurcating branch. By integrating (1.6a) from s to 1, we obtain $(EI)(s)\theta'(s) = \lambda \int_s^1 \sin\theta(s) \, ds$, which yields

$$\max_s |\theta'(s)| \leq \frac{\lambda}{\min_s (EI)(s)}. \tag{5.24}$$

Since branches $\mathcal{S}(\lambda_k^0)$ are unbounded, it follows that λ must be unbounded along each branch, for if not, θ and θ' would also have to be bounded in the C^0-norm, by (5.20) and (5.24), yielding a contradiction. Similar results can be obtained in the situations covered by Exs. 5.7 and 5.8. Sharper estimates can be derived by the use of the more delicate techniques discussed in Chap. VI.

Thus, the disposition of branches is at least qualitatively like that shown in Fig. 1.15. Since we know what every solution looks like on every bifurcating branch, we have obtained with little effort a qualitative description of all bifurcating solutions comparable in detail to that delivered by the exact solutions in terms of elliptic functions for problems with EI constant. The next exercise and the results of the next section show the effects caused by nonlinear material response.

5.25. Exercise. Derive inequalities like (5.24) under the conditions of Ex. 5.7 when the boundary conditions are $\theta(0) = 0 = \theta(1)$.

The boundary-value problem (1.6) possesses nonbifurcating branches corresponding to the noninflectional configurations like that illustrated in Fig. IV.2.15b. These configurations, which occur for negative values of λ, have θ's with values outside $[-\pi, \pi]$, so that the right-hand side of (5.21) is positive. These solutions can be found explicitly when $EI = B$ (const.). To show that there are such solutions when EI is not constant, we can consider the one-parameter family of problems with bending stiffness $\alpha B + (1-\alpha)EI(\cdot)$, $\alpha \in [0,1]$ and use a multiparameter continuation theory analogous to the Multiparameter Bifurcation Theorem 4.23. The approach is analogous to the use of degree theory in Sec. III.3.

This section is roughly based upon Crandall & Rabinowitz (1970) and Antman & Rosenfeld (1978). For generalizations of its analysis to extensible and shearable rods, which offer a rich collection of new phenomena, see Secs. VI.5–VI.10.

6. Perturbation Methods

The formal perturbation methods described in Sec. II.8 carry over to the bifurcation problems we are now studying. Here we describe some refinements applicable to one-parameter problems satisfying the hypotheses of Corollary 4.18, which fully justifies the perturbation approach. In particular, when its hypotheses hold, Corollary 4.18 implies that bifurcating solution pairs of (3.1) lie on branches of the form

$$\varepsilon \mapsto \lambda(\varepsilon) = \sum_{k=0}^{r-1} \frac{\lambda_{(k)}\varepsilon^k}{k!} + O(\varepsilon^r), \tag{6.1a}$$

$$\varepsilon \mapsto \mathbf{u}(\cdot,\varepsilon) = \sum_{k=1}^{r-1} \frac{\mathbf{u}_{(k)}(\cdot)\varepsilon^k}{k!} + O(\varepsilon^r) \tag{6.1b}$$

for sufficiently small ε. The parameter ε may be regarded as an amplitude of the solution $\mathbf{u}$ on the branch. We surround the indices on λ and $\mathbf{u}$ with parentheses to distinguish them from indices identifying the branches.

The equations for $\{\lambda_{(k-1)}, \mathbf{u}_{(k)}\}$ are obtained by substituting (6.1) into (3.1), differentiating the resulting equation k times with respect to ε, and then setting ε equal to 0. In this way we obtain

$$\text{(6.2)} \qquad \mathbf{f}[\lambda_{(0)}, \mathbf{0}] = \mathbf{0},$$

$$\text{(6.3)} \qquad (\partial \mathbf{f}/\partial \mathbf{u})^0 \cdot \mathbf{u}_{(1)} = \mathbf{0},$$

$$\begin{aligned} \text{(6.4)} \qquad (\partial \mathbf{f}/\partial \mathbf{u})^0 \cdot \mathbf{u}_{(2)} = &- (\partial^2 \mathbf{f}/\partial \mathbf{u} \partial \mathbf{u})^0 : \mathbf{u}_{(1)} \mathbf{u}_{(1)} \\ &- 2\lambda_{(1)} (\partial^2 \mathbf{f}/\partial \lambda \partial \mathbf{u})^0 : \mathbf{u}_{(1)}, \end{aligned}$$

$$\begin{aligned} \text{(6.5)} \qquad (\partial \mathbf{f}/\partial \mathbf{u})^0 \cdot \mathbf{u}_{(3)} = &- (\partial^3 \mathbf{f}/\partial \mathbf{u} \partial \mathbf{u} \partial \mathbf{u})^0 \mathrel{\therefore} \mathbf{u}_{(1)} \mathbf{u}_{(1)} \mathbf{u}_{(1)} \\ &- 3 (\partial^2 \mathbf{f}/\partial \mathbf{u} \partial \mathbf{u})^0 : \mathbf{u}_{(1)} \mathbf{u}_{(2)} \\ &- 3\lambda_{(1)} (\partial^3 \mathbf{f}/\partial \lambda \partial \mathbf{u} \partial \mathbf{u})^0 : \mathbf{u}_{(1)} \mathbf{u}_{(1)} \\ &- 3{\lambda_{(1)}}^2 (\partial^3 \mathbf{f}/\partial \lambda \partial \lambda \partial \mathbf{u})^0 : \mathbf{u}_{(1)} \\ &- 3\lambda_{(1)} (\partial^2 \mathbf{f}/\partial \lambda \partial \mathbf{u})^0 : \mathbf{u}_{(2)} \\ &- 3\lambda_{(2)} (\partial^2 \mathbf{f}/\partial \lambda \partial \mathbf{u})^0 : \mathbf{u}_{(1)}, \quad \text{etc.,} \end{aligned}$$

where $(\partial \mathbf{f}/\partial \mathbf{u})^0 \equiv (\partial \mathbf{f}/\partial \mathbf{u})[\lambda_{(0)}, \mathbf{0}]$, etc. The quadratic form $(\partial^2 \mathbf{f}/\partial \mathbf{u} \partial \mathbf{u})^0 : \mathbf{u}_{(1)} \mathbf{u}_{(1)}$ may be defined as $\frac{d}{d\varepsilon} (\partial \mathbf{f}/\partial \mathbf{u})[\lambda_{(0)}, \varepsilon \mathbf{u}_{(1)}] \cdot \mathbf{u}_{(1)} \big|_{\varepsilon=0}$, etc. Note that (6.2) is an identity.

If $\lambda_{(0)}$ is not an eigenvalue of (6.3), then its only solution is $\mathbf{0}$. It follows from (6.4), (6.5), and their successors that $\mathbf{u}_{(1)} = \mathbf{u}_{(2)} = \cdots = \mathbf{u}_{(r)} = \mathbf{0}$, in consonance with Theorem 4.1. We accordingly limit our attention to the case in which $\lambda_{(0)}$ is an eigenvalue of (6.3).

Each of the equations (6.4), (6.5), and their successors has the form of the inhomogeneous linear equation

$$\text{(6.6)} \qquad (\partial \mathbf{f}/\partial \mathbf{u})[\lambda_{(0)}, \mathbf{0}] \cdot \mathbf{v} = \mathbf{h}.$$

Since $\lambda_{(0)}$ is an eigenvalue, $(\partial \mathbf{f}/\partial \mathbf{u})[\lambda_{(0)}, \mathbf{0}]$ is singular, i.e., it fails to have an inverse. Consequently, (6.6) cannot be solved for all $\mathbf{h}$. The right-hand side of (6.4), however, contains the parameter $\lambda_{(0)}$ as well as a collection of parameters equal to the dimension n of the null space $\ker (\partial \mathbf{f}/\partial \mathbf{u})[\lambda_{(0)}, \mathbf{0}]$ in which $\mathbf{u}_{(1)}$ is a typical element. Thus we might be able to ensure the solvability of (6.4), etc., by adjusting these parameters appropriately. Lemma XVII.2.13, the simple version of the Alternative Theorem, implies that if (6.6) has a solution, then $\mathbf{h}$ annihilates the null space $\ker (\partial \mathbf{f}/\partial \mathbf{u}) \big[(\lambda_{(0)}, \mathbf{0}\big]^*$ of the adjoint:

$$\text{(6.7)} \qquad \langle \mathbf{h}, \mathbf{v}^* \rangle = 0 \quad \forall \mathbf{v}^* \in \ker (\partial \mathbf{f}/\partial \mathbf{u})[\lambda_{(0)}, \mathbf{0}]^*.$$

(This is just a generalized orthogonality condition.) Concrete applications of this principle are given in the remarks following (II.8.30) and below. If n^* is the dimension of $\ker (\partial \mathbf{f}/\partial \mathbf{u})[\lambda_{(0)}, \mathbf{0}]^*$, then (6.7) provides n^* equations that restrict the behavior of the parameters appearing in $\mathbf{h}$. In particular, if $\mathbf{I} - (\partial \mathbf{f}/\partial \mathbf{u})[\lambda_{(0)}, \mathbf{0}]$ (which is $\mathbf{L}(\lambda_{(0)})$ by (4.5)) is compact, then $n^* = n$ and (6.7) is also sufficient for (6.6) to have a solution (see Sec. XVII.2).

We now apply the perturbation method to (1.8), which is equivalent to (1.6). To handle (6.7) it is most convenient to seek solutions θ of (1.8) in the real Hilbert space $L_2(0,1)$, which has the inner product

$$\langle u, v\rangle \equiv \int_0^1 u(s)v(s)\,ds. \tag{6.8}$$

(It is easy to see that solutions of (1.8) in $L_2(0,1)$ are classical solutions of (1.6).) Equations (6.3)–(6.5) corresponding to (1.8) are

$$\theta_{(k)}(s) - \lambda_{(0)}\int_0^1 G(s,\xi)\theta_{(k)}(\xi)\,d\xi = h_{(k)}(s) \equiv \int_0^1 G(s,\xi)j_{(k)}(\xi)\,d\xi \tag{6.9}$$

where $j_{(1)} = 0$, $j_{(2)} = 2\lambda_{(1)}\theta_{(1)}$, $j_{(3)} = -\lambda_{(0)}{\theta_{(1)}}^3 + 3\lambda_{(1)}\theta_{(2)} + 3\lambda_{(2)}\theta_{(1)}$. These problems are equivalent to

$$\tfrac{d}{ds}[(EI)(s)\theta'_{(k)}(s)] + \lambda_{(0)}\theta_{(k)}(s) = -j_{(k)}(s), \tag{6.10a}$$

$$\theta_{(k)}(0) = 0 = \theta'_{(k)}(1). \tag{6.10b,c}$$

Problem (6.10) for $k=1$ has a nontrivial solution, which has the form

$$\theta_{(1)} = A_{(1)}\psi_n, \quad \text{if } \lambda_{(0)} = \lambda_n^0 \tag{6.11}$$

where $A_{(1)}$ is an arbitrary real constant and where we have used the notation introduced after (5.16). We restrict our attention to the $\lambda_{(0)}$ given by (6.11). In this case, we denote $h_{(k)}$, $j_{(k)}$, $\lambda_{(k)}$, and $\theta_{(k)}$ by $h_{(k)n}$, $j_{(k)n}$, $\lambda_{(k)n}$, and $\theta_{(k)n}$. Since the operator on the left-hand side of (6.9) is self-adjoint on $L_2(0,1)$, its null space equals the null space of its adjoint, which is span$\{\psi_n\}$. Thus (6.9) has a solution if and only if

$$\langle h_{(k)n}, \psi_n\rangle = 0. \tag{6.12}$$

6.13. Exercise. Prove that (6.12) is equivalent to

$$\langle j_{(k)n}, \psi_n\rangle = 0. \tag{6.14}$$

(This equation can be obtained by multiplying (6.10a) by ψ_n and integrating the resulting equation by parts subject to (6.10b,c). But this simple derivation of (6.14) yields only a necessary condition for solvability.)

A direct study of (1.6) in a suitable subspace of $L_2(0,1)$ introduces technical difficulties that are at bottom resolved by showing the equivalence of (1.6) to (1.8) (see Naimark (1969) or Stakgold (1979)). If (1.6) is to be studied directly, it is more convenient to do so in the Sobolev space $W_2^1(0,1)$. In this case, we would actually analyze the weak form of this problem, which corresponds to the Principle of Virtual Power. Condition (6.14) is still valid (see Nečas (1967)). If (1.8) were studied in other than a Hilbert space (as was done in Sec. 5), then a variety of technical obstacles must be overcome to construct the adjoint (see Taylor & Lay (1980)), but (6.14) still holds. We could study the equivalent problem (1.5) for $(\theta, M) \in L_2(0,1)$. Its linearization misses being self-adjoint, but the requisite solvability condition again reduces to (6.14).

To make our computations completely explicit, let us assume that $EI =$ const., in which case

$$\psi_n = \sqrt{2}\sin\tfrac{(2n-1)\pi}{2}s, \quad \lambda_n^0 = \tfrac{(2n-1)^2\pi^2}{4}EI. \tag{6.15}$$

We substitute (6.11) and (6.15) into (6.14) for $k = 2$ to obtain $\lambda_{(1)n}A_{(1)} = 0$. Since we seek nontrivial solutions, we take $A_{(1)} \neq 0$, whence

$$\lambda_{(1)n} = 0. \tag{6.16}$$

This condition significantly simplifies (6.10) for $k = 2$, which has a solution of the form

$$\theta_{(2)n}(s) = A_{(2)}\sin\tfrac{(2n-1)\pi}{2}s. \tag{6.17}$$

We now substitute (6.11) and (6.15)–(6.17) into (6.14) for $k = 3$ to obtain the most important consequence of our perturbation procedure:

$$\lambda_{(2)n} = \frac{\lambda_n^0\int_0^1 \theta_{(1)n}(s)^4\,ds}{3\int_0^1 \theta_{(1)n}(s)^2\,ds} = \tfrac{(2n-1)^2\pi^2}{16}EI{A_{(1)}}^2. \tag{6.18}$$

The solution of (6.10) for $k = 3$ is

$$\theta_{(3)n}(s) = A_{(3)}\sin\tfrac{(2n-1)\pi}{2}s + \frac{{A_{(1)}}^3}{32}\sin\tfrac{3(2n-1)\pi}{2}s. \tag{6.19}$$

Thus

$$\lambda(\varepsilon) = \tfrac{(2n-1)^2\pi^2}{4}EI + \varepsilon^2{A_{(1)}}^2\left[\tfrac{(2n-1)^2\pi^2}{32}\right]EI + O(\varepsilon^4), \tag{6.20a}$$

$$\begin{aligned}\theta(s,\varepsilon) = {} & \left[\varepsilon A_{(1)} + \frac{\varepsilon^2 A_{(2)}}{2} + \frac{\varepsilon^3 A_{(3)}}{6}\right]\sin\tfrac{(2n-1)\pi}{2}s \\ & + \frac{\varepsilon^3{A_{(1)}}^3}{192}\sin\tfrac{3(2n-1)\pi}{2}s + O(\varepsilon^4).\end{aligned} \tag{6.20b}$$

The error term $O(\varepsilon^4)$ rather than $O(\varepsilon^3)$ appears in (6.20a) because the oddness of (1.8) in θ implies that λ must be even in ε. The positivity of $\lambda_{(2)n}$ ensures that the bifurcation is supercritical, as shown in Fig. 1.15. The physical significance of supercriticality is discussed in Sec. 2. Note that the positivity of $\lambda_{(2)n}$ follows from (6.18) even when (6.15) does not hold.

We could simplify the form of (6.20b) by replacing $\varepsilon A_{(1)} + \frac{\varepsilon^2 A_{(2)}}{2} + \frac{\varepsilon^3 A_{(3)}}{6}$ by another small parameter η, the transformation from ε to η being invertible for small enough ε. A far more elegant way to effect this simplification is to define the amplitude ε as the projection of θ onto the normalized eigenfunction $\psi_n = \sqrt{2}\sin\frac{(2n-1)\pi}{2}s$:

$$\varepsilon \equiv \langle\theta(\cdot,\varepsilon), \psi_n\rangle. \tag{6.21}$$

If we differentiate this equation k times with respect to ε and then set $\varepsilon = 0$, we obtain

$$(6.22) \qquad \langle \theta_{(1)n}, \psi_n \rangle = 1, \quad \langle \theta_{(k)n}, \psi_n \rangle = 0 \quad \text{for } k = 2, 3, \dots.$$

The substitution of (6.17) and (6.19) into (6.22) yields $A_{(1)} = \sqrt{2}$, $A_{(2)} = 0 = A_{(3)}$, which produce the desired simplification of (6.20b).

The following exercise is strongly recommended.

6.23. Exercise. Suppose that the linear constitutive equation $M = EI\theta'$ is replaced by the nonlinear equation $M(s) = \hat{M}(\theta'(s))$ for a uniform rod. Suppose that $\hat{M}$ is an odd, strictly increasing, four-times continuously differentiable function. Use (6.21) to derive representations analogous to (6.20) for the modified version of problem (1.6) (cf. Exs. 5.7 and 5.9). Find conditions on $\hat{M}$ for $\lambda_{(2)}$ to be negative. That there are such conditions says that there is a nonlinear material response that can produce the dangerous subcritical bifurcation.

6.24. Exercise. Carry out formal perturbation expansions up to order 3 in ε of (3.14), of (3.23) with $\alpha = 0$, and of (3.25) with $\alpha = 1$.

The treatment of this section is roughly based on that of J. B. Keller (1968) (see Iooss & Joseph (1990)).

7. Dynamics and Stability

As this chapter has shown, nonlinearly elastic (and other kinds of) bodies may have several distinct equilibrium configurations under the same conditions. To determine whether a given equilibrium configuration is observed, we must analyze its stability. Since an equilibrium configuration is just a special motion that occurs for special initial conditions, we define the stability of a given motion.

Let a body have a motion $\mathbf{p}_0$ for a given system $\mathbf{f}_0$ of initial conditions, boundary conditions, applied loads, and other data. (For a material with memory, the initial data may include the entire history before the initial time.) $\mathbf{p}_0$ is a function of spatial variables and time. Let us denote the motion corresponding to the data $\mathbf{f}$ by $t \mapsto \mathbf{p}[\mathbf{f}](\cdot, t)$. (Here the dot occupies the slot for the spatial variables.) Thus $\mathbf{p}_0 \equiv \mathbf{p}[\mathbf{f}_0]$. Without loss of generality, we assume that initial data are prescribed at time 0.

It is of great practical importance to know whether $\mathbf{p}$ remains close to $\mathbf{p}_0$ when $\mathbf{f}$ remains close to $\mathbf{f}_0$. We can make this question mathematically precise by defining a distance $\pi[\mathbf{p}(\cdot, t), \mathbf{p}_0(\cdot, t)]$ between $\mathbf{p}(\cdot, t)$ and $\mathbf{p}_0(\cdot, t)$ and a distance $\phi[\mathbf{f}, \mathbf{f}_0]$ between $\mathbf{f}$ and $\mathbf{f}_0$. (It is by no means obvious which of many possible choices of these distances is both physically meaningful and mathematically convenient.) Given prescribed positive numbers ε and δ, we may pose the mathematically precise question of whether $\sup_{t \geq 0} \pi[\mathbf{p}(\cdot, t), \mathbf{p}_0(\cdot, t)] < \varepsilon$ if $\phi[\mathbf{f}, \mathbf{f}_0] < \delta$. It is mathematically easier to ask whether $\mathbf{p}_0$ is Lyapunov-stable:

7.1. Definition. The motion $[0, \infty) \ni t \mapsto \mathbf{p}_0(\cdot, t)$ is *stable (in the sense of Lyapunov)* iff for arbitrary $\varepsilon > 0$ there is a number $\delta > 0$ such that

$$\sup_{t \geq 0} \pi[\mathbf{p}(\cdot, t), \mathbf{p}_0(\cdot, t)] < \varepsilon \quad \text{if} \quad \phi[\mathbf{f}, \mathbf{f}_0] < \delta.$$

This motion is *asymptotically stable* (*in the sense of Lyapunov*) iff it is stable and if $\pi[\mathbf{p}(\cdot,t),\mathbf{p}_0(\cdot,t)] \to 0$ as $t \to \infty$.

The theory of Lyapunov stability for systems of ordinary differential equations (which describe discrete mechanical systems) is a highly developed subject. The study of the stability of solutions of partial differential equations of continuum mechanics is in a far more primitive state. *Elastic stability theory* primarily treats the study of the stability of equilibrium configurations of elastic bodies. In view of the technical difficulties presented by a rigorous analysis of equilibrium solutions of nonlinear evolution equations, elastic stability theory has traditionally employed simplified criteria of stability, logically distinct from Definition 7.1. The history of its relation to bifurcation theory has been a history of slow and painful mastery over entrenched misconception and imprecision. Indeed, the distinction between stability and equilibrium exercised the ancients:

> Anaximander thought that the earth floats freely, and is not supported on anything. Aristotle [*De Caelo*, 295^b], who often rejected the best hypotheses of his time, objected to the theory of Anaximander, that the earth, being at the centre, remained immovable because there was no reason for moving in one direction rather than another. If this were valid, he said, a man placed at the centre of a circle with food at various points of the circumference would starve to death for lack of reason to choose one portion of food rather than another. This argument reappears in scholastic philosophy, not in connection with astronomy, but with free will. It reappears in the form of "Buridan's ass," which was unable to choose between two bundles of hay placed at equal distances to right and left, and therefore died of hunger.*

Until the 1960's, a large part of elastic stability theory was generally understood to consist in the analysis of linearized equations like (6.10) for $k = 1$ and in the conventional interpretation of results along the lines of Sec. 2 (see Timoshenko & Gere (1961), e.g.). In this case, the trivial solution is said to be unstable by the *criterion of adjacent equilibrium*. Perturbation analyses of the sort leading to (6.20) are called analyses of *post-buckling behavior* in the engineering literature. This term also accounts for the study of imperfections.

If the equilibrium problem possesses a potential-energy functional that has a local minimum at equilibrium state $\mathbf{p}_0$, then $\mathbf{p}_0$ is said to be stable according to the *energy criterion* introduced by Lagrange. This criterion is readily justified for a class of discrete mechanical systems governed by ordinary differential equations. It is a far more delicate matter to justify it for continuous systems (see Ball, Knops, & Marsden (1978), Beevers & Craine (1988), Beevers & Silhavy (1988), Browne (1979), Caflisch & Maddocks (1984), and Potier-Ferry (1981, 1982)).

If the linearization of the equations of motion about $\mathbf{p}_0$ has only bounded solutions (for all initial conditions), then $\mathbf{p}_0$ is said to be *infinitesimally stable*. (Of course, boundedness must be defined in terms of a suitable metric.) Within the theory of ordinary differential equations, there are many

*Quoted from Russell (1945, pp. 212–213). It is interesting to note that Buridan, who lived in the fourteenth century, contributed to the development of the basic ideas of mechanics.

examples, particularly relevant to elasticity theory, in which infinitesimally stable equilibrium solutions are not stable. Again, the justification for partial differential equations of cases in which infinitesimal stability implies stability is delicate.

It is now common practice in engineering to study dynamical behavior of a continuum by studying the behavior of a discretization with a small number of degrees of freedom. Mathematical justifications of these approaches are difficult. See the comments at the end of Sec. II.11.

These various approaches should be used with circumspection, not merely because they have not been fully justified on mathematical grounds, but because they can be physically misleading. A problem illustrating the weakness of these criteria is that of the buckling of a rod under certain kinds of nonconservative torsion. Nikolai (1928, 1929) observed that the only equilibrium state is trivial, yet the rod becomes unstable by undergoing motions. A similar phenomenon occurs in Beck's (1952) problem for the buckling of a rod under a follower force. We discuss generalizations of these problems in Secs. VI.11 and IX.4. The need for careful treatments of stability questions has been emphasized by Bolotin (1961), Herrmann (1967a), Leipholz (1968,1978a), Panovko & Gubanova (1979), Ziegler (1977). The need for mathematically rigorous treatments has been emphasized by Knops & Wilkes (1973).

8. Bibliographical Notes

There is an extensive body of work on mathematical aspects of bifurcation theory. Here we list texts and survey articles that are intrinsically valuable, or have extensive bibliographies, or have potential applications to nonlinear elasticity.

Texts containing important mathematical material for general aspects of steady-state bifurcation theory, of the sort described in Sec. 4, include Berger (1977), Chow & Hale (1982), Deimling (1985), Fučik, Nečas, Souček, & Souček (1973), Golubitsky & Schaeffer (1985), Golubitsky, Stewart & Schaeffer (1988), Ize (1976), Krasnosel'skiĭ (1956, 1962), Krasnosel'skiĭ & Zabreĭko (1975), Loginov (1985), Mawhin (1979), Nirenberg (1974), Pimbley (1969), Rabinowitz (1975), Sattinger (1979, 1983), Vaĭnberg & Aizengendler (1966), Vaĭnberg & Trenogin (1969), Vanderbauwhede (1982), Zeidler (1986). Useful expository papers are those of Alexander (1981), Marsden (1978), Prodi (1974), Rabinowitz (1973a), and Stakgold (1971). Of all these works, Golubitsky, Stewart & Schaeffer (1988), Loginov (1985), Sattinger (1979, 1983), and Vanderbauwhede (1982) treat bifurcation problems invariant under groups of transformations. See Rheinboldt (1986) for a discussion of numerical methods for bifurcation problems.

Koiter (1945) developed variational methods for treating imperfection sensitivity in elastic structures. (See Koiter (1963), Hutchinson & Koiter (1970), and Thompson & Hunt (1984).) The question of determining how many different imperfection parameters are needed to describe essentially all kinds of behavior can be handled for certain (typically variational problems) by catastrophe theory (see Poston & Stewart (1978)) and more generally by singularity theory (see Golubitsky & Schaeffer (1985) and Golubitsky, Stewart, & Schaeffer (1988)).

For dynamical problems of bifurcation theory, see Carr (1981), Chow & Hale (1982), Guckenheimer & Holmes (1983), Hassard, Kazarinoff, & Wan (1981), Henry (1981), Iooss (1979), Iooss & Joseph (1990), Knops & Wilkes (1973), Marsden & McCracken (1976), Sattinger (1973), and Wiggins (1990). Useful compendia of papers directed toward chaotic motions are those of Holmes (1980) and Salam & Levi (1988).

CHAPTER VI

Global Bifurcation Problems for Strings and Rods

In this chapter we apply the theory of Chap. V to specific bifurcation problems for strings and rods. Here we confront certain singular boundary-value problems (like those described in Ex. V.5.6). We concentrate on effects caused by the nature of constitutive response. We develop techniques by which we can convert given boundary-value problems into mathematical forms suitable for global analysis, and techniques by which we can extract very detailed information about specific classes of problems. We begin by studying the multiplicity of steady rotating states of elastic strings.

1. The Equations for the Steady Whirling of Strings

We employ the notation of Chap. II. Let $\{\boldsymbol{i}, \boldsymbol{j}, \boldsymbol{k}\}$ be a fixed right-handed orthonormal basis for $\mathbb{E}^3$ with $\boldsymbol{k}$ pointing downward. Let ω be a fixed number and set

$$\boldsymbol{e}(t) \equiv \cos \omega t \boldsymbol{i} + \sin \omega t \boldsymbol{j}. \tag{1.1}$$

We study steady rotations of a string about $\boldsymbol{k}$, for which $\boldsymbol{r}$ has the form

$$\boldsymbol{r}(s,t) = x_1(s)\boldsymbol{e}(t) + x_2(s)\boldsymbol{k} \times \boldsymbol{e}(t) + z(s)\boldsymbol{k}. \tag{1.2}$$

Let s denote the arc-length parameter of the string in its natural state. We assume that $[0,1] \ni s \mapsto \boldsymbol{r}(s,t)$ is absolutely continuous for each t. We set

$$\nu(s) \equiv |\boldsymbol{r}_s(s,t)| \tag{1.3}$$

when (1.2) holds.

Suppose that the end $s = 1$ of the string is fixed at the origin:

$$x_1(1) = 0 = x_2(1), \quad z(1) = 0 \tag{1.4}$$

and suppose that a material point of weight μ is attached to the end $s = 0$. The only body force on the string is the force of gravity acting in the $\boldsymbol{k}$-direction. (Centrifugal forces, which are just misplaced acceleration terms, are automatically accounted for by the computation of acceleration from (1.2).) In this case, the integral form of the equations of motion for elastic strings (II.2.23), which adapts (II.2.7), (II.2.10b), and (II.2.23) to the boundary conditions at 0, is

$$(1.5) \quad \frac{N(s)x'_\alpha(s)}{\hat{\nu}(N(s),s)} + \omega^2\left[\int_0^s (\rho A)(\xi)x_\alpha(\xi)\,d\xi + \frac{\mu x_\alpha(0)}{g}\right] = 0, \quad \alpha = 1,2,$$

$$(1.6) \quad \frac{N(s)z'(s)}{\hat{\nu}(N(s),s)} = -\left[g\int_0^s (\rho A)(\xi)\,d\xi + \mu\right] \equiv -[s\gamma(s)+\mu].$$

(These equations are equivalent to a suitable version of the Linear Momentum Principle (II.3.1) for steady motions adapted to the boundary conditions in the manner of Sec. VIII.13.) We assume that ρA is everywhere positive and continuous and that $\hat{\nu}$ is continuous. As in Chap. III, we further require for each s that $\hat{\nu}(\cdot,s)$ be strictly increasing, $\hat{\nu}(N,s) \to \infty$ as $N \to \infty$, $\hat{\nu}(N,s) \to 0$ as $N \to -\infty$, and $\hat{\nu}(0,s) = 1$ (see Fig. III. 3.7). Our first result is both physically important and mathematically useful:

1.7. Proposition. *Let $\mathbf{r}$ and N satisfy (1.3)–(1.6) with $\mathbf{r}(\cdot,t)$ absolutely continuous and with N continuous. Then for each t, $\mathbf{r}(\cdot,t)$ is a plane curve.*

Proof. We set

$$(1.8) \qquad u_\alpha(s) \equiv N(s)x'_\alpha(s)/\hat{\nu}(N(s),s).$$

Then (1.5) implies that the u_α are continuously differentiable. It follows from (1.4) and (1.5) that (u_1,x_1) and (u_2,x_2) each satisfy the boundary-value problem

$$(1.9\text{a,b}) \qquad u' + \omega^2\rho A x = 0, \quad x' = \frac{u\hat{\nu}(N(s),s)}{N(s)},$$

$$(1.9\text{c,d}) \qquad gu(0) = -\omega^2\mu x(0), \quad x(1) = 0.$$

Since we are assuming that (1.3)–(1.6) has a solution, the function $s \mapsto N(s)/\hat{\nu}(N(s),s)$ may be regarded as known and the problem (1.9) may be regarded as linear in (u,x). Equation (1.9b) is singular wherever N vanishes. Equations (1.6) and (1.8) imply that

$$(1.10) \qquad N(s)^2 = u_1(s)^2 + u_2(s)^2 + [s\gamma(s)+\mu]^2.$$

Thus N vanishes nowhere if $\mu > 0$. If $\mu = 0$, then we deduce from (1.5) and (1.6) that N vanishes at 0 and only at 0. (Thus the problem with $\mu = 0$ is singular.)

The second-order linear system (1.9a,b) has two independent solutions and has one independent solution satisfying (1.9d). Let us fix one such independent solution and denote it by (u,v). Since (u_1,x_1) and (u_2,x_2) each satisfy (1.9), it follows that there are constants β_1 and β_2 such that

$(u_\alpha, x_\alpha) = \beta_\alpha(u, x)$. If $\beta_1^2 + \beta_2^2 = 0$, then $\boldsymbol{r}$ reduces to $z\boldsymbol{k}$, and we are done. If not, then $\boldsymbol{r}$ has the form

$$\text{(1.11)} \qquad \begin{aligned} \boldsymbol{r}(s,t) &= \sqrt{\beta_1^2 + \beta_2^2}\, x(s)\boldsymbol{e}^\sharp(t) + z(s)\boldsymbol{k}, \\ \boldsymbol{e}^\sharp(t) &\equiv \frac{\beta_1 \boldsymbol{e}(t) + \beta_2 \boldsymbol{k} \times \boldsymbol{e}(t)}{\sqrt{\beta_1^2 + \beta_2^2}}, \end{aligned}$$

which says that $\boldsymbol{r}(\cdot, t)$ is confined to the $\{\boldsymbol{e}^\sharp(t), \boldsymbol{k}\}$-plane. □

Clearly, this proof applies to (1.9a,b) subject to other boundary conditions, provided that N, which can vanish at most once, does not vanish on $(0,1)$. A more delicate analysis is required when N vanishes on $(0,1)$; see Ex. 3.6 below.

By readjusting the origin on the time-axis, we can identify $\boldsymbol{e}^\sharp$ with $\boldsymbol{e}$, and accordingly set $x_1 = x$, $x_2 = 0$, $u_1 = u$, and $u_2 = 0$. Thus the problem (1.3)–(1.6) is equivalent to (1.4), (1.6), (1.9), and

$$\text{(1.12}^\pm\text{)} \qquad N(s) = \pm\sqrt{u(s)^2 + [s\gamma(s) + \mu]^2}.$$

We could combine (1.9) and (1.12) to obtain

$$\text{(1.13}^\pm\text{)} \qquad \frac{d}{ds}\left[\frac{u'}{(\rho A)(s)}\right] \pm \frac{\omega^2 u \hat{\nu}(\pm\sqrt{u(s)^2 + [s\gamma(s) + \mu]^2}, s)}{\sqrt{u(s)^2 + [s\gamma(s) + \mu]^2}} = 0,$$

which is subject to the boundary conditions

$$\text{(1.14a,b)} \qquad g(\rho A)(0)u(0) = \mu u'(0), \quad u'(1) = 0.$$

If ρA is not differentiable, then the interpretation of (1.13) is given by (1.9) and (1.12). It proves more convenient to work with this system.

We now resolve the ambiguity of sign in (1.12$^\pm$). Since $\gamma > 0$ everywhere, Eq. (1.6) says that z' has the sign opposite to that of N. Thus the string must lie entirely above the support at the origin if $N \leq 0$ and must lie entirely below the support if $N \geq 0$. We now show that the former states are of little interest.

1.15. Proposition. *The only classical solution of* (1.9), (1.12$^-$) *is the trivial solution, which, in view of* (1.6), *describes the string balanced vertically above the support.*

Proof. If $\omega = 0$, the conclusion follows immediately from (1.9), so we henceforth assume that $\omega^2 > 0$. We first show that $u \leq 0$. Suppose not. Then the continuity of u would imply that u has a positive maximum $u(\sigma) > 0$ on $[0,1]$. Now σ cannot be 0, for if so, (1.9c) would imply that $x(0) < 0$, and (1.9a) would then imply that $u'(0) > 0$, which is incompatible with the maximality of $u(0)$. If σ is in $(0,1)$, then $u'(\sigma) = 0$ by elementary calculus; if $\sigma = 1$, then (1.9a,b) also implies that $u'(\sigma) = 0$. It would then follow from (1.9a,b), (1.12$^-$) that $x(\sigma) = 0$ and $x'(\sigma) <$

0. The continuity of x would imply that $x(\xi) > 0$ for all ξ lying in a sufficiently small neighborhood to the left of σ. Thus (1.9a) would imply that $u'(\xi) < 0$ for all such ξ, in contradiction to the assumption that $u(\sigma)$ is a maximum. Thus $u \le 0$. Similarly, we show that $u \ge 0$, so that $u = 0$. That $x = 0$ follows from (1.9a). (This argument is a primitive application of the maximum principle. A simpler and more traditional approach can be applied to (1.13^-) when ρA is continuously differentiable.) □

This result likewise extends to other boundary conditions.

1.16. Exercise. Transform (1.3), (1.5), (1.6), (1.12), and (1.13) by introducing the change of variables

$$s_* \equiv \frac{\int_0^s (\rho A)(\xi)\,d\xi}{\int_0^1 (\rho A)(\xi)\,d\xi}, \quad x_*(s_*) \equiv x(s), \quad z_*(s_*) \equiv z(s), \quad N_*(s_*) \equiv N(s),$$

$$\hat{\nu}_*(N_*(s_*), s_*) \equiv \frac{g\left[\int_0^1 (\rho A)(\xi)\,d\xi\right]^2}{(\rho A)(s)} \hat{\nu}(N(S), s), \quad u_*(s_*) \equiv \frac{N_*(s_*)x_*'(s_*)}{\hat{\nu}_*(N_*(s_*), s_*)}, \tag{1.17}$$

$$\omega_*^2 \equiv \frac{\omega^2}{g}, \quad \mu_* \equiv \frac{\mu}{g\int_0^1 (\rho A)(\xi)\,d\xi}.$$

We do not employ this mathematically convenient transformation, which has the effect of replacing ρA by 1 and g by 1 (at the expense of complicating $\hat{\nu}$), because it would force us to abandon the basic physical and geometrical variables we have become accustomed to.

2. Kolodner's Problem

In view of Proposition 1.15, we limit our study to (1.9) and (1.12^+). We treat just the difficult case in which $\mu = 0$. We show that we can formulate this problem so that it meets the hypotheses of the Global Bifurcation Theorem V.4.19.

We replace (1.9) and (1.12^+) with their associated integral equations

$$u(s) = -\omega^2 \int_0^s (\rho A)(\xi) x(\xi)\,d\xi, \tag{2.1a}$$

$$x(s) = -\int_s^1 \frac{u(\xi)\hat{\nu}\left(\sqrt{u(\xi)^2 + \xi^2\gamma(\xi)^2},\, \xi\right)}{\sqrt{u(\xi)^2 + \xi^2\gamma(\xi)^2}}\,d\xi, \tag{2.1b}$$

which correspond to (1.5). A straightforward application of the Arzelà-Ascoli Theorem V.5.3 implies that the integral operators on the right-hand sides of (2.1a,b) generate a compact mapping taking pairs (u, x) of continuous functions into pairs of continuous functions. Unfortunately, the linearization of (2.1), the second component of which is

$$x(s) = -\int_s^1 \frac{u(\xi)\hat{\nu}\left(\xi\gamma(\xi),\, \xi\right)}{\xi\gamma(\xi)}\,d\xi, \tag{2.2}$$

has an integral operator that does not generate such a compact mapping. In fact, there are continuous functions u vanishing at 0 such that the integral in (2.2) does not converge for $s = 0$ (e.g., take $u(s) = (\log s)^{-1}$.)

This difficulty is a direct consequence of the absence of μ. In fact, one way to remedy it would be to penalize the problem by introducing a sequence of μ's approaching 0. Each penalized problem satisfies the hypotheses of the Global Bifurcation Theorem. Theorem V.4.21 could then be used to show that the limiting problem has solution branches having the same properties as their approximating branches. We shall describe this approach in more detail in Sec. 3 in the context of a related problem, for which alternative approaches are not evident. Here we have a convenient alternative:

We set

$$u(s) = \sqrt{s}w(s), \quad h(w,s) \equiv \frac{\hat{\nu}\left(\sqrt{sw^2 + s^2\gamma(s)^2},\, s\right)}{\sqrt{w^2 + s\gamma(s)^2}}, \tag{2.3a,b}$$

which converts (2.1) to

$$w(s) = -\frac{\omega^2}{\sqrt{s}} \int_0^s (\rho A)(\xi) x(\xi)\, d\xi, \tag{2.4a}$$

$$x(s) = -\int_s^1 w(\xi) h(w(\xi), \xi)\, d\xi, \tag{2.4b}$$

and converts (2.2) to

$$x(s) = -\int_s^1 w(\xi) h(0,\xi)\, d\xi = -\int_s^1 \frac{w(\xi)\hat{\nu}\left(\xi\gamma(\xi),\, \xi\right)}{\sqrt{\xi}\gamma(\xi)}\, d\xi. \tag{2.5}$$

We write (2.4) as

$$(w,x) = \mathsf{L}(\omega^2)\cdot(w,x) + \mathsf{g}[\omega^2, w, x]. \tag{2.6}$$

The essential compactness properties of (2.6) are given in the following exercise.

2.7. Exercise. Let $\mathcal{X} \equiv C^0[0,1] \times C^0[0,1]$. Prove that

$$[0,\infty) \times \mathcal{X} \ni (\omega^2, w, x) \mapsto \mathsf{L}(\omega^2)\cdot(w,x), \quad \mathsf{g}[\omega^2, w, x] \in \mathcal{X}$$

are continuous and compact.

We now study the eigenvalues of the linear problem. The compactness of $\mathsf{L}(\omega^2)$, which is affine in ω^2, ensures that the eigenvalues have finite multiplicity. (This is a standard, useful property of compact linear mapping (see Riesz & Sz. Nagy (1955) or Stakgold (1979) among others). We, however, give a more direct proof of a much sharper result by the methods proceeding from (2.10) below. We readily show that the linearized problem is equivalent to the linearization of (1.9), namely,

$$u' + \omega^2 \rho A x = 0, \quad x' = \frac{u\hat{\nu}(s\gamma(s), s)}{s\gamma(s)}, \quad u(0) = 0, \quad x(1) = 0. \tag{2.8a,b,c,d}$$

From (2.8) we obtain

$$\begin{aligned}\omega^2 \int_0^1 \rho A x^2\, dx &= -\int_0^1 x u'\, ds = \int_0^1 u x'\, ds \\ &= \int_0^1 \frac{u^2 \hat{\nu}(s\gamma(s), s)}{s\gamma(s)}\, ds = \int_0^1 \frac{w^2 \hat{\nu}(s\gamma(s), s)}{\gamma(s)}\, ds,\end{aligned} \tag{2.9}$$

which tells us that the eigenvalues ω^2 are real and positive.

We could have replaced $\sqrt{s}$ in (2.3) with s^α for any $\alpha \in (0,1)$. The conclusion of Ex. 2.7 would still be valid and the corresponding rightmost integral of (2.9) would converge for any continuous w. The form of our actual (2.9) suggests that $\alpha = \frac{1}{2}$ is particularly attractive. This choice is actually motivated by the treatment of Bessel equations in the theory of linear operators.

To determine the properties of the set of eigenvalues of (2.8), which is a singular Sturm-Liouville system, we introduce polar coordinates for the (u, x)-plane by a Prüfer transformation

$$u = r \sin \phi, \quad -\omega x = r \cos \phi. \tag{2.10}$$

The substitution of (2.10) into (2.8a,b) yields

$$\phi' = \omega \left[\rho A \cos^2 \phi + \frac{\hat{\nu}(s\gamma(s), s)}{s\gamma(s)} \sin^2 \phi \right]. \tag{2.11}$$

Since this equation is singular at $s = 0$, it is not covered by the standard existence and uniqueness theory. To show that this equation has a unique solution satisfying the initial condition $\phi(0) = 0$, we may set $\phi(s) = \sqrt{s}\psi(s)$ and write this expression as the integral from 0 to s of the right-hand side of (2.11). We then use the Contraction Mapping Principle XVIII.13 as in Ex. XVIII.1.22 to show that this integral equation for ψ has a solution near 0. It is then an easy exercise to show that the solution does not blow up and therefore exists on [0,1]. Equation (2.11) is valid where r does not vanish. The proof of Lemma 2.16 below shows that if r should vanish anywhere, then it vanishes everywhere.

It is not difficult to see that there is a nontrivial solution of (2.8) for each positive integer k for which an ω^2 can be found such that

$$\phi(0) = 0, \quad \phi(1) = \frac{(2k-1)\pi}{2} \tag{2.12a,b}$$

(see Coddington & Levinson (1955, Chap. 8)). Since the solution ϕ of the initial-value problem (2.11), (2.12a) strictly increases pointwise with ω^2 and since $\phi(1)$ for this solution must be less than $\pi/2$ for small positive ω^2, we deduce

2.13. Sturmian Theorem. *The eigenvalues $\omega^2 = \lambda_1^0, \lambda_2^0, \dots$ of (2.8) are each simple and they can be ordered thus: $0 < \lambda_1^0 < \lambda_2^0 < \cdots$ with $\lambda_k^0 \to \infty$ as $k \to \infty$. The functions $u_{(1)k}$ and $x_{(1)k}$ constituting the eigenvector* *corresponding to* λ_k^0 *each have exactly* k *zeros on* $[0,1]$. *The zeros of* $u_{(1)k}$ *are distinct from those of* $x_{(1)k}$.

The only statement that does not follow from a careful study of the geometry of the Prüfer transformation is that $u_{(1)k}$ and $x_{(1)k}$ cannot vanish simultaneously at $s = 0$. A proof of it is an easy consequence of the proof of Lemma 2.16 below. Note that (2.8) and the last statement of this theorem imply that all the zeros of the components of the eigenfunctions are simple, save possibly the zero of $u_{(1)k}$ at $s = 0$.

By virtue of Ex. 2.7 and Theorem 2.13, we can apply Theorem V.4.19 to conclude that from each point $(\lambda_k^0, 0, 0)$ of $[0,\infty) \times \mathcal{X}$ there bifurcates a branch $\mathcal{S}(\lambda_k^0)$ of solutions of (2.6) in $\mathbb{R} \times \mathcal{X}$. (The space $\mathcal{X}$ is defined in Ex. 2.7.) We now show that u and x on this branch inherit the nodal properties of the eigenfunctions and preserve them globally. Moreover, $\mathcal{S}(\lambda_k^0)$ is unbounded and does not join up with any other branch. To do this, we need only refine the nodal theory of Sec. V.5 to account for the singular behavior at $s = 0$.

We define $\mathcal{Z}_j$ to be the set of all $(w, x) \in \mathcal{X}$ for which (i) w has exactly $j - 1$ zeros on $(0,1)$, at which x does not vanish, (ii) x has exactly $j - 1$ zeros on $(0,1)$, (iii) $x(0) \neq 0$, and (iv) $w(1) \neq 0$, $x(1) = 0$. Thus $\mathcal{Z}_j$ is open in $\mathcal{Z}$. Its boundary belongs to the set of $(w, x) \in \mathcal{X}$ for which w and x vanish at the same point in $(0,1]$ or for which $x(0) = 0$.

We can now use Theorem V.4.14 to show that if a nontrivial solution pair $(\omega^2, (w, x))$ in $\mathcal{S}(\lambda_k^0)$ lies sufficiently close to $(\lambda_k^0, (0,0))$, then $(w, x) \in \mathcal{Z}_k$. We use the argument on connectedness given in Sec. V.5 to show that this conclusion holds everywhere on $\mathcal{S}(\lambda_k^0)$ by showing that if $(u, x) \in \mathcal{S}(\lambda_k^0) \cap \partial\mathcal{Z}_k$, then $(w, z) = (0,0)$. Standard uniqueness theory implies this conclusion if w and x have a simultaneous positive zero. We need only prove that $(w, z) = (0,0)$ if $x(0) = 0$.

We require a preliminary result:

2.14. Exercise. Let x be absolutely continuous on $(0,1)$, say, so that x has a derivative x' a.e. on $(0,1]$ and $x(s) = x(a) + \int_a^s x'(\xi)\,d\xi$ for $a \in (0,1)$. Thus if $s > a$, then $|x(s)| \leq |x(a)| + \int_a^s |x'(\xi)|\,d\xi$. Now $|x|$ is also absolutely continuous, so that it is the indefinite integral of its derivative (which exists a.e.): $|x(s)| = |x(a)| + \int_a^s |x(\xi)|'\,d\xi$. Prove that

$$|x(s)|' \leq |x'(s)| \quad \text{a.e.} \tag{2.15}$$

2.16. Lemma. *Let* $(\omega^2, (w, x))$ *satisfy* (2.4) *and the additional condition that* $x(0) = 0$. *Then* $(w, z) = (0,0)$.

Proof. We define u by (2.3) and observe that (u, x) satisfies (1.9a,b) and consequently the initial conditions $x(0) = u(0) = u'(0) = 0$. Using (2.15), we obtain from (1.9) that

$$|u|' \leq C|x|, \quad |x|' \leq |x'| \leq K\frac{|u|}{s} \tag{2.17a,b}$$

where $C \equiv \omega^2 \max_s (\rho A)(s)$ and $K \equiv \max_s \hat{\nu}(\sqrt{u(s)^2 + s^2\gamma(s)^2}, s)/\min_s \gamma(s)$. It follows from (2.17b) that $x'(0) = 0$. We now set $x \equiv 2Ky$ and $v \equiv$

$|u| + |y|$. Then

$$v' \leq \frac{v}{2s} + 2KCv, \quad v(0) = 0, \quad \frac{v(s)}{s} \to 0 \quad \text{as } s \to 0. \tag{2.18a,b,c}$$

We multiply (2.18a) by its integrating factor to convert it to the form

$$\frac{d}{ds}\left[\frac{e^{-2KCs}}{\sqrt{s}} v(s)\right] \leq 0 \tag{2.19}$$

from which we obtain

$$\frac{e^{-2KCs}}{\sqrt{s}} v(s) \leq \lim_{a \to 0} \frac{v(a)}{\sqrt{a}} e^{-2KCa} = 0, \tag{2.20}$$

by virtue of the initial conditions. □

This proof represents a self-contained exposition of an important part of the theory of differential inequalities (see Hale (1969) or Hartman (1964)).

Let us summarize what we can prove by exploiting the results obtained so far by the methods of Sec. V.5:

2.21. Theorem. *For $k = 1, 2, \ldots$, the solution branch $\mathcal{S}(\lambda_k^0)$ of the system (2.4) bifurcates from the trivial branch at $(\lambda_k^0, (0,0))$, does not touch $\mathcal{S}(\lambda_l^0)$ for $l \neq k$, lies in $(0, \infty) \times \mathcal{X}$, and is unbounded there. On each such branch, (w, x) is a classical solution of the corresponding boundary-value problem. If $(\omega^2, (w, x)) \in \mathcal{S}(\lambda_k^0)$, then $(\omega^2, (-w, -x)) \in \mathcal{S}(\lambda_k^0)$ and $(w, x) \in \mathcal{Z}_k$.*

These global properties of branches were proved by Kolodner (1955) for uniform, inextensible strings by using classical methods of analysis and by Stuart (1976) by using Theorem V.4.19. We now study other properties that depend crucially on the nature of material response. For any continuous function v, we set

$$\|v\| \equiv \|v, C^0[0,1]\| \equiv \max_s |v(s)|. \tag{2.22}$$

We use this notation in this and the next section.

We examine solutions on the level

$$\|w\| + \|x\| = R. \tag{2.23}$$

Let $(\omega^2, (w, x)) \in \mathcal{S}(\lambda_k^0)$. Then (w, x) may be regarded as the eigenvector corresponding to the kth eigenvalue ω^2 of the *linear* Sturm-Liouville problem (2.4) in which the function $h(w(\cdot), \cdot)$ is regarded as given. If (w, x) also satisfies (2.23), then $h(w(s), s) \geq 1/\sqrt{R^2 + (\max_s \gamma(s))^2}$. Consider the modification of (2.4) obtained by replacing $h(w(s), s)$ by this lower bound and by replacing $(\rho A)(s)$ by $\min_s (\rho A)(s)$. We apply the Prüfer transformation (2.10) to our original problem (regarded as linear) and to the modified problem (which is truly linear). The inequalities subsisting between the coefficients for (2.11) and for its analog for the modified problem enable us to deduce that the kth eigenvalue ω^2 of the actual problem must be less than the kth eigenvalue of the modified problem:

$$\omega^2 \leq \frac{j_k^2 \sqrt{R^2 + \|\gamma\|^2}}{4 \min_s (\rho A)(s)} \equiv \lambda_k^+(R) \tag{2.24}$$

where j_k is the kth positive zero of the Bessel function J_0.

2.25. Exercise. Prove that the kth eigenvalue of the modified problem is given by the right-hand side of (2.24).

The argument we have just sketched is the basis of a proof of the *Sturmian Comparison Theorem* specialized to our problem. Another way to prove this and other versions, which is even more powerful, is to use minimax characterizations of eigenvalues of linear problems (see Courant & Hilbert (1953) or Stakgold (1979), e.g.).

Since each branch $\mathcal{S}(\lambda_l^0)$ is unbounded, we deduce from (2.23) and (2.24) that $\|w\| + \|x\|$ is unbounded on each such branch. Indeed, (2.24) says that in the bifurcation diagram in $(\omega^2, \|w\| + \|x\|)$-space, the image of the kth branch lies to the left of the curve whose equation is $\omega^2 = \lambda_k^+(\|w\| + \|x\|)$. The function λ_k^+ is asymptotically linear for large values of its argument.

Suppose that the string is strong in tension in the sense that there is a number M such that

$$\hat{\nu}(N, s) \leq M(1 + N) \quad \text{for } N \geq 0 \quad \text{and for all } s \in [0, 1]. \tag{2.26}$$

For such strings we have the upper bound

$$h(w(s), s) \leq M\left(\sqrt{s} + \frac{1}{\sqrt{s}\min_s \gamma(s)}\right). \tag{2.27}$$

The Comparison Theorem then implies that on $\mathcal{S}(\lambda_k^0)$, ω^2 has a positive lower bound:

$$\omega^2 \geq \frac{j_k^2}{M\|\rho A\| \left[1 + \frac{1}{\min_s \gamma(s)}\right]}. \tag{2.28}$$

Suppose now that the string is very strong in tension in the sense that

$$\frac{\hat{\nu}(N, s)}{N} \to 0 \quad \text{as } N \to \infty. \tag{2.29}$$

2.30. Lemma. *If (2.29) holds, then there is a real-valued function f such that if $(\omega^2, (w, x))$ is a solution pair of (2.4), then*

$$\|w\| + \|x\| \leq f(\omega^2). \tag{2.31}$$

Inequality (2.31) implies that $\omega^2 \to \infty$ on each bifurcating branch.

Proof. System (2.4) implies that

$$\|w\| \le \omega^2 \|\rho A\|\, \|x\|, \quad \|x\| \le \max_s \hat{\nu}\left(\sqrt{\|w\|^2 + \|\gamma\|^2}, s\right), \tag{2.32}$$

so that

$$\frac{\|w\| + \|x\|}{\max_s \hat{\nu}\left(\sqrt{(\|w\| + \|x\|)^2 + \|\gamma\|^2}, s\right)} \le 1 + \omega^2 \|\rho A\|. \tag{2.33}$$

It follows from (2.29) and (2.33) that $\|w\| + \|x\|$ must be bounded for fixed ω^2. Thus such a function f must exist. Indeed, if the function of $\|w\| + \|x\|$ defined by the left-hand side of (2.33) is strictly increasing, we define f to be its inverse. Otherwise, we define f to be the inverse of a strictly increasing lower bound for the left-hand side of (2.33), which always exists. □

We combine the results of our development beginning with (2.26) with Theorem 2.21 to obtain

2.34. Theorem. *Let (2.29) hold. For each ω^2 in $(\lambda_k^0, \lambda_{k+1}^0)$, and for each $j = 1, \dots, k$ there is at least one pair (w, x) such that $(\omega^2, (\pm w, \pm x)) \in \mathcal{S}(\lambda_j^0)$. On each branch $\mathcal{S}(\lambda_k^0)$, ω^2 has a lower bound and both ω^2 and $\|w\| + \|x\|$ are unbounded.*

In his study of the uniform inextensible string, Kolodner (1955) proved a sharper result in which the 'at least' in the second sentence of Theorem 2.34 is replaced with 'exactly'. We do not obtain such a conclusion here because it is not true, as we indicate below in Ex. 2.53. Nevertheless, this theorem says that the global bifurcation for very strong strings is substantially the same as that for inextensible strings.

2.35. Exercise. Use the classical methods of Kolodner to prove Theorem 2.34. This lengthy analysis shows both the beauty of a clever combination of classical methods and the limitations of such methods when assumptions such as (2.29) are removed.

We now study a class of strings for which the bifurcation is significantly different from that described by Theorem 2.34. Suppose that a string is weak in tension in the sense that there is a number $m > 0$ such that

$$\hat{\nu}(N, s) \ge mN \quad \text{for } N \ge 0 \quad \text{and for all } s \in [0, 1]. \tag{2.36}$$

Since (2.36) implies that $h(w, s) \ge m\sqrt{s}$, we can use the Comparison Theorem to prove

2.37. Theorem. *Let (2.36) hold. On $\mathcal{S}(\lambda_k^0)$, ω^2 is bounded above:*

$$\omega^2 \le \frac{\zeta_k^2}{m \min_s (\rho A)(s)} \equiv \eta_k \tag{2.38}$$

where ζ_k is the kth zero of the Bessel function $J_{(-1/2)}$.

Note that it is possible for a uniform string to satisfy neither (2.26) nor (2.36). It is also possible for a string to satisfy both of these conditions. For such strings, ω^2 is bounded above and has a positive lower bound on each bifurcating branch. Among such strings are those for which $\hat{\nu}(\cdot, s)$ is *asymptotically linear*, i.e., those for which there is a positive-valued function μ such that $N^{-1}\hat{\nu}(N, s) \to \mu(s)$ as $N \to \infty$. The corresponding bifurcation problem is termed asymptotically linear. There is an extensive and beautiful theory for such problems (see Dancer (1974b), Krasnosel'skiĭ (1956), Pimbley (1962, 1963), Rabinowitz (1973b), Stuart (1973b), and Toland (1973)). In this case, it can be shown that

$$\omega^2 \to \Lambda_k \quad \text{as } \|w\| + \|x\| \to \infty \text{ on } \mathcal{S}(\lambda_k^0) \tag{2.39}$$

where the Λ_k are the eigenvalues of the obvious linear problem at infinity. Dancer (1974b) showed that $\mathcal{S}(\lambda_k^0)$ is a curve for large $\|w\| + \|x\|$. In the context of our problem for whirling strings, the sharpness of these results is counterbalanced by the rarity of asymptotically linear functions within the class of physically admissible $\hat{\nu}$'s.

For strings that are very weak in tension in the sense that

$$\frac{\hat{\nu}(N, s)}{N} \to \infty \quad \text{as } N \to \infty, \tag{2.40}$$

we have

2.41. Theorem. *Let $u(s) = \sqrt{s}w(s) \equiv sy(s)$. If (2.40) holds, then*

$$\omega^2 \to 0 \quad \text{as } \|y\| \to \infty \text{ on } \mathcal{S}(\lambda_k^0) \text{ for } k = 1, 2, \dots. \tag{2.42}$$

Note that $\|w\| \to \infty$ and $\|u\| \to \infty$ as $\|y\| \to \infty$ by definition, and that $\|x\| \to \infty$ as $\|y\| \to \infty$ by (2.1a) and Theorem 2.37.

Proof for $k = 1$. We substitute (2.4b) into (2.4a) and integrate the resulting expression by parts to obtain

$$\begin{aligned} \frac{gy(s)}{\omega^2} = \frac{1}{s}\int_0^s \frac{\xi\gamma(\xi)y(\xi)\hat{\nu}\left(\xi\sqrt{y(\xi)^2+\gamma(\xi)^2}, \xi\right)}{\sqrt{y(\xi)^2+\gamma(\xi)^2}}\, d\xi \\ + \gamma(s)\int_s^1 \frac{y(\xi)\hat{\nu}\left(\xi\sqrt{y(\xi)^2+\gamma(\xi)^2}, \xi\right)}{\sqrt{y(\xi)^2+\gamma(\xi)^2}}\, d\xi. \end{aligned} \tag{2.43}$$

(We could replace (2.4) with (2.43) and show that the right-hand side of (2.43) defines a compact mapping from $C^1[0,1]$ into itself. The rest of the preceding analysis could be carried out in this setting, at the expense of somewhat more technical difficulty; cf. Antman (1980a).) On the first branch, w does not vanish on $(0, 1]$. Let us take it to be positive here. Then (2.43) yields

$$\begin{aligned} y(s) &\geq \frac{\omega^2}{g}\min_\xi \gamma(\xi)\int_0^1 \frac{\xi y(\xi)\hat{\nu}\left(\xi\sqrt{y(\xi)^2+\gamma(\xi)^2}, \xi\right)}{\sqrt{y(\xi)^2+\gamma(\xi)^2}}\, d\xi \\ &\geq y(1)\frac{\min_\xi \gamma(\xi)}{\|\gamma\|} \end{aligned} \tag{2.44}$$

together with an analogous upper bound for gy/ω^2, from which we obtain

$$y(1)\frac{\min_\xi \gamma(\xi)}{\|\gamma\|} \leq y(s) \leq y(0)\frac{\|\gamma\|}{\|\gamma(0)\|}. \tag{2.45}$$

These inequalities together with (2.40) imply that

$$\frac{sy(s)\hat{\nu}\left(s\sqrt{y(s)^2+\gamma(s)^2},\, s\right)}{\sqrt{y(s)^2+\gamma(s)^2}} \to \infty \text{ as } |y(1)| \to \infty \text{ for } s > 0. \tag{2.46}$$

The Comparison Theorem thus implies that

$$\omega^2 \to 0 \quad \text{as } |y(1)| \to \infty \tag{2.47}$$

on the first branch. We must obtain (2.42) from (2.47). Just as we deduced (2.44) from (2.43) we find that

$$u(s) \leq \frac{\|\gamma\|}{\min_s \gamma(s)} u(1) = \frac{\|\gamma\|}{\min_s \gamma(s)} y(1). \tag{2.48}$$

Thus (2.47) implies that

$$\omega^2 \to 0 \quad \text{as} \quad \|u\| \to \infty \tag{2.49}$$

on the first branch. Equation (2.1b) implies that

$$\|x\| \leq \int_0^1 \hat{\nu}\left(\sqrt{\|u\|^2+\|\gamma\|^2},\, s\right) ds,$$

so that

$$\omega^2 \to 0 \quad \text{as } \|x\| \to \infty \tag{2.50}$$

on the first branch. Now suppose that $\|y\| \to \infty$ on the first branch. Then (2.45) implies that $|y(0)| \to \infty$. Since (2.1a) and the definition of y imply that $y(0) = -\omega^2(\rho A)(0)x(0)$, we obtain from (2.38) that $|x(0)| \to \infty$ as $|y(0)| \to \infty$. Therefore, (2.50) implies that $\omega^2 \to 0$ as $\|y\| \to \infty$ on the first branch, which together with (2.50) implies (2.42). □

2.51. Exercise. By making suitable estimates and by using the Comparison Theorem, obtain a specific function φ with $\varphi(y) \to 0$ as $y \to \infty$ so that $\omega^2 \leq \varphi(y(1))$ on the first branch when there are numbers $m > 0$ and $p > 1$ such that $\hat{\nu}(N, s) \geq 1 + mN^p$ for $N \geq 0$. Thus we have an estimate for the rate at which the limit in (2.47) is attained.

The basic idea of the proof for arbitrary k is to apply the Prüfer transformation (2.10) to the system of ordinary differential equations corresponding to (2.1), obtain the analog of (2.11) with $\cos\phi$ and $\sin\phi$ expressed in terms of u and x, and then impose (2.12). Using (2.38), we find that

$$\begin{aligned}
\frac{(2k-1)\pi}{2\omega} &= \int_0^1 \frac{(\rho A)\omega^2 x^2 + u^2\hat{\nu}\left(\sqrt{u^2+s^2\gamma^2},\, s\right)/\sqrt{u^2+s^2\gamma^2}}{u^2+\omega^2 x^2}\, ds \\
&\geq \int_0^1 \frac{sy^2\hat{\nu}\left(s\sqrt{y^2+s^2\gamma^2},\, s\right)}{(s^2y^2+\eta_k x^2)\sqrt{y^2+s^2\gamma^2}}\, ds
\end{aligned} \tag{2.52}$$

for $(\omega^2, (w, x)) \in \mathcal{S}(\lambda_k^0)$. To prove the theorem, we need only use (2.40) to show that the right-hand side of (2.52) becomes infinite with $\|y\|$. To do this, we must ensure that the pointwise growth of y is not confined to spikes so narrow that the integrals in (2.52) remain bounded. We must therefore examine the detailed form of the solution (y, x) between their zeros. For this purpose, we can combine the Maximum Principle with very careful analysis. The details for a slightly different formulation are given by Antman (1980a). His methods were inspired by the work of Wolkowisky (1969). Related approaches are given by Rabinowitz (1973a) and Turner (1973).

We can illustrate the comment following Theorem 2.34 by showing that under its hypotheses there can be subcritical bifurcations. For this purpose, we may assume that the string is weak for small tension.

2.53. Exercise. Let ρA be constant and let there be numbers $m > 0$ and $p > 1$ such that $\hat{\nu}(N, s) = 1 + mN^p$ for $N \in [0, \rho A]$. Use the perturbation methods of Sec. V.6 to prove that the bifurcation from the trivial branch is subcritical if m is large enough.

2.54. Exercise. Carry out a complete analysis of the boundary-value problem (1.9) and (1.12$^+$) when $\mu > 0$. In particular, state and prove an analog of every result given in this section. (Note that this problem can be formulated entirely in terms of the variable u; the variable w is neither needed nor appropriate.)

2.55. Research Problem. Carry out a rigorous stability analysis of these whirling motions in the sense of Sec. V.7. It is probably necessary to assume that the material is dissipative, e.g., by assuming that (II.2.14), (II.2.21) hold. (See Toland (1979a,b) for a discussion of energy criteria of stability.)

3. Other Problems for Whirling Strings. Bibliographical Notes

Reeken's problem. Let us now suppose that each end of the string is fixed to the axis of rotation:

$$x_\alpha(0) = 0 = x_\alpha(1), \quad \alpha = 1, 2, \tag{3.1}$$

$$z(0) = 0, \quad z(1) = a \tag{3.2}$$

where a is a prescribed number, which without loss of generality we take to be nonnegative. (To have a meaningful problem for an inextensible string, it is necessary that $a < 1$.) This ostensibly innocuous variant of the boundary conditions of Secs. 1 and 2 leads to some serious technical difficulties.

In place of (1.5), (1.6), we have

(3.3)
$$\left.\frac{N(\xi)x'_\alpha(\xi)}{\hat{\nu}(N(\xi), \xi)}\right|_\eta^s + \omega^2 \int_\eta^s (\rho A)(\xi) x_\alpha(\xi)\, d\xi = 0 \quad \forall \eta, s \in [0, 1], \quad \alpha = 1, 2,$$

$$\frac{N(s)z'(s)}{\hat{\nu}(N(s), s)} = b\gamma(b) - s\gamma(s) \tag{3.4}$$

where b is a constant of integration. We retain (1.8). In place of (1.10), we have

$$N(s)^2 = u_1(s)^2 + u_2(s)^2 + [b\gamma(b) - s\gamma(s)]^2. \tag{3.5}$$

3.6. Exercise. Carry out the following steps to prove that if $\boldsymbol{r}$ and N satisfy (1.3) and (3.1)–(3.4) with $\boldsymbol{r}(\cdot,t)$ absolutely continuous and with N continuous, then for each t, $\boldsymbol{r}(\cdot,t)$ is a plane curve. If N does not vanish in $(0,1)$, then the proof of Proposition 1.7 goes through. Otherwise, if N vanishes on $(0,1)$, then (3.5) implies that it vanishes at $b \in (0,1)$, and that $u_1(b) = 0 = u_2(b)$. Thus (u_1, x_1) and (u_2, x_2) each satisfy this condition and (1.9a,b), (3.1), and (3.2). For given N, there is a linear multiple $\alpha^0(u^0, x^0)$ of nontrivial solutions (u^0, x^0) satisfying $x^0(0) = 0$ and satisfying (1.9a,b) on $[0,b)$, and there is a linear multiple $\alpha^1(u^1, x^1)$ of nontrivial solutions (u^1, x^1) satisfying $x^1(1) = 0$ and (1.9a,b) on $(b,1]$. The continuity of x implies that $\alpha^0 x^0(b) = \alpha^1 x^1(b)$. If these values are not 0, then this equation relates α^0 to α^1, giving a one-parameter family of solutions to our boundary-value problem. Thus we can apply the reasoning used in the proof of Proposition 1.7. If $x(b) = 0$, (use the theory of differential inequalities as in the proof of Lemma 2.16 to) prove that (1.9a,b) subject to the initial condition $u(b) = 0 = x(b)$ has the unique solution $(u, x) = (0, 0)$.

By virtue of this result, we may take $x_1 = x$, $x_2 = 0$, $u_1 = u$, $u_2 = 0$, just as in Sec. 1 . We say that a solution is trivial if and only if $u = 0$.

3.7. Exercise. Let (u, x) generate a nontrivial solution to (3.1)–(3.4) with x and z absolutely continuous and with N continuous. Prove that u is continuously differentiable and is twice continuously differentiable where N does not vanish and that u satisfies

$$\frac{d}{ds}\left[\frac{u'}{(\rho A)(s)}\right] + \frac{\omega^2 u \hat{\nu}(N(s), s)}{N(s)} = 0 \quad \text{where } N \text{ does not vanish,} \tag{3.8}$$

$$u'(0) = 0 = u'(1). \tag{3.9a,b}$$

3.10. Exercise. Prove the following results: Let (u, x) generate a nontrivial solution to (3.1)–(3.4) with x and z absolutely continuous and with N continuous. Then N is positive except possibly at a single point b. If $N(0) = 0$ or if $N(1) = 0$, then the solution is trivial. If $b \leq 0$, then the solution is trivial.

Thus N can be defined unambiguously in terms of u from (3.5):

$$N(s) = \sqrt{u(s)^2 + [b\gamma(b) - s\gamma(s)]^2}. \tag{3.11}$$

First suppose that $b \in (0,1)$ and that $N(b) \neq 0$. Then N is everywhere positive and continuously differentiable with $N(b) = |u(b)|$. It then follows from (1.8) and (3.4) that x' and z' are continuous with $z'(s) < 0$ for $s < b$ and $z'(s) > b$ for $s > b$. Thus, b is the lowest point on the string, and (3.8) holds everywhere.

Now suppose that $b \in (0,1)$ and that $N(b) = 0$, so that $u(b) = 0$. Then (3.3) implies that

$$u(s) = \omega^2 \int_s^b (\rho A)(\xi) x(\xi)\, d\xi \equiv [b\gamma(b) - s\gamma(s)] w(s), \tag{3.12}$$

where w is continuous on $[0,1]$ and is twice continuously differentiable on $[0,1] \setminus \{b\}$. If $w(b)$ were 0, then $u'(b)$ would be 0, and the solution would be trivial. Representation (3.12) reduces (3.11) to

$$N(s) = |b\gamma(b) - s\gamma(s)|\sqrt{1 + w(s)^2}. \tag{3.13}$$

It then follows from (1.8) and (3.4) that

$$(3.14)\qquad x'(b+) = -x'(b-) = -\frac{\hat{\nu}(0,b)w(b)}{\sqrt{1+w(b)^2}}, \quad z'(b+) = -z'(b-) = -\frac{\hat{\nu}(0,b)}{\sqrt{1+w(b)^2}}.$$

Thus, the string is folded at b, which is again the lowest point of the string by (3.4).

The treatment of the trivial solution reduces to that for the catenary in the special case that one support is above the other (see Sec. III.3). We find that there are at most three different kinds of solutions (meeting our regularity requirements) that are possible for a given a: a straight state (which is always possible), a state with a single fold below the lower support, and a state with a single fold above the upper support. There is exactly one purely tensile solution (either straight or folded) for each $a \geq 0$. By Ex. 3.10, these tensile configurations correspond to the only trivial branches from which bifurcation can take place.

We now sketch the treatment of solution branches for which N is positive everywhere on $[0,1]$. In this case, all solutions are classical. Our first objective is to choose b of (3.4) so that (3.2) is satisfied. Equations (3.4) and (3.11) imply that (3.2) holds if and only if

$$(3.15)\qquad \int_0^1 \frac{[b\gamma(b) - s\gamma(s)]\hat{\nu}(N(s),s)}{N(s)}\,ds = a$$

where $N(s)$ is given by (3.11). It is evident that if u is a nontrivial solution of our boundary-value problem, then we can solve (3.15) uniquely for b in terms of u and a, because the positivity of N prevents the integrand in (3.15) from being singular. This result is inadequate since we should like to solve (3.15) for b as a continuously differentiable functional of u and replace b in (3.11) and elsewhere with this representation. This cannot be done because the left-hand side of (3.15) does not define a differentiable functional of u on C^0. To circumvent this difficulty, we use an approximation method of the sort described in the remarks preceding Theorem V.4.21: Let m be a positive integer. We study the approximate problem in which $N(s)$ of (3.11) is replaced with $\sqrt{m^{-2} + u(s)^2 + [b\gamma(b) - s\gamma(s)]^2}$.

3.16. Exercise. Substitute this expression into the left-hand side of (3.15). Use the properties of $\hat{\nu}$ to prove that the resulting equation can be solved uniquely for b in terms of U, m, a:

$$(3.17)\qquad b = \beta[u,m]$$

where β is positive-valued and where $\beta[\cdot,m]$ is continuously differentiable on $C^0[0,1]$. Here the dependence of β on a is suppressed.

Now we replace the $N(s)$ of (3.8) with

$$(3.18)\qquad N[u,m](s) \equiv \sqrt{m^{-2} + u(s)^2 + \{\beta[u,m]\gamma(\beta[u,m]) - s\gamma(s)\}^2},$$

integrate the resulting equation twice with respect to x, and use (3.9a) to obtain

$$(3.19)\qquad u(s) = u(0) - \omega^2 \int_0^s (\rho A)(\xi) \int_0^\xi \frac{\hat{\nu}(N[u,m](\eta),\eta)u(\eta)}{N[u,m](\eta)}\,d\eta\,d\xi.$$

If $\omega \neq 0$, then a solution of the approximate problem satisfies (3.9b) if and only if

$$(3.20)\qquad \int_0^1 \frac{\hat{\nu}(N[u,m](s),s)u(s)}{N[u,m](s)}\,ds = 0.$$

Now we replace the only u in (3.20) that is not an argument of N with (3.19) and solve the resulting equation for $u(0)$. We substitute this value of $u(0)$ into (3.19) to get an integral equation for u, which we denote by

$$(3.21)\qquad u = \omega^2 T[u,m].$$

3.22. Exercise. Prove

3.23. Theorem. *Let $\mu_0^m, \mu_1^m, \dots$ be the eigenvalues of the linearization of* (3.21) *about the trivial solution. For each $m = 1, 2, \dots$, $k = 0, 1, 2, \dots$,* (3.21) *has a nontrivial branch $\mathcal{S}(\mu_k^m) \equiv \{((\omega^2)_k^m, u_k^m)\}$ in $[0, \infty) \times C^1[0,1]$ that is unbounded, bifurcates from the trivial branch at $(\mu_k^m, 0)$, and does not touch $\mathcal{S}(\mu_l^m)$ for $l \neq k$. Each solution pair $((\omega^2)_k^m, u_k^m)$ is a classical solution of the approximate version of* (3.8), (3.9) *with $(\omega^2)_0^m = 0$, $u_0^m = \text{const}$. Each nonzero u_k^m has exactly k zeros on $[0,1]$, which are simple, and $(\omega^2)_k^m < (\omega^2)_{k+1}^m$. If $((\omega^2)_k^m, u_k^m) \in \mathcal{S}(\mu_k^m)$, then $((\omega^2)_k^m, -u_k^m) \in \mathcal{S}(\mu_k^m)$.*

By following Sec. 2, we can determine the disposition of the branches $\mathcal{S}(\mu_k^m)$ for each $\hat{\nu}$. To construct the solutions of the actual problem we need two simple estimates given in the next exercise.

3.24. Exercise. Let $\mathcal{B}(R) \equiv \{u \in C^1[0,1] : \|u, C^1[0,1]\| \leq R\}$. Prove that there are numbers $\beta^+(R)$ and $\lambda_k(R)$ that are independent of m such that

$$\sup\{\beta[u,m] : u \in \mathcal{B}(R)\} \leq \beta^+(R), \tag{3.25}$$

$$(\omega^2)_k^m \leq \lambda_k(R) \quad \text{if } ((\omega^2)_k^m, u_k^m) \in \mathcal{S}(\mu_k^m) \cap [\mathbb{R} \times \mathcal{B}(R)]. \tag{3.26}$$

We now use Theorem V.4.21 to construct solutions of our actual problem as subsequential limits of solutions of our approximate problems. For this purpose, we introduce

$$\mathcal{G}_k \equiv \{(\omega^2, u) \in [0,\infty) \times C^1[0,1] : \omega^2 \leq \lambda_k(\|u, C^1[0,1]\|)\} \cup \{\infty\} \tag{3.27}$$

and define its topology by taking a neighborhood basis of ∞ to be $\{(\omega^2, u) : \|u, C^1[0,1]\| > j\}$, $j = 1, 2, \dots$. Let

$$\bar{\mathcal{S}}_k^m \equiv \mathcal{S}(\mu_k^m) \cup ([0, \lambda_k(0)] \times \{0\}) \cup \{\infty\}. \tag{3.28}$$

Then $\bar{\mathcal{S}}_k^m$ is compact in $\mathcal{G}_k$, and Theorem 3.23 implies that $[0, \lambda_k(0)] \times \{0\}$ and ∞ are not separated in $\bar{\mathcal{S}}_k^m$. We define the solution branches of our actual problem:

$$\begin{aligned} \mathcal{S}_k \equiv \{(\omega_k^2, u_k) : u_k = \omega_k^2 T[u_k, \infty], \\ u_k \text{ has exactly } k \text{ zeros on } [0,1], \text{ which are simple}\}, \end{aligned} \tag{3.29}$$

$$\bar{\mathcal{S}}_k \equiv \mathcal{S}_k \cup ([0, \lambda_k(0)] \times \{0\}) \cup \{\infty\}. \tag{3.30}$$

It is conceivable that $\mathcal{S}_k$ is empty. Our basic result says that it is not:

3.31. Theorem. *$[0, \lambda_k(0)] \times \{0\}$ and $\{\infty\}$ are not separated in $\mathcal{S}_k$, which consequently contains a connected subset bifurcating from $(\mu_k, 0)$ on which $\|u_k\|$ is unbounded. Here the μ_k are the eigenvalues of the linearization of $u = \omega^2 T[u, \infty]$ about the trivial solution. Moreover, $\mathcal{S}_k \cap \mathcal{S}_l = \emptyset$ if $l \neq k$.*

Consequently, the solution branches of the actual problem enjoy all the properties of the branches of the approximate problem given in Theorem 3.23. Theorem V.4.21 implies that Theorem 3.31 follows from a proof of the properties

(i) Let $\{m\}$ be any subsequence of the positive integers. If $((\omega^2)_k^m, u_k^m) \in \bar{\mathcal{S}}_k^m$, then $\{((\omega^2)_k^m, u_k^m)\}$ has a convergent subsequence.

(ii) If $((\omega^2)_k^m, u_k^m) \in \bar{\mathcal{S}}_k^m$ and if $((\omega^2)_k^m, u_k^m) \to ((\omega^2)_k, u_k)$ in $\mathcal{G}_k$ as $m \to \infty$, then $((\omega^2)_k, u_k) \in \bar{\mathcal{S}}_k$.

3.32. Exercise. Use the compactness inherent in the approximate problem to prove Theorem 3.31.

Note that $\beta[0] = \beta[0, \infty]$ is just the b for which there is a unique trivial tensile solution. If $\beta[0] \in (0,1)$, then $\beta[u_k] = \beta[u_k, \infty]$ also lies in $(0,1)$ for each u_k such that $(\omega_k^2, u_k) \in \mathcal{S}_k$, because if $\beta[u_k]$ were to equal either 0 or 1, then the solution would be trivial by Ex. 3.10. Thus, if the trivial tensile solution sags, then every bifurcating nontrivial solution sags; if the trivial tensile solution is taut, then every bifurcating nontrivial solution is taut.

Let us now briefly outline the main ideas of the difficult case in which $b \in (0,1)$ and $N(b) = 0$ so that $u(0) = 0$. It can be shown that if we try to imitate the development above, which begins with a representation for b as a functional of u, then we should encounter a host of apparently insurmountable analytic difficulties. We avoid these difficulties by letting b play the role of a free parameter with a status like that of ω^2. (This policy is equivalent to letting a play the role of a free parameter.)

3.33. Exercise. Prove that w, defined by (3.12), satisfies

$$w(s) = \omega^2 T[w,b](s)$$

$$\equiv \omega^2 \begin{cases} \frac{1}{b\gamma(b)-s\gamma(s)} \int_s^b \int_0^\xi \frac{[b\gamma(b)-\eta\gamma(\eta)]\hat{\nu}(N(\eta),\eta)w(\eta)}{N(\eta)}\, d\eta\, d\xi & \text{for } s \in (0,b), \\ \frac{1}{b\gamma(b)-s\gamma(s)} \int_b^s \int_\xi^1 \frac{[b\gamma(b)-\eta\gamma(\eta)]\hat{\nu}(N(\eta),\eta)w(\eta)}{N(\eta)}\, d\eta\, d\xi & \text{for } s \in (b,1) \end{cases} \tag{3.34}$$

where N is given by (3.13).

For u to be continuously differentiable on $(0,1)$, it is necessary that

$$[\![w]\!](b) = 0. \tag{3.35}$$

If we compute $[\![T[w,b]]\!](b)$ from (3.34), we obtain an expression that does not vanish in general, so that $T[\cdot, b]$ does not take $C^0[0,1]$ into itself. For this reason, we let b vary in the hope that it could be chosen to satisfy (3.35). The physical explanation is simple: Since $N(b) = 0$, the problem consisting of (3.8) on $(0,b)$ subject to (3.9a), and the problem consisting of (3.8) on $(b,1)$ subject to (3.9b) are nothing other than Kolodner problems. If ω^2 and a can be chosen so that the ends $s = b$ of each string touch (without communicating force to each other), then together they form a configuration corresponding to a solution of (3.8), (3.9), (3.35). The conditions for touching at b are that $[\![x]\!](b) = 0 = [\![z]\!](b)$.

To obtain solutions, we construct the families of nontrivial solution sheets for the two Kolodner problems with ω^2 and b as parameters. These sheets intersect on a countable family of curve-like continua in (ω^2, b, w)-space, which are the solution branches for our actual problem. It is remarkable that the eigenvalues at which bifurcation takes place each have multiplicity 2. It can be shown that each solution branch is globally characterized by the nodal properties of u. For details, see Alexander, Antman, & Deng (1983).

Stuart's problem. A technically simple problem for a string of length l, related to the nonsingular version of Kolodner's problem, is obtained by fixing the end $s = 0$ to the origin, attaching a ring of weight μ to the end $s = l$, and confining the ring to slide without friction on the axis of rotation. Then the boundary conditions are

$$x(0) = 0 = z(0), \quad x(1) = 0, \quad u'(1) = 0, \quad \frac{N(l)z'(l)}{\hat{\nu}(N(l),l)} = \mu. \tag{3.36}$$

A simple analysis yields results virtually identical to those of Sec. 1. In particular, the dependence of the disposition of branches on the strength of the string is the same.

In Stuart's problem, the end $s = 0$ of a string of length l is also fixed at the origin, while the end $s = l$ is attached to the positive z-axis at $z = 1$ and is subjected to a prescribed positive tension α, so that

$$x(0) = 0 = z(0), \quad x(l) = 0, \quad z(l) = 1, \quad N(l) = \alpha > 0. \tag{3.37}$$

The natural length l of the string is left unspecified to accommodate the prescription of $N(l)$.

It is not hard to see that the problem corresponding to (3.36) and Stuart's problem are intimately related to each other: In the former, the length is prescribed, but $N(1)$ is not; in the latter, $N(l)$ is prescribed, but l is not.

Despite this duality, a formulation like those used above has not proved convenient for Stuart's problem. In particular, the fundamental unknown used for Stuart's problem has been the function $[0,1] \ni z \mapsto X(z)$, which gives the graph of the deformed string. It is striking that the bifurcating branches for Stuart's problem have ω^2 bounded above and below on each bifurcating branch, no matter what the material response of the string. To explain this phenomenon, let us replace the $N(l)$ in (3.36) with α. Then (3.36), (3.37), (1.8), and the Chain Rule yield

$$\begin{gathered} \alpha z'(l) = \mu\hat{\nu}(\alpha, l), \quad \alpha = \mu\sqrt{1 + X'(1)^2}, \\ u(l) = \frac{\alpha X'(1)}{\sqrt{1 + X'(1)^2}} = \mu X'(1). \end{gathered} \tag{3.38}$$

Thus $|u(l)| \to \alpha$ as $|X'(1)| \to \infty$. This fact can be shown to imply that when the natural norm of X along a branch for Stuart's problem becomes infinite, the natural norm for u remains bounded. The physical source of this phenomenon is that in Stuart's problem, N is forced to be bounded, whereas it is not in the complementary problem based on (3.36).

Bibliographical notes. Kolodner (1955) gave a complete analysis of the problem in Sec. 2 for inextensible strings, by a beautiful combination of the shooting method with Sturmian theory. Stuart (1976) applied the global bifurcation theory of Chap. V to this problem, thereby furnishing a simple, alternative treatment. The work of Kolodner inspired further work on related problems by Reeken (1977, 1979, 1980, 1984b), Toland (1979a,b), and Wu (1972). The treatment in Sec. 2 for nonlinearly elastic strings is adapted from that of Antman (1980a) by kind permission of the Royal Society of Edinburgh.

The problem with boundary conditions (3.1), (3.2) was studied by Reeken (1980) for inextensible strings. He constructed the first branch of (smooth) solutions, not by continuing it from the eigenvalue of the linearization about the trivial solution, but by continuing it back from the asymptotic state at infinite angular velocity. Part of the treatment of Reeken's problem given in this section is adapted from that of Alexander, Antman, & Deng (1983) by kind permission of the Royal Society of Edinburgh.

Stuart's (1975) treatment of the problem with boundary conditions (3.37) for inextensible strings represents the first application of the global bifurcation theory of Chap. V to a physical problem. The treatment of it and of the dual problem with boundary conditions (3.36) is based on that of Antman (1980a). Stuart's problem is a primitive model for the process by which string is both drawn and whirled. A more refined model is analyzed in the next section.

4. The Drawing and Whirling of Strings

In this section we study the steady motion of a string that is simultaneously being drawn and whirled. This motion, which models the process by which fibers are manufactured, has a Coriolis acceleration, which makes the problem more complicated than those treated in the preceding sections. Our first goal is to give a careful formulation of the governing equations, because this problem exemplifies those technologically important problems in which the material points undergoing the motion change with time. Our second goal is to examine superficially the exceptionally rich structure of solutions and the mathematical methods for determining it. We adhere to the notation of Chap. II.

Kinematics. Let $\{\boldsymbol{i}, \boldsymbol{j}, \boldsymbol{k}\}$ be a fixed right-handed orthonormal basis for $\mathbb{E}^3$. Gravity is taken to act in either the $\boldsymbol{k}$ or $-\boldsymbol{k}$ direction. Let $a \geq 0$. We study the combined whirling and drawing motion under gravity of that part of a string, called the *active part*, lying between $\mathbf{0}$ and $a\boldsymbol{k}$ when it is being fed through an *inlet* at $\mathbf{0}$ and is being withdrawn through an *outlet* at $a\boldsymbol{k}$.

Let s be an arc length parameter for a natural reference configuration of the string. Let $s_0(\tau)$ and $s_a(\tau)$ denote the material points of the string that respectively pass through $\mathbf{0}$ and $a\boldsymbol{k}$ at time τ:

$$\boldsymbol{r}(s_0(t), t) = \mathbf{0}, \quad \boldsymbol{r}(s_a(t), t) = a\boldsymbol{k}. \tag{4.1}$$

As usual, we assume that $\boldsymbol{r}(\cdot, t)$ is absolutely continuous for all t.

We assume that the string is fed in at $\mathbf{0}$ and withdrawn at $a\boldsymbol{k}$ at the same constant rate γ:

$$s_0'(\tau) = s_a'(\tau) = -\gamma < 0, \tag{4.2}$$

so that

$$s_0(\tau) = s_0(0) - \gamma\tau, \quad s_a(\tau) = s_a(0) - \gamma\tau. \tag{4.3}$$

Thus

$$l \equiv s_a(\tau) - s_0(\tau) = s_a(0) - s_0(0) \tag{4.4}$$

is independent of τ. We assume that l is given. For uniform strings, this prescription is equivalent to the prescription of the amount of material in the active part of the string. We set

$$\sigma \equiv s_0(\tau) - s_0(t) = \gamma(t - \tau). \tag{4.5}$$

It is the amount of material that has passed through the inlet between times τ and t. If $s_0(t) \leq s_0(\tau) \leq s_a(t)$, then $\sigma \in [0, l]$. We could normalize our variables by taking $s_0(0) = 0$. The problem we study below is independent of t. Thus σ is just a parameter for the active part of the string.

We require that the motion of the string be steady by requiring that its active part lie on a steadily rotating rigid space curve $\boldsymbol{p}(\cdot, t)$, which (without loss of generality) must have the form

$$\begin{aligned} \boldsymbol{p}(\sigma,t) &= x_1(\sigma)\boldsymbol{e}(t) + x_2(\sigma)[\boldsymbol{k}\times\boldsymbol{e}(t)] + z(\sigma)\boldsymbol{k}, \\ \boldsymbol{e}(t) &\equiv \cos\omega t\boldsymbol{i} + \sin\omega t\boldsymbol{j}. \end{aligned} \tag{4.6}$$

We accordingly set

$$\begin{aligned} \boldsymbol{r}(s_0(\tau),t) &\equiv \boldsymbol{r}(s_0(t)+\sigma,t) \equiv \boldsymbol{r}(s_0(0)-\gamma t+\sigma,t) \\ &= \boldsymbol{p}(\sigma,t) \equiv \boldsymbol{p}(\gamma(t-\tau),t). \end{aligned} \tag{4.7}$$

For $\boldsymbol{r}(\cdot,t)$ to be absolutely continuous, it is necessary that $\boldsymbol{p}(\cdot,t)$ be absolutely continuous. In adopting (4.7), we are actually requiring that the motion of the string be a travelling wave with σ representing the fixed phase. It follows from (4.7) that

$$\boldsymbol{r}_s(s_0(\tau),t) = \boldsymbol{p}_\sigma(\sigma,t), \tag{4.8}$$

$$\nu(s_0(\tau),t) = |\boldsymbol{p}_\sigma(\sigma,t)| = \sqrt{x_1'(\sigma)^2 + x_2'(\sigma)^2 + z'(\sigma)^2} \equiv \tilde{\nu}(\sigma), \tag{4.9}$$

$$\boldsymbol{r}_t(s_0(\tau),t) = \gamma\boldsymbol{p}_\sigma(\sigma,t) + \omega\boldsymbol{k}\times\boldsymbol{p}(\sigma,t), \tag{4.10}$$

$$\begin{aligned} \boldsymbol{r}_{tt}(s_0(\tau),t) &= \gamma^2\boldsymbol{p}_{\sigma\sigma}(\sigma,t) + 2\omega\gamma\boldsymbol{k}\times\boldsymbol{p}_\sigma(\sigma,t) \\ &\quad - \omega^2[x_1(\sigma)\boldsymbol{e}(t) + x_2(\sigma)(\boldsymbol{k}\times\boldsymbol{e}(t))]. \end{aligned} \tag{4.11}$$

Note that the speeds at $\mathbf{0}$ and at $a\boldsymbol{k}$ of the material points occupying them are

$$|\boldsymbol{r}_t(s_0(t),t)| = \gamma\tilde{\nu}(0), \quad |\boldsymbol{r}_t(s_a(t),t)| = \gamma\tilde{\nu}(l). \tag{4.12}$$

These are not necessarily equal. For a uniform string, (4.2) can be interpreted as saying that the mass fluxes at $\mathbf{0}$ and $a\boldsymbol{k}$ are equal. This requirement, rather than the equality of the entrance and exit speeds, is essential for steady motions.

Mechanics. To ensure that the string can admit motions (4.7) under the sole body force of gravity, we require that the string be uniform, so that ρA be constant and so that its constitutive functions be independent of s. Otherwise, (4.7) would represent a constraint, which would have to be maintained by artificial (time-varying) constraint forces. Since ρA is constant, Eq. (4.4) implies that the total mass of the active part of the string is independent of t.

Following (4.7) we set

$$N(s_0(\tau),t) \equiv N(s_0(0)-\gamma t+\sigma,t) \equiv n(\sigma,t). \tag{4.13}$$

As part of our definition of steady motions we require that n be independent of t. If the string is inextensible, then

$$\tilde{\nu} = 1. \tag{4.14}$$

If the string is elastic and uniform, then (II.2.11) and (4.13) yield the constitutive equation

$$n(\sigma) = \hat{N}(\tilde{\nu}(\sigma)). \tag{4.15}$$

In view of the irregularities exhibited by solutions of string problems treated in Chaps. II and III and in the preceding sections of this chapter, and in view of the possibility that our dynamic problem could have shocks, we formulate our problem as a version of the Impulse-Momentum Law (II.3.1):

$$\int_{t_1}^{t_2} \frac{N(s,t)\boldsymbol{r}_s(s,t)}{\nu(s,t)}\bigg|_{s=s_0(\tau_1)}^{s=s_0(\tau_2)} dt + \varepsilon\rho A g[s_0(\tau_2) - s_0(\tau_1)](t_2 - t_1)\boldsymbol{k} = \rho A \int_{s_0(\tau_1)}^{s_0(\tau_2)} \boldsymbol{r}_t(s,t)\bigg|_{t=t_1}^{t=t_2} ds \tag{4.16a}$$

where $\varepsilon = \pm 1$ to account for gravity either assisting or opposing the drawing motion. We now use (4.2), (4.5), (4.8)–(4.10) to convert (4.16a) to

$$\int_{t_1}^{t_2} \frac{n(\gamma(t-\tau))\boldsymbol{p}_\sigma(\gamma(t-\tau),t)}{\rho A\tilde{\nu}(\gamma(t-\tau))}\bigg|_{\tau=\tau_1}^{\tau=\tau_2} dt - \varepsilon g\gamma(\tau_2 - \tau_1)(t_2 - t_1)\boldsymbol{k} = \int_{\gamma(t-\tau_1)}^{\gamma(t-\tau_2)} [\gamma\boldsymbol{p}_\sigma(\sigma,t) + \omega\boldsymbol{k}\times\boldsymbol{p}(\sigma,t)]\, d\sigma\bigg|_{t=t_1}^{t=t_2}. \tag{4.16b}$$

Since $\boldsymbol{p}(\cdot,t)$ is absolutely continuous and since $\boldsymbol{p}(\sigma,\cdot)$ is analytic, we can differentiate (4.16b) with respect to t_2 a.e., set $\sigma_2 = \gamma(t_2 - \tau_2)$ and $\sigma_1 = \gamma(t_2 - \tau_1)$, and replace t_2 by t to obtain

$$m(\sigma)\boldsymbol{p}_\sigma(\sigma,t)\big|_{\sigma_1}^{\sigma_2} + \varepsilon g(\sigma_2 - \sigma_1)\boldsymbol{k} = 2\omega\gamma\boldsymbol{k}\times\boldsymbol{p}(\sigma,t)\big|_{\sigma_1}^{\sigma_2} + \omega^2\int_{\sigma_1}^{\sigma_2} \boldsymbol{k}\times[\boldsymbol{k}\times\boldsymbol{p}(\sigma,t)]\, d\sigma \tag{4.17}$$

where

$$m(\sigma) \equiv \frac{n(\sigma)}{\rho A\tilde{\nu}(\sigma)} - \gamma^2. \tag{4.18}$$

If we introduce the complex horizontal displacement

$$x(\sigma) = x_1(\sigma) + ix_2(\sigma), \tag{4.19}$$

then we can rewrite (4.17) and (4.1) as the boundary-value problem

$$m(\xi)x'(\xi)\big|_c^\sigma - 2i\omega\gamma x(\xi)\big|_c^\sigma + \omega^2 \int_c^\sigma x(\xi)\,d\xi = 0, \tag{4.20}$$

$$x(0) = 0 = x(l), \tag{4.21}$$

$$m(\sigma)z'(\sigma) = \varepsilon g(b-\sigma), \tag{4.22}$$

$$z(0) = 0, \quad z(l) = a \tag{4.23}$$

where (4.20) is to hold for all $[c,\sigma] \subset [0,l]$ and where b is a constant of integration. These equations are to be supplemented with one of the constitutive equations (4.14) or (4.15) and with the definition of $\tilde{\nu}$ given in (4.9). If (4.20) has a solution with x an absolutely continuous function, then the first term of (4.20) must also be absolutely continuous. We can then differentiate (4.20) a.e. with respect to σ, obtaining a complex second-order ordinary differential equation. Observe that for elastic strings m could vanish at various points because this equation describes travelling waves for a hyperbolic system.

Standing in marked contrast to Proposition 1.7 is

4.24. Proposition. *If $\gamma > 0$, then the only solution of (4.20)–(4.23) corresponding to a string confined to a plane rotating about the* $\mathbf{k}$*-axis is the trivial solution, in which the string lies along the* $\mathbf{k}$*-axis.*

Proof. If the configuration of the string is confined to such a rotating plane, then we can choose x to be real. In this case, (4.20) and (4.21) imply that $\omega\gamma x = 0$. If $\gamma > 0$, then either the solution is trivial: $x = 0$, or else $\omega = 0$. In the latter case, (4.20) implies that

$$m(\sigma)x'(\sigma) = C \quad \text{(const.)}. \tag{4.25}$$

By Rolle's Theorem, the boundary conditions (4.21) imply that x' must vanish on $(0,l)$. Therefore $C = 0$. Equations (4.6), (4.22), and (4.25) thus imply that $m(\sigma)^2 = g^2(b-\sigma)^2$, so that m can vanish at most once. Thus $x' = 0$. It follows from (4.21) that $x = 0$. □

4.26. Exercise. Suppose that there is a number $\beta > 1$ such that

$$\hat{N}(\nu) = \alpha^2 \rho A \nu \quad \text{for } \nu \geq \beta \tag{4.27}$$

(see (II.7.1)). Find all the nontrivial solutions of (4.9), (4.20)–(4.23), and (4.27) when the data are such that $\tilde{\nu} \geq \beta$, and give explicit conditions for the data to meet this condition. Sketch the bifurcation diagram and describe a typical configuration.

We now give a very brief sketch of the bifurcation analysis of (4.20)–(4.23). The essential equation governing the bifurcation is (4.20). It is a complex second-order ordinary differential equation, which is equivalent to a pair of real second-order equations. As Ex. 4.26 suggests, neither x_1 nor x_2 preserves nodal properties along sheets of solution pairs $((\omega,\gamma),x)$. We introduce polar coordinates for the complex x-plane by

$$x(s) = v(s)e^{i\phi(s)}, \tag{4.28}$$

but we allow v to assume negative values so that it has a good chance of being smooth. A delicate regularity theory shows that if m and v are continuous and if (x, z) is an absolutely continuous solution of (4.20)–(4.23) for $\omega\gamma \neq 0$, then ϕ is continuously differentiable and v is twice continuously differentiable. These functions satisfy

$$(mv')' + \omega^2 \left(1 + \frac{\gamma^2}{m}\right) v = 0, \quad m\phi' = \omega\gamma \tag{4.29a,b}$$

where the prime denotes differentiation with respect to σ.

We now focus on inextensible strings, which are not susceptible of shocks. Here we require that $a < l$. Then (4.9) and (4.14) give

$$z' = \pm\sqrt{1 - |x'|^2}. \tag{4.30}$$

Equation (4.22) says that z' can vanish at most once. We get different representations for z' and m in terms of v from (4.29b) and (4.30) for the cases that z' is positive on $(0, l)$ or changes sign at $b \in (0, l)$. The boundary conditions (4.23) lead to an equation that can be solved uniquely for b in terms of a, v, ω, and γ for each case. The substitution of this b into (4.29a) yields a single ordinary functional differential equation for v for each case. The linearization of this equation is a Sturm-Liouville problem whose solution can be expressed in terms of the Bessel functions $J_{\pm 2i\omega\gamma/g}$, which behave like the sin and cos of the logarithm. Consequently, the eigenfunction of the linearized problem lacks both continuity and nodal structure.

To compensate for this difficulty, we regularize the problem as in (3.18) by suitably introducing a penalization $1/j$. We use the Multiparameter Global Bifurcation Theorem 4.23 together with standard uniqueness theory to prove that for each positive integer j there are two-dimensional sheets $\mathcal{S}(\mathcal{E}_k^j)$, $k = 1, 2, \ldots$, of solution pairs of the regularized problem bifurcating from the distinct eigencurves $\mathcal{E}_k^j$ of its linearization. If $((\omega, \gamma), v) \in \mathcal{S}(\mathcal{E}_k^j)$ and if $v \neq 0$, then v has exactly $k - 1$ zeros on $(0, 1)$, which are simple. Thus, it is the nodal structure of $|x|$ rather than the nodal structure of either x_1 or x_2 that is preserved globally on each sheet. See Fig. 4.31.

We now let $j \to \infty$ and use the Comparison Theorem to determine how the sheets move in this limit. We then use the Connectivity Theorem V.4.21 to show that the sheets of the regularized problems converge to two-dimensional connected sets and use the compactness inherent in the regularized problems to show that the limiting connected sets are connected sets of solution pairs of the actual problem preserving the nodal properties of the approximating sheets $\mathcal{S}(\mathcal{E}_k^j)$. This process shows that all the sheets bifurcate from the lines $\omega = 0$ and $\gamma = 0$ in the trivial plane. (These lines are the edges of the continuous spectra corresponding to the parameters ω and γ for the linearized problems.) The bifurcating sheets, having distinctive nodal properties, clearly do not inherit them from solutions of the linearized problem, which have no nodal structure. A typical bifurcation diagram is shown in Fig. 4.32.

Fig. 4.31. A typical configuration of a string that is being whirled and drawn.

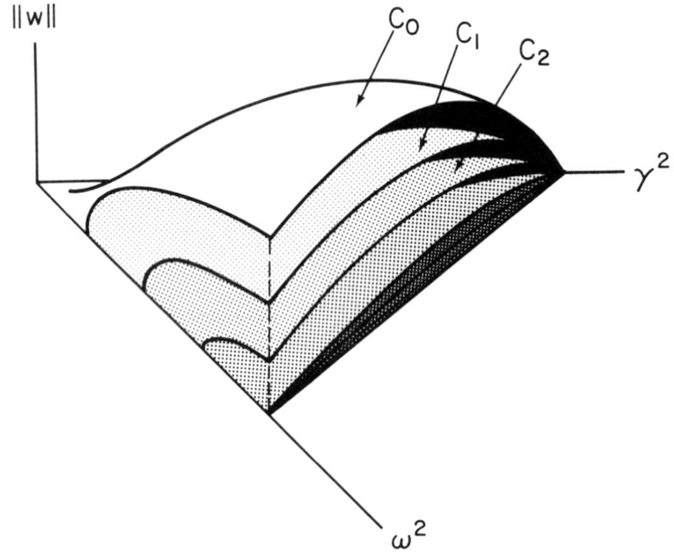

Fig. 4.32. A typical bifurcation diagram for the problem of this section.

The work in this section is based upon that of Antman & Reeken (1987, ©Society for Industrial and Applied Mathematics). Figures 4.31 and 4.32 come from this source.

4.33. Research Problem. Carry out a rigorous stability analysis of these motions in the sense of Sec. V.7. It is probably necessary to assume that the material is dissipative, e.g., by assuming that (II.2.14), (II.2.21) hold. Even a formal stability analysis would be useful.

5. Planar Buckling of Rods. Global Theory

We now extend the analysis of Sec. V.5 for the elastica to Cosserat rods that can suffer not only flexure, but also extension and shear, and that are governed by general nonlinear constitutive equations. We follow the notation and assumptions of Secs. IV.1, V.1, and V.5. In particular, we assume that the symmetry conditions (IV.2.9) hold. We further require that our constitutive functions be thrice continuously differentiable. We now suppose that the constitutive equations depend upon an additional parameter α, which measures the stiffness of the rod and which satisfies $0 < \alpha_* \leq \alpha \leq \infty$, where α_* is given. α may be related to the thickness of the rod; its presence does not affect the symmetry properties. We assume as in (V.1.3) and (V.1.4) that the reference configuration is defined by

(5.1) $$\boldsymbol{r}(s) = s\boldsymbol{i}, \quad s \in [0,1], \quad \boldsymbol{a} = \boldsymbol{i} \quad \text{so that } \theta = 0.$$

We assume that the end $s = 0$ of the rod is welded at the origin to a fixed rigid wall with normal $\boldsymbol{i}$ and that the end $s = 1$ is welded to a rigid wall with normal $-\boldsymbol{i}$, which is free to move in any direction having no $\boldsymbol{k}$-component. Thus the boundary conditions are

(5.2a,b) $$\theta(0) = 0 = \theta(1),$$

(5.2c,d) $$\boldsymbol{r}(0) = \boldsymbol{0}, \quad \boldsymbol{n}(1) = -\lambda\boldsymbol{i}.$$

(It is not hard to design a simple device that maintains these conditions.) We assume that no body force or couple is applied to the rod. Then the integral forms of the equilibrium equations, corresponding to (IV.2.10) with $\gamma = \theta$ and to (5.2a,d), reduce to

(5.3) $$\theta(s) = \int_0^s \hat{\mu}\,(-\lambda\cos\theta(\xi), \lambda\sin\theta(\xi), M(\xi), \alpha, \xi)\; d\xi,$$

(5.4) $$M(s) = M(0) - \lambda \int_0^s [\hat{\nu}\sin\theta(\xi) + \hat{\eta}\cos\theta(\xi)]\, d\xi$$

where the arguments of $\hat{\nu}$ and $\hat{\eta}$ in (5.4) are the same as those of $\hat{\mu}$ in (5.3).

To show that $M(0)$ can be chosen to make $\theta(1) = 0$, we substitute (5.4) into (5.3) evaluated at $s = 1$ and seek an $M(0)$ to satisfy

(5.5) $$\int_0^1 \hat{\mu}\left(-\lambda\cos\theta(s), \lambda\sin\theta(s), M(0) - \lambda\int_0^s \cdots\, d\xi, \alpha, s\right) ds = 0.$$

The M's that are arguments of $\hat{\nu}$ and $\hat{\eta}$ in (5.4) are not replaced with the right-hand side of (5.4). Thus the visible $M(0)$ in (5.5) is the only $M(0)$ that appears in (5.5). Conditions (1.23) and (2.9) imply that

(5.6) $$M \mapsto \hat{\mu}(N, H, M, \alpha, s) \quad \text{is odd and increasing.}$$

Hence for fixed $(\theta, M, \lambda, \alpha)$, (5.5) is positive for sufficiently large $M(0)$ and negative for sufficiently small $M(0)$. Thus the Intermediate Value Theorem implies that for fixed $(\theta, M, \lambda, \alpha)$, (5.5) has at least one solution for $M(0)$. Since $\mu_M > 0$ by (IV.1.23), this solution, denoted

$$M(0) = B[\theta, M, \lambda, \alpha], \tag{5.7}$$

is unique. Since our constitutive functions are smooth, the inequality $\mu_M > 0$ also enables us to deduce from the Implicit Function Theorem XVIII.1.27 in Banach Space that B is a smooth functional of its arguments on $C^0 \times C^0 \times [0, \infty] \times [\alpha_*, \infty]$. We replace $M(0)$ in (5.4) with $B[\theta, M, \lambda, \alpha]$. We set $\mathbf{u} \equiv (\theta, M)$ and denote the resulting form of (5.3) and (5.4) by

$$\mathbf{u} = \mathbf{L}(\lambda, \alpha) \cdot \mathbf{u} + \mathbf{g}[\lambda, \alpha, \mathbf{u}]. \tag{5.8}$$

We treat (λ, α) as a pair of eigenvalue parameters. We set $\mathcal{X} \equiv C^0[0,1] \times C^0[0,1]$. A straightforward application of the Arzelà-Ascoli and Bolzano-Weierstrass Theorems shows that

$$[0, \infty) \times [\alpha_*, \infty) \times \mathcal{X} \ni (\lambda, \alpha, \mathbf{u}) \mapsto \mathbf{L}(\lambda, \alpha) \cdot \mathbf{u}, \; \mathbf{g}[\lambda, \alpha, \mathbf{u}]$$

are compact and continuous. We therefore need only verify that $\mathbf{L}$ has the appropriate spectral properties in order to apply the local and global bifurcation theorems of Sec. V.4. The arguments of Sec. V.5 show that nodal properties of θ are preserved globally on any connected set of solution pairs not intersecting the trivial plane.

In view of these comments, we turn to the local study of bifurcation. Here we treat several novel issues associated with imperfections and singularities before returning to global questions.

Let $\hat{\omega}$ stand for any constitutive function, such as $\hat{\nu}$ or $\hat{\mu}_M$. Then we define

$$\omega^0(\lambda, \alpha) \equiv \hat{\omega}(-\lambda, 0, 0, \alpha). \tag{5.9}$$

To make our local study completely explicit, we assume that the material is uniform, so that the constitutive functions are independent of s. In this case, we find that

$$\begin{aligned} &[\mathbf{L}(\lambda, \alpha) \cdot \mathbf{u}](s) \\ &= \left(\mu_M^0 \int_0^s M(\xi)\, d\xi, \quad \lambda(\nu^0 + \lambda\eta_H^0) \left[\int_s^1 \theta(\xi)\, d\xi - \int_0^1 \xi\theta(\xi)\, d\xi \right] \right). \end{aligned} \tag{5.10}$$

Then the linearization of (5.8) about the trivial plane is equivalent to

$$\theta''_{(1)} + q(\lambda, \alpha)\theta_{(1)} = 0, \qquad \theta_{(1)}(0) = 0 = \theta_{(1)}(1), \tag{5.11}$$

where

$$q(\lambda, \alpha) \equiv \lambda\mu_M^0(\lambda, \alpha)[\nu^0(\lambda, \alpha) + \lambda\eta_H^0(\lambda, \alpha)]. \tag{5.12}$$

Problem (5.11) has a nontrivial solution, given by $\theta_{(1)}(s) =$ const. $\sin n\pi s$, if and only if there is a positive integer n such that

$$q(\lambda, \alpha) = n^2\pi^2. \tag{5.13}$$

We refer to the set of points (λ, α) satisfying (5.13) as the *nth eigencurve* (even though these points may not actually constitute one or even several curves). Note that our constitutive assumptions imply that

$$q(\lambda, \alpha) > 0 \quad \text{for } \lambda > 0, \qquad q(0, \alpha) = 0, \qquad q_\lambda(0, \alpha) > 0. \tag{5.14}$$

Simplicity of eigenvalues. We now carry out computations to check whether the hypotheses about simple eigenvalues of the Local and Global Multiparameter Bifurcation Theorems are met.

It is an elementary exercise to compute the operator $\mathbf{L}_\lambda$ from (5.10). Let $\mathbf{u}_{(1)} \equiv (\theta_{(1)}, M_{(1)}) = \text{const.}\big((n\pi)^{-1}\mu_M^0 \sin n\pi s, \cos n\pi s\big)$ denote the eigenvector corresponding to an eigenpair (λ, α) satisfying (5.13). Let $\langle\cdot,\cdot\rangle$ denote the L_2-inner product. Then the linear operator $\mathbf{P}(\lambda, \alpha)$ defined by

$$(5.15) \qquad \mathbf{P}(\lambda,\alpha)\cdot\mathbf{u} \equiv \frac{\langle\theta,\theta_{(1)}\rangle + \langle M, M_{(1)}\rangle}{\langle\theta_{(1)},\theta_{(1)}\rangle + \langle M_{(1)}, M_{(1)}\rangle}\mathbf{u}_{(1)} \equiv J(\mathbf{u},\lambda,\alpha)\mathbf{u}_{(1)}$$

projects $C^0 \times C^0$ onto span $\{\mathbf{u}_{(1)}\}$. Here $J(\cdot, \lambda, \alpha)$ is a linear functional.

Let us fix $\alpha = \alpha^\sharp$. Let $q(\lambda^\sharp, \alpha^\sharp) = n^2\pi^2$. Then according to (V.4.10) and (V.4.11), $\lambda^\sharp$ is a simple eigenvalue of the problem

$$(5.16) \qquad \mathbf{u}_{(1)} = \mathbf{L}(\lambda, \alpha^\sharp)\cdot\mathbf{u}_{(1)}$$

iff

$$(5.17) \qquad \dim\ker\big(\mathbf{I} - \mathbf{L}(\lambda^\sharp, \alpha^\sharp)\big) = 1,$$

$$(5.18) \qquad \dim\ker\big(\mathbf{I} - \mathbf{L}(\lambda^\sharp, \alpha^\sharp) - \mathbf{L}_\lambda(\lambda^\sharp, \alpha^\sharp)\cdot\mathbf{P}(\lambda^\sharp, \alpha^\sharp)\big) = 0.$$

$\mathbf{P}$ can be replaced with any projection onto span $\mathbf{u}_{(1)}$. The solution of (3.2) shows that (5.17) holds if $q(\lambda^\sharp, \alpha^\sharp) = n^2\pi^2$. A straightforward calculation shows that (θ, M) is in the null space given in (5.18) if and only if θ satisfies

$$(5.19) \qquad \theta'' + q^0\theta = (n\pi)^{-1}\big\{[(n\pi)^2 - q^0]\mu_{NM}^0 - \mu_M^0 q_\lambda^0\big\} J \sin n\pi s$$

and boundary conditions (5.2a,b). Here the superscript 0 denotes evaluation at $(\lambda^\sharp, \alpha^\sharp)$. We readily find (e.g., by using the alternative theorem and the definition of J) that the boundary-value problem for (5.19) has a nontrivial solution if and only if the coefficient of $J \sin n\pi s$ in (5.19) vanishes. It cannot do so if $q_\lambda(\lambda^\sharp, \alpha^\sharp) \neq 0$. Hence,

5.20. Lemma. *Let $q(\lambda^\sharp, \alpha^\sharp) = n^2\pi^2$. Then $\lambda^\sharp$ is a simple eigenvalue of (5.11) if and only if $q_\lambda(\lambda^\sharp, \alpha^\sharp) \neq 0$.*

Let us now fix $\lambda^\sharp$ and again let $q(\lambda^\sharp, \alpha^\sharp) = n^2\pi^2$. Then $\alpha^\sharp$ is a simple eigenvalue of the problem

$$(5.21) \qquad \mathbf{u}_{(1)} = \mathbf{L}(\lambda^\sharp, \alpha^\sharp)\cdot\mathbf{u}_{(1)}$$

if and only if (5.17) and (5.18) hold with $\mathbf{L}_\lambda$ replaced with $\mathbf{L}_\alpha$. By methods similar to those leading to Lemma 5.20 we obtain

5.22. Lemma. *Let $q(\lambda^\sharp, \alpha^\sharp) = n^2\pi^2$. Then $\alpha^\sharp$ is a simple eigenvalue of (5.21) if and only if $\mu_M^0 q_\alpha \neq 0$ at $(\lambda^\sharp, \alpha^\sharp)$.*

By using elementary operator theory or by direct computation we find that $(\theta, M) \in \ker\big((\mathbf{I} - \mathbf{L})^2\big)$ if and only if

$$(5.23) \qquad \theta'''' + 2q\theta'' + q^2\theta = 0, \quad \theta(0) = \theta''(0) = 0 = \theta(1) = \theta''(1).$$

A straightforward computation then yields

5.24. Lemma. *Let* $q(\lambda^\sharp, \alpha^\sharp) = n^2\pi^2$. *Then*

$$\ker\left([\mathsf{I} - \mathsf{L}(\lambda^\sharp, \alpha^\sharp)]^2\right) = \ker\left(\mathsf{I} - \mathsf{L}(\lambda^\sharp, \alpha^\sharp)\right). \tag{5.25}$$

These results enable us to apply the local and global bifurcation theory to branches bifurcating from simple eigenvalues. We discuss bifurcation from double eigenvalues in the following sections.

5.26. Problem. For nonuniform rods, develop a complete and detailed spectral theory analogous to that leading to Lemmas 5.20, 5.22, and 5.24.

6. Planar Buckling of Rods. Imperfection Sensitivity via Singularity Theory

To study the detailed local behavior of solution pairs near bifurcation points, it is useful to reduce our problem to a finite-dimensional problem by the Poincaré shooting method:

Let $\big(\theta(\cdot, \lambda, \alpha, m), M(\cdot, \lambda, \alpha, m)\big)$ denote the solution to the initial-value problem consisting of

$$\begin{aligned} \theta' &= \hat{\mu}(-\lambda\cos\theta, \Lambda\sin\theta, M, \alpha,), \\ M' &= -\lambda[\hat{\nu}(-\lambda\cos\theta, \Lambda\sin\theta, M, \alpha,)\sin\theta \\ &\qquad + \hat{\eta}(-\lambda\cos\theta, \Lambda\sin\theta, M, \alpha,)\cos\theta], \\ \theta(0) &= 0, \qquad M(0) = m. \end{aligned} \tag{6.1}$$

(θ, M) is a solution of our boundary-value problem if and only if it is a solution of (6.1) defined for $s \in [0, 1]$ that satisfies

$$\theta(1, \lambda, \alpha, m) = 0. \tag{6.2}$$

Since (6.1) admits the trivial solution for $m = 0$, standard perturbation theory for ordinary differential equations ensures that (6.1) has a solution defined for $s \in [0, 1]$ if $|m|$ is sufficiently small and that this solution depends smoothly on λ, α, and m (see Sec. XVIII.2).

We now study the finite-dimensional equation (6.2) in a neighborhood of $(\lambda, \alpha, m) = (\lambda_0, \alpha_0, 0)$ where $(\lambda_0, \alpha_0, 0)$ satisfies (5.13). If we can determine how m depends on λ and α, then we know how the solution of (5.8) depends on these variables.

We first observe that (IV.2.9) implies that the solution of (6.1) satisfies the oddness condition:

$$\theta(\cdot, \lambda, \alpha, m) = -\theta(\cdot, \lambda, \alpha, -m), \quad M(\cdot, \lambda, \alpha, m) = -M(\cdot, \lambda, \alpha, -m). \tag{6.3}$$

In carrying out the computations underlying the following results, we use the fact that the only derivatives up to order three of the constitutive functions at the trivial state that are not forced to vanish by the symmetry conditions (IV.2.9) are

(6.4)
$\nu^0,$
$\nu^0_N,\ \eta^0_H,\ \mu^0_M,$
$\nu^0_{NN},\ \nu^0_{HH},\ \nu^0_{MM},\ \eta^0_{NH},\ \mu^0_{NM},$
$\nu^0_{NNN},\ \nu^0_{NHH},\ \nu^0_{NMM},\ \eta^0_{NNH},\ \eta^0_{HHH},\ \eta^0_{HMM},\ \mu^0_{NNM},\ \mu^0_{HHM},\ \mu^0_{MMM}.$

The theory of ordinary differential equations implies that $\big(\theta_m(\cdot,\lambda,\alpha,0), M_m(\cdot,\lambda,\alpha,0)\big)$ satisfies the linear initial-value problem

$$\tilde{\theta}' = \mu^0_M \tilde{M}, \quad \tilde{M}' = -\lambda(\nu^0 + \lambda\eta^0_H)\tilde{\theta}, \quad \tilde{\theta}(0) = 0, \quad \tilde{M}(0) = 1. \tag{6.5}$$

We readily find that

$$\begin{aligned} \theta_m(s,\lambda,\alpha,0) &= \frac{\mu^0_M(\lambda,\alpha)}{\sqrt{q(\lambda,\alpha)}} \sin\sqrt{q(\lambda,\alpha)}s, \\ M_m(s,\lambda,\alpha,0) &= \cos\sqrt{q(\lambda,\alpha)}s, \end{aligned} \tag{6.6}$$

which (5.13) reduces to

$$\theta_m(s,\lambda_0,\alpha_0,0) = \frac{\mu^0_M(\lambda_0,\alpha_0)}{n\pi}\sin n\pi s, \quad M_m(s,\lambda_0,\alpha_0,0) = \cos n\pi s, \tag{6.7}$$

whence

$$\theta_m(1,\lambda_0,\alpha_0,0) = 0. \tag{6.8}$$

The symmetry condition (6.3) implies that

$$\begin{aligned} \theta_{mm}(\cdot,\lambda_0,\alpha_0,0) &= 0 = M_{mm}(\cdot,\lambda_0,\alpha_0,0), \\ \theta_\lambda(\cdot,\lambda_0,\alpha_0,0) &= 0 = M_\lambda(\cdot,\lambda_0,\alpha_0,0), \\ \theta_{\lambda\lambda}(\cdot,\lambda_0,\alpha_0,0) &= 0 = M_{\lambda\lambda}(\cdot,\lambda_0,\alpha_0,0), \\ \theta_\alpha(\cdot,\lambda_0,\alpha_0,0) &= 0 = M_\alpha(\cdot,\lambda_0,\alpha_0,0). \end{aligned} \tag{6.9}$$

From (6.6) (or else from a perturbation analysis of (6.1)), we find

$$\begin{aligned} \theta_{m\lambda}(s,\lambda,\alpha,0) &= -\left(\mu^0_{MN} q^{-1/2} + \tfrac{1}{2}\mu^0_M q^{-3/2} q_\lambda\right)\sin q^{1/2}s \\ &\qquad + \tfrac{1}{2}\mu^0_M q^{-1} q_\lambda s \cos q^{1/2}s, \\ M_{m\lambda}(s,\lambda,\alpha,0) &= -\tfrac{1}{2} q^{-1/2} q_\lambda s \sin q^{1/2}s, \end{aligned} \tag{6.10}$$

which, by (5.13), yields

$$\theta_{m\lambda}(1,\lambda_0,\alpha_0,0) = \frac{(-1)^n \mu^0_M q_\lambda}{2n^2\pi^2}. \tag{6.11}$$

We readily compute $\theta_{m\lambda\lambda}(1, \lambda_0, \alpha_0, 0)$ and $\theta_{m\alpha}(1, \lambda_0, \alpha_0, 0)$ from (6.6). A straightforward but lengthy perturbation process yields an expression for $\theta_{mmm}(1, \lambda_0, \alpha_0, 0)$. It depends on a complicated combination of derivatives of the constitutive functions up to order three. Thus we have no general constitutive restrictions to tell us about this variable. It is therefore highly probable that $\theta_{mmm}(1, \lambda_0, \alpha_0, 0) \neq 0$. The question of which derivatives of $\theta(1, \lambda, \alpha, 0)$ are needed at (λ_0, α_0) to give an accurate description of local behavior is answered by singularity theory (see Golubitsky & Schaeffer (1985, Chaps. II, III)):

Let α_0 be fixed and let (5.13) hold. Let $\varepsilon = \pm 1$, $\delta = \pm 1$. If

$$\text{(6.12a,b)} \qquad \theta(1, \lambda_0, \alpha_0, 0) = 0, \qquad \theta_m(1, \lambda_0, \alpha_0, 0) = 0,$$

$$\text{(6.12c,d)} \qquad \theta_{mm}(1, \lambda_0, \alpha_0, 0) = 0, \qquad \theta_\lambda(1, \lambda_0, \alpha_0, 0) = 0,$$

$$\text{(6.12e,f)} \quad \operatorname{sign} \theta_{mmm}(1, \lambda_0, \alpha_0, 0) = \varepsilon, \quad \operatorname{sign} \theta_{m\lambda}(1, \lambda_0, \alpha_0, 0) = \delta,$$

then this local problem for (6.2) is *strongly equivalent* to the problem

$$\text{(6.13)} \qquad (\varepsilon x^2 + \delta\omega)x = 0,$$

i.e., there exist smooth functions $(x, \omega) \mapsto F(x, \omega)$, $X(x, \omega)$ defined on a neighborhood of the origin $(0,0)$ in $\mathbb{R}^2$ with $F(0,0) \neq 0$, and $X_x(0,0) > 0$ such that

$$\text{(6.14)} \qquad (\varepsilon x^2 + \delta\omega)x = F(x, \omega)\theta\big(1, \lambda_0 + \omega, \alpha_0, X(x, \omega)\big).$$

Problem (6.13) is said to have a *normal form*. (Note that the function of (6.14) is the simplest function having the properties of (6.12).)

Problem (6.13) has two two-parameter *universal unfoldings* of the form

$$\text{(6.15a,b)} \qquad \varepsilon x^3 + \delta\omega x + \beta + \gamma x^2 = 0 \quad \text{or} \quad \varepsilon x^3 + \delta\omega x + \beta + \gamma\omega = 0.$$

These equations capture in a mathematically precise sense the effects of all sufficiently smooth small perturbations (imperfections) of (6.13) when (6.12) holds. If it happens that $\theta_{mmm}(1, \lambda_0, \alpha_0, 0) = 0$, then (6.2) is equivalent to some other problem (with a more complicated normal form).

There is no end to the number of ways we can introduce physical or geometrical imperfections into our problem: The natural reference configuration need not be straight, the end load need not be applied in the $-\boldsymbol{i}$ direction, the end load could be offset so that it is equipollent to a force and a moment, there can be a transverse load, etc. Singularity theory says that the variety of local mathematical response is not increased by introducing more than the two unfolding parameters and that the use of fewer parameters could very well result in missing important phenomena.

Suppose that we want to introduce a specific pair (β, γ) of imperfection parameters. Thus we want to study a problem of the form $G(\lambda, \alpha_0, m, \beta, \gamma) = 0$ for (λ, m) near $(\lambda_0, 0)$ when $G(\lambda, \alpha_0, m, 0, 0) = \theta(1, \lambda_0, \alpha_0, m)$ and when (5.13) holds. (α is playing a passive role here.) Then $G(\lambda, \alpha_0, m, \beta, \gamma)$ is a universal unfolding of $\theta(1, \lambda_0, \alpha_0, m)$ if and only if it meets the following simple test (see Golubitsky & Schaeffer (1985, Prop. III.4.4)):

$$\text{(6.16)} \qquad \det \begin{pmatrix} 0 & 0 & G_{m\lambda} & G_{mmm} \\ 0 & G_{\lambda m} & G_{\lambda\lambda} & G_{\lambda mm} \\ G_\beta & G_{\beta m} & G_{\beta\lambda} & G_{\beta mm} \\ G_\gamma & G_{\gamma m} & G_{\gamma\lambda} & G_{\gamma mm} \end{pmatrix} \neq 0 \quad \text{at } (\lambda_0, \alpha_0, 0, 0, 0).$$

In this case G is strongly equivalent to (6.15).

Suppose that $\varepsilon = 1$ and $\delta = -1$. A careful study of (6.15a) shows that if $\beta > \max\{0, \gamma^3/27\}$, then the solution branches have the form shown in Fig. 6.17a, and if $0 < \beta < \gamma^3/27$, then they have the form shown in Fig. 6.17b (cf. Fig. V.1.17). The figures for $\beta < 0$ are analogous. By the strong equivalence of (6.15) to the universal unfolding of (6.2), we know that the solution branches of the latter have the qualitative behavior of Fig. 6.17 near their bifurcation points. In particular, Fig. 6.17b indicates that a snapping phenomenon is likely to occur. The original study of this two-parameter universal unfolding for an elastica was carried out by Golubitsky & Schaeffer (1979).

Singularity theory is a rich and useful mathematical theory for studying solutions of finite-dimensional problems. Some procedure, such as the Lyapunov-Schmidt method or the Poincaré shooting method, based upon the Implicit Function Theorem (see Sec. XVIII.2), is needed to reduce infinite-dimensional problems to finite-dimensional problems to which the theory can be applied. Singularity theory is somewhat more general than catastrophe theory (see Poston & Stewart (1978)). The mathematical foundations of singularity theory consist in a complicated amalgamation of elementary analytic techniques, based on the Implicit Function Theorem and Taylor's Theorem, with algebraic techniques for systematizing computations. Group theory enters in an essential way for problems with symmetries (see Golubitsky, Stewart, & Schaeffer (1988)). The results of the theory are easy to apply because Golubitsky & Schaeffer (1985) have extensive tables of normal forms and their universal unfoldings for a variety of problems, including ours. For applications of catastrophe theory to problems of imperfection-sensitivity for discrete mechanical systems, see Thompson & Hunt (1984).

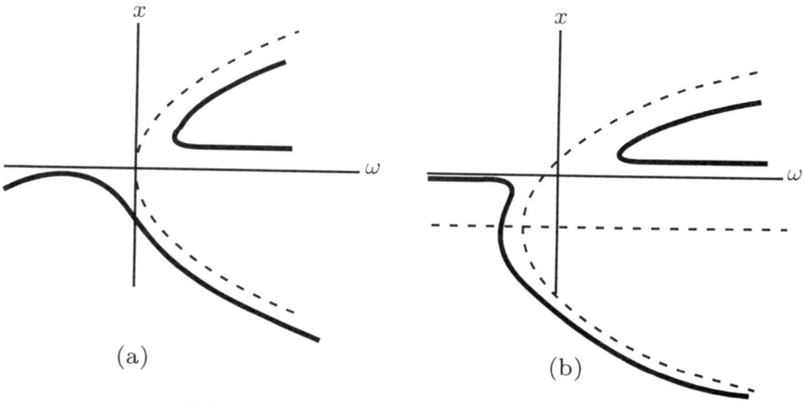

Fig. 6.17. Solution branches of (6.15).

7. Planar Buckling of Rods. Constitutive Assumptions

For the elastica treated in Sec. V.5, $q(\lambda,\alpha) = \lambda/EI$. (Here α can be related to the stiffness EI.) For this q, all the preceding computations become particularly simple. We now show that compressibility can produce q's for which

$$q(\lambda,\alpha) \to 0 \quad \text{as } \lambda \to \infty. \tag{7.1}$$

In view of (5.14) and (7.1), the function $q(\cdot,\alpha)$ has a positive maximum at a positive value of λ. Thus $q(\cdot,\alpha)$ may well have the form shown in Fig. 7.2. We shall show that for reasonable constitutive functions it is possible to vary the height of $q(\cdot,\alpha)$ by varying α. Thus, for appropriate α we can ensure that the maximum of this function occurs at $n^2\pi^2$, in which case the hypotheses supporting the universal unfolding just described are not valid. We can in fact force this maximum to lie below π^2, in which case there are no bifurcating branches. There are, however, nonbifurcating branches. The physical interpretation of such branches is straightforward: If a column, not too thin, composed of a relatively soft material is subjected to a gradually increasing end thrust, it shortens, thickens, but does not lose its straightness by buckling. There are no nearby equilibrium states. But if the thrust is large enough, then a large transverse force can move the compressed straight column into a stable large bowed configuration, where it will remain after the transverse force is removed. We wish to investigate these nonbifurcating branches.

We interrupt our mathematical analysis to show that (7.1) is plausible and to produce some concrete examples of constitutive functions for rods. Our approach is a primitive version of that to be conducted in Chap. XIV. Let us interpret the rod as a three-dimensional body (see Fig. IV.1.1) whose reference configuration is a cylinder, which for simplicity is taken to have the rectangular cross section $\mathcal{A} \equiv \{(x,y)\colon |x| \le h,\ 0 \le y \le 1\}$ of height $2h$ and of unit width. We assume that the deformed position $\tilde{\boldsymbol{p}}(x,y,s)$ of the material point (x,y,s) is constrained to have the form

$$\tilde{\boldsymbol{p}}(x,y,s) = \boldsymbol{r}(s) + x\boldsymbol{b}(s) + y\boldsymbol{k}, \quad \boldsymbol{r}'(s) = \nu(s)\boldsymbol{a}(s) \tag{7.3}$$

where $\boldsymbol{a}$ and $\boldsymbol{b}$ are defined in (IV.1.3). Thus the stretch of the longitudinal fiber through (x,y) is $\hat{\boldsymbol{p}}_s(x,y,s)\cdot\boldsymbol{a}(s) = \nu(s) - x\theta'(s)$. (Its positivity everywhere leads to the inequality $\nu > h|\theta'|$, which defines the domain of the constitutive functions; cf. (IV.3.22).) Constraint (7.3) ensures that $\eta = 0$.

In a deformed configuration of this rod, we let $T(x,y,s)$ denote the component of force per unit reference area (i.e., a *stress*) acting across the material cross section at s in the direction $\boldsymbol{a}(s)$. Then

$$N(s) = \int_0^1 dy \int_{-h}^{h} T(x,y,s)\,dx, \quad M(s) = -\int_0^1 dy \int_{-h}^{h} xT(x,y,s)\,dx. \tag{7.4}$$

Let us decompose T into a part T_{L} that makes no contribution to the integrals of (7.4) and a complementary part T_{A}. (In Chap. XIV we show that T_{L} can be identified with a component of the Lagrange multiplier maintaining the constraint (7.3).) For a uniform elastic material, we assume that $T_{\mathrm{A}}(x,y,s)$ depends only on the stretch of the longitudinal fiber through (x,y):

$$T_{\mathrm{A}}(x,y,s) = \hat{T}\big(\nu(s) - x\theta'(s)\big). \tag{7.5}$$

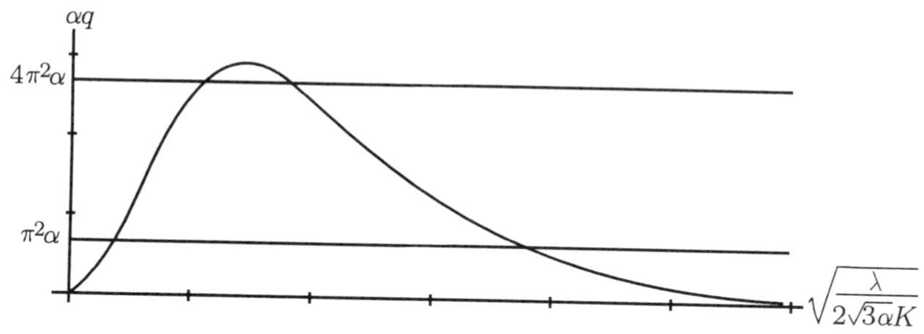

Fig. 7.2. Eigenvalues of (5.11) are those values of λ at which the graph of $q(\cdot, \alpha)$ intersects the horizontal lines with ordinates $n^2\pi^2$. The graph shown here, with an appropriate scaling, corresponds to (7.16) with $\beta = 1$.

We now define our constitutive equations for the planar deformation of a uniform unshearable elastic rod in terms of (7.5) by

(7.6) $$\hat{N}(\nu, \mu, \alpha) = \int_{-h}^{h} \hat{T}(\nu - x\mu)\, dx, \quad \hat{M}(\nu, \mu, \alpha) = -\int_{-h}^{h} x\hat{T}(\nu - x\mu)\, dx$$

where

(7.7) $$\alpha = h^2/3.$$

We can define the stored energy function W for the rod in terms of the corresponding function Ω for the constrained three-dimensional body by setting

(7.8) $$W(\nu, \mu, \alpha) \equiv \int_{-h}^{h} \Omega(\nu - x\mu)\, dx, \quad \Omega(\delta) \equiv \int_{1}^{\delta} T(u)\, du.$$

Then we can write (7.6) as

(7.9) $$N(\nu, \mu, \alpha) = W_\nu(\nu, \mu, \alpha), \quad M(\nu, \mu, \alpha) = W_\mu(\nu, \mu, \alpha).$$

We can readily represent derivatives of (7.6) with respect to ν and μ in terms of integrals of the derivatives of $\hat{T}$. If $\hat{B}(\nu, \mu, \alpha)$ stands for any such derivative of (7.6), then we set $B^*(\nu, \alpha) \equiv \hat{B}(\nu, 0, \alpha)$. Using these representations we find that all derivatives of (7.6) at $\mu = 0$ can be expressed in terms of derivatives of N^* with respect to ν alone:

(7.10)
$$\begin{aligned}
&N^*_\mu = M^*_\nu = M^*_\alpha = 0, \; M^*_\mu = \alpha N^*_\nu, \; N^*_\alpha = N^*/2\alpha, \\
&N^*_{\nu\mu} = N^*_{\mu\alpha} = M^*_{\nu\nu} = M^*_{\nu\alpha} = M^*_{\mu\mu} = M^*_{\alpha\alpha} = 0, \; N^*_{\mu\mu} = M^*_{\nu\mu} = \alpha N^*_{\nu\nu}, \\
&N^*_{\nu\alpha} = N^*_\nu/2\alpha, \; M^*_{\mu\alpha} = \tfrac{3}{2} N^*_\nu, \; N^*_{\alpha\alpha} = -N^*/4\alpha^2, \\
&N^*_{\nu\nu\mu} = N^*_{\mu\mu\mu} = M^*_{\nu\mu\mu} = M^*_{\nu\nu\nu} = 0, \; N^*_{\nu\mu\mu} = M^*_{\nu\nu\mu} = \alpha N^*_{\nu\nu\nu}, \; M^*_{\mu\mu\mu} = \tfrac{9}{5}\alpha^2 N^*_{\nu\nu\nu}.
\end{aligned}$$

Since $(N, M) \mapsto (\hat{\nu}(N, M, \alpha), \hat{\mu}(N, M, \alpha))$ is the inverse of the pair of constitutive functions defined by (7.6), we can represent those members of (6.4) not involving η or H in terms of these ν-derivatives of N^*. Since $\lambda = -N^*(\nu^0(\lambda, \alpha), \alpha)$ is an identity, we find from (5.12) and (7.10) that

$$(7.11) \qquad q(\lambda) = -\frac{N^*\left(\nu^0(\lambda, \alpha), \alpha\right) \nu^0(\lambda, \alpha)}{M^*_\mu\left(\nu^0(\lambda, \alpha), \alpha\right)} = -\frac{N^*\left(\nu^0(\lambda, \alpha), \alpha\right) \nu^0(\lambda, \alpha)}{\alpha N^*_\nu\left(\nu^0(\lambda, \alpha), \alpha\right)}$$

for an unshearable rod. Under our standard constitutive assumption that $N^*(\nu, \alpha) \to -\infty$ as $\nu \to 0$, we then obtain

$$(7.12) \qquad \lim_{\lambda \to \infty} q(\lambda) = -\lim_{\nu \to 0} \frac{N^*(\nu, \alpha)\nu}{\alpha N^*_\nu(\nu, \alpha)}.$$

If, furthermore, $N^*(\cdot, \alpha) \in C^2(0, 1)$, say, and if $N^*(\nu, \alpha)$ 'behaves like' $-C\nu^{-\beta}$ for ν small, where C and β are positive constants, then (7.1) holds. More generally, if we supplement the standard requirements that $N^*(\nu, \alpha) \to -\infty$ as $\nu \to 0$ and $N^*_\nu(\nu, \alpha) > 0$ with the requirement that $N^*_{\nu\nu}(\nu, \alpha)$ is defined and negative for sufficiently small and positive ν, then (7.1) holds.

7.13. Exercise. Prove this last statement and show that it need not be true if the condition on the negativity of $N^*_{\nu\nu}(\nu, \alpha)$ is not imposed.

We now study a specific family of constitutive functions:

$$(7.14\text{a,b}) \qquad N^*(\nu, \alpha) = 2hK(\nu^\gamma - \nu^{-\beta}) \text{ or, equivalently, } \hat{T}(\nu) = K(\nu^\gamma - \nu^{-\beta}),$$

where $K > 0$, $\beta > 0$, $\gamma \geq 0$ are given numbers. These functions, which meet our standard requirements, have the additional virtue that they permit the integrals in (7.6) to be evaluated explicitly. Equation (7.14) with $\gamma = 0$ is perfectly reasonable when $\nu \leq 1$, i.e., when the material is under compression. We shall accordingly use this version of (7.14) to illustrate local bifurcation from a trivial solution, which represents a uniformly compressed state. (It is physically unrealistic to take $\gamma = 0$ for global bifurcation because for very large λ, parts of the column can be in tension.) For $\gamma = 0$, (7.14a) is equivalent to

$$(7.15) \qquad \nu^0(\lambda, \alpha) = \left(\frac{2hK}{\lambda + 2hK}\right)^{1/\beta}.$$

Then by using our representations for (6.4) in terms of (7.10), we find

$$(7.16) \qquad q(\lambda, \alpha) = \frac{\lambda}{2hK\alpha\beta}\left(\frac{2hK}{\lambda + 2hK}\right)^{1+2/\beta}.$$

The graph of this function for $\beta = 1$ is shown in Fig. 7.2. Note that it has a unique maximum at $\lambda = hK\beta$ and that it satisfies (5.14) and (7.1). The corresponding eigencurves in the (λ, α)-plane are defined by substituting (7.7) and (7.16) into (5.12).

Equation (7.14) is typical of a wide range of one-dimensional elastic constitutive equations for compression. The combination of any such law with (7.10) would produce comparable results. The basic assumption underlying the derivation of (7.14) from (7.3) and (7.6) is that (7.3) in the form used makes no provision for transverse expansions and for shear. Preliminary studies show that accounting for transverse expansions in (7.3) seems to preserve much of the essential character of (7.10) and the ensuing analysis. But the treatment of shear introduces new features: The representation for H analogous to (7.4) involves a shear component of the stress besides the normal component T. Thus the resulting constitutive relation for H is independent of those for N and M, which are related by (7.10). In the absence of both a mathematical doctrine and a useful body of experimental results saying how H depends on elongations for large compressions, we must entertain the possibility, inherent in (5.12), that shearability could cause (7.1) to fail.

For $\beta = 3$ and $\gamma = 3$ in (7.14), we find that (7.6) reduces to (7.9) with

$$W = Kh\left\{\tfrac{1}{10}[5\nu^4 + 10\nu^2(h\mu)^2 + (h\mu)^4] + (\nu^2 - h^2\mu^2)^{-1}\right\}. \tag{7.17}$$

Notice how nicely (7.17) accounts for the condition that $\nu > h|\mu|$.

8. Planar Buckling of Rods. Nonbifurcating Branches

We assume that (7.1) holds, whether or not the rod is unshearable. Thus $q(\cdot, \alpha)$ has a positive maximum at a positive value of λ. In this section we study solutions of (6.2) for (λ, α) lying near a point (λ_0, α_0) for which

$$q(\lambda_0, \alpha_0) = n^2\pi^2, \quad q_\lambda(\lambda_0, \alpha_0) = 0, \quad q_{\lambda\lambda}(\lambda_0, \alpha_0) < 0. \tag{8.1a,b,c}$$

Conditions (8.1b) and (6.11) imply that

$$\theta_{m\lambda}(1, \lambda_0, \alpha_0, 0) = 0, \tag{8.2}$$

so that (6.12f) does not hold. We want to study the effect of varying only the parameter α on the bifurcation diagram. Such variations preserve the '$\mathbb{Z}_2$-symmetry' in (6.3). We do not introduce imperfection parameters as in Sec. 6. If we adopt (7.16), then from (8.1) and a lengthy computation we find that

(8.3)
$$\theta = 0, \quad \theta_m = 0, \quad \theta_{mm} = 0, \quad \theta_\lambda = 0, \quad \theta_{\lambda\lambda} = 0, \quad \theta_{m\lambda} = 0, \quad \theta_\alpha = 0,$$
$$(-1)^n\theta_{mmm} > 0, \quad (-1)^n\theta_{m\lambda\lambda} < 0, \quad (-1)^n\theta_{m\alpha} < 0$$

where the arguments of these derivatives of θ are $(1, \lambda_0, \alpha_0, 0)$. In particular, we have $(-1)^n 2n^2\pi^2\theta_{m\lambda\lambda} = \mu_M^0 q_{\lambda\lambda}$ and $(-1)^n 2n^2\pi^2\theta_{m\alpha} = \mu_M^0 q_\alpha$ at this point.

The $\mathbb{Z}_2$-singularity theory of Golubitsky & Schaeffer (1985, Chap. VI) says that if $\alpha = \alpha_0$ is fixed and if (8.3) holds, then problem (6.2) is *strongly $\mathbb{Z}_2$-equivalent* to the problem

$$(x^2 - \omega^2)x = 0, \tag{8.4}$$

i.e., there exist smooth functions $(x, \omega) \mapsto F(x, \omega)$, $X(x, \omega)$ defined on a neighborhood of the origin $(0, 0)$ in $\mathbb{R}^2$ with $F(0, 0) \neq 0$, $X_x(0, 0) > 0$, $F(\cdot, \omega)$ even, and $X(\cdot, \omega)$ odd such that

$$(x^2 - \omega^2)x = F(x, \omega)\theta\big(1, \lambda_0 + \omega, \alpha_0, X(x, \omega)\big). \tag{8.5}$$

Problem (8.4) has a one-parameter $\mathbb{Z}_2$-universal unfolding of the form

$$(x^2 - \omega^2 - \gamma)x = 0. \tag{8.6}$$

This equation captures the effects of all sufficiently smooth small perturbations of (8.4) that preserve the symmetry (6.3).

By using (8.5), which relates (6.2), (8.4), and (8.6), together with Golubitsky & Schaeffer's (1985) Definition VI.3.1 of a $\mathbb{Z}_2$-universal unfolding, we can readily show that $(\lambda, \alpha, m) \mapsto \theta(1, \lambda, \alpha, m)$ is a $\mathbb{Z}_2$-universal unfolding of $(\lambda, m) \mapsto \theta(1, \lambda, \alpha_0, m)$ if $\theta_{m\alpha}(1, \lambda_0, \alpha_0, 0) \neq 0$, an inequality which is ensured by the last inequality of (8.3). (This observation represents a solution of the *recognition problem* for these universal unfoldings.) Problem (8.6) bears essentially the same relation to (6.2) as (8.4) bears to (6.2) with α fixed at α_0. The variables (ω, γ, x) of (8.6) correspond to the variables $(\lambda - \lambda_0, \alpha - \alpha_0, m)$ of (6.2).

The graph of (8.6), a saddle surface, is easily constructed. A global version of it is shown in Fig. 9.33. In summary, we have

8.7. Theorem. *Let (λ_0, α_0) satisfy* (8.1) *and* (8.3). *Then for (λ, α, m) near $(\lambda_0, \alpha_0, 0)$, the bifurcation diagram for* (6.2) *is strongly equivalent to that for* (8.6) *with $(\lambda - \lambda_0, \alpha - \alpha_0, m)$ corresponding to (ω, γ, x). The number of solutions m of* (6.2) *for fixed (λ, α) with (λ, α, m) near $(\lambda_0, \alpha_0, 0)$ equals the number of intersections of the line parallel to the x-axis that passes through $(\omega, \gamma, 0)$ with the graph of* (8.6) *with (ω, γ) corresponding to the fixed (λ, α) under the equivalence generating* (8.5). *In particular, for fixed $\alpha < \alpha_0$, there are nontrivial branches that bifurcate from the trivial plane, while for fixed $\alpha > \alpha_0$, there are nontrivial branches that do not bifurcate from the trivial plane.*

If θ_{mmm} and $\theta_{m\lambda\lambda}$ have the same sign at $(1, \lambda_0, \alpha_0, 0)$, then (6.2) has a one-parameter $\mathbb{Z}_2$-universal unfolding of the form

$$(x^2 + \omega^2 \pm \gamma)x = 0, \tag{8.8}$$

the graph of which is readily constructed. Our considerations of the last section suggest that this situation is not so common.

Theorem 8.7 has an immediate global consequence. It asserts that nonbifurcating branches for fixed α are connected to bifurcating branches for a different fixed value of α by a two-dimensional surface of solution pairs. Therefore, the nonbifurcating branches inherit their nodal properties from those of the bifurcating branches, because these two branches can be joined by a branch containing no trivial solution pairs. Thus, we have exploited the availability of the parameter α to provide a connectivity method by which we can give a detailed qualitative picture of nonbifurcating branches.

9. Global Disposition of Solution Sheets

We now determine how the bifurcating sheets are disposed in $C^0 \times C^0 \times [0, \infty) \times [\alpha_*, \infty)$. Antman & Rosenfeld (1978) have shown by sophisticated versions of the techniques of Sec. V.5 that under very mild constitutive restrictions (which we do not spell out here, but which are met by (7.17)), there is a continuous positive-valued function f defined on $[0, \infty) \times [\alpha_*, \infty)$ with the property that if $((\lambda, \alpha), \mathbf{u})$ is a solution pair of (5.8), then $\|\mathbf{u}\| < f(\lambda, \alpha)$. (Here $\|\cdot\|$ denotes a norm for $\mathcal{X}$.) This fact means that

$\mathbf{u}$ cannot blow up unless λ or α approaches ∞. (Since Antman & Rosenfeld obtained their result for a whole family of materials, no parameter such as α intervenes in their analysis. By examining the role of h in (7.17) we can readily show that the estimate improves as h or α grows.) By using the Sturmian theory we can show that for bounded α the bifurcating sheets are bounded away from the plane $\lambda = 0$. (For refined results of this sort, see Olmstead (1977).) We can likewise show that under mild constitutive conditions there is a continuous positive-valued function g defined on an unbounded interval in $[0, \infty)$ with the property that if $(\mathbf{u}, \lambda, \alpha)$ is a *nontrivial* solution pair of (5.8), then $\alpha < g(\lambda)$ (see Antman & Pierce (1990)).

We now show how to analyze the behavior of branches for large λ. For simplicity, we just treat unshearable columns. Throughout our analysis, we fix α and consequently suppress its appearance. We shall show that the bifurcation diagrams for compressible columns can differ strikingly from that for the incompressible elastica shown in Fig. V.1.15. We begin with general considerations and then for simplicity restrict our attention to the material (7.17).

Let us assume that (7.9) holds and that the coercivity conditions (IV.1.24)–(IV.1.26) hold. Then our governing equations (for unshearable rods, which are versions of the differential equations corresponding to (5.3) and (5.4)) admit the first integral

$$\theta' W_\mu(\nu, \theta') + \nu W_\nu(\nu, \theta') - W(\nu, \theta') = c \text{ (const.)}, \tag{9.1a}$$

where ν is defined to be the solution of

$$-\lambda \cos\theta = W_\nu(\nu, \theta'). \tag{9.1b}$$

By Ex. IV.2.17, this integral can be written in the form

$$\theta' W^\sharp_\mu(-\lambda\cos\theta, \theta') - W^\sharp(-\lambda\cos\theta, \theta') = c, \tag{9.2}$$

which defines a family of trajectories in the (θ, θ')-phase plane. If $\theta'(0) > 0$, then any trajectory satisfying (5.2a,b) must first cross the positive θ-axis at some value $\theta = \psi$. In this case, we can evaluate c in (9.2) at $(\theta, \theta') = (\psi, 0)$ and find that c depends on ψ and λ. Exercise IV.2.17 also shows that (9.2) is equivalent to an equation of the form

$$h^2(\theta')^2 = F(\theta, \psi, \lambda) \tag{9.3}$$

where $F(\cdot, \psi, \lambda)$ has period 2π. Here the h^2, which appears throughout Sec. 7, is introduced for future algebraic convenience. The Implicit Function Theorem implies that F is continuously differentiable when W is twice continuously differentiable, etc.

The form of (9.3) says that ψ is the maximum value of θ for a solution of our boundary-value problem and that

$$F(\psi, \psi, \lambda) = 0. \tag{9.4a}$$

From the analysis of Sec. IV.2 we readily find that the only singular points in the phase portrait of (9.2) are centers at $(2k\pi, 0)$ and saddles at $((2k+1)\pi, 0)$, $k = 0, \pm 1, \pm 2, \dots$. Since there can be no singular points between $\theta = 0$ and $\theta = \pi$ for all values of the parameters, we find that

$$F(\theta, \psi, \lambda) > 0 \quad \text{for} \ -\psi < \theta < \psi. \tag{9.4b}$$

Since $(\theta, \theta') = (0, 0)$, $(\pi, 0)$ are singular points, F must satisfy

$$F_\theta(0, 0, \lambda) = 0, \quad F_\theta(\pi, \pi, \lambda) = 0. \tag{9.4c,d}$$

An examination of the linearizations of the differential equations corresponding to (5.3) and (5.4) about the saddle point $(\pi, 0)$, which reduces to the study of a simple eigenvalue problem, shows that the separatrices approach it along directions that are not parallel to the coordinate axes. Therefore,

$$F_{\theta\theta}(\pi, \pi, \lambda) > 0. \tag{9.4e}$$

Thus (9.2) (or (9.3)) has a phase portrait qualitatively the same as that for the elastica, (which is identical to that for the pendulum). But the dependence of our portrait on λ is radically different.

We now state and prove our basic multiplicity result for the material of (7.17). Afterwards we easily abstract it so that it applies to a wide class of materials.

9.5. Theorem. *Let (7.17) hold. with $\beta = 3 = \gamma$. There is a number Λ_n such that if $\lambda > \Lambda_n$, then (5.8) has (at least) two pairs of nontrivial solutions (reflections of each other) having exactly $n+1$ zeros on $[0,1]$.*

Proof. We obtain a solution of (5.8) if exactly one unit of the independent variable s is used up on a phase-plane trajectory that begins and ends on the θ'-axis. In particular, (9.3) then implies that there is a solution for which θ has exactly $n+1$ zeros on $[0,1]$ if λ and ψ satisfy

$$J(\psi, \lambda) \equiv \int_0^\psi \frac{d\theta}{\sqrt{F(\theta, \psi, \lambda)}} = \frac{1}{2hn}. \tag{9.6}$$

A solution pair (ψ, λ) of (9.6) determines a unique phase-plane trajectory for (9.3) and a unique solution θ of (5.8), which is defined implicitly by

$$s = h \int_0^\theta \frac{d\phi}{\sqrt{F(\phi, \psi, \lambda)}} \quad \text{for } 0 \le s \le \frac{1}{2n}. \tag{9.7}$$

ψ plays the same role in (9.6) as m plays in (6.2). Indeed, there is a unique $m \ge 0$ corresponding through (9.2) or (9.3) to each $\psi \in [0, \pi)$.

In view of (9.4a), it is not clear that J is continuously differentiable on $\{(\psi, \lambda): 0 \le \psi < \pi, \lambda > 0\}$. To show that it is, we introduce the changes of variables

$$u = \frac{\sin(\theta/2)}{\sin(\psi/2)}, \ \chi = \sin(\psi/2), \ F^\sharp(u, \chi, \lambda) = F(\theta, \psi, \lambda), \ J^\sharp(\chi, \lambda) = J(\psi, \lambda), \tag{9.8a}$$

so that

$$J^\sharp(\chi, \lambda) = 2\chi \int_0^1 \frac{du}{\sqrt{1 - \chi^2 u^2}\sqrt{F^\sharp(u, \chi, \lambda)}}. \tag{9.8b}$$

$J^\sharp$ is readily differentiated for $\chi < 1$. Indeed, J has as much differentiability on its domain as F has.

We must determine the behavior of $J(\cdot, \lambda)$ as $\lambda \to \infty$. We prove this theorem by showing that J has the patently nonuniform behavior:

$$J(\pi, \lambda) = \infty, \quad J(\psi, \lambda) \to \left\{ \begin{matrix} \infty & \text{for } 0 \le \psi < \pi/2 \\ 0 & \text{for } \pi/2 \le \psi < \pi \end{matrix} \right\} \quad \text{as } \lambda \to \infty. \tag{9.9a,b,c}$$

Conditions (9.4a,d) ensure that $F(\cdot, \pi, \lambda)$ is not integrable on $(0, \pi)$, so that (9.9a) holds.

The substitution of (7.17) into (9.1) shows that the corresponding function F of (9.3) is defined implicitly as the solution of the system

$$\frac{F^2}{10} + F\nu^2 + \frac{\nu^4}{2} - \frac{1}{\delta^2} = \frac{\xi^4}{2} - \frac{1}{\xi^2}, \tag{9.10a}$$

$$\nu(\nu^2 + F - \delta^{-4}) = -\lambda \cos\theta \tag{9.10b}$$

where

$$\delta^2 \equiv \nu^2 - F, \quad \xi^3 - \xi^{-3} \equiv -\lambda \cos\psi. \tag{9.10c,d}$$

We note that (9.10b) can be solved uniquely for ν as a function of the other variables appearing in this equation, in consonance with the development in Secs. IV.1 and IV.2 supporting (9.2).

We now turn to the proof of (9.9b). We let '$\sim$' designate asymptotic equivalence as $\lambda \to \infty$ and let 'O' and 'o' designate the Landau order symbols as $\lambda \to \infty$. We take ψ to be fixed in $[0, \pi/2)$, take θ to be fixed in $[0, \psi]$, and take λ large. (Let us emphasize that our results are not uniform in ψ and θ. Uniform results are not needed to prove (9.9).) Then (9.10d) implies that

$$\xi \sim (\lambda \cos \psi)^{-1/3}. \tag{9.11}$$

From (9.10a) we thus find that

$$\delta^2 \leq (\lambda \cos \psi)^{-2/3} = O(\lambda^{-2/3}), \tag{9.12a}$$

whence

$$\nu^2 = F + O(\lambda^{-2/3}). \tag{9.12b}$$

We solve (9.10a) for δ^{-2} and then substitute the resulting expression and (9.12b) into (9.10b) to obtain

$$\sqrt{F + O(\lambda^{-2/3})}\left\{\left[O(\lambda^{2/3}) + \tfrac{8}{5}F^2 + 2FO(\lambda^{-2/3})\right]^2 - 2F\right\} = \lambda \cos \theta, \tag{9.13}$$

whence

$$F = O(\lambda^{-2/3}), \quad \nu^2 = O(\lambda^{-2/3}), \tag{9.14}$$

the second equality following from (6.21b).

Now we substitute (9.14) back into (9.10a), from which we obtain the following refinement of (9.12a):

$$\delta^{-2} \sim (\lambda \cos \psi)^{2/3}. \tag{9.15}$$

We substitute (9.14) and (9.15) into (9.10b) and use (9.10c) to obtain

$$\lambda \cos \theta \sim \nu \delta^{-4} \sim \nu (\lambda \cos \psi)^{4/3}, \tag{9.16a}$$

$$F = \nu^2 - \delta^2 \sim \lambda^{-2/3} (\cos \psi)^{-8/3} [\cos^2 \theta - \cos^2 \psi]. \tag{9.16b}$$

Thus

$$J(\psi, \lambda) \sim \lambda^{1/3} (\cos \psi)^{4/3} \int_0^{\psi} \frac{d\theta}{\sqrt{\cos^2 \theta - \cos^2 \psi}}. \tag{9.17}$$

In particular, $J(0, \lambda) \sim \lambda^{1/3}\pi/2$. Condition (9.9b) follows from (9.17).

We now prove (9.9c). Let ψ be fixed in $(\pi/2, \pi)$. Then (9.10d) implies that

$$\xi \sim \lambda^{1/3}(-\cos \psi)^{1/3}. \tag{9.18}$$

For each $\theta \in (0, \psi)$, Eqs. (9.10a,c) imply that either (i) $F \to \infty$ and $\nu \to \infty$ as $\lambda \to \infty$ or (ii) $\nu \to \infty$ while F remains bounded as $\lambda \to \infty$. We prove (9.9c) by showing that (ii) cannot occur.

First let $\theta < \pi/2$. Then (9.10b,c) imply that $\delta^{-4}\nu = \delta^{-4}\sqrt{F + \delta^2} \to \infty$ as $\lambda \to \infty$. Were F not to approach ∞ as $\lambda \to \infty$, then δ would approach 0, in contradiction to the limit $\nu \to \infty$, which holds in (ii). For $\theta = \pi/2$, we find that (9.10b) implies that $\delta \to 0$, whence $F \to \infty$.

Now let $\theta \in (\pi/2, \psi)$. Were F not to approach ∞ as $\lambda \to \infty$, then (9.10b) would imply that $\nu \sim (-\lambda \cos \theta)^{1/3}$. We substitute this expression and the corresponding representation for ξ into (9.10a) to obtain an asymptotic quadratic equation for F. The positive solution of this equation is

$$\tfrac{1}{5}F \sim \lambda^{2/3}\left\{(-\cos\theta)^{2/3} + \sqrt{(-\cos\theta)^{4/3} + \tfrac{1}{5}[(-\cos\psi)^{4/3} - (-\cos\theta)^{4/3}]}\right\}, \tag{9.19}$$

which contradicts our false hypothesis.

Finally we treat the case that $\psi = \pi/2$. Here (9.10d) implies that $\xi = 1$, (9.10a) implies that $\delta^2 < 2$, whence (9.10c) implies that $F < \nu^2 < F + 2$. Thus either ν and F both approach ∞ or both stay bounded as $\lambda \to \infty$. In the former case, (9.10a) implies that $\delta \to 0$, and in the latter case, (9.10b) implies that $\delta \to 0$ when $\theta < \pi/2$. We thus conclude from (9.10c) that $\nu^2 \sim F$ when $\theta < \pi/2$. Equation (9.10a) then implies that ν^2 and $F \to \infty$ as $\lambda \to \infty$ and that $\delta^{-2} \sim \frac{8}{5}F^2$. Thus (9.10b) yields

$$F(\theta, \pi/2, \lambda) \sim \left(\tfrac{25}{64}\lambda \cos\theta\right)^{2/9}, \tag{9.20}$$

which yields (9.9c) for $\psi = \pi/2$. □

It is clear from this proof that it is not difficult to verify the hypotheses of the following generalization of Theorem 9.5.

9.21. Theorem. *If* (9.4) *and* (9.9) *hold, then the conclusion of Theorem* 9.5 *holds.*

We can obtain sharp asymptotic results for F for $\psi > \pi/2$ to complement those for $\psi < \pi/2$ obtained in the proof of Theorem 9.5. Let $\psi > \pi/2$ and let λ be large. We first study the case in which $\theta < \pi/2$. It then follows from (9.10b) that $\delta^{-4} > \nu^2 + F$. In the proof of Theorem 9.5, we showed that $\nu \to \infty$ and $F \to \infty$ as $\lambda \to \infty$. Thus $\delta \to 0$ as $\lambda \to \infty$ so that

$$\nu^2 \sim F. \tag{9.22}$$

We now solve (9.10a) for δ^{-2}, and substitute the resulting expression and (9.21) into (9.10b) to obtain

$$\lambda \cos\theta \sim \sqrt{F}\left\{\left[\tfrac{8}{5}F^2 - \tfrac{1}{2}(-\lambda\cos\psi)^{4/3}\right]^2 - 2F\right\}. \tag{9.23}$$

In view of (9.18), the only way this equation can hold is for

$$F \sim \tfrac{\sqrt{5}}{4}\xi^2 \sim \tfrac{\sqrt{5}}{4}(-\lambda\cos\psi)^{2/3} \sim \nu^2. \tag{9.24}$$

We now treat the case in which $\theta > \pi/2$. From (9.10b,c) we find that

$$\nu > \left(-\frac{\lambda}{2}\cos\theta\right)^{1/3}. \tag{9.25}$$

By combining (9.10a,b), we find that

$$\sqrt{F + \nu^2} > \frac{F^2}{10} + F\nu^2 + \frac{\nu^4}{2} - \frac{\xi^4}{2} \tag{9.26}$$

for sufficiently large λ. This inequality is possible only if

$$\frac{F^2}{10} + F\nu^2 + \frac{\nu^4}{2} \sim \frac{\xi^4}{2}, \quad \text{so that} \quad \nu^2 \sim -F + \sqrt{\tfrac{4}{5}F^2 + \xi^4}. \tag{9.27a,b}$$

Were δ^{-4} negligible with respect to $\nu^2 + F$, then (9.10a,b) would imply that

$$\nu(F + \nu^2) \sim -\lambda\cos\theta. \tag{9.28}$$

We substitute (9.27b) into (9.29) to obtain

$$\left\{(\lambda\cos\theta)^2 + F\left[\tfrac{4}{5}F^2 + \xi^4\right]\right\}^2 \sim \left[\tfrac{4}{5}F^2 + \xi^4\right]^3, \tag{9.29}$$

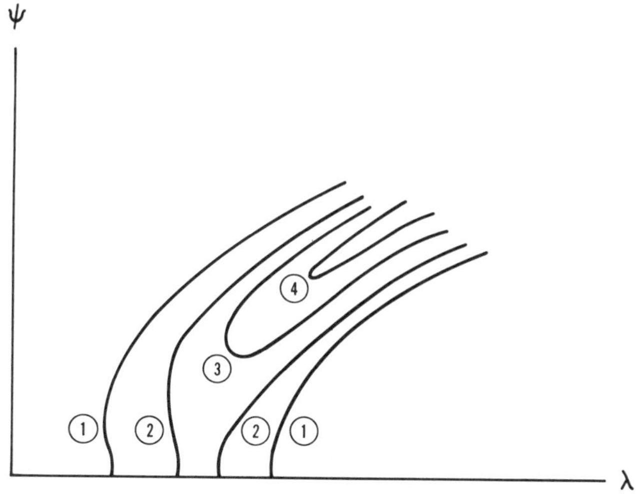

Fig. 9.32. Typical bifurcation diagram reflecting Theorem 9.30. Here we suppose that q has the form shown in Fig. 7.2. The number identifying each branch is the value of n introduced in (5.12) and is 1 less than the number of zeros of θ that characterizes each branch.

from which we would deduce that $F \sim \sqrt{5}\xi^2$. Then (9.27b) would imply that $\nu^2 = o(\lambda^{2/3})$, in contradiction to (9.25). Thus δ^{-4} must be comparable to $\nu^2 + F$. It then follows from (9.25) that δ^2 must be negligible relative to ν^2. Thus (9.10b) implies (9.22), and we recover (9.24) from (9.27). Therefore (9.24) holds for $\psi \in (\pi/2, \pi)$.

Property (9.9) implies that (9.6) has one solution ψ slightly less than $\pi/2$. In this solution, $|\theta| < \pi/2$, so that the column is everywhere compressed. The deformed shapes of the column corresponding to solutions of (9.6) with ψ slightly less than π can be determined by the techniques used by O'Malley (1976). They have the intuitively obvious form. For example, if $n = 1$, then the column is everywhere in tension except at boundary layers at its ends, where θ flips through an angle π. (In these solutions, θ stays near ψ, which is near π for all s away from the ends of its interval $[0, 1]$.) Equation (9.24) could be used as a starting point for an asymptotic representation of the solution as $\lambda \to \infty$. Note that it implies the surprising result that ν is large at the ends of the column where there is a large compressive thrust. This thrust is absorbed not by the reduction of ν, but by the increase in F and in the reduction of δ. In other words, ν does not exhibit boundary layer behavior. This fact, which is a consequence of our adoption of a refined constitutive model accounting for the preservation of orientation, would simplify a full asymptotic analysis.

To get detailed information about the disposition of solution branches corresponding to Theorem 9.21, we could undertake a further analysis of J, with the aim of refining (9.9), perhaps by showing that for λ large, the function $J_\lambda(\cdot, \lambda)$ vanishes exactly once. In this case, (9.6) would have exactly two solutions ψ for each large λ. By a suitable use of the regularity part of the Implicit Function Theorem, we could conclude that solution branches of (9.6) and thus solution branches of our boundary-value problem form two distinct C^1-curves going out to $\lambda = \infty$. Unfortunately, the requisite computations, based on the use of (9.10) to effect the differentiation of (9.8b), become exceedingly complicated as a consequence of the very nonuniformity that is reflected in (9.9). In place of such an analysis, we content ourselves with results in the same spirit, which are easily proved and which compensate for their deficiencies in detail with great generality:

9.30. Theorem. *Let (9.4) and (9.9) hold and let F be twice continuously differentiable. For each $\lambda > \Lambda_n$, (9.6) has (at least) four solutions $\pm\psi_1(\lambda)$, $\pm\psi_2(\lambda)$ with $\psi_1(\lambda) \to \pi/2$, $\psi_2(\lambda) \to \pi$ as $\lambda \to \infty$. Corresponding to each such solution is a solution pair of (5.8). Moreover, for almost all $h > 0$ the solution pairs $(\psi_1(\lambda), \lambda)$, $(\psi_2(\lambda), \lambda)$ of (9.6) lie on distinct C^1-manifolds (which are unions of C^1-curves).*

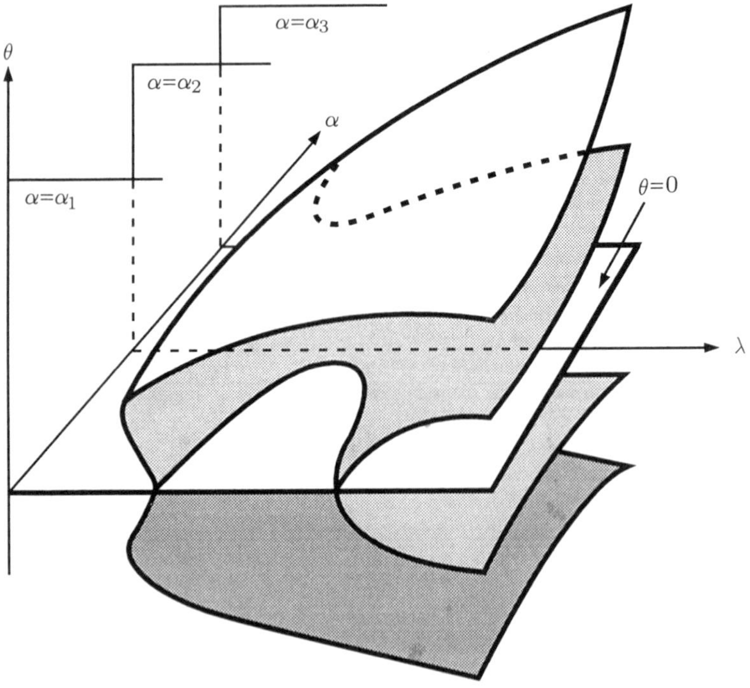

Fig. 9.33. Bifurcating sheet of solutions over the (λ, α)-plane. There are infinitely many such sheets, all nested.*

* This figure comes from Antman (1989) with permission of the American Society of Mechanical Engineers.

Proof. All these statements save the last are summaries of results obtained above. Since F is twice continuously differentiable, J is also. We can therefore obtain the last statement by invoking Sard's Theorem and results about C^1-manifolds (see Abraham & Robbin (1967, Thm. 15.1 and Cor. 17.2)). □

We can sacrifice a little more detail to get a result valid for all h. J. C. Alexander (in a private communication) has given an elementary topological proof of

9.31. Theorem. *Let (9.4) and (9.9) hold and let F be continuous. For each $\lambda > \Lambda_n$, (9.6) has (at least) four solutions $\pm\psi_1(\lambda)$, $\pm\psi_2(\lambda)$ with $\psi_1(\lambda) \to \pi/2$, $\psi_2(\lambda) \to \pi$ as $\lambda \to \infty$, and with $\{(\lambda, \pm\psi_1(\lambda)) : \lambda > \Lambda_n\}$ and $\{(\lambda, \pm\psi_2(\lambda)) : \lambda > \Lambda_n\}$ connected.*

We illustrate typical bifurcation diagrams reflecting these theorems in Figs. 9.32 and 9.33.

9.34. Problems. Carry out the analysis of this section when (7.14) holds with $\beta > 2$ and $\gamma > 1$. What can be said about the global disposition of solutions when $\beta > 0$ and $\gamma > 0$? Carry out the analysis for nonuniform rods (see Kath (1985)). Carry out the analysis for general classes of W.

The development of Secs. 5–9 is adapted from Antman & Pierce (1990, ©Society for Industrial and Applied Mathematics). Figures 6.17, 7.2, and 9.32 come from this source.

10. Other Planar Buckling Problems for Straight Rods

We can analyze a great many other problems for the planar buckling of rods by similar techniques. The essential ingredient in such studies is that the reaction $\boldsymbol{n}(1) \cdot \boldsymbol{j}$ to any kinematic boundary conditions imposed at the end $s = 1$ be zero. For example, consider the problem in which we replace the boundary conditions (5.2) with

$$\text{(10.1a–e)}\quad M(0) = 0 = M(1), \quad \boldsymbol{r}(0) = \boldsymbol{0}, \quad \boldsymbol{r}(1) \cdot \boldsymbol{j} = 0, \quad \boldsymbol{n}(1) \cdot \boldsymbol{i} = -\lambda.$$

These conditions describe a rod with with its end $s = 0$ hinged at $\boldsymbol{0}$ and with its end $s = 1$ hinged to a point that can slide freely along the $\boldsymbol{i}$-axis. From (IV.1.15) (cf. (5.4)) we then obtain

$$\text{(10.2)}\quad 0 = M(1) - M(0) = \int_0^1 [\boldsymbol{k} \times \boldsymbol{n}(1)] \cdot \boldsymbol{r}'(s)\, ds = -[\boldsymbol{r}(1) \cdot \boldsymbol{i}][\boldsymbol{n}(1) \cdot \boldsymbol{j}],$$

so that $\boldsymbol{n}(1) \cdot \boldsymbol{j} = 0$ as long as $\boldsymbol{r}(1) \neq \boldsymbol{0}$. It is thus clear that the entire apparatus of Secs. 5–9 can handle this problem with but cosmetic changes. The same is not true for the problem in which (5.2a,b) and (10.1c–e) hold, because the equation for $\boldsymbol{r}$ cannot be uncoupled from those for θ and M. In this case, we can express the reaction $\boldsymbol{n}(1) \cdot \boldsymbol{j}$ as a functional of (θ, M, λ). The analysis of nodal properties is technically much harder than that for (10.1). See Antman & Rosenfeld (1978). (In the terminology of structural mechanics, this problem is said to be *statically indeterminate* in the sense that the reactions at the end cannot be found a priori, as they can for the *statically determinate* problem for (10.2).)

10.3. Exercise. Prove that the bifurcating branches for the buckling of a rod under its own weight globally preserve the distinctive nodal patterns they inherit from the

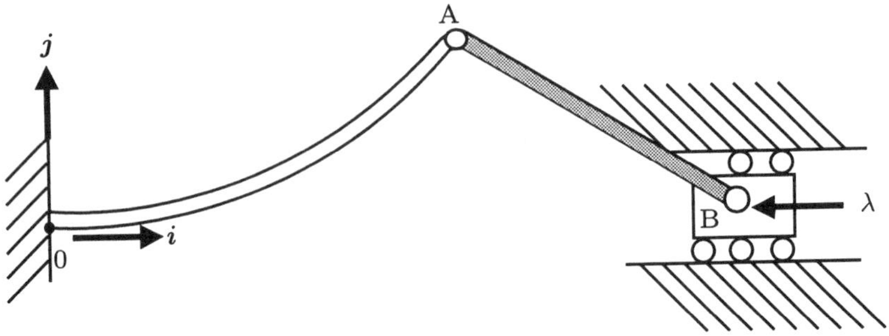

Fig. 10.5. Geometry of Problem 10.4.

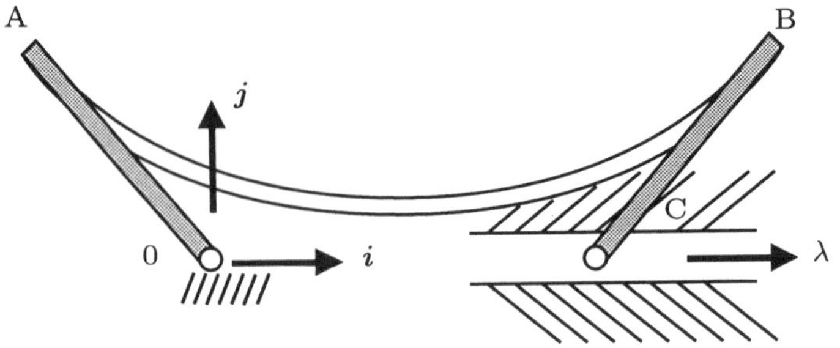

Fig. 10.7. Geometry of Problem 10.6.

eigenfunctions of the linearized problem corresponding to simple eigenvalues. Specifically, analyze the problem (IV.1.3), (IV.1.4), (IV.1.13), (IV.1.18), (IV.1.30) with $\boldsymbol{f}(s) = -(\rho A)(s)g\boldsymbol{i}$ subject to the boundary conditions $\theta(0) = 0$, $\boldsymbol{r}(0) = \boldsymbol{0}$, $M(1) = 0$, $\boldsymbol{n}(1) = \boldsymbol{0}$.

10.4. Problem. Consider the buckling of the structure illustrated in Fig. 10.5. The curve OA represents the $\boldsymbol{r}$ for a nonlinearly elastic rod of unit reference length, and the line AB represents a rigid shaft of length l. The rod is welded to a rigid wall at O, taken to be the origin $\boldsymbol{0}$, and hinged to the shaft at A. The shaft is hinged at point B, which is constrained to move along the $\boldsymbol{i}$-axis. An external horizontal force $-\lambda\boldsymbol{i}$ is applied at B. Carry out a global bifurcation analysis for this problem. Repeat the analysis when AB is replaced by a nonlinearly elastic rod. (The first problem for the elastica was solved by Stern (1979) by using elliptic functions.)

The second part of this problem deals with a primitive framework. There has been very little work done on nonlinear problems for frameworks, although there is an extensive linear theory. (See Bleich (1952), e.g., among the many references listed in Sec. V.2.) It would be interesting to determine whether there are buckling problems for frameworks that are amenable to global qualitative analysis in terms of nodal properties.

10.6. Problem. Consider the buckling of the structure illustrated in Fig. 10.7. The curve AB represents the $\boldsymbol{r}$ for a nonlinearly elastic rod of unit reference length. The lines OA and BC represent rigid shafts of length l. These rigid shafts are attached to the rod at A and B so that either (i) OA and BC have the same directions as $\boldsymbol{a}(\theta(0))$

and $-\boldsymbol{a}(\theta(1))$, respectively, or (ii) OA and BC have the same directions as $\boldsymbol{r}'(0)$ and $-\boldsymbol{r}'(1)$, respectively. The end O of the shaft OA is hinged to a fixed point, taken to be the origin $\mathbf{0}$, and the end C of the shaft BC is hinged to a point constrained to move along the $\boldsymbol{i}$-axis. An external force $\lambda\boldsymbol{i}$ is applied at C. Carry out a global bifurcation analysis of this problem. (For a study of the linearized problem, see Biezeno & Grammel (1953, Sec. IV16b).)

For an interesting collection of conservative loads for buckling problems see Gajewski & Życzkowski (1970).

11. Follower Loads

We now study an illuminating class of problems in which the end load $\boldsymbol{n}(1)$ depends on the configuration. In this case, $\boldsymbol{n}(1)$ is said to be a *follower load.* In particular, we assume that

(11.1) $$\boldsymbol{n}(1) = -\lambda\boldsymbol{c}, \quad \boldsymbol{c} = \cos\psi\boldsymbol{i} + \sin\psi\boldsymbol{j},$$

where $\boldsymbol{c}$ and ψ depend on the configuration of the end $s = 1$. We assume that $\lambda > 0$. Our remaining boundary conditions are

(11.2a,b,c) $$\theta(0) = 0, \quad \boldsymbol{r}(0) = \mathbf{0}, \quad M(1) = 0.$$

These problems are artificial. The force (11.1) may be realized by a feedback control. The one natural problem that can be modelled by follower loads is that for a hose ejecting water, the reaction of the ejected water at the nozzle being taken as tangential. The sinuous motion of hoses is typical of the dynamics that can be produced. We set

(11.3) $$\gamma \equiv \theta - \psi.$$

Then from (IV.2.10) we obtain

(11.4a) $$\gamma' = \hat{\mu}(-\lambda\cos\gamma, \lambda\sin\gamma, M, s),$$

(11.4b)
$$M' + \lambda[\hat{\nu}(-\lambda\cos\gamma, \lambda\sin\gamma, M, s)\sin\gamma + \hat{\eta}(-\lambda\cos\gamma, \lambda\sin\gamma, M, s)\cos\gamma] = 0.$$

We assume that the monotonicity condition (IV.1.23) holds. For simplicity we retain the symmetry conditions (IV.2.9). We now show that for various choices of (11.1) the only solutions of (11.1)–(11.4) are trivial.

11.5. Proposition. *Let* $\boldsymbol{c} = \boldsymbol{a}(1)$ *(so that* $\psi = \theta(1)$*). For* $\lambda > 0$*, the only solution of* (11.1)–(11.4) *is* $(\theta, M) = (0, 0)$.

Proof. Equation (11.3) implies that $\gamma(1) = 0$. The initial-value problem for (11.4) subject to this condition and to (11.2c) has a unique solution, which we recognize to be $(\gamma, M) = (\theta - \psi, M) = (0, 0)$. We use (11.2a) to prove that $\theta = \psi = 0$. □

Let us now set

(11.6) $$\boldsymbol{r}' = |\boldsymbol{r}'|(\cos\phi\boldsymbol{i} + \sin\phi\boldsymbol{j}), \quad \phi = \theta + \beta$$

so that

(11.7) $$\beta = \arctan\frac{\hat{\eta}(-\lambda\cos\gamma, \lambda\sin\gamma, M, s)}{\nu(-\lambda\cos\gamma, \lambda\sin\gamma, M, s)}$$

(cf. (IV.3.45) and (IV.3.46)).

11.8. Proposition. *Let* $\boldsymbol{c} = \boldsymbol{r}'(1)/|\boldsymbol{r}'(1)|$ *(so that* $\psi = \phi(1)$*). For* $\lambda > 0$ *and for* $|\gamma| < \pi/2$ *the only solution of* (11.1)–(11.4) *is* $(\theta, M) = (0, 0)$.

Proof. Equation (11.7), the monotonicity conditions, and the symmetry conditions (IV.2.9) imply that $\beta(1)$ has the same sign as $\gamma(1) = \theta(1) - \phi(1) = -\beta(1)$. Thus $\beta(1) = 0$ and $\psi = \theta(1)$. The proof is thus reduced to that of Proposition 11.5. □

11.9. Exercise. Let the rod be uniform. Use the phase portraits of Fig. IV.2.14 to show that if

$$\left[\psi - \frac{\theta(1)}{2}\right] \operatorname{sign} \theta(1) \geq 0, \tag{11.10}$$

then the only solution of (11.1)–(11.4) is $(\theta, M) = (0, 0)$.

These results do not mean that the straight states are stable. What happens is that their stability is lost through dynamical processes. The following exercise gives a formal computation of the buckling load. For a rigorous treatment, see Carr & Malhardeen (1979) and M.-S. Chen (1987). The latter paper shows that under some circumstances at least, stability is lost by a Hopf bifurcation.

11.11. Exercise. Linearize about the trivial state the equations of motion for a uniform rod subject to (11.1) and (11.2) with $\psi = \theta(1)$. Show that there is a positive number Λ, such that if $\lambda > \Lambda$, then these equations have solutions that grow without bound as $t \to \infty$.

Note that the nonlinear dynamical equations are not conservative because the follower load is not conservative. Thus these equations do not have a variational structure, i.e., they cannot be characterized by Hamilton's Principle. Moreover, the linearized equilibrium equations are consequently not self-adjoint. (This lack of self-adjointness is manifested in the imbalance in the boundary conditions, which we exploited to prove uniqueness of equilibrium states.)

There is an extensive engineering literature on nonconservative problems of elastic stability. See the references listed at the end of Sec. V.7. For treatments of problems for hoses, see Bajaj (1988), Bajaj & Sethna (1984), and Bajaj, Sethna, & Lundgren (1980.) For more complicated variants of the problems treated here, see Kordas & Życzkowski (1963) and Życzkowski & Gajewski (1971).

12. Buckling of Arches

We now treat the buckling of circular arches by the methods of Chap. V. A slight variant of our treatment can handle the closely related problem of the buckling of circular rings, which was treated in Sec. IV.3 by elementary methods. We use the same assumptions and notation as used in Sec. IV.3. In particular, we employ the monotonicity condition (IV.1.23), the growth conditions (IV.1.24)–(IV.1.26), and the symmetry condition (IV.1.27).

We take $-\alpha \leq s \leq \alpha$ with $0 < \alpha < \pi$. The reference configuration is defined by

$$\mathring{\boldsymbol{r}}(s) = \sin s \boldsymbol{i} - \cos s \boldsymbol{j}, \quad \mathring{\theta}(s) = s, \tag{12.1}$$

so that $\mathring{\boldsymbol{r}}$ describes a circular arc of radius 1. We assume that the ends of the arch are welded to blocks that slide in frictionless radial grooves. See Fig. 12.2.

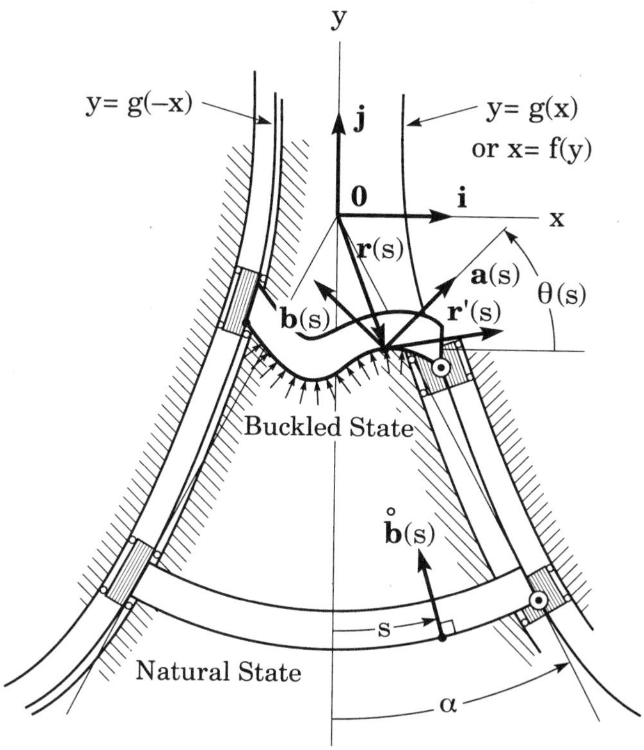

Fig. 12.2. Geometry of the natural reference configuration and of a buckled state of an arch subjected to hydrostatic pressure. The reference state of the arch is taken to be bowed down and the pressure p is taken to act upward in order to simplify sign conventions in the governing equations.

We assume that the material of the arch is uniform. Then the governing equations (IV.3.36), (IV.1.30), and (IV.1.4) can be reduced to

(12.3) $$N' = H\hat{\mu}(N,H,M) + p\hat{\eta}(N,H,M),$$

(12.4) $$-H' = N\hat{\mu}(N,H,M) + p\hat{\nu}(N,H,M),$$

(12.5) $$M' = N\hat{\eta}(N,H,M) - H\hat{\nu}(N,H,M),$$

(12.6) $$\theta' = \hat{\mu}(N,H,M),$$

(12.7) $$\boldsymbol{r}' = \hat{\nu}(N,H,M)\boldsymbol{a}(\theta) + \hat{\eta}(N,H,M)\boldsymbol{b}(\theta).$$

From (12.3)–(12.5) we obtain

(12.8) $$H'' + (N\hat{\mu}_H + p\hat{\nu}_H)H' + (\hat{\mu} + N\hat{\mu}_N + p\hat{\nu}_N)(H\hat{\mu} + p\hat{\eta}) - (N\hat{\mu}_M + p\hat{\nu}_M)(H\hat{\nu} - N\hat{\eta}) = 0,$$

$$\theta'' = \hat{\mu}_N(H\hat{\mu} + p\hat{\eta}) + \hat{\mu}_H H' + \hat{\mu}_M(N\hat{\eta} - H\hat{\nu}). \tag{12.9}$$

The boundary conditions corresponding to Fig. 12.2 are

$$\theta(\pm\alpha) = \pm\alpha, \quad H(\pm\alpha) = 0, \quad x(\pm\alpha) = \pm y(\pm\alpha)\tan\alpha. \tag{12.10a,b,c}$$

Here $x \equiv \boldsymbol{r}\cdot\boldsymbol{i}$, $y \equiv \boldsymbol{r}\cdot\boldsymbol{j}$. Condition (12.10b) is a consequence of (IV.1.27). The presence of frictionless grooves in Fig. 12.2 ensures that our extensible arches admit a one-parameter family of circular (trivial equilibrium states).

Our boundary-value problem consists of (12.3)–(12.7), (12.10).

We say that an equilibrium configuration is *trivial* iff there is no shear and if $\boldsymbol{r}$ describes a circle. Thus, in such a state $\eta = 0$ or, equivalently, $H = 0$. We limit our attention to positive values of p. Setting $H = 0$ in (12.3)–(12.5), we obtain

$$N = \text{const.}, \quad M = \text{const.}, \tag{12.11a,b}$$

$$N\hat{\mu}(N, 0, M) = -p\hat{\nu}(N, 0, M). \tag{12.12}$$

From (12.11) it follows that ν and μ are constant for trivial solutions, so that the curvature reduces to the constant μ/ν. That $\boldsymbol{r}$ is circular in the trivial state is thus a consequence of the assumption that $H = 0$. From (12.10a) we obtain

$$\hat{\mu}(N, 0, M) = 1. \tag{12.13}$$

The system (12.12), (12.13) has a unique solution for N and M when $p \geq 0$, as a consequence of the monotonicity and growth conditions. Note that (12.13) says that there is no change in the bending strain μ. (If $\hat{M}$ has the property that $\hat{M}(\nu, 0, \nu) = 0$, which is reasonable when $\boldsymbol{r}$ can be identified with the locus of centroids of the cross sections of three-dimensional rod, then we can identify this solution as $N = -p$, $M = 0$.)

Let $(N_0(p), M_0(p))$ denote the trivial solution. For any constitutive function such as $\hat{\nu}_N$, we set $\nu_N^0(p) \equiv \hat{\nu}_N(N_0(p), 0, M_0(p))$, etc., and usually suppress the argument p. The linearization of our boundary-value problem about this trivial solution uncouples, with the part corresponding to the linearization of (12.8) having the form

$$H_1'' + q(p)^2 H_1 = 0, \quad H_1(\pm\alpha) = 0 \tag{12.14}$$

where

$$q^2 \equiv (\mu^0 + p\eta_H^0)\left[\mu^0 + p\frac{(\mu^0)^2\nu_N^0 - \nu^0\mu^0(\mu_N^0 + \nu_M^0) + (\nu^0)^2\mu_M^0}{(\mu^0)^2}\right]. \tag{12.15}$$

Problem (12.14) has asymmetric nontrivial solutions

$$H_1(s) = \text{const.}\ \sin\tfrac{(2j+1)\pi}{2\alpha}s \quad \text{when} \quad q(p_0) = \frac{(2j+1)\pi}{2\alpha} \tag{12.16a,b}$$

where j is an integer, andhas symmetric nontrivial solutions

$$H_1(s) = \text{const. } \sin \tfrac{l\pi}{\alpha} s \quad \text{when} \quad q(p_0) = \frac{l\pi}{\alpha} \tag{12.16c,d}$$

where l is a nonzero integer. (The symmetry is taken with respect to the $\boldsymbol{j}$-axis.) The simplicity of eigenvalues p, which satisfy the characteristic equations (12.16b,d), is treated just as in Sec. 5.

We now study the global behavior of buckled states. We assume that the constitutive functions $\hat{\nu}, \hat{\eta}, \hat{\mu}$ are at least three times continuously differentiable. Let us define a *(solution) branch* of our boundary-value problem to be any connected set of $\big((N, H, M, \theta, \boldsymbol{r}), p\big)$ satisfying integral equations corresponding to the boundary-value problem with $(N, H, M, \theta, \boldsymbol{r})$ continuous. In view of our smoothness assumption on the constitutive functions, it follows from (12.3)–(12.7) that such a quintuple $(N, H, M, \theta, \boldsymbol{r})$ is twice continuously differentiable, so that (12.8) is satisfied in a classical sense. The existence of a global branch emanating from an eigenvalue of odd algebraic multiplicity of the linearization of the boundary-value problem about the trivial solution is assured by the Global Bifurcation Theorem V.4.19.

It follows from the symmetry condition (IV.1.27) that (12.8) admits the solution $H = 0$ no matter what functions N and M and what parameters p appear there. Since N and M are continuously differentiable functions along branches, we may regard (12.8) as a second-order ordinary differential equation for H alone. It has the property that if H were to have a double zero, then $H = 0$. It follows from the development of Sec. V.5 that H preserves the number of its simple zeros along any branch of solutions not containing a trivial solution. Thus, we can globally distinguish different nontrivial branches of solutions of this boundary-value problem by the nodal properties of the function H.

12.17. Exercise. Give a full justification, in the manner of Sec. V.5, of the global bifurcation results just described.

12.18. Exercise. Carry out a perturbation analysis of the bifurcation problem described above, showing that the bifurcation from simple eigenvalues corresponding to both (12.16b) and (12.16d) is of pitchfork type. This is to be expected for the asymmetric solutions corresponding to (12.16a) because these solutions come in mirror-image pairs, but it is somewhat surprising for the symmetric solutions, which do not enjoy any such pairing.

12.19. Exercise. For an inextensible, unshearable arch, with a constitutive equation of the form $\mu = \hat{\mu}(M)$, carry out a perturbation analysis of the problems for a doubly welded arch and for a doubly hinged arch with fixed ends. Their respective boundary conditions are

$$\theta(\pm\alpha) = \pm\alpha, \quad x(\pm\alpha) = \pm \sin\alpha, \quad y(\pm\alpha) = \cos\alpha, \tag{12.20}$$

$$M(\pm\alpha) = 0, \quad x(\pm\alpha) = \pm \sin\alpha, \quad y(\pm\alpha) = \cos\alpha. \tag{12.21}$$

Determine which bifurcations are transcritical.

The treatment of boundary conditions other than those of (12.10) can be very difficult for the general constitutive equations we are considering. For example, if the ends of the arch are hinged to blocks sliding in frictionless radial grooves, then for general constitutive functions there are no trivial solutions of the form described above for a range of p's:

12.22. Exercise. Let the ends of the arch be hinged to blocks sliding in frictionless grooves of the form $y = g(x)$, $y = g(-x)$ where g is a decreasing function defined on a subinterval of $(0,1)$. Then the boundary condition are

$$M(\pm\alpha) = 0, \quad y(\pm\alpha) = g(\pm x(\pm\alpha)), \quad \boldsymbol{n}(\pm\alpha) \cdot [\boldsymbol{i} \pm g'(\pm x(\pm\alpha))\boldsymbol{j}] = 0. \tag{12.23}$$

Suppose that the arch has the constitutive property that $\hat{\mu}(N,0,0) = 1$ for all N (which is unrealistic for thick arches). In this case, show that the problem with boundary condition (12.23) admits a family of trivial solutions with $y(\alpha) = y(-\alpha)$ when the grooves are radial: $g(x) = -x\cot\alpha$. Next suppose that $\hat{M}(\nu,0,\nu) = 0$ and show that the problem with boundary condition (12.23) admits a family of trivial solutions with $y(\alpha) = y(-\alpha)$ when the grooves have the form

$$g(x) = -\ln x + \ln\left(1 + \sqrt{1-x^2}\right) - \sqrt{1-x^2} + \text{const.} \quad \text{for } 0 < x < 1. \tag{12.24}$$

It is noteworthy that (12.24) does not depend on α. (In general, one can find a differential equation for g, which depends on α.)

The treatment of this section is adapted from Antman & Marlow (1993) (with kind permission from Elsevier Science Ltd, The Boulevard, Langford Lane, Kidlington OX5 1GB, UK. Figure 12.2 comes from this paper.) See Antman & Marlow (1993) for technically complicated variants of the problems of this section. Their work extends that of Antman & Dunn (1980).

12.25. Exercise. Adopt the methods of this section to study the buckling of a circular ring under hydrostatic pressure (see Sec. IV.3).

12.26. Problem. Suppose that the hydrostatic pressure on a circular arch or ring is replaced by (i) a constant normal force per unit reference arc length, (ii) a constant normal force depending on a suitably defined area enclosed by the ring or arch, (iii) a constant radial force per unit actual arc length, (iv) a constant radial force per unit reference arc length. Formulate and analyze the global bifurcation problems. (See Sills & Budiansky (1978) for a perturbation analysis, and Dickey & Roseman (1993) for a global analysis. For a discussion of the conservativeness of such loadings, see Fisher (1987, 1988, 1989).)

13. Buckling of Whirling Rods

We first formulate and analyze the buckling problem corresponding to Fig. V.2.2a. To preserve the symmetry of the problem, we assume that the rod is weightless. Let the radius of the rigid ring be R and let the length of the attached rod be 1. We define $\boldsymbol{e}(t)$ by (1.1):

$$\boldsymbol{e}(t) = \cos\omega t\boldsymbol{i} + \sin\omega t\boldsymbol{j}. \tag{13.1}$$

We replace the basis vectors $\boldsymbol{i}, \boldsymbol{j}, \boldsymbol{k}$ used in Chap. IV with $\boldsymbol{e}(t), \boldsymbol{k}, \boldsymbol{e} \times \boldsymbol{k}$, respectively. We seek steady solutions of the equations of motion (I.1.45), (IV.1.46) of the form

$$\begin{aligned} \boldsymbol{r}(s,t) &= r(s)\boldsymbol{e}(t) + z(s)\boldsymbol{k}, \\ \boldsymbol{a}(s,t) &= \cos\theta(s)\boldsymbol{e}(t) + \sin\theta(s)\boldsymbol{k}, \\ \boldsymbol{b}(s,t) &= -\sin\theta(s)\boldsymbol{e}(t) + \cos\theta(s)\boldsymbol{k}. \end{aligned} \tag{13.2}$$

These solutions correspond to steady rotations of the rod about $\boldsymbol{k}$ with angular velocity ω. We assume that the inertia term $\rho I = 0$. Then the equations of motion reduce to

$$\boldsymbol{n}_s + \omega^2 \rho A r\, \boldsymbol{e} = \boldsymbol{0}, \tag{13.3}$$

$$\boldsymbol{m}_s + \boldsymbol{r}_s \times \boldsymbol{n} = \omega^2 \rho J \cos\theta \sin\theta\, \boldsymbol{k} \times \boldsymbol{e}. \tag{13.4}$$

We take $s = 0$ to be the free end of the rod and $s = 1$ to be the end of the rod welded to the rigid ring. Then our boundary condition are

$$\boldsymbol{n}(0,t) = \boldsymbol{0}, \quad M(0) = 0, \quad \boldsymbol{r}(1,t) = R\boldsymbol{e}(t), \quad \theta(1) = 0. \tag{13.5}$$

We integrate (13.3), (13.4) subject to (13.5), and then use the definition (IV.1.4) of ν and η and the constitutive equations (IV.1.30) to convert our problem to the following system of integral equations for N, H, M, θ:

$$N(s) = -\,\omega^2 \left[\int_0^s (\rho A)(\xi) r(\xi)\, d\xi\right] \cos\theta(s), \tag{13.6a}$$

$$H(s) = \omega^2 \left[\int_0^s (\rho A)(\xi) r(\xi)\, d\xi\right] \sin\theta(s), \tag{13.6b}$$

$$\begin{aligned} M(s) = -\int_0^s &[H(\xi)\hat{\nu}\,(N(\xi), H(\xi), M(\xi), \xi) \\ &- N(\xi)\hat{\eta}\,(N(\xi), H(\xi), M(\xi), \xi)]\, d\xi \\ &- \omega^2 \int_0^s (\rho J)(\xi) \cos\theta(\xi) \sin\theta(\xi)\, d\xi, \end{aligned} \tag{13.6c}$$

$$\theta(s) = \int_s^1 \hat{\mu}\,(N(\xi), H(\xi), M(\xi), \xi)\; d\xi, \tag{13.6d}$$

where

$$\begin{aligned} r(s) = R - \int_s^1 &[\hat{\nu}\,(N(\xi), H(\xi), M(\xi), \xi) \cos\theta(\xi) \\ &- \hat{\eta}\,(N(\xi), H(\xi), M(\xi), \xi) \sin\theta(\xi)]\, d\xi. \end{aligned} \tag{13.6e}$$

Note that the trivial solutions of (13.6), defined by $\theta = 0$, were studied in Sec. IV.5.

13.7. Exercise. Carry out a detailed derivation of (13.6). Prove that (13.6) satisfies all the hypotheses of the Global Branching Theorem V.4.19. Prove that bifurcating branches are distinguished by the nodal properties of θ and that these nodal properties are inherited from the eigenfunctions corresponding to simple eigenvalues for the problem linearized about the trivial solution.

We now turn to the buckling problem for a heavy rod illustrated in Fig. V.2.1. This classical problem, which is the analog of Kolodner's whirling string problem treated in Sec. 2, provides some novel technical challenges in determining the nodal properties of solution branches. We retain (13.1). We now replace the basis vectors $\boldsymbol{i}, \boldsymbol{j}, \boldsymbol{k}$ used in Chap. IV with $\boldsymbol{k}, \boldsymbol{e}(t), \boldsymbol{k} \times \boldsymbol{e}$, respectively. We take $\boldsymbol{k}$ to point downward. We seek steady solutions of the equations of motion (IV.1.43) and (IV.1.44) of the form

$$\begin{aligned} \boldsymbol{r}(s,t) &= r(s)\boldsymbol{e}(t) + z(s)\boldsymbol{k}, \\ \boldsymbol{a}(s,t) &= \cos\theta(s)\boldsymbol{k} + \sin\theta(s)\boldsymbol{e}(t), \\ \boldsymbol{b}(s,t) &= -\sin\theta(s)\boldsymbol{k} + \cos\theta(s)\boldsymbol{e}(t). \end{aligned} \tag{13.8}$$

We take as the body force intensity the weight per unit reference length $g(\rho A)(s)\boldsymbol{k}$. We assume that the rod has unit length. We assume that the end $s = 0$ is welded at the origin to the plane with normal $\boldsymbol{k}$ and that the end $s = 1$ of the rod is free. Thus

$$r(0) = 0 = \theta(0), \quad \boldsymbol{n}(1,t) = \boldsymbol{0}, \quad M(1) = 0. \tag{13.9}$$

Let us set

$$Q(s) \equiv -\boldsymbol{n}(s,t) \cdot \boldsymbol{e}(t), \quad W(s) = g\int_s^1 (\rho A)(\xi)\, d\xi. \tag{13.10a,b}$$

Then the equations for steady motion can be put into the form

(13.11a) $Q' = \omega^2 \rho A r,$

(13.11b) $r' = \hat{\nu}(N,H,M)\sin\theta + \hat{\eta}(N,H,M)\cos\theta,$

(13.11c) $M' = -\hat{\nu}(N,H,M)H + \hat{\eta}(N,H,M)N + \omega^2 \rho J \cos\theta \sin\theta,$

(13.11d) $\theta' = \hat{\mu}(N,H,M)$

where

$$N = W\cos\theta - Q\sin\theta, \quad H = -W\sin\theta - Q\cos\theta. \tag{13.11e,f}$$

Therefore (13.9) reduces to

$$r(0) = 0 = \theta(0), \quad Q(1) = 0 = M(1). \tag{13.12}$$

13.13. Exercise. Derive (13.11) and (13.12) from the general equations of Chap. IV.

That the integral equations corresponding to (13.11) and (13.12) satisfy all the hypotheses of the Global Branching Theorem V.4.19 is demonstrated just as for (13.6). We now turn to the more delicate issue of nodal properties. In order to treat these we assume that the rod is unshearable, although we allow the constitutive functions for ν and μ to depend on H. We assume that the symmetry conditions (IV.2.9) hold.

13.14. Theorem. *Let the rod be unshearable. On each solution branch of (13.11), (13.12) on which $|\theta| < \frac{\pi}{2}$ and $(Q,r,M,\theta) \neq (0,0,0,0)$, the number of simple zeros of θ and of Q are equal and constant.*

We reduce the proof to a series of lemmas:

13.15. Lemma. *Let (Q, r, M, θ) satisfy (13.11), (13.12) for an unshearable rod, let $|\theta| < \frac{\pi}{2}$, and let $a \in [0, 1)$. Let $Q(a) \geq 0$, $r(a) \geq 0$, $M(a) \geq 0$, $\theta(a) \geq 0$, with $Q(a) + r(a) + M(a) + \theta(a) > 0$. Then each of the functions Q, r, M, θ is everywhere positive on $(a, 1]$.*

Proof. The hypotheses of this lemma say that $(Q(a), r(a), M(a), \theta(a))$ lies in the closure of the positive orthant in the four-dimensional phase space of (Q, r, M, θ), and does not coincide with the origin. Since $\hat{\eta} = 0$ for an unshearable rod, and since (IV.2.9) implies that θ' has the same sign as M, the differential equations (13.11) carry any such initial point into the interior of this orthant. □

An analogous argument (or the change of variables $\sigma = 1 - s$) yields

13.16. Lemma. *Let (Q, r, M, θ) satisfy (13.11), (13.12) for an unshearable rod, let $|\theta| < \frac{\pi}{2}$, and let $a \in (0, 1]$. Let $Q(a) \geq 0$, $r(a) \leq 0$, $M(a) \leq 0$, $\theta(a) \geq 0$, with $Q(a) - r(a) - M(a) + \theta(a) > 0$. Then each of the functions $Q, -r, -M, \theta$ is everywhere positive on $[0, a)$.*

13.17. Lemma. *Let (Q, r, M, θ) be a nontrivial solution of (13.11), (13.12) for an unshearable rod, let $|\theta| < \frac{\pi}{2}$, and let $a \in (0, 1)$. Suppose that either*

$$Q(a) = 0, \quad r(a) = 0 \tag{13.18a}$$

or

$$M(a) = 0, \quad \theta(a) = 0. \tag{13.18b}$$

Then in at least one of the intervals $[0, a)$ or $(a, 1]$, none of the four functions Q, r, M, θ vanishes anywhere.

Proof. The four conditions of (13.18) cannot hold simultaneously, because the solution would be trivial. Therefore, if (Q, r, M, θ) satisfies either (13.18a) or (13.18b), then (Q, r, M, θ) or $(-Q, -r, -M, -\theta)$ must satisfy the hypotheses of one of the Lemmas 13.16 or 13.17 (because (IV.2.9) implies that (13.11) is invariant under the transformation $(Q, r, M, \theta) \to (-Q, -r, -M, -\theta)$). □

Proof of Theorem 13.14. Suppose that there were an $a \in [0, 1]$ such that (13.18b), say, were to hold for a nontrivial solution. If $a \in (0, 1)$, then the conclusion of Lemma 13.17 would be incompatible with (13.12); if $a = 0$, then the conclusion of Lemma 13.15 would be incompatible with (13.12) and the invariance; if $a = 1$, then the conclusion of Lemma 13.16 would be incompatible with (13.12) and the invariance. Thus, a change in the nodal structure of θ or Q, signalled by a double zero, can only occur at a trivial solution. To prove that the number of simple zeros of θ and Q are equal, we observe that these numbers are inherited from those of the eigenfunctions of the linear problem as a consequence of the Local Bifurcation Theorem V.4.14. We leave the determination of the nodal properties of these eigenfunctions as the following exercise. □

13.19. Exercise. Determine the nodal properties of the solutions of the linearization about the trivial solution.

13.20. Research Problems. (i) Determine detailed qualitative properties of the solution branches for shearable rods. (ii) Suppose that W is replaced with $-W$. The resulting problem describes the rotation of the rod whose natural state lies above the support. Determine detailed qualitative properties of the solution branches for unshearable or shearable rods. (iii) Determine detailed qualitative properties of the solution branches for the problem corresponding to Fig. V.2.2b.

The treatment of the problem (13.6) is adapted from Antman & Nachman (1980) (with kind permission from Elsevier Science Ltd, The Boulevard, Langford Lane, Kidlington OX5 1GB, UK). The boundary-value problem (13.11) for weightless, inextensible, unshearable rods was formulated and analyzed by Odeh & Tadjbakhsh (1965).

Further analyses of their problem were carried out by Bazley & Zwahlen (1968), Parter (1970), Clément & Descloux (1984, 1991), Brezzi, Descloux, Rappaz, & Zwahlen (1984). The treatment given here generalizes and simplifies those of Parter (1970) and Clément & Descloux (1984).

An interesting variant of problem (13.11) was treated by Healey (1992) by group-theoretic methods of bifurcation theory. Here the base of the rod is welded to a fixed, rather than a rotating, horizontal plane. If the rod is transversely isotropic (see Sec. VIII.9), then there are branches of nontrivial solutions, which describe steady motions in which the axis lies in a fixed plane rotating about the vertical. The nodal properties of these branches have yet to be determined.

CHAPTER VII

Variational Methods

1. Introduction

In this chapter we describe a few simple applications of the calculus of variations to one-dimensional problems of nonlinear elasticity, with the purpose of showing how some very useful physical insights can be easily extracted from the basic theory. The reader should review Secs. II.10 and III.3. We require a mathematical setting somewhat different from that used in bifurcation theory; the technical aspects of this setting are given below in small type.

In Chap. VIII we generalize Hamilton's Principle for strings, stated in Sec. II.10, to hyperelastic rods moving in space under conservative loads. For static problems, there is no kinetic energy, and the Hamiltonian reduces to the potential energy, which consists of the stored energy and of the potential energy of the forces applied to the rod. In light of Hamilton's Principle, we expect that the Euler-Lagrange equations for the potential-energy functional are equilibrium equations.

To show that there actually are equilibrium states, we shall limit our attention to those states that minimize the potential-energy functional under suitable subsidiary conditions. These equilibrium states can be reckoned as stable in certain senses. (See Sec. V.7). The focus of this chapter is a description of methods for showing that the potential-energy functional is minimized and that the minimizer actually satisfies the Euler-Lagrange equations. In Sec. 2 we show how to obtain Euler-Lagrange equations for problems subject to isoperimetric constraints. Our goal in this section is to develop some simple and powerful techniques for deriving useful physical insights. Examples of such insights are given in Theorem 5.6 and in the developments of Secs. 6 and 7.

To fix ideas, let us study the equilibrium configurations of a hyperelastic rod under hydrostatic pressure p. We use the notation of Chap. IV. We allow the rod to have any planar reference shape. When p is positive, the pressure acts in the direction $\boldsymbol{k} \times \boldsymbol{r}'$. To be specific, let us assume that the rod has natural length L and that its ends are welded to vertical walls at $\mathbf{0}$ and $\boldsymbol{i}$. Thus

$$\text{(1.1a,b,c,d)} \qquad \boldsymbol{r}(0) = \mathbf{0}, \quad \boldsymbol{r}(L) = \boldsymbol{i}, \quad \theta(0) = 0 = \theta(L).$$

We wish to characterize the governing equations, described in Secs. IV.3 and VI.12 as the Euler-Lagrange equations for the potential-energy functional.

1.2. Exercise. Prove that the potential-energy functional for the hydrostatic pressure of Sec. IV.3 when applied to a rod satisfying boundary conditions (1.1) is the pressure p times the signed area enclosed by the $\mathring{\boldsymbol{r}}$ and $\boldsymbol{r}$. Use a version of Green's Theorem to prove that this area differs by a constant from

$$\frac{1}{2}\int_0^L \boldsymbol{k}\cdot[\boldsymbol{r}(s)\times\boldsymbol{r}'(s)]\,ds. \tag{1.3}$$

Thus, the potential energy for the hyperelastic rod under hydrostatic pressure p is

$$\int_0^L W(\nu(s),\eta(s),\theta'(s),s)\,ds+\frac{p}{2}\int_0^L \boldsymbol{k}\cdot[\boldsymbol{r}(s)\times\boldsymbol{r}'(s)]\,ds, \tag{1.4}$$

where W is the stored energy function for the rod. Note that $\boldsymbol{r}$ is related to the strains ν,η by (IV.1.4):

$$\boldsymbol{r}'=\nu\boldsymbol{a}(\theta)+\eta\boldsymbol{b}(\theta) \tag{1.5}$$

where

$$\boldsymbol{a}(\theta)=\cos\theta\boldsymbol{i}+\sin\theta\boldsymbol{j},\quad \boldsymbol{b}(\theta)=-\sin\theta\boldsymbol{i}+\cos\theta\boldsymbol{j}. \tag{1.6}$$

We shall accordingly study the minimization of (1.4) over a suitable class of functions $\nu,\eta,\theta,\boldsymbol{r}$ satisfying (1.1) and (1.5). The form of the differential constraint (1.5) is technically inconvenient. One way to avoid dealing with (1.5) is simply to replace $\boldsymbol{r}'$ and $\boldsymbol{r}$ in (1.4) respectively with (1.5) and

$$\boldsymbol{r}(s)=\int_0^s[\nu(\xi)\boldsymbol{a}(\theta(\xi))+\eta(\xi)\boldsymbol{b}(\theta(\xi))]\,d\xi\equiv\boldsymbol{r}[\nu,\eta,\theta](s). \tag{1.7}$$

Note that (1.7) satisfies (1.1a). We require that (1.7) satisfy (1.1b) by imposing the side condition that

$$\boldsymbol{r}[\nu,\eta,\theta](L)=\boldsymbol{i}. \tag{1.8}$$

It is technically easier to handle this integral constraint (called an *isoperimetric constraint*) than to handle the differential constraint (1.5). We can now define the potential-energy functional

$$\varPi[\nu,\eta,\theta]\equiv\varPsi[\nu,\eta,\theta]+p\varOmega[\nu,\eta,\theta] \tag{1.9a}$$

where

$$\varPsi[\nu,\eta,\theta]\equiv\int_0^L W(\nu(s),\eta(s),\theta'(s),s)\,ds, \tag{1.9b}$$

$$\varOmega[\nu,\eta,\theta]\equiv\frac{1}{2}\int_0^L \boldsymbol{k}\cdot\{\boldsymbol{r}[\nu,\eta,\theta](s)\times\boldsymbol{r}[\nu,\eta,\theta]'(s)\}\,ds. \tag{1.9c}$$

Suppose that p is prescribed. We seek to minimize Π over an *admissible* class of ν, η, θ that satisfy certain regularity conditions, the requirement (IV.1.9) (defining the domain of W to consist of orientation-preserving deformations), the boundary conditions (1.1c,d), and the constraint (1.8), and then to demonstrate that the minimizer satisfies appropriate Euler-Lagrange equations. Thus we need to develop a rule for determining the Euler-Lagrange equations for such constrained variational problems.

An alternative way to dispense with (1.5) is to use it as a definition of ν and η. In this case, we define the potential-energy functional

$$(1.10a) \qquad \Pi[\boldsymbol{r}, \theta] \equiv \Psi[\boldsymbol{r}, \theta] + p\Omega[\boldsymbol{r}]$$

where

$$(1.10b) \qquad \Psi[\boldsymbol{r}, \theta] \equiv \int_0^L W\big(\boldsymbol{r}'(s)\cdot\boldsymbol{a}(\theta(s)),\, \boldsymbol{r}'(s)\cdot\boldsymbol{b}(\theta(s)),\, \theta'(s),\, s\big)\, ds,$$

$$(1.10c) \qquad \Omega[\boldsymbol{r}] \equiv \tfrac{1}{2}\int_0^L \boldsymbol{k}\cdot\{\boldsymbol{r}(s)\times\boldsymbol{r}'(s)\}\, ds.$$

For prescribed p we seek to minimize Π over an admissible class of $\boldsymbol{r}, \theta$ that merely satisfy certain regularity conditions and the boundary conditions (1.1), and then to demonstrate that the minimizer satisfies appropriate Euler-Lagrange equations.

It is easy to construct other variants of these problems. For example, in (1.9) we could set $\mu = \theta'$ and then replace $\theta(s)$ by

$$(1.11) \qquad \theta[\mu](s) \equiv \int_0^s \mu(\xi)\, d\xi,$$

which accounts for (1.1c). Thus we regard μ, rather than θ, as a fundamental unknown; we accordingly replace θ with μ in the list of arguments of the energy functionals. In this case, we add to the side conditions the isoperimetric constraint that

$$(1.12) \qquad \theta[\mu](L) = 0,$$

which acxcounts for (1.1d) and which is analogous to (1.8). (Such formulations at best are generally inconvenient for partial differential equations.)

The use of (1.7) or (1.11) is contingent on the presence of a boundary condition for $\boldsymbol{r}$ or θ. If there were no such boundary conditions, we would have to introduce arbitrary constants of integration into (1.7) or (1.11). These unknown constants could be regarded as arguments of the energy functionals. We generally do not have to worry about such questions since in most of our applications of the calculus of variations to elasticity the arguments of the energies are geometric quantities and the arbitrary constants of integration merely fix a deformation to within a rigid displacement. Thus we are free to prescribe these constants or, equivalently, to prescribe corresponding subsidiary conditions for these geometric quantities when they are not otherwise prescribed. For example, consider the deformation of a ring, in which there are only 'periodicity' conditions (see Sec. IV.3)). In this case, we can fix the rigid displacement by prescribing (1.1a,c). Alternatively, we could prescribe the mean translation and mean rotation (to be zero):

$$\int_0^L \boldsymbol{r}(s)\,ds = \mathbf{0}, \quad \int_0^L \theta(s)\,ds = 0. \tag{1.13a,b}$$

To get a formula analogous to (1.7) when (1.13a) holds in place of a standard boundary condition, we integrate the identity $\xi\boldsymbol{r}'(\xi) = [\xi\boldsymbol{r}(\xi)]' - \boldsymbol{r}(\xi)$ over $(0,s)$ and integrate the identity $(L-\xi)\boldsymbol{r}'(\xi) = [(L-\xi)\boldsymbol{r}(\xi)]' + \boldsymbol{r}(\xi)$ over (s,L). Subtracting the resulting integrals, we obtain

$$L\boldsymbol{r}(s) = \int_0^L \boldsymbol{r}(\xi)\,d\xi + \int_0^s \xi\boldsymbol{r}'(\xi)\,d\xi - \int_s^L (L-\xi)\boldsymbol{r}'(\xi)\,d\xi. \tag{1.14}$$

Condition (1.13a) causes the first integral in (1.14) to vanish. We can now replace $\boldsymbol{r}'$ in (1.14) with (1.5), and define the resulting version of (1.14) as $\boldsymbol{r}[\nu,\eta,\theta](s)$. Clearly, the same approach can be used to replace (1.11) when (1.13b) holds.

To fix ideas, let us concentrate on the formulation (1.9). Instead of prescribing p, we could prescribe the size of solution, measured by the area change, by setting

$$\Omega[\nu,\eta,\theta] = \Omega_\sharp \tag{1.15}$$

where $\Omega_\sharp$ is given. In this case, we seek to minimize Ψ subject to the boundary conditions (1.1c,d) and the two constraints (1.8) and (1.15) and then to determine the Euler-Lagrange equations. Alternatively, we could leave p free, but prescribe the size of solution, measured by the stored energy, by setting

$$\Psi[\nu,\eta,\theta] = \Psi_\sharp \tag{1.16}$$

where $\Psi_\sharp$ is given. We now seek to extremize Ω subject to the boundary conditions (1.1c,d) and the two constraints (1.8) and (1.16) and then to determine the Euler-Lagrange equations.

The domain of $W(\cdot,\cdot,\cdot,s)$ is an open convex proper subset $\mathcal{V}(s)$ of $\mathbb{R}^3$ defined by (IV.1.22). One of the central difficulties we face in studying Ψ is caused by the reasonable requirements (IV.1.24), (IV.1.26) that the force $N = W_\nu$ or the couple $M = W_\mu$ become infinite in a total compression. A natural condition ensuring this (when W is convex in its first three arguments) is that the stored-energy density become infinite in a total compression, i.e., that $W(\cdot,\cdot,\cdot,s)$ approach ∞ as its arguments approach $\partial\mathcal{V}(s)$.

2. The Multiplier Rule

We first formulate the classical Multiplier Rule for constrained variational problems in $\mathbb{R}^n$, and then apply it to the the calculus of variations.

Let $\mathcal{A}$ be a domain in $\mathbb{R}^n$ and let

$$(x_1,\dots,x_n) \equiv \mathsf{x} \mapsto \varphi_0(\mathsf{x}), \varphi_1(\mathsf{x}), \dots, \varphi_k(\mathsf{x})$$

be continuously differentiable real-valued functions on $\text{cl}\,\mathcal{A}$. We consider the problem of minimizing φ_0 on $\text{cl}\,\mathcal{A}$ subject to the side conditions

$$\varphi_j = 0, \quad j = 1, \ldots, k \tag{2.1}$$

with $k < n$. We provisionally assume that each constraint $\varphi_j = 0$ defines an $(n-1)$-dimensional surface in $\mathbb{R}^n$ and that these surfaces intersect in an $(n-k)$-dimensional surface. Let us suppose that this problem has a minimizer $\bar{\mathsf{x}}$ that lies in $\mathcal{A}$. We want to characterize $\bar{\mathsf{x}}$ as a point at which the gradient of a certain function vanishes.

If we could solve the constraint equations (2.1) for k of the unknown components of x in terms of the rest, then we could substitute this solution into the argument of φ_0 and require that the gradient of the resulting composite function vanish. We eschew this procedure because (i) it is often impossible to get an explicit solution, (ii) system (2.1) might not admit a global implicit function theorem, (iii) such a procedure might destroy symmetries enjoyed by the original problem, and (iv) such a procedure might suppress the most natural description of a physical problem.

Let us first examine a special case, which illustrates the geometric content of the general technique. We take $n = 3$, $k = 2$. We assume that φ_0 has a local minimum, which necessarily occurs on the intersection of the surfaces defined by (2.1). We make the restrictive assumption that this minimum occurs at an interior point of the C^1-curve at which these surfaces intersect. We parametrize this curve by $s \mapsto \boldsymbol{\xi}(s)$ and assume that the tangent vector field $\boldsymbol{\xi}'$ vanishes nowhere. (If these surfaces do not intersect, then there can be no minimizer. If they intersect at isolated points, then a more refined analysis, carried out below, is required.)

By elementary calculus, the minimum of φ_0 must thus occur at a point $\bar{\mathsf{x}} = \boldsymbol{\xi}(\bar{s})$, at which

$$\left[\partial\varphi_j\left(\boldsymbol{\xi}(\bar{s})\right)/\partial\mathsf{x}\right] \cdot \boldsymbol{\xi}'(\bar{s}) = 0 \tag{2.2}$$

for $j = 0$. Now the $\partial\varphi_j(\mathsf{x})/\partial\mathsf{x}$ are respectively perpendicular to the surfaces $\varphi_j = 0$ at x. Thus (2.2) holds for $j = 1, 2$. Therefore $\boldsymbol{\xi}'(\bar{s})$ is perpendicular to each of the gradients $\partial\varphi_j\left(\boldsymbol{\xi}(\bar{s})\right)/\partial\mathsf{x}$, $j = 0, 1, 2$. If these three gradients were independent, then $\boldsymbol{\xi}'(\bar{s})$ would vanish, a contradiction. Since these gradients must be dependent, there are numbers $\lambda_0, \lambda_1, \lambda_2$, not all zero, such that

$$\partial[\lambda_0\varphi_0(\bar{\mathsf{x}}) + \lambda_1\varphi_1(\bar{\mathsf{x}}) + \cdots + \lambda_k\varphi_k(\bar{\mathsf{x}})]/\partial\mathsf{x} = \mathbf{0} \tag{2.3}$$

for $k = 2$. The numbers $\lambda_0, \ldots, \lambda_k$ are called the *Lagrange multipliers*. They are unknowns of the problem; they can clearly be subjected to some normalizing condition. If we wish to find candidates for minimizers by solving (2.3), we must also find the multipliers. We thus must solve the formally determinate system consisting of (2.1), (2.3), and the normalizing condition.

The following analytic generalization of this result handles the atypical cases omitted in the development culminating in (2.3).

2.4. Lemma. *Let $\mathcal{A}$ be a domain in $\mathbb{R}^n$ and let $\varphi_0, \varphi_1, \ldots, \varphi_k \in C^1(\operatorname{cl}\mathcal{A})$. If φ_0 has a local minimum at $\bar{\mathbf{x}} \in \mathcal{A}$ subject to the side conditions* (2.1), *then*

$$\operatorname{rank}\frac{\partial(\varphi_0, \varphi_1, \ldots, \varphi_k)}{\partial(x_1, x_2, \ldots, x_n)}(\bar{\mathbf{x}}) \equiv \operatorname{rank}\begin{pmatrix} \partial\varphi_0(\bar{\mathbf{x}})/\partial\mathbf{x} \\ \partial\varphi_1(\bar{\mathbf{x}})/\partial\mathbf{x} \\ \vdots \\ \partial\varphi_k(\bar{\mathbf{x}})/\partial\mathbf{x} \end{pmatrix} < k+1. \tag{2.5}$$

Proof. Suppose that (2.5) were not true, so that

$$\operatorname{rank}\frac{\partial(\varphi_0, \varphi_1, \ldots, \varphi_k)}{\partial(x_1, x_2, \ldots, x_n)}(\bar{\mathbf{x}}) = k+1, \tag{2.6}$$

which is the maximal rank possible for a $(k+1) \times n$ matrix with $k < n$. Now consider the underdetermined system

$$\begin{aligned} h + \varphi_0(\mathbf{x}) &= \varphi_0(\bar{\mathbf{x}}), \\ \varphi_1(\mathbf{x}) &= \varphi_1(\bar{\mathbf{x}}), \\ \vdots \quad & \quad \vdots \\ \varphi_k(\mathbf{x}) &= \varphi_k(\bar{\mathbf{x}}) \end{aligned} \tag{2.7}$$

for $(h, \mathbf{x})$, which is satisfied when $(h, \mathbf{x}) = (0, \bar{\mathbf{x}})$. Condition (2.6) would enable us to use the Implicit Function Theorem to solve for $k+1$ of the variables $x_1, \ldots, x_n$ in terms of h and of the remaining $n-(k+1)$ of the variables $x_1, \ldots, x_n$ whenever $(h, \mathbf{x})$ is near $(0, \bar{\mathbf{x}})$. Consequently, for each small $h > 0$, there would be an $\mathbf{x}$ such that $\varphi_0(\mathbf{x}) = \varphi_0(\bar{\mathbf{x}}) - h < \varphi_0(\bar{\mathbf{x}})$, so that $\bar{\mathbf{x}}$ could not be a minimizer. □

We immediately conclude from this lemma that the vectors $\partial\varphi_0(\bar{\mathbf{x}})/\partial\mathbf{x}$, $\partial\varphi_1(\bar{\mathbf{x}})/\partial\mathbf{x}, \ldots, \partial\varphi_k(\bar{\mathbf{x}})/\partial\mathbf{x}$ must be dependent, so that there exist real numbers $\lambda_0, \lambda_1, \ldots, \lambda_k$, not all zero, such that (2.3) holds. Condition (2.3) is the *Multiplier Rule* for $\mathbb{R}^n$. It says that a necessary condition for an interior minimum in a constrained problem can be reduced to the vanishing of a gradient of an auxiliary function.

If λ_0 can be shown not to vanish, the variational problem is called *normal.* In this case, we can take $\lambda_0 = 1$ as a normalization.

2.8. Exercise. Prove that the variational problem is normal if and only if

$$\operatorname{rank}\frac{\partial(\varphi_1, \ldots, \varphi_k)}{\partial(x_1, \ldots, x_n)}(\bar{\mathbf{x}}) = k. \tag{2.9}$$

We now turn to the general isoperimetric problem of minimizing a functional φ_0 subject to constraints $\varphi_j = 0$, $j = 1, \ldots, k$, where $\mathbf{u} \mapsto \varphi_0[\mathbf{u}], \varphi_1[\mathbf{u}], \ldots, \varphi_k[\mathbf{u}]$ are defined on a common domain of definition. Our intended applications motivate us to choose this domain of definition to consist of all vectors of the form $\mathbf{u}_0 + \mathbf{v}$ where $\mathbf{u}_0$ is fixed and $\mathbf{v}$ lies in a subset $\mathcal{A}$ of a Banach space $\mathcal{B}$. (See Chap. XVII for notation and fundamental concepts about Banach spaces.) The domain of definition of the φ_j is denoted $\mathbf{u}_0 + \mathcal{A}$. For example, consider the problem of minimizing Ψ of (1.10b) over all functions $(\boldsymbol{r}, \theta)$ that satisfy (i) the inequality $\int_0^L (|\boldsymbol{r}'|^\alpha + |\theta'|^\gamma)\, ds < \infty$ (where α and γ are given numbers greater than 1, which characterize the material response), (ii) the boundary conditions (1.1), (iii) the isoperimetric constraint $\Omega[\boldsymbol{r}] = \Omega_\sharp$, and (iv) the condition (IV.1.9) of orientation-preservation a.e. We identify $\mathbf{u}_0$ with $(\boldsymbol{r}_0, \theta_0)$ where $\boldsymbol{r}_0(s) = s\boldsymbol{i}/L$, $\theta_0 = 0$, we identify $\mathcal{B}$ with the space of functions $(\overset{\vartriangle}{\boldsymbol{r}}, \overset{\vartriangle}{\theta})$ satisfying $\int_0^L \big(|\overset{\vartriangle}{\boldsymbol{r}}'|^\alpha + |\overset{\vartriangle}{\theta}'|^\gamma\big)\, ds < \infty$ and the homogeneous boundary conditions $\overset{\vartriangle}{\boldsymbol{r}}(0) = \mathbf{0} = \overset{\vartriangle}{\boldsymbol{r}}(L)$, $\overset{\vartriangle}{\theta}(0) = 0 = \overset{\vartriangle}{\theta}(L)$, and we identify $\mathcal{A}$ with that subset of elements $(\overset{\vartriangle}{\boldsymbol{r}}, \overset{\vartriangle}{\theta})$ of $\mathcal{B}$ for which $(\boldsymbol{r}_0, \theta_0) + (\overset{\vartriangle}{\boldsymbol{r}}, \overset{\vartriangle}{\theta})$ preserves orientation a.e. The set $\mathcal{A}$ is introduced because its elements satisfy homogeneous boundary conditions, so that linear combinations of its elements also satisfy these same conditions.

Suppose that φ_0 subject to the constraints $\varphi_j = 0$, $j = 1, \dots, k$, has a local minimum at $\bar{\mathbf{u}} = \mathbf{u}_0 + \bar{\mathbf{v}}$, where $\bar{\mathbf{v}}$ is an interior point of $\mathcal{A}$. Let $\mathbf{v}_0, \dots, \mathbf{v}_n$ be arbitrary independent elements of $\mathcal{B}$ and let $\boldsymbol{\epsilon} \equiv (\varepsilon_0, \varepsilon_1, \dots, \varepsilon_k)$ be a collection of real numbers. Since $\bar{\mathbf{v}}$ is an interior point of $\mathcal{A}$, it follows that $\bar{\mathbf{u}} + \sum_0^k \varepsilon_j \mathbf{v}_j \in \mathcal{A}$ when $|\boldsymbol{\epsilon}|$ is sufficiently small. By definition of a local minimum,

$$\varphi_0[\bar{\mathbf{u}}] \le \varphi_0\left[\bar{\mathbf{u}} + \textstyle\sum_0^k \varepsilon_l \mathbf{v}_l\right] \tag{2.10a}$$

for all sufficiently small $\boldsymbol{\epsilon}$ satisfying

$$\varphi_j\left[\bar{\mathbf{u}} + \textstyle\sum_0^k \varepsilon_l \mathbf{v}_l\right] = 0, \quad j = 1, \dots, k. \tag{2.10b}$$

Let us suppose that the functions $\boldsymbol{\epsilon} \mapsto \varphi_j[\bar{\mathbf{u}} + \sum_0^k \varepsilon_l \mathbf{v}_l]$ are continuously differentiable at $\boldsymbol{\epsilon} = \mathbf{0}$, where $\boldsymbol{\epsilon} \mapsto \varphi_0[\bar{\mathbf{u}} + \sum_0^k \varepsilon_l \mathbf{v}_l]$ has a local minimum subject to the constraints. We identify $\boldsymbol{\epsilon}$ with $\mathbf{x}$ of Lemma 2.4. We note that

$$\begin{aligned} \frac{\partial \varphi_j[\bar{\mathbf{u}} + \sum_0^k \varepsilon_l \mathbf{v}_l]}{\partial \boldsymbol{\epsilon}} &= \left(\frac{\partial \varphi_j[\bar{\mathbf{u}} + \sum_0^k \varepsilon_l \mathbf{v}_l]}{\partial \varepsilon_0}, \dots, \frac{\partial \varphi_j[\bar{\mathbf{u}} + \sum_0^k \varepsilon_l \mathbf{v}_l]}{\partial \varepsilon_k} \right) \\ &= \frac{\partial \varphi_j[\bar{\mathbf{u}} + \sum_0^k \varepsilon_l \mathbf{v}_l]}{\partial \mathbf{u}} (\mathbf{v}_0, \dots, \mathbf{v}_k). \end{aligned} \tag{2.11}$$

Applying the Multiplier Rule to our present problem, we thus find that there exist real numbers $\lambda_0, \lambda_1, \dots, \lambda_k$, not all zero, such that

$$\sum_0^k \lambda_j \left.\frac{\partial \varphi_j[\bar{\mathbf{u}} + \sum_0^k \varepsilon_l \mathbf{v}_l]}{\partial \boldsymbol{\epsilon}}\right|_{\varepsilon=0} = 0. \tag{2.12}$$

In view of (2.11), we find that

$$\text{(2.13a,b)} \qquad \sum_0^k \lambda_j \frac{\partial \varphi_j[\bar{\mathbf{u}}]}{\partial \mathbf{u}} \cdot \mathbf{v} = 0 \quad \forall \mathbf{v} \in \mathcal{B}, \quad \text{i.e.,} \quad \sum_0^k \lambda_j \frac{\partial \varphi_j[\bar{\mathbf{u}}]}{\partial \mathbf{u}} = 0.$$

This is the *Multiplier Rule for Isoperimetric Problems.* We readily find that λ_0 can be taken equal to 1 (i.e., the problem is normal) if and only if the vectors $\partial\varphi_1[\bar{\mathbf{u}}]/\partial\mathbf{u}, \dots, \partial\varphi_k[\bar{\mathbf{u}}]/\partial\mathbf{u}$ (which belong to the dual space of $\mathcal{B}$) are independent.

2.14. Exercise. Formally apply the Multiplier Rule to obtain the Euler-Lagrange equations for the variational problems posed in the paragraphs following (1.9). For the problems based on (1.9), carry out a (rather lengthy) computation showing that these equations are equivalent to those for the boundary-value problem posed in Sec. IV.3.

For multiplier rules for side conditions like (1.5), consult standard references like Bliss (1946).

3. Direct Methods of the Calculus of Variations

The direct methods of the calculus of variations yield sufficient conditions ensuring that various minimization problems have solutions. We shall see that certain aspects of the theory are especially illuminating of the underlying physics. The basic existence theory is a generalization of Weierstrass's theorem that a continuous real-valued function on a closed and bounded (i.e., a compact) subset $\mathcal{A}$ of $\mathbb{R}^n$ attains its minimum there. We shall not start with the most natural abstract generalization of this theorem, but rather with a concrete version of it that will be immediately useful for one-dimensional problems of elasticity. Then we shall show how this concrete result can be deduced from the generalization of Weierstrass's theorem.

We study functionals of the form

$$\text{(3.1)} \qquad \varphi[\boldsymbol{\eta}, \boldsymbol{\zeta}] = \int_a^b f\big(\boldsymbol{\kappa}[\boldsymbol{\eta}, \boldsymbol{\zeta}](s), \boldsymbol{\zeta}(s), \boldsymbol{\eta}(s), \boldsymbol{\zeta}'(s), s\big)\, ds$$

where $\boldsymbol{\eta} \equiv (\eta_1, \dots, \eta_l)$ and $\boldsymbol{\zeta} \equiv (\zeta_1, \dots, \zeta_m)$ are collections of real-valued functions, and $\boldsymbol{\kappa}[\boldsymbol{\eta}, \boldsymbol{\zeta}] \equiv (\kappa_1[\boldsymbol{\eta}, \boldsymbol{\zeta}], \dots, \kappa_q[\boldsymbol{\eta}, \boldsymbol{\zeta}])$ is a collection of real-valued functions determined by $\boldsymbol{\eta}$ and $\boldsymbol{\zeta}$. There is a certain arbitrariness and redundancy in this notation, as the remarks following (1.10) show. For example, for the functionals in (1.9) we may identify $\boldsymbol{\eta} = (\nu, \eta)$, $\boldsymbol{\zeta} = \theta$, $\boldsymbol{\kappa}[\boldsymbol{\eta}, \boldsymbol{\zeta}] = \boldsymbol{r}[\nu, \eta, \theta]$, and for the functions of (1.10) we may identify $\boldsymbol{\zeta} = (\boldsymbol{r}, \theta)$ and have no need for $\boldsymbol{\eta}$. Thus without essential loss of generality for our one-dimensional problems of elasticity, we could drop $\boldsymbol{\zeta}$ and $\boldsymbol{\zeta}'$ as arguments of f in (3.1), or else we could drop $\boldsymbol{\eta}$ and $\boldsymbol{\kappa}$. We assume throughout this section that $[a, b]$ is a bounded interval.

Let $f_0, \dots, f_{k+1}$ be given real-valued functions on

$$\text{(3.2)} \qquad \{(\boldsymbol{\kappa}, \boldsymbol{\zeta}, \boldsymbol{\eta}, \boldsymbol{\zeta}', s) : \boldsymbol{\kappa} \in \mathbb{R}^q,\ \boldsymbol{\zeta} \in \mathbb{R}^m,\ (\boldsymbol{\eta}, \boldsymbol{\zeta}') \in \mathcal{V}(s),\ s \in [a, b]\}$$

where $\mathcal{V}(s)$ is an open convex subset of $\mathbb{R}^l \times \mathbb{R}^m$. We may assume that $f_0, \dots, f_{k+1}$ are extended real-valued functions on the closure of (3.2). Let $\psi_0, \dots, \psi_{k+1}$ be given real-valued functions and let $\mathbf{b}$ be a given $\mathbb{R}^p$-valued function on $\mathbb{R}^m \times \mathbb{R}^m$.

Our *minimization problem* is to minimize the functional φ_0 of the form

$$(3.3)\quad \varphi_0[\boldsymbol{\eta}, \boldsymbol{\zeta}] \equiv \psi_0\big(\boldsymbol{\zeta}(a), \boldsymbol{\zeta}(b)\big) + \int_a^b f_0\big(\boldsymbol{\kappa}[\boldsymbol{\eta}, \boldsymbol{\zeta}](s), \boldsymbol{\zeta}(s), \boldsymbol{\eta}(s), \boldsymbol{\zeta}'(s), s\big)\, ds$$

over a class $\mathcal{A}$ of admissible functions for which

$$(3.4)\qquad \mathbf{b}\big(\boldsymbol{\zeta}(a), \boldsymbol{\zeta}(b)\big) = \mathbf{0},$$

$$(3.5)\qquad \begin{aligned} \varphi_j[\boldsymbol{\eta}, \boldsymbol{\zeta}] &\equiv \psi_j\big(\boldsymbol{\zeta}(a), \boldsymbol{\zeta}(b)\big) \\ &\quad + \int_a^b f_j\big(\boldsymbol{\kappa}[\boldsymbol{\eta}, \boldsymbol{\zeta}](s), \boldsymbol{\zeta}(s), \boldsymbol{\eta}(s), \boldsymbol{\zeta}'(s), s\big)\, ds = 0, \quad j = 1, \dots, k, \end{aligned}$$

$$(3.6)\qquad \begin{aligned} \varphi_{k+1}[\boldsymbol{\eta}, \boldsymbol{\zeta}] &\equiv \psi_{k+1}\big(\boldsymbol{\zeta}(a), \boldsymbol{\zeta}(b)\big) \\ &\quad + \int_a^b f_{k+1}\big(\boldsymbol{\kappa}[\boldsymbol{\eta}, \boldsymbol{\zeta}](s), \boldsymbol{\zeta}(s), \boldsymbol{\eta}(s), \boldsymbol{\zeta}'(s), s\big)\, ds \le 0. \end{aligned}$$

We assume that there are reflexive Banach spaces $\boldsymbol{L}$ and $\boldsymbol{W}$ such that

$$(3.7)\qquad \text{if } (\boldsymbol{\eta}, \boldsymbol{\zeta}) \in \mathcal{A}, \text{ then } \boldsymbol{\eta} \in \boldsymbol{L} \text{ and } \boldsymbol{\zeta} \in \boldsymbol{W},$$

and such that

$$(3.8)\qquad \varphi_0[\boldsymbol{\eta}, \boldsymbol{\zeta}] \to \infty \quad \text{as } \|\boldsymbol{\eta}, \boldsymbol{L}\| + \|\boldsymbol{\zeta}, \boldsymbol{W}\| \to \infty \quad \text{for } (\boldsymbol{\eta}, \boldsymbol{\zeta}) \in \mathcal{A}.$$

Our choice of $\boldsymbol{L}$ and $\boldsymbol{W}$ is dictated by the form of f_0. For example, suppose that f_0 satisfies the following *coercivity* condition: Let $|\boldsymbol{\kappa}| + |\boldsymbol{\zeta}| \le R$. For each nonnegative R there is a positive number $C(R)$, there are continuous weight functions $\rho_1, \dots, \rho_l, \sigma_1, \dots, \sigma_m$ independent of R that are positive on (a, b), there are continuous functions $\alpha_1, \dots, \alpha_l, \beta_1, \dots, \beta_m$ independent of R and bounded below by a number exceeding 1, and there is an integrable function $\gamma(\cdot, R)$ such that

$$(3.9)\qquad \begin{aligned} &f_0(\boldsymbol{\kappa}, \boldsymbol{\zeta}, \boldsymbol{\eta}, \boldsymbol{\zeta}', s) \\ &\qquad \ge C(R)\left[\sum_1^l \rho_j(s)|\eta_j|^{\alpha_j(s)} + \sum_1^m \sigma_j(s)|\zeta_j'|^{\beta_j(s)}\right] + \gamma(s, R). \end{aligned}$$

For the simplest mathematical problems, we may assume that the weight functions ρ_j and σ_j are each 1, and that the exponents α_j and β_j are each constants. For our problems of elasticity, condition (3.9) says how the stored-energy function grows with the strains for large strains. The weight functions arise in problems for tapered rods (see Sec. VIII.10) and in problems in polar coordinates (see Sec. X.2).

When (3.9) holds we choose $\boldsymbol{L}$ to consist of all functions $\boldsymbol{\eta}$ for which the first sum in (3.9) is integrable on (a, b) and choose $\boldsymbol{W}$ to consist of all functions $\boldsymbol{\zeta}$ that are indefinite integrals of functions $\boldsymbol{\zeta}'$ for which the second sum in (3.9) is integrable on (a, b). If the α_j and β_j are constants, then $\boldsymbol{L}$ is a classical weighted Lebesgue space and $\boldsymbol{W}$ is a classical weighted Sobolev space, with their standard norms. In this case, if $\boldsymbol{\zeta}$ satisfies suitable boundary conditions (3.4) or if (3.9) is strengthened by suitably accounting for the dependence of f_0 on $\boldsymbol{\zeta}$, then (3.8) is a consequence of (the strengthened) (3.9). Condition (3.8) likewise follows from (3.9) when the α_j and β_j are not constants by defining the norms of $\boldsymbol{L}$ and $\boldsymbol{W}$ in the manner described below.

We are tacitly assuming that the minimization problem is well-defined on $\mathcal{A}$, which lies in $\boldsymbol{L} \times \boldsymbol{W}$. In particular, we are assuming that the weight functions are such that the boundary values of $\boldsymbol{\zeta}$ that appear in (3.3)–(3.6) are well defined for $\boldsymbol{\zeta} \in \boldsymbol{W}$. As we see below, this is ensured when the weight functions are constants.

Our basic result is

3.10. Theorem. *Let $\mathcal{A}$ be nonempty. Let the functions ψ_j, b, and f_j be continuous on their domains, let $f_j(\boldsymbol{\kappa}, \boldsymbol{\zeta}, \cdot, \cdot, s)$ be continuously differentiable, let f_0 and f_{k+1} be convex on $\mathcal{V}(s)$ and satisfy* (3.8)*, let ψ_0 and ψ_{k+1} be bounded below, let the f_j, $j = 1, \dots, k$, have the form*

$$f_j(\boldsymbol{\kappa}, \boldsymbol{\zeta}, \boldsymbol{\eta}, \boldsymbol{\zeta}', s) = \mathsf{g}_j[\boldsymbol{\eta}, \boldsymbol{\zeta}](s) \cdot \boldsymbol{\eta} + \mathsf{h}_j[\boldsymbol{\eta}, \boldsymbol{\zeta}](s) \cdot \boldsymbol{\zeta}' + \omega_j[\boldsymbol{\eta}, \boldsymbol{\zeta}](s) \tag{3.11}$$

with g_j, h_j, ω_j are compact mappings from $\boldsymbol{L} \times \boldsymbol{W}$ to $C^0[a, b]$. Let $\boldsymbol{\kappa}$ be a compact mapping from $\boldsymbol{L} \times \boldsymbol{W}$ to $C^0[a, b]$. Let the mapping taking $\boldsymbol{\zeta}'$ into $s \mapsto \int_c^s \boldsymbol{\zeta}'(\xi)\, d\xi$, where c is any given number in $[a, b]$, be compact from $\boldsymbol{W}$ to $C^0[a, b]$. Let (3.8) *hold. Then φ_0 has a minimum on $\mathcal{A}$.*

We apply this theorem to specific problems in Secs. 5 and 6, which the reader may wish to consult before proceeding further in this section. There the simple condition (3.8), which is a central hypothesis of Theorem (5.11), plays a critical role.

We now prove Theorem 3.10 in the context of the general theory.

Our fundamental abstract Minimization Theorem 3.21, stated below, is just a fancy version of the basic minimization theorem of Weierstrass; its simple proof is virtually identical once we marshal some technical preliminaries. We seek minimizers $(\boldsymbol{\eta}, \boldsymbol{\zeta}) \in \boldsymbol{L} \times \boldsymbol{W}$, rather than in $[C^0[a, b]]^l \times [C^1[a, b]]^m$, say, because the former space is mathematically and physically natural for variational problems, and because it is larger than the latter, so we have a greater chance of finding minimizers there. After showing that our problems have minimizers, we demonstrate (in the next section) that mild further assumptions ensure that these minimizers have far more smoothness. (The technical advantage of $\boldsymbol{L} \times \boldsymbol{W}$ is that it admits a generalization of the Bolzano-Weierstrass Theorem.)

Lower semicontinuity. Consider the discontinuous function

$$[0, \infty) \ni x \mapsto f(x) \equiv \begin{cases} a & \text{if } x = 0, \\ \frac{1}{x+1} \sin \frac{1}{x} & \text{if } x > 0. \end{cases} \tag{3.12}$$

If $a > -1$, then f has the infimum -1, which it does not attain. If $a \le -1$, then f has a minimum, a, at 0. In the latter case, f is lower semicontinuous on $[0, \infty)$: An extended real-valued function $\mathbb{R}^n \supset \Omega \ni \mathsf{x} \mapsto \varphi(\mathsf{x}) \in (-\infty, \infty]$ is *lower semicontinuous* at $\mathsf{x}_0 \in \Omega$ iff

$$\varphi(\mathsf{x}_0) \le \liminf \varphi(\mathsf{y}) \quad \text{as } \mathsf{y} \to \mathsf{x}_0, \quad \mathsf{y} \in \Omega. \tag{3.13}$$

φ is lower semicontinuous on Ω iff it is lower semicontinuous at every point of Ω. Many of the functionals of the calculus of variations are not continuous, but enjoy an analog of this property.

The following facts are consequences of (3.13).

3.14. Proposition. *Let $\varphi : \Omega \to (-\infty, \infty]$ with $\Omega \subset \mathbb{R}^n$.*

(i) *φ is lower semicontinuous at x_0 if and only if for arbitrary $\varepsilon > 0$ there exists a $\delta(\varepsilon) > 0$ such that $\varphi(\mathsf{x}_0) \le \varphi(\mathsf{x}) + \varepsilon$ for all $\mathsf{x} \in \Omega$ for which $|\mathsf{x} - \mathsf{x}_0| < \delta$.*
(ii) *φ is continuous at x_0 if and only if both φ and $-\varphi$ are lower semicontinuous.*
(iii) *φ is lower semicontinuous on Ω if and only if for each real λ the set $\mathcal{E}_\lambda \equiv \{\mathsf{x} \in \Omega : \varphi(\mathsf{x}) \le \lambda\}$ has the form $\mathcal{E}_\lambda = \mathcal{F} \cap \Omega$ where $\mathcal{F}$ is a closed set in $\mathbb{R}^n$.*
(iv) *If φ is lower semicontinuous on Ω and if Ω is compact (and nonempty), then φ is bounded below on Ω and attains its minimum there.*

Weak convergence. Only statement (iv) of Proposition 3.14, which is Weierstrass's Theorem, is not an immediate consequence of (3.13). A proof of it relies on the Bolzano-Weierstrass Theorem: *A bounded infinite set of points in $\mathbb{R}^n$ has a point of accumulation* or, equivalently, *a bounded sequence in $\mathbb{R}^n$ has a convergent subsequence.* (This property can be phrased as: *A closed and bounded set in $\mathbb{R}^n$ is (sequentially) compact.*) To carry the proof of statement (iv) over to a real-valued function defined on part of an infinite-dimensional space, we need only extend the Bolzano-Weierstrass Theorem to such spaces. This is not so easy:

Consider an orthonormal sequence $\{u_k\}$ for a real Hilbert space $\mathcal{H}$ with a (symmetric) inner product $\langle \cdot, \cdot \rangle$ and with norm $\| \cdot \|$ defined by $\|u\| \equiv \sqrt{\langle u, u \rangle}$. Thus $\langle u_k, u_l \rangle = \delta_{kl}$ where δ_{kl} is the Kronecker delta. (For example, $u_k(s) = \sqrt{2} \sin k\pi s$ defines an orthonormal sequence in the Hilbert space $L_2(0,1)$, whose inner product is defined by $\langle u, v \rangle \equiv \int_0^1 u(s)v(s)\,ds$.) But for $k \ne l$, we readily find that

$$\|u_k - u_l\|^2 = \langle u_k - u_l, u_k - u_l \rangle = \langle u_k, u_k \rangle + \langle u_l, u_l \rangle = 2, \tag{3.15}$$

so that no subsequence of $\{u_k\}$ can converge.

On the other hand, if $\{u_k\}$ is a complete orthonormal sequence, then any element f of $\mathcal{H}$ admits a Fourier expansion $f = \sum \langle f, u_k \rangle u_k$, which converges in the norm of the space. It follows that $\|f\|^2 = \sum |\langle f, u_k \rangle|^2$ (for each f). Since this infinite series of real numbers converges, its terms must converge to 0, i.e., $\langle f, u_k \rangle \to 0$ for each f. Since this limit holds for each f, it represents a sort of convergence for the sequence $\{u_k\}$. We say that a sequence v_k in $\mathcal{H}$ *converges weakly* to v iff $\langle f, v_k \rangle \equiv \langle v_k, f \rangle \to \langle v, f \rangle$ for every f in $\mathcal{H}$. It is this mode of convergence, when adapted for other Banach spaces, that furnishes the appropriate setting for the generalization of Weierstrass's Theorem. We now carry out the requisite adaptation.

Let $\mathcal{B}$ be a real Banach space, e.g., $L_\alpha(\Omega)$. A functional $u^* : \mathcal{B} \to \mathbb{R}$ is *linear* iff $u^*[\alpha_1 v_1 + \alpha_2 v_2] = \alpha_1 u^*[v_1] + \alpha_2 u^*[v_2]$ for all $\alpha_1, \alpha_2 \in \mathbb{R}$ and for all $v_1, v_2 \in \mathcal{B}$. A linear functional u^* is *bounded* (or, equivalently, *continuous*) iff for all $v \in \mathcal{B}$ there is a number K such that $|u^*[v]| \le K\|v, \mathcal{B}\|$. If u^* is a bounded linear functional on $\mathcal{B}$, it is convenient to replace $u^*[v]$ with $\langle v, u^* \rangle$. It can be shown that the set of all bounded linear functionals on $\mathcal{B}$ forms a real Banach space $\mathcal{B}^*$, called the *space dual to $\mathcal{B}$*, which has the norm

$$\|u^*, \mathcal{B}^*\| \equiv \sup\left\{ \frac{|\langle u, u^* \rangle|}{\|u, \mathcal{B}\|} : 0 \ne u \in \mathcal{B} \right\} \equiv \sup\{|\langle u, u^* \rangle| : \|u, \mathcal{B}\| = 1\}. \tag{3.16}$$

In particular, if $\alpha \in (1, \infty)$, then $[L_\alpha(\Omega)]^*$ can be identified with $[L_{\alpha^*}(\Omega)]$, where $\alpha^* \equiv \alpha/(\alpha - 1)$, with

$$\langle u, u^* \rangle \equiv \int_\Omega u(\mathsf{x})u^*(\mathsf{x})\, dv(\mathsf{x}) \quad \text{for } u \in L_\alpha(\Omega),\ u^* \in L_{\alpha^*}(\Omega),$$

which converges because the Hölder inequality (I.8.2) implies that $|\langle u, u^* \rangle| \leq \|u, L_\alpha\| \, \|u^*, L_{\alpha^*}\|$. The same results hold if we replace the differential volume $dv(\mathsf{x})$ with $\rho(\mathsf{x})dv(\mathsf{x})$ where ρ is positive on Ω. When the norm $\|u^*, \mathcal{B}^*\|$ is known, we can define $\|u, \mathcal{B}\|$ to be $\sup\{|\langle u, u^* \rangle| : \|u^*, \mathcal{B}^*\| = 1\}$ (which is not perfectly analogous to (3.16)). Such a definition is useful for the spaces $\boldsymbol{L}$ and $\boldsymbol{W}$ introduced in the remarks following (3.9).

A Banach space $\mathcal{B}$ is *reflexive* iff its second dual $\mathcal{B}^{**} \equiv [\mathcal{B}^*]^*$ can be identified with $\mathcal{B}$. We immediately see that L_α is reflexive for $\alpha \in (1, \infty)$. It can be shown that the Sobolev spaces W_α^1 are reflexive for $\alpha \in (1, \infty)$ and that the spaces $\boldsymbol{L}$ and $\boldsymbol{W}$ are reflexive as a consequence of the restrictions imposed on the exponents.

We now define weak convergence in an arbitrary Banach space. Recall that a sequence $\{u_k\}$ in Banach space $\mathcal{B}$ *converges* (*strongly*) to u in $\mathcal{B}$ iff $\|u_k - u, \mathcal{B}\| \to 0$ as $k \to \infty$. A sequence $\{u_k\}$ in Banach space $\mathcal{B}$ *converges weakly* to u in $\mathcal{B}$ iff

$$\langle u_k - u, v^* \rangle \to 0 \quad \forall\, v^* \in \mathcal{B}^*. \tag{3.17}$$

It follows from (3.16) that a strongly convergent sequence is weakly convergent, but the discussion centered on (3.15) shows that the converse is not true. (The converse is true in finite-dimensional spaces.) Thus, there are more weakly convergent sequences than strongly convergent sequences. It can be shown that *a weakly convergent sequence is a sequence bounded in the norm of the Banach space.*

To exploit this concept we need the following definitions: A functional φ on a subset $\mathcal{A}$ of a Banach space $\mathcal{B}$ is *sequentially weakly continuous* on $\mathcal{A}$ iff

$$\varphi[u] = \lim \varphi[u_k] \quad \text{as } k \to \infty \quad \forall\, u \in \mathcal{A} \text{ and } \forall\, \{u_k\} \subset \mathcal{A} \text{ converging weakly to } u. \tag{3.18a}$$

A functional φ on a subset $\mathcal{A}$ of a Banach space $\mathcal{B}$ is *sequentially weakly lower semicontinuous* on $\mathcal{A}$ iff

$$\varphi[u] \leq \liminf \varphi[u_k] \quad \text{as } k \to \infty \quad \forall\, u \in \mathcal{A} \text{ and } \forall\, \{u_k\} \subset \mathcal{A} \text{ converging weakly to } u. \tag{3.18b}$$

(Sequential weak lower semicontinuity is a much sharper restriction on a functional than is sequential strong lower semicontinuity because there are more weakly convergent sequences than strongly convergent sequences.) A set $\mathcal{E}$ in a Banach space $\mathcal{B}$ is *sequentially weakly closed* iff it contains the weak limits of all weakly convergent sequences in $\mathcal{E}$.

One important class of sequentially weakly closed sets is given by Mazur's Theorem: *A strongly closed convex set in a Banach space is sequentially weakly closed.* We shall actually employ the following illuminating result:

3.19. Proposition. *If φ is sequentially weakly lower semicontinuous on a Banach space $\mathcal{B}$, then for each $\lambda \in \mathbb{R}$, the set $\mathcal{E}_\lambda \equiv \{u \in \mathcal{B} : \varphi[u] \leq \lambda\}$ is sequentially weakly closed.*

Proof. Let $u_k \in \mathcal{E}_\lambda$ and let u_k converge weakly to u in $\mathcal{B}$. Then $\varphi[u] \leq \liminf \varphi[u_k] \leq \lambda$, so that $u \in \mathcal{E}_\lambda$. □

An essential tool in our analysis is the following analog of the Bolzano-Weierstrass Theorem:

3.20. Proposition. *A Banach space is reflexive if and only if every bounded sequence in it has a weakly convergent subsequence.*

The Minimization Theorem. We can now prove the analog of Weierstrass's Theorem:

3.21. Minimization Theorem. *A sequentially weakly lower semicontinuous functional φ (with values in $(-\infty, \infty]$) on a sequentially weakly closed nonempty subset $\mathcal{A}$ of a reflexive Banach space $\mathcal{B}$ is bounded below on $\mathcal{A}$ and has a minimum on $\mathcal{A}$ if either* (i) *$\mathcal{A}$ is bounded, or* (ii) *$\varphi[u] \to \infty$ as $\|u, \mathcal{B}\| \to \infty$ with $u \in \mathcal{A}$.*

Proof. If $\mathcal{A}$ has just a finite number of points, the result is immediate, so we assume that $\mathcal{A}$ has an infinite number of points. Let us first assume that $\mathcal{A}$ is bounded. We show that φ is bounded below on $\mathcal{A}$. If not, there would be a sequence $\{u_k\}$ in $\mathcal{A}$ with $\varphi[u_k] \leq -k$. Since $\mathcal{A}$ is a bounded set in a reflexive Banach space, Proposition 3.20 would imply that $\{u_k\}$ has a subsequence, which we also denote by $\{u_k\}$, that would converge weakly to some limit u in $\mathcal{B}$. Since $\mathcal{A}$ is sequentially weakly closed, it would follow that $u \in \mathcal{A}$. The nature of the range of φ implies that $\varphi[u] > -\infty$. Inequality (3.18b) would then imply that $\varphi[u] \leq \liminf \varphi[u_k]$, which is impossible because $\varphi[u_k] \to -\infty$.

We suppose that (i) holds, and show that φ attains a minimum on $\mathcal{A}$. Since φ is bounded below on $\mathcal{A}$, it has an infimum μ. By definition of infimum, there exists a sequence $\{u_k\} \subset \mathcal{A}$ such that $\mu = \lim \varphi[u_k]$. As above, this sequence has a subsequence, denoted the same way, that converges weakly to some limit $\bar{u}$ in $\mathcal{A}$. Thus $\varphi[\bar{u}] = \mu$ because $\varphi[\bar{u}] \geq \mu$ and because (3.18b) implies that

$$\varphi[\bar{u}] \leq \liminf \varphi[u_k] = \lim \varphi[u_k] = \mu. \tag{3.22}$$

Now suppose that (ii) holds. We choose a fixed point $u_\sharp$ in $\mathcal{A}$ and define $\mathcal{A}_\sharp \equiv \{u \in \mathcal{A} : \varphi[u] \leq \varphi[u_\sharp]\}$. It follows from Proposition 3.19 that $\mathcal{A}_\sharp$ is sequentially weakly closed. Moreover, $\mathcal{A}_\sharp$ is bounded, for if not, there would be a sequence $\{u_j\}$ in $\mathcal{A}_\sharp$ with $\|u_j, \mathcal{B}\| \to \infty$. But then alternative (ii) would imply that $\varphi[u_j] \to \infty$ in contradiction to the definition of $\mathcal{A}_\sharp$. We now apply the proof for case (i) to $\mathcal{A}_\sharp$ noting that φ has a minimum on $\mathcal{A}$ if and only if it has a minimum on $\mathcal{A}_\sharp$. □

Compact embeddings. Before proving Theorem 3.10 let us discuss its hypothesis about compact embeddings. We begin with

3.23. Compact Embedding Theorem. *Let $-\infty < a < b < \infty$ and $\alpha \in (1, \infty)$. An element of $W^1_\alpha(a,b)$ is (Hölder) continuous. A bounded sequence $\{u_k\}$ in $W^1_\alpha(a,b)$ has a uniformly convergent subsequence.*

Proof. The Hölder inequality (I.8.2) implies that if $u \in W^1_\alpha(a,b)$, then

$$|u(x) - u(y)| \leq |y - x|^{1/\alpha^*} \|u', L_\alpha(a,b)\|, \tag{3.24a}$$

which says that u is Hölder continuous with exponent $1/\alpha^*$. We are assuming that there is a number K such that $\|u_k, W^1_\alpha(a,b)\| \leq K$. Thus (3.24a) implies that

$$|u_k(x) - u_k(y)| \leq K|y - x|^{1/\alpha^*}, \tag{3.24b}$$

which implies that $\{u_k\}$ is equicontinuous. If the u_k are prescribed to have a single value at one point of $[a,b]$, or if the integrals of the u_k are prescribed to have a single value, then (3.24b) implies that the $\{u_k\}$ are bounded in $C^0[a,b]$. Even without these restrictions, the $\{u_k\}$ are bounded in $C^0[a,b]$, for if not, there would be a subsequence of the k's such that $|u_k(x_k)| \geq k$, whence (3.24b) would imply that $|u_k(x)| \geq |u_k(x_k)| - K|b-a|^{1/\alpha^*} \geq k - K|b-a|^{1/\alpha^*}$ for all x. But this inequality would imply that $\|u_k, L_\alpha(a,b)\| \to \infty$, a contradiction. The conclusion follows from the Arzelà-Ascoli Theorem (V.5.3). □

3.25. Exercise. Consider the set of functions u on $[0,1]$ whose (distributional) derivatives satisfy $\int_0^1 s^\omega |u'(s)|^\alpha ds < \infty$ where $\alpha \in (1, \infty)$. For what range of ω are these u's continuous on $[0,1]$?

Clearly, the same kind of result holds for elements of the Sobolev spaces $\boldsymbol{W}$ with appropriate weight functions. Such results support the hypotheses on compact embeddings of Theorem 3.10.

Proof of Theorem 3.10. To apply Theorem 3.21 to the proof of Theorem 3.10, we merely need to show the weak lower semicontinuity or the weak continuity of the functionals appearing there.

We first turn to the study of φ_0, introduced in (3.3). We extend its integrand f_0 from (3.2) to $\mathbb{R}^q \times \mathbb{R}^m \times \mathbb{R}^l \times \mathbb{R}^m \times [a,b]$ by defining f_0 for $(\eta, \zeta') \notin \mathcal{V}(s)$ to be ∞. Our main technical result is

3.26. Theorem. *Let φ_0 satisfy the hypotheses of Theorem 3.10, except possibly (3.8) Let there be an integrable function γ such that $f_0(\zeta, \zeta', s) \geq \gamma(s)$ (cf. (3.9)). Then φ_0 is sequentially weakly lower semicontinuous on $\boldsymbol{L} \times \boldsymbol{W}$ and a fortiori on $\mathcal{A}$.*

Proof. For notational simplicity, we assume that f_0 depends neither on $\boldsymbol{\kappa}$ nor $\boldsymbol{\eta}$. (In view of the development of Sec. 1 following (1.9), there is little loss of generality in making this restriction for problems in elasticity.) We further assume (without loss of generality for the ensuing development) that γ is independent of R. Let $\boldsymbol{\zeta}_i$ converge weakly in $\boldsymbol{W}$ to $\boldsymbol{\zeta}$, this convergence being defined in the obvious componentwise way. We recall that this sequence is accordingly bounded in this Banach space.

For any representative of $\boldsymbol{\zeta}$ and any real number M, we define

$$\mathcal{E}(M) \equiv \{s \in [a,b] : |\zeta(s)'| \leq M\}. \tag{3.27}$$

Now

$$\begin{aligned} f_0(\zeta_i(s), \zeta_i'(s), s) - \gamma(s) = f_0(\zeta(s), \zeta'(s), s) &- \gamma(s) \\ + f_0(\zeta_i(s), \zeta_i'(s), s) &- f_0(\zeta_i(s), \zeta'(s), s) \\ + f_0(\zeta_i(s), \zeta'(s), s) &- f_0(\zeta(s), \zeta'(s), s). \end{aligned} \tag{3.28}$$

Since $\boldsymbol{\zeta}_i \to \boldsymbol{\zeta}$ in $C^0[a,b]$ by virtue of the hypothesis on compactness of Theorem 3.10, there is a positive number $K(\varepsilon)$ such that

$$|f_0(\zeta_i(s), \zeta'(s), s) - f_0(\zeta(s), \zeta'(s), s)| < \varepsilon \quad \text{when } i \geq K(\varepsilon) \text{ for } s \in \mathcal{E}(M). \tag{3.29}$$

From (3.28) and (3.29) it follows that

$$\begin{aligned} \int_{\mathcal{E}(M)} [f_0(\zeta_i, \zeta_i', s) - \gamma]\, ds \geq & \int_{\mathcal{E}(M)} [f_0(\zeta, \zeta', s) - \gamma]\, ds \\ & + \int_{\mathcal{E}(M)} [f_0(\zeta_i, \zeta_i', s) - f_0(\zeta_i, \zeta', s)]\, ds - \varepsilon(b-a) \end{aligned} \tag{3.30}$$

when $i \geq K(\varepsilon)$. The convexity of $f_0(\zeta, \cdot, s)$ implies that

$$f_0(\zeta_i(s), \zeta_i'(s), s) - f_0(\zeta_i(s), \zeta'(s), s) \geq (\zeta_i' - \zeta') \cdot \frac{\partial f_0}{\partial \zeta'}(\zeta_i(s), \zeta'(s), s). \tag{3.31}$$

Since $f_0(\zeta_i(s), \zeta_i'(s), s) - \gamma(s) \geq 0$, we obtain from (3.30) and (3.31) that

(3.32)

$$\begin{aligned} &\int_a^b [f_0(\zeta_i, \zeta_i', s) - \gamma]\, ds \\ &\quad \geq \int_{\mathcal{E}(M)} [f_0(\zeta_i, \zeta_i', s) - \gamma]\, ds \\ &\quad \geq \int_{\mathcal{E}(M)} [f_0(\zeta, \zeta', s) - \gamma]\, ds + \int_{\mathcal{E}(M)} \left[(\zeta_i' - \zeta') \cdot \frac{\partial f_0}{\partial \zeta'}(\zeta_i, \zeta', s)\right] ds - \varepsilon(b-a) \\ &\quad = \int_{\mathcal{E}(M)} [f_0(\zeta, \zeta', s) - \gamma]\, ds \\ &\quad\quad + \int_{\mathcal{E}(M)} \left\{(\zeta_i' - \zeta') \cdot \left[\frac{\partial f_0}{\partial \zeta'}(\zeta_i, \zeta', s) - \frac{\partial f_0}{\partial \zeta'}(\zeta, \zeta', s)\right]\right\} ds \\ &\quad\quad + \int_{\mathcal{E}(M)} \left[(\zeta_i' - \zeta') \cdot \frac{\partial f_0}{\partial \zeta'}(\zeta, \zeta', s)\right] ds - \varepsilon(b-a) \end{aligned}$$

when $i \geq K(\varepsilon)$. Now we take the lim inf of each side of (3.32) as $i \to \infty$. We write the last integral as

$$\int_a^b (\boldsymbol{\zeta}_i' - \boldsymbol{\zeta}') \cdot \mathbf{m}\, ds \quad \text{where} \quad \mathbf{m} \equiv \begin{cases} \frac{\partial f_0}{\partial \boldsymbol{\zeta}'}(\boldsymbol{\zeta}, \boldsymbol{\zeta}', s) & \text{for } s \in \mathcal{E}(M), \\ 0 & \text{for } s \notin \mathcal{E}(M). \end{cases} \tag{3.33}$$

Since $\mathbf{m}$ is bounded (by the definition of $\mathcal{E}(M)$), it follows from the weak convergence of $\boldsymbol{\zeta}_i$ to $\boldsymbol{\zeta}$ that (3.33) approaches 0 as $i \to \infty$. Let β be any number such that $1 < \beta \leq \min\{\beta_j\}$. We find that there is a positive number K such that

$$\begin{aligned} &\int_{\mathcal{E}(M)} (\boldsymbol{\zeta}_i' - \boldsymbol{\zeta}') \cdot \left[\frac{\partial f_0}{\partial \boldsymbol{\zeta}'}(\boldsymbol{\zeta}_i, \boldsymbol{\zeta}', s) - \frac{\partial f_0}{\partial \boldsymbol{\zeta}'}(\boldsymbol{\zeta}, \boldsymbol{\zeta}', s) \right] ds \\ &\leq \left[\int_a^b |\boldsymbol{\zeta}_i' - \boldsymbol{\zeta}'|^\beta \, ds \right]^{1/\beta} \left[\int_{\mathcal{E}(M)} \left| \frac{\partial f_0}{\partial \boldsymbol{\zeta}'}(\boldsymbol{\zeta}_i, \boldsymbol{\zeta}', s) - \frac{\partial f_0}{\partial \boldsymbol{\zeta}'}(\boldsymbol{\zeta}, \boldsymbol{\zeta}', s) \right|^{\beta^*} ds \right]^{1/\beta^*} \\ &\leq K \left[\int_{\mathcal{E}(M)} \left| \frac{\partial f_0}{\partial \boldsymbol{\zeta}'}(\boldsymbol{\zeta}_i, \boldsymbol{\zeta}', s) - \frac{\partial f_0}{\partial \boldsymbol{\zeta}'}(\boldsymbol{\zeta}, \boldsymbol{\zeta}', s) \right|^{\beta^*} ds \right]^{1/\beta^*}, \end{aligned} \tag{3.34}$$

the first inequality coming from the Hölder inequality and the second from the boundedness of weakly convergent sequences. The last integrand of (3.34) converges uniformly to 0 as $i \to \infty$. Thus

$$\liminf \int_a^b [f_0(\boldsymbol{\zeta}_i, \boldsymbol{\zeta}_i', s) - \gamma]\, ds \geq \int_{\mathcal{E}(M)} [f_0(\boldsymbol{\zeta}, \boldsymbol{\zeta}', s) - \gamma]\; ds - \varepsilon(b-a) \tag{3.35}$$

for all $\varepsilon > 0$ and $M \geq 0$. This arbitrariness enables us to drop $\varepsilon(b-a)$ from the right-hand side of (3.35) and to replace the integral over $\mathcal{E}(M)$ with the integral over $[a, b]$. Thus the integral term in φ_0 is sequentially weakly lower semicontinuous. It is easy to show that $\psi_0(\boldsymbol{\zeta}_i(a), \boldsymbol{\zeta}_i(b)) \to \psi_0(\boldsymbol{\zeta}(a), \boldsymbol{\zeta}(b))$. □

We supplement this result with

3.36. Proposition. *$\mathcal{A}$ is sequentially weakly closed.*

Proof. Theorem 3.26 implies that φ_{k+1} is sequentially weakly lower semicontinuous, so that the set of $(\boldsymbol{\eta}, \boldsymbol{\zeta})$ satisfying (3.6) is sequentially weakly closed, by Proposition 3.19. Since a weakly convergent sequence of $\boldsymbol{\zeta}$'s is uniformly convergent, it follows that $\boldsymbol{\zeta} \mapsto \mathbf{b}(\boldsymbol{\zeta}(a), \boldsymbol{\zeta}(b))$ is sequentially weakly continuous, and thus the functions satisfying (3.4) are sequentially weakly closed. Likewise, the functional $\boldsymbol{\zeta} \mapsto \psi_j(\boldsymbol{\zeta}(a), \boldsymbol{\zeta}(b))$ in (3.5) is sequentially weakly continuous. By virtue of (3.11), the $f_j(\boldsymbol{\kappa}, \boldsymbol{\zeta}, \cdot, \cdot, s)$, $j = 1, \dots, k$, are linear; thus both $f_j(\boldsymbol{\kappa}, \boldsymbol{\zeta}, \cdot, \cdot, s)$ and $-f_j(\boldsymbol{\kappa}, \boldsymbol{\zeta}, \cdot, \cdot, s)$ are convex. We can apply Theorem 3.26 to the functionals defined by the integrals in (3.5) to conclude that they are sequentially weakly continuous. Thus, the set of functions satisfying (3.5) is sequentially weakly closed. □

Theorem 3.26, Proposition 3.36, and these last remarks ensure that the hypotheses of Theorem 3.21 are met. Thus Theorem 3.10 is proved.

Virtually all the concrete results we have stated in this section are valid under weaker hypotheses. For example, we could carry out the proof of an analog of Theorem 3.10 without the hypotheses about compact embeddings. In that case, we would use a compact embedding of $\boldsymbol{L} \times \boldsymbol{W}$ into a space larger than $C^0[a, b]$. See the references in Sec. 8. The abstract Theorem 3.21 is applicable to partial differential equations, but in this case, the analog of Theorem 3.26 for a single second-order partial differential equation is more difficult to prove. As we mention in Sec. XIII.2, the analog for systems of partial differential equations of the sort that arise in the three-dimensional theory of nonlinear elasticity has only been obtained in recent years.

4. The Bootstrap Method

In preceding chapters and in Sec. 2, we have treated Euler-Lagrange equations formally. For example, in the remarks following (2.10b) we have assumed that the functionals are continuously Gâteaux-differentiable at the minimizer. We now show that this differentiability and the fact that the minimizer satisfies the classical version of the Euler-Lagrange equations are natural consequences of further mild restrictions on the integrands of the functionals appearing in the minimization problem. This procedure is called the *bootstrap method.*

In the minimization problem posed in Sec. 3, we further assume that the functions ψ_j, b, and $f_j(\cdot,\cdot,\cdot,\cdot,s)$ are continuously differentiable, that the mapping $(\boldsymbol{\eta},\boldsymbol{\zeta}) \mapsto \boldsymbol{\kappa}[\boldsymbol{\eta},\boldsymbol{\zeta}]$ is continuously (Fréchet-) differentiable from $\boldsymbol{L} \times \boldsymbol{W}$ to $C^0[a,b]$, and that its derivative is compact.

To illustrate the essential idea of this method without some of the inherent difficulties of problems from elasticity, we study a functional φ_0 meeting the hypotheses placed on it above and in Theorem 3.10, but with $\mathcal{V}(s) = \mathbb{R}^l \times \mathbb{R}^m$. (this is the standard approach. In the next section we develop a more powerful method suitable for one-dimensional problems of elasticity.) We complement (3.9) with

4.1. Hypothesis. *There are mappings $\boldsymbol{L} \times \boldsymbol{W} \ni (\boldsymbol{\eta},\boldsymbol{\zeta}) \mapsto \Gamma[\boldsymbol{\eta},\boldsymbol{\zeta}](\cdot)$, $\gamma^+[\boldsymbol{\eta},\boldsymbol{\zeta}](\cdot)$ with $s \mapsto \Gamma[\boldsymbol{\eta},\boldsymbol{\zeta}](s)$ a bounded measurable real-valued function and with $s \mapsto \gamma^+[\boldsymbol{\eta},\boldsymbol{\zeta}](s)$ an integrable real-valued function such that*

$$
\begin{aligned}
(4.2)\quad & \left|\frac{\partial f_0}{\partial \boldsymbol{\kappa}}(\boldsymbol{\kappa},\boldsymbol{\zeta},\boldsymbol{\eta},\boldsymbol{\zeta}',s)\right| + \left|\frac{\partial f_0}{\partial \boldsymbol{\eta}}(\boldsymbol{\kappa},\boldsymbol{\zeta},\boldsymbol{\eta},\boldsymbol{\zeta}',s)\right| \\
& \quad + \left|\frac{\partial f_0}{\partial \boldsymbol{\zeta}}(\boldsymbol{\kappa},\boldsymbol{\zeta},\boldsymbol{\eta},\boldsymbol{\zeta}',s)\right| + \left|\frac{\partial f_0}{\partial \boldsymbol{\zeta}'}(\boldsymbol{\kappa},\boldsymbol{\zeta},\boldsymbol{\eta},\boldsymbol{\zeta}',s)\right| \\
& \leq \Gamma[\boldsymbol{\eta},\boldsymbol{\zeta}](s)\left[\sum_1^l \rho_j(s)|\eta_j|^{\alpha_j(s)} + \sum_1^m \sigma_j(s)|\zeta_j'|^{\beta_j(s)}\right] + \gamma^+[\boldsymbol{\eta},\boldsymbol{\zeta}](s).
\end{aligned}
$$

We now show that we can take the Gâteaux derivative of φ_0 at any $(\boldsymbol{\eta},\boldsymbol{\zeta}) \in \boldsymbol{L} \times \boldsymbol{W}$ in suitable directions. We use this result in the Multiplier Rule (2.13). For simplicity of exposition, let us assume that $\psi_0 = 0$ and that f_0 is independent of $\boldsymbol{\kappa}$ and $\boldsymbol{\eta}$. Then the Mean Value Theorem (I.4.3) implies that

(4.3)

$$
\begin{aligned}
\frac{\varphi_0[\boldsymbol{\zeta}+\varepsilon\overset{\vartriangle}{\boldsymbol{\zeta}}]-\varphi_0[\boldsymbol{\zeta}]}{\varepsilon} &= \int_a^b \left\{\left[\int_0^1 \frac{\partial f_0}{\partial \boldsymbol{\zeta}}(\boldsymbol{\zeta}+t\varepsilon\overset{\vartriangle}{\boldsymbol{\zeta}},\boldsymbol{\zeta}'+t\varepsilon\overset{\vartriangle}{\boldsymbol{\zeta}}{}',s)\,dt\right]\cdot\overset{\vartriangle}{\boldsymbol{\zeta}}\right. \\
&\quad \left. + \left[\int_0^1 \frac{\partial f_0}{\partial \boldsymbol{\zeta}'}(\boldsymbol{\zeta}+t\varepsilon\overset{\vartriangle}{\boldsymbol{\zeta}},\boldsymbol{\zeta}'+t\varepsilon\overset{\vartriangle}{\boldsymbol{\zeta}}{}',s)\,dt\right]\cdot\overset{\vartriangle}{\boldsymbol{\zeta}}{}'\right\} ds.
\end{aligned}
$$

We take $0 < \varepsilon < 1$. Now we choose $\overset{\vartriangle}{\boldsymbol{\zeta}}$ to be piecewise continuously differentiable. Since $\boldsymbol{\zeta}'$ merely belongs to $\boldsymbol{L}$, we cannot expect uniform convergence in (4.3) as $\varepsilon \to 0$. But (4.2) implies that the absolute values of the two integrals with respect to t over $[0,1]$ in (4.3) are dominated by

$$(4.4) \quad \int_0^1 \left\{ \Gamma[\boldsymbol{\zeta} + t\varepsilon\overset{\vartriangle}{\boldsymbol{\zeta}}](s) \sum_1^m \sigma_j(s)|\zeta_j' + t\varepsilon\overset{\vartriangle}{\zeta}{}_j'|^{\beta_j(s)} + \gamma^+[\boldsymbol{\zeta} + t\varepsilon\overset{\vartriangle}{\boldsymbol{\zeta}}](s) \right\} dt.$$

Now recall that a convex function h has the property that $h(\frac{1}{2}\zeta_1 + \frac{1}{2}\zeta_2) \leq \frac{1}{2}h(\zeta_1) + \frac{1}{2}h(\zeta_2)$. Since $\zeta \mapsto |\zeta|^\beta$ is convex, we thus find that

$$(4.5) \qquad |\zeta_j' + t\varepsilon\overset{\vartriangle}{\zeta}{}_j'|^{\beta_j(s)} \leq \tfrac{1}{2}|2\zeta_j'|^{\beta_j(s)} + \tfrac{1}{2}|2t\varepsilon\overset{\vartriangle}{\zeta}{}_j'|^{\beta_j(s)}.$$

It then follows from the definition of $\boldsymbol{W}$ that (4.4) is bounded above by an integrable function of s for all ε. The integrand in (4.3) has an obvious pointwise limit. Thus we can invoke the Lebesgue Dominated Convergence Theorem to show that φ_0 has a Gâteaux differential at $\boldsymbol{\zeta}$ in the direction of any piecewise continuously differentiable function $\overset{\vartriangle}{\boldsymbol{\zeta}}$, which is given by

$$(4.6) \qquad \varphi_0'[\boldsymbol{\zeta}] \cdot \overset{\vartriangle}{\boldsymbol{\zeta}} = \int_a^b \left\{ \frac{\partial f_0}{\partial \boldsymbol{\zeta}}(\boldsymbol{\zeta}, \boldsymbol{\zeta}', s) \cdot \overset{\vartriangle}{\boldsymbol{\zeta}} + \frac{\partial f_0}{\partial \boldsymbol{\zeta}'}(\boldsymbol{\zeta}, \boldsymbol{\zeta}', s) \cdot \overset{\vartriangle}{\boldsymbol{\zeta}}{}' \right\} ds.$$

To show that φ_0 has a Gâteaux differential at $\boldsymbol{\zeta}$ in the direction of any element of $\boldsymbol{L} \times \boldsymbol{W}$ we replace the bounds on $\partial f_0/\partial \boldsymbol{\eta}$ and $\partial f_0/\partial \boldsymbol{\zeta}'$ given in (4.2) with more restrictive bounds in which the α_j and β_j are replaced with $\alpha_j - 1$ and $\beta_j - 1$. We would then use the Hölder inequality. There is no advantage in carrying out this procedure for variational problems associated with ordinary differential equations, however.

The treatment of the Gâteaux differentiability of φ_j, $j = 1, \dots k$, follows easily from (3.11), which obviates the need for an hypothesis on the corresponding f_j comparable to (4.2).

Let us now assume that the φ_0 of the Minimization Problem of Sec. 3 when $f_0(\boldsymbol{\kappa}[\cdot,\cdot],\cdot,\cdot,\cdot,s)$ is confined to $\mathcal{V}(s)$ has the same Gâteaux differentiability as we have just obtained under simplifying assumptions. (We confront this question below for specific problems of elasticity.) We can apply the Multiplier Rule to the Minimization Problem provided that either the unilateral constraint (3.6) is not present or else it can be shown that any minimizer satisfying (3.6) must necessarily satisfy the corresponding equality. Later we show how this can be done. Let us assume that one of these conditions holds and furthermore, if the second condition holds, then f_0 satisfies (3.11). In this case, we denote φ_0 by a φ_j with j taken from $1, \dots, k$ and we denote φ_{k+1} by φ_0. (These conditions are typical of problems in elasticity. See Sec. 5.) Let us denote the minimizer by $(\bar{\boldsymbol{\eta}}, \bar{\boldsymbol{\zeta}})$. Then in view of our results on Gâteaux differentiability, the Multiplier Rule says that there are numbers $\lambda_0, \lambda_1, \dots, \lambda_k$ such that

$$(4.7) \qquad \varphi \equiv \sum_0^k \lambda_j \varphi_j$$

satisfies

$$\frac{\partial\varphi}{\partial\boldsymbol{\eta}}[\bar{\boldsymbol{\eta}},\bar{\boldsymbol{\zeta}}]\cdot\overset{\vartriangle}{\boldsymbol{\eta}}+\frac{\partial\varphi}{\partial\boldsymbol{\zeta}}[\bar{\boldsymbol{\eta}},\bar{\boldsymbol{\zeta}}]\cdot\overset{\vartriangle}{\boldsymbol{\zeta}}=0 \tag{4.8}$$

for all piecewise continuous $\overset{\vartriangle}{\boldsymbol{\eta}}$ and piecewise continuously differentiable $\overset{\vartriangle}{\boldsymbol{\zeta}}$. This is the weak form of the Euler-Lagrange equations for the Minimization Problem. For one-dimensional problems of elasticity, these equations correspond to the Principle of Virtual Power.

Let us denote the integrand of φ by f. To proceed further, we assume that f is independent of $\boldsymbol{\kappa}$ and $\boldsymbol{\eta}$. This is a substantive restriction; to avoid it we should have to impose further restrictions on $\boldsymbol{\kappa}$. Again, such a restriction is not crucial for our elasticity problems. Let ε be a small positive number and let c and s be any numbers satisfying $a < c < c+\varepsilon < s-\varepsilon < s < b$. We define the piecewise continuously differentiable function h by

$$h(\xi,s,c)\equiv\begin{cases}0 & \text{for } a\le\xi\le c,\\ \frac{\xi-c}{\varepsilon} & \text{for } c\le\xi\le c+\varepsilon,\\ 1 & \text{for } c+\varepsilon\le\xi\le s-\varepsilon,\\ \frac{s-\xi}{\varepsilon} & \text{for } s-\varepsilon\le\xi\le s,\\ 0 & \text{for } s\le\xi\le b.\end{cases} \tag{4.9}$$

Let $\mathbf{e}$ be a fixed but arbitrary element of $\mathbb{R}^m$. We change the dummy variable of integration in (4.8) to ξ and then choose $\overset{\vartriangle}{\boldsymbol{\zeta}}(\xi)=h(\xi,s,c)\mathbf{e}$. We substitute these variables into (4.8) to obtain

$$\begin{aligned}\frac{1}{\varepsilon}\int_c^{c+\varepsilon}\frac{\partial f}{\partial\boldsymbol{\zeta}'}(\bar{\boldsymbol{\zeta}}(\xi),\bar{\boldsymbol{\zeta}}'(\xi),\xi)\cdot\mathbf{e}\,d\xi-\frac{1}{\varepsilon}\int_{s-\varepsilon}^{s}\frac{\partial f}{\partial\boldsymbol{\zeta}'}(\bar{\boldsymbol{\zeta}}(\xi),\bar{\boldsymbol{\zeta}}'(\xi),\xi)\cdot\mathbf{e}\,d\xi&\\+\int_c^s\frac{\partial f}{\partial\boldsymbol{\zeta}}(\bar{\boldsymbol{\zeta}}(\xi),\bar{\boldsymbol{\zeta}}'(\xi),\xi)\cdot h(\xi,s,c)\mathbf{e}\,d\xi=0.&\end{aligned} \tag{4.10}$$

Since the first two terms in (4.10) are quotients, their limits as $\varepsilon\to0$ can be formally obtained by l'Hôpital's Rule wherever the derivatives of the integrals exist at $\varepsilon=0$. Since the integrands in (4.10) are integrable (by virtue of (4.2)), these integrals are absolutely continuous functions of ε; their derivatives with respect to ε at $\varepsilon=0$ exist for almost all c and s. The Lebesgue Dominate Convergence Theorem enables us to show that the limit of the last integral is given by this integral with h replaced with 1. Using the arbitrariness of $\mathbf{e}$, we thus obtain

$$\frac{\partial f}{\partial\boldsymbol{\zeta}'}(\bar{\boldsymbol{\zeta}}(s),\bar{\boldsymbol{\zeta}}'(s),s)=\frac{\partial f}{\partial\boldsymbol{\zeta}'}(\bar{\boldsymbol{\zeta}}(c),\bar{\boldsymbol{\zeta}}'(c),c)+\int_c^s\frac{\partial f}{\partial\boldsymbol{\zeta}}(\bar{\boldsymbol{\zeta}}(\xi),\bar{\boldsymbol{\zeta}}'(\xi),\xi)\,d\xi \tag{4.11}$$

for almost all c and s in $[a,b]$. For one-dimensional problems of elasticity, these equations correspond to the integral form of the equilibrium equations. That the integral form (4.8) implies the local form (4.11) is just a version of the *Fundamental Lemma of the Calculus of Variations.*

Let us fix c to be any value for which (4.11) holds. The right-hand side of (4.11) defines an absolutely continuous function of s. Since we know that there is a minimizer $(\bar{\boldsymbol{\zeta}}, \bar{\boldsymbol{\zeta}}')$, the left-hand side also defines an absolutely continuous function of s. We can therefore differentiate (4.11) with respect to s, obtaining the classical form of the Euler-Lagrange equations

$$\frac{d}{ds}\left[\frac{\partial f}{\partial \boldsymbol{\zeta}'}(\bar{\boldsymbol{\zeta}}(s), \bar{\boldsymbol{\zeta}}'(s), s)\right] = \frac{\partial f}{\partial \boldsymbol{\zeta}}(\bar{\boldsymbol{\zeta}}(s), \bar{\boldsymbol{\zeta}}'(s), s) \quad \text{a.e.} \tag{4.12}$$

To proceed further, we assume that this minimization problem is normal, so that we can take $\lambda_0 = 1$. (We are adhering to the renumbering of functionals described in the remarks preceding (4.7).) Then f is a convex function of $\boldsymbol{\zeta}'$. We now strengthen this assumption by requiring that f_0, and hence f, be a strictly convex function of $\boldsymbol{\zeta}'$, so that $\partial f/\partial \boldsymbol{\zeta}'$ is a strictly monotone function of $\boldsymbol{\zeta}'$. It follows that $\partial f(\bar{\boldsymbol{\zeta}}(s), \cdot, s)/\partial \boldsymbol{\zeta}'$ has an inverse $\boldsymbol{\omega}(\bar{\boldsymbol{\zeta}}(s), \cdot, s)$ on its range. Since we know that (4.11) has a solution $\bar{\boldsymbol{\zeta}}$, we know that the right-hand side of (4.11) is in this range. Therefore (4.11) is equivalent to

$$\bar{\boldsymbol{\zeta}}'(s) = \boldsymbol{\omega}\left(\bar{\boldsymbol{\zeta}}(s), \frac{\partial f}{\partial \boldsymbol{\zeta}'}(\bar{\boldsymbol{\zeta}}(c), \bar{\boldsymbol{\zeta}}'(c), c) + \int_c^s \frac{\partial f}{\partial \boldsymbol{\zeta}}(\bar{\boldsymbol{\zeta}}(\xi), \bar{\boldsymbol{\zeta}}'(\xi), \xi)\, d\xi, s\right). \tag{4.13}$$

Let us now further assume that f is twice continuously differentiable in $(\bar{\boldsymbol{\zeta}}, \bar{\boldsymbol{\zeta}}')$ and continuously differentiable in s. Then the classical Implicit Function Theorem implies that $\boldsymbol{\omega}$ is continuously differentiable. Since its arguments in (4.13) are continuous functions of s, it follows from (4.13) that $\bar{\boldsymbol{\zeta}}'$ is continuous. Consequently, the arguments of $\boldsymbol{\omega}$ in (4.13) are continuously differentiable functions of s. Therefore, (4.13) implies that $\bar{\boldsymbol{\zeta}}$ is twice continuously differentiable. In this case, we can carry out the differentiations everywhere, so that $\bar{\boldsymbol{\zeta}}$ is a classical solution of the Euler-Lagrange equations.

5. Inflation Problems

Existence. We now study extremum problems for the functionals introduced in (1.9). These problems describe the equilibrium of nonlinearly elastic arches subjected to hydrostatic pressure. We relegate the alternative treatment of these problems by the formulation (1.10) to an exercise.

We take the domain of $W(\cdot, \cdot, \cdot, s)$ to consist of those (ν, η, μ) satisfying (IV.1.22):

$$\nu > V(\mu, s). \tag{5.1}$$

For simplicity of exposition, we assume that there are continuous functions h_1 and h_2 with $h_1(s) \leq 0 \leq h_2(s)$ for all s such that

$$V(\mu, s) = \begin{cases} h_1(s)\mu & \text{for} \quad \mu < 0, \\ h_2(s)\mu & \text{for} \quad \mu > 0. \end{cases} \tag{5.2}$$

This form of V, mentioned in (IV.3.12), is particularly natural (see Sec. VIII.6).

In accord with (IV.1.23) and (IV.1.32), we assume that $W(\cdot,\cdot,\cdot,s)$ is twice continuously differentiable and has a positive-definite Hessian matrix. We require that W satisfy a special form of (3.9): There is a positive number C, there are continuous functions $\alpha_1, \alpha_2, \beta_1$ bounded below by a number exceeding 1, and there is an integrable function γ such that

$$(5.3) \qquad W(\nu,\eta,\mu,s) \geq C\left[\nu^{\alpha_1(s)} + |\eta|^{\alpha_2(s)} + |\mu|^{\beta_1(s)}\right] + \gamma(s).$$

(We do not include weight functions in (5.3) because their presence can add technical complications to our analysis; see Antman (1976b). Such weight functions are needed to treat rods whose thickness approaches zero at an end; see Sec. VIII.10.) For a uniform rod, in which the constitutive functions are independent of s, the exponents $\alpha_1, \alpha_2, \beta_1, \gamma$ are constants, and the resulting analysis becomes closer to the typical. In each problem we study, we take our admissible class of functions (ν, η, θ) to be a subset of the set $\mathcal{A}_0$ that consists of those functions that satisfy

$$(5.4) \qquad \int_0^L \left[\nu(s)^{\alpha_1(s)} + |\eta(s)|^{\alpha_2(s)} + |\theta'(s)|^{\beta_1(s)}\right] ds < \infty,$$

$$(5.5) \qquad \nu(s) \geq V(\theta'(s), s) \quad \text{a.e.},$$

together with (1.1c,d) and (1.8). Alternatively, we could replace (5.4) in the definition of $\mathcal{A}_0$ with the requirement that $\Psi[\nu,\eta,\theta'] < \infty$, which implies (5.4), by virtue of (5.3). We take our variables to lie in the appropriate specialization of the spaces of $\boldsymbol{L}$ and $\boldsymbol{W}$.

We assert that $\mathcal{A}_0$ is sequentially weakly closed. Indeed, since the functional of (5.4) is sequentially weakly lower semicontinuous by Theorem 3.26, it follows from Proposition 3.36 that the functions satisfying (5.4) form a weakly closed set. To show that the functions further satisfying (5.5) form a weakly closed set, we first show that they form a strongly closed convex set and then use Mazur's Theorem, stated before Proposition 3.19, to deduce that they form a sequentially weakly closed set. That the set is convex is immediate. To show that it is strongly closed, we need only use the fact that a strongly convergent sequence in a space L_α, $\alpha \geq 1$, has a subsequence that converges a.e.

Our major result about the three variational problems for the functionals of (1.9) is the following

5.6. Theorem. *Let the functionals Π, Ψ, Ω be defined by* (1.9).

(i) *Let p be prescribed. If $\alpha_1(s) > 2$ and $\alpha_2(s) > 2$ for all s, then the functional $\Pi = \Psi + p\Omega$ has a minimum on $\mathcal{A}_0$.*

(ii) *Let $\Omega_\sharp$ be prescribed. The functional Ψ has a minimum over the elements of $\mathcal{A}_0$ for which* (1.15) *holds.*

(iii) *Let $\Psi_\sharp$ be prescribed. The functional Ω has a minimum and a maximum over the elements of $\mathcal{A}_0$ for which $\Psi[\nu,\eta,\theta] \leq \Psi_\sharp$.*

Proof. These results follow directly from Theorem 3.10 or alternatively from the abstract theory developed in Sec. 3. (The functional Ψ

is sequentially weakly lower semicontinuous and the functionals Ω and $(\nu, \eta, \theta) \mapsto \boldsymbol{r}[\nu, \eta, \theta]$ are sequentially weakly continuous.) The only point that is worthy of comment is that Ω is quadratic in ν and η and has no definite sign. Consequently, in the proof of (i), to ensure that $\Pi = \Psi + p\Omega$ approaches ∞ as the norm of (ν, η, θ) approaches ∞ (see (3.8)), we must take $\alpha_1(s) > 2$ and $\alpha_2(s) > 2$ for all s. (See Ex. 5.7(i).) In this case, the internal energy dominates the potential energy of the applied force system. This issue does not arise in cases (ii) and (iii), because the two potentials are not competing with each other. □

We refine statement (iii) of Theorem 5.6 below.

5.7. Exercise. (i) Prove that $|\Omega[\nu, \eta, \theta]| \leq \int_0^L (\nu^2 + \eta^2)\, ds$. (ii) Carry out the proof of Theorem 5.6 for the functionals of (1.10).

Let us interpret these minimizers as very weak solutions of the Euler-Lagrange equations for an arch under hydrostatic pressure. (Later we show that under further assumptions, these minimizers are classical equilibrium states.) The physical content of this theorem is that an arch admits equilibrium states for all pressures provided that the material is strong enough in tension and shear. On the other hand, an arch always admits equilibrium states of all sizes when the size is measured by the area change or by the internal energy. These observations are in complete agreement with the pure inflation problem for a circular ring or arch, which has much the same form as the pure inflation problem for a string (cf. Secs. III.5, IV.3, and VI.12).

Regularity. We now assume that the stored-energy density becomes infinite in a total compression:

$$W(\nu, \eta, \mu, s) \to \infty \quad \text{as } \nu - V(\mu, s) \to 0. \tag{5.8}$$

Let (ν, η, θ) be an extremizer of one of the variational problems for (1.9) (see Theorem 5.6). Since Ψ is finite at this minimizer, it follows from (5.8) that the set of s's at which $\nu(s) - V(\theta'(s), s) = 0$ has measure zero. We wish to show that this set is empty and to show that the minimizer defines a regular solution of the boundary-value problem for the Euler-Lagrange equations. This objective is much harder to achieve for our problems of elasticity than for the problems treated in Sec. 4, because the derivative of W grows faster than W when ν is near $V(\theta', s)$, so that hypotheses like (4.2) are untenable. (Consider an energy for longitudinal motion: $W(\nu) = A/\nu + B\nu^{\alpha_1}$.) To carry out an effective proof of regularity, we must exploit the strong repulsion inherent in (5.8). To minimize technicalities with side conditions, we study the minimizer of (1.10a) satisfying (1.1).

5.9. Theorem. *Let $W(\cdot, \cdot, \cdot, s)$ be continuously differentiable on its domain, and let W be continuous on its domain. Let (5.3) and (5.8) hold. Let $(\boldsymbol{r}, \theta)$ minimize (1.10a) on $\mathcal{A}_0$, which consists of all $(\boldsymbol{r}, \theta)$ satisfying the inequality $\Psi[\nu, \eta, \theta'] < \infty$ and the boundary conditions (1.1), with ν and η defined by (1.5). (Elements of $\mathcal{A}_0$ satisfy (5.4) and (5.5) in view* *of (5.3) and (5.8).) Then $(\boldsymbol{r}, \theta)$ satisfies the Euler-Lagrange equations a.e., and $s \mapsto \nu(s) - V(\theta'(s), s)$ is everywhere positive.*

Proof. We must show that Π is Gâteaux-differentiable in suitable directions. Then we can employ the bootstrap method. To avoid the difficulties with (5.8), we construct variations $(\overset{\triangle}{\boldsymbol{r}}, \overset{\triangle}{\theta})$ that vanish where the minimizer is close to violating (5.5): For any positive integer n, let

$$G_n(s) \equiv \sup\{|W_\nu(\tilde{\nu}, \tilde{\eta}, \tilde{\mu}, s)| + |W_\eta(\tilde{\nu}, \tilde{\eta}, \tilde{\mu}, s)| + |W_\mu(\tilde{\nu}, \tilde{\eta}, \tilde{\mu}, s)| : |\tilde{\nu} - \nu(s)| + |\tilde{\eta} - \eta(s)| + |\tilde{\mu} - \theta'(s)| < \tfrac{1}{n}\}, \tag{5.10a}$$

$$\mathcal{E}_n \equiv \{s \in [0, L] : G_n(s) \le n\}, \tag{5.10b}$$

$$\chi_n(s) \equiv \begin{cases} 1 & \text{if } s \in \mathcal{E}_n, \\ 0 & \text{if } s \notin \mathcal{E}_n. \end{cases} \tag{5.10c}$$

Our hypotheses on W ensure that $\mathcal{E}_n$ is measurable. Clearly, $\mathcal{E}_n \subset \mathcal{E}_{n+1}$ and $[0, L] \setminus \cup_{n=1}^\infty \mathcal{E}_n$ has measure zero.

We first choose an arbitrary bounded function $s \mapsto \mu_n^\sharp(s)$ and define

$$\theta_n(s; t) \equiv \theta(s) + t \int_0^s \chi_n(\xi) \mu_n^\sharp(\xi)\, d\xi. \tag{5.10d}$$

In order for θ_n to satisfy (1.1c,d), we require that $\mu_n^\sharp$ satisfy

$$\varphi_1[\theta_n(\cdot; t)] \equiv \theta_n(L; t) = t \int_{\mathcal{E}_n} \mu_n^\sharp\, ds = 0. \tag{5.10e}$$

Next we introduce arbitrary functions $(s, t) \mapsto \nu_n^\sharp(s, t)$, $\eta_n^\sharp(s, t)$ bounded in s and continuously differentiable in t for $|t|$ small. We define

$$\nu_n(s; t) \equiv \nu(s) + t\chi_n(s)\nu_n^\sharp(s, t), \quad \eta_n(s; t) \equiv \eta(s) + t\chi_n(s)\eta_n^\sharp(s, t), \tag{5.11a,b}$$

$$\begin{aligned} \boldsymbol{r}_n(s; t) &\equiv \int_0^s [\nu_n(\xi; t)\boldsymbol{a}(\theta_n(\xi; t)) + \eta_n(\xi; t)\boldsymbol{b}(\theta_n(\xi; t))]\, d\xi \\ &\equiv \boldsymbol{r}(s) + \int_0^s \{\nu(\xi)[\boldsymbol{a}(\theta_n(\xi; t)) - \boldsymbol{a}(\theta(\xi))] + \eta(\xi)[\boldsymbol{b}(\theta_n(\xi; t)) - \boldsymbol{b}(\theta(\xi))]\}\, d\xi \\ &\quad + t \int_0^s \chi_n(\xi)[\nu_n^\sharp(\xi; t)\boldsymbol{a}(\theta_n(\xi; t)) + \eta_n^\sharp(\xi; t)\boldsymbol{b}(\theta_n(\xi; t))]\, d\xi. \end{aligned} \tag{5.11c}$$

The functions $\boldsymbol{r}_n$ and θ_n generate ν_n, η_n of (5.11a,b) via (1.5). To ensure that $\boldsymbol{r}_n$ satisfies (1.1a,b), we now choose $(\nu_n^\sharp, \eta_n^\sharp)$ so that

$$\begin{aligned} 0 &= \frac{1}{t}\boldsymbol{\phi}[\boldsymbol{r}_n(\cdot; t), \theta_n(\cdot; t)] \\ &\equiv \frac{1}{t}\int_0^L \{\nu(s)[\boldsymbol{a}(\theta_n(s; t)) - \boldsymbol{a}(\theta(s))] + \eta(s)[\boldsymbol{b}(\theta_n(s; t)) - \boldsymbol{b}(\theta(s))]\}\, ds \\ &\quad + \int_{\mathcal{E}_n} [\nu_n^\sharp(s; t)\boldsymbol{a}(\theta_n(s; t)) + \eta_n^\sharp(s; t)\boldsymbol{b}(\theta_n(s; t))]\, ds \end{aligned} \tag{5.11d}$$

for all small $|t|$, i.e., we now choose any vector $(\nu_n^\sharp, \eta_n^\sharp)$ whose projections on the two independent vectors $(\cos\theta_n, \sin\theta_n)$ and $(-\sin\theta_n, \cos\theta_n)$ are prescribed according to (5.11d). Note that there is no difficulty at $t = 0$. ($\nu_n^\sharp$ and $\eta_n^\sharp$ depend on t so that they can satisfy (5.11d).)

Since

$$\text{(5.12)}\qquad \begin{aligned} &\Pi[\boldsymbol{r}_n(\cdot;t),\theta_n(\cdot;t)]-\Pi[\boldsymbol{r},\theta]\\ &=\int_{\mathcal{E}_n}\left[W(\nu_n(s;t),\eta_n(s;t),\theta_n'(s;t),s)-W(\nu(s),\eta(s),\theta'(s),s)\right]ds\\ &\qquad+\frac{p}{2}\boldsymbol{k}\cdot\int_0^L\left[\boldsymbol{r}_n(s;t)\times\boldsymbol{r}_n'(s;t)-\boldsymbol{r}(s)\times\boldsymbol{r}'(s)\right]ds,\end{aligned}$$

we find that $\Pi[\boldsymbol{r}_n(\cdot;t),\theta_n(\cdot;t)]<\infty$ for $|t|$ sufficiently small, so that $(\boldsymbol{r}_n(\cdot,t),\theta_n(\cdot,t))\in\mathcal{A}_0$ for $|t|$ sufficiently small. Since the minimum of $t\mapsto\Pi[\boldsymbol{r}_n(\cdot;t),\theta_n(\cdot;t)]$ is attained at $t=0$ and since ν_n,η_n,θ_n satisfy the side conditions (5.10e) and (5.11d), we imitate the argument leading to the Multiplier Rule (2.13) to conclude that there are constants λ_0, λ_1, $\boldsymbol{\lambda}$, not all zero, such that

$$\text{(5.13a)}\qquad \tfrac{d}{dt}\{\lambda_0\Psi[\boldsymbol{r}_n(\cdot;t),\theta_n(\cdot;t)]+\lambda_0 p\Omega[\boldsymbol{r}_n(\cdot;t),\theta_n(\cdot;t)] +\lambda_1\varphi_1[\theta_n(\cdot;t)]+\boldsymbol{\lambda}\cdot\boldsymbol{\phi}[\boldsymbol{r}_n(\cdot;t),\theta_n(\cdot;t)]\}|_{t=0}=0$$

for all $\nu_n^\sharp(\cdot;0),\eta_n^\sharp(\cdot;0),\mu_n^\sharp$. By integrating by parts and by using (5.10e) we find that

(5.13b)

$$\begin{aligned}&\frac{d}{dt}\Psi[\boldsymbol{r}_n(\cdot;t),\theta_n(\cdot;t)]\Big|_{t=0}\\ &=\int_{\mathcal{E}_n}\Big[W_\nu(\nu(s),\eta(s),\theta'(s),s)\nu_n^\sharp(s;0)+W_\eta(\nu(s),\eta(s),\theta'(s),s)\eta_n^\sharp(s;0)\\ &\qquad+W_\mu(\nu(s),\eta(s),\theta'(s),s)\mu_n^\sharp(s)\Big]ds\end{aligned}$$

(5.13c)

$$\begin{aligned}&\frac{d}{dt}\Omega[\boldsymbol{r}_n(\cdot;t),\theta_n(\cdot;t)]\Big|_{t=0}\\ &=\int_{\mathcal{E}_n}\Big\{-\tfrac{1}{2}|\boldsymbol{r}(s)|^2\mu_n^\sharp(s)+\boldsymbol{r}(s)\cdot[-\nu_n^\sharp(s;0)\boldsymbol{b}(\theta(s))+\eta_n^\sharp(s;0)\boldsymbol{a}(\theta(s))]\Big\}ds,\end{aligned}$$

(5.13d)

$$\frac{d}{dt}\varphi_1[\theta_n(\cdot;t)]\Big|_{t=0}=\int_{\mathcal{E}_n}\mu_n^\sharp(s)\,ds,$$

(5.13e)

$$\begin{aligned}&\frac{d}{dt}\boldsymbol{\phi}[\boldsymbol{r}_n(\cdot;t),\theta_n(\cdot;t)]\Big|_{t=0}\\ &=\int_{\mathcal{E}_n}\Big\{-[\boldsymbol{k}\times\boldsymbol{r}(s)]\mu_n^\sharp(s)+\nu_n^\sharp(s;0)\boldsymbol{a}(\theta(s))+\eta_n^\sharp(s;0)\boldsymbol{b}(\theta(s))\Big\}ds.\end{aligned}$$

Using the arbitrariness of $\nu_n^\sharp(\cdot;0)$, $\eta_n^\sharp(\cdot;0)$, $\mu_n^\sharp$, we immediately deduce from (5.13) that

$$\text{(5.14a)}\qquad \lambda_0W_\nu(\nu(s),\eta(s),\theta'(s),s)=\lambda_0p\boldsymbol{r}(s)\cdot\boldsymbol{b}(\theta(s))-\boldsymbol{\lambda}\cdot\boldsymbol{a}(\theta(s)),$$

$$\text{(5.14b)}\qquad \lambda_0W_\eta(\nu(s),\eta(s),\theta'(s),s)=-\lambda_0p\boldsymbol{r}(s)\cdot\boldsymbol{a}(\theta(s))-\boldsymbol{\lambda}\cdot\boldsymbol{b}(\theta(s)),$$

$$\text{(5.14c)}\qquad \lambda_0W_\mu(\nu(s),\eta(s),\theta'(s),s)=\lambda_0\tfrac{1}{2}p|\boldsymbol{r}(s)|^2-\lambda_1+\boldsymbol{\lambda}\cdot[\boldsymbol{k}\times\boldsymbol{r}(s)]$$

a.e. on $\mathcal{E}_n$. It easily follows from (5.14) that if $\lambda_0=0$, then $\lambda_1=0$, $\boldsymbol{\lambda}=\boldsymbol{0}$, in contradiction to the Multiplier Rule. We accordingly take $\lambda_0=1$. Since $\mathcal{E}_n\subset\mathcal{E}_{n+1}$ it follows that λ_1 and $\boldsymbol{\lambda}$ are independent of n. The properties of $\mathcal{E}_n$ show that (5.14) holds a.e. on $[0,L]$. Since the right-hand side of (5.14) is continuous on $[0,L]$, we can assume that (5.14) holds everywhere on $[0, L]$, and we deduce from (5.8) that $s \mapsto \nu(s) - V(\theta'(s), s)$ is everywhere positive. Since the right-hand side of (5.14) is absolutely continuous on $[0, L]$, we can differentiate (5.14) a.e., and as in Sec. 4, thus find that the Euler-Lagrange equations hold a.e. □

Note that this theorem does not require the hypothesis that $W(\cdot, \cdot, \cdot, s)$ be convex. It suffices that the variational problem have a minimizer. Convexity is used in the next theorem.

5.15. Exercise. Show that (5.14) is equivalent to the appropriate specialization of the equations derived in Sec. IV.1.

5.16. Theorem. *Let the hypotheses of Theorem 5.9 hold and let $W(\cdot, \cdot, \cdot, s)$ be uniformly convex. Then $(\boldsymbol{r}, \theta)$ is a classical solution of the Euler-Lagrange equations.*

Proof. From (5.14) we obtain

$$\begin{aligned} \text{(5.17a)} \quad \hat{N}(\nu(s), \eta(s), \theta'(s), s) &\equiv W_\nu(\nu(s), \eta(s), \theta'(s), s) \\ &= p\boldsymbol{r}(s) \cdot \boldsymbol{b}(\theta(s)) - \boldsymbol{\lambda} \cdot \boldsymbol{a}(\theta(s)), \quad \text{etc.} \end{aligned}$$

The uniform convexity of W implies that (5.17a) is equivalent to a system of the form

$$\text{(5.17b)} \qquad \nu(s) = \hat{\nu}\big(p\boldsymbol{r}(s) \cdot \boldsymbol{b}(\theta(s)) - \boldsymbol{\lambda} \cdot \boldsymbol{a}(\theta(s)), \dots\big), \quad \text{etc.}$$

Since the first three arguments of $\hat{\nu}, \dots$ are continuous on $[0, L]$, it follows that $\nu, \dots$ are continuous, so that the solution is classical. □

5.18. Exercise. Prove the analogs of Theorems 5.9 and 5.16 for the three problems treated in Theorem 5.6.

5.19. Problem. Determine the regularity of equilibrium states when the energy W is not required to satisfy (5.8). Is the minimizer a weak solution of the Euler-Lagrange equation? Suppose that V has the form (5.2) and that

(5.20)

$$\hat{N}(\nu, \eta, \mu, s) \to -\infty, \quad \hat{M}(\nu, \eta, \mu, s) \to \begin{Bmatrix} \infty & \text{for } \mu > 0, \\ -\infty & \text{for } \mu < 0 \end{Bmatrix} \quad \text{as } \nu - V(\mu, s) \to 0.$$

Is it then possible to show that for a weak solution $\nu - V(\theta', s) = 0$ nowhere?

Normality for the problem with prescribed strain energy. We now refine Theorem 5.6(iii):

5.21. Theorem. *Let the functionals Π, Ψ, Ω be defined by (1.9) or (1.10). Let $\Psi_\sharp$ be prescribed. The functional Ω has a minimum and a maximum over the elements of $\mathcal{A}_0$ for which $\Psi[\nu, \eta, \theta] = \Psi_\sharp$.*

Sketch of proof. For simplicity, we use Exercise 5.7 to show that Theorem 5.6(iii) holds for the functionals of (1.10). Suppose for contradiction that the extremizer $(\boldsymbol{r}, \theta)$ were to satisfy $\Psi[\boldsymbol{r}, \theta] < \Psi_\sharp$. This inequality suggests that Ω has an unconstrained minimum on $\mathcal{A}_0$ and that its Gâteaux derivative must vanish. But small variations of the minimizer can violate (5.5). It is consequently not obvious that Ω is Gâteaux-differentiable at the extremizer. We avoid this difficulty by constructing variations by the methods used in the proof of Theorem 5.9. In place of (5.15), we accordingly find that

$$\text{(5.22)} \qquad p\boldsymbol{k} \times \boldsymbol{r}(s) = \boldsymbol{c},$$

which implies that $\boldsymbol{r}' = \boldsymbol{0}$, in violation of (5.5). □

In our treatment of regularity, we have avoided confronting problems with side conditions. An ad hoc procedure for doing so would be to refine the variations $\nu_n^\sharp$, $\eta_n^\sharp$, $\mu_n^\sharp$ used in the proof of Theorem 5.9 so that they are consistent with the isoperimetric constraints. For this purpose, we can use the ideas developed in Sec. 2. Alternatively, we could formulate a Multiplier Rule applicable to the original functionals, rather than to their Gâteaux derivatives (see Cannon, Cullum, & Polak (1970)). We omit the details of such constructions, assuming that versions of (2.13) can be established just like (5.14) and (5.15).

Let us consider the problems (for (1.10)) in which (1.15) or (1.16) holds. We should like to normalize the Lagrange multipliers so that the multiplier associated with Ψ is 1. In this case, the multiplier associated with Ω can be identified as the pressure necessary to maintain the constraint. For showing that the multiplier associated with Ψ can be taken to be 1, we need only show that this multiplier does not vanish. Were it to do so, then the Gâteaux derivative of Ω would have to vanish, leading to (5.22) and the consequent contradiction.

5.23. Exercise. Show that the multiplier associated with Ψ can be taken to be 1 when (1.9) is used and either (1.15) or (1.16) holds. In this case, (1.8) must be used. It has its own multiplier.

6. Problems for Whirling Rods

We now study variational aspects of the boundary-value problem (IV.5.10):

$$N'' + \lambda\hat{\nu}(N) = 0, \quad N(0) = 0, \quad N'(1) = -\lambda R, \tag{6.1a,b,c}$$

which describes the radial steady states of a rotating uniform elastic rod of length 1, with its end $s = 0$ free, and with its end $s = 1$ attached to a rigid ring of radius R rotating about its center with angular speed $\sqrt{\lambda/\rho A}$. N is the tension in the rod and $\hat{\nu}(N)$ is the stretch it produces. We introduce the conjugate W^* of the stored-energy function; it has the property that

$$\hat{\nu}(N) = W_N^*(N). \tag{6.2}$$

The usual properties of $\hat{\nu}$ yield

$$\begin{gathered} W^*(0) = 0, \quad W_N^*(0) = 1, \quad W_N^*(N) > 0, \quad W_{NN}^*(N) > 0, \\ W_N^*(N) \to 0 \quad \text{as } N \to -\infty, \quad W_N^*(N) \to \infty \quad \text{as } N \to \infty. \end{gathered} \tag{6.3}$$

Let $\lambda > 0$ be fixed. We consider the problem of minimizing the functional

$$\Phi[N] \equiv \int_0^1 \left[\tfrac{1}{2}N'(s)^2 - \lambda W^*(N(s))\right] ds + \lambda R N(1) \tag{6.4}$$

over the closed convex set

$$\mathcal{K} \equiv \{N \in W_2^1(0,1) : N(0) = 0, N(s) \le 0 \text{ for } 0 \le s \le 1\}. \tag{6.5}$$

Since the constraint that $N \le 0$ in (6.5) is an a priori mathematical restriction, and not a consequence of fundamental mechanics, we shall have to impose further restrictions to ensure that minimizers of (6.5) correspond to solutions of (6.1).

6.6. Exercise. Noting that (6.3) implies that there is a constant $C > 0$ such that $-CN \leq W^*(N) \leq 0$ for $N \leq 0$, prove that $\Phi[N] \to \infty$ as $\|N, W_2^1\| \to \infty$ and consequently that Φ is minimized at a function $\bar{N} \in \mathcal{K}$.

Thus

(6.7) $$\tfrac{1}{t}\left[\Phi(\bar{N} + tQ) - \Phi(\bar{N})\right] \geq 0 \quad \forall\, Q \in \mathcal{K}, \quad t > 0.$$

The Lebesgue Dominated Convergence Theorem implies that the limit of (6.7) exists as $t \searrow 0$ and is given by

(6.8) $$\int_0^1 \left[\bar{N}'Q' - \lambda W_N^*(\bar{N})Q\right] ds + \lambda R Q(1) \geq 0.$$

Let $\sigma \in (0,1)$, let ε be small and positive, and choose

(6.9) $$Q(s) = \begin{cases} 0 & \text{for } 0 \leq s \leq \sigma, \\ (\sigma - s)/\varepsilon & \text{for } \sigma \leq s \leq \sigma + \varepsilon, \\ -1 & \text{for } \sigma + \varepsilon \leq s \leq 1. \end{cases}$$

Substituting (6.9) into (6.8) and taking the limit as $\varepsilon \to 0$, we obtain

(6.10) $$\bar{N}'(\sigma) \leq \lambda \int_\sigma^1 W_N^*(\bar{N}(\xi))\, d\xi - \lambda R.$$

Since $\bar{N} \in \mathcal{K}$, (6.3) and (6.10) imply that

(6.11) $$\bar{N}'(\sigma) \leq \lambda[1 - \sigma - R]$$

(for all such σ). To proceed further and to eliminate unphysical minimizers of Φ on $\mathcal{K}$, we assume that $R \geq 1$. Then $\bar{N}'(s) < 0$ for $s > 0$. Integrating (6.11) with respect to σ from 0 to s and using the condition that $\bar{N}(0) = 0$, we find that $\bar{N}(s) < 0$ for $s > 0$. Now let α be an arbitrary number in $(0,1)$ and let Q be given by (6.9). If $\sigma \geq \alpha$ and if t is sufficiently small, then $\bar{N} + t(-Q) \in \mathcal{K}$. We can therefore replace Q in (6.7) with $-Q$ and accordingly replace (6.8) and (6.10) with the corresponding equations, the latter holding for $\sigma \in [\alpha, 1)$. Since α is arbitrary, the equation corresponding to (6.10) holds for all σ. Taking its limit as $\sigma \to 1$, we recover (6.1c), and differentiating it with respect to σ, we recover (6.1a). Thus

6.12. Theorem. *Let λ, R be given with $R \geq 1$. Then (6.1) has a solution $\bar{N}$, which minimizes Φ on $\mathcal{K}$ and which satisfies*

(6.13) $$-\lambda R s < \bar{N}(s) < -Rs + s - \tfrac{1}{2}s^2 \quad \textit{and} \quad \bar{N}'(s) < 0 \quad \textit{for} \quad s > 0.$$

We can relax the restrictions caused by the membership of N in $\mathcal{K}$ if we take the material to be sufficiently strong in resisting extension. The following exercise supports the basic calculations in the proof of the next theorem. We set

(6.14) $$\mathcal{W} \equiv \{N \in W_2^1(0,1) : N(0) = 0\}.$$

6.15. Exercise. (i) Use the Cauchy-Bunyakovskiĭ-Schwarz inequality to prove that if $N \in \mathcal{W}$, then

(6.16) $$N(s)^2 = 2\int_0^s N(t)N'(t)\, dt \leq 2\left\{\int_0^1 N^2\, dt\right\}^{1/2} \left\{\int_0^1 [N']^2\, dt\right\}^{1/2},$$

so that

(6.17) $$\|N, L_2(0,1)\|^2 \leq 4\|N', L_2(0,1)\|^2.$$

This last inequality can be considerably sharpened by using variational arguments: (ii) Prove that the functional $N \mapsto \|N', L_2(0,1)\|^2$ has a minimum on the subset of $\mathcal{W}$ consisting of those N's for which $\|N, L_2(0,1)\|^2 = 1$ and show that this minimum is $\pi^2/4$. Thus (6.17) can be replaced with

(6.18) $$\|N, L_2(0,1)\|^2 \leq \frac{4}{\pi^2}\|N', L_2(0,1)\|^2.$$

6.19. Theorem. *Let λ and R be fixed. If there are numbers $\alpha \in (0, \pi^2/8)$ and $K > 0$ such that*

$$\lambda W^*(N) \leq \alpha N^2 + K \quad \text{for } N > 0, \tag{6.20}$$

then (6.1) *has a solution, which minimizes Φ on $\mathcal{W}$.*

Proof. Inequalities (6.18) and (6.20) imply that

$$\begin{aligned} \Phi(N) - \lambda R N(1) &\geq \int_0^1 \tfrac{1}{2} N'(s)^2 \, ds - \lambda \int_{\{s : N(s) > 0\}} W^*(N(s)) \, ds \\ &\geq -K + \left[\frac{1}{2} - \frac{4\alpha}{\pi^2}\right] \int_0^1 N'(s)^2 \, ds. \end{aligned} \tag{6.21}$$

Thus $\Phi(N) \to \infty$ as $\|N, W_2^1\| \to \infty$. The proof that Φ is minimized at a function $\bar{N} \in \mathcal{W}$ and that this minimizer is a classical solution of (6.1) follows from a straightforward application of the theory of Secs. 3 and 4. □

7. The Second Variation. Bifurcation Problems

Let us consider a bifurcation problem for a circular arch or ring. It admits a trivial circular equilibrium state. The mere availability of an existence theorem provides us with no new information, because we already know a solution. In this case, we can, however, use the results of Sec. 5 to show that there is a nontrivial solution for certain ranges of pressure.

To be specific, we know the existence of a (classical) minimizer and of a trivial solution. To show the existence of a nontrivial solution, we merely need to show that the trivial solution is not a minimizer. We do this by showing that it does not have a nonnegative-definite second variation, and therefore fails to satisfy a necessary condition for minimization.

Under the conditions prevailing in Secs. 4 and 6, we know that the minimizers of the variational functionals over the admissible sets, which are subsets of Sobolev spaces, are classical solutions of the Euler-Lagrange equations. These minimizers are a fortiori minimizers over subsets of spaces of smooth functions. Consequently, the study of necessary conditions can be carried out in a completely classical setting. We now outline the theory of the second variation in the simplest case, which corresponds to minimizing (1.10a) over continuously differentiable $(\boldsymbol{r}, \theta)$ that satisfy (1.1) (and (5.1)).

We use the notation introduced in the paragraph following (2.9). Let φ_0 have an unconstrained local minimum at a point $\bar{\mathbf{u}} = \mathbf{u}_0 + \bar{\mathbf{v}}$ in $\mathbf{u}_0 + \mathcal{A}$. Let $\mathbf{v}$ be an arbitrary element of the Banach space $\mathcal{B}$ containing $\mathcal{A}$. If $\varepsilon \mapsto \varphi_0[\bar{\mathbf{u}} + \varepsilon \mathbf{v}]$ is twice continuously differentiable for $|\varepsilon|$ small, then

$$\frac{\partial \varphi_0[\bar{\mathbf{u}}]}{\partial \mathbf{u}} \cdot \mathbf{v} \equiv \left. \frac{\partial \varphi_0[\bar{\mathbf{u}} + \varepsilon \mathbf{v}]}{\partial \varepsilon} \right|_{\varepsilon = 0} = 0, \tag{7.1}$$

as before, and

$$\frac{\partial^2 \varphi_0[\bar{\mathbf{u}}]}{\partial \mathbf{u} \partial \mathbf{u}} : \mathbf{v}\mathbf{v} \equiv \left. \frac{\partial^2 \varphi_0[\bar{\mathbf{u}} + \varepsilon \mathbf{v}]}{\partial \varepsilon^2} \right|_{\varepsilon = 0} \geq 0. \tag{7.2}$$

(The left-hand side of (7.2) is the *second variation* of φ_0 at $\bar{\mathbf{u}}$.) If $\bar{\mathbf{u}}$ satisfies (7.1) for all $\mathbf{v}$ in a dense subset of $\mathcal{B}$, then it is a (weak) solution of the Euler-Lagrange equations, but if there is a single $\mathbf{v}$ in $\mathcal{B}$ for which it does *not* satisfy (7.2), then $\bar{\mathbf{u}}$ cannot be a minimizer of φ_0. We apply this principle to a trivial solution of a bifurcation problem to show that it cannot be a minimizer, so that the minimizer, known to exist from other considerations, must be another function. Therefore, there must be multiple solutions.

Let us apply this method to the buckling of a circular ring, treated in Sec IV.3 (cf. Sec. VI.12). We take the reference configuration of the ring to be defined by

$$\mathring{\boldsymbol{r}}(s) = \sin s\boldsymbol{i} - \cos s\boldsymbol{j}, \quad \mathring{\theta}(s) = s. \tag{7.3}$$

We assume that W is independent of s. In place of (1.1), we assume that $\boldsymbol{r}$ and $s \mapsto \theta(s) - s$ when extended to the whole real line have period 2π. We could fix the rigid displacement of the ring by introducing the isoperimetric constraints

$$\int_{-\pi}^{\pi} \boldsymbol{r}(s)\,ds = \mathbf{0}, \quad \int_{-\pi}^{\pi} \theta(s)\,ds = 0, \tag{7.4}$$

but refrain from doing so because they complicate the analysis slightly. The methods of Secs. 3–5 ensure that (1.10a) (with $L = 2\pi$) has a classical minimizer satisfying (5.4) and (5.5), and these periodicity conditions. The smoothness of this minimizer enables us to choose $\mathcal{B}$ to consist of $(\hat{\boldsymbol{r}}, \hat{\theta}) \in C^1[0, 2\pi]$ having period 2π.

Let the first variation of Π of (1.10a) vanish at $(\bar{\boldsymbol{r}}, \bar{\theta})$. Then (5.14), (5.15) imply that the second variation here is the integral over $(-\pi, \pi)$ of

$$\begin{aligned} &\bar{W}_{\nu\nu}x^2 + \bar{W}_{\eta\eta}y^2 + \bar{W}_{\mu\mu}(\hat{\theta}')^2 + 2\bar{W}_{\nu\eta}xy + 2\bar{W}_{\nu\mu}x\hat{\theta}' + 2\bar{W}_{\eta\mu}y\hat{\theta}' \\ &+ p\boldsymbol{k}\cdot(\hat{\boldsymbol{r}} \times \hat{\boldsymbol{r}}') + 2\hat{\theta}[p\bar{\boldsymbol{r}}\cdot\hat{\boldsymbol{r}}' + \boldsymbol{k}\cdot(\boldsymbol{c} \times \hat{\boldsymbol{r}}')] + (\hat{\theta})^2[p\boldsymbol{k}\cdot(\bar{\boldsymbol{r}} \times \bar{\boldsymbol{r}}') - \boldsymbol{c} \times \bar{\boldsymbol{r}}'] \end{aligned} \tag{7.5}$$

where $x = \hat{\boldsymbol{r}}' \cdot \bar{\boldsymbol{a}} + \hat{\theta}\bar{\boldsymbol{r}}' \cdot \bar{\boldsymbol{b}}$, $y = \hat{\boldsymbol{r}}' \cdot \bar{\boldsymbol{b}} - \hat{\theta}\bar{\boldsymbol{r}}' \cdot \bar{\boldsymbol{a}}$, where $\bar{\boldsymbol{a}} \equiv \boldsymbol{a}(\bar{\theta})$, etc., and where $\bar{W}_{\nu\nu} \equiv W_{\nu\nu}(\bar{\boldsymbol{r}}\cdot\bar{\boldsymbol{a}}, \bar{\boldsymbol{r}}\cdot\bar{\boldsymbol{b}}, \bar{\theta}')$, etc. We assume that W is convex, so that the quadratic form of second derivatives of W in (7.5) is positive.

We study trivial solutions of the form

$$\bar{\boldsymbol{r}}(s) = -\bar{\nu}\boldsymbol{b}(\bar{\theta}(s)), \quad \bar{\theta}(s) = s \tag{7.6}$$

where $\bar{\nu}$ is constant. Thus $\bar{\eta} = 0$. Our standard assumption that $W_\eta(\nu, 0, \mu) = 0$ enables us to deduce from (5.15) that $\boldsymbol{c} = \mathbf{0}$ and that $\bar{\nu}$ is the (unique) solution of

$$W_\nu(\bar{\nu}, 0, 1) = -p\bar{\nu}. \tag{7.7}$$

We suppress the dependence of $\bar{\nu}$ on p. For this trivial solution, (7.5) reduces to

$$(7.8)\quad \bar{W}_{\nu\nu}(\overset{\vartriangle}{\boldsymbol{r}}{}' \cdot \bar{\boldsymbol{a}})^2 + \bar{W}_{\eta\eta}(\overset{\vartriangle}{\boldsymbol{r}}{}' \cdot \bar{\boldsymbol{b}} - \overset{\vartriangle}{\theta}\bar{\nu})^2 + \bar{W}_{\mu\mu}(\overset{\vartriangle}{\theta}{}')^2 + 2\bar{W}_{\nu\mu}\overset{\vartriangle}{\boldsymbol{r}}{}' \cdot \bar{\boldsymbol{a}}\overset{\vartriangle}{\theta}{}' \\ + p\left[\boldsymbol{k} \cdot (\overset{\vartriangle}{\boldsymbol{r}} \times \overset{\vartriangle}{\boldsymbol{r}}{}') - 2\overset{\vartriangle}{\theta}\bar{\nu}\bar{\boldsymbol{b}} \cdot \overset{\vartriangle}{\boldsymbol{r}}{}' + (\overset{\vartriangle}{\theta})^2\bar{\nu}^2\right].$$

We take the test functions $(\overset{\vartriangle}{\boldsymbol{r}}, \overset{\vartriangle}{\theta})$ to be generated by the eigenfunctions of the equilibrium problem linearized about the trivial state. This linearization includes

$$(7.9)\qquad \overset{\vartriangle}{\eta}{}'' + q(p)^2\overset{\vartriangle}{\eta} = 0, \quad \overset{\vartriangle}{\eta} \text{ has period } 2\pi$$

where

$$(7.10)\qquad q(p)^2 \equiv \left(1 + \frac{p}{\bar{W}_{\eta\eta}}\right)\left(1 + p\frac{\bar{\nu}^2\bar{W}_{\nu\nu} + 2\bar{\nu}\bar{W}_{\nu\mu} + \bar{W}_{\mu\mu}}{\bar{W}_{\nu\nu}\bar{W}_{\mu\mu} - \bar{W}_{\nu\mu}^2}\right).$$

Note that the convexity of W ensures that $q(p)^2 > 1$ when $p > 0$. The eigenvalues p are solutions of

$$(7.11)\qquad q(p)^2 = n^2$$

where n is an integer ≥ 2 (cf. (VI.12.18), (VI.12.19)). For a given n, (7.11) may have none, one, several, or even infinitely many solutions. To each n for which (7.11) has a solution corresponds an eigenfunction of the form

$$(7.12)\qquad \begin{aligned} \overset{\vartriangle}{\eta}(s) &= B\sin(ns - \omega), \\ \overset{\vartriangle}{\nu}(s) &= -B\frac{A}{n}\cos(ns - \omega) \\ &\equiv -\frac{(p + \bar{W}_{\eta\eta})(\bar{W}_{\mu\mu} + \bar{\nu}\bar{W}_{\nu\mu})}{\bar{W}_{\nu\nu}\bar{W}_{\mu\mu} - \bar{W}_{\nu\mu}^2}\frac{B}{n}\cos(ns - \omega), \\ \overset{\vartriangle}{\theta}(s) &= B\frac{C}{n^2}\sin(ns - \omega) + E \\ &\equiv \frac{(p + \bar{W}_{\eta\eta})(\bar{\nu}\bar{W}_{\nu\nu} + \bar{W}_{\nu\mu})}{\bar{W}_{\nu\nu}\bar{W}_{\mu\mu} - \bar{W}_{\nu\mu}^2}\frac{B}{n^2}\sin(ns - \omega) + E, \\ \overset{\vartriangle}{\boldsymbol{r}}{}'(s) &= \overset{\vartriangle}{\nu}\bar{\boldsymbol{a}} + (\bar{\nu}\overset{\vartriangle}{\theta} + \overset{\vartriangle}{\eta})\bar{\boldsymbol{b}} \end{aligned}$$

where B, E, ω are arbitrary constants. The substitution of (7.12) into (7.8) and the (lengthy) integration of the resulting expression over $(-\pi, \pi)$ yields the second variation, which is πB^2 times

$$(7.13)\quad \bar{W}_{\nu\nu}\frac{A^2}{n^2} + \bar{W}_{\eta\eta} + \bar{W}_{\mu\mu}\frac{A^2}{n^2} - 2\bar{W}_{\nu\mu}\frac{AC}{n^2} \\ -p\left[\frac{1}{n^2 - 1}\left(\frac{A}{n}\right)^2 + \frac{2n}{n^2 + 1}\left(\frac{A}{n}\right)\left(1 + \frac{\bar{\nu}C}{n^2}\right) + \frac{n^2}{n^2 - 1}\left(1 + \frac{\bar{\nu}C}{n^2}\right)^2 - 1\right].$$

Note that the first line of (7.13) is positive, by the convexity of W, and that the second line is negative when p is positive. Thus we conclude that *if there is an* $n \geq 2$ (*typically* $n = 2$) *and a* p *for which* (7.13) *is negative, then the trivial solution* (7.6) *does not minimize* Π, *so that the minimum is achieved elsewhere.* For simpler problems, such as that of Sec. V.5, the analog of (7.13) becomes correspondingly simple. See Ex. 7.15.

Since an equilibrium state that minimizes the energy can often be regarded as stable, and since a state at which the energy does not have a local minimum can be regarded as unstable, results of the kind we have just obtained may be interpreted as giving a range of values for which the trivial state is unstable. See the comments in Sec. V.7.

7.14. Exercise. Derive (7.5)–(7.13).

7.15. Exercise. Formulate the buckling problems of Secs. V.5 and VI.5 for hyperelastic rods as a variational problem, and use the methods of this section to determine a range of the loading parameter λ for which there are nontrivial solutions (which minimize the potential-energy functional).

7.16. Exercise. Linearize the equations of motion for a circular elastic ring under hydrostatic pressure about a trivial equilibrium state. The resulting linear equations of motion are autonomous in t (and s). The solutions are sums of functions that are exponentials in t. The equilibrium state is linearly stable if none of these exponentials grows without bound as a function of t. Otherwise, this state is linearly unstable. Discuss the relation of this kind of stability to the positive-definiteness of the second variation. What happens if the inertia terms depend on s? Study these questions for the dynamical analog of the problem of Sec. V.5.

8. Notes

The proof of Theorem 5.9 is modelled on that of Ball (1981a), which uses weaker hypotheses. It similar to that of Antman (1976b), which uses slightly different hypotheses. Alternative proofs of similar theorems are given by Antman (1970b, 1971, 1972), Antman & Brezis (1978), and Seidman & Wolfe (1988). A related approach for nonvariational problems using the method of variational inequalities is given by Antman (1983a). Section 6 is adapted from Antman & Nachman (1980). (with kind permission from Elsevier Science Ltd, The Boulevard, Langford Lane, Kidlington OX5 1GB, UK. For further applications and refinements of the theory of Sec. 7, see Maddocks (1984, 1987, 1994). For a brief discussion of the implications of energy minimization on stability, see Sec. V.7.

Most of the classical books on the calculus of variations, such as those of Bliss (1946), Gel'fand & Fomin (1961), and many of the more modern books, such as that of Hestenes (1966), do not treat the question of existence, but are concerned with so-called necessary and sufficient conditions for minima: The former give conditions on the variational functional that must hold at an minimizer, such as the vanishing of the first variation of the functional at an 'interior' minimizer. The latter give conditions ensuring that a (weak) solution of the Euler-Lagrange equations is a minimizer. The texts of Akhiezer (1955), Cesari (1983), Ewing (1969), Ioffe & Tihomirov (1974), among others treat the existence question for one-dimensional problems whose Euler-Lagrange equations are ordinary differential equations.

There is now an extensive literature on minimax methods which give the existence not only of a minimizer, but also of families of 'saddle points' of the functional. See Berger (1977), Deimling (1985), Rabinowitz (1986), Vaĭnberg (1956), and Zeidler (1985). These methods have seldom been brought to bear on problems of nonlinear elasticity, even those governed by ordinary differential equations.

In recent years there has been a major effort to treat problems with nonconvex energies, which are used to describe phase changes. See the texts of Dacorogna (1989), Ekeland & Temam (1972), and Struwe (1990) for general considerations, and see Fosdick & James (1981), Gurtin & Temam (1981), and James (1981) for specific applications to elasticity. For references to the extensive recent work on the much more complicated variational problems of three-dimensional nonlinear elasticity, which necessarily have nonconvex energies, some of which may model phase changes, see the discussion in Sec. XIII.2.

Static contact problems for elastic bodies lead to variational inequalities, for which there is a rich literature: See Baiocchi & Capelo (1978), Duvaut & Lions (1972), Fichera (1972b), Hlaváček, Haslinger, Nečas, & Lovišek (1983), Kinderlehrer & Stampacchia (1980), Panagiotopoulos (1985). But the methods developed so far seem inadequate to treat the kinds of problems formulated in this book.

CHAPTER VIII

The Special Cosserat Theory of Rods

1. Introduction

In this chapter we generalize the development of Chap. IV by formulating the general dynamical theory of rods that can undergo large deformations in space by suffering flexure, torsion, extension, and shear. We call the resulting geometrically exact theory the *special Cosserat theory of rods.* In Sec. 2 we outline an honest derivation of the governing equations for elastic and viscoelastic rods. Here we scarcely pause for motivation, interpretation, and justification of our results. The purpose of this presentation is twofold: to establish a framework for the ensuing careful treatment in subsequent sections and to demonstrate that there is a short and pleasant path leading from fundamental physical principles to the governing equations. Armed with these results, the reader interested in the treatment of concrete problems is ready to begin the following chapter. A full appreciation of the theory, however, requires a study of the topics covered in the remainder of the present chapter, which also serves as an easy entrée into ideas important for three-dimensional theories of solids.

Throughout this and the ensuing chapters, we adhere to the following conventions (unless there is a statement to the contrary). Lower-case Latin subscripts, other than s and t, range over $1, 2, 3$. Expressions containing twice-repeated lower-case Latin indices are summed from 1 to 3. Thus

$$u_k v_k = u_1 v_1 + u_2 v_2 + u_3 v_3. \tag{1.1}$$

Lower-case Greek subscripts range over 1 and 2. Expressions containing twice-repeated lower-case Greek indices are summed from 1 to 2. Thus

$$u_\alpha v_\alpha = u_1 v_1 + u_2 v_2. \tag{1.2}$$

The *Kronecker delta* δ_{ij} and the *alternating symbol* ε_{ijk} are defined by

(1.3)

$$\delta_{ij} \equiv \begin{cases} 1 & \text{if } i = j, \\ 0 & \text{if } i \neq j, \end{cases} \quad \varepsilon_{ijk} \equiv \begin{cases} 1 \text{ if } (i,j,k) = (1,2,3),\ (2,3,1),\ (3,1,2), \\ -1 \text{ if } (i,j,k) = (3,2,1),\ (1,3,2),\ (2,1,3), \\ 0 \text{ otherwise.} \end{cases}$$

If $\{\boldsymbol{e}_k\}$ is a right-handed orthonormal basis for $\mathbb{E}^3$, then

$$\delta_{ij} = \boldsymbol{e}_i \cdot \boldsymbol{e}_j, \quad \varepsilon_{ijk} = (\boldsymbol{e}_i \times \boldsymbol{e}_j) \cdot \boldsymbol{e}_k. \tag{1.4}$$

We set

$$\varepsilon_{\alpha\beta} \equiv \varepsilon_{\alpha\beta 3}. \tag{1.5}$$

2. Outline of the Essential Theory

Suppose that the reference configuration of a slender solid body is a solid right circular cylinder of length $s_2 - s_1$. The *material points* or *particles* of this body are identified with their positions in this configuration. The arc-length parameter s of the axis of the cylinder identifies a *material (cross) section* of the body, which consists of all material points whose reference positions are on the plane perpendicular to the axis at s.

The forms of the governing equations we obtain are unaffected by assuming that the reference configuration is a cylinder. This assumption actually represents no loss of generality for any body that can be continuously deformed into such a cylindrical segment, because we do not assume that this configuration is natural. But for a body whose natural configuration is not cylindrical, it may be very inconvenient to exploit this observation. We treat the general case in the next section.

We now construct the simple model of a Cosserat rod to describe the motion of our slender three-dimensional solid body. The *motion of a special Cosserat rod* is defined by three vector-valued functions

$$[s_1, s_2] \times \mathbb{R} \ni (s,t) \mapsto \boldsymbol{r}(s,t),\ \boldsymbol{d}_1(s,t),\ \boldsymbol{d}_2(s,t) \in \mathbb{E}^3 \tag{2.1}$$

with $\boldsymbol{d}_1(s,t)$ and $\boldsymbol{d}_2(s,t)$ orthonormal. $\boldsymbol{r}(\cdot,t)$ may be interpreted as the configuration of the axis at time t. $\boldsymbol{d}_1(\cdot,t)$ and $\boldsymbol{d}_2(\cdot,t)$ may be interpreted as characterizing the configuration of the material section (at) s at time t. In particular, $\boldsymbol{d}_1(s,t)$ and $\boldsymbol{d}_2(s,t)$ may be regarded as characterizing the configurations of a pair of orthogonal material lines of the section s. Our task is to furnish equations for the determination of (2.1). These equations should reflect the geometry inherent in (2.1), basic mechanical principles, and reasonable constitutive assumptions.

Throughout this and the next two chapters, we take the word *rod* to be synonymous with *special Cosserat rod*, unless there is a specific statement to the contrary. We have defined the motion of a rod, but have not defined a rod itself. (See the comments following (3.5).)

Geometry of deformation. We set

$$\boldsymbol{d}_3 \equiv \boldsymbol{d}_1 \times \boldsymbol{d}_2. \tag{2.2}$$

Thus $\{\boldsymbol{d}_k(s,t)\}$ for each (s,t) is a right-handed orthonormal basis for $\mathbb{E}^3$. These vectors are called *directors*. Thus there are vector-valued functions $\boldsymbol{u}$ and $\boldsymbol{w}$ such that

$$\partial_s \boldsymbol{d}_k = \boldsymbol{u} \times \boldsymbol{d}_k, \quad \partial_t \boldsymbol{d}_k = \boldsymbol{w} \times \boldsymbol{d}_k. \tag{2.3a,b}$$

Since the basis $\{\boldsymbol{d}_k\}$ is natural for the intrinsic description of deformation, we decompose relevant vector-valued functions with respect to it:

$$\boldsymbol{u} = u_k \boldsymbol{d}_k, \quad \boldsymbol{r}_s = v_k \boldsymbol{d}_k, \quad \boldsymbol{w} = w_k \boldsymbol{d}_k. \tag{2.4a,b,c}$$

2.5. Exercise. Prove (2.3). Obtain representations for $\{u_i, w_i\}$ in terms of $\{\boldsymbol{d}_k, \partial_s \boldsymbol{d}_k, \partial_t \boldsymbol{d}_k\}$.

The triples

$$\mathsf{u} \equiv (u_1, u_2, u_3), \quad \mathsf{v} \equiv (v_1, v_2, v_3) \tag{2.6}$$

are the *strain* variables corresponding to the motion (2.1). For fixed t, the functions $\mathsf{u}(\cdot, t)$ and $\mathsf{v}(\cdot, t)$ determine $\boldsymbol{r}(\cdot, t)$, $\boldsymbol{d}_1(\cdot, t)$, $\boldsymbol{d}_2(\cdot, t)$ (the *configuration at time* t) to within a rigid motion and thus account for change of shape. (We distinguish between a vector $\boldsymbol{u}$ and a triple u; see Sec. I.4.) u_1 and u_2 measure flexure, u_3 measures torsion, v_1 and v_2 measure shear, and v_3 measures dilatation.

A primitive condition ensuring that orientation is preserved is that

$$v_3 \equiv \boldsymbol{r}_s \cdot \boldsymbol{d}_3 > 0. \tag{2.7}$$

This condition implies that (i) $|\boldsymbol{r}_s| > 0$, so that the local ratio of deformed to reference length of the axis cannot be reduced to zero, and (ii) a typical section s cannot undergo a total shear in which the plane determined by $\boldsymbol{d}_1(s,t)$ and $\boldsymbol{d}_2(s,t)$ is tangent to the curve $\boldsymbol{r}(\cdot, t)$ at $\boldsymbol{r}(s,t)$.

Mechanics. Let $s_1 < a < b < s_2$. In the configuration at time t, the material of $(b, s_2]$ exerts a *resultant contact force* $\boldsymbol{n}^+(b,t)$ and a *resultant contact torque* $\boldsymbol{r}(b,t) \times \boldsymbol{n}^+(b,t) + \boldsymbol{m}^+(b,t)$ *about* $\mathbf{0}$ on the material of $[a, b]$. $\boldsymbol{m}^+(b,t)$ is the corresponding *resultant contact couple*. Just as in Sec. II.2, $\boldsymbol{n}^+(\cdot, t)$ and $\boldsymbol{m}^+(\cdot, t)$ depend only on the section separating the two bodies and are independent of any other property of the bodies. The resultant contact force and contact torque about $\mathbf{0}$ exerted on $[a, b]$ by $[s_1, a)$ are denoted by $-\boldsymbol{n}^-(a,t)$ and $-\boldsymbol{r}(a,t) \times \boldsymbol{n}^-(a,t) - \boldsymbol{m}^-(a,t)$. The resultant of all other forces acting on $[a, b]$ in this configuration is assumed to have the form

$$\int_a^b \boldsymbol{f}(s,t)\, ds \tag{2.8}$$

and the resultant of all other torques is assumed to have the form

$$\int_a^b [\boldsymbol{r}(s,t) \times \boldsymbol{f}(s,t) + \boldsymbol{l}(s,t)]\, ds. \tag{2.9}$$

$\boldsymbol{f}(s,t)$ and $\boldsymbol{l}(s,t)$ are respectively the *body force* and *body couple per unit reference length at* (s,t).

We denote the linear and angular momenta of $[a, b]$ at time t by

$$\int_a^b [(\rho A)(s) \boldsymbol{r}_t(s,t) + \boldsymbol{q}_t(s,t)]\, ds, \tag{2.10}$$

$$\begin{aligned} \int_a^b [(\rho A)(s) \boldsymbol{r}(s,t) \times \boldsymbol{r}_t(s,t) &+ \boldsymbol{r}(s,t) \times \boldsymbol{q}_t(s,t) \\ &+ \boldsymbol{q}(s,t) \times \boldsymbol{r}_t(s,t) + \boldsymbol{h}(s,t)]\, ds. \end{aligned} \tag{2.11}$$

ρA may be interpreted as the mass density per reference length. The functions $\boldsymbol{q}$ and $\boldsymbol{h}$ depend on $\boldsymbol{r}$, $\boldsymbol{d}_1$, $\boldsymbol{d}_2$ and their time derivatives. Physically reasonable choices for these functions seem to demand inspiration from the three-dimensional theory (or else from some abstract mathematical characterization of dynamical equations of mechanical systems; see DeSilva & Whitman (1971)). In Secs. 3 and 4 we treat this question from a three-dimensional viewpoint and show that the simplest reasonable forms for $\boldsymbol{q}$ and $\boldsymbol{h}$ are

$$\text{(2.12a,b)} \qquad \boldsymbol{q} = \boldsymbol{0}, \quad \boldsymbol{h}(s,t) = (\rho J_{pq})(s) w_q(s,t) \boldsymbol{d}_p(s,t).$$

Here the $(\rho J_{\gamma\delta})(s)$ are the components of a positive-definite symmetric 2×2 matrix, which may be thought of as a matrix of mass-moments of inertia of the section s, $\rho J_{\gamma 3} = \rho J_{3\gamma} = 0$, $\rho J_{33} = \rho J_{\gamma\gamma}$. Of course, the issue of adopting a reasonable form for $\boldsymbol{q}$ and $\boldsymbol{h}$ does not arise for statical problems.

By requiring that the resultant force acting on the material of $[a,b]$ equal the time derivative of the linear momentum, we obtain the integral form of the equations of motion, which differs from (II.2.7) only by the presence of $\boldsymbol{q}$. More generally, by requiring that the impulse of forces acting on $[a,s]$ over the time interval $[0,\tau]$ equal the change in linear momentum of this body over this time interval, we obtain the *Linear Impulse-Momentum Law*:

$$\text{(2.13)} \quad \int_0^\tau \left[\boldsymbol{n}^+(s,t) - \boldsymbol{n}^-(a,t)\right] dt + \int_0^\tau \int_a^s \boldsymbol{f}(\xi,t)\, d\xi\, dt$$
$$= \int_a^s \{(\rho A)(\xi)[\boldsymbol{r}_t(\xi,\tau) - \mathbf{r}_t(\xi,0)] + \boldsymbol{q}_t(\xi,\tau) - \boldsymbol{q}_t(\xi,0)\}\, d\xi$$

for all a and s satisfying $s_1 < a < s < s_2$. Similarly, by requiring that the impulse of torques about $\boldsymbol{0}$ equal the change in angular momentum about $\boldsymbol{0}$, we obtain the *Angular Impulse-Momentum Law*:

(2.14)

$$\int_0^\tau \left[\boldsymbol{m}^+(s,t) - \boldsymbol{m}^-(a,t) + \boldsymbol{r}(s,t) \times \boldsymbol{n}^+(s,t) - \boldsymbol{r}(a,t) \times \boldsymbol{n}^-(a,t)\right] dt$$
$$+ \int_0^\tau \int_a^s [\boldsymbol{r}(\xi,t) \times \boldsymbol{f}(\xi,t) + \boldsymbol{l}(\xi,t)]\, d\xi\, dt$$
$$= \int_a^s \{\boldsymbol{r}(\xi,t) \times [(\rho A)(\xi)\boldsymbol{r}_t(\xi,t) + \boldsymbol{q}_t(\xi,t)]$$
$$+ \boldsymbol{q}(\xi,t) \times \boldsymbol{r}_t(\xi,t) + \boldsymbol{h}(\xi,t)\}\, d\xi \Big|_{t=0}^{t=\tau}.$$

The extension of these laws to intervals for which $a = s_1$ or $s = s_2$ is straightforward: Prescribed forces and couples acting at the ends of the rod must be accounted for. (See Sec. 13. No such forces are treated in Sec. IV.1.)

By imitating the procedures of Secs. II.2 and II.3, we can obtain the obvious analogs of the results found there. In particular, we find that the superscripts '±' are superfluous. Assuming that all the functions appearing in (2.13) and (2.14) are sufficiently regular, we can differentiate these equations with respect to s and τ to obtain

$$\boldsymbol{n}_s + \boldsymbol{f} = \rho A \boldsymbol{r}_{tt} + \boldsymbol{q}_{tt}, \tag{2.15}$$

$$\boldsymbol{m}_s + (\boldsymbol{r} \times \boldsymbol{n})_s + \boldsymbol{r} \times \boldsymbol{f} + \boldsymbol{l} = \rho A \boldsymbol{r} \times \boldsymbol{r}_{tt} + \boldsymbol{r} \times \boldsymbol{q}_{tt} + \boldsymbol{q} \times \boldsymbol{r}_{tt} + \boldsymbol{h}_t. \tag{2.16}$$

The substitution of (2.15) into (2.16) reduces the latter to

$$\boldsymbol{m}_s + \boldsymbol{r}_s \times \boldsymbol{n} + \boldsymbol{l} = \boldsymbol{q} \times \boldsymbol{r}_{tt} + \boldsymbol{h}_t. \tag{2.17}$$

Equations (2.15) and (2.17) are the classical forms of the *equations of motion for (the special theory of Cosserat) rods.*

By integrating (2.17) over $[a, s] \times [0, \tau]$, we obtain a reduced version of (2.14):

$$\int_0^\tau \left\{ \boldsymbol{m}(s,t) - \boldsymbol{m}(a,t) + \int_a^s [\boldsymbol{r}_s(\xi,t) \times \boldsymbol{n}(\xi,t) + \boldsymbol{l}(\xi,t)]\, d\xi \right\} dt = \int_a^s [\mathbf{c}(\xi,\tau) - \mathbf{c}(\xi,0)]\, d\xi \tag{2.18a}$$

where

$$\mathbf{c}(s,t) \equiv \boldsymbol{q}(s,t) \times \boldsymbol{r}_t(s,t) + \int_0^t \boldsymbol{r}_t(s,\eta) \times \boldsymbol{q}_t(s,\eta)\, d\eta + \boldsymbol{h}(s,t). \tag{2.18b}$$

By imposing mild conditions on the functions entering (2.13) and (2.14), we could deduce (2.18) directly from (2.13) and (2.14) without resorting to the classical equations. Thus the Impulse-Momentum Laws (2.13) and (2.18) are equivalent to (2.13) and (2.14).

2.19. Exercise. Carry out this direct derivation of (2.18) under the assumption that the functions entering the integrals are sufficiently smooth. (An approximation argument can be used to extend the validity of this derivation to the most general class of functions for which these integrals make sense as Lebesgue integrals. Alternatively, the derivation can be effected directly by the careful use of standard tools of real analysis, such as Fubini's Theorem.)

Let

$$\mathsf{m} \equiv (m_1, m_2, m_3), \quad \mathsf{n} \equiv (n_1, n_2, n_3) \quad \text{where} \quad m_k \equiv \boldsymbol{m} \cdot \boldsymbol{d}_k, \quad n_k \equiv \boldsymbol{n} \cdot \boldsymbol{d}_k. \tag{2.20}$$

We may call m_1 and m_2 the *bending couples* (or *bending moments*), m_3 the *twisting couple* (or *twisting moment*), n_1 and n_2 the *shear forces*, and $\boldsymbol{n} \cdot \boldsymbol{r}_s / |\boldsymbol{r}_s|$ the *tension*. These terms are not strictly analogous to those used in structural mechanics, in which it is usually assumed that $\boldsymbol{d}_3 = \boldsymbol{r}_s / |\boldsymbol{r}_s|$.

Constitutive equations. The rod is called *elastic* if there are constitutive functions $\hat{\mathsf{m}}$ and $\hat{\mathsf{n}}$ such that

$$\mathsf{m}(s,t) = \hat{\mathsf{m}}(\mathsf{u}(s,t), \mathsf{v}(s,t), s), \quad \mathsf{n}(s,t) = \hat{\mathsf{n}}(\mathsf{u}(s,t), \mathsf{v}(s,t), s). \tag{2.21}$$

For each fixed s, the common domain of these constitutive functions is a subset of u's and v's describing deformations that preserve orientation. At the least, the domain consists of u's and v's satisfying (2.7). The rod is called *viscoelastic of differential type of complexity 1* if there are functions $\hat{\mathsf{m}}$ and $\hat{\mathsf{n}}$ such that

$$\begin{aligned} \mathsf{m}(s,t) &= \hat{\mathsf{m}}\big(\mathsf{u}(s,t), \mathsf{v}(s,t), \mathsf{u}_t(s,t), \mathsf{v}_t(s,t), s\big), \\ \mathsf{n}(s,t) &= \hat{\mathsf{n}}(\mathsf{u}(s,t), \mathsf{v}(s,t), \mathsf{u}_t(s,t), \mathsf{v}_t(s,t), s). \end{aligned} \tag{2.22}$$

More generally, we could allow $\hat{\mathsf{m}}$ and $\hat{\mathsf{n}}$ to be determined by the past history of u and v.

If representations for $\boldsymbol{q}$ and $\boldsymbol{h}$ in terms of $\boldsymbol{r}$, $\boldsymbol{d}_1$, $\boldsymbol{d}_2$ are given, then the substitution of (2.21) into (2.15) and (2.17) yields a twelfth-order quasilinear system of partial differential equations for $\boldsymbol{r}$, $\boldsymbol{d}_1$, $\boldsymbol{d}_2$. The study of concrete problems for this system is the object of the next chapter.

Our brief development of these equations raises a number of important questions:

(i) What are specific interpretations for $\boldsymbol{r}$, $\boldsymbol{d}_1$, $\boldsymbol{d}_2$ for rods whose reference configurations need not be cylindrical?

(ii) How do u and v characterize the change of shape?

(iii) What are suitable generalizations of (2.7)?

(iv) What are suitable choices for $\boldsymbol{q}$ and $\boldsymbol{h}$?

(v) How can $\{\boldsymbol{d}_k\}$ be effectively represented, so as to exploit its orthonormality?

(vi) What are appropriate boundary conditions? How can these be incorporated naturally into impulse-momentum laws and into a Principle of Virtual Power?

(vii) Why must the constitutive functions have just the forms shown in (2.21) and (2.22)? Are these forms the most general in their classes? Do they ensure that material behavior is unaffected by rigid motions? What are reasonable physical restrictions to impose on these functions? How do such restrictions affect the classifications of the governing systems of partial differential equations?

We address these and other pertinent questions in the rest of this chapter.

We can now explain the condition (II.2.10) that states that $\boldsymbol{n}$ is tangent to $\boldsymbol{r}$ in a string. We define a string to be a rod that (i) has no bending or torsional stiffness, so that $\hat{\mathsf{m}} = \mathbf{0}$, (ii) has no relative linear momentum $\boldsymbol{q}_t$, and (iii) has no relative angular momentum $\boldsymbol{h}$. If $\boldsymbol{l} = \mathbf{0}$, then (2.17) reduces to (II.2.10).

Readers wishing to progress as quickly as possible to the analysis of concrete problems can do so by reading Sec. 5 through (5.8), reading enough of Sec. 4 to appreciate the basic ideas of Sec. 6, getting a thorough understanding of Sec. 8, and appreciating the ideas of Sec. 15. When necessary, they can return to Secs. 9, 11, and 16 for requisite background.

3. The Exact Equations of Motion

We now develop the exact equations of motion for rods from a three-dimensional viewpoint. Since we merely seek representations for the acceleration terms, we do not require anything resembling the full apparatus of three-dimensional continuum mechanics. Our approach has the virtue of giving a preview both of some of the geometric foundations of the three-dimensional theory and of methods for constructing more sophisticated rod theories.

A *body* is a set, whose elements are called *material points*, that has mass, can sustain forces, and can occupy regions of Euclidean 3-space $\mathbb{E}^3$. (We do not pause to assign these terms mathematically precise interpretations by means of measure theory; see Truesdell (1991a).) That a body can occupy regions of $\mathbb{E}^3$ means that its material points can be put into a one-to-one correspondence with the points of such regions. We distinguish one such correspondence and call it the *reference configuration.* We denote the region occupied by the body in its reference configuration by $\mathcal{B}$. Note that the reference configuration is not merely $\mathcal{B}$: It is the identification of each material point with its position $\boldsymbol{z}$ in $\mathcal{B}$. We nevertheless refer to the body itself as $\mathcal{B}$. We assume that $\mathcal{B}$ is an open set.

Suppose that there is a continuously differentiable invertible mapping

$$\operatorname{cl}\mathcal{B} \ni \boldsymbol{z} \mapsto \tilde{\mathbf{x}}(\boldsymbol{z}) \in \mathbb{R}^3$$

such that the Jacobian

$$\det \frac{\partial \tilde{\mathbf{x}}}{\partial \boldsymbol{z}}(\boldsymbol{z}) > 0 \quad \forall\, \boldsymbol{z} \in \operatorname{cl}\mathcal{B}. \tag{3.1}$$

The function $\tilde{\mathbf{x}}$ assigns a triple of curvilinear coordinates $\mathbf{x} \equiv (x^1, x^2, x^3 \equiv s)$ to each $\boldsymbol{z}$ in $\operatorname{cl}\mathcal{B}$. See Fig. 3.2. We denote the inverse of $\tilde{\mathbf{x}}$ by $\tilde{\boldsymbol{z}}$. By virtue of (3.1), the Inverse Function Theorem implies that $\tilde{\boldsymbol{z}}$ is continuously differentiable and that

$$j(\mathbf{x}) \equiv \det \frac{\partial \tilde{\boldsymbol{z}}}{\partial \mathbf{x}}(\mathbf{x}) > 0 \quad \forall\, \mathbf{x} \in \tilde{\mathbf{x}}(\operatorname{cl}\mathcal{B}). \tag{3.3}$$

Our first minimal restriction on the shape of $\mathcal{B}$ to make it rod-like is that the coordinate s range over an interval (s_1, s_2) as $\mathbf{x}$ ranges over $\tilde{\mathbf{x}}(\mathcal{B})$, i.e., $(s_1, s_2) = \{s \colon \mathbf{x} \in \tilde{\mathbf{x}}(\mathcal{B})\}$. The *material section* $\mathcal{B}(a)$ is the set of all material points of $\mathcal{B}$ whose coordinates have the form (x^1, x^2, a), i.e.,

$$\mathcal{B}(a) \equiv \{\boldsymbol{z} \in \mathcal{B} \colon \tilde{\mathbf{x}}(\boldsymbol{z}) = (x^1, x^2, a)\}. \tag{3.4}$$

Our second minimal restriction on $\mathcal{B}$ is that $\mathcal{B}(s)$ be bounded for each s in $[s_1, s_2]$. For any subinterval $\mathcal{I}$ of $[s_1, s_2]$, we define

$$\mathcal{B}(\mathcal{I}) \equiv \bigcup_{s \in \mathcal{I}} \mathcal{B}(s) \equiv \{\boldsymbol{z} \in \mathcal{B} \colon \tilde{\mathbf{x}}(\boldsymbol{z}) = (x^1, x^2, a),\ a \in \mathcal{I}\}. \tag{3.5}$$

Thus $\mathcal{B} = \mathcal{B}((s_1, s_2))$.

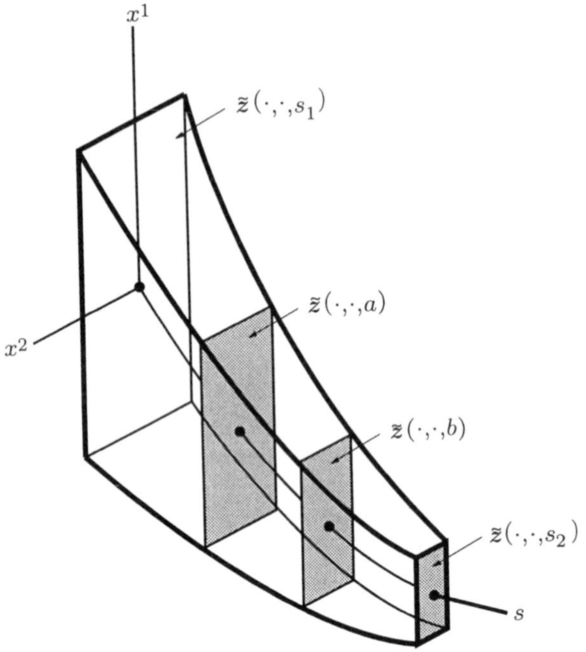

Fig. 3.2. Curvilinear coordinates for the reference configuration of $\mathcal{B}$.

We have defined a theory of rods and a motion of a Cosserat rod, but we have not defined a *rod* itself. We could declare any body $\mathcal{B}$ meeting these two minimal restrictions to be a rod, but such a body need not be slender. Since there is no simple, obvious, universally acceptable mathematical characterization of slenderness, we avoid defining *rod*, but continue to apply the term informally to the bodies we study in this chapter.

Let $\boldsymbol{p}(\boldsymbol{z}, t)$ be the position of material point $\boldsymbol{z}$ at time t. We assume that the domain of $\boldsymbol{p}(\cdot, t)$ is $\operatorname{cl}\mathcal{B}$. Then $\boldsymbol{p}(\cdot, t)$ is the configuration of the closure $\operatorname{cl}\mathcal{B}$ of the body at time t and $\boldsymbol{p}(\operatorname{cl}\mathcal{B}, t)$ is the region occupied by $\operatorname{cl}\mathcal{B}$ at time t. We define $\tilde{\boldsymbol{p}}(\mathbf{x}, t) \equiv \boldsymbol{p}(\tilde{\boldsymbol{z}}(\mathbf{x}), t)$. Then $\tilde{\boldsymbol{p}}(\cdot, \cdot, s, t)$ is the configuration of the section $\mathcal{B}(s)$ at time t. See Fig. 3.6.

We are now ready to formulate the equations of motion. For $s_1 < a < b < s_2$, let $\boldsymbol{n}^+(b, t)$ denote the resultant contact force exerted on $\mathcal{B}([s_1, b])$ by $\mathcal{B}((b, s_2])$. We define other forces and torques by analogous modifications of the definitions given in Sec. 2. Let $\rho(\boldsymbol{z})$ be the mass density at $\boldsymbol{z}$ in $\mathcal{B}$ in the reference configuration. (We assume that it is well defined.) We set $\tilde{\rho}(\mathbf{x}) \equiv \rho(\tilde{\boldsymbol{z}}(\mathbf{x}))$ and $\mathcal{A}(s) \equiv \tilde{\mathbf{x}}(\mathcal{B}(s))$. In mechanics, the linear momentum of a body is defined to be the integral over the body of its velocity times its differential mass. Accordingly, the *linear momentum* of $\mathcal{B}([a, b])$ at time t is

$$\int_a^b \left\{ \int_{\mathcal{A}(s)} \tilde{\rho}(\mathbf{x}) \tilde{\boldsymbol{p}}_t(\mathbf{x}, t) \jmath(\mathbf{x})\, dx^1\, dx^2 \right\} ds. \tag{3.7}$$

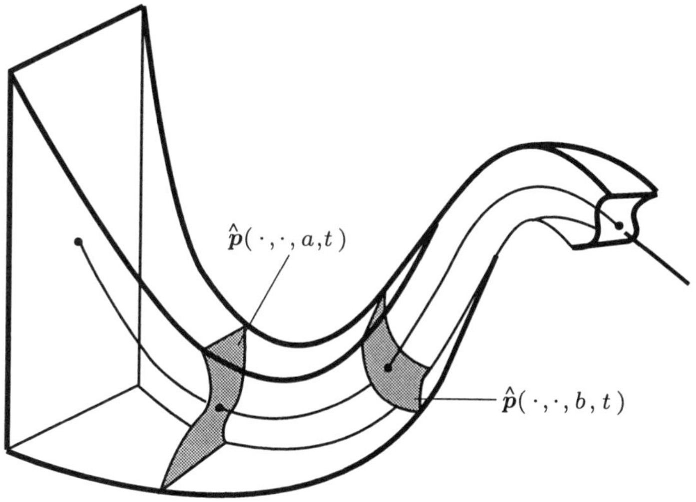

Fig. 3.6. Configuration $\tilde{\boldsymbol{p}}(\cdot,\cdot,s,t)$ of the section $\mathcal{B}(s)$ at time t.

The *Linear Impulse-Momentum Law* for $\mathcal{B}([a,b])$ is

$$\text{(3.8)}\quad \int_0^\tau \left[\boldsymbol{n}^+(b,t) - \boldsymbol{n}^-(a,t)\right] dt + \int_0^\tau \int_a^b \mathbf{f}(s,t)\, d\xi\, dt = \int_a^b \left\{ \int_{\mathcal{A}(s)} \tilde{\rho}(\mathbf{x})[\tilde{\boldsymbol{p}}_t(\mathbf{x},\tau) - \tilde{\boldsymbol{p}}_t(\mathbf{x},0)]\, \jmath(\mathbf{x})\, dx^1\, dx^2 \right\} ds.$$

Under mild conditions like those discussed in Secs. II.2 and II.3, we can delete the superscripts $\pm$ from $\boldsymbol{n}$.

To convert (3.8) to a form useful for rod theories we introduce a special position vector $\boldsymbol{r}(s,t)$ determined from $\tilde{\boldsymbol{p}}(\cdot,\cdot,s,t)$ in any convenient manner. For example, we could set

$$\text{(3.9)}\qquad (\rho A)(s)\boldsymbol{r}(s,t) \equiv \int_{\mathcal{A}(s)} \tilde{\rho}(\mathbf{x})\tilde{\boldsymbol{p}}(\mathbf{x},t)\jmath(\mathbf{x})\, dx^1\, dx^2$$

where

$$\text{(3.10)}\qquad (\rho A)(s) \equiv \int_{\mathcal{A}(s)} \tilde{\rho}(\mathbf{x})\jmath(\mathbf{x})\, dx^1\, dx^2$$

is the mass density in the reference configuration per unit of s. In this case, (3.7) reduces to

$$\text{(3.11)}\qquad \int_a^b (\rho A)(s)\boldsymbol{r}_t(s,t)\, ds.$$

Alternatively, if $\tilde{\boldsymbol{z}}(0,0,s) \in \operatorname{cl}\mathcal{B}$ for each s, then we could take

$$\boldsymbol{r}(s,t) \equiv \tilde{\boldsymbol{p}}(0,0,s,t). \tag{3.12}$$

No matter what choice we make for $\boldsymbol{r}$, the linear momentum (3.7) always has the form

$$\int_a^b [(\rho A)(s)\boldsymbol{r}_t(s,t) + \boldsymbol{q}_t(s,t)]\,ds \tag{3.13}$$

where

$$\boldsymbol{q}(s,t) \equiv \int_{\mathcal{A}(s)} \tilde{\rho}(\mathbf{x})[\tilde{\boldsymbol{p}}(\mathbf{x},t) - \boldsymbol{r}(s,t)]\,j(\mathbf{x})\,dx^1\,dx^2. \tag{3.14}$$

$\boldsymbol{q}_t$ is the *linear momentum relative to* $\boldsymbol{r}(s,t)$. Thus (3.8) reduces to (2.13).

The *angular momentum* of $\mathcal{B}([a,b])$ about $\mathbf{0}$ at time t is

$$\int_a^b \left[\int_{\mathcal{A}(s)} \tilde{\rho}(\mathbf{x})\tilde{\boldsymbol{p}}(\mathbf{x},t) \times \tilde{\boldsymbol{p}}_t(\mathbf{x},t) j(\mathbf{x})\,dx^1\,dx^2\right] ds, \tag{3.15}$$

which we readily decompose into the form

$$\begin{aligned}\int_a^b [(\rho A)(s)\boldsymbol{r}(s,t) \times \boldsymbol{r}_t(s,t) &+ \boldsymbol{r}(s,t) \times \boldsymbol{q}_t(s,t) \\ &+ \boldsymbol{q}(s,t) \times \boldsymbol{r}_t(s,t) + \boldsymbol{h}(s,t)]\,ds\end{aligned} \tag{3.16}$$

where $\boldsymbol{q}$ is given by (3.14) and where

$$\boldsymbol{h}(s,t) \equiv \int_{\mathcal{A}(s)} \tilde{\rho}(\mathbf{x})[\tilde{\boldsymbol{p}}(\mathbf{x},t)-\boldsymbol{r}(s,t)]\times[\tilde{\boldsymbol{p}}_t(\mathbf{x},t)-\boldsymbol{r}_t(s,t)]\,j(\mathbf{x})\,dx^1\,dx^2 \tag{3.17}$$

is the *angular momentum relative to* $\boldsymbol{r}(s,t)$. The *Angular Impulse-Momentum Law* for $\mathcal{B}([a,b])$ is

(3.18)

$$\begin{aligned}&\int_0^{\mathcal{T}} \left[\boldsymbol{m}^+(b,t) - \boldsymbol{m}^-(a,t) + \boldsymbol{r}(b,t) \times \boldsymbol{n}^+(b,t) - \boldsymbol{r}(a,t) \times \boldsymbol{n}^-(a,t)\right] dt \\ &+ \int_0^{\mathcal{T}}\int_a^b [\boldsymbol{r}(s,t) \times \mathbf{f}(s,t) + \boldsymbol{l}(s,t)]\,ds\,dt \\ &\quad = \int_a^b [(\rho A)(s)\boldsymbol{r}(s,t) \times \boldsymbol{r}_t(s,t) + \boldsymbol{r}(s,t) \times \boldsymbol{q}_t(s,t) \\ &\qquad\qquad + \boldsymbol{q}(s,t) \times \boldsymbol{r}_t(s,t) + \boldsymbol{h}(s,t)]\,ds\Big|_0^{\mathcal{T}}.\end{aligned}$$

Under mild conditions we can delete the superscripts $\pm$ from $\boldsymbol{m}$. Definitions (3.14) and (3.17) reduce (3.18) to (2.14). If $\boldsymbol{q} = \mathbf{0}$, which happens if (3.9) holds, then (3.18) becomes much simpler.

We emphasize that (3.8) and (3.18) are exact laws of motion. No approximations or constraints have been introduced. Since the function $\tilde{\boldsymbol{p}}$ depends on $(\mathbf{x}, t)$ and not merely on (s, t), these formulations are not appropriate for a theory in which the motion is described solely by functions of (s, t). We use (3.8) and (3.18) to construct such a theory in the next section.

4. The Equations of Constrained Motion

We generate a rod theory from the results of the last section by replacing the exact representations (3.14) and (3.17) for $\boldsymbol{q}$ and $\boldsymbol{h}$ with approximate expressions depending only on $\boldsymbol{r}$, $\boldsymbol{d}_1$, and $\boldsymbol{d}_2$. We interpret the approximation as coming from a constraint that restricts the position field $\tilde{\boldsymbol{p}}$ to have the form

$$\tilde{\boldsymbol{p}}(\mathbf{x}, t) = \boldsymbol{r}(s, t) + \boldsymbol{g}\big(\boldsymbol{r}(s, t), \boldsymbol{d}_1(s, t), \boldsymbol{d}_2(s, t), \mathbf{x}, t\big) \tag{4.1}$$

no matter what force system is applied to the body. Here $\boldsymbol{g}$ is a prescribed function. (In the terminology of rigid-body mechanics, (4.1) is a holonomic, rheonomic constraint.) The form of (4.1) assigns a specific interpretation to $\boldsymbol{r}$, $\boldsymbol{d}_1$, and $\boldsymbol{d}_2$.

There are a variety of specializations of (4.1) that are reasonable:

(4.2a) $\boldsymbol{g}$ is independent of t (i.e., the constraint is scleronomic).
(4.2b) $\boldsymbol{g}$ is independent of $\boldsymbol{r}$.
(4.2c) $\boldsymbol{g}$ is independent of s.
(4.2d) $\boldsymbol{g}$ is linear in $\boldsymbol{r}$, $\boldsymbol{d}_1$, and $\boldsymbol{d}_2$.
(4.2e) $\boldsymbol{g}$ is a linear combination of $\boldsymbol{r}$, $\boldsymbol{d}_1$, and $\boldsymbol{d}_2$.
(4.2f) In the reference configuration, $\boldsymbol{g}$ reduces to a linear combination of $\boldsymbol{r}$, $\boldsymbol{d}_1$, and $\boldsymbol{d}_2$.
(4.2g) $\boldsymbol{g}(\boldsymbol{r}, \boldsymbol{d}_1, \boldsymbol{d}_2, \mathbf{x}, t) = x^1\boldsymbol{d}_1 + x^2\boldsymbol{d}_2$.

This last form, so special, is also the most traditional. It enjoys all the properties (4.2a–f).

Suppose for the moment that x^1 and x^2 are rectangular Cartesian coordinates for each section $\mathcal{B}(s)$. Let (4.2g) hold. Then (4.1) describes a deformation that takes the plane reference section $\mathcal{B}(s)$ spanned by the reference values $\mathring{\boldsymbol{d}}_1(s)$, $\mathring{\boldsymbol{d}}_2(s)$ of $\boldsymbol{d}_1(s, \cdot)$, $\boldsymbol{d}_2(s, \cdot)$ into an undeformed plane spanned by the actual values $\boldsymbol{d}_1(s, t)$, $\boldsymbol{d}_2(s, t)$ at time t. The vectors $\mathring{\boldsymbol{d}}_1(s)$, $\mathring{\boldsymbol{d}}_2(s)$ could be taken as the principal axes of inertia of the section $\mathcal{B}(s)$.

We now obtain simple explicit formulas for $\boldsymbol{q}$ and $\boldsymbol{h}$ for $\boldsymbol{g}$'s satisfying properties (4.2a,b,e). In this case, (4.1) has the form

$$\tilde{\boldsymbol{p}}(\mathbf{x}, t) = \boldsymbol{r}(s, t) + \varphi_\gamma(\mathbf{x})\boldsymbol{d}_\gamma(s, t). \tag{4.3a}$$

We suppose that the φ_γ are continuously differentiable and that $\det(\partial\varphi_\gamma/\partial x^\mu) > 0$ on $\mathcal{A}(s)$. Thus $\varphi_1(\cdot, \cdot, s)$ and $\varphi_2(\cdot, \cdot, s)$ are independent for each s. When (4.3a) holds, we assume that the reference configuration is correspondingly defined by

$$\tilde{\boldsymbol{z}}(\mathbf{x}) = \mathring{\boldsymbol{r}}(s) + \varphi_\alpha(\mathbf{x})\mathring{\boldsymbol{d}}_\alpha(s) \tag{4.3b}$$

where $\{\mathring{\boldsymbol{d}}_1, \mathring{\boldsymbol{d}}_2, \mathring{\boldsymbol{d}}_3 \equiv \partial_s\mathring{\boldsymbol{r}}\}$ is a right-handed orthonormal basis.

If (4.3a) holds, we obtain from (3.14) and (3.17) that

$$\boldsymbol{q}(s,t) = (\rho I_\gamma)(s)\boldsymbol{d}_\gamma(s,t), \tag{4.4}$$

$$\begin{aligned} \boldsymbol{h}(s,t) &= \varepsilon_{\gamma\mu}\varepsilon_{\delta\nu}(\rho J_{\mu\nu})(s)\boldsymbol{d}_\gamma(s,t) \times \partial_t\boldsymbol{d}_\delta(s,t), \\ &\equiv (\rho J_{\mu\nu})(s)w_\nu(s,t)\boldsymbol{d}_\mu(s,t) + (\rho J_{\gamma\gamma})(s)w_3(s,t)\boldsymbol{d}_3(s,t) \end{aligned} \tag{4.5}$$

where

$$(\rho I_\gamma)(s) \equiv \int_{\mathcal{A}(s)} \tilde{\rho}(\mathbf{x})\varphi_\gamma(\mathbf{x})j(\mathbf{x})\,dx^1\,dx^2, \tag{4.6}$$

$$(\rho J_{\gamma\delta})(s) \equiv \varepsilon_{\gamma\mu}\varepsilon_{\delta\nu} \int_{\mathcal{A}(s)} \tilde{\rho}(\mathbf{x})\varphi_\mu(\mathbf{x})\varphi_\nu(\mathbf{x})j(\mathbf{x})\,dx^1\,dx^2. \tag{4.7}$$

In evaluating the rightmost term of (4.5), we have used (2.3b). The $(\rho I_\gamma)(s)$ are generalized first moments of mass and the $(\rho J_{\gamma\delta})(s)$ are generalized second moments of mass. (The alternating symbols are introduced so that the indices on ρJ correspond to the axes about which these moments are defined. The consequent complexity here leads to simplifications elsewhere.) We can make $\boldsymbol{q} = \mathbf{0}$ by choosing the φ_γ to make (4.6) vanish. See Ex. 4.8. Since the φ_γ are associated with the reference configuration, we could always choose the reference values of $\boldsymbol{d}_\alpha$ so as to diagonalize the matrix with components $\rho J_{\gamma\delta}$.

4.8. Exercise. (i) Find j from (4.3b). Note that it depends on the $\lambda_\alpha \equiv \mathring{\boldsymbol{d}}_3 \cdot \partial_s\mathring{\boldsymbol{d}}_\alpha$. (ii) Suppose that the $\lambda_\alpha(s) = 0$. Give a geometric interpretation of this restriction. Show that if (4.2g) is used and that if $\tilde{\boldsymbol{z}}(0,0,\cdot) \equiv \mathring{\boldsymbol{r}}$ is the curve of centroids of mass of the sections of $\mathcal{B}$, then (4.6) vanishes, so that $\boldsymbol{q} = \mathbf{0}$. (iii) Suppose that the $\lambda_\alpha(s)$ are not both 0. Show that if (4.2g) is used and that if $\tilde{\boldsymbol{z}}(0,0,\cdot) \equiv \mathring{\boldsymbol{r}}$ is defined so that $\int_{\mathcal{A}(s)} x^\alpha\tilde{\rho}(\mathbf{x})\,dx^1\,dx^2 = 0$, then in general (4.6) does not vanish, even if $\tilde{\rho}$ is independent of x^1 and x^2. (iv) Prove that if $\mathcal{A}(s)$ is symmetric about the x^1 and x^2 axes and if $\tilde{\rho}$ is even in x^1 and x^2, then there are functions a^α such that the choice of $\varphi_\alpha(\mathbf{x}) = x^\alpha - a^\alpha(s)$ causes (4.6) to vanish provided that

$$[\lambda_1(s)]^2 \int_{\mathcal{A}(s)} (x^1)^2\tilde{\rho}(\mathbf{x})\,dx^1\,dx^2 + [\lambda_2(s)]^2 \int_{\mathcal{A}(s)} (x^2)^2\tilde{\rho}(\mathbf{x})\,dx^1\,dx^2 < \tfrac{1}{4}(\rho A)(s). \tag{4.9}$$

(v) Let $\mathcal{A}(s)$ and $\tilde{\rho}$ be arbitrary. Use the Implicit Function Theorem to prove that if the $\mathring{\boldsymbol{d}}_3(s) \cdot \partial_s\mathring{\boldsymbol{d}}_\alpha(s)$ are sufficiently close to 0, then there are functions a and b as in (iv) such that (4.6) vanishes.

The key to appreciating the significance of Ex. 4.8 is that the Jacobian j appears in the integrand of (3.10), which defines the mass per unit length. To see what j does, consider a typical annular sector defined in polar coordinates (r, θ) by $a-h \leq r \leq a+h$, $0 \leq \theta \leq \bar{\theta}$. Here $0 < h < a$. The circle $r = a$ goes through the centroid of each cross section. But the area of the region defined by $a \leq r \leq a+h$, $0 \leq \theta \leq \bar{\theta}$ exceeds that of the region defined by $a-h \leq r \leq a$, $0 \leq \theta \leq \bar{\theta}$. The circle that divides the sector into equal areas has radius $\sqrt{a^2+h^2}$. There is an interesting open problem of characterizing those bodies $\mathcal{B}$ that admit a coordinate system defined by (4.3b) with (4.6) vanishing.

4.10. Exercise. Let $\tilde{\rho}$ be independent of x^1 and x^2. Show that the integrand in (4.6) can be written as a two-dimensional divergence over $\mathcal{A}(s)$. Use the Divergence Theorem to get representations for (4.6) as integrals over $\partial\mathcal{A}(s)$.

4.11. Exercise. Use the Cauchy-Bunyakovskiĭ-Schwarz inequality to prove that the matrix with components $\rho J_{\mu\nu}$ is positive-definite.

We now show that the expression (4.3) is scarcely more general than that given by (4.2g) when $\overset{\circ}{\boldsymbol{r}}$ is straight. In this case, $j = \det(\partial\varphi_\gamma/\partial x^\mu) > 0$. For any fixed point $(\bar{x}^1, \bar{x}^2)$ in the domain of $(\varphi_1(\cdot,\cdot,s), \varphi_2(\cdot,\cdot,s))$, let the mapping defined by

$$y_\alpha = \varphi_\alpha(\mathbf{x}) - \varphi_\alpha(\bar{x}^1, \bar{x}^2, s) \tag{4.12a}$$

be an invertible mapping taking $\operatorname{cl}\mathcal{A}(s)$ into a closed and bounded set $\operatorname{cl}\mathcal{A}^\sharp(s)$ in (y_1, y_2)-space. Let us set $\mathbf{y} \equiv (y_1, y_2, s)$. Thus (4.12a) is equivalent to a system of the form

$$x^\alpha = \psi^\alpha(\mathbf{y}). \tag{4.12b}$$

If we substitute (4.12) into (4.6), we obtain

$$(\rho I_\gamma)(s) = \int_{\mathcal{A}^\sharp(s)} \tilde{\rho}(\psi^1(\mathbf{y}), \psi^1(\mathbf{y}), s) y_\gamma \, dy_1 \, dy_2 \tag{4.13}$$

We get a similar simplification for (4.7). This procedure could break down if the Jacobian of (4.12a) vanishes, say on $\partial\mathcal{A}(s)$. Choices of the φ_α to be other than x^α are sometimes useful for treating special kinds of boundary behavior (see Antman (1972, Sec. 5)).

We shall often take $\boldsymbol{q} = \mathbf{0}$ and take $\boldsymbol{h}$ to have the form (4.5), in which case (2.15) and (2.17) reduce to

$$\boldsymbol{n}_s + \boldsymbol{f} = \rho A \boldsymbol{r}_{tt}, \tag{4.14}$$

$$\boldsymbol{m}_s + \boldsymbol{r}_s \times \boldsymbol{n} + \boldsymbol{l} = \boldsymbol{h}_t = \partial_t[\varepsilon_{\gamma\mu}\varepsilon_{\delta\nu}(\rho J_{\mu\nu})(s)\boldsymbol{d}_\gamma(s,t) \times \partial_t \boldsymbol{d}_\delta(s,t)]. \tag{4.15}$$

The right-hand side of (4.15) can be expressed in terms of $\boldsymbol{w}$ and $\boldsymbol{w}_t$ via (4.5).

For certain problems in which it is important to account for detailed effects of a force system applied to the lateral boundary of a two- or three-dimensional body, it may be convenient to interpret $\boldsymbol{r}$ as a material curve on this boundary. In particular, consider a two-dimensional problem in which the rod may be interpreted as a cross section of a cylindrical shell. If the outer surface of the shell is exposed to the effects of moving fluids or is in contact with another body, then it is useful to take $\boldsymbol{r}$ to correspond to this surface. In this case, $\boldsymbol{q} \neq \mathbf{0}$. (Cf. Lanza de Cristoforis & Antman (1991).)

5. The Strains and the Strain Rates

In this section we discuss the geometrical significance of u and v, we show that u, v, u_t, v_t are invariant under rigid motions, and we show that u and v determine a configuration to within a rigid motion. These strains are defined by (2.3)–(2.6). By taking the dot product of (2.3a) with $\boldsymbol{d}_j \times \boldsymbol{d}_k$, we obtain

$$u_j = \tfrac{1}{2}\varepsilon_{jkl}(\partial_s \boldsymbol{d}_k) \cdot \boldsymbol{d}_l. \tag{5.1}$$

To interpret the significance of u and v, we study some especially simple deformations. Let $\{\boldsymbol{i}, \boldsymbol{j}, \boldsymbol{k}\}$ be a fixed, right-handed, orthonormal basis for $\mathbb{E}^3$. We take $[s_1, s_2] = [0,1]$ and assume that the reference configuration, which corresponds to a straight prismatic three-dimensional body, is defined by

$$\mathring{\boldsymbol{r}}(s) = s\boldsymbol{k}, \quad \mathring{\boldsymbol{d}}_1(s) = \boldsymbol{i}, \quad \mathring{\boldsymbol{d}}_2(s) = \boldsymbol{j}, \quad \mathring{\boldsymbol{d}}_3(s) = \boldsymbol{k}. \tag{5.2}$$

We first study the pure elongation in which the deformed configuration is given by

$$\boldsymbol{r}(s) = \lambda s\boldsymbol{k}, \quad \boldsymbol{d}_1(s) = \boldsymbol{i}, \quad \boldsymbol{d}_2(s) = \boldsymbol{j}, \quad \boldsymbol{d}_3(s) = \boldsymbol{k}. \tag{5.3}$$

Then $v_3 = \lambda$, the ratio of deformed to reference length. All other strains are zero. This interpretation of v_3 as a *stretch* is overly simplistic. Equations (2.2) and (2.4) imply that in general

$$v_3 = \boldsymbol{r}_s \cdot (\boldsymbol{d}_1 \times \boldsymbol{d}_2). \tag{5.4}$$

Thus v_3 is the volume of the parallelepiped with sides $\boldsymbol{r}_s$, $\boldsymbol{d}_1$, $\boldsymbol{d}_2$, and therefore represents the ratio of deformed to reference volume. That this ratio never be reduced to zero is embodied in the inequality (2.7). In general, the stretch of the axis is $|\boldsymbol{r}_s| = \sqrt{v_k v_k}$.

The pure shear deformation defined by

$$\boldsymbol{r}(s) = s[\alpha\boldsymbol{i} + \boldsymbol{k}], \quad \boldsymbol{d}_1(s) = \boldsymbol{i}, \quad \boldsymbol{d}_2(s) = \boldsymbol{j}, \quad \boldsymbol{d}_3(s) = \boldsymbol{k} \tag{5.5}$$

yields $v_1 = \alpha$, $v_3 = 1$, with all other strains equal to zero. This v_1 describes the shear of $\boldsymbol{d}_1$ with respect to $\boldsymbol{r}_s$. This interpretation is likewise simplistic because stretching can contribute to v_1. In general, a measure of pure shear is

$$\frac{v_1}{\sqrt{v_k v_k}} = \frac{\boldsymbol{d}_1 \cdot \boldsymbol{r}_s}{|\boldsymbol{r}_s|}. \tag{5.6}$$

The torsional deformation defined by

$$\begin{gathered} \boldsymbol{r}(s) = s\boldsymbol{k}, \\ \boldsymbol{d}_1(s) = \cos(\mu s)\boldsymbol{i} + \sin(\mu s)\boldsymbol{j}, \quad \boldsymbol{d}_2(s) = \boldsymbol{k} \times \boldsymbol{d}_1(s), \quad \boldsymbol{d}_3(s) = \boldsymbol{k} \end{gathered} \tag{5.7}$$

yields $v_3 = 1$, $u_3 = \mu$, with all other strains equal to zero. Thus u_3 measures the amount of twist.

The deformation of the configuration (5.2) into the circular configuration given by

$$\begin{gathered} \boldsymbol{r}(s) = \lambda[\sin(\omega s)\boldsymbol{k} - \cos(\omega s)\boldsymbol{i}], \\ \boldsymbol{d}_3(s) = \cos(\omega s)\boldsymbol{k} + \sin(\omega s)\boldsymbol{i}, \quad \boldsymbol{d}_1(s) = \boldsymbol{j} \times \boldsymbol{d}_3(s), \quad \boldsymbol{d}_2(s) = \boldsymbol{j} \end{gathered} \tag{5.8}$$

yields $v_3 = \lambda\omega$, $u_2 = \omega$, with all other strains equal to zero. Thus u_2 measures the amount of flexure. Note that u_2 is not the curvature $1/\lambda$ of the circle $\boldsymbol{r}$: The strain u_2 is the derivative with respect to the reference arc length s of the tangent angle ωs to the curve $\boldsymbol{r}$ at $\boldsymbol{r}(s)$, whereas the curvature is the derivative of this angle with respect to the deformed arc length. Thus u_2 is not influenced by those changes in the curvature due to the inflation of the circular configuration, caused by varying λ.

5.9. Exercise. Find the most general deformation of configuration (5.2) for which u and v are constant.

We now show that our strains and strain rates are invariant under rigid motions. Thus these quantities can only give information about the change of shape. We assume that the $\boldsymbol{d}_\alpha$ are each differences of two position vectors. It follows that if $\{\boldsymbol{r}^\sharp, \boldsymbol{d}_\alpha^\sharp\}$ differ from $\{\boldsymbol{r}, \boldsymbol{d}_\alpha\}$ by a rigid motion, then there are functions $t \mapsto \boldsymbol{c}(t)$ and $t \mapsto \boldsymbol{Q}(t)$ where $\boldsymbol{Q}(t)$ is a proper orthogonal tensor such that

$$\boldsymbol{r}^\sharp = \boldsymbol{c} + \boldsymbol{Q}\cdot\boldsymbol{r}, \quad \boldsymbol{d}_\alpha^\sharp = \boldsymbol{Q}\cdot\boldsymbol{d}_\alpha. \tag{5.10}$$

$\boldsymbol{c}(t)$ represents a translation and $\boldsymbol{Q}(t)$ a rotation. It follows from (2.4), (5.10), and the orthogonality of $\boldsymbol{Q}$ that

$$v_k^\sharp = \boldsymbol{r}_s^\sharp \cdot \boldsymbol{d}_k^\sharp = (\boldsymbol{Q}\cdot\boldsymbol{r}_s)\cdot(\boldsymbol{Q}\cdot\boldsymbol{d}_k) = \boldsymbol{r}_s\cdot\boldsymbol{d}_k = v_k. \tag{5.11}$$

Thus v is invariant under rigid motions.

5.12. Exercise. Prove that u, u_t, v_t are invariant under rigid motions.

We now show that u and v determine a configuration to within a rigid motion. It follows that these quantities entirely characterize the change of shape. Our development, which generalizes a standard method of differential geometry, is important also because it gives a coordinate-free treatment of the directors, which supports effective numerical methods for their computation (see Simo & Vu-Quoc (1988)).

For our analysis, we regard t as fixed and accordingly suppress its appearance. Let u be a given function of s, which for simplicity we take to be continuous. Then (2.3a) is a ninth-order *linear* system of ordinary differential equations for (the components of) $\{\boldsymbol{d}_k\}$, which we can write in a form lacking the symbol for the cross-product:

$$\partial_s \boldsymbol{d}_k = u_l \varepsilon_{lkm} \boldsymbol{d}_m. \tag{5.13}$$

Let s_0 be fixed in $[s_1, s_2]$. Then the standard theory for linear systems of ordinary differential equations (see Coddington & Levinson (1955, Chap. 3)) says that each initial-value problem for (5.13) has a unique solution defined on the whole interval $[s_1, s_2]$. In particular, there are continuously differentiable tensor-valued functions $s \mapsto \boldsymbol{D}_{kl}(s)$ with $\boldsymbol{D}_{kl}(s_0) = \delta_{kl}\boldsymbol{I}$ such that each solution of (5.13) has the form

$$\boldsymbol{d}_k(s) = \boldsymbol{D}_{kl}(s) \cdot \boldsymbol{d}_l(s_0). \tag{5.14}$$

The tensors $\boldsymbol{D}_{kl}$ generate the fundamental matrix solution of (5.13).

We adopt the normalization that $\{\boldsymbol{d}_k(s_0)\}$ is a right-handed orthonormal basis for $\mathbb{E}^3$. Now (5.13) admits the seven integrals

$$\begin{aligned} \boldsymbol{d}_k \cdot \boldsymbol{d}_l &= \boldsymbol{d}_k(s_0) \cdot \boldsymbol{d}_l(s_0) = \delta_{kl}, \\ (\boldsymbol{d}_1 \times \boldsymbol{d}_2) \cdot \boldsymbol{d}_3 &= [\boldsymbol{d}_1(s_0) \times \boldsymbol{d}_2(s_0)] \cdot \boldsymbol{d}_3(s_0) = 1 \end{aligned} \tag{5.15}$$

because (2.3a) implies that

$$\begin{aligned} \partial_s(\boldsymbol{d}_k \cdot \boldsymbol{d}_l) &= (\boldsymbol{u} \times \boldsymbol{d}_k) \cdot \boldsymbol{d}_l + \boldsymbol{d}_k \cdot (\boldsymbol{u} \times \boldsymbol{d}_l) = 0, \\ \partial_s[\boldsymbol{d}_1 \cdot (\boldsymbol{d}_2 \times \boldsymbol{d}_3)] &= (\boldsymbol{u} \times \boldsymbol{d}_1) \cdot (\boldsymbol{d}_2 \times \boldsymbol{d}_3) + \cdots = 0. \end{aligned} \tag{5.16}$$

The last integral of (5.15) is not completely independent of the first six, which imply that $(\boldsymbol{d}_1 \times \boldsymbol{d}_2) \cdot \boldsymbol{d}_3 = \pm 1$. Since this triple scalar product is continuous, it must equal 1 or -1 everywhere. It therefore inherits its sign from its value at s_0.

Let $\{\boldsymbol{d}_k^\sharp\}$ be any other solution of (5.13). That a fundamental matrix is independent of initial data implies that $\{\boldsymbol{d}_k^\sharp\}$ is expressed in terms of its initial data $\{\boldsymbol{d}_k^\sharp(s_0)\}$ by formula (5.14):

$$\boldsymbol{d}_k^\sharp(s) = \boldsymbol{D}_{kl}(s) \cdot \boldsymbol{d}_l^\sharp(s_0). \tag{5.17}$$

Since $\{\boldsymbol{d}_k(s_0)\}$ and $\{\boldsymbol{d}_k^\sharp(s_0)\}$ are each right-handed and orthonormal, there is a proper orthogonal tensor $\boldsymbol{Q}$ such that

$$\boldsymbol{d}_l^\sharp(s_0) = \boldsymbol{Q} \cdot \boldsymbol{d}_l(s_0). \tag{5.18}$$

Since $\{\boldsymbol{d}_l(s_0)\}$ is an orthonormal basis, we can express $\boldsymbol{d}_k$ by

$$\boldsymbol{d}_k = [\boldsymbol{d}_k \cdot \boldsymbol{d}_l(s_0)]\boldsymbol{d}_l(s_0). \tag{5.19}$$

By comparing this equation with (5.14) and (5.17), we find that

$$\boldsymbol{D}_{kl}(s) = [\boldsymbol{d}_k(s) \cdot \boldsymbol{d}_l(s_0)]\boldsymbol{I} = [\boldsymbol{d}_k^\sharp(s) \cdot \boldsymbol{d}_l^\sharp(s_0)]\boldsymbol{I}. \tag{5.20a,b}$$

By substituting (5.20a) into (5.17) and using (5.18), we find that

$$\begin{aligned} \boldsymbol{d}_k^\sharp(s) &= \boldsymbol{d}_k(s) \cdot \boldsymbol{d}_l(s_0)\boldsymbol{Q} \cdot \boldsymbol{d}_l(s_0) \\ &= \boldsymbol{Q} \cdot \{[\boldsymbol{d}_k(s) \cdot \boldsymbol{d}_l(s_0)]\boldsymbol{d}_l(s_0)\} = \boldsymbol{Q} \cdot \boldsymbol{d}_k(s). \end{aligned} \tag{5.21}$$

This says that two sets of director fields corresponding to the same $\mathbf{u}$ differ by a rigid motion. That (5.13) always has a solution of the form (5.14) ensures that $\mathbf{u}$ determines $\{\boldsymbol{d}_k\}$ uniquely to within a rigid motion.

Since $\{\boldsymbol{d}_k(s)\}$ and $\{\boldsymbol{d}_k(s_0)\}$ are each right-handed and orthonormal, there is a proper orthogonal tensor $\boldsymbol{D}$ such that

$$\boldsymbol{d}_k(s) = \boldsymbol{D} \cdot \boldsymbol{d}_k(s_0). \tag{5.22}$$

$\boldsymbol{D}$ can be found by comparing (5.22) with (5.14) or (5.20a). Similarly, there is a proper orthogonal tensor $\boldsymbol{D}^\sharp$ such that

$$\boldsymbol{d}_k^\sharp(s) = \boldsymbol{D}^\sharp \cdot \boldsymbol{d}_k^\sharp(s_0). \tag{5.23}$$

Despite the presence of the same $\boldsymbol{D}_{kl}$ in (5.14) and (5.17), the tensors $\boldsymbol{D}$ and $\boldsymbol{D}^\sharp$ are not equal in general. Useful representations of $\boldsymbol{D}$ as a solution of (5.13) can be obtained in terms of the formula of Rodrigues (see Gibbs & Wilson (1901), Goldstein (1980), Simo & Vu-Quoc (1988), e.g.).

5.24. Exercise. Prove that $\boldsymbol{D}^\sharp = \boldsymbol{Q} \cdot \boldsymbol{D} \cdot \boldsymbol{Q}^*$, where $\boldsymbol{Q}$ is defined by (5.18).

Let $\mathbf{u}$ and $\mathbf{v}$ be given and let $\{\boldsymbol{d}_k\}$ be determined from $\mathbf{u}$ by (5.14). We find $\boldsymbol{r}$ by integrating (2.4):

$$\boldsymbol{r}(s) = \boldsymbol{r}(s_0) + \int_{s_0}^{s} v_k(\xi)\boldsymbol{d}_k(\xi)\, d\xi. \tag{5.25}$$

If $\{\boldsymbol{r}^\sharp, \boldsymbol{d}_k^\sharp\}$ is another configuration likewise determined from $\mathbf{u}$ and $\mathbf{v}$, then we can use (5.21) to obtain

$$\begin{aligned} \boldsymbol{r}^\sharp(s) &= \boldsymbol{r}^\sharp(s_0) + \int_{s_0}^{s} v_k(\xi)\boldsymbol{d}_k^\sharp(\xi)\, d\xi \\ &= \boldsymbol{r}^\sharp(s_0) + \boldsymbol{Q} \cdot \int_{s_0}^{s} v_k(\xi)\boldsymbol{d}_k(\xi)\, d\xi \\ &= [\boldsymbol{r}^\sharp(s_0) - \boldsymbol{Q} \cdot \boldsymbol{r}(s_0)] + \boldsymbol{Q} \cdot \boldsymbol{r}(s), \end{aligned} \tag{5.26}$$

which has the form (5.10). Thus $\mathbf{u}$ and $\mathbf{v}$ (subject to the normalization that $\{\boldsymbol{d}_k(s_0)\}$ be right-handed and orthonormal) determine $\{\boldsymbol{r}, \boldsymbol{d}_k\}$ uniquely to within a rigid motion.

Let $\mathring{\boldsymbol{r}}$, $\mathring{\boldsymbol{d}}_k$, $\mathring{\mathbf{u}}$, $\mathring{\mathbf{v}}$ be the values of $\boldsymbol{r}$, $\boldsymbol{d}_k$, $\mathbf{u}$, $\mathbf{v}$ in the reference configuration. We always take

$$\mathring{\boldsymbol{r}}_s = \mathring{\boldsymbol{d}}_3 \tag{5.27}$$

so that

$$\mathring{v}_1 = 0 = \mathring{v}_2, \quad \mathring{v}_3 = 1. \tag{5.28}$$

As Ex. 5.29 shows, $\mathring{\mathbf{u}} = \mathbf{0}$ only for rods having a straight untwisted reference configuration. Moreover, as Ex. 5.9 shows, the class of reference configurations for which $\mathring{\mathbf{u}}$ is constant is small.

5.29. Exercise. Let $\mathring{\boldsymbol{r}}$ have curvature $\mathring{\kappa}$ and torsion $\mathring{\tau}$. Let $\mathring{\boldsymbol{d}}_1$ be the unit normal field to $\mathring{\boldsymbol{r}}$, let $\mathring{\boldsymbol{d}}_2$ be the unit binormal field to $\mathring{\boldsymbol{r}}$, and let $\mathring{\boldsymbol{d}}_3$ be the unit tangent field to $\mathring{\boldsymbol{r}}$. Use the Frenet-Serret formulas to show that

$$\mathring{u}_1 = 0, \quad \mathring{u}_2 = \mathring{\kappa}, \quad \mathring{u}_3 = \mathring{\tau}. \tag{5.30}$$

We call v_1 and v_2 the *shear strains*, v_3 the *dilatation*, u_1 and u_2 the *flexural strains*, and u_3 the *torsional strain*. $|\boldsymbol{r}_s| = \sqrt{v_k v_k}$ is the *stretch*.

5.31. Exercise. Prove that

$$\partial_s u_k \equiv \partial_s(\boldsymbol{u} \cdot \boldsymbol{d}_k) = \boldsymbol{u}_s \cdot \boldsymbol{d}_k, \quad \partial_t w_k \equiv \partial_t(\boldsymbol{w} \cdot \boldsymbol{d}_k) = \boldsymbol{w}_t \cdot \boldsymbol{d}_k. \tag{5.32}$$

6. The Preservation of Orientation

A configuration of a three-dimensional body should be one-to-one (or *injective*) so that no two material points simultaneously occupy the same point in space. Injectivity is a global restriction on configurations and accordingly poses serious difficulties for analysis. A far more modest restriction is that the deformation preserve orientation, which ensures that it be locally injective. If the configuration $\boldsymbol{p}(\cdot, t)$ is differentiable, then it preserves orientation if its Jacobian is positive:

$$0 < \det \frac{\partial \boldsymbol{p}}{\partial \boldsymbol{z}}(\tilde{\boldsymbol{z}}(\mathbf{x}), t) = j(\mathbf{x})^{-1} \det \frac{\partial \tilde{\mathbf{p}}}{\partial \mathbf{x}}(\mathbf{x}, t) \quad \text{for } \mathbf{x} \in \tilde{\mathbf{x}}(\operatorname{cl} \mathcal{B}). \tag{6.1}$$

Here we have used the notation of Sec. 3. The analog of (6.1) for rods, which strengthens (2.7), is introduced in

6.2. Theorem. (i) *The requirement that (6.1) hold for all $\tilde{\mathbf{p}}$ of the form (4.3) with the φ_γ continuous on $\tilde{\mathbf{x}}(\operatorname{cl} \mathcal{B})$ implies that there is a function $(u_1, u_2, s) \mapsto V(u_1, u_2, s)$ for which $V(0, 0, s) = 0$ and $V(\cdot, \cdot, s)$ is convex and homogeneous of degree 1 such that*

$$v_3 > V(u_1, u_2, s). \tag{6.3}$$

(ii) *Furthermore, if there is a point $(x_0^1, x_0^2) \in \operatorname{cl} \mathcal{A}(s)$ such that*

$$\varphi_1(x_0^1, x_0^2, s) = 0 = \varphi_2(x_0^1, x_0^2, s) \tag{6.4}$$

or if there is a function $\omega \geq 0$ such that

$$\int_{\mathcal{A}(s)} \omega(\mathbf{x}) \, dx^1 \, dx^2 > 0 \quad \text{but} \quad \int_{\mathcal{A}(s)} \omega(\mathbf{x}) \varphi_\alpha(\mathbf{x}) \, dx^1 \, dx^2 = 0, \tag{6.5}$$

then $V \geq 0$. (iii) *Finally, if the φ_γ are continuously differentiable and satisfy*

$$\det \left(\partial \varphi_\gamma / \partial x^\mu \right) > 0 \tag{6.6}$$

everywhere on a domain containing $\tilde{\mathbf{x}}(\operatorname{cl} \mathcal{B})$, and if for each u_1, u_2, s, any curve in the (x^1, x^2)-plane defined by

$$\varphi_1(x^1, x^2, s) u_2 - \varphi_2(x^1, x^2, s) u_1 = 0 \tag{6.7}$$

that touches $\partial \mathcal{A}(s)$ has a part that lies in $\mathcal{A}(s)$, then $V(u_1, u_2, s) > 0$ for $u_\alpha u_\alpha > 0$.

We say that a deformation of a special Cosserat rod *preserves orientation* iff (6.3) holds for all $s \in [s_1, s_2]$. Most of the works that acknowledge the need for a condition like (6.3) choose the much weaker (2.7) for this purpose. Condition (2.7) arises in a purely one-dimensional context, whereas

(6.3) depends upon concepts from the three-dimensional theory, just as the expressions for the momenta do.

To interpret the significance of (6.3), we consider the two-dimensional deformation $s\boldsymbol{k} + x\boldsymbol{i} \mapsto (1 - x)(\sin s\boldsymbol{k} - \cos s\boldsymbol{i})$ for $0 \leq s \leq \pi/2$, $|x| \leq h$, which takes lines parallel to the $\boldsymbol{k}$-axis into concentric circles. If $h \geq 1$, then the line $x = 1$ is deformed into the point $\mathbf{0}$ and the lines $x = a$ with $a > h$ have their orientations reversed. Condition (6.3), which prevents such total compressions and reversals of orientation, says that this deformation for $h \geq 1$ is physically inadmissible. The hypotheses of statement (iii) relate the geometry of $\mathcal{A}(s)$ to the functions φ_α. In particular, suppose that $\varphi_\alpha(\mathbf{x}) = x^\alpha$. Then (6.7) defines a family of rays in the (x^1, x^2)-plane. If $\partial\mathcal{A}(s)$ is nowhere radial (which depends not only upon the shape of $\mathcal{A}(s)$ but also upon its placement relative to the origin in (x^1, x^2)-space), then these hypotheses are met. See Ex. 6.18.

It might seem that any realistic theory of materials need not be concerned with extreme deformations that violate (6.1) or (6.3). We require (6.1) and (6.3) in order to prescribe constitutive restrictions that penalize and often prevent the violation of these inequalities by solutions of the governing equations. Without such a penalization, one could, for example, inadvertently carry out a numerical study of configurations that are have reversed orientation. Static problems for bodies having corners typically have solutions with severe singularities at the corners. A correct analysis of the behavior of the solution in the vicinity of such singular point, which is often the most important aspect of a problem for such a body, depends crucially on constitutive equations that correctly account for extreme strains. The same remarks apply to bodies undergoing shocks.

Proof of Theorem 6.2. The substitution of (4.3) into (6.1) yields

$$v_3(s) > V^\sharp(u_1(s), u_2(s), \mathbf{x}) \quad \forall\, (x^1, x^2) \in \operatorname{cl} \mathcal{A}(s) \tag{6.8}$$

for all $s \in [s_1, s_2]$ where

$$V^\sharp(u_1, u_2, \mathbf{x}) \equiv \varphi_1(\mathbf{x})u_2 - \varphi_2(\mathbf{x})u_1. \tag{6.9}$$

(Note that if (6.4) or (6.5) holds, then (6.8) implies (2.7).) The set of strains (u_1, u_2, v_3) satisfying (6.8), denoted $\mathcal{V}(s)$, consists of exactly those satisfying

$$v_3(s) > V(u_1(s), u_2(s), s) \tag{6.10}$$

for all $s \in [s_1, s_2]$ where

$$V(u_1, u_2, s) \equiv \max\left\{V^\sharp(u_1, u_2, \mathbf{x}) : (x^1, x^2) \in \operatorname{cl} \mathcal{A}(s)\right\}. \tag{6.11}$$

To determine the properties of V, we let

$$\mathcal{H}(\mathbf{x}) \equiv \{(u_1, u_2, v_3) \colon v_3(s) > V^\sharp(u_1(s), u_2(s), \mathbf{x})\}. \tag{6.12}$$

Since $V^\sharp(\cdot, \cdot, \mathbf{x})$ is linear, $\mathcal{H}(\mathbf{x})$ is an open half-space in (u_1, u_2, v_3)-space. Since the intersection of half-spaces is a convex set, it follows that

$$\mathcal{V}(s) \equiv \cap\{\mathcal{H}(\mathbf{x}) \colon (x^1, x^2) \in \operatorname{cl} \mathcal{A}(s)\} \tag{6.13}$$

is a convex set in (u_1, u_2, v_3)-space consisting exactly of those strains satisfying (6.10). Thus the function $V(\cdot,\cdot,s)$ must be convex and therefore continuous, from which it follows that $\mathcal{V}(s)$ is open. The linearity of $V^\sharp(\cdot,\cdot,\mathbf{x})$ implies that if $(u_1,u_2,v_3) \in \mathcal{V}(s)$, then $(\alpha u_1, \alpha u_2, \alpha v_3) \in \mathcal{V}(s)$ for $\alpha > 0$. This result says that $\mathcal{V}(s)$ is a solid cone in $\mathbb{R}^3$ and implies that $V(\cdot,\cdot,s)$ is homogeneous of degree 1. This completes the proof of statement (i).

Since (6.11) implies that $V(u_1,u_2,s) \geq V^\sharp(u_1,u_2,x_0^1,x_0^2,s)$ and that

$$V(u_1,u_2,s) \geq \int_{\mathcal{A}(s)} V^\sharp(u_1,u_2,\mathbf{x})\,dx^1\,dx^2/|\mathcal{A}(s)|,$$

statement (ii) follows immediately from (6.4) and(6.5).

To prove statement (iii), which says that the cone $\partial\mathcal{V}(s)$ has no generators in the plane $v_3 = 0$, we assume for contradiction that there is a pair $(\tilde{u}_1,\tilde{u}_2) \neq (0,0)$ and a pair $(\tilde{x}^1,\tilde{x}^2) \in \operatorname{cl}\mathcal{A}(s)$ such that

$$V(\tilde{u}_1,\tilde{u}_2,s) \equiv \max\{V^\sharp(\tilde{u}_1,\tilde{u}_2,\mathbf{x}) : (x^1,x^2 \in \operatorname{cl}\mathcal{A}(s)\} \equiv V^\sharp(\tilde{u}_1,\tilde{u}_2,\tilde{x}^1,\tilde{x}^2,s). \tag{6.14}$$

Let us first suppose that the maximizer $(\tilde{x}^1,\tilde{x}^2)$ lies in the open set $\mathcal{A}(s)$. Then the gradient of $(x^1,x^2) \mapsto V^\sharp(\tilde{u}_1,\tilde{u}_2,\mathbf{x})$ must vanish at $(\tilde{x}^1,\tilde{x}^2)$:

$$\begin{pmatrix} \partial\varphi_1(\tilde{x}^1,\tilde{x}^2,s)/\partial x^1 & \partial\varphi_2(\tilde{x}^1,\tilde{x}^2,s)/\partial x^1 \\ \partial\varphi_1(\tilde{x}^1,\tilde{x}^2,s)/\partial x^2 & \partial\varphi_2(\tilde{x}^1,\tilde{x}^2,s)/\partial x^2 \end{pmatrix} \begin{pmatrix} -\tilde{u}_2 \\ \tilde{u}_1 \end{pmatrix} = \begin{pmatrix} 0 \\ 0 \end{pmatrix}. \tag{6.15}$$

In view of (6.6), system (6.15) implies that $\tilde{u}_1 = 0 = \tilde{u}_2$, in contradiction to our assumption.

Now let us suppose that $(\tilde{x}^1,\tilde{x}^2) \in \partial\mathcal{A}(s)$. Then (6.6) and (6.14) enable us to use the Implicit-Function Theorem to deduce that the equation $V^\sharp(\tilde{u}_1,\tilde{u}_2,\mathbf{x}) = 0$ has a curve of solutions (x^1,x^2) in the (x^1,x^2)-plane that contains $(\tilde{x}^1,\tilde{x}^2)$. If this curve penetrates $\mathcal{A}(s)$, then $V^\sharp$ is maximized at an interior point, which we have just shown leads to a contradiction. □

We note that much of the structure of $\mathcal{V}(s)$ is preserved for the general constraint (4.1). See Sec. XIV.2.

We now examine three methods by which $V(s)$ can be explicitly found when $(x^1,x^2) \mapsto \delta(u_1,u_2,v_3,\mathbf{x})$ is known to have its minimum on $\partial\mathcal{A}(s)$.

(i) If $(x^1,x^2) \mapsto \varphi_1(\mathbf{x}), \varphi_1(\mathbf{x})$ are harmonic (i.e., solutions of Laplace's equation) or, more generally, are solutions of certain uniformly elliptic partial differential equations, then these functions assume their extrema on $\partial\mathcal{A}(s)$, whence it follows that $(x^1,x^2) \mapsto V^\sharp(u_1,u_2,\mathbf{x})$ is maximized here for each (u_1,u_2). Under these conditions, $\partial\mathcal{V}(s)$ is the envelope of $\{\mathcal{H}(\mathbf{x}) \colon (x^1,x^2) \in \partial\mathcal{A}(s)\}$. For simplicity, let us assume that $\mathcal{A}(s)$ is simply-connected. Let $\partial\mathcal{A}(s)$ be parametrized by the curve $[\xi_1,\xi_2] \ni \xi \mapsto \big(y^1(\xi,s), y^2(\xi,s)\big)$ whose values at ξ_1 and ξ_2 agree. $\partial\mathcal{A}(s)$ is said to be continuously differentiable iff it admits such a parametrization in which $\big(y^1(\cdot,s), y^2(\cdot,s)\big)$ is continuously differentiable and its derivative with respect to ξ vanishes nowhere. In this case, the envelope of $\{\mathcal{H}(y^1(\xi,s), y^2(\xi,s), s) \colon \xi \in [\xi_1,\xi_2]\}$ is obtained by solving

$$\frac{\partial}{\partial\xi} V^\sharp\big(u_1,u_2,v_3,y^1(\xi,s),y^2(\xi,s),s\big) = 0 \tag{6.16}$$

for ξ and substituting it into

$$v^3 = V^\sharp\big(u_1,u_2,y^1(\xi,s),y^2(\xi,s),s\big) = 0, \tag{6.17}$$

from which we read off V. We recognize that (6.16) is just a necessary condition for the maximization of $(x^1,x^2) \mapsto V^\sharp(u_1,u_2,\mathbf{x})$. If $\partial\mathcal{A}(s)$ is only piecewise continuously differentiable, then the process of envelope formation or, equivalently, the process of minimization must be carried out separately on each segment where $\partial\mathcal{A}(s)$ is continuously differentiable. When minimization is performed, the values of $V^\sharp$ at the ends of each segment must be accounted for.

6.18. Exercise. Let $\varphi_\alpha(\mathbf{x}) = x^\alpha$ and $\mathcal{A}(s) = \{(x^1, x^2) : 0 \leq a(s) \leq (x^1)^2 + (x^2)^2 \leq h(s)\}$. Show that $\mathcal{V}(s)$ is a right circular cone and

$$V(s) = h(s)\sqrt{(u_1)^2 + (u_2)^2}. \tag{6.19}$$

Show that if $\mathcal{A}(s) = \{(x^1, x^2) : |x^1| \leq h^1(s), |x^2| \leq h^2(s)\}$, then $\mathcal{V}(s)$ is a cone with a rectangular cross section and

$$V(s) = h^2(s)|u_1| + h^1(s)|u_2|. \tag{6.20}$$

Find V for $\mathcal{A}(s) = \{(x^1, x^2) : |x^1| \leq h^1(s), 0 \leq x^2 \leq h^2(s)\}$. Interpret these V's in light of Theorem 6.2(iii).

(ii) Let the φ_α be harmonic. If $\mathcal{A}(s)$ is simply-connected, then $\mathcal{A}(s)$ can be conformally mapped onto a disk. The compositions of the φ_α with the inverse conformal mapping are harmonic on the disk and therefore assume their extrema on the circular boundary. In this case, the problem of determining $\mathcal{V}(s)$ and V reduces to an equivalent problem on the disk. (But it is important to note that if $\partial\mathcal{A}(s)$ is not sufficiently smooth, then the composite functions need not be well behaved on the boundary of the disk.) Analogous results hold if $\mathcal{A}(s)$ is doubly-connected, for then $\mathcal{A}(s)$ can be conformally mapped onto a circular annulus.

6.21. Problem. Find formulas for V in these cases.

(iii) Let $(\tilde{x}^1, \tilde{x}^2)$ be any fixed point in $\mathcal{A}(s)$. Let $\mathcal{A}(s)$ be convex and let the matrix $(\partial\varphi_\alpha/\partial x^\beta)$ be positive-definite on $\operatorname{cl}\mathcal{A}(s)$. Thus the mapping (4.12) is *uniformly strictly monotone*. By the theory of Sec. 8, it is an invertible mapping taking $\operatorname{cl}\mathcal{A}(s)$ into a closed convex set $\operatorname{cl}\mathcal{A}^\sharp(s)$ in (y_1, y_2)-space, which has $(0,0)$ in its interior. Thus we have to minimize the affine function

$$(y_1, y_2) \mapsto v_3 - y_1 u_2 + y_2 u_1 \tag{6.22}$$

(which is a special case of that defined in (6.9)) on $\operatorname{cl}\mathcal{A}^\sharp(s)$. Since it is affine, this function assumes its minimum on $\partial\mathcal{A}^\sharp(s)$. If $\mathcal{A}^\sharp(s)$ is strictly convex, there is a unique minimizer, which is readily found from the geometric construction of Fig. 6.23 by the following argument. The function (6.22) has a planar graph over $\operatorname{cl}\mathcal{A}^\sharp(s)$. This graph has the normal $(u_2, -u_1, 1)$. The direction of steepest descent lies along the intersection of this graph with the plane containing the normal and the vertical vector $(0,0,1)$. This direction, defined wherever $u_\alpha u_\alpha \neq 0$, is $-(-u_2, u_1, u_\alpha u_\alpha)$. The projection of this direction onto the (y_1, y_2)-plane is $(u_2, -u_1)$. The minimizer $(\bar{y}_1, \bar{y}_2)$ of (6.22) on $\operatorname{cl}\mathcal{A}^\sharp(s)$ is at the intersection of $\partial\mathcal{A}^\sharp(s)$ with the half-ray from $(0,0)$ in the direction $(u_2, -u_1)$.

Let us describe $\partial\mathcal{A}^\sharp(s)$ by polar coordinates:

$$\sqrt{y_\alpha y_\alpha} = f(\theta, s), \quad \tan\theta = y_2/y_1. \tag{6.24}$$

Then the construction of Fig. 6.23 yields

$$\begin{pmatrix} \bar{y}_1 \\ \bar{y}_2 \end{pmatrix} = \frac{f(-\operatorname{sign}(u_1)\arctan(u_1/u_2), s)}{\sqrt{u_\alpha u_\alpha}} \begin{pmatrix} u_2 \\ -u_1 \end{pmatrix}. \tag{6.25}$$

The substitution of (6.25) into the analog of (6.10) yields

$$V(s) = f(-\operatorname{sign}(u_1)\arctan(u_1/u_2), s)\sqrt{u_\alpha u_\alpha} \tag{6.26}$$

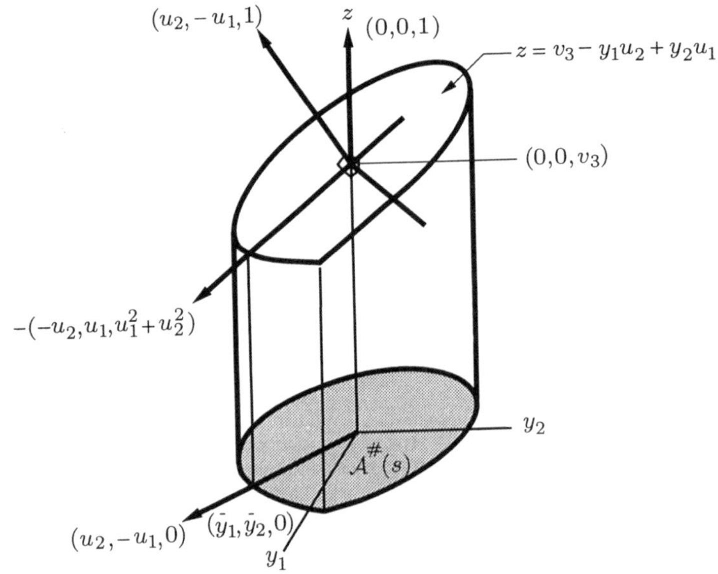

Fig. 6.23. Construction locating the minimizer of (6.22).

by virtue of the equivalence of (6.3) and (6.10). The results reported here extend those of Antman (1976a, Sec. 4).

6.27. Exercise. Derive (6.19) and (6.20) from (6.26).

7. Constitutive Equations Invariant under Rigid Motions

In this section we examine why the constitutive equations should have the restricted forms (2.21) and (2.22). We begin by studying constitutive equations in far more general forms. A special Cosserat rod of a *simple elastic material* is governed by constitutive equations of the form

$$(7.1) \qquad \begin{aligned} \boldsymbol{n}(s,t) &= \hat{\boldsymbol{n}}\big(\boldsymbol{r}(s,t), \boldsymbol{r}_s(s,t), \boldsymbol{d}_\alpha(s,t), \partial_s\boldsymbol{d}_\alpha(s,t), s\big), \\ \boldsymbol{m}(s,t) &= \hat{\boldsymbol{m}}\big(\boldsymbol{r}(s,t), \boldsymbol{r}_s(s,t), \boldsymbol{d}_\alpha(s,t), \partial_s\boldsymbol{d}_\alpha(s,t), s\big). \end{aligned}$$

The common domain of $\hat{\boldsymbol{n}}$ and $\hat{\boldsymbol{m}}$ consists of those values of its arguments satisfying (6.3). We do not treat non-simple materials, for which $\hat{\boldsymbol{n}}$ and $\hat{\boldsymbol{m}}$ depend on higher s-derivatives of $\boldsymbol{r}$ and $\boldsymbol{d}_\alpha$. (See the remarks following (I.2.12).) We do not allow $\hat{\boldsymbol{n}}$ and $\hat{\boldsymbol{m}}$ to depend on the absolute time t for the reasons stated after (I.2.11).

We require that these constitutive equations be *frame-indifferent* (or *objective*), i.e., be invariant under rigid motions. We now determine the restrictions imposed on (7.1) by this fundamental principle.

Let $\{\boldsymbol{r}, \boldsymbol{d}_\alpha\}$ define a configuration and let $\boldsymbol{n}$ and $\boldsymbol{m}$ be the resultants acting in this configuration. Suppose, as in Sec. 5, that the $\boldsymbol{d}_\alpha$ are differences of position vectors (cf. (4.2g) and (4.3)). Let $\{\boldsymbol{r}^\sharp, \boldsymbol{d}_\alpha^\sharp\}$ be another configuration that differs from $\{\boldsymbol{r}, \boldsymbol{d}_\alpha\}$ by a rigid motion. Let $\boldsymbol{n}^\sharp$ and $\boldsymbol{m}^\sharp$ be the resultants acting in the configuration $\{\boldsymbol{r}^\sharp, \boldsymbol{d}_\alpha^\sharp\}$. Then there is a vector-valued function $\boldsymbol{c}$ and a proper orthogonal tensor-valued function $\boldsymbol{Q}$ such that

$$\begin{aligned} \boldsymbol{r}^\sharp(s,t) - \boldsymbol{c}(t) = \boldsymbol{Q}(t)\cdot\boldsymbol{r}(s,t), \quad \boldsymbol{d}_\alpha^\sharp(s,t) = \boldsymbol{Q}(t)\cdot\boldsymbol{d}_\alpha(s,t), \\ \boldsymbol{n}^\sharp(s,t) = \boldsymbol{Q}(t)\cdot\boldsymbol{n}(s,t), \quad \boldsymbol{m}^\sharp(s,t) = \boldsymbol{Q}(t)\cdot\boldsymbol{m}(s,t). \end{aligned} \tag{7.2}$$

We require that the constitutive functions $\tilde{\boldsymbol{n}}$ and $\tilde{\boldsymbol{m}}$ assigning resultants to the geometric variables be unaffected by the rigid motion defined by $\boldsymbol{c}$ and $\boldsymbol{Q}$, i.e., $\boldsymbol{n}^\sharp$ and $\boldsymbol{m}^\sharp$ must be delivered by the same constitutive functions $\tilde{\boldsymbol{n}}$ and $\tilde{\boldsymbol{m}}$ of $\{\boldsymbol{r}^\sharp, \boldsymbol{r}_s^\sharp, \boldsymbol{d}_\alpha^\sharp, \partial_s\boldsymbol{d}_\alpha^\sharp\}$ as those that deliver $\boldsymbol{n}$ and $\boldsymbol{m}$ from $\{\boldsymbol{r}, \boldsymbol{r}_s, \boldsymbol{d}_\alpha, \partial_s\boldsymbol{d}_\alpha\}$. Thus

$$\boldsymbol{n}^\sharp = \tilde{\boldsymbol{n}}(\boldsymbol{r}^\sharp, \boldsymbol{r}_s^\sharp, \boldsymbol{d}_\alpha^\sharp, \partial_s\boldsymbol{d}_\alpha^\sharp), \quad \text{etc.} \tag{7.3}$$

We substitute (7.2) into (7.3) to obtain

$$\boldsymbol{Q}\cdot\boldsymbol{n} = \tilde{\boldsymbol{n}}(\boldsymbol{c} + \boldsymbol{Q}(t)\cdot\boldsymbol{r},\ \boldsymbol{Q}(t)\cdot\boldsymbol{r}_s,\ \boldsymbol{Q}(t)\cdot\boldsymbol{d}_\alpha,\ \boldsymbol{Q}(t)\cdot\partial_s\boldsymbol{d}_\alpha,\ s), \quad \text{etc.} \tag{7.4}$$

We substitute (7.1) into (7.4) to get

$$\begin{aligned} &\boldsymbol{Q}\cdot\tilde{\boldsymbol{n}}(\boldsymbol{r}, \boldsymbol{r}_s, \boldsymbol{d}_\alpha, \partial_s\boldsymbol{d}_\alpha, s) \\ &\quad = \tilde{\boldsymbol{n}}(\boldsymbol{c} + \boldsymbol{Q}(t)\cdot\boldsymbol{r}, \boldsymbol{Q}(t)\cdot\boldsymbol{r}_s, \boldsymbol{Q}(t)\cdot\boldsymbol{d}_\alpha, \boldsymbol{Q}(t)\cdot\partial_s\boldsymbol{d}_\alpha, s), \quad \text{etc.,} \end{aligned} \tag{7.5}$$

which must hold for all $\boldsymbol{c}$ and for all proper orthogonal $\boldsymbol{Q}$. We now confront the specific mathematical problem of determining the form of those functions $\tilde{\boldsymbol{n}}$ and $\tilde{\boldsymbol{m}}$ satisfying (7.5) for all $\boldsymbol{c}$ and $\boldsymbol{Q}$.

Let us first choose $\boldsymbol{Q} = \boldsymbol{I}$. Then (7.5) implies that $\tilde{\boldsymbol{n}}$ and $\tilde{\boldsymbol{m}}$ are independent of their first argument $\boldsymbol{r}$, which we accordingly drop. We next set

$$\tilde{\boldsymbol{n}}(\boldsymbol{r}_s, \boldsymbol{d}_\alpha, \partial_s\boldsymbol{d}_\alpha, s) = \tilde{n}_k(\boldsymbol{r}_s, \boldsymbol{d}_\alpha, \partial_s\boldsymbol{d}_\alpha, s)\boldsymbol{d}_k, \quad \text{etc.} \tag{7.6}$$

The substitution of (7.6) into (7.5) yields

$$\tilde{n}_k(\boldsymbol{r}_s, \boldsymbol{d}_\alpha, \partial_s\boldsymbol{d}_\alpha, s) = \tilde{n}_k(\boldsymbol{Q}(t)\cdot\boldsymbol{r}_s,\ \boldsymbol{Q}(t)\cdot\boldsymbol{d}_\alpha,\ \boldsymbol{Q}(t)\cdot\partial_s\boldsymbol{d}_\alpha,\ s), \quad \text{etc.,} \tag{7.7}$$

for all proper orthogonal $\boldsymbol{Q}$. Equation (7.7) says that each of the functions $\tilde{n}_k$ is a *hemitropic* scalar-valued function (i.e., a scalar-valued function invariant under the proper orthogonal group). We determine the form of such functions from

7.8. Cauchy's Representation Theorem (1850). *Let*

$$\begin{aligned} S(\boldsymbol{y}_1, \dots, \boldsymbol{y}_N) &\equiv \{\boldsymbol{y}_a\cdot\boldsymbol{y}_b,\ 1 \le a \le b \le N\}, \\ T(\boldsymbol{y}_1, \dots, \boldsymbol{y}_N) &\equiv \{(\boldsymbol{y}_a\times\boldsymbol{y}_b)\cdot\boldsymbol{y}_c,\ 1 \le a < b < c \le N\}, \\ I(\boldsymbol{y}_1, \dots, \boldsymbol{y}_N) &\equiv S(\boldsymbol{y}_1, \dots, \boldsymbol{y}_N) \cup T(\boldsymbol{y}_1, \dots, \boldsymbol{y}_N) \end{aligned} \tag{7.9}$$

denote various ordered sets of invariants of the vectors $\boldsymbol{y}_1, \dots, \boldsymbol{y}_N$. *Let* $\mathcal{E}$ *consist of those vectors* $\boldsymbol{y}_1, \dots, \boldsymbol{y}_N$ *satisfying a collection of relations of the forms*

$$\psi(I(\boldsymbol{y}_1, \dots, \boldsymbol{y}_N)) \prec 0 \tag{7.10}$$

where $\prec$ *stands for any of the relations* $<$, $\leq$, $=$. *If* γ *is a hemitropic scalar-valued function on* $\mathcal{E}$, *i.e., if*

$$\gamma(\boldsymbol{Q} \cdot \boldsymbol{y}_1, \dots, \boldsymbol{Q} \cdot \boldsymbol{y}_N) = \gamma(\boldsymbol{y}_1, \dots, \boldsymbol{y}_N) \tag{7.11}$$

for all $(\boldsymbol{y}_1, \dots, \boldsymbol{y}_n) \in \mathcal{E}$ *and for all proper orthogonal* $\boldsymbol{Q}$, *then* γ *depends only on* $I(\boldsymbol{y}_1, \dots, \boldsymbol{y}_N)$, *and conversely. If the functions in* (7.10) *depend only on* $S(\boldsymbol{y}_1, \dots, \boldsymbol{y}_N)$ *and if* γ *is a isotropic scalar-valued function on* $\mathcal{E}$, *so that* (7.11) *holds for all* $(\boldsymbol{y}_1, \dots, \boldsymbol{y}_n) \in \mathcal{E}$ *and for all orthogonal* $\boldsymbol{Q}$, *then* γ *depends only on* $S(\boldsymbol{y}_1, \dots, \boldsymbol{y}_N)$, *and conversely.*

The proof of this theorem is given below. We identify γ with $\tilde{n}_k$, identify $(\boldsymbol{y}_1, \dots, \boldsymbol{y}_N)$ with $(\boldsymbol{r}_s, \boldsymbol{d}_1, \boldsymbol{d}_2, \partial_s \boldsymbol{d}_1, \partial_s \boldsymbol{d}_2)$, and identify the relations of the form (7.10) with

$$v_3 > V(u_1, u_2, s), \quad \boldsymbol{d}_\alpha \cdot \boldsymbol{d}_\beta = \delta_{\alpha\beta}, \quad \partial_s \boldsymbol{d}_\alpha \cdot \boldsymbol{d}_\beta + \boldsymbol{d}_\alpha \cdot \partial_s \boldsymbol{d}_\beta = 0. \tag{7.12}$$

We readily compute $I(\boldsymbol{r}_s, \boldsymbol{d}_1, \boldsymbol{d}_2, \partial_s \boldsymbol{d}_1, \partial_s \boldsymbol{d}_2)$ and find that it depends on (u, v). Conversely, given $S(\boldsymbol{r}_s, \boldsymbol{d}_1, \boldsymbol{d}_2, \partial_s \boldsymbol{d}_1, \partial_s \boldsymbol{d}_2)$ and given (6.3) (which is included in (7.12)), we can uniquely determine (u, v). Thus any function of $I(\boldsymbol{r}_s, \boldsymbol{d}_1, \boldsymbol{d}_2, \partial_s \boldsymbol{d}_1, \partial_s \boldsymbol{d}_2)$ is a function of (u, v) and vice versa. We therefore obtain

7.13. Theorem. *The most general frame-indifferent constitutive functions of the form* (7.1) *are* (2.21):

$$\mathsf{m}(s,t) = \hat{\mathsf{m}}\big(\mathsf{u}(s,t), \mathsf{v}(s,t), s\big), \quad \mathsf{n}(s,t) = \hat{\mathsf{n}}\big(\mathsf{u}(s,t), \mathsf{v}(s,t), s\big). \tag{7.14}$$

The common domain of definition of $\hat{\mathsf{m}}$ and $\hat{\mathsf{n}}$ is

$$\begin{aligned} \mathcal{V} &\equiv \{(\mathsf{u}, \mathsf{v}, s) \colon (\mathsf{u}, \mathsf{v}) \in \mathcal{V}(s),\ s \in [s_1, s_2]\} \\ &= \cap\{\mathcal{V}(s) \times \{s\} \colon s \in [s_1, s_2]\}. \end{aligned} \tag{7.15}$$

The rod is called *hyperelastic* iff there exists a function $W : \mathcal{V} \to \mathbb{R}$ such that

$$\hat{\mathsf{m}}(\mathsf{u}, \mathsf{v}, s) = \partial W(\mathsf{u}, \mathsf{v}, s)/\partial \mathsf{u}, \quad \hat{\mathsf{n}}(\mathsf{u}, \mathsf{v}, s) = \partial W(\mathsf{u}, \mathsf{v}, s)/\partial \mathsf{v}. \tag{7.16}$$

If $\hat{\mathsf{m}}(\cdot, \cdot, s)$ and $\hat{\mathsf{m}}(\cdot, \cdot, s)$ are continuously differentiable, then the simple-connectedness of $\mathcal{V}(s)$ implies that a necessary and sufficient condition for (7.16) to hold is that the 6×6 matrix

$$\begin{pmatrix} \partial \hat{\mathsf{m}}/\partial \mathsf{u} & \partial \hat{\mathsf{m}}/\partial \mathsf{v} \\ \partial \hat{\mathsf{n}}/\partial \mathsf{u} & \partial \hat{\mathsf{n}}/\partial \mathsf{v} \end{pmatrix} \tag{7.17}$$

be symmetric. W is called the *stored-energy function.* Had we introduced W independently of (7.16), then we could have taken its arguments to be those of (7.1) and used Cauchy's Representation Theorem to show that frame-indifference requires W to depend only on the arguments shown in (7.16). There are compelling thermodynamic reasons, discussed in Sec. XII.14, for assuming that an elastic body is hyperelastic. We refrain from making this assumption universal because we can analyze most of our steady-state problems without it and because certain natural approximation schemes can destroy hyperelasticity.

A special Cosserat rod of a *simple viscoelastic material of differential type with complexity 1* is governed by constitutive equations of the form

$$\begin{aligned}(7.18)\quad \boldsymbol{n}(s,t) = \tilde{\boldsymbol{n}}\big(&\boldsymbol{r}_s(s,t), \boldsymbol{d}_\alpha(s,t), \partial_s\boldsymbol{d}_\alpha(s,t), \boldsymbol{r}_{st}(s,t),\\ &\partial_t\boldsymbol{d}_\alpha(s,t), \partial_{ts}\boldsymbol{d}_\alpha(s,t), s\big), \quad \text{etc.}\end{aligned}$$

Defining the $\tilde{n}_k$ as in (7.6), we follow the development for elastic rods to show that frame-indifferent constitutive functions must satisfy the following generalization of (7.7):

$$\begin{aligned}(7.19)\quad &\tilde{n}_k(\boldsymbol{r}_s, \boldsymbol{d}_\alpha, \partial_s\boldsymbol{d}_\alpha, \boldsymbol{r}_{st}, \partial_t\boldsymbol{d}_\alpha, \partial_{ts}\boldsymbol{d}_\alpha, s)\\ &= \tilde{n}_k(\boldsymbol{Q}\cdot\boldsymbol{r}_s,\ \boldsymbol{Q}\cdot\boldsymbol{d}_\alpha,\ \boldsymbol{Q}\cdot\partial_s\boldsymbol{d}_\alpha,\ \boldsymbol{Q}_t\cdot\boldsymbol{r}_s+\boldsymbol{Q}\cdot\boldsymbol{r}_{st},\\ &\qquad \boldsymbol{Q}_t\cdot\boldsymbol{d}_\alpha+\boldsymbol{Q}(t)\cdot\partial_t\boldsymbol{d}_\alpha,\ \boldsymbol{Q}_t\cdot\partial_s\boldsymbol{d}_\alpha+\boldsymbol{Q}\cdot\partial_{ts}\boldsymbol{d}_\alpha,\ s), \quad \text{etc.,}\end{aligned}$$

for all proper orthogonal and differentiable $\boldsymbol{Q}$. The orthogonality of $\boldsymbol{Q}$ implies that

$$(7.20)\qquad \boldsymbol{Q}_t = -\boldsymbol{Q}\cdot\boldsymbol{Q}_t^*\cdot\boldsymbol{Q}$$

and that $\boldsymbol{Q}_t^*\cdot\boldsymbol{Q}$ is antisymmetric. Thus we may choose

$$(7.21)\qquad \boldsymbol{Q}_t^*\cdot\boldsymbol{Q} = \boldsymbol{w}\times$$

where $\boldsymbol{w}$ is defined in (2.3b). The substitution of (7.20) and (7.21) into the right-hand side of (7.19) reduces it to

$$\begin{aligned}(7.22)\quad \tilde{n}_k\big(&\boldsymbol{Q}\cdot\boldsymbol{r}_s,\ \boldsymbol{Q}\cdot\boldsymbol{d}_\alpha,\ \boldsymbol{Q}\cdot\partial_s\boldsymbol{d}_\alpha,\ \boldsymbol{Q}\cdot(\boldsymbol{r}_{st}-\boldsymbol{w}\times\boldsymbol{r}_s),\\ &\boldsymbol{Q}\cdot(\partial_t\boldsymbol{d}_\alpha-\boldsymbol{w}\times\boldsymbol{d}_\alpha),\ \boldsymbol{Q}\cdot(\partial_{ts}\boldsymbol{d}_\alpha-\boldsymbol{w}\times\partial_s\boldsymbol{d}_\alpha),\ s\big).\end{aligned}$$

Now (2.3b) and (2.4b) imply that

$$(7.23)\qquad \boldsymbol{r}_{st}-\boldsymbol{w}\times\boldsymbol{r}_s = (\partial_t v_k)\boldsymbol{d}_k + v_k\boldsymbol{w}\times\boldsymbol{d}_k - \boldsymbol{w}\times\boldsymbol{r}_s = (\partial_t v_k)\boldsymbol{d}_k.$$

Similarly,

$$(7.24)\qquad \partial_{ts}\boldsymbol{d}_\alpha - \boldsymbol{w}\times\partial_s\boldsymbol{d}_\alpha = \partial_s(\boldsymbol{w}\times\boldsymbol{d}_\alpha) - \boldsymbol{w}\times(\boldsymbol{u}\times\boldsymbol{d}_\alpha) = \boldsymbol{w}_s\times\boldsymbol{d}_\alpha.$$

Since $\partial_t \partial_s \boldsymbol{d}_k = \partial_s \partial_t \boldsymbol{d}_k$, we obtain

$$\partial_t(\boldsymbol{u} \times \boldsymbol{d}_k) = \partial_s(\boldsymbol{w} \times \boldsymbol{d}_k), \tag{7.25}$$

which implies that

$$(\boldsymbol{w}_s - \boldsymbol{u}_t) \times \boldsymbol{d}_k = u_k \boldsymbol{w} - w_k \boldsymbol{u}. \tag{7.26}$$

We take the dot product of (7.26) with $\boldsymbol{d}_l \times \boldsymbol{d}_k$ to obtain

$$\boldsymbol{w}_s = \boldsymbol{u}_t + \boldsymbol{u} \times \boldsymbol{w} = (\partial_t u_k)\boldsymbol{d}_k. \tag{7.27}$$

We now use (2.3), (7.23), (7.24), and (7.27) to write (7.22) as

$$\tilde{n}_k\big(\boldsymbol{Q}\cdot v_l \boldsymbol{d}_l,\ \boldsymbol{Q}\cdot\boldsymbol{d}_\alpha,\ \boldsymbol{Q}\cdot(\boldsymbol{u}\times\boldsymbol{d}_\alpha),\ (\partial_t v_l)\boldsymbol{Q}\cdot\boldsymbol{d}_l,\ \boldsymbol{0},\ (\partial_t u_l)\boldsymbol{Q}\cdot(\boldsymbol{d}_l\times\boldsymbol{d}_\alpha),\ s\big), \tag{7.28}$$

which must equal the left-hand side of (7.19) for all proper orthogonal $\boldsymbol{Q}$. Choosing $\boldsymbol{Q} = \boldsymbol{I}$, we find that $\tilde{n}_k$ must be independent of its fifth argument. In this case, the equality of the left-hand side of (7.19) with (7.28) has exactly the form (7.11). Thus Cauchy's Representation Theorem 7.8 implies that the $\tilde{n}_k$ each depend on the dot and triple scalar products of

$$\{v_l \boldsymbol{d}_l,\ \boldsymbol{d}_\alpha,\ \boldsymbol{u} \times \boldsymbol{d}_\alpha,\ (\partial_t v_l)\boldsymbol{d}_l,\ (\partial_t u_l)(\boldsymbol{d}_l \times \boldsymbol{d}_\alpha)\}. \tag{7.29}$$

7.30. Theorem. *The most general frame-indifferent constitutive functions of the form* (7.18) *are* (2.22)*:*

$$\begin{aligned} \mathsf{m}(s,t) &= \hat{\mathsf{m}}\big(\mathsf{u}(s,t), \mathsf{v}(s,t), \mathsf{u}_t(s,t), \mathsf{v}_t(s,t), s\big), \\ \mathsf{n}(s,t) &= \hat{\mathsf{n}}\big(\mathsf{u}(s,t), \mathsf{v}(s,t), \mathsf{u}_t(s,t), \mathsf{v}_t(s,t), s\big). \end{aligned} \tag{7.31}$$

Proof. We have just proved that if (7.18) is invariant under rigid motions, then (7.31) holds. In this process, we made the special choice (7.21). To show that this choice causes no loss of generality, we need only verify that (7.31) is invariant under rigid motions. □

We now prove Cauchy's Representation Theorem for hemitropic functions. The proof for isotropic functions is analogous. We are given that

$$\gamma(\boldsymbol{z}_1, \dots, \boldsymbol{z}_N) = \gamma(\boldsymbol{y}_1, \dots, \boldsymbol{y}_N) \tag{7.32}$$

for all $(\boldsymbol{y}_1, \dots, \boldsymbol{y}_N) \in \mathcal{E}$ and for all $(\boldsymbol{z}_1, \dots, \boldsymbol{z}_N)$ of the form $\boldsymbol{z}_a = \boldsymbol{Q} \cdot \boldsymbol{y}_a$ for all proper orthogonal $\boldsymbol{Q}$. We now reduce the proof to the proof of

7.33. Lemma. *Let*

$$I(\boldsymbol{y}_1, \dots, \boldsymbol{y}_N) = I(\boldsymbol{z}_1, \dots, \boldsymbol{z}_N) \tag{7.34}$$

for all $(\boldsymbol{y}_1, \dots, \boldsymbol{y}_N)$ *and* $(\boldsymbol{z}_1, \dots, \boldsymbol{z}_N)$ *in* $\mathcal{E}$. *If* span $\mathcal{E}$ *has dimension* = 3, *then there exists a proper orthogonal* $\boldsymbol{Q}$ *such that*

$$\boldsymbol{z}_a = \boldsymbol{Q} \cdot \boldsymbol{y}_a \tag{7.35}$$

for $a = 1, \dots, N$. If span $\mathcal{E}$ *has dimension* < 3, *then there exists an orthogonal* $\boldsymbol{Q}$ *such that* (7.35) *holds for* $a = 1, \dots, N$.

Note that (7.35) implies (7.34). Furthermore, if span $\mathcal{E}$ has dimension < 3, then $T(\boldsymbol{y}_1, \dots, \boldsymbol{y}_N)$ is trivial.

Lemma 7.33 implies that (7.32) must hold for all $(\boldsymbol{y}_1, \dots, \boldsymbol{y}_N)$ and $(\boldsymbol{z}_1, \dots, \boldsymbol{z}_N)$ in $\mathcal{E}$ satisfying (7.34). It follows that $\gamma(\boldsymbol{y}_1, \dots, \boldsymbol{y}_N)$ must depend only on $I(\boldsymbol{y}_1, \dots, \boldsymbol{y}_N) = I(\boldsymbol{z}_1, \dots, \boldsymbol{z}_N)$, which are the only variables that the left- and right-hand sides of (7.32) have in common.

Proof of Lemma 7.33. Let (7.34) hold. By renumbering the vectors $\boldsymbol{y}_1, \dots, \boldsymbol{y}_N$ if necessary, we may assume that $\{\boldsymbol{y}_1, \dots, \boldsymbol{y}_P\}$ is a basis for span$\{\boldsymbol{y}_1, \dots, \boldsymbol{y}_N\}$. Thus $P \le 3$ (because the $\boldsymbol{y}_a \in \mathbb{E}^3$) and $P \le N$. Since a necessary and sufficient condition for the independence of $\{\boldsymbol{y}_1, \dots, \boldsymbol{y}_P\}$ is that the Gram determinant

$$\det\big(\boldsymbol{y}_a \cdot \boldsymbol{y}_b\big)_{a,b=1,\dots,P} \neq 0, \tag{7.36}$$

condition (7.34) implies that $\{\boldsymbol{z}_1, \dots, \boldsymbol{z}_P\}$ is a basis for span$\{\boldsymbol{z}_1, \dots, \boldsymbol{z}_N\}$. Thus there is a unique linear transformation $\boldsymbol{Q}\colon$ span$\{\boldsymbol{y}_1, \dots, \boldsymbol{y}_P\} \to$ span$\{\boldsymbol{z}_1, \dots, \boldsymbol{z}_P\}$ such that (7.35) holds for $a = 1, \dots, P$ and such that $\boldsymbol{Q}$ is the identity on the orthogonal complement of span$\{\boldsymbol{y}_1, \dots, \boldsymbol{y}_P\}$.

We now show that (7.35) holds for all a. Let $P < a \le N$. Since $\{\boldsymbol{y}_1, \dots, \boldsymbol{y}_P\}$ and $\{\boldsymbol{z}_1, \dots, \boldsymbol{z}_P\}$ respectively are bases for span$\{\boldsymbol{y}_1, \dots, \boldsymbol{y}_N\}$ and span$\{\boldsymbol{z}_1, \dots, \boldsymbol{z}_N\}$, there are uniquely determined sets of numbers $\{\alpha_{ba}, b = 1, \dots, P\}$ and $\{\beta_{ba}, b = 1, \dots, P\}$ such that

$$\boldsymbol{y}_a = \sum_{b=1}^{P} \alpha_{ba} \boldsymbol{y}_b, \quad \boldsymbol{z}_a = \sum_{b=1}^{P} \beta_{ba} \boldsymbol{z}_b \quad \text{for } P < a \le N. \tag{7.37}$$

Now (7.34) and (7.35) for $a = 1, \dots, P$ imply that

$$\sum_{b=1}^{P} \beta_{ba} \boldsymbol{z}_b \cdot \boldsymbol{z}_c = \boldsymbol{z}_a \cdot \boldsymbol{z}_c = \boldsymbol{y}_a \cdot \boldsymbol{y}_c = \sum_{b=1}^{P} \alpha_{ba} \boldsymbol{y}_b \cdot \boldsymbol{y}_c = \sum_{b=1}^{P} \alpha_{ba} \boldsymbol{z}_b \cdot \boldsymbol{z}_c, \quad c = 1, \dots, P. \tag{7.38}$$

For each $a > P$, this is a homogeneous linear system for $\{\beta_{ba} - \alpha_{ba}\}$. Condition (7.36) ensures that this system has the unique solution 0, so that $\beta_{ba} = \alpha_{ba}$. Thus (7.35) for $a = 1, \dots, P$ and (7.36) yield

$$\boldsymbol{Q} \cdot \boldsymbol{y}_a = \sum_{b=1}^{P} \alpha_{ba} \boldsymbol{Q} \cdot \boldsymbol{y}_b = \sum_{b=1}^{P} \alpha_{ba} \boldsymbol{z}_b = \boldsymbol{z}_a. \tag{7.39}$$

To show that $\boldsymbol{Q}$ is orthogonal, we use (7.34) and (7.35) to obtain

$$(\boldsymbol{Q} \cdot \boldsymbol{y}_a) \cdot (\boldsymbol{Q} \cdot \boldsymbol{y}_b) = \boldsymbol{z}_a \cdot \boldsymbol{z}_b = \boldsymbol{y}_a \cdot \boldsymbol{y}_b \quad \text{for } 1 \le a \le b \le P. \tag{7.40}$$

Since $\boldsymbol{Q}$ is the identity on the orthogonal complement of span$\{\boldsymbol{y}_1, \dots, \boldsymbol{y}_P\}$, it follows from (7.40) that $\boldsymbol{Q}$ is orthogonal. If $P = 3$, then we likewise obtain

$$(\det \boldsymbol{Q})[(\boldsymbol{y}_1 \times \boldsymbol{y}_2) \cdot \boldsymbol{y}_3] \equiv [(\boldsymbol{Q} \cdot \boldsymbol{y}_1) \times (\boldsymbol{Q} \cdot \boldsymbol{y}_2)] \cdot (\boldsymbol{Q} \cdot \boldsymbol{y}_3) = (\boldsymbol{z}_1 \times \boldsymbol{z}_2) \cdot \boldsymbol{z}_3 = (\boldsymbol{y}_1 \times \boldsymbol{y}_2) \cdot \boldsymbol{y}_3. \tag{7.41}$$

Thus $\det \boldsymbol{Q} = 1$, so that $\boldsymbol{Q}$ is a proper orthogonal tensor. (This proof is modelled on that of Truesdell & Noll (1965, Sec. 11).) □

8. Monotonicity and Growth Conditions

If the constitutive equations (7.14) for elastic rods are to conform to our intuition and experience, they should meet some minimal physical restrictions. These might ensure that an increase in the tension $\boldsymbol{n} \cdot \boldsymbol{r}_s / |\boldsymbol{r}_s|$ accompany an increase in the stretch $|\boldsymbol{r}_s|$, an increase in a shear component n_α of the contact force accompany an increase in the shear strain v_α, an increase in a bending couple m_α accompany an increase in the flexure u_α, and an increase in twisting couple m_3 accompany an increase in the twist u_3. If $\hat{\mathsf{m}}(\cdot,\cdot,s)$ and $\hat{\mathsf{n}}(\cdot,\cdot,s)$ are differentiable, then these conditions are ensured by the requirement:

$$\text{(8.1)} \qquad \begin{pmatrix} \partial\hat{\mathsf{m}}/\partial\mathsf{u} & \partial\hat{\mathsf{m}}/\partial\mathsf{v} \\ \partial\hat{\mathsf{n}}/\partial\mathsf{u} & \partial\hat{\mathsf{n}}/\partial\mathsf{v} \end{pmatrix} \quad \text{is positive-definite.}$$

(This condition is equivalent to the positive-definiteness of the symmetric part of this matrix.) The generalization of (8.1) to functions $\hat{\mathsf{m}}(\cdot,\cdot,s)$ and $\hat{\mathsf{n}}(\cdot,\cdot,s)$ that are not differentiable is that $\big(\hat{\mathsf{m}}(\cdot,\cdot,s), \hat{\mathsf{n}}(\cdot,\cdot,s)\big)$ be *uniformly monotone on compact subsets* of its domain $\mathcal{V}(s)$, i.e., for each compact set $\mathcal{K}(s) \subset \mathcal{V}(s)$ there is a number $c(\mathcal{K}(s)) > 0$ such that

$$\text{(8.2)} \qquad \begin{aligned} \Phi(\mathsf{u}_1, \mathsf{v}_1, \mathsf{u}_2, \mathsf{v}_2, s) &\equiv [\hat{\mathsf{m}}(\mathsf{u}_1, \mathsf{v}_1, s) - \hat{\mathsf{m}}(\mathsf{u}_2, \mathsf{v}_2, s)] \cdot (\mathsf{u}_1 - \mathsf{u}_2) \\ &\quad + [\hat{\mathsf{n}}(\mathsf{u}_1, \mathsf{v}_1, s) - \hat{\mathsf{n}}(\mathsf{u}_2, \mathsf{v}_2, s)] \cdot (\mathsf{v}_1 - \mathsf{v}_2) \\ &> c(\mathcal{K}(s))\big(|\mathsf{u}_1 - \mathsf{u}_2|^2 + |\mathsf{v}_1 - \mathsf{v}_2|^2\big) \\ &\qquad \forall\, (\mathsf{u}_1, \mathsf{v}_1),\ (\mathsf{u}_2, \mathsf{v}_2) \in \mathcal{K}(s). \end{aligned}$$

Here the dot product $\mathsf{m} \cdot \mathsf{u}$ of elements of $\mathbb{R}^3$ is defined to be $m_k u_k$. (Since m and u are components of $\boldsymbol{m}$ and $\boldsymbol{u}$ with respect to an orthonormal basis, it follows that $\mathsf{m} \cdot \mathsf{u} = \boldsymbol{m} \cdot \boldsymbol{u}$.) Conditions (8.1) and (8.2) are equivalent when $\hat{\mathsf{m}}(\cdot,\cdot,s)$ and $\hat{\mathsf{n}}(\cdot,\cdot,s)$ are differentiable. A useful generalization of (8.2) is that $\big(\hat{\mathsf{m}}(\cdot,\cdot,s), \hat{\mathsf{n}}(\cdot,\cdot,s)\big)$ be *strictly monotone* on $\mathcal{V}(s)$:

$$\text{(8.3)} \qquad \Phi(\mathsf{u}_1, \mathsf{v}_1, \mathsf{u}_2, \mathsf{v}_2, s) > 0 \qquad \forall\, (\mathsf{u}_1, \mathsf{v}_1),\ (\mathsf{u}_2, \mathsf{v}_2) \in \mathcal{V}(s) \ \text{ with } \ (\mathsf{u}_1, \mathsf{v}_1) \neq (\mathsf{u}_2, \mathsf{v}_2).$$

A weaker and less useful restriction is that $\big(\hat{\mathsf{m}}(\cdot,\cdot,s), \hat{\mathsf{n}}(\cdot,\cdot,s)\big)$ be *monotone* on $\mathcal{V}(s)$:

$$\text{(8.4)} \qquad \Phi(\mathsf{u}_1, \mathsf{v}_1, \mathsf{u}_2, \mathsf{v}_2, s) \geq 0 \quad \forall\, (\mathsf{u}_1, \mathsf{v}_1),\ (\mathsf{u}_2, \mathsf{v}_2) \in \mathcal{V}(s).$$

Condition (8.4) is equivalent to the nonnegative-definiteness of the matrix of (8.1) when $\hat{\mathsf{m}}(\cdot,\cdot,s)$ and $\hat{\mathsf{n}}(\cdot,\cdot,s)$ are differentiable. It is easy to see that a function that is uniformly monotone on compact sets is strictly monotone and that a strictly monotone function is monotone. The distinction between these classes of functions is best seen in the setting of functions from $\mathbb{R}$ to itself: The function

$$x \mapsto \begin{cases} x & \text{for } \ x \leq 0, \\ 0 & \text{for } \ 0 \leq x \leq 1, \\ x - 1 & \text{for } \ 1 \leq x \end{cases}$$

is monotone, but not strictly monotone. The function $x \mapsto x^3$ is strictly monotone, but not uniformly monotone on compact sets. The function $x \mapsto x$ is uniformly monotone everywhere and the function $x \mapsto \arctan x$ is uniformly monotone on compact sets.

8.5. Exercise. Prove that (8.1) and (8.2) are equivalent when $\hat{\mathsf{m}}(\cdot, \cdot, s)$ and $\hat{\mathsf{n}}(\cdot, \cdot, s)$ are differentiable.

Conditions (8.1)–(8.3) each ensure not only the minimal restrictions listed above, but also that the coupling of effects, associated with the off-diagonal elements of the matrix of (8.1), are subservient to the direct effects, associated with the diagonal elements. We adopt (8.3) as our basic constitutive assumption for elastic rods.

Let us describe the $\boldsymbol{d}_k$ by Euler angles (see Sec. 11). The substitution of the constitutive equations (7.16) into the equations of motion (2.15) and (2.16) yields a quasilinear system of six second-order partial differential equations for $\boldsymbol{r}$ and the Euler angles. The requirement that this system be totally hyperbolic (see Courant & Hilbert (1961)) is exactly (8.1). The physical import of this requirement is that the rod can undergo a complete range of wave-like behavior. In particular, the linearization of the equations about any state has six real wave speeds. Condition (8.1) may be termed the *strong ellipticity condition* because it ensures that the static part of the equation, which is obtained by setting all time derivatives equal to 0 and which depends on the constitutive functions, is strongly elliptic. Condition (8.3) supports a complete and natural existence and regularity theory, without preventing the appearance of buckling and other kinds of instabilities, whose study constitutes one of our main objectives.

Ericksen (1975, 1977b, 1980a) has emphasized that weaker constitutive hypotheses, which can violate (8.3) locally, may describe important physical phenomena in crystals not traditionally associated with elasticity. See Ball & James (1987) Carr, Gurtin, & Slemrod (1984), Gurtin & Temam (1981), James (1979, 1980) for further developments along these lines. Antman & Carbone (1977), in their study of a general Cosserat theory of rods (in which $\boldsymbol{d}_1$, $\boldsymbol{d}_2$, and $\boldsymbol{r}_s$ need merely be independent), showed that the corresponding generalization of (8.3), by itself, permits instabilities of a kind that have traditionally been associated with plasticity (see Sec.XIV.6). Coleman (1985) and N. C. Owen (1987) have shown how these two viewpoints can be reconciled. By adhering to (8.3) I do not mean to suggest that the question of determining suitable constitutive restrictions for elastic rods is settled.

Let $(\boldsymbol{a}, \boldsymbol{b})$ be a unit vector in $\mathbb{R}^3 \times \mathbb{R}^3$. The line in $\mathbb{R}^3 \times \mathbb{R}^3$ through $(\boldsymbol{u}_0, \boldsymbol{v}_0)$ parallel to $(\boldsymbol{a}, \boldsymbol{b})$ is

$$\mathbb{R} \ni \sigma \mapsto (\boldsymbol{u}_0, \boldsymbol{v}_0) + (\boldsymbol{a}, \boldsymbol{b})\sigma. \tag{8.6}$$

Then (8.3) implies that

$$\sigma \mapsto \hat{\mathsf{m}}(\boldsymbol{u}_0 + \sigma\boldsymbol{a}, \boldsymbol{v}_0 + \sigma\boldsymbol{b}) \cdot \boldsymbol{a} + \hat{\mathsf{n}}(\boldsymbol{u}_0 + \sigma\boldsymbol{a}, \boldsymbol{v}_0 + \sigma\boldsymbol{b}) \cdot \boldsymbol{b} \tag{8.7}$$

is strictly increasing. We use this observation in constructing growth conditions.

We might expect that an infinite elongation of a rod must be maintained by an infinite tensile force, that infinite shear strains must be maintained by infinite shear forces, that an infinite twist must be maintained by an infinite twisting couple, etc. The restrictions on shearing and twisting are expressed by

$$\hat{n}_1(\mathbf{u},\mathbf{v},s) \to \pm\infty \quad \text{as } v_1 \to \pm\infty, \tag{8.8a}$$

$$\hat{n}_2(\mathbf{u},\mathbf{v},s) \to \pm\infty \quad \text{as } v_2 \to \pm\infty, \tag{8.8b}$$

$$\hat{m}_3(\mathbf{u},\mathbf{v},s) \to \pm\infty \quad \text{as } u_3 \to \pm\infty. \tag{8.8c}$$

The formulation of comparable restrictions on $\hat{m}_1$, $\hat{m}_2$, and $\hat{n}_3$ requires more care because the corresponding strains u_1, u_2, v_3 are confined to the open convex cone $\mathcal{V}(s)$ (see Sec. 6). We say that the strain $(\mathbf{u},\mathbf{v})$ is *extreme* iff either $|(\mathbf{u},\mathbf{v})| \equiv \sqrt{|\mathbf{u}|^2+|\mathbf{v}|^2} = \infty$ or $(\mathbf{u},\mathbf{v}) \in \partial\mathcal{V}(s)$. We adopt the general constitutive principle that extreme strains must be accompanied by an infinite resultant, i.e., if $(\mathbf{u},\mathbf{v})$ is extreme, then $|(\hat{\mathbf{m}}(\mathbf{u},\mathbf{v},s),\hat{\mathbf{n}}(\mathbf{u},\mathbf{v},s))| = \infty$. Condition (8.3) specifies how this unboundedness is to be attained: The line (8.6) intersects the convex set $\mathcal{V}(s)$ either nowhere, or on a bounded open interval, or on an open half-line, or on a whole line. We impose the

8.9. Coercivity Hypothesis. *The function* (8.7) *approaches* $\pm\infty$ *as* σ *approaches an extreme of its interval, with the sign on* ∞ *dictated by the monotonicity of* (8.7).

This condition includes (8.8). Let us fix $(\mathbf{u}_0,\mathbf{v}_0)$ in $\mathcal{V}(s)$. An hypothesis inspired directly by (8.3) and related to Hypothesis 8.9 is the coercivity condition

$$\frac{\Phi(\mathbf{u},\mathbf{v},\mathbf{u}_0,\mathbf{v}_0,s)}{\sqrt{|\mathbf{u}-\mathbf{u}_0|^2+|\mathbf{v}-\mathbf{v}_0|^2}} \to \infty \quad \text{as } (\mathbf{u},\mathbf{v}) \text{ becomes extreme.} \tag{8.10}$$

In Theorem XIX.2.30, we show that (8.3) and either Hypothesis 8.9 or (8.10) imply that if $\mathbf{m},\mathbf{n},s$ are given, then the algebraic equations

$$\mathbf{m} = \hat{\mathbf{m}}(\mathbf{u},\mathbf{v},s), \quad \mathbf{n} = \hat{\mathbf{n}}(\mathbf{u},\mathbf{v},s) \tag{8.11}$$

can be uniquely solved for $(\mathbf{u},\mathbf{v})$. The uniqueness is an immediate consequence of (8.3): If (8.11) had two solutions $(\mathbf{u}_1,\mathbf{v}_1)$ and $(\mathbf{u}_2,\mathbf{v}_2)$, then the left-hand side of (8.3) would vanish, forcing these two solutions to be identical. (We prove the existence statement specialized for hyperelastic rods at the end of this section.) The solution, denoted by

$$\mathbf{u} = \hat{\mathbf{u}}(\mathbf{m},\mathbf{n},s), \quad \mathbf{v} = \hat{\mathbf{v}}(\mathbf{m},\mathbf{n},s), \tag{8.12}$$

generates a set of constitutive equations equivalent to (7.14). Moreover, if $\hat{\mathbf{m}}(\cdot,\cdot,s)$ and $\hat{\mathbf{n}}(\cdot,\cdot,s)$ are continuously differentiable and satisfy the slightly stronger requirement (8.1), then the classical inverse function theorem implies that $\hat{\mathbf{u}}(\cdot,\cdot,s)$ and $\hat{\mathbf{v}}(\cdot,\cdot,s)$ are continuously differentiable. In this case, Proposition 8.23, proved below, implies that

$$\begin{pmatrix} \partial\hat{\mathbf{u}}/\partial\mathbf{m} & \partial\hat{\mathbf{u}}/\partial\mathbf{n} \\ \partial\hat{\mathbf{v}}/\partial\mathbf{m} & \partial\hat{\mathbf{v}}/\partial\mathbf{n} \end{pmatrix} \quad \text{is positive-definite.} \tag{8.13}$$

The following variant of these results is also useful. Condition (8.3) and either Hypothesis 8.9 or (8.10) imply that if $u_1, u_2, m_3, \mathsf{n}, s$ are given, then the algebraic equations

$$m_3 = \hat{m_3}(\mathsf{u}, \mathsf{v}, s), \quad \mathsf{n} = \hat{\mathsf{n}}(\mathsf{u}, \mathsf{v}, s) \tag{8.14}$$

have a unique solution

$$u_3 = u_3^\sharp(u_1, u_2, m_3, \mathsf{n}, s), \quad \mathsf{v} = \mathsf{v}^\sharp(u_1, u_2, m_3, \mathsf{n}, s). \tag{8.15}$$

Define

$$m_\beta^\sharp(u_1, u_2, m_3, \mathsf{n}, s) \equiv \hat{m_\beta}\big(u_1, u_2, u_3^\sharp(u_1, u_2, m_3, \mathsf{n}, s), \mathsf{v}^\sharp(u_1, u_2, m_3, \mathsf{n}, s), s\big). \tag{8.16}$$

Then (8.15) and

$$m_\beta = m_\beta^\sharp(u_1, u_2, m_3, \mathsf{n}, s) \tag{8.17}$$

generate a set of constitutive equations equivalent to (7.14). Moreover, if $\hat{\mathsf{m}}(\cdot, \cdot, s)$ and $\hat{\mathsf{n}}(\cdot, \cdot, s)$ are continuously differentiable and satisfy (8.1), then the classical Inverse-Function Theorem implies that the functions of (8.15) and (8.17) are continuously differentiable and that

$$\begin{pmatrix} \partial m_1^\sharp/\partial u_1 & \partial m_1^\sharp/\partial u_2 & \partial m_1^\sharp/\partial m_3 & \partial m_1^\sharp/\partial \mathsf{n} \\ \partial m_2^\sharp/\partial u_1 & \partial m_2^\sharp/\partial u_2 & \partial m_2^\sharp/\partial m_3 & \partial m_2^\sharp/\partial \mathsf{n} \\ \partial u_3^\sharp/\partial u_1 & \partial u_3^\sharp/\partial u_2 & \partial u_3^\sharp/\partial m_3 & \partial u_3^\sharp/\partial \mathsf{n} \\ \partial \mathsf{v}^\sharp/\partial u_1 & \partial \mathsf{v}^\sharp/\partial u_2 & \partial \mathsf{v}^\sharp/\partial m_3 & \partial \mathsf{v}^\sharp/\partial \mathsf{n} \end{pmatrix} \quad \text{is positive-definite.} \tag{8.18}$$

We now furnish a general proof of (8.13) and (8.18). Let a and b be non-negative integers. Let $\mathcal{V}$ be an open convex subset of $\mathbb{R}^a \times \mathbb{R}^b$. (We define $\mathbb{R}^0$ to be the empty set.) Let $\mathcal{V} \ni (\mathsf{x}, \mathsf{y}) \mapsto \big(\hat{\mathsf{f}}(\mathsf{x}, \mathsf{y}), \hat{\mathsf{g}}(\mathsf{x}, \mathsf{y})\big) \in \mathbb{R}^a \times \mathbb{R}^b$ be continuously differentiable and uniformly monotone so that

$$\mathsf{a} \cdot \frac{\partial \hat{\mathsf{f}}}{\partial \mathsf{x}} \cdot \mathsf{a} + \mathsf{a} \cdot \frac{\partial \hat{\mathsf{f}}}{\partial \mathsf{y}} \cdot \mathsf{b} + \mathsf{b} \cdot \frac{\partial \hat{\mathsf{g}}}{\partial \mathsf{x}} \cdot \mathsf{a} + \mathsf{b} \cdot \frac{\partial \hat{\mathsf{g}}}{\partial \mathsf{y}} \cdot \mathsf{b} > 0 \quad \forall\, (\mathsf{a}, \mathsf{b}) \in \mathbb{R}^a \times \mathbb{R}^b \quad \text{with} \quad \mathsf{a} \cdot \mathsf{a} + \mathsf{b} \cdot \mathsf{b} \neq 0. \tag{8.19}$$

For each f in the range of $\hat{\mathsf{f}}(\cdot, \mathsf{y})$, the algebraic equation

$$\mathsf{f} = \hat{\mathsf{f}}(\mathsf{x}, \mathsf{y}) \tag{8.20}$$

has a unique solution for x:

$$\mathsf{x} = \mathsf{x}^\sharp(\mathsf{f}, \mathsf{y}). \tag{8.21}$$

We now define

$$\mathsf{g}^\sharp(\mathsf{f}, \mathsf{y}) \equiv \hat{\mathsf{g}}\big(\mathsf{x}^\sharp(\mathsf{f}, \mathsf{y}), \mathsf{y}\big). \tag{8.22}$$

8.23. Proposition. *Under these conditions,*

$$\begin{pmatrix} \partial \mathsf{x}^\sharp/\partial \mathsf{f} & \partial \mathsf{x}^\sharp/\partial \mathsf{y} \\ \partial \mathsf{g}^\sharp/\partial \mathsf{f} & \partial \mathsf{g}^\sharp/\partial \mathsf{y} \end{pmatrix} \quad \textit{is positive-definite.} \tag{8.24}$$

Proof. The definitions of $\mathsf{x}^\sharp$ and $\mathsf{g}^\sharp$ yield the identities

$$\mathsf{f} = \hat{\mathsf{f}}\big(\mathsf{x}^\sharp(\mathsf{f}, \mathsf{y}), \mathsf{y}\big), \quad \hat{\mathsf{g}}(\mathsf{x}, \mathsf{y}) = \mathsf{g}^\sharp\big(\hat{\mathsf{f}}(\mathsf{x}, \mathsf{y}), \mathsf{y}\big), \tag{8.25}$$

which imply that

$$(8.26)\qquad \begin{aligned} \mathsf{I} &= (\partial\hat{\mathsf{f}}/\partial\mathsf{x})\cdot(\partial\mathsf{x}^\sharp/\partial\mathsf{f}), & \mathsf{O} &= (\partial\hat{\mathsf{f}}/\partial\mathsf{x})\cdot(\partial\mathsf{x}^\sharp/\partial\mathsf{y}) + \partial\hat{\mathsf{f}}/\partial\mathsf{y},\\ \partial\hat{\mathsf{g}}/\partial\mathsf{x} &= (\partial\mathsf{g}^\sharp/\partial\mathsf{f})\cdot(\partial\hat{\mathsf{f}}/\partial\mathsf{x}), & \partial\hat{\mathsf{g}}/\partial\mathsf{y} &= (\partial\mathsf{g}^\sharp/\partial\mathsf{f})\cdot(\partial\hat{\mathsf{f}}/\partial\mathsf{y}) + \partial\mathsf{g}^\sharp/\partial\mathsf{y}.\end{aligned}$$

(Here I represents the identity operator on $\mathbb{R}^a$ and O represents the zero operator from $\mathbb{R}^b$ to $\mathbb{R}^a$.) We substitute

$$(8.27)\qquad \mathsf{a} = (\partial\mathsf{x}^\sharp/\partial\mathsf{y})\cdot\mathsf{b} + (\partial\mathsf{x}^\sharp/\partial\mathsf{f})\cdot\mathsf{c}$$

into (8.19) and use (8.26) to obtain

$$(8.28)\qquad \mathsf{c}\cdot\frac{\partial\mathsf{x}^\sharp}{\partial\mathsf{f}}\cdot\mathsf{c} + \mathsf{c}\cdot\frac{\partial\mathsf{x}^\sharp}{\partial\mathsf{y}}\cdot\mathsf{b} + \mathsf{b}\cdot\frac{\partial\mathsf{g}^\sharp}{\partial\mathsf{f}}\cdot\mathsf{c} + \mathsf{b}\cdot\frac{\partial\mathsf{g}^\sharp}{\partial\mathsf{y}}\cdot\mathsf{b} > 0 \quad \forall\,(\mathsf{c},\mathsf{b})\in\mathbb{R}^a\times\mathbb{R}^b,$$
$$\text{with}\quad \mathsf{c}\cdot\mathsf{c}+\mathsf{b}\cdot\mathsf{b}\neq 0. \quad \square$$

This proof simplifies that of Antman (1974b). Condition (8.13) follows from the identifications $\mathsf{f} = (\mathsf{m},\mathsf{n})$ and $\mathsf{x} = (\mathsf{u},\mathsf{v})$ with g and y vacuous. Condition (8.18) follows from the identifications $\mathsf{f} = (m_3,\mathsf{n})$, $\mathsf{g} = (m_1,m_2)$, $\mathsf{x} = (u_3,\mathsf{v})$, $\mathsf{y} = (u_1,u_2)$.

If the rod is hyperelastic, so that (7.16) holds, then the constitutive functions of (8.12) and of (8.15) and (8.17) can each be expressed as the gradient of a scalar-valued function, as we now show for the variables entering Proposition 8.22. Suppose that there is a scalar-valued function $(\mathsf{x},\mathsf{y})\mapsto W(\mathsf{x},\mathsf{y})$ such that

$$(8.29)\qquad \hat{\mathsf{f}} = \partial W/\partial\mathsf{x}, \quad \hat{\mathsf{g}} = \partial W/\partial\mathsf{y}$$

and such that the hypotheses of Proposition 8.23 hold. We make the Legendre transformation

$$(8.30)\qquad W^\sharp(\mathsf{f},\mathsf{y}) \equiv \mathsf{f}\cdot\mathsf{x}^\sharp(\mathsf{f},\mathsf{y}) - W\big(\mathsf{x}^\sharp(\mathsf{f},\mathsf{y}),\mathsf{y}\big).$$

Thus

$$(8.31)\qquad \mathsf{x}^\sharp = \partial W^\sharp/\partial\mathsf{f}, \quad \mathsf{g}^\sharp = -\partial W^\sharp/\partial\mathsf{y}.$$

In particular, if (7.16) holds, then the constitutive functions of (8.15) and (8.17) can be put into the form

$$(8.32)\qquad m_\alpha^\sharp = -\partial W^\sharp/\partial u_\alpha, \quad u_3^\sharp = \partial W^\sharp/\partial m_3, \quad \mathsf{v}^\sharp = \partial W^\sharp/\partial\mathsf{n}$$

where

$$(8.33)\qquad \begin{aligned} W^\sharp(u_1,u_2,m_3,\mathsf{n},s) \equiv{}& m_3 u_3^\sharp(u_1,u_2,m_3,\mathsf{n},s) + \mathsf{n}\cdot\mathsf{v}^\sharp(u_1,u_2,m_3,\mathsf{n},s)\\ &- W\big(u_1,u_2,u_3^\sharp(u_1,u_2,m_3,\mathsf{n},s),\mathsf{v}^\sharp(u_1,u_2,m_3,\mathsf{n},s),s\big).\end{aligned}$$

8.34. Exercise. Show that (7.16) and (8.1) imply that W is strictly convex when $\hat{\mathsf{m}}$ and $\hat{\mathsf{n}}$ are differentiable. Discuss the convexity of $\pm W^\sharp$ of (8.30) when $(\hat{\mathsf{f}},\hat{\mathsf{g}})$ is strictly convex and differentiable.

A useful restriction on constitutive equations (7.31) for viscoelastic rods ensuring that frictional effects increase with the strain rates is that:

$$(8.35)\qquad (\dot{\mathsf{u}},\dot{\mathsf{v}}) \mapsto \big(\hat{\mathsf{m}}(\mathsf{u},\mathsf{v},\dot{\mathsf{u}},\dot{\mathsf{v}},s),\hat{\mathsf{n}}(\mathsf{u},\mathsf{v},\dot{\mathsf{u}},\dot{\mathsf{v}},s)\big) \text{ is uniformly monotone.}$$

This condition ensures that the equations of motion have a parabolic character, so that many effective mathematical tools are available for their analysis. When (8.35) holds, the monotonicity restrictions on $\hat{\mathbf{m}}(\cdot,\cdot,\dot{\mathbf{u}},\dot{\mathbf{v}},s)$, $\hat{\mathbf{n}}(\cdot,\cdot,\dot{\mathbf{u}},\dot{\mathbf{v}},s)$ are not so crucial (see Dafermos (1969), Andrews & Ball (1982), and Pego (1987)).

We now prove the equivalence of (8.11) and (8.12) for hyperelastic rods, for which (7.16) holds. We assume that $W(\cdot,\cdot,s)$ is continuously differentiable and satisfies the strengthened coercivity condition that for fixed $(\mathbf{u}_0,\mathbf{v}_0)$ in $\mathcal{V}(s)$,

$$(8.36)\qquad \frac{W(\mathbf{u},\mathbf{v},s)}{\sqrt{|\mathbf{u}-\mathbf{u}_0|^2+|\mathbf{v}-\mathbf{v}_0|^2}}\to\infty \quad \text{as } (\mathbf{u},\mathbf{v}) \text{ becomes extreme.}$$

(This condition says that $W(\cdot,\cdot,s)$ has superlinear growth at infinity.) We further assume that (8.3) holds. It says that $W(\cdot,\cdot,s)$ is strictly convex on $\mathcal{V}(s)$. Let $\mathbf{m}$, $\mathbf{n}$, s be given. Then (7.16) implies that (8.11) is equivalent to the vanishing of the derivative of

$$(8.37)\qquad (\mathbf{u},\mathbf{v})\mapsto \Omega(\mathbf{u},\mathbf{v},s)\equiv W(\mathbf{u},\mathbf{v},s)-\mathbf{m}\cdot\mathbf{u}-\mathbf{n}\cdot\mathbf{u}.$$

We prove that (8.11) can be solved for $(\mathbf{u},\mathbf{v})$ by showing that (8.37) has a minimum on the open set $\mathcal{V}(s)$. We write

$$(8.38)\qquad \Omega(\mathbf{u},\mathbf{v},s)=\sqrt{|\mathbf{u}-\mathbf{u}_0|^2+|\mathbf{v}-\mathbf{v}_0|^2}\left\{\frac{W(\mathbf{u},\mathbf{v},,s)}{\sqrt{|\mathbf{u}-\mathbf{u}_0|^2+|\mathbf{v}-\mathbf{v}_0|^2}}\right.$$
$$\left.-\frac{\mathbf{m}\cdot\mathbf{u}}{\sqrt{|\mathbf{u}-\mathbf{u}_0|^2+|\mathbf{v}-\mathbf{v}_0|^2}}-\frac{\mathbf{n}\cdot\mathbf{v}}{\sqrt{|\mathbf{u}-\mathbf{u}_0|^2+|\mathbf{v}-\mathbf{v}_0|^2}}\right\}.$$

Condition (8.36) implies that the first term in braces approaches ∞, while the other two terms stay bounded, as $(\mathbf{u},\mathbf{v})$ becomes extreme. Thus

$$(8.39)\qquad \Omega(\mathbf{u},\mathbf{v},s)\to\infty \quad \text{as } (\mathbf{u},\mathbf{v}) \text{ becomes extreme.}$$

We seek to minimize (8.37) on the set $\{(\mathbf{u},\mathbf{v})\colon \Omega(\mathbf{u},\mathbf{v},s)\le\Omega(\mathbf{u}_0,\mathbf{v}_0,s)\}$. This set is closed because (8.37) is continuous and it is bounded because if not, (8.39) would be violated. Thus (8.37) has an absolute minimizer, which lies in this set. Since this minimizer is necessarily an interior point of (the open set) $\mathcal{V}(s)$, the derivative of (8.37) must vanish at the minimizer. Thus (8.11) has a solution. The uniqueness follows from the proof given above.

We conclude this section with a useful and elementary result, which is not well known. It implies that the principal subdeterminants of the derivative of a uniformly monotone function are each positive-definite. Thus, if (8.1) holds, then the principal subdeterminants of the matrix of (8.1) are each positive.

8.40. Theorem. *Let* $\mathbf{A}$ *be a real* $n\times n$ *matrix that is positive-definite (but not necessarily symmetric). Then the principal subdeterminants of* $\mathbf{A}$ *are each positive.*

This result is standard for symmetric, positive-definite matrices. We use this fact in the ensuing proof.

Proof. It is enough to show that $\det \mathsf{A} > 0$, because the same argument can be applied to each principal subdeterminant. Observe that the symmetric tensor $\mathsf{A}^* \cdot \mathsf{A}$ is positive-definite, for if not, there would be a $\mathbf{u} \neq \mathbf{0}$ such that $0 \geq \mathbf{u} \cdot \mathsf{A}^* \cdot \mathsf{A} \cdot \mathbf{u} \equiv |\mathsf{A} \cdot \mathbf{u}|^2$, which implies that $\mathbf{u} \cdot \mathsf{A} \cdot \mathbf{u} = 0$, a contradiction. Therefore, $[\det \mathsf{A}]^2 = \det(\mathsf{A}^* \cdot \mathsf{A}) > 0$, by results for symmetric matrices. Now let $2\mathsf{B} = \mathsf{A} + \mathsf{A}^*$, $2\mathsf{C} = \mathsf{A} - \mathsf{A}^*$. We know that $\det \mathsf{A} \equiv \det(\mathsf{B} + \mathsf{C}) \neq 0$, that $\det \mathsf{B} > 0$, and that $\mathsf{C} \mapsto \det(\mathsf{B} + \mathsf{C}) \neq 0$ is continuous. Thus $\det(\mathsf{B} + \mathsf{C}) > 0$. □

An alternative proof can be based on the observation that the equation $\det(\mathsf{B} + \mathsf{C}) = 0$, which says that the eigenvalues of the skew C with respect to the symmetric, positive-definite B are each equal to -1, is inconsistent with the fact that such eigenvalues must be pure imaginary numbers.

9. Transverse Isotropy

We shall see that the equations for special Cosserat rods often readily yield detailed information about solutions if the rods are transversely isotropic. In this section we define this concept and obtain a convenient mathematical characterization of it. Although the mathematical theory is closely related to that of frame-indifference, the concepts are quite distinct.

Many materials, such as rubber or steel (at least on the macroscopic level), having no preferred material directions are called *isotropic*. Other materials, such as wood and animal muscle, having pronounced grain patterns, are called *aeolotropic* or *anisotropic*. (We employ the former unwieldy term because the Greek prefix *an* of the latter term is easily confused with the English article *an* in speech.) Aeolotropic materials are readily manufactured by imbedding reinforcing fibers in an isotropic material. To test whether a material is isotropic, we may subject a small ball of it to a given system of forces and observe the consequent deformation (possibly time-dependent) it suffers. Next we rotate the ball before applying the same system of forces to it and observing the deformation. (See Fig. 9.1.) If for each fixed force system, these deformations are independent of rotations and reflections, then the material is *isotropic*. Otherwise, it is *aeolotropic*. An essentially equivalent test is obtained by subjecting rotated states of the ball to a fixed deformation and asking whether the force system needed to maintain each such deformation is independent of the rotation. (The first test is associated with constitutive functions giving strains in terms of force intensities (stresses), whereas the second is associated with constitutive functions giving force intensities in terms of strains. Constitutive functions of the latter type are mathematically more convenient. In general, they are not invertible.) Instead of rotating the body in these tests, we could alternatively hold the body fixed and rotate the testing apparatus. We could also characterize isotropy by the requirement that the experiment in Fig. 9.1 be unaffected by the choice of markings on the ball. If the response of the material is not the same for all orthogonal transformations, but only for those that leave a given material direction fixed, then the material is called *transversely isotropic*. Note that these

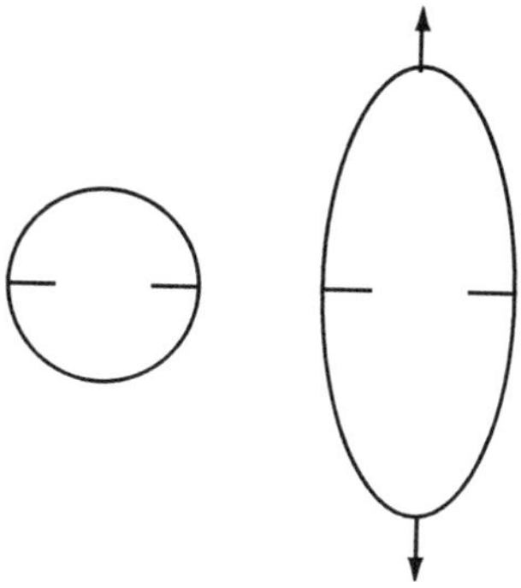

Fig. 9.1a. Deformation of an aeolotropic ball under a given force system.

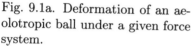

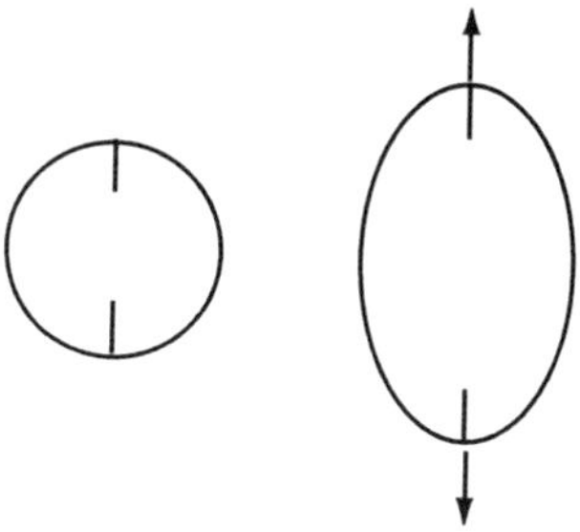

Fig. 9.1b. Deformation of the rotated ball by the same force system.

notions of isotropy are purely local in that they describe the behavior at material points.

In the theory of Cosserat rods, the directors may be regarded as identifying material lines. Thus they play a role similar to that of the markings on the ball in Fig. 9.1: They are objects that can be rotated in experiments testing for isotropy. Alternatively, they can register the rotation of testing apparatus. Since the role of $\boldsymbol{d}_3$ is so distinct from those of $\boldsymbol{d}_1$ and $\boldsymbol{d}_2$, there is no point in letting $\boldsymbol{d}_3$ vary in testing for isotropy. Thus the kind of isotropy relevant for rod theories is transverse isotropy.

In a rod theory, the constitutive equations characterize not only the material of the three-dimensional body, but also the shape of the body. In the classical linear theories and in Kirchhoff's Kinetic Analogy (1859), which we discuss in Sec. 16, the shape of the cross section enters only through the moments of inertia of area and mass of the cross section (see (4.6), (4.7)). The larger the principal moments of inertia of the cross section, the greater the stiffness with which these rods resist both bending and twisting. In these classical theories, a straight rod having a common line of centroids for both mass and area has transverse isotropy if and only if the principal area and mass moments of inertia of the cross section about the line of centroids are respectively equal. (The principal mass moments of inertia are the eigenvalues of $(\rho J_{\alpha\beta})$ when $\varphi_\alpha = x^a$.) See Fig. 9.2. Since the response of a nonlinearly elastic rod cannot be characterized solely by these moments of inertia, we must characterize transverse isotropy by a mathematical statement that reflects the physical notions discussed above.

Buzano, Geymonat, & Poston (1985) have studied the buckling in space of a cylindrical rod with an equilateral triangular cross section. Its principal moments of inertia are equal, so that it meets the classical criterion for transverse isotropy. But the theory they employ has enough sophistication to distinguish between a circular cross section and an equilateral triangular cross section. The buckling patterns they obtain exhibit some striking differences from those for rods with circular cross sections. See Antman & Marlow (1992) for a simpler example within the Cosserat theory.

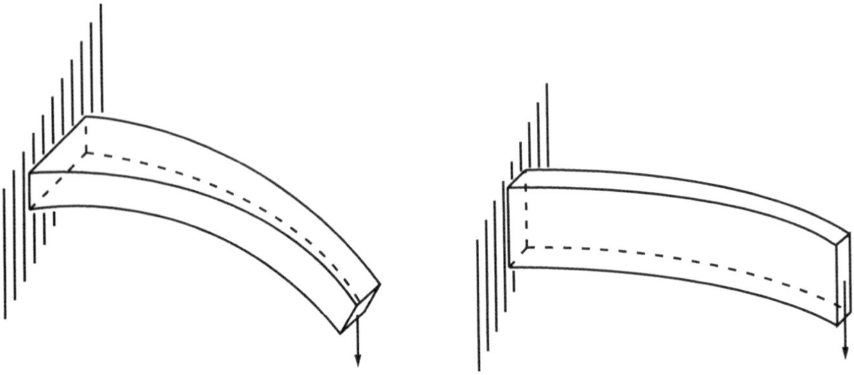

Fig. 9.2. Different deflections caused in the same rod by the same force system applied to configurations differing by a rotation. As the shape of the cross section suggests, the different response indicates that the rod is not transversely isotropic.

Now we obtain precise and useful mathematical characterizations of transverse isotropy for elastic Cosserat rods. Let $\{\mathring{\boldsymbol{d}}_k\}$ be the values of $\{\boldsymbol{d}_k\}$ in the reference configuration. Let $\boldsymbol{A}(s)$ be an orthogonal tensor that leaves $\mathring{\boldsymbol{d}}_3(s)$ unchanged. We introduce another pair of directors $\{\mathring{\boldsymbol{d}}^\sharp_\alpha(s)\}$ in $\operatorname{span}\{\mathring{\boldsymbol{d}}_\alpha(s)\}$ by

$$(9.3)\qquad \mathring{\boldsymbol{d}}^\sharp_k(s) = \boldsymbol{A}^*(s)\cdot\mathring{\boldsymbol{d}}_k(s) \quad\text{with}\quad \mathring{\boldsymbol{d}}^\sharp_3(s) = \boldsymbol{A}^*(s)\cdot\mathring{\boldsymbol{d}}_3(s) = \mathring{\boldsymbol{d}}_3(s).$$

We want to study the invariance of constitutive equations under transformations (9.3). But the directors $\mathring{\boldsymbol{d}}_k$ do not intervene in (7.14). We accordingly formulate this invariance under an induced transformation of the directors $\boldsymbol{d}_k$.

We know that there is a proper orthogonal $\boldsymbol{B}(s,t)$ such that

$$(9.4)\qquad \boldsymbol{d}_k(s,t) = \boldsymbol{B}(s,t)\cdot\mathring{\boldsymbol{d}}_k(s).$$

We define

$$(9.5)\qquad \boldsymbol{d}^\sharp_k(s,t) = \boldsymbol{B}(s,t)\cdot\mathring{\boldsymbol{d}}^\sharp_k(s).$$

Then

$$(9.6)\quad \boldsymbol{d}^\sharp_k(s,t) = [\boldsymbol{B}(s,t)\cdot\boldsymbol{A}^*(s)\cdot\boldsymbol{B}^*(s,t)]\cdot\boldsymbol{d}_k(s,t) \equiv \boldsymbol{Q}^*(s,t)\cdot\boldsymbol{d}_k(s,t),$$

so that

$$(9.7)\qquad \boldsymbol{d}_3 = \boldsymbol{d}^\sharp_3 = \boldsymbol{Q}^*\cdot\boldsymbol{d}_3.$$

Equations (9.6) and (9.7) say that $\boldsymbol{Q}$ is an orthogonal tensor that leaves $\boldsymbol{d}_3$ invariant. These considerations show that any statement about invariance under orthogonal transformations $\boldsymbol{A}^*$ of $\{\mathring{\boldsymbol{d}}_k\}$ that leave $\mathring{\boldsymbol{d}}_3$ unchanged is equivalent to a statement about invariance under orthogonal transformations $\boldsymbol{Q}^*$ of $\{\boldsymbol{d}_k\}$ that leave $\boldsymbol{d}_3$ unchanged.

We now characterize constitutive equations for transversely isotropic elastic Cosserat rods as those that are invariant under the choice of $\{\mathring{\boldsymbol{d}}_k\}$ (i.e., under the choice of markings). These equations are therefore invariant under $\boldsymbol{A}^*$ and $\boldsymbol{Q}^*$. Let a deformation $(\boldsymbol{r}_s, \boldsymbol{u})$ be given. By virtue of (7.14), this deformation produces the resultants

$$\hat{m}_k(\boldsymbol{u}\cdot\boldsymbol{d}_1,\dots)\boldsymbol{d}_k, \quad \text{etc.} \tag{9.8}$$

We can also describe this deformation with respect to the basis $\{\boldsymbol{d}_k^\sharp\}$ and obtain resultants

$$\hat{m}_k^\sharp(\boldsymbol{u}\cdot\boldsymbol{d}_1^\sharp,\dots)\boldsymbol{d}_k^\sharp, \quad \text{etc.,} \tag{9.9}$$

which equal (9.8). There is a sharp sign on the constitutive functions in (9.9) because we have no assurance that if the arguments are the same, then the corresponding components of the resultants are the same. The material is *transversely isotropic* iff $\hat{m}_k^\sharp = \hat{m}_k$. In this case, the equality of (9.8) and (9.9) yields

$$\hat{m}_k(\boldsymbol{u}\cdot\boldsymbol{d}_1^\sharp,\dots) = \boldsymbol{d}_k^\sharp\cdot\boldsymbol{d}_l\,\hat{m}_l(\boldsymbol{u}\cdot\boldsymbol{d}_1,\dots), \quad \text{etc.,} \tag{9.10a}$$

or, equivalently,

$$\hat{m}_k(\boldsymbol{u}\cdot\boldsymbol{Q}^*\cdot\boldsymbol{d}_1,\dots) = \boldsymbol{d}_k\cdot\boldsymbol{Q}\cdot\boldsymbol{d}_l\,\hat{m}_l(\boldsymbol{u}\cdot\boldsymbol{d}_1,\dots), \quad \text{etc.} \tag{9.10b}$$

(We can regard the arguments $\boldsymbol{u}\cdot\boldsymbol{Q}^*\cdot\boldsymbol{d}_1,\dots$ of $\hat{m}_k$ on the left-hand side of (9.10) as describing a rotation taking $\boldsymbol{d}_k$ to $\boldsymbol{Q}^*\cdot\boldsymbol{d}_k$ followed by the deformation $(\boldsymbol{r}_s, \boldsymbol{u})$. Thus (9.10b) conforms to our second interpretation of a test for isotropy.) Let

$$Q_{ij}^* = Q_{ji} \equiv \boldsymbol{d}_i\cdot\boldsymbol{Q}^*\cdot\boldsymbol{d}_j = \boldsymbol{d}_i\cdot\boldsymbol{d}_j^\sharp \tag{9.11}$$

denote the components of $\boldsymbol{Q}^*$ with respect to the basis $\{\boldsymbol{d}_k\}$ (so that $\boldsymbol{Q} = \boldsymbol{d}_k\boldsymbol{d}_k^\sharp$). Then we can write (9.10b) as

(9.12a)
$$\hat{m}_\alpha(Q_{1\nu}u_\nu, Q_{2\nu}u_\nu, u_3, Q_{1\nu}v_\nu, Q_{2\nu}v_\nu, v_3, s) = Q_{\alpha\beta}\hat{m}_\beta(u_1,u_2,u_3,v_1,v_2,v_3,s),$$
(9.12b)
$$\hat{m}_3(Q_{1\nu}u_\nu, Q_{2\nu}u_\nu, u_3, Q_{1\nu}v_\nu, Q_{2\nu}v_\nu, v_3, s) = \hat{m}_3(u_1,u_2,u_3,v_1,v_2,v_3,s), \text{etc.,}$$

or, equivalently,

$$\hat{\mathsf{m}}(\mathsf{Q}\cdot\mathsf{u},\, \mathsf{Q}\cdot\mathsf{v},\, s) = \mathsf{Q}\cdot\hat{\mathsf{m}}(\mathsf{u},\mathsf{v},s), \quad \text{etc.,} \tag{9.13}$$

where

$$\mathbf{Q} \equiv \begin{pmatrix} Q_{11} & Q_{12} & 0 \\ Q_{21} & Q_{22} & 0 \\ 0 & 0 & 1 \end{pmatrix}. \tag{9.14}$$

9.14. Exercise. By making a judicious choice of $\boldsymbol{Q}$, use (9.12a) to prove that

$$\hat{m}_1(\mathsf{u}, \mathsf{v}, s) = 0 = \hat{n}_1(\mathsf{u}, \mathsf{v}, s) \quad \text{if} \quad u_1 = 0 = v_1, \quad \text{etc.,} \tag{9.15}$$

for a transversely isotropic rod.

9.16. Theorem. *The constitutive functions for a transversely isotropic elastic Cosserat rod have the form*

$$\hat{m}_\alpha(\mathsf{u}, \mathsf{v}, s) = m\big(J(\mathsf{u}, \mathsf{v}), s\big)u_\alpha + m^\times\big(J(\mathsf{u}, \mathsf{v}), s\big)v_\alpha, \tag{9.17a}$$

$$\hat{n}_\alpha(\mathsf{u}, \mathsf{v}, s) = n^\times\big(J(\mathsf{u}, \mathsf{v}), s\big)u_\alpha + n\big(J(\mathsf{u}, \mathsf{v}), s\big)v_\alpha, \tag{9.17b}$$

$$\hat{m}_3(\mathsf{u}, \mathsf{v}, s) = \check{m}_3\big(J(\mathsf{u}, \mathsf{v}), s\big), \tag{9.17c}$$

$$\hat{n}_3(\mathsf{u}, \mathsf{v}, s) = \check{n}_3\big(J(\mathsf{u}, \mathsf{v}), s\big) \tag{9.17d}$$

where

$$J(\mathsf{u}, \mathsf{v}) \equiv (u_\alpha u_\alpha,\ u_\alpha v_\alpha,\ v_\alpha v_\alpha,\ u_3,\ v_3). \tag{9.18}$$

The stored-energy function W for a transversely isotropic hyperelastic Cosserat rod depends only on $J(\mathsf{u}, \mathsf{v})$ and s.

Proof. Representations (9.17c,d) and that for the stored-energy function W follow from (9.12b) by a version of Cauchy's Representation Theorem 7.8 appropriate for $\mathbb{R}^2$. Since this theorem applies only to scalar-valued functions, we must reduce the proof of (9.17a,b) to that for such functions. Toward this end, we calculate that

$$(\mathsf{u}, \mathsf{v}) \mapsto \varepsilon_{\alpha\beta}\hat{m}_\alpha(\mathsf{u}, \mathsf{v}, s)u_\beta, \quad \varepsilon_{\alpha\beta}\hat{m}_\alpha(\mathsf{u}, \mathsf{v}, s)v_\beta \tag{9.19a,b}$$

have the same invariance as $\hat{m}_3$ in (9.12b) under *proper* orthogonal transformations. By Cauchy's Representation Theorem for hemitropic functions, (9.19) can be represented as functions of $\big(J(\mathsf{u}, \mathsf{v}), s\big)$. (Note that $T(\mathsf{u}, \mathsf{v})$, defined in (7.9), is trivial because $N = 2$.) If $u_1v_2 - u_2v_1 \neq 0$, then we can use Cramer's rule to solve for $\hat{m}_1$ and $\hat{m}_2$ as functions of the expressions on the right-hand side of (9.19), thereby obtaining (9.17a).

Now suppose that $u_1v_2 - u_2v_1 = 0$. Then u and v must have the form

$$\begin{pmatrix} u_1 \\ u_2 \end{pmatrix} = u\begin{pmatrix} \cos\theta \\ \sin\theta \end{pmatrix}, \quad \begin{pmatrix} v_1 \\ v_2 \end{pmatrix} = v\begin{pmatrix} \cos\theta \\ \sin\theta \end{pmatrix}. \tag{9.20}$$

When (9.20) holds, we can identify $\hat{m}_\alpha(\mathsf{u}, \mathsf{v}, s)$ with the left-hand side of (9.12a) by choosing

$$(9.21) \qquad \begin{pmatrix} u_1 \\ u_2 \end{pmatrix} = \begin{pmatrix} u \\ 0 \end{pmatrix}, \quad \begin{pmatrix} v_1 \\ v_2 \end{pmatrix} = \begin{pmatrix} v \\ 0 \end{pmatrix}, \quad \mathsf{Q} = \begin{pmatrix} \cos\theta & -\sin\theta \\ \sin\theta & \cos\theta \end{pmatrix}.$$

Thus (9.12a) and Ex. 9.14 imply that

(9.22)

$$\begin{aligned} \hat{m}_1(u\cos\theta, u\sin\theta, u_3, v\cos\theta, v\sin\theta, v_3, s) &= \cos\theta\hat{m}_1(u, 0, u_3, v, 0, v_3, s), \\ \hat{m}_2(u\cos\theta, u\sin\theta, u_3, v\cos\theta, v\sin\theta, v_3, s) &= \sin\theta\hat{m}_1(u, 0, u_3, v, 0, v_3, s), \end{aligned}$$

which has the form of (9.17a), although there are no unique representations for m and $m^\times$ in terms of $\hat{m}_1(u, 0, u_3, v, 0, v_3, s)$. The proof of (9.17b) is the same. □

9.23. Exercise. Prove that transversely isotropic viscoelastic materials (7.31) have representations analogous to (9.17) of the form

$$(9.24) \qquad \begin{aligned} \hat{m}_\alpha(\mathsf{u}, \mathsf{v}, \dot{\mathsf{u}}, \dot{\mathsf{v}}, s) &= m(J(\mathsf{u}, \mathsf{v}, \dot{\mathsf{u}}, \dot{\mathsf{v}}), s)u_\alpha + m^\times(J(\mathsf{u}, \mathsf{v}, \dot{\mathsf{u}}, \dot{\mathsf{v}}), s)v_\alpha \\ &\quad + \mu(J(\mathsf{u}, \mathsf{v}, \dot{\mathsf{u}}, \dot{\mathsf{v}}), s)\dot{u}_\alpha + \mu^\times(J(\mathsf{u}, \mathsf{v}, \dot{\mathsf{u}}, \dot{\mathsf{v}}), s)\dot{v}_\alpha, \end{aligned}$$

etc., where

(9.25)

$$\begin{aligned} &J(\mathsf{u}, \mathsf{v}, \dot{\mathsf{u}}, \dot{\mathsf{v}}) \\ &\equiv u_\alpha u_\alpha,\ u_\alpha v_\alpha,\ v_\alpha v_\alpha,\ u_\alpha \dot{u}_\alpha,\ u_\alpha \dot{v}_\alpha,\ v_\alpha \dot{u}_\alpha,\ v_\alpha \dot{v}_\alpha,\ \dot{u}_\alpha \dot{u}_\alpha,\ \dot{u}_\alpha \dot{v}_\alpha,\ \dot{v}_\alpha \dot{v}_\alpha,\ u_3,\ v_3,\ \dot{u}_3,\ \dot{v}_3. \end{aligned}$$

It is tempting to use a continuity argument to deduce the nonunique representation (9.17a) when $u_1 v_2 - u_2 v_1 = 0$ from that for the generic case that $u_1 v_2 - u_2 v_1 \neq 0$. The representation obtained by Cramer's rule in the latter case shows the difficulty with such an approach. A similar difficulty with regularity is apparent in (9.22) with $v = 0$, for then we deduce that

$$(9.26) \qquad m(J(\mathsf{u}, 0, 0, v_3), s) = \frac{\hat{m}_1(\sqrt{u_\alpha u_\alpha}, 0, u_3, 0, 0, v_3, s)}{\sqrt{u_\alpha u_\alpha}},$$

which could be ill behaved for $u_\alpha u_\alpha = 0$. Such troubles with regularity and uniqueness are typical of analyses involving invariants.

For a rod to be transversely isotropic, the inertia terms must obey invariance requirements like those imposed on the constitutive functions. Let (4.4) and (4.5) hold. Then the inertial terms are invariant under orthogonal $\boldsymbol{Q}$ that leaves $\boldsymbol{d}_3$ invariant if and only if

$$(9.27a) \qquad \rho I_1 = 0 = \rho I_2$$

and there is a scalar-valued function ρJ such that

$$(9.27b) \qquad \begin{pmatrix} \rho J_{11} & \rho J_{12} \\ \rho J_{21} & \rho J_{22} \end{pmatrix} = \rho J \begin{pmatrix} 1 & 0 \\ 0 & 1 \end{pmatrix}.$$

9.28. Exercise. Prove this statement.

We could include in our definition of transverse isotropy the reasonable requirement that

$$(9.29)\quad u_3 \mapsto \hat{m}_3(\mathbf{u}, \mathbf{v}, s) \quad \text{is odd,} \quad u_3 \mapsto \hat{m}_\alpha(\mathbf{u}, \mathbf{v}, s),\ \hat{\mathbf{n}}(\mathbf{u}, \mathbf{v}, s) \quad \text{are even.}$$

In this case, we replace u_3 in (9.18) with ${u_3}^2$ and multiply the right-hand side of (9.17c) by u_3.

We now show that *if the monotonicity condition* (8.1) *holds, then a transversely isotropic elastic rod must have a straight reference state.* Consider (9.17a) in the natural reference state:

$$(9.30)\qquad m(\mathring{u}_\alpha \mathring{u}_\alpha, 0, 0, \mathring{u}_3, 1, s)\mathring{u}_\alpha = 0.$$

Either $\mathring{u}_1 = 0 = \mathring{u}_2$ or $m(\mathring{u}_\alpha \mathring{u}_\alpha, 0, 0, \mathring{u}_3, 1, s) = 0$. Suppose that the first alternative does not hold, say that $\mathring{u}_1 \neq 0$. It follows that $u_1 \mapsto m(u_1^2 + \mathring{u}_2^2, 0, 0, \mathring{u}_3, 1, s)$ vanishes at $\pm\mathring{u}_1$. Thus $u_1 \mapsto \hat{m}_1(u_1, \mathring{u}_2, \mathring{u}_3, \mathring{\mathbf{v}}, s) = m(u_1^2 + \mathring{u}_2^2, 0, 0, \mathring{u}_3, 1, s)u_1$ cannot have a positive derivative, so that (8.1) cannot hold.

Note that representation (9.17) is not convenient for expressing monotonicity and growth conditions. As we shall see elsewhere in this book, such representations never are.

10. Uniform Rods. Singular Problems

An elastic special Cosserat rod is called *constitutively uniform* iff the functions $\hat{\mathbf{m}}$ and $\hat{\mathbf{n}}$ of (7.14) are independent of s. An analogous definition holds for a viscoelastic rod. If (4.14) and (4.15) hold, a special Cosserat rod is called *inertially uniform* iff ρA and the $\rho J_{\alpha\beta}$ are independent of s. A rod is called *uniform* iff it is both constitutively and inertially uniform. A real three-dimensional rod may fail to be uniform because (i) its material properties change along its length (i.e., the body in inhomogeneous), or (ii) its cross-sectional size and shape vary with s, or (iii) $\mathring{\mathbf{u}}$ (which is the only set of strains that need not be constant functions in the reference configuration) depends on s.

Let us show that (iii) causes nonuniformity for elastic rods. Suppose for simplicity that $\mathring{\mathbf{u}}$ is continuously differentiable and that $\mathring{\mathbf{u}}_s \neq \mathbf{0}$ (see Ex. 5.9). Note that (5.28) ensures that $\mathring{\mathbf{v}}$ is constant. Let the reference configuration be natural, so that

$$(10.1a)\qquad \hat{\mathbf{m}}(\mathring{\mathbf{u}}(s), \mathring{\mathbf{v}}, s) = \mathbf{0} \quad \forall\, s,$$

whence

$$(10.1b)\qquad \frac{\partial \hat{\mathbf{m}}}{\partial \mathbf{u}}(\mathring{\mathbf{u}}(s), \mathring{\mathbf{v}}, s) \cdot \mathring{\mathbf{u}}_s + \frac{\partial \hat{\mathbf{m}}}{\partial s}(\mathring{\mathbf{u}}(s), \mathring{\mathbf{v}}, s) = \mathbf{0} \quad \forall\, s.$$

Were the rod uniform, then (10.1b) would imply that $[\partial \hat{\mathbf{m}}(\mathring{\mathbf{u}}, \mathring{\mathbf{v}})/\partial \mathbf{u}] \cdot \mathring{\mathbf{u}}_s = \mathbf{0}$. But the monotonicity condition (8.1) implies that the matrix $\partial \hat{\mathbf{m}}/\partial \mathbf{u}$ is nonsingular, so that (10.1) would imply that $\mathring{\mathbf{u}}_s = \mathbf{0}$, a contradiction.

The situation (i) is rare for rods manufactured for technological purposes. Consequently, it is reasonable to suppose that the functions ρA and the $\rho J_{\alpha\beta}$ are in fact products of a constant ρ, interpreted as the mass density per unit reference volume, with functions A and $J_{\alpha\beta}$, interpreted as generalized cross-sectional areas and moments of inertia of area of the reference configuration. In particular (cf. (3.10) and (4.7)), we can take

(10.2)
$$A(s) \equiv \int_{\mathcal{A}(s)} j(\mathbf{x})\, dx^1\, dx^2, \quad J_{\gamma\delta}(s) \equiv \varepsilon_{\gamma\mu}\varepsilon_{\delta\nu} \int_{\mathcal{A}(s)} \varphi_\mu(\mathbf{x})\varphi_\nu(\mathbf{x}) j(\mathbf{x})\, dx^1\, dx^2.$$

In the same spirit, we may take

$$(10.3) \qquad \begin{aligned} \hat{\mathbf{n}}(\mathbf{u}, \mathbf{v}, s) &= A(s)\bar{\mathbf{n}}(\mathbf{u}, \mathbf{v}, s), \\ \hat{m}_\alpha(\mathbf{u}, \mathbf{v}, s) &= J_{\alpha\beta}(s)\overline{m}_\beta(\mathbf{u}, \mathbf{v}, s), \\ \hat{m}_3(\mathbf{u}, \mathbf{v}, s) &= J_{\alpha\alpha}(s)\overline{m}_3(\mathbf{u}, \mathbf{v}, s). \end{aligned}$$

Even if (ii) is the only source of nonuniformity, the functions $\bar{\mathbf{n}}$, $\overline{m}_\beta$, and $\overline{m}_3$ cannot in general be taken to be independent of s. Indeed, if the rod is hyperelastic, then the symmetry of (7.17) would force severe restrictions on the constitutive functions. A fully convincing demonstration of the naturalness of (10.3) must await the derivation in Chap. XIV of rod equations from the three-dimensional equations. Such a derivation is entirely in the spirit of Sec. 4, but depends upon the three-dimensional notion of stress and upon three-dimensional constitutive equations.

In the Kinetic Analogy of Kirchhoff (1859), which is a special case of our Cosserat theory based on the constraint that $\boldsymbol{r}_s = \boldsymbol{d}_3$, the constitutive equations have the linear form

$$(10.4) \qquad \hat{m}_\alpha(\mathbf{u}, \mathbf{v}, s) = EJ_{\alpha\beta}(s)[u_\beta - \mathring{u}_\beta], \quad \hat{m}_3(\mathbf{u}, \mathbf{v}, s) = EJ_{\alpha\alpha}(s)u_3$$

where E is the (effective) elastic modulus. In linear theories of extensible, shearable elastic rods, such as that of Timoshenko (1921), the constitutive equations for $\mathbf{n}$ have the form

$$(10.5) \qquad \hat{n}_\alpha(\mathbf{u}, \mathbf{v}, s) = GA(s)v_\alpha, \quad \hat{n}_3(\mathbf{u}, \mathbf{v}, s) = EA(s)[v_3 - 1]$$

where G is the (effective) shear modulus. In these linear theories, the v_k are approximated by linear combinations of s-derivatives of the displacements.

Our purpose in using (10.3) is not to force our nonlinear constitutive functions into the mold of classical theories, but to characterize in an explicit and convenient form the dependence of these functions on s. Such a characterization proves to be very useful for the treatment of singular problems in which $\hat{\mathbf{m}}(\cdot, \cdot, s) \to \mathbf{0}$ and $\hat{\mathbf{n}}(\cdot, \cdot, s) \to \mathbf{0}$ as $s \to s_1$ or s_2. We accordingly assume that the coefficients of the barred functions in (10.3) exactly capture the limit process. Thus $\bar{m}_1(\cdot, \cdot, s) \not\to 0$ as $s \to s_1$ or s_2, etc.

The considerations of this section apply to rods of any material.

10.6 Exercise. Suppose that $\mathring{\mathsf{u}}_s \neq \mathbf{0}$. Prove that constitutive equations satisfying the monotonicity condition (8.1) cannot have the form

$$\hat{m}_\alpha(\mathsf{u}, \mathsf{v}, s) = J_{\alpha\beta}\bar{m}_\beta(\mathsf{u}, \mathsf{v}), \quad \hat{m}_3(\mathsf{u}, \mathsf{v}, s) = J_{\alpha\alpha}\bar{m}_3(\mathsf{u}, \mathsf{v}). \tag{10.7}$$

11. Representations for the Directors in Terms of Euler Angles

The use of any six components of the directors $\boldsymbol{d}_1$ and $\boldsymbol{d}_2$ to define a configuration of a special Cosserat rod is awkward because these six components are subject to three constraints expressing the orthonormality of $\boldsymbol{d}_1$ and $\boldsymbol{d}_2$. In this section we show how $\boldsymbol{d}_1$ and $\boldsymbol{d}_2$ can be determined from just three Euler angles θ, ϕ, ψ in such a way that the constraint of orthonormality is automatically satisfied (at the cost of introducing a polar singularity that can be the source of serious analytical difficulty).

Let $\{\boldsymbol{j}_l\}$ be a fixed orthonormal basis for $\mathbb{E}^3$. We describe $\boldsymbol{d}_3$ by spherical coordinates:

$$\boldsymbol{d}_3 = \sin\theta\,\boldsymbol{h} + \cos\theta\,\boldsymbol{j}_3, \quad \boldsymbol{h} \equiv \cos\phi\,\boldsymbol{j}_1 + \sin\phi\,\boldsymbol{j}_2. \tag{11.1}$$

$\boldsymbol{d}_1$ and $\boldsymbol{d}_2$ are confined to the plane perpendicular to $\boldsymbol{d}_3$. If we can identify a distinguished line in this plane, then we can fix the locations of $\boldsymbol{d}_1$ and $\boldsymbol{d}_2$ by specifying the angle that one of them makes with respect to this line. Our development is based on this simple observation and on the corresponding construction shown in Fig. 11.2.

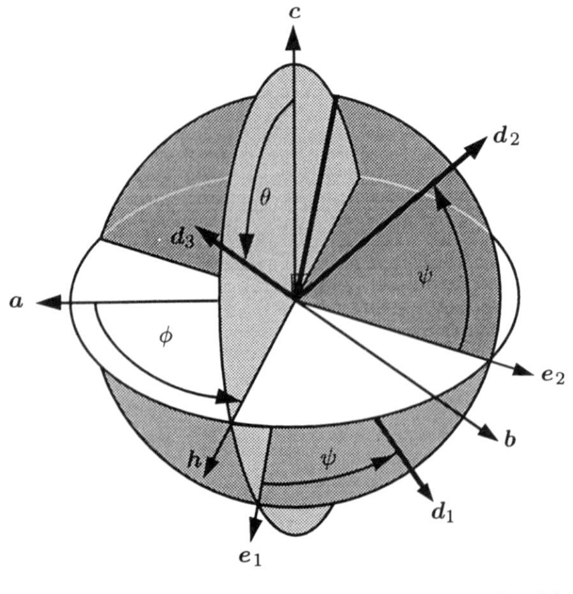

Fig. 11.2. The relationship of the directors to the fixed basis via the Euler angles.

If $\sin\theta \neq 0$, then the plane perpendicular to $\boldsymbol{d}_3$ intersects the $\{\boldsymbol{j}_1, \boldsymbol{j}_2\}$-plane along the line perpendicular to both $\boldsymbol{j}_3$ and $\boldsymbol{d}_3$. A unit vector along this *line of nodes* is

$$\boldsymbol{k}_2 \equiv \frac{\boldsymbol{j}_3 \times \boldsymbol{d}_3}{|\boldsymbol{j}_3 \times \boldsymbol{d}_3|} = \boldsymbol{j}_3 \times \boldsymbol{h} = -\sin\phi\,\boldsymbol{j}_1 + \cos\phi\,\boldsymbol{j}_2. \tag{11.3}$$

Set

$$\boldsymbol{k}_1 \equiv \boldsymbol{k}_2 \times \boldsymbol{d}_3 = \cos\theta\,\boldsymbol{h} - \sin\theta\,\boldsymbol{j}_3, \quad \boldsymbol{k}_3 = \boldsymbol{d}_3. \tag{11.4}$$

$\boldsymbol{k}_1$ and $\boldsymbol{k}_2$ form an orthonormal basis for the plane perpendicular to $\boldsymbol{d}_3$. Let ψ be the angle from $\boldsymbol{k}_1$ to $\boldsymbol{d}_1$. Then

$$\boldsymbol{d}_1 = \cos\psi\,\boldsymbol{k}_1 + \sin\psi\,\boldsymbol{k}_2, \quad \boldsymbol{d}_2 = -\sin\psi\,\boldsymbol{k}_1 + \cos\psi\,\boldsymbol{k}_2, \tag{11.5}$$

together with (11.1), (11.3), (11.4), give the $\boldsymbol{d}_k$ in terms of the Euler angles. In the terminology of rigid-body mechanics, θ is the *nutation angle*, ϕ is the *precession angle*, and ψ is the *spin angle.*

Using the relations

$$\boldsymbol{h}_s = \phi_s \boldsymbol{k}_2, \quad \partial_s \boldsymbol{k}_2 = -\phi_s \boldsymbol{h}, \quad \partial_s \boldsymbol{k}_1 = -\theta_s \boldsymbol{d}_3 + \cos\theta\,\phi_s \boldsymbol{k}_2 \tag{11.6}$$

we compute the $\partial_s \boldsymbol{d}_k$ and then use (5.1) to obtain

$$\begin{aligned} u_1 &= \theta_s \sin\psi - \phi_s \cos\psi \sin\theta, \\ u_2 &= \theta_s \cos\psi + \phi_s \sin\psi \sin\theta, \\ u_3 &= \psi_s + \phi_s \cos\theta. \end{aligned} \tag{11.7}$$

The corresponding expressions for $\boldsymbol{w}$ are obtained by replacing the s-derivatives with t-derivatives.

There is no standard notation for Euler angles. Although the Euler angles uniquely determine the configuration of the $\boldsymbol{d}_k$, they are not uniquely determined by the $\boldsymbol{d}_k$. Some of this ambiguity is evident in (11.7): If $\cos\theta = 1$, then only the sum $\psi_s + \phi_s$ can be determined from u. This difficulty, associated with the polar singularity introduced in (11.1), can be avoided by using the invariant characterization of u and w inherent in (2.3) and developed in Sec. 5, or by using alternative representations of proper orthogonal transformations in $\mathbb{E}^3$, as is done in rigid-body mechanics (see Simo & Vu-Quoc (1988)).

12. Boundary Conditions

In this section we describe a whole range of physically reasonable boundary conditions and then show how they can be expressed by a compact formalism. The motivation for our careful treatment arises not from a compulsive desire for completeness, but from the observation that global behavior and stability of solutions for rod problems depends crucially on subtle distinctions in boundary conditions; see Antman & Kenney (1981). In this section we exhibit only the classical versions of the boundary conditions in their conventional forms (see (II.2.3)) deferring to the next section a discussion of their weak forms.

The configuration at time t of the end section s_1 of a special Cosserat rod is $\{\boldsymbol{r}(s_1,t), \boldsymbol{d}_1(s_1,t), \boldsymbol{d}_2(s_1,t)\}$. (The inequality $v_3 \equiv \boldsymbol{r}_s \cdot \boldsymbol{d}_3 > 0$ of (2.7) determines the side of the plane $\operatorname{span}\{\boldsymbol{d}_1(s_1,t), \boldsymbol{d}_2(s_1,t)\}$ on which lies the material of the rod near s_1.) The prescription of the motion of an end of the rod is equivalent to welding the end to a rigid plane whose motion is prescribed. We can prescribe the entire motion of an end or merely part of it together with complementary mechanical conditions.

Whether the welding of an end leads to a well-posed mathematical problem when the end is tapered to zero thickness depends on whether the boundary values of the $\boldsymbol{d}_\alpha$ are assumed in the sense of trace. The answer to this question in turn depends on the functional setting of the problem, which is strongly influenced by the form of the functions $J_{\alpha\beta}$ of Sec. 10 and by the behavior of the constitutive functions for large strains. The crudeness of our model permits welding under some circumstances in which the end is tapered to zero thickness; see Antman (1976b, Sec. 2).

Kinematic boundary conditions fixing translation. We could prescribe the position $\boldsymbol{r}(s_1,\cdot)$ of the end s_1 or we could allow it to move on a curve or surface undergoing a prescribed motion or we could allow it to move freely in space. If $\boldsymbol{r}(s_1,t)$ is confined to a moving surface whose equation is $\gamma(\boldsymbol{z},t)=0$, then

$$\gamma(\boldsymbol{r}(s_1,t),t)=0. \tag{12.1}$$

If $\boldsymbol{r}(s_1,t)$ is confined to a moving curve, then it moves on the intersection of two moving surfaces and therefore satisfies two equations like (12.1).

We now develop a single compact formula to describe the four kinds of boundary conditions on $\boldsymbol{r}(s_1,t)$. If $\boldsymbol{r}(s_1,\cdot)$ is prescribed, then there is a given (continuous) function $\boldsymbol{r}^1$ such that

$$\boldsymbol{r}(s_1,t)=\boldsymbol{r}^1(t). \tag{12.2}$$

If $\boldsymbol{r}(s_1,t)$ is confined to a moving curve defined parametrically by $\mathcal{I} \ni a \mapsto \boldsymbol{r}^1(a,t)$ where $\mathcal{I}$ is an interval, then

$$\boldsymbol{r}(s_1,t)=\boldsymbol{r}^1(a(t),t). \tag{12.3}$$

$a(t)$ is a *generalized coordinate* locating $\boldsymbol{r}(s_1,t)$ on the curve $\boldsymbol{r}^1(\cdot,t)$. If $\boldsymbol{r}(s_1,t)$ is confined to a moving surface defined parametrically by $\mathcal{A} \ni \mathsf{a} \mapsto \boldsymbol{r}^1(\mathsf{a},t)$ where $\mathcal{A}$ is a domain in $\mathbb{R}^2$ with possibly part of its boundary, then

$$\boldsymbol{r}(s_1,t)=\boldsymbol{r}^1(\mathsf{a}(t),t). \tag{12.4}$$

$\mathsf{a}(t)$ is a pair of generalized coordinates locating $\boldsymbol{r}(s_1,t)$ on the surface $\boldsymbol{r}^1(\cdot,t)$.

We now unify and generalize these considerations. Let $\mathcal{T}$ denote the time interval on which the motion takes place. Let $\mathcal{A}^1$ denote a connected set in $\mathbb{R}^3\times\mathcal{T}$ having the property that each of its sections $\mathcal{A}^1(\tau)\equiv\{(\mathsf{a},t)\in\mathcal{A}^1 : t=\tau\}$ is nonempty. We assume that there is a continuous function $\mathcal{A}^1 \ni (\mathsf{a},t)\mapsto \boldsymbol{r}^1(\mathsf{a},t)\in\mathbb{E}^3$ such that

$$\boldsymbol{r}(s_1,t)=\boldsymbol{r}^1(\mathsf{a}(t),t). \tag{12.5}$$

This means that $\boldsymbol{r}(s_1, t)$ is specified by whatever part of the triple $\mathsf{a}(t)$ of generalized coordinates appears in the right-hand side of (12.5). If $\mathcal{A}^1(t)$ is a point for all t, then (12.5) prescribes $\boldsymbol{r}(s_1, t)$ and is therefore equivalent to (12.2). If $\mathcal{A}^1(t)$ is a segment of a curve for all t, then (12.5) restricts $\boldsymbol{r}(s_1, t)$ to a curve in $\mathbb{E}^3$, so that (12.5) is equivalent to (12.3). Similarly, we recover (12.4) when $\mathcal{A}^1(t)$ is a two-dimensional region for all t and we recover the unconstrained case when $\mathcal{A}^1(t)$ is a three-dimensional region for all t.

Let us study the second case more carefully. If $\mathcal{A}^1(t)$ is a segment of a curve of finite length that is not a loop, then it may or may not contain its end points. For illustration, suppose that the constraint curve in $\mathbb{E}^3$ is the half-open segment $\boldsymbol{r}^1(\cdot, t)$ defined by $\boldsymbol{r}^1(\mathsf{a}, t) = a_1 \boldsymbol{i}$ for $0 < a_1 \leq l$. We interpret $\boldsymbol{r}^1(\cdot, t)$ as the groove illustrated in Fig. 12.6. The end of the rod can touch the end $l\boldsymbol{i}$ of the groove, in which case the end of the groove exerts a reactive force on the end of the rod. The end of the rod cannot touch the end $\mathbf{0}$ of the groove. The placement of a spring in the groove between this end of the groove and the end of the rod would prevent this contact (provided that an infinite force is required to compress this spring to zero length).

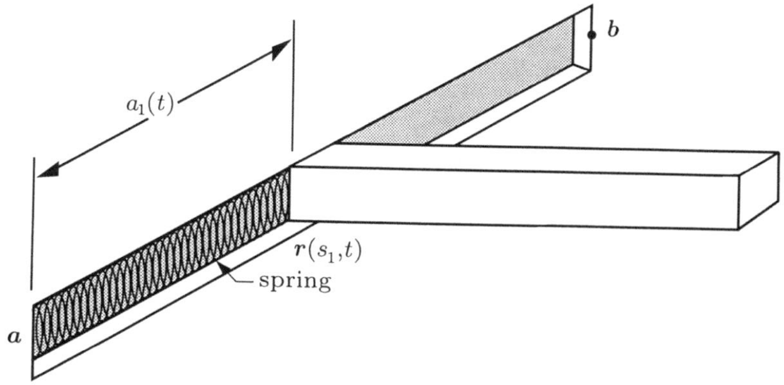

Fig. 12.6. Configuration at time t of a rod whose end s_1 is constrained to lie on a half-open segment.

Suppose that $\mathsf{a}_0 \in \mathcal{A}^1(\tau)$ and that the directions in which $\boldsymbol{r}^1(\cdot, \tau)$ has a linear Gâteaux differential at a_0 form a linear space of dimension $\rho^1(\mathsf{a}_0, \tau)$, which can be 0, 1, 2, or 3. In these cases, which are not exhaustive (because the differential might be defined only on a proper subset of such a linear space), we denote the Gâteaux derivative by $\partial \boldsymbol{r}^1(\mathsf{a}_0, \tau)/\partial \mathsf{a}$ and can identify $\rho^1(\mathsf{a}_0, \tau)$ with the rank of this transformation. $\rho^1(\mathsf{a}_0, \tau)$ is the number of *translational degrees of freedom* of the end s_1 at (a_0, τ). When there are no unilateral contacts included in the constraints, we can subsume the special cases of $\mathcal{A}^1$ described in preceding paragraphs under this formalism.

The use of generalized coordinates a in (12.5) means that (12.5) might afford merely a local description of boundary conditions. A global description, in the spirit of (12.1), can be readily framed by using concepts of manifold theory (see Abraham & Marsden (1979) or Arnol'd (1974)). For the treatment of unilateral conditions, see the references mentioned in the last paragraph of Sec. VII.8.

Kinematic boundary conditions fixing rotation. To describe boundary conditions fixing rotation, we imitate the development beginning with (12.5). Let $\mathcal{B}^1$ denote a connected set in $\mathbb{R}^3 \times \mathcal{T}$ having the property that each of its sections $\mathcal{B}^1(\tau) \equiv \{(\mathsf{b}, t) \in \mathcal{B}^1 : t = \tau\}$ is nonempty. We assume that there are continuous functions $\mathcal{B}^1 \ni (\mathsf{b}, t) \mapsto \boldsymbol{d}_k^1(\mathsf{b}, t) \in \mathbb{E}^3$, where the $\boldsymbol{d}_k^1(\mathsf{b}, t)$ form a right-handed orthonormal system such that

$$\boldsymbol{d}_k(s_1, t) = \boldsymbol{d}_k^1(\mathsf{b}(t), t). \tag{12.7}$$

To avoid technical difficulties with unilateral constraints, let us suppose $\mathsf{b}_0 \in \mathcal{B}^1(\tau)$ and that the directions in which the $\boldsymbol{d}_k^1(\cdot, \tau)$ have linear Gâteaux differentials at b_0 form a linear space. Then in analogy with our treatment of $\partial \boldsymbol{r}^1(\mathsf{a}_0, \tau)/\partial \mathsf{a}$ and with (2.3), we can write

$$\partial \boldsymbol{d}_k^1(\mathsf{b}_0, \tau)/\partial \mathsf{b} = \mathsf{Z}^1(\mathsf{b}_0, \tau) \times \boldsymbol{d}_k^1(\mathsf{b}_0, \tau). \tag{12.8}$$

Let the rank of $\mathsf{Z}^1(\mathsf{b}_0, \tau)$ be $\delta^1(\mathsf{b}_0, \tau)$, which is the number of *rotational degrees of freedom* of the end s_1 at (b_0, τ). If $\delta^1(\mathsf{b}_0, \tau) = 0$, then the $\boldsymbol{d}_k$ are fixed by (12.7). If $\delta^1 = 1$, then $\mathsf{Z}^1(\mathsf{b}_0, \tau)$ has the same effect as a vector $\boldsymbol{z}^1(\mathsf{b}_0, \tau)$, namely, (12.7) constrains the end s_1 to rotate instantaneously about the axis $\boldsymbol{z}^1(\mathsf{b}_0, \tau)$. An important special case of this condition is that in which a material direction $\alpha_k \boldsymbol{d}_k(s_1, t)$ is attached to an axis (hinge) having a prescribed alignment $t \mapsto \boldsymbol{e}_3(t)$:

$$\alpha_k \boldsymbol{d}_k(s_1, t) = \boldsymbol{e}_3(t). \tag{12.9}$$

Here the α_k are assumed to be constant Cartesian components of a unit vector and $\boldsymbol{e}_3$ is a unit vector. To put this condition into the parametric form (12.7), we prescribe $\boldsymbol{e}_1$ and $\boldsymbol{e}_2$ in any convenient way so that $\{\boldsymbol{e}_k\}$ forms a right-handed orthonormal triad. Since $\alpha_k \boldsymbol{d}_k$ is a material vector, we introduce another material basis $\{\boldsymbol{d}_k^\sharp\}$ by $\boldsymbol{d}_k^\sharp(s_1, t) = \boldsymbol{Q}^* \cdot \boldsymbol{d}_k(s_1, t)$ where $\boldsymbol{Q}$ is any constant proper orthogonal tensor such that $\boldsymbol{d}_3^\sharp(s_1, t) = \alpha_k \boldsymbol{d}_k(s_1, t)$. Then (12.9) can be put into the parametric form (12.7) by

$$\begin{aligned} \boldsymbol{d}_1(s_1, t) = \boldsymbol{d}_1^1(b_1(t), t) &\equiv \cos b_1(t) \boldsymbol{Q} \cdot \boldsymbol{e}_1(t) + \sin b_1(t) \boldsymbol{Q} \cdot \boldsymbol{e}_2(t), \\ \boldsymbol{d}_2(s_1, t) = \boldsymbol{d}_2^1(b_1(t), t) &\equiv -\sin b_1(t) \boldsymbol{Q} \cdot \boldsymbol{e}_1(t) + \cos b_1(t) \boldsymbol{Q} \cdot \boldsymbol{e}_2(t), \\ \boldsymbol{d}_3(s_1, t) = \boldsymbol{d}_3^1(b_1(t), t) &\equiv \boldsymbol{Q} \cdot \boldsymbol{e}_3(t). \end{aligned} \tag{12.10}$$

12.11. Exercise. Compute $\boldsymbol{z}^1$ and $\boldsymbol{w}(s_1, t)$ from (12.10).

If $\delta^1 = 2$, then (12.8) says that there is no rotation about one time-dependent direction. The easiest way to produce examples of (12.7) is to identify b with the Euler angles θ, ϕ, ψ and constrain these to lie in a two-dimensional surface in their space. For this purpose, let $\boldsymbol{Q}$ be a constant proper orthogonal tensor, and identify a special material triad $\boldsymbol{d}_k^\sharp(s_1, t)$ by $\boldsymbol{d}_k^\sharp(s_1, t) = \boldsymbol{Q}^* \cdot \boldsymbol{d}_k(s_1, t)$. Let $t \mapsto \boldsymbol{e}_1(t), \boldsymbol{e}_2(t), \boldsymbol{e}_3(t)$ be a prescribed right-handed orthonormal moving basis. We now identify $\{\boldsymbol{e}_1(t), \boldsymbol{e}_2(t), \boldsymbol{e}_3(t), \boldsymbol{d}_k^\sharp\}$ with the variables $\{\boldsymbol{\jmath}_1, \boldsymbol{\jmath}_2, \boldsymbol{\jmath}_3, \boldsymbol{d}_k\}$ of Sec. 11. We could then impose the constraint that there be no spin about $\boldsymbol{e}_3^\sharp$ relative to the frame $\{\boldsymbol{e}_k\}$ simply by fixing ψ while leaving the other Euler angles free.

We get an important example by constraining the orientation of the end so that

$$0 = \boldsymbol{d}_1^\sharp \cdot \boldsymbol{e}_2(t) = \cos\psi\cos\theta\sin\phi + \sin\psi\cos\phi. \tag{12.12}$$

This boundary condition is maintained by a (*Cardan* or *Hooke*) *universal joint*, illustrated in Fig. 12.13, whose outer gimbal has a prescribed motion. Then (12.7) has the form

$$\begin{aligned}
\boldsymbol{d}_1^\sharp(s_1,t) &= \cos\psi(t)[\cos\theta(t)\cos\phi(t)\boldsymbol{e}_1(t) - \sin\theta(t)\boldsymbol{e}_3(t)] - \sin\psi(t)\phi(t)\boldsymbol{e}_1(t),\\
\boldsymbol{d}_2^\sharp(s_1,t) &= -\sin\psi(t)[\cos\theta(t)\,(\cos\phi(t)\boldsymbol{e}_1(t) + \sin\phi(t)\boldsymbol{e}_2(t)) - \sin\theta(t)\boldsymbol{e}_3(t)]\\
&\quad + \cos\psi(t)[-\sin\phi(t)\boldsymbol{e}_1(t) + \cos\phi(t)\boldsymbol{e}_2(t)],\\
\boldsymbol{d}_3^\sharp(s_1,t) &= \sin\theta(t)[\cos\phi(t)\boldsymbol{e}_1(t) + \sin\phi(t)\boldsymbol{e}_2(t)] + \cos\theta(t)\boldsymbol{e}_3(t),
\end{aligned} \tag{12.14}$$

where the Euler angles are subject to (12.12).

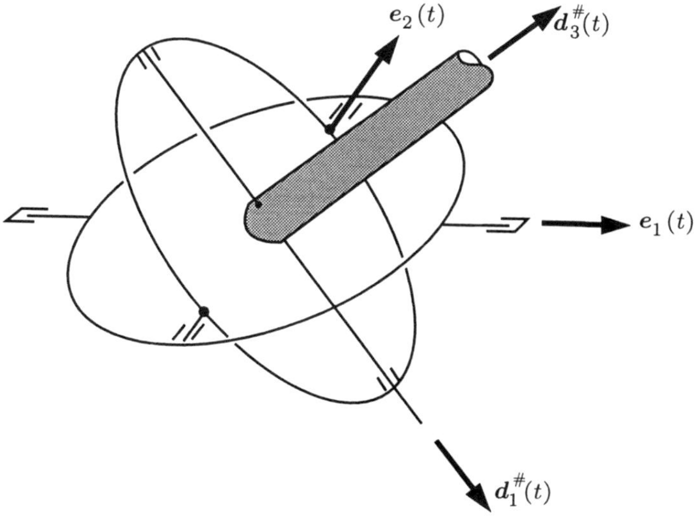

Fig. 12.13. A universal joint. For the support defined by (12.12), the motion of $\boldsymbol{e}_2$ is prescribed.

12.15. Exercise. Compute $\mathsf{Z}^1(\mathsf{b}_0,\tau)$ from (12.12) and (12.14).

If $\delta^1 = 3$, then the orientation of the section s_1 is free. If the position $\boldsymbol{r}(s_1)$ is prescribed, then the end can be supported in a ball-and-socket joint. If $\boldsymbol{r}(s_1)$ is confined to a curve, then the socket can be replaced by a tube. If $\boldsymbol{r}(s_1)$ is confined to a surface, there does not appear to be a simple mechanical device that prevents $\boldsymbol{r}(s_1)$ from leaving the surface. In such cases, there is usually some force, for example, coming from gravity or from an arrangement of springs that presses the end s_1 into the surface. Strictly speaking, the boundary conditions are *unilateral constraints*.

12.16. Exercise. Let

$$\partial_t \boldsymbol{d}_k^1(\mathsf{b},t) = \boldsymbol{w}^1(\mathsf{b},t) \times \boldsymbol{d}_k^1(\mathsf{b},t). \tag{12.17}$$

Show that (12.7) implies that

$$\boldsymbol{w}(s_1,t) = [\mathsf{Z}^1(\mathsf{b}(t),t) \times \boldsymbol{d}_k^1(\mathsf{b}(t),t)] \cdot \partial_t \mathsf{b}(t) + \boldsymbol{w}^1(\mathsf{b}(t),t) \times \boldsymbol{d}_k^1(\mathsf{b}(t),t). \tag{12.18}$$

In place of (12.7), we could have prescribed (12.18). Making the angular velocity $\boldsymbol{w}(s_1, t)$ the fundamental kinematical variable is analogous to using quasi-coordinates in rigid body mechanics (see Whittaker (1937)).

The generalized coordinates a and b may have components in common. For example, a hinge with axis $\boldsymbol{d}_1(s_1,t)$ may be attached to the end s_1 with this hinge constrained to slide in the (moving) track to which $\boldsymbol{r}(s_1,t)$ is constrained with the axis of the hinge tangent to the track. We model this track by a moving space curve $\mathcal{I} \ni a \mapsto \boldsymbol{r}^1(a,t)$ and take

$$\boldsymbol{r}(s_1,t) = \boldsymbol{r}^1(a(t),t), \quad \boldsymbol{d}_1(s_1,t) = \frac{\partial \boldsymbol{r}^1(a(t),t)/\partial a}{|\partial \boldsymbol{r}^1(a(t),t)/\partial a|} \tag{12.19}$$

(cf. (12.3)). In view of this example, we can generalize (12.5) and (12.7) by allowing $\boldsymbol{r}^1(\cdot)$ and $\boldsymbol{d}_k^1(\cdot)$ to have a nonempty subset $\mathcal{E}^1$ of points $(\mathsf{a}, \mathsf{b}, t)$ in $\mathbb{R}^6 \times \mathcal{T}$ as their common domain of definition.

Of course, all the comments about end s_1 apply to end s_2, the behavior of which is described with the analogous notation. In view of the last paragraph, we take as our kinematic boundary conditions

$$\boldsymbol{r}(s_\alpha,t) = \boldsymbol{r}^\alpha(\mathsf{a}^\alpha(t), \mathsf{b}^\alpha(t), t), \quad \boldsymbol{d}_k(s_\alpha,t) = \boldsymbol{d}_k^\alpha(\mathsf{a}^\alpha(t), \mathsf{b}^\alpha(t), t), \tag{12.20a,b}$$

$\alpha = 1, 2$, where $\boldsymbol{r}^\alpha(\cdot)$ and $\boldsymbol{d}_k^\alpha(\cdot)$ are defined on $\mathcal{E}^\alpha \subset \mathbb{R}^6 \times \mathcal{T}$.

Mechanical boundary conditions. The constraints imposed on the motion of the ends by the kinematic boundary conditions (12.20) are enforced by components of resultants at the ends that are not known a priori. In general, they cannot be determined until a full solution of the governing equations and their side conditions is obtained. The values of these constraint components at the ends are whatever are necessary to maintain the constraint. On the other hand, components of the resultants that tend to move the ends in ways permitted by the kinematic boundary conditions may be prescribed. In fact, for well-set problems these components must be prescribed. (Often they are prescribed to vanish.) We first illustrate these notions by example and then incorporate our findings into a rule for the prescription of mechanical boundary conditions that complement (12.20).

If $\boldsymbol{r}(s_1,t)$ is prescribed, then $\boldsymbol{n}(s_1,t)$, the reaction necessary to enforce the constraint, is not prescribed. If $\boldsymbol{r}(s_1,t)$ is confined to a curve $\mathcal{I} \ni a \mapsto \boldsymbol{r}^1(a,t)$ undergoing a prescribed motion, then only the component of $\boldsymbol{n}(s_1,t)$ instantaneously tangent to the curve can be prescribed; the components of $\boldsymbol{n}(s_1,t)$ orthogonal to the curve maintain the constraint. Since a tangent to $\boldsymbol{r}^1(\cdot,t)$ at $\boldsymbol{r}(s_1,t)$ is $\partial\boldsymbol{r}^1(a(t),t)/\partial a$, we accordingly prescribe $t \mapsto \boldsymbol{n}(s_1,t)\cdot\partial\boldsymbol{r}^1(a(t),t)/\partial a$. Note that the actual velocity of the end $\boldsymbol{r}_t(s_1,t) = [\partial\boldsymbol{r}^1(a(t),t)/\partial a]a'(t) + \partial\boldsymbol{r}^1(a(t),t)/\partial t$ is not necessarily tangent to $\boldsymbol{r}^1(\cdot,t)$ at $\boldsymbol{r}(s_1,t)$. (It is if the curve is fixed in space.)

Similarly, if $\boldsymbol{r}(s_1,t)$ is confined to a moving surface $\mathcal{A} \ni \mathsf{a} \mapsto \boldsymbol{r}^1(\mathsf{a},t)$ (see (12.4)), then we prescribe only $t \mapsto \boldsymbol{n}(s_1,t)\cdot\partial\boldsymbol{r}^1(\mathsf{a}(t),t)/\partial\mathsf{a}$. If $\boldsymbol{r}(s_1,t)$ is free, then (12.5) holds with $\rho^1 = 3$. In this case, we prescribe $t \mapsto \boldsymbol{n}(s_1,t)\cdot\partial\boldsymbol{r}^1(\mathsf{a}(t),t)/\partial\mathsf{a}$, which is the same as prescribing $\boldsymbol{n}(s_1,t)$ itself.

The prescribed values of the components of $\boldsymbol{n}(s_1,t)$ may depend on the position, velocity, and even the past history of the end. A dependence on the position of the end may account for spring supports, a dependence on the velocity may account for friction or for internal dissipation in supporting springs, and the dependence on past history may account for more general viscoelastic processes. For simplicity of exposition, we assume that the prescribed components of $\boldsymbol{n}(s_\alpha, t)$ depend only on the position, velocity, orientation, and angular velocity of the end s_α. Similar considerations apply to the specification of couples at the ends. Under the assumption that $(\mathsf{a}^\alpha, \mathsf{b}^\alpha) \mapsto \boldsymbol{r}^\alpha(\mathsf{a}^\alpha, \mathsf{b}^\alpha, t), \boldsymbol{d}_k^\alpha(\mathsf{a}^\alpha, \mathsf{b}^\alpha, t)$ are differentiable, all these cases of mechanical boundary conditions are embodied in the requirement that there are functions $(\boldsymbol{r}, \dot{\boldsymbol{r}}, \boldsymbol{d}_k, \boldsymbol{w}, t) \mapsto \boldsymbol{n}^\alpha(\boldsymbol{r}, \dot{\boldsymbol{r}}, \boldsymbol{d}_k, \boldsymbol{w}, t), \boldsymbol{m}^\alpha(\boldsymbol{r}, \dot{\boldsymbol{r}}, \boldsymbol{d}_k, \boldsymbol{w}, t)$ such that

(12.21a)
$$\big[\boldsymbol{n}(s_\alpha, t) - \boldsymbol{n}^\alpha\big(\boldsymbol{r}(s_\alpha, t), \boldsymbol{r}_t(s_\alpha, t),\ \boldsymbol{d}_k(s_\alpha, t), \boldsymbol{w}(s_\alpha, t), t\big)\big] \cdot \left\{ \begin{array}{c} \partial \boldsymbol{r}^\alpha(\mathsf{a}^\alpha(t), \mathsf{b}^\alpha(t), t)/\partial \mathsf{a}^\alpha \\ \partial \boldsymbol{r}^\alpha(\mathsf{a}^\alpha(t), \mathsf{b}^\alpha(t), t)/\partial \mathsf{b}^\alpha \end{array} \right\} = \mathbf{0},$$

(12.21b)
$$\big[\boldsymbol{m}(s_\alpha, t) - \boldsymbol{m}^\alpha\big(\boldsymbol{r}(s_\alpha, t), \boldsymbol{r}_t(s_\alpha, t),\ \boldsymbol{d}_k(s_\alpha, t), \boldsymbol{w}(s_\alpha, t), t\big)\big] \cdot \left\{ \begin{array}{c} \mathsf{Y}^\alpha(\mathsf{a}^\alpha(t), \mathsf{b}^\alpha(t), t) \\ \mathsf{Z}^\alpha(\mathsf{a}^\alpha(t), \mathsf{b}^\alpha(t), t) \end{array} \right\} = \mathbf{0}$$

where

$$\begin{aligned} \partial \boldsymbol{d}_k^\alpha(\mathsf{a}, \mathsf{b}, t)/\partial \mathsf{a} &= \mathsf{Y}^\alpha(\mathsf{a}, \mathsf{b}, t) \times \boldsymbol{d}_k^\alpha(\mathsf{a}, \mathsf{b}, t), \\ \partial \boldsymbol{d}_k^\alpha(\mathsf{a}, \mathsf{b}, t)/\partial \mathsf{b} &= \mathsf{Z}^\alpha(\mathsf{a}, \mathsf{b}, t) \times \boldsymbol{d}_k^\alpha(\mathsf{a}, \mathsf{b}, t). \end{aligned} \tag{12.21c}$$

The domain of $\boldsymbol{n}^\alpha(\cdot, \dot{\boldsymbol{r}}, \boldsymbol{d}_k, \boldsymbol{w}, t)$ and $\boldsymbol{m}^\alpha(\cdot, \dot{\boldsymbol{r}}, \boldsymbol{d}_k, \boldsymbol{w}, t)$ is the range of $\boldsymbol{r}^\alpha(\cdot, \cdot, t)$, and the domain of $\boldsymbol{n}^\alpha(\boldsymbol{r}, \cdot, \boldsymbol{d}_k, \boldsymbol{w}, t)$ and $\boldsymbol{m}^\alpha(\boldsymbol{r}, \cdot, \boldsymbol{d}_k, \boldsymbol{w}, t)$ is the range of

$$(\mathsf{a}, \dot{\mathsf{a}}, \mathsf{b}, \dot{\mathsf{b}}) \mapsto [\partial \boldsymbol{r}^\alpha(\mathsf{a}, \mathsf{b}, t)/\partial \mathsf{a}] \cdot \dot{\mathsf{a}} + [\partial \boldsymbol{r}^\alpha(\mathsf{a}, \mathsf{b}, t)/\partial \mathsf{b}] \cdot \dot{\mathsf{b}} + \partial \boldsymbol{r}^\alpha(\mathsf{a}, \mathsf{b}, t)/\partial t,$$

etc.

For static problems mechanical boundary conditions cannot be prescribed arbitrarily. They must be consistent with the requirement that the resultant force and moment must vanish. Such conditions are necessary conditions for the existence of solutions to equilibrium problems.

Twist. Consider a statical problem in which $\delta^1 = 0 = \delta^2$ so that $\boldsymbol{d}_1$ and $\boldsymbol{d}_2$ are prescribed at each end of the rod. If the end s_2 of the rod is twisted about $\boldsymbol{d}_3(s_2)$ through an angle equal to an integral multiple of 2π, then $\boldsymbol{d}_1(s_2)$ and $\boldsymbol{d}_2(s_2)$ remain unchanged. Thus it is likely that equilibrium problems for the rod under the boundary conditions specifying $\boldsymbol{d}_1$ and $\boldsymbol{d}_2$ at the ends have multiple solutions. This multiplicity is unsatisfactory because the boundary conditions ignore the physical process of twisting. (There is no such ambiguity for dynamical problems because the $\boldsymbol{d}_k(s_\alpha)$ can be expected to evolve continuously from their initial states.) It is possible to prescribe the amount of twist in a way that is geometrically and analytically satisfactory, the most effective way involving the linking number of Gauss. This number is a topological invariant, the use of which clarifies a very rich and complicated geometrical situation. For details, see Alexander & Antman (1982).

13. Impulse-Momentum Laws and the Principle of Virtual Power

In this section we extend (2.13) and (2.14) to intervals (a, b) containing the end points, at which we must account for boundary conditions. We then show that these impulse-momentum laws are equivalent to a precisely formulated Principle of Virtual Power. We concentrate on the treatment of boundary conditions and on the Angular Impulse-Momentum Law, which present novelties not encompassed in the development of Secs. II.3 and II.4. For simplicity of exposition, we assume that (2.12) holds.

We first extend the Linear Impulse-Momentum Law to the whole interval $[s_1, s_2]$. The classical forms of the initial conditions are conventionally denoted

$$\text{(13.1a,b)} \qquad \boldsymbol{r}(s,0) = \boldsymbol{g}_0(s), \quad \boldsymbol{r}_t(s,0) = \boldsymbol{g}_1(s)$$

(see (II.2.4)). We assume that $\boldsymbol{g}_0$ and $\boldsymbol{g}_1$ are integrable on $[s_1, s_2]$. We assume that (13.1b) holds in the sense of trace (see (II.3.4)) and, just as in Sec. II.3, incorporate it into the Linear Impulse-Momentum Law by replacing $\boldsymbol{r}_t(\xi, 0)$ in (2.13) with $\boldsymbol{g}_1(\xi)$. Then just as in (II.3.7), we find that the Linear Impulse-Momentum Law implies that (13.1b) holds in the sense of trace. To fix ideas, let us first suppose that $\rho^1 = 3 = \rho^2$. In this case, the conventional classical forms of the boundary conditions (12.21a) reduce to

$$\text{(13.2)} \qquad \boldsymbol{n}(s_\alpha, t) = \boldsymbol{n}^\alpha\big(\boldsymbol{r}(s_\alpha, t), \boldsymbol{r}_t(s_\alpha, t), \boldsymbol{d}_k(s_\alpha, t), \boldsymbol{w}(s_\alpha, t), t\big).$$

We assume that the functions $t \mapsto \boldsymbol{n}^\alpha\big(\boldsymbol{r}(s_\alpha, t), \boldsymbol{r}_t(s_\alpha, t), \boldsymbol{d}_k(s_\alpha, t), \boldsymbol{w}(s_\alpha, t), t\big)$ are locally integrable on $[0, \infty)$. We extend (2.13) to the whole interval $[s_1, s_2]$ and incorporate (13.2) into the resulting equation by replacing $\boldsymbol{n}^-(a, t) \equiv \boldsymbol{n}(a, t)$ with $\boldsymbol{n}^\alpha\big(\boldsymbol{r}(s_\alpha, t), \boldsymbol{r}_t(s_\alpha, t), \boldsymbol{d}_k(s_\alpha, t), \boldsymbol{w}(s_\alpha, t), t\big)$ if $a = s_1$ and by making the corresponding adjustment if $s = s_2$. If $a = s_1$ and if $s < s_2$, then with the modifications just introduced (2.13) becomes

$$\text{(13.3)} \qquad \begin{aligned} &\int_0^T \left[\boldsymbol{n}(s,t) - \boldsymbol{n}^1\big(\boldsymbol{r}(s_1, t), \boldsymbol{r}_t(s_1, t), \boldsymbol{d}_k(s_1, t), \boldsymbol{w}(s_1, t), t\big) + \int_{s_1}^s \boldsymbol{f}(\xi, t)\, d\xi\right] dt \\ &\qquad = \int_{s_1}^s (\rho A)(\xi)[\boldsymbol{r}_t(\xi, t) - \boldsymbol{g}_1(\xi)]\, d\xi. \end{aligned}$$

By following the proof of (II.3.7), we readily find that (13.2) holds in the sense of trace, i.e., there are sets $\mathcal{A}$ and $\mathcal{B}$ with Lebesgue measure $|\mathcal{A}| = s_2 - s_1$ and $|\mathcal{B} \cap [0, T]| = T$ for all $T \geq 0$ such that

$$\text{(13.4)} \qquad \lim_{\mathcal{A} \ni s \to s_\alpha} \int_{t_1}^{t_2} [\boldsymbol{n}(s,t) - \boldsymbol{n}^\alpha\big(\boldsymbol{r}(s_\alpha, t), \boldsymbol{r}_t(s_\alpha, t), \boldsymbol{d}_k(s_\alpha, t), \boldsymbol{w}(s_\alpha, t), t\big)]\, dt = 0$$

for all $t_1, t_2 \in \mathcal{B}$.

We can now derive the Principle of Virtual Power corresponding to our generalization of the Linear Impulse-Momentum Law by modifying the approach of Sec. II.4. There we introduced a test function $\boldsymbol{y}$ with support in $(s_1, s_2)\times[0,\tau)$ (where (s_1, s_2) was taken to equal $(0,1)$). The requirement that this $\boldsymbol{y}$ vanish at s_1 and s_2 reflects the boundary conditions (2.3) or (3.3). We suspend this requirement in our present development. Since the boundary of $[s_1, s_2]$ is disconnected, it is convenient to treat the boundary conditions at each end separately by the contrivance of writing the test function $\boldsymbol{y}$ as the difference

$$\boldsymbol{y} = \boldsymbol{y}^1 - \boldsymbol{y}^2 \tag{13.5}$$

where $\boldsymbol{y}^1$ and $\boldsymbol{y}^2$ are twice continuously differentiable, have compact support on $[s_1, s_2]\times[0,\infty)$, and satisfy the boundary conditions $\boldsymbol{y}^1(s_2) = \boldsymbol{0} = \boldsymbol{y}^2(s_1)$. We operate on (13.3) with $\int_0^\infty d\tau \int_{s_1}^{s_2} db\, \boldsymbol{y}^1_{st}(b,\tau)\cdot$, operate on the corresponding version of (2.13) for $b = s_2$, namely, $\int_0^\tau [\boldsymbol{n}^2(\boldsymbol{r}(s_2,t),\dots) - \boldsymbol{n}(a,t) + \cdots$, with $\int_0^\infty d\tau \int_{s_1}^{s_2} da\, \boldsymbol{y}^1_{st}(a,\tau)\cdot$, add the resulting equations, and follow (II.4.1) and (II.4.2) to obtain the *Principle of Virtual Power corresponding to the Linear Impulse-Momentum Law*:

$$\begin{aligned}\int_0^\infty \int_{s_1}^{s_2} [\boldsymbol{n}(s,t)\cdot\boldsymbol{y}_s(s,t) - \boldsymbol{f}(s,t)\cdot\boldsymbol{y}(s,t)]\,ds\,dt \\ - \int_0^\infty [\boldsymbol{n}^2\big(\boldsymbol{r}(s_2,t),\boldsymbol{r}_t(s_2,t),\boldsymbol{d}_k(s_2,t),\boldsymbol{w}(s_2,t),t\big)\cdot\boldsymbol{y}(s_2,t) \\ - \boldsymbol{n}^1\big(\boldsymbol{r}(s_1,t),\boldsymbol{r}_t(s_1,t),\boldsymbol{d}_k(s_1,t),\boldsymbol{w}(s_1,t),t\big)\cdot\boldsymbol{y}(s_1,t)]\,dt \\ = \int_0^\infty \int_{s_1}^{s_2} (\rho A)(s)[\boldsymbol{r}_t(s,t) - \boldsymbol{g}_1(s)]\cdot\boldsymbol{y}_t(s,t)\,ds\,dt\end{aligned} \tag{13.6}$$

for all $\boldsymbol{y}$'s in $C^2\big([s_1, s_2]\times[0,\infty)\big)$ with compact support and, more generally, for all $\boldsymbol{y}$'s in $W_1^\infty\big([s_1, s_2]\times[0,\infty)\big)$ with compact support.

13.7. Exercise. Follow the pattern of Sec. II.4 to show that (13.6) implies the generalized Linear Impulse-Momentum Law consisting in (13.3) and its analog for s_2.

We now turn to the treatment of (12.20) and (12.21). Since their formulation was specifically contrived to fit the Principle of Virtual Power, we first formulate a version of it capable of handling these boundary conditions. We assume that the composite functions of t whose values are given by the right-hand sides of (12.21) are locally integrable on $[0,\infty)$. We use a variant $\overset{\vartriangle}{\boldsymbol{r}}$ of the traditional notation $\delta\boldsymbol{r}$ for the test function (virtual velocity) $\boldsymbol{y}$. Suppose that the functions $\boldsymbol{r}^\alpha(\cdot,\cdot,t)$ and $\boldsymbol{d}_k^\alpha(\cdot,\cdot,t)$ of (12.20) are continuously differentiable. Then the *Principle of Virtual Power corresponding to the Linear Impulse-Momentum Law* states that (13.6) holds for all piecewise continuously differentiable $\overset{\vartriangle}{\boldsymbol{r}}$'s with compact support on $[s_1, s_2]\times[0,\infty)$ that have the form

$$\overset{\vartriangle}{\boldsymbol{r}}(s_\alpha,t) = \frac{\partial\boldsymbol{r}^\alpha}{\partial\mathsf{a}^\alpha}(\mathsf{a}^\alpha(t),\mathsf{b}^\alpha(t),t)\cdot\overset{\vartriangle}{\mathsf{a}}^\alpha(t) + \frac{\partial\boldsymbol{r}^\alpha}{\partial\mathsf{b}^\alpha}(\mathsf{a}^\alpha(t),\mathsf{b}^\alpha(t),t)\cdot\overset{\vartriangle}{\mathsf{b}}^\alpha(t) \tag{13.8}$$

with $\overset{\vartriangle}{\mathsf{a}}^\alpha$ and $\overset{\vartriangle}{\mathsf{b}}^\alpha$ piecewise continuously differentiable with compact support. Condition (13.8) says that the test function $\overset{\vartriangle}{\boldsymbol{r}}$ is tangent to the manifolds defined by (12.20).

In particular, for fixed t, a curve on the manifold (12.20a) with $\alpha = 1$ has the form $\varepsilon \mapsto \boldsymbol{r}^1(\tilde{\mathsf{a}}^1(\varepsilon), \tilde{\mathsf{b}}^1(\varepsilon), t)$. Therefore, a tangent to this manifold is the derivative of this function with respect to ε evaluated at $\varepsilon = 0$. The one-parameter family of position functions that lie on this curve for $s = s_1$ is $\boldsymbol{r}(s, t, \varepsilon)$ with $\boldsymbol{r}(s_1, t, \varepsilon) = \boldsymbol{r}^1(\tilde{\mathsf{a}}^1(\varepsilon), \tilde{\mathsf{b}}^1(\varepsilon), t)$. The differentiation of this equation with respect to ε yields (13.8), provided we denote $\partial \mathsf{u}(\cdot, \varepsilon)/\partial \varepsilon|_{\varepsilon=0}$ by $\overset{\vartriangle}{\mathsf{u}}$ for any function u.

We now show that this principle implies that the boundary conditions (12.21a) are assumed in the sense of trace. We tacitly suppose that all the integrals we exhibit make sense as Lebesgue integrals. (For a discussion of the technical details that we now suppress, see Secs. II.3 and II.4.) Let $\sigma \in (s_1, s_2)$, let $\varepsilon \in (0, s_2 - \sigma)$, and set

$$\phi(s; \sigma, \varepsilon) \equiv \begin{cases} 1 & \text{for} \quad s_1 \le s \le \sigma, \\ -\frac{s-(\sigma+\varepsilon)}{\varepsilon} & \text{for} \quad \sigma \le s \le \sigma + \varepsilon, \\ 0 & \text{for} \quad \sigma + \varepsilon \le s \le s_2. \end{cases} \tag{13.9}$$

Let $0 \le t_1 < t_2$, let $2\varepsilon < t_2 - t_1$, and set

$$\psi(t; \varepsilon) \equiv \begin{cases} 0 & \text{for} \quad 0 \le t \le t_1, \\ \frac{t-t_1}{\varepsilon} & \text{for} \quad t_1 \le t \le t_1 + \varepsilon, \\ 1 & \text{for} \quad t_1 + \varepsilon \le t \le t_2 - \varepsilon, \\ \frac{t_2 - t}{\varepsilon} & \text{for} \quad t_2 - \varepsilon \le t \le t_2, \\ 0 & \text{for} \quad t_2 \le t. \end{cases} \tag{13.10}$$

Let e be an arbitrary triple of real numbers. Let $\overset{\vartriangle}{\boldsymbol{r}}(s_1, t)$ be given by (13.8) with $\overset{\vartriangle}{\mathsf{a}}{}^1(t) = \psi(t; \varepsilon)\mathsf{e}$ and with $\overset{\vartriangle}{\mathsf{b}}{}^1 = \mathbf{0}$. We substitute

$$\overset{\vartriangle}{\boldsymbol{r}}(s, t) = \phi(s; \sigma, \varepsilon)\overset{\vartriangle}{\boldsymbol{r}}(s_1, t) \tag{13.11}$$

into (13.6) and let $\varepsilon \to 0$. In this process, we find that

$$\lim_{\varepsilon \to 0} \int_0^\infty \int_{s_1}^{s_2} \boldsymbol{n}(s, t) \cdot \overset{\vartriangle}{\boldsymbol{r}}_s(s, t)\, ds\, dt = -\int_{t_1}^{t_2} \boldsymbol{n}(\sigma, t) \cdot \frac{\partial \boldsymbol{r}^1}{\partial \mathsf{a}^1}(\mathsf{a}^1(t), \mathsf{b}^1(t), t) \cdot \mathsf{e}\, dt. \tag{13.12}$$

If we now let $\sigma \to s_1$ (through an appropriate dense set) in the resulting form of (13.6) and if we use the arbitrariness of e, we obtain

$$\lim_{s \to s_1} \int_{t_1}^{t_2} \left[\boldsymbol{n}(\sigma, t) - \boldsymbol{n}^1(\boldsymbol{r}(s_1, t), \boldsymbol{r}_t(s_1, t), \boldsymbol{d}_k(s_1, t), \boldsymbol{w}(s_1, t), t)\right] \cdot \frac{\partial \boldsymbol{r}^1}{\partial \mathsf{a}^1}(\mathsf{a}^1(t), \mathsf{b}^1(t), t)\, dt = \mathbf{0} \tag{13.13}$$

(for almost every t_1 and t_2) just as in (13.4). The other boundary conditions of (12.21a) are treated analogously. Thus the Principle of Virtual Power implies that (12.21) holds in the sense of trace.

13.14. Exercise. Prove the equivalence of our general version of the Principle of Virtual Power corresponding to the Linear Impulse-Momentum Law with a generalized version of the Linear Impulse-Momentum Law when for each α the $\boldsymbol{r}(s_\alpha, t)$ are confined to moving points, lines, or planes, or are unrestrained for all time. In this case, (12.20a) and (12.21a) hold.

We now derive the Principle of Virtual Power corresponding to the Angular Impulse-Momentum law in the simple special case that $\delta^1 = 3 = \delta^2$, which is analogous to (13.2). In this case, the classical boundary conditions, which come from (12.21b), have the form

$$\boldsymbol{m}(s_\alpha,t)=\boldsymbol{m}^1\big(\boldsymbol{r}(s_\alpha,t),\boldsymbol{r}_t(s_\alpha,t),\boldsymbol{d}_k(s_\alpha,t),\boldsymbol{w}(s_1,t),t\big). \tag{13.15}$$

The classical forms of the initial conditions are

$$\boldsymbol{d}_k(s,0)=\boldsymbol{d}_k^0(s),\quad \boldsymbol{h}(s,0)=\boldsymbol{h}^0(s). \tag{13.16a,b}$$

We tacitly assume that all data are such that all integrals make sense as Lebesgue integrals. We assume that (13.16a) holds in the sense of trace.

We extend the reduced Angular Impulse-Momentum Law (2.18) to the interval $[s_1,s_2]$ by incorporating the boundary condition (13.15) just as in (13.3). We identify

$$\overset{\vartriangle}{\boldsymbol{d}}_k=\boldsymbol{z}\times\boldsymbol{d}_k. \tag{13.17}$$

We introduce a test function $\boldsymbol{z}$ as in (13.5) and follow the procedure leading to (13.6) to obtain the *Principle of Virtual Power corresponding to the Angular Impulse-Momentum Law*:

(13.18)

$$\begin{aligned}\int_0^\infty\int_{s_1}^{s_2}&[\boldsymbol{m}(s,t)\cdot\boldsymbol{z}_s(s,t)-[\boldsymbol{r}_s(s,t)\times\boldsymbol{n}(s,t)+\boldsymbol{l}(s,t)]\cdot\boldsymbol{z}(s,t)]\,ds\,dt\\ &-\int_0^\infty[\boldsymbol{m}^2\big(\boldsymbol{r}(s_2,t),\boldsymbol{r}_t(s_2,t),\boldsymbol{d}_k(s_2,t),\boldsymbol{w}(s_2,t),t\big)\cdot\boldsymbol{z}(s_2,t)\\ &\qquad-\boldsymbol{m}^1\big(\boldsymbol{r}(s_1,t),\boldsymbol{r}_t(s_1,t),\boldsymbol{d}_k(s_1,t),\boldsymbol{w}(s_1,t),t\big)\cdot\boldsymbol{z}(s_1,t)]\,dt\\ &=\int_0^\infty\int_{s_1}^{s_2}[\boldsymbol{h}(s,t)-\boldsymbol{h}^0(s)]\cdot\boldsymbol{z}_t(s,t)\,ds\,dt\end{aligned}$$

for all $\boldsymbol{z}$'s in $C^2\big([s_1,s_2]\times[0,\infty)\big)$ with compact support and, more generally, for all $\boldsymbol{z}$'s in $W_1^\infty\big([s_1,s_2]\times[0,\infty)\big)$ with compact support.

As with (13.6), we generalize (13.18) to handle general boundary conditions (12.20) and (12.21) by requiring that (13.18) hold for $\boldsymbol{z}$'s meeting obvious smoothness requirements and having the form

$$\boldsymbol{z}(s_\alpha,t)=\overset{\vartriangle}{\mathsf{a}}{}^\alpha(t)\cdot\mathsf{Y}^\alpha\big(\mathsf{a}^\alpha(t),\mathsf{b}^\alpha(t),t\big)+\overset{\vartriangle}{\mathsf{b}}{}^\alpha(t)\cdot\mathsf{Z}^\alpha\big(\mathsf{a}^\alpha(t),\mathsf{b}^\alpha(t),t\big) \tag{13.19}$$

where Y^α and Z^α are defined after (12.21b).

We can unify our two Principles of Virtual Power by simply adding (13.6) and (13.18). To obtain a useful variant of this combined form, which illuminates the development given above, we observe that our kinematical variables $\boldsymbol{r}$ and $\boldsymbol{d}_k$ are constrained by

$$\boldsymbol{d}_k\cdot\boldsymbol{d}_l=\delta_{kl}, \tag{13.20}$$

by the boundary conditions (12.20), and by the initial conditions (13.1a) and (13.16a). These constraints may be regarded as restricting $\{\boldsymbol{r}, \boldsymbol{d}_k\}$ to a certain manifold $\mathcal{M}$ in the function space of $\{\boldsymbol{r}, \boldsymbol{d}_k\}$. We denote a typical tangent vector to this manifold at $(\boldsymbol{r}, \boldsymbol{d}_k)$ by $(\overset{\vartriangle}{\boldsymbol{r}}, \overset{\vartriangle}{\boldsymbol{d}}_k)$. To compute $(\overset{\vartriangle}{\boldsymbol{r}}, \overset{\vartriangle}{\boldsymbol{d}}_k)$ at a particular element $(\boldsymbol{r}, \boldsymbol{d}_k)$ of the function space, we choose a C^1-curve $[-1,1] \ni \gamma \mapsto (\boldsymbol{r}(\cdot,\cdot,\gamma), \boldsymbol{d}_k(\cdot,\cdot,\gamma))$ in $\mathcal{M}$ that passes through $(\boldsymbol{r}, \boldsymbol{d}_k)$ at $\gamma = 0$. Then a tangent vector to this curve, which is of course tangent to $\mathcal{M}$ at $(\boldsymbol{r}, \boldsymbol{d}_k)$, is $(\overset{\vartriangle}{\boldsymbol{r}}, \overset{\vartriangle}{\boldsymbol{d}}_k) = (\partial_\gamma \boldsymbol{r}(\cdot,\cdot,0), \partial_\gamma \boldsymbol{d}_k(\cdot,\cdot,0))$. The arbitrariness of the choice of curve is reflected in the arbitrariness in $(\overset{\vartriangle}{\boldsymbol{r}}, \overset{\vartriangle}{\boldsymbol{d}}_k)$. We justify the choice (13.8) by requiring the curve to satisfy (12.20a):

$$\gamma \mapsto \boldsymbol{r}(s_\alpha, \cdot, \gamma) = \boldsymbol{r}^\alpha(\mathsf{a}^\alpha(\cdot,\gamma), \mathsf{b}^\alpha(\cdot,\gamma), \cdot). \tag{13.21}$$

Then (13.8) is given by

$$\overset{\vartriangle}{\boldsymbol{r}}(s_\alpha, t) = \partial_\gamma \boldsymbol{r}(s_\alpha, t, 0) \quad \text{with} \quad \overset{\vartriangle}{\mathsf{a}}{}^\alpha = \partial_\gamma \mathsf{a}^\alpha(\cdot, 0), \quad \overset{\vartriangle}{\mathsf{b}}{}^\alpha = \partial_\gamma \mathsf{b}^\alpha(\cdot, 0). \tag{13.22}$$

The choice (13.19) is justified the same way.

Now (2.4b) and (13.17) imply that

$$\overset{\vartriangle}{\boldsymbol{r}}_s = (\overset{\vartriangle}{v}_k)\boldsymbol{d}_k + v_k(\boldsymbol{z} \times \boldsymbol{d}_k) = (\overset{\vartriangle}{v}_k)\boldsymbol{d}_k + \boldsymbol{z} \times \boldsymbol{r}_s, \quad \boldsymbol{z}_s = (\overset{\vartriangle}{u}_k)\boldsymbol{d}_k. \tag{13.23}$$

Hence the combined form of the Principle of Virtual Power is equivalent to

$$\begin{aligned} &\int_0^\infty \int_{s_1}^{s_2} [\mathsf{n} \cdot \overset{\vartriangle}{\mathsf{v}} + \mathsf{m} \cdot \overset{\vartriangle}{\mathsf{u}} - \boldsymbol{f} \cdot \overset{\vartriangle}{\boldsymbol{r}} - \boldsymbol{l} \cdot \boldsymbol{z}]\, ds\, dt \\ &\quad - \int_0^\infty [\boldsymbol{n}^2 \cdot \overset{\vartriangle}{\boldsymbol{r}}(s_2,t) - \boldsymbol{n}^1 \cdot \overset{\vartriangle}{\boldsymbol{r}}(s_1,t) + \boldsymbol{m}^2 \cdot \boldsymbol{z}(s_2,t) - \boldsymbol{m}^1 \cdot \boldsymbol{z}(s_1,t)]\, dt \\ &\quad = \int_0^\infty \int_{s_1}^{s_2} \{(\rho A)(s)[\boldsymbol{r}_t(s,t) - \boldsymbol{g}_1(s)] \cdot \overset{\vartriangle}{\boldsymbol{r}}_t + [\boldsymbol{h}(s,t) - \boldsymbol{h}^0(s)] \cdot \boldsymbol{z}_t(s,t)\}\, ds\, dt \end{aligned} \tag{13.24}$$

for all $\overset{\vartriangle}{\boldsymbol{r}}$'s satisfying (13.8) and for all $\boldsymbol{z}$'s satisfying (13.19).

13.25. Exercise. Carry out the development in this section when (2.12) does not hold.

13.26. Exercise. Follow the pattern of Sec. II.5 to determine jump conditions at curves of discontinuity in the (s,t)-plane.

14. Hamilton's Principle for Hyperelastic Rods

We now carry out the analog of Sec. II.10 for rods. We do not pause to spell out the obvious regularity assumptions needed to make sense of our various expressions. We assume that the rod is hyperelastic, so that (7.16) holds. The *stored energy* in the rod at time t is

$$\Psi[\boldsymbol{r}, \boldsymbol{d}_k](t) \equiv \int_{s_1}^{s_2} W(\mathsf{u}(s,t), \mathsf{v}(s,t), s)\, ds. \tag{14.1}$$

Suppose that there is a scalar-valued function ω such that the applied force and couple densities have the form

$$\begin{aligned} \boldsymbol{f}(s,t) &= -\frac{\partial \omega}{\partial \boldsymbol{r}}(\boldsymbol{r}(s,t), \boldsymbol{d}_k(s,t), s), \\ \boldsymbol{l}(s,t) &= -\boldsymbol{d}_k(s,t) \times \frac{\partial \omega}{\partial \boldsymbol{d}_k}(\boldsymbol{r}(s,t), \boldsymbol{d}_k(s,t), s). \end{aligned} \tag{14.2}$$

ω is the *potential-energy density* of the body loads. Suppose that there are scalar-valued functions ω^α such that the end loads $\boldsymbol{n}^\alpha$ and $\boldsymbol{m}^a$ of (12.21) have the forms

(14.3)

$$\boldsymbol{n}^\alpha\big(\boldsymbol{r}(s_\alpha,t),\boldsymbol{d}_k(s_\alpha,t)\big) = (-1)^{\alpha+1}\frac{\partial\omega^\alpha}{\partial\boldsymbol{r}}\big(\boldsymbol{r}(s_\alpha,t),\boldsymbol{d}_k(s_\alpha,t)\big),$$

$$\boldsymbol{m}^\alpha\big(\boldsymbol{r}(s_\alpha,t),\boldsymbol{d}_k(s_\alpha,t)\big) = (-1)^{\alpha+1}\boldsymbol{d}_k(s_\alpha,t)\times\frac{\partial\omega^\alpha}{\partial\boldsymbol{d}_k}\big(\boldsymbol{r}(s_\alpha,t),\boldsymbol{d}_k(s_\alpha,t)\big).$$

The ω^α are the *potential energies* of the end loads. We set

(14.4)

$$\Omega[\boldsymbol{r},\boldsymbol{d}_k](t) \equiv \int_{s_1}^{s_2}\omega\big(\boldsymbol{r}(s,t),\boldsymbol{d}_k(s,t),s\big)\,ds + \sum_{\alpha=1}^{2}\omega^\alpha\big(\boldsymbol{r}(s_\alpha,t),\boldsymbol{d}_k(s_\alpha,t)\big).$$

$\Psi+\Omega$ is the *potential-energy functional* for the special Cosserat rod. The *kinetic energy* of the rod at time t is

$$(14.5)\quad K[\boldsymbol{r},\boldsymbol{d}_k](t) \equiv \tfrac{1}{2}\int_{s_1}^{s_2}[(\rho A)(s)\boldsymbol{r}_t(s,t)\cdot\boldsymbol{r}_t(s,t) + 2\rho I_\beta(\boldsymbol{r}_t\times\boldsymbol{w})\cdot\boldsymbol{d}_\beta + (\rho J_{\beta\gamma})(s)w_\beta(s,t)w_\gamma(s,t) + (\rho J_{\gamma\gamma})(s)w_3(s,t)^2]\,ds.$$

(This kinetic energy can be constructed directly from (2.10), (2.11), or else from the integral over the volume of $\frac{1}{2}\tilde{\boldsymbol{p}}_t\cdot\tilde{\boldsymbol{p}}_t$ with $\tilde{\boldsymbol{p}}$ given by (4.3a).) The *Lagrangian functional* of the rod for $\{\boldsymbol{r},\boldsymbol{d}_k\}$ having period T in t is

$$(14.6)\qquad \Lambda[\boldsymbol{r},\boldsymbol{d}_k] \equiv \int_0^T\{K[\boldsymbol{r},\boldsymbol{d}_k](t) - \Psi[\boldsymbol{r},\boldsymbol{d}_k](t) - \Omega[\boldsymbol{r},\boldsymbol{d}_k](t)\}\,dt.$$

Hamilton's Principle states that the weak form of the governing equations for hyperelastic rods subject to conservative loads and to appropriate initial and final conditions can be characterized by the vanishing of the Gâteaux differential of the Lagrangian functional in every direction compatible with the boundary conditions.

14.7. Exercise. Demonstrate the validity of Hamilton's Principle by showing that the vanishing of the Gâteaux derivative of Λ for $\{\boldsymbol{r},\boldsymbol{d}_k\}$ having period T in t and satisfying (12.20) is just the Principle of Virtual Power (13.24).

14.8. Exercise. Consider a viscoelastic rod (2.22) satisfying (8.35). Suppose that there is a scalar-valued function W such that

$$(14.9)\qquad \hat{\mathsf{m}}(\mathsf{u},\mathsf{v},\mathbf{0},\mathbf{0},s) = \partial W(\mathsf{u},\mathsf{v},s)/\partial\mathsf{u},\quad \hat{\mathsf{n}}(\mathsf{u},\mathsf{v},\mathbf{0},\mathbf{0},s) = \partial W(\mathsf{u},\mathsf{v},s)/\partial\mathsf{v}.$$

Let (14.1)–(14.5) hold. Prove that the total energy is nowhere increasing:

$$(14.10)\qquad \tfrac{d}{dt}\{K[\boldsymbol{r},\boldsymbol{d}_k](t) + \Psi[\boldsymbol{r},\boldsymbol{d}_k](t) + \Omega[\boldsymbol{r},\boldsymbol{d}_k](t)\} \le 0$$

(cf. Ex. II.2.29).

15. Material Constraints

In this section we study *material* (or *internal*) *constraints*, which restrict the kinds of deformations that a special Cosserat rod can suffer. Typical material constraints are those requiring the rod to be unshearable and inextensible, no matter what the resultant contact force and couple are. A different kind of constraint, termed *external*, is exemplified by the requirement that $\boldsymbol{r}$, $\boldsymbol{d}_1$, and $\boldsymbol{d}_3$ be coplanar, no matter what the applied forces are. Such constraints are maintained by devices external to the rod. They are discussed in the next section.

Material constraints are introduced because they simplify the treatment of the governing equations and because they are observed to be approximately true for important classes of deformations of real rods. For example, the shears and the extension are typically very small in a rod subject to purely terminal loads of 'moderate' size. A rational description of the relationship between problems with small shears and extensions and those in which these strains are constrained to be zero requires some care in its formulation. We address this question after we lay down the general theory. We note that the Special Cosserat Theory of Rods is itself a constrained version of the General Cosserat Theory, described in Sec. 17. Indeed, each theory of rods belongs to a hierarchy of such theories with the simpler theories representing constrained versions of the more complicated theories.

The simplest and most important material constraints for a special Cosserat rod are those that prevent it from suffering shear and extension:

$$\mathsf{v} = (0, 0, 1). \tag{15.1}$$

We assume that (15.1) holds no matter what loading system is applied to the rod and therefore no matter what values are assumed by m and n. Thus (15.1) is a constitutive restriction. Traditional doctrine asserts that the constitutive equations for an unshearable, inextensible elastic rod consist of (15.1) and

$$\mathsf{m}(s,t) = \hat{\mathsf{m}}\big(\mathsf{u}(s,t), s\big), \tag{15.2}$$

with n not prescribed as a constitutive function so that it remains free to assume whatever values are necessary to maintain the constraint (15.1). A generalization of (15.2) is

$$\mathsf{m}(s,t) = \mathsf{m}^\flat\big(\mathsf{u}(s,t), \mathsf{n}(s,t), s\big). \tag{15.3}$$

In either case, n, and consequently $\boldsymbol{n}$, plays the role of a fundamental unknown in the governing equations. The presence of (15.1) ensures that the number of equations equals the number of unknowns.

To motivate (15.3), we first study an unconstrained elastic rod for which (8.3) and either Hypothesis 8.9 or property (8.10) hold. Then by the results of Sec. 8, the constitutive equations (2.21) are equivalent to equations of the form

$$\mathsf{v}(s,t) = \mathsf{v}^\flat\big(\mathsf{u}(s,t), \mathsf{n}(s,t), s\big), \quad \mathsf{m}(s,t) = \mathsf{m}^\flat\big(\mathsf{u}(s,t), \mathsf{n}(s,t), s\big). \tag{15.4}$$

We regard the constraint (15.1) as a limiting case of nearly constrained materials with constitutive equations of the form (15.4). We accordingly generate (15.1) and (15.3) by the simple device of taking the constitutive function $\mathbf{v}^\flat = (0,0,1)$. In this procedure, we impose no restriction on $\mathbf{n}$. That $\mathbf{n}$ should be unrestricted is consistent with the Principle of Virtual Power (13.24), which is independent of $\mathbf{n}$ when (15.1) holds.

We obtain analogous results for viscoelastic rods (of differential type of complexity 1). If (8.35) holds, then the constitutive equations (7.31) are equivalent to equations of the form

$$
\begin{aligned}
\mathbf{v}_t(s,t) &= \dot{\mathbf{v}}^\flat\big(\mathbf{u}(s,t), \mathbf{u}_t(s,t), \mathbf{v}(s,t), \mathbf{n}(s,t), s\big), \\
\mathbf{m}(s,t) &= \mathbf{m}^\flat\big(\mathbf{u}(s,t), \mathbf{u}_t(s,t), \mathbf{v}(s,t), \mathbf{n}(s,t), s\big).
\end{aligned}
\tag{15.5}
$$

By defining $\dot{\mathbf{v}}^\flat$ to equal $\mathbf{0}$, (or, more generally, by defining $\dot{\mathbf{v}}^\flat(\mathbf{u}, \dot{\mathbf{u}}, \mathbf{0}, \mathbf{n}, s) = \mathbf{0}$) and by assuming that there is a time t_0 such that $\mathbf{v}(t_0) = (0,0,1)$ we reduce (15.5) to (15.1) and an equation of the form

$$
\mathbf{m}(s,t) = \mathbf{m}^\flat\big(\mathbf{u}(s,t), \mathbf{u}_t(s,t), \mathbf{n}(s,t), s\big). \tag{15.6}
$$

A hyperelastic rod subject to constraint (15.1) under a conservative loading can be described by Hamilton's Principle. We use the Multiplier Rule (see Bliss (1946), e.g.) to formulate an equivalent principle in which we replace the integrand $W(\mathbf{u}(s,t), \mathbf{v}(s,t), s)$ of the unconstrained problem with $W(\mathbf{u}(s,t), s) + \mathbf{n}(s,t) \cdot \mathbf{v}(s,t)$ for the constrained problem. Here $\mathbf{n}$ is a triple of Lagrange multipliers. We then find that the governing Euler-Lagrange equations correspond to the special choice (15.2), rather than to the more general (15.3).

To explain this phenomenon in the context in which we obtained (15.3), we introduce a Legendre transformation $W^\flat$ as in (8.30) so that the constitutive functions of (15.4) have the form

$$
\mathbf{v}^\flat(\mathbf{u}, \mathbf{n}, s) = \partial W^\flat(\mathbf{u}, \mathbf{n}, s)/\partial \mathbf{n}, \quad \mathbf{m}^\flat(\mathbf{u}, \mathbf{n}, s) = -\partial W^\flat(\mathbf{u}, \mathbf{n}, s)/\partial \mathbf{u}. \tag{15.7}
$$

Thus the matrix

$$
\begin{pmatrix} \partial \mathbf{m}^\flat/\partial \mathbf{u} & \partial \mathbf{m}^\flat/\partial \mathbf{n} \\ -\partial \mathbf{v}^\flat/\partial \mathbf{u} & -\partial \mathbf{v}^\flat/\partial \mathbf{n} \end{pmatrix} \tag{15.8}
$$

is symmetric. Now in any limit process in which $\mathbf{v}^\flat \to (0,0,1)$, the symmetry of (15.8) is destroyed unless $\partial \mathbf{m}^\flat/\partial \mathbf{n}$ also goes to $\mathbf{O}$. Thus for hyperelastic rods, (15.3) reduces to (15.2).

The foregoing considerations can be extended to any consistent set of K material constraints of the frame-indifferent form

$$
\boldsymbol{\kappa}(\mathbf{u}, \mathbf{v}, s) = 0. \tag{15.9}
$$

We take $K \leq 5$. If $K = 6$, then the deformation is rigid. We regard these constraints as defining a $(6-K)$-dimensional surface in the six-dimensional space of $(\mathbf{u}, \mathbf{v})$. At any point of this surface, we introduce a local set of curvilinear coordinates in the surface and a complementary set normal to the surface. The coordinates in the surface are treated like $\mathbf{u}$ and those normal to the surface like $\mathbf{v}$. The projection of $(\mathbf{m}, \mathbf{n})$ onto the surface is defined constitutively; its projection normal to the surface is not so defined (and plays the role of $\mathbf{n}$ described above). The details of this approach are given by Antman (1982). For a full treatment in the setting of three-dimensional continuum mechanics, see Sec. XII.12.

We now develop a formal procedure that yields representations for solutions of the equations for nearly constrained elastic materials. These procedures are particularly useful to treat problems whose constrained versions reduce to the Kirchhoff Kinetic Analogy (see (10.4)) because many of these can be solved in closed form in terms of elliptic functions. (See the references in Chap. V.) Our perturbation methods yield corrections for slight extensibility and shearability, which enable their effect to be assessed. Let ε be a nonnegative parameter that measures the discrepancy between $\mathbf{v}$ and its constrained value $(0,0,1)$. We introduce a one-parameter family of materials of the form (15.4):

$$\text{(15.10a,b)} \quad \mathbf{v}^\flat(\mathbf{u}, \mathbf{n}, s) = (0,0,1) + \varepsilon \mathbf{v}^\star(\mathbf{u}, \mathbf{n}, s, \varepsilon), \quad \mathbf{m}^\flat(\mathbf{u}, \mathbf{n}, s) = \mathbf{m}^\star(\mathbf{u}, \mathbf{n}, s, \varepsilon)$$

where $\mathbf{v}^\star$ and $\mathbf{m}^\star$ are assumed to be $(p+1)$-times continuously differentiable functions of $\mathbf{u}$, $\mathbf{n}$, and ε. We assume that all the variables of the governing equations have Taylor expansions up to the pth power in ε:

$$\text{(15.11)} \qquad \mathbf{u}(s,t,\varepsilon) = \sum_{a=0}^{p} \mathbf{u}_a(s,t) \frac{\varepsilon^a}{a!} + O(\varepsilon^{p+1}), \quad \text{etc.}$$

We find the unknown coefficients by differentiating the equations repeatedly with respect to ε and then setting $\varepsilon = 0$ (cf. Sec. II.8). Clearly, the problem obtained by simply setting $\varepsilon = 0$ in the governing equations is that for the material satisfying the constraint (15.1).

15.12. Exercise. Carry out this perturbation process to obtain the first correction to the equations of motion for elastic materials satisfying the constraint (15.1). Note that all corrections are governed by linear equations.

There is a simple variant of this procedure that does not require that we express our constitutive equations explicitly in the form (15.4) or (15.10). Inspired by (15.10a), we introduce a new strain variable $\mathbf{y}$ by

$$\text{(15.13)} \qquad \mathbf{v} = (0,0,1) + \varepsilon \mathbf{y}$$

and adopt constitutive equations in the form

(15.14a,b)

$$\mathbf{m}(s,t) = \mathbf{m}^+(\mathbf{u}(s,t), \mathbf{y}(s,t), s, \varepsilon), \quad \mathbf{n}(s,t) = \mathbf{n}^+(\mathbf{u}(s,t), \mathbf{y}(s,t), s, \varepsilon),$$

where $\mathbf{m}^+$ and $\mathbf{n}^+$ are assumed to be $(p+1)$-times continuously differentiable functions of $\mathbf{u}$, $\mathbf{y}$, and ε. We assume that $\mathbf{y}$ and all the other variables of the governing equations have the form (15.11).

15.15. Exercise. If $\mathbf{n}^{+}(\mathbf{u}, \cdot, s, \varepsilon)$ is invertible, show that the equations for the leading term (for $\varepsilon = 0$) of the governing equations correspond to (15.1) and (15.3). Show that if furthermore (15.14a) is specialized to

$$\mathbf{m}(s,t) = \bar{\mathbf{m}}(\mathbf{u}(s,t), s) + \varepsilon \check{\mathbf{m}}(\mathbf{u}(s,t), \mathbf{y}(s,t), s, \varepsilon), \tag{15.16}$$

then the equations for the leading term correspond to (15.1) and (15.2). Find the equations governing the first-order corrections. (After $\mathbf{n}_0$ and $\mathbf{u}_0$ are found, then $\mathbf{y}_0$ can be found from (15.14b).) Show that the hyperelastic specialization of (15.16) and (15.14b) is obtained by taking the stored-energy function W to have the form

$$W(\mathbf{u}, \mathbf{v}, s) = \bar{W}(\mathbf{u}, s) + \varepsilon \check{W}(\mathbf{u}, \mathbf{y}, s, \varepsilon). \tag{15.17}$$

There remains the substantial analytic problem of justifying the validity of the formal expansions (15.11).

Most of this section is based on Antman (1982). A formal theory for the relaxation of the constraint of incompressibility was developed by Spencer (1964). A refinement of his approach in the setting of planar rod theories was given by Antman (1968); it is generalized in Ex. 15.15. For mathematical justifications of such methods, see Schochet (1985).

16. External Constraints, Planar Problems, Classical Theories

External constraints enforcing planarity. Suppose that the natural reference configuration of a rod, regarded for the moment as a three-dimensional body, is a bent and twisted prism with a rectangular cross section. That is, this reference configuration is generated by translating the center of a rectangle along a fixed C^1-space curve with the plane of the rectangle perpendicular to the curve and with the rectangle allowed to rotate about its center in some prescribed way. We could force this rod to undergo purely 'planar' deformations by flattening out the reference curve, untwisting the cross sections so that a pair of opposite sides of the rectangle now generate a pair of parallel planes, and then confining the rod to move between a pair of lubricated rigid planes containing these parallel planes. See Fig. 16.1.

Fig. 16.1a. Natural reference configuration.

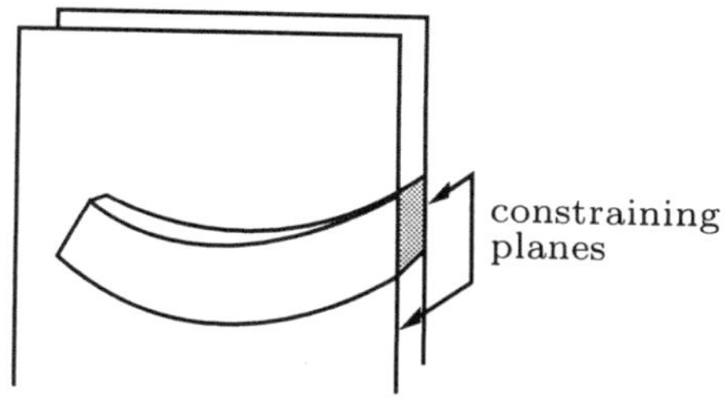

Fig. 16.1b. Constrained configuration.

Let $\{\boldsymbol{i}, \boldsymbol{j}, \boldsymbol{k}\}$ be a fixed orthonormal basis for $\mathbb{E}^3$. Let the constraining planes be perpendicular to $\boldsymbol{k}$. In terms of the special Cosserat theory, these constraints can be expressed by

$$\boldsymbol{r} \cdot \boldsymbol{k} = 0, \quad \boldsymbol{d}_2 = \boldsymbol{k}. \tag{16.2}$$

Following the ideas developed in Secs. 12, 13, and 15, we conclude from (16.2) that the virtual velocities $\overset{\vartriangle}{\boldsymbol{r}}$ and $\boldsymbol{z}$ must satisfy

$$\overset{\vartriangle}{\boldsymbol{r}} \cdot \boldsymbol{k} = 0, \quad \overset{\vartriangle}{\boldsymbol{d}}_2 = \boldsymbol{z} \times \boldsymbol{d}_2 = \boldsymbol{0}. \tag{16.3}$$

Thus the expressions

$$\boldsymbol{f} \cdot \boldsymbol{k} \equiv \boldsymbol{f} \cdot \boldsymbol{d}_2, \quad \boldsymbol{l} \cdot \boldsymbol{d}_1, \quad \boldsymbol{l} \cdot \boldsymbol{d}_3 \tag{16.4}$$

do not appear in the Principle of Virtual Power (13.24) because their coefficients vanish by virtue of (16.3). In their places are the constraint forces and couples (Lagrange multipliers) maintaining (16.2).

These considerations readily handle any constraints like (16.2). Such constraints are termed *external* because they are enforced by the specific environment in which the rod is placed, whereas a material (or *internal*) constraint operates in every environment and characterizes the material response. In particular, (16.2) implies that $v_2 = 0$. The force n_2 nevertheless is determined from the constitutive equations, which give the value of this component necessary to enforce the constraint. If the constraining planes were removed, then v_2 could assume any value. If $v_2 = 0$ were a material constraint, then it would always be in effect and n_2 would not be determined constitutively.

Our three-dimensional interpretation of (16.2) is equivalent to the prescription of certain kinds of position boundary conditions on the lateral boundary of the rod. This equivalence is the underlying source of the similarity between our treatments of boundary conditions in Sec. 12 and of external constraints here. The boundary conditions (12.20) may be regarded as external constraints.

Naturally planar motions. A rod may have configurations satisfying (16.2) without being forced to do so by constraints. Suppose that the natural reference configuration of a rod satisfies (16.2), that $\boldsymbol{f} \cdot \boldsymbol{k} = 0$, that $\boldsymbol{k} \times \boldsymbol{l} = \boldsymbol{0}$, that $\rho J_{12} = 0$, and that the constitutive equations are such that $\boldsymbol{n} \cdot \boldsymbol{k} = 0$ and $\boldsymbol{k} \times \boldsymbol{m} = \boldsymbol{0}$ when $\boldsymbol{r}_s \cdot \boldsymbol{k} \equiv v_2 = 0$ and $\boldsymbol{k} \times \boldsymbol{u} = \boldsymbol{0}$. Then the equations of motion may be expected to admit planar solutions (satisfying (16.2)). (These solutions, however, need not be stable; see Fig. V.2.3.)

When we study planar motions of a rod, we usually take the plane of motion to be the $\{\boldsymbol{i}, \boldsymbol{j}\}$-plane. We then adopt the following notation:

$$\boldsymbol{d}_3 \equiv \boldsymbol{a} = \cos\theta \boldsymbol{i} + \sin\theta \boldsymbol{j}, \quad \boldsymbol{d}_1 \equiv \boldsymbol{b} = -\sin\theta \boldsymbol{i} + \cos\theta \boldsymbol{j}, \quad \boldsymbol{d}_2 = \boldsymbol{k}, \tag{16.5}$$

$$\boldsymbol{r}_s = \nu \boldsymbol{a} + \eta \boldsymbol{b}, \tag{16.6}$$

$$\boldsymbol{m} = M\boldsymbol{k}, \quad \boldsymbol{n} = N\boldsymbol{a} + H\boldsymbol{b}, \tag{16.7}$$

so that

$$u_1 = 0, \quad u_2 = \theta_s, \quad u_3 = 0, \tag{16.8}$$

$$v_1 = \eta, \quad v_2 = 0, \quad v_3 = \nu, \tag{16.9}$$

$$m_1 = 0, \quad m_2 = M, \quad m_3 = 0, \quad n_1 = H, \quad n_2 = 0, \quad n_3 = N. \tag{16.10}$$

Classical nonlinear theories. The classical elastic rod theory of Kirchhoff (1859), called the *kinetic analogy*, is based on the material constraints (15.1) and on linear constitutive equations of the form

$$m_\alpha(s,t) = (EJ_{\alpha\beta})(s)[u_\beta(s,t) - \overset{\circ}{u}_\beta(s)], \quad m_3(s,t) = D(s)\, u_3(s,t). \tag{16.11}$$

(In this case, the governing equilibrium equations for a terminally loaded uniform rod are the same as the equations of motion of a heavy rigid body.) The functions $EJ_{\alpha\beta}$ are defined as the $\rho J_{\alpha\beta}$ of (4.7) (see Sec. 10). The torsional stiffness is usulally found by solving the St. Venant torsion problem of linear elasticity; it reduces to $EJ_{\alpha\alpha}$ for isotropic rods with a circular cross section.) The further requirement that the motion be naturally planar reduces (16.11) to the *Bernoulli-Euler* constitutive equation for the *elastica*

$$M(s,t) = (EJ_{22})(s)[\theta_s(s,t) - \overset{\circ}{\theta}_s(s)] \tag{16.12}$$

due to Jas. Bernoulli (1694), Euler (1727), D. Bernoulli (1732), and Euler (1732) (see Truesdell (1960)).

16.13. Exercise. Find the equations of motion for the elastica when it is subject to zero body loads and specialize them to equilibrium equations.

Many authors ignore 'rotatory inertia' in their formulation of the equations of motion for rods by setting $\rho J_{\alpha\beta} = 0$ (thus causing the angular momentum to vanish). For thin rods, these inertias are small. Nevertheless, we shall not introduce this assumption, which simplifies the governing equations and changes their type, because it does not necessarily simplify the analysis (see Caflisch & Maddocks (1984)). Indeed, a careful analysis of the role of small rotatory inertia seems to require a delicate asymptotic treatment of initial layers.

Classical linear theories. We now linearize the equations of motion of nonlinearly elastic rods about their reference configurations. We then show how the imposition of further restrictions leads to a variety of theories for elastic rods standard in the engineering literature. For simplicity of exposition, we restrict our attention to the naturally planar motions just discussed. In view of (16.5)–(16.10), we can write the governing equations of motion (2.15) and (2.17) under the assumption (2.12) in the form

$$N_s - H\theta_s + \boldsymbol{f} \cdot \boldsymbol{a} = \rho A \boldsymbol{r}_{tt} \cdot \boldsymbol{a}, \tag{16.14}$$

$$H_s + N\theta_s + \boldsymbol{f} \cdot \boldsymbol{b} = \rho A \boldsymbol{r}_{tt} \cdot \boldsymbol{b}, \tag{16.15}$$

$$M_s + \nu H - \eta N + l = \rho J \theta_{tt} \tag{16.16}$$

where $l = \boldsymbol{l} \cdot \boldsymbol{k}$ and $\rho J = \rho J_{22}$. The constitutive equations (2.21) for unconstrained elastic rods reduce to (IV.1.21):

$$\begin{aligned} N(s,t) &= \hat{N}(\nu(s,t), \eta(s,t), \theta_s(s,t), s), \\ H(s,t) &= \hat{H}(\nu(s,t), \eta(s,t), \theta_s(s,t), s), \\ M(s,t) &= \hat{M}(\nu(s,t), \eta(s,t), \theta_s(s,t), s). \end{aligned} \tag{16.17}$$

We assume that $\hat{N}$ and $\hat{M}$ are even in η and that $\hat{H}$ is odd in η. We denote the third argument of these constitutive functions by μ. We assume that these functions vanish in the reference configuration.

We follow the procedures of Sec. II.9. We let ε be a small parameter and take the body forces and the initial data to have the form

$$\boldsymbol{f} = \varepsilon \boldsymbol{f}_1, \quad l = \varepsilon l_1, \tag{16.18a,b}$$

$$\boldsymbol{r}(s,0) = \mathring{\boldsymbol{r}}(s) + \varepsilon \boldsymbol{u}_1, \quad \boldsymbol{r}_t(s,0) = \varepsilon \boldsymbol{v}_1. \tag{16.18c,d}$$

We assume that the boundary data are perturbations of data valid for the reference configuration; we do not bother to spell out the possible forms, which are analogous to (16.18c,d).

Let us first study the case in which the reference configuration is straight with sections perpendicular to the axis. In this case, we set

$$\boldsymbol{r}(s,t) = x(s,t)\boldsymbol{i} + y(s,t)\boldsymbol{j}. \tag{16.19}$$

The reference configuration is then characterized by

$$\mathring{\boldsymbol{r}}(s) = s\boldsymbol{i}, \quad \mathring{x}(s) = s, \quad \mathring{y}(s) = 0, \quad \mathring{\boldsymbol{a}}(s) = \boldsymbol{i}, \quad \mathring{\boldsymbol{b}}(s) = \boldsymbol{j}. \tag{16.20}$$

For naturally straight rods, we also require that $\hat{N}$ and $\hat{H}$ be even in μ and that $\hat{M}$ be odd in μ.

We denote the variable of the first perturbation by subscripts 1. By the procedures of Sec. II.9 we readily find that (16.5), (16.6), (16.14)–(16.17) yield the first perturbation

$$\begin{aligned} \boldsymbol{a}_1 &= \theta_1 \boldsymbol{j}, \quad \boldsymbol{b}_1 = -\theta_1 \boldsymbol{i}, \\ \partial_s x_1 &= \nu_1, \quad \partial_s y_1 = \theta_1 + \eta_1, \\ \partial_s N_1 + \boldsymbol{f}_1 \cdot \boldsymbol{i} &= \rho A \partial_{tt} x_1, \\ \partial_s H_1 + \boldsymbol{f}_1 \cdot \boldsymbol{j} &= \rho A \partial_{tt} y_1, \\ \partial_s M_1 + H_1 + l_1 &= \rho J \partial_{tt} \theta_1, \\ N_1 &= N_\nu^0(s)\nu_1 \equiv N_\nu(1,0,0,s)\nu_1, \\ H_1 &= H_\eta^0(s)\eta_1 \equiv H_\eta(1,0,0,s)\eta_1, \\ M_1 &= M_\mu^0(s)\partial_s\theta_1 \equiv M_\mu(1,0,0,s)\partial_s\theta_1 \end{aligned} \tag{16.21}$$

with the form of the linearized constitutive equations reflecting the symmetries of (16.17) in η and μ. The monotonicity condition (8.3) implies that the moduli $N_\nu^0, H_\eta^0, M_\mu^0$ are each positive. From (16.21) we readily obtain the linear hyperbolic system

$$\partial_s[N_\nu^0 \partial_s x_1] + \boldsymbol{f}_1 \cdot \boldsymbol{i} = \rho A \partial_{tt} x_1, \tag{16.22}$$

$$\partial_s[H_\eta^0(\partial_s y_1 - \theta_1)] + \boldsymbol{f}_1 \cdot \boldsymbol{j} = \rho A \partial_{tt} y_1, \tag{16.23a}$$

$$\partial_s[M_\mu^0 \partial_s \theta_1] + H_\eta^0(\partial_s y_1 - \theta_1) + l_1 = \rho J \partial_{tt} \theta_1. \tag{16.23b}$$

Notice that (16.22) is uncoupled from (16.23). For both mathematical convenience and physical interpretation, it seems best to retain (16.23) in the present form as a system of two second-order equations for y_1 and θ_1. It is nevertheless traditional in structural mechanics to convert (16.23) to a single fourth-order equation. In the rest of this section we drop the subscript 1 from the variables of (16.23).

16.24. Exercise. Derive the identities

$$M_{ss} + \rho A y_{tt} - \boldsymbol{f} \cdot \boldsymbol{j} + l_s = (\rho J \theta_{tt})_s, \tag{16.25a}$$

$$y_s = \theta + \left(H_\eta^0\right)^{-1} \left[\rho J \theta_{tt} - (M_\mu^0 \theta_s)_s - l\right] \tag{16.25b}$$

to reduce (16.23) to

$$\left[\frac{(M_\mu^0 \theta_s)_{ss}}{\rho A}\right]_s + \theta_{tt} + \frac{\rho J \theta_{tttt}}{(H_\eta^0)} - \frac{(M_\mu^0 \theta_{tts})_s}{(H_\eta^0)} - \left[\frac{(\rho J \theta_{tt})_s}{\rho A}\right]_s = \frac{l_{tt}}{H_\eta^0} + \left(\frac{l_s}{\rho A}\right)_s - \left(\frac{\boldsymbol{f} \cdot \boldsymbol{j}}{\rho A}\right)_s. \tag{16.26}$$

Note that (16.26) contains the fourth-order derivatives $\theta_{tttt}, \theta_{sstt}, \theta_{ssss}$. It is common to neglect rotatory inertia by setting $\rho J = 0$. (For thin rods, ρJ can be reckoned as negligible with respect to ρA.) In this case, the mathematical structure of (16.23) or (16.26) (which resides in their classification as partial differential equations) changes significantly. One can expect that differences in solutions in the resulting equations would be most pronounced in an initial layer.

For unshearable materials, the linearized variable η is constrained to vanish, so that

$$\theta = y_s, \tag{16.27}$$

and H is not delivered by a constitutive equation. The substitution of (16.27) into (16.25a), which is still valid, then yields

$$\left(M_\mu^0 y_{ss}\right)_{ss} + \rho A y_{tt} - \boldsymbol{f} \cdot \boldsymbol{j} + l_s = \left(\rho J y_{stt}\right)_s. \tag{16.28}$$

(If we divide (16.28) by ρA and then differentiate it with respect to s, we get an equation equivalent to (16.26) with H_η^0 formally set equal to ∞.) The most commonly used version of the equation for the flexural motion of a rod is obtained by setting $\rho J = 0$ in (16.28).

Because of the artificiality of the equations (16.26) and (16.28), it is sometimes necessary to carry out a variety of contortions to handle boundary conditions. For example, suppose that H is required to vanish at an end. From (16.21) we find that the appropriate boundary condition for (16.26) is that the complicated expression

$$\left(M_\mu^0 \theta_s\right)_s + l - \rho J \theta_{tt}$$

vanish at that end.

If the reference configuration is not straight, then the uncoupling between the extension in (16.22) and the flexure and shear in (16.23) is lost. The following exercise illustrates this.

16.29. Exercise. Represent $\boldsymbol{r}$ in polar coordinates by

$$\begin{gathered} \boldsymbol{r}(s,t) = r(s,t)\boldsymbol{e}_1(s,t), \\ \boldsymbol{e}_1(s,t) = \cos\phi(s,t)\boldsymbol{i} + \sin\phi(s,t)\boldsymbol{j}, \quad \boldsymbol{e}_2(s,t) = -\sin\phi(s,t)\boldsymbol{i} + \cos\phi(s,t)\boldsymbol{j}. \end{gathered} \tag{16.30}$$

(The coordinates (r, ϕ) of (16.29) replace (x, y) of (16.19).) Obtain the equations of motion analogous to (16.22) and (16.23).

17. General Theories of Cosserat Rods

In Sec. 1 we defined the motion of a special Cosserat rod to be specified by the three functions $\boldsymbol{r}$, $\boldsymbol{d}_1$, and $\boldsymbol{d}_2$ where $\{\boldsymbol{d}_1(s,t), \boldsymbol{d}_2(s,t)\}$ is orthonormal for each (s,t). The directors $\boldsymbol{d}_1(s,t)$ and $\boldsymbol{d}_2(s,t)$ fix a plane through $\boldsymbol{r}(s,t)$ and a line in this plane through $\boldsymbol{r}(s,t)$. This information is interpreted as characterizing the gross behavior of the material section s at time t. An obvious generalization of this theory is obtained by suspending the requirement that the directors be orthonormal. The resulting generalization, called the *full 2-director Cosserat theory*, has enough versatility to describe stretching and shearing within cross sections. We can retain the interpretations of $\boldsymbol{d}_1$ and $\boldsymbol{d}_2$ implicit in (4.1) and (4.3).

More general theories can be constructed by increasing the number of directors from two to any finite number. The additional directors allow additional three-dimensional features of of deformed cross sections, such as curvatures, to be described. For example, we could interpret five (unconstrained) directors $\boldsymbol{d}_1, \boldsymbol{d}_2, \boldsymbol{d}_{11}, \boldsymbol{d}_{12}, \boldsymbol{d}_{22}$ by a generalization of (4.2g):

$$\tag{17.1} \begin{aligned} \tilde{\boldsymbol{p}}(\mathbf{x},t) &= \boldsymbol{r}(s,t) + x^1\boldsymbol{d}_1(s,t) + x^2\boldsymbol{d}_2(s,t) \\ &\quad + \tfrac{1}{2}(x^1)^2\boldsymbol{d}_{11}(s,t) + x^1x^2\boldsymbol{d}_{12}(s,t) + (x^2)^2\boldsymbol{d}_{22}(s,t). \end{aligned}$$

The treatment of the geometry of deformation for any of these theories can be carried out in imitation of the development of Secs. 1–8. In doing so we confront the technical difficulty that the failure of $\{\boldsymbol{d}_1, \boldsymbol{d}_2\}$ to be orthonormal deprives the analysis of a physically natural and mathematically convenient orthonormal basis $\{\boldsymbol{d}_k\}$ and prevents the appearance of the vector $\boldsymbol{u}$, which played such an important role in this chapter. But if $\boldsymbol{r}$, $\boldsymbol{d}_1$, and $\boldsymbol{d}_2$ essentially retain the interpretations assigned to them in Sec. 2, then we can require them to satisfy

$$\tag{17.2} \boldsymbol{r}_s \cdot (\boldsymbol{d}_1 \times \boldsymbol{d}_2) > 0,$$

which generalizes (2.7). (Under mild assumptions on the representation of $\tilde{\boldsymbol{p}}$, it can be shown by the methods of Sec. 6 that (17.2) is a consequence of the three-dimensional requirement that $\tilde{\boldsymbol{p}}$ preserve orientation.) Thus $\{\boldsymbol{d}_1, \boldsymbol{d}_2, \boldsymbol{r}_s\}$ is a basis at each (s,t). It, together with its dual basis, can be used to give a full treatment of the geometry of deformation.

The basic difficulty in constructing general Cosserat theories is to introduce a full set of equations of motion. In particular, since the configuration of the full 2-director Cosserat theory is described by three unconstrained vector-valued functions, we need three vector equations of motion to have as many equations as unknowns. Two of these vector equations should be the impulse-momentum laws developed in Secs. 2, 4, and 13. What is the third? There are two approaches by which we could answer this question. In the first, we would replace the development of Secs. 3 and 4 with a far more sophisticated analysis, tantamount to the use of the full three-dimensional theory of continuum mechanics. This approach, developed in Chap. XIV, while not intrinsically one-dimensional, yields a completely satisfactory physical interpretation of the resulting equations. In the second approach, we construct the requisite equations by postulating them in the framework of a suitable principle of virtual power. In doing so, we would tacitly rely on the doctrine that all equations of continuum mechanics have such characterizations. This approach would yield the governing equations with little effort, but at the sacrifice of a concrete physical intuition based upon explicit representations for the additional generalized force resultants. (For conservative problems, this approach is formally equivalent to generating the governing equations by Hamilton's Principle.) Thus it is only the special Cosserat theory that admits an intuitively natural, intrinsic formulation, at least for statics. But even here, a deeper probing of the basic concepts inexorably leads to the doors of the three-dimensional theory of solids. The virtue of the special Cosserat theory is that the meaning of the resultants $\boldsymbol{n}$ and $\boldsymbol{m}$ is plain, so that only elementary aspects of the full three-dimensional theory are needed to illuminate the basic notions of the special Cosserat theory, as Secs. 3, 4, and 6 show. For intrinsic developments of general Cosserat theories, see Sec. XIV.7 and the references cited at the end of the next section.

18. Historical and Bibliographical Notes

A theory of rods may be viewed as a theory of one-dimensional bodies that can move in $\mathbb{E}^3$ or as an approximation to the theory of certain slender three-dimensional bodies. In this chapter we have examined in detail the nature of a single intrinsically one-dimensional theory, the special Cosserat theory. In Sec. 17 we give a cursory description of more general *intrinsic* or *direct* theories and in Sec. XIV.7 we give a more detailed account. (The derivation of rod theories from the three-dimensional theory is carried out in Chap. XIV.) Our presentation, especially that of Secs. 3, 4, 6, and 17, shows that although it is logically possible to construct intrinsically one-dimensional theories, at least for static problems, some inspiration, however tacit, from the three-dimensional theory is needed to prescribe the form of acceleration terms, to express the requirement that the deformation preserve orientation, and to interpret the variables of more complicated models. Here we outline the development of the main ideas that have led to modern direct theories of rods.

The first major steps in creating a theory of rods were taken by Jas. Bernoulli (1694). He gave the equation of moment balance for the planar deformation of an (inextensible, unshearable) elastica and also gave a separate *nonlinear* constitutive equation relating the curvature and the bending couple. To obtain this constitutive equation, he regarded the rod as behaving like a bundle of noninteracting fibers, each of which can be stretched (or compressed). Euler (1727) derived the classical linear relation between bending couple and change in curvature by assuming the elongation to vary linearly across each cross section, by assuming each fiber to obey Hooke's law, and by integrating the moment of the corresponding 'stress' across the section to get the constitutive equation for the bending couple. In this last step, Euler followed the method introduced by Leibniz (1684). Euler's work was long unpublished. The definitive form of this constitutive equation, the *Bernoulli-Euler law of bending* (16.12), was obtained by D. Bernoulli (1732). A full and coherent exposition of the theory of the elastica under an arbitrary prescribed system of planar loads, based upon the moment balance, was given by Euler (1727). The power and beauty of this theory became fully evident in Euler's (1744) analysis of all equilibrium states possible for a uniform, initially straight elastica under terminal loads. The full theory for the planar motion of elasticae, incorporating the equations for both force and moment balance, was given by Euler (1771, 1774). Although certain statically determinate problems can be handled with the theory expounded in Euler's paper of 1727, the treatment of statically indeterminate problems, the treatment of the statics of rods that can suffer extension and shear, and the treatment of all dynamical problems requires the full set of equations he derived in 1771 and 1774. In this work he introduced the notion of shear force.

Euler (1775) went far toward establishing the equations for the motion of rods in space. In this process, he developed much of the apparatus of the differential geometry of curves in space. (A full account of the researches described so far can be found in Truesdell (1960).) Euler, however, lacked a theory of strain for rods that can deform in space. St. Venant (1843, 1845) introduced the notion of twist (*gauchissement*). Kirchhoff (1859), Clebsch (1862), Thomson & Tait (1867), and Love (1893) successively refined the notion of strain and thereby established Kirchhoff's kinetic analogy, a completely satisfactory, geometrically exact theory for the deformation of rods in space. It is the natural generalization of the elastica theory (see (16.11) and (16.12)); see Dill (1992).

By using a variational characterization of the governing equations, the Cosserats (1907, 1909) introduced as the generalization of Kirchhoff's theory what we term the special Cosserat theory. Many subsequent independent refinements of rod theory, such as Timoshenko's (1921) treatment of shear deformation, can be subsumed under the

Cosserat theory. Ericksen & Truesdell (1958) revived the Cosserats' theory, extended it to the full 3-director theory, and gave a complete and precise analysis of strain. Alexander & Antman (1982) gave a global treatment of strain, resolving certain paradoxes in the specification of boundary conditions.

Ericksen & Truesdell formulated the classical equilibrium equations for the contact force $\boldsymbol{n}$ and contact couple $\boldsymbol{m}$. They did not, however, discuss the additional resultants corresponding to the additional kinematical variables through the Principle of Virtual Power, the additional equilibrium equations in which these resultants appear, and constitutive equations relating the resultants to the full set of strains. Their work inspired new interest in the formulation of geometrically exact rod theories more general than Kirchhoff's.

Cohen (1966) used a variational method to construct a theory for the statics of a 3-director theory of hyperelastic rods. Green & Laws (1966) formulated a thermodynamical 2-director theory of rods, which is marred only by the artificiality of their introduction of the equations of motion for the very resultants not considered by Ericksen & Truesdell (see Antman (1972, p. 668)). DeSilva & Whitman (1969, 1971) refined and elaborated these formulations, giving a natural postulational development for thermodynamical 3-director theories. Cohen & Wang (1989) gave a coordinate-free treatment of the full Cosserat theory. All these authors used frame-indifference to simplify the forms of the governing equations. (There is no obvious need that is met by theories with exactly three directors. Indeed, (17.1) suggests that if a 2-director theory is inadequate, then it should be replaced with a 5-director theory.) A summary of this work, together with additional references, is given by Antman (1972). Many of the details of the special Cosserat theory presented in this chapter were introduced by Antman (1974b).

CHAPTER IX

Spatial Problems for Cosserat Rods

1. Summary of the Governing Equations

In this chapter we study spatial deformations for nonlinearly elastic rods. We collect here the governing partial differential equations for transversely isotropic rods from the preceding chapter. Equations (VIII.2.3) and (VIII.2.4) yield

(1.1a,b,c) $$\boldsymbol{r}_s = v_k \boldsymbol{d}_k, \quad \partial_s \boldsymbol{d}_k = \boldsymbol{u} \times \boldsymbol{d}_k, \quad \boldsymbol{u} = u_k \boldsymbol{d}_k.$$

Under the assumptions (VIII.2.12), which are appropriate for transversely isotropic rods, we reduce (VIII.2.15) and (VIII.2.17) for zero body loads to

(1.2a,b) $$\boldsymbol{n}_s = \rho A \boldsymbol{r}_{tt}, \quad \boldsymbol{m}_s + \boldsymbol{r}_s \times \boldsymbol{n} = \varepsilon_{\alpha\gamma}\varepsilon_{\beta\delta}\rho J_{\gamma\delta}\partial_t\{\boldsymbol{d}_\alpha \times \partial_t \boldsymbol{d}_\beta\}.$$

We assume that the constitutive functions (VIII. 2.21) satisfy the monotonicity condition (VIII.8.3) and the coercivity condition (VIII.8.10), so that they can be inverted to yield (VIII.8.12). We further assume that the material is transversely isotropic so that (VIII.9.27) holds and so that these constitutive equations have a form corresponding to that of (VIII.9.17):

(1.3a) $$\hat{u}_\alpha(\mathsf{m},\mathsf{n}) = u\big(J(\mathsf{m},\mathsf{n}),s\big)m_\alpha + u^\times\big(J(\mathsf{m},\mathsf{n}),s\big)n_\alpha,$$

(1.3b) $$\hat{v}_\alpha(\mathsf{m},\mathsf{n}) = v^\times\big(J(\mathsf{m},\mathsf{n}),s\big)m_\alpha + v\big(J(\mathsf{m},\mathsf{n}),s\big)n_\alpha,$$

(1.3c,d) $$\hat{u}_3(\mathsf{m},\mathsf{n}) = \check{u}_3\big(J(\mathsf{m},\mathsf{n}),s\big), \quad \hat{v}_3(\mathsf{m},\mathsf{n}) = \check{v}_3\big(J(\mathsf{m},\mathsf{n}),s\big),$$

(1.3e) $$J(\mathsf{m},\mathsf{n}) \equiv (m_\alpha m_\alpha,\, m_\alpha n_\alpha,\, n_\alpha n_\alpha,\, m_3,\, n_3).$$

We assume that the constitutive functions u, $u^\times$, $v^\times$, v, $\check{u}_3$, and $\check{v}_3$ are continuously differentiable.

1.4. Exercise. For an isotropic, hyperelastic material, prove that

(1.5) $$u^\times = v^\times.$$

Let $\{\boldsymbol{i}_l\}$ be a fixed right-handed orthonormal basis, to be determined from the data of our problems. Identifying this basis with $\{\boldsymbol{j}_l\}$ of Sec.

VIII.11 and using the auxiliary basis $\{\boldsymbol{k}_l\}$ introduced there, we obtain representations for the directors and for $\mathbf{u}$ in terms of the Euler angles:

(1.6)
$$\begin{aligned} \boldsymbol{d}_1 &= \cos\psi(\cos\theta\boldsymbol{h} - \sin\theta\boldsymbol{i}_3) + \sin\psi\,\boldsymbol{i}_3 \times \boldsymbol{h} = \cos\psi\boldsymbol{k}_1 + \sin\psi\boldsymbol{k}_2, \\ \boldsymbol{d}_2 &= -\sin\psi(\cos\theta\boldsymbol{h} - \sin\theta\boldsymbol{i}_3) + \cos\psi\,\boldsymbol{i}_3 \times \boldsymbol{h} = -\sin\psi\boldsymbol{k}_1 + \cos\psi\boldsymbol{k}_2, \\ \boldsymbol{d}_3 &= \sin\theta\boldsymbol{h} + \cos\theta\boldsymbol{i}_3 = \boldsymbol{k}_3, \\ \boldsymbol{h} &\equiv \cos\phi\boldsymbol{i}_1 + \sin\phi\boldsymbol{i}_2 = \cos\theta\boldsymbol{k}_1 + \sin\theta\boldsymbol{k}_3, \end{aligned}$$

from which we find

(1.7a) $$u_1 = \theta_s \sin\psi - \phi_s \sin\theta\cos\psi,$$

(1.7b) $$u_2 = \theta_s \cos\psi + \phi_s \sin\theta\sin\psi,$$

(1.7c) $$u_3 = \psi_s + \phi_s \cos\theta.$$

We denote the components of $\boldsymbol{m}$, $\boldsymbol{n}$, $\boldsymbol{u}$, $\boldsymbol{v}$ with respect to $\{\boldsymbol{k}_l\}$ by capital letters:

(1.8a) $$\boldsymbol{m} = m_l\boldsymbol{d}_l = M_l\boldsymbol{k}_l, \quad \text{etc.},$$

so that

(1.8b) $$\begin{aligned} M_1 &= m_1\cos\psi - m_2\sin\psi, \\ M_2 &= m_1\sin\psi + m_2\cos\psi, \\ M_3 &= m_3, \quad \text{etc.} \end{aligned}$$

In particular,

(1.9) $$U_1 = -\phi_s\sin\theta, \quad U_2 = \theta_s.$$

(The use of these components emphasizes the fundamental role played by M_2, the component of $\boldsymbol{m}$ about the line of nodes $\boldsymbol{k}_2$, which is most responsible for changes in θ.)

Equilibrium problems. We set the acceleration terms in (1.2) equal to zero. We take the scaled arc length s of the reference configuration to be confined to $[0,1]$. We assume that $\boldsymbol{n}(1)$ and $\boldsymbol{m}(1)$ are prescribed. Then (1.2a) implies that $\boldsymbol{n}$ is constant. We choose the basis $\{\boldsymbol{i}_k\}$ so that

(1.10a) $$\boldsymbol{n}(s) = \boldsymbol{n}(1) = -\Lambda\boldsymbol{i}_3.$$

From (1.6) we then find

(1.10b) $$\begin{aligned} n_1 = \Lambda\sin\theta\cos\psi, \quad n_2 = -\Lambda\sin\theta\sin\psi, \quad n_3 = -\Lambda\cos\theta, \\ N_1 = \Lambda\sin\theta, \quad N_2 = 0. \end{aligned}$$

From the components of (1.2b) in the $\boldsymbol{k}_2, \boldsymbol{h}, \boldsymbol{i}_3, \boldsymbol{d}_3$-directions and from (1.3), (1.7) we now obtain

$$\begin{aligned}
&(1.11a) \quad M_2' + (M_1 \cos\theta + m_3 \sin\theta)\phi' \\
&\qquad\qquad + \Lambda(\Lambda v \sin\theta\cos\theta + v^{\times} M_1 \cos\theta + \check{v}_3 \sin\theta) = 0, \\
&(1.11b) \quad (M_1 \cos\theta + m_3 \sin\theta)' - M_2(\phi' + \Lambda v^{\times}) = 0, \\
&(1.11c) \quad (\boldsymbol{m} \cdot \boldsymbol{i}_3) \equiv -M_1 \sin\theta + m_3 \cos\theta = \alpha \quad \text{(const.)}, \\
&(1.11d) \quad m_3' = (u^{\times} - v^{\times})\Lambda \sin\theta M_2, \\
&(1.11e) \quad \theta' = u M_2, \\
&(1.11f) \quad -\phi' \sin\theta = u M_1 + \Lambda \sin\theta u^{\times}, \\
&(1.11g) \quad \psi' = \check{u}_3 + [u M_1 + \Lambda \sin\theta u^{\times}] \cot\theta,
\end{aligned}$$

where the arguments of the constitutive functions $u, u^{\times}, v^{\times}, v, \check{u}_3, \check{v}_3$ are s and

$$(1.12) \qquad J(\mathsf{m}, \mathsf{n}) = \big(M_1^2 + M_2^2, \Lambda M_1 \sin\theta, \Lambda^2 \sin^2\theta, m_3, -\Lambda\cos\theta\big).$$

A straightforward study of the integral laws of equilibrium shows that all solutions are as regular as the constitutive functions permit.

2. Kirchhoff's Problem for Helical Equilibrium States

We begin our analysis of (1.1) and (1.11) by seeking solutions with

$$(2.1) \qquad \theta = \text{const.}$$

Since u must be positive by the strict monotonicity condition, it follows from (1.11b,d,e) that

$$(2.2\text{a,b,c}) \quad M_2 = 0, \quad M_1 \cos\theta + m_3 \sin\theta = \text{(const.)}, \quad m_3 = \beta \quad \text{(const.)}.$$

Conditions (1.11c) and (2.2b,c) imply that M_1 is constant. Thus (1.12) is constant, so that all the constitutive functions for the strains assume constant values. If $\sin\theta \neq 0$, then from (1.11c) and (2.2) we obtain

$$(2.3) \qquad -M_1 = \frac{\alpha - \beta\cos\theta}{\sin\theta}.$$

If $\sin\theta \neq 0$, then (1.11f,g) imply that ϕ' and ψ' are constants uniquely determined from θ, Λ, α, and β. Thus all the variables are uniquely determined from θ and the loads. (But θ need not be uniquely determined from the data, as we shall see below.) If $\sin\theta = 0$, then all we can conclude from (1.7) is that $\psi' \pm \phi' = \check{u}_3(M_1^2, 0, 0, \beta, \mp\Lambda)$. This ambiguity is a consequence of the polar singularity in the description of the $\boldsymbol{d}_k$ by Euler angles; it leads to no ambiguity in the description of the actual configuration.

Let $\sigma(s)$ be the actual arc length from 0 to s. Then

$$\sigma' = \sqrt{\hat{v}_\alpha \hat{v}_\alpha + \hat{v}_3^2} = \text{const.} \tag{2.4}$$

Let $\boldsymbol{r} \equiv x_k \boldsymbol{i}_k$. The substitution of (1.6) and (1.3b) into (1.1a) then yields

$$\frac{dx_1}{d\sigma} = a\cos\phi, \quad \frac{dx_2}{d\sigma} = a\sin\phi, \quad \frac{dx_3}{d\sigma} = b, \tag{2.5}$$

where

$$a\sigma' \equiv \quad (v\Lambda \sin\theta + M_1 v^\times)\cos\theta + \check{v}_3 \sin\theta, \tag{2.6a}$$

$$b\sigma' \equiv -(v\Lambda \sin\theta + M_1 v^\times)\sin\theta + \check{v}_3 \cos\theta. \tag{2.6b}$$

If $\sin\theta \neq 0$, then ϕ' is a constant, so that

$$\phi = \frac{\phi'}{\sigma'}\sigma + \text{const.} \tag{2.6c}$$

Thus (2.5) is the equation of a right circular helix (possibly degenerate) of radius $|a\sigma'/\phi'|$ and pitch b. If $\sin\theta = 0$, then $u_1 = 0 = u_2$ by (1.7) and $n_1 = 0 = n_2$ by (1.10b). Since u is positive by the strict monotonicity condition, it follows that $m_1 = 0 = m_2$. Equation (1.3b) then implies that $v_1 = 0 = v_2$. Thus (2.6) implies that $a = 0$, so that (2.5) represents a straight line along $\boldsymbol{i}_3$. Thus, in either case, if the governing equations admit solutions with θ constant, then the axis $\boldsymbol{r}$ of the corresponding configuration must be helical. (The study of helical solutions is *Kirchhoff's problem.*) The constants a and b cannot vanish simultaneously because $\check{v}_3 > 0$.

Note that (1.6), (1.10a), and (2.5) yield

$$\begin{aligned} \boldsymbol{n} \times \frac{\boldsymbol{r}'}{\sigma'} &= -\Lambda a(-\sin\phi \boldsymbol{i}_1 + \cos\phi \boldsymbol{i}_2) = -\Lambda a(\sin\psi \boldsymbol{d}_1 + \cos\psi \boldsymbol{d}_2), \\ \boldsymbol{n} \cdot \frac{\boldsymbol{r}'}{\sigma'} &= -\Lambda b. \end{aligned} \tag{2.7}$$

Thus Λa is the magnitude of the component of the resultant contact force in the plane perpendicular to $\boldsymbol{r}'$, and Λb is the component of the resultant force along $\boldsymbol{r}'$.

To verify that there actually exist solutions of this kind, we must find θ, ϕ, ψ to satisfy the equilibrium equations, which by virtue of (2.2) reduce to

$$\begin{aligned} &(M_1 \cos\theta + \beta \sin\theta)\phi' \\ &\qquad = \Lambda[\check{v}_3 \sin\theta + (v\Lambda \sin\theta + M_1 v^\times)\cos\theta] = \Lambda a \sigma', \end{aligned} \tag{2.8}$$

$$-M_1 \sin\theta + \beta\cos\theta = \alpha, \tag{2.9}$$

$$-\phi'\sin\theta = M_1 u + \Lambda u^\times \sin\theta, \tag{2.10}$$

$$\psi' = \check{u}_3 - \phi'\cos\theta. \tag{2.11}$$

Here the constitutive functions depend on (1.12), which reduces to

$$(2.12)\qquad \left(M_1^2, \Lambda M_1 \sin\theta, \Lambda^2 \sin^2\theta, \beta, -\Lambda\cos\theta\right).$$

If $\sin\theta \neq 0$, then (2.8)–(2.10) are easily reduced to a single equation for θ. If $\sin\theta = 0$, then the results of the paragraph following (2.6) imply that (2.8) and (2.10) are identically satisfied, that (2.9) reduces to $\alpha = \pm\beta$, and that (2.11) reduces to

$$(2.13)\qquad \psi' \pm \phi' = \check{u}_3(0,0,0,\alpha,\Lambda).$$

The existence and properties of solutions are treated in the exercises below.

A helical deformation, if it exists for given Λ and $\boldsymbol{m}(1)$, can always be maintained by a *wrench*, which is a statically equivalent load system consisting of the force $-\Lambda\boldsymbol{i}_3$ acting along the $\boldsymbol{i}_3$-axis (rather than being applied at $\boldsymbol{r}(1)$) and of a couple with axis parallel to $\boldsymbol{i}_3$. The force and couple may be communicated to the end $\boldsymbol{r}(1)$ of the rod by a rigid shaft joining it to the $\boldsymbol{i}_3$-axis. To show that there is such a wrench, we first observe that our actual terminal loading is statically equivalent to the force $-\Lambda\boldsymbol{i}_3$ acting along the $\boldsymbol{i}_3$-axis and the couple $\boldsymbol{m}(1) + \boldsymbol{r}(1) \times (-\Lambda\boldsymbol{i}_3)$. This load system is a wrench if the components of the couple in the $\boldsymbol{i}_1$- and $\boldsymbol{i}_2$-directions vanish. An evaluation of these components using (2.1)–(2.6) shows that these components vanish if and only if (2.8)–(2.12) hold (see Love (1927, Sec. 270)).

Let us now consider deformations with a straight axis, which occur only if $a\phi' = 0$, in consequence of (2.5). Equation (2.8) implies that $\Lambda a = 0$ if $\phi' = 0$. If $\Lambda = 0$, then we can choose $\boldsymbol{i}_3$ arbitrarily. By taking it to lie along $\boldsymbol{r}$, we effectively cause a to vanish. Thus deformations with a straight axis occur only if $a = 0$, i.e., only if

$$(2.14)\qquad -(M_1 v^{\times} + v\Lambda\sin\theta)\cos\theta = \check{v}_3 \sin\theta.$$

We first seek straight configurations with $\sin\theta = 0$. By the comments following (2.12) we see that (2.8)–(2.10) are identically satisfied if $\alpha = \beta\cos\theta = \pm\beta$. In this case, the stretch and twist are given by

$$(2.15)\qquad v_3 = \check{v}_3(0,0,0,\beta,-\Lambda\cos\theta), \quad \psi' + \phi'\cos\theta = \check{u}_3(0,0,0,\beta,-\Lambda\cos\theta).$$

If the reference configuration is natural, then $\check{v}_3(0,0,0,0,0) = 1$ and $\check{u}_3(0,0,0,0,0) = 0$. The *Poynting effect*, observed experimentally and deduced analytically in the three-dimensional theory of nonlinear elastic materials (see Truesdell & Noll (1965, Sec. 66)) is that a twist can produce a change in length. Thus there is no reason to require that $\check{v}_3(0,0,0,\beta,0) = 1$ for $\beta \neq 0$. Assumption (VIII.9.4) would, however, imply that $\partial\hat{v}^3/\partial m_3 = 0$ when $m_3 = 0$, so that the Poynting effect is a second-order effect for small twist.

2.16. Exercise. Discuss the existence of configurations with straight axes when $a = 0, \sin\theta \neq 0, \phi' = 0$, and when $a = 0, \sin\theta \neq 0, \phi' \neq 0$.

Equations (2.5) imply that the axis is circular if and only if $b = 0$ and $\phi' \neq 0$. It follows from (2.6) that $b = 0$ if and only if

$$(2.17)\qquad (-M_1 v^{\times} + v\Lambda\sin\theta)\sin\theta = -\check{v}_3\cos\theta.$$

$a \neq 0$ because $b = 0$. Moreover, $\sin\theta \neq 0$ because if so, condition (2.17) would cause $\check{v}_3$ to vanish.

2.18. Exercise. Discuss the existence of configurations with circular axes when $\theta = \pi/2$.

2.19. Exercise. Discuss the existence of configurations with a nondegenerate helical axis.

Kirchhoff (1859) found helical configurations for his kinetic analogue. (See Love (1927, Secs. 270–272).) This work was extended to shearable and extensible rods with linear constitutive laws by Whitman & DeSilva (1974). The treatment of this section represents a modification of that of Antman (1974b), and appears with the kind permission of the *Quarterly of Applied Mathematics.*

3. General Solutions for Equilibria

We now study general equilibrium states of (1.1)–(1.3) for transversely isotropic rods. We assume that (1.5) holds, so that (2.2c) holds. In this case, Eqs. (1.10), (1.11c), (2.2c) constitute a set of five integrals for our twelfth-order (six degree-of-freedom) system. We use these integrals to reduce (1.11a,e) to

$$\theta' = uM_2, \tag{3.1}$$

$$\begin{aligned} M_2' = &- \left[\frac{\alpha - \beta\cos\theta}{\sin^2\theta} u - \Lambda u^\times\right]\left[\frac{\beta - \alpha\cos\theta}{\sin\theta}\right] \\ &- \Lambda\left\{\left[v\Lambda\sin\theta - v^\times\left(\frac{\alpha-\beta\cos\theta}{\sin\theta}\right)\right]\cos\theta + \check{v}_3\sin\theta\right\} \end{aligned} \tag{3.2}$$

where the constitutive functions depend on s and

$$\left(M_2^2 + \left(\frac{\alpha - \beta\cos\theta}{\sin\theta}\right)^2, -\Lambda(\alpha - \beta\cos\theta), \Lambda^2\sin^2\theta, \beta, -\Lambda\cos\theta\right). \tag{3.3}$$

Let us assume that the material is uniform (which is crucial for the ensuing study) and is hyperelastic (which simplifies the ensuing study). Thus the constitutive functions may be written in the form

$$\mathsf{u} = \frac{\partial W^*}{\partial \mathsf{m}}(\mathsf{m},\mathsf{n}), \quad \mathsf{v} = \frac{\partial W^*}{\partial \mathsf{n}}(\mathsf{m},\mathsf{n}) \tag{3.4}$$

where W^* depends only on (3.3) because of transverse isotropy.

The equilibrium version of (1.2) then yields

(3.5)

$$\begin{aligned} 0 &= \boldsymbol{r}'\cdot\boldsymbol{n}' + \boldsymbol{u}\cdot(\boldsymbol{m}' + \boldsymbol{r}'\times\boldsymbol{n}) \\ &= \boldsymbol{r}'\cdot(n_k'\boldsymbol{d}_k + n_k\boldsymbol{u}\times\boldsymbol{d}_k) + \boldsymbol{u}\cdot(m_k'\boldsymbol{d}_k + m_k\boldsymbol{u}\times\boldsymbol{d}_k + \boldsymbol{r}'\times\boldsymbol{n}) \\ &= \mathsf{v}\cdot\mathsf{n}' + \boldsymbol{r}'\cdot(\boldsymbol{u}\times\boldsymbol{n}) + \mathsf{u}\cdot\mathsf{m}' + \boldsymbol{u}\cdot(\boldsymbol{r}'\times\boldsymbol{n}) = \frac{\partial W^*}{\partial\mathsf{n}}\cdot\mathsf{n}' + \frac{\partial W^*}{\partial\mathsf{m}}\cdot\mathsf{m}' \\ &= \frac{d}{ds}W^*. \end{aligned}$$

Thus our system for hyperelastic rods has the ‘energy’ integral

$$W^*(\mathbf{m}(s), \mathbf{n}(s)) = \text{const.}, \tag{3.6}$$

which, when combined with the integrals previously obtained, gives a family of trajectories for the autonomous version of (3.1), (3.2) in the (θ, M_2)-phase plane.

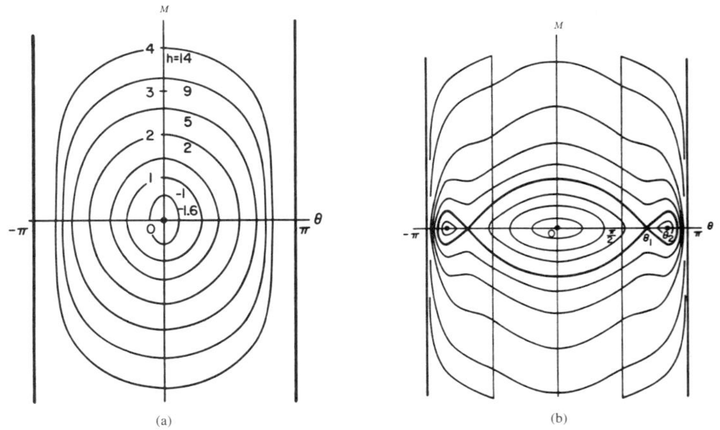

Figs. 3.7a,b. Typical phase portraits for (3.1) and (3.2) for $\alpha = \beta \neq 0$ and $\Lambda > 0$. A shear instability is manifested in the portrait at the right.

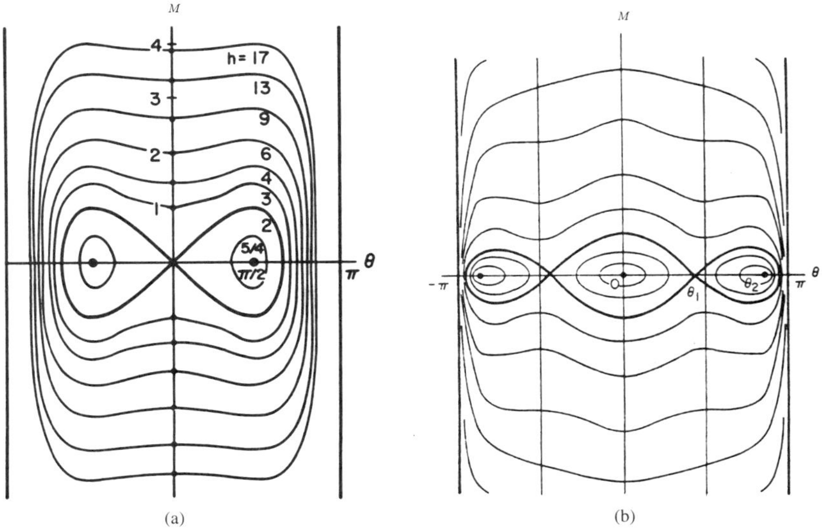

Figs. 3.8a,b. Typical phase portraits for (3.1) and (3.2) for $\alpha = \beta \neq 0$ and $\Lambda < 0$. A shear instability is manifested in the portrait at the right.

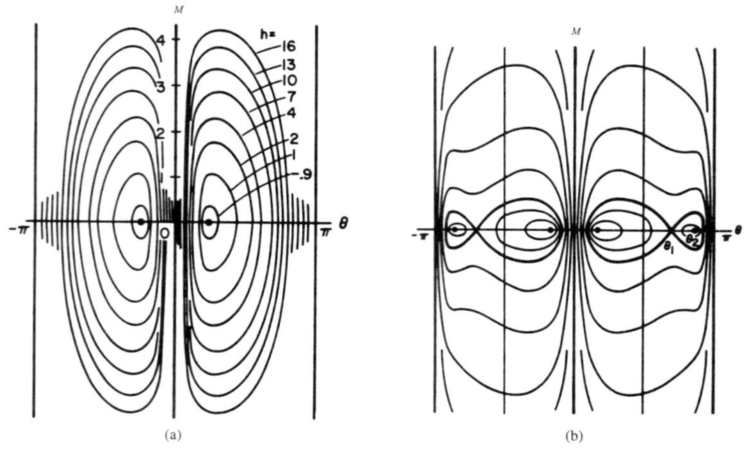

Fig. 3.9a,b. Typical phase portraits for (3.1) and (3.2) for $\alpha > 0$, $\beta > 0$, $\alpha \neq \beta$ and $\Lambda > 0$. A shear instability is manifested in the portrait at the right. The portraits for other cases in which $\alpha \neq \pm\beta$ are similar.

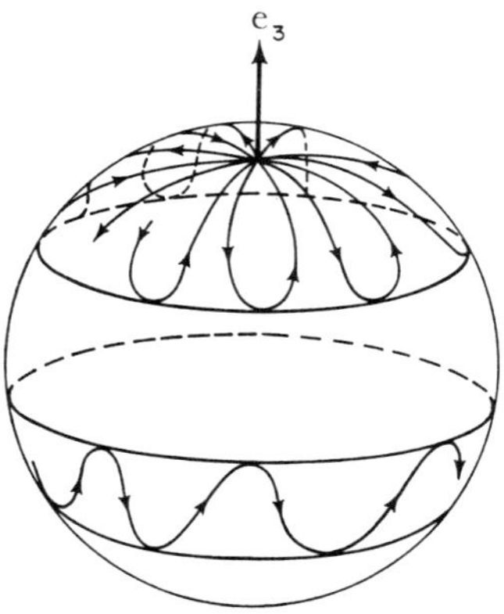

Fig. 3.11. Trajectories of $\boldsymbol{d}_3$ on the unit sphere when $\alpha = \beta \neq 0$ and ϕ' is everywhere positive. The petal-shaped trajectory through the north pole corresponds to a closed orbit of Fig. 3.7 or Fig. 3.8 that encloses the center at the origin. The banded trajectory corresponds to a closed orbit of these figures that encloses a center other than the origin.

The symmetries inherent in (3.3) ensure that the phase portrait of (3.1) and (3.2) is symmetric about the θ-axis, is symmetric about the M-axis, and has period 2π in θ. We can readily sketch the phase portraits for different parameter ranges. For $\alpha = 0 = \beta$ and for $\Lambda > 0$, the phase portraits are given by Fig. IV.2.14 with γ replaced by θ. If $\Lambda < 0$, the phase

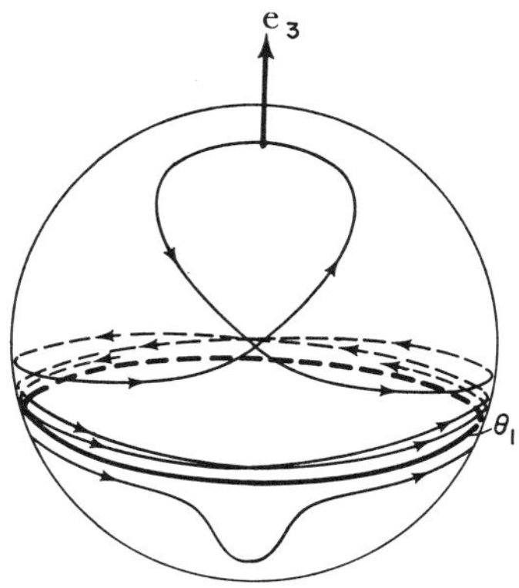

Fig. 3.12. Trajectories of $\boldsymbol{d}_3$ on the unit sphere when $\alpha = \beta \neq 0$ and ϕ' is everywhere positive. The trajectory above the circle $\theta = \theta_1$ corresponds to separatrix arc from θ_1 to $-\theta_1$ of Fig. 3.7 or Fig. 3.8b. The trajectory below the circle $\theta = \theta_1$ corresponds to the separatrix arc enclosing the center at θ_1 of Fig. 3.7.

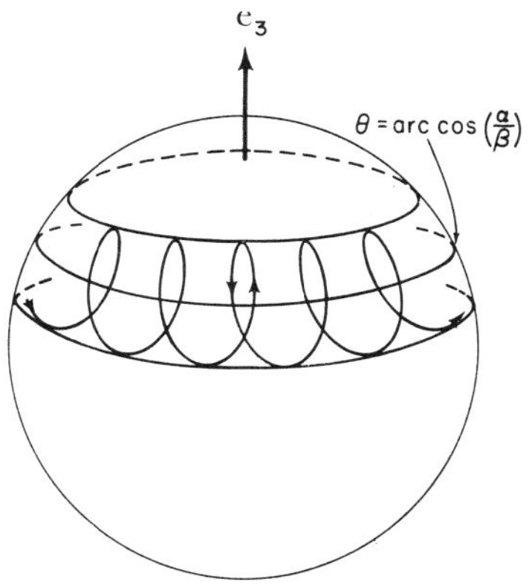

Fig. 3.13. Typical trajectory of $\boldsymbol{d}_3$ on the unit sphere when $0 < \alpha < \beta$ corresponding to a closed orbit of Fig. 3.9 on which ϕ' changes sign.

portrait is translated horizontally by π. Other cases are sketched in Figs. 3.7–3.9. Note that if $\alpha = \beta$, then we can use the identity $\frac{1-\cos\theta}{\sin\theta} = \frac{\sin\theta}{1+\cos\theta}$.

These portraits show how θ behaves as a function of s. Since θ and ϕ are spherical coordinates of $\boldsymbol{d}_3$, we can determine the behavior of ϕ and thus of $\boldsymbol{d}_3$ by combining our information on θ with the specialization of (1.11f). In particular, if $\alpha = \beta$, then (1.11f) reduces to

(3.10) $$\phi' = \frac{\beta u}{1 + \cos\theta} - \Lambda u^\times .$$

The first term on the right has the sign of β; we have no grounds to choose a sign for the second term on the right. It is not unreasonable to take $u^{\times} = 0$, at least for illustrative purposes. In Figs. 3.11–3.13 we sketch some typical trajectories of $\boldsymbol{d}_3$ on the unit sphere that correspond to certain trajectories of Figs. 3.7–3.9. Using (1.1) and (1.11g) we could determine the qualitative behavior of $\{\boldsymbol{r}, \boldsymbol{d}_k\}$ from the information at hand. But since we can expect the shear strains to be small, the behavior of $\boldsymbol{d}_3$ effectively indicates that of $\boldsymbol{r}'$.

There have been many treatments of Kirchhoff's kinetic analogue for uniform, transversely isotropic rods by elliptic functions. Among the classical treatments are those of Hess (1884, 1885), Born (1906), and Nikolai (1916); see Ilyukhin (1979). The results of this section represent a generalization of some of those of Antman & Jordan (1975). This material together with all the figures shown here is reproduced by kind permission of the Royal Society of Edinburgh.

4. Travelling Waves in Straight Rods

We now assume that the rod is infinitely long so that $s \in (-\infty, \infty)$. Let f stand for a typical dependent variable appearing in the system (1.1), (1.2), (VIII.9.17) for transversely isotropic, uniform rods. We seek *travelling wave solutions* of this system, for which

$$f(s,t) = \bar{f}(s - ct) \tag{4.1}$$

where the real number c is called the *wave speed.* If we substitute representations of the form (4.1) into our system, set

$$\xi \equiv s - ct, \tag{4.2}$$

drop the bars over functions of ξ, and denote differentiation with respect to ξ by a prime, then we can use (VIII.9.27) to reduce (1.2) to

$$\boldsymbol{n}' = c^2 \rho A \boldsymbol{r}'', \quad \boldsymbol{m}' + \boldsymbol{r}' \times \boldsymbol{n} = c^2 \rho J (\boldsymbol{d}_\alpha \times \boldsymbol{d}_\alpha')'. \tag{4.3a,b}$$

Without loss of generality, we denote the integral of (4.3a) by

$$\boldsymbol{n} - c^2 \rho A \boldsymbol{r}' = -\Lambda \boldsymbol{i}_3 \quad \text{(const.)} \tag{4.4}$$

where $\{\boldsymbol{i}_k\}$ is a right-handed orthonormal basis, as in the preceding two sections. Replacing the components of $\boldsymbol{n}$ by their constitutive representations, we use (1.6) to obtain the following generalization of (1.10b):

$$\hat{n}_1(\mathsf{u}, \mathsf{v}) - c^2 \rho A v_1 = \Lambda \sin\theta \cos\psi, \tag{4.5a}$$

$$\hat{n}_2(\mathsf{u}, \mathsf{v}) - c^2 \rho A v_2 = -\Lambda \sin\theta \sin\psi, \tag{4.5b}$$

$$\hat{n}_3(\mathsf{u}, \mathsf{v}) - c^2 \rho A v_3 = -\Lambda \cos\theta. \tag{4.5c}$$

Using the constitutive equations (VIII.9.17) together with (1.7) and (1.8) we obtain from (4.5a,b) that

$$n^\times \theta' + (n - c^2\rho A)V_2 = 0, \tag{4.6a}$$

$$n^\times \phi' \sin\theta - (n - c^2\rho A)V_1 = -\Lambda \sin\theta, \tag{4.6b}$$

From (4.3b), (4.4), and the constitutive equations we obtain the following generalization of (1.11c):

$$\begin{aligned}(4.7)\quad &\boldsymbol{m}\cdot\boldsymbol{i}_3 - c^2\rho J(\boldsymbol{d}_\gamma \times \boldsymbol{d}_\gamma')\cdot\boldsymbol{i}_3 \\ &\equiv [-M_1 + c^2\rho J U_1]\sin\theta + [\hat{m}_3(\mathbf{u},\mathbf{v}) - 2c^2\rho J u_3]\cos\theta = \alpha \text{ (const.)}.\end{aligned}$$

The substitution of (VIII.9.17) into the $\boldsymbol{d}_3$-component of (4.3b) and the use of the constitutive restriction that $m^\times = n^\times$ yields the following generalization of (1.11c):

$$\hat{m}_3(\mathbf{u},\mathbf{v}) - 2c^2\rho J u_3 = \beta \quad \text{(const.)}. \tag{4.8}$$

4.9. Exercise. Show that if the material is hyperelastic, so that (VIII.3.16) holds, and is uniform, then (4.3) and (4.4) have the integral

$$\mathbf{u}\cdot\frac{\partial W}{\partial \mathbf{u}} + \mathbf{v}\cdot\frac{\partial W}{\partial \mathbf{v}} - W - \frac{c^2}{2}[\rho A\mathbf{v}\cdot\mathbf{v} + \rho J(u_\gamma u_\gamma + 2{u_3}^2)] = \text{const.} \tag{4.10}$$

For materials that are not hyperelastic, we must replace (4.10) by an equation analogous to (3.1) and (3.2). By virtue of (VIII.9.17) and (1.7) the constitutive equation for M_2 is

$$M_2 = m\theta' + m^\times V_2. \tag{4.11}$$

To get the equation for M_2 we dot (4.3b) with $\boldsymbol{k}_2$ and use (1.1), (1.7), (4.4), and (4.8) to obtain

(4.12)

$$\begin{aligned} M_2' - c^2\rho J\theta'' &= -\,\phi'(\boldsymbol{m} - c^2\rho J\boldsymbol{d}_\alpha \times \boldsymbol{d}_\alpha')\cdot\boldsymbol{h} + (\boldsymbol{n}\cdot\boldsymbol{i}_3)(\boldsymbol{r}'\cdot\boldsymbol{h}) \\ &= -\,\phi'[(M_1 + c^2\rho J\phi'\sin\theta)\cos\theta + \beta\sin\theta] \\ &\quad - [\Lambda - c^2\rho A(-V_1\sin\theta + v_3\cos\theta)][V_1\cos\theta + v_3\sin\theta] = 0.\end{aligned}$$

We now discuss the qualitative behavior of (4.5)–(4.8), and (4.12). The presence of the terms with c^2 may destroy the monotonicity of the left-hand sides of (4.5), (4.7), and (4.8) with respect to $v_1, v_2, v_3, \phi'\sin\theta, u_3$. In this case, we cannot reduce (4.12) to a single second-order differential equation for θ. Suppose we interpret (4.5), (4.7), and (4.8) as defining $v_1, v_2, v_3, \phi'\sin\theta, u_3$ as multivalued functions of (θ, θ') and the parameters

Λ, α, β. We can substitute these multivalued representations into (4.12), which thereby has a family of phase portraits for each fixed set of parameters. In particular, if (4.10) holds, then the substitution of these representations into the left-hand side of (4.10) converts it into a multivalued function of (θ, θ'). The phase portrait for each fixed set of parameters corresponds to the level curves of its multivalued graph.

By seeking travelling waves, we have reduced our original system of hyperbolic partial differential equations to a system of ordinary differential equations. But the inherent lack of invertibility of hyperbolic equations still remains for our system of ordinary differential equations. Indeed, the characteristic speeds are those values of c for which the matrix of partial derivatives of the principal parts of the ordinary differential equations for travelling waves with respect to $(\mathbf{u}, \mathbf{v})$ is singular. At such characteristic speeds, the uniqueness theory for initial-value problems for ordinary differential equations breaks down.

In Fig. 4.13 we show a phase portrait that arises when c is characteristic. Since the smooth curves $ABCD$ and $EBGH$ are tangent at B, an actual trajectory may switch there (and at D) or not at each passage. (This tangency follows from the fact that c is characteristic; the switching is permitted because of the absence of a uniqueness theory.) Because of this switching, there is an uncountable family of oscillatory travelling waves with $\theta \in C^2$, a typical member of which is sketched in Fig. 4.14.

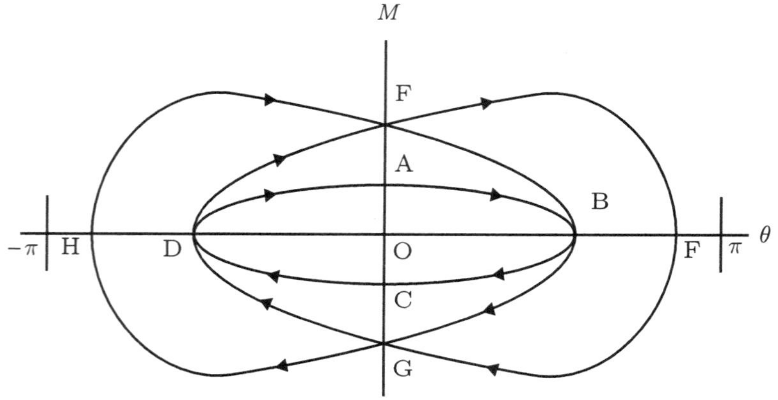

Fig. 4.13. A phase portrait for (4.5), (4.7), (4.8), and (4.12) when c is a characteristic speed.

We have used (VIII.9.17) in favor of (1.3) because the latter form of the constitutive equations is of very limited utility in the study of partial differential equations for rods. A deeper investigation of travelling waves, directed toward such questions as stability, would inevitably lead to the partial differential equations. The work of this section generalizes part of that of Antman & Liu (1979), who also examined the possibility of solutions with discontinuities compatible with the Rankine-Hugoniot jump conditions and with entropy conditions. This material, together with Figs. 4.13 and 4.14, appears with the kind permission of the *Quarterly of Applied Mathematics*.

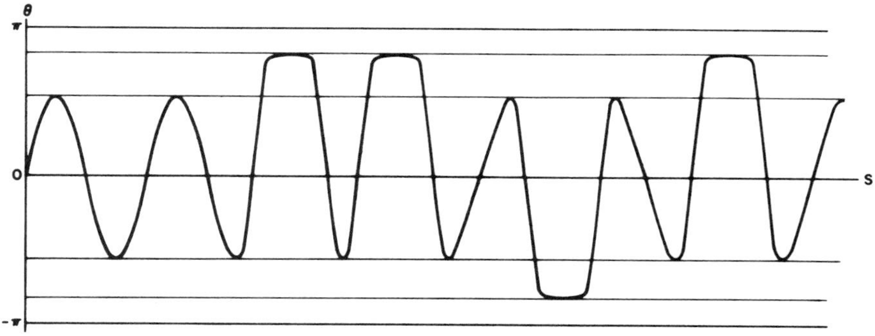

Fig. 4.14. A typical member of the uncountable family of oscillatory travelling waves corresponding to Fig. 4.13.

4.15. Exercise. Formulate the corresponding problem for travelling waves for transversely isotropic, uniform, viscoelastic rods having the form indicated in Ex. VIII.9.23 and satisfying (VIII.8.35). The governing system of ordinary differential equations is now well-posed. Their analysis, which would illuminate the nature of shock structure for elastic rods, is an open research problem.

There are several works devoted to the treatment of large displacements for the kinetic analogue, often by the use of elliptic functions. They include Born (1906), Frisch-Fay (1962), Funk (1970), Ilyukhin (1979), Love (1927), Nikolai (1955), Popov (1948), and Saalschütz (1880). These books contain extensive bibliographies. Moreover, much of value for the spatial deformation of rods can be found in books on rigid-body mechanics in consequence of the kinetic analogue. There are descriptions of large rotations that do not employ the Euler angles and thereby avoid the concomitant difficulties with polar singularities.

Mielke & Holmes (1988) have demonstrated chaotic spatial deformations for infinitely long rods that are not transversely isotropic. Such phenomena might be expected, even for hyperelastic rods, because the governing equations are not totally integrable.

5. Buckling under Terminal Thrust and Torque

We now study the buckling in space of a transversely isotropic rod subjected to terminal thrust and torque. We assume that (1.5) holds, so that we can use (2.2c). Our governing equations are (3.1)–(3.3).

We assume that the end $s = 0$ of the rod is welded to the $\{\boldsymbol{i}, \boldsymbol{j}\}$-plane at the origin and that the end $s = 1$ is welded to a plane with normal $-\boldsymbol{k}$ that is otherwise free to move in any manner. The end $s = 1$ is subjected to a compressive thrust of magnitude λ acting in the $-\boldsymbol{k}$ direction and to a torque whose component about $\boldsymbol{k}$ is μ. Thus we require that

$$\text{(5.1a–d)} \qquad \boldsymbol{r}(0) = \boldsymbol{0}, \quad \boldsymbol{d}_1(0) = \boldsymbol{i}, \quad \boldsymbol{d}_2(0) = \boldsymbol{j}, \quad \boldsymbol{d}_3(0) = \boldsymbol{k},$$

$$\text{(5.2a–c)} \qquad \boldsymbol{n}(1) = -\lambda \boldsymbol{k}, \quad \boldsymbol{m}(1) \cdot \boldsymbol{k} = \mu, \quad \boldsymbol{d}_3(1) = \boldsymbol{k}.$$

In view of (5.2a), we identify $\{\boldsymbol{i}_1, \boldsymbol{i}_2, \boldsymbol{i}_3\}$ with $\{\boldsymbol{i}, \boldsymbol{j}, \boldsymbol{k}\}$ (see (1.10a)). Conditions (5.1b–d) imply that

$$\text{(5.3)} \qquad \theta(0) = 0 = \theta(1).$$

Combining the integrals (1.10), (1.11c), and (2.2c) with (5.1d,f), we obtain

$$\Lambda = \lambda, \quad \alpha = \beta = \mu. \tag{5.4}$$

Thus the quantity $\frac{\alpha - \beta\cos\theta}{\sin\theta}$, which portends singular behavior in (3.2) and (3.3), is reduced to the innocuous $\frac{\mu\sin\theta}{1+\cos\theta}$, and (3.1)–(3.3) become

$$\theta' = uM_2, \tag{5.5}$$

$$\begin{aligned} M_2' = {} & \mu(\check{u}_3 - \mu u + v^\times \lambda\cos\theta)\frac{\sin\theta}{1+\cos\theta} \\ & - \lambda\sin\theta(\check{v}_3 - \mu u^\times + v\lambda\cos\theta), \end{aligned} \tag{5.6}$$

$$\left(M_2^2 + \left(\frac{\mu\sin\theta}{1+\cos\theta}\right)^2, -\lambda\mu(1-\cos\theta), \lambda^2\sin^2\theta, \mu, -\lambda\cos\theta, s \right). \tag{5.7}$$

We can immediately convert our boundary-value problem (5.1)–(5.3), (5.5)–(5.7) to the form (VI.5.8) with α replaced by μ. Thus we can imitate the local and global bifurcation analysis of Secs. VI.5–VI.9. In particular, if the rod is uniform, then the linearization of the problem about the trivial (λ, μ)-plane is equivalent to (VI.5.11) with

$$q(\lambda,\mu) \equiv u^0(\lambda,\mu)\{\lambda[v_3^0(\lambda,\mu) + \lambda v^0(\lambda,\mu)] + \mu^2 u^0(\lambda,\mu)\}. \tag{5.8}$$

The eigenvalues are pairs (λ, μ) for which there is a positive integer n satisfying

$$q(\lambda,\mu) = n^2\pi^2. \tag{5.9}$$

For Kirchhoff's kinetic analogue (see (VIII.10.4)), for which the rod is inextensible and unshearable, so that $v = 0$, $v^\times = 0$, $\check{v}_3 = 1$, and for which the constitutive equations for the couples are linear, Eq. (5.9) reduces to

$$u^0(\lambda + \mu^2 u^0) = n^2\pi^2. \tag{5.10}$$

Here u^0 is a positive constant. Thus (5.10) describes a family of parabolas. For $\lambda = 0$, we find that the the smallest buckling torque is $\mu = \pm n\pi/\sqrt{u^0}$. When (5.10) holds we can make the structure safer, i.e., raise the lowest buckling torque, by making λ negative, i.e., by subjecting the rod to tension. When (5.8) holds, however, it is quite possible to induce instabilities by making λ large and negative because (5.8) has a term with λ^2 multiplied by a positive-valued function. Thus the presence of shearability leads to hitherto unexpected instabilities.

Now let us study a three-dimensional version of the follower-load problem of Sec. VI.10. We retain (5.1) and replace (5.2) with

$$\boldsymbol{n}(1) = -\lambda \boldsymbol{d}_3(1), \quad \boldsymbol{m}(1) = \mu \boldsymbol{d}_3(1). \tag{5.11}$$

We accordingly take $\boldsymbol{i}_3$ to equal $\boldsymbol{d}_3(1)$ so that (1.6) implies $\theta(1) = 0 \mod 2\pi$. It follows from (1.8) that $M(1) = 0$. We again find from our integrals that (5.4) holds. Thus our differential equations are (5.5)–(5.7), which are nonsingular. The initial-value problem for these equations subject to $\theta(1) = 0 = M(1)$ has the unique solution $\theta = 0 = M$. Note that this result holds even if $\lambda = 0$. This fact does not indicate that the rod is stable. It loses stability by dynamical processes. One source of this phenomenon is that the torque in (5.11) is not conservative. (It cannot be applied by a simple mechanical device using weights and pulleys; see Ziegler (1977).)

5.12. Exercise. Suppose that a rod has all the properties used in this section. Let its end $s = 0$ be welded to the $\{\boldsymbol{i}, \boldsymbol{j}\}$-plane so that (5.1) holds. Let the end $s = 1$ be held in a universal joint (see Fig. VIII.12.13) and be subjected to (5.2a,b). Formulate the remaining boundary condition and determine the nodal properties of bifurcating solution branches.

5.13. Exercise. For $\lambda = 0$, study the linearized equations of motion for the nonconservative (5.11) and determine the linearized stability of the trivial state.

The account in this section is based upon Antman & Kenney (1981), who treat a wide variety of boundary conditions, which produce a wide variety of responses. Interesting examples of secondary bifurcation have been analyzed by Kovári (1969) for Kirchhoff's kinetic analogue and by Maddocks (1984) for an inextensible, unshearable, transversely isotropic rod with general nonlinear constitutive equations for flexure and torsion. It seems evident that his methods could carry over to the class of materials treated in this section.

6. Lateral Instability

A rod having a plane of symmetry subjected to a loading with the same symmetry suffers a *lateral instability* when a solution loses that symmetry. (See Fig. V.2.3.) We formulate the problem in general for a special Cosserat rod and then subject a special case to a global bifurcation analysis.

We assume that the $\{\boldsymbol{i}, \boldsymbol{j}\}$-plane is the plane of symmetry. In the reference configuration, let

$$(6.1) \quad \mathring{\boldsymbol{r}}' = \mathring{\boldsymbol{d}}_3, \quad \mathring{\boldsymbol{d}}_1 = -\sin\mathring{\theta}\boldsymbol{i} + \cos\mathring{\theta}\boldsymbol{j}, \quad \mathring{\boldsymbol{d}}_2 = \boldsymbol{k}, \quad \mathring{\boldsymbol{d}}_3 = \cos\mathring{\theta}\boldsymbol{i} + \sin\mathring{\theta}\boldsymbol{j}.$$

We assume that the constitutive functions have the form

$$(6.2) \quad \begin{aligned} &u_1 = \sigma_1(\iota, s)m_1, \quad u_2 = \hat{u}_2(\iota, s), \quad u_3 = \sigma_{32}(\iota, s)n_2 + \sigma_3(\iota, s)m_3, \\ &v_1 = \tau_1(\iota, s)n_1, \quad v_2 = \tau_2(\iota, s)n_2 + \tau_{23}(\iota, s)m_3, \quad v_3 = \hat{v}_3(\iota, s) \end{aligned}$$

where $\iota \equiv (n_1^2, n_3, m_1^2, m_2, n_2^2 + m_3^2, n_2m_3)$. That (6.2) captures the symmetry of the material response of the rod with respect to the $\{\boldsymbol{i}, \boldsymbol{j}\}$-plane is demonstrated in Ex. XIV.5.34. We assume that the body force and couple preserve the symmetry:

$$(6.3) \quad \boldsymbol{f} \cdot \boldsymbol{k} = 0, \quad \boldsymbol{k} \times \boldsymbol{l} = \boldsymbol{0}.$$

Then the equilibrium equation for the force balance is

$$(6.4) \quad \boldsymbol{n}' + \boldsymbol{f} = \boldsymbol{0},$$

and the componential forms of the equilibrium equations of the moment balance are

(6.5)

$$
\begin{aligned}
&m_1' - (\sigma_{32}n_2 + \sigma_3 m_3)m_2 + u_2 m_3 + (\tau_2 n_2 + \tau_{23} m_3)n_3 - v_3 n_2 + \boldsymbol{l}\cdot\boldsymbol{d}_1 = 0,\\
&m_2' + (\sigma_{32}n_2 + \sigma_3 m_3)m_1 - \sigma_1 m_3 m_1 - \tau_1 n_1 n_3 + v_3 n_1 + \boldsymbol{l}\cdot\boldsymbol{d}_2 = 0,\\
&m_3' - m_1 u_2 + \sigma_1 m_1 m_2 + \tau_1 n_1 n_2 - (\tau_2 n_2 + \tau_{23} m_3)n_1 + \boldsymbol{l}\cdot\boldsymbol{d}_3 = 0.
\end{aligned}
$$

Boundary conditions. We study a particular set of boundary conditions. We assume that the end $s = 0$ is welded at the origin to a fixed rigid wall lying in the $\{\boldsymbol{j}, \boldsymbol{k}\}$-plane, so that

(6.6) $$\boldsymbol{r}(0) = \boldsymbol{0}, \quad \boldsymbol{d}_1(0) = \boldsymbol{j}, \quad \boldsymbol{d}_2(0) = \boldsymbol{k}, \quad \boldsymbol{d}_3(0) = \boldsymbol{i}.$$

We assume that the force $\boldsymbol{n}(1)$ at the end $s = 1$ is prescribed with

(6.7) $$\boldsymbol{n}(1)\cdot\boldsymbol{k} = 0.$$

We finally assume that the end $s = 1$ is freely hinged about the material fiber $\boldsymbol{d}_1(1)$, which is constrained to be parallel to the $\{\boldsymbol{i}, \boldsymbol{j}\}$-plane, and that this end is subject to a couple whose $\boldsymbol{k}$-component is the prescribed number μ. Thus

(6.8a,b,c) $$\boldsymbol{d}_1(1)\cdot\boldsymbol{k} = 0, \quad m_1(1) = 0, \quad \boldsymbol{m}(1)\cdot\boldsymbol{k} = \mu.$$

Since $\boldsymbol{n}(1)$ is prescribed, it follows from (6.3), (6.4), and (6.7) that

(6.9) $$\boldsymbol{n}(s) = \boldsymbol{n}(1) - \int_s^1 \boldsymbol{f}(\xi)\,d\xi, \quad \boldsymbol{n}\cdot\boldsymbol{k} = 0.$$

We can substitute (6.9) into (6.5).

We now simplify our problem in order to get one that has a simple nodal pattern. We assume that $\boldsymbol{f} = \boldsymbol{0}$, $\boldsymbol{n}(1) = \boldsymbol{0}$, $\boldsymbol{l} = 0$. We further require that the reference configuration of the rod be straight with a doubly symmetric cross section. In this case, Ex. XIII.5.34 shows that

(6.10) $$u_2 = \sigma_2(\iota, s)m_2$$

where ι here and in (6.2) has m_2 replaced with m_2^2. (For the deep beams that occur in traditional engineering studies of lateral instabilities, $\sigma_1 > \sigma_2$.) Under these conditions, (6.5) reduces to

(6.11a) $$m_1' + (\sigma_2 - \sigma_3)m_2 m_3 = 0,$$
(6.11b) $$m_2' + (\sigma_3 - \sigma_1)m_3 m_1 = 0,$$
(6.11c) $$m_3' + (\sigma_1 - \sigma_2)m_1 m_2 = 0.$$

Conditions (6.8) imply that there is a reactive couple β such that

(6.12) $$\boldsymbol{m}(1) = \mu\boldsymbol{k} + \beta\boldsymbol{d}_1(1)\times\boldsymbol{k}.$$

Since the balance of moments now reduces to $\boldsymbol{m}' = \boldsymbol{0}$, it follows that $\boldsymbol{m}(s) = \boldsymbol{m}(1)$, so that

(6.13a,b,c)
$$m_1(0) = \boldsymbol{m}(0)\cdot\boldsymbol{j} = -\beta\boldsymbol{d}_1(1)\cdot\boldsymbol{i}, \quad m_2(0) = \mu, \quad m_3(0) = \beta\boldsymbol{d}_1(1)\cdot\boldsymbol{j}.$$

Our boundary-value problem consists of (1.1), (6.2), and (6.9)–(6.11) subject to (6.6) and (6.8).

6.14. Exercise. Assume that the rod is uniform. Find the trivial solution in which $\boldsymbol{r}$ describes a circle. Solve the linearization of this boundary-value problem about the trivial solution. Recast the nonlinear boundary-value problem as a system of integral equations in a form suitable for the application of the Global Bifurcation Theorem V.4.19.

Let us now study the nodal properties of m_1. In view of (6.8b), we can use the methods of Sec. V.5 to show that the number of simple zeros of m_1 on $(0,1)$ are preserved wherever the following property holds: If m_1 has a double zero on $[0,1]$ or merely vanishes at $s=0$, then $m_1 = 0$. Let us accordingly first suppose that m_1 has a double zero at s_0 in $[0,1]$. Then (6.11a) implies that

$$m_1 = 0 \quad \text{and} \quad (\sigma_2 - \sigma_3)m_2 m_3 = 0 \quad \text{at } s_0. \tag{6.15}$$

If $m_3(s_0) = 0$, then the initial-value problem for (6.11a,c) (with m_2 regarded as given) subject to $m_1(s_0) = 0 = m_3(s_0)$ has the unique solution $m_1 = 0 = m_3$. If $m_2(s_0) = 0$, then the initial-value problem for (6.11a,b) subject to $m_1(s_0) = 0 = m_2(s_0)$ has the unique solution $m_1 = 0 = m_2$, which is incompatible with (6.8c) for $\mu \neq 0$. Thus, if $\sigma_2 - \sigma_3$ vanishes nowhere or everywhere, then the possession of a double zero by m_1 implies that $m_1 = 0 = m_3$, and the solution is trivial.

Now suppose that $m_1(0) = 0$. Then (6.13a) implies that either $\beta = 0$ or $\boldsymbol{d}_1(1)\cdot\boldsymbol{i} = 0$. In the former case, (6.13c) implies that $m_3(0) = 0$, and the solution is trivial. The latter condition can occur only if $\boldsymbol{d}_1(1)$ is rotated through an angle $\frac{\pi}{2}$ from its reference value. Thus

6.16. Theorem. *Let $\boldsymbol{n} = \boldsymbol{0} = \boldsymbol{l}$. The number of simple zeros of m_1 on $(0,1]$ are preserved along any connected family $\mathcal{S}$ of solution pairs for the boundary-value problem under study, provided that $\mathcal{S}$ does not contain a trivial solution pair, that $\sigma_2 - \sigma_3$ vanishes nowhere on $(0,1]$ for any solution pair in $\mathcal{S}$, and that $\boldsymbol{d}_1(1)\cdot\boldsymbol{i} \neq 0$ for every solution pair in $\mathcal{S}$.*

6.17. Exercise. Prove an analog of Theorem 6.16 for m_3.

6.18. Exercise. Let (6.6) hold and consider the two boundary-value problems obtained by replacing (6.8) by each of the following two alternatives:

$$\boldsymbol{m}(1) = \mu\boldsymbol{k}, \tag{6.19}$$
$$\boldsymbol{m}(1) = \mu\boldsymbol{d}_2, \tag{6.20}$$

Let $\boldsymbol{n} = \boldsymbol{0}$, but allow $\boldsymbol{l} \neq \boldsymbol{0}$. Prove that the boundary-value problems consisting of (1.1), (6.2), (6.5), (6.6), (6.10) and either (6.19) or (6.20) have only trivial (equilibrium) solutions.

6.21. Exercise. Study the stability of the trivial solutions of Ex. 6.18 when the rod is uniform by linearizing the equations of motion about the trivial equilibrium state.

6.22. Problem. Analyze equilibrium problems for which (6.8) is replaced with each of the following sets of boundary conditions.

$$\boldsymbol{d}_2(1)\cdot\boldsymbol{j}=0,\quad \boldsymbol{m}(1)\cdot\boldsymbol{j}=0,\quad m_2(1)=\mu, \tag{6.23}$$

$$\boldsymbol{d}_2(1)=\boldsymbol{k},\quad m_2(1)=\mu. \tag{6.24}$$

(Interpret these conditions in terms of hinges and their dispositions.)

6.25. Research Problem. Analyze the global qualitative behavior of buckled states of boundary-value problems for (6.2) and (6.5).

The work of this section is based on that of Antman (1984). For accounts of the extensive engineering literature on linearized instability, see the references listed in Sec. V.2. For more recent work along this line, see Reissner (1989). The kinetic analogue corresponding to Kirchhoff's equations (for inextensible unshearable rods with linear relations between u and m) when $\boldsymbol{n}=\boldsymbol{0}=\boldsymbol{l}$ is the force-free motion of a rigid body with a fixed point. This theory has a geometric description in terms of the rolling of Poinsot's ellipsoids. Presumably, our theory with $\boldsymbol{n}=\boldsymbol{0}=\boldsymbol{l}$ has an analogous description (of limited utility for boundary-value problems). The analysis of these equations for rods in terms of elliptic functions was first carried out by Hess (1884).

CHAPTER X

Axisymmetric Equilibria of Cosserat Shells

1. Formulation of the Governing Equations

In this section we show how easy it is to formulate geometrically exact theories for the axisymmetric deformation of shells that can suffer flexure, mid-surface extension, and shear. These are the equations we shall analyze in the next few sections. We limit our attention here to statics, deferring to Chap. XIV the general formulation of dynamical equations for problems lacking symmetry. Our approach imitates that of Sec. IV.1. We define a *plate* to be a shell with a flat natural configuration. From the viewpoint of large deformations, there is little to distinguish the two theories.

There is a voluminous and contentious literature on the derivation of various approximate theories of shells, both linear and nonlinear, such as the von Kármán theory, discussed in Sec. XIV.15. Most of these theories were originally devised by imposing ad hoc truncations on the equations describing the deformation and by using linear stress-strain laws. Some of these theories can now be obtained as formal asymptotic limits of the equations of the three-dimensional theory as a thickness parameter goes to zero, provided that the data in the form of applied forces and boundary conditions have special scalings in terms of the thickness parameter (see Sec. XIV.15).

There are quite a few rigorous mathematical studies of boundary-value problems for these nonlinear approximate theories (see Vorovich (1989)). I believe that the analog in the exact theory of virtually any rigorous result for an axisymmetric equilibrium problem for an approximate theory can be demonstrated with available mathematical tools. (The same is not yet true of problems governed by partial differential equations.) The rest of this chapter exemplifies this point. Moreover, not only is the treatment of the exact theory often mathematically and mechanically more transparent, but it also often yields a far richer description of physical effects.

Geometry of Deformation. Let $\{\boldsymbol{i}, \boldsymbol{j}, \boldsymbol{k}\}$ be a fixed right-handed orthonormal basis for Euclidean 3-space. For each real number ϕ, we set

$$(1.1) \qquad \boldsymbol{e}_1(\phi) = \cos\phi\boldsymbol{i} + \sin\phi\boldsymbol{j}, \quad \boldsymbol{e}_2(\phi) = -\sin\phi\boldsymbol{i} + \cos\phi\boldsymbol{j}, \quad \boldsymbol{e}_3 = \boldsymbol{k}.$$

The *axisymmetric configuration of a special Cosserat shell*, which can suffer flexure, mid-surface extension, and shear is determined by a pair of vector-valued functions $\boldsymbol{r}$ and $\boldsymbol{b}$ of s and ϕ of the form

$$(1.2) \quad \boldsymbol{r}(s,\phi) = r(s)\boldsymbol{e}_1(\phi) + z(s)\boldsymbol{k}, \quad \boldsymbol{b}(s,\phi) = -\sin\theta(s)\boldsymbol{e}_1(\phi) + \cos\theta(s)\boldsymbol{k}.$$

The variable s lies in an interval of the form $[s_1, s_2]$ and ϕ lies in $[0, 2\pi]$. We define

$$(1.3) \qquad \boldsymbol{a}(s,\phi) = \boldsymbol{e}_2(\phi) \times \boldsymbol{b}(s,\phi) = \cos\theta(s)\boldsymbol{e}_1(\phi) + \sin\theta(s)\boldsymbol{k}.$$

The function $\boldsymbol{r}$ defines an axisymmetric surface in $\mathbb{E}^3$. The values of these and other geometric variables in the reference configuration are designated by the superscript $\circ$. We assume that s is the arc-length parameter of each curve $\mathring{\boldsymbol{r}}(\cdot, \phi)$ of longitude (so that $|\mathring{\boldsymbol{r}}_s(s, \phi)| = 1$) and that $\mathring{\boldsymbol{b}} = -\boldsymbol{e}_2 \times \partial_s \mathring{\boldsymbol{r}}$.

To interpret the variables of (1.2) and (1.3), we assume that the reference configuration of an axisymmetric shell, regarded as a thin three-dimensional body, has an axisymmetric material reference surface, called the *material base surface*, defined by the function $\mathring{\boldsymbol{r}}$. The surface $\mathring{\boldsymbol{r}}$ is often taken to be the mid-surface. The vector $\mathring{\boldsymbol{r}}(s, \phi)$, or simply the pair (s, ϕ), identifies a typical material point on this base surface. ϕ is an azimuthal coordinate for this surface. The vector $\boldsymbol{r}(s, \phi)$ is interpreted as the deformed position of the material point (s, ϕ). The vector $\boldsymbol{b}(s, \phi)$ is interpreted as characterizing the deformed configuration of the material fiber whose reference configuration is on the normal to the base surface through $\mathring{\boldsymbol{r}}(s, \phi)$.

The axisymmetry ensures that the section of the shell with a typical $\{\boldsymbol{e}_1(\phi), \boldsymbol{k}\}$-plane behaves like every other section. The deformation of such a section can be described by a planar rod theory. The shell, however, has additional strains that influence its constitutive response.

We set

$$\boldsymbol{r}_s(s, \phi) \equiv \nu(s)\boldsymbol{a}(s, \phi) + \eta(s)\boldsymbol{b}(s, \phi),$$

$$\tau \equiv \frac{r}{\mathring{r}}, \qquad \sigma \equiv \frac{\sin\theta}{\mathring{r}}, \qquad \mu \equiv \theta'. \tag{1.4}$$

The strains ν, η, μ have the same meanings as for the planar deformation of rods. τ, the stretch in the azimuthal direction, is the ratio of deformed to reference length of a circle of latitude. σ measures the amount of bending about $\boldsymbol{a}$. To appreciate its significance, consider the deformation of a planar annulus into the frustum of a cone. The material is clearly bent. The strain μ equals its reference value of 0, but σ changes, accounting for this mode of flexure. This choice of σ for a flexural strain, while geometrically reasonable, might seem to be pulled out of thin air. In Sec. XIV.13 we construct a general theory for Cosserat shells in which natural flexural strains arise. When these are specialized to axisymmetric deformations, these strains reduce to σ and μ.

The strain variables for our theory are

$$\mathsf{q} \equiv (\tau, \nu, \eta, \sigma, \mu). \tag{1.5}$$

To ensure that these variables correspond to deformations that preserve orientation, we require that admissible strains satisfy a certain system of inequalities, which correspond to the requirement that the Jacobian of a three-dimensional deformation be positive.

1.6. Exercise. Suppose that the reference configuration of the three-dimensional body is

$$\overline{\mathcal{B}} \equiv \{\boldsymbol{z} = \mathring{\boldsymbol{r}}(s, \phi) + x^3 \mathring{\boldsymbol{b}}(s, \phi) : s_1 \le s \le s_2,\ 0 \le \phi \le 2\pi,\ -h^- \le x^3 \le h^+\}$$

where $0 \leq h^-, h^+$. Let the deformed position of the material point with coordinates (s, ϕ, x^3) be constrained to have the form $\hat{\boldsymbol{p}}(s, \phi, x^3) = \boldsymbol{r}(s, \phi) + x^3 \boldsymbol{b}(s, \phi)$. Follow Sec. VIII.6 to show that orientation is preserved if and only if

$$\nu > \max\{h^-, h^+\}|\mu|, \quad \tau > \max\{h^-, h^+\}|\sigma|. \tag{1.7}$$

Stress resultants and the equilibrium equations. Let $\boldsymbol{n}_1(s_0, \phi)$ and $\boldsymbol{m}_1(s_0, \phi)$ denote the resultant contact force and contact couple per unit reference length of the circle $\phi \mapsto \mathring{\boldsymbol{r}}(s_0, \phi)$ of radius $\mathring{r}(s_0)$ that are exerted across this section at $\mathring{\boldsymbol{r}}(s_0, \phi)$ by the material with $s \geq s_0$ on the material with $s < s_0$. Let $\boldsymbol{n}_2(s, \phi)$ and $\boldsymbol{m}_2(s, \phi)$ denote the resultant contact force and contact couple per unit reference length of the curve $s \mapsto \mathring{\boldsymbol{r}}(s, \phi_0)$ that are exerted across this section at $\mathring{\boldsymbol{r}}(s, \phi_0)$ by the material with $\phi \in [\phi_0, \phi_0 + \varepsilon]$ on the material with $\phi \in [\phi_0 - \varepsilon, \phi_0)$. Here ε is a small positive number. By a variant of the procedure used in Sec. IV.1 we find that the reactions to these resultants have the same values with signs reversed. Since we seek axisymmetric configurations, we require that these resultants have the form

$$\boldsymbol{n}_1(s, \phi) = N(s)\boldsymbol{a}(s, \phi) + H(s)\boldsymbol{b}(s, \phi), \quad \boldsymbol{n}_2(s, \phi) = T(s)\boldsymbol{e}_2(\phi), \tag{1.8a,b}$$

$$\boldsymbol{m}_1(s, \phi) = -M(s)\boldsymbol{e}_2(\phi), \qquad \boldsymbol{m}_2(s, \phi) = \Sigma(s)\boldsymbol{a}(s, \phi). \tag{1.8c,d}$$

Let us assume that the shell is subjected to an axisymmetric body force of intensity $\boldsymbol{f}(s, \phi) = f_1(s)\boldsymbol{e}_1(\phi) + f_3(s)\boldsymbol{k}$ and a body couple of intensity $-l(s)\boldsymbol{e}_2(s, \phi)$ per unit reference area of $\mathring{\boldsymbol{r}}$ at (s, ϕ). For example, a hydrostatic pressure of intensity p per unit actual area of $\boldsymbol{r}$ has an $\boldsymbol{f}$ of the form

$$\boldsymbol{f} = p \frac{\boldsymbol{r}_s \times \boldsymbol{r}_\phi}{|\mathring{\boldsymbol{r}}_s \times \mathring{\boldsymbol{r}}_\phi|} = p \frac{r}{\mathring{r}}(\nu \boldsymbol{a} + \eta \boldsymbol{b}) \times \boldsymbol{e}_2 = p\tau(\nu \boldsymbol{b} - \eta \boldsymbol{a}). \tag{1.9}$$

Then summing forces and moments on the segment consisting of all material points with coordinates (ξ, ψ) with $s_1 \leq \xi \leq s$, $0 \leq \psi \leq \phi$, we obtain

(1.10a)

$$\begin{aligned} &\int_0^\phi [N(s)\boldsymbol{a}(s, \psi) + H(s)\boldsymbol{b}(s, \psi)]\mathring{r}(s)\, d\psi \\ &- \int_0^\phi [N(s_1)\boldsymbol{a}(s_1, \psi) + H(s_1)\boldsymbol{b}(s_1, \psi)]\mathring{r}(s_1)\, d\psi \\ &+ \int_{s_1}^s T(\xi)\boldsymbol{e}_2(\phi)\, d\xi - \int_{s_1}^s T(\xi)\boldsymbol{e}_2(0)\, d\xi + \int_{s_1}^s \int_0^\phi \boldsymbol{f}(\xi, \psi)\mathring{r}(\xi)\, d\psi\, d\xi = \boldsymbol{0}, \end{aligned}$$

(1.10b)

$$\int_0^\phi \{-M(s)\boldsymbol{e}_2(\psi) + \boldsymbol{r}(s,\psi) \times [N(s)\boldsymbol{a}(s,\psi) + H(s)\boldsymbol{b}(s,\psi)]\}\mathring{r}(s)\,d\psi$$

$$-\int_0^\phi \{-M(s_1)\boldsymbol{e}_2(\psi) + \boldsymbol{r}(s_1,\psi)\times[N(s_1)\boldsymbol{a}(s_1,\psi) + H(s_1)\mathbf{b}(s_1,\psi)]\}\mathring{r}(s_1)\,d\psi$$

$$+\int_{s_1}^s [\Sigma(\xi)\boldsymbol{a}(\xi,\phi) + \boldsymbol{r}(\xi,\phi) \times T(\xi)\boldsymbol{e}_2(\phi)]\,d\xi$$

$$-\int_{s_1}^s [\Sigma(\xi)\boldsymbol{a}(\xi,0) + \boldsymbol{r}(\xi,0) \times T(\xi)\boldsymbol{e}_2(0)]\,d\xi$$

$$+\int_{s_1}^s\int_0^\phi [\boldsymbol{r}(\xi,\psi) \times \boldsymbol{f}(\xi,\psi) - l(\xi)\boldsymbol{e}_2(\phi)]\mathring{r}(\xi)\,d\psi\,d\xi = \mathbf{0}.$$

Differentiating (1.10) with respect to s and ϕ, we obtain the classical form of the equilibrium equations:

$$\tag{1.11} \tfrac{\partial}{\partial s}[\mathring{r}(N\boldsymbol{a} + H\boldsymbol{b})] - T\boldsymbol{e}_1 + \mathring{r}\boldsymbol{f} = \mathbf{0},$$

$$\tag{1.12} \tfrac{d}{ds}[\mathring{r}(s)M] - \Sigma\cos\theta + \mathring{r}(\nu H - \eta N) + \mathring{r}l = 0.$$

Constitutive relations. The shell is *nonlinearly elastic* iff there are functions $\hat{T}$, $\hat{N}$, $\hat{H}$, $\hat{\Sigma}$, $\hat{M}$ of $\mathbf{q}$ and s such that

$$\tag{1.13} T(s) = \hat{T}(\mathbf{q}(s), s), \quad \text{etc.}$$

The common domain of definition of these functions corresponds to $\mathbf{q}$'s that preserve orientation. In particular, we can take this domain to be defined by (1.7). We assume that the constitutive functions are continuously differentiable and are thrice continuously differentiable in $\mathbf{q}$. We require that these functions satisfy the *monotonicity conditions*: The matrices

$$\tag{1.14a,b} \frac{\partial(\hat{N},\hat{H},\hat{M})}{\partial(\nu,\eta,\mu)}, \quad \frac{\partial(\hat{T},\hat{\Sigma})}{\partial(\tau,\sigma)} \quad \text{are positive-definite.}$$

Condition (1.14a) ensures that an increase in the bending strain μ is accompanied by an increase in the bending couple M, etc. We also require that compatible coercivity conditions hold:

$$\tag{1.15a} \begin{Bmatrix} \hat{T}(\mathbf{q},s) \\ \hat{N}(\mathbf{q},s) \end{Bmatrix} \to -\infty \quad \text{as} \quad \begin{Bmatrix} \tau \searrow \max\{h^-,h^+\}|\sigma| \\ \nu \searrow \max\{h^-,h^+\}|\mu| \end{Bmatrix}$$

if $\begin{Bmatrix} \nu \\ \tau \end{Bmatrix}$ is bounded above and if η,σ,μ are bounded,

$$\tag{1.15b} \begin{Bmatrix} \hat{T}(\mathbf{q},s) \\ \hat{N}(\mathbf{q},s) \end{Bmatrix} \to \infty \quad \text{as} \quad \begin{Bmatrix} \tau \\ \nu \end{Bmatrix} \to \infty$$

if $\begin{Bmatrix} \nu - \max\{h^-,h^+\}|\mu| \\ \tau - \max\{h^-,h^+\}|\sigma| \end{Bmatrix}$ has a positive lower bound

and if η,σ,μ are bounded,

(1.15c) $$\hat{H}(\mathbf{q}, s) \to \pm\infty \quad \text{as } \eta \to \pm\infty$$

if (τ, ν, σ, μ) lies in a compact subset of

$$\{(\tau, \nu, \sigma, \mu) : \tau > \max\{h^-, h^+\}|\sigma|,\ \nu > \max\{h^-, h^+\}|\mu|\},$$

(1.15d) $$\left\{ \begin{array}{c} \hat{\Sigma}(\mathbf{q}, s) \\ \hat{M}(\mathbf{q}, s) \end{array} \right\} \to \pm\infty \quad \text{as} \quad \left\{ \begin{array}{c} \sigma \\ \mu \end{array} \right\} \to \pm \left\{ \begin{array}{c} \tau/\max\{h^-, h^+\} \\ \nu/\max\{h^-, h^+\} \end{array} \right\}$$

if $\left\{ \begin{array}{c} (\nu, \eta, \mu) \\ (\tau, \eta, \mu) \end{array} \right\}$ lies in a compact subset of

$$\left\{ \begin{array}{c} \{(\nu, \eta, \mu) : \nu > \max\{h^-, h^+\}|\mu|\} \\ \{(\tau, \eta, \mu) : \tau > \max\{h^-, h^+\}|\sigma|\} \end{array} \right\}.$$

These conditions are more complicated than those of Sec. IV.1. To appreciate their significance, we focus on the first condition of (1.15b), and for simplicity we suppose that $(\eta, \sigma, \mu) = (0, 0, 0)$. We consider a specimen of material whose reference configuration is a small square with τ measuring the stretch in the horizontal direction and ν, in the vertical direction. The first condition of (1.15b) says that if the horizontal stretch τ becomes infinite while the vertical stretch ν has a positive lower bound, then the horizontal tension T must become infinite to maintain such a state. If the lower bound on ν were not present, then it would seem reasonable to expect that τ could be forced to ∞ with bounded T by squeezing ν down to 0. It is this interaction between effects in different directions that gives elasticity its richness and complexity. (See the discussion following (XII.2.30). These interactions are absent in the problems discussed so far in this book.)

That conditions such as (1.14) and (1.15) are regarded as physically reasonable and worthy of study does not imply that alternative conditions are necessarily unreasonable or unworthy of study.

In view of (1.14) and (1.15), we can invoke the Global Implicit Function Theorem XIX.2.30 to show that the system of nonlinear algebraic equations

(1.16) $$\hat{N}(\mathbf{q}, s) = n, \quad \hat{H}(\mathbf{q}, s) = h, \quad \hat{M}(\mathbf{q}, s) = m$$

can be uniquely solved for (ν, η, μ) in terms of the other variables appearing in this equation. We denote this solution by

(1.17a) $$\nu = \nu^\sharp(\tau, n, h, \sigma, m, s), \text{ etc.}$$

We then define $T^\sharp$ and $\Sigma^\sharp$ by

(1.17b) $$T^\sharp(, \tau, n, h, \sigma, \mu, s) \equiv \hat{T}(\tau, \nu^\sharp, \eta^\sharp, \sigma, \mu^\sharp, s), \quad \text{etc.},$$

where the arguments of $\nu^\sharp$, etc., are shown in (1.17a).

For simplicity, we assume that the natural state is stress-free, so that

(1.18) $$\hat{T}(\mathbf{q}, s) = \hat{N}(\mathbf{q}, s) = \hat{H}(\mathbf{q}, s) = \hat{\Sigma}(\mathbf{q}, s) = \hat{M}(\mathbf{q}, s) = 0$$

$$\text{for } \mathbf{q} = (1, 1, 0, \mathring{\sigma}, \mathring{\theta}').$$

2. Buckling of a Transversely Isotropic Circular Plate

We specialize the theory of Sec. 1 to circular plates of unit radius by taking $s_1 = 0$, $s_2 = 1$, and

$$\mathring{r}(s) = s, \quad \mathring{\theta}(s) = 0. \tag{2.1}$$

We assume that there is neither applied force nor couple, so that $\boldsymbol{f} = \mathbf{0}$, $l = 0$. The axisymmetry implies that for a classical solution, the following boundary conditions should hold at the center of the plate:

$$r(0) = 0, \quad \theta(0) = 0, \quad \eta(0) = 0. \tag{2.2a,b,c}$$

We assume that the edge $s = 1$ is constrained to be parallel to $\boldsymbol{k}$:

$$\theta(1) = 0. \tag{2.3}$$

We assume that a normal pressure of intensity $\lambda g(\rho(1))$ units of force per reference length is applied to the edge $s = 1$ of the plate, so that

$$N(1) = -\lambda g(r(1)), \quad H(1) = 0. \tag{2.4a,b}$$

If the edge pressure has intensity λ units of force per *deformed* length, then $g(r) = r$, whereas if it has intensity λ units of force per *reference* length, then $g(r) = 1$. For simplicity, we assume that the latter is the case.

Replacing (s_1, s) in (1.10a) with $(s, 1)$ and using (2.4), we obtain

$$-sN(s) = \left[\lambda + \int_s^1 T(\xi)\, d\xi\right] \cos\theta(s), \tag{2.5a}$$

$$sH(s) = \left[\lambda + \int_s^1 T(\xi)\, d\xi\right] \sin\theta(s). \tag{2.5b}$$

By differentiating these equations, we recover the componential forms of (1.11):

$$\tfrac{d}{ds}[sN(s)] = sH(s)\theta'(s) + T(s)\cos\theta(s), \tag{2.6a}$$

$$-\tfrac{d}{ds}[sH(s)] = sN(s)\theta'(s) - T(s)\sin\theta(s). \tag{2.6b}$$

Substituting (2.5) into (1.12) we obtain

$$\tfrac{d}{ds}[sM(s)] - \Sigma(s)\cos\theta(s) + \left[\lambda + \int_s^1 T(\xi)\, d\xi\right][\nu(s)\sin\theta(s) + \eta(s)\cos\theta(s)] = 0, \tag{2.7}$$

which resembles (VI.6.1). From (2.5) we readily obtain

$$H\cos\theta = -N\sin\theta, \quad (sN/\cos\theta)' = T, \tag{2.8a,b}$$

which is equivalent to (2.6). (We can get (2.8a) directly by taking the dot product of (1.11) with $\boldsymbol{k}$.)

We require that the material meet the following minimal restrictions on its symmetry:

$$\hat{T}, \hat{N}, \hat{\Sigma}, \hat{M} \text{ are even in } \eta, \quad \hat{H} \text{ is odd in } \eta, \tag{2.9a}$$

$$\hat{T}, \hat{N}, \hat{H} \text{ are unchanged under } (\sigma, \mu) \mapsto (-\sigma, -\mu), \tag{2.9b}$$

$$\hat{\Sigma}, \hat{M} \text{ change sign under } (\sigma, \mu) \mapsto (-\sigma, -\mu). \tag{2.9c}$$

These conditions ensure that deformed states come in mirror images. Condition (2.9c) is not as simple as one might hope. The source of these constitutive assumptions is the three-dimensional theory; see Sec. XIV.11.

We assume that the constitutive functions are independent of s (for simplicity of exposition) and satisfy the restricted *isotropy* conditions:

$$\hat{N}(\tau, \nu, 0, \sigma, \mu) = \hat{T}(\nu, \tau, 0, \mu, \sigma), \quad \hat{M}(\tau, \nu, 0, \sigma, \mu) = \hat{\Sigma}(\nu, \tau, 0, \mu, \sigma). \tag{2.10}$$

These conditions are termed restricted because η is taken to be 0 here. If (2.10) holds, we call the plate *transversely isotropic*. A gross but useful oversimplification of the import of (2.10) is that it says that $\hat{N}$ depends on ν the same way that $\hat{T}$ depends on τ, etc. The general isotropy condition, which we study in a three-dimensional context in Sec. XII.13, says that the material response is unaffected by rotations of material fibers in the reference configuration (cf. Sec. VIII.9). For axisymmetric deformations, we need only require a suitable invariance under rotations of $\pi/2$ that take longitudinal fibers into azimuthal fibers, or vice versa. It is precisely such rotations that are accounted for in (2.10).

Unbuckled states. We first seek unbuckled states for which $\eta = 0 = \theta$. Then all the equilibrium equations are identically satisfied except for (2.5a). We substitute the constitutive equations (1.13) into (2.5a) and seek solutions of the form

$$r(s) = r_0(s) = ks \tag{2.11}$$

where k is a constant to be determined. The isotropy condition reduces the resulting form of (2.5a) to

$$T(k, k, 0, 0, 0) = -\lambda. \tag{2.12}$$

We augment (1.14) and (1.15) with the mild assumption that $T(k, k, 0, 0, 0)$ strictly decreases from 0 to $-\infty$ as k decreases from 1 to 0 and is positive for $k > 1$. Then for each $\lambda \geq 0$, Eq. (2.12) has a unique solution for k, which we denote by $k(\lambda)$.

The linearization. Let R represent any constitutive function from (1.13) or a derivative of such a function. Then we set

$$R^0(\lambda) \equiv \hat{R}(k(\lambda), k(\lambda), 0, 0, 0). \tag{2.13}$$

The linearization of (2.7) and (1.13) about the unbuckled state defined by $r_0(s, \lambda) = k(\lambda)s$ is the Bessel equation

$$(s\theta'_{(1)})' - s^{-1}\theta_{(1)} + \gamma(\lambda)^2 s\theta_{(1)} = 0 \tag{2.14}$$

where

$$\gamma(\lambda)^2 \equiv \frac{\lambda}{M_\mu^0(\lambda)}\left[k(\lambda) + \frac{\lambda}{H_\eta^0(\lambda)}\right] = -\frac{N^0(\lambda)}{M_\mu^0(\lambda)}\left[k(\lambda) - \frac{N^0(\lambda)}{H_\eta^0(\lambda)}\right]. \tag{2.15}$$

The eigenvalues λ^0 of (2.14) subject to (2.2b) and (2.3) are solutions of

$$J_1(\gamma(\lambda)) = 0 \tag{2.16}$$

where J_1 is the Bessel function of order 1. The corresponding eigenfunction is $s \mapsto J_1(\gamma(\lambda^0)s)$. In particular, if j_p is the $(p+1)$st zero of J_1 and if there is a λ^0 such that $\gamma(\lambda^0) = j_p$, then $J_1(j_p s)$ is the corresponding eigenfunction, which has exactly p zeros on $(0, 1)$, each simple.

2.17. Exercise. Linearize the entire boundary-value problem about the trivial branch, obtaining (2.14) and (2.15).

We can expect γ^2 to look like the q of Sec. VI.7. We could then carry out a local analysis along the lines of Sec. VI.8. To carry out a global analysis, we require that the boundary-value problem can be cast in the form required by the One-Parameter Global Bifurcation Theorem V.4.19.

Compactness. We now wish to convert our problem to the form (V.3.1), (V.4.5) with $(\lambda, \mathbf{u}) \mapsto \mathbf{L}(\lambda)\mathbf{u}$, $\mathbf{g}[\lambda, \mathbf{u}]$ compact and continuous. Our inclination is to follow the scheme begun in Sec. VI.5 for rod problems and use the natural integral equations of equilibrium as our starting point. But (2.5) suggests that we shall encounter technical problems, at least in dealing with the polar singularity at $s = 0$. It might be possible to resolve the difficulties by adopting suitable weightings on the unknowns, but a study of (2.5) and (2.7) suggests that the efficacy of weightings depends crucially on the details of the constitutive response.

The approach we adopt is suggested by the linearized problem, which is governed by a Bessel equation. We know from elementary spectral theory (see Stakgold (1979)) that many eigenvalue problems involving Bessel equations can be transformed into integral equations of the form $[\mathbf{I} - \mathbf{L}(\lambda)] \cdot \mathbf{u} = \mathbf{0}$ with $\mathbf{L}(\lambda)$ compact. Our plan is to force our boundary-value problem into a form suggested by the linear problem with the hope that the constitutive properties will enable us to prove that the nonlinear problem has the same compactness properties as the linear problem. We merely sketch the salient ideas, omitting a lot of computation.

We substitute (1.13) into (2.6) and (2.7) and then carry out the differentiations on the left-hand sides of the resulting equations. Condition (1.14a) enables us to use Cramer's rule to solve these equations for $s\nu'$, $s\eta'$, $s\mu'$. We then use (1.2) to force these equations into a mold suggested by the linearization:

$$L_1\rho \equiv \frac{d}{ds}[s\rho'] - \frac{\rho}{s} = f_1, \tag{2.18a}$$

$$L_2 z \equiv \frac{d}{ds}[s z'] = f_2, \tag{2.18b}$$

$$L_3\theta \equiv \frac{d}{ds}[s\theta'] - \frac{\theta}{s} = f_3 \tag{2.18c}$$

where $\rho \equiv r - r_0$ and where the $f's$ are complicated expressions involving all the geometric variables and the parameter λ. The variables ρ, z, and θ satisfy the boundary conditions

$$\rho(0) = 0, \tag{2.19a}$$

$$N_\nu^0\rho'(1) + N_\tau^0\rho(1) = -\hat{N}(\mathbf{q}(1)) - \lambda + N_\nu^0\rho'(1) + N_\tau^0\rho(1) \equiv b, \tag{2.19b}$$

$$z(0) = 0 = z'(1), \tag{2.19c}$$

$$\theta(0) = 0 = \theta(1). \tag{2.19d}$$

We express f_1, f_2, f_3 in terms of ρ, z, θ by using (1.4).

Motivated by the treatment of linear problems involving Bessel's operator, we define $\mathbf{u} \equiv (u_1, u_2, u_3)$ by

$$\sqrt{s}u_1(s) \equiv (L_1\rho)(s), \quad \sqrt{s}u_2(s) \equiv (L_2 z)(s), \quad \sqrt{s}u_3(s) \equiv (L_3\theta)(s). \tag{2.20}$$

We integrate these equations subject to the boundary condition (2.19) to express (ρ, z, θ) as integral operators acting on (u_1, u_2, u_3). In particular,

$$\theta[u_3](s) = \int_0^1 K_3(s,\xi)\sqrt{\xi}u_3(\xi)\,d\xi \tag{2.21}$$

where K_3, Green's function for L_3 subject to (2.19d), is given by

$$K_3(s,\xi) = \begin{cases} \frac{1}{2}\left(s - \frac{1}{s}\right)\xi & \text{for } \xi < s, \\ \frac{1}{2}\left(\xi - \frac{1}{\xi}\right)s & \text{for } s < \xi. \end{cases} \tag{2.22}$$

We now replace every expression in f_1, f_2, f_3 by representations of the form (2.21). We denote the resulting f_i by $f_i[\lambda, \mathbf{u}]$. Using (2.18) and (2.20), we can convert our boundary-value problem to

$$u_i(s) = \frac{f_i[\lambda, \mathbf{u}](s)}{\sqrt{s}}. \tag{2.23}$$

We want to show that the right-hand side of (2.23) defines a mapping from $\mathbb{R} \times [C^0[0,1]]^3$ to $[C^0[0,1]]^3$ that is continuous and compact.

Now let us examine carefully some crucial steps in the process by which (2.23) was constructed. From (2.7) and (1.13) we obtain

$$s(\hat{M}_\sigma \sigma' + \hat{M}_\mu \theta'' + \cdots) = \hat{\Sigma}\cos\theta - \hat{M} + \cdots, \tag{2.24}$$

which we can rewrite as

$$\hat{M}_\mu L_3 \theta = \hat{\Sigma} - \hat{M} - \left(\frac{\theta}{s} - \theta'\right)\hat{M}_\mu - \left(\frac{\theta}{s} - \theta'\right)\hat{M}_\sigma + \cdots. \tag{2.25}$$

Here the ellipsis contains terms like $(\cos\theta - 1)\hat{\Sigma}$, which prove to be innocuous.

Thus we can expect the right-hand sides of (2.23) to contain an expression of the form (2.25) divided by $\sqrt{s}$. Since $\hat{\Sigma}$ and $\hat{M}$ depend on $s^{-1}\sin\theta$ and θ', which (2.21) does not require to be particularly well behaved at $s = 0$, this term could well cause trouble. It is not evident that (2.25) divided by $\sqrt{s}$ is even continuous. We now show how isotropy extricates us from the threatened difficulty.

For simplicity of exposition, let us assume that $\hat{\Sigma}$ and $\hat{M}$ depend only on σ and μ and let us replace $\sigma(s)$ with $\theta(s)/s$. Then in place of (2.25) we accordingly study $s \mapsto \Omega[\theta](s)/\sqrt{s}$ where

$$\begin{aligned}\Omega[\theta](s) \equiv{}& \hat{\Sigma}\left(\frac{\theta(s)}{s}, \theta'(s)\right) - \hat{M}\left(\frac{\theta(s)}{s}, \theta'(s)\right) \\ &- \left(\frac{\theta(s)}{s} - \theta'(s)\right)\left[\hat{M}_\mu\left(\frac{\theta(s)}{s}, \theta'(s)\right) + \hat{M}_\sigma\left(\frac{\theta(s)}{s}, \theta'(s)\right)\right]\end{aligned} \tag{2.26}$$

when θ is given by (2.21) and u_3 is continuous. It is still not evident that (2.26) is even continuous.

Now (2.21) implies that

(2.27a)
$$2\theta[u_3](s) = \int_0^s \left(s - \frac{1}{s}\right)\xi^{3/2} u_3(\xi)\, d\xi + \int_s^1 s\left(\xi^{3/2} - \xi^{-1/2}\right) u_3(\xi)\, d\xi,$$
(2.27b)
$$2\theta[u_3]'(s) = \int_0^s \left(1 + \frac{1}{s^2}\right)\xi^{3/2} u_3(\xi)\, d\xi + \int_s^1 \left(\xi^{3/2} - \xi^{-1/2}\right) u_3(\xi)\, d\xi,$$
(2.27c)
$$2\frac{\theta[u_3](s)}{s} = \int_0^s \left(1 - \frac{1}{s^2}\right)\xi^{3/2} u_3(\xi)\, d\xi + \int_s^1 \left(\xi^{3/2} - \xi^{-1/2}\right) u_3(\xi)\, d\xi.$$

The Arzelà-Ascoli Theorem V.5.3 implies that the right-hand sides of (2.27) define compact mappings taking C^0 into itself.

Now the isotropy condition that $\hat{\Sigma}(\sigma, \mu) = \hat{M}(\mu, \sigma)$, which mimics (2.10), implies that $\hat{\Sigma}_\sigma(\sigma, \mu) = \hat{M}_\mu(\mu, \sigma)$ and $\hat{\Sigma}_\mu(\sigma, \mu) = \hat{M}_\sigma(\mu, \sigma)$. Here

$\hat{M}_\mu(\mu, \sigma)$ denotes the derivative of $\hat{M}$ with respect to its *second argument*, which here is occupied by σ. Thus

$$\begin{aligned}(2.28)\quad \Omega[\theta](s) &= \hat{\Sigma}\left(\frac{\theta(s)}{s}, \theta'(s)\right) - \hat{\Sigma}\left(\theta'(s), \frac{\theta(s)}{s}\right) \\ &- \left(\frac{\theta(s)}{s} - \theta'(s)\right)\left[\hat{\Sigma}_\sigma\left(\theta'(s), \frac{\theta(s)}{s}\right) + \hat{\Sigma}_\mu\left(\theta'(s), \frac{\theta(s)}{s}\right)\right],\end{aligned}$$

which we recognize as the difference between $\hat{\Sigma}\left(\frac{\theta(s)}{s}, \theta'(s)\right)$ and its linear approximation at $\left(\theta'(s), \frac{\theta(s)}{s}\right)$. Assuming that $\hat{\Sigma}$ is twice continuously differentiable, we use Taylor's theorem to write (2.28) as

$$(2.29)\qquad \alpha\left(\theta[u_3]'(s), \frac{\theta[u_3](s)}{s}\right)\left[\theta[u_3]'(s) - \frac{\theta[u_3](s)}{s}\right]^2$$

where α is a continuous function.

2.30. Exercise. Obtain an explicit representation for α as an integral of derivatives of constitutive functions that shows that α is continuous when the constitutive functions are twice continuously differentiable.

It follows from the definition of a compact mapping, given before the statement of Theorem VI.4.19, that the composition of a continuous mapping with a compact mapping (in either order) is compact and that the product of compact mappings from C^0 to C^0 is compact. Therefore, to prove the compactness of the mapping taking u_3 into (2.26), we need only prove that the mapping taking u_3 to

$$(2.31)\qquad s \mapsto s^{-1/4}\left[\theta[u_3]'(s) - \frac{\theta[u_3](s)}{s}\right] = s^{-9/4}\int_0^s \xi^{3/2} u_3(\xi)\, d\xi$$

is compact. (The equality in (2.31) follows from (2.27).)

2.32. Exercise. Prove that the mapping taking u_3 into (2.31) is compact, but that the mappings taking u_3 into $s \mapsto s^{-1/4}\theta[u_3]'(s)$ and $s \mapsto s^{-1/4}\theta[u_3](s)/s$ are not compact.

Thus we can apply the Global Bifurcation Theorem VI.4.19 to branches of our problem that bifurcate from eigenvalues of odd algebraic multiplicity.

Nodal properties. Just as in Sec. V.5 and Chap. VI, a proof that the nodal properties are preserved on any connected set of solution pairs not containing a trivial solution devolves on a uniqueness theorem for initial-value problems for the governing equations. Just as in Sec. VI.2, the proof of this uniqueness theorem is routine except for the case of zero initial data at the singular point $s = 0$. We again merely sketch the main ideas:

We suppose that a solution of (2.23) generates via (2.21) a θ (which is necessarily continuously differentiable) that has a double zero at 0. Now ρ and z have representations like (2.21), from which we conclude that $\rho(0) =$

$0 = \rho'(0)$ and $\eta(0) = 0$. It then follows from the definition of ρ that $\hat{N}(\mathbf{q}(s)) \to N^0 = -\lambda$ as $s \to 0$. Then (2.5a) implies that

$$\Gamma(s, \lambda) \equiv \frac{1}{s}\left[\lambda + \int_s^1 \hat{T}(\mathbf{q}(\xi))\, d\xi\right] \tag{2.33}$$

defines a continuous function.

The positivity of $\hat{H}_\eta$, ensured by (1.14a), and the coercivity condition (1.15c) imply that $\eta \mapsto \hat{H}(\mathbf{q})$ has an inverse $H \mapsto \tilde{\eta}(\tau, \nu, H, \sigma, \mu)$. Since $\tilde{\eta}(\tau, \nu, 0, \sigma, \mu) = 0$ by (2.9a), we obtain from (2.5b) and (2.33) that

$$\begin{aligned} \tilde{\eta}(\tau, \nu, \hat{H}(\mathbf{q}), \sigma, \mu) &= \left[\int_0^1 \tilde{\eta}_H(\tau, \nu, t\hat{H}(\mathbf{q}), \sigma, \mu)\, dt\right] \hat{H}(\mathbf{q}) \\ &\equiv \chi(\mathbf{q})\hat{H}(\mathbf{q}) = \chi(\mathbf{q})\Gamma \sin\theta. \end{aligned} \tag{2.34}$$

We replace the η in (2.7) with (2.34) and again use the ideas leading to (2.25) and (2.29) to express (2.7) as

$$\hat{M}_\mu L_3 \theta = \alpha\left[\theta' - \frac{\theta}{s}\right]^2 - s\Gamma[\nu + \chi\Gamma]\sin\theta + \cdots . \tag{2.35a}$$

Since we have an existence theory, we can regard (2.35) as an equation for θ alone. It is clear from the form of this equation that if θ has a double zero in $(0, 1]$, then $\theta = 0$. We now deduce the same conclusion when θ has a double zero at 0.

Now we set $\psi \equiv s\theta' + \theta$, $\omega \equiv s\theta' - \theta$ and rewrite (2.35a) as the system

$$\begin{Bmatrix} \psi' \\ \omega' \end{Bmatrix} = \begin{Bmatrix} \psi/s \\ -\omega/s \end{Bmatrix} + \frac{\alpha}{\hat{M}_\mu}\frac{\omega^2}{s^2} - \frac{s}{\hat{M}_\mu}\Gamma[\nu + \chi\Gamma]\sin\tfrac{1}{2}(\psi - \omega) + \cdots . \tag{2.35b}$$

We want to show that the only solution of this equation when θ has a double zero at 0 is the trivial solution.

From (2.31) it follows that $s \mapsto s^{-1/2}[\theta'(s) - \theta(s)/s] = \omega s^{-3/2}$ is bounded. Since we are restricting our attention to a specific solution, it then follows that there is a positive number C such that the second term on the right-hand side of (2.35b) is bounded in absolute value by $C\omega s^{-1/2}$. Let $v \equiv |\psi| + |\omega|$ (cf. Ex. VI.2.14). In light of the preceding remarks, (2.35b) implies that there is another positive number C such that

$$v' \leq \frac{v}{s} + \frac{Cv}{\sqrt{s}}, \quad \frac{v(s)}{s} \to 0 \quad \text{as } s \to 0, \tag{2.36}$$

the limit condition corresponding to the double zero of θ. The only solution of (2.36) is $v = 0$. This fact ensures the uniqueness that supports the nodal theory for our problem. (If the term v/s on the right-hand side of (2.36) were multiplied by a constant greater than 1, then the solution of the resulting inequality need not be 0. That this constant is exactly 1 is a consequence of isotropy.)

2.37. Exercise. Prove all the statements in the last paragraph.

This section is based upon Antman (1978a).

3. Remarkable Trivial States of Aeolotropic Circular Plates

We now study trivial states of the problem of Sec. 2 for uniform plates when (2.10) does not hold. These have a surprisingly rich and complicated structure. Since (2.9) holds, the governing equations for $\theta = 0 = \eta$ reduce to

$$\text{(3.1)} \qquad \tfrac{d}{ds}[s\hat{N}(r(s)/s, r'(s))] = \hat{T}(r(s)/s, r'(s)),$$

which comes from (2.6a), (1.4), and (1.13). Here and below we set $\hat{N}(\tau, \nu) = \hat{N}(\tau, \nu, 0, 0, 0)$, etc. The boundary conditions come from (2.2a) and (2.4a,b):

$$\text{(3.2)} \qquad r(0) = 0, \quad \hat{N}(r(1), r'(1)) = -\lambda.$$

To avoid tedious technicalities in our analysis, we supplement the constitutive restriction (1.14) with some auxiliary requirements. The first of these is: The matrix

$$\text{(3.3)} \qquad \frac{\partial(\hat{T}, \hat{N})}{\partial(\tau, \nu)} \quad \text{is positive-definite.}$$

This condition roughly says that a change in τ has more effect on $\hat{T}$ than it does on $\hat{N}$.

3.4. Exercise. Let (3.3) hold and let $\lambda \geq 0$. Prove that the boundary-value problem (3.1), (3.2) has at most one solution $r \in C^2(0, 1]$ for which $s \mapsto s\hat{N}(r(s)/s, r'(s))$ is bounded.

We further require that

$$\text{(3.5a,b)} \qquad \hat{T}_\nu > 0, \quad \hat{N}_\tau > 0.$$

To appreciate the significance of (3.5), consider the deformation of a rectangular block with edges parallel to the x, y, z-axes. If we fix the length in the x-direction, increase the length in the y-direction, and apply zero force to the faces perpendicular to the z-axis, we might expect the tensions in both the x- and y-directions to increase. If we identify ν and τ with the stretches in the x- and y-directions respectively, then this argument yields (3.5b).

Let $\nu^\sharp(\tau, n)$ denote the unique solution of the equation $\hat{N}(\tau, \nu) = n$; its existence is ensured by (1.14a) and a corresponding growth condition. We set $T^\sharp(\tau, n) \equiv \hat{T}(\tau, \nu^\sharp(\tau, n))$. Then the boundary-value problem (3.1), (3.2) is equivalent to the semilinear problem

$$\text{(3.6a,b)} \qquad \tfrac{d}{ds}(s\breve{\tau}) = \nu^\sharp(\breve{\tau}, \breve{n}), \quad \tfrac{d}{ds}(s\breve{n}) = T^\sharp(\breve{\tau}, \breve{n}),$$

$$\text{(3.6c)} \qquad s\breve{\tau}(s) \to 0 \quad \text{as } s \to 0,$$

$$\text{(3.6d,e)} \qquad \breve{n}(1) = -\lambda, \quad \breve{\tau}(s) > 0 \quad \text{for } s > 0.$$

3.7. Exercise. Convert the specializations of (1.14), (1.15a,b), and (1.18) for $\eta = \sigma = \mu = 0$ and the conditions (3.3) and (3.5) to equivalent restrictions on $\nu^\sharp$ and $T^\sharp$.

Let

$$s = e^{\xi - 1}, \quad \breve{\tau}(s) = \tau(\xi), \quad \breve{n}(s) = n(\xi). \tag{3.8}$$

Then (3.6) is equivalent to the autonomous problem

$$\tfrac{d}{d\xi}\tau = \nu^\sharp(\tau, n) - \tau, \quad \tfrac{d}{d\xi}n = T^\sharp(\tau, n) - n, \quad -\infty < \xi < 1, \tag{3.9a,b}$$

$$e^\xi \tau(\xi) \to 0 \quad \text{as} \quad \xi \to -\infty, \tag{3.9c}$$

$$n(1) = -\lambda, \quad \tau(\xi) > 0 \quad \text{for} \quad \xi > -\infty. \tag{3.9d,e}$$

Thus we can study (3.9) by phase-plane methods.

The vertical isoclines of (3.9) consist of those points (τ, n) with $\tau \geq 0$ for which

$$\tau = \nu^\sharp(\tau, n) \quad \text{or, equivalently,} \quad n = \hat{N}(\tau, \tau) \tag{3.10}$$

and the horizontal isoclines consist of those points (τ, n) for which

$$n = T^\sharp(\tau, n) \quad \text{or, equivalently,} \quad n = \hat{N}(\tau, \nu), \quad n = \hat{T}(\tau, \nu). \tag{3.11}$$

Conditions (3.3), (3.5b), (1.15a,b), and (1.18) imply that (3.10) is equivalent to an equation of the form

$$\tau = v(n) \tag{3.12}$$

with v increasing from 0 to 1 to ∞ as n increases from $-\infty$ to 0 to ∞, and that (3.11) is equivalent to an equation of the form

$$\tau = h(n) \tag{3.13}$$

with $h(n) \to 0$ as $n \to -\infty$, $h(0) = 0$. The functions v and h are thrice continuously differentiable because $\hat{T}$ and $\hat{N}$ are.

Let

$$\begin{aligned} \mathcal{U} &\equiv \{(\tau, n) : 0 < \tau, \ -\infty < n < \infty\}, \\ \overline{\mathcal{U}} &\equiv \{(\tau, n) : 0 \leq \tau \leq \infty, \ -\infty \leq n \leq \infty\}. \end{aligned} \tag{3.14}$$

$\overline{\mathcal{U}}$ is the phase space for our problem. We introduce the open quadrants

$$\begin{aligned} &\mathcal{Q}_1 \equiv \{(\tau, n) : 1 < \tau, \ 0 < n\}, \quad \mathcal{Q}_2 \equiv \{(\tau, n) : 0 < \tau < 1, \ 0 < n\}, \\ &\mathcal{Q}_3 \equiv \{(\tau, n) : 0 < \tau < 1, \ n < 0\}, \quad \mathcal{Q}_4 \equiv \{(\tau, n) : 1 < \tau, \ n < 0\}. \end{aligned} \tag{3.15}$$

The singular points of (3.9) in $\mathcal{U}$ are points at which the horizontal and vertical isoclines intersect and thus are points (τ, n) for which

$$n = \hat{T}(\tau, \tau) = \hat{N}(\tau, \tau). \tag{3.16}$$

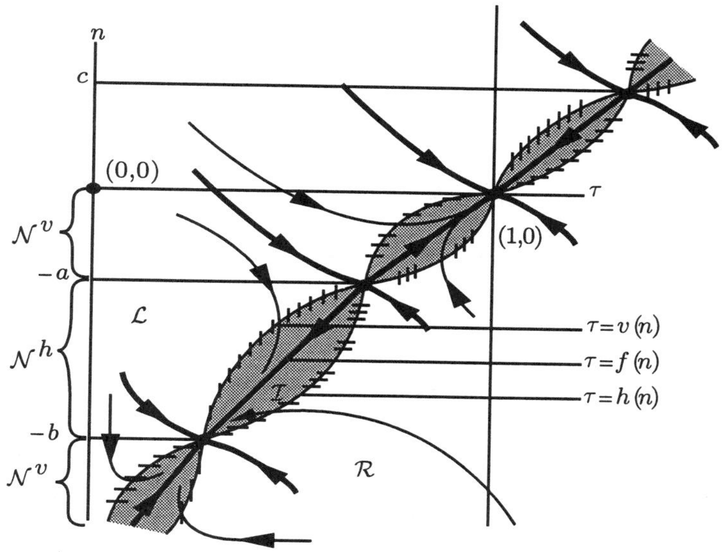

Fig. 3.17. Typical phase portrait of (3.9) showing the isoclines, the invariant regions $\mathcal{I}$, $\mathcal{L}$, and $\mathcal{R}$, and the trajectories on the curve $\tau = f(n)$. Here the horizontal and vertical isoclines intersect transversally at $n = 0, a, b$. The singular points $(1, 0)$ and $(-b, f(-b))$ are attractive nodes and the singular points $(-a, f(-a))$ and $(c, f(c))$ are saddle points.

(Other candidates for singular points in $\overline{\mathcal{U}}$, at which the direction field is not defined, are points on its boundary $\tau = 0$ and points at infinity. These points require separate study.)

Let $\mathcal{I}$ be the (disconnected) open region lying between the isoclines. Let $\mathcal{L}$ be the open region lying to the left of both isoclines and let $\mathcal{R}$ be the open region lying to the right of both isoclines. We illustrate these regions in Fig. 3.17.

It is instructive to examine the isoclines (3.10) and (3.11) for an isotropic material, for which (2.10) implies that

$$\hat{T}(\tau, \nu) = \hat{N}(\nu, \tau). \tag{3.18}$$

In this case, the equation

$$\hat{N}(\tau, \nu) = \hat{T}(\tau, \nu) \equiv \hat{N}(\nu, \tau), \tag{3.19}$$

which comes from (3.11), has a solution for ν, namely, $\nu = \tau$. The corresponding horizontal isoclines defined by (3.11) are then given by $n = \hat{N}(\tau, \tau)$, which is the same equation as (3.10). Thus the vertical isoclines are also horizontal isoclines and are therefore curves of singular points. If (3.19) admits solutions for ν other than $\nu = \tau$, then there are horizontal isoclines not coincident with the vertical isoclines, but these contain no critical points in $\mathcal{U}$. In any event, all other trajectories impinge transversally from the left and the right onto the curve of singular points.

If we make the constitutive assumption that

$$\frac{\partial}{\partial \nu}\left[\hat{N}(\tau,\nu) - \hat{T}(\tau,\nu)\right]\bigg|_{\nu=\tau} > 0 \quad \forall \tau, \tag{3.20}$$

then the solution $\nu = \tau$ of (3.19) is unique, and the horizontal and vertical isoclines coincide. Condition (3.20) is ensured by (3.3) and (3.18).

Now conditions (1.14) and (1.18) ensure via Ex. 3.7 that

$$[\nu^\sharp(1,n) - 1]n > 0 \quad \text{for } n \neq 0. \tag{3.21}$$

Thus (3.9a) implies that trajectories touching the line $\{(1,n) : n > 0\}$ cross it transversally from $\mathcal{Q}_2$ to $\mathcal{Q}_1$ and that trajectories touching the line $\{(1,n) : n < 0\}$ cross it transversally from $\mathcal{Q}_4$ to $\mathcal{Q}_3$. Similarly, we find that

$$(\tau - 1)T^\sharp(\tau, 0) > 0 \quad \text{for } \tau \neq 1. \tag{3.22}$$

Thus (3.9b) implies that trajectories touching the segment $\{(\tau,0) : 0 < \tau < 1\}$ cross it transversally from $\mathcal{Q}_2$ to $\mathcal{Q}_3$ and that trajectories touching the line $\{(\tau,0) : 1 < \tau\}$ cross it transversally from $\mathcal{Q}_4$ to $\mathcal{Q}_1$. Conditions (3.21) and (3.22) imply that $\mathcal{Q}_1$ and $\mathcal{Q}_3$ are *positively invariant regions*, i.e., trajectories entering them never leave. The disposition of the vector fields in $\mathcal{I}$, $\mathcal{L}$, and $\mathcal{R}$ shows that $\mathcal{I}$ is positively invariant and that $\mathcal{L}$ and $\mathcal{R}$ are negatively invariant regions. (A *negatively invariant region* is one with the property that if a trajectory is ever in it, then it must have been in it for all smaller values of the independent variable.)

Since (3.9a,b) is autonomous, it is invariant under shifts of ξ. Therefore, if a trajectory originates at a point, such as a singular point, for which ξ may be chosen to equal $-\infty$, then any point on the trajectory with ξ finite may be regarded as corresponding to $\xi = 1$. A solution of (3.9) accordingly corresponds to a trajectory in $\mathcal{U}$ that originates at a singular point or possibly a point on the line $\tau = 0$ or a point at ∞, that exhausts an infinite amount of independent variable ξ, that terminates on the line $n = -\lambda$, and that satisfies (3.9c) (which is automatic if τ is bounded on the trajectory). Our fundamental result is

3.23. Lemma. *Let* (1.15a,b), (1.18), (3.3), *and* (3.5) *hold. A trajectory having a single point in* $\mathcal{L}$ *or in* $\mathcal{R}$ *cannot correspond to a solution of* (3.9). *A trajectory corresponding to a solution must therefore begin at a singular point and remain thereafter in a connected component of* $\mathcal{I}$.

Proof. Our constitutive assumptions imply that

$$\tfrac{d}{d\xi}\tau > 0, \quad \tfrac{d}{d\xi}n < 0, \quad \tfrac{d}{d\xi}\nu^\sharp(\tau(\xi), n(\xi)) < 0 \quad \text{in } \mathcal{L}. \tag{3.24}$$

Since every point in $\mathcal{L}$ lies above the vertical isocline, it follows that

$$n > \hat{N}(\tau,\tau) \quad \text{or, equivalently,} \quad \nu^\sharp(\tau,n) > \tau \quad \text{for } (\tau,n) \in \mathcal{L}. \tag{3.25}$$

Let $-\infty < \omega < 1$ and let $(\tau(\omega), n(\omega)) \in \mathcal{L}$. Since $\mathcal{L}$ is negatively invariant, the trajectory terminating at $(\tau(\omega), n(\omega)) \in \mathcal{L}$ lies entirely in $\mathcal{L}$. Inequalities (3.24) and (3.25) imply that on this trajectory

$$\nu^\sharp(\tau(\xi), n(\xi)) \geq \nu^\sharp(\tau(\omega), n(\omega)) > \tau(\omega) \quad \text{for } \xi \leq \omega. \tag{3.26a,b}$$

Substituting (3.26) into (3.9a), we obtain

$$\tfrac{d}{d\xi}\tau \geq \nu^\sharp(\tau(\omega), n(\omega)) - \tau, \tag{3.27a}$$

or, equivalently,

$$\tfrac{d}{d\xi}(e^\xi \tau) \geq \nu^\sharp(\tau(\omega), n(\omega)) e^\xi. \tag{3.27b}$$

Integrating (3.27b) from χ to ω, we obtain

$$e^\chi \tau(\chi) \leq e^\omega[\tau(\omega) - \nu^\sharp(\tau(\omega), n(\omega))] + \nu^\sharp(\tau(\omega), n(\omega)) e^\chi \quad \text{for} \quad \chi \leq \omega \tag{3.28}$$

on this trajectory. Since the first term on the right-hand side of (3.28) is negative by (3.26b), the entire right-hand side becomes negative as $\chi \to -\infty$. Thus there is a finite negative value of χ, depending on $(\tau(\omega), n(\omega))$, at which this trajectory touches the line $\tau = 0$. Therefore this trajectory cannot correspond to a solution of (3.9), and our assertion about $\mathcal{L}$ is proved.

Now let us study trajectories terminating in $\mathcal{R}$, in which all the inequalities of (3.24) and (3.25) are reversed. We then prove the reverse of (3.28), which says that $e^\chi \tau(\chi)$ has a positive lower bound, so that (3.9c) cannot hold. Thus the last statement of the lemma must hold. □

Exercise 3.4 implies that a trajectory beginning at a singular point and necessarily remaining thereafter in a connected component of $\mathcal{I}$ must be unique. Thus each such trajectory must lie on a curve defined by

$$\tau = f(n), \tag{3.29}$$

which lies between the horizontal and vertical isoclines. Let

$$\begin{aligned} \mathcal{N}^h &\equiv \{n \in (-\infty, 0) : h(n) > v(n)\}, \\ \mathcal{N} &\equiv \{n \in (-\infty, 0] : h(n) = v(n)\}, \\ \mathcal{N}^v &\equiv \{n \in (-\infty, 0) : h(n) < v(n)\}. \end{aligned} \tag{3.30}$$

Note that $0 \in \mathcal{N}$. Since v and h are continuous, $\mathcal{N}$ is closed while $\mathcal{N}^h$ and $\mathcal{N}^v$ are open. $\mathcal{N}^h$ and $\mathcal{N}^v$, either or both of which could be empty, can therefore be decomposed as unions of a countable number of disjoint open intervals. The singular points lying in $\mathcal{Q}_3$ are points of $\{(\tau, n) : n \in \mathcal{N},\ \tau = h(n)\}$. A study of the phase portrait Fig. 3.17 in light of Lemma 3.23 leads to

3.31. Theorem. *Let (1.15a,b), (1.18), (3.3), and (3.5) hold. Then for $\lambda \geq 0$, problem (3.1), (3.2) has a unique solution $r_0(\cdot, \lambda) \in C^0[0,1] \cap C^2(0,1]$ and, equivalently, problem (3.6) has a unique solution $(\tau_0(\cdot,\lambda), N^0(\cdot,\lambda)) \in C^1(0,1]$. If $-\lambda \in \mathcal{N}$, then $r_0(s,\lambda) = sf(-\lambda)$. If $-\lambda \in \mathcal{N}^h$, then $-\lambda$ belongs to a component open interval $(-b,-a)$ of $\mathcal{N}^h$. The function $N^0(\cdot,\lambda)$ strictly decreases from $N^0(0,\lambda) = -a$ to $N^0(1,\lambda) = -\lambda$, and $r_0(s,\lambda) = sf(N^0(s,\lambda))$. If $-\lambda \in \mathcal{N}^v$, then $-\lambda$ belongs to a component open interval $(-d,-c)$ of $\mathcal{N}^v$. The function $N^0(\cdot,\lambda)$ strictly increases from $N^0(0,\lambda) = -d$ to $N^0(1,\lambda) = -\lambda$, and $r_0(s,\lambda) = sf(N^0(s,\lambda))$. (See Fig. 3.17.)*

In particular, if $\mathcal{N}^h$ has a component open interval of the form $(-b,0)$, then $N^0(0,\lambda) = 0$ for all $\lambda \in (0,b)$. Thus the center of the plate is stress-free for a range of boundary pressures. If $\mathcal{N}^h = (-\infty, 0)$, then the center of the plate is stress-free for all pressures λ. If $\mathcal{N}^v$ has a component open interval of the form $(-d,0)$, then $N^0(0,\lambda) = -d$ for all $\lambda \in (0,d]$. Thus the smallest amount of pressure on the boundary causes the stress at the center to jump from 0 to a nonzero value, at which it remains while λ increases to d. If $\mathcal{N}^v = (-\infty,0)$, then the normal stresses T and N at the center of the plate equal $-\infty$ for all pressures λ. If $\mathcal{N}^h$ has a component open interval of the form $(-\infty,-b)$, then for large enough λ, the normal stresses at the center of the plate equal $-\infty$. Thus if $\mathcal{N}^h \neq (-\infty,0)$, then the solutions of (3.6) do not depend continuously on $\lambda \in [0,\infty)$.

For the analysis of the buckling problem in the next section, we must determine the detailed behavior of its trivial solutions $(\tau_0(\cdot,\lambda), N^0(\cdot,\lambda))$ near $s = 0$. This behavior, described by the next theorem and the subsequent discussion, is obtained by perturbing (3.9) about its singular points in $\mathcal{U}$. Using a notation like that of (2.13), we set

$$A_1(s,\lambda) \equiv \frac{T^0_\nu(s,\lambda) - N^0_\tau(s,\lambda)}{2N^0_\nu(s,\lambda)}, \quad B_1(s,\lambda) \equiv \sqrt{A_1(s,\lambda)^2 + \frac{T^0_\tau(s,\lambda)}{N^0_\nu(s,\lambda)}}, \tag{3.32}$$

$$\alpha_1(\lambda) \equiv A_1(0,\lambda), \quad \beta_1(\lambda) \equiv B_1(0,\lambda).$$

Note that $A_1 = 0$ for hyperelastic plates.

3.33. Theorem. *Let the hypotheses of Theorem 3.31 hold. If $-\infty < N^0(0,\lambda) < 0$, then the limits of $\tau_0(s,\lambda)$ and $\nu_0(s,\lambda)$ as $s \to 0$ exist and are equal and positive. Moreover,*

$$\alpha_1(\lambda) + \beta_1(\lambda) \geq 1 \tag{3.34a}$$

or, equivalently,

$$T^0_\tau(0,\lambda) + T^0_\nu(0,\lambda) \geq N^0_\tau(0,\lambda) + N^0_\nu(0,\lambda). \tag{3.34b}$$

There are numbers $A(\lambda)$, $B(\lambda)$, $C(\lambda)$, and $D(\lambda)$ such that

$$\begin{Bmatrix} r_0(s,\lambda) \\ N^0(s,\lambda) \end{Bmatrix} = \begin{Bmatrix} s\tau_0(0,\lambda) \\ N^0(0,\lambda) \end{Bmatrix} + \begin{Bmatrix} A(\lambda) \\ B(\lambda) \end{Bmatrix} s^{\alpha_1(\lambda)+\beta_1(\lambda)} + \begin{Bmatrix} C(\lambda) \\ D(\lambda) \end{Bmatrix} s^{2[\alpha_1(\lambda)+\beta_1(\lambda)]} + o\big(s^{2[\alpha_1(\lambda)+\beta_1(\lambda)]}\big) \quad \text{as } s \to 0. \tag{3.35}$$

The expansions for r_0' and r_0'' are given by the formal derivatives of (3.35). *If* $\alpha_1(\lambda) + \beta_1(\lambda) = 1$, *then all the terms on the right-hand side of* (3.35), *except the first, vanish.*

If the material is hyperelastic, then (3.34) reduces to $T_\tau^0(0, \lambda) \geq N_\nu^0(0, \lambda)$. When the strict inequality in (3.34) holds, we may say that the plate is circularly reinforced at the center. Note that $(\tau_0(0, \lambda), N^0(0, \lambda))$ is a saddle point for (3.9a,b).

3.36. Exercise. Prove Theorem 3.33.

For problems for which $(\tau_0(0, \lambda), N^0(0, \lambda)) \to (0, -\infty)$ as $s \to 0$, we have that

$$\nu_0(s, \lambda) \to 0, \quad T^0(s, \lambda) \to -\infty \quad \text{as } s \to 0, \tag{3.37a,b}$$

for if (3.37a) were not to hold, then ν_0 would have a positive lower bound a, so that (3.6) would imply the contradiction that $\tau_0 \geq a$. The limit (3.37b) then follows from that for τ_0, from (3.37a), and from (1.15a). For these problems the behavior of solutions near $s = 0$ devolves upon the fine structure of the constitutive functions. The behavior is typified by that for the artificial example of the *Taylor plate*, which has no tensile or flexural strength in the azimuthal direction. (It generalizes G. I. Taylor's (1919) model for a parachute.) Its constitutive functions for trivial solutions have the defining property that

$$\hat{N}_\tau = 0, \quad \hat{T} = 0. \tag{3.38}$$

3.39. Exercise. Solve (3.9) in closed form when (3.38) holds. Sketch the phase portrait for (3.9).

The work in this section is based on that of Antman & Negrón-Marrero (1987) and Negrón-Marrero & Antman (1990). (Much of the exposition is adapted from the former reference, ©Martinus Nijhoff Publishers, Dordrecht, and is reprinted by permission of Kluwer Academic Publishers.) Transformations like (3.9) have been used for elasticity problems by Biot (1976), Callegari, Reiss, & Keller (1971), Sivaloganathan (1986), Stuart (1985), and Szeri (1990). See Sec. XII.7.

4. Buckling of Aeolotropic Plates

Here we sketch how the buckling problem for aeolotropic plates differs from that for isotropic plates treated in Sec. 2. We first study the linearization of our boundary-value problem about the trivial solution described in Sec. 3. The linearization of (2.7) and (1.13) is

$$\begin{aligned}
&[sM_\mu^0(s, \lambda)\theta']' - 2M_\mu^0(s, \lambda)A_3(s, \lambda)\theta' \\
&+ M_\sigma^0(s, \lambda)'\theta + M_\mu^0(s, \lambda)\left[A_3(s, \lambda)^2 - B_3(s, \lambda)^2\right]\theta/s \\
&+ sN^0(s, \lambda)\left[N^0(s, \lambda)/H_\eta^0(s, \lambda) - r_0'(s, \lambda)\right]\theta = 0
\end{aligned} \tag{4.1}$$

where

$$A_3(s, \lambda) \equiv \frac{\Sigma_\mu^0(s, \lambda) - M_\sigma^0(s, \lambda)}{2M_\mu^0(s, \lambda)}, \quad B_3(s, \lambda) \equiv \sqrt{A_3(s, \lambda)^2 + \frac{\Sigma_\sigma^0(s, \lambda)}{M_\mu^0(s, \lambda)}}. \tag{4.2}$$

Here $M_\mu^0(s, \lambda) = M_\mu(r_0(s, \lambda)/s, r_0'(s, \lambda), 0, 0, 0)$ is the value of M_μ at the trivial state, etc. Note that it depends on s and, in fact, can inherit any singular behavior of r_0 in s or discontinuous behavior of it in λ. Consequently, the analysis of (4.1) is a far more challenging exercise than that provided by (2.14). Representations (3.35) can be used to support an analysis (using comparison theory) to determine the nature of eigenvalues and eigenfunctions of (4.1) subject to (2.2b) and (2.3).

If $(\tau_0(0,\lambda), N^0(0,\lambda)) \to (0, -\infty)$ as $s \to 0$, so that (3.37) holds, then (3.35) is not valid. In this case, an inkling of the complexity of the solutions of the linearized boundary-value problem is offered by the degenerate problem of the Taylor plate, defined by the constitutive restrictions that $\hat{T} = 0 = \hat{\Sigma}$ and that $\hat{N}$, $\hat{H}$, $\hat{M}$ are each independent of τ and σ. The Taylor plate may thus be regarded as consisting of an infinite array of radially disposed rods.

4.3. Exercise. Let a Taylor plate have constitutive functions of the form

$$\nu^\sharp(n) = (1 - Kn)^{-k} \quad \text{for } n \le 0, \tag{4.4a}$$

$$M^0_\mu(s,\lambda) = \frac{L}{\nu_0(s,\lambda)^l} N^0_\nu(s,\lambda) \quad \text{for } \nu_0(s,\lambda) \le 1, \tag{4.4b}$$

where $K, L, k > 0$; $l \ge 0$. Analyze the resulting boundary-value problem for (4.1) by the Prüfer transformation in the manner of Sec. VI.2. For an unshearable plate, prove that if the positive integer j is sufficiently large, then there is no eigenfunction having more than j zeros, and that if there is an eigenfunction with exactly j zeros, then the corresponding eigenvalues cannot accumulate at $\lambda = 0$. For a shearable plate, let $H^0_\eta(s,\lambda)$ be a positive constant. Prove that if $k(1+l) < 1$, then there is a sequence of eigenvalues $\{\lambda_j\}$ going to ∞ with corresponding eigenfunctions having exactly $j + 1$ zeros.

The conversion of our boundary-value problem to the form (V.3.1), (V.4.5) is like that for the transversely isotropic plate. But here we replace (2.20) with an expression suggested by (4.1) for s near 0. Surprisingly, the presence of aeolotropy actually simplifies the analysis, as we shall see, by affording ways to replace the troublesome factor s^{-1} with $s^{1-\varepsilon}$ for a positive ε. In certain respects, the isotropic case thus represents an extreme of technical difficulty.

In many cases, nontrivial solutions satisfy the initial conditions $\theta(0) = 0 = \theta'(0)$. These conditions would seem to suggest difficulties in reducing a proof that nodal properties are preserved to a proof of the uniqueness of trivial solutions for θ when it has a double zero. Fortunately, a change of variables of the form $\theta(s) = s^\alpha \psi(s)$, where $\alpha \equiv A(0,\lambda) + B(0,\lambda) - 1$ accounts for the amount of anisotropy, leads to problems for the new variable ψ when $\alpha \ge 1$ in which a solution is trivial if and only if $\psi(0) = 0 = \psi'(0)$. The requisite uniqueness theory leads to an inequality like (2.36), but with vs^{-1} replaced with $Cvs^{\varepsilon-1}$ where ε is a positive number measuring the amount of anisotropy. We no longer have a control on C, but this fact causes no difficulties because the presence of ε significantly reduces the strength of the singularity at $s = 0$.

The product of this global bifurcation analysis is that the number of zeros of the associated function ψ are inherited from the eigenfunctions of the linearized problem corresponding to simple eigenvalues and are preserved globally, except possibly across planes of the form $\lambda = \lambda^\sharp$ where $\lambda^\sharp$ is a value of λ at which the solution of the trivial problem jumps. Consequently, we can expect the bifurcation diagram for a given branch to have the form shown in Fig. 4.5. As λ is slowly raised through the value $\lambda^\sharp$, the nontrivial solution along the branch in Fig. 4.5 suddenly jumps. The effect is similar to that of snap-buckling, but the mathematical nature of the process is entirely different. A snap-buckling would correspond to an S-shaped bifurcation diagram, which does not occur in traditional theories for our problem for transversely isotropic plates (but could result for such plates from the nonlinear material response, as Fig. VI.9.33 suggests).

We now ask if it is possible to determine nodal properties of the disconnected upper branch in Fig. 4.5. Since such disconnectedness does not occur for the transversely isotropic plate, we could show that the upper branch has the same nodal properties as the lower if we could embed our problem into a family of problems parametrized by a number β, which measures the amount of aeolotropy, in such a way that the resulting problem with two parameters λ and β has a sheet of solution triples joining

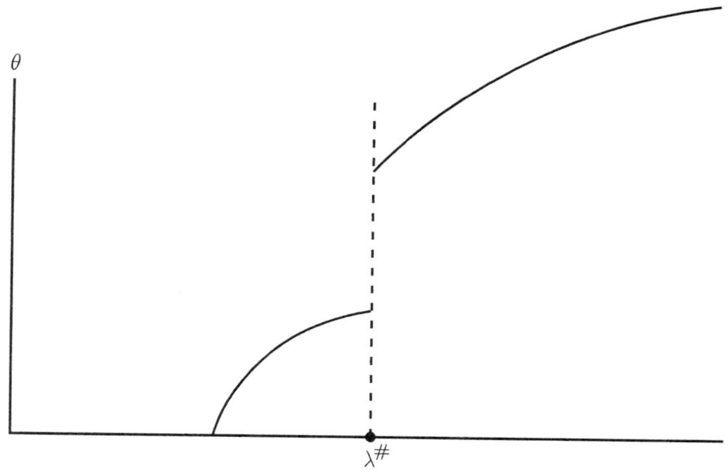

Fig. 4.5. Bifurcation diagram showing a disconnected branch for the buckling of an anisotropic plate.

the disconnected branches of Fig. 4.5 with the connected branches for the isotropic plate. Were it possible to construct such a sheet of solution triples, we could find a connected branch of solutions lying in this sheet, joining the lower branch of Fig. 4.5 (for $\beta = 1$) to the connected branch for the isotropic plate (for $\beta = 0$), and then going from this branch to the upper branch of Fig. 4.5 (for $\beta = 1$).

The obvious family of constitutive functions parametrized by β has the form

$$\mathbf{q} \mapsto \beta\hat{N}(\mathbf{q}) + (1-\beta)\bar{N}(\mathbf{q}), \quad \text{etc.,} \tag{4.6}$$

where $\hat{N}$ is the actual constitutive function for the aeolotropic material and $\bar{N}$ is the constitutive function for an isotropic material. Our strategy fails miserably when functions (4.6) are used: An analysis of the phase portrait Fig. 3.17 for the trivial state shows that the isoclines enclosing the separatrix (3.29), which determines solutions, move closer together as a $\beta \to 0$, but the singular points, which are the intersections of the isoclines, scarcely move at all. Thus the gaps between the singular points, which are responsible for the jumps in the trivial solutions, are scarcely moved for $\beta > 0$. For $\beta = 0$, however, the isoclines coincide and all jumps disappear. Therefore, the solutions generated by (4.6) depend discontinuously on β, having a jump at $\beta = 0$. Fortunately, the same phase portrait suggests how to construct a much subtler embedding that does work: Here, for β near 1, the isoclines are squeezed into coincidence near the singular points. As β is reduced to 0, the isoclines are progressively squeezed into coincidence over the intervals between the original singular points. Thus nodal properties of branches are preserved across gaps. This account of the buckling of anisotropic plates is based on the work of Negrón-Marrero & Antman (1990).

5. Buckling of Spherical Shells

We now specialize the theory of Section 1 to uniform, transversely isotropic, spherical shells of unit radius by taking $s_1 = 0$, $s_2 = \pi$, and

$$\overset{\circ}{r}(s) = \sin s, \quad \overset{\circ}{\theta}(s) = s. \tag{5.1}$$

Note that s measures the arc length along circles of longitude from the *south* pole. Our boundary conditions require that the deformation be regular at the poles:

$$r(0) = 0 = r(\pi), \quad \eta(0) = 0 = \eta(\pi), \quad \theta(0) = 0 = \theta(\pi). \tag{5.2}$$

We assume that the applied force is an externally applied hydrostatic pressure (1.9) and we assume that there is no applied couple. Then equations (1.11) and (1.12) reduce to

$$[\sin s\hat{N}(s)]' - \hat{T}(s)\cos\theta(s) - \sin s\hat{H}(s)\theta'(s) - pr(s)\eta(s) = 0, \tag{5.3}$$

$$[\sin s\hat{H}(s)]' + \hat{T}(s)\sin\theta(s) + \sin s\hat{N}(s)\theta'(s) + pr(s)\nu(s) = 0, \tag{5.4}$$

$$[\sin s\hat{M}(s)]' - \hat{\Sigma}(s)\cos\theta(s) + \sin s[\nu(s)\hat{H}(s) - \eta(s)\hat{N}(s)] = 0. \tag{5.5}$$

As in (2.8a), we can combine (5.3) and (5.4) to obtain

$$\sin s[\hat{N}(s)\sin\theta(s) + \hat{H}(s)\cos\theta(s)] + \tfrac{1}{2}pr(s)^2 = 0. \tag{5.6}$$

We assume that the material of the shell is homogeneous and isotropic in the restricted sense of (2.10). We retain (2.9a), but reject (2.9b,c). See the discussion of these conditions in Sec. XIV.11.

Our boundary-value problem consists of (1.4), (1.13), (5.2)–(5.5). Our constitutive assumptions ensure that it admits the trivial solution (unbuckled state) in which the shell remains spherical, unsheared, and uniformly compressed, so that $\nu = \tau = k$ (const.), $\eta = 0$, $\theta(s) = s$, $\sigma = 1$, $\mu = 1$ if and only if k satisfies

$$N(k, k, 0, 1, 1) \equiv T(k, k, 0, 1, 1) = -pk^2/2. \tag{5.7}$$

We adopt very mild constitutive restrictions sufficient to ensure that this equation has a unique solution, which we denote by $k = k(p)$.

An examination of the boundary-value problem does not disclose any obvious candidate for a variable whose number of zeros might be preserved globally on branches. The equations seem too highly coupled for that. We nevertheless carry out the linearization of the boundary-value problem about the trivial solution to see if it exposes any interesting relations that we can exploit.

5.8. Exercise. Use the isotropy condition (2.10) to show that this linearization has the form

(5.9a,b) $$\frac{d}{ds}(\tau_1 \sin s) = \nu_1 \cos s - (\eta_1 + k\theta_1)\sin s, \quad \sigma_1 \sin s = \theta_1 \cos s,$$

(5.9c) $$N_\nu^0 \xi + N_\mu^0 L\theta_1 = \left[(N_\tau^0 + H_\eta^0 + pk)\eta_1 + \left(kN_\tau^0 + N_\sigma - N_\mu^0 + \tfrac{1}{2}pk^2\right)\theta_1\right]\sin s,$$

(5.9d) $$H_\eta^0 \frac{d}{ds}(\eta_1 \sin s) + \left(N_\sigma + N_\mu^0 - \tfrac{1}{2}pk^2\right)\frac{d}{ds}(\theta_1 \sin s) = -\left(N_\nu^0 + N_\tau^0 + pk\right)(\tau_1 + \nu_1)\sin s,$$

(5.9e) $$M_\nu^0 \xi + M_\mu^0 L\theta_1 = \left[(M_\tau^0 - kH_\eta^0 - \tfrac{1}{2}pk^2)\eta_1 + \left(kM_\tau^0 + M_\sigma - M_\mu^0 - M^0\right)\theta_1\right]\sin s,$$

(5.9f) $$\eta_1(0) = 0 = \eta_1(\pi), \quad \theta_1(0) = 0 = \theta_1(\pi)$$

where

(5.9g) $$\xi \equiv \frac{d}{ds}(\nu_1 \sin s) - \tau_1 \cos s,$$

(5.9h) $$(Lu)(s) \equiv [u'(s)\sin s]' - u(s)/\sin s.$$

Differentiate (5.9d) with respect to s and use (5.9a,c,d) to eliminate the terms involving τ_1 and ν_1 from the differentiated form of (5.9d). Show that the resulting equation is given by the second row of the following matrix equation, in which the first and third rows are just (5.9c) and (5.9e).

(5.10a) $$\begin{pmatrix} N_\nu^0 & 0 & N_\mu^0 \\ 0 & H_\eta^0 & N_\sigma^0 - \frac{1}{2}pk^2 - N_\mu^0\left(\frac{N_\tau^0 + pk}{N_\nu^0}\right) \\ M_\nu^0 & 0 & M_\mu^0 \end{pmatrix} \begin{pmatrix} \xi \\ L\eta_1 \\ L\theta_1 \end{pmatrix}$$
$$= \eta_1 \sin s \begin{pmatrix} N_\tau^0 + H_\eta^0 + pk \\ \frac{(N_\nu^0 - N_\tau^0 - H_\eta^0 - pk)(N_\nu^0 + N_\tau^0 + pk)}{N_\nu^0} \\ M_\tau^0 - kH_\eta^0 - \frac{1}{2}pk^2 \end{pmatrix}$$
$$+ \theta_1 \sin s \begin{pmatrix} kN_\tau^0 + N_\sigma^0 - N_\mu^0 + \frac{1}{2}pk^2 \\ \frac{(kN_\nu^0 - N_\sigma^0 + N_\mu^0 - \frac{1}{2}pk^2)(N_\nu^0 + N_\tau^0 + pk)}{N_\nu^0} \\ kM_\tau^0 + M_\sigma^0 - M_\mu^0 - M^0 \end{pmatrix}.$$

Use Cramer's rule to solve (5.10a) for $L\eta_1$ and $L\theta_1$, obtaining

(5.10b) $$(L\eta_1)(s) + A(p)\eta_1(s)\sin s = -a(p)\theta_1(s)\sin s,$$

(5.10c) $$(L\theta_1)(s) + B(p)\theta_1(s)\sin s = -b(p)\eta_1(s)\sin s$$

where the constants A, B, a, b can be read off from (5.10a).

Thus the linearized problem leads to the coupled pair (5.10b,c) of second-order ordinary differential equations. We can choose complex numbers α, β, C such that $\omega = \alpha\eta_1 + \beta\theta_1$ satisfies the Legendre equation

(5.11) $$(L\omega)(s) + C\omega \sin s = 0, \quad \omega(0) = 0 = \omega(\pi)$$

if and only if (η_1, θ_1) satisfy (5.10b,c). Problem (5.11) has a nontrivial solution if and only if

$$C = n(n+1), \quad n = 1, 2, \ldots, \tag{5.12}$$

in which case,

$$\omega(s) = \omega_n(s) = \text{const.}\, P_n^1(\cos s) \tag{5.13}$$

where P_n^1 is the associated Legendre function of the first kind of degree n and order 1. Equation (5.12) is equivalent to

$$g(p; n) \equiv [A(p) - n(n+1)][B(p) - n(n+1)] - a(p)b(p) = 0. \tag{5.14}$$

It is not hard to show that the eigenfunctions corresponding to the eigenvalues p, which satisfy (5.14), have the form

$$(\eta(s), \theta(s)) = ([n(n+1) - B]/b,\ 1)\, P_n^1(\cos s). \tag{5.15}$$

The function $s \mapsto P_n^1(\cos s)$ has exactly $n+1$ zeros on $[0, \pi]$, including those at 0 and π, each of which is simple. Thus *every nontrivial solution of problem* (5.8) *is characterized by the fact that* η *and* θ *have exactly the same zeros.*

The treatment of the global bifurcation theory leads to a somewhat more complicated version of that for the plate in Sec. 2. We omit the details. We now turn to the most interesting aspect of the analysis, the treatment of nodal properties. Our results for the linearized problem together with perturbation results, which we do not describe, motivate us to study whether the number of *simultaneous* zeros of η and $\theta - s$ is fixed on each branch.

The justification of such a conclusion, like that of Sec. VII.4, is much trickier than that for the nodal theory for plates, which relies on two mathematical facts: (i) If g is a given continuously differentiable real-valued function having exactly $n+1$ zeros on a closed interval, each of which is simple, then any continuously differentiable function f sufficiently close to g (in the C^1-norm) has exactly the same number of zeros. (Thus the set $\mathcal{S}_n$ of such functions g is open.) (ii) Roughly speaking, if such a function g is perturbed enough to change the number of its zeros, then along the way it must evolve through a function having a double zero. (As we note in Sec. V.5, this remark is not strictly correct because branches of solution pairs are not necessarily curves, but are merely connected sets. What is correct and useful is that the boundary of $\mathcal{S}_n$ consists of functions with double zeros.) In other words, the specification of g as a function of s gives the graph of a plane curve, small perturbations of which leave its nodal properties unchanged. The change of these nodal properties is signalled by the appearance of a double zero.

Now suppose that we consider two continuously differentiable functions η and $\theta - s$ of s. Their graph is a curve in the three-dimensional space with coordinates $s, \eta, \theta - s$. The functions η and $\theta - s$ have simultaneous zeros when this curve intersects the s-axis. But typically the slightest perturbation of this curve would destroy such intersections. Thus the nodal properties of pairs of continuously differentiable functions differ markedly from those for a single continuously differentiable function. To get useful information about nodal properties of η and $\theta - s$ we must exploit the fact that they are not merely arbitrary continuously differentiable functions, but are also solutions of our boundary-value problem.

For this purpose, we introduce polar coordinates ξ and χ by the Prüfer transformation

$$\eta(s) = \xi(s)\cos\chi(s), \quad \theta(s) - s = \xi(s)\sin\chi(s). \tag{5.16}$$

A zero of ξ is a simultaneous zero of η and $\theta - s$. We now confront a second polar singularity caused by the transformations (5.16). A rather delicate analysis is needed to show that ξ satisfies a well-defined second-order equation. One unpleasant, but not surprising, feature of this equation is that the mid-surface stretches are inextricably coupled with the shearing and bending terms. A careful analysis of the relationship between the equation for ξ and those of our boundary-value problem shows that the number of zeros of ξ is preserved globally on branches (except if they cross certain small, precisely defined regions of solution-parameter space). It can also be shown that the nodal property of preserving the number of simultaneous zeros of η and $\theta - s$ is not vacuous: There are nontrivial branches with this property. The complexity of this result vis-à-vis those for the plate may be regarded as a nonlinear analog of the fact that the linear equations for the equilibrium of an elastic plate uncouple into equations of lower order than those for shells. The treatment of nodal properties for solutions of problems of this section is a complicated analog of that of Sec. VI.4.

These nodal properties illuminate the interesting numerical work of Bauer, Reiss, & Keller (1970) on the axisymmetric buckling of spherical shells. Using a theory of von Kármán type, they found that all the bifurcating branches they computed are connected. This demonstration required the very expensive computation of large buckled states. Our nodal properties show that such connectedness cannot occur for our geometrically exact theory. The discrepancy occurs because theories of von Kármán type are based on a number of hypotheses that are not valid for large deformations. A detailed treatment of the results of this section is given by Shih & Antman (1986). The sketch just given generalizes their work by suspending their special symmetry requirements, which the treatment of Sec. XIV.11 suggests is not enforced by the underlying three-dimensional physical interpretation.

6. Buckling of Cylindrical Shells

We now formulate the problem of the axisymmetric buckling of a circular cylindrical tube under a compressive thrust applied to its edges. We scale the problem so that the shell has length $2l$ and radius 1. The reference configuration is characterized by

$$\mathring{\theta} = \tfrac{\pi}{2}, \quad \mathring{\boldsymbol{r}}(s,\phi) = \boldsymbol{e}_1(\phi) + s\boldsymbol{k}, \quad \mathring{r} = 1, \quad \mathring{\boldsymbol{a}} = \boldsymbol{k}, \quad \mathring{\boldsymbol{b}} = -\boldsymbol{e}_1 \tag{6.1}$$

for $-l \le s \le l$. Thus (1.4) is simplified.

We assume that there is no body force applied to the shell. We assume that the edges are kept parallel and subject to a vertical thrust of intensity λ, so that

$$\theta(\pm l) = \tfrac{\pi}{2}, \quad \boldsymbol{n}_1(\pm l, \phi) = -\lambda \boldsymbol{k}, \quad H(\pm l) = 0, \quad N(\pm l) = -\lambda. \tag{6.2a,b,c,d}$$

We do not bother to impose a condition on z that would fix a rigid translation. We set

$$\psi \equiv \frac{\pi}{2} - \theta. \tag{6.3}$$

From (1.10)–(1.12) we then can write the equilibrium equations in the form

$$\begin{aligned} N(s) &= -\lambda \cos\psi(s) + \left[\int_{-l}^{s} T(t)\,dt\right] \sin\psi(s), \\ H(s) &= -\lambda \sin\psi(s) - \left[\int_{-l}^{s} T(t)\,dt\right] \cos\psi(s), \end{aligned} \tag{6.4}$$

$$M' - \Sigma \sin\psi + \nu H - \eta N = 0. \tag{6.5}$$

Note that the boundary conditions imply that

$$\int_{-l}^{l} T(t)\,dt = 0. \tag{6.6}$$

From (6.4) we obtain the integral

$$N(s)\cos\psi + H(s)\sin\psi = -\lambda. \tag{6.7}$$

We assume that the shell is nonlinearly elastic and is uniform, so that s does not appear explicitly as an argument of the constitutive functions. Using (1.17) and (6.7) we can write the governing system of ordinary differential equations from (6.5), (1.11), (1.4) as the following semilinear system for M, H, ψ, r:

$$M' = \Sigma^\sharp \sin\psi - \nu^\sharp H - (\lambda \sec\psi + H \tan\psi)\eta^\sharp, \tag{6.8a}$$

$$H' = -(\lambda \sec\psi + H \tan\psi)\mu^\sharp - T^\sharp \cos\psi, \tag{6.8b}$$

$$\psi' = \mu^\sharp, \tag{6.8c}$$

$$r' = \nu^\sharp \sin\psi - \eta^\sharp \cos\psi \tag{6.8d}$$

where the arguments of the functions with sharp signs are

$$r, \quad -(\lambda \sec\psi + H\tan\psi), \quad H, \quad \cos\psi, \quad M. \tag{6.8e}$$

Our boundary-value problem is (6.8) subject to (6.2a,c).

It is simple to put this boundary-value problem into a form to which we can apply the Global Bifurcation Theorem V.4.19. This problem does not admit any obvious nodal structure.

Let us assume that $\hat{H}, \hat{\Sigma}, \hat{M}$ vanish when $(\eta, \sigma, \mu) = (0, 1, 0)$. Then the boundary-value problem admits a trivial solution with $\psi = 0 = \eta$, with the radius (hoop stretch) r and longitudinal stretch $\nu = z'$ determined by

$$\hat{N}(r, \nu, 0, 1, 0) = -\lambda, \quad \hat{T}(r, \nu, 0, 1, 0) = 0. \tag{6.9}$$

Conditions (1.15) can be shown to ensure the existence of at least one solution (r, ν) of (6.9) (see Chap. XIX), but (1.14) is insufficient to ensure the uniqueness of the solution. In contrast to the trivial solution for the compressed plate studied in Sec. 2, the hoop stress T vanishes here, although (6.9) indicates that the radius varies with the thrust. To see that T must vanish here, imagine that the shell is sliced in half by a plane through $\boldsymbol{k}$. If T were not zero, then each half would be subject to a net force and would consequently accelerate.

6.10. Problem. Carry out a perturbation analysis of (6.8), (6.2a,c), determining whether stability is lost by a transcritical bifurcation. (The problem does not have enough symmetry to preclude transcritical bifurcations a priori.)

6.11. Research Problem. Characterize global bifurcation by means of nodal structure.

7. Asymptotic Shape of Inflated Shells

Consider a closed axisymmetric shell (a spheroidal shell) with an arbitrary axisymmetric reference shape. It is subjected to an internal hydrostatic pressure. We study the asymptotic shape of the shell when the enclosed volume becomes large. This is the analog of the problem treated in Sec. IV.4. (Since there are spherical shells that can enclose an arbitrarily large volume at finite pressure, we take the volume rather than the pressure to be the prescribed large parameter.) Isaacson (1965) initiated the study of such asymptotic states for membranes. See Wu (1979) for a systematic survey of work on this problem for membranes.

Let ε be a small positive parameter. The large deformation of the shell is characterized by the requirement that the volume enclosed by it be that of a ball of radius $1/\varepsilon$:

$$-\frac{2\pi}{3}\int_0^\pi \rho(\boldsymbol{r}\times\boldsymbol{r}')\cdot\boldsymbol{e}_2\,ds = \frac{4\pi}{3\varepsilon^3}. \tag{7.1}$$

We accordingly scale the strains by

$$\nu = \varepsilon^{-1}\nu^\sharp, \quad \tau = \varepsilon^{-1}\tau^\sharp, \quad \boldsymbol{r} = \varepsilon^{-1}\boldsymbol{r}^\sharp. \tag{7.2}$$

The substitution of (7.2) into (1.13) introduces the small parameter ε into these constitutive equations. We rescale these equations in a way that reflects the underlying mechanical response for large ν and τ. For example, we take

$$N(\mathsf{q}, s) = \alpha(\varepsilon)\sum_{k=0}^{K}\alpha_k(\varepsilon)N^k(\mathsf{q}^\sharp, s) + o(\alpha_K(\varepsilon)) \tag{7.3}$$

where α is a positive decreasing function on $(0,1]$ with $\alpha(\varepsilon)\to\infty$ as $\varepsilon\to 0$, where $\alpha_0 = 1$, where $\{\alpha_k\}$ is an asymptotic sequence, and where $\mathsf{q}^\sharp$ equals q with ν and τ replaced with $\nu^\sharp$ and $\tau^\sharp$. (It is interesting to note that even if the original theory is hyperelastic, i.e., if the constitutive functions are derivatives of a scalar strain-energy function, the leading terms of (7.3), etc., do not correspond to a hyperelastic material.)

Under these assumptions one can prove that the solution can be represented as an asymptotic series in ε. The leading term corresponds to a solution of a hydrostatically loaded membrane. It can be shown that if the constitutive functions are isotropic (but not necessarily homogeneous) and if area changes dominate length changes in affecting the response of the tensions T and N for large τ and ν, then the leading term describes a spherical shell (of very large radius). In particular, if for very large τ and ν the isotropic strain energy function behaves like $A(\tau^2+\nu^2)^a + B(\tau\nu)^b$, where A, B, a, b are positive numbers, and if $b > a$, then the second hypothesis is met. These results are based on the work of Antman & Calderer (1985b), part of whose exposition is reproduced here with the kind permission of Cambridge University Press.

8. Membranes

We obtain the equations for membranes from those for shells by assuming that $l = 0$ and by making the constitutive assumptions that the membrane cannot sustain couples, so that the contact force is tangent to the surface $\boldsymbol{r}$:

$$\hat{M} = 0 = \hat{\Sigma}, \quad \boldsymbol{r}_s \times [\hat{N}\boldsymbol{a} + \hat{H}\boldsymbol{b}] = \boldsymbol{0}. \tag{8.1}$$

Thus $\boldsymbol{n}_1 = N\boldsymbol{a} + H\boldsymbol{b}$ has the form $K\boldsymbol{r}_s/|\boldsymbol{r}_s|$. If we assume that the longitudinal and azimuthal tensions K and T have constitutive equations independent of η, σ, μ, then the governing equations for an elastic membrane are exactly the same as (1.11) in which the shear strain $\eta = 0$, provided we identify K with N:

$$\tfrac{d}{ds}(\mathring{r}N\cos\theta) - T + \mathring{r}\boldsymbol{f}\cdot\boldsymbol{e}_1 = 0, \tag{8.2}$$

$$\tfrac{d}{ds}(\mathring{r}N\sin\theta) + \mathring{r}\boldsymbol{f}\cdot\boldsymbol{k} = 0, \tag{8.3}$$

with

$$\boldsymbol{r}_s = r'\boldsymbol{e}_1 + z'\boldsymbol{k} = \nu(\cos\theta\boldsymbol{e}_1 + \sin\theta\boldsymbol{k}), \quad \tau = r/\overset{\circ}{r}, \tag{8.4}$$

$$T(s) = \hat{T}(\tau(s), \nu(s), s), \quad N(s) = \hat{N}(\tau(s), \nu(s), s). \tag{8.5}$$

In place of (1.14) and (1.15), we merely require that

$$\hat{N}_\nu > 0, \quad \hat{N}(\tau, \nu, s) \to \begin{Bmatrix} +\infty \\ -\infty \end{Bmatrix} \quad \text{as } \nu \to \begin{Bmatrix} \infty \\ 0 \end{Bmatrix} \text{ for fixed } \tau, s. \tag{8.6}$$

We want to obtain a convenient set of differential equations. From (8.2) and (8.3) or from (1.11), we immediately obtain

$$\tfrac{d}{ds}(\overset{\circ}{r}N) - T\cos\theta + \overset{\circ}{r}\boldsymbol{f}\cdot\boldsymbol{a} = 0. \tag{8.7}$$

Let $F'(s) = -\overset{\circ}{r}(s)\boldsymbol{f}(s)\cdot\boldsymbol{k}$. Then (8.3) admits the integral

$$\overset{\circ}{r}N\sin\theta = F + \text{const.} \tag{8.9}$$

For example, if the body force is due solely to hydrostatic pressure (1.9), then $F(s) = \frac{1}{2}pr(s)^2$. The constant is determined by boundary conditions.

Now as in the paragraph containing (3.6), we let $\nu^\sharp(\tau, n, s)$ denote the unique solution of the equation $\hat{N}(\tau, \nu, s) = n$; its existence is ensured by (8.6). We set $T^\sharp(\tau, n, s) \equiv \hat{T}(\tau, \nu^\sharp(\tau, n, s), s)$. Then (8.2)–(8.5) is equivalent to the semilinear system

$$\tfrac{d}{ds}(\overset{\circ}{r}n) = T^\sharp(\tau, n, s)\cos\theta - \overset{\circ}{r}\boldsymbol{f}\cdot\boldsymbol{a}, \tag{8.10}$$

$$\tfrac{d}{ds}(\overset{\circ}{r}\tau) = \nu^\sharp(\tau, n, s)\cos\theta \tag{8.11}$$

for (τ, n), where $\cos\theta$ can be expressed in terms of n and τ by (8.9) with n replacing N (with different representations of $\cos\theta$ having to be patched together where θ crosses odd multiples of $\frac{\pi}{2}$. Thus the governing equations can be reduced to a semilinear second-order system. Variants of these equations were developed by Dickey (1983).

The exact equations for membranes are treated in Green & Adkins (1970) and in the references given in the last section. Many membrane problems have interesting singularities that cause mathematical difficulties. These have almost exclusively been treated for approximate theories. Thus many fascinating problems, such as that of Callegari, Reiss, & Keller (1971), have not been given a full treatment within the exact theory. (Preliminary results for this problem are described by Antman (1981).) For a survey of nonlinear problems for approximate theories, see Weinitschke (1987) and Weinitschke & Grabmüller (1992).

CHAPTER XI

Tensors

1. Tensor Algebra

We now give a deeper treatment of the material introduced in Secs. I.4 and IV.1. Much of our exposition consists of assertions of standard results, the proofs of which are given in the references cited at the end of Sec. 2.

Vectors. A *vector* is an element of $\mathbb{E}^3$, which is abstract real three-dimensional inner-product space. Vectors are denoted by lower-case boldface symbols. The inner product of vectors $\boldsymbol{a}$ and $\boldsymbol{b}$ is denoted $\boldsymbol{a} \cdot \boldsymbol{b}$. If $\boldsymbol{a} \cdot \boldsymbol{b} = \boldsymbol{0}$ for all $\boldsymbol{b}$, then $\boldsymbol{a} = \boldsymbol{0}$. The cross product of $\boldsymbol{a}$ and $\boldsymbol{b}$ is denoted $\boldsymbol{a} \times \boldsymbol{b}$.

Let $\{\boldsymbol{a}_1, \boldsymbol{a}_2, \boldsymbol{a}_3\}$ be a basis for $\mathbb{E}^3$. Then any $\boldsymbol{u} \in \mathbb{E}^3$ can be represented as a linear combination of these base vectors:

$$\boldsymbol{u} = \sum_{k=1}^{3} u^k \boldsymbol{a}_k. \tag{1.1a}$$

We adopt the conventions that Latin indices have range $1, 2, 3$, that Greek indices have range $1, 2$, and that twice-repeated indices are summed over their range. Thus (1.1a) can be abbreviated as

$$\boldsymbol{u} = u^k \boldsymbol{a}_k. \tag{1.1b}$$

To find the components u^k we introduce the dual basis: By the Riesz Representation Theorem (see Stakgold (1979, p. 293)), or by a direct computation, we can show that there exists a unique basis $\{\boldsymbol{a}^k\}$, called the *basis dual to* $\{\boldsymbol{a}_k\}$, such that

$$\boldsymbol{a}^k \cdot \boldsymbol{a}_l = \delta_l^k \equiv \begin{cases} 1 & \text{if } k = l, \\ 0 & \text{if } k \neq l. \end{cases} \tag{1.2a}$$

Indeed, we readily find that

$$\boldsymbol{a}^3 = \frac{\boldsymbol{a}_1 \times \boldsymbol{a}_2}{(\boldsymbol{a}_1 \times \boldsymbol{a}_2) \cdot \boldsymbol{a}_3}, \quad \text{etc.} \tag{1.2b}$$

Then

$$u^k = \boldsymbol{u} \cdot \boldsymbol{a}^k \quad \text{so that} \quad \boldsymbol{u} = (\boldsymbol{u} \cdot \boldsymbol{a}^k)\boldsymbol{a}_k. \tag{1.3a}$$

Similarly,

$$(1.3b)\qquad \boldsymbol{u} = u_k \boldsymbol{a}^k \quad \text{where} \quad u_k = \boldsymbol{u}\cdot\boldsymbol{a}_k \quad \text{so that} \quad \boldsymbol{u} = (\boldsymbol{u}\cdot\boldsymbol{a}_k)\boldsymbol{a}^k.$$

If $\boldsymbol{u} = u^k\boldsymbol{a}_k$ and $\boldsymbol{v} = v_k\boldsymbol{a}^k$, then

$$(1.4)\qquad \boldsymbol{u}\cdot\boldsymbol{v} = u^k v_k.$$

A basis $\{\boldsymbol{a}_k\}$ is called *orthonormal* iff $\boldsymbol{a}_k\cdot\boldsymbol{a}_l = \delta_{kl} \equiv \delta_l^k$. For such a basis, $\boldsymbol{a}^k = \boldsymbol{a}_k$.

Second-order tensors. Elements of the nine-dimensional space Lin $\equiv \mathcal{L}(\mathbb{E}^3, \mathbb{E}^3)$ of linear transformations of $\mathbb{E}^3$ to itself are called (*second-order*) *tensors*. We may call vectors *first-order tensors* and scalars (i.e., real numbers) *zeroth-order tensors*. We denote second-order tensors by boldface upper-case symbols. The value of a tensor $\boldsymbol{A}$ at vector $\boldsymbol{u}$ is the vector denoted by $\boldsymbol{A}\cdot\boldsymbol{u}$ (rather than by the more common $\boldsymbol{Au}$.)

The *product* $\boldsymbol{A}\cdot\boldsymbol{B}$ of tensors $\boldsymbol{A}$ and $\boldsymbol{B}$ is the tensor defined by

$$(1.5)\qquad (\boldsymbol{A}\cdot\boldsymbol{B})\cdot\boldsymbol{u} = \boldsymbol{A}\cdot(\boldsymbol{B}\cdot\boldsymbol{u}) \quad \forall\, \boldsymbol{u}\in\mathbb{E}^3.$$

(The usual notation for this product is $\boldsymbol{AB}$.) We define $\boldsymbol{A}^2 \equiv \boldsymbol{A}\cdot\boldsymbol{A}$, $\boldsymbol{A}^3 \equiv \boldsymbol{A}\cdot\boldsymbol{A}\cdot\boldsymbol{A}$, etc.

The *transpose* (or *adjoint*) $\boldsymbol{A}^*$ of $\boldsymbol{A}$ is the tensor defined by

$$(1.6a)\qquad \boldsymbol{v}\cdot(\boldsymbol{A}\cdot\boldsymbol{u}) = (\boldsymbol{A}^*\cdot\boldsymbol{v})\cdot\boldsymbol{u} \quad \forall\, \boldsymbol{u},\boldsymbol{v}\in\mathbb{E}^3.$$

The existence of the transpose follows from the Riesz Representation Theorem. In light of (1.6a), we set

$$(1.6b)\qquad \boldsymbol{v}\cdot\boldsymbol{A} = \boldsymbol{A}^*\cdot\boldsymbol{v}.$$

Thus we can drop the parentheses on the left-hand side of (1.6b). It is a simple exercise to show that

$$(1.7a,b)\qquad (\alpha\boldsymbol{A}+\beta\boldsymbol{B})^* = \alpha\boldsymbol{A}^*+\beta\boldsymbol{B}^* \quad \forall\,\alpha,\beta\in\mathbb{R}, \qquad (\boldsymbol{A}\cdot\boldsymbol{B})^* = \boldsymbol{B}^*\cdot\boldsymbol{A}^*.$$

$\boldsymbol{A}$ is called *symmetric* iff $\boldsymbol{A} = \boldsymbol{A}^*$, and is called *antisymmetric* or *skew* iff $\boldsymbol{A} = -\boldsymbol{A}^*$. Every tensor $\boldsymbol{A}$ has a unique decomposition as a sum of a symmetric tensor and a skew tensor:

$$(1.8)\qquad \boldsymbol{A} = \tfrac{1}{2}(\boldsymbol{A}+\boldsymbol{A}^*) + \tfrac{1}{2}(\boldsymbol{A}-\boldsymbol{A}^*).$$

Every skew tensor $\boldsymbol{W}$ has associated with it a unique vector $\boldsymbol{w}$, called the *axial vector*, corresponding to $\boldsymbol{W}$ such that

$$(1.9a)\qquad \boldsymbol{W}\cdot\boldsymbol{u} = \boldsymbol{w}\times\boldsymbol{u} \quad \forall\,\boldsymbol{u}$$

or, alternatively,

$$\boldsymbol{v}\cdot\boldsymbol{W}\cdot\boldsymbol{u}=\boldsymbol{w}\cdot(\boldsymbol{u}\times\boldsymbol{v})\quad\forall\boldsymbol{u},\boldsymbol{v}. \tag{1.9b}$$

The six-dimensional subspace of Lin consisting of the symmetric tensors is denoted Sym and the three-dimensional subspace of Lin consisting of the skew tensors is denoted Skw.

The identity tensor $\boldsymbol{I}$ and the zero tensor $\boldsymbol{O}$ are defined by

$$\boldsymbol{I}\cdot\boldsymbol{u}=\boldsymbol{u},\quad \boldsymbol{O}\cdot\boldsymbol{u}=\boldsymbol{0}\quad\forall\boldsymbol{u}. \tag{1.10}$$

The function $\boldsymbol{u}\mapsto\boldsymbol{u}\cdot\boldsymbol{A}\cdot\boldsymbol{u}$ is called the *quadratic form of* $\boldsymbol{A}$. Only the symmetric part of $\boldsymbol{A}$ enters it. A tensor $\boldsymbol{A}$ (not necessarily symmetric) is called *positive-definite* iff $\boldsymbol{u}\cdot\boldsymbol{A}\cdot\boldsymbol{u}>0$ for all $\boldsymbol{u}\neq\boldsymbol{0}$. It is called *positive-semidefinite* iff $\boldsymbol{u}\cdot\boldsymbol{A}\cdot\boldsymbol{u}\geq 0$. It is called *negative-definite* or *negative-semidefinite* iff $-\boldsymbol{A}$ is positive-definite or positive-semidefinite, respectively. $\boldsymbol{A}$ is called *indefinite* iff it is neither positive- nor negative-semidefinite.

1.11. Alternative Theorem. *The equation* $\boldsymbol{A}\cdot\boldsymbol{u}=\boldsymbol{b}$ *has a solution* $\boldsymbol{u}$ *if and only if* $\boldsymbol{b}$ *is orthogonal to the null space of* $\boldsymbol{A}^*$, *i.e., if and only if* $\boldsymbol{b}\cdot\boldsymbol{z}=0$ *for all* $\boldsymbol{z}$ *satisfying* $\boldsymbol{A}^*\cdot\boldsymbol{z}\equiv\boldsymbol{z}\cdot\boldsymbol{A}=\boldsymbol{0}$.

$\boldsymbol{A}$ is *invertible* (or *nonsingular*) iff it is a one-to-one mapping of $\mathbb{E}^3$ onto itself, i.e., iff the equation $\boldsymbol{A}\cdot\boldsymbol{u}=\boldsymbol{b}$ has a solution $\boldsymbol{u}$ for each $\boldsymbol{b}$.

1.12. Proposition. $\boldsymbol{A}$ *is invertible if and only if the only solution of* $\boldsymbol{A}\cdot\boldsymbol{u}=\boldsymbol{0}$ *is* $\boldsymbol{u}=\boldsymbol{0}$.

If $\boldsymbol{A}$ is invertible, then there exists a unique $\boldsymbol{A}^{-1}$ in Lin, called the *inverse of* $\boldsymbol{A}$, such that if $\boldsymbol{A}\cdot\boldsymbol{u}=\boldsymbol{v}$, then $\boldsymbol{u}=\boldsymbol{A}^{-1}\cdot\boldsymbol{v}$. It then follows that

$$\boldsymbol{A}\cdot\boldsymbol{A}^{-1}=\boldsymbol{A}^{-1}\cdot\boldsymbol{A}=\boldsymbol{I}. \tag{1.13}$$

If $\boldsymbol{A}$ and $\boldsymbol{B}$ are invertible, then

$$(\boldsymbol{A}\cdot\boldsymbol{B})^{-1}=\boldsymbol{B}^{-1}\cdot\boldsymbol{A}^{-1},\quad (\alpha\boldsymbol{A})^{-1}=\frac{1}{\alpha}\boldsymbol{A}^{-1}\quad\forall\alpha\in\mathbb{R}\setminus\{0\}. \tag{1.14}$$

We define $\boldsymbol{A}^{-2}\equiv\boldsymbol{A}^{-1}\cdot\boldsymbol{A}^{-1}$, etc. The operations of inverse and transpose commute so that $(\boldsymbol{A}^{-1})^*=(\boldsymbol{A}^*)^{-1}$, which may be denoted by $\boldsymbol{A}^{-*}$. The invertible tensors form the *general linear group* GL(3). The subgroup of GL(3) consisting of tensors with a positive determinant is denoted Lin^+. (The determinant is defined below.)

A complex number λ is an *eigenvalue* of $\boldsymbol{A}$ iff there is a nonzero vector $\boldsymbol{v}$ such that $\boldsymbol{A}\cdot\boldsymbol{v}=\lambda\boldsymbol{v}$. Such a vector $\boldsymbol{v}$ is an *eigenvector* corresponding to λ. If $\boldsymbol{A}\in\mathrm{Sym}$, then its eigenvectors corresponding to distinct eigenvalues are orthogonal. Moreover, $\mathbb{E}^3$ has an orthonormal basis consisting of eigenvectors of $\boldsymbol{A}$.

A tensor $\boldsymbol{Q}$ is said to be *orthogonal* iff $|\boldsymbol{Q}\cdot\boldsymbol{u}|=|\boldsymbol{u}|\quad\forall\boldsymbol{u}$. $\boldsymbol{Q}$ is orthogonal if and only if $\boldsymbol{Q}^*=\boldsymbol{Q}^{-1}$. The orthogonal tensors form the *orthogonal group*

O(3). If $\boldsymbol{Q}$ is orthogonal, then $\det \boldsymbol{Q} = \pm 1$. $\boldsymbol{Q}$ is said to be a *rotation* or *proper-orthogonal* iff it is orthogonal and $\det \boldsymbol{Q} = 1$. Proper orthogonal tensors describe rigid rotations. They form the *special orthogonal group* SO(3).

The *dyadic product* $\boldsymbol{uv}$ of vectors $\boldsymbol{u}$ and $\boldsymbol{v}$ (often denoted elsewhere by $\boldsymbol{u} \otimes \boldsymbol{v}$) is the tensor defined by

(1.15a) $$(\boldsymbol{uv}) \cdot \boldsymbol{w} = (\boldsymbol{v} \cdot \boldsymbol{w})\boldsymbol{u} \quad \forall \boldsymbol{w}.$$

This formula shows that it is not necessary to use the parentheses. Note that (1.5) and (1.15a) imply that

(1.15b) $$(\boldsymbol{ab}) \cdot (\boldsymbol{cd}) = (\boldsymbol{b} \cdot \boldsymbol{c})\boldsymbol{ad}.$$

A dyadic product is called a *dyad.*

1.16. Proposition. *If* $\{\boldsymbol{a}_k\}$ *and* $\{\boldsymbol{b}_l\}$ *are bases for* $\mathbb{E}^3$, *then* $\{\boldsymbol{a}_k\boldsymbol{b}_l\}$ *is a basis for* Lin.

This result implies that many operations on tensors can be defined by giving their effects on dyads. In particular, $(\boldsymbol{ab})^* = \boldsymbol{ba}$. It follows from (1.3) that if $\{\boldsymbol{a}_k\}$ is a basis, then

(1.17) $$\boldsymbol{I} = \boldsymbol{a}^k\boldsymbol{a}_k = \boldsymbol{a}_k\boldsymbol{a}^k.$$

1.18. Exercise. Given a dyadic basis for Lin, construct one for Sym.

On Lin, we define the function that assigns the scalar $\operatorname{tr} \boldsymbol{A}$, called the *trace* of $\boldsymbol{A}$, to any tensor $\boldsymbol{A}$ by defining it for dyads:

(1.19) $$\operatorname{tr}(\boldsymbol{ab}) \equiv \boldsymbol{a} \cdot \boldsymbol{b}.$$

On Lin, we similarly define an *inner product* ':', which assigns the real number $\boldsymbol{A} : \boldsymbol{B}$ to any pair of tensors $\boldsymbol{A}$ and $\boldsymbol{B}$, by defining it for dyads:

(1.20a) $$(\boldsymbol{ab}) : (\boldsymbol{cd}) \equiv (\boldsymbol{a} \cdot \boldsymbol{c})(\boldsymbol{b} \cdot \boldsymbol{d}).$$

(Some authors define the operator ':' to be something other than an inner product.) Compare (1.20a) with (1.15b). It follows that

(1.20b) $$\boldsymbol{A} : (\boldsymbol{ab}) = \boldsymbol{a} \cdot \boldsymbol{A} \cdot \boldsymbol{b} = (\boldsymbol{ab}) : \boldsymbol{A}, \quad \text{so that} \quad \boldsymbol{A} : \boldsymbol{B} = \boldsymbol{B} : \boldsymbol{A}.$$

We define

(1.20c) $$|\boldsymbol{A}| \equiv \sqrt{\boldsymbol{A} : \boldsymbol{A}}.$$

The trace satisfies

(1.21a) $$\operatorname{tr}(\boldsymbol{A} \cdot \boldsymbol{B}^*) = \boldsymbol{A} : \boldsymbol{B}, \quad \operatorname{tr} \boldsymbol{A} = \boldsymbol{A} : \boldsymbol{I} = \boldsymbol{I} : \boldsymbol{A}.$$

Note that (1.20c) and (1.21a) imply that

(1.21b) $$\operatorname{tr}(\boldsymbol{A}^* \cdot \boldsymbol{A}) = \boldsymbol{A} : \boldsymbol{A} = |\boldsymbol{A}|^2.$$

1.22. Exercise. Prove that

(1.23a) If $\boldsymbol{A} : \boldsymbol{B} = 0$ for all $\boldsymbol{B} \in$ Lin, then $\boldsymbol{A} = \boldsymbol{O}$.

(1.23b) $\boldsymbol{A} : \boldsymbol{B} = \boldsymbol{A}^* : \boldsymbol{B}^*$.

(1.23c) If $\boldsymbol{A} \in$ Skw and $\boldsymbol{B} \in$ Sym, then $\boldsymbol{A} : \boldsymbol{B} = 0$.

(1.23d) If $\boldsymbol{A} : \boldsymbol{B} = 0 \quad \forall \boldsymbol{B} \in$ Sym, then $\boldsymbol{A} \in$ Skw.

(1.23e) If $\boldsymbol{A} : \boldsymbol{B} = 0 \quad \forall \boldsymbol{B} \in$ Skw, then $\boldsymbol{A} \in$ Sym.

Tensors of higher order. A typical element of the space $\mathcal{L}(\mathbb{E}^3, \text{Lin})$ of linear transformations that assign a second-order tensor to a vector is denoted $\boldsymbol{u} \mapsto \mathsf{M} \cdot \boldsymbol{u}$. We define the triadic product $\boldsymbol{abc}$ to be an element of this space by setting

$$(\boldsymbol{abc}) \cdot \boldsymbol{u} = (\boldsymbol{c} \cdot \boldsymbol{u})\boldsymbol{ab} \quad \forall \boldsymbol{u}. \tag{1.24a}$$

It can be shown that if $\{\boldsymbol{a}_k\}$, $\{\boldsymbol{b}_l\}$, and $\{\boldsymbol{c}_m\}$ are bases for $\mathbb{E}^3$, then $\{\boldsymbol{a}_k\boldsymbol{b}_l\boldsymbol{c}_m\}$ is a basis for $\mathcal{L}(\mathbb{E}^3, \text{Lin})$.

A typical element of the space $\mathcal{L}(\text{Lin}, \mathbb{E}^3)$ is denoted $\boldsymbol{A} \mapsto \mathsf{M} : \boldsymbol{A}$. We define the triadic product $\boldsymbol{abc}$ to be an element of this space by setting

$$(\boldsymbol{abc}) : \boldsymbol{uv} = (\boldsymbol{b} \cdot \boldsymbol{u})(\boldsymbol{c} \cdot \boldsymbol{v})\boldsymbol{a} \quad \forall \boldsymbol{u}, \boldsymbol{v}. \tag{1.24b}$$

This space likewise has a triadic basis.

We identify $\mathcal{L}(\mathbb{E}^3, \text{Lin})$ with $\mathcal{L}(\text{Lin}, \mathbb{E}^3)$ and call their elements *third-order tensors*. (In this process, we tacitly use the inner-product structure of our spaces to shortcut the use of duality theory in these definitions.)

In this way, we can construct tensors of any order and develop a formalism for the use of dots. For example, a fourth-order tensor $\mathcal{D}$, regarded as a member of $\mathcal{L}(\text{Lin}, \text{Lin})$, has the value $\mathcal{D} : \boldsymbol{A}$ at the second-order tensor $\boldsymbol{A}$. In our work, we shall have little need for tensors of order higher than 2. Fourth-order tensors typically arise as Fréchet derivatives of a nonlinear mapping of Lin into itself. We shall often refer to tensors like $\boldsymbol{abc}$, $\boldsymbol{abcd}$, etc., merely as dyads, and not bother with a terminology that indicates their order.

Components. Let $\{\boldsymbol{a}_k\boldsymbol{b}_l\}$ be a basis for Lin. (Such bases occur frequently in continuum mechanics.) By definition, there is a 3×3 matrix (A^{kl}), called the *matrix of* $\boldsymbol{A}$ *with respect to* $\{\boldsymbol{a}_k\boldsymbol{b}_l\}$, such that

$$\boldsymbol{A} = A^{kl}\boldsymbol{a}_k\boldsymbol{b}_l. \tag{1.25}$$

The entries of the matrix are called the *components of* $\boldsymbol{A}$ *with respect to* $\{\boldsymbol{a}_k\boldsymbol{b}_l\}$. By taking the inner product of (1.25) with $\boldsymbol{a}^p\boldsymbol{b}^q$, we find that

$$A^{pq} = \boldsymbol{a}^p\boldsymbol{b}^q : \boldsymbol{A} = \boldsymbol{a}^p \cdot \boldsymbol{A} \cdot \boldsymbol{b}^q = \boldsymbol{A} : \boldsymbol{a}^p\boldsymbol{b}^q. \tag{1.26}$$

In particular,

$$(\boldsymbol{uv}) = (\boldsymbol{a}^k \cdot \boldsymbol{u})(\boldsymbol{b}^l \cdot \boldsymbol{v})\boldsymbol{a}_k\boldsymbol{b}_l, \quad \boldsymbol{I} = (\boldsymbol{a}^k \cdot \boldsymbol{b}^l)\boldsymbol{a}_k\boldsymbol{b}_l. \tag{1.27a,b}$$

If $\boldsymbol{A}$ is given by (1.25), then

$$\boldsymbol{A}^* = A^{kl}\boldsymbol{b}_l\boldsymbol{a}_k. \tag{1.28a}$$

If, furthermore, $\boldsymbol{a}_k = \boldsymbol{b}_k$, then

$$\boldsymbol{A} = A^{kl}\boldsymbol{a}_k\boldsymbol{a}_l, \quad \boldsymbol{A}^* = A^{lk}\boldsymbol{a}_k\boldsymbol{a}_l. \tag{1.28b}$$

Let $\boldsymbol{v} = \boldsymbol{A} \cdot \boldsymbol{u}$. Let $\boldsymbol{A}$ be given by (1.25), $\boldsymbol{v} = v^k \boldsymbol{a}_k$, and $\boldsymbol{u} = u_m \boldsymbol{c}^m$. Then

(1.29a) $$v^k = A^{kl} \boldsymbol{b}_l \cdot \boldsymbol{c}^m u_m,$$

which reduces to the usual

(1.29b) $$v^k = A^{km} u_m$$

when $\boldsymbol{c}^m = \boldsymbol{b}^m$.

1.30. Exercise. Find the components of $\boldsymbol{A} \cdot \boldsymbol{B}$. If $\boldsymbol{A}$ is given by (1.25), how should the basis for $\boldsymbol{B}$ be chosen to make the formula for $\boldsymbol{A} \cdot \boldsymbol{B}$ as simple as possible?

From (1.20a) we obtain

(1.31) $$\boldsymbol{A} : \boldsymbol{B} = (A^{kl} \boldsymbol{a}_k \boldsymbol{b}_l) : (B_{mn} \boldsymbol{a}^m \boldsymbol{b}^n) = A^{ml} B_{ml},$$

which confirms that ':' is an inner product.

In most books on linear algebra, the components of a tensor $\boldsymbol{A}$ are introduced thus: Let $\{\boldsymbol{c}_k\}$ be a basis for $\mathbb{E}^3$. Since $\boldsymbol{A} \cdot \boldsymbol{c}_k \in \mathbb{E}^3$, it is a linear combination of the $\boldsymbol{c}_m$:

(1.32) $$\boldsymbol{A} \cdot \boldsymbol{c}_k = A^m{}_k \boldsymbol{c}_m.$$

The $A^m{}_k$ are the components of $\boldsymbol{A}$ generated by the basis $\{\boldsymbol{c}_k\}$. Since

(1.33) $$\boldsymbol{c}^m \cdot \boldsymbol{A} \cdot \boldsymbol{c}_k = A^m{}_k,$$

the $A^m{}_k$ are therefore the components of $\boldsymbol{A}$ with respect to the basis $\{\boldsymbol{c}_j \boldsymbol{c}^l\}$.

Let $\boldsymbol{A}$ be symmetric. Let $\{\boldsymbol{e}_k\}$ be an orthonormal basis of eigenvectors corresponding to the eigenvalues λ_k of $\boldsymbol{A}$. Condition (1.26) then yields the *spectral representation* of $\boldsymbol{A}$:

(1.34) $$\boldsymbol{A} = \sum_{k=1}^{3} \lambda_k \boldsymbol{e}_k \boldsymbol{e}_k,$$

which says that $\boldsymbol{A}$ has a diagonal matrix with respect to its eigenvector basis. For $\alpha \in \mathbb{R}$, we define

(1.35) $$\boldsymbol{A}^\alpha = \sum_{k=1}^{3} \lambda_k^\alpha \boldsymbol{e}_k \boldsymbol{e}_k$$

whenever the right-hand side makes sense. This definition is consistent with those for integral values of α given previously. If $\boldsymbol{A}$ is positive-definite (so that all its eigenvalues are positive), then (1.35) implies that $\boldsymbol{A}^\alpha$ is also, provided that λ_k^α is chosen to be positive when α is not an integer. In particular, under this convention, if n is a positive integer and $\boldsymbol{A}$ is positive-definite, then (1.35) with $\alpha = 1/n$ can be shown to define the unique positive-definite nth root of $\boldsymbol{A}$ satisfying $\boldsymbol{A}^{1/n} \cdots \boldsymbol{A}^{1/n} = \boldsymbol{A}$. (Here $\boldsymbol{A}^{1/n}$ appears n times.)

An alternative notation. The following notational scheme is both convenient for handwritten manuscript and contains some conceptual advantages to compensate for its ugliness. A vector is written as a symbol over a single tilde, a second-order tensor as a symbol over a double tilde, a third-order tensor as a symbol over a triple tilde, etc. Thus each tensor symbol has as many tildes as its order. Moreover, each dot cancels two tildes, as we see in the following examples:

$$\underset{\sim}{\mathrm{v}} = \underset{\approx}{\mathrm{A}} \cdot \underset{\sim}{\mathrm{u}}, \quad \underset{\approx}{\mathrm{V}} = \underset{\approx\approx}{\mathrm{A}} : \underset{\approx}{\mathrm{U}}, \tag{1.36}$$

which correspond to $\boldsymbol{v} = \boldsymbol{A} \cdot \boldsymbol{u}$, $\boldsymbol{V} = \mathcal{A} : \boldsymbol{U}$.

Determinants. Let $\boldsymbol{u}, \boldsymbol{v}, \boldsymbol{w}$ be independent and let $\boldsymbol{A} \in \mathrm{Lin}$. We define

$$\det \boldsymbol{A} \equiv \frac{(\boldsymbol{A} \cdot \boldsymbol{u}) \cdot [(\boldsymbol{A} \cdot \boldsymbol{v}) \times (\boldsymbol{A} \cdot \boldsymbol{w})]}{\boldsymbol{u} \cdot (\boldsymbol{v} \times \boldsymbol{w})}. \tag{1.37}$$

It can be shown that $\det \boldsymbol{A}$ is independent of $\boldsymbol{u}, \boldsymbol{v}, \boldsymbol{w}$. Since $\boldsymbol{u} \cdot (\boldsymbol{v} \times \boldsymbol{w})$ is the volume of a parallelipiped $\mathcal{P}$ with sides $\boldsymbol{u}, \boldsymbol{v}, \boldsymbol{w}$, it follows that $(\boldsymbol{A} \cdot \boldsymbol{u}) \cdot [(\boldsymbol{A} \cdot \boldsymbol{v}) \times (\boldsymbol{A} \cdot \boldsymbol{w})]$ is the volume of the deformed parallelipiped $\boldsymbol{A} \cdot \mathcal{P}$ in which each point of $\mathcal{P}$ has been subjected to the linear transformation $\boldsymbol{A}$. Thus $\det \boldsymbol{A}$ is the ratio of these volumes. The product of two tensors satisfies the important identities

$$\det(\boldsymbol{A} \cdot \boldsymbol{B}) = \det \boldsymbol{A} \det \boldsymbol{B}, \quad \det \boldsymbol{A} = \det \boldsymbol{A}^*, \tag{1.38}$$

which follow from (1.37).

$\boldsymbol{A}$ is invertible if and only if $\det \boldsymbol{A} \neq 0$. In particular, if λ is an eigenvalue of $\boldsymbol{A}$, then it satisfies the *characteristic equation* for $\boldsymbol{A}$:

$$\det(\boldsymbol{A} - \lambda \boldsymbol{I}) = 0. \tag{1.39}$$

For any $\lambda \in \mathbb{C}$, definition (1.37) implies that $\det(\boldsymbol{A} - \lambda \boldsymbol{I})$ is a cubic in λ:

$$\det(\boldsymbol{A} - \lambda \boldsymbol{I}) = -\lambda^3 + \mathrm{I}_{\boldsymbol{A}} \lambda^2 - \mathrm{II}_{\boldsymbol{A}} \lambda + \mathrm{III}_{\boldsymbol{A}}. \tag{1.40}$$

The coefficients $\mathrm{I}_{\boldsymbol{A}}$, $\mathrm{II}_{\boldsymbol{A}}$, $\mathrm{III}_{\boldsymbol{A}}$ are the *principal invariants* of $\boldsymbol{A}$. They are defined by (1.37) and (1.40). They satisfy

$$\mathrm{I}_{\boldsymbol{A}} = \operatorname{tr} \boldsymbol{A}, \quad \mathrm{II}_{\boldsymbol{A}} = \tfrac{1}{2}[(\operatorname{tr} \boldsymbol{A})^2 - \operatorname{tr}(\boldsymbol{A}^2)], \quad \mathrm{III}_{\boldsymbol{A}} = \det \boldsymbol{A}. \tag{1.41}$$

It can be shown that

$$\operatorname{tr}(\boldsymbol{A}^3) = \mathrm{I}_{\boldsymbol{A}}^3 - 3\mathrm{I}_{\boldsymbol{A}} \mathrm{II}_{\boldsymbol{A}} + 3\mathrm{III}_{\boldsymbol{A}}. \tag{1.42}$$

Thus (1.41) and (1.42) imply that $\operatorname{tr} \boldsymbol{A}$, $\operatorname{tr}(\boldsymbol{A}^2)$, and $\operatorname{tr}(\boldsymbol{A}^3)$ form a set of invariants equivalent to the principal invariants. If $\lambda_1, \lambda_2, \lambda_3$ are the eigenvalues of $\boldsymbol{A}$, then

$$\mathrm{I}_{\boldsymbol{A}} = \lambda_1 + \lambda_2 + \lambda_3, \quad \mathrm{II}_{\boldsymbol{A}} = \lambda_1\lambda_2 + \lambda_2\lambda_3 + \lambda_3\lambda_1, \quad \mathrm{III}_{\boldsymbol{A}} = \lambda_1\lambda_2\lambda_3. \tag{1.43}$$

Equations (1.39) and (1.40) imply that any real tensor $\boldsymbol{A}$ has at least one real eigenvalue.

1.44. Cayley-Hamilton Theorem. *Every tensor $\boldsymbol{A}$ satisfies its own characteristic equation:*

$$-\boldsymbol{A}^3 + \mathrm{I}_{\boldsymbol{A}} \boldsymbol{A}^2 - \mathrm{II}_{\boldsymbol{A}} \boldsymbol{A} + \mathrm{III}_{\boldsymbol{A}} \boldsymbol{I} = \boldsymbol{O}. \tag{1.45}$$

2. Tensor Calculus

Consider a smooth function $\mathbb{E}^2 \ni \boldsymbol{z} \mapsto \varphi(\boldsymbol{z}) \in \mathbb{R}$, whose graph can be visualized as a surface over the plane $\mathbb{E}^2$. This surface has a geometry independent of the coordinates we use to describe it. In particular, the tangent plane and the curvature at each point of the surface, which can be determined by partial derivatives of first and second order, have an invariant character that can be described by tensors of first and second order. In this section we refine slightly the coordinate-free approach of defining derivatives introduced in Sec. I.4.

A *tensor field* is a function that assigns a tensor to each point of its domain of definition. For example, the velocity field of a body at a fixed time assigns the velocity vector to each material point of the body. Consider the following mappings, which are typical of those we shall encounter in solid mechanics:

$$\mathbb{E}^3 \ni \boldsymbol{u} \mapsto \varphi(\boldsymbol{u}) \in \mathbb{R}, \tag{2.1a}$$

$$\mathrm{Lin} \ni \boldsymbol{U} \mapsto \psi(\boldsymbol{U}) \in \mathbb{R}, \tag{2.1b}$$

$$\mathrm{Sym} \ni \boldsymbol{U} \mapsto \vartheta(\boldsymbol{U}) \in \mathbb{R}, \tag{2.1c}$$

$$\mathbb{E}^3 \ni \boldsymbol{u} \mapsto \boldsymbol{f}(\boldsymbol{u}) \in \mathbb{E}^3, \tag{2.1d}$$

$$\mathrm{Lin} \ni \boldsymbol{U} \mapsto \boldsymbol{F}(\boldsymbol{U}) \in \mathrm{Lin}. \tag{2.1e}$$

If these functions are Fréchet-differentiable (or even just Gâteaux-differentiable with a differential linear in the variation), then their derivatives can be found by the procedure of Sec. I.4. In particular, the derivative $\frac{\partial \vartheta}{\partial \boldsymbol{U}}(\boldsymbol{A}) \equiv \partial\vartheta(\boldsymbol{A})/\partial\boldsymbol{U} \equiv \vartheta_{\boldsymbol{U}}(\boldsymbol{A})$ of (2.1c) at $\boldsymbol{A}$ is a second-order tensor, which can be found from

$$\frac{\partial}{\partial \varepsilon}\vartheta(\boldsymbol{A}+\varepsilon\boldsymbol{B})\bigg|_{\varepsilon=0} = \frac{\partial \vartheta}{\partial \boldsymbol{U}}(\boldsymbol{A}) : \boldsymbol{B} \quad \forall\, \boldsymbol{B} \in \mathrm{Sym}. \tag{2.2}$$

The derivative of (2.1e) at $\boldsymbol{A}$ is a fourth-order tensor denoted by

$$\frac{\partial \boldsymbol{F}}{\partial \boldsymbol{U}}(\boldsymbol{A}) \equiv \partial\boldsymbol{F}(\boldsymbol{A})/\partial\boldsymbol{U} \equiv \boldsymbol{F}_{\boldsymbol{U}}(\boldsymbol{A}).$$

The last two forms have the virtue that the ordering of the boldface symbols $\boldsymbol{F}$ and $\boldsymbol{U}$ conforms to the dyadic ordering of these symbols.

We thus find that the derivative of $\mathbb{E}^3 \ni \boldsymbol{u} \mapsto \varphi(\boldsymbol{u}) \equiv \boldsymbol{c}\cdot\boldsymbol{u}$ is $\boldsymbol{c}$ since

$$\frac{\partial \varphi(\boldsymbol{a})}{\partial \boldsymbol{u}}\cdot\boldsymbol{v} \equiv \frac{\partial}{\partial\varepsilon}\boldsymbol{c}\cdot(\boldsymbol{a}+\varepsilon\boldsymbol{v})\bigg|_{\varepsilon=0} = \boldsymbol{c}\cdot\boldsymbol{v}. \tag{2.3}$$

Similarly,

$$\frac{\partial}{\partial \boldsymbol{u}}|\boldsymbol{u}|^2 = \frac{\partial}{\partial \boldsymbol{u}}\boldsymbol{u}\cdot\boldsymbol{u} = 2\boldsymbol{u}, \quad \frac{\partial}{\partial \boldsymbol{u}}|\boldsymbol{u}| = \frac{\partial}{\partial \boldsymbol{u}}\sqrt{\boldsymbol{u}\cdot\boldsymbol{u}} = \frac{\boldsymbol{u}}{|\boldsymbol{u}|}. \tag{2.4}$$

The computation of $\partial \det \boldsymbol{A}/\partial \boldsymbol{A}$ involves a couple of tricks. Let $\{\boldsymbol{a}_k\}$ be an arbitrary basis. From (1.37) and (1.20) we obtain

(2.5a)
$$\begin{aligned}
&\boldsymbol{a}_1 \cdot (\boldsymbol{a}_2 \times \boldsymbol{a}_3)(\partial \det \boldsymbol{A}/\partial \boldsymbol{A}) : \boldsymbol{B} \\
&= (\boldsymbol{B} \cdot \boldsymbol{a}_1) \cdot [(\boldsymbol{A} \cdot \boldsymbol{a}_2) \times (\boldsymbol{A} \cdot \boldsymbol{a}_3)] \\
&\quad + (\boldsymbol{B} \cdot \boldsymbol{a}_2) \cdot [(\boldsymbol{A} \cdot \boldsymbol{a}_3) \times (\boldsymbol{A} \cdot \boldsymbol{a}_1)] + (\boldsymbol{B} \cdot \boldsymbol{a}_3) \cdot [(\boldsymbol{A} \cdot \boldsymbol{a}_1) \times (\boldsymbol{A} \cdot \boldsymbol{a}_2)] \\
&\equiv \boldsymbol{a}_1 \cdot (\boldsymbol{a}_2 \times \boldsymbol{a}_3)(\operatorname{cof} \boldsymbol{A}) : \boldsymbol{B}
\end{aligned}$$

where the *cofactor* tensor $\operatorname{cof} \boldsymbol{A}$ is defined by

(2.5b)
$$\operatorname{cof} \boldsymbol{A} \equiv \frac{[(\boldsymbol{A} \cdot \boldsymbol{a}_2)\times(\boldsymbol{A} \cdot \boldsymbol{a}_3)]\boldsymbol{a}_1 + [(\boldsymbol{A} \cdot \boldsymbol{a}_3)\times(\boldsymbol{A} \cdot \boldsymbol{a}_1)]\boldsymbol{a}_2 + [(\boldsymbol{A} \cdot \boldsymbol{a}_1)\times(\boldsymbol{A} \cdot \boldsymbol{a}_2)]\boldsymbol{a}_3}{\boldsymbol{a}_1 \cdot (\boldsymbol{a}_2 \times \boldsymbol{a}_3)}.$$

This definition is likewise independent of the basis $\{\boldsymbol{a}_k\}$. If $\boldsymbol{A}$ is singular, then there is a nontrivial vector in its null space, which we can take to be $\boldsymbol{a}_3$, say. In this case, (2.5) simplifies considerably. Indeed, if $\boldsymbol{A}$ has a two-dimensional null space, then $\partial \det \boldsymbol{A}/\partial \boldsymbol{A} = \boldsymbol{O}$. We now obtain a more elegant representation for the derivative when $\boldsymbol{A}$ is nonsingular. We first obtain a useful identity from (1.37), which we rewrite as

$$\text{(2.6a)} \qquad \boldsymbol{u} \cdot \{(\det \boldsymbol{A})(\boldsymbol{v} \times \boldsymbol{w}) - \boldsymbol{A}^* \cdot [(\boldsymbol{A} \cdot \boldsymbol{v}) \times (\boldsymbol{A} \cdot \boldsymbol{w})]\} = 0.$$

Since this relation holds for all $\boldsymbol{u}$, it follows that

$$\text{(2.6b)} \qquad \boldsymbol{A}^* \cdot [(\boldsymbol{A} \cdot \boldsymbol{v}) \times (\boldsymbol{A} \cdot \boldsymbol{w})] = (\det \boldsymbol{A})(\boldsymbol{v} \times \boldsymbol{w}).$$

When $\boldsymbol{A}$ is nonsingular, we use (2.6b), (1.2), (1.17) to simplify (2.5):

(2.7a)
$$\begin{aligned}
&(\partial \det \boldsymbol{A}/\partial \boldsymbol{A}) : \boldsymbol{B} \\
&= (\det \boldsymbol{A})(\boldsymbol{A}^*)^{-1} \cdot \left[\frac{(\boldsymbol{a}_2 \times \boldsymbol{a}_3)\boldsymbol{a}_1}{(\boldsymbol{a}_2 \times \boldsymbol{a}_3) \cdot \boldsymbol{a}_1} + \frac{(\boldsymbol{a}_3 \times \boldsymbol{a}_1)\boldsymbol{a}_2}{(\boldsymbol{a}_3 \times \boldsymbol{a}_1) \cdot \boldsymbol{a}_2} + \frac{(\boldsymbol{a}_1 \times \boldsymbol{a}_2)\boldsymbol{a}_3}{(\boldsymbol{a}_1 \times \boldsymbol{a}_2) \cdot \boldsymbol{a}_3}\right] : \boldsymbol{B} \\
&= (\det \boldsymbol{A})(\boldsymbol{A}^*)^{-1} : \boldsymbol{B},
\end{aligned}$$

so that

$$\text{(2.7b)} \qquad \partial \det \boldsymbol{A}/\partial \boldsymbol{A} = \det \boldsymbol{A}\boldsymbol{A}^{-*}.$$

By interchanging $\boldsymbol{A}$ and $\boldsymbol{A}^*$ in (2.7b), we get a representation for $\boldsymbol{A}^{-1}$, which is Cramer's Rule.

An alternative proof of (2.7) uses (1.38) and (1.40):

$$\text{(2.8a)} \qquad \begin{aligned}
\det(\boldsymbol{A} + \varepsilon \boldsymbol{B}) &= \varepsilon^3 \det[(\varepsilon^{-1}\boldsymbol{I} + \boldsymbol{B} \cdot \boldsymbol{A}^{-1}) \cdot \boldsymbol{A}] \\
&= (\det \boldsymbol{A})\varepsilon^3 \left(\frac{1}{\varepsilon^3} + \frac{1}{\varepsilon^2}\mathrm{I}_{\boldsymbol{B}\cdot\boldsymbol{A}^{-1}} + \frac{1}{\varepsilon}\mathrm{II}_{\boldsymbol{B}\cdot\boldsymbol{A}^{-1}} + \mathrm{III}_{\boldsymbol{B}\cdot\boldsymbol{A}^{-1}}\right) \\
&= (\det \boldsymbol{A})[1 + \varepsilon \operatorname{tr}(\boldsymbol{B} \cdot \boldsymbol{A}^{-1}) + \cdots],
\end{aligned}$$

so that

$$(\partial \det \boldsymbol{A}/\partial \boldsymbol{A}) : \boldsymbol{B} = (\det \boldsymbol{A})\operatorname{tr}(\boldsymbol{B} \cdot \boldsymbol{A}^{-1}), \tag{2.8b}$$

which yields (2.7) by virtue of (1.21a).

2.9. Exercise. Prove that

$$(\partial \boldsymbol{U}/\partial \boldsymbol{U}) : \boldsymbol{B} = \boldsymbol{B}, \quad (\partial \boldsymbol{U}^*/\partial \boldsymbol{U}) : \boldsymbol{B} = \boldsymbol{B}^*,$$
$$[\partial(\boldsymbol{A} \cdot \boldsymbol{U})/\partial \boldsymbol{U}] : \boldsymbol{B} = \boldsymbol{A} : \boldsymbol{B}, \quad [\partial(\boldsymbol{U} \cdot \boldsymbol{A})/\partial \boldsymbol{U}] : \boldsymbol{B} = \boldsymbol{A} : \boldsymbol{B}, \tag{2.10}$$
$$[\partial(\boldsymbol{U}^* \cdot \boldsymbol{U})/\partial \boldsymbol{U}] : \boldsymbol{B} = \boldsymbol{U}^* : \boldsymbol{B} + \boldsymbol{B}^* : \boldsymbol{U}.$$

Let $\boldsymbol{U}$ be invertible. Prove that

$$(\partial \boldsymbol{U}^{-1}/\partial \boldsymbol{U}) : \boldsymbol{B} = -\boldsymbol{U}^{-1} \cdot \boldsymbol{B} \cdot \boldsymbol{U}^{-1}. \tag{2.11a}$$

Let $\boldsymbol{C}$ be positive-definite and let $\boldsymbol{U}(\boldsymbol{C})$ be its positive-definite square root. Prove that

$$\boldsymbol{U} \cdot \partial \boldsymbol{U}/\partial \boldsymbol{C} + (\boldsymbol{U} \cdot \partial \boldsymbol{U}/\partial \boldsymbol{C})^* = \boldsymbol{I}. \tag{2.11b}$$

(To get an explicit representation for $\partial \boldsymbol{U}/\partial \boldsymbol{C}$, see Sidoroff (1978).)

Let $\boldsymbol{u} \mapsto \boldsymbol{f}(\boldsymbol{u})$ and $\boldsymbol{v} \mapsto \boldsymbol{g}(\boldsymbol{v})$ be differentiable. Then the derivative of their composition satisfies the *Chain Rule*:

$$\partial \boldsymbol{f}(\boldsymbol{g}(\boldsymbol{v}))/\partial \boldsymbol{v} = [\partial \boldsymbol{f}(\boldsymbol{g}(\boldsymbol{v}))/\partial \boldsymbol{u}] \cdot \partial \boldsymbol{g}(\boldsymbol{v})/\partial \boldsymbol{v}. \tag{2.12}$$

Example. Let $\boldsymbol{u} \mapsto \boldsymbol{f}(\boldsymbol{u})$ be differentiable and let $\{\boldsymbol{a}_k\}$ and $\{\boldsymbol{b}_l\}$ be bases for $\mathbb{E}^3$ independent of $\boldsymbol{u}$. Let $\boldsymbol{u} = u^l \boldsymbol{a}_l$ and $\boldsymbol{f}(\boldsymbol{u}) = f^k(u^l \boldsymbol{a}_l)\boldsymbol{b}_k$. Then

$$\frac{\partial f^k}{\partial u^l} = \frac{\partial \boldsymbol{f} \cdot \boldsymbol{b}^k}{\partial u^l} = \boldsymbol{b}^k \cdot \frac{\partial \boldsymbol{f}}{\partial \boldsymbol{u}} \cdot \frac{\partial \boldsymbol{u}}{\partial u^l} = \boldsymbol{b}^k \cdot \frac{\partial \boldsymbol{f}}{\partial \boldsymbol{u}} \cdot \boldsymbol{a}_l. \tag{2.13}$$

Thus the $\frac{\partial f^k}{\partial u^l}$ are the components of the tensor $\frac{\partial \boldsymbol{f}}{\partial \boldsymbol{u}}$ with respect to the basis $\{\boldsymbol{b}_k \boldsymbol{a}^l\}$.

Gradients and related operators. The notation for the derivative $\partial \boldsymbol{f}/\partial \boldsymbol{u}$ is ideally suited for the Chain Rule because it captures the correct ordering for the underlying dyadic structure. We now show how it is related to the standard notations of vector analysis.

The *gradient* $\nabla \psi$ of $\mathbb{E}^3 \ni \boldsymbol{z} \mapsto \psi(\boldsymbol{z}) \in \mathbb{R}$ is defined by

$$\nabla \psi \equiv \partial \psi/\partial \boldsymbol{z}. \tag{2.14}$$

The *gradient* $\nabla \boldsymbol{f}$ of $\mathbb{E}^3 \ni \boldsymbol{z} \mapsto \boldsymbol{f}(\boldsymbol{z}) \in \mathbb{E}^3$ is defined by

$$\nabla \boldsymbol{f} \equiv (\partial \boldsymbol{f}/\partial \boldsymbol{z})^*. \tag{2.15}$$

The transpose is used to make the notation conform to classical usage. Indeed, let $\{\boldsymbol{a}_k\}$ be a basis independent of $\boldsymbol{z}$ and set $\boldsymbol{z} = z^k \boldsymbol{a}_k$ with $z^k = \boldsymbol{z} \cdot \boldsymbol{a}^k$. Using techniques like those of (2.13), we find that

$$\partial \boldsymbol{f}/\partial \boldsymbol{z} = (\partial \boldsymbol{f}/\partial z^k)\boldsymbol{a}^k, \tag{2.16}$$

so that

$$(\partial \boldsymbol{f}/\partial \boldsymbol{z})^* = \boldsymbol{a}^k(\partial \boldsymbol{f}/\partial z^k) = \left(\boldsymbol{a}^k \frac{\partial}{\partial z^k}\right) \boldsymbol{f} = \nabla \boldsymbol{f}. \tag{2.17}$$

Thus we make the classical identification

$$\nabla \equiv \boldsymbol{a}^k \frac{\partial}{\partial z^k}. \tag{2.18}$$

We define

$$\operatorname{div} \boldsymbol{f} \equiv \nabla \cdot \boldsymbol{f} \equiv \operatorname{tr}(\partial \boldsymbol{f}/\partial \boldsymbol{z}) = \boldsymbol{I} : (\partial \boldsymbol{f}/\partial \boldsymbol{z}) = (\partial \boldsymbol{f}/\partial \boldsymbol{z}) : \boldsymbol{I}, \tag{2.19a}$$

$$\operatorname{div} \boldsymbol{T} \equiv \nabla \cdot \boldsymbol{T}^* \equiv (\partial \boldsymbol{T}/\partial \boldsymbol{z}) : \boldsymbol{I}, \tag{2.19b}$$

$$\operatorname{curl} \boldsymbol{f} \equiv \nabla \times \boldsymbol{f}. \tag{2.19c}$$

An alternative characterization of $\operatorname{div} \boldsymbol{T}$ is that

$$(\operatorname{div} \boldsymbol{T}) \cdot \boldsymbol{c} = \nabla \cdot (\boldsymbol{T}^* \cdot \boldsymbol{c}) = \operatorname{div}(\boldsymbol{T}^* \cdot \boldsymbol{c}) = \operatorname{tr}[\partial(\boldsymbol{T}^* \cdot \boldsymbol{c})/\partial \boldsymbol{z}] \quad \forall\, \boldsymbol{c} \in \mathbb{E}^3. \tag{2.20}$$

We shall reserve the symbol $\boldsymbol{z}$ to represent material points of a body and shall apply operators involving ∇ only to functions depending on $\boldsymbol{z}$.

2.21. Exercise. Find the components of the tensors of (2.19). For treating the curl, use the alternating symbol introduced in Sec. VIII.1.

Integral theorems. Let $\boldsymbol{\Psi} \in C^1(\Omega) \cap C^0(\operatorname{cl}\Omega)$ be a tensor-valued function of any order defined on a bounded domain Ω of $\mathbb{E}^3$. Let $\partial\Omega$ be a regular surface in the sense of Kellogg (1929). Then

$$\int_\Omega \partial \boldsymbol{\Psi}(\boldsymbol{z})/\partial \boldsymbol{z}\, dv = \int_{\partial\Omega} \boldsymbol{\Psi}(\boldsymbol{z}) \boldsymbol{\nu}(\boldsymbol{z})\, da(\boldsymbol{z}) \tag{2.22}$$

where dv is the differential volume element, $da(\boldsymbol{z})$ is the differential surface element of $\partial\Omega$ at $\boldsymbol{z}$, and $\boldsymbol{\nu}(\boldsymbol{z})$ is the outer unit normal to $\partial\Omega$ at $\boldsymbol{z}$. In particular,

$$\int_\Omega \nabla \varphi(\boldsymbol{z})\, dv = \int_{\partial\Omega} \boldsymbol{\nu}(\boldsymbol{z}) \varphi(\boldsymbol{z})\, da(\boldsymbol{z}), \tag{2.23a}$$

$$\int_\Omega \nabla \boldsymbol{f}(\boldsymbol{z})\, dv = \int_{\partial\Omega} \boldsymbol{\nu}(\boldsymbol{z}) \boldsymbol{f}(\boldsymbol{z})\, da(\boldsymbol{z}), \tag{2.23b}$$

$$\int_\Omega \nabla \cdot \boldsymbol{f}(\boldsymbol{z})\, dv = \int_{\partial\Omega} \boldsymbol{\nu}(\boldsymbol{z}) \cdot \boldsymbol{f}(\boldsymbol{z})\, da(\boldsymbol{z}), \tag{2.23c}$$

$$\int_\Omega \nabla \times \boldsymbol{f}(\boldsymbol{z})\, dv = \int_{\partial\Omega} \boldsymbol{\nu}(\boldsymbol{z}) \times \boldsymbol{f}(\boldsymbol{z})\, da(\boldsymbol{z}), \tag{2.23d}$$

$$\begin{aligned} \int_\Omega \operatorname{div} \boldsymbol{T}(\boldsymbol{z})\, dv &\equiv \int_\Omega \nabla \cdot \boldsymbol{T}^*(\boldsymbol{z})\, dv = \int_{\partial\Omega} \boldsymbol{\nu}(\boldsymbol{z}) \cdot \boldsymbol{T}^*(\boldsymbol{z})\, da(\boldsymbol{z}) \\ &= \int_{\partial\Omega} \boldsymbol{T}(\boldsymbol{z}) \cdot \boldsymbol{\nu}(\boldsymbol{z})\, da(\boldsymbol{z}). \end{aligned} \tag{2.23e}$$

The divergence of a tensor was defined to yield (2.23e) in which the rightmost term has $\boldsymbol{T}$ itself acting on $\boldsymbol{\nu}$. Identity (2.23a) is the Green-Gauss-Ostrogradskiĭ Theorem. Identity (2.23c) is the Divergence Theorem.

2.24. Exercise. Derive (2.23a) under favorable regularity assumptions (by elementary calculus) and then derive (2.23c,d,e) from it.

Let $\mathcal{S}$ be a one-sided surface in $\mathbb{E}^3$ that is regular in the sense of Kellogg. Let it be bounded by a compatibly oriented curve $\mathcal{C} = \partial\mathcal{S}$ regular in the sense of Kellogg. We define $\mathcal{S}$ so that it does not contain any point of $\mathcal{C}$. Let $\boldsymbol{\nu}(\boldsymbol{z})$ and $da(\boldsymbol{z})$ denote the unit normal vector to $\mathcal{S}$ at $\boldsymbol{z}$ and the differential surface area of $\mathcal{S}$ at $\boldsymbol{z}$. Let $\boldsymbol{t}(\boldsymbol{z})$ and $ds(\boldsymbol{z})$ denote the unit tangent vector to $\mathcal{C}$ at $\boldsymbol{z}$ and the differential arc length of $\mathcal{C}$ at $\boldsymbol{z}$. Let $\boldsymbol{f} \in C^1(\mathcal{S}) \cap C^0(\mathcal{S} \cup \mathcal{C})$. Then two forms of the Kelvin (Stokes) Theorem are

$$\int_{\mathcal{S}} \boldsymbol{\nu}(\boldsymbol{z}) \cdot \operatorname{curl} \boldsymbol{f}(\boldsymbol{z})\, da(\boldsymbol{z}) = \int_{\mathcal{C}} \boldsymbol{t}(\boldsymbol{z}) \cdot \boldsymbol{f}(\boldsymbol{z})\, ds(\boldsymbol{z}), \tag{2.25a}$$

$$\int_{\mathcal{S}} \boldsymbol{\nu}(\boldsymbol{z}) \times \operatorname{curl} \boldsymbol{f}(\boldsymbol{z})\, da(\boldsymbol{z}) = \int_{\mathcal{C}} \boldsymbol{t}(\boldsymbol{z}) \times \boldsymbol{f}(\boldsymbol{z})\, ds(\boldsymbol{z}). \tag{2.25b}$$

Proofs of all these integral theorems under favorable regularity assumptions are found in standard texts on calculus and on vector analysis. Proofs under weak classical assumptions are given by Kellogg (1929). Proofs under very weak conditions (which are important in the modern mathematical analysis of problems of continuum mechanics) use the ideas of geometric measure theory (see Evans & Gariepy (1991), Federer (1969), Morgan (1988), and Ziemer (1989)).

The notational scheme introduced in this and the preceding section represents a marriage of the dyadic system of Gibbs (see Gibbs & Wilson (1901)) with a modern coordinate-free approach to linear algebra and to analysis in Euclidean space (see Bowen & Wang (1976), Dieudonné (1960), Halmos (1958), Nickerson, Spencer, & Steenrod (1959), Noll (1987), among others.) Our scheme is similar in approach, though not in detail, to those used by Gurtin (1981a) and Truesdell & Noll (1965).

3. Indicial Notation

In the first half of the twentieth century, partly in response to the theory of relativity, a very detailed indicial notation was developed for Riemannian geometry. In the 1950's (just when it was being extirpated from geometry), this notation was adopted in continuum mechanics, where it flourished for about a decade and was then superseded by more modern notations. We outline this notation because we use a variant of it in the shell theory in Chap. XIV and because important works on continuum mechanics and elasticity employ it consistently.

Let $\mathbf{x} = (x^1, x^2, x^3)$ be curvilinear coordinates for a region $\mathcal{B} \subset \mathbb{E}^3$. Let $\hat{\boldsymbol{z}}(\mathbf{x})$ assign the position of the point with coordinates $\mathbf{x}$. Then the vectors

$$\boldsymbol{g}_k(\mathbf{x}) \equiv \frac{\partial \tilde{\boldsymbol{z}}}{\partial x^k}(\mathbf{x}) \equiv \tilde{\boldsymbol{z}}_{,k}(\mathbf{x}) \tag{3.1}$$

are tangent to the coordinate curves. We assume that

$$0 < \det \frac{\partial \tilde{\boldsymbol{z}}}{\partial \mathbf{x}} \equiv [\boldsymbol{g}_1(\mathbf{x}) \times \boldsymbol{g}_2(\mathbf{x})] \cdot \boldsymbol{g}_3(\mathbf{x}) \quad \text{for } \boldsymbol{z} \in \mathcal{B} \tag{3.2}$$

so that the $\boldsymbol{g}_k$ are independent. Let $\{\boldsymbol{g}^k\}$ be the basis dual to $\{\boldsymbol{g}_k\}$. We can use (3.1) to verify that $\boldsymbol{g}^k(\mathbf{x}) = \frac{\partial \tilde{x}^k}{\partial \boldsymbol{z}}(\tilde{\boldsymbol{z}}(\mathbf{x}))$. Then we can represent any vector field $\boldsymbol{w}$ by

$$\boldsymbol{w}(\mathbf{x}) = w^k(\mathbf{x})\boldsymbol{g}_k(\mathbf{x}) = w_k(\mathbf{x})\boldsymbol{g}^k(\mathbf{x}). \tag{3.3}$$

The w^k are the *contravariant components* of $\boldsymbol{w}$ and the w_k are the *covariant components* of $\boldsymbol{w}$. Similarly, we can represent any tensor field with respect to a basis consisting of dyadic products of the $\boldsymbol{g}_k$ and the $\boldsymbol{g}^k$. For example,

$$\boldsymbol{T} = T^{kl}\boldsymbol{g}_k\boldsymbol{g}_l = T^k{}_l\boldsymbol{g}_k\boldsymbol{g}^l = T_k{}^l\boldsymbol{g}^k\boldsymbol{g}_l = T_{kl}\boldsymbol{g}^k\boldsymbol{g}^l. \tag{3.4}$$

The T^{kl} are the *contravariant components* of $\boldsymbol{T}$, the T_{kl} are the *covariant components* of $\boldsymbol{T}$, and the $T^k{}_l$ and the $T_k{}^l$ are *mixed components* of $\boldsymbol{T}$. In particular, (1.27b) implies that

$$\boldsymbol{I} = g^{kl}\boldsymbol{g}_k\boldsymbol{g}_l = g_l^k\boldsymbol{g}_k\boldsymbol{g}^l = g_{kl}\boldsymbol{g}^k\boldsymbol{g}^l \tag{3.5}$$

where

$$g^{kl} \equiv \boldsymbol{g}^k \cdot \boldsymbol{g}^l, \quad g_l^k \equiv \delta_l^k, \quad g_{kl} \equiv \boldsymbol{g}_k \cdot \boldsymbol{g}_l. \tag{3.6}$$

The g^{kl} are the *contravariant components*, the g_{kl} are the *covariant components*, and the g_l^k are the *mixed components* of the identity or *metric* tensor $\boldsymbol{I}$. (That the quadratic form of $\boldsymbol{I}$ is a sum of squares corresponds to the fact that the differential arc length squared is a sum of squares in Euclidean space, so $\boldsymbol{I}$ is the metric tensor for $\mathbb{E}^3$.) We readily find that

$$\boldsymbol{g}^k = g^{kl}\boldsymbol{g}_l, \quad \boldsymbol{g}_k = g_{kl}\boldsymbol{g}^l. \tag{3.7}$$

For any tensor $\boldsymbol{\Psi}$, we set

$$\frac{\partial \boldsymbol{\Psi}}{\partial x^k} \equiv \boldsymbol{\Psi}_{,k}. \tag{3.8}$$

Thus (3.3), (3.4), and (3.8) imply that

$$\boldsymbol{w}_{,m} = w^k_{,m}\boldsymbol{g}_k + w^k\boldsymbol{g}_{k,m}, \tag{3.9a}$$

$$\boldsymbol{T}_{,m} = T^{kl}_{,m}\boldsymbol{g}_k\boldsymbol{g}_l + T^{kl}\boldsymbol{g}_{k,m}\boldsymbol{g}_l + T^{kl}\boldsymbol{g}_k\boldsymbol{g}_{l,m}. \tag{3.9b}$$

Since $\boldsymbol{g}_{k,m}$ is a vector, it can be represented as a linear combination of base vectors:

$$\boldsymbol{g}_{k,m} = \Gamma^p_{km}\boldsymbol{g}_p = \Gamma_{kmp}\boldsymbol{g}^p. \tag{3.10}$$

Thus we can write (3.9) as

$$\boldsymbol{w}_{,m} = (w^p_{,m}\boldsymbol{g}_k + w^k\Gamma^p_{km})\boldsymbol{g}_p \equiv w^p{}_{|m}\boldsymbol{g}_p, \tag{3.11a}$$

$$\boldsymbol{T}_{,m} = (T^{pq}_{,m} + T^{iq}\Gamma^p_{im} + T^{pj}\Gamma^q_{jm})\boldsymbol{g}_p\boldsymbol{g}_q \equiv T^{pq}{}_{|m}\boldsymbol{g}_p\boldsymbol{g}_q. \tag{3.11b}$$

$w^p{}_{|m}$ is the *covariant derivative* of w^p and $T^{pq}{}_{|m}$ is the *covariant derivative* of T^{pq}. Γ^p_{km} and Γ_{kmp} are *Christoffel symbols*. All formulas involving components with respect to the bases consisting of dyadic products of $\{\boldsymbol{g}_k\}$ and $\{\boldsymbol{g}^k\}$ are valid when any index is systematically raised or lowered.

Let $\{\boldsymbol{i}_k\}$ be an orthonormal basis and let $\boldsymbol{z} = z^k\boldsymbol{i}_k$. By using (2.18) and the Chain Rule, we obtain

$$\nabla = \boldsymbol{i}^k\frac{\partial}{\partial z^k} = \boldsymbol{i}^k\frac{\partial \tilde{x}^l}{\partial z^k}\frac{\partial}{\partial x^l} = \boldsymbol{g}^l\frac{\partial}{\partial x^l}. \tag{3.12}$$

3.13. Exercise. Prove that

$$\Gamma^k_{km} = \frac{1}{\sqrt{\det g_{ij}}}\frac{\partial\sqrt{\det g_{ij}}}{\partial x^m}. \tag{3.14}$$

In general, the $\boldsymbol{g}_k$ and the $\boldsymbol{g}^k$ are not unit vectors, even when they are orthogonal (as for cylindrical and spherical coordinates). They typically have different dimensions. Thus, components with respect to such bases are unsuitable for the estimations usually required in the analysis of differential equations and unsuitable for physical interpretation. Consequently, physical components of tensors are introduced. Such components can be introduced more directly by the formalism developed in Secs. 1 and 2. For fuller developments of the tensorial apparatus described here with an orientation towards continuum mechanics, see Doyle & Ericksen (1956), Ericksen (1960), Eringen (1962), Green & Zerna (1968), and Truesdell & Toupin (1960).

CHAPTER XII

Three-Dimensional Continuum Mechanics

In this chapter we present a formulation of continuum mechanics directed toward the treatment of the behavior of solids. It is based on the material (or *Lagrangian*) description of motion. Virtually all the problems we treat will be cast in this formulation. At the conclusion of the chapter, we describe the modifications needed for the spatial (or *Eulerian*) formulation, which is essential for fluid mechanics and which has been used in solid mechanics. References for standard material of this chapter expressed in a related style are Chadwick (1976), Gurtin (1981a), and Truesdell & Noll (1965). The last work has extensive historical notes. References for specialized topics in this chapter are given in the individual sections.

1. Kinematics

For the purposes of the classical mechanics we employ, a (*three-dimensional*) *body* can be informally defined to be a set that can occupy regions of $\mathbb{E}^3$, that has volume, that has mass, and that can sustain forces. (For more general problems of continuum physics, we could ask that a body sustain heat flux, have entropy and energy, have electric charge, and sustain electromagnetic fields.) The elements of a body are called *material points* or *particles*. The use of the latter term does not imply any association between material points and the discrete particles of modern physics.

To make these concepts absolutely precise, it is necessary to employ modern mathematical concepts. For example, the requirement that a body be able to occupy regions of $\mathbb{E}^3$ is expressed by: A body is a topological space that can be mapped homeomorphically onto a region of $\mathbb{E}^3$. (A *homeomorphism* is a continuous mapping having a continuous inverse.) Each such mapping defines a *configuration* of the body. We distinguish one configuration of the body and call it the *reference configuration*. This configuration might be one occupied by the body at a certain (initial) instant of time, or it might be a natural (stress-free) configuration, or it might be some ideal configuration, unlikely to be occupied by the body. For example, in the study of water waves, a nice reference configuration might be one in which the free surface is flat, and therefore unlikely ever to form the actual free boundary of a large body of water on earth. The requirements that a body have volume, have mass, and can sustain forces can be expressed by saying that the body is measurable in a suitable sense, which we do not bother to make precise (cf. Secs. 6 and 7).

We identify a body with the region $\mathcal{B}$ it occupies in its reference configuration, and we identify the material points of the body with the positions

z they occupy in $\mathcal{B}$. The bodies we study are connected open sets together with any part of their boundaries.

For very careful discussions of mathematical models for bodies, see Noll (1966, 1993), Noll & Virga (1988), and Truesdell (1991a). They treat refined notions of bodies that can account for what happens to material points that lie on a surface interior to a body when that body is fractured along that surface.

A *motion* of body $\mathcal{B}$ is a one-parameter family $\{\boldsymbol{p}(\cdot, t),\ t \in \mathcal{I}\}$ of its configurations $\mathcal{B} \ni \boldsymbol{z} \mapsto \boldsymbol{p}(\boldsymbol{z}, t) \in \mathbb{E}^3$. Here $\mathcal{I}$ is an interval of $\mathbb{R}$. $\boldsymbol{p}(\boldsymbol{z}, t)$ is the *position* of material point $\boldsymbol{z}$ at time t.

In the absence of a statement to the contrary, we assume that $\boldsymbol{p}$ has as many continuous derivatives as are exhibited. Problems of existence are typically posed in larger spaces of functions that do not possess classical derivatives of the orders appearing in the equations, because there is more likelihood of finding a solution in such a space. We comment on such questions in Sec. XIII.2.

We require that for (almost) every t the actual position field $\boldsymbol{p}(\cdot, t)$ of a body be injective (i.e., one-to-one) so that two distinct material points cannot simultaneously occupy the same position in space. This requirement is the *Principle of Impenetrability of Matter.* This principle is a global restriction on $\boldsymbol{p}(\cdot, t)$. A related local requirement is that $\boldsymbol{p}(\cdot, t)$ *preserve orientation*, i.e., that

$$(1.1) \qquad \det \boldsymbol{p}_{\boldsymbol{z}}(\boldsymbol{z}, t) > 0$$

for (almost) all $\boldsymbol{z} \in \mathcal{B}$ and for (almost) all t. By the remarks of Sec. XI.1, this condition ensures that the local ratio of actual to reference volume never be reduced to zero.

1.2. Exercise. Find a smooth mapping $\boldsymbol{z} \mapsto \boldsymbol{p}(\boldsymbol{z})$ for which (1.1) holds everywhere, but is not injective.

A very effective version of the impenetrability of matter was introduced by Ciarlet & Nečas (1987):

$$(1.3) \qquad v(\boldsymbol{p}(\mathcal{B}, t)) \geq \int_{\mathcal{B}} |\det \boldsymbol{p}_{\boldsymbol{z}}(\boldsymbol{z}, t)| \, dv.$$

Here v denotes the volume (measure). To appreciate (1.3), suppose that $\boldsymbol{p}_{\boldsymbol{z}}(\cdot, t)$ were not injective, so that parts of the deformed body would overlap. Then the right-hand side would locally account for all the contributions to the volume of the deformed body and accordingly overestimate it. Condition (1.3) says that the actual volume cannot be less than this overestimate and thus that the cause of the overestimation, namely, the lack of injectivity, must be absent.

The class of position fields satisfying (1.1) is very complicated: It is not closed in $C^1(\operatorname{cl} B)$, it is not convex, and it may have an infinite number of connected components that are not convex. Indeed, consider the deformation of a toroid that is effected by slicing it so as to make it simply-connected, by rotating one sliced face through an integral multiple of 2π relative to the other, and then by gluing the faces together so that material points originally contiguous are again so. Each such deformation preserves orientation, but deformation of one family cannot be continuously deformed into another. For a discussion of these and related topological questions, see Alexander & Antman (1982), Antman (1976a), and Pierce & Whitman (1980).

The *velocity* and *acceleration* of material point $\boldsymbol{z}$ at time t are $\boldsymbol{p}_t(\boldsymbol{z}, t)$ and $\boldsymbol{p}_{tt}(\boldsymbol{z}, t)$. We use the abbreviation

$$\boldsymbol{F} \equiv \boldsymbol{p}_{\boldsymbol{z}}. \tag{1.4}$$

It is the transpose of the gradient of the position field. In the literature, it is simply called the *deformation gradient.*

A formulation of continuum mechanics in which the material point $\boldsymbol{z}$ is the independent spatial variable is called a *material formulation* (or ahistorically, a *Lagrangian formulation*). A formulation of continuum mechanics in which the fixed point in space is the independent spatial variable is called a *spatial formulation* (or ahistorically, an *Eulerian formulation*). We discuss the latter in Sec. 15.

1.5. Exercise. Prove that the set of continuously differentiable $\boldsymbol{F}$'s satisfying (1.1) is not convex. Prove that they form a cone with vertex $\boldsymbol{O}$. (A *cone* $\mathcal{C}$ *with vertex* $\boldsymbol{A}$ is a set in a linear space with the property that if $\boldsymbol{F} \in \mathcal{C}$, then $\alpha(\boldsymbol{F} - \boldsymbol{A}) \in \mathcal{C}$ for all $\alpha > 0$.)

2. Strain

We now compare the geometry of an arbitrary deformed configuration with that of the reference configuration in order to develop convenient quantities to measure the change of shape.

Let $[a, b] \ni s \mapsto \breve{\boldsymbol{z}}(s) \in \mathcal{B}$ be a continuously differentiable curve of material points. We assume that s is the arc-length parameter, so that $|\breve{\boldsymbol{z}}'| = 1$ and so that $\breve{\boldsymbol{z}}'(s)$ is the unit tangent vector to $\breve{\boldsymbol{z}}$ at $\breve{\boldsymbol{z}}(s)$.

Suppose that the body undergoes a motion $\boldsymbol{p}$. Then at time t the material points forming $\breve{\boldsymbol{z}}$ lie along the curve $[a, b] \ni s \mapsto \boldsymbol{p}(\breve{\boldsymbol{z}}(s), t) \in \boldsymbol{p}(\mathcal{B}, t)$ (see Fig. 2.1). A tangent to this curve at $\boldsymbol{p}(\breve{\boldsymbol{z}}(s), t)$ is $\boldsymbol{p}_{\boldsymbol{z}}(\breve{\boldsymbol{z}}(s), t) \cdot \breve{\boldsymbol{z}}'(s) \equiv \boldsymbol{F}(\breve{\boldsymbol{z}}(s), t) \cdot \breve{\boldsymbol{z}}'(s)$. The length of $\boldsymbol{p}(\breve{\boldsymbol{z}}(\cdot), t)$ is

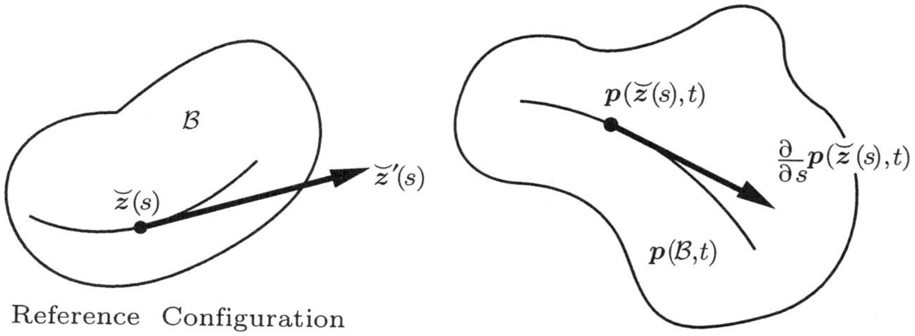

Fig. 2.1. The material curve $\breve{\boldsymbol{z}}$ and its deformed image.

$$(2.2)\quad \int_a^b \sqrt{[\boldsymbol{F}(\breve{\boldsymbol{z}}(s),t)\cdot\breve{\boldsymbol{z}}'(s)]\cdot[\boldsymbol{F}(\breve{\boldsymbol{z}}(s),t)\cdot\breve{\boldsymbol{z}}'(s)]}\,ds = \int_a^b \sqrt{\breve{\boldsymbol{z}}'(s)\cdot\boldsymbol{C}(\breve{\boldsymbol{z}}(s),t)\cdot\breve{\boldsymbol{z}}'(s)}\,ds$$

where

$$(2.3)\quad \boldsymbol{C}(\boldsymbol{z},t) \equiv \boldsymbol{F}(\boldsymbol{z},t)^*\cdot\boldsymbol{F}(\boldsymbol{z},t)$$

is the (*right Cauchy-*) *Green deformation tensor* at $(\boldsymbol{z},t)$. It is symmetric and is positive-definite where $\boldsymbol{F}$ is nonsingular.

We now show that $\boldsymbol{C}$ completely describes change of shape locally. In particular, we apply the rules of l'Hôpital and Leibniz to (2.2) to obtain

$$(2.4)\quad \lim_{b\searrow a}\frac{\text{length}\,(\boldsymbol{p}(\breve{\boldsymbol{z}}(\cdot),t))}{\text{length}\,(\breve{\boldsymbol{z}})} = \sqrt{\breve{\boldsymbol{z}}'(a)\cdot\boldsymbol{C}(\breve{\boldsymbol{z}}(a),t)\cdot\breve{\boldsymbol{z}}'(a)}.$$

We now regard $\breve{\boldsymbol{z}}(a)$ as a typical material point in the interior of $\mathcal{B}$ and regard $\breve{\boldsymbol{z}}'(a)$ as a typical material direction or *fiber* at $\breve{\boldsymbol{z}}(a)$. We accordingly set $\boldsymbol{z}=\breve{\boldsymbol{z}}(a)$ and $\boldsymbol{n}=\breve{\boldsymbol{z}}'(a)$. Thus our earlier observations say that the material fiber $\boldsymbol{n}$ at $\boldsymbol{z}$ is taken by the motion $\boldsymbol{p}$ to $\boldsymbol{p}_{\boldsymbol{z}}(\boldsymbol{z},t)\cdot\boldsymbol{n}$. The ratio of deformed to reference length of the fiber $\boldsymbol{n}$ is

$$(2.5)\quad |\boldsymbol{F}(\boldsymbol{z},t)\cdot\boldsymbol{n}| \equiv \sqrt{\boldsymbol{n}\cdot\boldsymbol{C}(\boldsymbol{z},t)\cdot\boldsymbol{n}}.$$

$|\boldsymbol{F}(\boldsymbol{z},t)\cdot\boldsymbol{n}|$ is the *stretch* at $(\boldsymbol{z},t)$ along the fiber $\boldsymbol{n}$. The fiber $\boldsymbol{n}$ at $\boldsymbol{z}$ is said to be *compressed, unstretched,* or *extended* according as $|\boldsymbol{F}(\boldsymbol{z},t)\cdot\boldsymbol{n}|<1$, $|\boldsymbol{F}(\boldsymbol{z},t)\cdot\boldsymbol{n}|=1$, or $|\boldsymbol{F}(\boldsymbol{z},t)\cdot\boldsymbol{n}|>1$. If $|\boldsymbol{F}(\boldsymbol{z},t)\cdot\boldsymbol{n}|=1$ for all $\boldsymbol{n}$, which happens if and only if

$$(2.6)\quad \boldsymbol{C}(\boldsymbol{z},t)=\boldsymbol{I},$$

then the deformation at $(\boldsymbol{z},t)$ is said to be *rigid.*

We now determine the geometrical significance of the components of $\boldsymbol{C}$. Let $\{\boldsymbol{a}^k\}$ be a basis for $\mathbb{E}^3$. Then

$$(2.7)\quad \boldsymbol{C}=C_{kl}\boldsymbol{a}^k\boldsymbol{a}^l=(\boldsymbol{a}_k\cdot\boldsymbol{C}\cdot\boldsymbol{a}_l)\boldsymbol{a}^k\boldsymbol{a}^l=(\boldsymbol{a}_k\cdot\boldsymbol{F}^*)\cdot(\boldsymbol{F}\cdot\boldsymbol{a}_l)\boldsymbol{a}^k\boldsymbol{a}^l.$$

Thus

$$(2.8)\quad \frac{C_{11}(\boldsymbol{z},t)}{\boldsymbol{a}_1(\boldsymbol{z})\cdot\boldsymbol{a}_1(\boldsymbol{z})}=\frac{\boldsymbol{a}_1(\boldsymbol{z})}{|\boldsymbol{a}_1(\boldsymbol{z})|}\cdot\boldsymbol{C}(\boldsymbol{z},t)\cdot\frac{\boldsymbol{a}_1(\boldsymbol{z})}{|\boldsymbol{a}_1(\boldsymbol{z})|}=\left|\boldsymbol{F}(\boldsymbol{z},t)\cdot\frac{\boldsymbol{a}_1(\boldsymbol{z})}{|\boldsymbol{a}_1(\boldsymbol{z})|}\right|^2$$

is the square of the stretch of the material fiber along $\boldsymbol{a}_1$.

To interpret the off-diagonal components of $\boldsymbol{C}$, we let $\boldsymbol{m}$ and $\boldsymbol{n}$ denote two unit fibers at $\boldsymbol{z}$ subtending an angle $\mathring{\theta}$. Thus $|\boldsymbol{m}|=1=|\boldsymbol{n}|$ and

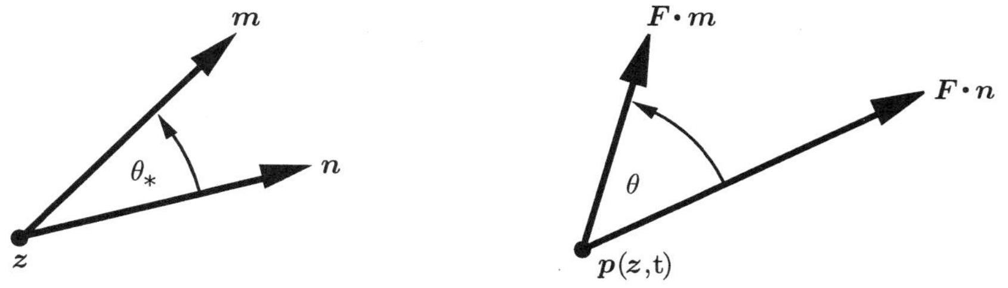

Fig. 2.9. Shear deformation.

$\boldsymbol{m}\cdot\boldsymbol{n} = \cos\overset{\circ}{\theta}$. (See Fig. 2.9.) The deformed images of these fibers subtend the angle θ defined by

$$\text{(2.10)}\qquad \cos\theta = \frac{\boldsymbol{F}\cdot\boldsymbol{m}}{|\boldsymbol{F}\cdot\boldsymbol{m}|}\cdot\frac{\boldsymbol{F}\cdot\boldsymbol{n}}{|\boldsymbol{F}\cdot\boldsymbol{n}|} = \frac{\boldsymbol{m}\cdot\boldsymbol{F}^*\cdot\boldsymbol{F}\cdot\boldsymbol{n}}{|\boldsymbol{F}\cdot\boldsymbol{m}|\,|\boldsymbol{F}\cdot\boldsymbol{n}|} = \frac{\boldsymbol{m}\cdot\boldsymbol{C}\cdot\boldsymbol{n}}{|\boldsymbol{F}\cdot\boldsymbol{m}|\,|\boldsymbol{F}\cdot\boldsymbol{n}|}.$$

The reduction $\gamma(\boldsymbol{m},\boldsymbol{n}) = \overset{\circ}{\theta} - \theta$ in angle between $\boldsymbol{m}$ and $\boldsymbol{n}$ is found from the two equations:

$$\text{(2.11)}\qquad \cos(\overset{\circ}{\theta} - \gamma) = \frac{\boldsymbol{m}\cdot\boldsymbol{C}\cdot\boldsymbol{n}}{|\boldsymbol{F}\cdot\boldsymbol{m}|\,|\boldsymbol{F}\cdot\boldsymbol{n}|},\quad \cos\overset{\circ}{\theta} = \boldsymbol{m}\cdot\boldsymbol{n}.$$

If $\overset{\circ}{\theta} = \pi/2$, then

$$\text{(2.12)}\qquad \sin\gamma = \frac{\boldsymbol{m}\cdot\boldsymbol{C}\cdot\boldsymbol{n}}{|\boldsymbol{F}\cdot\boldsymbol{m}|\,|\boldsymbol{F}\cdot\boldsymbol{n}|},$$

and γ is then called the *orthogonal shear*. If $\boldsymbol{n} = \frac{\boldsymbol{a}_1(\boldsymbol{z})}{|\boldsymbol{a}_1(\boldsymbol{z})|}$ and $\boldsymbol{m} = \frac{\boldsymbol{a}_2(\boldsymbol{z})}{|\boldsymbol{a}_2(\boldsymbol{z})|}$, then

$$\text{(2.13)}\qquad \cos(\overset{\circ}{\theta} - \gamma) = \frac{C_{12}}{\sqrt{C_{11}}\sqrt{C_{22}}}.$$

Thus the component $\boldsymbol{a}_1\cdot\boldsymbol{C}\cdot\boldsymbol{a}_2$, together with the stretches, determines the shear between fibers originally along vectors $\boldsymbol{a}_1$ and $\boldsymbol{a}_2$.

Let $\boldsymbol{p}$ be prescribed so that $\boldsymbol{C}$ is known. We fix t. The question of determining which fibers $\boldsymbol{n}$ at a given material point $\boldsymbol{z}$ suffer the greatest and least stretches is equivalent to determining which $\boldsymbol{n}$'s extremize $\boldsymbol{n}\cdot\boldsymbol{C}\cdot\boldsymbol{n}$ subject to $\boldsymbol{n}\cdot\boldsymbol{n} = 1$. By the method of Lagrange multipliers, there is a number C such that the extremizer $\boldsymbol{n}$ satisfies

$$\text{(2.14a)}\qquad \partial[\boldsymbol{n}\cdot(\boldsymbol{C} - C\boldsymbol{I})\cdot\boldsymbol{n}]/\partial\boldsymbol{n} = \boldsymbol{0}.$$

Using the methods of Sec. XI.2, we readily show that (2.14a) is equivalent to the eigenvalue problem:

$$(\boldsymbol{C} - C\boldsymbol{I}) \cdot \boldsymbol{n} = \boldsymbol{0}, \quad \boldsymbol{n} \cdot \boldsymbol{n} = 1. \tag{2.14b}$$

This problem has solutions, necessarily nontrivial, if and only if

$$\det(\boldsymbol{C} - C\boldsymbol{I}) = 0. \tag{2.15}$$

The solutions C_1, C_2, C_3 of (2.15) are the eigenvalues of $\boldsymbol{C}$ and the corresponding solutions $\boldsymbol{n}_1$, $\boldsymbol{n}_2$, $\boldsymbol{n}_3$ are the corresponding eigenvectors. Since $\boldsymbol{C}$ is symmetric and positive-definite, the eigenvalues are real and positive and the eigenvectors can be chosen to be orthonormal. The eigenvectors of $\boldsymbol{C}(\boldsymbol{z}, t)$ determine the *principal axes of strain* at $(\boldsymbol{z}, t)$, which are the material directions that suffer extreme stretch under the motion $\boldsymbol{p}$. The eigenvalues are the squares of the *principal stretches.*

Let us set

$$\boldsymbol{a} \equiv \frac{\boldsymbol{m}}{|\boldsymbol{F} \cdot \boldsymbol{m}|}, \quad \boldsymbol{b} \equiv \frac{\boldsymbol{n}}{|\boldsymbol{F} \cdot \boldsymbol{n}|}. \tag{2.16}$$

Then the problem of finding orthogonal pairs of directions $\boldsymbol{m}$ and $\boldsymbol{n}$ for which the orthogonal shear, defined by (2.12), is extremized is equivalent to extremizing $\sin\gamma = \boldsymbol{a} \cdot \boldsymbol{C} \cdot \boldsymbol{b}$ subject to the constraints

$$\boldsymbol{a} \cdot \boldsymbol{C} \cdot \boldsymbol{a} = 1 = \boldsymbol{b} \cdot \boldsymbol{C} \cdot \boldsymbol{b}, \quad \boldsymbol{a} \cdot \boldsymbol{b} = 0. \tag{2.17}$$

2.18. Exercise. Use the method of Lagrange multipliers to prove: There are three pairs of orthogonal directions that extremize $\sin\gamma$. Each such pair lies in a plane spanned by a pair of eigenvectors of $\boldsymbol{C}$, and each vector of the pair makes an angle of $\pm\frac{\pi}{4}$ mod $\frac{\pi}{2}$ with the eigenvectors in the plane. The extreme values of $\sin\gamma$ are $\frac{C_k - C_l}{C_k + C_l}$.

The planes spanned by $\{\boldsymbol{n}_1 + \boldsymbol{n}_2, \boldsymbol{n}_3\}$, $\{\boldsymbol{n}_1 - \boldsymbol{n}_2, \boldsymbol{n}_3\}$, etc., are the *principal planes of shear.*

Any invertible tensor-valued function of $\boldsymbol{C}$ can be used as a measure of strain in place of $\boldsymbol{C}$. Three measures widely used are $\boldsymbol{C}$, which is algebraically convenient, $\boldsymbol{C}^{1/2}$, defined by (XI.1.35), which is geometrically convenient (see Sec. 4), and the *material strain tensor*

$$\boldsymbol{E} \equiv \tfrac{1}{2}(\boldsymbol{C} - \boldsymbol{I}), \tag{2.19}$$

which has the virtue that it vanishes in the reference configuration and in rigid motions.

The *displacement of* $\boldsymbol{z}$ at time t is

$$\boldsymbol{u}(\boldsymbol{z}, t) \equiv \boldsymbol{p}(\boldsymbol{z}, t) - \boldsymbol{z}. \tag{2.20}$$

Its has the virtue that it vanishes in the reference configuration and the disadvantage that its use complicates virtually every expression in which it appears. For example, in terms of $\boldsymbol{u}$ the simple (1.1) becomes $\det(\boldsymbol{I} + \boldsymbol{u}_{\boldsymbol{z}}(\boldsymbol{z}, t)) > 0$.

2.21. Exercise. Find $\boldsymbol{E}$ in terms of $\boldsymbol{u}$.

There are many strain tensors, e.g., $\frac{1}{\alpha}(\boldsymbol{C}^{\alpha/2} - \boldsymbol{I})$, $\alpha > 0$, that agree with $\boldsymbol{E}$ when $\boldsymbol{E}$ is small, but disagree markedly when $\boldsymbol{E}$ is large. Thus the use of constitutive functions delivering the stress as a linear function of some strain tensor typically gives the same kind of results for small strain as any other such constitutive function, but the resulting behavior for large strains can be significantly different.

3. Compatibility

If $\boldsymbol{p}$ is given, then we can compute $\boldsymbol{C}$ from (1.4) and (2.3):

$$\boldsymbol{C} = \boldsymbol{p}_{\boldsymbol{z}}^{*} \cdot \boldsymbol{p}_{\boldsymbol{z}}. \tag{3.1}$$

On the other hand, if $\boldsymbol{C}$ is given, then (3.1) represents an overdetermined system of six equations for the three components of $\boldsymbol{p}$. Here we give *compatibility* conditions on $\boldsymbol{C}$ ensuring that $\boldsymbol{p}$ can be found. Such conditions are useful in linear elasticity because an important class of static problems can be posed entirely in terms of the strain $\boldsymbol{E}$. For a solution $\boldsymbol{E}$ of such a problem to be meaningful, it must correspond to a position field $\boldsymbol{p}$. No such formulations have yet succeeded in nonlinear elasticity (although the compatibility equations have been exploited with great effect in the work of Ericksen (1954)). Nevertheless, the question remains interesting because the set of positive-definite $\boldsymbol{C}$'s is a convex set of tensor-valued functions, whereas $\boldsymbol{p}$'s satisfying (1.1) do not form a convex set of vector-valued functions (see Ex. 1.5).

We begin by giving a careful treatment of the much simpler problem for the existence of a potential, and then outline the related ideas needed to determine those $\boldsymbol{C}$'s for which (3.1) has a solution $\boldsymbol{p}$.

Let $\mathcal{B}$ be a domain in $\mathbb{E}^3$.

3.2. Proposition. *Let $\boldsymbol{f}$ be given on $\mathcal{B}$. If*

$$\nabla \omega \equiv \omega_{\boldsymbol{z}} = \boldsymbol{f} \tag{3.3}$$

has a solution $\omega \in C^2(\mathcal{B})$, then $\boldsymbol{f} \in C^1(\mathcal{B})$ and

$$\nabla \times \boldsymbol{f} = \boldsymbol{0} \quad \text{or, equivalently, } \boldsymbol{f}_{\boldsymbol{z}} \text{ is symmetric.} \tag{3.4}$$

Proof. Take the curl of (3.3). □

3.5. Theorem. *If $\boldsymbol{f} \in C^1(\mathcal{B})$ is given and satisfies (3.4), and if $\mathcal{B}$ is simply-connected, then (3.3) has a solution $\omega \in C^2(\mathcal{B})$ unique to within an additive constant.*

Proof. Let $\boldsymbol{a}$ be a fixed point in $\mathcal{B}$ and let $\boldsymbol{b}$ be an arbitrary point in $\mathcal{B}$. Let $\mathcal{C}_1$ and $\mathcal{C}_2$ be piecewise smooth curves in $\mathcal{B}$ joining $\boldsymbol{a}$ to $\boldsymbol{b}$. We set

$$\omega_\alpha(\boldsymbol{b}) = \chi + \int_{\mathcal{C}_\alpha} \boldsymbol{f}(\boldsymbol{z}) \cdot d\boldsymbol{z} \tag{3.6a}$$

where χ is a given number and where $d\boldsymbol{z} = \boldsymbol{t}(\boldsymbol{z})\, ds(\boldsymbol{z})$ in the notation of (XI.2.25). Since $\mathcal{B}$ is simply-connected, there is a smooth surface $\mathcal{S}$ spanning the closed circuit $\mathcal{C}_1 - \mathcal{C}_2$ that lies entirely in $\mathcal{B}$. It follows from (XI.2.25) and (3.4) that

$$\omega_1(\boldsymbol{b}) - \omega_2(\boldsymbol{b}) = \int_{\mathcal{C}_1 - \mathcal{C}_2} \boldsymbol{f}(\boldsymbol{z}) \cdot d\boldsymbol{z} = \int_{\mathcal{S}} \boldsymbol{\nu} \cdot (\nabla \times \boldsymbol{f})\, da = 0. \tag{3.6b}$$

Thus the line integral in (3.6a) is independent of path. If $\mathcal{C}(\boldsymbol{y})$ is any continuously differentiable path in $\mathcal{B}$ from $\boldsymbol{a}$ to $\boldsymbol{y}$, then it follows that

$$\omega(\boldsymbol{y}) = \omega(\boldsymbol{a}) + \int_{\mathcal{C}(\boldsymbol{y})} \boldsymbol{f}(\boldsymbol{z}) \cdot d\boldsymbol{z} \tag{3.6c}$$

defines a continuously differentiable function on $\mathcal{B}$ for given $\omega(\boldsymbol{a})$.

We now show that this ω satisfies (3.3). Let $\boldsymbol{z}$ be an arbitrary point of $\mathcal{B}$, which we can take to lie on an intermediate point of $\mathcal{C}(\boldsymbol{b})$. We parametrize $\mathcal{C}(\boldsymbol{b})$ by $[a, b] \ni s \mapsto \breve{\boldsymbol{z}}(s)$ with $\breve{\boldsymbol{z}}(a) = \boldsymbol{a}$ and $\breve{\boldsymbol{z}}(b) = y$. Thus (3.6c) implies that

$$\omega(\breve{\boldsymbol{z}}(s)) = \omega(\boldsymbol{a}) + \int_a^s \boldsymbol{f}(\breve{\boldsymbol{z}}(\xi)) \cdot \breve{\boldsymbol{z}}'(\xi)\, d\xi, \tag{3.7}$$

whence

$$\omega_{\boldsymbol{z}}(\breve{\boldsymbol{z}}(s)) \cdot \breve{\boldsymbol{z}}'(s) = \boldsymbol{f}(\breve{\boldsymbol{z}}(s)) \cdot \breve{\boldsymbol{z}}'(s). \tag{3.8}$$

The arbitrariness of $\boldsymbol{z}$ and of $\mathcal{C}(\boldsymbol{b})$ means that we can take $\breve{\boldsymbol{z}}(s) = \boldsymbol{z}$ and take $\breve{\boldsymbol{z}}'(s)$ arbitrary. We can accordingly cancel $\breve{\boldsymbol{z}}'(s)$ from (3.8) to get (3.3). □

We immediately obtain

3.9. Corollary. *Let $\mathcal{B}$ be multiply-connected, but otherwise let all the hypotheses of Theorem 3.5 hold. Let each hole in $\mathcal{B}$ be encircled by a simple closed (irreducible) piecewise smooth curve C_k lying entirely in $\mathcal{B}$, $k = 1, \dots, K$, where K is the number of holes. If*

$$\int_{C_k} \boldsymbol{f} \cdot d\boldsymbol{z} = 0, \quad k = 1, \dots, K, \tag{3.10}$$

then the conclusion of Theorem 3.5 still holds.

These results are readily extended for $\boldsymbol{z}$'s in higher-dimensional spaces by replacing our version of the Kelvin (Stokes) Theorem with its natural generalization in terms of differential forms. Generalizations of these classical results for much weaker regularity assumptions are given by Ladyzhenskaia (1969), L. Schwartz (1966, Thm. II.VI), and Temam (1977). These theorems suffice to treat the linearization of (3.1) about the reference configuration (see Sokolnikoff (1956)) but are insufficient for (3.1) itself. For these we outline a classical approach, which represents a modification of that of Eisenhart (1926).

We first consider a nonlinear multidimensional version of (3.3):

$$w^\nu_{\boldsymbol{z}} = \boldsymbol{f}^\nu(\mathbf{w}, \boldsymbol{z}), \quad \nu = 1, \dots, N, \quad \mathbf{w} \equiv (w^1, \dots, w^N). \tag{3.11}$$

Here we adopt the summation convention that twice-repeated Greek indices are summed from 1 to N.

Just as in the proof of Proposition 3.2, we obtain

3.12. Proposition. *If $\boldsymbol{f}^\nu \in C^1(\mathbb{R}^N \times \mathcal{B})$ and if* (3.11) *has a solution $\mathbf{w} \in C^2(\mathcal{B})$, then*

$$\frac{\partial \boldsymbol{f}^\nu}{\partial w^\mu}(\mathbf{w}(\boldsymbol{z}), \boldsymbol{z}) \frac{\partial w^\mu}{\partial \boldsymbol{z}}(\boldsymbol{z}) + \frac{\partial \boldsymbol{f}^\nu}{\partial \boldsymbol{z}}(\mathbf{w}(\boldsymbol{z}), \boldsymbol{z}) \quad \textit{is symmetric} \tag{3.13}$$

or, equivalently,

$$\frac{\partial \boldsymbol{f}^\nu}{\partial w^\mu}(\mathbf{w}(\boldsymbol{z}), \boldsymbol{z}) \boldsymbol{f}^\mu(\mathbf{w}(\boldsymbol{z}), \boldsymbol{z}) + \frac{\partial \boldsymbol{f}^\nu}{\partial \boldsymbol{z}}(\mathbf{w}(\boldsymbol{z}), \boldsymbol{z}) \quad \textit{is symmetric.} \tag{3.14}$$

The analog of Theorem 3.5 is

3.15. Theorem. *If $f^\nu \in C^1(\mathbb{R}^N \times \mathcal{B})$, if*

$$\frac{\partial \boldsymbol{f}^\nu}{\partial w^\mu} \boldsymbol{f}^\mu(\mathbf{w}(\boldsymbol{z}), \boldsymbol{z}) + \frac{\partial \boldsymbol{f}^\nu}{\partial \boldsymbol{z}} \quad \text{is symmetric,} \tag{3.16}$$

and if $\mathcal{B}$ is simply-connected, then the problem consisting of (3.11) *and the condition that*

$$\mathbf{w}(\boldsymbol{a}) \quad \text{is prescribed} \tag{3.17}$$

for some $\boldsymbol{a} \in \mathcal{B}$ has a unique solution on a neighborhood of $\boldsymbol{a}$ that is twice continuously differentiable there. If the $\boldsymbol{f}^\nu(\cdot, \boldsymbol{z})$ are linear, then this solution exists on all of $\mathcal{B}$.

Sketch of Proof. Let $\mathcal{C}_0$ and $\mathcal{C}_1$ be simple, smooth curves in $\mathcal{B}$ joining $\boldsymbol{a}$ to $\boldsymbol{b}$ given by $[a, b] \ni s \mapsto \check{\boldsymbol{z}}(s, 0)$, $\check{\boldsymbol{z}}(s, 1)$ with $\check{\boldsymbol{z}}(a, 0) = \boldsymbol{a} = \check{\boldsymbol{z}}(a, 1)$ and $\check{\boldsymbol{z}}(b, 0) = \boldsymbol{b} = \check{\boldsymbol{z}}(b, 1)$. Since $\mathcal{B}$ is simply-connected, there is a continuously differentiable surface $\mathcal{S}$ in $\mathcal{B}$ that spans the circuit $\mathcal{C}_0 - \mathcal{C}_1$, which can be parametrized thus: $[a, b] \times [0, 1] \ni (s, t) \mapsto \check{\boldsymbol{z}}(s, t)$.

Standard theory of ordinary differential equations says that the initial-value problem consisting of

$$\frac{dw^\nu}{ds} = \boldsymbol{f}^\nu(\mathbf{w}, \check{\boldsymbol{z}}(s, t)) \cdot \check{\boldsymbol{z}}_s(s, t), \quad \nu = 1, \ldots, N, \tag{3.18}$$

and (3.17) has a unique solution $s \mapsto \check{\mathbf{w}}(s, t)$ on some neighborhood of $s = a$, which is a continuously differentiable function of (s, t). Moreover, if the $\boldsymbol{f}^\nu(\cdot, \boldsymbol{z})$ are linear, then this solution exists on all of $[a, b] \times [0, 1]$. If $\boldsymbol{b}$ is close enough to $\boldsymbol{a}$, then even for nonlinear equations, we can choose $\check{\boldsymbol{z}}$ to ensure that this solution exists on all of $[a, b] \times [0, 1]$.

A computation based on (3.16) and (XI.2.25), like that of the proof of Theorem 3.5, then shows that $\check{\mathbf{w}}(b, 0) = \boldsymbol{b} = \check{\mathbf{w}}(b, 1)$. By an argument like that leading to (3.6c), we find that the solution of (3.18) and (3.17) generates a function $\mathbf{w}$ defined on a neighborhood of $\boldsymbol{a}$. By an argument like that involving (3.8), we show that this function is in fact a solution of (3.11). □

For multiply-connected domains $\mathcal{B}$, conditions like (3.10) must be imposed. These are restrictions not merely on the $\boldsymbol{f}^\nu$, but also on the boundary values of $\mathbf{w}$ on these internal boundaries.

We now turn to (3.1), which is not even quasilinear. We obtain a quasilinear system by differentiating (3.1) with respect to $\boldsymbol{z}$. The integration of this differentiated equation produces (3.1) modified by the presence of an arbitrary constant symmetric tensor $\boldsymbol{A}$ of integration. Thus (3.1) is equivalent to the differentiated equation and the further condition that

$$\boldsymbol{A} = \boldsymbol{O}. \tag{3.19}$$

A considerable amount of algebraic manipulation converts the differentiated equation to a *linear* system of form

$$\boldsymbol{p}_z = \boldsymbol{F}, \quad \boldsymbol{F}_z = \boldsymbol{F} \cdot \boldsymbol{C}^{-1} \cdot \mathbf{H} \tag{3.20a,b}$$

where $\mathbf{H}$ is a third-order tensor (related to the Christoffel symbols) that depends on $\boldsymbol{C}_z$ (which is given). We identify (3.20) with (3.11). Condition (3.16) for (3.20a) is automatically satisfied by virtue of (3.20b). Condition (3.16) for (3.20b) is equivalent to the vanishing of the (fourth-order) Riemann-Christoffel curvature tensor $\mathcal{R}$ for $\boldsymbol{C}$:

$$\mathcal{R} = \mathcal{O}. \tag{3.21a}$$

These equations are called the *compatibility equations.* In the notation of Section XI.3, the components of $\mathcal{R}$ are

$$R_{hijk} \equiv \tfrac{1}{2}(C_{hk,ij} + C_{ij,hk} - C_{hj,ik} - C_{ik,hj}) + (C^{-1})^{lm}(\Gamma_{ijm}\Gamma_{hkl} - \Gamma_{ikm}\Gamma_{hjl}) \tag{3.21b}$$

where

$$\Gamma_{ijk} = \tfrac{1}{2}(C_{ik,j} + C_{jk,i} - C_{ij,k}). \tag{3.21c}$$

Of the 81 components R_{hijk} of $\mathcal{R}$, only six are independent. These can be taken to be $R_{2323}, R_{3131}, R_{1212}, R_{3112}, R_{1223}, R_{2331}$.

The system (3.20) has twelve arbitrary constants of integration. It can be shown that these suffice to satisfy (3.19) and to fix the rigid motion of $\boldsymbol{p}$. Hence,

3.22. Compatibility Theorem. *Let $\boldsymbol{p} \in C^3(\mathcal{B})$ be given. Then $\boldsymbol{C} \in C^2(\mathcal{B})$ and the Riemann-Christoffel curvature tensor for $\boldsymbol{C}$ vanishes. If $\boldsymbol{C}$ is a given function in $C^2(\mathcal{B})$, if $\mathcal{B}$ is simply-connected, and if the Riemann-Christoffel curvature tensor for $\boldsymbol{C}$ vanishes, then* (3.1) *has a solution $\boldsymbol{p} \in C^3(\mathcal{B})$, unique to within a rigid motion.*

If $\mathcal{B}$ is not simply-connected, then conditions like (3.10) must be adduced to ensure that the second statement of the Compatibility Theorem holds. As an example of what happens in the absence of conditions like (3.10), consider the dislocation (see Ex. XIII.4.14) given in cylindrical coordinates (s, ϕ, z) by $\boldsymbol{z} \mapsto \boldsymbol{z} + \phi \boldsymbol{k}$. This deformation is discontinuous on a circular cylindrical tube with axis along $\boldsymbol{k}$. By the methods of Sec. 5 below, we find that $\boldsymbol{F} = \boldsymbol{I} + s^{-1}\boldsymbol{k}(-\sin\phi \boldsymbol{i} + \cos\phi \boldsymbol{j})$. Thus $\boldsymbol{F}$ and consequently $\boldsymbol{C}$ are continuous.

The linearization of (3.1) about the reference configuration possesses a weak formulation (see T. W. Ting (1974)). It is not known whether (3.1) itself admits a weak form appropriate for the kinds of function spaces used in the existence theory of nonlinear elasticity.

4. Rotation

The main tool in analyzing deformation is

4.1. Cauchy's Polar Decomposition Theorem. *Any nonsingular tensor $\boldsymbol{F}$ can be written uniquely in the forms*

$$\boldsymbol{F} = \boldsymbol{R} \cdot \boldsymbol{U} = \boldsymbol{V} \cdot \boldsymbol{R} \tag{4.2a,b}$$

where $\boldsymbol{R}$ is orthogonal and $\boldsymbol{U}$ and $\boldsymbol{V}$ are symmetric and positive-definite.

Proof. Set

$$\boldsymbol{U} \equiv (\boldsymbol{F}^* \cdot \boldsymbol{F})^{1/2}, \quad \boldsymbol{R} \equiv \boldsymbol{F} \cdot \boldsymbol{U}^{-1}. \tag{4.3a,b}$$

Thus $\boldsymbol{U}$ is positive-definite. (If (1.4) holds, then we recognize that $\boldsymbol{U}$ is the positive-definite square root of $\boldsymbol{C}$.) To justify formula (4.2a), we need only show that $\boldsymbol{R}$ is orthogonal:

$$\boldsymbol{R}^* \cdot \boldsymbol{R} = [\boldsymbol{U}^{-1} \cdot \boldsymbol{F}^*] \cdot [\boldsymbol{F} \cdot \boldsymbol{U}^{-1}] = \boldsymbol{U}^{-1} \cdot \boldsymbol{U}^2 \cdot \boldsymbol{U}^{-1} = \boldsymbol{I}. \tag{4.4}$$

The proof of (4.2b) is analogous. We find that

$$\boldsymbol{V} = (\boldsymbol{F} \cdot \boldsymbol{F}^*)^{1/2}. \tag{4.5}$$

To show uniqueness of (4.2a), suppose that $\boldsymbol{F} = \tilde{\boldsymbol{R}} \cdot \tilde{\boldsymbol{U}}$ also. Then $\boldsymbol{F}^* \cdot \boldsymbol{F} = \tilde{\boldsymbol{U}} \cdot \tilde{\boldsymbol{U}}$. But $\boldsymbol{F}^* \cdot \boldsymbol{F}$ has a unique positive-definite square root so that $\boldsymbol{U} = \tilde{\boldsymbol{U}}$, whence $\tilde{\boldsymbol{R}} = \boldsymbol{R}$. □

4.6. Corollary. *If the hypotheses of Theorem* 4.1 *hold and if furthermore* $\det \boldsymbol{F} > 0$, *then* $\det \boldsymbol{R} = 1$, *so that $\boldsymbol{R}$ is a proper-orthogonal tensor, called the* rotation *tensor corresponding to $\boldsymbol{F}$.*

Now we make the intended identification (1.4), and show that $\boldsymbol{R}$ rotates the principal axes of strain. Indeed, these principal axes, which are the eigenvectors $\boldsymbol{n}_k$ of $\boldsymbol{C}$, are also the eigenvectors of $\boldsymbol{U} = \boldsymbol{C}^{1/2}$. Thus the

$$\boldsymbol{F} \cdot \boldsymbol{n}_k = \boldsymbol{R} \cdot \boldsymbol{U} \cdot \boldsymbol{n}_k = \sqrt{C_k} \boldsymbol{R} \cdot \boldsymbol{n}_k \quad \text{(no sum)} \tag{4.7a}$$

are orthogonal. This result gives the following representation for

$$\boldsymbol{R} = \boldsymbol{R} \cdot \boldsymbol{n}_k \boldsymbol{n}^k = \sum_{k=1}^{3} \frac{\boldsymbol{F} \cdot \boldsymbol{n}_k \boldsymbol{n}^k}{\sqrt{C_k}} \tag{4.7b}$$

and supports the

4.8. Cauchy Decomposition Theorem. *The deformation about material point $\boldsymbol{z}$ may be regarded as resulting from a translation of $\boldsymbol{z}$, a rigid rotation of the principal axes, and stretches along these axes. The translation, rotation, and stretch may occur in any order, their representations depending on the order.*

A deformation exhibiting these concepts is given by the planar deformation of a rectangle in which the material is compressed in the horizontal direction to half its length and is then rotated through $\pi/2$. The same final configuration is obtained by first rotating the body and then compressing it in the vertical direction. If $\boldsymbol{i}_1$ and $\boldsymbol{i}_2$ are unit vectors in the horizontal and vertical directions, then the first procedure takes $z^\alpha \boldsymbol{i}_\alpha$ to $\xi^\alpha \boldsymbol{i}_\alpha$ and then takes $\xi^\alpha \boldsymbol{i}_\alpha$ to $y^\alpha \boldsymbol{i}_\alpha$ where

$$\begin{pmatrix} y^1 \\ y^2 \end{pmatrix} = \begin{pmatrix} 0 & -1 \\ 1 & 0 \end{pmatrix} \begin{pmatrix} \xi^1 \\ \xi^2 \end{pmatrix}, \quad \begin{pmatrix} \xi^1 \\ \xi^2 \end{pmatrix} = \begin{pmatrix} \frac{1}{2} & 0 \\ 0 & 1 \end{pmatrix} \begin{pmatrix} z^1 \\ z^2 \end{pmatrix}. \tag{4.9a}$$

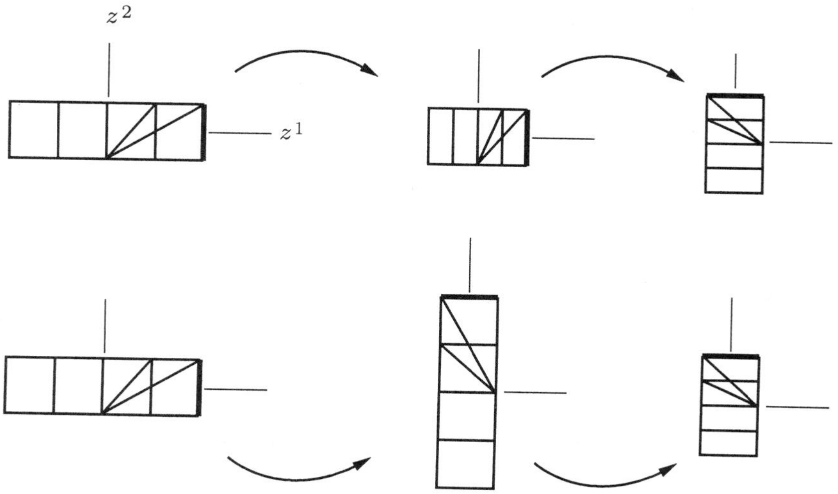

Fig. 4.10. The deformations (4.9). Notice what happens to each straight line.

The second procedure takes $z^\alpha \boldsymbol{i}_\alpha$ to $\eta^\alpha \boldsymbol{i}_\alpha$ and then takes $\eta^\alpha \boldsymbol{i}_\alpha$ to $y^\alpha \boldsymbol{i}_\alpha$ where

$$(4.9b)\qquad \begin{pmatrix} y^1 \\ y^2 \end{pmatrix} = \begin{pmatrix} 1 & 0 \\ 0 & \frac{1}{2} \end{pmatrix} \begin{pmatrix} \eta^1 \\ \eta^2 \end{pmatrix}, \quad \begin{pmatrix} \eta^1 \\ \eta^2 \end{pmatrix} = \begin{pmatrix} 0 & -1 \\ 1 & 0 \end{pmatrix} \begin{pmatrix} z^1 \\ z^2 \end{pmatrix}.$$

These deformations are illustrated in Fig. 4.10.

For a discussion of mean rotation, the specification of which can be used to fix the rigid rotation of a deformable body, see Truesdell & Toupin (1960, Sec. B.I.c).

5. Examples

Let $\{\boldsymbol{i}_k = \boldsymbol{i}^k\}$ be a fixed right-handed orthonormal basis for $\mathbb{E}^3$. We set $\boldsymbol{z} = z^k \boldsymbol{i}_k$ so that the z^k are Cartesian coordinates of $\boldsymbol{z}$. Let K be a positive number. *Simple shear* is a deformation $\boldsymbol{p}$ of the form

$$(5.1)\qquad \boldsymbol{p}(\boldsymbol{z}) = (z^1 + Kz^2)\boldsymbol{i}_1 + z^2\boldsymbol{i}_2 + z^3\boldsymbol{i}_3,$$

which is illustrated in Fig. 5.2. Thus

$$(5.3)\qquad \frac{\partial \boldsymbol{p}}{\partial \boldsymbol{z}} = \boldsymbol{F} = \frac{\partial \boldsymbol{p}}{\partial z^k}\frac{\partial z^k}{\partial \boldsymbol{z}} = \frac{\partial \boldsymbol{p}}{\partial z^k}\boldsymbol{i}^k$$

$$= \boldsymbol{i}_1\boldsymbol{i}^1 + (K\boldsymbol{i}_1 + \boldsymbol{i}_2)\boldsymbol{i}^2 + \boldsymbol{i}_3\boldsymbol{i}^3,$$

$$(5.4)\qquad \left(\frac{\partial \boldsymbol{p}}{\partial \boldsymbol{z}}\right)^* = \boldsymbol{F}^* = \boldsymbol{i}^1\boldsymbol{i}_1 + \boldsymbol{i}^2(K\boldsymbol{i}_1 + \boldsymbol{i}_2) + \boldsymbol{i}^3\boldsymbol{i}_3,$$

(5.5)

$$\boldsymbol{C} = \left(\frac{\partial \boldsymbol{p}}{\partial \boldsymbol{z}}\right)^* \cdot \left(\frac{\partial \boldsymbol{p}}{\partial \boldsymbol{z}}\right) = \boldsymbol{F}^* \cdot \boldsymbol{F}$$

$$= \boldsymbol{i}^1\boldsymbol{i}^1 + K(\boldsymbol{i}^1\boldsymbol{i}^2 + \boldsymbol{i}^2\boldsymbol{i}^1) + (1+K^2)\boldsymbol{i}^2\boldsymbol{i}^2 + \boldsymbol{i}^3\boldsymbol{i}^3.$$

The matrix of $\boldsymbol{C}$ with respect to the basis $\{\boldsymbol{i}^k\boldsymbol{i}^l\}$ is

$$(5.6)\qquad (C_{kl}) = \begin{pmatrix} 1 & K & 0 \\ K & 1+K^2 & 0 \\ 0 & 0 & 1 \end{pmatrix}.$$

The invariants of $\boldsymbol{C}$, which can be read off from (5.6), are

$$(5.7a,b,c)\qquad \mathrm{I}_C = 3 + K^2, \quad \mathrm{II}_C = 3 + K^2, \quad \mathrm{III}_C = 1.$$

Since $\mathrm{III}_C = \det \boldsymbol{C} = \det(\boldsymbol{F}^* \cdot \boldsymbol{F}) = (\det \boldsymbol{F})^2$, it follows from (5.7c) that the deformation (5.1) is locally volume-preserving, which is geometrically obvious.

5.8. Exercise. Find the principal axes, the principal stretches, the rotation tensor, and the images of the principal axes for the deformation (5.1). (See Truesdell & Toupin (1960, Sec. 45).)

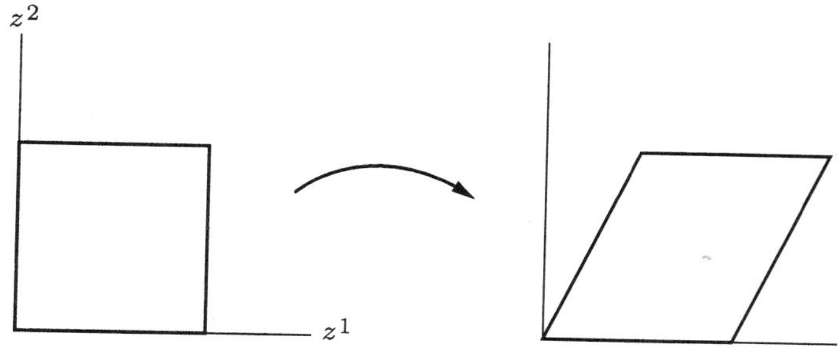

Fig. 5.2. Simple shear.

This example of simple shear can easily be treated in Cartesian coordinates. In the following example, the presence of curvilinear coordinates puts a premium on an accurate formalism to produce $\boldsymbol{F}$ and $\boldsymbol{C}$. For this purpose, we use the tensorial methods developed in Chap. XI.

Let f, g, h be given real-valued functions defined on an interval of $[0, \infty)$. Let $\alpha, \beta, \gamma, \delta$ be given real numbers. We study deformations that take the material point with cylindrical coordinates (s, ϕ, z) to the position with cylindrical coordinates $(f(s), g(s) + \alpha\phi + \beta z, h(s) + \gamma\phi + \delta z)$. We postpone describing the geometry of this class of deformations so we can concentrate on the practical question of computing $\boldsymbol{F}$ and $\boldsymbol{C}$.

We set $\mathbf{x} \equiv (s, \phi, z)$. The function $\tilde{\boldsymbol{z}}$ that assigns the reference position to $\mathbf{x}$ is defined by

$$\tilde{\boldsymbol{z}}(\mathbf{x}) = s\boldsymbol{j}_1(\phi) + z\boldsymbol{j}_3 \tag{5.9}$$

where

$$\begin{aligned} \boldsymbol{j}_1(\phi) &\equiv \boldsymbol{j}^1(\phi) \equiv \cos\phi\, \boldsymbol{i}_1 + \sin\phi\, \boldsymbol{i}_2, \\ \boldsymbol{j}_2(\phi) &\equiv \boldsymbol{j}^2(\phi) \equiv -\sin\phi\, \boldsymbol{i}_1 + \cos\phi\, \boldsymbol{i}_2, \\ \boldsymbol{j}_3 &\equiv \boldsymbol{j}^3 \equiv \boldsymbol{i}_3. \end{aligned} \tag{5.10}$$

We now set

$$\tilde{\boldsymbol{p}}(\mathbf{x}) \equiv \boldsymbol{p}(\tilde{\boldsymbol{z}}(\mathbf{x})), \quad \omega(\mathbf{x}) \equiv g(s) + \alpha\phi + \beta z. \tag{5.11a,b}$$

Then our deformation is defined by

$$\tilde{\boldsymbol{p}}(\mathbf{x}) = f(s)\boldsymbol{k}_1(\mathbf{x}) + [h(s) + \gamma\phi + \delta z]\boldsymbol{k}_3, \tag{5.12}$$

where

$$\begin{aligned} \boldsymbol{k}_1(\mathbf{x}) &\equiv \boldsymbol{k}^1(\mathbf{x}) \equiv \cos\omega(\mathbf{x})\boldsymbol{i}_1 + \sin\omega(\mathbf{x})\boldsymbol{i}_2, \\ \boldsymbol{k}_2(\mathbf{x}) &\equiv \boldsymbol{k}^2(\mathbf{x}) \equiv -\sin\omega(\mathbf{x})\boldsymbol{i}_1 + \cos\omega(\mathbf{x})\boldsymbol{i}_2, \\ \boldsymbol{k}_3 &\equiv \boldsymbol{k}^3 \equiv \boldsymbol{i}_3. \end{aligned} \tag{5.13}$$

Since the bases introduced in (5.10) and (5.13) are orthonormal, they are their own duals. Nevertheless, we position the indices to conform to standard notations for general bases. (Our notation therefore is compatible with that of Sec. XI.3.) In particular, it is conventional in differential geometry to put the indices on coordinates as superscripts, as in x^k. In expressions such as $\frac{\partial \psi}{\partial x^k}$, the index is reckoned to be a subscript. The indices on bases are designed to conform to the diagonal summation convention. Thus we define the components of $\boldsymbol{F}$ in (5.17) below to be the coefficients of $\boldsymbol{k}^i \boldsymbol{j}_l$. The subscript l corresponds to a derivative. When we take components of stress in Chap. XIII we assign their indices so that the indicial form of the inner product $\boldsymbol{T} : \boldsymbol{F}_t$, which is the stress power, conforms to the diagonal summation convention. These policies, which promote efficiency in bookkeeping and ease in translating our results to those found in the literature, may be ignored without causing any damage.

Let $\tilde{\mathbf{x}}$ be the inverse of $\tilde{\boldsymbol{z}}$. By the Chain Rule,

$$(5.14a,b) \qquad \frac{\partial \boldsymbol{p}}{\partial \boldsymbol{z}}(\tilde{\boldsymbol{z}}(\mathbf{x})) = \frac{\partial \tilde{\boldsymbol{p}}}{\partial \mathbf{x}}(\mathbf{x}) \cdot \frac{\partial \tilde{\mathbf{x}}}{\partial \boldsymbol{z}}(\tilde{\boldsymbol{z}}(\mathbf{x})) = \frac{\partial \tilde{\boldsymbol{p}}}{\partial x^k}(\mathbf{x}) \frac{\partial \tilde{x}^k}{\partial \boldsymbol{z}}(\tilde{\boldsymbol{z}}(\mathbf{x})).$$

To find $\partial \tilde{\mathbf{x}} / \partial \boldsymbol{z}$ we also use the Chain Rule:

$$(5.15a,b) \qquad \boldsymbol{I} = \frac{\partial \tilde{\boldsymbol{z}}}{\partial \mathbf{x}}(\mathbf{x}) \cdot \frac{\partial \tilde{\mathbf{x}}}{\partial \boldsymbol{z}}(\tilde{\boldsymbol{z}}(\mathbf{x})) = \frac{\partial \tilde{\boldsymbol{z}}}{\partial x^k}(\mathbf{x}) \frac{\partial \tilde{x}^k}{\partial \boldsymbol{z}}(\tilde{\boldsymbol{z}}(\mathbf{x})).$$

We readily compute $(\partial \tilde{\boldsymbol{z}} / \partial x^k)(\mathbf{x})$ from (5.9), substitute it into (5.15b), and then take the dot product of the resulting expression with the $\boldsymbol{j}^l$ to obtain

$$(5.16) \qquad \begin{aligned} \frac{\partial \tilde{s}}{\partial \boldsymbol{z}}(\tilde{\boldsymbol{z}}(\mathbf{x})) &= \frac{\partial \tilde{x}^1}{\partial \boldsymbol{z}}(\tilde{\boldsymbol{z}}(\mathbf{x})) = \boldsymbol{j}^1(\phi), \\ \frac{\partial \tilde{\phi}}{\partial \boldsymbol{z}}(\tilde{\boldsymbol{z}}(\mathbf{x})) &= \frac{\partial \tilde{x}^2}{\partial \boldsymbol{z}}(\tilde{\boldsymbol{z}}(\mathbf{x})) = \frac{1}{s}\boldsymbol{j}^2(\phi), \\ \frac{\partial \tilde{z}}{\partial \boldsymbol{z}}(\tilde{\boldsymbol{z}}(\mathbf{x})) &= \frac{\partial \tilde{x}^3}{\partial \boldsymbol{z}}(\tilde{\boldsymbol{z}}(\mathbf{x})) = \boldsymbol{j}^3. \end{aligned}$$

We readily compute $\partial \tilde{\boldsymbol{p}} / \partial x^k$ from (5.12) and (5.13). We substitute these expressions and (5.16) into (5.14b) to obtain

$$(5.17) \quad \begin{aligned} \boldsymbol{F}(\tilde{\boldsymbol{z}}(\mathbf{x})) = {} & [f'(s)\boldsymbol{k}_1(\mathbf{x}) + f(s)g'(s)\boldsymbol{k}_2(\mathbf{x}) + h'(s)\boldsymbol{k}_3]\boldsymbol{j}^1(\phi) \\ & + \frac{1}{s}[\alpha f(s)\boldsymbol{k}_2(\mathbf{x}) + \gamma \boldsymbol{k}_3]\boldsymbol{j}^2(\phi) + [\beta f(s)\boldsymbol{k}_2(\mathbf{x}) + \delta \boldsymbol{k}_3]\boldsymbol{j}^3, \end{aligned}$$

$$(5.18) \quad \begin{aligned} \boldsymbol{C} &\equiv \boldsymbol{F}^* \cdot \boldsymbol{F} \\ &= \{\boldsymbol{j}^1[f'\boldsymbol{k}_1 + fg'\boldsymbol{k}_2 + h'\boldsymbol{k}_3] + \cdots\} \cdot \{[f'\boldsymbol{k}_1 + fg'\boldsymbol{k}_2 + h'\boldsymbol{k}_3]\boldsymbol{j}^1 + \cdots\} \\ &= [(f')^2 + (fg')^2 + (h')^2]\boldsymbol{j}^1\boldsymbol{j}^1 + \frac{1}{s}[\alpha f^2 g' + \gamma h'](\boldsymbol{j}^1\boldsymbol{j}^2 + \boldsymbol{j}^2\boldsymbol{j}^1) \\ &\quad + [\beta f^2 g' + \delta h'](\boldsymbol{j}^1\boldsymbol{j}^3 + \boldsymbol{j}^3\boldsymbol{j}^1) + \frac{1}{s^2}[(\alpha f)^2 + \gamma^2]\boldsymbol{j}^2\boldsymbol{j}^2 \\ &\quad + \frac{1}{s}[\alpha\beta f^2 + \gamma\delta](\boldsymbol{j}^2\boldsymbol{j}^3 + \boldsymbol{j}^3\boldsymbol{j}^2) + [(\beta f)^2 + \delta^2]\boldsymbol{j}^3\boldsymbol{j}^3, \end{aligned}$$

$$\text{(5.19a)} \qquad \mathrm{I}_{\boldsymbol{C}} = (f')^2 + (fg')^2 + (h')^2 + \left(\frac{\alpha f}{s}\right)^2 + \left(\frac{\gamma}{s}\right)^2 + (\beta f)^2 + \delta^2,$$

$$\text{(5.19b)} \qquad \mathrm{II}_{\boldsymbol{C}} = (f')^2 \left[\left(\frac{\alpha f}{s}\right)^2 + \left(\frac{\gamma}{s}\right)^2 + (\beta f)^2 + \delta^2\right] + [\delta f g' - \beta f h']^2$$
$$+ \left[\left(\frac{\gamma}{s}\right) fg' - \left(\frac{\alpha f}{s}\right) h'\right]^2 + \left[\left(\frac{\alpha f}{s}\right)\delta - \left(\frac{\gamma}{s}\right)\beta f\right]^2,$$

$$\text{(5.19c)} \qquad \mathrm{III}_{\boldsymbol{C}} = \left[(\alpha\delta - \beta\gamma)\left(\frac{f}{s}\right) f'\right]^2,$$

$$\text{(5.19d)} \qquad \det \boldsymbol{F} = (\alpha\delta - \beta\gamma)\left(\frac{f}{s}\right) f'.$$

Note that the reciprocal of s occurs in (5.17)–(5.19). If the domain of $\boldsymbol{p}$ includes the origin, then $\boldsymbol{F}$ could be singular there. This singularity is at the least a source of analytic difficulty and may give rise to solutions with singularities (see Sec. X.3). Thus it is essential that the disposition of the s's in all the formulas be absolutely accurate.

The deformation (5.12) is natural for bodies whose boundaries are coordinate surfaces for cylindrical coordinates. If $g = 0 = h$, $\alpha = 1 = \delta$, and $\gamma = 0$, then (5.12) represents pure torsion. In general, α accounts for circumferential stretch and δ for longitudinal stretch. β and γ are shear parameters. The three functions f, g, h determine a space curve. The deformed image of the circular cylinder $s =$ const. is a circular cylinder. The deformed images of the planes $\phi =$ const. and $z =$ const. are helicoidal surfaces generated by subjecting the space curve determined by f, g, h to screw motions, consisting in simultaneous translations along the z-axis and rotations about it of the space curve, with the rates determined by the parameters β and γ.

5.20. Exercise. Let (s, θ, ϕ) be spherical coordinates for $\mathbb{E}^3$. Consider the deformation

$$\text{(5.21)} \qquad \tilde{\boldsymbol{p}}(s, \theta, \phi) = f(s)[\sin\theta(\cos\phi\, \boldsymbol{i}_1 + \sin\phi\, \boldsymbol{i}_2) + \cos\theta\, \boldsymbol{i}_3].$$

Find $\boldsymbol{F}$ for it.

5.22. Exercise. Let $\mathcal{B}$ be the hemispherical shell consisting of those material points whose spherical polar coordinates s, θ, ϕ satisfy $1 - \epsilon \le s \le 1 + \epsilon$, $0 \le \theta \le \pi/2$, $0 \le \phi \le 2\pi$. (Here ϵ is a small positive number.) Give a geometrical description of the deformation that takes the material point in $\mathcal{B}$ with spherical polar coordinates s, θ, ϕ to the point in space with cylindrical polar coordinates r, ϕ, z with $r = 2\theta/\pi$, $\phi = \phi$, $z = s - 1$, and find the deformation tensor $\mathbf{F} = \partial \mathbf{p}/\partial \mathbf{z}$.

5.23. Exercise. Let (x, y, z) be Cartesian coordinates relative to the basis $\{\boldsymbol{i}, \boldsymbol{j}, \boldsymbol{k}\}$. Find $\boldsymbol{F}$ for the deformation

$$\text{(5.24)} \quad \boldsymbol{p}(x, y, z) = f(x)[\cos(g(x) + \alpha y + \beta z)\boldsymbol{i} + \sin(g(x) + \alpha y + \beta z)\boldsymbol{j}] + [h(x) + \gamma y + \delta z]\boldsymbol{k}.$$

6. Mass and Density

The mass of a body is commonly defined as the amount of material in the body. This physical definition leaves undefined *amount* and *material.* To employ mass in our work, we must replace this intuitive but vague physical notion with a precise mathematical notion: First of all, the *mass* of a body $\mathcal{B}$ is a nonnegative number $m(\mathcal{B})$ associated with $\mathcal{B}$. We refine our notion of *body* introduced in Sec. 1 by requiring that a body must have a mass associated with it that satisfies the following requirements: (i) The empty body $\emptyset$ has zero mass: $m(\emptyset) = 0$. (ii) If $\mathcal{A}$ and $\mathcal{B}$ are bodies with $\mathcal{A} \subset \mathcal{B}$, then $m(\mathcal{A}) \leq m(\mathcal{B})$. If the $\mathcal{B}_k$ form a countable sequence of disjoint bodies, then $m(\cup \mathcal{B}_k) = \sum m(\mathcal{B}_k)$. In mathematical parlance, m is therefore a *measure.* Thus, if we accept the axiom of choice, there are very irregular bodies for which mass cannot be defined. For our work, we take bodies to be connected open sets of material points together with any parts of their boundaries. We thereby encounter no measure-theoretic difficulties. We write

$$m(\mathcal{B}) = \int_{\mathcal{B}} dm(\boldsymbol{z}), \tag{6.1}$$

implicit in which is the presence of a compatible notion of integral. The interpretation of (6.1) is that the total mass of a body is additively determined by its distribution over parts of the body. (In the parlance of physics, mass is an *extensive* quantity.)

By having m depend only on $\mathcal{B}$, we are effectively positing the *Principle of Conservation of Mass*: Mass is an intrinsic property of a body and does not depend on the configuration it occupies. Therefore,

$$m(\mathcal{B}) = m(\boldsymbol{p}(\mathcal{B}, t)) = \int_{\boldsymbol{p}(\mathcal{B},t)} dm(\boldsymbol{p}(\boldsymbol{z}, t)) \tag{6.2}$$

for all (continuous injections) $\boldsymbol{p}(\cdot, t)$, for all t, and for all bodies $\mathcal{B}$. Condition (6.2) implies that

$$\frac{d}{dt} \int_{\boldsymbol{p}(\mathcal{B},t)} dm(\boldsymbol{p}(\boldsymbol{z}, t)) = 0. \tag{6.3}$$

Let us now suppose that if any part (subbody) $\mathcal{P}$ of $\mathcal{B}$ has zero volume, then it has zero mass. (Thus we are ostensibly dismissing from consideration the particles treated in the discipline of particle mechanics. In Sec. 7 we show that a viable particle mechanics survives this assumption about $\mathcal{B}$.) In this case, (the Radon-Nikodym Theorem implies that) there exists a (locally Lebesgue-integrable) *reference density* $\rho \geq 0$ such that

$$m(\mathcal{B}) = \int_{\mathcal{B}} \rho(\boldsymbol{z})\, dv(\boldsymbol{z}). \tag{6.4}$$

ρ is the local ratio at $\boldsymbol{z}$ of mass to volume in the reference configuration. Similarly, the Principle of Conservation of Mass implies that there exists a (locally Lebesgue-integrable) density function $\boldsymbol{p}(\mathcal{B},t) \times \mathbb{R} \ni (\boldsymbol{y},t) \mapsto \breve{\rho}(\boldsymbol{y},t) \in [0,\infty]$ such that

$$m(\mathcal{B}) = m(\boldsymbol{p}(\mathcal{B},t)) = \int_{\boldsymbol{p}(\mathcal{B},t)} \breve{\rho}(\boldsymbol{y},t)\, dv(\boldsymbol{y}). \tag{6.5}$$

Here $dv(\boldsymbol{y})$ is the differential volume at position $\boldsymbol{y}$. The density of the material point occupying position $\boldsymbol{y}$ at time t is $\breve{\rho}(\boldsymbol{y},t)$. The density of material point $\boldsymbol{z}$ at time t is $\breve{\rho}(\boldsymbol{p}(\boldsymbol{z},t),t)$. The substitution of (6.4) and (6.5) into (6.2) and the use of the formula for change of variables yields

$$\begin{aligned} \int_{\mathcal{B}} \rho(\boldsymbol{z})\, dv(\boldsymbol{z}) &= \int_{\boldsymbol{p}(\mathcal{B},t)} \breve{\rho}(\boldsymbol{y},t)\, dv(\boldsymbol{y}) \\ &= \int_{\mathcal{B}} \breve{\rho}(\boldsymbol{p}(\boldsymbol{z},t),t) \det \boldsymbol{F}(\boldsymbol{z},t)\, dv(\boldsymbol{z}). \end{aligned} \tag{6.6}$$

But $\mathcal{B}$ is arbitrary since each (measurable) part of a body is a body. Hence (6.6) yields the *local form of the Principle of Conservation of Mass*:

$$\rho(\boldsymbol{z}) = \breve{\rho}(\boldsymbol{p}(\boldsymbol{z},t),t) \det \boldsymbol{F}(\boldsymbol{z},t) \quad \text{a.e.} \tag{6.7}$$

This equation is called the *material form of the continuity equation*. It tells how the density $\breve{\rho}(\boldsymbol{p}(\boldsymbol{z},t),t)$ is determined from the datum ρ and from the tensor $\boldsymbol{F}$. The availability of this equation means that the unknown $\breve{\rho}$ can always be found after the motion is determined. In the spatial formulation of mechanics (discussed in Sec. 15), the density is a fundamental unknown.

7. Stress and the Equations of Motion

Force. Stress is the intensity of force per unit area. It, rather than force itself, is responsible for significant deformation of materials. However unsatisfactory the intuitive definition of mass is, the intuitive definitions of force are even worse. Noll (1959) (see Gurtin (1974), Noll (1966), and Truesdell (1991a)) created a very elegant postulational treatment of force exerted by one body on another. His theory adapts the measure-theoretic ideas underlying the concepts of mass and density to the far richer physical setting of force. Rather than giving a superficial outline of this work, we content ourselves with assuming that the *resultant force* $\mathbf{g}(t;\mathcal{B})$ *on body* $\mathcal{B}$ *at time* t is a vector that is additively determined from its distribution over parts of $\mathcal{B}$ (cf. (6.1)). We accordingly write

$$\boldsymbol{g}(t;\mathcal{B}) = \int_{\mathcal{B}} d\boldsymbol{g}(\boldsymbol{z},t;\mathcal{B}). \tag{7.1}$$

We further assume that the *resultant couple* $\mathbf{c}(t;\mathcal{B})$ *on* $\mathcal{B}$ *at time* t is another vector that is additively determined from its distribution over parts of $\mathcal{B}$ and has a representation like (7.1). (A *couple* is a sort of pure torque.)

The *resultant torque or resultant moment* $\boldsymbol{m}(t, \boldsymbol{a}(t); \mathcal{B})$ *on* $\mathcal{B}$ *about* $\boldsymbol{a}(t)$ *at time* t is defined to be

$$\boldsymbol{m}(t, \boldsymbol{a}(t); \mathcal{B}) \equiv \int_{\mathcal{B}} [\boldsymbol{p}(\boldsymbol{z}, t) - \boldsymbol{a}(t)] \times d\boldsymbol{g}(\boldsymbol{z}, t; \mathcal{B}) + \boldsymbol{c}(t; \mathcal{B}). \tag{7.2}$$

The *Newton-Euler Laws of Motion* consist of the *Balance of Linear Momentum*

$$\boldsymbol{g}(t; \mathcal{B}) = \frac{d}{dt} \int_{\mathcal{B}} \boldsymbol{p}_t(\boldsymbol{z}, t)\, dm(\boldsymbol{z}) \quad \forall\, \mathcal{B} \tag{7.3}$$

and the *Balance of Angular Momentum*

$$\boldsymbol{m}(t, \boldsymbol{0}; \mathcal{B}) = \frac{d}{dt} \int_{\mathcal{B}} \boldsymbol{p}(\boldsymbol{z}, t) \times \boldsymbol{p}_t(\boldsymbol{z}, t)\, dm(\boldsymbol{z}) \quad \forall\, \mathcal{B}. \tag{7.4}$$

The integrals in (7.3) and (7.4) are respectively the *Linear Momentum* and the *Angular Momentum*. It is an easy matter to generalize (7.3) and (7.4) to impulse-momentum laws, which are discussed in Sec. 9.

Many books on particle and rigid-body mechanics contain a proof that (7.3) implies (7.4) when the body $\mathcal{B}$ consists of a *finite number of unconstrained mass points exerting central forces on each other.* Their authors, making no further comment on the law (7.4), tacitly seem to assume that they have established (7.4) in all reasonable generality. There are no general proofs that (7.3) implies (7.4) when any of the special assumptions is suspended. We avoid a logical impasse by using Euler's simple device of postulating (7.4), secure in the knowledge that its validity in classical mechanics has been amply justified by experience.

The *mass center* $\boldsymbol{p}_{\mathcal{B}}(t)$ *of* $\mathcal{B}$ is defined by

$$\boldsymbol{p}_{\mathcal{B}}(t) \equiv \frac{1}{m(\mathcal{B})} \int_{\mathcal{B}} \boldsymbol{p}(\boldsymbol{z}, t)\, dm(\boldsymbol{z}). \tag{7.5}$$

Then (7.3) yields the *Law of Motion of the Mass Center*:

$$\boldsymbol{g}(t; \mathcal{B}) = m(\mathcal{B}) \frac{d^2}{dt^2} \boldsymbol{p}_{\mathcal{B}}(t). \tag{7.6}$$

This is the same law as that governing the Newtonian particle. Rather than defining a Newtonian particle as a body with zero volume but positive mass, we can merely define it to be the mass center of a body. Thus we need not go through a song and dance to explain why the earth may be regarded as having zero volume for certain problems and having large volume for other problems.

7.7. Exercise. Prove that (7.3) and (7.4) imply that

$$\boldsymbol{m}(t, \boldsymbol{a}(t); \mathcal{B}) = \frac{d}{dt} \int_{\mathcal{B}} [\boldsymbol{p}(\boldsymbol{z}, t) - \boldsymbol{a}(t)] \times \frac{\partial}{\partial t} [\boldsymbol{p}(\boldsymbol{z}, t) - \boldsymbol{a}(t)]\, dm(\boldsymbol{z}) \tag{7.8}$$

when $\boldsymbol{a}(t)$ is a constant vector or when $\boldsymbol{a}(t) = \boldsymbol{p}_{\mathcal{B}}(t)$.

We assume that $\boldsymbol{g}(t; \mathcal{B})$ has the following decomposition:

$$\boldsymbol{g}(t; \mathcal{B}) = \int_{\mathcal{B}} \boldsymbol{f}(\boldsymbol{z}, t)\, dv(\boldsymbol{z}) + \int_{\partial \mathcal{B}} \boldsymbol{t}(\boldsymbol{z}, t; \partial \mathcal{B})\, da(\boldsymbol{z}). \tag{7.9}$$

$\boldsymbol{f}(\boldsymbol{z},t)$ is the *body force (intensity) per unit reference volume at* $(\boldsymbol{z},t)$ and $\boldsymbol{t}(\boldsymbol{z},t;\partial\mathcal{B})$ is the *surface traction per unit reference area of* $\partial\mathcal{B}$ *at* $(\boldsymbol{z},t)$. As we show below, $\boldsymbol{f}$ at $(\boldsymbol{z},t)$ could depend on the position of each material point of $\mathcal{B}$ at time t. We tacitly understand that $\boldsymbol{f}$ should thus properly be expressed as a composite function of its arguments by $\boldsymbol{f}(\boldsymbol{z},t) = \hat{\boldsymbol{f}}(\boldsymbol{z},t,\boldsymbol{p}(\cdot,t))$. The same remarks apply to the boundary values for the traction, which are discussed below in Sec. 8.

For thermomechanical problems, which are the only ones treated in this book, we take $\boldsymbol{c}(t;\mathcal{B}) = \boldsymbol{0}$. Then the resultant torque on $\mathcal{B}$ about $\boldsymbol{0}$ is the moment about $\boldsymbol{0}$ of the force distribution (7.9):

(7.10)
$$\boldsymbol{m}(t,\boldsymbol{0};\mathcal{B}) = \int_{\mathcal{B}} \boldsymbol{p}(\boldsymbol{z},t)\times\boldsymbol{f}(\boldsymbol{z},t)\,dv(\boldsymbol{z}) + \int_{\partial\mathcal{B}} \boldsymbol{p}(\boldsymbol{z},t)\times\boldsymbol{t}(\boldsymbol{z},t;\partial\mathcal{B})\,da(\boldsymbol{z}).$$

Examples. The terrestrial weight of a body $\mathcal{B}$ is the magnitude of the (total) body force

$$-\int_{\mathcal{B}} \rho(\boldsymbol{z})g\boldsymbol{k}\,dv(\boldsymbol{z}) \tag{7.11}$$

where g is the acceleration of gravity and $\boldsymbol{k}$ points upward. Here $\boldsymbol{f}(\boldsymbol{z},t) = -\rho g\boldsymbol{k}$. More generally, the gravitational attraction of a spherical earth on a body $\mathcal{B}$ outside the earth is

$$-MG\int_{\mathcal{B}} \frac{\rho(\boldsymbol{z})\boldsymbol{p}(\boldsymbol{z},t)}{|\boldsymbol{p}(\boldsymbol{z},t)|^3}\,dv(\boldsymbol{z}) \tag{7.12}$$

where M is the mass of the earth, G is the universal gravitational constant, and $\boldsymbol{0}$ is at the earth's center. Thus the body force intensity

$$\boldsymbol{f}(\boldsymbol{z},t) = -MG\rho(\boldsymbol{z})\boldsymbol{p}(\boldsymbol{z},t)/|\boldsymbol{p}(\boldsymbol{z},t)|^3$$

depends on the unknown $\boldsymbol{p}(\boldsymbol{z},t)$. If $\mathcal{B}$ were part of a self-gravitating body, then the contribution of the rest of $\mathcal{B}$ to its body force intensity $\boldsymbol{f}(\boldsymbol{z},t)$ at $\boldsymbol{z}$ would depend not merely on $\boldsymbol{p}(\boldsymbol{z},t)$, but on $\boldsymbol{p}(\cdot,t)$.

About the only other kinds of body forces found in nature are 'reversed effective forces', such as centrifugal and Coriolis forces, which are due to the acceleration of the reference frame used, and forces of electromagnetic origin. These can have the character of the self-gravitating forces just described.

In the rest of our work on continuum mechanics, we assume that the mass density ρ is defined a.e. for the bodies under study, so that (6.4) holds.

Stress. Let $\mathcal{S}$ be a smooth surface forming the common part of the boundary of bodies $\mathcal{A}$ and $\mathcal{B}$ that otherwise have no intersection. Let $\boldsymbol{\nu}(\boldsymbol{z})$ denote the unit normal field at $\boldsymbol{z}$ in $\mathcal{S}$ pointing from $\mathcal{B}$ to $\mathcal{A}$. Note that $\boldsymbol{\nu}$ merely orients a material surface in the reference configuration. It has no further geometrical significance. Then the *contact force exerted by $\mathcal{A}$ on $\mathcal{B}$ at time t* is

$$\int_S \boldsymbol{t}(\boldsymbol{z}, t; \partial\mathcal{B})\, da(\boldsymbol{z}). \tag{7.13}$$

We intend that this expression be independent of the choice of $\mathcal{A}$ and $\mathcal{B}$, provided that they meet the restrictions just stated. Using a precisely delimited version of the postulational system alluded to above, Noll (1959) (see Noll (1966) and Truesdell (1991a)) proved that the balance of angular momentum implies *Cauchy's postulate*: $\boldsymbol{t}(\boldsymbol{z}, t; \partial\mathcal{B})$ depends on ∂B only through $\boldsymbol{\nu}(\boldsymbol{z})$. We assume this and henceforth replace $\boldsymbol{t}(\boldsymbol{z}, t; \partial\mathcal{B})$ with $\boldsymbol{t}(\boldsymbol{z}, t, \boldsymbol{\nu}(\boldsymbol{z}))$. The fundamental result of this section is

7.14. Cauchy's Stress Theorem. *If $\boldsymbol{t}(\cdot, t, \boldsymbol{\nu}(\boldsymbol{z}))$ and $\rho\boldsymbol{p}_{tt}(\cdot, t) - \boldsymbol{f}(\cdot, t)$ are continuous on $\mathcal{B}$, then there exists a second-order tensor field $\boldsymbol{T}(\cdot, t)$, called the* first Piola-Kirchhoff stress tensor *field, such that*

$$\boldsymbol{t}(\boldsymbol{z}, t, \boldsymbol{\nu}) = \boldsymbol{T}(\boldsymbol{z}, t) \cdot \boldsymbol{\nu}. \tag{7.15}$$

Proof. We fix t. First we assume that $\boldsymbol{y}$ is an interior point of $\mathcal{B}$. Let $\{\boldsymbol{e}_k\}$ be an orthonormal basis and let $\boldsymbol{\nu}$ be any unit vector satisfying $\boldsymbol{\nu} \cdot \boldsymbol{e}_k > 0$. If h is a sufficiently small positive number, then $\mathcal{B}$ contains the tetrahedron $\mathcal{B}(h)$ bounded by the plane through $\boldsymbol{y} + h\boldsymbol{\nu}$ with normal $\boldsymbol{\nu}$ and by the three planes through $\boldsymbol{y}$ with normals $\boldsymbol{e}_1, \boldsymbol{e}_2, \boldsymbol{e}_3$. If the area of the face $\Pi(h)$ with normal $\boldsymbol{\nu}$ is Ah^2, then the areas of the three other faces $\Pi_1(h), \Pi_2(h), \Pi_3(h)$ are $Ah^2\boldsymbol{\nu} \cdot \boldsymbol{e}_1, Ah^2\boldsymbol{\nu} \cdot \boldsymbol{e}_2, Ah^2\boldsymbol{\nu} \cdot \boldsymbol{e}_3$, and the volume of the tetrahedron is $\frac{1}{3}Ah^3$. Under assumption (7.9), the Balance of Linear Momentum for the tetrahedron reduces to

$$\begin{aligned} \int_{\Pi(h)} \boldsymbol{t}(\boldsymbol{z}, t, \boldsymbol{\nu})\, da(\boldsymbol{z}) + \sum_{k=1}^{3} \int_{\Pi_k(h)} \boldsymbol{t}(\boldsymbol{z}, t, -\boldsymbol{e}_k)\, da(\boldsymbol{z}) \\ = \int_{\mathcal{B}(h)} [\rho\boldsymbol{p}_{tt}(\boldsymbol{z}, t) - \boldsymbol{f}(\boldsymbol{z}, t)]\, dv(\boldsymbol{z}). \end{aligned} \tag{7.16}$$

We divide this equation by Ah^2 and let $h \to 0$. Since the integrand on the right-hand side of (7.16) is continuous, its integral is bounded by a constant times $\frac{1}{3}Ah^3$. By using the Mean-Value Theorem (applied to components of (7.16)) and the continuity of $\boldsymbol{t}(\cdot, t, \boldsymbol{\nu})$ we thus obtain

$$\boldsymbol{t}(\boldsymbol{y}, t, \boldsymbol{\nu}) = -\sum_{k=1}^{3} \boldsymbol{t}(\boldsymbol{y}, t, -\boldsymbol{e}_k)\boldsymbol{e}_k \cdot \boldsymbol{\nu} \tag{7.17}$$

for every $\boldsymbol{y} \in \operatorname{int} \mathcal{B}$ and for every $\boldsymbol{\nu}$ lying in the interior of the first octant.

Since (7.17) holds for all orthonormal bases $\{\boldsymbol{e}_k\}$, it follows that it holds for all unit vectors $\boldsymbol{\nu}$. Since $\boldsymbol{t}(\cdot, t, \boldsymbol{\nu})$ is continuous and since $\mathcal{B}$ is an open set together with part of its boundary, Eq. (7.17) holds for all $\boldsymbol{y} \in \mathcal{B}$. We now define

$$T(\boldsymbol{z}, t) \equiv -\sum_{k=1}^{3} \boldsymbol{t}(\boldsymbol{z}, t, -\boldsymbol{e}_k)\boldsymbol{e}_k, \tag{7.18}$$

which with (7.17) yields (7.15). □

It immediately follows from (7.15) that

$$\boldsymbol{t}(\boldsymbol{z}, t, \boldsymbol{\nu}) = -\boldsymbol{t}(\boldsymbol{z}, t, -\boldsymbol{\nu}) \tag{7.19}$$

for all unit vectors $\boldsymbol{\nu}$. This is *Newton's Law of Action and Reaction*, derived as a consequence of the Balance of Linear Momentum and the representation (7.9). We can use (7.19) to simplify the right-hand side of (7.18) by removing the minus signs.

This exposition is modelled on Gurtin's (1981a). For a discussion of variants, see Truesdell (1991a). This theorem is true under far weaker regularity assumptions. See Gurtin & Martins (1976), Ziemer (1983), and Šilhavý (1985, 1991).

Let us now substitute (7.15) into (7.9), which we then substitute into (7.3) with $\mathcal{B}$ replaced by a typical subbody $\mathcal{A}$. The linear dependence of the traction on $\boldsymbol{\nu}$ enables us to apply the Divergence Theorem in the form (XI.2.23e) to the resulting equation to convert it to

$$\int_{\mathcal{A}} [\nabla \cdot \boldsymbol{T}^* + \boldsymbol{f} - \rho \boldsymbol{p}_{tt}] \, dv = \boldsymbol{0} \tag{7.20}$$

for all $\mathcal{A} \subset \mathcal{B}$. If the integrand of (7.20) is continuous on $\mathcal{B}$, assumed to be an open set together with any part of its boundary, then the arbitrariness of $\mathcal{A}$ ensures that

$$\nabla \cdot \boldsymbol{T}^* + \boldsymbol{f} = \rho \boldsymbol{p}_{tt} \tag{7.21}$$

everywhere on $\mathcal{B}$. Equation (7.21) is the material form of the *classical equations of motion*. It is the local version of (7.3).

7.22. Exercise. Prove that (7.20) implies (7.21) under the indicated assumptions.

Let us now determine the local consequences of the Balance of Angular Momentum. Substituting (7.15) and (7.10) into (7.4), we obtain

$$\int_{\partial\mathcal{A}} \boldsymbol{p} \times [\boldsymbol{T} \cdot \boldsymbol{\nu}] \, da + \int_{\mathcal{A}} [\boldsymbol{p} \times \boldsymbol{f} - \rho \boldsymbol{p} \times \boldsymbol{p}_{tt}] \, dv = \boldsymbol{0}. \tag{7.23}$$

We can apply the Divergence Theorem to the first integral in (7.23) to convert it to the integral over $\mathcal{A}$ of the unappetizing $\nabla \cdot (\boldsymbol{p} \times \boldsymbol{T})^*$. This expression could readily be evaluated by introducing components. The first step of a coordinate-free procedure is to take the dot product of (7.23) with $(\boldsymbol{a} \times \boldsymbol{b})$ where $\boldsymbol{a}$ and $\boldsymbol{b}$ are arbitrary constant vectors. Then the first integrand becomes

$$\begin{aligned}(7.24)\qquad (\boldsymbol{a}\cdot\boldsymbol{p})[\boldsymbol{b}\cdot(\boldsymbol{T}\cdot\boldsymbol{\nu})]-&[\boldsymbol{a}\cdot(\boldsymbol{T}\cdot\boldsymbol{\nu})](\boldsymbol{b}\cdot\boldsymbol{p})\\ &= \boldsymbol{a}\cdot[\boldsymbol{p}(\boldsymbol{T}\cdot\boldsymbol{\nu})-(\boldsymbol{T}\cdot\boldsymbol{\nu})\boldsymbol{p}]\cdot\boldsymbol{b}\\ &= \boldsymbol{a}\cdot[\boldsymbol{p}(\boldsymbol{\nu}\cdot\boldsymbol{T}^*)-(\boldsymbol{\nu}\cdot\boldsymbol{T}^*)\boldsymbol{p}]\cdot\boldsymbol{b}\\ &= [(\boldsymbol{\nu}\cdot\boldsymbol{T}^*)\boldsymbol{p}]:(\boldsymbol{ba}-\boldsymbol{ab}).\end{aligned}$$

The Divergence Theorem implies that the integral over $\partial\mathcal{A}$ of $[\boldsymbol{\nu}\cdot(\boldsymbol{T}^*\boldsymbol{p})]:\boldsymbol{ba}$ equals the integral over $\mathcal{A}$ of

$$\begin{aligned}(7.25)\qquad [\nabla\cdot\boldsymbol{T}^*\boldsymbol{p}]:\boldsymbol{ba} &= \nabla\cdot[(\boldsymbol{T}^*\cdot\boldsymbol{b})(\boldsymbol{p}\cdot\boldsymbol{a})]\\ &= [\nabla\cdot(\boldsymbol{T}^*\cdot\boldsymbol{b})](\boldsymbol{p}\cdot\boldsymbol{a})+(\boldsymbol{T}^*\cdot\boldsymbol{b})\cdot(\nabla\boldsymbol{p}\cdot\boldsymbol{a})\\ &= \boldsymbol{b}\cdot[(\nabla\cdot\boldsymbol{T}^*)\boldsymbol{p}+(\boldsymbol{T}\cdot\boldsymbol{F}^*)]\cdot\boldsymbol{a}\\ &= \boldsymbol{ba}:[(\nabla\cdot\boldsymbol{T}^*)\boldsymbol{p}+(\boldsymbol{T}\cdot\boldsymbol{F}^*)].\end{aligned}$$

The second equality in (7.25) is the standard identity for the divergence of the product of a scalar- and a vector-valued function. It is a consequence of (XI.2.19a). We can thus reduce (7.23) to

$$(7.26)\qquad (\boldsymbol{ba}-\boldsymbol{ab}):\int_{\mathcal{A}}\{\boldsymbol{T}\cdot\boldsymbol{F}^*+[(\nabla\cdot\boldsymbol{T}^*)+\boldsymbol{f}-\rho\boldsymbol{p}_{tt}]\,\boldsymbol{p}\}\,dv=0.$$

Since the bracketed term vanishes by (7.21) and since $\mathcal{A}$ is arbitrary, we deduce from (7.26) that $(\boldsymbol{ba}-\boldsymbol{ab}):\boldsymbol{T}\cdot\boldsymbol{F}^*=0$. Since Skw has a basis of elements of the form $\boldsymbol{ba}-\boldsymbol{ab}$ and since $\boldsymbol{a}$ and $\boldsymbol{b}$ are arbitrary, we conclude from (XI.1.23e) that

$$(7.27\text{a,b})\qquad \boldsymbol{T}\cdot\boldsymbol{F}^* \ \text{ is symmetric}: \quad \boldsymbol{F}\cdot\boldsymbol{T}^*=\boldsymbol{T}\cdot\boldsymbol{F}^*.$$

This is the local version of (7.4).

7.28. Exercise. Suppose that

$$(7.29)\qquad \boldsymbol{c}(\mathcal{B},t)=\int_{\partial\mathcal{B}}\boldsymbol{\Gamma}\cdot\boldsymbol{\nu}\,da$$

(cf. (7.9) and (7.15)) is not $\boldsymbol{0}$. Determine the resulting generalization of (7.27). For the basic concepts of couple stress, see Truesdell & Toupin (1960, Part D).

The lack of symmetry of $\boldsymbol{T}$ prompts us to introduce another material stress tensor, the *second Piola-Kirchhoff stress tensor* $\boldsymbol{S}$, by

$$(7.30)\qquad \boldsymbol{T}=\boldsymbol{F}\cdot\boldsymbol{S}.$$

By premultiplying (7.27b) by $\boldsymbol{F}^{-1}$ and postmultiplying it by $\boldsymbol{F}^{-*}$, we find that $\boldsymbol{S}$ is symmetric.

To interpret the Piola-Kirchhoff stress tensors, we let $\{\boldsymbol{e}_k\}$ be an orthonormal basis. The deformation $\boldsymbol{p}(\cdot, t)$ carries the material plane through $\boldsymbol{z}$ with normal $\boldsymbol{e}_3$ to the surface with tangent plane at $\boldsymbol{p}(\boldsymbol{z}, t)$ spanned by $\boldsymbol{F}(\boldsymbol{z}, t) \cdot \boldsymbol{e}_1$ and $\boldsymbol{F}(\boldsymbol{z}, t) \cdot \boldsymbol{e}_2$. Using (XI.2.6b), we find that the unit normal to this surface at $\boldsymbol{p}(\boldsymbol{z}, t)$ is

$$(7.31) \qquad \tilde{\boldsymbol{\xi}}_3(\boldsymbol{z}, t) = \frac{[\boldsymbol{F}(\boldsymbol{z}, t) \cdot \boldsymbol{e}_1] \times [\boldsymbol{F}(\boldsymbol{z}, t) \cdot \boldsymbol{e}_2]}{|[\boldsymbol{F}(\boldsymbol{z}, t) \cdot \boldsymbol{e}_1] \times [\boldsymbol{F}(\boldsymbol{z}, t) \cdot \boldsymbol{e}_2]|} = \frac{\boldsymbol{F}^{-*} \cdot \boldsymbol{e}_3}{\sqrt{\boldsymbol{e}_3 \cdot \boldsymbol{C}^{-1} \cdot \boldsymbol{e}_3}}.$$

We illustrate $\boldsymbol{T}(\boldsymbol{z}, t) \cdot \boldsymbol{e}_3$ in Fig. 7.32. Even though $\boldsymbol{T}(\boldsymbol{z}, t) \cdot \boldsymbol{e}_3$ acts across the deformed image of the material plane through $\boldsymbol{z}$ with normal $\boldsymbol{e}_3$, its intensity is measured per unit reference area. In Sec. 15 we define the Cauchy stress, which is the intensity of force per actual area of the surface across which it acts. To compute the average first Piola-Kirchhoff stress across a material surface, we need only measure the total force across the surface because the reference area of the surface is a piece of data. To compute an average stress that is a force intensity per actual area, we must also measure the actual area. The Piola-Kirchhoff stress has been invidiously termed the *engineering stress* with the tacit suggestion that its definition rests upon some unjustified approximation. In fact, it is precisely formulated, physically natural, and especially well suited for solid mechanics.

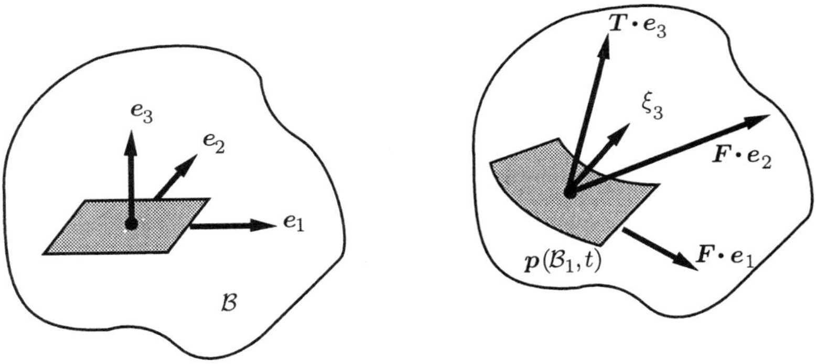

Fig. 7.32. The Piola-Kirchhoff traction vector $\boldsymbol{T}(\boldsymbol{z}, t) \cdot \boldsymbol{e}_3$ across the material plane through $\boldsymbol{z}$ with normal $\boldsymbol{e}_3$.

The vector $\boldsymbol{T}(\boldsymbol{z}, t) \cdot \boldsymbol{e}_3$ can be decomposed in any convenient way. It may be represented as a linear combination of the base vectors tangent to curvilinear coordinate curves in the reference configuration. It is also convenient to represent $\boldsymbol{T}(\boldsymbol{z}, t) \cdot \boldsymbol{e}_3$ in terms of a basis associated with the deformed image of the reference plane. In particular, (7.30) and (7.31) imply that the component of $\boldsymbol{T}(\boldsymbol{z},t)\cdot\boldsymbol{e}_3$ normal to this surface is

$$[\boldsymbol{T}(\boldsymbol{z},t)\cdot\boldsymbol{e}_3]\cdot\tilde{\boldsymbol{\xi}}_3(\boldsymbol{z},t)=\frac{\boldsymbol{e}_3\cdot\boldsymbol{S}(\boldsymbol{z},t)\cdot\boldsymbol{e}_3}{\sqrt{\boldsymbol{e}_3\cdot\boldsymbol{C}^{-1}\cdot\boldsymbol{e}_3}}. \tag{7.33}$$

Thus the diagonal component $\boldsymbol{e}_3\cdot\boldsymbol{S}(\boldsymbol{z},t)\cdot\boldsymbol{e}_3$ of $\boldsymbol{S}$ accounts for the normal component of traction across the deformed image of the material plane through $\boldsymbol{z}$ with normal $\boldsymbol{e}_3$. The full meaning of $\boldsymbol{S}$ is apparent when we represent $\boldsymbol{T}(\boldsymbol{z},t)\cdot\boldsymbol{e}_3$ as a linear combination of the elements of the basis $\{\boldsymbol{F}\cdot\boldsymbol{e}_k\}$. By (XI.1.2a) and (XI.2.6b), the dual basis is $\{\boldsymbol{F}^{-*}\cdot\boldsymbol{e}_k\}$. Thus

$$\boldsymbol{T}\cdot\boldsymbol{e}_3=(\boldsymbol{T}\cdot\boldsymbol{e}_3)\cdot(\boldsymbol{F}^{-*}\cdot\boldsymbol{e}_k)\boldsymbol{F}\cdot\boldsymbol{e}_k=(\boldsymbol{e}_3\cdot\boldsymbol{S}\cdot\boldsymbol{e}_k)\boldsymbol{F}\cdot\boldsymbol{e}_k. \tag{7.34}$$

Thus the components of $\boldsymbol{T}\cdot\boldsymbol{e}_3$ associated with the deformed image of the reference plane are the components of $\boldsymbol{S}\cdot\boldsymbol{e}_3$ relative to the basis in the reference configuration.

We shall find that certain parts of solid mechanics are most easily described in terms of $\boldsymbol{T}$, while others are most easily described in terms of $\boldsymbol{S}$.

8. Boundary and Initial Conditions

We now give a general description of boundary and initial conditions for three-dimensional bodies in the spirit of Sec. VIII.12. Our development is actually simpler than that of Sec. VIII.12 because we do not have to worry about rotation and torque. Consequently, we merely sketch the main ideas.

We assume that $\partial\mathcal{B}$ is sufficiently regular that it has a unit normal vector defined a.e. We decompose $\partial\mathcal{B}$ as the union

$$\partial\mathcal{B}=\bigcup_{c=0}^{3}\mathcal{S}_c(t) \tag{8.1}$$

of disjoint sets $\mathcal{S}_c(t)$, on which different kinds of boundary conditions hold:

If material point $\boldsymbol{z}\in\mathcal{S}_0(t)$, then its position is prescribed at time t. Thus there is a function $\bar{\boldsymbol{p}}$ defined on $\{(\boldsymbol{z},t):\boldsymbol{z}\in\mathcal{S}_0(t),\ t\in[0,\infty)\}$ (for initial-boundary-value problems) such that

$$\boldsymbol{p}(\boldsymbol{z},t)=\bar{\boldsymbol{p}}(\boldsymbol{z},t). \tag{8.2}$$

The material point $\boldsymbol{z}$ has 0 degrees of freedom. In this case, the traction at $(\boldsymbol{z},t)$ is not prescribed. (Components of the traction that are not prescribed on the boundary are whatever are necessary to make the body $\mathcal{B}$ move subject to the prescribed boundary and initial conditions; they are found a posteriori from the solution of a well-set problem.)

At the other extreme, if material point $\boldsymbol{z}\in\mathcal{S}_3(t)$, then its traction is prescribed at time t. Thus there is a function $\bar{\boldsymbol{\tau}}$, whose domain is defined just like that of $\bar{\boldsymbol{p}}$, such that

$$\boldsymbol{T}(\boldsymbol{z},t)\cdot\boldsymbol{\nu}(\boldsymbol{z})=\bar{\boldsymbol{\tau}}(\boldsymbol{z},t). \tag{8.3}$$

The position of $\boldsymbol{z} \in \mathcal{S}_3(t)$ at time t is not restricted in any way, so that it can go wherever it is pushed. The material point $\boldsymbol{z}$ has 3 degrees of freedom.

Here and below, it is tacitly understood that $\bar{\boldsymbol{\tau}}$ may depend, say, on the restriction of $\boldsymbol{p}(\cdot, t)$ to ∂B, although we suppress an explicit notation indicating such a dependence.

In a situation intermediate between the two preceding cases, if a material point $\boldsymbol{z} \in \mathcal{S}_2(t)$, then it slides on a moving surface $\gamma(\cdot, \boldsymbol{z}, t)$ (of class C^1) so that

$$\gamma(\boldsymbol{p}(\boldsymbol{z}, t), \boldsymbol{z}, t) = 0. \tag{8.4}$$

The material point $\boldsymbol{z}$ has 2 degrees of freedom. In this case, we can prescribe tractions tangent to the surface. Let $\boldsymbol{\xi}$ be the unit normal to this surface:

$$\boldsymbol{\xi}(\boldsymbol{p}, \boldsymbol{z}, t) = \frac{\partial\gamma(\boldsymbol{p}(\boldsymbol{z}, t), \boldsymbol{z}, t)/\partial \boldsymbol{p}}{|\partial\gamma(\boldsymbol{p}(\boldsymbol{z}, t), \boldsymbol{z}, t)/\partial \boldsymbol{p}|}. \tag{8.5}$$

We complement (8.4) with

$$[\boldsymbol{I} - \boldsymbol{\xi}(\boldsymbol{p}(\boldsymbol{z}, t), \boldsymbol{z}, t)\boldsymbol{\xi}(\boldsymbol{p}(\boldsymbol{z}, t), \boldsymbol{z}, t)] \cdot [\boldsymbol{T}(\boldsymbol{z}, t) \cdot \boldsymbol{\nu}(\boldsymbol{z}) - \bar{\boldsymbol{\tau}}(\boldsymbol{z}, t)] = \boldsymbol{0} \tag{8.6}$$

where $(\boldsymbol{z}, t) \mapsto [\boldsymbol{I} - \boldsymbol{\xi}(\boldsymbol{p}(\boldsymbol{z}, t), \boldsymbol{z}, t)\boldsymbol{\xi}(\boldsymbol{p}(\boldsymbol{z}, t), \boldsymbol{z}, t)] \cdot \bar{\boldsymbol{\tau}}(\boldsymbol{z}, t)$ is a prescribed function with values in $\mathbb{E}^3$.

In the remaining intermediate case, if $\boldsymbol{z} \in \mathcal{S}_1(t)$, then it is confined to a moving curve (of class C^1), taken to be the intersection of the moving surfaces $\gamma_\alpha(\cdot, \boldsymbol{z}, t)$, $\alpha = 1, 2$ so that

$$\gamma_\alpha(\boldsymbol{p}(\boldsymbol{z}, t), \boldsymbol{z}, t) = 0, \qquad \alpha = 1, 2. \tag{8.7}$$

The material point $\boldsymbol{z}$ has 1 degree of freedom. Let $\boldsymbol{e}(\cdot, \boldsymbol{z}, t)$ be the unit tangent vector field to the curve. We prescribe the traction tangent to the curve by

$$\boldsymbol{e}(\boldsymbol{p}(\boldsymbol{z}, t), \boldsymbol{z}, t)\boldsymbol{e}(\boldsymbol{p}(\boldsymbol{z}, t), \boldsymbol{z}, t) \cdot [\boldsymbol{T}(\boldsymbol{z}, t) \cdot \boldsymbol{\nu}(\boldsymbol{z}) - \bar{\boldsymbol{\tau}}(\boldsymbol{z}, t)] = \boldsymbol{0} \tag{8.8}$$

where $\boldsymbol{e}(\boldsymbol{p}(\boldsymbol{z}, t), \boldsymbol{z}, t)\boldsymbol{e}(\boldsymbol{p}(\boldsymbol{z}, t), \boldsymbol{z}, t) \cdot \bar{\boldsymbol{\tau}}(\boldsymbol{z}, t)$ is given.

For the few technical aspects of Sec. 9, which we do not emphasize, it suffices that $\partial\mathcal{B}$ be a locally Lipschitz-continuous surface and that the $\partial\mathcal{S}_c(t)$ be formed of locally Lipschitz-continuous curves.

Following Sec. VIII.12, we subsume all these cases under a compact formalism: We assume that there are given functions

$$\partial\mathcal{B} \times [0, \infty) \times \mathbb{R}^3 \ni (\boldsymbol{z}, t, \mathsf{q}) \mapsto (\bar{\boldsymbol{p}}(\boldsymbol{z}, t, \mathsf{q}), \bar{\boldsymbol{\tau}}(\boldsymbol{z}, t, \mathsf{q})) \in \mathbb{E}^3 \times \mathbb{E}^3 \tag{8.9}$$

such that a complementary combination of components of position and traction are specified in terms of an unknown function

$$[0, \infty) \ni t \mapsto \mathsf{q}(\boldsymbol{z}, t) \in \mathbb{R}^3 \tag{8.10}$$

via

$$p(z,t) = \bar{p}(z,t,\mathsf{q}(z,t)), \tag{8.11}$$

$$[\nu(z,t)\cdot T^*(z,t) - \bar{\tau}(z,t,\mathsf{q}(z,t))]\cdot \frac{\partial \bar{p}}{\partial \mathsf{q}}(z,t,\mathsf{q}(z,t)) = \mathbf{0} \tag{8.12}$$

for $(z,t) \in \partial\mathcal{B} \times [0,\infty)$. For each z, the function $\mathsf{q}(z,\cdot)$ is an unknown of the problem. (For example, if the material point z is restricted to the unit sphere centered at the origin, then we take $\mathsf{q} = (\theta,\phi)$ and $\bar{p}(z,t,\mathsf{q}) = \sin\theta(\cos\phi i + \sin\phi j) + \cos\theta k$.) In (8.12), $\bar{\tau}$ has a meaning slightly different from that of the $\bar{\tau}$'s that appeared in such previous expressions as (8.6). The rank of $[\partial\bar{p}/\partial\mathsf{q}](z,t,\mathsf{q}(z,t))$ is the number of degrees of freedom of the material point z at time t. We thus find that

$$\mathcal{S}_c(t) = \{z \in \partial\mathcal{B} : \operatorname{rank}[\partial\bar{p}/\partial\mathsf{q}](z,t,\mathsf{q}(z,t)) = c\}. \tag{8.13}$$

We assume that p satisfies initial conditions of the form

$$p(z,0) = p_0(z), \quad p_t(z,0) = p_1(z). \tag{8.14a,b}$$

Let us now study the weak form of the the boundary conditions for position. Such versions are to hold when the solutions are not classical. We do not pause, however, to spell out the function spaces in which such solutions are sought. We assume that $\partial\mathcal{B}$ is sufficiently smooth (uniformly Lipschitz-continuous). Then given any material point z_0 in $\partial\mathcal{B}$, we can find a neighborhood $\mathcal{N}(z_0)$ of z_0 on the surface $\partial\mathcal{B}$ and a unit vector $k(z_0)$ such that $\{z + \zeta k(z_0) : z \in \mathcal{N}(z_0), 0 \le \zeta \le \varepsilon\} \subset \operatorname{cl}\mathcal{B}$ for a sufficiently small positive number ε. See Fig. 8.15. Moreover, if $\mathcal{B}$ is bounded, then there are a finite number of material points $z_1,\dots,z_N$ in $\partial\mathcal{B}$ such that $\cup_{b=1}^N \mathcal{N}(z_b)$ covers $\partial\mathcal{B}$. (If $\mathcal{B}$ is unbounded, then there is a locally finite set of such points $z_1,\dots$.)

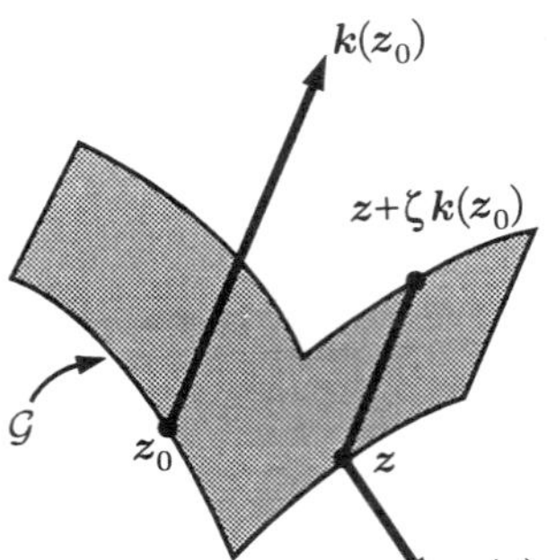

Fig. 8.15. The nature of $\mathcal{B}$ near a material point z_0 on its boundary.

We now assume that (8.11) holds in the sense of trace:

$$\lim_{\zeta\searrow 0}\int_{t_1}^{t_2}\int_{\mathcal{G}} [p(z+\zeta k(z_b),t) - \bar{p}(z,t,\mathsf{q}(z,t))]\, da(z)\, dt = \mathbf{0} \tag{8.16}$$

for each b, for 'almost every' time interval $[t_1, t_2]$, and for 'almost every' measurable subset $\mathcal{G}$ of $\mathcal{N}(\boldsymbol{z}_b)$. (We explain what 'almost every' means in Sec. 9.) The traction boundary conditions (8.12) have an analogous generalization, which is implicit in the Impulse-Momentum Laws and the Principle of Virtual Power formulated in the next section.

We likewise assume that the initial condition (8.14a) holds in the sense of trace:

$$\lim_{t\searrow 0} \int_{\mathcal{P}} \rho(\boldsymbol{z})[\boldsymbol{p}(\boldsymbol{z},t) - \boldsymbol{p}_0(\boldsymbol{z})]\, dv(\boldsymbol{z}) = \boldsymbol{0} \tag{8.17}$$

for 'almost every' $\mathcal{P} \subset \mathcal{B}$. For well-behaved ρ, condition (8.17) is unaffected by its removal. The initial condition (8.14b) is implicit in the integral laws formulated in the next section.

The material of this section is largely based on Antman & Osborn (1979).

9. Impulse-Momentum Laws and the Principle of Virtual Power

We now carry over the development of Secs. II.3 and II.4 to three-dimensional continuum mechanics. The reader should review these sections, especially the discussions of motivations. Let us first suppose that $\mathcal{S}_1$ and $\mathcal{S}_2$ are empty. The straightforward generalization of the Balance of Linear Momentum (7.3), (7.9), (7.15) to allow for less regularity in the dependence on t and to account explicitly for traction boundary conditions is the *Linear Impulse-Momentum Law*:

$$\begin{aligned} \int_0^\tau \int_{\partial\mathcal{P}\setminus\mathcal{S}_3(t)} \boldsymbol{T}(\boldsymbol{z},t)\cdot\boldsymbol{\nu}(\boldsymbol{z})\, da(\boldsymbol{z})\, dt + \int_0^\tau \int_{\partial\mathcal{P}\cap\mathcal{S}_3(t)} \bar{\boldsymbol{\tau}}(\boldsymbol{z},t)\, da(\boldsymbol{z})\, dt \\ + \int_0^\tau \int_{\mathcal{P}} \boldsymbol{f}(\boldsymbol{z},t)\, dv(\boldsymbol{z})\, dt = \int_{\mathcal{P}} \rho(\boldsymbol{z})[\boldsymbol{p}_t(\boldsymbol{z},\tau) - \boldsymbol{p}_1(\boldsymbol{z})]\, dv(\boldsymbol{z}) \end{aligned} \tag{9.1}$$

for almost all (sufficiently nice) $\mathcal{P} \subset \mathcal{B}$ and for almost all $\tau > 0$.

9.2. Exercise. Prove that (9.1) yields generalizations of (8.3) and (8.14b) like (8.16) and (8.17).

We can easily generalize (9.1) to the case in which $\mathcal{S}_1$ and $\mathcal{S}_2$ need not be empty. We merely replace the region of integration $\partial\mathcal{P}\setminus\mathcal{S}_3(t)$ in the first integral of (9.1) with $\partial\mathcal{P} \setminus [\mathcal{S}_1(t) \cup \mathcal{S}_2(t) \cup \mathcal{S}_3(t)]$ and add to the left-hand side of (9.1) the terms

$$\begin{aligned} &\int_0^\tau \int_{\partial\mathcal{P}\cap\mathcal{S}_1(t)} \bar{\boldsymbol{\tau}}(\boldsymbol{z},t)\cdot\boldsymbol{e}(\boldsymbol{p}(\boldsymbol{z},t),\boldsymbol{z},t)\boldsymbol{e}(\boldsymbol{p}(\boldsymbol{z},t),\boldsymbol{z},t)\, da(\boldsymbol{z})\, dt \\ &+ \int_0^\tau \int_{\partial\mathcal{P}\cap\mathcal{S}_1(t)} [\boldsymbol{I} - \boldsymbol{e}(\boldsymbol{p}(\boldsymbol{z},t),\boldsymbol{z},t)\boldsymbol{e}(\boldsymbol{p}(\boldsymbol{z},t),\boldsymbol{z},t)]\cdot\boldsymbol{T}(\boldsymbol{z},t)\cdot\boldsymbol{\nu}(\boldsymbol{z})\, da(\boldsymbol{z})\, dt \\ &+ \int_0^\tau \int_{\partial\mathcal{P}\cap\mathcal{S}_2(t)} [\boldsymbol{I} - \boldsymbol{\xi}(\boldsymbol{p}(\boldsymbol{z},t),\boldsymbol{z},t)\boldsymbol{\xi}(\boldsymbol{p}(\boldsymbol{z},t),\boldsymbol{z},t)]\cdot\bar{\boldsymbol{\tau}}(\boldsymbol{z},t)\, da(\boldsymbol{z})\, dt \\ &+ \int_0^\tau \int_{\partial\mathcal{P}\cap\mathcal{S}_2(t)} \boldsymbol{\xi}(\boldsymbol{p}(\boldsymbol{z},t),\boldsymbol{z},t)\boldsymbol{\xi}(\boldsymbol{p}(\boldsymbol{z},t),\boldsymbol{z},t)\cdot\boldsymbol{T}(\boldsymbol{z},t)\cdot\boldsymbol{\nu}(\boldsymbol{z})\, da(\boldsymbol{z})\, dt \end{aligned} \tag{9.3}$$

coming from (8.6) and (8.8).

To write the resulting form of the Linear Impulse-Momentum Law in a form that utilizes the compact notation (8.12), we merely have to replace $\partial\bar{\mathsf{p}}/\partial\mathsf{q}$ of (8.12) with the tensor $\boldsymbol{\Pi}$ that is the orthogonal projector onto the tangent space at $\boldsymbol{p}(\boldsymbol{z},t)$ of the manifold (8.11). (If (8.11) corresponds to (8.4), then $\boldsymbol{\Pi} = \boldsymbol{I} - \boldsymbol{\xi}\boldsymbol{\xi}$.) Note that $\boldsymbol{\Pi}$ is symmetric. In this case, we can write the Linear Impulse-Momentum Law as

$$\begin{aligned}
&\int_0^\tau \int_{\partial\mathcal{P}\setminus\partial\mathcal{B}} \boldsymbol{T}(\boldsymbol{z},t)\cdot\boldsymbol{\nu}(\boldsymbol{z})\,da(\boldsymbol{z})\,dt + \int_0^\tau\int_{\mathcal{P}} \boldsymbol{f}(\boldsymbol{z},t)\,dv(\boldsymbol{z})\,dt \\
&\quad + \int_0^\tau \int_{\partial\mathcal{P}\cap\partial\mathcal{B}} \bar{\boldsymbol{\tau}}(\boldsymbol{z},t,\mathsf{q}(\boldsymbol{z},t))\cdot\boldsymbol{\Pi}(\boldsymbol{z},t,\mathsf{q}(\boldsymbol{z},t))\,da(\boldsymbol{z})\,dt \\
&\quad + \int_0^\tau \int_{\partial\mathcal{P}\cap\partial\mathcal{B}} \boldsymbol{\nu}(\boldsymbol{z})\cdot\boldsymbol{T}(\boldsymbol{z},t)^*\cdot[\boldsymbol{I}-\boldsymbol{\Pi}(\boldsymbol{z},t,\mathsf{q}(\boldsymbol{z},t))]\,da(\boldsymbol{z})\,dt \\
&\qquad = \int_{\mathcal{P}} \rho(\boldsymbol{z})[\boldsymbol{p}_t(\boldsymbol{z},\tau)-\boldsymbol{p}_1(\boldsymbol{z})]\,dv(\boldsymbol{z})
\end{aligned} \tag{9.4}$$

for almost all (sufficiently nice) $\mathcal{P}\subset\mathcal{B}$ and for almost all $\tau > 0$.

Since we can treat the symmetry condition (7.27) as a constitutive hypothesis (cf. Sec. 10), we do not pause to generalize the Balance of Angular Momentum to an Angular Impulse-Momentum Law. We likewise do not examine the corresponding Principle of Virtual Power. For details of these laws, see Antman & Osborn (1979).

The *Principle of Virtual Power* for forces or, equivalently, the *weak form of the equations of motion* (7.21) subject to boundary conditions (8.11) and (8.12) and to initial conditions (8.14) is

$$\begin{aligned}
&\int_0^\infty \int_{\mathcal{B}} \left[\boldsymbol{T} : \frac{\partial\boldsymbol{\eta}}{\partial\boldsymbol{z}} - \boldsymbol{f}\cdot\boldsymbol{\eta} - \rho\boldsymbol{p}_t\cdot\boldsymbol{\eta}_t\right] dv(\boldsymbol{z})\,dt \\
&-\int_0^\infty\int_{\partial\mathcal{B}} \bar{\boldsymbol{\tau}}(\boldsymbol{z},t,\mathsf{q}(\boldsymbol{z},t))\cdot\boldsymbol{\eta}(\boldsymbol{z},t)\,da(\boldsymbol{z})\,dt - \int_{\mathcal{B}} \rho(\boldsymbol{z})\boldsymbol{p}_1(\boldsymbol{z})\cdot\boldsymbol{\eta}(\boldsymbol{z},0)\,dv(\boldsymbol{z}) = 0
\end{aligned} \tag{9.5}$$

for all (nice enough) $\boldsymbol{\eta}$ with compact support in $(\operatorname{cl} B)\times[0,\infty)$ that satisfy the boundary conditions

$$\boldsymbol{\eta}(\boldsymbol{z},t) = \frac{\partial\bar{\boldsymbol{p}}}{\partial\mathsf{q}}(\boldsymbol{z},t,\mathsf{q}(\boldsymbol{z},t))\cdot\mathsf{r}(\boldsymbol{z},t) \quad \text{for} \quad (\boldsymbol{z},t)\in\partial\mathcal{B}\times[0,\infty) \tag{9.6}$$

for all (nice enough) r.

9.7. Exercise. Suppose that (7.21), (8.11), and (8.12) hold in the classical sense. Derive (9.5) and (9.6) by taking the inner product of (7.21) with $\boldsymbol{\eta}$ and then by integrating the resulting expression by parts while using (8.12). (This is the traditional derivation of the Principle of Virtual Power. Its deficiencies are described in Sec. II.4.)

The fundamental result of this section can be informally stated thus:

9.8. Theorem (Antman & Osborn (1979)). *The Impulse-Momentum Law* (9.4) *and the Principle of Virtual Power* (9.5), (9.6) *are equivalent* *when the fields entering these laws are sufficiently well behaved for all the integrals that appear to make sense as Lebesgue integrals.*

A typical requirement on the fields is that $\boldsymbol{T}$ be locally integrable. In this case, Fubini's theorem must be used to show that the traction $\boldsymbol{T} \cdot \boldsymbol{\nu}$ is integrable on 'almost all' surfaces $\partial\mathcal{P}$ in $\mathcal{B}$. The full proof of this theorem requires measure-theoretic constructions that put it outside the scope of this book. (The proof is technically much harder than that of the corresponding result in Sec. II.4.) Nevertheless, the conceptual foundation of the proof is so simple that it should be used to replace Ex. 9.7 as the natural motivation for the Principle of Virtual Power. The essential feature of the proof is that no field is ever left without the protective clothing of an integral. Thus it is never required to have a pointwise significance (and it is never subjected to the abuse of being multiplied by some function of the unknowns with the consequent adverse effects described in the third paragraph of Sec. II.4).

Derivation of the Principle of Virtual Power from the Impulse-Momentum Law. Let $\boldsymbol{z} = z^k \boldsymbol{i}_k$ where $\{\boldsymbol{i}_k\}$ is a fixed orthonormal basis for $\mathbb{E}^3$. For simplicity, we assume that $\mathcal{S}_3 \subset \{\boldsymbol{z} : z^3 = 0\}$. We take $\mathcal{P}$ to be the cell

$$\mathcal{Q}(\boldsymbol{u}) \equiv \{\boldsymbol{z}\colon z^1 \in (-u^1, u^1),\ z^2 \in (-u^2, u^2),\ z^3 \in (0, u^3)\} \tag{9.9}$$

where $\boldsymbol{u} = u^k \boldsymbol{i}_k$ with $u^1, u^2, u^3 > 0$. We assume that $\mathcal{Q}(\boldsymbol{u}) \subset \mathcal{B}$ and that $\partial\mathcal{Q}(\boldsymbol{u}) \cap \partial\mathcal{B} \subset \mathcal{S}_3$. The specialization of (9.4) to $\mathcal{P} = \mathcal{Q}(\boldsymbol{y}) \subset \mathcal{Q}(\boldsymbol{u})$, which has the form (9.1), reduces to

$$\begin{aligned} &\int_0^{\mathcal{T}} \int_0^{y^3} \int_{-y^2}^{y^2} \boldsymbol{T}(\boldsymbol{z}, t) \cdot \boldsymbol{i}^1 \Big|_{z^1=-y^1}^{z^1=y^1} dz^2\, dz^3\, dt \\ &+ \int_0^{\mathcal{T}} \int_{-y^1}^{y^1} \int_0^{y^3} \boldsymbol{T}(\boldsymbol{z}, t) \cdot \boldsymbol{i}^2 \Big|_{z^2=-y^2}^{z^2=y^2} dz^3\, dz^1\, dt \\ &+ \int_0^{\mathcal{T}} \int_{-y^2}^{y^2} \int_{-y^1}^{y^1} \boldsymbol{T}(\boldsymbol{z}, t) \cdot \boldsymbol{i}^3 \Big|^{z^3=y^3} dz^1\, dz^2\, dt \\ &+ \int_0^{\mathcal{T}} \int_{-y^2}^{y^2} \int_{-y^1}^{y^1} \bar{\boldsymbol{\tau}}(z^1 \boldsymbol{i}_1 + z^2 \boldsymbol{i}_2, t)\, dz^1\, dz^2\, dt \\ &+ \int_0^{\mathcal{T}} \int_0^{y^3} \int_{-y^2}^{y^2} \int_{-y^1}^{y^1} \boldsymbol{f}(\boldsymbol{z}, t)\, dz^1\, dz^2\, dz^3\, dt \\ &= \int_0^{y^3} \int_{-y^2}^{y^2} \int_{-y^1}^{y^1} \rho(\boldsymbol{z})[\boldsymbol{p}_t(\boldsymbol{z}, t) - \boldsymbol{p}_1(\boldsymbol{z})]\, dz^1\, dz^2\, dz^3. \end{aligned} \tag{9.10}$$

Let ω, u^1, u^2, u^3 be positive. We introduce functions ψ_0, ψ_1, ψ_2, ψ_3 with properties:

$$(9.11)\qquad \begin{array}{lll} \psi_0 & \text{has support in } [0,\omega), & \\ \psi_1 & \text{has support in } (-u^1,u^1), & \psi_1(z^1)=\psi_1(-z^1), \\ \psi_2 & \text{has support in } (-u^2,u^2), & \psi_2(z^2)=\psi_2(-z^2), \\ \psi_3 & \text{has support in } [0,u^3). & \end{array}$$

Let $\boldsymbol{c}$ be a constant vector. We operate on (9.10) with

$$(9.12)\qquad \frac{1}{4}\int_0^\omega\int_0^{u^3}\int_{-u^2}^{u^2}\int_{-u^1}^{u^1} dy^1\,dy^2\,dy^3\,d\tau\psi_0'(\tau)\psi_1'(y^1)\psi_2'(y^2)\psi_3'(y^3)\boldsymbol{c}\cdot$$

(noting that $\int_0^{u^3}\int_{-u^2}^{u^2}\int_{-u^1}^{u^1}=\int_{\mathcal{Q}(\boldsymbol{u})}$). Since

(9.13)

$$\int_0^\omega d\tau\psi_0'(\tau)\int_0^\tau \varphi(t)\,dt=-\int_0^\omega \psi_0(\tau)\varphi(\tau)\,d\tau,$$

$$\int_{-u^1}^{u^1} dy^1\psi_1'(y^1)\int_{-y^1}^{y^1}\varphi(z^1)\,dz^1=-\int_{-u^1}^{u^1}\psi_1(y^1)[\varphi(y^1)+\varphi(-y^1)]\,dy^1$$
$$=-2\int_{-u^1}^{u^1}\psi_1(y^1)\varphi(y^1)\,dy^1,$$

etc., we thus obtain

(9.14)

$$\begin{aligned}
&-\int_0^\omega\int_{\mathcal{Q}(\boldsymbol{u})}\psi_0(\tau)\psi_1'(y^1)\psi_2(y^2)\psi_3(y^3)\boldsymbol{c}\cdot\boldsymbol{T}(\boldsymbol{z},t)\cdot\boldsymbol{i}^1\,dv(\boldsymbol{y})\,d\tau\\
&-\int_0^\omega\int_{\mathcal{Q}(\boldsymbol{u})}\psi_0(\tau)\psi_1(y^1)\psi_2'(y^2)\psi_3(y^3)\boldsymbol{c}\cdot\boldsymbol{T}(\boldsymbol{z},t)\cdot\boldsymbol{i}^2\,dv(\boldsymbol{y})\,d\tau\\
&-\int_0^\omega\int_{\mathcal{Q}(\boldsymbol{u})}\psi_0(\tau)\psi_1(y^1)\psi_2(y^2)\psi_3'(y^3)\boldsymbol{c}\cdot\boldsymbol{T}(\boldsymbol{z},t)\cdot\boldsymbol{i}^3\,dv(\boldsymbol{y})\,d\tau\\
&+\psi_3(0)\int_0^\omega\int_{-u^2}^{u^2}\int_{-u^1}^{u^1}\psi_0(\tau)\psi_1(y^1)\psi_2(y^2)\boldsymbol{c}\cdot\bar{\boldsymbol{\tau}}(y^1\boldsymbol{i}_1+y^2\boldsymbol{i}_2,t)\,dy^1\,dy^2\,d\tau\\
&+\int_0^\omega\int_{\mathcal{Q}(\boldsymbol{u})}\psi_0(\tau)\psi_1(y^1)\psi_2(y^2)\psi_3(y^3)\boldsymbol{c}\cdot\boldsymbol{f}(\boldsymbol{z},t)\,dv(\boldsymbol{y})\,d\tau\\
&\quad=-\int_0^\omega\int_{\mathcal{Q}(\boldsymbol{u})}\psi_0'(\tau)\psi_1(y^1)\psi_2(y^2)\psi_3(y^3)\rho(\boldsymbol{y})\boldsymbol{c}\cdot\boldsymbol{p}_t(\boldsymbol{y},t)\,dv(\boldsymbol{y})\,d\tau\\
&\quad\quad+\psi_0(0)\int_{\mathcal{Q}(\boldsymbol{u})}\psi_1(y^1)\psi_2(y^2)\psi_3(y^3)\rho(\boldsymbol{y})\boldsymbol{c}\cdot\boldsymbol{p}_1(\boldsymbol{y})\,dv(\boldsymbol{y})\,d\tau.
\end{aligned}$$

Let us now set

$$(9.15)\qquad \boldsymbol{\eta}(\boldsymbol{z},t)=\psi_0(t)\psi_1(z^1)\psi_2(z^2)\psi_3(z^3)\boldsymbol{c}.$$

Since (9.11) implies that this $\boldsymbol{\eta}$ vanishes outside $\mathcal{Q}(\boldsymbol{u}) \times [0, \omega]$, Eq. (9.14) reduces to (9.5).

Simpler constructions work when $\mathcal{Q}(\boldsymbol{u}) \cap \partial\mathcal{B} \subset \mathcal{S}_0 \subset \{\boldsymbol{z} : z^3 = 0\}$ and when $\mathcal{Q}(\boldsymbol{u}) \cap \partial\mathcal{B} = \emptyset$. All such constructions can be extended to the generalized versions of (8.11) and (8.12). Thus (9.5) holds for a special class of $\boldsymbol{\eta}$'s like those of (9.11) and (9.15). Now if $\partial\mathcal{B}$ is uniformly Lipschitz-continuous, then essentially by definition there is an obvious Lipschitz-continuous mapping that takes a typical region

$$\{\boldsymbol{z} + \zeta\boldsymbol{k}(\boldsymbol{z}_0) : \boldsymbol{z} \in \mathcal{N}(\boldsymbol{z}_0), 0 \leq \zeta \leq \varepsilon\}$$

abutting the boundary into a cell of the form $\mathcal{Q}(\boldsymbol{u})$. (See Fig. 8.15.) This mapping has a Lipschitz-continuous inverse. Thus we can carry out our construction for any such region merely by inserting into our integrals the Jacobian of the Lipschitz-continuous mapping, which exists a.e. Thus we can partition $\mathcal{B}$ into a locally finite number of Lipschitz-continuous images of cells like $\mathcal{Q}(\boldsymbol{u})$, each cell having a finite number of intersections with other such cells, such that (9.5) holds for an $\boldsymbol{\eta}$ that is supported on such an image of a cell and that is a transformed version of a $\boldsymbol{\eta}$ like that of (9.11) and (9.15). But linear combinations of such $\boldsymbol{\eta}$'s supported on images of cells are dense in appropriate Sobolev spaces (with one derivative) corresponding to the regularity of the ψ's, so that (9.5) holds for all $\boldsymbol{\eta}$'s in these Sobolev spaces.

Now we can define what the phrase 'for almost every $\mathcal{P} \subset \mathcal{B}$' means. Let $\mathcal{Q} \equiv \mathcal{Q}(\boldsymbol{u})$ with $u^1 = u^2 = u^3 = 1$. Let $\boldsymbol{g} : \mathcal{Q} \to \mathbb{E}^3$ be a one-to-one Lipschitz-continuous function that has a Lipschitz continuous inverse and that takes each face of the cell $\mathcal{Q}$ into the graph of a Lipschitz-continuous function. A property is said to hold *for almost every* $\mathcal{P} \subset \mathcal{B}$ iff for each such $\boldsymbol{g}$ the property holds on $\boldsymbol{g}(\mathcal{Q}(\boldsymbol{y})) \cap \operatorname{cl}\mathcal{B}$ for almost all $\boldsymbol{y} \in [0,1]^3$ in the sense of Lebesgue measure on $\mathbb{R}^3$.

Derivation of the Impulse-Momentum Law from the Principle of Virtual Power. Using the notions associated with Lipschitz-continuous mappings discussed in the last two paragraphs, we can reduce the derivation of the Impulse-Momentum Law for arbitrary bodies to its derivation for cells. We may accordingly restrict our attention to the case described in the paragraph containing (9.9). Let

$$\text{(9.16)} \quad \mathcal{Q}(\boldsymbol{w}) \subset \mathcal{Q}(\boldsymbol{y}) \subset \mathcal{Q}(\boldsymbol{u}) \subset \mathcal{B}, \quad \partial\mathcal{Q}(\boldsymbol{u}) \cap \partial\mathcal{B} \subset \mathcal{S}_3 \subset \{\boldsymbol{z} \colon z^3 = 0\}.$$

Then $\mathcal{Q}(\boldsymbol{y})$ is the union of $\mathcal{Q}(\boldsymbol{w})$ and five truncated pyramids, which we denote by $\operatorname{cl}\mathcal{K}_1^+$, $\operatorname{cl}\mathcal{K}_2^+$, $\operatorname{cl}\mathcal{K}_3^+$, $\operatorname{cl}\mathcal{K}_1^-$, $\operatorname{cl}\mathcal{K}_2^-$, where the $\mathcal{K}_k^+$ lie between the planes $z^k = w^k$ and $z^k = y^k$ for $k = 1, 2, 3$ and the $\mathcal{K}_k^-$ lie between the planes $z^k = -w^k$ and $z^k = -y^k$ for $k = 1, 2$. Let

(9.17)
$$\alpha(\boldsymbol{z}) = \begin{cases} 1 & \text{for } \boldsymbol{z} \in \operatorname{cl} \mathcal{Q}(\boldsymbol{w}), \\ \frac{y^k - z^k}{y^k - w^k} & \text{for } \boldsymbol{z} \in \operatorname{cl} \mathcal{K}_k^+, \quad k = 1, 2, 3, \\ \frac{y^k + z^k}{y^k - w^k} & \text{for } \boldsymbol{z} \in \operatorname{cl} \mathcal{K}_k^-, \quad k = 1, 2, \\ 0 & \text{for } \boldsymbol{z} \notin \operatorname{cl} \mathcal{Q}(\boldsymbol{y}), \end{cases}$$

(9.18)
$$\beta(t) = \begin{cases} 1 & \text{for } 0 \le t \le \tau, \\ \frac{\omega - t}{\omega - \tau} & \text{for } \tau \le t \le \omega, \\ 0 & \text{for } \omega \le t. \end{cases}$$

Let $\boldsymbol{c}$ be an arbitrary fixed vector. We set

(9.19)
$$\boldsymbol{\eta}(\boldsymbol{z}, t) = \alpha(\boldsymbol{z})\beta(t)\boldsymbol{c}.$$

(We may note in passing that $\boldsymbol{\eta}$ has the form (9.15).) We substitute (9.19) into (9.5) to obtain

(9.20)
$$\begin{aligned} &\boldsymbol{c} \cdot \int_0^\omega \beta(t) \sum_{k=1}^3 \frac{1}{y^k - w^k} \left(\int_{\mathcal{K}_k^+} + \int_{\mathcal{K}_k^-} \right) \boldsymbol{T}(\boldsymbol{z}, t) \cdot \boldsymbol{i}^k \, dv(\boldsymbol{z}) \, dt \\ &- \boldsymbol{c} \cdot \int_0^\omega \beta(t) \int_{\mathcal{Q}(\boldsymbol{y})} \alpha(\boldsymbol{z}) \boldsymbol{f}(\boldsymbol{z}, t) \, dv(\boldsymbol{z}) \, dt \\ &+ \frac{\boldsymbol{c}}{\omega - \tau} \cdot \int_\tau^\omega \int_{\mathcal{Q}(\boldsymbol{y})} \alpha(\boldsymbol{z}) \rho(\boldsymbol{z}) \boldsymbol{p}_t(\boldsymbol{z}, t) \, dv(\boldsymbol{z}) \, dt \\ &- \boldsymbol{c} \cdot \int_0^\omega \beta(t) \int_{\{\boldsymbol{z} \in \mathcal{Q}(\boldsymbol{y}) : z^3 = 0\}} \alpha(\boldsymbol{z}) \bar{\boldsymbol{\tau}}(\boldsymbol{z}, t) \, da(\boldsymbol{z}) \, dt \\ &- \boldsymbol{c} \cdot \int_{\mathcal{Q}(\boldsymbol{y})} \alpha(\boldsymbol{z}) \rho(\boldsymbol{z}) \boldsymbol{p}_1(\boldsymbol{z}) \, dv(\boldsymbol{z}) = 0. \end{aligned}$$

The arbitrariness of $\boldsymbol{c}$ enables us to cancel it from (9.20).

Let

(9.21)
$$\mathcal{E}_k^\pm(z^k) = \{\boldsymbol{v} \in \mathcal{K}_k^\pm : v^k = z^k\}.$$

Let us substitute the identities

(9.22)
$$\int_{\mathcal{K}_k^+} dv(\boldsymbol{z}) = \int_{w^k}^{y^k} \int_{\mathcal{E}_k^+(z^k)} da(\boldsymbol{z}) \, dz^k,$$

etc., into (9.20). If $\boldsymbol{T}$ were continuous, then we could use the Fundamental Theorem of Calculus and the rule for changing the order of integration

(Fubini's Theorem) to show that as $(\boldsymbol{y}, \omega) \to (\boldsymbol{w}, \tau)$, the first integral of (9.20) approaches

$$(9.23) \qquad -\int_0^\tau \sum_{k=1}^3 \int_{\mathcal{E}_k^+(w^k)} \boldsymbol{T}(\boldsymbol{z}, t) \cdot \boldsymbol{i}^k \, da(\boldsymbol{z}) \, dt$$

for all $(\boldsymbol{w}, \tau)$. Similarly, the fourth integral of (9.20) would approach

$$(9.24) \qquad \int_{\mathcal{Q}(\boldsymbol{w})} \rho(\boldsymbol{z}) \boldsymbol{p}_t(\boldsymbol{z}, t) \, dv(\boldsymbol{z}).$$

Under these conditions, the limit of (9.20) as $(\boldsymbol{y}, \omega) \to (\boldsymbol{w}, \tau)$ is the Impulse-Momentum Law (9.4). A surprisingly delicate measure-theoretic argument, which we omit, is needed to justify the same conclusion when $\boldsymbol{T}$ is merely integrable.

The presentation of Secs. 8 and 9 is based upon the work of Antman & Osborn (1979), which may be consulted for the full details of the proofs sketched here. For further refinements, see Šilhavý (1991). The proof of a generalization of Theorem 9.8 suitable for the kinds of function spaces used in the modern theory of quasilinear hyperbolic equations remains an open problem.

10. Constitutive Equations of Mechanics

Our description of kinematics and strain and our equations of motion apply to any material body. The equations of motion (7.21) and (7.27) represent six scalar restrictions on the twelve components of $\boldsymbol{p}$ and $\boldsymbol{T}$. We simultaneously obtain a determinate system and describe the material properties of bodies by relating $\boldsymbol{p}$ and $\boldsymbol{T}$ by *constitutive equations*. We now study these from a general viewpoint. In later sections we treat specific classes of materials.

For any function $\mathbb{R} \ni t \mapsto \boldsymbol{\Psi}(\boldsymbol{z}, t)$, we define the *history of* $\boldsymbol{\Psi}(\boldsymbol{z}, \cdot)$ *up to time* t, i.e., the *past history of* $\boldsymbol{\Psi}(\boldsymbol{z}, \cdot)$, to be the function

$$(10.1) \qquad [0, \infty) \ni \tau \mapsto \boldsymbol{\Psi}^t(\boldsymbol{z}, \tau) \equiv \boldsymbol{\Psi}(\boldsymbol{z}, t - \tau).$$

Thus τ measures time backwards from t.

There are two classes of mechanical constitutive equations: In the first class, the stress $\boldsymbol{T}(\boldsymbol{z}, t)$ at material point $\boldsymbol{z}$ at time t of body $\mathcal{B}$ is completely determined by the history up to time t of $\boldsymbol{p}$ over the whole body:

$$(10.2) \qquad \boldsymbol{T}(\boldsymbol{z}, t) = \hat{\boldsymbol{T}}(\boldsymbol{p}^t(\cdot, \cdot), \boldsymbol{z}, t).$$

In the second class, there are material constraints restricting the kinds of motion a body can sustain. For example, if a body is *rigid*, then it cannot change its shape no matter what system of forces is applied to it. This material constraint of rigidity is characterized by the constitutive restriction $\boldsymbol{C} = \boldsymbol{I}$. If a body is *incompressible*, then no part of it can change its volume no matter what system of forces is applied to it. This material constraint of incompressibility is characterized by the constitutive restriction $\det \boldsymbol{C} = 1$ or, equivalently, $\det \boldsymbol{F} = 1$. Such material constraints are accompanied by adaptations of (10.2) in which $\hat{\boldsymbol{T}}$ is restricted to histories satisfying the material constraints. We discuss material constraints in Sec. 12. We now focus on (10.2).

There is no evidence that real materials require the level of generality contemplated in (10.2). In particular, we may assume that $\boldsymbol{T}(\boldsymbol{z}, t)$ does not depend on the history of motion of material points outside an arbitrary neighborhood of $\boldsymbol{z}$. This assumption permits constitutive equations of the form

$$\boldsymbol{T}(\boldsymbol{z}, t) = \hat{\boldsymbol{T}}(\boldsymbol{p}^t(\boldsymbol{z}, \cdot), \boldsymbol{p}^t_{\boldsymbol{z}}(\boldsymbol{z}, \cdot), \boldsymbol{p}^t_{\boldsymbol{z}\boldsymbol{z}}(\boldsymbol{z}, \cdot), \dots, \boldsymbol{z}, t). \tag{10.3}$$

There is a considerable body of research on special forms of (10.3) when $\boldsymbol{p}^t_{\boldsymbol{z}\boldsymbol{z}}(\boldsymbol{z}, \cdot)$, at least, is present as an argument. In brief, this work has indicated that the effect of the presence of $\boldsymbol{z}$-derivatives of $\boldsymbol{p}$ of order higher than 1 is slight, accounting for phenomena associated with surface tension, except where the solution would otherwise be irregular, e.g., at shocks. Recent work has indicated that constitutive equations with such higher derivatives help resolve very difficult mathematical and mechanical problems. See Truesdell & Noll (1965) for an account of early work and Ball & James (1987), Carr, Gurtin & Slemrod (1984), Dunn & Serrin (1985), Hagan & Slemrod (1983), among others, for a discussion of important advances.

In this book we restrict our attention to *simple materials*, which have constitutive relations of the form

$$\boldsymbol{T}(\boldsymbol{z}, t) = \hat{\boldsymbol{T}}(\boldsymbol{p}^t(\boldsymbol{z}, \cdot), \boldsymbol{p}^t_{\boldsymbol{z}}(\boldsymbol{z}, \cdot), \boldsymbol{z}, t). \tag{10.4}$$

The second argument of $\hat{\boldsymbol{T}}$ is restricted to histories of $\boldsymbol{F}(\boldsymbol{z}, \cdot)$ for which $\det \boldsymbol{F} > 0$. We assume that $\hat{\boldsymbol{T}}$ identically satisfies the symmetry condition (7.27). Thus we have no need to concern ourselves with it any more. In the next section we show that $\hat{\boldsymbol{T}}$ should not depend on its first argument. Our forthcoming definitions are made in light of this observation.

An *elastic* material has a constitutive equation of the form

$$\boldsymbol{T}(\boldsymbol{z}, t) = \hat{\boldsymbol{T}}(\boldsymbol{p}_{\boldsymbol{z}}(\boldsymbol{z}, t), \boldsymbol{z}), \tag{10.5}$$

so that the stress is determined by the actual deformation gradient and is otherwise independent of its past history. If the $\hat{\boldsymbol{T}}$ of (10.4) is independent of its first argument, then its restriction to histories for which $\boldsymbol{F}(\boldsymbol{z}, \cdot)$ is constant has the form (10.5). In this sense, the equilibrium of simple materials reduces to the equilibrium of elastic materials.

A *viscoelastic material of differential type of complexity 1* has a constitutive equation of the form

$$\boldsymbol{T}(\boldsymbol{z}, t) = \hat{\boldsymbol{T}}(\boldsymbol{p}_{\boldsymbol{z}}(\boldsymbol{z}, t), \boldsymbol{p}_{\boldsymbol{z}t}(\boldsymbol{z}, t), \boldsymbol{z}). \tag{10.6}$$

The substitution of (10.5) or (10.6) into (7.21) yields a quasilinear system of partial differential equations, which present major challenges to modern analysis. The substitution of (10.4) into (7.21) typically yields even more difficult partial-functional differential equations.

In general, a material is said to be of *differential type of complexity n* iff it has a constitutive equation of the form

$$(10.7) \qquad \boldsymbol{T}(\boldsymbol{z},t) = \hat{\boldsymbol{T}}(\boldsymbol{p_z}(\boldsymbol{z},t), \partial_t \boldsymbol{p_z}(\boldsymbol{z},t), \dots, \partial_t^n \boldsymbol{p_z}(\boldsymbol{z},t), \boldsymbol{z}).$$

Here n is a positive integer. Many constitutive equations in this class with $n \geq 2$ apparently do not describe real materials. A material is said to be of *rate type* iff it has a constitutive equation of the form

$$(10.8) \quad \partial_t^m \boldsymbol{T}(\boldsymbol{z},t) = \hat{\boldsymbol{G}}(\boldsymbol{T}(\boldsymbol{z},t), \partial_t \boldsymbol{T}(\boldsymbol{z},t), \dots, \partial_t^{m-1} \boldsymbol{T}(\boldsymbol{z},t), \boldsymbol{p_z}(\boldsymbol{z},t), \partial_t \boldsymbol{p_z}(\boldsymbol{z},t), \dots \partial_t^n \boldsymbol{p_z}(\boldsymbol{z},t), \boldsymbol{z}).$$

Here m and n are positive integers. For (10.8), which is an ordinary differential equation for $\boldsymbol{T}(\boldsymbol{z},\cdot)$, to define a constitutive equation of the form (10.4) it must have a globally defined solution. Even in this case, the material properties depend on the initial conditions for $\boldsymbol{T}$. A material is said to be of *integral type* iff it has a constitutive equation in which $\boldsymbol{T}(\boldsymbol{z},t)$ is specified as an integral operator on $\boldsymbol{F}^t(\boldsymbol{z},\cdot)$. A material is said to be of *internal-variable type* iff it has a constitutive of the form

$$(10.9) \qquad \boldsymbol{T}(\boldsymbol{z},t) = \hat{\boldsymbol{T}}(\boldsymbol{p_z}(\boldsymbol{z},t), \partial_t \boldsymbol{p_z}(\boldsymbol{z},t), \dots \partial_t^n \boldsymbol{p_z}(\boldsymbol{z},t), \boldsymbol{\Psi}(\boldsymbol{z},t), \boldsymbol{z})$$

where $\boldsymbol{\Psi}$ is a function satisfying some auxiliary equation, typically an ordinary differential equation of the form

$$(10.10) \qquad \partial_t \boldsymbol{\Psi}(\boldsymbol{z},t) = \boldsymbol{\Gamma}(\boldsymbol{p_z}(\boldsymbol{z},t), \partial_t \boldsymbol{p_z}(\boldsymbol{z},t), \dots \partial_t^n \boldsymbol{p_z}(\boldsymbol{z},t), \boldsymbol{\Psi}(\boldsymbol{z},t), \boldsymbol{z}).$$

The caveats for materials of rate type also apply to this class. Clearly, these classes of materials overlap and are not exhaustive.

We say that the material described by (10.4) is *homogeneous* iff $\hat{\boldsymbol{T}}$ and ρ do not depend explicitly on $\boldsymbol{z}$. (The concept of homogeneity is actually somewhat simpler than the concept of uniformity introduced in Sec. VIII.10. But there is a three-dimensional concept of uniformity that is much deeper; see Truesdell & Noll (1965, Sec. 27)).

In the next few sections we discuss general restrictions on (10.4). We then specialize our studies to problems for specific classes of materials.

There is a challenging open question related to the classification of constitutive equations. If we are presented with a specimen of material and know that it has a specific kind of constitutive equation, e.g., (10.5) or (10.6), then we could devise sensible experiments (see Rivlin & Saunders (1951)) to determine the constitutive function $\hat{\boldsymbol{T}}$. On the other hand, if we do not know the class of constitutive equations to which the material belongs, then there is no obvious procedure to determine it from (a finite number of) experiments. Mathematically, the question of determining material properties from experiments is equivalent to determining a class of equations from information about its solutions.

11. Invariance under Rigid Motions

In this section we determine the restrictions imposed on the general constitutive equation (10.4) by the requirement that it be unaffected by rigid motions.

Suppose that in a laboratory we suspend a variety of weights from a weightless spring of natural length l and conclude that in static deformations the weight necessary to produce a change in length δ is $k\delta$ where k is a positive constant. Next we now cause the spring to oscillate in the vertical direction and determine indirectly that in such dynamical deformations the weight necessary to produce a change in length δ is also $k\delta$. Finally we confine the spring to a frictionless radial groove on a horizontal turntable with one end of the spring fixed at the center and the other end attached to a small body of mass m. Let the polar coordinates of the tip mass at time t be $(r(t), \phi(t))$. We spin the turntable, so that ϕ is given. What are the equations of motion for the radial coordinate r of the tip mass? The radial component of Newton's equation of motion is $f = m(\ddot{r} - r\dot{\phi}^2)$ where f is the radial component of force on the tip mass. What is f? If we give the expected answer that $f = -k(r - l)$, then we are tacitly assuming that the material properties of the spring are unchanged by the rotation, even though the form of the equations is significantly affected by the rotation. If the response were affected by the rigid motion, then we could not set up the problem and could not pose it to students of elementary mechanics.

Now suppose we take the spring to the top of a mountain in Nepal or to the bottom of a mine in West Virginia to measure the acceleration of gravity g by the formula $g = k\delta/m$. This formula would be useless were k to change with position (This experiment was proposed by Hooke.)

We could conduct any of these experiments at different times. If they are to be valid, we must know how the material properties of the spring vary with time. Our experiments would not be valid if the material properties were to depend on an absolute time, which we do not know, rather than with a time lapse, since the time of manufacture, say.

To avoid the difficulties we have just alluded to, we adopt the *Principle of Frame-Indifference* or *Principle of Objectivity*: Constitutive functions are invariant under rigid motions and time shifts. We now turn this principle into a precise mathematical statement, which can be effectively applied to specialize the form of our constitutive equations. (Cf. Sec. VIII.7.)

If a body undergoes a motion $\boldsymbol{p}$, then a motion differing from $\boldsymbol{p}$ by a rigid motion relative to a different clock is given by

$$\boldsymbol{p}_\sharp(\boldsymbol{z}, \sigma) = \boldsymbol{c}(t) + \boldsymbol{Q}(t) \cdot \boldsymbol{p}(\boldsymbol{z}, t), \quad \sigma = t + a \tag{11.1}$$

where $\boldsymbol{Q}$ is a proper-orthogonal tensor. Therefore,

$$\boldsymbol{p}_\sharp(\boldsymbol{z}, t + a - \tau) = \boldsymbol{c}(t - \tau) + \boldsymbol{Q}(t - \tau) \cdot \boldsymbol{p}(\boldsymbol{z}, t - \tau), \tag{11.2}$$

$$\boldsymbol{p}_\sharp^\sigma(\boldsymbol{z}, \tau) = \boldsymbol{c}^t(\tau) + \boldsymbol{Q}^t(\tau) \cdot \boldsymbol{p}^t(\boldsymbol{z}, \tau). \tag{11.3}$$

Thus

$$\frac{\partial \boldsymbol{p}_\sharp^\sigma}{\partial \boldsymbol{z}} = \boldsymbol{Q}^t \cdot \frac{\partial \boldsymbol{p}^t}{\partial \boldsymbol{z}}. \tag{11.4}$$

We apply the Polar Decomposition Theorem to (11.4) to write it as

$$\boldsymbol{R}_\sharp^\sigma \cdot \boldsymbol{U}_\sharp^\sigma = \boldsymbol{Q}^t \cdot \boldsymbol{R}^t \cdot \boldsymbol{U}^t. \tag{11.5}$$

Since the polar decomposition is unique and since $\boldsymbol{Q}^t \cdot \boldsymbol{R}^t$ is proper-orthogonal, we conclude from (11.5) that

$$\boldsymbol{R}_\sharp^\sigma = \boldsymbol{Q}^t \cdot \boldsymbol{R}^t, \quad \boldsymbol{U}_\sharp^\sigma = \boldsymbol{U}^t. \tag{11.6}$$

Now let $\boldsymbol{T} \cdot \boldsymbol{\nu}$ be the force per unit reference area acting across the material surface with unit normal $\boldsymbol{\nu}$ when the body undergoes the motion $\boldsymbol{p}$. We expect that if a body undergoes the motion $\boldsymbol{p}_\sharp$ defined by (11.1), then the corresponding traction field $\boldsymbol{T}_\sharp \cdot \boldsymbol{\nu}$ should just be the rotation of $\boldsymbol{T} \cdot \boldsymbol{\nu}$:

$$\boldsymbol{T}_\sharp(\boldsymbol{z}, \sigma) \cdot \boldsymbol{\nu} = \boldsymbol{Q}(t) \cdot \boldsymbol{T}(\boldsymbol{z}, t) \cdot \boldsymbol{\nu} \tag{11.7}$$

so that

$$\boldsymbol{T}_\sharp(\boldsymbol{z}, \sigma) = \boldsymbol{Q}(t) \cdot \boldsymbol{T}(\boldsymbol{z}, t). \tag{11.8}$$

This is just an informal statement of the requirement that the constitutive equations be invariant under rigid motions. A formal mathematical statement, the *Principle of Frame-Indifference*, is: If for fixed $\boldsymbol{z}$ a set of constitutive equations relating $\boldsymbol{p}$ and $\boldsymbol{T}$ hold, then they must hold for $\boldsymbol{p}_\sharp$ and $\boldsymbol{T}_\sharp$ for all $\boldsymbol{c}, \boldsymbol{Q}, a$. Thus if $\boldsymbol{p}$ and $\boldsymbol{T}$ satisfy (10.4):

$$\boldsymbol{T}(\boldsymbol{z}, t) = \hat{\boldsymbol{T}}(\boldsymbol{p}^t(\boldsymbol{z}, \cdot), \boldsymbol{p}_{\boldsymbol{z}}^t(\boldsymbol{z}, \cdot), \boldsymbol{z}, t), \tag{11.9}$$

then $\boldsymbol{p}_\sharp$ and $\boldsymbol{T}_\sharp$ also satisfy it:

$$\boldsymbol{T}_\sharp(\boldsymbol{z}, \sigma) = \hat{\boldsymbol{T}}(\boldsymbol{p}_\sharp^\sigma(\boldsymbol{z}, \cdot), (\boldsymbol{p}_\sharp)_{\boldsymbol{z}}^\sigma(\boldsymbol{z}, \cdot), \boldsymbol{z}, \sigma). \tag{11.10}$$

We now substitute (11.1) and (11.8) into (11.10) to convert it to

$$\boldsymbol{Q}(t) \cdot \boldsymbol{T}(\boldsymbol{z}, t) = \hat{\boldsymbol{T}}(\boldsymbol{c}^t(\cdot) + \boldsymbol{Q}^t(\cdot) \cdot \boldsymbol{p}^t(\boldsymbol{z}, \cdot), \boldsymbol{Q}^t(\cdot) \cdot \frac{\partial \boldsymbol{p}^t}{\partial \boldsymbol{z}}(\boldsymbol{z}, \cdot), \boldsymbol{z}, t + a). \tag{11.11}$$

We now replace the left-hand side of (11.11) with (11.9) to obtain

$$\begin{aligned} &\boldsymbol{Q}(t) \cdot \hat{\boldsymbol{T}}(\boldsymbol{p}^t(\boldsymbol{z}, \cdot), \boldsymbol{p}_{\boldsymbol{z}}^t(\boldsymbol{z}, \cdot), \boldsymbol{z}, t) \\ &\qquad = \hat{\boldsymbol{T}}(\boldsymbol{c}^t(\cdot) + \boldsymbol{Q}^t(\cdot) \cdot \boldsymbol{p}^t(\boldsymbol{z}, \cdot), \boldsymbol{Q}^t(\cdot) \cdot \frac{\partial \boldsymbol{p}^t}{\partial \boldsymbol{z}}(\boldsymbol{z}, \cdot), \boldsymbol{z}, t + a), \end{aligned} \tag{11.12}$$

which must hold for all $\boldsymbol{c} \in \mathbb{E}^3$, $\boldsymbol{Q} \in \mathrm{SO}(3)$, $a \in \mathbb{R}$ and for all $\boldsymbol{p}(\boldsymbol{z}, \cdot)$, $\boldsymbol{F}(\boldsymbol{z}, \cdot)$ with $\det \boldsymbol{F}(\boldsymbol{z}, \cdot) > 0$.

Our basic result is the elegant theorem of Noll (1958):

11.13. Theorem. *Constitutive equation* (11.9) *is invariant under rigid motions and time shifts (i.e., satisfies* (11.12) *for all* $\mathbf{c}, \boldsymbol{Q}, a$) *if and only if* $\hat{\boldsymbol{T}}$ *is independent of* $\mathbf{p}^t$ *and of* t *and has the form*

$$\hat{\boldsymbol{T}}(\boldsymbol{F}^t(\boldsymbol{z},\cdot),\boldsymbol{z}) = \boldsymbol{R}(\boldsymbol{z},t)\cdot\hat{\boldsymbol{T}}(\boldsymbol{U}^t(\boldsymbol{z},\cdot),\boldsymbol{z}). \tag{11.14}$$

(*Here the notation of* (4.2) *is used.*)

Proof. Let us set $\boldsymbol{Q} = \boldsymbol{I}$ and $a = 0$ in (11.12), which immediately implies that $\hat{\boldsymbol{T}}$ is independent of its first argument. Similarly, we show that $\hat{\boldsymbol{T}}$ is independent of its last argument.

Now the Polar Decomposition Theorem enables us to write (11.12) in the form

$$\boldsymbol{Q}(t)\cdot\hat{\boldsymbol{T}}(\boldsymbol{F}^t(\boldsymbol{z},\cdot),\boldsymbol{z}) = \hat{\boldsymbol{T}}(\boldsymbol{Q}^t(\cdot)\cdot\boldsymbol{R}^t(\boldsymbol{z},\cdot)\cdot\boldsymbol{U}^t(\boldsymbol{z},\cdot),\boldsymbol{z}). \tag{11.15}$$

Since $\boldsymbol{z}$ is being held fixed, we can choose the arbitrary $\boldsymbol{Q}(t)$ to equal $\boldsymbol{R}(\boldsymbol{z},t)^*$, so that (11.15) becomes

$$\boldsymbol{R}(\boldsymbol{z},t)^*\cdot\hat{\boldsymbol{T}}(\boldsymbol{F}^t(\boldsymbol{z},\cdot),\boldsymbol{z}) = \hat{\boldsymbol{T}}(\boldsymbol{U}^t(\boldsymbol{z},\cdot),\boldsymbol{z}), \tag{11.16}$$

which is equivalent to (11.14). We now show that (11.14) implies (11.12). Equation (11.14) implies that

$$\boldsymbol{Q}(t)\cdot\hat{\boldsymbol{T}}(\boldsymbol{F}^t(\boldsymbol{z},\cdot),\boldsymbol{z}) = \boldsymbol{Q}(t)\cdot\boldsymbol{R}(\boldsymbol{z},t)\cdot\hat{\boldsymbol{T}}(\boldsymbol{U}^t(\boldsymbol{z},\cdot),\boldsymbol{z}). \tag{11.17}$$

Since (11.14) is an identity in $\boldsymbol{F}$, we can replace $\boldsymbol{F}$ with $\boldsymbol{Q}\cdot\boldsymbol{F} = \boldsymbol{Q}\cdot\boldsymbol{R}\cdot\boldsymbol{U}$ and thus replace $\boldsymbol{R}$ with $\boldsymbol{Q}\cdot\boldsymbol{R}$ to get

$$\hat{\boldsymbol{T}}(\boldsymbol{Q}^t\cdot\boldsymbol{F}^t(\boldsymbol{z},\cdot),\boldsymbol{z}) = \boldsymbol{Q}(t)\cdot\boldsymbol{R}(\boldsymbol{z},t)\cdot\hat{\boldsymbol{T}}(\boldsymbol{U}^t(\boldsymbol{z},\cdot),\boldsymbol{z}), \tag{11.18}$$

which equals $\boldsymbol{Q}(t)\cdot\hat{\boldsymbol{T}}(\boldsymbol{F}^t(\boldsymbol{z},\cdot),\boldsymbol{z})$ by (11.17). □

We shall use a somewhat more convenient form of (11.14) obtained by combining it with (7.30):

$$\begin{aligned} \boldsymbol{S}(\boldsymbol{z},t) &= \boldsymbol{F}(\boldsymbol{z},t)^{-1}\cdot\boldsymbol{T}(\boldsymbol{z},t) \\ &= \boldsymbol{U}(\boldsymbol{z},t)^{-1}\cdot\boldsymbol{R}(\boldsymbol{z},t)^{-1}\cdot\boldsymbol{R}(\boldsymbol{z},t)\cdot\hat{\boldsymbol{T}}(\boldsymbol{U}^t(\boldsymbol{z},\cdot),\boldsymbol{z}) \\ &= \boldsymbol{U}(\boldsymbol{z},t)^{-1}\cdot\hat{\boldsymbol{T}}(\boldsymbol{U}^t(\boldsymbol{z},\cdot),\boldsymbol{z}) \\ &\equiv \hat{\boldsymbol{S}}(\boldsymbol{C}^t(\boldsymbol{z},\cdot),\boldsymbol{z}). \end{aligned} \tag{11.19}$$

This result shows that the frame-indifferent versions of constitutive equations (10.7) for materials of differential type and (10.8) for materials of rate type have the forms

$$\boldsymbol{S}(\boldsymbol{z},t) = \hat{\boldsymbol{S}}(\boldsymbol{C}(\boldsymbol{z},t),\partial_t\boldsymbol{C}(\boldsymbol{z},t),\dots,\partial_t^n\boldsymbol{C}(\boldsymbol{z},t),\boldsymbol{z}), \tag{11.20}$$

$$\begin{aligned} \partial_t^m\boldsymbol{S}(\boldsymbol{z},t) = \hat{\boldsymbol{H}}(&\boldsymbol{S}(\boldsymbol{z},t),\partial_t\boldsymbol{S}(\boldsymbol{z},t),\dots,\partial_t^{m-1}\boldsymbol{S}(\boldsymbol{z},t),\\ &\boldsymbol{C}(\boldsymbol{z},t),\partial_t\boldsymbol{C}(\boldsymbol{z},t),\dots,\partial_t^n\boldsymbol{C}(\boldsymbol{z},t),\boldsymbol{z}). \end{aligned} \tag{11.21}$$

The treatment of (10.9) and (10.10) is similar.

12. Material Constraints

Liquids and rubbers are nearly incompressible. The fibers that reinforce rubber tires make them nearly inextensible in certain directions. These are examples of materials subject to *material* or (*internal*) *constraints*, which are constitutive restrictions on the kinds of deformations a body can suffer. The presence of material constraints can lead to great simplifications in the governing equations. In this section, we extend the analysis of Sec. VIII.15 to three-dimensional problems. In doing so, we confront some difficulties with compatibility that are not present in problems with but one independent spatial variable.

A material point $\boldsymbol{z}$ of a body is subjected to a *simple material constraint* if there is a real-valued function $\gamma(\cdot, \boldsymbol{z})$ such that

$$\gamma(\boldsymbol{F}(\boldsymbol{z}, t), \boldsymbol{z}) = 0 \tag{12.1}$$

for any motion of the body. Since such a constraint is a constitutive restriction, it must be frame-indifferent, i.e., invariant under rigid motions. Thus the constitutive function γ must satisfy

$$\gamma(\boldsymbol{Q} \cdot \boldsymbol{F}, \boldsymbol{z}) = \gamma(\boldsymbol{F}, \boldsymbol{z}) \tag{12.2}$$

for every proper-orthogonal $\boldsymbol{Q}$ and for every $\boldsymbol{F}$ satisfying (12.1) with $\det \boldsymbol{F} > 0$. For simplicity of exposition, we assume that (12.1) is equivalent to an equation of the form

$$\kappa(\boldsymbol{C}(\boldsymbol{z}, t), \boldsymbol{z}) = 0. \tag{12.3}$$

(Frame-indifference is not central to our analysis. The general case falls under the treatment given at the end of this section.)

The material constraint most important for continuum mechanics is that for an *incompressible material*, which is characterized by the following version of (12.3):

$$\det \boldsymbol{C} = 1. \tag{12.4}$$

A material that is inextensible in the direction $\boldsymbol{a}$, where $\boldsymbol{a}$ is a unit vector, is characterized by the constraint

$$\boldsymbol{a} \cdot \boldsymbol{C} \cdot \boldsymbol{a} = 1. \tag{12.5}$$

Ericksen (1986) has shown that the material constraint

$$\operatorname{tr} \boldsymbol{C} = 3, \tag{12.6}$$

where tr denotes trace, arises in the description of the behavior of crystals. Bell (1985) has observed that the material constraint

$$\operatorname{tr} \boldsymbol{U} = 3 \tag{12.7}$$

where $\boldsymbol{U}$ is the positive-definite square root of $\boldsymbol{C}$ closely describes experimental data for the plastic deformation of metals. (A comprehensive study of Bell's constraint in nonlinear elasticity has been carried out by Podio-Guidugli (1990) and by Beatty & Hayes (1992).) A *rigid material* is characterized by the constraint that

$$\boldsymbol{C} = \boldsymbol{I}. \tag{12.8}$$

The details of the application of our theory to rigid materials is left as an exercise.

It is interesting to note that (12.4) and (12.6) imply that $\boldsymbol{C} = \boldsymbol{I}$. For a proof of this fact, which seems to be known, see Marlow (1992). Thus the two constraints (12.4) and (12.6) are equivalent to the six constraints of rigidity. Likewise, (12.4) and (12.7) imply that $\boldsymbol{C} = \boldsymbol{I}$ (see Beatty & Hayes (1992)).

We shall also study the *Kirchhoff* constraints

$$C_{i3} \equiv \tilde{\boldsymbol{p}}_{,i} \cdot \tilde{\boldsymbol{p}}_{,3} = \delta_{i3}, \tag{12.9}$$

where the C_{kl} are the components of $\boldsymbol{C}$ defined in Sec. XI.3 and where $\tilde{\boldsymbol{p}}(\mathbf{x}, t) \equiv \boldsymbol{p}(\tilde{\boldsymbol{z}}(\mathbf{x}), t)$. (To follow the development given in this section, it suffices to take $\mathbf{x}$ to be a triple of Cartesian coordinates.) The constraints (12.9) arise in various simple models for shells. We shall see that these constraints have a character quite different from that of the other constraints just described.

The $\boldsymbol{C}$'s that enter constraints like (12.3) not only must take values in the cone of positive-definite symmetric tensors, but also must satisfy the compatibility conditions of Sec. 3. We begin the study of constraints by developing a naive theory that ignores questions of compatibility. We then refine this theory, showing that the naive theory is valid for constraints (12.4)–(12.8), but not for (12.9) and not for an important class of constraints that arise in theories for rods and shells (see Chap. XIV). Our basic approach consists in regarding a constrained material as a limit of a large class of unconstrained materials.

Let us study a material point $\boldsymbol{z}$ that is subjected to a system of K (independent and consistent) simple material constraints of the form

$$\kappa_a(\boldsymbol{C}, \boldsymbol{z}) = 0, \quad a = 1, \ldots, K < 6, \tag{12.10}$$

which restrict $\boldsymbol{C}$ to a $(6-K)$-dimensional manifold $\mathcal{K}(\boldsymbol{z})$ in the six-dimensional inner-product space Sym of symmetric tensors. In particular, for $K = 1$, which is the most important case, $\mathcal{K}(\boldsymbol{z})$ corresponding to (12.3) is a five-dimensional manifold in the six-dimensional inner-product space Sym of $\boldsymbol{C}$. If in this case we assume that $\kappa_{\boldsymbol{C}}(\boldsymbol{C}, \boldsymbol{z})$ is not $\boldsymbol{O}$ for $\boldsymbol{C} \in \mathcal{K}(\boldsymbol{z})$ and is therefore normal to $\mathcal{K}(\boldsymbol{z})$ at $\boldsymbol{C}$, then $\mathcal{K}(\boldsymbol{z})$ admits a five-dimensional tangent plane at each point.

We use (XI.2.7b) to show that

$$\kappa_{\boldsymbol{C}}(\boldsymbol{C}, \boldsymbol{z}) = \boldsymbol{C}^{-1} \tag{12.11}$$

when $\kappa(\boldsymbol{C}, \boldsymbol{z}) = \det \boldsymbol{C} - 1$ and when (12.4) is satisfied. Similarly,

$$\kappa_{\boldsymbol{C}}(\boldsymbol{C}, \boldsymbol{z}) = \boldsymbol{a}\boldsymbol{a} \tag{12.12}$$

when $\kappa(\boldsymbol{C}, \boldsymbol{z}) = \boldsymbol{a} \cdot \boldsymbol{C} \cdot \boldsymbol{a} - 1$ and when (12.5) is satisfied, and

$$\kappa_{\boldsymbol{C}}(\boldsymbol{C}, \boldsymbol{z}) = \boldsymbol{I} \tag{12.13}$$

when (12.6) holds.

12.14. Exercise. Prove that

$$\kappa_{\boldsymbol{C}}(\boldsymbol{C}, \boldsymbol{z}) = (\det \boldsymbol{U})\boldsymbol{U}^{-1} \tag{12.15}$$

when (12.7) holds. (This is a difficult computation, based upon the Cayley-Hamilton Theorem XI.1.44; see Beatty & Hayes (1992), Podio-Guidugli (1990), Sidoroff (1978), and T. C. T. Ting (1985).)

More generally, we assume that the rank of $\big(\partial\kappa_1/\partial\boldsymbol{C}, \dots, \partial\kappa_K/\partial\boldsymbol{C}\big)$ is K. Then $\mathcal{K}(\boldsymbol{z})$ can be locally described by

$$\boldsymbol{C} = \check{\boldsymbol{C}}(\boldsymbol{\Gamma}, \boldsymbol{O}) \tag{12.16}$$

where $\boldsymbol{\Gamma} \equiv (\Gamma_1, \dots \Gamma_{6-K}) \in \mathbb{R}^{6-K}$ is a set of independent curvilinear coordinates for $\mathcal{K}(\boldsymbol{z})$ and where $\boldsymbol{O}$ is a collection of K zeros.

For an unconstrained material, we can introduce curvilinear coordinates $(\boldsymbol{\Gamma}, \boldsymbol{\Delta})$ for Sym in a neighborhood of any point $\boldsymbol{C}_0$ of $\mathcal{K}$ by

$$\boldsymbol{C} = \check{\boldsymbol{C}}(\boldsymbol{\Gamma}, \boldsymbol{\Delta}) \tag{12.17}$$

where $\boldsymbol{\Delta} \equiv (\Delta_1, \dots \Delta_K) \in \mathbb{R}^K$. Thus $\mathcal{K}$ is characterized by $\boldsymbol{\Delta} = \boldsymbol{O}$, and $\boldsymbol{\Gamma}$ is a set of curvilinear coordinates for $\mathcal{K}$. We assume that $\check{\boldsymbol{C}}$ is continuously differentiable. The strain $\boldsymbol{C}(\boldsymbol{z}, t)$, provided it is near $\boldsymbol{C}_0 \in \mathcal{K}$, is thus determined by $(\boldsymbol{\Gamma}(\boldsymbol{z}, t), \boldsymbol{\Delta}(\boldsymbol{z}, t))$ through (12.17). A basis for any point in this neighborhood of $\boldsymbol{C}_0$ is

$$\frac{\partial \check{\boldsymbol{C}}}{\partial \Gamma^1}(\boldsymbol{\Gamma}, \boldsymbol{\Delta}), \dots, \frac{\partial \check{\boldsymbol{C}}}{\partial \Gamma^{6-K}}(\boldsymbol{\Gamma}, \boldsymbol{\Delta}), \frac{\partial \check{\boldsymbol{C}}}{\partial \Delta^1}(\boldsymbol{\Gamma}, \boldsymbol{\Delta}), \dots, \frac{\partial \check{\boldsymbol{C}}}{\partial \Delta^K}(\boldsymbol{\Gamma}, \boldsymbol{\Delta}). \tag{12.18}$$

When (12.4) or (12.5) hold, we can take $\boldsymbol{\Delta}$ to be a scalar Δ with $\Delta = \det \boldsymbol{C} - 1$ or $\Delta = \boldsymbol{a} \cdot \boldsymbol{C} \cdot \boldsymbol{a} - 1$, respectively. In the former case, we define $\boldsymbol{\Gamma}$ by $\boldsymbol{C} = (1 + \Delta)^{\frac{1}{3}} \boldsymbol{\Gamma}$, which is of the form (12.17). Thus $\boldsymbol{\Gamma} = \boldsymbol{C}/(\det \boldsymbol{C})^{\frac{1}{3}}$.

We conceive of the constitutive equations for constrained materials as being the limits of those for (all possible) slightly constrained materials. To convert this statement into a precise mathematical criterion, we express (11.19), the general frame-indifferent constitutive equation for unconstrained materials, in terms of the strain variables $(\boldsymbol{\Gamma}, \boldsymbol{\Delta})$.

Whenever $\boldsymbol{C}(\boldsymbol{z}, t)$ is close enough to $\mathcal{K}$ for $(\boldsymbol{\Gamma}(\boldsymbol{z}, t), \boldsymbol{\Delta}(\boldsymbol{z}, t))$ to be uniquely defined, we can decompose the second Piola-Kirchhoff stress $\boldsymbol{S}(\boldsymbol{z}, t)$ into the form

$$\boldsymbol{S}(\boldsymbol{z}, t) = \boldsymbol{S}_{\mathrm{L}}(\boldsymbol{z}, t) + \boldsymbol{S}_{\mathrm{A}}(\boldsymbol{z}, t) \tag{12.19}$$

where $\boldsymbol{S}_{\text{L}}(\boldsymbol{z},t)$ is orthogonal to $\mathcal{K}$ at $\check{\boldsymbol{C}}(\boldsymbol{\Gamma}(\boldsymbol{z},t),\boldsymbol{O})$, and $\boldsymbol{S}_{\text{A}}(\boldsymbol{z},t)$ belongs to a space complementary to that of $\boldsymbol{S}_{\text{L}}(\boldsymbol{z},t)$. In particular, we can take $\boldsymbol{S}_{\text{A}}(\boldsymbol{z},t)$ to be tangent to $\mathcal{K}$ at $\check{\boldsymbol{C}}(\boldsymbol{\Gamma}(\boldsymbol{z},t),\boldsymbol{O})$. This choice of space for $\boldsymbol{S}_{\text{A}}(\boldsymbol{z},t)$ is not always the most convenient, as we shall see. Both $\boldsymbol{S}_{\text{L}}$ and $\boldsymbol{S}_{\text{A}}$ are required to be symmetric.

Then the constitutive equation (11.19):

$$\boldsymbol{S}(\boldsymbol{z},t) = \hat{\boldsymbol{S}}(\boldsymbol{C}^t(\boldsymbol{z},\cdot),\boldsymbol{z}) \tag{12.20}$$

is equivalent to equations of the form

$$\boldsymbol{S}_{\text{L}}(\boldsymbol{z},t) = \check{\boldsymbol{S}}_{\text{L}}(\boldsymbol{\Gamma}^t(\boldsymbol{z},\cdot),\boldsymbol{\Delta}^t(\boldsymbol{z},\cdot),\boldsymbol{z}), \tag{12.21}$$

$$\boldsymbol{S}_{\text{A}}(\boldsymbol{z},t) = \check{\boldsymbol{S}}_{\text{A}}(\boldsymbol{\Gamma}^t(\boldsymbol{z},\cdot),\boldsymbol{\Delta}^t(\boldsymbol{z},\cdot),\boldsymbol{z}). \tag{12.22}$$

Let us assume that if $\boldsymbol{\Delta}$ is small, then (12.21) can be solved for $\boldsymbol{\Delta}$, so that it is equivalent to an equation of the form

$$\boldsymbol{\Delta}(\boldsymbol{z},t) = \boldsymbol{\Delta}^\sharp(\boldsymbol{\Gamma}^t(\boldsymbol{z},\cdot),\boldsymbol{S}^t_{\text{L}}(\boldsymbol{z},\cdot),\boldsymbol{z}). \tag{12.23}$$

We now substitute (12.23) into (12.22) to convert it to an equation of the form

$$\boldsymbol{S}_{\text{A}}(\boldsymbol{z},t) = \boldsymbol{S}^\sharp_{\text{A}}(\boldsymbol{\Gamma}^t(\boldsymbol{z},\cdot),\boldsymbol{S}^t_{\text{L}}(\boldsymbol{z},\cdot),\boldsymbol{z}). \tag{12.24}$$

Thus for $\boldsymbol{C}$ sufficiently close to $\mathcal{K}$, the constitutive equation (12.20), for unconstrained materials, is equivalent to the system (12.23), (12.24). We could, of course, shortcut this argument by assuming at the outset that (12.20) is equivalent to system (12.23), (12.24).

Note that the material satisfying the constraint (12.16) is naturally embedded in the family of materials (12.23), (12.24). We now obtain the constitutive equations for the constrained material as a limit of those for unconstrained materials by the simple device of letting $\boldsymbol{\Delta}^\sharp \to \boldsymbol{O}$ (while $\boldsymbol{S}^\sharp_{\text{A}}$ can vary in any manner). In this process, (12.23) is replaced by $\boldsymbol{\Delta} = \boldsymbol{O}$ and the form of (12.24) is unchanged. Thus the constitutive equations for a constrained simple material have the form

$$\boldsymbol{\Delta} = \boldsymbol{O}, \tag{12.25a}$$

$$\boldsymbol{S}(\boldsymbol{z},t) = \boldsymbol{S}_{\text{L}}(\boldsymbol{z},t) + \boldsymbol{S}^\sharp_{\text{A}}(\boldsymbol{\Gamma}^t(\boldsymbol{z},\cdot),\boldsymbol{S}^t_{\text{L}}(\boldsymbol{z},\cdot),\boldsymbol{z}) \tag{12.25b}$$

where

$$\boldsymbol{S}_{\text{L}}(\boldsymbol{z},t) : \overset{\vartriangle}{\boldsymbol{C}}(\boldsymbol{z},t) = 0 \text{ with } \overset{\vartriangle}{\boldsymbol{C}}(\boldsymbol{z},t) \in \operatorname{span}\left\{\frac{\partial \tilde{\boldsymbol{C}}}{\partial \Gamma_a}(\boldsymbol{\Gamma}(\boldsymbol{z},t),\boldsymbol{O})\right\} \tag{12.26a}$$

or, equivalently,

$$\boldsymbol{S}_{\text{L}}(\boldsymbol{z},t) = \sum_a \lambda_a(\boldsymbol{z},t)\frac{\partial \kappa_a}{\partial \boldsymbol{C}}(\tilde{\boldsymbol{C}}(\boldsymbol{\Gamma}(\boldsymbol{z},t),\boldsymbol{O})) \tag{12.26b}$$

where the λ_a are scalar-valued functions. If we take $\boldsymbol{S}_{\mathrm{A}}(\boldsymbol{z},t)$ to be tangent to $\mathcal{K}$, then

$$(12.27) \qquad \boldsymbol{S}_{\mathrm{A}}^{\sharp} : \frac{\partial \kappa_a}{\partial \boldsymbol{C}}(\tilde{\boldsymbol{C}}(\boldsymbol{\Gamma}, \boldsymbol{O})) = 0, \quad a = 1, \ldots, K.$$

Note that (12.26b) forces $\boldsymbol{S}_{\mathrm{L}}$ to be symmetric. Moreover, it says that $\boldsymbol{S}_{\mathrm{L}}$ is determined by $\boldsymbol{C}$ and by a collection $\boldsymbol{\Lambda}$ of scalar fields λ_a. We indicate this fact by setting

$$(12.28a) \qquad \boldsymbol{S}_{\mathrm{L}}(\boldsymbol{z},t) = \hat{\boldsymbol{S}}_{\mathrm{L}}(\boldsymbol{C}(\boldsymbol{z},t), \boldsymbol{\Lambda}(\boldsymbol{z},t), \boldsymbol{z})$$

and writing (12.25b) as

$$(12.28b) \qquad \boldsymbol{S}(\boldsymbol{z},t) = \hat{\boldsymbol{S}}_{\mathrm{L}}(\boldsymbol{C}(\boldsymbol{z},t), \boldsymbol{\Lambda}(\boldsymbol{z},t), \boldsymbol{z}) + \hat{\boldsymbol{S}}_{\mathrm{A}}(\boldsymbol{C}^t(\boldsymbol{z},\cdot), \boldsymbol{\Lambda}^t(\boldsymbol{z},\cdot), \boldsymbol{z}).$$

Note that $\boldsymbol{S}_{\mathrm{L}}$ or, equivalently, the *Lagrange multiplier* $\boldsymbol{\Lambda}$, is not specified as a constitutive function of other variables and is consequently an unknown of any problem in which it appears. The *extra stress* $\hat{\boldsymbol{S}}_{\mathrm{A}}$ is defined only for histories $\boldsymbol{C}^t$ that satisfy the constraint (12.10) (or (12.16)).

Motivated by this development, we are led to adopt the *Local Constraint Principle* (*Principle of Determinism*) of Truesdell & Noll (1965), (see Antman (1982)) which consists in the requirement that (12.28) hold subject to (12.26), so that the *reactive Piola-Kirchhoff stress* $\boldsymbol{S}_{\mathrm{L}}$ (associated with the Lagrange multiplier) must be orthogonal to the constraint manifold (12.10). Thus it locally does no virtual work in any deformation tangent to the constraint manifold. Antman (1982) observed that the derivation just given allows $\boldsymbol{S}_{\mathrm{A}}$ to depend upon $\boldsymbol{S}_{\mathrm{L}}$, as is assumed in a number of special theories (see Richmond & Spitzig (1980) and Renardy (1986)), and he showed that $\boldsymbol{S}_{\mathrm{A}}$ must be independent of $\boldsymbol{S}_{\mathrm{L}}$ for hyperelastic materials (see Exercise 12.31 below). The subscript A signifies *active* and the subscript L signifies *Lagrange* or *latent.*

Let us reexamine the limit process leading to (12.25). For simplicity of notation, we assume that the material is elastic, so that we can dispense with memory effects. From (12.22)–(12.24) we find that

$$(12.29) \qquad \boldsymbol{S}_{\mathrm{A}}^{\sharp}(\boldsymbol{\Gamma}, \boldsymbol{S}_{\mathrm{L}}, \boldsymbol{z}) = \check{\boldsymbol{S}}_{\mathrm{A}}(\boldsymbol{\Gamma}, \boldsymbol{\Delta}^{\sharp}(\boldsymbol{\Gamma}, \boldsymbol{S}_{\mathrm{L}}, \boldsymbol{z}), \boldsymbol{z}).$$

Were we to let $\boldsymbol{\Delta}^{\sharp} \to \boldsymbol{O}$ *while preventing* $\check{\boldsymbol{S}}_{\mathrm{A}}$ *from changing,* then $\boldsymbol{S}_{\mathrm{A}}^{\sharp}$ would be independent of $\boldsymbol{S}_{\mathrm{L}}$ in the limit. But we are considering *any* family of materials for which $\boldsymbol{\Delta}^{\sharp} \to \boldsymbol{O}$, so that $\check{\boldsymbol{S}}_{\mathrm{A}}$ may indeed change. To see that $\boldsymbol{S}_{\mathrm{L}}$ (or $\boldsymbol{\Lambda}$) could survive such a limit process as an argument of $\boldsymbol{S}_{\mathrm{A}}^{\sharp}$ (or $\check{\boldsymbol{S}}_{\mathrm{A}}$), we consider the following simple analog of the theory we have just developed. Corresponding to (12.21), (12.22) is the system

$$(12.30a) \qquad \lambda = \check{\lambda}(\gamma, \delta, \varepsilon) \equiv \frac{\varepsilon h(\gamma)}{\delta}, \quad \alpha = \check{\alpha}(\gamma, \delta, \varepsilon) \equiv \frac{\varepsilon f(\gamma)}{\delta}$$

where ε is a small positive number. This system is equivalent to

$$(12.30b) \qquad \delta = \delta^{\sharp}(\gamma, \lambda, \varepsilon) \equiv \frac{\varepsilon h(\gamma)}{\lambda}, \quad \alpha = \alpha^{\sharp}(\gamma, \lambda, \varepsilon) \equiv \frac{\varepsilon \lambda f(\gamma)}{\varepsilon h(\gamma)}.$$

In the limit as $\varepsilon \searrow 0$, $\delta^\sharp \to 0$ and $\alpha^\sharp$ depends on λ.

The equations of three-dimensional hyperelasticity can be given a variational characterization, in which the role of the Lagrange multiplier is determined by the standard processes of the Calculus of Variations. In this setting, it does not appear in $\hat{\boldsymbol{S}}_{\mathrm{A}}$. That it should not so appear is also a consequence of our construction:

12.31. Exercise. A material is *hyperelastic* iff there is a stored-energy function $\mathrm{Sym} \times \mathcal{B} \ni (\boldsymbol{C}, \boldsymbol{z}) \mapsto W(\boldsymbol{C}, \boldsymbol{z}) \in \mathbb{R}$ such that

$$\boldsymbol{S}(\boldsymbol{z}, t) = \hat{\boldsymbol{S}}(\boldsymbol{C}(\boldsymbol{z}, t), \boldsymbol{z}) = 2W_{\boldsymbol{C}}(\boldsymbol{C}(\boldsymbol{z}, t), \boldsymbol{z}). \tag{12.32}$$

Show that for such a material the process leading from (12.23) and (12.24) to (12.25)–(12.28) leads to an $\hat{\boldsymbol{S}}_{\mathrm{A}}$ independent of $\boldsymbol{\Lambda}$. (See the argument surrounding (VIII.15.7) and (VIII.15.8).)

If we apply this theory to incompressible materials, for which (12.4) holds, then we can use the identity that $\partial \det \boldsymbol{F}/\partial \boldsymbol{F} = (\boldsymbol{F}^*)^{-1} \det \boldsymbol{F}$ for nonsingular $\boldsymbol{F}$ to find that the constitutive equations (12.23)–(12.28) reduce to equations of the form

$$\det \boldsymbol{C}(\boldsymbol{z}, t) = 1, \tag{12.33a}$$

$$\boldsymbol{S}(\boldsymbol{z}, t) = -p(\boldsymbol{z}, t)\boldsymbol{C}^{-1}(\boldsymbol{z}, t) + \hat{\boldsymbol{S}}_{\mathrm{A}}(\boldsymbol{C}^t(\boldsymbol{z}, \cdot), p^t(\boldsymbol{z}, \cdot), \boldsymbol{z}). \tag{12.33b}$$

Here $\boldsymbol{\Lambda}$ reduces to the scalar p, called the *pressure*. If we adopt (12.27), then

$$\hat{\boldsymbol{S}}_{\mathrm{A}} : \boldsymbol{C}^{-1} = 0. \tag{12.33c}$$

It is more illuminating to define $\hat{\boldsymbol{S}}_{\mathrm{A}}$ by the alternative condition

$$\hat{\boldsymbol{S}}_{\mathrm{A}} : \boldsymbol{C} = 0. \tag{12.33d}$$

(Note that the span of $\boldsymbol{C}^{-1}$ and the orthogonal complement of $\boldsymbol{C}$ are complementary, for if not, $\boldsymbol{C}^{-1}$ would lie in this orthogonal complement, so that $\boldsymbol{C}^{-1} : \boldsymbol{C}$, which equals 3, would have to vanish.) In this case, (12.33b,d) say that the pressure $p = -\frac{1}{3}\boldsymbol{S} : C$ is the negative of the mean normal Cauchy stress. (See Sec. 15. A reader accustomed to the spatial formulation of incompressibility, as used in fluid dynamics, might find the form of (12.33) to be strange. It is a consequence of our use of the second Piola-Kirchhoff stress instead of the Cauchy stress.)

It follows from (12.13) and (12.15) that

$$\boldsymbol{S}_{\mathrm{L}} = \mu \boldsymbol{I}, \quad \boldsymbol{S}_{\mathrm{L}} = \lambda (\det \boldsymbol{U}) \boldsymbol{U}^{-1}, \tag{12.34a,b}$$

respectively, when (12.6) and (12.7) hold. Here μ and λ are scalars. If (with slight loss of generality) we choose $\boldsymbol{a} = \boldsymbol{g}_3$ (with $\boldsymbol{g}_3$ a unit vector) in (12.5), then it reduces to $C_{33} = 1$. In this case, the x^3-coordinate curves are inextensible and x^3 is the arc-length parameter for every configuration. It follows from (12.26) that

$$\begin{aligned} &S_{\mathrm{L}}^{33} \equiv \boldsymbol{g}^3 \cdot \boldsymbol{S}_{\mathrm{L}} \cdot \boldsymbol{g}^3 \quad \text{is arbitrary}, \\ &S_{\mathrm{L}}^{kl} \equiv \boldsymbol{g}^k \cdot \boldsymbol{S}_{\mathrm{L}} \cdot \boldsymbol{g}^l = 0 \quad \text{for } \{k, l\} \neq \{3, 3\}. \end{aligned} \tag{12.35}$$

Similarly, for (12.9) we find that

(12.36a,b) $$S_{\mathrm{L}}^{3k} \text{ is arbitrary}, \quad S_{\mathrm{L}}^{\alpha\beta} = 0.$$

We shall argue that of all the consequences we have derived from the Local Constraint Principle, only (12.36b) is unsatisfactory.

Kirchhoff constraints. The Kirchhoff constraints (12.9), which we rewrite as

(12.37) $$C_{3k} \equiv \tilde{\boldsymbol{p}}_{,3} \cdot \tilde{\boldsymbol{p}}_{,k} = \delta_{3k},$$

form the starting point of the derivations of many shell theories, both linear and nonlinear. The material surface defined by $x^3 = 0$ is usually taken as the midsurface of the shell (under the tacit assumption that this surface exists). We now examine these and related systems of constraints.

We differentiate (12.37) to obtain

(12.38) $$\tilde{\boldsymbol{p}}_{,33} \cdot \tilde{\boldsymbol{p}}_{,3} = 0, \quad \tilde{\boldsymbol{p}}_{,3\alpha} \cdot \tilde{\boldsymbol{p}}_{,3} = 0, \quad \tilde{\boldsymbol{p}}_{,33} \cdot \tilde{\boldsymbol{p}}_{,\alpha} = 0.$$

Thus $\tilde{\boldsymbol{p}}_{,33}$ is orthogonal to each member of the basis $\{\tilde{\boldsymbol{p}}_{,k}\}$ and is therefore $\mathbf{0}$. Hence $\tilde{\boldsymbol{p}}$ must have the form

(12.39) $$\tilde{\boldsymbol{p}}(\mathbf{x}, t) = \boldsymbol{r}(x^1, x^2, t) + x^3 \boldsymbol{d}(x^1, x^2, t)$$

where

(12.40a,b) $$\boldsymbol{r}_{,\alpha} \cdot \boldsymbol{d} = 0, \quad |\boldsymbol{d}| = 1$$

as a consequence of (12.37). We now compute from (12.39) that

(12.41a) $$C_{\alpha\beta} = (\boldsymbol{r}_{,\alpha} + x^3 \boldsymbol{d}_{,\alpha}) \cdot (\boldsymbol{r}_{,\beta} + x^3 \boldsymbol{d}_{,\beta}).$$

Thus $C_{\alpha\beta}$ is quadratic in x^3:

(12.41b) $$C_{\alpha\beta,333} = 0.$$

In deducing (12.36), we assumed that for each $\boldsymbol{z}$ we could prescribe $C_{\alpha\beta}$ arbitrarily. But if we regard $\boldsymbol{C}$ as a function on $\mathcal{B}$ everywhere subjected to (12.37), then the $C_{\alpha\beta}$ are not arbitrary, but are themselves subject to the nonsimple constraints (12.41). In general, *the geometrically admissible functions $\boldsymbol{C}$ are not merely arbitrary functions taking values in the cone of positive-definite symmetric tensor-valued functions, but are rather arbitrary functions taking values in the subset of this cone consisting of compatible $\boldsymbol{C}$'s for which there exist injective position fields $\mathbf{p}$ satisfying* (1.1), (1.4), *and* (2.3). If $\boldsymbol{C}$ is twice continuously differentiable on $\mathcal{B}$ and if $\mathcal{B}$ is simply-connected, then a necessary and sufficient condition for $\boldsymbol{C}$ to be compatible is that the Riemann-Christoffel curvature tensor for it vanish. (See Sec. 3.)

12.42. Exercise. Show that if $\boldsymbol{C}$ satisfies (12.37), then three of the compatibility equations (3.21a) form a system of ordinary differential equations for the dependence of the $C_{\alpha\beta}$ on x^3.

We could of course replace the Kirchhoff constraints (12.37) with the equivalent set of constraints on the position field $\boldsymbol{p}$ requiring it to have the form (12.39), (12.40). In this case, we need not concern ourselves with compatibility. We then require an appropriate principle for constraints on the position field. Let us note that (12.39) and (12.40) cause no difficulty in meeting the requirements of frame-indifference because (12.39) and (12.40) define an equivalence class of position fields differing by rigid motions.

In treating more complicated analogs of (12.37), it is convenient to generalize (12.39) and (12.40) rather than to generalize (12.37) directly, because the geometrical significance of the constraints on the position field is usually obvious, while that of the corresponding constraints on the strains is usually not. For example, suppose we constrain $\boldsymbol{p}$ by requiring that (12.39) hold subject only to the requirement that $|\boldsymbol{d}| = 1$. We readily deduce from (1.4) and (2.3) that $C_{33} = 1$, $C_{\alpha 3,3} = 0$. Note that these restrictions, which are local, fail to be simple because they involve a partial derivative and therefore they do not fall under the control of the Local Constraint Principle. These restrictions are equivalent to the original restrictions on $\boldsymbol{p}$, as we can readily check by integrating these equations for the strains to get the requisite $\boldsymbol{p}$.

The constraint system (12.39) and (12.40) says that for each fixed (x^1, x^2) the position field $\tilde{\boldsymbol{p}}(x^1, x^2, \cdot, t)$ is constrained to lie in the six-dimensional linear subspace of its function space spanned by the functions $x^3 \mapsto \boldsymbol{e}_k, x^3\boldsymbol{e}_k$, $k = 1, 2, 3$, where $\{\boldsymbol{e}_k\}$ is a basis for $\mathbb{E}^3$. The corresponding strains $C_{\alpha\beta}(x^1, x^2, \cdot, t)$ lie in the nine-dimensional linear subspace of its function space spanned by products of the functions $x^3 \mapsto 1, x^3, (x^3)^2$ with basis elements for the three-dimensional space of symmetric tensors. We expect the corresponding Lagrange multipliers $S_{\mathrm{L}}^{\alpha\beta}$ to lie in some dual function space (i.e., to be 'orthogonal' to these position fields in a suitable sense). If so, there should be a very large collection of such multipliers. But (12.36) says that the collection of such multipliers consists exactly of the zero multipliers. The global generalization of the Constraint Principle framed below is designed to be physically natural and to remedy this and more fundamental defects.

The five constraints (12.3)–(12.8), which are used almost exclusively in two- and three-dimensional elasticity, seem to have no clear-cut effect on the compatibility equations (4.7), i.e., they seem to produce no clear-cut restrictions on the position field. Thus they do not seem to suffer from the complications we have just described for (12.37). We now demonstrate this fact by an indirect approach, which relies on the fine structure of Lagrange multipliers.

The Global Constraint Principle. We now study position fields on $\mathcal{B} \times \mathcal{I}$ where $\mathcal{I}$ is some time interval on the real line. We construct a global principle of determinism by imitating the development of the local principle sketched in Section 3. In much of our analysis, t is an inessential parameter; we retain it for completeness.

We consider position fields $\boldsymbol{p}(\cdot, t)$ lying in an admissible class $\mathfrak{P}$ of continuous one-to-one functions on $\mathcal{B}$ for which the generalized derivative $\boldsymbol{p}_{\boldsymbol{z}}(\cdot, t)$ is a locally Lebesgue-integrable function satisfying (1.1) a.e. for almost every t. For simplicity of exposition, we refrain from allowing the derivative of $\boldsymbol{p}$ to be a measure.) We assume that $\mathfrak{P}$ inherits its topology from a normed space $\mathfrak{W}$ in which it lies and that $\mathfrak{P}$ is an open subset of $\mathfrak{W}$. We furthermore assume that $\mathfrak{W}$ has the property that if $\boldsymbol{p}_{\boldsymbol{z}}^*(\cdot, t) \cdot \boldsymbol{p}_{\boldsymbol{z}}(\cdot, t) = \boldsymbol{I}$, then $\boldsymbol{p}(\boldsymbol{z}, t) = \boldsymbol{z}$ for all $\boldsymbol{z}$. We thus take $\mathfrak{W}$ to be a quotient space in which rigid motions are factored out. This factoring can be explicitly effected by requiring mean translations and rotations to vanish. (A natural example of $\mathfrak{W}$ is a suitable quotient space of the Sobolev space W_p^1 of vector-valued functions whose first derivatives are Lebesgue-integrable to a power $p \geq 1$.)

For each $\boldsymbol{\eta} \in \mathfrak{W}$ (and each $\boldsymbol{p} \in \mathfrak{P}$), we define the linear form $\langle \cdot, \boldsymbol{\eta}_{\boldsymbol{z}} \rangle$ by

$$\langle \boldsymbol{T}, \boldsymbol{\eta}_{\boldsymbol{z}} \rangle \equiv \int_{\mathcal{B}} \boldsymbol{T}(\boldsymbol{z}) : \boldsymbol{\eta}_{\boldsymbol{z}}(\boldsymbol{z})\, dv(\boldsymbol{z}) = \tfrac{1}{2} \int_{\mathcal{B}} \boldsymbol{S}(\boldsymbol{z}) : \overset{\vartriangle}{\boldsymbol{C}}(\boldsymbol{z})\, dv(\boldsymbol{z}) \tag{12.43a}$$

where we have used (7.30) and the definition

$$\overset{\vartriangle}{\boldsymbol{C}} = (\overset{\vartriangle}{\boldsymbol{p}}_{\boldsymbol{z}})^* \cdot \boldsymbol{p}_{\boldsymbol{z}} + (\boldsymbol{p}_{\boldsymbol{z}})^* \cdot \overset{\vartriangle}{\boldsymbol{p}}_{\boldsymbol{z}}. \tag{12.43b}$$

The set of $\boldsymbol{T}$'s for which (12.43) is defined is the dual space $\mathfrak{T}$ of the space of derivatives of elements of $\mathfrak{W}$.

We study general constraints of the form

$$\gamma(\boldsymbol{p}(\cdot, t)) = \boldsymbol{0}. \tag{12.44a}$$

Let $\mathfrak{K}$ be the subset of $\mathfrak{P}$ consisting of all $\boldsymbol{p}(\cdot,t)$ satisfying (12.44a). We assume that (12.44a) is frame-indifferent; in particular, $\mathfrak{K}$ has the property that if $\boldsymbol{p}(\cdot,t) \in \mathfrak{K}$, then so is every position field that differs from $\boldsymbol{p}$ by a rigid motion, i.e., if $\boldsymbol{p}(\cdot,t) \in \mathfrak{K}$, then so is $\boldsymbol{c}(t) + \boldsymbol{Q}(t) \cdot \boldsymbol{p}(\cdot,t)$ for every vector $\boldsymbol{c}(t)$ and for every proper orthogonal tensor $\boldsymbol{Q}(t)$. The linearization of (12.44a) about $\boldsymbol{p}(\cdot,t)$ in $\mathfrak{K}$ is denoted

$$\boldsymbol{\Gamma}(\boldsymbol{p}(\cdot,t)) \cdot \boldsymbol{\eta}(\cdot,t) = \mathbf{0}. \tag{12.44b}$$

The solutions $\boldsymbol{\eta}(\cdot,t)$ of (12.44b) are called *vectors tangent to* $\mathfrak{K}$ *at* $\boldsymbol{p}(\cdot,t)$; they form the *tangent space to* $\mathfrak{K}$ *at* $\boldsymbol{p}(\cdot,t)$.

We first study 'nearly constrained' materials, those for which $\boldsymbol{p}$ is 'close' to $\mathfrak{K}$. We can represent any such $\boldsymbol{p}$ as a sum

$$\boldsymbol{p} = \boldsymbol{\pi} + \boldsymbol{q} \tag{12.45}$$

where $\boldsymbol{\pi} \in \mathfrak{K}$ and $\boldsymbol{q}$ is not tangent to $\mathfrak{K}$ at $\boldsymbol{\pi}$. We adopt any convenient rule for this decomposition. (We have in mind constraints for which $\boldsymbol{\pi}$ is defined by the right-hand side of (12.39).)

For given $\boldsymbol{\pi} \in \mathfrak{K}$, we can decompose the first Piola-Kirchhoff stress tensor $\boldsymbol{T}$, assumed to lie in $\mathfrak{T}$, into the sum

$$\boldsymbol{T} = \boldsymbol{T}_{\mathrm{L}} + \boldsymbol{T}_{\mathrm{A}} \tag{12.46a}$$

where $\boldsymbol{T}_{\mathrm{L}}$ is in the annihilator of the tangent space to $\mathfrak{K}$ at $\boldsymbol{\pi}$ in the sense that

$$\langle \boldsymbol{T}_{\mathrm{L}}(\cdot,t), \boldsymbol{\eta}_{\boldsymbol{z}}(\cdot,t) \rangle = 0 \tag{12.46b}$$

for all tangent vectors $\boldsymbol{\eta}$ to $\mathfrak{K}$ at $\boldsymbol{\pi}$ and where $\boldsymbol{T}_{\mathrm{A}}$ lies in a space complementary to that of $\boldsymbol{T}_{\mathrm{L}}$. This decomposition induces an analogous decomposition of the second Piola-Kirchhoff stress tensor

$$\boldsymbol{S} = \boldsymbol{S}_{\mathrm{L}} + \boldsymbol{S}_{\mathrm{A}} \quad \text{where} \quad \boldsymbol{S}_{\mathrm{L}} = \boldsymbol{F}^{-1} \cdot \boldsymbol{T}_{\mathrm{L}}, \quad \boldsymbol{S}_{\mathrm{A}} = \boldsymbol{F}^{-1} \cdot \boldsymbol{T}_{\mathrm{A}}. \tag{12.46c}$$

The difference between (12.46c) and (12.19) is that (12.46c) cannot use a pointwise orthogonality condition.

We now imitate the approach leading to the Local Constraint Principle to obtain a constitutive theory for constrained simple materials. Let $\boldsymbol{q}_{\boldsymbol{z}}$ have the polar decomposition $\boldsymbol{Q} \cdot \boldsymbol{M}$ where $\boldsymbol{Q}$ is a rotation and $\boldsymbol{M}$ is symmetric. The constitutive equations (11.19) for an unconstrained simple material are equivalent to the specification of $\boldsymbol{T}_{\mathrm{L}}(\boldsymbol{z},t)$ and $\boldsymbol{T}_{\mathrm{A}}(\boldsymbol{z},t)$ as functionals of the past history of

$$[\boldsymbol{\pi}_{\boldsymbol{z}}]^* \cdot [\boldsymbol{\pi}_{\boldsymbol{z}} + \boldsymbol{Q} \cdot \boldsymbol{M}] + \boldsymbol{M} \cdot \boldsymbol{Q}^* \cdot \boldsymbol{\pi}_{\boldsymbol{z}} + \boldsymbol{M} \cdot \boldsymbol{M}. \tag{12.47}$$

We suppose that these constitutive relations are equivalent to equations of the form

$$\begin{aligned} \boldsymbol{M}(\boldsymbol{z},t) &= \boldsymbol{M}^{\sharp}\left(\boldsymbol{\pi}_{\boldsymbol{z}}^t(\boldsymbol{z},\cdot), \boldsymbol{T}_{\mathrm{L}}^t(\boldsymbol{z},\cdot), \boldsymbol{z}\right), \\ \boldsymbol{T}_{\mathrm{A}}(\boldsymbol{z},t) &= \boldsymbol{T}_{\mathrm{A}}^{\sharp}\left(\boldsymbol{\pi}_{\boldsymbol{z}}^t(\boldsymbol{z},\cdot), \boldsymbol{T}_{\mathrm{L}}^t(\boldsymbol{z},\cdot), \boldsymbol{z}\right) \end{aligned} \tag{12.48}$$

when $\boldsymbol{M}$ is 'small', for $\boldsymbol{z} \in \mathcal{B}$. By setting $\boldsymbol{M}^{\sharp} = \boldsymbol{O}$, we obtain the equations for the constrained material for which $\boldsymbol{p} \in \mathfrak{K}$:

$$\boldsymbol{q}_{\boldsymbol{z}}(\boldsymbol{z},t) = \boldsymbol{O}, \tag{12.49}$$

$$\boldsymbol{T}(\boldsymbol{z},t) = \boldsymbol{T}_{\mathrm{L}}(\boldsymbol{z},t) + \boldsymbol{T}_{\mathrm{A}}^{\sharp}\left(\boldsymbol{\pi}_{\boldsymbol{z}}^t(\boldsymbol{z},\cdot), \boldsymbol{T}_{\mathrm{L}}^t(\boldsymbol{z},\cdot), \boldsymbol{z}\right) \tag{12.50a}$$

where

$$\langle \boldsymbol{T}_{\mathrm{L}}(\cdot,t), \boldsymbol{\eta}_{\boldsymbol{z}}(\cdot,t) \rangle = 0 \quad \forall \boldsymbol{\eta} \quad \text{tangent to } \mathfrak{K}. \tag{12.50b}$$

and where $\boldsymbol{T}^\sharp_{\mathrm{A}}$ takes values in a space complementary to that of $\boldsymbol{T}_{\mathrm{L}}$.

As in Sec. 3 it is convenient to set

$$\boldsymbol{T}_{\mathrm{L}}(\boldsymbol{z},t) = \hat{\boldsymbol{T}}_{\mathrm{L}}\left(\boldsymbol{\pi_z}(\boldsymbol{z},t), \boldsymbol{\Lambda}(\boldsymbol{z},t), \boldsymbol{z}\right), \tag{12.51a}$$

$$\boldsymbol{T}^\sharp_{\mathrm{A}}\left(\boldsymbol{\pi}^t_{\boldsymbol{z}}(\boldsymbol{z},\cdot), \boldsymbol{T}^t_{\mathrm{L}}(\boldsymbol{z},\cdot), \boldsymbol{z}\right) = \hat{\boldsymbol{T}}_{\mathrm{A}}\left(\boldsymbol{\pi}^t_{\boldsymbol{z}}(\boldsymbol{z},\cdot), \boldsymbol{\Lambda}^t(\boldsymbol{z},\cdot), \boldsymbol{z}\right). \tag{12.51b}$$

Here $\boldsymbol{\Lambda}$ represents a collection of unknown functions that together with the motion determine $\boldsymbol{T}_{\mathrm{L}}$ (cf. (12.28)). The constitutive functions we employ must be frame-indifferent, but here we do not use a notation that makes this requirement explicit.

The *Global Constraint Principle* consists in the structure of the constitutive restrictions (12.49)– (12.51) on $\boldsymbol{T}_{\mathrm{L}}$. *Since this principle is a constitutive restriction, $\boldsymbol{\eta}$ need not satisfy any boundary conditions, which merely characterize the environment in which the body is placed. In contrast, the Principle of Virtual Power* (9.5), *a fundamental law of mechanics, holds for all $\boldsymbol{\eta}$ satisfying* (9.6); *these $\boldsymbol{\eta}$'s need not be tangent to $\mathfrak{K}$. If, however, the $\boldsymbol{\eta}$ in* (9.5) *is taken to be tangent to $\mathfrak{K}$, then* (12.50b) *implies that* (9.5) *is independent of $\boldsymbol{T}_{\mathrm{L}}$. Equations for $\boldsymbol{T}_{\mathrm{L}}$ are obtained by using* (9.5) *for $\boldsymbol{\eta}$'s not tangent to $\mathfrak{K}$.*

The three differences between the Local and Global Constraint Principles are that for the latter: (i) We use position fields and their gradients, so that the issue of compatibility is explicit. (ii) We no longer necessarily have an inner product, so we sacrifice some of the detail, such as (12.27), given by the Local Constraint Principle. (iii) Our approach is global in that the bilinear form in (12.50b) is defined on infinite-dimensional spaces.

Since each open subset of $\mathcal{B}$ is itself a body with the same properties as $\mathcal{B}$, one might argue that the Global Constraint Principle can be thereby localized. This argument is specious because some constraints, notably the Kirchhoff constraint, depend on $\mathcal{B}$. In particular, the representation (12.39) involves a distinguished material surface, namely, the reference configuration of $\boldsymbol{r}$. As our preceding development shows, it is not prima facie clear that (12.37) can be localized. Thus we cannot blithely assume that we can replace the region $\mathcal{B}$ of integration for (12.50a) by arbitrary nice subsets of $\mathcal{B}$. (In particular, in Chap. XIV we could identify certain kinds of bodies as rods or shells. But not all subsets of these bodies are rods or shells.) Thus in the Global Constraint Principle the only way we admit of reducing the region of integration for (12.50a) is by reducing the support of $\boldsymbol{\eta}$. As we shall see, this cannot be easily done (because of compatibility issues). Indeed, if (12.39) holds, then $\overset{\vartriangle}{\boldsymbol{p}}$ has the same form as $\tilde{\boldsymbol{p}}$, and it is impossible to reduce the support of $\overset{\vartriangle}{\boldsymbol{p}}$ in the x^3-direction without making $\overset{\vartriangle}{\boldsymbol{d}}$ vanish on a corresponding set.

We now show that the Global Constraint Principle implies the consequences of the Local Constraint Principle for the single constraints (12.4)–(12.7). If a single simple constraint in the form (12.1) holds, then $\boldsymbol{\eta}$ is tangent to $\mathfrak{K}$ if and only if it satisfies the single underdetermined first-order linear partial differential equation

$$\frac{\partial \gamma}{\partial \boldsymbol{F}} : \frac{\partial \boldsymbol{\eta}}{\partial \boldsymbol{z}} = 0. \tag{12.52}$$

We want to deduce from the Global Constraint Principle that

$$\boldsymbol{T}_{\mathrm{L}} = \lambda \frac{\partial \gamma}{\partial \boldsymbol{F}}, \tag{12.53}$$

which is equivalent to (12.26b), for (12.4)–(12.7). To this end, we could try to localize (12.50b) by constructing a rich enough class of $\boldsymbol{\eta}$'s with compact support satisfying (12.52). Since we are faced with the construction of a field $\boldsymbol{\eta}$ from restrictions on its derivatives, we are again confronting the question of compatibility, albeit in a mild form. Alternatively, we could set $\boldsymbol{T}_{\mathrm{L}} = \lambda \frac{\partial \gamma}{\partial \boldsymbol{F}} + \boldsymbol{H}$ where $\frac{\partial \gamma}{\partial \boldsymbol{F}} : \boldsymbol{H} = 0$. If we could choose the gradient $\partial \boldsymbol{\eta}/\partial \boldsymbol{z}$ to equal $\boldsymbol{H}$, then (12.50) would immediately yield the desired result. Since we have no assurance that $\boldsymbol{H}$ is a gradient, we must use another approach.

We shall demonstrate that (12.53) essentially holds for (12.4)–(12.7) by invoking a multiplier rule. We now introduce the notation needed to state this rule: Let $\mathfrak{X}$ and $\mathfrak{Y}$ be Banach spaces and let $\mathfrak{X}^*$ and $\mathfrak{Y}^*$ be their duals. (See Chap. XVII for definitions.) Let the value of the functional x^* (in $\mathfrak{X}^*$) at x in $\mathfrak{X}$ be denoted $\langle x^*, x\rangle$ and let the value of the functional y^* (in $\mathfrak{Y}^*$) at y in $\mathfrak{Y}$ be denoted $\langle y^*, y\rangle$. We denote the space of all continuous linear transformations from $\mathfrak{X}$ to $\mathfrak{Y}$ by $\mathfrak{L}(\mathfrak{X}, \mathfrak{Y})$. If $G \in \mathfrak{L}(\mathfrak{X}, \mathfrak{Y})$, then its adjoint, which is in $\mathfrak{L}(\mathfrak{Y}^*, \mathfrak{X}^*)$, is denoted G^*.

12.54. Multiplier Rule. *Let $\tau^* \in \mathfrak{X}^*$ and let $G \in \mathfrak{L}(\mathfrak{X}, \mathfrak{Y})$. If*

$$\langle \tau^*, x\rangle = 0 \quad \forall x \in \mathfrak{X} \quad \textit{such that } Gx = 0 \tag{12.55}$$

and if G maps $\mathfrak{X}$ onto $\mathfrak{Y}$, then there is an element $\lambda^* \in \mathfrak{Y}^*$ such that

$$\langle \tau^*, x\rangle + \langle \lambda^*, Gx\rangle = 0 \quad \forall x \in \mathfrak{X}. \tag{12.56}$$

We shall identify x with $\boldsymbol{\eta}$, λ^* with the λ of (12.26b), the equation $\langle \tau^*, x\rangle = 0$ with (12.50), and the equation $Gx = 0$ with (12.52). The simple proof of this theorem, which is the same as Theorem XVII.2.24, is given in Sec. XVII.2.

To apply this theorem, we need only verify the surjectivity of G, i.e., we need only show that the nonhomogeneous version

$$\frac{\partial \gamma}{\partial \boldsymbol{F}} : \frac{\partial \boldsymbol{\eta}}{\partial \boldsymbol{z}} = \omega \tag{12.57}$$

of (12.52) has a solution $\boldsymbol{\eta}$ for all ω. In the examples to be analyzed here and at the end of Sec. 15, we examine (12.57) formally before discussing suitable function spaces for it.

We first treat (12.5), because it is technically the easiest constraint. Choosing $\boldsymbol{a} = \boldsymbol{g}_3$, we convert (12.5) to

$$\tilde{\boldsymbol{p}}_{,3} \cdot \tilde{\boldsymbol{p}}_{,3} = 1, \tag{12.58}$$

so that (12.44b) is equivalent to

$$\tilde{\boldsymbol{p}}_{,3} \cdot \overset{\vartriangle}{\boldsymbol{p}}_{,3} = 0. \tag{12.59}$$

In view of (12.43a), condition (12.50b) has the form

$$\int_{\tilde{\mathbf{x}}(\mathcal{B})} \boldsymbol{\tau}_{\mathrm{L}}^{k}(\mathbf{x}, t) \cdot \overset{\vartriangle}{\boldsymbol{p}}_{,k}(\mathbf{x}, t)\, dv(\mathbf{x}) = 0 \tag{12.60}$$

for all $\overset{\vartriangle}{\boldsymbol{p}}$ satisfying (12.59). Equation (12.57) reduces to

$$\tilde{\boldsymbol{p}}_{,3} \cdot \overset{\vartriangle}{\boldsymbol{p}}_{,3} = \omega. \tag{12.61}$$

Let us take

$$\overset{\vartriangle}{\boldsymbol{p}}_{,3} = \omega \tilde{\boldsymbol{p}}_{,3}, \tag{12.62}$$

which satisfies (12.61) by virtue of (12.58). We integrate (12.62) with respect to x^3 to find a $\overset{\vartriangle}{\boldsymbol{p}}$ that satisfies (12.61).

Now (12.58) implies that $\tilde{\boldsymbol{p}}_{,3}(\cdot, t) \in L_\infty(\tilde{\mathbf{x}}(\mathcal{B}))$. If we take $\omega \in L_q(\tilde{\mathbf{x}}(\mathcal{B}))$ where $q \geq 1$, then (12.62) further implies that $\overset{\vartriangle}{\boldsymbol{p}}_{,3}(\cdot, t) \in L_q(\tilde{\mathbf{x}}(\mathcal{B}))$. We choose function spaces for ω and $\overset{\vartriangle}{\boldsymbol{p}}$ to be consistent with this requirement.

Invoking the Multiplier Rule, we find that there are functions $\tilde{\lambda}$ and λ such that

$$\begin{aligned} 0 &= \int_{\tilde{\mathbf{x}}(\mathcal{B})} \left[\boldsymbol{\tau}_{\mathrm{L}}^{k}(\mathbf{x}, t) - \tilde{\lambda}(\mathbf{x}, t)\tilde{\boldsymbol{p}}_{,3}(\mathbf{x}, t)\delta_3^k\right] \cdot \overset{\vartriangle}{\boldsymbol{p}}_{,k}(\mathbf{x}, t)\, j(\mathbf{x})\, dx^1\, dx^2\, dx^3 \\ &\equiv \int_{\mathcal{B}} [\boldsymbol{T}_{\mathrm{L}}(\boldsymbol{z}, t) - \lambda(\boldsymbol{z}, t)\tilde{\boldsymbol{p}}_{,3}(\tilde{\mathbf{x}}(\boldsymbol{z}), t)\boldsymbol{g}_3(\tilde{\mathbf{x}}(\boldsymbol{z}))] : \boldsymbol{\eta}_{\boldsymbol{z}}(\boldsymbol{z}, t)\, dv(\boldsymbol{z}) \end{aligned} \tag{12.63}$$

for all $\overset{\vartriangle}{\boldsymbol{p}}$ or, equivalently, for all $\boldsymbol{\eta}$. This equation implies that

$$\text{(12.64a)} \qquad \left[j\left(\boldsymbol{\tau}_{\mathrm{L}}^{k} - \tilde{\lambda}\tilde{\boldsymbol{p}}_{,3}\delta_{3}^{k}\right)\right]_{,k} = \mathbf{0}$$

in the sense of distributions on $\tilde{\mathbf{x}}(\mathcal{B})$ or, equivalently, that

$$\text{(12.64b)} \qquad \nabla \cdot [\boldsymbol{T}_{\mathrm{L}} - \lambda\tilde{\boldsymbol{p}}_{,3}\boldsymbol{g}_3]^* = \mathbf{0}$$

in the sense of distributions on $\mathcal{B}$ (cf. (7.21)). Thus we conclude that

$$\text{(12.65a)} \qquad \boldsymbol{\tau}_{\mathrm{L}}^{k} = \tilde{\lambda}\tilde{\boldsymbol{p}}_{,3}\delta_{3}^{k} + \boldsymbol{n}^{k}$$

or, equivalently, that

$$\text{(12.65b)} \qquad \boldsymbol{T}_{\mathrm{L}} = \lambda\tilde{\boldsymbol{p}}_{,3}\boldsymbol{g}_3 + \boldsymbol{N}$$

where $(j\boldsymbol{n}^k)_{,k} = \mathbf{0}$ or $\nabla \cdot \boldsymbol{N}^* = \mathbf{0}$ in the sense of distributions, i.e.,

$$\text{(12.66)} \qquad \int_{\mathcal{B}} \boldsymbol{N} : \boldsymbol{\eta}_{\mathbf{z}}\, dv = 0 \quad \forall\, \boldsymbol{\eta} \in \mathfrak{D}(\mathcal{B}).$$

Here $\mathfrak{D}(\mathcal{B})$ is the space of infinitely differentiable functions with compact support in $\mathcal{B}$. But any such distribution $\boldsymbol{N}$ can enter neither into (12.50b) nor into the Principle of Virtual Power (9.5). Accordingly, we can set $\boldsymbol{N}$ equal to $\boldsymbol{O}$. In this case, (12.65) is equivalent to (12.35).

One of the virtues of the theory of stress developed by Noll (1966) on the basis of forces acting between bodies (see Truesdell (1991a)) is that it leads to a uniquely defined stress, unpolluted by stresses $\boldsymbol{N}$ having zero divergences.

The use of the Multiplier Rule obviates the need for constructing functions $\overset{\vartriangle}{\boldsymbol{p}}$ with compact support that satisfy (12.52). This construction is easy for (12.5), which reduces (12.52) to (12.59), but difficult for (12.4), (12.6), and (12.7). We avoid this difficult construction by treating these three constraints by the Multiplier Rule in the setting of the spatial formulation in Sec. 15.

We now study the Kirchhoff constraints (12.37) or (12.39). For these constraints, we can write the system corresponding to (12.57) as

$$\text{(12.67a,b)} \qquad \tilde{\boldsymbol{p}}_{,\alpha} \cdot \overset{\vartriangle}{\boldsymbol{p}}_{,3} + \tilde{\boldsymbol{p}}_{,3} \cdot \overset{\vartriangle}{\boldsymbol{p}}_{,\alpha} = \omega_\alpha, \quad \tilde{\boldsymbol{p}}_{,3} \cdot \overset{\vartriangle}{\boldsymbol{p}}_{,3} = \omega_3$$

where $\tilde{\boldsymbol{p}}$ is given by (12.39). Let $\{\boldsymbol{h}^1, \boldsymbol{h}^2, \boldsymbol{d}\}$ be the basis dual to $\{\tilde{\boldsymbol{p}}_{,k}\}$. (It can be found explicitly.) By differentiating both equations of (12.67) with respect to x^3, differentiating (12.67b) with respect to x^α, and using (12.39), we obtain the representation

$$\text{(12.68)} \qquad \overset{\vartriangle}{\boldsymbol{p}}_{,33} = (\omega_{\alpha,3} - \omega_{3,\alpha})\,\boldsymbol{h}^\alpha + \omega_{3,3}\boldsymbol{d}.$$

If we integrate this equation twice with respect to x^3, substitute the resulting expression for $\overset{\vartriangle}{\boldsymbol{p}}$ into (12.67a), and then differentiate this modified version of (12.67a) twice with respect to x^3, we obtain the compatibility equations

$$\text{(12.69)} \qquad \left(\omega_{\beta,3} - \omega_{3,\beta}\right)_{,3} = \boldsymbol{h}^\alpha_{,\beta} \cdot \boldsymbol{d}\,(\omega_{\alpha,3} - \omega_{3,\alpha}),$$

which is a linear system of ordinary differential equations for the $\omega_{\beta,3} - \omega_{3,\beta}$. It is clear that we can find ω_k's that do not satisfy this equation. Thus it is impossible to fulfill the hypotheses of the Multiplier Rule.

All these considerations allow us to enunciate the following informally stated consequence of the Multiplier Rule:

12.70. Theorem. *Let body $\mathcal{B}$ be subjected to a system of K simple material constraints of the form*

$$\gamma_a(\partial \boldsymbol{p}/\partial \boldsymbol{z}, \boldsymbol{z}) = 0, \quad a = 1, \dots, K < 6. \tag{12.71}$$

If the linear system

$$\frac{\partial \gamma_a}{\partial \boldsymbol{F}}\left(\frac{\partial \boldsymbol{p}}{\partial \boldsymbol{z}}, \boldsymbol{z}\right) : \frac{\partial \boldsymbol{\eta}}{\partial \boldsymbol{z}} = \omega_a, \quad a = 1, \dots, K < 6, \tag{12.72}$$

has a solution $\boldsymbol{\eta}$ for all ω_a in a suitable function space on $\mathcal{B}$, then the Global Constraint Principle in a suitable function-space setting reduces to the Local Constraint Principle in that (12.26b) *holds (to within a divergence-free stress).*

It is easy to generalize this statement to arbitrary constraints of the form (12.44a).

We now examine the reactive stress for a system of constraints somewhat weaker than the Kirchhoff constraints (12.37). We generalize these by adopting (12.39) without the restrictions (12.40). Then the virtual positions (displacements, velocities), tangent to the constraint manifold (12.39), are

$$\overset{\vartriangle}{\boldsymbol{p}}(\mathbf{x}) = \overset{\vartriangle}{\boldsymbol{r}}(x^1, x^2) + x^3 \overset{\vartriangle}{\boldsymbol{d}}(x^1, x^2) \tag{12.73}$$

where

$$\overset{\vartriangle}{\boldsymbol{r}} \text{ and } \overset{\vartriangle}{\boldsymbol{d}} \text{ are arbitrary.} \tag{12.74}$$

In view of (12.43a), condition (12.50b) reduces to

$$\int_{\tilde{\mathbf{x}}(\mathcal{B})} \left[\boldsymbol{\tau}_{\mathrm{L}}^{\alpha} \cdot \left(\overset{\vartriangle}{\boldsymbol{r}}_{,\alpha} + x^3 \overset{\vartriangle}{\boldsymbol{d}}_{,\alpha}\right) + \boldsymbol{\tau}_{\mathrm{L}}^{3} \cdot \overset{\vartriangle}{\boldsymbol{d}}\right] dv(\mathbf{x}) = 0 \tag{12.75}$$

for all $\overset{\vartriangle}{\boldsymbol{r}}$ and $\overset{\vartriangle}{\boldsymbol{d}}$, where

$$\boldsymbol{\tau}_{\mathrm{L}}^{k} = \boldsymbol{T}_{\mathrm{L}} \cdot \boldsymbol{g}^k = S_{\mathrm{L}}^{kl} \boldsymbol{p}_{,l} \tag{12.76}$$

with $\boldsymbol{p}$ given by (12.39) and with $\boldsymbol{S}_{\mathrm{L}}$ symmetric.

Let us assume that

$$\tilde{\mathbf{x}}(\mathcal{B}) = \{\mathbf{x} \in \mathbb{R}^3 : (x^1, x^2) \in \mathcal{M} \subset \mathbb{R}^2,\ h^-(x^1, x^2) < x^3 < h^+(x^1, x^2)\}. \tag{12.77}$$

Then (12.75) reduces to

$$\int_{\mathcal{M}} \left[\left(\int_{h^-}^{h^+} \boldsymbol{\tau}_{\mathrm{L}}^{\alpha}\, j\, dx^3\right) \cdot \overset{\vartriangle}{\boldsymbol{r}}_{,\alpha} + \left(\int_{h^-}^{h^+} x^3 \boldsymbol{\tau}_{\mathrm{L}}^{\alpha}\, j\, dx^3\right) \cdot \overset{\vartriangle}{\boldsymbol{d}}_{,\alpha} + \left(\int_{h^-}^{h^+} \boldsymbol{\tau}_{\mathrm{L}}^{3}\, j\, dx^3\right) \cdot \overset{\vartriangle}{\boldsymbol{d}}\right] dx^1\, dx^2 = 0. \tag{12.78}$$

The arbitrariness of $\overset{\vartriangle}{\boldsymbol{r}}$ and $\overset{\vartriangle}{\boldsymbol{d}}$ implies that (12.78) is equivalent to

$$\left(\int_{h^-}^{h^+} \boldsymbol{\tau}_{\mathrm{L}}^{\alpha}\, j\, dx^3\right)_{,\alpha} = \mathbf{0}, \tag{12.79}$$

$$\left(\int_{h^-}^{h^+} x^3 \boldsymbol{\tau}_{\mathrm{L}}^{\alpha}\, j\, dx^3\right)_{,\alpha} - \left(\int_{h^-}^{h^+} \boldsymbol{\tau}_{\mathrm{L}}^{3}\, j\, dx^3\right) = \mathbf{0} \tag{12.80}$$

in the sense of distributions. Thus the Global Constraint Principle implies that the Lagrange multipliers $\boldsymbol{\tau}_{\mathrm{L}}^k$ for (12.39) belong to the infinite-dimensional space of functions satisfying (12.79) and (12.80). The integrals appearing in (12.79) and (12.80) are stress resultants. Their significance is discussed in Chap. XIV.

In addition to (12.79) and (12.80), these reactive stresses satisfy the underdetermined problem

$$(12.81a) \qquad \left(j\boldsymbol{\tau}_{\mathrm{L}}^k\right)_{,k} = -\left(j\boldsymbol{\tau}_{\mathrm{A}}^k\right)_{,k} - j\boldsymbol{f} + \tilde{\rho} j \partial_{tt}\boldsymbol{\pi},$$

$$(12.81b) \qquad \boldsymbol{\tau}_{\mathrm{L}}^k \nu_k = -\boldsymbol{\tau}_{\mathrm{A}}^k \nu_k + \overline{\boldsymbol{\tau}} \quad \text{for } x^3 = h^{\pm}(x^1, x^2),$$

in the sense of distributions. Note that the terms on the right-hand sides of (12.81) are typically known from the solution of the primary problem. (Representations for the Lagrange multipliers can be obtained in terms of Betti-Gwyther-Finzi-Maxwell-Morera potentials for the stress, which allow for the arbitrariness inherent in (12.81); see Truesdell & Toupin (1960, Chap. D.IV). These representations simplify greatly for two-dimensional problems.)

To treat the Kirchhoff constraints (12.39) and (12.40), we use (12.73) with (12.74) replaced by the requirement that $\overset{\vartriangle}{\boldsymbol{r}}$ and $\overset{\vartriangle}{\boldsymbol{d}}$ are arbitrary functions satisfying

$$(12.82) \qquad \boldsymbol{d} \cdot \overset{\vartriangle}{\boldsymbol{d}} = 0, \quad \boldsymbol{d} \cdot \overset{\vartriangle}{\boldsymbol{r}}_{,\alpha} + \boldsymbol{r}_{,\alpha} \cdot \overset{\vartriangle}{\boldsymbol{d}} = 0.$$

The details of the treatment for these constraints are given in Sec. XIV.10.

The theory of material constraints (for three-dimensional continuum mechanics) was developed and refined by Ericksen & Rivlin (1954), Truesdell & Noll (1965), Antman (1982), and Antman & Marlow (1991). Most of this section is based on this last work. Other pertinent references are Green, Naghdi, & Trapp (1970), Gurtin & Podio-Guidugli (1973), Podio-Guidugli (1990), Podio-Guidugli & Vianello (1989), and Vianello (1990). A formal theory for the relaxation of the constraint of incompressibility was developed by Spencer (1964). A refinement of his approach in the setting of planar rod theories was given by Antman (1968); it is generalized in Ex. VIII.15.15. For mathematical justifications of such methods see Le Dret (1985), Le Tallec & Oden (1981), Rostamian (1978, 1981), and Schochet (1985).

13. Isotropy

We now characterize materials whose mechanical properties have no preferred directions. Our treatment happens to be much simpler and more direct than that given for rods in Sec. VIII.9, which we refer to for motivation. Of the three testing scenarios described at the beginning of that section, we adopt the following: We deform a small ball of material and measure the stress needed to maintain the deformation. Now we repeat the experiment, keeping the reference configuration of the ball the same, but rotating the testing apparatus. If the stress necessary to maintain the rotated deformation is just the rotated version of the stress needed to maintain the original deformation, and if this property holds for all rotations, then the material is said to be *hemitropic*; otherwise it is called *aeolotropic*. Our concept of hemitropy is always associated with a given reference configuration. We now convert this verbiage to precise mathematics.

We consider simple materials with constitutive equations of the general form

$$(13.1) \qquad \boldsymbol{T}(\boldsymbol{z}, t) = \hat{\boldsymbol{T}}(\boldsymbol{F}^t(\boldsymbol{z}, \cdot), \boldsymbol{z}).$$

We specialize our results to frame-indifferent versions of (13.1) shortly. We fix $(\boldsymbol{z}, t)$ and suppress their appearance as arguments of $\boldsymbol{T}$ and $\boldsymbol{F}$.

Let $\{\boldsymbol{a}_k \boldsymbol{b}_l\}$ be a dyadic basis for Lin. In the first experiment, the body suffers the deformation given by $\boldsymbol{F} = F^{kl}\boldsymbol{a}_k\boldsymbol{b}_l$. Let $\boldsymbol{Q}$ be a proper-orthogonal tensor and let $\boldsymbol{a}_k^\sharp = \boldsymbol{Q}\cdot\boldsymbol{a}_k$ and $\boldsymbol{b}_l^\sharp = \boldsymbol{Q}\cdot\boldsymbol{b}_l$. In the second experiment, the body suffers the deformation given by $\boldsymbol{F}_\sharp = F^{kl}\boldsymbol{a}_k^\sharp\boldsymbol{b}_l^\sharp$, which has the *same* components as $\boldsymbol{F}$. Thus

$$\boldsymbol{F}_\sharp = F^{kl}\boldsymbol{Q}\cdot\boldsymbol{a}_k\boldsymbol{Q}\cdot\boldsymbol{b}_l = \boldsymbol{Q}\cdot\boldsymbol{F}\cdot\boldsymbol{Q}^*. \tag{13.2}$$

Let $\boldsymbol{T} = T^{kl}\boldsymbol{a}_k\boldsymbol{b}_l$ be the stress corresponding to $\boldsymbol{F}$ and let $\boldsymbol{T}_\sharp = T_\sharp^{kl}\boldsymbol{a}_k^\sharp\boldsymbol{b}_l^\sharp$ be the stress corresponding to $\boldsymbol{F}_\sharp$. If $T^{kl} = T_\sharp^{kl}$, then the material (at $\boldsymbol{z}$) is *hemitropic*, in which case we find as in (13.2) that

$$\boldsymbol{T}_\sharp = \boldsymbol{Q}\cdot\boldsymbol{T}\cdot\boldsymbol{Q}^*. \tag{13.3}$$

We now substitute (13.1) into (13.3) to obtain

$$\hat{\boldsymbol{T}}(\boldsymbol{Q}\cdot\boldsymbol{F}^t(\boldsymbol{z},\cdot)\cdot\boldsymbol{Q}^*, \boldsymbol{z}) \equiv \hat{\boldsymbol{T}}(\boldsymbol{F}_\sharp^t(\boldsymbol{z},\cdot),\boldsymbol{z}) = \boldsymbol{Q}\cdot\hat{\boldsymbol{T}}(\boldsymbol{F}^t(\boldsymbol{z},\cdot),\boldsymbol{z})\cdot\boldsymbol{Q}^*. \tag{13.4}$$

Thus the material (13.1) is *hemitropic* iff (13.4) holds for all proper-orthogonal $\boldsymbol{Q}$. Notice that this condition is not the same as (11.12) for frame-indifference. There we did not rotate material fibers. We say that the material (13.1) is *isotropic* iff (13.4) holds for all orthogonal $\boldsymbol{Q}$.

13.5. Proposition. *If* (13.1) *is hemitropic, then it is isotropic.*

The proof is an immediate consequence of the fact that any orthogonal tensor is either proper-orthogonal or the negative of a proper-orthogonal tensor $\boldsymbol{Q}$ and that $(-\boldsymbol{Q})\cdot\boldsymbol{A}\cdot(-\boldsymbol{Q})^* = \boldsymbol{Q}\cdot\boldsymbol{A}\cdot\boldsymbol{Q}^*$ for all $\boldsymbol{A}\in$ Lin. This proposition does not hold for constitutive functions involving tensors of odd rank, as happens if vectors are either the independent or dependent variables in the constitutive equation.

Even for aeolotropic materials, Eq. (13.4) can be satisfied for certain $\boldsymbol{Q}$'s. In particular, it is satisfied for $\boldsymbol{Q} = \boldsymbol{I}$. Any $\boldsymbol{Q}$ for which (13.4) is satisfied is called a *symmetry transformation*. The set of all such transformations forms a group, the *symmetry group of the material*, which is a subgroup of the special (proper) orthogonal group. The symmetry group determines the nature of the aeolotropy. A material is thus hemitropic iff its symmetry group is the special orthogonal group.

To get the isotropic version of the frame-indifferent constitutive equation (11.14), we merely replace $\boldsymbol{F}$ in (13.4) with $\boldsymbol{U}$. To get the isotropic version of (11.19), we first observe that

$$\hat{\boldsymbol{T}}(\boldsymbol{F}^t(\boldsymbol{z},\cdot),\boldsymbol{z}) = \boldsymbol{F}(\boldsymbol{z},t)\cdot\hat{\boldsymbol{S}}(\boldsymbol{C}^t(\boldsymbol{z},\cdot),\boldsymbol{z}). \tag{13.6}$$

We substitute (13.6) into (13.4) to obtain the following alternative characterization of frame-indifferent hemitropic simple materials:

$$\hat{\boldsymbol{S}}(\boldsymbol{Q}\cdot\boldsymbol{C}^t(\boldsymbol{z},\cdot)\cdot\boldsymbol{Q}^*,\boldsymbol{z}) = \boldsymbol{Q}\cdot\hat{\boldsymbol{S}}(\boldsymbol{C}^t(\boldsymbol{z},\cdot),\boldsymbol{z})\cdot\boldsymbol{Q}^*. \tag{13.7}$$

For special classes of materials, we can get very specific representations of hemitropic materials. Let us first consider the frame-indifferent version of an elastic material (defined by (10.5) and (11.19)):

$$\boldsymbol{S}(\boldsymbol{z},t)=\hat{\boldsymbol{S}}(\boldsymbol{C}(\boldsymbol{z},t),\boldsymbol{z}), \tag{13.8}$$

with $\hat{\boldsymbol{S}}$ as symmetric tensor-valued function of the symmetric tensor $\boldsymbol{C}$. To it we can apply the

13.9. Representation Theorem for Hemitropic Tensor-Valued Functions. *Let $\mathcal{A}$ be a subset of* Sym *that is invariant under rotations. (Thus if $\boldsymbol{C}\in\mathcal{A}$, then $\boldsymbol{Q}\cdot\boldsymbol{C}\cdot\boldsymbol{Q}^*\in\mathcal{A}$ for all proper-orthogonal $\boldsymbol{Q}$.) A function*

$$\mathcal{A}\ni\boldsymbol{C}\mapsto\hat{\boldsymbol{S}}(\boldsymbol{C},\boldsymbol{z})\in\mathrm{Sym} \tag{13.10}$$

is hemitropic (and therefore isotropic) if and only if there are scalar-valued functions φ_0, φ_1, φ_2 of the invariants of $\boldsymbol{C}$ and of $\boldsymbol{z}$ such that

$$\hat{\boldsymbol{S}}(\boldsymbol{C},\boldsymbol{z})=\varphi_0(\iota(\boldsymbol{C}),\boldsymbol{z})\boldsymbol{I}+\varphi_1(\iota(\boldsymbol{C}),\boldsymbol{z})\boldsymbol{C}+\varphi_2(\iota(\boldsymbol{C}),\boldsymbol{z})\boldsymbol{C}^2 \tag{13.11}$$

where

$$\iota(\boldsymbol{C})\equiv(\mathrm{I}_{\boldsymbol{C}},\mathrm{II}_{\boldsymbol{C}},\mathrm{III}_{\boldsymbol{C}}). \tag{13.12}$$

By using the Cayley-Hamilton Theorem XI.1.44, we readily find that (13.11) is equivalent to

$$\hat{\boldsymbol{S}}(\boldsymbol{C},\boldsymbol{z})=\psi_{-1}(\iota(\boldsymbol{C}),\boldsymbol{z})\boldsymbol{C}^{-1}+\psi_0(\iota(\boldsymbol{C}),\boldsymbol{z})\boldsymbol{I}+\psi_1(\iota(\boldsymbol{C}),\boldsymbol{z})\boldsymbol{C}. \tag{13.13}$$

If $\tilde{\boldsymbol{S}}(\cdot,\boldsymbol{z})$ is both hemitropic and linear, then it has the form

$$\tilde{\boldsymbol{S}}(\boldsymbol{E},\boldsymbol{z})=\lambda(\boldsymbol{z})\mathrm{I}_{\boldsymbol{E}}\boldsymbol{I}+2\mu(\boldsymbol{z})\boldsymbol{E}. \tag{13.14}$$

The constitutive functions of linear elasticity and of Newtonian fluids have this character.

We follow Gurtin (1981a) in reducing the proof of Theorem 13.9 to the proofs of a sequence of simpler results, which have intrinsic interest. We let $\mathcal{A}$ be as in the statement of the theorem.

A scalar-valued function $\mathcal{A}\ni\boldsymbol{C}\mapsto\varphi(\boldsymbol{C})\in\mathbb{R}$ is *hemitropic* iff $\varphi(\boldsymbol{C})=\varphi(\boldsymbol{Q}\cdot\boldsymbol{C}\cdot\boldsymbol{Q}^*)$ for all proper-orthogonal $\boldsymbol{Q}$. This function is *isotropic* iff this relation holds for all orthogonal $\boldsymbol{Q}$. Thus a hemitropic scalar-valued function of a symmetric tensor is isotropic. The principal invariants of a tensor are therefore isotropic scalar-valued functions. Consequently the function $\mathcal{A}\ni\boldsymbol{C}\mapsto\psi(\iota(\boldsymbol{C}))$ is isotropic. The converse of this observation is also true:

13.15. Representation Theorem for Hemitropic Scalar-Valued Functions. *If $\mathcal{A} \ni \boldsymbol{C} \mapsto \varphi(\boldsymbol{C})$ is hemitropic, then there is a function $\iota(\boldsymbol{C}) \mapsto \tilde{\varphi}(\iota(\boldsymbol{C}))$ such that*

$$\varphi(\boldsymbol{C}) = \tilde{\varphi}(\iota(\boldsymbol{C})). \tag{13.16}$$

Proof. The hemitropy of φ implies that

$$\varphi(\boldsymbol{C}) = \varphi(\boldsymbol{D}) \tag{13.17}$$

for all $\boldsymbol{D}$ of the form $\boldsymbol{Q} \cdot \boldsymbol{C} \cdot \boldsymbol{Q}^*$ with $\boldsymbol{Q}$ orthogonal. Note that $\boldsymbol{C}$ and $\boldsymbol{D}$ have the same invariants. We prove the theorem by showing that

$$\varphi(\boldsymbol{A}) = \varphi(\boldsymbol{B}) \tag{13.18a}$$

whenever

$$\iota(\boldsymbol{A}) = \iota(\boldsymbol{B}) \tag{13.18b}$$

for hemitropic φ. Indeed, (13.18) says that φ cannot depend on anything other than the common invariants in (13.18b), i.e., (13.16) holds. Since $\boldsymbol{C}$ and $\boldsymbol{D}$ have common invariants, we can identify them with $\boldsymbol{A}$ and $\boldsymbol{B}$ of (13.18a), and thus deduce (13.18b) and (13.16).

We now prove (13.18). Let (13.18b) hold. Then (XI.1.39) and (XI.1.40) imply that $\boldsymbol{A}$ and $\boldsymbol{B}$ have the same eigenvalues $\{C_k\}$. By the spectral representation (XI.1.34), $\boldsymbol{A}$ and $\boldsymbol{B}$ have orthonormal bases of eigenvectors $\{\boldsymbol{a}_k\}$ and $\{\boldsymbol{b}_k\}$ such that

$$\boldsymbol{A} = \sum_{k=1}^{3} C_k \boldsymbol{a}_k \boldsymbol{a}_k, \quad \boldsymbol{B} = \sum_{k=1}^{3} C_k \boldsymbol{b}_k \boldsymbol{b}_k. \tag{13.19a,b}$$

Let $\boldsymbol{Q}$ be defined by $\boldsymbol{Q} \cdot \boldsymbol{a}_k = \boldsymbol{b}_k$. Therefore (13.19) implies that $\boldsymbol{Q} \cdot \boldsymbol{A} \cdot \boldsymbol{Q}^* = \boldsymbol{B}$. Since φ is hemitropic, $\varphi(\boldsymbol{A}) = \varphi(\boldsymbol{Q} \cdot \boldsymbol{A} \cdot \boldsymbol{Q}^*) = \varphi(\boldsymbol{B})$, which is (13.18a). □

13.20. Theorem. *Let $\mathcal{A} \ni \boldsymbol{C} \mapsto \boldsymbol{G}(\boldsymbol{C}) \in$ Sym be hemitropic. Then $\boldsymbol{C}$ and $\boldsymbol{G}(\boldsymbol{C})$ have the same eigenvectors.*

Proof. Let $\boldsymbol{c}$ be an eigenvector of $\boldsymbol{C}$ in $\mathcal{A}$ and let $\boldsymbol{Q}$, which is orthogonal but not proper, be the reflection across the plane perpendicular to $\boldsymbol{c}$:

$$\boldsymbol{Q} \cdot \boldsymbol{c} = -\boldsymbol{c}, \quad \boldsymbol{Q} \cdot \boldsymbol{a} = \boldsymbol{a} \quad \text{if } \boldsymbol{a} \cdot \boldsymbol{c} = 0. \tag{13.21}$$

Using the spectral representation (XI.1.34) for $\boldsymbol{C}$, we find that $\boldsymbol{Q} \cdot \boldsymbol{C} \cdot \boldsymbol{Q}^* = \boldsymbol{C}$. Since $\boldsymbol{G}$ is isotropic, we then find that $\boldsymbol{Q} \cdot \boldsymbol{G}(\boldsymbol{C}) \cdot \boldsymbol{Q}^* = \boldsymbol{G}(\boldsymbol{Q} \cdot \boldsymbol{C} \cdot \boldsymbol{Q}^*) = \boldsymbol{G}(\boldsymbol{C})$, so that $\boldsymbol{Q} \cdot \boldsymbol{G}(\boldsymbol{C}) = \boldsymbol{G}(\boldsymbol{C}) \cdot \boldsymbol{Q}$. Thus

$$\boldsymbol{Q} \cdot \boldsymbol{G}(\boldsymbol{C}) \cdot \boldsymbol{c} = \boldsymbol{G}(\boldsymbol{C}) \cdot \boldsymbol{Q} \cdot \boldsymbol{c} = -\boldsymbol{G}(\boldsymbol{C}) \cdot \boldsymbol{c} \tag{13.22}$$

so that $\boldsymbol{Q}$ takes $\boldsymbol{G}(\boldsymbol{C}) \cdot \boldsymbol{c}$ to its negative. But (13.21) implies that this can happen only if $\boldsymbol{G}(\boldsymbol{C}) \cdot \boldsymbol{c}$ is parallel to $\boldsymbol{c}$, i.e., only if there is a number λ such that $\boldsymbol{G}(\boldsymbol{C}) \cdot \boldsymbol{c} = \lambda \boldsymbol{c}$. Thus $\boldsymbol{c}$ is an eigenvector of $\boldsymbol{G}(\boldsymbol{C})$. □

13.23. Lemma. *Let $\boldsymbol{C} \in$ Sym have distinct eigenvalues $\{C_k\}$ and corresponding orthonormal eigenvectors $\{\boldsymbol{c}_k\}$. Then*

$$\operatorname{span}\{\boldsymbol{I}, \boldsymbol{C}, \boldsymbol{C}^2\} = \operatorname{span}\{\boldsymbol{c}_1\boldsymbol{c}_1, \boldsymbol{c}_2\boldsymbol{c}_2, \boldsymbol{c}_3\boldsymbol{c}_3\} \quad \textit{in } \text{Lin}. \tag{13.24}$$

Proof. We first show that $\{\boldsymbol{I}, \boldsymbol{C}, \boldsymbol{C}^2\}$ is linearly independent in Lin. Thus we must show that if there are numbers α, β, γ such that

$$\alpha \boldsymbol{C}^2 + \beta \boldsymbol{C} + \gamma \boldsymbol{I} = \boldsymbol{O}, \tag{13.25}$$

then $\alpha = \beta = \gamma = 0$. We let (13.25) operate on the eigenvectors $\{\boldsymbol{c}_k\}$ to obtain $(\alpha C_k^2 + \beta C_k + \gamma)\boldsymbol{c}_k = \boldsymbol{0}$ (no sum) so that $\alpha C_k^2 + \beta C_k + \gamma = 0$. Thus the three distinct eigenvalues are solutions of the same quadratic equation, which is possible only if the quadratic equation is completely degenerate: $\alpha = \beta = \gamma = 0$. The spectral representation (13.19a) implies that $\{\boldsymbol{I}, \boldsymbol{C}, \boldsymbol{C}^2\}$ lies in $\operatorname{span}\{\boldsymbol{c}_1\boldsymbol{c}_1, \boldsymbol{c}_2\boldsymbol{c}_2, \boldsymbol{c}_3\boldsymbol{c}_3\}$, so that (13.24) must hold. □

A simple proof yields

13.26. Lemma. *Let $C \in \mathrm{Sym}$ have exactly two distinct eigenvalues: C_1 corresponding to eigenvector c_1 and $C_2 = C_3$ whose eigenspace is the orthogonal complement of c_1. Then*

$$\text{span}\{I, C\} = \text{span}\{c_1c_1, I - c_1c_1\} \quad \text{in Lin}. \tag{13.27}$$

Proof of Theorem 13.9. First suppose that C has three distinct eigenvalues and accordingly has a spectral representation of the form (13.19a). Then Theorem 13.20 implies that $\hat{S}$ has a corresponding representation

$$\hat{S}(C, z) = \sum_{k=1}^{3} \sigma_k(z) c_k c_k. \tag{13.28}$$

It then follows from (13.24) that there are scalar-valued functions α_0, α_1, α_2 such that

$$\hat{S}(C, z) = \alpha_0(C, z)I + \alpha_1(C, z)C + \alpha_2(C, z)C^2. \tag{13.29}$$

If C has exactly two distinct eigenvalues, we can similarly use (13.27) to show that (13.29) holds with $\alpha_2 = 0$. If all the eigenvalues of C are equal, then C is proportional to I and again (13.29) holds.

To complete the proof, we need only show that $\alpha_0(\cdot, z), \alpha_1(\cdot, z), \alpha_2(\cdot, z)$ are isotropic, for then we can invoke Theorem 13.15. Since $\hat{S}(\cdot, z)$ is isotropic, it follows that

$$\hat{S}(C, z) - Q^* \cdot \hat{S}(Q \cdot C \cdot Q^*, z) \cdot Q = O. \tag{13.30}$$

We substitute (13.29) into (13.30) to obtain

$$\begin{aligned}[]
[\alpha_0(C, z) - \alpha_0(Q \cdot C \cdot Q^*, z)]I + [\alpha_1(C, z) - \alpha_1(Q \cdot C \cdot Q^*, z)]C& \\
+ [\alpha_2(C, z) - \alpha_2(Q \cdot C \cdot Q^*, z)]C^2 &= O.
\end{aligned} \tag{13.31}$$

If the eigenvalues of C are distinct, then the independence of $\{I, C, C^2\}$, implied by (13.24), forces the coefficients in (13.31) to vanish, so that $\alpha_0(\cdot, z), \alpha_1(\cdot, z), \alpha_2(\cdot, z)$ are isotropic. If C has exactly two distinct eigenvalues, then the vanishing of α_2 enables us to use (13.27) to show the isotropy of $\alpha_0(\cdot, z)$ and $\alpha_1(\cdot, z)$. If the eigenvalues of C are equal, then $\alpha_1 = 0 = \alpha_2$ and the isotropy of α_0 follows immediately from (13.31). □

The following result is useful for constitutive functions like those of (11.20) and (11.21) in which the constitutive function has a finite number of tensor arguments and for more general constitutive functions of thermomechanics in which vectors also appear as independent and dependent variables in the constitutive equations.

13.32. General Representation Theorem. *Let γ be an isotropic scalar-valued function, g be an isotropic vector-valued function, and G be an isotropic tensor-valued function of symmetric tensors $U_1, \ldots, U_M$ and vectors $v_1, \ldots, v_N$ so that*

$$\begin{aligned}
\gamma(U_1, \ldots, v_1, \ldots) &= \gamma(Q \cdot U_1 \cdot Q^*, \ldots, Q \cdot v_1, \ldots), \\
Q \cdot g(U_1, \ldots, v_1, \ldots) &= g(Q \cdot U_1 \cdot Q^*, \ldots, Q \cdot v_1, \ldots), \\
Q \cdot G(U_1, \ldots, v_1, \ldots) \cdot Q^* &= G(Q \cdot U_1 \cdot Q^*, \ldots, Q \cdot v_1, \ldots)
\end{aligned} \tag{13.33}$$

for all orthogonal Q. (It is understood that these vectors and tensors are defined for $\mathbb{E}^3$.) Let $\iota(U_1, \ldots, v_1, \ldots)$ denote a functionally independent set of all the invariants obtained from the following list by replacing A, B, C by all possible choices of $U_1, \ldots, U_M$, by replacing a, b by all possible choices of $v_1, \ldots, v_N$, and by including duplications unless there is a statement to the contrary:

$$\begin{gathered}
\text{tr}\, A,\ \text{tr}\,(A \cdot B),\ \text{tr}\,(A \cdot B \cdot C),\ a \cdot b,\ a \cdot A \cdot b,\ a \cdot A \cdot B \cdot b, \\
\text{tr}\,(A^2 \cdot B^2) \quad \text{for } A \neq B.
\end{gathered} \tag{13.34}$$

Then γ has the form

$$\gamma(\boldsymbol{U}_1,\ldots,\boldsymbol{v}_1,\ldots)=\tilde{\gamma}(\iota(\boldsymbol{U}_1,\ldots,\boldsymbol{v}_1,\ldots)). \tag{13.35}$$

$\boldsymbol{g}$ has the form

$$\boldsymbol{g}(\boldsymbol{U}_1,\ldots,\boldsymbol{v}_1,\ldots)=\sum_{\mu}\gamma^{\mu}(\iota(\boldsymbol{U}_1,\ldots,\boldsymbol{v}_1,\ldots))\boldsymbol{g}_{\mu}(\boldsymbol{U}_1,\ldots,\boldsymbol{v}_1,\ldots) \tag{13.36}$$

where the $\boldsymbol{g}_\mu$ form a functionally independent set of all the vectors obtained from the following list by replacing $\boldsymbol{A},\boldsymbol{B}$ by all possible choices of $\boldsymbol{U}_1,\ldots,\boldsymbol{U}_M$, by replacing $\boldsymbol{a}$ by all possible choices of $\boldsymbol{v}_1,\ldots,\boldsymbol{v}_N$, and by including duplications:

$$\boldsymbol{a},\ \boldsymbol{A}\cdot\boldsymbol{a},\ \boldsymbol{A}\cdot\boldsymbol{B}\cdot\boldsymbol{a}. \tag{13.37}$$

$\boldsymbol{G}$ has the form

$$\boldsymbol{G}(\boldsymbol{U}_1,\ldots,\boldsymbol{v}_1,\ldots)=\sum_{\nu}\gamma^{\nu}(\iota(\boldsymbol{U}_1,\ldots,\boldsymbol{v}_1,\ldots))\boldsymbol{G}_{\nu}(\boldsymbol{U}_1,\ldots,\boldsymbol{v}_1,\ldots) \tag{13.38}$$

where the $\boldsymbol{G}_\nu$ form a functionally independent set of all the symmetric tensors obtained from the following list by replacing $\boldsymbol{A},\boldsymbol{B}$ by all possible choices of $\boldsymbol{U}_1,\ldots,\boldsymbol{U}_M$, by replacing $\boldsymbol{a},\boldsymbol{b}$ by all possible choices of $\boldsymbol{v}_1,\ldots,\boldsymbol{v}_N$, and by including duplications unless there is a statement to the contrary:

(13.39)

$$\begin{gathered}\boldsymbol{I},\ \boldsymbol{A},\ \boldsymbol{A}\cdot\boldsymbol{B}+\boldsymbol{B}\cdot\boldsymbol{A},\\ \boldsymbol{A}^2\cdot\boldsymbol{B}+\boldsymbol{B}\cdot\boldsymbol{A}^2\quad\text{for } \boldsymbol{A}\neq\boldsymbol{B},\\ \boldsymbol{ab}+\boldsymbol{ba},\ \boldsymbol{a}(\boldsymbol{A}\cdot\boldsymbol{b})+(\boldsymbol{A}\cdot\boldsymbol{b})\boldsymbol{a},\ \boldsymbol{a}(\boldsymbol{A}^2\cdot\boldsymbol{b})+(\boldsymbol{A}^2\cdot\boldsymbol{b})\boldsymbol{a},\\ (\boldsymbol{A}\cdot\boldsymbol{a})(\boldsymbol{B}\cdot\boldsymbol{a})-(\boldsymbol{B}\cdot\boldsymbol{a})(\boldsymbol{A}\cdot\boldsymbol{a})\quad\text{for } \boldsymbol{A}\neq\boldsymbol{B},\quad \boldsymbol{A}\cdot\boldsymbol{B}\cdot\boldsymbol{aa}-\boldsymbol{aA}\cdot\boldsymbol{B}\cdot\boldsymbol{a}\quad\text{for } \boldsymbol{A}\neq\boldsymbol{B}.\end{gathered}$$

There is an extensive literature on representation theorems for scalar-, vector-, and tensor-valued functions invariant under subgroups of the orthogonal group. See Green & Adkins (1970), Pipkin & Rivlin (1966), and Truesdell & Noll (1965, Chap. BII) for summaries of results and extensive bibliographies. Theorem 13.9 is due to Rivlin & Ericksen (1955). An incomplete proof was given by Richter (1948). Our proof uses ideas of Serrin (1959, Sec. 59). Theorem 13.32 is due to G. F. Smith (1971) and Wang (1970b).

14. Thermomechanics

Consider two scenarios:
(i) At an initial time a block and a table have the same temperature. The block is slid across the table. It comes to rest with the temperatures of the block and table each higher than their initial values.
(ii) The block is at rest on the table, each at the same fixed temperature. Each body supplies thermal energy to their common interface, causing the block to accelerate and reducing the temperature of each body.

Scenario (ii) is virtually the reverse of scenario (i). Scenario (i) conforms to our experience; (ii) does not. It is a goal of thermodynamics to establish a sound basis of laws explaining why (ii) cannot occur. We seek a thermodynamics rich enough to encompass the behavior of heat-conducting continua undergoing large deformations.

The most familiar concepts of thermodynamics are temperature and heat. Temperature measures hotness, which is not the same thing as heat.

We 'know' what it means for one object to be hotter than another, but we are hard-pressed to say what it means for one object to be twice as hot as another. We know what heat is: We pay for it. But there is a qualitative difference between heats at various temperatures, as we can discover by trying to heat a house to 20°C with a million calories at 0°C. Thus temperature, hotness, and heat present subtle obstacles to our understanding. A notion such as entropy, about which we have little intuition, presents substantial difficulties.

In this section we outline a rational theory of continuum thermodynamics and apply it to constitutive equations of thermoviscoelasticity of differential type to determine the restrictions it imposes on these equations. We then specialize these results to thermoelastic media. (In Sec. XV.3 we use the same methods to treat thermoplasticity.)

Fundamental concepts. The *kinetic energy* of body $\mathcal{B}$ at time t is

$$K[\boldsymbol{p};\mathcal{B}](t) \equiv \tfrac{1}{2}\int_{\mathcal{B}} \rho(\boldsymbol{z})|\boldsymbol{p}_t(\boldsymbol{z},t)|^2\,dv(\boldsymbol{z}). \tag{14.1}$$

The (*mechanical*) *power* at time t of the forces acting on body $\mathcal{B}$ is

$$\begin{aligned} P[\boldsymbol{f},\boldsymbol{T}\cdot\boldsymbol{\nu};\mathcal{B}](t) \equiv \int_{\mathcal{B}} &\boldsymbol{f}(\boldsymbol{z},t)\cdot\boldsymbol{p}_t(\boldsymbol{z},t)\,dv(\boldsymbol{z}) \\ &+ \int_{\partial\mathcal{B}} [\boldsymbol{T}(\boldsymbol{z},t)\cdot\boldsymbol{\nu}(\boldsymbol{z})]\cdot\boldsymbol{p}_t(\boldsymbol{z},t)\,da(\boldsymbol{z}). \end{aligned} \tag{14.2}$$

The integral

$$\int_{t_1}^{t_2} P[\boldsymbol{f},\boldsymbol{T}\cdot\boldsymbol{\nu};\mathcal{B}](t)\,dt \tag{14.3}$$

is the *work done on* $\mathcal{B}$ *in the time interval* (t_1,t_2).

Now performing work on a body can make it move (i.e., can cause an increase in its kinetic energy) or can make it hotter. Heating a body can make it hotter or can cause it to do work. Alongside the mechanical power we accordingly introduce the nonmechanical power Q. Since we are not considering electromagnetic effects, we call this power the *heat power*. We assume that it has the form

$$Q[r,\boldsymbol{q};\mathcal{B}](t) \equiv \int_{\mathcal{B}} r(\boldsymbol{z},t)\,dv(\boldsymbol{z}) + \int_{\partial\mathcal{B}} \boldsymbol{q}(\boldsymbol{z},t)\cdot\boldsymbol{\nu}(\boldsymbol{z})\,da(\boldsymbol{z}). \tag{14.4}$$

r is the rate at which heat is generated at $(\boldsymbol{z},t)$ per unit volume of the reference configuration. $\boldsymbol{q}$ is the *heat flux vector*; $\boldsymbol{q}\cdot\boldsymbol{\nu}$ is rate per unit reference area at which heat enters the surface with outer normal $\boldsymbol{\nu}$. (Some authors switch the sign of $\boldsymbol{q}$.)

Doing work on a body and adding heat to it produces motion and causes the storage of energy, which is recoverable to do work. This energy, called the *internal energy*, is denoted

$$E[\varepsilon;\mathcal{B}](t) \equiv \int_{\mathcal{B}} \varepsilon(\boldsymbol{z},t)\,dv(\boldsymbol{z}). \tag{14.5}$$

ε is the *internal energy density per unit reference volume.* $K + E$ is the *total energy.*

We postulate the *Energy Balance* (*First Law of Thermodynamics*):

$$\frac{d}{dt}K + \frac{d}{dt}E = P + Q. \tag{14.6}$$

We substitute (14.1), (14.2), (14.4), (14.5) into (14.6) and use the Divergence Theorem to convert the surface integrals to volume integrals. Since $\mathcal{B}$ is arbitrary, we obtain

$$\rho \boldsymbol{p}_{tt} \cdot \boldsymbol{p}_t + \varepsilon_t = \boldsymbol{f} \cdot \boldsymbol{p}_t + \nabla \cdot (\boldsymbol{T}^* \cdot \boldsymbol{p}_t) + \nabla \cdot \boldsymbol{q} + r. \tag{14.7}$$

The introduction of components shows that

$$\nabla \cdot (\boldsymbol{T}^* \cdot \boldsymbol{p}_t) = (\nabla \cdot \boldsymbol{T}^*) \cdot \boldsymbol{p}_t + \boldsymbol{T} : \boldsymbol{p}_{zt}. \tag{14.8}$$

We substitute (14.8) into (14.7) and use (7.21) to obtain the *energy equation:*

$$\varepsilon_t = \boldsymbol{T} : \boldsymbol{F}_t + \nabla \cdot \boldsymbol{q} + r. \tag{14.9}$$

To get a coordinate-free proof of (14.8), we use (XI.2.19a) to obtain

$$\nabla \cdot (\boldsymbol{T}^* \cdot \boldsymbol{p}_t) = \nabla \cdot (\boldsymbol{p}_t \cdot \boldsymbol{T}) = [\partial(\boldsymbol{p}_t \cdot \boldsymbol{T})/\partial \boldsymbol{z}] : \boldsymbol{I}. \tag{14.10}$$

We compute $\partial(\boldsymbol{p}_t \cdot \boldsymbol{T})/\partial \boldsymbol{z}$ by the process of Sec. I.4:

$$\begin{aligned}
[\partial(\boldsymbol{p}_t \cdot \boldsymbol{T})/\partial \boldsymbol{z}] \cdot \boldsymbol{y} &= \frac{\partial}{\partial \varepsilon}[\boldsymbol{p}_t(\boldsymbol{z} + \varepsilon \boldsymbol{y}, t) \cdot \boldsymbol{T}(\boldsymbol{z} + \varepsilon \boldsymbol{y}, t)]\bigg|_{\varepsilon=0} \\
&= [(\partial \boldsymbol{p}_t/\partial \boldsymbol{z}) \cdot \boldsymbol{y}] \cdot \boldsymbol{T} + \boldsymbol{p}_t \cdot [(\partial \boldsymbol{T}/\partial \boldsymbol{z}) \cdot \boldsymbol{y}] \\
&= [\boldsymbol{T}^* \cdot (\partial \boldsymbol{p}_t/\partial \boldsymbol{z}) + \boldsymbol{p}_t \cdot (\partial \boldsymbol{T}/\partial \boldsymbol{z})] \cdot \boldsymbol{y}.
\end{aligned} \tag{14.11}$$

Therefore, (XI.1.21a) and (XI.2.19) imply that

$$\begin{aligned}
\nabla \cdot (\boldsymbol{T}^* \cdot \boldsymbol{p}_t) &= [\boldsymbol{T}^* \cdot (\partial \boldsymbol{p}_t/\partial \boldsymbol{z})] : \boldsymbol{I} + \boldsymbol{p}_t \cdot (\partial \boldsymbol{T}/\partial \boldsymbol{z}) : \boldsymbol{I} \\
&= \operatorname{tr}[\boldsymbol{T}^* \cdot (\partial \boldsymbol{p}_t/\partial \boldsymbol{z})] + \boldsymbol{p}_t \cdot (\nabla \cdot \boldsymbol{T}^*) \\
&= \boldsymbol{T} : \boldsymbol{F}_t + \boldsymbol{p}_t \cdot (\nabla \cdot \boldsymbol{T}^*).
\end{aligned} \tag{14.12}$$

It is useful to get an alternative representation for the stress power $\boldsymbol{T} : \boldsymbol{F}_t$, which appears in (14.9). From the dyadic identity

$$(\boldsymbol{a}_1\boldsymbol{a}_2 \cdot \boldsymbol{b}_1\boldsymbol{b}_2) : (\boldsymbol{c}_1\boldsymbol{c}_2) = (\boldsymbol{b}_1\boldsymbol{b}_2) : (\boldsymbol{a}_2\boldsymbol{a}_1 \cdot \boldsymbol{c}_1\boldsymbol{c}_2), \tag{14.13a}$$

we obtain the general identity

$$(\boldsymbol{A} \cdot \boldsymbol{B}) : \boldsymbol{C} = \boldsymbol{B} : (\boldsymbol{A}^* \cdot \boldsymbol{C}). \tag{14.13b}$$

By using the definition (7.30) of $\boldsymbol{S}$, (14.13b), the symmetry of $\boldsymbol{S}$, and the definition (2.3) of $\boldsymbol{C}$ we obtain

$$\begin{aligned}
\boldsymbol{T} : \boldsymbol{F}_t = (\boldsymbol{F} \cdot \boldsymbol{S}) : \boldsymbol{F}_t = \boldsymbol{S} : (\boldsymbol{F}^* \cdot \boldsymbol{F}_t) &= \tfrac{1}{2}\boldsymbol{S} : (\boldsymbol{F}^* \cdot \boldsymbol{F}_t + \boldsymbol{F}_t^* \cdot \boldsymbol{F}) \\
&= \tfrac{1}{2}\boldsymbol{S} : \boldsymbol{C}_t.
\end{aligned} \tag{14.14}$$

(This identity can be obtained very easily by using components.)

The *absolute or Kelvin temperature* $\theta(\boldsymbol{z}, t)$ of $\boldsymbol{z}$ at time t is a scalar-valued function. We postulate that

$$\theta(\boldsymbol{z}, t) > 0 \quad \forall (\boldsymbol{z}, t). \tag{14.15}$$

Entropy. The Energy Balance expresses the equivalence of power and the rate of energy production, but does not prescribe the manner in which conversion can take place. Thus it says nothing about irreversibility and nothing about dissipation. Entropy is the notion associated with irreversibility.

Consider a homogeneous body $\mathcal{B}$ that has a uniform temperature $\theta(t)$. We take as a fundamental postulate that the rate $Q[\mathcal{B}](t)$ at which heat can be added to a body is bounded above by a function $\theta(t)\frac{d}{dt}H[\mathcal{B}](t)$, depending on the state of the body and on its constitution:

$$Q \leq \theta \tfrac{d}{dt} H[\mathcal{B}]. \tag{14.16}$$

$H[\mathcal{B}](t_2) - H[\mathcal{B}](t_1)$ is the *entropy* produced in the body in the time interval (t_1, t_2). Condition (14.16) is the *Clausius-Planck Form of the Second Law of Thermodynamics* for homogeneous processes.

Now we consider materials that need not have uniform fields at each instant of time. We introduce the *entropy* of $\mathcal{B}$ at time t:

$$H[\eta, \mathcal{B}](t) = \int_{\mathcal{B}} \eta(\boldsymbol{z}, t)\, dv(\boldsymbol{z}). \tag{14.17}$$

η is the *entropy per unit reference volume.* A generalization of (14.16) to continua without uniform fields is the *Clausius-Duhem Form of the Second Law of Thermodynamics* or the *Clausius-Duhem Inequality*:

$$\int_{\mathcal{B}} \frac{r(\boldsymbol{z},t)}{\theta(\boldsymbol{z},t)}\, dv(\boldsymbol{z}) + \int_{\partial\mathcal{B}} \frac{\boldsymbol{q}(\boldsymbol{z},t)\cdot\boldsymbol{\nu}(\boldsymbol{z})}{\theta(\boldsymbol{z},t)}\, da(\boldsymbol{z}) \leq \frac{d}{dt}\int_{\mathcal{B}} \eta(\boldsymbol{z},t)\, dv(\boldsymbol{z}). \tag{14.18}$$

Its local form is

$$r + \theta \nabla \cdot \left(\frac{\boldsymbol{q}}{\theta}\right) \leq \theta \eta_t. \tag{14.19}$$

Let

$$\boldsymbol{g} \equiv \nabla\theta. \tag{14.20}$$

The substitution of (14.9) and (14.14) into (14.19) yields

$$\theta\eta_t - \varepsilon_t + \tfrac{1}{2}\boldsymbol{S} : \boldsymbol{C}_t + \frac{\boldsymbol{q}\cdot\boldsymbol{g}}{\theta} \geq 0. \tag{14.21}$$

If we introduce the *Helmholtz free-energy density* ψ *per unit reference volume* by

$$\psi \equiv \varepsilon - \eta\theta, \tag{14.22}$$

then (14.21) reduces to

$$-\psi_t - \eta\theta_t + \tfrac{1}{2}\boldsymbol{S} : \boldsymbol{C}_t + \frac{\boldsymbol{q}\cdot\boldsymbol{g}}{\theta} \geq 0, \tag{14.23}$$

which is the local form of the Clausius-Duhem Inequality that we employ.

Another generalization of (14.16) to continua is the *Clausius-Planck Inequality*:

$$r + \nabla \cdot \boldsymbol{q} \leq \theta \eta_t. \tag{14.24}$$

It does not seem to have a natural integral form.

We may adopt the Clausius-Duhem Inequality as a universal law of nature or merely as an interesting hypothesis worthy of study. Rather than trying to make the plausible hypothesis (14.16) compelling and then trying to justify (14.18) as its natural generalization (which can be done in the setting of gas dynamics (see Truesdell & Bharatha (1977)), we shall be content with determining consequences of (14.19) when the fields are smooth. Our methods can be applied to any inequality replacing (14.19).

Constitutive restrictions. We define a *simple thermomechanical material* to be one with constitutive equations of the form

$$\begin{aligned}
\boldsymbol{T}(\boldsymbol{z}, t) &= \hat{\boldsymbol{T}}(\boldsymbol{F}^t(\boldsymbol{z}, \cdot), \theta^t(\boldsymbol{z}, \cdot), \boldsymbol{g}^t(\boldsymbol{z}, \cdot), \boldsymbol{z}), \\
\boldsymbol{q}(\boldsymbol{z}, t) &= \hat{\boldsymbol{q}}(\boldsymbol{F}^t(\boldsymbol{z}, \cdot), \theta^t(\boldsymbol{z}, \cdot), \boldsymbol{g}^t(\boldsymbol{z}, \cdot), \boldsymbol{z}), \\
\eta(\boldsymbol{z}, t) &= \hat{\eta}(\boldsymbol{F}^t(\boldsymbol{z}, \cdot), \theta^t(\boldsymbol{z}, \cdot), \boldsymbol{g}^t(\boldsymbol{z}, \cdot), \boldsymbol{z}), \\
\psi(\boldsymbol{z}, t) &= \hat{\psi}(\boldsymbol{F}^t(\boldsymbol{z}, \cdot), \theta^t(\boldsymbol{z}, \cdot), \boldsymbol{g}^t(\boldsymbol{z}, \cdot), \boldsymbol{z}),
\end{aligned} \tag{14.25}$$

Note that these equations permit a very general class of couplings between thermal and mechanical effects. We shall show how the Clausius-Duhem inequality can be used to specialize the form of these equations.

14.26. Exercise. Put (14.25) into frame-indifferent form.

A *thermomechanical process* for a body consists of the fields $\boldsymbol{p}, \theta, \boldsymbol{T}, \boldsymbol{q}, \eta$, ψ that satisfy the inequalities $\det \boldsymbol{p}_z > 0$ and $\theta > 0$, the balance of linear momentum, the balance of angular momentum, and the balance of energy. We can vary these processes by varying the body force $\boldsymbol{f}$ and the rate of heat supply r.

We may interpret the Clausius-Duhem inequality (or any entropy inequality we adopt as a statement of the Second Law of Thermodynamics) as a requirement that a real material can never behave so that the entropy inequality is ever violated. Thus we adopt the

14.27. Coleman-Noll Entropy Principle. *The constitutive functions* (14.25) *must satisfy the entropy inequality for all thermomechanical processes.*

Since $\boldsymbol{f}$ and r are at our disposal, we need not worry about the fields constituting a thermomechanical process satisfying the balance of linear momentum or the balance of energy. The balance of angular momentum is automatically ensured by the requirement that the constitutive function $\hat{\boldsymbol{T}}$ satisfy it identically. Thus Principle 14.27 has the

14.28. Corollary. *The constitutive functions* (14.25) *must satisfy the entropy inequality identically.*

We now apply this corollary to the class of thermoviscoelastic materials of differential type having constitutive equations of the form

$$\boldsymbol{T}(\boldsymbol{z},t)=\hat{\boldsymbol{T}}(\boldsymbol{F}(\boldsymbol{z},t),\boldsymbol{F}_t(\boldsymbol{z},t),\theta(\boldsymbol{z},t),\boldsymbol{g}(\boldsymbol{z},t),\boldsymbol{z}),\quad \text{etc.} \tag{14.29}$$

We assume that these functions are continuously differentiable. Corollary 14.28 applies to 'all' processes. These could be defined as the most general processes for which the integral balance laws make sense (cf. Sec. 9). Here we merely determine the effect of Principle 14.27 when it is restricted to smooth processes. We accordingly substitute (14.29) into the classical version (14.23) of the Clausius-Duhem inequality and use the Chain Rule to obtain

$$\left[\hat{\boldsymbol{T}}-\frac{\partial\hat{\psi}}{\partial\boldsymbol{F}}\right]:\boldsymbol{F}_t-\frac{\partial\hat{\psi}}{\partial\boldsymbol{F}_t}:\boldsymbol{F}_{tt}-\left[\hat{\eta}+\frac{\partial\hat{\psi}}{\partial\theta}\right]\theta_t-\frac{\partial\hat{\psi}}{\partial\boldsymbol{g}}\cdot\boldsymbol{g}_t+\frac{\hat{\boldsymbol{q}}\cdot\boldsymbol{g}}{\theta}\geq 0 \tag{14.30}$$

for all thermomechanical processes. We limit our attention to a fixed $(\boldsymbol{z},t)$. Then by varying the process, we can vary $\boldsymbol{F},\boldsymbol{F}_t,\boldsymbol{F}_{tt},\theta,\theta_t,\boldsymbol{g},\boldsymbol{g}_t$ *independently* at $(\boldsymbol{z},t)$. Note that $\boldsymbol{F}_{tt},\theta_t,\boldsymbol{g}_t$ are *not* arguments of the constitutive functions. Thus the values of the constitutive functions are unaltered when we vary these three functions at $(\boldsymbol{z},t)$.

Condition (14.30) implies that

$$\frac{\partial\hat{\psi}}{\partial\boldsymbol{g}}=\boldsymbol{0}, \tag{14.31}$$

for if not, we could take the arbitrary $\boldsymbol{g}_t$ to be a large positive multiple of $\frac{\partial\hat{\psi}}{\partial\boldsymbol{g}}$ and thereby cause (14.30) to be violated. Similarly, we find that

$$\frac{\partial\hat{\psi}}{\partial\boldsymbol{F}_t}=\boldsymbol{O} \tag{14.32}$$

and that the coefficient of θ_t in (14.30) must vanish. Thus the constitutive equation for the free-energy density ψ in (14.29) reduces to

$$\psi(\boldsymbol{z},t)=\hat{\psi}(\boldsymbol{F}(\boldsymbol{z},t),\theta(\boldsymbol{z},t),\boldsymbol{z}) \tag{14.33}$$

and the constitutive function for the entropy density must have the special form

$$\hat{\eta}(\boldsymbol{F}(\boldsymbol{z},t),\theta(\boldsymbol{z},t),\boldsymbol{z})=-\frac{\partial\hat{\psi}}{\partial\theta}(\boldsymbol{F}(\boldsymbol{z},t),\theta(\boldsymbol{z},t),\boldsymbol{z}). \tag{14.34}$$

Conditions (14.32)–(14.34) reduce (14.30) to

$$\left[\hat{\boldsymbol{T}}-\frac{\partial\hat{\psi}}{\partial\boldsymbol{F}}\right]:\boldsymbol{F}_t+\frac{\hat{\boldsymbol{q}}\cdot\boldsymbol{g}}{\theta}\geq 0\quad\forall\,\boldsymbol{F},\boldsymbol{F}_t,\theta,\boldsymbol{g}. \tag{14.35}$$

Let us replace $\boldsymbol{F}_t$ by $\alpha \boldsymbol{F}_t$ and $\boldsymbol{g}$ by $\beta \boldsymbol{g}$ where α and β are real numbers. We obtain

$$(14.36)\quad \alpha\left[\hat{\boldsymbol{T}}(\boldsymbol{F}, \alpha\boldsymbol{F}_t, \theta, \beta\boldsymbol{g}, \boldsymbol{z}) - \frac{\partial\hat{\psi}}{\partial\boldsymbol{F}}(\boldsymbol{F}, \theta, \boldsymbol{z})\right] : \boldsymbol{F}_t + \beta\frac{\hat{\boldsymbol{q}}(\boldsymbol{F}, \alpha\boldsymbol{F}_t, \theta, \beta\boldsymbol{g}, \boldsymbol{z})\cdot\boldsymbol{g}}{\theta} \geq 0 \quad \forall\, \boldsymbol{F}, \boldsymbol{F}_t, \theta, \boldsymbol{g}.$$

Now the left-hand side of (14.36), regarded as a function of α and β, is ≥ 0 and vanishes for $\alpha = 0 = \beta$. It therefore has an interior minimum at $\alpha = 0 = \beta$, so that its derivatives with respect to α and β must vanish here:

$$(14.37\text{a})\qquad \left[\hat{\boldsymbol{T}}(\boldsymbol{F}, \boldsymbol{O}, \theta, \boldsymbol{0}, \boldsymbol{z}) - \frac{\partial\hat{\psi}}{\partial\boldsymbol{F}}(\boldsymbol{F}, \theta, \boldsymbol{z})\right] : \boldsymbol{F}_t = 0,$$

$$(14.37\text{b})\qquad \hat{\boldsymbol{q}}(\boldsymbol{F}, \boldsymbol{O}, \theta, \boldsymbol{0}, \boldsymbol{z})\cdot\boldsymbol{g} = 0.$$

Since $\boldsymbol{F}_t$ and $\boldsymbol{g}$ are arbitrary, (14.37) reduces to

$$(14.38)\qquad \hat{\boldsymbol{T}}(\boldsymbol{F}, \boldsymbol{O}, \theta, \boldsymbol{0}, \boldsymbol{z}) = \frac{\partial\hat{\psi}}{\partial\boldsymbol{F}}(\boldsymbol{F}, \theta, \boldsymbol{z}),$$

$$(14.39)\qquad \hat{\boldsymbol{q}}(\boldsymbol{F}, \boldsymbol{O}, \theta, \boldsymbol{0}, \boldsymbol{z}) = \boldsymbol{0}.$$

Equation (14.38) says that the equilibrium mechanical response at constant temperature is determined by the derivative of the free-energy density with respect to $\boldsymbol{F}$. Equation (14.39) says that heat does not flow when $\boldsymbol{F}_t = \boldsymbol{O}$ and $\boldsymbol{g} = \boldsymbol{0}$.

In a purely mechanical theory, a material is elastic iff the only field that the stress depends on is a deformation tensor (see (10.5)). This material is *hyperelastic* iff the stress is the derivative of a scalar field with respect to a suitable deformation tensor. Equation (14.38) shows that the equilibrium response of a thermoelastic material may be regarded as hyperelastic if the stress is independent of θ. In *isothermal* deformations, in which θ is constant, the same conclusion holds.

Our use of the entropy inequality implies that there are uncouplings in the constitutive equations, which need not be postulated a priori. Further uncouplings typically obtain when the material response is invariant under some symmetry groups.

Let us introduce the *dissipative part of the stress*:

$$(14.40)\qquad \boldsymbol{T}^{\mathrm{D}}(\boldsymbol{F}, \boldsymbol{F}_t, \theta, \boldsymbol{g}, \boldsymbol{z}) \equiv \hat{\boldsymbol{T}}(\boldsymbol{F}, \boldsymbol{F}_t, \theta, \boldsymbol{g}, \boldsymbol{z}) - \hat{\boldsymbol{T}}(\boldsymbol{F}, \boldsymbol{O}, \theta, \boldsymbol{0}, \boldsymbol{z}),$$

so that (14.38) implies that the constitutive function for the stress has the form

$$(14.41)\qquad \hat{\boldsymbol{T}}(\boldsymbol{F}, \boldsymbol{F}_t, \theta, \boldsymbol{g}, \boldsymbol{z}) = \frac{\partial\hat{\psi}}{\partial\boldsymbol{F}}(\boldsymbol{F}, \theta, \boldsymbol{z}) + \boldsymbol{T}^{\mathrm{D}}(\boldsymbol{F}, \boldsymbol{F}_t, \theta, \boldsymbol{g}, \boldsymbol{z}).$$

The substitution of (14.41) into (14.35) yields the *dissipation inequality*:

$$\boldsymbol{T}^{\mathrm{D}} : \boldsymbol{F}_t + \frac{\hat{\boldsymbol{q}} \cdot \boldsymbol{g}}{\theta} \geq 0 \quad \forall \boldsymbol{F}, \boldsymbol{F}_t, \theta, \boldsymbol{g}, \tag{14.42}$$

from which we obtain the *mechanical dissipation inequality*:

$$\boldsymbol{T}^{\mathrm{D}}(\boldsymbol{F}, \boldsymbol{F}_t, \theta, \boldsymbol{0}, \boldsymbol{z}) : \boldsymbol{F}_t \geq 0 \quad \forall \boldsymbol{F}, \boldsymbol{F}_t, \theta \tag{14.43}$$

and the *Fourier heat conduction inequality*:

$$\hat{\boldsymbol{q}}(\boldsymbol{F}, \boldsymbol{O}, \theta, \boldsymbol{g}, \boldsymbol{z}) \cdot \boldsymbol{g} \geq 0 \quad \forall \boldsymbol{F}, \theta, \boldsymbol{g}, \tag{14.44}$$

which says that heat flows in a direction 'opposed' to that of the temperature gradient. The *Fourier heat law* says that there is a positive-definite tensor $\boldsymbol{K}(\boldsymbol{z})$ such that $\hat{\boldsymbol{q}}(\boldsymbol{F}, \boldsymbol{O}, \theta, \boldsymbol{g}, \boldsymbol{z}) = \boldsymbol{K}(\boldsymbol{z}) \cdot \boldsymbol{g}$. For an isotropic material, $\boldsymbol{K}$ reduces to a scalar multiple of the identity.

We could easily work backwards to show that these results are also sufficient to ensure that the Clausius-Duhem inequality holds for (14.29).

For the analysis of the partial differential equations obtained by substituting (14.29) into the balance of linear momentum and the balance of energy, it is mathematically convenient and it might be physically reasonable to strengthen (14.43) and (14.44) by requiring that $\boldsymbol{T}^{\mathrm{D}}(\boldsymbol{F}, \cdot, \theta, \cdot, \boldsymbol{z})$ and $\hat{\boldsymbol{q}}(\boldsymbol{F}, \cdot, \theta, \cdot, \boldsymbol{z})$ satisfy the *strict monotonicity condition*:

$$\begin{aligned} &[\boldsymbol{T}^{\mathrm{D}}(\boldsymbol{F}, \boldsymbol{F}_t^2, \theta, \boldsymbol{g}^2, \boldsymbol{z}) - \boldsymbol{T}^{\mathrm{D}}(\boldsymbol{F}, \boldsymbol{F}_t^1, \theta, \boldsymbol{g}^1, \boldsymbol{z})] : [\boldsymbol{F}_t^2 - \boldsymbol{F}_t^1] \\ &\quad + \frac{1}{\theta}[\hat{\boldsymbol{q}}(\boldsymbol{F}, \boldsymbol{F}_t^2, \theta, \boldsymbol{g}^2, \boldsymbol{z}) - \hat{\boldsymbol{q}}(\boldsymbol{F}, \boldsymbol{F}_t^1, \theta, \boldsymbol{g}^1, \boldsymbol{z})] \cdot [\boldsymbol{g}^2 - \boldsymbol{g}^1] > 0 \\ &\qquad \text{for} \quad [\boldsymbol{F}_t^2 - \boldsymbol{F}_t^1] : [\boldsymbol{F}_t^2 - \boldsymbol{F}_t^1] + [\boldsymbol{g}^2 - \boldsymbol{g}^1] \cdot [\boldsymbol{g}^2 - \boldsymbol{g}^1] > 0. \end{aligned} \tag{14.45}$$

This condition gives the governing partial differential equations a strong parabolic character. I suspect, however, that the weaker restriction in which $\boldsymbol{T}^{\mathrm{D}}$ is a strongly elliptic function of $\boldsymbol{F}_t$ (see Sec. XIII.2) might be more natural.

For a *thermoelastic material* the constitutive functions of (14.29) are independent of $\boldsymbol{F}_t$. In this case, a straightforward specialization of the development beginning with (14.30) allows us to replace (14.41) with

$$\hat{\boldsymbol{T}}(\boldsymbol{F}, \theta, \boldsymbol{g}, \boldsymbol{z}) = \frac{\partial \hat{\psi}}{\partial \boldsymbol{F}}(\boldsymbol{F}, \theta, \boldsymbol{z}) \tag{14.46}$$

and to simplify the other relations. If we set

$$\tilde{\psi}(\boldsymbol{C}, \theta, \boldsymbol{z}) \equiv \hat{\psi}(\boldsymbol{F}, \theta, \boldsymbol{z}), \tag{14.47}$$

then (14.46) is equivalent to

$$\hat{\boldsymbol{S}}(\boldsymbol{C}, \theta, \boldsymbol{z}) = 2\frac{\partial \tilde{\psi}}{\partial \boldsymbol{C}}(\boldsymbol{C}, \theta, \boldsymbol{z}). \tag{14.48}$$

14.49. Exercise. Replace (14.29) with constitutive functions of the form

$$\begin{aligned} \boldsymbol{T}(\boldsymbol{z},t) &= \check{\boldsymbol{T}}(\boldsymbol{F}(\boldsymbol{z},t),\boldsymbol{F}_t(\boldsymbol{z},t),\eta(\boldsymbol{z},t),\boldsymbol{g}(\boldsymbol{z},t),\boldsymbol{z}), \\ \boldsymbol{q}(\boldsymbol{z},t) &= \check{\boldsymbol{q}}(\boldsymbol{F}(\boldsymbol{z},t),\boldsymbol{F}_t(\boldsymbol{z},t),\eta(\boldsymbol{z},t),\boldsymbol{g}(\boldsymbol{z},t),\boldsymbol{z}), \\ \theta(\boldsymbol{z},t) &= \check{\theta}(\boldsymbol{F}(\boldsymbol{z},t),\boldsymbol{F}_t(\boldsymbol{z},t),\eta(\boldsymbol{z},t),\boldsymbol{g}(\boldsymbol{z},t),\boldsymbol{z}), \\ \varepsilon(\boldsymbol{z},t) &= \check{\varepsilon}(\boldsymbol{F}(\boldsymbol{z},t),\boldsymbol{F}_t(\boldsymbol{z},t),\eta(\boldsymbol{z},t),\boldsymbol{g}(\boldsymbol{z},t),\boldsymbol{z}). \end{aligned} \tag{14.50}$$

Imitate the analysis beginning with (14.30) to determine necessary and sufficient conditions for (14.50) to satisfy the Clausius-Duhem inequality in the form (14.21). Use the reduced constitutive equations to show that the equilibrium mechanical response is hyperelastic if the deformation is *isentropic*, i.e., if η is constant. These conclusions also follow from the Chain Rule from those for (14.29) if we assume that (14.29) has sufficient invertibility for it to be equivalent to (14.50). A similar analysis can be carried out when $\boldsymbol{F}, \boldsymbol{F}_t, \varepsilon, \boldsymbol{g}$ are the independent variables and $\boldsymbol{T}, \boldsymbol{q}, 1/\theta, \eta$ are the dependent variables of the constitutive equations. Show that the equilibrium mechanical response is generally not hyperelastic for *isoenergetic* deformations, i.e., deformations for which ε is constant.

In the nineteenth century, continuum mechanics was founded under the leadership of Cauchy, electromagnetism, under the leadership of Maxwell, and thermodynamics, under the leadership of Carnot and Clausius. The mathematical skills of the last two were vastly inferior to those of the first two. The deficiencies in substance and precision that typify the vast literature on thermodynamics since the time of Clausius may be attributed to this inferiority. Although thermodynamics has been fashioned into a serviceable tool for gas dynamics, chemistry, and some other fields, until recently it (in painful contrast to continuum mechanics and electromagnetism) has lacked a formulation capable of describing fields varying with space and time and governed by nonlinear constitutive equations. Scathing critiques of the inadequacies of traditional thermodynamics have been given by Bridgman (1941) and by Truesdell in a long series of writings culminating in Truesdell's books of 1980 and 1984, which give lively opinionated histories of thermodynamics. Logical developments of the ideas of Carnot are given by Truesdell & Bharatha (1977) and D. R. Owen (1984).

The development leading from (14.16) to (14.23) is a reduction of an argument introduced by Truesdell (1984). The Entropy Principle was introduced by Coleman & Noll (1963). The treatment following (14.29) is based on this work and on that of Coleman & Mizel (1964). It was extended to a large class of materials of the form (14.25) having *fading memory* by Coleman (1964) (see Gurtin (1968) and Truesdell (1984)).

The constitutive restrictions imposed by the Clausius-Duhem inequality on a linearly heat-conducting compressible Newtonian fluid (governed by the Navier-Stokes equations for compressible fluids and by Fourier's heat law) are in complete accord with those universally accepted for such materials. On the other hand, as our comments surrounding (14.45) suggest, the consequences of the Clausius-Duhem inequality are not strong enough to deliver certain mathematically attractive and natural restrictions, while some of its consequences, such as uncouplings, are of limited mathematical advantage.

Entropy inequalities are important in the study of shocks, for which the integral balance laws, possibly interpreted in a very general form, must be used in favor of their classical forms. The precise role that entropy inequalities, such as that of Clausius and Duhem, should play in the theory of shocks is not yet clear (see Dafermos (1983, 1984, 1985), Ericksen (1991), and the discussion in Chap. XVI). For these reasons, we cannot unequivocably embrace the Clausius-Duhem inequality as a universally valid characterization of irreversibility.

For investigations of how entropy inequalities are founded on more primitive axioms, see Day (1972) and the deep and systematic work of Coleman & Owen summarized in their article of 1974. The entire logical structure of thermodynamics has been critically examined by Serrin (1986), who suggest grounds on which the Energy Balance should be replaced by an energy inequality. (In this connection, it is interesting to note that weak solutions of the Navier-Stokes equations are only known to satisfy such an inequality.)

15. The Spatial Formulation

Kinematics. $\boldsymbol{p}(\boldsymbol{z},t)$ is the position of material point $\boldsymbol{z}$ at time t. Under our assumption that $\boldsymbol{p}(\cdot,t)$ is one-to-one for all t, it has an inverse $\boldsymbol{p}(\mathcal{B},t) \ni \boldsymbol{y} \mapsto \boldsymbol{q}(\boldsymbol{y},t) \in \mathcal{B}$. Thus $\boldsymbol{q}(\boldsymbol{y},t)$ is the material point occupying position $\boldsymbol{y}$ at time t. We can replace $\boldsymbol{z}$ with $\boldsymbol{q}(\boldsymbol{y},t)$ in all the functions of continuum mechanics we have introduced above. For example, if $\boldsymbol{u}(\boldsymbol{z},t) \equiv \boldsymbol{p}(\boldsymbol{z},t) - \boldsymbol{z}$ is the displacement of the material point $\boldsymbol{z}$, then we can define the displacement, velocity, and acceleration of the material point occupying position $\boldsymbol{y}$ at time t by

$$\begin{gathered}\breve{\boldsymbol{u}}(\boldsymbol{y},t) \equiv \boldsymbol{u}(\boldsymbol{q}(\boldsymbol{y},t),t) = \boldsymbol{y} - \boldsymbol{q}(\boldsymbol{y},t), \\ \breve{\boldsymbol{v}}(\boldsymbol{y},t) \equiv \boldsymbol{p}_t(\boldsymbol{q}(\boldsymbol{y},t),t), \quad \breve{\boldsymbol{a}}(\boldsymbol{y},t) \equiv \boldsymbol{p}_{tt}(\boldsymbol{q}(\boldsymbol{y},t),t)\end{gathered} \tag{15.1}$$

or, equivalently, by

$$\begin{gathered}\breve{\boldsymbol{u}}(\boldsymbol{p}(\boldsymbol{z},t),t) \equiv \boldsymbol{u}(\boldsymbol{z},t), \\ \breve{\boldsymbol{v}}(\boldsymbol{p}(\boldsymbol{z},t),t) \equiv \boldsymbol{p}_t(\boldsymbol{z},t), \quad \breve{\boldsymbol{a}}(\boldsymbol{p}(\boldsymbol{z},t),t) \equiv \boldsymbol{p}_{tt}(\boldsymbol{z},t).\end{gathered} \tag{15.2}$$

A formulation of continuum physics in which the independent variables are $(\boldsymbol{y},t)$ is called a *spatial* (or *Eulerian*) formulation (which was apparently introduced by d'Alembert). From the viewpoint of physics, the use of spatial descriptions is artificial since attention is shifted from the physical entity of a material point, the seat of physical activity, to the purely geometric entity of a point in space. But for many important problems of fluid mechanics and for a few special problems of solid mechanics there are compelling mathematical advantages in using the spatial formulation. We outline a formulation of continuum mechanics in spatial coordinates primarily to make the literature on solid mechanics couched in such formulations accessible to readers and to furnish readers knowledgeable in fluid dynamics or electromagnetism a means to bring their knowledge to bear on solid mechanics.

Let us first determine how $\breve{\boldsymbol{u}}, \breve{\boldsymbol{v}}, \breve{\boldsymbol{a}}$ are related. From (15.1) and (15.2) we obtain

$$\begin{aligned}\breve{\boldsymbol{v}}(\boldsymbol{p}(\boldsymbol{z},t),t) &= \partial_t \boldsymbol{p}(\boldsymbol{z},t) = \partial_t \boldsymbol{u}(\boldsymbol{z},t) = \partial_t \breve{\boldsymbol{u}}(\boldsymbol{p}(\boldsymbol{z},t),t) \\ &= \breve{\boldsymbol{u}}_{\boldsymbol{y}}(\boldsymbol{p}(\boldsymbol{z},t),t) \cdot \boldsymbol{p}_t(\boldsymbol{z},t) + \breve{\boldsymbol{u}}_t(\boldsymbol{p}(\boldsymbol{z},t),t) \\ &= \breve{\boldsymbol{u}}_{\boldsymbol{y}}(\boldsymbol{p}(\boldsymbol{z},t),t) \cdot \breve{\boldsymbol{v}}(\boldsymbol{p}(\boldsymbol{z},t),t) + \breve{\boldsymbol{u}}_t(\boldsymbol{p}(\boldsymbol{z},t),t)\end{aligned} \tag{15.3a}$$

or, equivalently,

$$\breve{\boldsymbol{v}}(\boldsymbol{y},t) = \breve{\boldsymbol{u}}_{\boldsymbol{y}}(\boldsymbol{y},t) \cdot \breve{\boldsymbol{v}}(\boldsymbol{y},t) + \breve{\boldsymbol{u}}_t(\boldsymbol{y},t). \tag{15.3b}$$

This is a linear equation for $\breve{\boldsymbol{v}}$.

15.4. Exercise. Show that (15.3b) can be uniquely solved for $\breve{\boldsymbol{v}}$.

By the method leading to (15.3b), we obtain

$$\breve{\boldsymbol{a}}(\boldsymbol{y},t) = \breve{\boldsymbol{v}}_{\boldsymbol{y}}(\boldsymbol{y},t) \cdot \breve{\boldsymbol{v}}(\boldsymbol{y},t) + \breve{\boldsymbol{v}}_t(\boldsymbol{y},t). \tag{15.5}$$

Thus the acceleration $\breve{\boldsymbol{a}}$ is nonlinear in the velocity $\breve{\boldsymbol{v}}$ and its derivatives! The spatial description must have many mathematical virtues to compensate for the disadvantage of introducing this *convective* nonlinearity into every equation of motion.

15.6. Exercise. Let $\boldsymbol{z} = z^k \boldsymbol{i}_k$ and $\boldsymbol{y} = y^k \boldsymbol{i}_k$ where $\{\boldsymbol{i}_k\}$ is a fixed orthonormal basis. Let $\boldsymbol{y} = \boldsymbol{p}(\boldsymbol{z}, t)$ be given by

$$\begin{aligned} y^1 &= \tfrac{1}{2}(z^1 + z^2)e^t + \tfrac{1}{2}(z^1 - z^2)e^{-t}, \\ y^2 &= \tfrac{1}{2}(z^1 + z^2)e^t - \tfrac{1}{2}(z^1 - z^2)e^{-t}, \\ y^3 &= z^3. \end{aligned} \tag{15.7}$$

Find $\boldsymbol{q}(\boldsymbol{y}, t)$, $\breve{\boldsymbol{u}}(\boldsymbol{y}, t)$; find $\breve{\boldsymbol{v}}(\boldsymbol{y}, t)$ both by (15.1) and by solving (15.3b).

In the material formulation, the independent variable $\boldsymbol{z}$ is confined to the fixed region $\mathcal{B}$, whereas in the spatial formulation, the independent variable $\boldsymbol{y}$ is confined to $\boldsymbol{p}(\mathcal{B}, t)$, which in general is unknown. For many problems of fluid mechanics, the fluids are confined to a prescribed, possibly time-dependent region of space, so that $\boldsymbol{p}(\mathcal{B}, t)$ is given. Otherwise, for free-surface problems, some methods of a material formulation are introduced to make the problem accessible. (These may include the use of streamlines as coordinates or the linearization of the governing equations about the reference configuration.) In solid mechanics, a spatial description is useful, not for initial-boundary-value problems, but for inverse problems in which the motion of the body or its boundary is specified, and other data, such as the force system or the reference shape or the constitutive functions, are to be determined.

When the spatial formulation is used, it is occasionally useful to determine whether a moving surface always consists of the same material points, i.e., is a *material surface*. A material surface may be specified by an equation of the form $\varphi(\boldsymbol{z}) = 0$. At time t, it occupies the surface consisting of those points $\boldsymbol{y}$ in space satisfying $\varphi(\boldsymbol{q}(\boldsymbol{y}, t)) = 0$.

15.8. Lagrange's Criterion. *The moving surface $\mathcal{S}$ in a deforming material defined by $\psi(\boldsymbol{y}, t) = 0$ is a material surface if and only if*

$$\psi_{\boldsymbol{y}} \cdot \breve{\boldsymbol{v}} + \psi_t = 0. \tag{15.9}$$

Proof. If $\mathcal{S}$ is a material surface, then there is a function φ such that $\varphi(\boldsymbol{z}) = \psi(\boldsymbol{p}(\boldsymbol{z}, t), t)$. Equation (15.9) follows from the equation $\varphi_t(\boldsymbol{z}) = 0$. Conversely, suppose that (15.9) holds. Set $\omega(\boldsymbol{z}, t) = \psi(\boldsymbol{p}(\boldsymbol{z}, t), t)$. (The equation $\omega(\boldsymbol{z}, t) = 0$ defines the reference positions of the material points that at time t are on $\mathcal{S}$.) Equation (15.9) implies that $\omega_t(\boldsymbol{z}, t) = 0$ so that ω is independent of t. □

Strain. Following the development leading from (2.2) to (2.4), we find that the local ratio of deformed to reference length of the material fiber at $(\boldsymbol{y}, t)$ along unit vector $\boldsymbol{a}$ is

$$\frac{1}{\sqrt{\boldsymbol{a} \cdot \boldsymbol{q}_{\boldsymbol{y}}(\boldsymbol{y}, t)^* \cdot \boldsymbol{q}_{\boldsymbol{y}}(\boldsymbol{y}, t) \cdot \boldsymbol{a}}}. \tag{15.10}$$

$\boldsymbol{q}_{\boldsymbol{y}}(\boldsymbol{y}, t)^* \cdot \boldsymbol{q}_{\boldsymbol{y}}(\boldsymbol{y}, t)$ has been termed the *Cauchy deformation tensor*. By the definition of $\boldsymbol{q}$, we have the equivalent relations

$$\boldsymbol{y} = \boldsymbol{p}(\boldsymbol{q}(\boldsymbol{y}, t), t), \quad \boldsymbol{z} = \boldsymbol{q}(\boldsymbol{p}(\boldsymbol{z}, t), t), \tag{15.11}$$

whence we obtain

$$(15.12)\qquad \boldsymbol{I} = \boldsymbol{p_z}(\boldsymbol{q}(\boldsymbol{y},t),t)\cdot \boldsymbol{q_y}(\boldsymbol{y},t), \quad \boldsymbol{I} = \boldsymbol{q_y}(\boldsymbol{p}(\boldsymbol{z},t),t)\cdot \boldsymbol{p_z}(\boldsymbol{z},t).$$

Thus we may introduce the *left Cauchy-Green deformation tensor* $\boldsymbol{B}$ by

$$(15.13)\quad \boldsymbol{B}(\boldsymbol{y},t) \equiv \boldsymbol{V}(\boldsymbol{q}(\boldsymbol{y},t),t)^2 \equiv \boldsymbol{F}(\boldsymbol{q}(\boldsymbol{y},t),t)\cdot \boldsymbol{F}(\boldsymbol{q}(\boldsymbol{y},t),t)^* = \boldsymbol{q_y}(\boldsymbol{y},t)^{-1}\cdot \boldsymbol{q_y}(\boldsymbol{y},t)^{-*}$$

(cf. (4.5)). Thus the Cauchy deformation tensor $\boldsymbol{q_y}(\boldsymbol{y},t)^*\cdot \boldsymbol{q_y}(\boldsymbol{y},t) = \boldsymbol{B}^{-1}$. We may introduce the *spatial strain tensor* $\breve{\boldsymbol{E}}$ (cf. (2.19)) by

$$(15.14)\quad 2\breve{\boldsymbol{E}}(\boldsymbol{y},t) \equiv \boldsymbol{I} - \boldsymbol{q_y}(\boldsymbol{y},t)^*\cdot \boldsymbol{q_y}(\boldsymbol{y},t) = \breve{\boldsymbol{u}}_{\boldsymbol{y}}(\boldsymbol{y},t) + \breve{\boldsymbol{u}}_{\boldsymbol{y}}(\boldsymbol{y},t)^* - \breve{\boldsymbol{u}}_{\boldsymbol{y}}(\boldsymbol{y},t)^*\cdot \breve{\boldsymbol{u}}_{\boldsymbol{y}}(\boldsymbol{y},t),$$

the second equality coming from (15.1). (Compare this result with Ex. 2.21.)

Rates. From (15.2) we obtain

$$(15.15)\qquad \boldsymbol{F}_t(\boldsymbol{z},t) = \boldsymbol{p}_{\boldsymbol{z}t}(\boldsymbol{z},t) = \frac{\partial}{\partial \boldsymbol{z}}\breve{\boldsymbol{v}}(\boldsymbol{p}(\boldsymbol{z},t),t) = \breve{\boldsymbol{v}}_{\boldsymbol{y}}(\boldsymbol{p}(\boldsymbol{z},t),t)\cdot \boldsymbol{F}(\boldsymbol{z},t).$$

On the other hand, the polar decomposition (4.2) yields

$$(15.16)\qquad \boldsymbol{F}_t = \boldsymbol{R}_t\cdot \boldsymbol{U} + \boldsymbol{R}\cdot \boldsymbol{U}_t.$$

Now we focus our attention at a particular time τ and choose $\boldsymbol{p}(\cdot,\tau)$ to be the reference configuration. Thus we identify $\boldsymbol{z} = \boldsymbol{p}(\boldsymbol{z},\tau) = \boldsymbol{y}$. Equating (15.15) to (15.16) for $t=\tau$, we obtain

$$(15.17)\qquad \breve{\boldsymbol{v}}_{\boldsymbol{y}}(\boldsymbol{y},\tau) = \boldsymbol{R}_t(\boldsymbol{y},\tau) + \boldsymbol{U}_t(\boldsymbol{y},\tau).$$

Since $\boldsymbol{U}$ is symmetric, so is $\boldsymbol{U}_t$. By differentiating the identity $\boldsymbol{R}(\boldsymbol{z},t)^*\cdot \boldsymbol{R}(\boldsymbol{z},t) = \boldsymbol{I}$ with respect to t and then setting $t=\tau$, we obtain

$$(15.18)\qquad \boldsymbol{R}_t(\boldsymbol{y},\tau)^* + \boldsymbol{R}_t(\boldsymbol{y},\tau) = \boldsymbol{O},$$

so that $\boldsymbol{R}_t(\boldsymbol{y},\tau)$ is the skew part of $\breve{\boldsymbol{v}}_{\boldsymbol{y}}(\boldsymbol{y},\tau)$. We can therefore define the *stretching tensor* $\boldsymbol{D}$ and the *spin tensor* $\boldsymbol{W}$ to be the symmetric and skew parts of $\nabla\breve{\boldsymbol{v}} \equiv \breve{\boldsymbol{v}}_{\boldsymbol{y}}^*$ so that at time τ, $\boldsymbol{D}(\boldsymbol{y},\tau) = \boldsymbol{U}_t(\boldsymbol{y},\tau)$ and $\boldsymbol{W}(\boldsymbol{y},\tau) = -\boldsymbol{R}_t(\boldsymbol{y},\tau)$. The axial vector $\boldsymbol{w}$ corresponding to $\boldsymbol{W}$ (see (XI.1.9)) is the instantaneous angular velocity of the principal axes of stretch relative to the present configuration. $2\boldsymbol{w}$ is the *vorticity*. The diagonal components of $\boldsymbol{D}$ measure the rates of length changes per unit instantaneous length along the instantaneous principal axes, and its off-diagonal components measure the rates of angle changes relative to these axes.

15.19. Exercise. Prove that

$$(15.20)\qquad 2\boldsymbol{w} = \operatorname{curl}\breve{\boldsymbol{v}}$$

where the curl is taken with respect to the variables $\boldsymbol{y}$.

From (XI.2.7b) and (15.15) we obtain

$$(15.21)\qquad \begin{aligned} \partial_t[\det \boldsymbol{F}(\boldsymbol{z},t)] &= \det \boldsymbol{F}(\boldsymbol{z},t)\operatorname{tr}[\boldsymbol{F}_t(\boldsymbol{z},t)\cdot \boldsymbol{F}(\boldsymbol{z},t)^{-1}] \\ &= \det \boldsymbol{F}(\boldsymbol{z},t)\operatorname{tr}[\breve{\boldsymbol{v}}_{\boldsymbol{y}}(\boldsymbol{p}(\boldsymbol{z},t),t)\cdot \boldsymbol{F}(\boldsymbol{z},t)\cdot \boldsymbol{F}(\boldsymbol{z},t)^{-1}] \\ &= \det \boldsymbol{F}(\boldsymbol{z},t)\operatorname{div}\breve{\boldsymbol{v}}(\boldsymbol{p}(\boldsymbol{z},t),t) \end{aligned}$$

where the divergence is taken with respect to $\boldsymbol{y}$. Thus $\operatorname{div}\breve{\boldsymbol{v}}(\boldsymbol{y},t)$ is the rate of change of volume at $(\boldsymbol{y},t)$ per unit actual volume. It follows that the requirement that a material undergo a locally volume-preserving deformation, which is specified by the nonlinear equation $\det \boldsymbol{F} = 1$ in the material formulation, is equivalently specified by the linear equation

$$\operatorname{div} \breve{\boldsymbol{v}} = 0 \tag{15.22}$$

in the spatial formulation. This fact is one of the virtues of the spatial formulation. It is particularly useful for the description of liquids, which are effectively incompressible. (Ebin & Saxton (1986) balanced (15.22) against the natural virtues of the material formulation to treat the existence of equilibrium states of incompressible nonlinearly elastic bodies.)

Let φ be an arbitrary real-valued function of $\boldsymbol{y}$ and t. Using (15.21) and (15.2) in conjunction with the rule for the change of variables in integrals, we obtain the very important identities

$$\begin{aligned}
&\frac{d}{dt}\int_{\boldsymbol{p}(\mathcal{B},t)} \varphi(\boldsymbol{y},t)\,dv(\boldsymbol{y}) \\
&\quad= \frac{d}{dt}\int_{\mathcal{B}} \varphi(\boldsymbol{p}(\boldsymbol{z},t),t)\det \boldsymbol{p}_{\boldsymbol{z}}(\boldsymbol{z},t)\,dv(\boldsymbol{z}) \\
&\quad= \int_{\mathcal{B}} [\varphi_t(\boldsymbol{p}(\boldsymbol{z},t),t) + \varphi_{\boldsymbol{y}}(\boldsymbol{p}(\boldsymbol{z},t),t)\cdot\breve{\boldsymbol{v}}(\boldsymbol{p}(\boldsymbol{z},t),t) \\
&\qquad\quad + \varphi(\boldsymbol{p}(\boldsymbol{z},t),t)\operatorname{div}\breve{\boldsymbol{v}}(\boldsymbol{p}(\boldsymbol{z},t),t)]\det \boldsymbol{F}(\boldsymbol{z},t)\,dv(\boldsymbol{z}) \\
&\quad= \int_{\boldsymbol{p}(\mathcal{B},t)} [\varphi_t(\boldsymbol{y},t) + \varphi_{\boldsymbol{y}}(\boldsymbol{y},t)\cdot\breve{\boldsymbol{v}}(\boldsymbol{y},t) + \varphi(\boldsymbol{y},t)\operatorname{div}\breve{\boldsymbol{v}}(\boldsymbol{y},t)]\,dv(\boldsymbol{y}) \\
&\quad= \int_{\boldsymbol{p}(\mathcal{B},t)} \{\varphi_t(\boldsymbol{y},t) + \operatorname{div}[\varphi(\boldsymbol{y},t)\breve{\boldsymbol{v}}(\boldsymbol{y},t)]\}\,dv(\boldsymbol{y}) \\
&\quad= \int_{\boldsymbol{p}(\mathcal{B},t)} \varphi_t(\boldsymbol{y},t)\,dv(\boldsymbol{y}) + \int_{\partial\boldsymbol{p}(\mathcal{B},t)} \varphi(\boldsymbol{y},t)\breve{\boldsymbol{v}}(\boldsymbol{y},t)\cdot\boldsymbol{\xi}(\boldsymbol{y},t)\,da(\boldsymbol{y}),
\end{aligned} \tag{15.23}$$

which constitute the *Transport Theorem*. In the last equation of (15.23), $\boldsymbol{\xi}(\boldsymbol{y},t)$ denotes the unit outer normal to $\partial\boldsymbol{p}(\mathcal{B},t)$ at $\boldsymbol{y}$.

Mass. If we take the time derivative of the local form (6.7) of mass conservation and use (15.21), we get the *spatial form of the continuity equation*:

$$\breve{\rho}_t + \operatorname{div}(\breve{\rho}\breve{\boldsymbol{v}}) = 0. \tag{15.24}$$

We can alternatively derive (15.24) by using (15.23) to take the time derivative of (6.5) and then by using the arbitrariness of $\boldsymbol{p}(\mathcal{B},t)$. Finally, we can get (15.24) by requiring that the rate at which the mass within any fixed region Ω of space increases equals the rate at which it enters through $\partial\Omega$:

$$\frac{d}{dt}\int_{\Omega} \breve{\rho}(\boldsymbol{y},t)\,dv(\boldsymbol{y}) = -\int_{\partial\Omega} \breve{\rho}(\boldsymbol{y},t)\breve{\boldsymbol{v}}(\boldsymbol{y},t)\cdot\boldsymbol{\xi}(\boldsymbol{y})\,da(\boldsymbol{y}), \tag{15.25}$$

and then by using (15.23). We can now use (15.24) in (15.23) to get the *Corollary to the Transport Theorem*:

$$\frac{d}{dt}\int_{\boldsymbol{p}(\mathcal{B},t)} \breve{\rho}(\boldsymbol{y},t)\varphi(\boldsymbol{y},t)\,dv(\boldsymbol{y}) = \int_{\boldsymbol{p}(\mathcal{B},t)} \breve{\rho}(\boldsymbol{y},t)[\varphi_t(\boldsymbol{y},t)+\varphi_{\boldsymbol{y}}(\boldsymbol{y},t)\cdot\breve{\boldsymbol{v}}(\boldsymbol{y},t)]\,dv(\boldsymbol{y}). \tag{15.26}$$

(This gives a mathematically precise interpretation of the statement that $\breve{\rho}\,dv = \text{const.}$)

One of the virtues of the spatial formulation for solid mechanics as well as fluid mechanics is that it allows the following definition to be made: A motion is *steady* iff $\breve{\boldsymbol{v}}_t = \boldsymbol{0}$ and $\breve{\rho}_t = 0$.

Stress. We assume that the resultant force $\boldsymbol{g}(t;\mathcal{B})$ on body $\mathcal{B}$ at time t has the decomposition

$$(15.27)\qquad \boldsymbol{g}(t;\mathcal{B}) = \int_{\boldsymbol{p}(\mathcal{B},t)} \breve{\boldsymbol{f}}(\boldsymbol{y},t)\,dv(\boldsymbol{y}) + \int_{\partial\boldsymbol{p}(\mathcal{B},t)} \boldsymbol{\sigma}(\boldsymbol{y},t;\partial\boldsymbol{p}(\mathcal{B},t))\,da(\boldsymbol{y}).$$

$\breve{\boldsymbol{f}}(\boldsymbol{y},t)$ is the body force per unit actual volume at $(\boldsymbol{y},t)$ and $\boldsymbol{\sigma}(\boldsymbol{y},t;\partial\boldsymbol{p}(\mathcal{B},t))$ is the surface traction per unit actual area at $(\boldsymbol{y},t)$ of $\partial\boldsymbol{p}(\mathcal{B},t)$. By assuming that $\boldsymbol{\sigma}(\boldsymbol{y},t;\partial\boldsymbol{p}(\mathcal{B},t))$ depends on $\partial\boldsymbol{p}(\mathcal{B},t)$ only through its unit outer normal vector $\boldsymbol{\xi}(\boldsymbol{y},t)$ at $\boldsymbol{y}$, we can imitate the proof of Cauchy's Stress Theorem 7.14 (by taking $\partial\boldsymbol{p}(\mathcal{B},t)$ to be a suitable tetrahedron) and conclude that there is a second-order tensor $\boldsymbol{\Sigma}$, called the *Cauchy* stress tensor (or simply *the* stress tensor), such that

$$(15.28)\qquad \boldsymbol{\sigma}(\boldsymbol{y},t;\partial\boldsymbol{p}(\mathcal{B},t)) = \boldsymbol{\Sigma}(\boldsymbol{y},t)\cdot\boldsymbol{\xi}(\boldsymbol{y},t)$$

for $\boldsymbol{y}\in\partial\boldsymbol{p}(\mathcal{B},t)$. We then use appropriate forms of the Balance of Linear and Angular Momentum together with (15.23) to obtain the classical equations of motion

$$(15.29)\qquad \nabla\cdot\boldsymbol{\Sigma}^* + \breve{\boldsymbol{f}} = \breve{\rho}\breve{\boldsymbol{a}},$$

$$(15.30)\qquad \boldsymbol{\Sigma}\quad\text{is symmetric},$$

which are the counterparts of (7.21) and (7.27). Here the divergence is taken with respect to the variable $\boldsymbol{y}$. (Condition (15.30) says that the asterisk in (15.29) is unnecessary.)

To find out how $\boldsymbol{\Sigma}$ is related to $\boldsymbol{T}$, we equate (7.9) to (15.27) to conclude that

$$(15.31)\qquad \int_{\partial\boldsymbol{p}(\mathcal{B},t)} \boldsymbol{\Sigma}(\boldsymbol{y},t)\cdot\boldsymbol{\xi}(\boldsymbol{y},t)\,da(\boldsymbol{y}) = \int_{\partial\mathcal{B}} \boldsymbol{T}(\boldsymbol{z},t)\cdot\boldsymbol{\nu}(\boldsymbol{z})\,da(\boldsymbol{z})$$

for all $\mathcal{B}$. We need to determine how $\partial\mathcal{B}$ is affected by the deformation. Let s^1 and s^2 be parameters for a surface patch of $\partial\mathcal{B}$ that is described by $(s^1,s^2)\mapsto\hat{\boldsymbol{z}}(s^1,s^2)$. On this patch

$$(15.32)\qquad \boldsymbol{\nu}(\boldsymbol{z})\,da(\boldsymbol{z}) = \frac{\partial\hat{\boldsymbol{z}}}{\partial s^1}\times\frac{\partial\hat{\boldsymbol{z}}}{\partial s^2}\,ds^1\,ds^2.$$

The image of this patch under $\boldsymbol{p}(\cdot,t)$ is $(s^1,s^2)\mapsto\boldsymbol{p}(\hat{\boldsymbol{z}}(s^1,s^2),t)$, for which

$$(15.33)\qquad \begin{aligned} \boldsymbol{\xi}(\boldsymbol{y},t)\,da(\boldsymbol{y}) &= \left(\boldsymbol{F}\cdot\frac{\partial\hat{\boldsymbol{z}}}{\partial s^1}\right)\times\left(\boldsymbol{F}\cdot\frac{\partial\hat{\boldsymbol{z}}}{\partial s^2}\right)ds^1\,ds^2 \\ &= (\det\boldsymbol{F})\boldsymbol{F}^{-*}\cdot\left(\frac{\partial\hat{\boldsymbol{z}}}{\partial s^1}\times\frac{\partial\hat{\boldsymbol{z}}}{\partial s^2}\right)ds^1\,ds^2 \\ &= (\det\boldsymbol{F})\boldsymbol{F}^{-*}\cdot\boldsymbol{\nu}(\boldsymbol{z})\,da(\boldsymbol{z}) \end{aligned}$$

by virtue of (XI.2.6b) and (15.32). Thus we conclude from (15.31) that

$$\begin{aligned}\boldsymbol{\Sigma}(\boldsymbol{y},t) &= [\det \boldsymbol{F}(\boldsymbol{z},t)]^{-1}\boldsymbol{T}(\boldsymbol{z},t)\cdot\boldsymbol{F}(\boldsymbol{z},t)^* \\ &= [\det \boldsymbol{F}(\boldsymbol{z},t)]^{-1}\boldsymbol{F}(\boldsymbol{z},t)\cdot\boldsymbol{S}(\boldsymbol{z},t)\cdot\boldsymbol{F}(\boldsymbol{z},t)^*.\end{aligned} \tag{15.34}$$

We use (15.34) to deduce frame-indifferent forms of constitutive equations for $\boldsymbol{\Sigma}$. Observe that the very simple form (11.21) for materials of the rate type becomes very complicated in terms of the Cauchy stress tensor: Both the spin and the vorticity tensors intervene when time derivatives of $\boldsymbol{S}$ are expressed in terms of time derivatives of $\boldsymbol{\Sigma}$. A complicated theory of invariant stress rates was made superfluous by the theory of Sec. 11. The stress $\boldsymbol{\Sigma}$ is particularly useful for Newtonian fluids, which have constitutive equations of the form $\boldsymbol{\Sigma} = -p\boldsymbol{I} + \mu\boldsymbol{D}$, where p and μ are scalar fields independent of $\boldsymbol{D}$.

Material constraints. We now complete the demonstration that the Global Constraint Principle reduces to the Local Constraint Principle for the constraints (12.4), (12.6), and (12.7).

We introduce spatial coordinates and reduce (12.57) to

$$\boldsymbol{A} : \frac{\partial \breve{\boldsymbol{\eta}}}{\partial \boldsymbol{y}} = \breve{\boldsymbol{\omega}} \tag{15.35}$$

where

$$\boldsymbol{A} \equiv \frac{\partial \gamma}{\partial \boldsymbol{F}} \cdot \boldsymbol{F}^*. \tag{15.36}$$

Since system (15.35) is underdetermined, we may seek solutions in the form

$$\breve{\boldsymbol{\eta}} = \frac{\partial \breve{\psi}}{\partial \boldsymbol{y}} \tag{15.37}$$

where $\breve{\psi}$ depends on $\boldsymbol{y}$ and t. If $\boldsymbol{A}$ is symmetric and positive-definite, then (15.35) and (15.37) form a linear elliptic equation for $\breve{\psi}$ on $\boldsymbol{p}(\mathcal{B},t)$. The regularity of $\boldsymbol{A}$ (which reduces to the identity for (12.4)) depends only on the regularity of $\boldsymbol{p}$. We can invoke the standard theories of linear elliptic equations to ensure existence of solutions in a whole scale of compatible choices of function spaces for $\breve{\boldsymbol{\omega}}$ and $\breve{\psi}$.

15.38. Exercise. Show that

$$\boldsymbol{A} = \boldsymbol{I}, \quad \boldsymbol{A} = 2\boldsymbol{F}\cdot\boldsymbol{F}^*, \quad \boldsymbol{A} = \boldsymbol{V} \tag{15.39}$$

for (12.4), (12.6), (12.7), respectively. Here $\boldsymbol{V}$ is the positive-definite square root of $\boldsymbol{B} \equiv \boldsymbol{F}\cdot\boldsymbol{F}^*$. Each of these $\boldsymbol{A}$'s is symmetric and positive-definite.

Had we substituted

$$\boldsymbol{\eta} = \frac{\partial \psi}{\partial \boldsymbol{z}} \tag{15.40}$$

into (12.52), we would have obtained

$$\frac{\partial \gamma}{\partial \boldsymbol{F}} : \frac{\partial \psi}{\partial \boldsymbol{z}} = \boldsymbol{\omega}. \tag{15.41}$$

For (12.6) and (12.7), the coefficient tensor $\partial\gamma/\partial\boldsymbol{F}$ is not positive-definite, and our argument breaks down.

CHAPTER XIII

Elasticity

1. Summary of the Governing Equations

We now collect all the results of Chap. XII that are pertinent to elasticity. $\boldsymbol{p}(\boldsymbol{z},t)$ is the position of material point $\boldsymbol{z}$ of body $\mathcal{B}$ at time t. We define

$$\boldsymbol{F} \equiv \boldsymbol{p}_{\boldsymbol{z}}, \quad \boldsymbol{C} \equiv \boldsymbol{F}^* \cdot \boldsymbol{F}. \tag{1.1}$$

The classical form of the balance of linear momentum is

$$\nabla \cdot \boldsymbol{T}^* + \boldsymbol{f} = \rho \boldsymbol{p}_{tt} \tag{1.2}$$

where $\boldsymbol{T}$ is the first Piola-Kirchhoff stress tensor, $\boldsymbol{f}$ is the prescribed body force intensity per unit reference volume, and ρ is the mass density per unit reference volume.

The material of $\mathcal{B}$ is *elastic* iff there is a function

$$\text{Lin}^+ \times \mathcal{B} \ni (\boldsymbol{F},\boldsymbol{z}) \mapsto \hat{\boldsymbol{T}}(\boldsymbol{F},\boldsymbol{z}) \equiv \boldsymbol{F} \cdot \hat{\boldsymbol{S}}(\boldsymbol{F}^* \cdot \boldsymbol{F},\boldsymbol{z}), \tag{1.3a}$$

where $\hat{\boldsymbol{S}}(\boldsymbol{F}^* \cdot \boldsymbol{F},\boldsymbol{z}) \in \text{Sym}$, such that

$$\boldsymbol{T}(\boldsymbol{z},t) = \hat{\boldsymbol{T}}(\boldsymbol{F}(\boldsymbol{z},t),\boldsymbol{z}). \tag{1.3b}$$

(The symmetry of $\hat{\boldsymbol{S}}$ ensures that the balance of angular momentum is automatically satisfied.) The substitution of (1.3b) into (1.2) converts it into the following quasilinear sixth-order system of partial differential equations (equivalent to three coupled second-order scalar partial differential equations):

$$\nabla \cdot \hat{\boldsymbol{T}}^*(\boldsymbol{p}_{\boldsymbol{z}}(\boldsymbol{z},t),\boldsymbol{z}) + \boldsymbol{f}(\boldsymbol{z},t) = \rho(\boldsymbol{z})\boldsymbol{p}_{tt}(\boldsymbol{z},t). \tag{1.4}$$

The body force $\boldsymbol{f}$ could be prescribed as a composite function depending on $(\boldsymbol{z},t)$ through $\boldsymbol{p}$, $\boldsymbol{p}_t$, etc.

For incompressible materials, we replace (1.3b) with

$$\det \boldsymbol{C} = 1, \tag{1.5}$$

$$\hat{\boldsymbol{T}}(\boldsymbol{F},p,\boldsymbol{z}) \equiv \boldsymbol{F} \cdot [-p\boldsymbol{C}^{-1} + \hat{\boldsymbol{S}}_{\text{A}}(\boldsymbol{C},p,\boldsymbol{z})], \tag{1.6a}$$

$$\hat{\boldsymbol{T}}_{\text{A}}(\boldsymbol{F},p,\boldsymbol{z}) \equiv \boldsymbol{F} \cdot \hat{\boldsymbol{S}}_{\text{A}}(\boldsymbol{C},p,\boldsymbol{z}) \tag{1.6b}$$

where the extra stress $\hat{\boldsymbol{S}}_{\mathrm{A}}$, which we take to satisfy $\hat{\boldsymbol{S}}_{\mathrm{A}} : \boldsymbol{C} = 0$ (see (XII.12.33d)), is defined only for $\boldsymbol{C}$'s satisfying (1.5).

For unconstrained isotropic materials, $\hat{\boldsymbol{S}}$ has the form

$$(1.7)\qquad \begin{aligned}\hat{\boldsymbol{S}}(\boldsymbol{C},\boldsymbol{z}) &= \alpha_0(\iota(\boldsymbol{C}),\boldsymbol{z})\boldsymbol{I} + \alpha_1(\iota(\boldsymbol{C}),\boldsymbol{z})\boldsymbol{C} + \alpha_2(\iota(\boldsymbol{C}),\boldsymbol{z})\boldsymbol{C}^2 \\ &= \beta_{-1}(\iota(\boldsymbol{C}),\boldsymbol{z})\boldsymbol{C}^{-1} + \beta_0(\iota(\boldsymbol{C}),\boldsymbol{z})\boldsymbol{I} + \beta_1(\iota(\boldsymbol{C}),\boldsymbol{z})\boldsymbol{C}\end{aligned}$$

where

$$(1.8)\qquad \iota(\boldsymbol{C}) \equiv (\mathrm{I}_C, \mathrm{II}_C, \mathrm{III}_C).$$

For incompressible isotropic materials, (1.6) reduces to

$$(1.9)\qquad \hat{\boldsymbol{T}}(\boldsymbol{F},p,\boldsymbol{z}) \equiv \boldsymbol{F}\cdot[-p\boldsymbol{C}^{-1} + \psi_0(\iota(\boldsymbol{C}),p,\boldsymbol{z})\boldsymbol{I} + \psi_1(\iota(\boldsymbol{C}),p,\boldsymbol{z})\boldsymbol{C}]$$

where now $\iota(\boldsymbol{C}) \equiv (\mathrm{I}_C, \mathrm{II}_C)$ since $\mathrm{III}_C = 1$.

The material defined by (1.3) is *hyperelastic* iff there exists a scalar-valued *stored-energy* or *strain-energy* (*density*) *function* W such that

$$(1.10\mathrm{a})\qquad \begin{aligned}&\hat{\boldsymbol{T}}(\boldsymbol{F},\boldsymbol{z}) = \partial W(\boldsymbol{F}^*\cdot\boldsymbol{F},\boldsymbol{z})/\partial\boldsymbol{F} \quad \text{or, equivalently,}\\ &\hat{\boldsymbol{S}}(\boldsymbol{C},\boldsymbol{z}) = 2\,\partial W(\boldsymbol{C},\boldsymbol{z})/\partial\boldsymbol{C}.\end{aligned}$$

The material defined by (1.5), (1.6) is *hyperelastic* iff there exists a stored-energy function W such that

$$(1.10\mathrm{b})\qquad \begin{aligned}&\hat{\boldsymbol{T}}_{\mathrm{A}}(\boldsymbol{F},\boldsymbol{z}) = \partial W(\boldsymbol{F}^*\cdot\boldsymbol{F},\boldsymbol{z})/\partial\boldsymbol{F} \quad \text{or, equivalently,}\\ &\hat{\boldsymbol{S}}_{\mathrm{A}}(\boldsymbol{C},\boldsymbol{z}) = 2\,\partial W(\boldsymbol{C},\boldsymbol{z})/\partial\boldsymbol{C}\end{aligned}$$

where the domain of $\hat{\boldsymbol{S}}_{\mathrm{A}}$ is restricted to those $\boldsymbol{C}$ satisfying (1.5). A hyperelastic material is sometimes termed *Green-elastic*, in which case the elastic material defined by (1.3) is termed *Cauchy-elastic*. In Sec. XII.14 we showed that the Coleman-Noll Entropy Principle implies that elastic materials should be hyperelastic. Less sophisticated arguments also support the same conclusion. Nevertheless, it is useful to preserve the generality of Cauchy elasticity because there are a variety of approximation schemes, numerical, perturbational, and analytic, that lead to worthy problems lacking the variational structure of hyperelasticity. In particular, the constitutive function for the leading term of the asymptotic analysis carried out in Sec. IV.4 is not hyperelastic, even if the constitutive function for the full problem is hyperelastic. As we have seen in Chaps. IV, VI, IX, and X, there are many *equilibrium* problems for which hyperelasticity is not crucial for an effective analysis.

We adopt boundary and initial conditions like those discussed in Sec. XII.8.

1.11. Exercise. Let the body force intensity $\boldsymbol{f}$ and the applied traction $\bar{\boldsymbol{\tau}}$ (see (XII.8.12)) be conservative, so that there are potential-energy functions Ω and ω such that

$$(1.12)\qquad \boldsymbol{f}(\boldsymbol{z},t) = -\frac{\partial\Omega}{\partial\boldsymbol{p}}(\boldsymbol{p}(\boldsymbol{z},t),\boldsymbol{z}),\quad \bar{\boldsymbol{\tau}}(\boldsymbol{z},t,\mathfrak{q}(\boldsymbol{z},t)) = -\frac{\partial\omega}{\partial\boldsymbol{p}}(\boldsymbol{p}(\boldsymbol{z},t),\boldsymbol{z}).$$

Let (XII.8.11) hold. Prove that the equations governing the motion of a hyperelastic body $\mathcal{B}$ under the action of the force system (1.12) are the Euler-Lagrange equations for the Lagrangian functional

(1.13)
$$L[\boldsymbol{p}] \equiv \int_{t_1}^{t_2} \left\{ \int_{\mathcal{B}} \left[\tfrac{1}{2}\rho \boldsymbol{p}_t \cdot \boldsymbol{p}_t - W(\boldsymbol{p}_z^* \cdot \boldsymbol{p}_z, \boldsymbol{z}) - \Omega(\boldsymbol{p}, \boldsymbol{z}) \right] dv(\boldsymbol{z}) - \int_{\partial\mathcal{B}} \omega(\boldsymbol{p}, \boldsymbol{z})\, da(\boldsymbol{z}) \right\} dt,$$

which is defined for all sufficiently well-behaved $\boldsymbol{p}$ satisfying (XII.8.11).

1.14. Exercise. The theory of Sec. XII.13 says that a hyperelastic isotropic material has a stored-energy function of the form $W^\dagger(\iota(\boldsymbol{C}), \boldsymbol{z})$. Use the results of Sec. XI.2 to show that (1.10a) reduces to

$$\hat{\boldsymbol{T}}(\boldsymbol{F}, \boldsymbol{z}) = \boldsymbol{F} \cdot \left[2\frac{\partial W^\dagger}{\partial \mathrm{I}_C}\boldsymbol{I} + 2\frac{\partial W^\dagger}{\partial \mathrm{II}_C}(\mathrm{I}_C \boldsymbol{I} - \boldsymbol{C}) + \frac{\partial W^\dagger}{\partial \mathrm{III}_C}\boldsymbol{C}^{-1} \right]. \tag{1.15}$$

The corresponding version for incompressible media is obtained by replacing $\partial W^\dagger / \partial \mathrm{III}_C$ with $-p$.

2. Constitutive Restrictions

Since we are using a material, rather than a spatial, formulation of the governing equations, the acceleration term is just the innocuous $\boldsymbol{p}_{tt}$ appearing on the right-hand side of (1.4). Thus the entire responsibility for the mathematical structure of this equation devolves upon the constitutive function $\hat{\boldsymbol{T}}$. In this section we discuss restrictions on $\hat{\boldsymbol{T}}$ that are both physically reasonable and mathematically convenient. We consider the following desiderata for $\hat{\boldsymbol{T}}$:

(i) An increase in a component of strain should be accompanied by an increase in a corresponding component of stress.
(ii) Extreme strains (those for which $|\boldsymbol{C}| = \infty$ or for which $\det \boldsymbol{C} = 0$) should be maintained by infinite stresses.
(iii) The equations of motion should admit solutions with wave-like behavior.
(iv) Well-set initial-boundary-value problems for the equations of motion should have solutions.
(v) For appropriate data, the equilibrium equations should admit multiple solutions, so that buckling can be described.
(vi) Solutions should have an appropriate level of regularity.

This list is notable for its intentional imprecision. To make (i) and (ii) precise, we must select suitable measures of deformation from among $\boldsymbol{F}, \boldsymbol{C}, \boldsymbol{U}$, etc., and suitable measures of stress from among $\boldsymbol{T}, \boldsymbol{S}, \boldsymbol{\Sigma}$, etc. To convert these conditions into reasonable mathematical statements, we must further account for the fact that an elongation in one direction is typically accompanied by contractions in the transverse directions. This coupling of effects is a source of the richness and difficulty of (1.4). We shall respond to (iii) by requiring that (1.4) be hyperbolic. To make (iv)–(vi) precise, we should have to define the class of functions in which solutions are to be sought.

Order-preservation. Let us turn to (i), which says that the mapping from strains to stresses is *order-preserving* in some as yet unspecified sense. The most mathematically attractive notion of order-preservation is that $\hat{\boldsymbol{T}}(\cdot, \boldsymbol{z})$ be *strictly monotone*:

$$(2.1) \qquad [\hat{\boldsymbol{T}}(\boldsymbol{G} + \alpha \boldsymbol{H}, \boldsymbol{z}) - \hat{\boldsymbol{T}}(\boldsymbol{G}, \boldsymbol{z})] : \boldsymbol{H} > 0 \quad \forall\, \boldsymbol{G} \in \mathrm{Lin}^+,$$

$$(2.2) \qquad \forall\, \boldsymbol{H} \neq \boldsymbol{O}, \ \ \forall \alpha \in (0,1] \text{ such that } \det(\boldsymbol{G} + \alpha \boldsymbol{H}) > 0.$$

(Lin^+ is defined in Sec. XI.1.) If $\hat{\boldsymbol{T}}(\cdot, \boldsymbol{z})$ is differentiable, then we can divide (2.1) by α and take the limit as $\alpha \searrow 0$ to obtain an inequality whose strict form is

$$(2.3) \qquad \boldsymbol{H} : [\partial \hat{\boldsymbol{T}}(\boldsymbol{F}, \boldsymbol{z}) / \partial \boldsymbol{F}] : \boldsymbol{H} > 0 \quad \forall\, \boldsymbol{F} \in \mathrm{Lin}^+,$$

$$(2.4) \qquad \forall\, \boldsymbol{H} \neq \boldsymbol{O}.$$

For a hyperelastic material, the analog of (2.1) and (2.2) is that $\boldsymbol{F} \mapsto W(\boldsymbol{F}^* \cdot \boldsymbol{F}, \boldsymbol{z})$ be strictly convex.

Were $\hat{\boldsymbol{T}}$ to be defined on all of Lin, to satisfy (2.1) and (2.2), and to satisfy mild growth conditions, then boundary-value problems for the equilibrium equation corresponding to (1.4) would have an extensive mathematical theory: Weak solutions exist. If the loading is dead, these solutions are unique to within a rigid displacement. The weak solutions possess a considerable amount of regularity. (See Giaquinta (1983).)

2.5. Exercise. Prove the statement about uniqueness. (Hill (1957) first observed this result, which is an extension of the Kirchhoff Uniqueness Theorem of linear elasticity.)

At first sight, these virtues seem irresistible. But the uniqueness means that an elastic rod could not buckle under thrust, however thin the rod and however large the thrust. One of the reasons for suffering the complexity of nonlinear elasticity is to be able to describe buckling. Somewhat weaker uniqueness results hold if the domain of $\hat{\boldsymbol{T}}$ is that given in (1.3) (see Gurtin & Spector (1979).)

Another reason for rejecting (2.1), (2.2) is given by

2.6. Exercise. Suppose that $\hat{\boldsymbol{T}}(\boldsymbol{I}, \boldsymbol{z}) = \boldsymbol{O}$. Prove that (2.1), (2.2) are incompatible with frame-indifference. (Hint: Take $\boldsymbol{G} = \boldsymbol{I}$ and $\boldsymbol{H} = \boldsymbol{Q} - \boldsymbol{I}$ in (2.1), (2.2) where $\boldsymbol{Q}$ is proper-orthogonal.)

A third reason for rejecting (2.1), (2.2) is connected with desideratum (ii). A standard assumption in monotone operator theory to achieve the requisite growth is that the left-hand side of (2.1) approach ∞ as $|\boldsymbol{G} + \boldsymbol{H}| \to \infty$ or as $\det(\boldsymbol{G} + \boldsymbol{H}) \to 0$ along any line containing $\boldsymbol{G}$. Since the domain Lin^+ of $\hat{\boldsymbol{T}}(\cdot, \boldsymbol{z})$ is not convex (see Ex. XII.1.5), there are line segments in the domain with interior points very close to the boundary of the domain. At such interior points of such line segments, the left-hand side of (2.1) is very large, so that (2.1), (2.2) cannot hold over the entire line segment. A related phenomenon holds for hyperelastic materials because a convex W cannot approach ∞ on the boundary of a nonconvex set.

The easiest mathematical way to allow nonuniqueness, while preserving most of the analytical advantages of (2.1), (2.2) is to allow $\hat{\boldsymbol{T}}$ to depend upon $\boldsymbol{p}$ in addition to $\boldsymbol{p_z}$. This ploy is prohibited by the Principle of Frame-Indifference.

In all of our studies of problems with but one spatial variable in Chaps. I–X, we required that the principal parts of the spatial differential operators (of elasticity) be strictly monotone in their highest derivatives. The presence of lower-order terms in most of these equations prevents the entire operator from being strictly monotone. Since the operator in the left-most term of (1.4) does not have lower-order terms, a requirement that its principal part be strictly monotone would ensure that the entire differential operator be strictly monotone, and thus be a condition far more restrictive than that imposed on our degenerate problems with only one spatial variable. In Chap. XIV we discuss the relationship between monotonicity conditions for elastic rod and shell theories and the Strong Ellipticity Condition, which we now introduce.

A condition weaker than (2.1), (2.2), yet having a well-established mathematical status, is (the strict form of) the *Strong Ellipticity Condition*: Inequality (2.1) holds

$$\forall\, \boldsymbol{H} \text{ of rank } 1,\ \forall \alpha \in (0,1] \text{ such that } \det(\boldsymbol{G}+\alpha\boldsymbol{H}) > 0. \tag{2.7}$$

Note that the set of tensors of rank 1 is exactly the set of dyads. If $\hat{\boldsymbol{T}}(\cdot,\boldsymbol{z})$ is differentiable, then a slightly stronger restriction is that (2.3) hold

$$\forall\, \boldsymbol{H} \text{ of rank } 1. \tag{2.8}$$

For hyperelastic materials, the strict form of the Strong Ellipticity Condition when generalized slightly by relaxing the differentiability of W yields the requirement that $W(\cdot,\boldsymbol{z})$ be *rank-one convex*:

$$W(\boldsymbol{G}+\alpha\boldsymbol{H},\boldsymbol{z}) \le (1-\alpha)W(\boldsymbol{G},\boldsymbol{z}) + \alpha W(\boldsymbol{G}+\boldsymbol{H},\boldsymbol{z}) \quad \forall\, \boldsymbol{G} \in \mathrm{Lin}^+ \tag{2.9}$$

and (2.7) holds. When $W(\cdot,\boldsymbol{z})$ is twice continuously differentiable, (2.3), (2.8) reduce to the *Strong Legendre-Hadamard Condition* of the Calculus of Variations.

We now obtain a useful mechanical interpretation of the Strong Ellipticity Condition. Let $\boldsymbol{a}$ and $\boldsymbol{b}$ be arbitrary unit vectors. Then $\boldsymbol{ab}$ is a unit dyad: $|\boldsymbol{ab}| = 1$. We can decompose $\boldsymbol{F}$ into its component along $\boldsymbol{ab}$ and its orthogonal complement:

$$\boldsymbol{F} = (\boldsymbol{a}\cdot\boldsymbol{F}\cdot\boldsymbol{b})\boldsymbol{ab} + [\boldsymbol{F} - (\boldsymbol{a}\cdot\boldsymbol{F}\cdot\boldsymbol{b})\boldsymbol{ab}]. \tag{2.10}$$

We can therefore study the variation of the first term of (2.10) while the second is held fixed. It follows from the definition (XI.1.37) of the determinant that if $\boldsymbol{a}$ and $\boldsymbol{b}$ are independent of $\boldsymbol{F}$, or more generally, if $\boldsymbol{a}$ and $\boldsymbol{b}$ are independent of $\boldsymbol{a}\cdot\boldsymbol{F}\cdot\boldsymbol{b}$, then

$$\boldsymbol{a}\cdot\boldsymbol{F}\cdot\boldsymbol{b} \mapsto \det\boldsymbol{F} \text{ is affine for fixed } \boldsymbol{F} - (\boldsymbol{a}\cdot\boldsymbol{F}\cdot\boldsymbol{b})\boldsymbol{ab}. \tag{2.11}$$

In this case, it follows that

$$\mathcal{D}(\boldsymbol{ab}) \equiv \{\boldsymbol{a}\cdot\boldsymbol{F}\cdot\boldsymbol{b} \in \mathbb{R} :\ \det\boldsymbol{F} > 0\} \tag{2.12}$$

is either an open half-line, or a line, or the empty set (even though Lin^+ is not convex). We are suppressing the dependence of $\mathcal{D}$ on $\boldsymbol{F}$. We accordingly set

$$\mathcal{D}(\boldsymbol{ab}) \equiv \big(l^-(\boldsymbol{ab}), l^+(\boldsymbol{ab})\big). \tag{2.13}$$

$l^-(\boldsymbol{ab})$ is either $-\infty$ or a finite number; $l^+(\boldsymbol{ab})$ is either ∞ or a finite number. If $\mathcal{D}(\boldsymbol{ab})$ is empty, we can simply equate l^- and l^+. By choosing $\boldsymbol{H} = \eta\boldsymbol{ab}$ in (2.1) and (2.8), we deduce from the Strong Ellipticity Condition that

$$\big(l^-(\boldsymbol{ab}), l^+(\boldsymbol{ab})\big) \ni \boldsymbol{a}\cdot\boldsymbol{F}\cdot\boldsymbol{b} \mapsto \boldsymbol{a}\cdot\hat{\boldsymbol{T}}(\boldsymbol{F},\boldsymbol{z})\cdot\boldsymbol{b} \text{ is strictly increasing.} \tag{2.14}$$

Conditions (2.3) and (2.8) imply that

$$\frac{\partial[\boldsymbol{a}\cdot\hat{\boldsymbol{T}}(\boldsymbol{F},\boldsymbol{z})\cdot\boldsymbol{b}]}{\partial(\boldsymbol{a}\cdot\boldsymbol{F}\cdot\boldsymbol{b})} > 0 \tag{2.15}$$

provided that $\boldsymbol{a}$ and $\boldsymbol{b}$ are independent of $\boldsymbol{a}\cdot\boldsymbol{F}\cdot\boldsymbol{b}$. This condition says that the $\boldsymbol{ab}$-component of the first Piola-Kirchhoff stress tensor is an increasing function of the corresponding component of $\boldsymbol{F}$.

Let us now examine the dynamical significance of the Strong Ellipticity Condition by classifying the type of (1.4) as a system of partial differential equations from the viewpoint of wave propagation (see Courant & Hilbert (1961)). Since such classifications are purely local, we study the behavior of a solution $\boldsymbol{p}$ for $(\boldsymbol{z}, t)$ near $(\boldsymbol{z}_0, t_0)$ by studying the linearization of (1.4) about an equilibrium state with constant deformation $\boldsymbol{F}(\boldsymbol{z}_0, t_0)$ for a homogeneous elastic body occupying all space $\mathbb{E}^3$ with constitutive function $\boldsymbol{F} \mapsto \hat{\boldsymbol{T}}(\boldsymbol{F}, \boldsymbol{z}_0)$ and constant density $\rho(\boldsymbol{z}_0)$. This linearization has the form

$$\nabla\cdot\left[\frac{\partial\hat{\boldsymbol{T}}}{\partial\boldsymbol{F}}(\boldsymbol{F}(\boldsymbol{z}_0,t_0),\boldsymbol{z}_0) : \frac{\partial\boldsymbol{u}}{\partial\boldsymbol{z}}\right]^* = \rho(\boldsymbol{z}_0)\frac{\partial^2\boldsymbol{u}}{\partial t^2}, \tag{2.16a}$$

which is most conveniently written in Cartesian components as

$$\frac{\partial\hat{T}_i{}^p}{\partial F^j{}_q}(\boldsymbol{F}(\boldsymbol{z}_0,t_0),\boldsymbol{z}_0)\frac{\partial^2 u^j}{\partial z^q\partial z^p} = \rho(\boldsymbol{z}_0)\delta_{ij}\frac{\partial^2 u^j}{\partial t^2}. \tag{2.16b}$$

We seek solutions of (2.16) in the form of plane waves travelling in direction $\boldsymbol{n}$ with speed c, i.e., solutions of the form

$$\boldsymbol{u}(\boldsymbol{z},t) = \boldsymbol{v}(\boldsymbol{n}\cdot\boldsymbol{z} - ct), \quad |\boldsymbol{n}| = 1. \tag{2.17}$$

The substitution of (2.17) into (2.16) yields

$$\left[\frac{\partial\hat{T}_i{}^p}{\partial F^j{}_q}n_p n_q - c^2\rho\delta_{ij}\right](v^j)'' = 0. \tag{2.18}$$

For a given $\boldsymbol{n}$, this equation has a nontrivial solution $\boldsymbol{v}''$, i.e., (2.16) admits a travelling wave in direction $\boldsymbol{n}$, if and only if the *acoustic tensor* $\boldsymbol{Q}(\boldsymbol{n})$ with components $Q_{ij}(\boldsymbol{n}) \equiv \left(\partial \hat{T}_i{}^p / \partial F^j{}_q\right) n_p n_q$ has a positive eigenvalue $c^2\rho$, which satisfies

$$\det\left(\boldsymbol{Q} - c^2\rho \boldsymbol{I}\right) = 0. \tag{2.19}$$

The left-hand side of (1.4) is said to be *elliptic* at any solution for which there is a positive solution $c^2\rho$ to (2.19). The Strong Ellipticity Condition ensures that $\boldsymbol{Q}$, which is not necessarily symmetric, is positive-definite. A fortiori, it ensures that $\boldsymbol{n} \cdot \boldsymbol{Q}(\boldsymbol{n}) \cdot \boldsymbol{n} > 0$. An application of Theorem XIX.2.32 shows that this last condition implies that there is a positive eigenvalue $c^2\rho$ with a corresponding eigenvector $\boldsymbol{v}'' = \boldsymbol{n}$ (see Truesdell (1966) or Wang & Truesdell (1973, Sec. VI.3)). Thus the Strong Ellipticity Condition ensures the existence of a longitudinal travelling wave. Equation (1.4) is *hyperbolic* iff for each $\boldsymbol{n}$, all the eigenvalues of (2.18) are positive and if the corresponding eigenvectors span $\mathbb{E}^3$. The Strong Ellipticity Condition ensures the positivity of the eigenvalues. If $\boldsymbol{Q}(\boldsymbol{n})$ is symmetric for all $\boldsymbol{n}$, then the Strong Ellipticity Condition also ensures the hyperbolicity of (1.4), which means that it admits the full range of wave-like behavior. (It can be shown that $\boldsymbol{Q}(\boldsymbol{n})$ is symmetric for all $\boldsymbol{F}$ and $\boldsymbol{n}$ if and only if the material is hyperelastic (see Wang & Truesdell (1973, Sec. VI.3)).

For further discussion of the Strong Ellipticity Condition for unconstrained materials, see P. J. Chen (1973), Y.-C. Chen (1991), Hayes (1969), Hayes & Rivlin (1961), Rosakis (1990), Rivlin & Sawyers (1977, 1978), Simpson & Spector (1983, 1987), Truesdell & Noll (1965), and Wang & Truesdell (1973).

2.20. Exercise. Convert (2.3) and (2.8) into an equivalent statement about the dependence of $\hat{\boldsymbol{S}}$ on $\boldsymbol{C}$.

The *Strong Coleman-Noll Condition* is that (2.1) hold

$$\forall\, \boldsymbol{H} \in \mathrm{Sym} \setminus \{\boldsymbol{O}\}. \tag{2.21}$$

This condition implies several plausible constitutive restrictions and is therefore more restrictive than any of its consequences. A detailed discussion of the relationship of these various constitutive assumptions was carried out by Truesdell & Toupin (1963), augmented versions of which have been given by Truesdell & Noll (1965) and by Wang & Truesdell (1973). We shall find that the Strong Coleman-Noll Condition is not as attractive as the Strong Ellipticity Condition for a number of our constructions.

Ericksen (1980a) and elsewhere has pointed out that (2.14), at least for certain components, fails to describe phenomena observed in phase transitions in pure crystals. Those who have studied such problems have typically relaxed conditions like (2.14) by requiring that they hold everywhere except on a bounded subset of Lin^+.

We adopt the Strong Ellipticity Condition as our basic constitutive hypothesis for nonlinear elasticity, without endorsing it as the sole avenue to mechanical truth. Sometimes it is appropriate to supplement it with additional restrictions, which may be consequences of the Strong Coleman-Noll Condition. Sometimes it is appropriate to weaken it to describe phase transitions.

Let us introduce the *adjugate tensor*

$$\boldsymbol{F}^{\times} \equiv [\partial(\det \boldsymbol{F})/\partial \boldsymbol{F}]^{*}, \tag{2.22}$$

which is the transposed cofactor tensor. For $\boldsymbol{F} \in \mathrm{Lin}^{+}$, Eq. (XI.2.7b) implies that

$$\boldsymbol{F}^{\times} = (\det \boldsymbol{F})\boldsymbol{F}^{-1}. \tag{2.23}$$

An important refinement of the Strong Legendre-Hadamard Condition is *Ball's Condition of Polyconvexity*: There is a convex function

$$\mathrm{Lin}^{+} \times \mathrm{Lin}^{+} \times (0,\infty) \ni (\boldsymbol{G}, \boldsymbol{H}, \delta) \mapsto \tilde{W}(\boldsymbol{G}, \boldsymbol{H}, \delta, \boldsymbol{z}) \in \mathbb{R} \tag{2.24}$$

such that

$$W(\boldsymbol{F}^{*} \cdot \boldsymbol{F}, \boldsymbol{z}) = \tilde{W}(\boldsymbol{F}, \boldsymbol{F}^{\times}, \det \boldsymbol{F}, \boldsymbol{z}). \tag{2.25}$$

Note that the dependence of W on $\boldsymbol{C} = \boldsymbol{F}^{*} \cdot \boldsymbol{F}$ further restricts the form of $\tilde{W}$. The form of the right-hand side of (2.25) implies that the domain of $\tilde{W}$ can be extended to $\mathrm{Lin}\times\mathrm{Lin}\times(0,\infty)$ without altering the significance of this equation. In this case, the domain of $\tilde{W}(\cdot,\cdot,\cdot,\boldsymbol{z})$ is an open half-space in the nineteen-dimensional space $\mathrm{Lin}\times\mathrm{Lin}\times\mathbb{R}$, whereas the domain of $\boldsymbol{F} \mapsto W(\boldsymbol{F}^{*} \cdot \boldsymbol{F}, \boldsymbol{z})$ is the nonconvex set Lin^{+}. (The domain of $\boldsymbol{C} \mapsto W(\boldsymbol{C}, \boldsymbol{z})$ is the open convex subset of positive-definite tensors in Sym.) The Polyconvexity Condition supports the only extant proofs of the existence of minimizers of the potential energy functional for conservative static problems,

Morrey (1952, 1966) introduced the concept of quasiconvexity, described by an integral inequality, and used it to obtain the existence of minimizers of variational problems corresponding to well-behaved quasilinear elliptic systems, which do not suffer from the peculiarities of the equations of elasticity associated with the requirement that $\det \boldsymbol{F} > 0$. It was shown by Ball (1977a,b) that polyconvexity implies quasiconvexity and it was essentially shown by Morrey that quasiconvexity implies rank-one convexity. Šverák (1991, 1992a) showed that rank-one convexity does not imply quasiconvexity, and produced an example of a frame-indifferent, isotropic, quasiconvex function bounded below that is not polyconvex. Also see Alibert & Dacorogna (1992), Rosakis & Simpson (1994), and Šverák (1992b).

Order-preservation for constrained materials. We now study analogs of the Strong Ellipticity Condition for elastic materials subject to a single material constraint of the form (XII.12.1):

$$\gamma(\boldsymbol{F}(\boldsymbol{z},t), \boldsymbol{z}) = 0 \tag{2.26}$$

for which the global constraint principle reduces to the local constraint principle. (Other cases, important for rod and shell theories, are treated in Chap. XIV.) It is convenient to refrain from putting (2.26) into frame-indifferent form. Corresponding to (XII.12.26)–(XII.12.28) are constitutive equations of the form $\boldsymbol{T}(\boldsymbol{z},t) = \hat{\boldsymbol{T}}(\boldsymbol{F}(\boldsymbol{z},t), \lambda(\boldsymbol{z},t), \boldsymbol{z})$ where

$$\hat{\boldsymbol{T}}(\boldsymbol{F},\lambda,\boldsymbol{z}) = \lambda \frac{\partial \gamma}{\partial \boldsymbol{F}}(\boldsymbol{F},\boldsymbol{z}) + \hat{\boldsymbol{T}}_{\mathrm{A}}(\boldsymbol{F},\lambda,\boldsymbol{z}). \tag{2.27}$$

We now extend the Strong Ellipticity Condition (2.1), (2.7) to such materials by requiring that

$$[\hat{\boldsymbol{T}}(\boldsymbol{G}+\alpha\boldsymbol{ab},\lambda,\boldsymbol{z}) - \hat{\boldsymbol{T}}(\boldsymbol{G},\lambda,\boldsymbol{z})] : \boldsymbol{ab} > 0 \tag{2.28}$$

for all $\boldsymbol{G} \in \mathrm{Lin}^+$, $\boldsymbol{ab} \neq \boldsymbol{O}$, $\alpha \in (0,1]$ such that

$$\det(\boldsymbol{G}+\alpha\boldsymbol{ab}) > 0, \quad \gamma(\boldsymbol{G}+\alpha\boldsymbol{ab}) = 0. \tag{2.29a,b}$$

The requirement that (2.29b) hold for all α implies that $\gamma_{\boldsymbol{F}}(\boldsymbol{G}+\alpha\boldsymbol{ab}) : \boldsymbol{ab} = 0$ for all α. We thus deduce from (2.28) and (2.27) that

$$[\hat{\boldsymbol{T}}_{\mathrm{A}}(\boldsymbol{G}+\alpha\boldsymbol{ab},\lambda,\boldsymbol{z}) - \hat{\boldsymbol{T}}_{\mathrm{A}}(\boldsymbol{G},\lambda,\boldsymbol{z})] : \boldsymbol{ab} > 0 \tag{2.30}$$

for all $\boldsymbol{G}$, $\boldsymbol{ab} \neq \boldsymbol{O}$, $\alpha \in (0,1]$ such that (2.29) holds. This is the desired analog of (2.1), (2.7). To get the analog of (2.3), (2.8) when $\hat{\boldsymbol{T}}_{\mathrm{A}}$ and $\gamma(\cdot,\boldsymbol{z})$ are differentiable, we divide (2.30) by α, let $\alpha \searrow 0$, and differentiate (2.29b) with respect to α at $\alpha = 0$ to obtain an inequality whose strict form is

$$\boldsymbol{ab} : \frac{\partial \hat{\boldsymbol{T}}_{\mathrm{A}}}{\partial \boldsymbol{F}} : \boldsymbol{ab} > 0 \quad \forall\, \boldsymbol{F} \in \mathrm{Lin}^+, \tag{2.31}$$

$$\forall \boldsymbol{ab} \neq \boldsymbol{O} \quad \text{such that} \quad \frac{\partial \gamma}{\partial \boldsymbol{F}}(\boldsymbol{F},\boldsymbol{z}) : \boldsymbol{ab} = 0. \tag{2.32}$$

(Note that (2.31) and (2.32) represent a restriction of (2.28) and (2.29) more radical than the restriction (2.3) and (2.8) of (2.1) and (2.7), because (2.32) allows $\boldsymbol{ab}$ to be a tangent vector to the constraint manifold, whereas (2.29) requires that $\boldsymbol{G}+\alpha\boldsymbol{ab}$ lie in the constraint manifold for all $\alpha \in [0,1]$.)

Let us now study strong ellipticity for constrained elastic materials from the viewpoint of classification, which we again approach by examining travelling waves for the linearization. We linearize the system (1.2), (2.26), (2.27) about an equilibrium state with constant deformation $\boldsymbol{F}(\boldsymbol{z}_0, t_0)$ and constant multiplier $\lambda(\boldsymbol{z}_0)$ for a homogeneous elastic body occupying all space $\mathbb{E}^3$ as above, obtaining a system somewhat more complicated than (2.16) for the linearized displacement $\boldsymbol{u}$ and the linearized multiplier $\overset{\vartriangle}{\lambda}$. We seek solutions of this system in the form of plane waves travelling in direction $\boldsymbol{n}$ with speed c, i.e., solutions of the form

$$\boldsymbol{u}(\boldsymbol{z},t) = \boldsymbol{v}(\boldsymbol{n}\cdot\boldsymbol{z} - ct), \quad \overset{\vartriangle}{\lambda}(\boldsymbol{z},t) = \mu(\boldsymbol{n}\cdot\boldsymbol{z} - ct), \quad |\boldsymbol{n}| = 1, \tag{2.33}$$

which satisfy

(2.34a)

$$\left\{ \left[\frac{\partial \hat{T}_{\mathrm{A}i}{}^{p}}{\partial F^{j}{}_{q}} + \lambda \frac{\partial^2 \gamma}{\partial F^{i}{}_{p} \partial F^{j}{}_{q}} \right] n_p n_q - c^2 \rho \delta_{ij} \right\} (v^j)'' + \left[\frac{\partial \hat{T}_{\mathrm{A}i}{}^{p}}{\partial \lambda} + \frac{\partial \gamma}{\partial F^{i}{}_{p}} \right] n_p \mu' = 0,$$

(2.34b)
$$\frac{\partial \gamma}{\partial F^{i}{}_{p}} n_p (v^i)'' = 0.$$

(Note that we have obtained (2.34b) by adding one extra derivative to v_i' in the linearization of (2.26).) For a given $\boldsymbol{n}$, this system has a nontrivial solution $\boldsymbol{v}'', \mu'$ if and only if (2.34) has a positive eigenvalue $c^2\rho$. In this case, the left-hand sides of (1.2) and (2.26) subject to (2.27) constitute an elliptic operator at the solution $\boldsymbol{p}, \lambda$. An examination of the characteristic equation for (2.34) shows that it is merely quadratic in $c^2\rho$ (because (2.34b) implies that eigenvectors $\boldsymbol{v}''$ must lie in the two-dimensional orthogonal complement of $\gamma_{\boldsymbol{F}} \cdot \boldsymbol{n}$). If the characteristic equation has two positive eigenvalues corresponding to two independent eigenvectors, then the left-hand sides of (1.2) and (2.26) subject to (2.27) constitute a strongly elliptic operator at the solution $\boldsymbol{p}, \lambda$.

Now let us specialize our attention to hyperelastic materials, for which $\hat{\boldsymbol{T}}_{\mathrm{A}}$ is independent of λ. We follow Scott & Hayes (1985). Let

(2.35a,b)
$$Q_{ij} \equiv \left[\frac{\partial \hat{T}_{\mathrm{A}i}{}^{p}}{\partial F^{j}{}_{q}} + \lambda \frac{\partial^2 \gamma}{\partial F^{i}{}_{p} \partial F^{j}{}_{q}} \right] n_p n_q, \quad \boldsymbol{m}(\boldsymbol{n}) \equiv \frac{\gamma_{\boldsymbol{F}} \cdot \boldsymbol{n}}{|\gamma_{\boldsymbol{F}} \cdot \boldsymbol{n}|}.$$

Notice the presence of the multiplier λ in the acoustic tensor $\boldsymbol{Q}(\boldsymbol{n})$, which is symmetric. We solve (2.34a) for μ' by taking the dot product of (2.34a) with $\boldsymbol{m}$, and substitute the result into (2.34) to obtain

(2.36a,b)
$$[(\boldsymbol{I} - \boldsymbol{m}\boldsymbol{m}) \cdot \boldsymbol{Q} - c^2 \rho \boldsymbol{I}] \cdot \boldsymbol{v}'' = \boldsymbol{0}, \quad \boldsymbol{m} \cdot \boldsymbol{v}'' = 0.$$

Note that $(\boldsymbol{I} - \boldsymbol{m}\boldsymbol{m}) \cdot \boldsymbol{Q}$ need not be symmetric even though $\boldsymbol{Q}$ is. Nevertheless, (2.36) has real eigenvalues, as we now show.

We introduce the modified tensor

(2.37)
$$\tilde{\boldsymbol{Q}} \equiv \boldsymbol{Q} - \boldsymbol{m}\boldsymbol{m} \cdot \boldsymbol{Q} - \boldsymbol{Q} \cdot \boldsymbol{m}\boldsymbol{m} + (\boldsymbol{m} \cdot \boldsymbol{Q} \cdot \boldsymbol{m}) \boldsymbol{m}\boldsymbol{m}$$

and consider the eigenvalue problem

(2.38a,b)
$$[\tilde{\boldsymbol{Q}} - c^2 \rho \boldsymbol{I}] \cdot \boldsymbol{v}'' = \boldsymbol{0}, \quad \boldsymbol{m} \cdot \boldsymbol{v}'' = 0.$$

It follows from (2.37) that $\tilde{\boldsymbol{Q}} \cdot \boldsymbol{m} = \boldsymbol{0}$, so that (2.38a) is a well-defined eigenvalue problem on the orthogonal complement of $\boldsymbol{m}$. Since $\tilde{\boldsymbol{Q}}$ is symmetric, it has two real eigenvalues ρc^2 (not necessarily distinct) with corresponding orthogonal eigenvectors. It also follows from (2.37) that if $(c^2\rho, \boldsymbol{v}'')$ is an eigenpair of (2.36) if and only if it is an eigenpair of (2.38). Thus (2.36) has two real eigenvalues with orthogonal eigenvectors.

Let us now take the dot product of (2.36a) with an eigenvector $\boldsymbol{v}''$ and use (2.36b) to obtain

(2.39)
$$\boldsymbol{v}'' \cdot \boldsymbol{Q} \cdot \boldsymbol{v}'' = c^2 \rho \boldsymbol{v}'' \cdot \boldsymbol{v}'',$$

from which we conclude that the eigenvalues are positive for all $\boldsymbol{n}$ if $\boldsymbol{Q}$ is positive-definite for all $\boldsymbol{n}$, i.e., if

(2.40)
$$\boldsymbol{a}\boldsymbol{b} : \left[\frac{\partial \hat{\boldsymbol{T}}_{\mathrm{A}}}{\partial \boldsymbol{F}} + \lambda \frac{\partial^2 \gamma}{\partial \boldsymbol{F} \partial \boldsymbol{F}} \right] : \boldsymbol{a}\boldsymbol{b} > 0 \quad \forall\, \boldsymbol{F} \in \mathrm{Lin}^+,$$

subject to (2.32). Thus (2.40) and (2.32), which are not the same as (2.31) and (2.32), say that the left-hand sides of (1.2) and (2.26) subject to (2.27) constitute a strongly elliptic operator at the solution $\boldsymbol{p}, \lambda$.

We now specialize our study to incompressible elastic media, for which (2.29a) is automatically satisfied and (2.29b) reduces to

$$\det(\boldsymbol{F} + \alpha \boldsymbol{ab}) = 1 \quad \forall\, \alpha \in [0,1], \tag{2.41}$$

which is essentially equivalent to

$$\frac{\partial}{\partial \alpha} \det(\boldsymbol{F} + \alpha \boldsymbol{ab}) \equiv (\boldsymbol{F} + \alpha \boldsymbol{ab})^{-*} : \boldsymbol{ab} \equiv \boldsymbol{b} \cdot (\boldsymbol{F} + \alpha \boldsymbol{ab})^{-*} \cdot \boldsymbol{a} = 0. \tag{2.42}$$

We set $\alpha = 0$ in (2.42) to get (2.32).

For a careful treatment of the Strong Ellipticity Condition for incompressible media, beginning from the foregoing analysis of travelling waves, see Zee & Sternberg (1983). Also see Ericksen (1953), Fosdick & MacSithigh (1986), and Scott (1975).

2.43. Exercise. Show that for hyperelastic incompressible materials, the multiplier $\lambda = -p$ can be eliminated from (2.34), so that (2.40) reduces to (2.31), and the Strong Ellipticity Condition, discussed for arbitrary constraints, can be readily imposed on the equations everywhere, rather than just at solutions.

In practice, the specializations of conditions (2.28), (2.29) and (2.31), (2.32) to incompressible materials apply to shear deformations. A straightforward and physically natural analysis of problems for incompressible bodies in which there is stretching requires more extensive alterations of the Strong Ellipticity Condition (see Wissmann (1991)). Note that Ball's condition is readily specialized to incompressible materials.

Growth conditions. We now complement the Strong Ellipticity Condition with compatible growth or *coercivity* conditions. We assume that if (2.14) holds, then

(2.44)
$$\boldsymbol{a} \cdot \hat{\boldsymbol{T}}(\boldsymbol{F}, \boldsymbol{z}) \cdot \boldsymbol{b} \to \pm\infty \quad \text{as } \boldsymbol{a} \cdot \boldsymbol{F} \cdot \boldsymbol{b} \to l^{\pm}(\boldsymbol{ab}) \text{ for fixed } \boldsymbol{F} - (\boldsymbol{a} \cdot \boldsymbol{F} \cdot \boldsymbol{b})\boldsymbol{ab}.$$

We assume analogous results for incompressible media for those $\boldsymbol{ab}$ satisfying (2.42).

For certain analytical work, it is useful to have a more refined description of the behavior of $\hat{\boldsymbol{T}}$ under extreme strains. In particular, we might need to study processes in which $\boldsymbol{a} \cdot \boldsymbol{F} \cdot \boldsymbol{b} \to l^{\pm}(\boldsymbol{ab})$, but for which $\boldsymbol{F} - (\boldsymbol{a} \cdot \boldsymbol{F} \cdot \boldsymbol{b})\boldsymbol{ab}$ is not fixed. To motivate further growth conditions, we examine the behavior of an isotropic material point. Let $\{\boldsymbol{e}_k\}$ be a fixed orthonormal basis of eigenvectors of $\boldsymbol{C}(\boldsymbol{z}, t)$ or, equivalently, of $\boldsymbol{U}(\boldsymbol{z}, t)$. We henceforth suppress the arguments $\boldsymbol{z}, t$. Let $\{U_i\}$ be the corresponding eigenvalues of $\boldsymbol{U}$. The isotropy implies that the $\{\boldsymbol{e}_k\}$ are also eigenvectors of $\boldsymbol{R}^{-1} \cdot \hat{\boldsymbol{T}}(\boldsymbol{F}, \boldsymbol{z}) = \boldsymbol{U} \cdot \hat{\boldsymbol{S}}(\boldsymbol{C}, \boldsymbol{z})$, whose eigenvalues are denoted $\{T_k\}$.

Condition (2.44) implies that if U_1 and U_2 are fixed, then $T_3 \to -\infty$ as $U_3 \to 0$ and $T_3 \to \infty$ as $U_3 \to \infty$. Under these conditions, it is also plausible that $T_1 \to -\infty$ and $T_2 \to -\infty$ as $U_3 \to 0$ and that $T_1 \to \infty$ and $T_2 \to \infty$ as $U_3 \to \infty$. (This property is an extreme version of the Poisson ratio effect of linear elasticity.) Similar effects are captured by the statements that if T_3 is fixed and if $U_3 \to 0$, then $U_1 \to \infty$ or $U_2 \to \infty$, and that if T_1 and T_2 are fixed and if $U_1 \to 0$ and $U_2 \to 0$, then $U_3 \to \infty$. The common feature of each of these effects is that extreme behavior associated with one direction must always be accompanied by extreme behavior associated with another direction. A general statement of this principle, couched in a form that complements (2.14), is given by Antman (1983a). For a comprehensive treatment of the behavior of stress under extreme strain, see Podio-Guidugli & Vergara Caffarelli (1991).

For hyperelastic materials, it is customary to strengthen (2.44) by requiring that

$$W(\boldsymbol{C}, \boldsymbol{z}) \to \infty \quad \text{as any eigenvalue of } \boldsymbol{C} \text{ approaches } 0 \text{ or } \infty, \tag{2.45}$$

which is equivalent to the requirement that

$$W(\boldsymbol{C}, \boldsymbol{z}) \to \infty \quad \text{as } \det \boldsymbol{C} \to 0 \text{ and as } \operatorname{tr} \boldsymbol{C} \to \infty. \tag{2.46}$$

Note that

$$\operatorname{tr} \boldsymbol{C} = |\boldsymbol{F}|^2 = |\boldsymbol{U}|^2. \tag{2.47}$$

It is often useful to require that there be numbers $a > 1$, $A > 0$, and an integrable function φ such that

$$W(\boldsymbol{C}, \boldsymbol{z}) \geq A(\operatorname{tr} \boldsymbol{C})^{a/2} - \varphi(\boldsymbol{z}). \tag{2.48}$$

This condition, which gives a lower bound for the rate of growth of W for large $\boldsymbol{C}$, typically plays an essential role in the applications of the direct methods of the Calculus of Variations to problems of elasticity. It is typically accompanied by a complementary condition giving upper bounds for $|\partial W/\partial \boldsymbol{C}|$. (There are circumstances in which (2.48) must be replaced by a much more sophisticated bound; see Antman (1976a, Sec. 3d)). Conditions like (2.48) have analogs for Cauchy-elastic materials.

Existence and regularity. Desiderata (iii)–(vi) are largely concerned with questions of existence and regularity. We take the requirement (iii) that any solutions of (1.4) should have wave-like behavior and the concomitant need for a precise definition of such behavior as being subsumed under the Strong Ellipticity Condition, which is intimately related to the hyperbolicity of (1.4). This condition supports the existence theories for initial-value problems and initial-boundary-value problems for (1.4) for short time of C. Chen & von Wahl (1982), Dafermos & Hrusa (1985), and Hughes, Kato, & Marsden (1977). Ebin & Saxton (1986), Ebin & Simanca (1992), and Hrusa & Renardy (1988) have treated the corresponding problem for incompressible media. Thus desideratum (iv) is met by the Strong Ellipticity Condition.

The question of existence for the equilibrium equations is somewhat less satisfactory. Of course, equilibrium solutions cannot exist if there is a net force or moment applied to the body. Ball (1977a,b) was able to make a significant extension of the techniques of Morrey (1966) to show that if (2.25) and suitable growth conditions hold, then the total energy functional has a minimum over a natural class of admissible functions in an appropriate Sobolev space. The Gâteaux-differentiability of the energy functional at the minimizer would imply that the minimizer is a weak solution of the Euler-Lagrange equations for the functional, i.e., that the minimizer satisfies the Principle of Virtual Power. The requirement that the deformation preserve orientation is one of the main obstacles that has so far prevented the proof of the requisite differentiability. There have been a number of major advances in determining the regularity properties of minimizers. Cf. Ball (1980, 1981b), Ball & Murat (1984), Bauman & Phillips (1994), Bauman, Owen, & Phillips (1992), Giaquinta, Modica, & Souček (1989), and Müller (1990), but this issue is difficult and far from settled. One interesting result is that of Zhang (1991), who showed that under certain circumstances the absolute minimizer is the solution given by the Implicit Function Theorem, and is therefore smooth. For a comprehensive treatment of results up to 1987, see Ciarlet (1988). Important advances have even been made for problems for crystals in which the restricted convexity of the Strong Legendre-Hadamard Condition is suspended on a bounded set. See Ball & James (1987, 1992), Chipot & Kinderlehrer (1988), Dacorogna (1981, 1989), Fonseca (1987, 1988, 1989a,b), Kinderlehrer & Pedregal (1991), and Modica (1987). These existence and regularity questions for nonlinear elasticity have stimulated a small, but very significant set of contributions to the theory of quasilinear partial differential equations. We comment on local existence theories based on the Implicit Function Theorem at the end of Sec. 11.

Special constitutive equations. It can be shown that the constitutive equation (1.9) for an incompressible isotropic elastic material can be put into the form

$$\boldsymbol{\Sigma} = -p\boldsymbol{I} + \gamma_1(\iota(\boldsymbol{B}), p, z)\boldsymbol{B} + \gamma_{-1}(\iota(\boldsymbol{B}), p, z)\boldsymbol{B}^{-1} \tag{2.49}$$

where $\boldsymbol{\Sigma}$ is the Cauchy stress, and $\boldsymbol{B} \equiv \boldsymbol{F} \cdot \boldsymbol{F}^*$, with ι defined after (1.9). In the special case that γ_1 and γ_{-1} are independent of ι, the material is hyperelastic so that γ_1 and γ_{-1} are also independent of p. In this case, (2.49) reduces to the constitutive equation for a *Mooney-Rivlin material*:

$$\boldsymbol{\Sigma} = -p\boldsymbol{I} + \mu(z)\left[\tfrac{1}{2} + \beta(z)\right]\boldsymbol{B} - \mu(z)\left[\tfrac{1}{2} - \beta(z)\right]\boldsymbol{B}^{-1}. \tag{2.50}$$

Here μ and β are given moduli. This constitutive equation arises in certain approximation schemes. It gives a reasonable approximation to some experimental data for rubber. Its primary virtue is that it can greatly simplify the analysis of many problems of incompressible elasticity, yielding explicit formulas in place of qualitative results. In the variables of (1.9), Eq. (2.50) becomes

$$\boldsymbol{T} = -p\boldsymbol{F}^{-*} + \mu\left[\tfrac{1}{2} + \beta\right]\boldsymbol{F} - \mu\left[\tfrac{1}{2} - \beta\right]\boldsymbol{F}^{-*} \cdot \boldsymbol{C}^{-1}. \tag{2.51}$$

The special case of (2.50) with $\beta = 0$ describes the *neo-Hookean material.*

There are a variety of special constitutive hypotheses based upon experimental evidence or on mathematical simplicity, which we do not pause to discuss. See Carroll (1988), Ogden (1984, Chap. 7), Truesdell (1952), and Truesdell & Noll (1965, Secs. 94, 95) for references and evaluations.

3. Semi-Inverse Problems of Equilibrium in Cylindrical Coordinates

We now embark on a study of restricted classes of solutions of several equilibrium problems for elastic bodies. The resulting problems, which enjoy a number of symmetries, are governed by equations with fewer independent variables and are consequently far easier to analyze than the original system of quasilinear partial differential equations. We can thus derive important physical insights into the behavior of nonlinearly elastic materials. We also obtain a collection of quite specific solutions about which we can perform illuminating perturbation analyses. We begin with problems that are invariant under changes of the cylindrical coordinates ϕ and z.

We introduce the cylindrical coordinates $\mathbf{x} \equiv (s, \phi, z)$ and study the deformations described by (XII.5.9)–(XII.5.19):

$$\tilde{\boldsymbol{p}}(\mathbf{x}) = f(s)\boldsymbol{k}_1(\mathbf{x}) + [h(s) + \gamma\phi + \delta z]\boldsymbol{k}_3, \tag{3.1}$$

$$\begin{aligned} \boldsymbol{F}(\tilde{\boldsymbol{z}}(\mathbf{x})) &= [f'(s)\boldsymbol{k}_1(\mathbf{x}) + f(s)g'(s)\boldsymbol{k}_2(\mathbf{x}) + h'(s)\boldsymbol{k}_3]\boldsymbol{j}^1(\phi) \\ &\quad + \frac{1}{s}[\alpha f(s)\boldsymbol{k}_2(\mathbf{x}) + \gamma\boldsymbol{k}_3]\boldsymbol{j}^2(\phi) + [\beta f(s)\boldsymbol{k}_2(\mathbf{x}) + \delta\boldsymbol{k}_3]\boldsymbol{j}^3 \\ &\equiv F^i{}_q(s)\boldsymbol{k}_i(\mathbf{x})\boldsymbol{j}^q(\phi), \end{aligned} \tag{3.2}$$

$$\begin{aligned} \boldsymbol{C} &= [(f')^2 + (fg')^2 + (h')^2]\boldsymbol{j}^1\boldsymbol{j}^1 + \frac{1}{s}[\alpha f^2 g' + \gamma h'](\boldsymbol{j}^1\boldsymbol{j}^2 + \boldsymbol{j}^2\boldsymbol{j}^1) \\ &\quad + [\beta f^2 g' + \delta h'](\boldsymbol{j}^1\boldsymbol{j}^3 + \boldsymbol{j}^3\boldsymbol{j}^1) + \frac{1}{s^2}[(\alpha f)^2 + \gamma^2]\boldsymbol{j}^2\boldsymbol{j}^2 \\ &\quad + \frac{1}{s}[\alpha\beta f^2 + \gamma\delta](\boldsymbol{j}^2\boldsymbol{j}^3 + \boldsymbol{j}^3\boldsymbol{j}^2) + [(\beta f)^2 + \delta^2]\boldsymbol{j}^3\boldsymbol{j}^3 \\ &\equiv C_{pq}(s)\boldsymbol{j}^p(\phi)\boldsymbol{j}^q(\phi) \end{aligned} \tag{3.3}$$

where

$$\begin{aligned} \boldsymbol{j}_1(\phi) &\equiv \boldsymbol{j}^1(\phi) \equiv \cos\phi\,\boldsymbol{i}_1 + \sin\phi\,\boldsymbol{i}_2, \\ \boldsymbol{j}_2(\phi) &\equiv \boldsymbol{j}^2(\phi) \equiv -\sin\phi\,\boldsymbol{i}_1 + \cos\phi\,\boldsymbol{i}_2, \\ \boldsymbol{j}_3 &\equiv \boldsymbol{j}^3 \equiv \boldsymbol{i}_3, \end{aligned} \tag{3.4}$$

$$\begin{aligned} \boldsymbol{k}_1(\mathbf{x}) &\equiv \boldsymbol{k}^1(\mathbf{x}) \equiv \cos\omega(\mathbf{x})\boldsymbol{i}_1 + \sin\omega(\mathbf{x})\boldsymbol{i}_2, \\ \boldsymbol{k}_2(\mathbf{x}) &\equiv \boldsymbol{k}^2(\mathbf{x}) \equiv -\sin\omega(\mathbf{x})\boldsymbol{i}_1 + \cos\omega(\mathbf{x})\boldsymbol{i}_2, \\ \boldsymbol{k}_3 &\equiv \boldsymbol{k}^3 \equiv \boldsymbol{i}_3, \\ \omega(\mathbf{x}) &\equiv g(s) + \alpha\phi + \beta z. \end{aligned} \tag{3.5}$$

Note that each component of $\boldsymbol{F}$ and $\boldsymbol{C}$ depends only on s.

Using (XII.5.16), we find that

$$\nabla \equiv \frac{\partial}{\partial \boldsymbol{z}} = \frac{\partial \tilde{x}^l}{\partial \boldsymbol{z}}(\tilde{\boldsymbol{z}}(\mathbf{x}))\frac{\partial}{\partial x^l} = \boldsymbol{j}^1(\phi)\frac{\partial}{\partial s} + \frac{1}{s}\boldsymbol{j}^2(\phi)\frac{\partial}{\partial \phi} + \boldsymbol{j}^3\frac{\partial}{\partial z}. \tag{3.6}$$

We shall apply the divergence operator based on (3.6) to the first Piola-Kirchhoff stress tensor to get the equations of equilibrium. Let us decompose $\boldsymbol{T}$ like $\boldsymbol{F}$:

$$\boldsymbol{T}(\tilde{\boldsymbol{z}}(\mathbf{x})) = T_i{}^q(\tilde{\boldsymbol{z}}(\mathbf{x}))\boldsymbol{k}^i(\mathbf{x})\boldsymbol{j}_q(\phi). \tag{3.7}$$

We first look at unconstrained media, assuming that the constitutive functions (1.3) are such that the components of $\boldsymbol{T}$ in (3.7) depend only on s.

3.8. Exercise. Show that if $\hat{\boldsymbol{S}}$ is independent of ϕ and z and if (1.7) holds, then $\hat{\boldsymbol{T}}$ has this property. (The isotropy condition is sufficient, but not necessary for this property to hold: There are many aeolotropic materials whose aeolotropy respects the translation-invariance of the problem under changes of ϕ and z. See Sec. X.3.)

3.9. Exercise. Show that if the components of $\boldsymbol{T}$ in (3.7) depend on $\mathbf{x}$ only through s, and if (3.1)–(3.5) hold, then the equations of equilibrium for zero body force, i.e., the components of $\nabla \cdot \boldsymbol{T}^* = \mathbf{0}$, reduce to the system of ordinary differential equations:

$$\frac{d}{ds}(sT_1{}^1) = K \equiv sT_2{}^1 g'(s) + \alpha T_2{}^2 + \beta s T_2{}^3, \tag{3.10}$$

$$\frac{d}{ds}(sT_2{}^1) + sT_1{}^1 g'(s) + \alpha T_1{}^2 + \beta s T_1{}^3 = 0, \tag{3.11}$$

$$\frac{d}{ds}(sT_3{}^1) = 0. \tag{3.12}$$

3.13. Exercise. Find the three nontrivial components of the symmetry condition (XII.7.27), one of which is

$$f\left[T_1{}^1 g' + \frac{\alpha}{s}T_1{}^2 + \beta T_1{}^3\right] = f' T_2{}^1. \tag{3.14}$$

The substitution of (3.14) into (3.11) yields

$$\frac{d}{ds}(sfT_2{}^1) = 0. \tag{3.15}$$

We may now substitute our constitutive equations for unconstrained materials into (3.10), (3.15), and (3.12) to get a system of three second-order ordinary differential equations for $f, g, h, \alpha, \beta, \gamma, \delta$. We shall append suitable subsidiary conditions to this system.

Incompressible media. The requirement that $\det \boldsymbol{F} = 1$, which characterizes incompressibility, is reduced by (3.2) to

$$(\alpha\delta - \beta\gamma)s^{-1} f(s) f'(s) = 1 \tag{3.16}$$

(see (XII.5.19)), whence we obtain

(3.17)
$$f(s) = \sqrt{\frac{s^2 + C}{\alpha\delta - \beta\gamma}}$$

where C is a constant of integration. Thus the form of f is quite explicit.

In line with our treatment of unconstrained materials, we assume that the constitutive function (1.6) implies that the components of $\boldsymbol{F} \cdot \hat{\boldsymbol{S}}_{\text{A}}$ with respect to the basis $\{\boldsymbol{k}^i \boldsymbol{j}_q\}$ depend only on s. Isotropy ensures this independence. The remaining contributor to the first Piola-Kirchhoff stress in (1.6) is the pressure term $-p(\mathbf{x})\boldsymbol{F}(\tilde{\boldsymbol{z}}(\mathbf{x}))^{-*}$. Since p is unknown, we cannot automatically assume that it is independent of ϕ and z. Perhaps the easiest way to obtain $\boldsymbol{F}^{-1}$ is to employ Cramer's Rule to find the inverse of the matrix of $\boldsymbol{F}$ with respect to the basis used in (3.2) and then to observe that this inverse is the matrix of $\boldsymbol{F}^{-1}$ with respect to the transposed dual $\{\boldsymbol{j}_q \boldsymbol{k}^i\}$ of the basis $\{\boldsymbol{k}_i \boldsymbol{j}^q\}$. (Since the bases $\{\boldsymbol{j}^p\}$ and $\{\boldsymbol{k}_i\}$ are orthonormal, they are their own duals. We retain the distinction in the positioning of indices for the reasons stated in the paragraph following (XII.5.13).) We thus find that

(3.18)
$$\begin{aligned}
&\boldsymbol{F}(\tilde{\boldsymbol{z}}(\mathbf{x}))^{-1}\\
&\quad= (\alpha\delta - \beta\gamma)\frac{f(s)}{s}\boldsymbol{j}_1(\phi)\boldsymbol{k}^1(\mathbf{x})\\
&\qquad+ \boldsymbol{j}_2(\phi)\{-f(s)[\delta g'(s) - \beta h'(s)]\boldsymbol{k}^1(\mathbf{x}) + \delta f'(s)\boldsymbol{k}^2(\mathbf{x}) - \beta f(s) f'(s)\boldsymbol{k}^3\}\\
&\qquad+ \boldsymbol{j}_3\left\{\frac{f(s)}{s}[\gamma g'(s) - \alpha h'(s)]\boldsymbol{k}^1(\mathbf{x}) - \frac{\gamma f'(s)}{s}\boldsymbol{k}^2(\mathbf{x}) + \alpha\frac{f(s)}{s} f'(s)\boldsymbol{k}^3\right\}
\end{aligned}$$

where (3.16) and (3.17) hold. It follows from (1.6) that

(3.19)
$$\begin{aligned}
&T_1{}^1 = \overset{+}{T}_1{}^1 - \frac{p}{f'}, && T_1{}^2 = \overset{+}{T}_1{}^2 + pf(\delta g' - \beta h'), && T_1{}^3 = \overset{+}{T}_1{}^3 - p\frac{f}{s}(\gamma g' - \alpha h'),\\
&T_2{}^1 = \overset{+}{T}_2{}^1, && T_2{}^2 = \overset{+}{T}_2{}^2 - p\delta f', && T_2{}^3 = \overset{+}{T}_2{}^3 + p\gamma\frac{f'}{s},\\
&T_3{}^1 = \overset{+}{T}_3{}^1, && T_3{}^2 = \overset{+}{T}_3{}^2 + \frac{p\beta s}{\alpha\delta - \beta\gamma}, && T_3{}^3 = \overset{+}{T}_3{}^3 - \frac{p\alpha}{\alpha\delta - \beta\gamma}
\end{aligned}$$

where the $\overset{+}{T}_i{}^q$ are the components of $\hat{\boldsymbol{T}}_{\text{A}} = \boldsymbol{F} \cdot \hat{\boldsymbol{S}}_{\text{A}}$ with respect to the basis shown in (3.7). For isotropic incompressible media, for which (1.9) holds, we obtain

(3.20)

$$\begin{aligned}
\overset{+}{T}_1{}^1 &= \psi_0 f' + \psi_1 f'[(f')^2 + (fg')^2 + (h')^2],\\
\overset{+}{T}_1{}^2 &= \psi_1 \frac{f'}{s}[\alpha f^2 g' + \gamma h'],
\end{aligned}$$

$$\overset{+}{T}_1{}^3 = \psi_1 f'[\beta f^2 g' + \delta h'],$$

$$\overset{+}{T}_2{}^1 = \psi_0 f g' + \psi_1 f \{g'[(f')^2 + (fg')^2 + (h')^2] + \frac{\alpha}{s^2}[\alpha f^2 g' + \gamma h'] + \beta[\beta f^2 g' + \delta h']\},$$

$$\overset{+}{T}_2{}^2 = \psi_0 \frac{\alpha}{s} f + \psi_1 \frac{f}{s}\left\{g'[\alpha f^2 g' + \gamma h'] + \frac{\alpha}{s^2}[(\alpha f)^2 + \gamma^2] + \beta[\alpha\beta f^2 + \gamma\delta]\right\},$$

$$\overset{+}{T}_2{}^3 = \psi_0 \beta f + \psi_1 f\left\{g'[\beta f^2 g' + \delta h'] + \frac{\alpha}{s^2}[\alpha\beta f^2 + \gamma\delta] + \beta[(\beta f)^2 + \delta^2]\right\},$$

$$\overset{+}{T}_3{}^1 = \psi_0 h' + \psi_1 \{h'[(f')^2 + (fg')^2 + (h')^2] + \frac{\gamma}{s^2}[\alpha f^2 g' + \gamma h'] + \delta[\beta f^2 g' + \delta h']\},$$

$$\overset{+}{T}_3{}^2 = \psi_0 \frac{\gamma}{s} + \frac{\psi_1}{s}\left\{h'[\alpha f^2 g' + \gamma h'] + \frac{\gamma}{s^2}[(\alpha f)^2 + \gamma^2] + \delta[\alpha\beta f^2 + \gamma\delta]\right\},$$

$$\overset{+}{T}_3{}^3 = \psi_0 \delta + \psi_1\left\{h'[\beta f^2 g' + \delta h'] + \frac{\gamma}{s^2}[\alpha\beta f^2 + \gamma\delta] + \delta[(\beta f)^2 + \delta^2]\right\}$$

(since (3.1)–(3.5), (3.16), and (3.17) hold). Here ψ_0 and ψ_1 depend only on I_C and II_C of (XII.5.19) (and possibly on p for a Cauchy-elastic material). We assume that ψ_0 and ψ_1 are independent of p.

In place of (3.10), (3.15), and (3.12), the equilibrium equations are

$$(3.21)\qquad -(\alpha\delta - \beta\gamma) f p_s + (\delta g' - \beta h') f p_\phi - (\gamma g' - \alpha h')\frac{f}{s} p_z + \frac{\partial}{\partial s}(s\overset{+}{T}_1{}^1) = K^+ \equiv s\overset{+}{T}_2{}^1 g' + \alpha \overset{+}{T}_2{}^2 + \beta s \overset{+}{T}_2{}^3,$$

$$(3.22)\qquad -\frac{\delta p_\phi - \gamma p_z}{\alpha\delta - \beta\gamma} s + \frac{\partial}{\partial s}(s f \overset{+}{T}_2{}^1) = 0,$$

$$(3.23)\qquad \frac{\beta p_\phi - \alpha p_z}{\alpha\delta - \beta\gamma} s + \frac{\partial}{\partial s}(s \overset{+}{T}_3{}^1) = 0.$$

From (3.22) and (3.23) we find that p_ϕ and p_z are independent of ϕ and z. Differentiating (3.21) with respect to ϕ and z, we find that $p_{\phi s} = 0 = p_{zs}$ so that p has the form

$$(3.24)\qquad p(s, \phi, z) = \bar{p}(s) + A\phi + Bz$$

where A and B are constants. Since the functional form of f is known, Eqs. (3.21)–(3.23) form a system of ordinary differential equations for $\bar{p}, g, h$ depending upon several parameters, which are either specified or else to be determined from additional conditions. Note that this system is linear in $\bar{p}$.

3.25. Exercise. Derive (3.21)–(3.23). (Since p and therefore $\boldsymbol{T}$ can depend on ϕ and z, these equations *cannot* be obtained by direct substitution of (3.19) into (3.10), (3.15), and (3.12). Instead, it is necessary to obtain the general form of $\nabla \cdot \boldsymbol{T}^* = \mathbf{0}$ from (3.6) and (3.7).)

4. Torsion and Other Problems in Cylindrical Coordinates for Incompressible Bodies

We assume that (3.1)–(3.5), (3.16), and (3.17) hold with ψ_0 and ψ_1 independent of ϕ, z, and p. Thus (3.19)–(3.24) hold. We first treat torsion problems for which $\gamma = 0$, $g = 0 = h$. Then (3.19) and (3.20) reduce to

$$\tag{4.1a}\begin{aligned}
T_1{}^1 &= -\frac{p}{f'} + \psi_0 f' + \psi_1 (f')^3,\\
T_1{}^2 &= 0, \quad T_1{}^3 = 0, \quad T_2{}^1 = 0,\\
T_2{}^2 &= -p\,\delta f' + \psi_0 \alpha \frac{f}{s} + \psi_1 \alpha \frac{f^3}{s^3}[\alpha^2 + \beta^2 s^2],\\
T_2{}^3 &= \beta f \left\{ \psi_0 + \psi_1 \left[\frac{\alpha^2 f^2}{s^2} + (\beta f)^2 + \delta^2 \right] \right\},\\
T_3{}^1 &= 0, \quad T_3{}^2 = \beta s \left(\frac{p}{\alpha\delta} + \psi_1 \frac{\alpha\delta f^2}{s^2} \right),\\
T_3{}^3 &= -\frac{p}{\delta} + \psi_0 \delta + \psi_1 \delta[(\beta f)^2 + \delta^2]
\end{aligned}$$

where ψ_0 and ψ_1 depend only on s and

$$\tag{4.1b}\begin{aligned}
\mathrm{I}_C &= (f')^2 + \left(\frac{\alpha f}{s} \right)^2 + (\beta f)^2 + \delta^2,\\
\mathrm{II}_C &= (f')^2 \left[\left(\frac{\alpha f}{s} \right)^2 + (\beta f)^2 + \delta^2 \right] + \left(\frac{\alpha\delta f}{s} \right)^2.
\end{aligned}$$

If $f(0)$ is required to be 0, then (4.1b) reduces to

$$\tag{4.1c} \mathrm{I}_C = \frac{1}{\alpha\delta} + \frac{\alpha}{\delta} + \frac{(\beta s)^2}{\alpha\delta} + \delta^2, \quad \mathrm{II}_C = \frac{1}{\alpha\delta}\left[\frac{\alpha}{\delta} + \frac{(\beta s)^2}{\alpha\delta} + \delta^2 \right] + \alpha\delta.$$

We define (*the generalized shear modulus*)

$$\tag{4.2} \mu(\mathrm{I}_C, \mathrm{II}_C) \equiv \psi_0 + \left[\frac{\alpha^2 f^2}{s^2} + (\beta f)^2 + \delta^2 \right] \psi_1.$$

For our class of deformations, condition (2.42) reduces to

$$\tag{4.3} (\boldsymbol{b}\cdot\boldsymbol{j}_1)\alpha\delta\frac{f}{s}(\boldsymbol{k}^1\cdot\boldsymbol{a}) + (\boldsymbol{b}\cdot\boldsymbol{j}_2)(\delta f'\boldsymbol{k}^2 - \beta f f'\boldsymbol{k}^3)\cdot\boldsymbol{a} + (\boldsymbol{b}\cdot\boldsymbol{j}_3)\alpha\frac{f}{s}f'(\boldsymbol{k}^3\cdot\boldsymbol{a}) = 0.$$

Using (3.2)–(3.5) and the Strong Ellipticity Condition (2.30), (2.31), we find that

$$\tag{4.4}\begin{aligned}
\frac{\partial}{\partial\beta}\overset{+}{T}_2{}^3 &= \frac{\partial}{\partial\beta}(\boldsymbol{k}_2 \cdot \hat{\boldsymbol{T}}_{\mathrm{A}} \cdot \boldsymbol{j}^3)\\
&= -z\boldsymbol{k}_1 \cdot \hat{\boldsymbol{T}}_{\mathrm{A}} \cdot \boldsymbol{j}^3 + \boldsymbol{k}_2\boldsymbol{j}^3 : \frac{\partial \hat{\boldsymbol{T}}_{\mathrm{A}}}{\partial \boldsymbol{F}} : \frac{\partial \boldsymbol{F}}{\partial \beta}\\
&= 0 + f\boldsymbol{k}_2\boldsymbol{j}^3 : \frac{\partial \hat{\boldsymbol{T}}_{\mathrm{A}}}{\partial \boldsymbol{F}} : \boldsymbol{k}_2\boldsymbol{j}^3 > 0.
\end{aligned}$$

This inequality together with the growth condition (2.44) implies that $\beta \mapsto \beta\mu(\mathrm{I}_C, \mathrm{II}_C)$ strictly increases from $-\infty$ to ∞ with β. Since $\beta\mu$ vanishes for $\beta = 0$, it follows that $\mu > 0$.

Pure torsion. We take $\mathcal{B}$ to be the segment of a right circular cylinder of unit radius with axis along $\boldsymbol{i}_3$ defined by

$$\{(s, \phi, z) : 0 \le s < 1, 0 \le \phi \le 2\pi, 0 \le z \le z_1\}. \tag{4.5a}$$

We study the pure torsion of $\mathcal{B}$, which is defined by (3.1)–(3.5) with $\alpha = 1$, $\gamma = 0$, $\delta = 1$, $g = 0$, $h = 0$, with β prescribed, and with the outer radius prevented from changing, so that $f(1) = 1$. Then

$$f(s) = s. \tag{4.5b}$$

(In place of the requirement that $f(1) = 1$, we could have merely required that the axis remain intact: $f(0) = 0$, which also ensures that $C = 0$ in (3.17), so that (4.5b) holds.)

The traction on the planar end, defined by $z = z_1$, is $\boldsymbol{T} \cdot \boldsymbol{j}^3 = T_i^{\ 3}\boldsymbol{k}^i$. Its component in the $\boldsymbol{k}^3$-direction is $T_3^{\ 3}$. The traction on the lateral cylindrical surface, defined by $s = 1$, is $\boldsymbol{T} \cdot \boldsymbol{j}^1 = T_i^{\ 1}\boldsymbol{k}^i = T_1^{\ 1}\boldsymbol{k}^1$. If we require that the lateral surface be traction-free, i.e., that

$$T_i^{\ 1} = 0 \quad \text{for } s = 1, \tag{4.6a}$$

then (3.19), (3.24), and (4.1) imply that $A = 0 = B$ so that $-\bar{p} + \psi_0 + \psi_1 = 0$ for $s = 1$. In this case, (3.22) and (3.23) are identically satisfied. The pressure $p = \bar{p}$ is determined by integrating (3.21) subject to (4.6a):

$$-p + \overset{+}{T}_1{}^1 = \int_s^1 \frac{\overset{+}{T}_1{}^1 - \overset{+}{T}_2{}^2 - \beta\xi \overset{+}{T}_2{}^3}{\xi}\, d\xi. \tag{4.6b}$$

Note that (4.1) and (4.5b) imply that this integral converges for $s = 0$. We use the expression for p given by (4.6b) to find that the normal component of force over the plane $z = z_1$ is

$$2\pi \int_0^1 (-p + \overset{+}{T}_3{}^3)\, s\, ds = \pi \int_0^1 [2\overset{+}{T}_3{}^3 - \overset{+}{T}_1{}^1 - \overset{+}{T}_2{}^2 - \beta s \overset{+}{T}_2{}^3]\, s\, ds. \tag{4.7}$$

Substituting (4.1) and (4.5b) into (4.7), we find that the bracketed term in its integrand is

$$-(\beta s)^2\{\psi_0 + \psi_1[1 + (\beta s)^2]\}. \tag{4.8}$$

Note that the term in braces is not the same as the specialization of (4.2), which is positive. Mild further restrictions, such as the E-inequalities (see Truesdell & Noll (1965, Sec. 53)), experimentally observed to hold for rubbers, ensure that (4.8) is negative. In this case, a compressive thrust is necessary to effect the pure torsion.

The resultant torque across any plane section with normal $\boldsymbol{j}^3$ is

(4.9)
$$\begin{aligned}\int_0^{2\pi}\int_0^1 \boldsymbol{p}\times\boldsymbol{T}\cdot\boldsymbol{j}^3 s\,ds\,d\phi &= \int_0^{2\pi}\int_0^1 (s\boldsymbol{k}_1 + z\boldsymbol{k}_3)\times({T_2}^3\boldsymbol{k}^2 + {T_3}^3\boldsymbol{k}^3)s\,ds\,d\phi \\ &= 2\pi\int_0^1 s^2{T_2}^3\,ds\,\boldsymbol{k}_3 \equiv T\boldsymbol{k}_3\end{aligned}$$

(since $\int_0^{2\pi}\boldsymbol{k}_1\,d\phi = \boldsymbol{0} = \int_0^{2\pi}\boldsymbol{k}_2\,d\phi$). This equation gives the torque as a nonlinear function of the twist β.

Let $\overset{+}{T}(\beta)$ be defined to be the expression obtained by replacing the integrand in the last integral in (4.9) with the constitutive function from (4.1). Were we to prescribe T and leave the twist β free, then we would have to solve the equation $\overset{+}{T}(\beta) = T$ for β as a function of T. Condition (4.4) together with the growth condition (2.44) implies that $\overset{+}{T}$ strictly increases from $-\infty$ to ∞ with β, so that β can be uniquely determined.

A variant of our torsion problem is obtained by prescribing the twist β, prescribing the traction ${T_1}^1$ on the lateral surface to be 0, and prescribing the normal force resultant on the ends to be 0: $\int_0^1 {T_3}^3\,s\,ds = 0$. In this case, δ is left free and (4.5b) is replaced with $\sqrt{\delta}f(s) = s$. In general, we find that $\delta \neq 1$, so that both the radius and the length of the cylinder are changed in producing the twisted state by applying twisting moments to the ends.

4.10. Exercise. Find an equation for δ for this problem. Compare conditions that would ensure that there exists a unique solution δ with the condition coming from (4.4).

We remark that all these solutions are valid for bodies of any shape, in particular, for parts of a full cylinder bounded by cylindrical coordinate surfaces, such as wedges and tubes.

Suppose that we prescribe a full boundary-value problem for the cylinder $\mathcal{B}$ with zero traction on the lateral surface and with prescribed position field

$$\tilde{\boldsymbol{p}}(\mathbf{x}) = s\boldsymbol{k}_1(\mathbf{x}) + z\boldsymbol{k}_3, \quad \omega(\mathbf{x}) \equiv \phi + \beta z \quad \text{for } z = 0, z_1. \tag{4.11}$$

Then the solution discussed above is *a* solution of this boundary-value problem; it need not be the only solution. On the basis of the analysis of Sec. IX.4, we may expect the body to suffer a torsional buckling at a critical value of the twist. Similar remarks apply to problems in which the cylinder is compressed.

Other problems. In the following exercises, we pose a variety of other problems for incompressible, isotropic elastic media that have deformations of the form (3.1).

4.12. Exercise. Axial compression and elongation. For the body $\mathcal{B}$ of (4.5a), let $\alpha = 1$, $\beta = 0 = \gamma$, $g = 0 = h$ and let δ be prescribed. Let the traction on the lateral surface be $\mathbf{0}$. Find an explicit representation for the traction on the ends in terms of the constitutive functions ψ_0 and ψ_1.

4.13. Exercise. Azimuthal dislocation. Let the body be the cylindrical segment with a wedge removed that is defined by $0 \leq s < 1$, $0 < \phi < \phi_1$, $0 < z < z_1$ with $0 < \phi_1 < 2\pi$. (For the sake of visualization, it is useful to take $\phi_1 > \pi$.) Consider the deformation obtained by welding the faces $\phi = 0, \phi_1$ together so that the deformed body is a complete circular cylindrical segment. Such a deformation can be put into the form (3.1) by taking $\alpha = 2\pi/\phi_1$, $\beta = 0 = \gamma$, and $g = 0 = h$. Suppose that the traction on the lateral surface is $\mathbf{0}$ and that the resultant force applied to the ends $z = 0, z_1$ is $\mathbf{0}$. Find an equation for the unknown δ. Discuss its solvability.

4.14. Exercise. Shearing dislocation. Let the body be the half-slit annular or cylindrical segment defined by $s_0 \leq s < 1$, $0 < \phi < 2\pi$, $0 < z < z_1$ with $0 < s_0 < 1$. Consider the deformation (3.1) with $\alpha = 1$, $\beta = 0$, $g = 0 = h$, and with γ a prescribed positive number. Sketch the deformed shape, which can be maintained by suitably welding the slit faces together. Suppose that the traction on the lateral surface is $\mathbf{0}$ and that the resultant force applied to the ends $z = 0, z_1$ is $\mathbf{0}$. Find an equation for the unknown δ. Discuss its solvability. (For the trickier case in which $s_0 = 0$, see the comments following (5.4) below.)

4.15. Exercise. Cavitation. For the full cylinder $\mathcal{B}$ described above, let $\alpha = 1 = \delta$, $\beta = 0 = \gamma$, $g = 0 = h$. (i) Let $f(1)$ be a prescribed number greater than 1. Show that the equilibrium equations have a solution with $f(0) > 0$. This solution describes a configuration in which the axis of the cylinder opens into a cylindrical cavity. (ii) Let ${T_1}^1$ be a prescribed positive number λ for $s = 1$. Show that there is a critical value λ^0 of λ such that the equilibrium problem has a solution of the form (3.1) with the center intact if $\lambda < \lambda^0$ and has a solution of the form (3.1) with cavitation if $\lambda > \lambda^0$. (See the discussion in Sec. 7.)

4.16. Exercise. Inflation. Let the body be the tube defined by $s_0 \leq s < 1$, $0 < \phi < 2\pi$, $0 < z < z_1$ with $0 < s_0 < 1$. Let $\alpha = 1 = \delta$, $\beta = 0 = \gamma$, $g = 0 = h$. Let the inner cylinder be subjected to a simple hydrostatic pressure of intensity λ units of force per *actual area* and let the outer cylinder be traction-free. Find the tractions at the ends necessary to maintain the equilibrium of the annular cylinder.

4.17. Exercise. Eversion. Let the body be the tube given in Ex. 4.16. For the deformation (3.1) defined by $\alpha = 1$, $\delta = -1$, $\beta = 0 = \gamma$, $g = 0 = h$, $f > 0$, which turns the tube inside out, find the tractions at the ends necessary to keep the tube in equilibrium with zero tractions on the cylindrical faces.

The Strong Ellipticity Condition is incapable of ensuring the uniqueness of solutions for several of these problems. A discussion of slightly stronger related conditions that do ensure uniqueness is given by Wissmann (1991).

In the next section we discuss problems in which g and h are not required to vanish.

4.18. Exercise. Give a geometric interpretation of the deformation

$$\tilde{\boldsymbol{p}}(\mathbf{x}) = \frac{1}{2}\alpha\beta^2 s^2 \boldsymbol{i} + \frac{\phi}{\alpha\beta}\boldsymbol{j} + \left[\frac{z}{\beta} + \frac{\gamma\phi}{\alpha\beta}\right]\boldsymbol{k}, \quad \alpha\beta \neq 0 \tag{4.19}$$

where $\mathbf{x}$ is defined just as in this section. Carry out an analysis of appropriate boundary-value problems along the lines of this section.

5. Problems in Cylindrical Coordinates for Compressible Bodies

For compressible bodies, we cannot use (3.16) and (3.17). Thus we cannot avoid confronting bona fide differential equations consisting of (3.10), (3.15), and (3.12) subject to suitable constitutive equations of the form (1.5). In general, we confront quasilinear boundary-value problems for (f, g, h). The existence and regularity theory for these problems (obtained by Antman (1983a)) relies crucially on the Strong Ellipticity Condition and the growth conditions. It is made difficult by the presence of the requirement that $\det \boldsymbol{F} > 0$. (By the same device as that leading to (6.15) below, we can show that the Strong Ellipticity Condition implies that the matrix $\partial(\hat{T}_1{}^1, f\hat{T}_2{}^1, \hat{T}_3{}^1)/\partial(f', g', h')$ is positive-definite. Consequently, the governing equations can be treated by powerful methods of pseudo-monotone operator theory, which we do not discuss here.)

In this section, we show how to extract from the growth conditions some useful information about solutions without solving for them. For simplicity of exposition, we assume that the material is hyperelastic. Here we define the body $\mathcal{B}$ by

$$\mathcal{B} \equiv \{\tilde{\boldsymbol{z}}(\mathbf{x}) : s_0 \le s < 1,\ 0 < \phi < \phi_1,\ 0 < z < z_1\}. \tag{5.1}$$

Let us now adopt the growth condition (2.48) and specialize it to the semi-inverse deformation (3.1):

(5.2)

$$W(\boldsymbol{C}, \boldsymbol{z}) \ge A(\operatorname{tr} \boldsymbol{C})^{a/2} - \varphi(\boldsymbol{z})$$
$$= A\left[(f')^2 + (fg')^2 + (h')^2 + \left(\frac{\alpha f}{s}\right)^2 + \left(\frac{\gamma}{s}\right)^2 + (\beta f)^2 + \delta^2\right]^{a/2} - \varphi(\boldsymbol{z}).$$

(See (XII.5.19a). Recall that we require that $A > 0$ and $a > 1$.) We have more chance of finding solutions of our semi-inverse boundary-value problems if we seek them in a large class of functions rather than in a small class defined by very nice regularity properties. (The singularity for $s = 0$, which caused singularities in some of the solutions of Sec. 4, makes us suspicious of too optimistic an assumption of regularity.) A reasonable, large class of *admissible* functions (f, g, h) and parameters $(\alpha, \beta, \gamma, \delta)$ are those that make the total stored energy bounded:

$$\int_{\mathcal{B}} W(\boldsymbol{C}(\boldsymbol{z}), \boldsymbol{z})\, dv(\boldsymbol{z}) = \phi_1 z_1 \int_{s_0}^1 sW(\boldsymbol{C}(\tilde{\boldsymbol{z}}(\mathbf{x}), s)\, ds < \infty. \tag{5.3}$$

It then follows from (5.2) that admissible functions and parameters satisfy

(5.4)

$$\int_{s_0}^1 s\left[(f')^2 + (fg')^2 + (h')^2 + \left(\frac{\alpha f}{s}\right)^2 + \left(\frac{\gamma}{s}\right)^2 + (\beta f)^2 + \delta^2\right]^{a/2} ds < \infty.$$

(We are tacitly assuming that the functions (f, g, h) are Lebesgue-measurable.)

We immediately deduce from (5.4) that if $s_0 = 0$ and if $a \geq 2$, then there is no admissible solution with $\gamma \neq 0$. (Of course, a full existence theory is required to show that such shearings with $\gamma \neq 0$ generate admissible solutions for $1 < a < 2$.) Under the usual assumption that $\alpha \neq 0$ we likewise deduce that if $s_0 = 0$ and if $a \geq 2$, then there is no admissible solution with $f(0) > 0$. Thus cavitation cannot occur if the material is strong in resisting tension.

If $A(\operatorname{tr}\boldsymbol{C})^{a/2}$ represents the asymptotic behavior of W for large $\boldsymbol{C}$, then $\hat{T}_1{}^1(\boldsymbol{F}, \boldsymbol{z})$ is asymptotic to $(F^1{}_1)^{a-1}$, etc. Thus the response is asymptotically superlinear, linear, or sublinear if $a > 2$, $= 2$, or < 2, respectively.

Now let us consider a problem for a wedge $\mathcal{B}$ for which $s_0 = 0$ and $\phi_1 < 2\pi$. We fix a vertical rigid translation by setting $h(1) = 0$. Let us assume that the tractions on the faces $\phi = 0, \phi_1$ are zero and that the traction $\boldsymbol{T}\cdot\boldsymbol{j}^1$ on the cylindrical face $s = 1$ is the prescribed constant vector $T\boldsymbol{k}^3$. The total vertical force applied to this cylindrical face is therefore $\phi_1 z_1 T\boldsymbol{k}^3$. We are not free to prescribe tractions arbitrarily on the end faces because they must be adjusted to ensure that the resultant force and torque on the body be zero. Thus there must in fact be a concentrated force along the axis, because the integral of (3.12) (which is a consequence of the Principle of Virtual Power) implies that

$$\lim_{s\to 0} sT_3{}^1 = T. \tag{5.5}$$

We ask whether $h(0)$ is bounded. The answer is furnished by the following embedding based on the Hölder inequality. We assume that (5.4) holds. Let $0 < x \leq y \leq 1$. Then for $a \neq 2$,

(5.6)

$$\begin{aligned}
|h(y) - h(x)| &\leq \int_x^y |h'(s)|\, ds = \int_x^y s^{-\frac{1}{a}} \left| s^{\frac{1}{a}} h'(s)\right| ds \\
&\leq \left\{\int_x^y s^{-\frac{1}{a}\frac{a}{a-1}}\, ds\right\}^{\frac{a-1}{a}} \left\{\int_x^y \left| s^{\frac{1}{a}} h'(s)\right|^a ds\right\}^{\frac{1}{a}} \\
&\leq \text{const.} \left\{\int_x^y s^{-\frac{1}{a-1}}\, ds\right\}^{\frac{a-1}{a}} = \text{const.} \left[y^{\frac{a-2}{a-1}} - x^{\frac{a-2}{a-1}}\right]^{\frac{a-1}{a}}.
\end{aligned}$$

Thus h is continuous on any compact subset of $(0, 1]$. If $a > 2$, then (5.6) implies that h is continuous on $[0, 1]$, which implies that $h(0)$ is bounded.

An analogous problem for a wedge is that in which the traction $\boldsymbol{T}\cdot\boldsymbol{j}^1$ on the cylindrical face $s = 1$ is the prescribed vector $T\boldsymbol{k}^2$. In this case, the integral of (3.15) implies that there is a concentrated torque applied to the axis. Setting $g(1) = 0$ to fix rigid rotations about the axis, we can ask whether $g(0)$ is bounded. Conditions ensuring that it is are more difficult because the analog of the computation (5.6) is complicated by the presence of the function f.

5.7. Exercise. Study this torsion problem for incompressible materials.

Note that in many of our examples, the rate of growth $a = 2$ is a threshold for qualitatively different behavior. A number of nonlinear constitutive equations for elastic response, closely inspired by linear elasticity, use constitutive equations corresponding to $a = 2$. Others use energies quadratic in $\boldsymbol{E}$, which correspond to $a = 4$. The study of a single constitutive equation may generate misleadingly narrow conclusions about the general nature of the behavior of solutions to a given problem.

5.8. Research Problem. Analyze the existence of solutions to suitable boundary-value problems for semi-inverse deformations of Sec. 3 for incompressible materials when g and h are not required to vanish (see Antman (1983a)).

5.9. Research Problem. Analyze the existence of solutions to suitable boundary-value problems for semi-inverse deformations of the form

$$\tilde{\boldsymbol{p}}(\mathbf{x}) = f(s)\boldsymbol{i} + [g(s) + \alpha\phi + \beta z]\boldsymbol{j} + [h(s) + \gamma\phi + \delta z]\boldsymbol{k}. \tag{5.10}$$

(If $g = 0 = h$ and if the deformation is volume-preserving, then (5.10) reduces to (4.19).)

6. Deformation of a Compressible Cube into an Annular Wedge and Related Deformations

We let $\boldsymbol{z} = z^k\boldsymbol{i}_k = x\boldsymbol{i} + y\boldsymbol{j} + z\boldsymbol{k}$ and study deformations of the form

$$\boldsymbol{p}(\boldsymbol{z}) = f(x)\boldsymbol{k}_1(\boldsymbol{z}) + [h(x) + \gamma y + \delta z]\boldsymbol{k}_3 \tag{6.1a}$$

where

$$\begin{aligned} \boldsymbol{k}_1(\boldsymbol{z}) &\equiv \boldsymbol{k}^1(\boldsymbol{z}) \equiv \cos\omega(\boldsymbol{z})\boldsymbol{i} + \sin\omega(\boldsymbol{z})\boldsymbol{j}, \\ \boldsymbol{k}_2(\boldsymbol{z}) &\equiv \boldsymbol{k}^2(\boldsymbol{z}) \equiv -\sin\omega(\boldsymbol{z})\boldsymbol{i} + \cos\omega(\boldsymbol{z})\boldsymbol{j}, \\ \boldsymbol{k}_3 &\equiv \boldsymbol{k}^3 \equiv \boldsymbol{k}, \end{aligned} \tag{6.1b}$$

$$\omega(\boldsymbol{z}) \equiv g(x) + \alpha y + \beta z. \tag{6.1c}$$

Here $\alpha, \beta, \gamma, \delta$ are constants and f, g, h are functions to be determined. A special case of this deformation takes a cube into an annular wedge. Semi-inverse problems for these deformations can be analyzed by more elementary methods than those used for the complete study of the problem discussed in Sec. 5.

6.2. Exercise. Describe and sketch the effect of the deformation (6.1) on a cube. (Hint: Break up the deformation into a composition of simpler deformations, starting with a pure bending.)

6.3. Exercise. Show that

$$\begin{aligned} \boldsymbol{F}(\boldsymbol{z}) = {} & [f'(x)\boldsymbol{k}_1(\boldsymbol{z}) + f(x)g'(x)\boldsymbol{k}_2(\boldsymbol{z}) + h'(x)\boldsymbol{k}_3]\boldsymbol{i}^1 \\ & + [\alpha f(x)\boldsymbol{k}_2(\boldsymbol{z}) + \gamma\boldsymbol{k}_3]\boldsymbol{i}^2 + [\beta f(x)\boldsymbol{k}_2(\boldsymbol{z}) + \delta\boldsymbol{k}_3]\boldsymbol{i}^3 \\ \equiv {} & F^i{}_q(x)\boldsymbol{k}_i(\boldsymbol{z})\boldsymbol{i}^q, \end{aligned} \tag{6.4}$$

$$(6.5)\qquad \begin{aligned} \boldsymbol{C} &= [(f')^2 + (fg')^2 + (h')^2]\boldsymbol{i}^1\boldsymbol{i}^1 + [\alpha f^2 g' + \gamma h'](\boldsymbol{i}^1\boldsymbol{i}^2 + \boldsymbol{i}^2\boldsymbol{i}^1) \\ &+ [\beta f^2 g' + \delta h'](\boldsymbol{i}^1\boldsymbol{i}^3 + \boldsymbol{i}^3\boldsymbol{i}^1) + [(\alpha f)^2 + \gamma^2]\boldsymbol{i}^2\boldsymbol{i}^2 \\ &+ [\alpha\beta f^2 + \gamma\delta](\boldsymbol{i}^2\boldsymbol{i}^3 + \boldsymbol{i}^3\boldsymbol{i}^2) + [(\beta f)^2 + \delta^2]\boldsymbol{i}^3\boldsymbol{i}^3 \\ &\equiv C_{pq}(s)\boldsymbol{i}^p\boldsymbol{i}^q. \end{aligned}$$

The requirement that $\det \boldsymbol{F} > 0$ reduces to

$$(6.6)\qquad \det \boldsymbol{F} = (\alpha\delta - \beta\gamma) f f' > 0.$$

By making obvious sign conventions, we may assume that each factor of the second term of (6.6) is positive.

We decompose $\boldsymbol{T}$ thus:

$$(6.7)\qquad \boldsymbol{T} = T_p{}^q(\boldsymbol{z})\boldsymbol{k}^p(x)\boldsymbol{i}_q.$$

The equilibrium equations for zero body forces have the form

$$(6.8)\qquad \frac{d}{dx}(T_1{}^1) = T_2{}^1 g'(x) + \alpha T_2{}^2 + \beta T_2{}^3,$$

$$(6.9)\qquad \frac{d}{dx}(T_2{}^1) + T_1{}^1 g'(x) + \alpha T_1{}^2 + \beta T_1{}^3 = 0,$$

$$(6.10)\qquad \frac{d}{dx}(T_3{}^1) = 0$$

(cf. (3.10)–(3.12)). The symmetry conditions (XII.7.27) imply that

$$(6.11)\qquad f[T_1{}^1 g' + \alpha T_1{}^2 + \beta T_1{}^3] = f' T_2{}^1,$$

the substitution of which into (6.9) yields

$$(6.12)\qquad \frac{d}{dx}(f T_2{}^1) = 0.$$

Equations (6.10) and (6.12) yield the integrals

$$(6.13\text{a,b})\qquad f T_2{}^1 = G \quad (\text{const.}), \quad T_3{}^1 = H \quad (\text{const.}).$$

The governing equations for our problem consist of (6.8), (6.13), and the constitutive equations (1.3b). We shall append suitable subsidiary conditions to this system. We assume that the constitutive functions are such that the components of $\boldsymbol{T}$ introduced in (6.7) are independent of y and z. Isotropy ensures this.

Without loss of generality, we suppose that $\mathcal{B}$ is the unit cube $\{\boldsymbol{z} : 0 < z^k < 1\}$. We may assume that a subset of the constants $\alpha, \beta, \gamma, \delta$ is prescribed and that a complementary set of subsidiary conditions (on appropriate stress resultants) is prescribed. We fix a rigid displacement by prescribing

$$(6.14)\qquad g(0) = 0, \quad h(0) = 0.$$

We require that any tractions prescribed on the faces $x = 0, 1$ be compatible with the integrals of (6.13). We prescribe either $g(1)$ or G and we prescribe either $h(1)$ or H. Note that $T_3{}^1$ is a dead shear load on the faces $x = 0, 1$, whereas $f T_2{}^1$ is a shear load that neither is dead nor corresponds to a constant Cauchy stress.

The effect of the Strong Ellipticity Condition. Let us assume for simplicity of exposition that $\hat{\boldsymbol{T}}$ is continuously differentiable. We now transform our equations further by bringing the Strong Ellipticity Condition and associated growth conditions to bear on them. Choosing $\boldsymbol{b} = \boldsymbol{i}$ in (2.15), we obtain

$$\boldsymbol{a} \cdot \frac{\partial(\hat{\boldsymbol{T}} \cdot \boldsymbol{i})}{\partial(\boldsymbol{F} \cdot \boldsymbol{i})} \cdot \boldsymbol{a} = \boldsymbol{a} \cdot \frac{\partial(\hat{\boldsymbol{T}} \cdot \boldsymbol{i})}{\partial \boldsymbol{p}_x} \cdot \boldsymbol{a} = a^l \frac{\partial \hat{T}_l{}^1}{\partial F^m{}_1} a^m > 0 \tag{6.15}$$

where $\boldsymbol{a} = a^l \boldsymbol{k}_l$. (Thus $\hat{\boldsymbol{T}} \cdot \boldsymbol{i}$ is a strictly monotone function of $\boldsymbol{p}_x$ for fixed values of its other arguments.)

In consonance with (6.15) we impose the following special cases of the growth conditions (2.44) when $\boldsymbol{F}$ has the form (6.4):

$$\begin{aligned} &\hat{T}_1{}^1 \to \pm\infty \quad \text{as } F^1{}_1 = f' \to \begin{cases} \infty, \\ 0, \end{cases} \\ &\hat{T}_2{}^1 \to \pm\infty \quad \text{as } F^2{}_1 = fg' \to \pm\infty, \\ &\hat{T}_3{}^1 \to \pm\infty \quad \text{as } F^3{}_1 = h' \to \pm\infty \end{aligned} \tag{6.16}$$

for fixed values of the other arguments.

Conditions (6.15) and (6.16) enable us to apply the Global Implicit Function Theorem XIII.2.30 to show that when $\boldsymbol{F}$ has the form (6.4), then the (algebraic) equations

$$\hat{T}_1{}^1(\boldsymbol{F}, x) = T, \tag{6.17a}$$

$$f\hat{T}_2{}^1(\boldsymbol{F}, x) = G, \tag{6.17b}$$

$$\hat{T}_3{}^1(\boldsymbol{F}, x) = H, \tag{6.17c}$$

can be uniquely solved for (f', g', h') in terms of the other variables, and are therefore equivalent to a system of the form

$$f' = \varphi(T, G/f, H, \alpha f, \gamma, \beta f, \delta, x), \tag{6.18a}$$

$$fg' = \zeta(T, G/f, H, \alpha f, \gamma, \beta f, \delta, x), \tag{6.18b}$$

$$h' = \eta(T, G/f, H, \alpha f, \gamma, \beta f, \delta, x), \tag{6.18c}$$

where φ is positive-valued. The Local Implicit Function Theorem implies that φ, ζ, η are continuously differentiable because $\hat{\boldsymbol{T}}$ is. Note that (6.17b,c) correspond to (6.13). Consequently, the governing equations (6.8), (6.13), and (1.3b) are equivalent to the semilinear system of ordinary differential equations consisting of (6.18) and

$$\begin{aligned} T' &= \Phi(T, G/f, H, \alpha f, \gamma, \beta f, \delta, \alpha, \beta, x) \\ &\equiv -\hat{T}_2{}^1(F^p{}_q \boldsymbol{k}_p \boldsymbol{i}^q, x)\zeta(T, G/f, H, \alpha f, \gamma, \beta f, \delta, x)/f \\ &\quad - \alpha \hat{T}_2{}^2(F^p{}_q \boldsymbol{k}_p \boldsymbol{i}^q, x) - \beta \hat{T}_2{}^3(F^p{}_q \boldsymbol{k}_p \boldsymbol{i}^q, x) \end{aligned} \tag{6.19}$$

where

$$F^1{}_1 = \varphi(T, G/f, H, \alpha f, \gamma, \beta f, \delta, x),$$
$$F^2{}_1 = \zeta(T, G/f, H, \alpha f, \gamma, \beta f, \delta, x),$$
$$F^3{}_1 = \eta(T, G/f, H, \alpha f, \gamma, \beta f, \delta, x),$$

and the remaining components of $\boldsymbol{F}$ are given by (6.4).

We readily find that (for fixed $f, \alpha, \beta, \gamma, \delta$) the quadratic form for the matrix

$$\frac{\partial(\hat{T}_1{}^1, f\hat{T}_2{}^1, \hat{T}_3{}^1)}{\partial(f', g', h')}$$

is the quadratic forms in (6.15) for suitable $\boldsymbol{a}$ and is therefore positive-definite. We can accordingly solve (6.17) for f', g', h' in terms of the other variables, where we can now replace G/f with G. The choice made above, however, proves more useful for the ensuing analysis.

Note that the fourth-order system (6.18), (6.19) is degenerate because the right-hand sides are independent of g and h. It is accordingly controlled by the second-order system (6.18a), (6.19). (A similar phenomenon occurs in the problems treated in Secs. VI.5, IX.4.) In particular, if $\alpha, \beta, \gamma, \delta, G, H$ are prescribed, then (6.18a) and (6.19) are uncoupled from (6.18b,c), which can be solved for g and h after f and T are found by simply integrating (6.18b,c):

(6.20)
$$g(x) = \int_0^x f(\xi)^{-1}\zeta(T(\xi), G/f(\xi), H, \alpha f(\xi), \gamma, \beta f(\xi), \delta, \xi)\, d\xi,$$
$$h(x) = \int_0^x \eta(T(\xi), G/f(\xi), H, \alpha f(\xi), \gamma, \beta f(\xi), \delta, \xi)\, d\xi.$$

On the other hand, if $g(1)$ and $h(1)$ are prescribed, then G and H must be found as solutions of the equations

(6.21a)
$$g(1) = \int_0^1 f(\xi)^{-1}\zeta(T(\xi), G/f(\xi), H, \alpha f(\xi), \gamma, \beta f(\xi), \delta, \xi)\, d\xi,$$

(6.21b)
$$h(1) = \int_0^1 \eta(T(\xi), G/f(\xi), H, \alpha f(\xi), \gamma, \beta f(\xi), \delta, \xi)\, d\xi,$$

which are coupled to (6.18a) and (6.19).

We shall analyze boundary-value problems for (6.18a), (6.19), (6.21) by continuing solutions from a distinguished special solution. In most problems, the natural choice for the special solution would be a trivial solution, which would correspond to the reference configuration for our elasticity problem. Unfortunately, the trivial state, which is a cube, is not in the form (6.1) of admissible solutions; it is rather a singular limit of such solutions. Indeed, the cubical reference configuration is obtained by setting $\beta = 0$, $\gamma = 0$, $\delta = 1$, $g = 0$, $h = 0$ and by taking the formal limits $\alpha \to 0$, $f \to \infty$, $\alpha f \to 1$. In this case, the family of solutions may be said to be *continued from infinity* (see the references in the paragraph containing

(VI.2.39)). Restricting our attention to α's that are positive, we can make a change of variables to render this technical difficulty innocuous: We set

$$q \equiv \alpha f, \quad \bar{\beta} \equiv \beta/\alpha, \quad \bar{g}_1 \equiv g(1)/\alpha, \quad \bar{G} \equiv \alpha G. \tag{6.22}$$

In this case, (6.18a), (6.19), (6.21) reduce to

$$q' = \alpha\varphi(T, \bar{G}/q, H, q, \gamma, \bar{\beta}q, \delta, x), \tag{6.23a}$$

$$T' = \Phi(T, \bar{G}/q, H, q, \gamma, \bar{\beta}q, \delta, \alpha, \alpha\bar{\beta}, x), \tag{6.23b}$$

$$\bar{g}_1 = \int_0^1 q(\xi)^{-1}\zeta(T(\xi), \bar{G}/q(\xi), H, q(\xi), \gamma, \bar{\beta}q(\xi), \delta, \xi)\, d\xi, \tag{6.24a}$$

$$h(1) = \int_0^1 \eta(T(\xi), \bar{G}/q(\xi), H, q(\xi), \gamma, \bar{\beta}q(\xi), \delta, \xi)\, d\xi. \tag{6.24b}$$

Let us suppose that

$$(q, T) \in \mathcal{E} \equiv \{(q, T) \in C^0(0,1) \times C^0(0,1) : q(0) > 0\}. \tag{6.25}$$

The Strong Ellipticity Condition and the growth conditions (6.16) ensure that the right-hand sides of (6.24) define a monotone and coercive mapping of $(\bar{G}, H)$ (see Proposition VIII.8.23 and the accompanying discussion), so that the Global Implicit Function Theorem XIX.2.30 implies that $\bar{G}$ and H can be found uniquely in terms of $\bar{g}_1$, $h(1)$, and other data of (6.24):

$$\begin{aligned} \bar{G} &= G^\sharp[T, q, \alpha, \bar{\beta}, \gamma, \delta, \bar{g}_1, h(1)], \\ H &= H^\sharp[T, f, \alpha, \bar{\beta}, \gamma, \delta, \bar{g}_1, h(1)]. \end{aligned} \tag{6.26}$$

The Local Implicit Function Theorem implies that $G^\sharp$ and $H^\sharp$ are continuously differentiable functions of their arguments with $(q, T) \in \mathcal{E}$. If $g(1)$ and H are prescribed, or if $h(1)$ and G are prescribed, a more elementary argument based on the Intermediate Value Theorem gives analogous conclusions.

Connected families of solutions of the boundary-value problem. Let us now suppose that α, $\bar{\beta}$, γ, δ, $\bar{g}_1$, $h(1)$ are prescribed with $\alpha\delta - \beta\gamma > 0$, $\alpha > 0$. We impose boundary conditions of the form

$$T(0) = T_0 \quad \text{or} \quad q(0) = q_0 > 0, \tag{6.27a,b}$$

$$T(1) = T_1 \quad \text{or} \quad q(1) = q_1 > 0 \tag{6.27c,d}$$

where q_0, T_0, q_1, T_1 are prescribed. Conditions more general than (6.27a,c) lead to no serious difficulties in the analysis. Since $\alpha, \bar{\beta}, \gamma, \delta$ are prescribed, the tractions on the faces $y = 0, 1$, $z = 0, 1$ ensure that the resultant force and moment on the body vanish when (6.27a,c) are prescribed. To be specific and to avoid minor problems with Neumann boundary conditions, we restrict our attention to (6.27a,d) (which do not render the requirement that f be everywhere positive, coming from (6.6), completely innocuous, as would (6.27a,c)). Then (6.23), (6.24), and(6.27a,d) are equivalent to

$$\text{(6.28a)}\qquad q(x) = q_1 - \alpha \int_x^1 \varphi(T(\xi), G^\sharp/q(\xi), H^\sharp, q(\xi), \gamma, \bar{\beta}q(\xi), \delta, \xi)\, d\xi,$$

$$\text{(6.28b)}\qquad T(x) = T_0 + \int_0^x \Phi(T(\xi), G^\sharp/q(\xi), H^\sharp, q, \gamma, \bar{\beta}q(\xi), \delta, \alpha, \alpha\bar{\beta}, \xi)\, d\xi$$

where the arguments of $G^\sharp$ and $H^\sharp$ are given in (6.26). We write (6.28) as

$$\text{(6.29a)}\qquad (q, T) = \boldsymbol{\kappa}[q, T, \boldsymbol{\mu}]$$

where

$$\text{(6.29b)}\qquad \begin{aligned} &\boldsymbol{\mu} \equiv (\alpha, \bar{\beta}, \gamma, \delta, q_1, T_0, \bar{g}_1, h(1)) \\ &\in \mathcal{M} \equiv \{\boldsymbol{\mu} \in \mathbb{R}^8 : \alpha > 0,\ \alpha\delta - \bar{\beta}\gamma > 0,\ q_1 > 0\}. \end{aligned}$$

6.30. Exercise. Prove

6.31. Proposition. *$\boldsymbol{\kappa}$ is a continuous and compact mapping from $\mathcal{E} \times \mathcal{M}$ to $\mathcal{E}$.*

We wish to study how solutions depend on the eight-dimensional parameter $\boldsymbol{\mu}$. If we restrict our attention to $\boldsymbol{\mu}$'s confined to a curve in $\mathcal{M}$, then we could use a continuation theorem like Theorem III.3.30 (see the proof of Theorem XIX.4.1). For multiparameter problems, we may appeal to the following analog of Theorem V.4.23. This theorem essentially says that if an equation like (6.29) has a special solution pair $((\mathring{q}, \mathring{T}), \mathring{\boldsymbol{\mu}})$ at which the linearization of (6.29) with respect to (q, T) is nonsingular, then (6.29) has a connected family of solution pairs containing $((\mathring{q}, \mathring{T}), \mathring{\boldsymbol{\mu}})$ each point of which has dimension at least equal to the number of parameters of $\boldsymbol{\mu}$. Moreover, this connected family cannot come to an abrupt stop.

6.32. Multiparameter Global Continuation Theorem. *Let $\mathcal{X}$ be a Banach space and let $\{\mathcal{O}(\epsilon), 0 < \epsilon < 1\}$ be a family of open sets (not necessarily bounded) in $\mathcal{X} \times \mathbb{R}^m$ for which*

$$(\mathring{\mathsf{u}}, \mathring{\boldsymbol{\lambda}}) \in \mathcal{O}(\epsilon), \qquad 0 < \epsilon < 1,$$

$$\operatorname{cl} \mathcal{O}(\epsilon_2) \subset \mathcal{O}(\epsilon_1) \quad \text{for } 0 < \epsilon_1 < \epsilon_2 < 1.$$

Let

$$\mathcal{O} \equiv \cup_{0<\epsilon<1} \mathcal{O}(\epsilon).$$

Let $\mathsf{f} : \mathcal{O} \to \mathcal{X}$ be continuous, let $\mathsf{f}[\mathring{\mathsf{u}}, \mathring{\boldsymbol{\lambda}}] = \mathring{\mathsf{u}}$, and let $\mathsf{f} : \mathcal{O}(\epsilon) \to \mathcal{X}$ be compact for $0 < \epsilon < 1$. Let I denote the identity operator on $\mathcal{X}$. Let the Fréchet derivative $\mathsf{I} - \partial\mathsf{f}[\mathring{\mathsf{u}}, \mathring{\boldsymbol{\lambda}}]/\partial\mathsf{u} : \mathcal{X} \to \mathcal{X}$ of $\mathsf{u} \mapsto \mathsf{u} - \mathsf{f}[\mathsf{u}, \boldsymbol{\lambda}]$ at $(\mathring{\mathsf{u}}, \mathring{\boldsymbol{\lambda}})$ exist and be invertible. Let $\mathcal{S} \equiv \{(\mathsf{u}, \boldsymbol{\lambda}) \in \mathcal{O} : \mathsf{u} = \mathsf{f}[\mathsf{u}, \boldsymbol{\lambda}]\}$ and let $\mathcal{S}_0$ be the connected component of $\mathcal{S}$ containing $(\mathring{\mathsf{u}}, \mathring{\boldsymbol{\lambda}})$. (In a neighborhood of $(\mathring{\mathsf{u}}, \mathring{\boldsymbol{\lambda}})$, $\mathcal{S}$ agrees with $\mathcal{S}_0$.) Then one of the following

statements is true:
(i) $\mathcal{S}_0$ *is bounded and there is an* $\epsilon^* \in (0,1)$ *such that* $\mathcal{S}_0 \subset \mathcal{O}(\epsilon^*)$. *There is an essential map (i.e., a continuous map not homotopic to a constant)* $\boldsymbol{\sigma}$ *from* $\mathcal{S}_0$ *onto the m-dimensional sphere* $\mathbb{S}^m$ *whose restriction to* $\mathcal{S}_0 \setminus \{(\overset{\circ}{\mathsf{u}}, \overset{\circ}{\boldsymbol{\lambda}})\}$ *is inessential. Moreover,* $\mathcal{S}_0$ *contains a connected subset* $\mathcal{S}_{00}$ *that contains* $(\overset{\circ}{\mathsf{u}}, \overset{\circ}{\boldsymbol{\lambda}})$, *that has the same properties as* $\mathcal{S}_0$ *with respect to* σ, *and that has the property that each point of it has Lebesgue dimension at least* m.
(ii) $\mathcal{S}_0 \setminus \mathcal{O}(\epsilon) \neq \emptyset$ *for all* $\epsilon \in (0,1)$ *or* $\mathcal{S}_0$ *is unbounded. For each* $\epsilon \in (0,1)$ *there is a modified equation* $\mathsf{u} = \varphi[\mathsf{u}, \boldsymbol{\lambda}, \epsilon]\mathsf{f}[\mathsf{u}, \boldsymbol{\lambda}]$ *defined on all of* $\mathcal{X} \times \mathbb{R}^m$ *that agrees with* $\mathsf{u} = \mathsf{f}[\mathsf{u}, \boldsymbol{\lambda}]$ *on* $\mathcal{O}(\epsilon)$. *The one-point compactification* $\mathcal{S}_0^+(\epsilon)$ *of the connected component* $\mathcal{S}_0(\epsilon)$ *containing* $(\overset{\circ}{\mathsf{u}}, \overset{\circ}{\boldsymbol{\lambda}})$ *of the set of solution pairs of the modified equation has the same properties as* $\mathcal{S}_0$ *in statement* (i).

The basic proof of this theorem is given by Alexander & Yorke (1976). This statement embodies refinements due to Alexander & Antman (1981) and Lanza & Antman (1991). Also see Alexander & Antman (1983), Fitzpatrick, Massabò, & Pejsachowicz (1983), and Ize, Massabò, Pejsachowicz, & Vignoli (1985).

We assume that

$$\hat{T}_2{}^1 = 0 = \hat{T}_3{}^1 \quad \text{when} \quad F^2{}_1 = 0 = F^3{}_1. \tag{6.33}$$

6.34. Exercise. Prove that if the material is isotropic, then (6.33) holds. Prove that if (6.33) holds and if $\bar{g}_1 = 0 = h(1)$, then $g = 0 = h$ and $\bar{G} = 0 = H$.

We also require that the reference configuration be a stress-free natural state and that

$$\boldsymbol{H} : [\partial\hat{\boldsymbol{T}}(\boldsymbol{I}, \boldsymbol{z})/\partial\boldsymbol{F}] : \boldsymbol{H} > 0 \quad \forall \boldsymbol{H} \in \mathrm{Sym} \setminus \{\boldsymbol{O}\}. \tag{6.35}$$

This is just a special case of the Strong Coleman-Noll Condition; it is slightly weaker than the requirement that this inequality hold for all $\boldsymbol{H} \neq \boldsymbol{O}$, which is standard in linear elasticity.

We show that (6.29) satisfies the hypotheses of Theorem 6.32 by virtue of Proposition 6.31 and the following two exercises:

6.36. Exercise. Use Ex. 6.34 to show that for $\alpha = \bar{\beta} = \gamma = 0$, $\delta = 1$, $\bar{g}_1 = 0 = h(1)$, Eq. (6.29) admits the trivial solution $(q, T) = (1, 0)$, which corresponds to the cubical reference configuration.

6.37. Exercise. Show that (6.35) implies that the linearization of (6.29) about the trivial state described by Ex. 6.36 is nonsingular.

Growth conditions like those discussed in the paragraphs following (2.44) ensure that $f(0) > 0$. To treat weaker coercivity conditions, one can use devices like those of Ex. V.5.10. Also see Antman (1978b). For a treatment of acceptable dual pairs of data when not all of $\alpha, \beta, \gamma, \delta$ are prescribed, see Antman (1983a).

Qualitative behavior of solutions. Let us assume that the material is homogeneous, so that the constitutive functions do not depend explicitly on x. We write (6.18a) and (6.19) in the form

$$f' = \psi(T, f, G, H, \alpha, \beta, \gamma, \delta), \tag{6.38a}$$

$$T' = \Psi(T, f, G, H, \alpha, \beta, \gamma, \delta). \tag{6.38b}$$

For simplicity of exposition we fix the parameters $G, H, \alpha, \beta, \gamma, \delta$, suppress their appearance, and study the phase portrait Fig. 6.39 for (6.38). Since $\psi > 0$, so that $f' > 0$, it follows that there are no singular points and therefore no closed orbits and that all trajectories move from left to right. For certain ranges of parameters, growth conditions like those discussed in the paragraphs following (2.44) imply that

$$\Psi(f,T) \to \begin{cases} +\infty \\ -\infty \end{cases} \quad \text{as} \quad f \to \begin{cases} \infty, \\ 0. \end{cases} \tag{6.40}$$

For a subset of such parameter ranges, the Strong Ellipticity Condition implies that $\partial\Psi/\partial f > 0$.

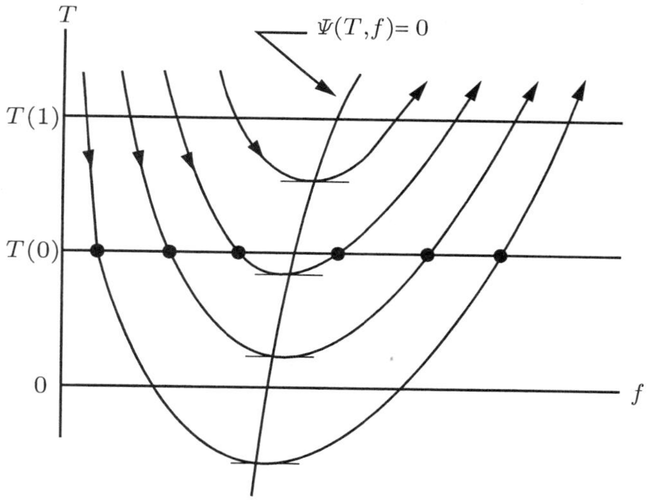

Fig. 6.39. Phase portrait of (6.38) when (6.40) holds and $\partial\Psi/\partial f > 0$.

6.41. Exercise. Prove that if $G = 0 = H$ and if (6.33) holds, then the Strong Ellipticity Condition implies that $\partial\Psi/\partial f > 0$.

If (6.40) holds and if $\partial\Psi/\partial f > 0$, then the phase portrait has the form shown in Fig. 6.39. Note that if $T(0)$ and $T(1)$ are prescribed, then candidates for solutions correspond to trajectories joining the two horizontal lines with ordinates $T(0)$ and $T(1)$. On such trajectories, T is either monotone or else has an interior minimum. Similar observations can be made for other pairs of boundary conditions. Such a trajectory actually corresponds to a solution of our boundary-value problem if exactly 1 unit of the independent variable x is used up in the traversal of the trajectory. In the next section we show how to obtain existence theorems by such an approach.

6.42. Problem. Use this approach to obtain existence theorems for boundary-value problems for (6.38).

6.43. Problem. Show that nonhomogeneous materials exhibit the same qualitative properties as those implicit in Fig. 6.39 (e.g., by using a Prüfer transformation; see (VI.2.10)).

The material of this section is adapted from Antman (1978b) by kind permission of Elsevier Science Publishers.

7. Dilatation, Cavitation, Inflation, and Eversion

We now study spherically symmetric deformations of compressible nonlinear elastic bodies. Let $\mathbf{x} \equiv (s, \theta, \phi)$ be spherical polar coordinates for $\mathbb{E}^3$, and let

$$\begin{aligned} \boldsymbol{a}_1(\theta,\phi) &\equiv \boldsymbol{a}^1(\theta,\phi) \equiv \sin\theta(\cos\phi\boldsymbol{i} + \sin\phi\boldsymbol{j}) + \cos\theta\boldsymbol{k}, \\ \boldsymbol{a}_2(\theta,\phi) &\equiv \boldsymbol{a}^2(\theta,\phi) \equiv \cos\theta(\cos\phi\boldsymbol{i} + \sin\phi\boldsymbol{j}) - \sin\theta\boldsymbol{k}, \\ \boldsymbol{a}_3(\phi) &\equiv \boldsymbol{a}^3(\phi) \equiv -\sin\phi\boldsymbol{i} + \cos\phi\boldsymbol{j} \end{aligned} \tag{7.1}$$

be the orthonormal basis associated with

$$\tilde{\boldsymbol{z}}(\mathbf{x}) \equiv s\boldsymbol{a}_1(\theta,\phi), \tag{7.2}$$

which assigns positions (in the reference configuration) to $\mathbf{x}$. Let $0 \le \alpha < 1$, $0 \le \beta < \gamma \le \pi$. We study bodies of the form

$$\mathcal{B} = \operatorname{int}\{\boldsymbol{z} = \tilde{\boldsymbol{z}}(\mathbf{x}) : \alpha \le s \le 1,\ \beta \le \theta \le \gamma,\ 0 \le \phi < 2\pi\}. \tag{7.3}$$

(For $0 < \alpha$, $0 < \beta < \gamma < \pi$, $\mathcal{B}$ is a zone of a spherical shell; for $0 < \alpha$, $\beta = 0$, $\gamma = \pi$, $\mathcal{B}$ is a full spherical shell; for $\alpha = 0$, $\beta = 0$, $\gamma = \pi$, $\mathcal{B}$ is a ball.)

For $\varepsilon = \pm$, we set

$$\begin{gathered} \boldsymbol{b}_1(\theta,\phi) \equiv \boldsymbol{b}^1(\theta,\phi) \equiv \varepsilon\boldsymbol{a}_1(\varepsilon\theta,\phi), \quad \boldsymbol{b}_2(\theta,\phi) \equiv \boldsymbol{b}^2(\theta,\phi) \equiv \boldsymbol{a}_2(\varepsilon\theta,\phi), \\ \boldsymbol{b}_3(\phi) \equiv \boldsymbol{b}^3(\phi) \equiv \varepsilon\boldsymbol{a}_3(\phi). \end{gathered} \tag{7.4}$$

Thus $\{\boldsymbol{b}_k\}$ is also an orthonormal basis for $\mathbb{E}^3$.

We study semi-inverse problems for which the deformation is restricted to have the form

$$\tilde{\boldsymbol{p}}(\mathbf{x}) = r(s)\boldsymbol{b}_1(\theta,\phi). \tag{7.5}$$

This deformation represents an inflation or dilatation if $\varepsilon = +$, and an eversion if $\varepsilon = -$.

We deduce from (7.5) that

(7.6)

$$\begin{aligned} \boldsymbol{F}(\tilde{\boldsymbol{z}}(\mathbf{x})) &= r'(s)\boldsymbol{b}_1(\theta,\phi)\boldsymbol{a}^1(\theta,\phi) + \frac{r(s)}{s}[\boldsymbol{b}_2(\theta,\phi)\boldsymbol{a}^2(\theta,\phi) + \varepsilon\boldsymbol{b}_3(\phi)\boldsymbol{a}^3(\phi)] \\ &\equiv F^k{}_l(s)\boldsymbol{b}_k(\theta,\phi)\boldsymbol{a}^l(\theta,\phi), \end{aligned}$$

$$\det \boldsymbol{F}(\tilde{\boldsymbol{z}}(\mathbf{x})) = \varepsilon r'(s)\left[\frac{r(s)}{s}\right]^2, \tag{7.7}$$

so that the requirement that $\det \boldsymbol{F} > 0$ everywhere essentially reduces to

$$\frac{r(s)}{s} > 0, \quad \varepsilon r'(s) > 0 \quad \forall\, s \in [\alpha, 1]. \tag{7.8}$$

Thus the right stretch tensor $\boldsymbol{U}$, which is the positive-definite square root of $\boldsymbol{C}$, is given by

(7.9)

$$\begin{aligned}\boldsymbol{U}(\tilde{\boldsymbol{z}}(\mathbf{x})) &= \varepsilon r'(s)\boldsymbol{a}^1(\theta,\phi)\boldsymbol{a}^1(\theta,\phi) + \frac{r(s)}{s}[\boldsymbol{a}^2(\theta,\phi)\boldsymbol{a}^2(\theta,\phi) + \boldsymbol{a}^3(\phi)\boldsymbol{a}^3(\phi)]\\ &\equiv U_{kl}(s)\boldsymbol{a}^k(\theta,\phi)\boldsymbol{a}^l(\theta,\phi).\end{aligned}$$

Its principal invariants are

$$\mathrm{I}_U = \varepsilon r' + 2\frac{r}{s}, \quad \mathrm{II}_U = \frac{r}{s}\left[2\varepsilon r' + \frac{r}{s}\right], \quad \mathrm{III}_U = \varepsilon r'\left[\frac{r}{s}\right]^2. \tag{7.10}$$

7.11. Exercise. Derive (7.6)–(7.10) (see Sec. XII.5).

For the problems we consider, it is marginally more convenient to replace the constitutive function (1.3a) delivering the second Piola-Kirchhoff stress tensor for a nonlinearly elastic material with one of the form

$$(\boldsymbol{U}, \boldsymbol{z}) \mapsto \check{\boldsymbol{S}}(\boldsymbol{U}, \boldsymbol{z}). \tag{7.12}$$

We require that the material be transversely isotropic in the sense that

$$\check{\boldsymbol{S}}(\boldsymbol{Q}(\boldsymbol{z}) \cdot \boldsymbol{U} \cdot \boldsymbol{Q}(\boldsymbol{z})^*, \boldsymbol{z}) = \boldsymbol{Q}(\boldsymbol{z}) \cdot \check{\boldsymbol{S}}(\boldsymbol{U}, \boldsymbol{z}) \cdot \boldsymbol{Q}(\boldsymbol{z})^* \tag{7.13}$$

for all orthogonal $\boldsymbol{Q}(\boldsymbol{z})$ that leave $\boldsymbol{a}^1(\theta,\phi)$ invariant:

$$\boldsymbol{Q}(\boldsymbol{z}) \cdot \boldsymbol{a}^1(\theta,\phi) = \boldsymbol{a}^1(\theta,\phi). \tag{7.14}$$

We assume that the inhomogeneity is spherically symmetric, so that the components of $\check{\boldsymbol{S}}$ with respect to the basis $\{\boldsymbol{a}_k\boldsymbol{a}_l\}$ depend explicitly on $\boldsymbol{z}$ only through s. For $\boldsymbol{U}$'s that are diagonal with respect to the basis $\{\boldsymbol{a}^k\boldsymbol{a}^l\}$, we set

$$\check{S}^{kl}(U_{11}, U_{22}, U_{33}, s) \equiv \boldsymbol{a}^k(\theta,\phi) \cdot \check{\boldsymbol{S}}(\boldsymbol{U}, \boldsymbol{z}) \cdot \boldsymbol{a}^l(\theta,\phi). \tag{7.15}$$

7.16. Exercise. Use (7.13) and (7.14) with $\boldsymbol{Q} \cdot \boldsymbol{a}^2 = \boldsymbol{a}^3$, $\boldsymbol{Q} \cdot \boldsymbol{a}^3 = -\boldsymbol{a}^2$ to prove that

$$\check{S}^{11}(U,V,W,s) = \check{S}^{11}(U,W,V,s), \tag{7.16a}$$

$$\check{S}^{33}(U,V,W,s) = \check{S}^{22}(U,W,V,s), \tag{7.16b}$$

$$\check{S}^{23}(U,V,V,s) = \check{S}^{31}(U,V,V,s) = \check{S}^{12}(U,V,V,s) = 0. \tag{7.16c}$$

The properties of isotropic materials can be found from (7.16a,b) by a simultaneous cyclic permutation of the indices 1,2,3 and the arguments U, V, W and from the observation in Sec. XII.13 that $(\check{S}^{ij})$ is diagonal when (U_{ij}) is diagonal. Alternatively, we could use representations like (1.7) for isotropic tensor-valued functions. (We could have likewise treated transverse isotropy by obtaining a suitable representation theorem.)

For $\boldsymbol{F}$'s that are diagonal with respect to the basis $\{\boldsymbol{b}_k\boldsymbol{a}^l\}$, so that the corresponding $\boldsymbol{U}$'s are diagonal with respect to the basis $\{\boldsymbol{a}^k\boldsymbol{a}^l\}$, and for $\boldsymbol{S}$'s that are diagonal with respect to the basis $\{\boldsymbol{a}_k\boldsymbol{a}_l\}$, the constitutive function delivering the first Piola-Kirchhoff stress tensor has the form

$$\hat{\boldsymbol{T}}(\boldsymbol{F},\boldsymbol{z}) = \hat{T}_k{}^l(F^1{}_1, F^2{}_2, F^3{}_3, s)\boldsymbol{b}^k\boldsymbol{a}_l \tag{7.17a}$$

where

$$\begin{aligned}\hat{T}_p{}^p(F^1{}_1, F^2{}_2, F^3{}_3, s) &= F^p{}_p\check{S}^{pp}(U_{11}, U_{22}, U_{33}, s)\\ &= F^p{}_p\check{S}^{pp}(|F^1{}_1|, |F^2{}_2|, |F^3{}_3|, s), \quad p = 1,2,3 \quad \text{not summed,}\end{aligned} \tag{7.17b}$$

and where the remaining components are each 0. Conditions (7.6), (7.7), and (7.9) give

$$\begin{aligned}\hat{T}_1{}^1(F^1{}_1, F^2{}_2, F^3{}_3, s) &= r'\check{S}^{11}\left(\varepsilon r', \frac{r}{s}, \frac{r}{s}, s\right),\\ \hat{T}_2{}^2(F^1{}_1, F^2{}_2, F^3{}_3, s) &= \frac{r}{s}\check{S}^{22}\left(\varepsilon r', \frac{r}{s}, \frac{r}{s}, s\right),\\ \hat{T}_3{}^3(F^1{}_1, F^2{}_2, F^3{}_3, s) &= \varepsilon\frac{r}{s}\check{S}^{33}\left(\varepsilon r', \frac{r}{s}, \frac{r}{s}, s\right).\end{aligned} \tag{7.17c}$$

To make our notation conform to that of Sec. X.3, whose methods we shall use to analyze our problems, we set

$$\begin{aligned}\hat{T}_1{}^1(\nu,\tau,\varepsilon\tau,s) &= \nu\check{S}^{11}(\varepsilon\nu,\tau,\tau,s) \equiv N^\varepsilon(\tau,\nu,s),\\ \hat{T}_2{}^2(\nu,\tau,\varepsilon\tau,s) &= \tau\check{S}^{22}(\varepsilon\nu,\tau,\tau,s) \equiv T^\varepsilon(\tau,\nu,s),\\ \hat{T}_3{}^3(\nu,\tau,\varepsilon\tau,s) &= \varepsilon\tau\check{S}^{33}(\varepsilon\nu,\tau,\tau,s) \equiv \varepsilon T^\varepsilon(\tau,\nu,s),\end{aligned} \tag{7.17d}$$

with the last two equations reflecting the transverse isotropy condition (7.16b).

7.18. Exercise. Prove that if the components of $\boldsymbol{T}$ with respect to basis $\{\boldsymbol{b}^k\boldsymbol{a}_l\}$ are diagonal and are independentof (θ,ϕ), then the only nontrivial equilibrium equation for zero body force reduces to

$$\frac{d}{ds}\left[s^2T_1{}^1\right] - sT_2{}^2 - \varepsilon sT_3{}^3 = 0. \tag{7.19a}$$

Prove that for a deformation of the form (7.5) the insertion of constitutive equations for a transversely isotropic material reduces these equations to the single quasilinear second-order ordinary differential equation

$$\frac{d}{ds}\left[s^2 N^\varepsilon\left(\frac{r}{s}, r', s\right)\right] - 2sT^\varepsilon\left(\frac{r}{s}, r', s\right) = 0, \tag{7.19b}$$

which is the object of our study. Note that this equation is very similar to (X.3.1). The development in the rest of this section should be compared with that of Sec. X.3.

Using (7.17), we readily find that the Strong Ellipticity Condition implies that

$$N^\varepsilon_\nu > 0. \tag{7.20}$$

7.21. Exercise. Prove that the Strong Ellipticity Condition does not imply that $T^\varepsilon_\tau > 0$. This is one respect in which our spherically symmetric problem differs from the axisymmetric problem treated in Sec. X.3.

Note that (7.8) implies that r' ranges over $(0, \infty)$ when $\varepsilon = +$ and ranges over $(-\infty, 0)$ when $\varepsilon = -$. In consonance with (2.44), we require for fixed τ and s that

$$N^+(\nu, \tau, s) \to \pm\infty, \quad N^-(\nu, \tau, s) \to \pm\infty \quad \text{as } \nu \text{ approaches its extremes,} \tag{7.22}$$

the signs taken to be compatible with (7.20). In this case, (7.20) and the Intermediate Value Theorem ensure that the equation $N^\pm(\tau, \nu, s) = n$ is equivalent to an equation of the form $\nu = \nu^\varepsilon(\tau, n, s)$. Setting $\tilde{T}^\varepsilon(\tau, n, s) \equiv T^\varepsilon(\tau, \nu^\varepsilon(\tau, n, s), s)$, we reduce (7.19b) to the semilinear system

$$\frac{d}{ds}(s\breve{\tau}) = \nu^\pm(\breve{\tau}, \breve{n}, s), \quad \frac{d}{ds}(s^2\breve{n}) = 2s\tilde{T}^\pm(\breve{\tau}, \breve{n}, s). \tag{7.23±}$$

Let us use the transformation

$$s = e^{\xi-1}, \quad \breve{\tau}(s) = \tau(\xi), \quad \breve{n}(s) = n(\xi), \tag{7.24}$$

which is the same as (X.3.8), to convert (7.23) to

$$\frac{d}{d\xi}\tau = \nu^\pm(\tau, n, e^{\xi-1}) - \tau, \quad \frac{d}{d\xi}n = 2[\tilde{T}^\pm(\tau, n, e^{\xi-1}) - n] \tag{7.25±}$$

for $1 + \ln\alpha < \xi < 1$.

Condition (7.20) implies that

$$\nu^\varepsilon_n > 0, \tag{7.26}$$

so that $n \mapsto \nu^+(\tau, n, s)$ strictly increases from 0 to ∞ and $n \mapsto \nu^-(\tau, n, s)$ strictly increases from $-\infty$ to 0 as n increases from $-\infty$ to ∞. We can follow Sec. X.3 to deduce useful consequences of hypotheses like

$$\frac{\partial(T^\pm, N^\pm)}{\partial(\tau, \nu)} \quad \text{is positive-definite.} \tag{7.27±}$$

7.28. Exercise. Use (7.17d) to prove that

$$\nu^-(\tau, n, s) = -\nu^+(\tau, -n, s), \quad \tilde{T}^-(\tau, n, s) = \tilde{T}^+(\tau, -n, s), \tag{7.29}$$

so that (7.25−) can be written as

$$\frac{d}{d\xi}\tau = -\nu^+(\tau, -n, e^{\xi-1}) - \tau, \quad \frac{d}{d\xi}n = 2[\tilde{T}^+(\tau, -n, e^{\xi-1}) - n] \tag{7.30−}$$

for $1 + \ln \alpha < \xi < 1$. These identities can be used to convert constitutive restrictions on T^+ and N^+ into constitutive restrictions on T^- and N^-: Prove that (7.27+) is equivalent to (7.26) and

$$\tilde{T}^-_\tau \nu^-_n - \tilde{T}^-_n \nu^-_\tau > 0, \quad \tilde{T}^-_\tau > 0, \tag{7.31a,b}$$

that when (7.20) holds, the conditions

$$T^+_\nu > 0, \quad N^+_\tau > 0 \tag{7.31c,d}$$

are equivalent to

$$\tilde{T}^-_n < 0, \quad \nu^-_\tau > 0, \tag{7.31e}$$

and that when (7.20) holds, the condition

$$N^+_\tau < N^+_\nu \tag{7.31f}$$

is equivalent to

$$\nu^-_\tau < 1. \tag{7.31g}$$

We now focus our attention on homogeneous materials, for which the constitutive functions are independent of s. Thus (7.25) is autonomous. The phase portrait of (7.25+) is constructed exactly as in Sec. X.3. The treatment of isotropy differs somewhat from that centered on (X.3.18)–(X.3.20) because (X.3.18) does not hold. For our spherical problems, we find that the equation $N^+(\tau, \nu) = T^+(\tau, \nu)$, associated with the horizontal isoclines (see (X.3.19)), nevertheless has a solution for ν given by $\nu = \tau$, by virtue of (7.17d) and the transverse isotropy.

7.32. Exercise. Show that when (7.26) and (7.31a,b,e,g) hold, the phase portrait for (7.25−) or, equivalently, (7.27−), has the character of that shown in Fig. 7.33. Show that the horizontal isoclines are defined by $n = T^+(\tau, \nu) = -N^+(\tau, \nu)$. Note that there is no vertical isocline because $\nu^- < 0$.

Compression of a ball. We now study the deformation of a ball subject to a uniform normal traction of intensity $-\lambda g(r(1))$ per unit reference area on its boundary, so that

$$n(1) = -\lambda g(\tau(1)). \tag{7.34}$$

Here λ is a given number, and g is a given function that is assumed to satisfy $g(r) > 0,\ g'(r) \geq 0$ for $r > 0$. For a dead load, $g(r) = 1$, and for a hydrostatic load (with a constant intensity of force per unit deformed area),

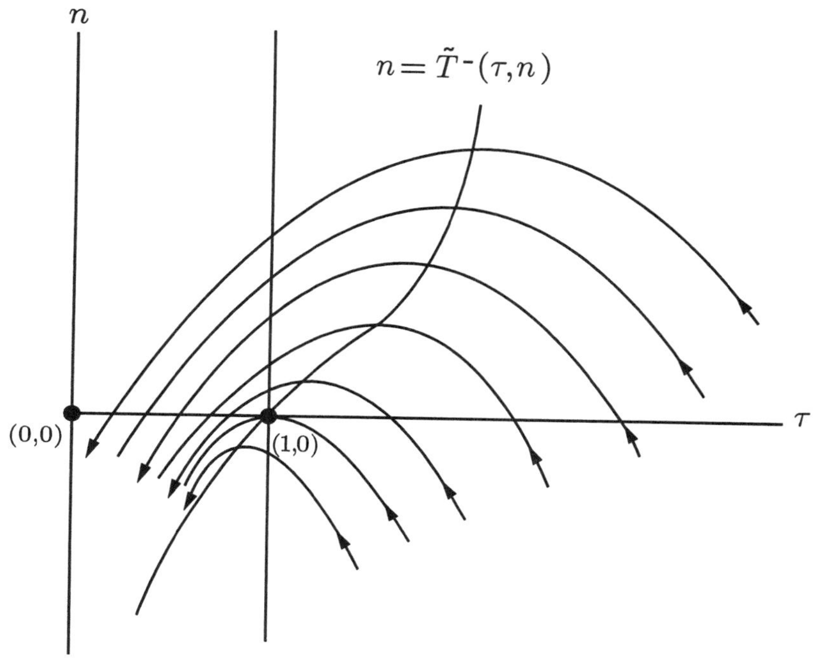

Fig. 7.33. Typical phase portrait of (7.25−) under the assumption that (7.26) and (7.31a,b,e,g) hold.

$g(r) = r^2$. These loads are pressures for $\lambda > 0$ and tensions for $\lambda < 0$. For problems in which $\lambda > 0$ we require that the center remain intact, so that $r(0) = 0$ or, equivalently,

(7.35) $$e^{\xi-1}\tau(\xi) \to 0 \quad \text{as } \xi \to -\infty.$$

For $\lambda < 0$, we assume that either (7.35) holds or that a spherical cavity forms at the center. In the latter case, $r(0) > 0$ and the normal stress $T_1{}^1(0) = 0$. Thus

(7.36a,b) $$\liminf_{\xi\to-\infty} e^{\xi-1}\tau(\xi) > 0, \qquad \lim_{\xi\to-\infty} n(\xi) = 0.$$

For all these problems we use (7.25+).

The phase-plane analysis of (7.25+), (7.35) for $\lambda > 0$ is identical to that carried out in Sec. X.3 for aeolotropic plates. The much simpler treatment in Sec. X.2 for isotropic plates carries over to isotropic balls.

7.37. Exercise. For an isotropic ball subject to a uniform normal pressure with $\lambda > 0$ and either $g(r) = 1$ or $g(r) = r^2$, discuss the existence and uniqueness of solutions of the form $r(s) = ks$, where k is a positive constant, in light of constitutive hypotheses of the sorts treated in this chapter.

Observe that the treatment of problems in which $r(1)$ or, equivalently, $\tau(1)$ is a prescribed number less than 1 is virtually identical.

Dilatation and cavitation of a ball. Let us first study the dilatation problem for a ball. Here λ is negative, and (7.34) and (7.35) hold. We can invoke the results of Sec. X.3 to conclude that under mild constitutive assumptions there is a unique solution of the boundary-value problem. It corresponds to a segment of the separatrix in Fig. X.3.17 (in the quadrant $\mathcal{Q}_1$) beginning at a saddle point and terminating on the curve $n = -\lambda g(\tau)$. The behavior of such solutions depends on the disposition of the horizontal and vertical isoclines in $\mathcal{Q}_1$, and may exhibit all the complexity of compressive states described by Theorem X.3.31. Under mild constitutive restrictions, the separatrix lies on the curve defined by $\tau = f(n)$ with the domain of f the whole n-axis. The problem for a dead load, for which $g = 1$, always has a unique solution. On the other hand, there are materials for which the curve $\tau = f(n)$ does not intersect the curve $n = -\lambda\tau^2$ for $-\lambda$ sufficiently large. Thus there are bodies for which there is no spherically symmetric equilibrium state with center intact under the action of a hydrostatic tension on its boundary. When no such equilibrium state exists, we might expect that at least one of the following alternatives occurs: There is a spherically symmetric equilibrium state with cavitation, there is an equilibrium state (possibly with cavitation) that is not spherically symmetric, or there is no equilibrium state and the body behaves dynamically. In fact, it follows from the ensuing development that there cannot be spherically symmetric equilibrium states with cavitation when there are no spherically symmetric equilibrium states without cavitation.

We now study cavitational solutions and compare them with solutions in which the center is intact. Stable cavitational solutions are of great importance in solid mechanics because they furnish a mechanism for spontaneous fracture. Most contemporary researches in fracture mechanics employ special constitutive criteria for the onset of fracture (in circumstances much more general than those we study here).

In $\mathcal{R}$ (see Fig. X.3.17), $\tau' < 0$ and $n' > 0$. Now (7.31c) implies that $\nu_\tau^+ < 0$. This condition and the inequality $\nu_n^+ > 0$, which follows from (7.20), imply that

$$\text{(7.38)} \qquad \frac{d}{d\xi}\nu^+(\tau(\xi), n(\xi)) > 0 \quad \text{when } (\tau(\xi), n(\xi)) \in \mathcal{R}.$$

7.39. Exercise. Imitate the proof of Lemma X.3.23 to demonstrate

7.40. Lemma. *Let* (7.38) *hold for trajectories lying in* $\mathcal{R}$. *Then* $e^{\xi-1}\tau(\xi)$ *has a positive lower bound on any trajectory in* $\mathcal{R}$.

Thus (7.36a) is automatically satisfied on such trajectories. Now it is not hard to show that any half-trajectory terminating in $\mathcal{R}$ is defined for $\xi \to -\infty$. It follows that on such half-trajectories $\tau(\xi) \to \infty$ as $\xi \to -\infty$. Thus for $\lambda < 0$, a (cavitating) solution of (7.25+) satisfying (7.34) and

(7.36) corresponds to a trajectory that is asymptotic to the positive τ-axis (as $\xi \to -\infty$) and that terminates on the curve $n = -\lambda g(\tau)$. Below we sketch how to show that there are materials for which there exists a trajectory $\mathcal{C}$ that is asymptotic to the positive τ-axis. For our immediate purposes, we simply assume that there exists such a trajectory and that it is unique.

It is illuminating to consider as an alternative to (7.34) the requirement that the outer radius be prescribed to exceed 1:

$$\tau(1) = \omega > 1. \tag{7.41}$$

For simplicity of exposition, we take the g in (7.34) to correspond to a dead load: $g = 1$. For other g's, there can be no spherically symmetric equilibrium state with cavitation when there is no spherically symmetric equilibrium state with center intact, because nonexistence in the latter case occurs when the curve $n = -\lambda g(\tau)$ lies to the left of the separatrix curve $\tau = f(n)$, while $\mathcal{C}$ must lie to the right of the separatrix curve.

The existence, multiplicity, and qualitative properties of solutions follow from an examination of the appropriate phase portraits. The stability of solutions does not (although one can make educated guesses about it). For hyperelastic bodies, we can characterize stable solutions formally as those that minimize the energy in suitable classes. Cavitation, however, presents some peculiarities about the class of admissible functions that are appropriate for stability. (See Ball (1982), Giaquinta, Modica, & Souček (1989), Müller & Spector (1995), Müller, Tang, & Yan (1995), and Sivaloganathan (1986)).

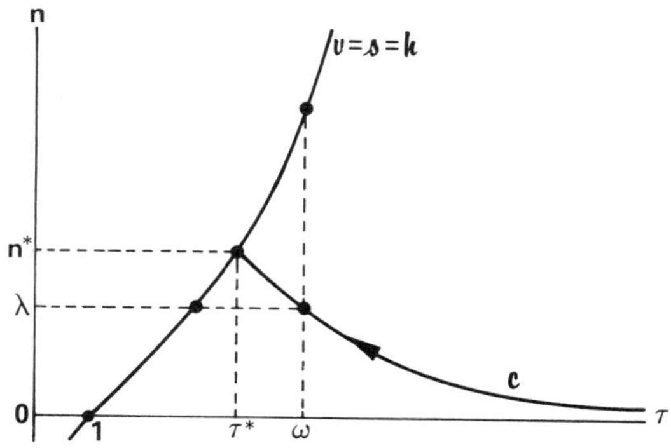

Fig. 7.42. Typical phase portrait of the quadrant $\mathcal{Q}_1$ for (7.25+) for an isotropic material.

In Figs. 7.42, 7.43, and 7.44 are exhibited typical phase portraits of the quadrant $\mathcal{Q}_1$ for (7.25+) for an isotropic material and for two kinds of aeolotropic materials. From Fig. 7.42, for isotropic materials, we find that for $0 < -\lambda \leq n^*$ there is exactly one equilibrium state with center intact

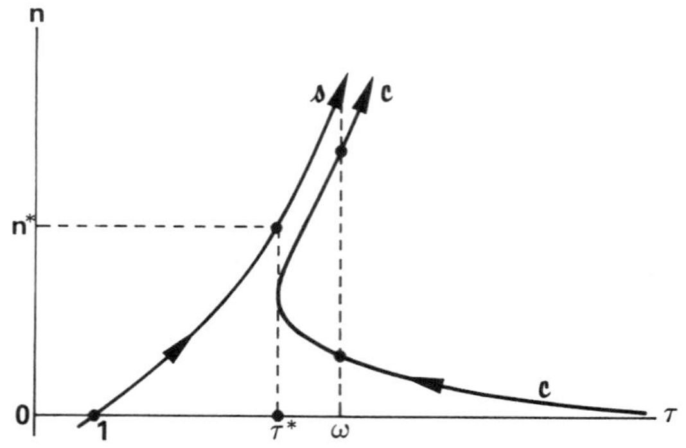

Fig. 7.43. Typical phase portrait of the quadrant $\mathcal{Q}_1$ for (7.25+) for an aeolotropic material when $(\tau, n) = (1, 0)$ is a saddle point and when $\mathcal{Q}_1$ is free of critical points.

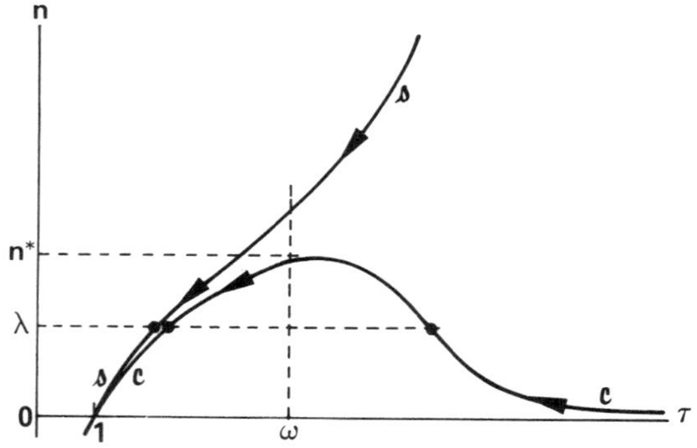

Fig. 7.44. Typical phase portrait of the quadrant $\mathcal{Q}_1$ for (7.25+) for an aeolotropic material when $(\tau, n) = (1, 0)$ is an attractive node and when $\mathcal{Q}_1$ is free of critical points.

satisfying (7.34) and exactly one cavitating equilibrium state satisfying (7.34). If $-\lambda > n^*$, there is exactly one equilibrium state, which is intact. (This state presumably is unstable: Cavitation may well occur, but in a dynamical process.) Note that the outer radius of this intact state is the same as that for a cavitating state, but the normal stress $-\lambda$ on the boundary for the intact state exceeds that for the cavitating state at the same radius. This normal stress for the cavitating state is presumably the most that can be borne in a stable equilibrium state with the prescribed outer radius.

From Fig. 7.42 we also find that for $1 \leq \omega < \tau^*$, there is exactly one equilibrium state satisfying (7.41), which is intact. For $\tau^* \leq \omega$ there are exactly two equilibrium states satisfying (7.41): one intact, the other not. For hyperelastic materials, the cavitating state is stable and the intact state is unstable according to the energy criterion (see Ball (1982)).

For phase portraits for aeolotropic materials of the form of Fig. 7.43, we find that for each $-\lambda > 0$, there is exactly one intact state and one cavitating state satisfying (7.34). For $1 \leq \omega < \tau^*$, there is exactly one equilibrium state satisfying (7.41), which is intact. For $\tau^* < \omega$, there is exactly one intact equilibrium state satisfying (7.34) and there are exactly two cavitating equilibrium states satisfying (7.34). Extrapolating from Ball's results about the stability of the isotropic body, we surmise that the only one of these three states that is stable is the cavitating state on the lower branch of $\mathcal{C}$. (n is apparently smallest on this branch.) As ω is increased past τ^*, the (presumably) stable equilibrium state snaps from one with center intact to a cavitating state. No such jumping occurs for isotropic materials. For the traction problem, we might conjecture that there is no stable equilibrium state for $\lambda \geq n^*$. The deductions from Fig. 7.44 are left as an informal exercise.

We now show that there are materials for which the phase portrait has a trajectory asymptotic to the positive τ-axis. Since $\tau'(\xi) < 0$ and $\tau(\xi) \to \infty$ as $\xi \to -\infty$ in $\mathcal{R}$, we can solve the equation $\tau(\xi) = \tau^\dagger$ for $\xi = \xi^\dagger(\tau^\dagger)$ when $\tau(\cdot)$ corresponds to a trajectory in $\mathcal{R}$. We then set $n^\dagger(\tau^\dagger) \equiv n(\xi^\dagger(\tau^\dagger))$ or, equivalently, $n^\dagger(\tau) \equiv n(\xi^\dagger(\tau))$. Then (7.25+) in $\mathcal{R}$ implies that $n^\dagger$ satisfies the equation

$$\frac{dn^\dagger}{d\tau} = \frac{2[\tilde{T}^+(\tau, n^\dagger) - n^\dagger]}{\nu^+(\tau, n^\dagger) - \tau}, \tag{7.45a}$$

which we rewrite as

$$\frac{d}{d\tau}\left(\frac{n^\dagger(\tau)}{\tau^2}\right) = \frac{2}{\tau^2(\nu^+(\tau, n^\dagger) - \tau)}\left[\tilde{T}^+(\tau, n^\dagger) - \frac{n^\dagger \nu^+(\tau, n^\dagger)}{\tau}\right] \equiv H(\tau, n^\dagger). \tag{7.45b}$$

Let ω be any positive number and let $\tau > \omega$. Then (7.45b) yields

$$\frac{n^\dagger(\tau)}{\tau^2} = \frac{n^\dagger(\omega)}{\omega^2} + \int_\omega^\tau H(\sigma, n^\dagger(\sigma))\, d\sigma. \tag{7.46}$$

If the constitutive functions are such that the integral in (7.46) defines a bounded, continuous function of ω and τ that is decreasing in τ, then, for any fixed $(\omega, n^\dagger(\omega))$, (7.46) has a limit as $\tau \to \infty$. For fixed ω, we can use the Intermediate Value Theorem to choose $n^\dagger(\omega)$ so that this limit is 0. With this choice, the solution of (7.46) gives the desired trajectory, provided that $(\omega, n^\dagger(\omega)) \in \mathcal{R}$.

We shall find that it is fairly easy to impose conditions on the constitutive functions ν^+ and $\tilde{T}^+$ that ensure these properties. But constitutive functions for nonlinearly elastic materials customarily deliver the stress as a given function of strain. We accordingly obtain sufficient conditions for constitutive equations in this form. Our procedure is reminiscent of that used in Sec. VI.9.

We assume that there are positive numbers $A_1, B_1, \dots, c_2, d_2$ such that

$$T^+(\tau, \nu) \sim -A_1\tau^{-a_1} + B_1\tau^{b_1} - Cc_1\tau^{-c_1-1}\nu^{-c_2} + Dd_1\tau^{d_1-1}\nu^{d_2}, \tag{7.47}$$

$$N^+(\tau, \nu) \sim -A_2\nu^{-a_2} + B_2\nu^{b_2} - Cc_2\tau^{-c_1}\nu^{-c_2-1} + Dd_2\tau^{d_1}\nu^{d_2-1} \tag{7.48}$$

as $\tau \to \infty$, $\nu \to 0$. By this we mean that T^+ and N^+ equal the terms on the right-hand sides of (7.47) and (7.48) and terms negligible with respect to these as as $\tau \to \infty$, $\nu \to 0$. These limits are chosen because under our general constitutive assumptions they correspond to large τ and bounded n. (The representations (7.47) and (7.48) are reasonable for all ν and τ.)

7.49. Exercise. Determine the ranges of the parameters in (7.47) and (7.48) for which the right-hand sides satisfy each of the constitutive restrictions (7.20), (7.27+), and (7.31c,e).

We seek corresponding asymptotic representations for $\tilde{T}^+$ and ν^+. We get the latter by seeking a representation for the solution ν of the equation

$$n = N^+(\tau, \nu) \tag{7.50}$$

when (7.48) holds in the form

$$\nu \sim \tau^{-\mu(n)}\chi(\tau, n) \tag{7.51}$$

with μ and χ positive-valued for each real n and with $\chi(\tau, n)$ having a positive limit, depending on n, as $\tau \to \infty$. We substitute (7.51) into (7.48) and then substitute the result into (7.50) to obtain an asymptotic equation involving a sum of powers of τ. For this equation to be satisfied, μ must be chosen so that the two highest powers of τ in (7.50) are equal (cf. Sec. IV.4). (The systematic exploitation of this observation when the exponents $a_1, \ldots, d_2$ are given specific numerical values constitutes the Newton Polygon Method; see Dieudonné (1949).) Then $\lim_{\tau\to\infty} \chi(\tau, n) \equiv M$ is determined by the requirement that the sum of terms with the highest powers of τ in the resulting version of (7.50) vanish in the limit as $\tau \to \infty$:

$$0 \sim -A_2\tau^{\mu a_2}M^{-a_2} - Cc_2\tau^{-c_1+\mu(c_2+1)}M^{-c_2-1} + Dd_2\tau^{d_1-\mu(d_2-1)}M^{d_2-1}. \tag{7.52}$$

Here we have retained all terms that cannot be immediately disqualified as negligible. The candidates for μ, obtained by the pairwise equating of the exponents on τ, are the positive numbers among

$$\frac{c_1}{c_2 + 1 - a_2}, \quad \frac{d_1}{a_2 + d_2 - 1}, \quad \frac{d_1 + c_2}{d_2 + c_2}. \tag{7.53}$$

(The last entry in (7.53) is the only one that is clearly positive.) We note that μ and M found by this procedure are independent of n. We obtain the asymptotic representation for $\tilde{T}^+$ by substituting that for ν^+ into T^+:

$$\tilde{T}^+(\tau, n) \sim B_1\tau^{b_1} - Cc_1M^{-c_2}\tau^{-c_1-1+\mu c_2} + Dd_1M^{d_2}\tau^{d_1-1-\mu d_2}. \tag{7.54}$$

Now we substitute (7.51) and (7.54) into (7.46) to obtain

(7.55)

$$\frac{n^\dagger(\tau)}{\tau^2} \sim \frac{n^\dagger(\omega)}{\omega^2} - \int_\omega^\tau (B_1\sigma^{b_1-3} - Cc_1M^{-c_2}\sigma^{-c_1-4+\mu c_2} + Dd_1M^{d_2}\sigma^{d_1-4-\mu d_2})\, d\sigma$$

as $\tau \to \infty$. Since we know that the general constitutive restrictions supporting the phase portraits of Figs. 7.43 and 7.44 force $n^\dagger$ to be decreasing, by the definition of $\mathcal{R}$, it follows that the C-term in (7.53) cannot be dominant as $\tau \to \infty$. This condition and the requirement that the integrand in (7.55) be integrable on (ω, ∞) are ensured by

$$-c_1 + \mu c_2 < b_1 + 1, \quad -c_1 + \mu c_2 < d_1 - \mu d_2, \quad b_1 < 2, \quad d_1 < 3 + \mu d_2. \tag{7.56}$$

In view of the argument following (7.46), conditions (7.56) imply the existence of a unique trajectory asymptotic to the positive τ-axis.

If necessary, we could ensure that $(\omega, n^\dagger(\omega)) \in \mathcal{R}$ by taking the constants B_1 and D sufficiently small. Whether it is necessary to make such an assumption depends on the nature of the boundary of $\mathcal{R}$, which is determined by the disposition of the horizontal and vertical isoclines in $\mathcal{Q}_1$. (This disposition may be found by a complementary asymptotic analysis involving other constants of (7.45).) Thus we conclude that there are cavitating solutions when (7.45) and (7.46) hold (and possibly when further minor restrictions are imposed). This asymptotic analysis corrects that of Antman & Negrón-Marrero (1987).

The details of the cavitation problem for cylinders are virtually identical to those for a ball. For a cylinder, however, hydrostatic tension is described by $g(r) = r$. The important distinction between cylinders and balls is that the nature of tensile solutions for certain materials depends crucially on the choice of g.

Inflation of a shell. Let us take $\alpha > 0$. We can consider any combination of traction boundary conditions and position boundary conditions for (7.25+) subject to the requirement that if we prescribe both inner and outer radii, then the former must be the smaller. To be specific, let us prescribe dead loads $n(1+\ln\alpha) = n_* < 0$ and $n(1) = n^* > n_*$ and let us suppose that the phase portrait has the form shown in Fig. 7.57, so that the only critical point is a saddle at $(1, 0)$. If $n^* \geq 0$, then the only candidates for solutions are the upward-moving trajectories to the right of the separatrices. If $n^* < 0$, then these trajectories, together with some of those in the region below both lower separatrices, are candidates.

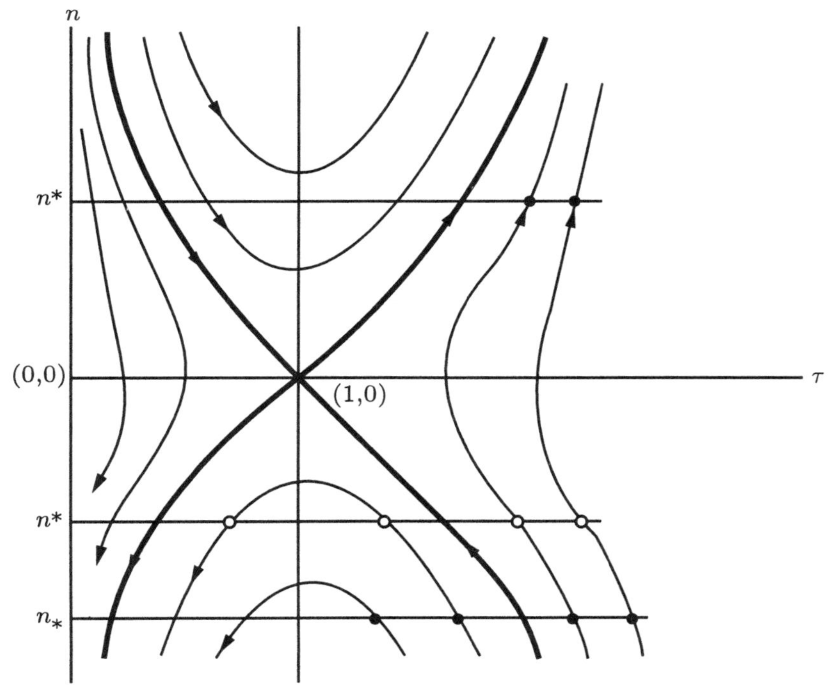

Fig. 7.57. Phase portrait for (7.25+) when its only critical point is a saddle at $(1, 0)$.

Since $dn/d\xi > 0$ in the region to the right of the separatrices we can define each trajectory there by an equation of the form $\tau = \tau^\dagger(n, \omega)$ where

ω is the value of τ at which the trajectory intersects the line $n = n_*$. Then $\tau^\dagger$ is the solution of the initial-value problem

$$\text{(7.58a)} \qquad \frac{d\tau}{dn} = \frac{\nu^+(\tau, n) - \tau}{2[\tilde{T}^+(\tau, n) - n]}, \qquad \tau(n_*) = \omega.$$

Such a trajectory generates a solution to our boundary-value problem whenever ω can be chosen so that ξ varies from $1 + \ln \alpha$ to 1 on this trajectory, i.e., so that

$$\text{(7.58b)} \qquad -\ln \alpha = \int_{1+\ln \alpha}^{1} d\xi = \int_{n_*}^{n^*} \frac{dn}{2[\tilde{T}^+(\tau^\dagger(n, \omega), n) - n]} \equiv \Phi(\omega).$$

$\Phi(\omega)$ is the amount of independent variable ξ used up on the segment of the trajectory through (ω, n_*) terminating on the line $n = n^*$. From the phase portrait (7.57), we find that $\tau^\dagger_\omega > 0$ and that $\tau^\dagger_\omega \nearrow \infty$ as $\omega \nearrow \infty$. Our growth conditions imply that $\tilde{T}^+(\tau, n) \to \infty$ as $\tau \to \infty$ for n in any bounded interval. Therefore, $\Phi(\omega) \to 0$ as a $\omega \to \infty$. If we further assume that $\tilde{T}^+_\tau > 0$, which is a consequence of (7.27+), we find that $\Phi'(\omega) < 0$ for all $\omega > \omega_0$ where ω_0 is the value of τ at which the line $n = n_*$ intersects the stable separatrix.

Let us first study the case in which $n^* \geq 0$. From standard properties of critical points, it follows that $\Phi(\omega_0) = \infty$. It then follows from the Intermediate Value Theorem that there is an ω satisfying (7.58b) and consequently a solution to our boundary-value problem, which is unique when Φ is strictly decreasing.

Now suppose that $n^* < 0$. We first examine solutions of our boundary-value problem for trajectories to the right of the separatrices. We know that $\Phi(\omega) \to 0$ as $\omega \to \infty$. Let us presume that $\Phi(\omega_0)$ can be computed. Then the Intermediate Value Theorem implies that there is a solution of our boundary-value problem corresponding to a trajectory to the right of the separatrices if $-\ln \alpha \leq \Phi(\omega_0)$, i.e., if α is sufficiently close to 1 or if n^* is sufficiently close to 0. There is at most one solution trajectory in this region if Φ is strictly decreasing. If Φ is strictly decreasing and if $-\ln \alpha > \Phi(\omega_0)$, there is no solution trajectory in this region.

For $n^* < 0$, we now study trajectories in the region below the lower separatrices. The graphs of these trajectories have the form $n = n^\dagger(\tau, \omega)$ where ω is the rightmost value of τ at which the trajectory intersects the line $n = n_*$. Then $n^\dagger(\cdot, \omega)$ is the solution of the initial-value problem

$$\text{(7.59a)} \qquad \frac{dn}{d\tau} = \frac{2[\tilde{T}^+(\tau, n) - n]}{\nu^+(\tau, n) - \tau}, \qquad n(\omega) = n_*.$$

Just as above, we find that $n^\dagger(\cdot, \omega)$ generates a solution of the boundary-value problem if ω satisfies

$$\text{(7.59b)} \qquad -\ln \alpha = \int_{1+\ln \alpha}^{1} d\xi = -\int_{\tau^*}^{\omega} \frac{d\tau}{\nu^+(\tau, n^\dagger(\tau, \omega)) - \tau} \equiv \Psi(\omega)$$

where τ^* is any solution of $n^\dagger(\tau, \omega) = n^*$.

Let $\bar{\omega}$ identify the trajectory satisfying (7.59a) that is tangent to the line $n = n^*$. For $\omega \in (\bar{\omega}, \omega_0)$, there are two values of τ^* such that $n^\dagger(\tau, \omega) = n^*$. Let us first choose τ^* to be the smaller of these values. Then for $\bar{\omega} \leq \omega \leq \omega_0$, $\Psi(\omega)$ ranges over an interval containing $[\Psi(\bar{\omega}), \infty]$. Thus, if $-\lambda\alpha \geq \Psi(\bar{\omega})$, i.e., if α is sufficiently small or if n^* is sufficiently large, then there is a solution of our boundary-value problem. In particular, if $-\lambda\alpha > \Psi(\bar{\omega})$, then for this solution n has an internal maximum. (Note that this range of α complements, at least roughly, the range corresponding to trajectories to the right of the separatrices.)

Now let $\omega \in (\bar{\omega}, \omega_0)$ and let τ^* be the larger of the two solutions of $n^\dagger(\tau, \omega) = n^*$. Then Ψ can be defined just like Φ above, and, under mild constitutive assumptions, can be assumed to satisfy $\Psi'(\omega) > 0$ for $\omega \in (\bar{\omega}, \omega_0)$. We find that if $\Psi(\bar{\omega}) \leq -\lambda\alpha \geq \Psi(\omega_0)$, then there is a solution of our boundary-value problem in which n is increasing.

There are a variety of more sophisticated methods (including those used in Sec. IV.5 and in Chap. VII) to show that this problem has solutions. The procedure used here has the virtue that it gives the qualitative properties of the solutions and multiplicity results, and the disadvantage that it does not give the stability of solutions.

7.60. Exercise. Analyze by the methods of this section all the qualitatively distinct possibilities for boundary-value problems corresponding to Fig. 7.57 and for the phase portrait for (7.25+) when its only critical point is an attractive node at $(1, 0)$.

7.61. Exercise. Use phase-plane methods to determine the existence and qualitative properties of these inflation problems for all the alternative specifications of position and traction boundary conditions for an isotropic material.

7.62. Exercise. For hyperelastic materials, use the direct methods of the calculus of variations to prove the existence of solutions to the problem of this section under all the alternative specifications of position and traction boundary conditions for an isotropic material.

Eversion of a shell. Half a tennis ball can easily be turned inside out and remain in an unloaded everted configuration. One can even contemplate the eversion of a whole tennis ball by slicing it, pulling it through the slit, and then gluing the slit together again. A finite length of rubber tube can be everted, possibly with some difficulty. (See the illustrated discussion of Truesdell (1978).) That such eversions can be effected suggests that equilibrium equations for such nonlinear elastic bodies admit at least two (stable) classical solutions (differing by more than a rigid displacement).

The nature of the everted states suggest that the analytic problems of demonstrating their existence and determining their properties are unambiguously global: They are impervious to attack by perturbation methods and by obvious global continuation methods, like those described in Sec. 6.

Here we treat the eversion problem for spherical shells by methods just like those we used for the inflation problem. We assume that (7.26) and (7.27+) hold, that the corresponding natural growth conditions hold, so that the phase portrait Fig. 7.33 is valid. Under these conditions, the horizontal isocline has the form $\tau = h(n)$ with $h(0) = 0$ and with h strictly increasing from 1 to ∞ as n increase from 0 to ∞. Since $\nu^- < 0$, by (7.26a), there are no vertical isoclines in Fig. 7.33 and therefore no singular points in $\mathcal{Q}_1$.

Let the graph of the solution of (7.25−) satisfying the initial conditions $\tau = h(\bar{n})$, $n = \bar{n}$ with $\bar{n} \geq 0$ be denoted $\tau \mapsto n^{\ddagger}(\tau, \bar{n})$. Under our constitutive assumptions, for each $\bar{n} > 0$ this graph has exactly two intersections with the line $n = 0$, which are denoted $\omega_\star(\bar{n})$ and $\omega^\star(\bar{n})$ with $\omega_\star(\bar{n}) < \omega^\star(\bar{n})$. An everted equilibrium configuration of the shell corresponds to a solution of (7.25−) satisfying the boundary conditions $n(1 + \ln \alpha) = 0 = n(1)$. Such a solution in turn corresponds to $n^{\ddagger}(\cdot, \bar{n})$ when $\bar{n}$ satisfies the following analog of (7.59b):

$$(7.63) \qquad -\ln \alpha = -\int_{\omega_\star(\bar{n})}^{\omega^\star(\bar{n})} \frac{d\tau}{\nu^-(\tau, n^{\ddagger}(\tau, \bar{n})) - \tau} \equiv X(\bar{n}).$$

Now $X(0) = 0$. Since $\omega_\star < 1$, since $\omega^\star(\bar{n}) \nearrow \infty$ as $\bar{n} \nearrow \infty$, since $\omega_\star(\bar{n}) \searrow 0$ as $\bar{n} \nearrow \infty$, and since $\nu^-(\tau, n) \nearrow 0$ as $n \to \infty$, it follows that $X(\bar{n}) \to \infty$ as $n \to \infty$. It follows from Leibniz's Rule that $X'(\bar{n}) > 0$ for all $\bar{n} > 0$. Therefore, (7.63) has a unique solution for each $\alpha \in (0, 1)$, so that there is a unique spherically symmetric everted state when (7.26), (7.27+), and compatible growth conditions hold. Uniqueness can be lost when (7.27+) is suspended. It is of course a straightforward matter to replace the zero traction conditions for the eversion problem with nonzero traction conditions. The methods are the same.

7.64. Exercise. Suppose that the shell is not complete, i.e., $\beta > 0$ or $\gamma < \pi$. Then the material edges of the everted shell, defined by $\theta = \beta, \gamma$, are constrained to lie on the cones defined by $\theta = -\beta, -\gamma$. Determine the traction on these edges, which is not zero, and prove that the resultant forces on these edges are zero.

These results explain why an everted hemispherical shell with zero traction over its entire boundary cannot have the everted spherical form given by (7.5) with $\varepsilon = -1$. Those everted states that look nearly spherical (see Truesdell (1978)) exhibit flaring near the edge. This flaring can be attributed to a boundary-layer effect corresponding to the discrepancy between zero traction on the edge and zero force on the edge.

7.65. Research Problem. Carry out an asymptotic analysis of this flaring with respect to the thickness parameter $1 - \alpha$.

7.66. Exercise. Formulate and fully analyze the eversion of cylindrical tubes of finite length.

7.67. Exercise. Work out the existence theory for the inflation and the eversion of incompressible shell.

7.68. Exercise. Study the cavitation problem for an incompressible elastic ball when the normal stress is prescribed and when the outer radius is prescribed. The latter problem gives a simple paradigm for the whole process of cavitation.

The solution for the inflation of an incompressible spherical shell was obtained by Green & Shield (1950) and the solution for eversion by Ericksen (1955a). For a treatment of existence for these problems in the spirit of Sec. 4, see Adeleke (1983). Studies of the corresponding problems for compressible media were carried out by Adkins (1955) and Green (1955). For full accounts of this early work, see Green & Adkins (1970), Green & Zerna (1968), Ogden (1984), and Truesdell & Noll (1965).

The material formulation given in the beginning of this section follows Antman (1979a). Figures 7.43–7.45 and much of the analysis in this section through (7.56) come from Antman & Negrón-Marrero (1987) (©Martinus Nijhoff Publishers, Dordrecht, reprinted by permission of Kluwer Academic Publishers). Ball (1982) initiated the study of cavitation (for isotropic hyperelastic materials). His work has been the inspiration of numerous studies: Giaquinta, Modica, & Souček (1989), Horgan & Abeyaratne (1986), Müller & Spector (1995), Müller, Tang, & Yan (1995)), Pericak-Spector & Spector (1988), Podio-Guidugli, Vergara Caffarelli, & Virga (1986), Sivaloganathan (1986), and Stuart (1985).

The analysis of the inflation of a shell given above is based on that of Sivaloganathan (1986); the analogous treatment of eversion is based on Szeri (1990). For materials that are not homogeneous, the phase-plane methods used throughout this section fail, although it might be possible to employ methods based on the Prüfer transformation. For such problems, one can use variational methods (see Ball (1982) and Sivaloganthan (1986) for hyperelastic problems), variational inequalities (see Antman (1979a)), and shooting methods (see Stuart (1985)). For problems of dilatation and inflation, one can use continuation methods like those of Gauss & Antman (1984) and Negrón-Marrero (1985).

8. Other Semi-Inverse Problems

Problems for prisms. In Sec. IX.2 we studied the qualitative behavior of transversely isotropic elastic rods subject only to terminal loads. We found that the qualitative behavior is dictated by phase portraits relating a bending couple to a suitable bending strain. Most of the orbits in these portraits are closed, and therefore correspond to periodic behavior of these two variables. The solutions as a whole, however, need not be periodic. Here we study a family of semi-inverse problems for the three-dimensional theory of elasticity that is designed to correspond to the solutions observed in the rod theory.

We assume that the body $\mathcal{B}$ is an infinite prism

$$\mathcal{B} = \Omega \times \operatorname{span}\{\boldsymbol{k}\} \tag{8.1}$$

where Ω is a domain in $\mathbb{E}^2$. We assume that the material properties of the body do not vary with the coordinate z^3 along $\boldsymbol{k}$. We seek deformations $\boldsymbol{p}$ for which there is a constant vector $\boldsymbol{a}$, a real number ζ, and a constant rotation tensor $\boldsymbol{R}$ such that

$$\boldsymbol{p}(\boldsymbol{z} + \zeta\boldsymbol{k}) = \boldsymbol{R} \cdot [\boldsymbol{p}(\boldsymbol{z}) + \boldsymbol{a}]. \tag{8.2}$$

This deformation has the property that $\boldsymbol{C}$ has period ζ in the $\boldsymbol{k}$-direction. An important special case of this deformation (which Ericksen (1980b) calls a deformation of *St. Venant type*) is that in which there are functions $\zeta \mapsto \boldsymbol{a}(\zeta), \boldsymbol{R}(\zeta)$ such that (8.2) holds for all ζ.

We assume that the body is subject to zero body force. Then the equilibrium equations imply that the resultant force $\boldsymbol{n}(z^3)$ and the resultant torque $\boldsymbol{h}(z^3)$ about any fixed point $\boldsymbol{c}$ acting across the section z^3 are independent of z^3:

$$\boldsymbol{n}(z^3) \equiv \int_\Omega \boldsymbol{T}(\boldsymbol{z}) \cdot \boldsymbol{k}\, dz^1\, dz^2 = \text{const.}, \tag{8.3a}$$

$$\boldsymbol{h}(z^3) \equiv \int_\Omega [\boldsymbol{p}(\boldsymbol{z}) - \boldsymbol{c}] \times \boldsymbol{T}(\boldsymbol{z}) \cdot \boldsymbol{k}\, dz^1\, dz^2 = \text{const.} \tag{8.3b}$$

8.4. Exercise. Make judicious choices of the test functions in the Principle of Virtual Power to give a direct proof of (8.3).

We want to determine some simple properties of deformations satisfying (8.2). First we suppose that $\boldsymbol{R} \neq \boldsymbol{I}$. Then (by Euler's Theorem on Rigid Motions) there is a unique line of vectors $\boldsymbol{b}$, forming the axis of rotation, that is invariant under $\boldsymbol{R}$, i.e., $\boldsymbol{R}\cdot\boldsymbol{b}=\boldsymbol{b}$. These vectors span the null space of $\boldsymbol{I}-\boldsymbol{R}$. The set of vectors orthogonal to the axis of rotation have the form $(\boldsymbol{R}^*-\boldsymbol{I})\cdot\boldsymbol{d}$ where $\boldsymbol{d}$ is an arbitrary vector, because by the Alternative Theorem XVII.2.20, the range of $\boldsymbol{R}^*-\boldsymbol{I}$ is the orthogonal complement of the null space of its adjoint $\boldsymbol{R}-\boldsymbol{I}$. Thus $\boldsymbol{a}$ admits the orthogonal decomposition $\boldsymbol{a}=\boldsymbol{b}+(\boldsymbol{R}^*-\boldsymbol{I})\cdot\boldsymbol{d}$, which we substitute into (8.2) to obtain

$$\boldsymbol{p}(\boldsymbol{z}+\zeta\boldsymbol{k})-\boldsymbol{d}=\boldsymbol{R}\cdot[\boldsymbol{p}(\boldsymbol{z})-\boldsymbol{d}]+\boldsymbol{b}. \tag{8.5}$$

Therefore $\boldsymbol{F}(\boldsymbol{z}+\zeta\boldsymbol{k})=\boldsymbol{R}\cdot\boldsymbol{F}(\boldsymbol{z})$. It follows from frame-indifference (see (XII.11.14)) that

$$\begin{aligned}\boldsymbol{T}(\boldsymbol{z}+\zeta\boldsymbol{k})&=\hat{\boldsymbol{T}}(\boldsymbol{F}(\boldsymbol{z}+\zeta\boldsymbol{k}),z^1,z^2)=\hat{\boldsymbol{T}}(\boldsymbol{R}\cdot\boldsymbol{F}(\boldsymbol{z}),z^1,z^2)\\&=\boldsymbol{R}\cdot\hat{\boldsymbol{T}}(\boldsymbol{F}(\boldsymbol{z}),z^1,z^2)=\boldsymbol{R}\cdot\boldsymbol{T}(\boldsymbol{z}).\end{aligned} \tag{8.6}$$

Thus $\boldsymbol{n}(z^3+\zeta)=\boldsymbol{R}\cdot\boldsymbol{n}(z^3)$. It then follows from (8.3a) that

$$\boldsymbol{n}(z^3)=\boldsymbol{R}\cdot\boldsymbol{n}(z^3)\quad\forall z^3, \tag{8.7}$$

which, together with (8.3a), implies that $\boldsymbol{n}$ is a constant vector parallel to the axis $\boldsymbol{b}$ of rotation. Let us now take $\boldsymbol{c}=\boldsymbol{d}$. Then we likewise find from (8.3b) that

$$\begin{aligned}\boldsymbol{h}(z^3)=\boldsymbol{h}(z^3+\zeta)&=\int_\Omega\{\boldsymbol{R}\cdot[\boldsymbol{p}(\boldsymbol{z})-\boldsymbol{d}]+\boldsymbol{b}\}\times\boldsymbol{R}\cdot\boldsymbol{T}(\boldsymbol{z})\cdot\boldsymbol{k}\,dz^1\,dz^2\\&=\boldsymbol{R}\cdot\boldsymbol{h}(z^3)+\boldsymbol{b}\times\boldsymbol{n}=\boldsymbol{R}\cdot\boldsymbol{h}(z^3)\quad\forall z^3.\end{aligned} \tag{8.8}$$

In summary, if $\boldsymbol{R}\neq\boldsymbol{I}$ and if $\boldsymbol{c}=\boldsymbol{d}$, then $\boldsymbol{n}$ and $\boldsymbol{h}$ are constant vectors parallel to the axis of rotation.

8.9. Exercise. Prove that if $\boldsymbol{R}=\boldsymbol{I}$ and if $\boldsymbol{n}\neq\boldsymbol{0}$, then $\boldsymbol{c}$ can be chosen so that $\boldsymbol{a}$, $\boldsymbol{h}$, and $\boldsymbol{n}$ are all parallel.

8.10. Research Problem. Suppose that Ω is a disk and that the material of $\mathcal{B}$ is transversely isotropic with respect to the axis $\boldsymbol{k}$. Determine the qualitative behavior of the material centerline in deformations of the form (8.2). (Cf. Sec. XIV.8.)

This material is based on the work of Ericksen (1980b). For analyses of the St. Venant deformation by using similar ideas, see Ball (1977b), Ericksen (1977a,b, 1983), Muncaster (1979), and the treatment in Sec. XIV.8.

Another problem in cylindrical coordinates. We now briefly discuss a variant of the deformations described in Sec. 3. Again we take $\mathbf{x}\equiv(s,\phi,z)$, but now we consider

$$\tilde{\boldsymbol{p}}(\mathbf{x})=sf(\phi)\boldsymbol{k}_1(\mathbf{x})+[sh(\phi)+\delta z]\boldsymbol{k}_3, \tag{8.11}$$

where

$$\begin{aligned}\boldsymbol{k}_1(\mathbf{x})&\equiv\boldsymbol{k}^1(\mathbf{x})\equiv\cos\omega(\mathbf{x})\boldsymbol{i}_1+\sin\omega(\mathbf{x})\boldsymbol{i}_2,\\\boldsymbol{k}_2(\mathbf{x})&\equiv\boldsymbol{k}^2(\mathbf{x})\equiv-\sin\omega(\mathbf{x})\boldsymbol{i}_1+\cos\omega(\mathbf{x})\boldsymbol{i}_2,\\\boldsymbol{k}_3&\equiv\boldsymbol{k}^3\equiv\boldsymbol{i}_3,\\\omega(\mathbf{x})&\equiv g(\phi)+\alpha\ln s.\end{aligned} \tag{8.12}$$

The reference state is characterized by $f=1$, $g(\phi)=\phi$, $h=0$, $\alpha=0$, $\delta=1$. Setting

$$\boldsymbol{j}_1(\phi)\equiv\boldsymbol{j}^1(\phi)\equiv\cos\phi\,\boldsymbol{i}_1+\sin\phi\,\boldsymbol{i}_2,\quad\boldsymbol{j}_2(\phi)\equiv\boldsymbol{j}^2(\phi)\equiv-\sin\phi\,\boldsymbol{i}_1+\cos\phi\,\boldsymbol{i}_2,\quad\boldsymbol{j}_3\equiv\boldsymbol{j}^3\equiv\boldsymbol{i}_3, \tag{8.13}$$

we readily find that

$$\begin{aligned}(8.14)\qquad \boldsymbol{F}(\tilde{\boldsymbol{z}}(\mathbf{x})) &= [f(\phi)\boldsymbol{k}_1(\mathbf{x}) + \alpha f(\phi)\boldsymbol{k}_2(\mathbf{x}) + h(\phi)\boldsymbol{k}_3]\boldsymbol{j}^1(\phi)\\ &\quad + [f'(\phi)\boldsymbol{k}_1(\mathbf{x}) + f(\phi)g'(\gamma)\boldsymbol{k}_2(\mathbf{x}) + h'(\phi)\boldsymbol{k}_3]\boldsymbol{j}^2(\phi) + +\delta\boldsymbol{k}_3\boldsymbol{j}^3\\ &\equiv F^i{}_q(s)\boldsymbol{k}_i(\mathbf{x})\boldsymbol{j}^q(\phi),\end{aligned}$$

$$\begin{aligned}(8.15)\qquad \boldsymbol{C} &= [(1+\alpha^2)f^2 + h^2]\boldsymbol{j}^1\boldsymbol{j}^1 + [ff' + \alpha f^2 g' + hh'](\boldsymbol{j}^1\boldsymbol{j}^2 + \boldsymbol{j}^2\boldsymbol{j}^1)\\ &\quad + [(f')^2 + (fg')^2 + (h')^2]\boldsymbol{j}^2\boldsymbol{j}^2 + \delta h(\boldsymbol{j}^1\boldsymbol{j}^3 + \boldsymbol{j}^3\boldsymbol{j}^1)\\ &\quad + \delta h'(\boldsymbol{j}^2\boldsymbol{j}^3 + \boldsymbol{j}^3\boldsymbol{j}^2) + \delta^2\boldsymbol{j}^3\boldsymbol{j}^3\\ &\equiv C_{pq}(s)\boldsymbol{j}^p(\phi)\boldsymbol{j}^q(\phi),\end{aligned}$$

$$(8.16)\qquad \det \boldsymbol{F} = \delta[fg' - \alpha f']f.$$

In view of (8.16), we naturally restrict our variables to satisfy

$$(8.17)\qquad f > 0, \quad \delta > 0, \quad fg' > \alpha f'.$$

8.18. Exercise. Describe geometrically the effect of (8.11) on a wedge. Derive (8.14)–(8.16). Find the principal invariants of $\boldsymbol{C}$.

We decompose $\boldsymbol{T}$ like $\boldsymbol{F}$:

$$(8.19)\qquad \boldsymbol{T}(\tilde{\boldsymbol{z}}(\mathbf{x})) = T_i{}^q(\tilde{\boldsymbol{z}}(\mathbf{x}))\boldsymbol{k}^i(\mathbf{x})\boldsymbol{j}_q(\phi).$$

We assume that the unconstrained constitutive functions (1.3) are such that the components of $\boldsymbol{T}$ in (8.19) depend only on ϕ. (Isotropy ensures this for a material whose components in (8.19) depend explicitly on $\mathbf{x}$ only through ϕ.) In this case, the equations of equilibrium for zero body force, i.e., the components of $\nabla \cdot \boldsymbol{T}^* = \mathbf{0}$, reduce to the system of ordinary differential equations:

$$(8.20)\qquad \frac{dT_1{}^2}{d\phi} + T_1{}^1 - \alpha T_2{}^1 - g'T_2{}^2 = 0,$$

$$(8.21)\qquad \frac{dT_2{}^2}{d\phi} + \alpha T_1{}^1 + T_2{}^1 + g'T_1{}^2 = 0,$$

$$(8.22)\qquad \frac{dT_3{}^2}{d\phi} + T_3{}^1 = 0.$$

8.23. Exercise. Derive (8.20)–(8.22) and find the three nontrivial components of the symmetry condition (XII.7.27).

We may now substitute our constitutive equations for unconstrained materials into (8.20)–(8.22) to get a system of three second-order ordinary differential equations for f, g, h, α, δ. If the constitutive functions do not depend explicitly on ϕ, then this system is autonomous. A large variety of suitable subsidiary conditions can be appended to this system (see Antman (1983a)). It seems likely that simplified versions of the methods of Antman (1983a) or of variational methods can be used to prove that suitable boundary-value problems have weak solutions and that these weak solutions are classical.

In view of (8.16), the defining requirement for incompressible media that $\det \boldsymbol{F} = 1$ reduces to an explicit expression for g' in terms of f, f', α, δ. Here, in contrast to the problems treated in Sec. 4, we confront a bona fide system of ordinary differential equations.

The deformation (8.11) is a generalization of that studied by Fu, Rajagopal, & Szeri (1990) and Rajagopal & Carroll (1992), who took $\alpha = 0, h = 0$.

We can also formulate an analogous deformation for a cone. In the notation of Sec. 7, we consider a deformation of the form

$$(8.24)\qquad \tilde{\boldsymbol{p}}(\mathbf{x}) = sf(\theta)\boldsymbol{a}_1(\theta, \phi).$$

8.25. Problem. Analyze deformations of the form (8.24).

9. Universal and Non-Universal Deformations

We have studied a variety of specific static problems in which the form of the deformation is strongly restricted. Now any given deformation is a solution of the equilibrium equations provided that the body force $\boldsymbol{f}$ be adjusted to accommodate it. The class of physically natural body forces, however, is quite small, encompassing merely gravitational forces, forces due to reversed acceleration, and forces of electromagnetic origin. We exclude deformations that must be maintained by artificial body forces by limiting our attention to those that can be maintained entirely by tractions on the boundary of the body. These deformations are termed *controllable.*

Among the controllable deformations are those that can occur for every material in a given class. Such deformations are called *universal* for that class. For example, consider the affine static deformations, for which $\boldsymbol{F}$ is constant, of homogeneous elastic bodies. Since $\hat{\boldsymbol{T}}(\boldsymbol{F})$ is consequently constant, the equilibrium equation $\nabla \cdot \hat{\boldsymbol{T}}^*(\boldsymbol{F}) = \boldsymbol{0}$ is automatically satisfied. Thus these affine deformations are universal for such materials. (In fact, they are the only universal deformations, as we show below.) In these deformations, the constitutive function $\hat{\boldsymbol{T}}$ enters only through the boundary conditions. It is this feature that makes universal deformations especially useful for determining material properties: One can subject a homogeneous elastic body to a given affine deformation $\boldsymbol{F}$ and measure the traction induced on the boundary. From the tractions, one can readily determine the stress corresponding to the (constant) $\boldsymbol{F}$. (This procedure was actually carried out with great success by Rivlin & Saunders (1951) for rubber, an incompressible homogeneous elastic material.) Clearly, there is no reason to expect that a non-universal deformation can be maintained by surface tractions alone, so that this experimental process could not be carried out for such a deformation. Alternative procedures relying on a prescription of surface tractions or other data would be tantamount to solving a boundary-value problem, with the attendant difficulties described in Sec. 2.

It is clear that if $\mathcal{U}(\mathcal{M})$ is the class of universal deformations of a class $\mathcal{M}$ of materials and if $\mathcal{N}$ is a subclass of $\mathcal{M}$: $\mathcal{N} \subset \mathcal{M}$, then $\mathcal{U}(\mathcal{M})$ is a subclass of $\mathcal{U}(\mathcal{N})$: $\mathcal{U}(\mathcal{M}) \subset \mathcal{U}(\mathcal{N})$. Thus, as we show below, the most powerful characterizations of universal deformations are those that apply to the most restricted class of materials. Let us remark that many of these comments readily carry over to dynamical problems.

We begin our study of the characterizations of universal deformations with the most elementary of theorems:

9.1. Theorem. *All universal static deformations of homogeneous elastic bodies are affine deformations.*

Proof. Let $\boldsymbol{F}$ denote such a deformation. Then the equilibrium equation in coordinates is

$$\frac{\partial \hat{T}_i{}^j}{\partial F^k{}_l}(\boldsymbol{F}) \frac{\partial F^k{}_l}{\partial xj} = 0, \tag{9.2a}$$

which is equivalent to

$$ab : \frac{\partial \hat{\boldsymbol{T}}}{\partial \boldsymbol{F}}(\boldsymbol{F}) : [\boldsymbol{F}_z \cdot \boldsymbol{b}] = 0 \quad \forall \boldsymbol{a}, \boldsymbol{b}. \tag{9.2b}$$

We write (9.2b) as $\mathsf{A} \therefore \boldsymbol{F}_z = 0$. The arbitrariness of the material ensures that the third-order tensor A is arbitrary. Hence it follows that $\boldsymbol{F}_z = \mathbf{O}$, so that $\boldsymbol{F}$ is constant. □

9.3. Problem. Find all universal static deformations of homogeneous elastic bodies that satisfy the Strong Ellipticity Condition.

In view of the remarks preceding its statement, Theorem 9.1 is a consequence of

9.4. Theorem. *All universal static deformations of homogeneous, isotropic elastic bodies are affine deformations.*

Sketch of proof. We substitute (1.7) into the equilibrium equation. The arbitrariness of the functions $\beta_{-1}, \beta_0, \beta_1$ imposes several restrictions on $\boldsymbol{F}$ and the invariants $\iota(\boldsymbol{C})$. These, together with the restrictions coming from the further requirement that $\boldsymbol{C}$ satisfy the compatibility equations (XII.3.21a), yield a set of conditions that force $\boldsymbol{F}$ to be constant. The details of the proof are given by Wang & Truesdell (1973, Sec. IV.2). □

This theorem is in turn a special case of the following more powerful theorem of Ericksen (1955b):

9.5. Theorem. *All universal static deformations of homogeneous, isotropic hyperelastic bodies are affine deformations.*

For incompressible materials, the arbitrariness of the pressure p allows a much richer class of deformations. We summarize the situation as it is now understood:

9.6. Theorem. *All universal static deformations of homogeneous, isotropic, incompressible, hyperelastic bodies belong to one of the following classes:*

$$\boldsymbol{F}_z = \mathbf{O}, \quad \det \boldsymbol{F} = 1, \tag{9.7}$$

$$\begin{aligned} \boldsymbol{p}(x\boldsymbol{i} + y\boldsymbol{j} + z\boldsymbol{k}) = \sqrt{2\kappa x}[\cos(\alpha y + \beta z)\boldsymbol{i} + \sin(\alpha y + \beta z)\boldsymbol{j}] \\ + (\gamma y + \delta z)\boldsymbol{k} \quad \textit{with } \kappa(\alpha\delta - \beta\gamma) = 1, \end{aligned} \tag{9.8}$$

$$\begin{aligned} \tilde{\boldsymbol{p}}(s, \phi, z) = \tfrac{1}{2}\kappa s^2 \boldsymbol{i} + (\alpha\phi + \beta z)\boldsymbol{j} + (\gamma\phi + \delta z)\boldsymbol{k} \\ \textit{with } \kappa(\alpha\delta - \beta\gamma) = 1, \end{aligned} \tag{9.9}$$

$$\begin{aligned} \tilde{\boldsymbol{p}}(s, \phi, z) = \sqrt{\kappa s^2 + \lambda}[\cos(\alpha\phi + \beta z)\boldsymbol{i} + \sin(\alpha\phi + \beta z)\boldsymbol{j}] \\ + (\gamma\phi + \delta z)\boldsymbol{k} \quad \textit{with } \kappa(\alpha\delta - \beta\gamma) = 1, \end{aligned} \tag{9.10}$$

$$\tilde{\boldsymbol{p}}(s, \theta, \phi) = \sqrt{\kappa \pm s^3}\{\sin(\pm\theta)[\cos\phi\boldsymbol{i} + \sin\phi\boldsymbol{j}] + \cos\theta\boldsymbol{k}\}, \tag{9.11}$$

$$\tilde{\boldsymbol{p}}(s,\phi,z) = \kappa s[\cos(\alpha \ln s + \beta\phi)\boldsymbol{i} + \sin(\alpha \ln s + \beta\phi)\boldsymbol{j}] + \delta\boldsymbol{k} \quad \text{with } \kappa^2\beta\delta = 1, \tag{9.12}$$

and deformations for which there is a functional relationship among the twelve quantities

$$C_m, \quad \boldsymbol{e}_k \cdot (\nabla \times \boldsymbol{e}_l) \tag{9.13}$$

where the C_m are the eigenvalues and the $\boldsymbol{e}_k$ the corresponding unit eigenvectors of $\boldsymbol{C}$.

We have studied many deformations of the forms (9.7)–(9.12). A concrete characterization of the last class, associated with (9.13), is not known. (It is conceivably empty.)

Most of Theorem 9.6 is due to Ericksen (1954). His analysis was a tour de force combining mechanics and geometry. He showed that universal deformations had to be of the form (9.7)–(9.11) or else satisfy certain geometric restrictions. Subsequent refinements of these restrictions, leading to class (9.12) and the last class, were carried out by Fosdick (1966, 1971), Fosdick & Schuler (1969), Kafadar (1972), Klingbeil & Shield (1966), Marris (1975, 1982), Marris & Shiau (1970), and Singh & Pipkin (1965). For a study of universal deformations in the dynamics of simple materials, see Carroll (1967).

One way to escape the straitjacket that Theorems 9.5 and 9.6 impose on possible deformations is to relax the restrictions on the deformation by allowing semi-inverse deformations, discussed in Secs. 5–8. The existence theorems for these problems assert that such solutions are possible for every material meeting the Strong Ellipticity Condition and associated growth conditions. A second method is to prescribe a class of deformations and restrict the material response to ensure that such deformations are possible. We now describe some examples. In the next section we describe an example that combines these two approaches.

Consider the deformation of a slab

$$\mathcal{B} \equiv \{\boldsymbol{z} = x\boldsymbol{i} + y\boldsymbol{j} + s\boldsymbol{k} : 0 < s < 1\} \tag{9.14}$$

in which each layer at height s stays at its own height and undergoes a rigid deformation. Such a deformation has the form

$$\boldsymbol{p}(x,y,z) = u(s)\boldsymbol{i} + v(s)\boldsymbol{j} + \boldsymbol{\Omega}(s)\cdot\boldsymbol{z} \tag{9.15}$$

where $\boldsymbol{\Omega}(s)$ is a rotation about the $\boldsymbol{k}$-axis through an angle $\omega(s)$. With respect to the basis $\{\boldsymbol{i},\boldsymbol{j},\boldsymbol{k}\}$, $\boldsymbol{\Omega}(s)$ has the matrix

$$[\boldsymbol{\Omega}] = \begin{pmatrix} \cos\omega & -\sin\omega & 0 \\ \sin\omega & \cos\omega & 0 \\ 0 & 0 & 1 \end{pmatrix}. \tag{9.16}$$

Without loss of generality, we may assume that $u(0) = v(0) = \omega(0) = 0$. We prescribe $u(1)$, $v(1)$, $\omega(1)$. We seek ordinary differential equations for u, v, ω.

$\boldsymbol{F}(\boldsymbol{z})$ has the matrix

$$\begin{pmatrix} \cos\omega(s) & -\sin\omega(s) & u'(s) - x\omega'(s)\sin\omega(s) - y\omega'(s)\cos\omega(s) \\ \sin\omega(s) & \cos\omega(s) & v'(s) + x\omega'(s)\cos\omega(s) - y\omega'(s)\sin\omega(s) \\ 0 & 0 & 1 \end{pmatrix}. \tag{9.17}$$

$C(z)$ has the invariants

$$\begin{aligned} \mathrm{I}_C(z) = \mathrm{II}_C(z) &= 3 + [u'(s) - x\omega'(s)\sin\omega(s) - y\omega'(s)\cos\omega(s)]^2 \\ &\quad + [v'(s) + x\omega'(s)\cos\omega(s) - y\omega'(s)\sin\omega(s)]^2, \\ \mathrm{III}_C &= 1. \end{aligned} \tag{9.18}$$

We assume that the slab is composed of an isotropic incompressible elastic material. Since F depends on x and y, we cannot expect the resulting equations of equilibrium to be independent of x and y. Thus, in general, it is impossible to obtain ordinary differential equations for u, v, ω. One way to avoid this difficulty is to restrict (9.15) further by requiring that $\omega' = 0$. (The dynamical problem in which $\omega = 0$ is treated in Sec. 13.) Alternatively, we can restrict our attention to materials for which x and y do not enter the equilibrium equations. Such a material is the Mooney-Rivlin material (2.50).

9.19. Exercise. Show that the equations of equilibrium for the Mooney-Rivlin material are independent of x and y. Obtain and solve the ordinary differential equations for u, v, ω.

The analysis of the deformation (9.14) for $\omega' = 0$ was carried out for arbitrary incompressible materials and for $\omega' \neq 0$ for Mooney-Rivlin materials by Rajagopal & Wineman (1984, 1985). This deformation was inspired by steady solutions of the Navier-Stokes equations. The degenerate character of the Mooney-Rivlin constitutive equation gives it a strong resemblance to the constitutive equation of a Newtonian fluid.

9.20. Problem. Carry out an analysis of deformations of the form

$$p(x, y, z) = f(z)(x\boldsymbol{i} + y\boldsymbol{j}) + h(z)\boldsymbol{k} \tag{9.21}$$

for compressible and incompressible isotropic materials. Show that there are materials for which the functions f and g can be found as solutions of ordinary differential equations. (See Currie & Hayes (1981).)

10. Antiplane Problems

As we have seen, the equilibrium of a nonlinearly elastic body is governed by a quasilinear system of three coupled second-order partial differential equations. Now the theory of quasilinear elliptic systems (in particular, the theory of variational problems) under the kinds of assumptions we need for nonlinear elasticity has not yet reached a level of completion comparable to that for the two degenerate cases in which there is but one independent variable and in which there is but one dependent variable. For the semi-inverse problems discussed in Secs. 3–8, the partial differential equations of elasticity reduce to ordinary differential equations and accordingly fall into the first case. A full arsenal of analytic methods, including some not emphasized in this book, are available for their analysis. Are there likewise interesting problems of elasticity governed by a single second-order partial differential equation? There is only one such class of problems of this sort, the antiplane deformations of an incompressible body, that has been extensively investigated. Although the theory has few problems that are physically illuminating or practically important, its relative simplicity makes it a very attractive laboratory in which to develop analytic techniques and insights. We now sketch the main features of the theory, which is equally useful for other branches of solid mechanics.

Let Ω be a domain in $\operatorname{span}\{\boldsymbol{i}_1, \boldsymbol{i}_2\}$ and let $\mathcal{B}$ be the cylindrical domain $\Omega \times \operatorname{span}\{\boldsymbol{k}\}$, with generators parallel to $\boldsymbol{k}$. An *antiplane (shearing) motion* of $\mathcal{B}$ has the form

$$p(z, t) = z^\alpha \boldsymbol{i}_\alpha + [z^3 + w(z^1, z^2, t)]\boldsymbol{k}. \tag{10.1}$$

Here repeated Greek indices are summed over 1,2. Thus

$$\text{(10.2a,b)} \qquad \boldsymbol{F} = \boldsymbol{I} + w_{,\alpha}\boldsymbol{k}\boldsymbol{i}^\alpha, \quad \boldsymbol{C} = \boldsymbol{I} + w_{,\alpha}w_{,\beta}\boldsymbol{i}^\alpha\boldsymbol{i}^\beta + w_{,\alpha}(\boldsymbol{k}\boldsymbol{i}^\alpha + \boldsymbol{i}^\alpha\boldsymbol{k}),$$

$$\text{(10.3)} \qquad \mathrm{I}_C = 3 + w_{,1}^2 + w_{,2}^2 = \mathrm{II}_C, \quad \mathrm{III}_C = 1.$$

(We could ostensibly generalize (10.1) by replacing it with

$$\boldsymbol{p}(\boldsymbol{z},t) = \lambda^{-1/2}z^\alpha\boldsymbol{i}_\alpha + [\lambda z^3 + w(z^1,z^2,t)]\boldsymbol{k}$$

where the constant $\lambda > 0$ accounts for stretches in the axial and transverse directions. But we can include such effects in (10.1) either by choosing the stretched configuration as the reference configuration or by redefining z^k.)

Let us assume that the body force acts only in the z^3-direction: $\boldsymbol{f} = f^3\boldsymbol{k}$, and that the prescribed density ρ and body force f^3 do not depend on z^3. We assume that the body consists of an incompressible elastic material (1.5), (1.6) in which $\hat{\boldsymbol{S}}_{\mathrm{A}}$ is independent of z^3. Then (1.4)–(1.6) reduce to

$$\text{(10.4)} \qquad -p_{,\alpha} + p_{,3}w_{,\alpha} + \overset{+}{T}{}_\alpha{}^\beta{}_{,\beta} = 0,$$

$$\text{(10.5)} \qquad -p_{,3} + \overset{+}{T}{}_3{}^\beta{}_{,\beta} + f^3 = \rho w_{tt}$$

where the $\overset{+}{T}{}_q{}^i$ are the components of $\hat{\boldsymbol{T}}_{\mathrm{A}}$.

10.6. Exercise. Show that (10.4) and (10.5) imply that $p_{,3} = c$, a constant.

Equations (10.4) and (10.5) form an overdetermined system of three equations for the two unknown functions w and p. If this system has a solution, then cross differentiation of (10.4) shows that

$$\text{(10.7)} \qquad \overset{+}{T}{}_1{}^\beta{}_{,\beta 2} = \overset{+}{T}{}_2{}^\beta{}_{,\beta 1}.$$

Here and throughout this section, we assume that the $\overset{+}{T}{}_\alpha{}^\beta$ and w have as many derivatives as appear in our analysis. Conversely, if (10.7) has a solution w (say, a solution to an initial-boundary-value problem) and if Ω is simply-connected, then (10.4) can be solved for p (cf. Sec. XII.3). (If Ω is not simply-connected, then suitable additional restrictions must be imposed on the boundary values of w.) We pose the question of determining conditions on the constitutive functions to ensure that (10.7) is identically satisfied for any regular solution of (10.5).

For example, suppose that the material is isotropic, in which case (1.9) holds. Then (10.7) reduces to

$$\text{(10.8)} \qquad [(\psi_1 w_{,1}w_{,1})_{,1} + (\psi_1 w_{,1}w_{,2})_{,2}]_{,2} = [(\psi_1 w_{,2}w_{,2})_{,2} + (\psi_1 w_{,1}w_{,2})_{,1}]_{,1}.$$

Clearly, (10.8) holds for the degenerate material for which $\psi_1 = 0$. If we seek those ψ_1's for which (10.8) is identically satisfied for all w, then we find only that $\psi_1 = 0$. We can expect to get a much stronger result by requiring that (10.8) hold only for those w's that satisfy (10.5). The basic result, for the equilibrium of hyperelastic materials (1.15), is

10.9. Theorem (Knowles (1976)). *Let $f^3 = 0$ and let $(\mathrm{I}_C, \mathrm{II}_C) \mapsto W^\dagger(\mathrm{I}_C, \mathrm{II}_C)$ be independent of $\boldsymbol{z}$ and satisfy the ellipticity condition*

$$\text{(10.10)} \qquad \frac{d}{d\gamma}\left\{\gamma\left[\frac{\partial W^\dagger}{\partial \mathrm{I}_C}(3+\gamma^2, 3+\gamma^2) + \frac{\partial W^\dagger}{\partial \mathrm{II}_C}(3+\gamma^2, 3+\gamma^2)\right]\right\} > 0 \quad \forall \gamma > 0.$$

Then the body admits nontrivial antiplane equilibrium states, i.e., condition (10.7) is identically satisfied for any equilibrium solution w of (10.5), if and only if there is a constant β such that $W^\dagger$ satisfies

$$\text{(10.11)} \qquad \beta\frac{\partial W^\dagger}{\partial \mathrm{I}_C}(3+\gamma^2, 3+\gamma^2) = (1-\beta)\frac{\partial W^\dagger}{\partial \mathrm{II}_C}(3+\gamma^2, 3+\gamma^2)$$

for all $\gamma \geq 0$.

Antiplane problems, a class of semi-inverse problems, were introduced into nonlinear elasticity by Adkins (1954). Theorem 10.9 ensures that the class of admissible materials is reasonably rich, so that from the viewpoint of material response, antiplane problems are worthwhile. See Knowles (1977a) for an analogous result for compressible media. For studies of fracture for antiplane problems, see Knowles (1977b), among others.

Papers on antiplane strain exploiting the special analytic virtue of having but a single dependent variable are those of Gurtin & Temam (1981) and Bauman & Phillips (1990). (They compensate for this simplicity by considering nonconvex stored energies.) Among related papers on dynamical problems for other materials are those of Engler (1989) and Antman & Szymczak (1995) (see Sec. XV.5). For these other materials, there are (as yet) no analogs of Theorem 10.9, so that the compatibility condition must be checked on an ad hoc basis.

10.12. Problem. Obtain the analog of Theorem 10.9 for the deformation

$$\tilde{\boldsymbol{p}}(r,\theta,z) = r[\cos(\theta+g(r,z))\boldsymbol{i} + \sin(\theta+g(r,z))\boldsymbol{j}] + z\boldsymbol{k} \tag{10.13}$$

(see Tao, Rajagopal, & Wineman (1990)).

11. Perturbation Methods

Perturbation methods, the same as those described in Secs. II.8 and V.6, can be applied in an illuminating way to many three-dimensional problems of elasticity. In particular, several of the solutions of the inverse and semi-inverse problems described above can serve as trivial solutions for bifurcation problems. Here we first formulate the perturbation process for the spatial buckling and necking of an incompressible, homogeneous, isotropic elastic cylinder under prescribed end displacement. For these problems, we restrict our attention to hyperelastic materials, not only because they are physically natural, but also because they ensure that the (principal parts of the) linearized equations are self-adjoint. We describe some of the details for the problems of necking and barrelling. We then describe the instabilities of an incompressible body under constant, dead, normal traction. We finally comment briefly on complications produced by pure traction boundary-value problems.

An incompressible, homogeneous, isotropic hyperelastic cylinder under prescribed end displacement. We assume that the lateral surface of the cylinder is traction-free and that the end planes are held horizontal and are lubricated, so that there is no horizontal component of the traction on these faces. We prescribe the distance $l\delta$ between the end planes. As in Sec. 3, we use cylindrical coordinates $\mathbf{x} = (s, \phi, z)$ and take the body to be the cylinder

$$\mathcal{B} \equiv \{\tilde{z}(\mathbf{x}) : 0 \leq s < l,\ 0 \leq \phi \leq 2\pi,\ 0 < z < l\}.$$

We define the basis $\{\boldsymbol{j}_q\}$ by (3.4), but do not yet introduce the base vectors $\boldsymbol{k}_i$ as a generalization of (3.5), however, because the polar coordinates implicit in their use unduly restrict the class of deformations. We denote the position of the material point with coordinates $\mathbf{x}$ by $\tilde{\boldsymbol{p}}(\mathbf{x})$. Then

$$\boldsymbol{F}(\tilde{z}(\mathbf{x})) = \tilde{\boldsymbol{p}}_s(\mathbf{x})\boldsymbol{j}_1(\phi) + \tfrac{1}{s}\tilde{\boldsymbol{p}}_\phi(\mathbf{x})\boldsymbol{j}_2(\phi) + \tilde{\boldsymbol{p}}_z(\mathbf{x})\boldsymbol{j}_3. \tag{11.1}$$

We likewise set

$$\boldsymbol{T} = \boldsymbol{\tau}^q \boldsymbol{j}_q. \tag{11.2}$$

Then the governing boundary-value problem consists of the equilibrium equations:

$$\nabla \cdot \boldsymbol{T}^* \equiv \left(\boldsymbol{j}^1 \frac{\partial}{\partial s} + \frac{1}{s} \boldsymbol{j}^2 \frac{\partial}{\partial \phi} + \boldsymbol{j}^3 \frac{\partial}{\partial z} \right) \cdot \boldsymbol{j}_q \boldsymbol{\tau}^q = \mathbf{0}, \quad \mathbf{x} \in \tilde{\mathbf{x}}(\mathcal{B}); \tag{11.3}$$

the constraint of incompressibility:

$$\det \boldsymbol{F} \equiv \tfrac{1}{s} \left(\tilde{\boldsymbol{p}}_s \times \tilde{\boldsymbol{p}}_\phi \right) \cdot \tilde{\boldsymbol{p}}_z = 1, \quad \mathbf{x} \in \tilde{\mathbf{x}}(\mathcal{B}); \tag{11.4}$$

the constitutive equation (see (1.9)) for homogeneous, isotropic materials:

$$\boldsymbol{F}^* \cdot \boldsymbol{T} = -p\boldsymbol{I} + \zeta(\iota(\boldsymbol{C}))\boldsymbol{C} + \eta(\iota(\boldsymbol{C}))\boldsymbol{C}^2, \quad \zeta \equiv \psi_0, \quad \eta \equiv \psi_1, \tag{11.5a}$$

which for hyperelastic materials (see (1.10b), (1.15)) reduces to

$$\begin{aligned} \boldsymbol{F}^* \cdot \boldsymbol{T} &= -p\boldsymbol{I} + 2\boldsymbol{C} \cdot \frac{\partial W}{\partial \boldsymbol{C}} \\ &= -p\boldsymbol{I} + 2\boldsymbol{C} \cdot \left[\left(\frac{\partial W^\dagger}{\partial \mathrm{I}_{\boldsymbol{C}}} + 2\frac{\partial W^\dagger}{\partial \mathrm{II}_{\boldsymbol{C}}} \mathrm{I}_{\boldsymbol{C}} \right) \boldsymbol{I} - \frac{\partial W^\dagger}{\partial \mathrm{II}_{\boldsymbol{C}}} \boldsymbol{C} \right]; \end{aligned} \tag{11.5b}$$

the requirement that the lateral surface be traction-free:

$$\boldsymbol{\tau}^1 = \boldsymbol{T} \cdot \boldsymbol{j}^1 = \mathbf{0} \quad \text{for } s = 1; \tag{11.6}$$

the requirement that the ends be horizontal and separated by the distance $l\delta$:

$$\tilde{\boldsymbol{p}}(s, \phi, 0) \cdot \boldsymbol{i}_3 = 0, \quad \tilde{\boldsymbol{p}}(s, \phi, l) \cdot \boldsymbol{i}_3 = l\delta; \tag{11.7}$$

the requirement that the horizontal components of the tractions on the end faces vanish:

$$\boldsymbol{\tau}^3 \cdot \boldsymbol{i}_1 = 0 = \boldsymbol{\tau}^3 \cdot \boldsymbol{i}_2 \quad \text{for } z = 0, l; \tag{11.8}$$

and a pair of restrictions that fix the rigid motions permitted by the requirements given above:

$$\tilde{\boldsymbol{p}}(0, \phi, 0) = \mathbf{0}, \quad \tilde{\boldsymbol{p}}(1, 0, 0) \cdot \boldsymbol{i}_2 = 0. \tag{11.9}$$

We allow δ to be any positive number, so that the cylinder can be either compressed or extended. (In the latter case, it would, however, be difficult to construct devices, such as lubricated magnets, that could maintain the prescribed boundary conditions at the ends.)

For each $\delta > 0$, this boundary-value problem admits the trivial solution defined by

$$\tilde{\boldsymbol{p}}_{(0)}(\mathbf{x}) = \delta^{-1/2} s \boldsymbol{j}_1(\phi) + \delta z \boldsymbol{j}_3, \tag{11.10}$$

$$\boldsymbol{F}_{(0)} = \delta^{-1/2}(\boldsymbol{j}_1\boldsymbol{j}^1 + \boldsymbol{j}_2\boldsymbol{j}^2) + \delta \boldsymbol{j}_3\boldsymbol{j}^3, \tag{11.11}$$

$$\boldsymbol{T}_{(0)} = [-p_{(0)}\delta^{-1} + \zeta_{(0)}\delta + \eta_{(0)}\delta^3]\boldsymbol{j}^3\boldsymbol{j}_3, \tag{11.12}$$

$$p_{(0)} = \zeta_{(0)}\delta^{-1} + \eta_{(0)}\delta^{-2} \tag{11.13}$$

where

$$\zeta_{(0)} = \zeta(2\delta^{-1} + \delta^2, \delta^{-2} + 2\delta), \quad \eta_{(0)} = \eta(2\delta^{-1} + \delta^2, \delta^{-2} + 2\delta). \tag{11.14}$$

This solution is just that studied in Ex. 4.12.

We treat δ as the bifurcation parameter. Following the procedure of Sec. V.6, we introduce a small parameter ε, which measures the magnitude of deviations from the trivial state, and then expand all the variables of our boundary-value problem as Taylor polynomials in ε to the extent permitted by the smoothness of the data, which here are the functions ζ and η:

(11.15)
$$\delta = \delta_{(0)} + \varepsilon\delta_{(1)} + \frac{\varepsilon^2}{2!}\delta_{(1)} + \cdots, \quad \tilde{\boldsymbol{p}} = \tilde{\boldsymbol{p}}_{(0)} + \varepsilon\tilde{\boldsymbol{p}}_{(1)} + \frac{\varepsilon^2}{2!}\tilde{\boldsymbol{p}}_{(1)} + \cdots, \quad \text{etc.}$$

We proceed formally in that we do not spell out the function spaces in which our variables are to lie and we do not justify the validity of the expansion. (See the discussion at the end of this section.) The first-order terms satisfy the boundary-value problem

$$\nabla \cdot \boldsymbol{T}^*_{(1)} = \mathbf{0}, \quad \mathbf{x} \in \tilde{\mathbf{x}}(\mathcal{B}), \tag{11.16}$$

$$\boldsymbol{F}^{-*}_{(0)} : \boldsymbol{F}_{(1)} = 0, \tag{11.17}$$

$$\begin{aligned} \boldsymbol{T}_{(1)} = &- p_{(1)}\boldsymbol{F}^{-*}_{(0)} - \boldsymbol{F}^{-*}_{(0)} \cdot \boldsymbol{F}^*_{(1)} \cdot \boldsymbol{T}_{(0)} + \zeta_{(1)}\boldsymbol{F}_{(0)} \\ &+ \zeta_{(0)}\boldsymbol{F}^{-*}_{(0)} \cdot \boldsymbol{C}_{(1)} + \eta_{(1)}\boldsymbol{F}_{(0)} \cdot \boldsymbol{C}_{(0)} \\ &+ \eta_{(0)}\boldsymbol{F}^{-*}_{(0)} \cdot \left(\boldsymbol{C}_{(0)} \cdot \boldsymbol{C}_{(1)} + \boldsymbol{C}_{(1)} \cdot \boldsymbol{C}_{(0)}\right), \end{aligned} \tag{11.18}$$

$$\boldsymbol{T}_{(1)} \cdot \boldsymbol{j}^1 = \mathbf{0} \quad \text{for } s = 1, \tag{11.19}$$

$$\tilde{\boldsymbol{p}}_{(1)}(s, \phi, 0) \cdot \boldsymbol{i}_3 = 0, \quad \tilde{\boldsymbol{p}}_{(1)}(s, \phi, l) \cdot \boldsymbol{i}_3 = \delta_{(1)} l, \tag{11.20}$$

$$\boldsymbol{j}_1 \cdot \boldsymbol{T}_{(1)} \cdot \boldsymbol{j}^3 = 0 = \boldsymbol{j}_2 \cdot \boldsymbol{T}_{(1)} \cdot \boldsymbol{j}^3 \quad \text{for } z = 0, l, \tag{11.21}$$

$$\tilde{\boldsymbol{p}}_{(1)}(0, \phi, 0) = \mathbf{0}, \quad \tilde{\boldsymbol{p}}_{(1)}(1, \phi, 0) \cdot \boldsymbol{i}_2 = 0, \tag{11.22}$$

where $\boldsymbol{C}_{(1)} = \boldsymbol{F}^*_{(0)} \cdot \boldsymbol{F}_{(1)} + \boldsymbol{F}^*_{(1)} \cdot \boldsymbol{F}_{(0)}$, where

$$\zeta_{(1)} = \frac{\partial \zeta}{\partial \mathrm{I}_{\boldsymbol{C}}}(\iota(\boldsymbol{C}_{(0)}))\,\mathrm{tr}\,\boldsymbol{C}_{(1)} + \frac{\partial \zeta}{\partial \mathrm{II}_{\boldsymbol{C}}}(\iota(\boldsymbol{C}_{(0)}))\left[\mathrm{tr}\,\boldsymbol{C}_{(0)}\mathrm{tr}\,\boldsymbol{C}_{(1)} - \boldsymbol{C}_{(0)} : \boldsymbol{C}_{(1)}\right],$$

etc., and where δ is replaced by $\delta_{(0)}$ in all expressions for $\boldsymbol{F}_{(0)}$. (The form of (11.5) enables us to avoid using the identity

$$\text{(11.23)} \qquad \frac{\partial \boldsymbol{F}^{-*}}{\partial \varepsilon} = -\boldsymbol{F}^{-*} \cdot \frac{\partial \boldsymbol{F}^*}{\partial \varepsilon} \cdot \boldsymbol{F}^{-*},$$

coming from (XI.2.10) and (XI.211a), to differentiate $\boldsymbol{F}^{-*}$.)

For $\delta_{(1)} = 0$, problem (11.16)–(11.22) is homogeneous and admits the zero solution generated by $(\boldsymbol{p}_{(1)}, p_{(1)}) = (\mathbf{0}, 0)$. Let us suppose that this homogeneous problem also admits a (smooth) nontrivial solution $(\bar{\boldsymbol{p}}, \bar{p})$ at some value of $\delta_{(0)}$, which is an eigenvalue of the homogeneous problem. Then (11.16)–(11.22) imply that

$$\text{(11.24)} \qquad \begin{aligned} 0 &= -\int_{\mathcal{B}} \bar{\boldsymbol{p}} \cdot \nabla \cdot \boldsymbol{T}^*_{(1)}\, dv = \int_{\mathcal{B}} \boldsymbol{T}_{(1)} : \bar{\boldsymbol{F}}\, dv \\ &= \int_{\mathcal{B}} \bar{\boldsymbol{T}} : \boldsymbol{F}_{(1)}\, dv = \int_{\partial\mathcal{B}} \tilde{\boldsymbol{p}}_{(1)} \cdot \bar{\boldsymbol{T}} \cdot \boldsymbol{\nu}\, da \\ &= \delta_{(1)} l \int_{\{z \in \partial\mathcal{B} : z = l\}} \boldsymbol{j}_3 \cdot \bar{\boldsymbol{T}} \cdot \boldsymbol{j}^3\, da \end{aligned}$$

where $\bar{\boldsymbol{C}}$ and $\bar{\boldsymbol{T}}$ are obtained from $\boldsymbol{C}_{(1)}$ and $\boldsymbol{T}_{(1)}$ by replacing $\boldsymbol{F}_{(1)}$ wherever it appears by $\bar{\boldsymbol{F}}$. (This entire computation, which relies on the hyperelasticity, should properly be carried out in the context of the Principle of Virtual Power.)

11.25. Exercise. Derive (11.24).

By an analysis like that of Secs. VI.5–VI.8, we can show that the last integral in (11.24) vanishes if the eigenvalue $\delta_{(0)}$ is multiple. Let us suppose that this integral does not vanish. Then (11.24) implies that $\delta_{(1)} = 0$. This means that $\bar{\boldsymbol{p}} = \boldsymbol{p}_{(1)}$, etc. We are now faced with the eigenvalue problem (11.16)–(11.22) in the cylinder $\mathcal{B}$. The eigenvalues are critical loads for various kinds of deformations. We study this problem in the special case that the deformation is restricted to be axisymmetric. We accordingly preclude the treatment of buckling, shear, and torsional instabilities.

Necking and barrelling instabilities. For axisymmetric deformations, $\tilde{\boldsymbol{p}}$ has the form

$$\text{(11.26)} \qquad \tilde{\boldsymbol{p}}(\mathbf{x}) = f(s,z)\boldsymbol{j}_1(\phi) + h(s,z)\boldsymbol{j}_3,$$

so that (11.1) and (11.4) reduce to

$$\text{(11.27)} \qquad \boldsymbol{F} = (f_s \boldsymbol{j}_1 + h_s \boldsymbol{j}_3)\boldsymbol{j}^1 + \frac{f}{s}\boldsymbol{j}_2\boldsymbol{j}^2 + (f_z \boldsymbol{j}_1 + h_z \boldsymbol{j}_3)\boldsymbol{j}^3,$$

$$\text{(11.28)} \qquad \frac{f}{s}(f_s h_z - f_z h_s) = 1,$$

and so that

$$\text{I}_C = f_s^2 + h_s^2 + \frac{f^2}{s^2} + f_z^2 + h_z^2,$$
$$\text{II}_C = \frac{f^2}{s^2}(f_s^2 + h_s^2 + f_z^2 + h_z^2) + \frac{s^2}{f^2}. \tag{11.29}$$

Set $\boldsymbol{T} = T_i{}^q \boldsymbol{j}^i \boldsymbol{j}_q$. The constitutive equations (11.5) then reduce to

$$T_1{}^1 = -p\frac{f}{s}h_z + \zeta f_s + \eta[f_s(f_s^2 + h_s^2) + f_z(f_s f_z + h_s h_z)],$$
$$T_1{}^3 = \ \ p\frac{f}{s}h_s + \zeta f_z + \eta[f_s(f_s f_z + h_s h_z) + f_z(f_z^2 + h_z^2)],$$
$$T_2{}^2 = -p\frac{s}{f} + \zeta\frac{f}{s} + \eta\frac{f^3}{s^3}, \tag{11.30}$$
$$T_3{}^1 = \ \ p\frac{f}{s}f_z + \zeta h_s + \eta[h_s(f_s^2 + h_s^2) + h_z(f_s f_z + h_s h_z)],$$
$$T_3{}^3 = -p\frac{f}{s}f_s + \zeta h_z + \eta[h_s(f_s f_z + h_s h_z) + h_z(f_z^2 + h_z^2)],$$

the other components of $\boldsymbol{T}$ vanishing. From (11.3) (and (11.30)) we obtain

$$\partial_s(sT_1{}^1) - T_2{}^2 + s\partial_z T_1{}^3 = 0, \tag{11.31a}$$
$$\partial_s(sT_3{}^1) + s\partial_z T_3{}^3 = 0, \tag{11.31b}$$
$$p_\phi = 0. \tag{11.32}$$

Conditions (11.6)–(11.8) reduce to

$$T_1{}^1 = 0 = T_3{}^1 \quad \text{for} \quad s = 1, \tag{11.33}$$
$$h(s,0) = 0, \quad h(s,l) = l\delta, \tag{11.34}$$
$$T_1{}^3 = 0 \quad \text{for } z = 0, l. \tag{11.35}$$

Axisymmetry replaces (11.9) with

$$f(0,z) = 0, \tag{11.36}$$

which says that the axis must remain intact.

11.37. Exercise. Derive (11.27)–(11.36).

We now linearize the boundary-value problem (11.28), (11.30)–(11.36) about the trivial state (11.10)–(11.14), which is described by

$$f_{(0)} = \delta^{-1/2}s, \quad h_{(0)} = \delta z. \tag{11.38}$$

The linearization of (11.28) is

$$\partial_s[sf_{(1)}] + \delta^{-3/2}s\partial_z h_{(1)} = 0. \tag{11.39}$$

The linearization of (11.30) is

$$
\begin{aligned}
(T_1{}^1)_{(1)} &= -\,\delta^{1/2}p_{(1)} - \delta p_{(0)}\frac{f_{(1)}}{s} - \delta^{-1/2}p_{(0)}\partial_z h_{(1)} \\
&\quad + (\delta^{-1/2}Z + \delta^{-3/2}H)\left[\frac{\delta^{-3/2}}{s}\partial_s(sf_{(1)}) + \partial_z h_{(1)}\right] \\
&\quad + (\zeta_{(0)} + 3\delta^{-1}\eta_{(0)})\partial_s f_{(1)}, \\
(T_1{}^3)_{(1)} &= [\delta^{-1/2}p_{(0)} + \delta^{1/2}\eta_{(0)}]\partial_s h_{(1)} + [\zeta_{(0)} + (\delta^{-1} + \delta^2)\eta_{(0)}]\partial_z f_{(1)}, \\
(T_2{}^2)_{(1)} &= -\,\delta^{1/2}p_{(1)} + (\delta p_{(0)} + \zeta_{(0)} + 3\delta^{-1}\eta_{(0)})\frac{f_{(1)}}{s} \\
&\quad + (\delta^{-1/2}Z + \delta^{-3/2}H)\left[\frac{\delta^{-3/2}}{s}\partial_s(sf_{(1)}) + \partial_z h_{(1)}\right], \\
(T_3{}^1)_{(1)} &= [\delta^{-1/2}p_{(0)} + \delta^{1/2}\eta_{(0)}]\partial_z f_{(1)} + [\zeta_{(0)} + (\delta^{-1} + \delta^2)\eta_{(0)}]\partial_s h_{(1)}, \\
(T_3{}^3)_{(1)} &= -\,\delta^{-1}p_{(1)} - \delta^{-1/2}p_{(0)}\frac{1}{s}\partial_s(sf_{(1)}) \\
&\quad + (\delta Z + \delta^3 H)\left[\frac{\delta^{-3/2}}{s}\partial_s(sf_{(1)}) + \partial_z h_{(1)}\right] + [\zeta_{(0)} + 3\delta^2\eta_{(0)}]\partial_z h_{(1)},
\end{aligned}
\tag{11.40}
$$

with

$$
Z \equiv 2\left[\delta\frac{\partial\zeta}{\partial\,\mathrm{I}_C} + \frac{\partial\zeta}{\partial\,\mathrm{II}_C}\right], \quad H \equiv 2\left[\delta\frac{\partial\eta}{\partial\,\mathrm{I}_C} + \frac{\partial\eta}{\partial\,\mathrm{II}_C}\right], \tag{11.41}
$$

where the arguments of the derivatives of ζ are those of (11.14). Since (11.31)–(11.36) are linear in the visible variables, their linearizations are obtained by appending the subscript (1) on each of their variables. In consonance with the discussion following (11.24), we take $\delta_{(1)} = 0$, so that corresponding to (11.34) are the boundary conditions

$$
h_{(1)}(s,0) = 0 = h_{(1)}(s,l). \tag{11.42}
$$

From (11.35), (11.40), and (11.42) we obtain

$$
\partial_z f_{(1)}(s,0) = 0 = \partial_z f_{(1)}(s,l) \tag{11.43}
$$

provided that $\zeta_{(0)} + (\delta_{(0)}^{-1} + \delta_{(0)}{}^2)\eta_{(0)} \neq 0$, which we assume. We seek solutions of our linear system in the form

$$
\begin{aligned}
f_{(1)}(s,z) &= \sum_{k=1}^{\infty} f_{1k}(s)\cos\frac{k\pi z}{l}, \\
h_{(1)}(s,z) &= \sum_{k=1}^{\infty} h_{1k}(s)\sin\frac{k\pi z}{l}, \\
p_{(1)}(s,z) &= \sum_{k=1}^{\infty} p_{1k}(s)\cos\frac{k\pi z}{l}
\end{aligned}
\tag{11.44}
$$

where $f_{1k}(s)$, $h_{1k}(s)$, $p_{1k}(s)$ are to be determined. The form of $f_{(1)}$ and $h_{(1)}$ is chosen to satisfy (11.42) and (11.43). The form of $p_{(1)}$ is then dictated by any of the partial differential equations in which it appears. The cosine series do not contain terms corresponding to $k = 0$ because the linear system forces $f_{10} = 0$, $p_{10} = 0$.

11.45. Exercise. Substitute (11.44) into the linear boundary-value problem and eliminate h_{1k} and p_{1k} to get a fourth-order linear equation for f_{1k} subject to appropriate boundary conditions. Use Bessel functions to solve this boundary-value problem and find an explicit equation for the eigenvalues corresponding to the eigenfunction f_{1k}. (There are two cases. The analysis, which is lengthy, requires care and precision.) Discuss the multiplicity and the disposition of the entire set of eigenvalues.

11.46. Problem. Determine criteria ensuring that the bifurcations from the eigenvalues are supercritical and subcritical.

The treatment just given is akin to those of Sensenig (1963) and Wesołowski (1962, 1963) and is typical of those for an extensive literature on perturbation methods in nonlinear elasticity, much of it under the names of 'small deformations upon large' and 'incremental' theories. Davis (1989) and Simpson & Spector (1985) gave careful treatments of barrelling, and Spector (1984) refined the bifurcation picture of Wesołowski. See Green & Adkins (1970), Green & Zerna (1968), Ogden (1984), and Truesdell & Noll (1965) for compendia of solutions and references.

The methods used here can be extended not only to problems for compressible media, but also for problems of thermoelasticity, etc. For problems in cylindrical coordinates, separated solutions like those of (11.44) are governed by systems of coupled Bessel-type equations. Criteria ensuring that such systems can be solved in terms of Bessel functions are given by Burridge (1984). For analogous results about Legendre functions for problems in spherical coordinates, see Burridge (1969).

Instability of an incompressible body under constant normal traction. We study the deformation of a homogeneous, isotropic, incompressible body $\mathcal{B}$ (e.g., a cube) under a constant dead normal traction τ

$$\boldsymbol{T}(\boldsymbol{z})\cdot\boldsymbol{\nu}(\boldsymbol{z}) = \tau\boldsymbol{\nu}(\boldsymbol{z}) \quad \text{for } \boldsymbol{z}\in\partial\mathcal{B}. \tag{11.47}$$

We limit our attention to linear deformations of $\mathcal{B}$, i.e., deformations of the form $\boldsymbol{z}\mapsto\boldsymbol{F}\cdot\boldsymbol{z}$ where $\boldsymbol{F}$ is a constant tensor satisfying (11.4).

Substituting (11.5) for such deformations into the equilibrium equations (11.3), we find that the pressure p is constant. Thus $\boldsymbol{T}$ is constant. We assume that the normal vectors $\{\boldsymbol{\nu}(\boldsymbol{z})\}$ on $\partial\mathcal{B}$ span $\mathbb{E}^3$. In this case, the substitution of (11.5) into (11.47) yields

$$-p\boldsymbol{F}^{-*} + \zeta\boldsymbol{F} + \eta\boldsymbol{F}\cdot\boldsymbol{C} = \tau\boldsymbol{I}, \tag{11.48a}$$

which the Polar Decomposition Theorem XII.4.1 converts to

$$-p\boldsymbol{U}^{-1} + \zeta\boldsymbol{U} + \eta\boldsymbol{U}^3 = \tau\boldsymbol{R}^*. \tag{11.48b}$$

Let the $U_k\equiv\sqrt{C_k}$ be the eigenvalues of $\boldsymbol{U}$, and let $\{\boldsymbol{e}_k\}$ be a corresponding right-handed orthonormal basis of eigenvectors. Giving $\boldsymbol{U}$ its spectral representation (XI.1.34), we deduce from (11.48b) that if $\tau\neq 0$, then the proper-orthogonal $\boldsymbol{R}$ has a diagonal matrix with respect to the basis $\{\boldsymbol{e}_k\}$, whose diagonal entries must accordingly be ±1 and whose determinant must equal 1. A suitable enumeration of the eigenvectors implies that either $\boldsymbol{R}=\boldsymbol{I}$ (so that $\boldsymbol{F}=\sum U_k\boldsymbol{e}_k\boldsymbol{e}_k$) or that $\boldsymbol{R}=-\boldsymbol{e}_1\boldsymbol{e}_1-\boldsymbol{e}_2\boldsymbol{e}_2+\boldsymbol{e}_3\boldsymbol{e}_3$ (so that $\boldsymbol{F}=-U_1\boldsymbol{e}_1\boldsymbol{e}_1-U_2\boldsymbol{e}_2\boldsymbol{e}_2+U_3\boldsymbol{e}_3\boldsymbol{e}_3$). Thus the components of (11.48b) with respect to the basis $\{\boldsymbol{e}_k\}$ can be written as

$$[\zeta U_k + \eta U_k^3 - \varepsilon(k)\tau]U_k = p \quad \text{(no summation)} \tag{11.48c}$$

where $\varepsilon(k) = -1$ if $k = 1, 2$ and $\boldsymbol{R} = -\boldsymbol{e}_1\boldsymbol{e}_1 - \boldsymbol{e}_2\boldsymbol{e}_2 + \boldsymbol{e}_3\boldsymbol{e}_3$, and $\varepsilon(k) = 1$ otherwise. Here ζ and η depend on the invariants $U_1+U_2+U_3$ and $U_1U_2+U_2U_3+U_3U_1$. The system, (11.48c) together with the incompressibility condition

$$U_1U_2U_3 = 1, \tag{11.49}$$

yields four equations for the four unknowns p and the U_k.

11.50. Exercise. Carry out in detail the preceding development.

Note that the problem with $\boldsymbol{R} = -\boldsymbol{e}_1\boldsymbol{e}_1 - \boldsymbol{e}_2\boldsymbol{e}_2 + \boldsymbol{e}_3\boldsymbol{e}_3$ describes an equilibrium in which the body has been rotated about the $\boldsymbol{e}_3$-axis through an angle of π while the directions of the tractions remain are unchanged. See Fig. 11.51. Thus this equilibrium is also the solution of a different problem in which the normal traction is not constant. We discuss problems like this below. For the time being, we limit our attention to the case that $\boldsymbol{R} = \boldsymbol{I}$. This problem admits the trivial (rigid) solution $U_1 = U_2 = U_3 = 1$. If we assume that the extra stress vanishes in the reference configuration, then we find that $p = -\tau$ for the trivial solution. We now study the bifurcation of nontrivial solutions from the trivial solution by using a refinement of the singularity theory described in Secs. VI.6 and VI.8.

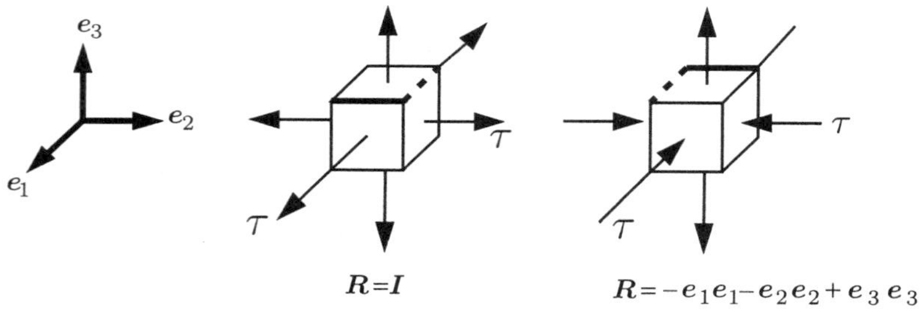

Fig. 11.51. Trivial equilibrium configuration of a cube with $\boldsymbol{R} = \boldsymbol{I}$ and a nontrivial configuration with $\boldsymbol{R} = -\boldsymbol{e}_1\boldsymbol{e}_1 - \boldsymbol{e}_2\boldsymbol{e}_2 + \boldsymbol{e}_3\boldsymbol{e}_3$.

Let us set $\mathsf{u} = (u_1, u_2, u_3)$ where $u_k \equiv \ln U_k$ and set $\mathsf{c} = (1, 1, 1)$. Then (11.48c), (11.49) becomes

$$\mathsf{h}(\mathsf{u}, \tau) = p\,\mathsf{c}, \quad \mathsf{u}\cdot\mathsf{c} \equiv u_1 + u_2 + u_3 = 0 \tag{11.52a,b}$$

where $\mathsf{h} \equiv (h_1, h_2, h_3)$ with $h_k(\mathsf{u}, \tau) \equiv [\zeta e^{u_k} + \eta e^{3u_k} - \tau]e^{u_k}$. Let $\boldsymbol{\Pi}$ be the orthogonal projection of $\mathbb{R}^3$ onto the plane defined by (11.52b):

$$\boldsymbol{\Pi}\cdot\mathsf{u} = \mathsf{u} - \tfrac{1}{3}(\mathsf{u}\cdot\mathsf{c})\,\mathsf{c}. \tag{11.53}$$

Thus u satisfies (11.52b) if and only if $\boldsymbol{\Pi}\cdot\mathsf{u} = \mathsf{u}$. Operating on (11.52a) with $\boldsymbol{\Pi}\cdot$ and with $(\mathsf{I} - \boldsymbol{\Pi})\cdot$ we reduce it to

$$\boldsymbol{\Pi}\cdot\mathsf{h}(\mathsf{u}, \tau) = \mathbf{0}, \quad (\mathsf{I} - \boldsymbol{\Pi})\cdot[p\,\mathsf{c} - \mathsf{h}(\mathsf{u}, \tau)] = \mathbf{0}.$$

Thus (11.52) has a solution $(\mathbf{u}, p)$ if and only if

$$\text{(11.54a,b,c)} \qquad \boldsymbol{\Pi} \cdot \mathbf{u} = \mathbf{u}, \quad \boldsymbol{\Pi} \cdot \mathbf{h}(\mathbf{u}, \tau) = \mathbf{0}, \quad p = \tfrac{1}{3}\mathbf{h}(\mathbf{u}, \tau) \cdot \mathbf{c}.$$

Note that (for $\varepsilon(k) = 1$) problem (11.48c), (11.49), and therefore (11.54) is invariant under permutations of the index k. In particular, let S_3 be the group of permutations of three symbols. If $\mathbf{G} \in S_3$, then $\mathbf{G} \cdot \mathbf{u}$ is a triple whose indices form a permutation of those of $\mathbf{u}$. Thus $\mathbf{h}(\mathbf{G}\cdot\mathbf{u}, \tau) = \mathbf{G}\cdot\mathbf{h}(\mathbf{u}, \tau)$, i.e., $\mathbf{h}(\cdot, \tau)$ is *equivariant under* S_3. Thus if $\mathbf{u}$ is a solution of (11.54), then so is $\mathbf{G} \cdot \mathbf{u}$.

Now we introduce a Cartesian coordinate system on the plane (11.52b) by methods like those used in the introduction of Euler angles in Sec. VIII.11: We introduce the standard basis $\mathbf{j}_1 \equiv (1, 0, 0)$, etc., for $\mathbb{R}^3$. Then the plane defined by (11.52b) is perpendicular to $\mathbf{c} \equiv \mathbf{j}_1 + \mathbf{j}_2 + \mathbf{j}_3$; it is therefore spanned by the orthonormal basis

$$\text{(11.55)} \qquad \begin{aligned} \boldsymbol{a} &\equiv \tfrac{1}{\sqrt{2}}(\mathbf{j}_1 + \mathbf{j}_2 + \mathbf{j}_3) \times \mathbf{j}_3 \equiv \tfrac{1}{\sqrt{2}}(\mathbf{j}_1 - \mathbf{j}_2), \\ \boldsymbol{b} &\equiv \tfrac{1}{\sqrt{6}}(\mathbf{j}_1 + \mathbf{j}_2 + \mathbf{j}_3) \times (\mathbf{j}_1 - \mathbf{j}_2) \equiv \tfrac{1}{\sqrt{6}}(\mathbf{j}_1 + \mathbf{j}_2 - 2\mathbf{j}_3). \end{aligned}$$

If $\mathbf{u}$ satisfies (11.52b), then it has the form $\mathbf{u} = x\boldsymbol{a} + y\boldsymbol{b}$ with

$$x = (u_1 - u_2)/\sqrt{2}, \quad y = (u_1 + u_2 - 2u_3)/\sqrt{6}.$$

We introduce the complex variable $z = x + iy$. The complex conjugate, real part, and imaginary part of any complex number w are denoted by $\bar{w}$, $\operatorname{Re} w$, and $\operatorname{Im} w$. We can express (11.54b) as an equivalent equation for z, which we write as

$$\text{(11.56)} \qquad g(z, \sigma) = 0$$

where g is complex-valued, $\sigma \equiv \tau - \tau_0$, and τ_0 is an eigenvalue of the linearization of our problem about the trivial solution. The equivariance of (11.54) under S_3 implies that (11.56) is invariant under all compositions of $z \mapsto e^{2\pi/3}z$ (rotations through $2\pi/3$) and $z \mapsto \bar{z}$ (reflections). (That is, the problem for z is equivariant under the *dihedral group* D_3 of all symmetries of the equilateral triangle. This is a manifestation of the intuitive fact that S_3 is isomorphic to D_3, which can be seen by labelling the vertices of an equilateral triangle with 1,2,3.)

To avoid technical difficulties, let us assume that $g \in C^\infty(\mathbb{C} \times \mathbb{R})$, where $\mathbb{C}$ denotes the complex plane. It can be shown that such a D_3-equivariant g has the form

$$\text{(11.57)} \qquad g(z, \sigma) = a(|z|^2, \operatorname{Re} z^3, \sigma)z + b(|z|^2, \operatorname{Re} z^3, \sigma)\bar{z}^2$$

where a and b are C^∞ functions. (Representation (11.57) is a complicated analog of the fact that an even, real-valued function f of a real variable has the form $f(x) = h(x^2)$.) Let us now study (11.57) in a neighborhood of a bifurcation point $(z, \sigma) = (0, 0)$. At such a point, $a(0,0,0) = 0$. It is shown by Golubitsky, Stewart, & Schaeffer (1988) that if $a(0,0,0) = 0$, $b(0,0,0) \neq 0$, $a_\sigma(0,0,0) \neq 0$, then (11.56) is D_3-*equivalent* to

$$\delta \sigma z + \bar{z}^2 = 0 \tag{11.58}$$

where δ =sign $[b(0,0,0)a_\sigma(0,0,0)]$: i.e., if $\mathbf{g} = (\operatorname{Re} g, \operatorname{Im} g)$, then there is a nonsingular matrix $\mathbf{S}(z,\sigma)$ and an infinitely differentiable one-to-one mapping (Z, Σ) of $\mathbb{C} \times \mathbb{R}$ onto itself such that

$$\begin{pmatrix} \operatorname{Re}\,[\delta\sigma z + \bar{z}^2] \\ \operatorname{Im}\,[\delta\sigma z + \bar{z}^2] \end{pmatrix} = \mathbf{S}(z,\sigma) \cdot \mathbf{g}\big(Z(z,\sigma), \Sigma(\sigma)\big) \tag{11.59}$$

where $\mathbf{S}, Z, \Sigma$ preserve the symmetry (cf. (VI.6.14)). Moreover, (11.58) is its own universal unfolding. Note that the hypotheses leading to (11.58) are generic.

We easily see that the solution set for (11.58) in (z, σ)-space consists of the trivial line $z = 0$ and the three nontrivial lines

$$z = -\delta\sigma, \quad z = -\delta\sigma e^{2\pi i/3}, \quad z = -\delta\sigma e^{-2\pi i/3}. \tag{11.60}$$

If we trace back the significance of each of these branches, we find that they correspond to states in which exactly two of the stretches λ_k are equal. Perhaps surprisingly, this group-theoretical analysis thus shows that there is no bifurcation to a state with three unequal stretches. A formal study of stability on the nontrivial branches, testing whether they provide local minima to the energy of a hyperelastic body restricted to affine deformations, shows that these branches are unstable (near the bifurcation point). Just as in Sec. VI.8, we could illuminate the global behavior of branches and their stability by studying a more degenerate problem, namely, that for which $b(0,0,0) = 0$. In physical terms, this condition is satisfied for a Mooney-Rivlin material.

The treatment just given is based on that of Ball & Schaeffer (1983). Also see Golubitsky, Stewart, & Schaeffer (1988). These references give detailed treatments of the degenerate problem for which $b(0,0,0) = 0$. The general problem of the dead-load traction of a cube was introduced by Rivlin (1948b). For further developments of this problem, see Rivlin (1974) and Sawyers (1976).

Comments. The treatment of the instability of the elastic cylinder under prescribed end displacement is purely formal: It has not yet been justified by an appropriate version of the Implicit Function Theorem. The difficulty, as pointed out by Ciarlet (1988) and Valent (1988), is that the amount of regularity that can be demonstrated for problems with mixed boundary conditions is typically insufficient for the hypotheses of available forms of the theorem. This difficulty does not arise in the problem with boundary condition (11.47) because the problem is finite-dimensional since we have restricted our attention to affine deformations. It would not arise in unrestricted traction boundary-value problems either, because the boundary conditions are not mixed.

Figure 11.51 illustrates another source of trouble: multiple solutions for the same dead loading. In distinguishing these solutions, the role of rotation is central.

A necessary condition for the solution of an equilibrium problem with traction boundary conditions is that the resultant force and torque on the body be zero:

$$\int_{\mathcal{B}} \boldsymbol{f}\, dv + \int_{\partial\mathcal{B}} \bar{\boldsymbol{\tau}}\, da = \mathbf{0}, \tag{11.61}$$

$$\int_{\mathcal{B}} \boldsymbol{p} \times \boldsymbol{f}\, dv + \int_{\partial\mathcal{B}} \boldsymbol{p} \times \bar{\boldsymbol{\tau}}\, da = \mathbf{0}. \tag{11.62}$$

Condition (11.62) is a source of difficulty, because it depends on the unknown $\boldsymbol{p}$. (Condition (11.61) would depend on $\boldsymbol{p}$ only if the load were not dead.) Systematic perturbation schemes for equilibrium problems must deal with this issue, which is related to the use of the Alternative Theorem. The basic method, a refinement of that outlined below in Sec. 15 in which the trivial configuration is the reference configuration, is due to Signorini (1930, 1949, 1955). Accounts of this method are given by Grioli (1962), Wang & Truesdell (1973), and Truesdell & Noll (1965). Refinements of the method are given by Bharatha & Levinson (1978), Capriz & Podio-Guidugli (1974, 1979, 1982), Green & Spratt (1954), Grioli (1983), and Rivlin & Topakoglu (1954). Standard perturbation methods work without difficulty for dynamical problems, for which (11.61) and (11.62) do not intervene.

11.63. Exercise. Expand solutions of the pure torsion problem for an incompressible cylinder, formulated in Sec. 4, in terms of the parameter β about $\beta = 0$ up to terms of order β^2. Interpret the second-order effects found.

The justification of Signorini's expansion about the reference configuration by means of the Implicit Function Theorem was carried out by Stoppelli (1954, 1955, 1957–1958) and Van Buren (1968). Stoppelli confronted the bifurcation problems associated with Fig. 11.50. Accounts of Stoppelli's work are given by Grioli (1962), Wang & Truesdell (1973), Truesdell & Noll (1965), and Valent (1988). See Ciarlet (1988) and especially Valent (1988) for careful expositions of the technical questions that underlie the use of the Implicit Function Theorem. A complete modern treatment of these bifurcation problems, using singularity theory and group theory, was carried out by Chillingworth, Marsden, & Wan (1982, 1983) and Wan & Marsden (1984). See Marsden & Hughes (1983) and Pierce (1989) for accounts of this work. The difficulties in justifying perturbation methods for dynamical problems are the same as those discussed in Chap. II.

12. Radial Motions of an Incompressible Tube

We now study a dynamic inverse problem, corresponding to the static problems formulated in Sec. 3, for a homogeneous, isotropic, incompressible infinite tube of inner radius s_0 and outer radius 1. We adopt the time-dependent versions of (3.1) and (3.16) with $\alpha = 1 = \delta$, $\beta = 0 = \gamma$, and $g = 0 = h$:

$$\tilde{\boldsymbol{p}}(\mathbf{x}, t) = f(s,t)\boldsymbol{j}_1(\phi) + z\boldsymbol{j}_3, \tag{12.1}$$

$$\boldsymbol{F} = f_s \boldsymbol{j}_1 \boldsymbol{j}^1 + s^{-1} f \boldsymbol{j}_2 \boldsymbol{j}^2 + \boldsymbol{j}_3 \boldsymbol{j}^3, \tag{12.2}$$

$$f(s,t) f_s(s,t) = s \quad \text{so that} \quad f(s,t)^2 = s^2 + r(t)^2 - s_0^2 \tag{12.3}$$

where

$$r(t) \equiv f(s_0, t). \tag{12.4}$$

These equations describe the purely radial motions of the tube. (Since we are fixing the longitudinal stretch $\delta = 1$, there is no such motion for a solid cylinder, for which $s_0 = 0$.)

Let us assume that the tractions on the two cylindrical surfaces are prescribed:

$$(12.5)\quad T_1{}^1(s_0,t) = -\frac{r(t)}{s_0}\pi_0(r(t),t), \quad T_1{}^1(1,t) = -f(1,t)\pi_1(f(1,t),t).$$

(π_0 and π_1 measure force per unit actual area.) Following Secs. 3 and 4, we find that the equations of motion reduce to

$$(12.6)\qquad [sT_1{}^1]_s - T_2{}^2 = s\rho f_{tt},$$

$$(12.7)\quad T_1{}^1 = -p\frac{f}{s} + \psi_0 f_s + \psi_1 (f_s)^3, \quad T_2{}^2 = -pf_s + \psi_0\frac{f}{s} + \psi_1\frac{f^3}{s^3}$$

where each argument of ψ_0 and ψ_1 is $(f_s)^2 + f^2/s^2 + 1$. We used (12.3) in obtaining (12.7). The remaining equations of motion and our boundary conditions show that p is independent of ϕ and z.

Using (12.3) and (3.19), we can write (12.6) as

$$(12.8)\qquad p_s = \frac{(s\overset{+}{T}_1{}^1)_s - \overset{+}{T}_2{}^2 - s\rho f_{tt}}{f}.$$

We integrate this equation from s_0 to 1, integrating the first term on the right-hand side by parts, and then use the boundary conditions (12.5). Using (12.3), we thus obtain the following ordinary differential equation for r:

$$(12.9a)\qquad \int_{s_0}^1 \left[\frac{s\rho f_{tt}}{f} + \frac{\overset{+}{T}_2{}^2}{f} - \frac{s^2\overset{+}{T}_1{}^1}{f^3}\right] ds = \pi_0 - \pi_1.$$

Explicitly integrating the first term in the integrand of (12.9a) and using (12.7), we obtain

$$(12.9b)\quad \frac{\rho}{2}\left\{[rr_{tt} + (r_t)^2]\ln\left(1+\frac{\kappa^2}{r^2}\right) - \frac{(r_t)^2\kappa^2}{r^2+\kappa^2}\right\} + \frac{\Phi'(r)}{4r} = \pi_0 - \pi_1$$

where $\kappa^2 \equiv 1 - s_0^2$ and

$$(12.9c)\qquad \Phi'(r) = 4r\int_{s_0}^1 \frac{1}{s}\left[\psi_0\left(1-\frac{s^4}{f^4}\right) + \psi_1\frac{s^2}{f^2}\left(\frac{f^4}{s^4} - \frac{s^4}{f^4}\right)\right] ds.$$

Equation (12.9b) can be written as

$$(12.9d)\qquad \frac{d}{dt}\left\{\rho\left[r^2(r_t)^2\ln\left(1+\frac{\kappa^2}{r^2}\right)\right] + \Phi(r)\right\} = 4rr_t(\pi_0 - \pi_1).$$

If $\pi_0 - \pi_1$ does not depend explicitly on t, then (12.9b) is autonomous and (12.9d) yields the energy integral for (12.9b), which gives the orbits of its phase portrait.

12.10. Exercise. Discuss how the qualitative behavior of solutions of (12.9) is influenced by the constitutive functions ψ_0 and ψ_1 when $\pi_0 - \pi_1 = 0$ and when $\pi_0 - \pi_1$ is a positive constant. Compare this problem with Ex. III.7.4.

12.11. Exercise. Formulate and analyze the purely radial motions of an incompressible, isotropic, elastic, spherical shell.

This work is based on Knowles (1960). Also see Guo & Solecki (1963). For a related analysis, see Knowles & Jakub (1965). Exercises 12.10 and 12.11 show that there are initial conditions for which solutions become unbounded. For a broad class of constitutive assumptions, Calderer (1983) went far beyond Exs. 12.10 and 12.11 in analyzing phase portraits like that generated by (12.9c) when $\pi_0 - \pi_1$ is independent of t. In particular, she exhaustively determined the qualitative behavior of all solutions, showing when solutions blow up in finite time. (For a general class of related problems, Ball (1978) showed that weak solutions do not exist for all time. For partial differential equations, in contrast to ordinary differential equations, this finding does not necessarily imply blowup. Consequently, Calderer's work illuminated Ball's analysis.) Calderer (1986) treated analogous problems for viscoelastic materials with memory.

These radial motions are universal. We discuss such motions in the next section.

13. Universal Motions of Incompressible Bodies

Recall that a motion $\boldsymbol{p}$ is controllable for an incompressible elastic body iff it is a solution of (1.4)–(1.6) with $\boldsymbol{f} = \boldsymbol{0}$, i.e., iff there is a scalar pressure field p such that

$$(13.1\text{a,b}) \qquad \nabla \cdot \left[-p \boldsymbol{p}_{\boldsymbol{z}}^{-1} + \hat{\boldsymbol{T}}_{\text{A}}(\boldsymbol{p}_{\boldsymbol{z}}, p, \cdot)^{*} \right] = \rho \boldsymbol{p}_{tt}, \quad \det \boldsymbol{p}_{\boldsymbol{z}} = 1.$$

A motion $\boldsymbol{p}$ is *universal* for (a class of) incompressible elastic bodies iff for each material in the class there is a pressure field p such that (13.1) holds.

We now use the freedom afforded by the arbitrariness of the pressure field p to construct universal motions from the universal deformations discussed in Sec. 9. We assume that the extra stress $\hat{\boldsymbol{T}}_{\text{A}}$ is independent of the pressure p. We say that a motion $\boldsymbol{p}$ of an incompressible body is *quasi-equilibrated* iff for each fixed time t the deformation $\boldsymbol{p}(\cdot, t)$ is a solution of an equilibrium problem with zero body force, i.e., iff there exists a scalar field $(\boldsymbol{z}, t) \mapsto \pi(\boldsymbol{z}, t)$ such that

$$(13.2\text{a,b}) \qquad \nabla \cdot \left[-\pi \boldsymbol{p}_{\boldsymbol{z}}^{-1} + \hat{\boldsymbol{T}}_{\text{A}}(\boldsymbol{p}_{\boldsymbol{z}}, \cdot)^{*} \right] = \boldsymbol{0}, \quad \det \boldsymbol{p}_{\boldsymbol{z}} = 1.$$

By subtracting (13.2a) from (13.1a), we find that if $\boldsymbol{p}$ is quasi-equilibrated and controllable, then there is a scalar field

$$(13.3\text{a}) \qquad \zeta = \pi - p$$

such that

$$(13.3\text{b}) \qquad \nabla \cdot \left[\zeta \boldsymbol{p}_{\boldsymbol{z}}^{-1} \right] = \rho \boldsymbol{p}_{tt}.$$

Conversely, if if $\boldsymbol{p}$ is quasi-equilibrated and if (13.3b) holds, then $\boldsymbol{p}$ is controllable. (Truesdell (1962) introduced the theory presented here.)

13.4. Exercise. Convert the left-hand side of (13.3b) to its spatial form. (For $\rho =$ const., there is an extensive literature on motions satisfying (13.3b); see Truesdell & Toupin (1960, Secs. 105–138).)

The following observation makes the class of quasi-equilibrated motions so useful.

13.5. Theorem (Wang (1970a)). *Let $\mathcal{M}$ denote any class of incompressible materials with the property that $\hat{\boldsymbol{T}}_{\mathrm{A}}$ is independent of the pressure. A motion $\boldsymbol{p}$ is universal for $\mathcal{M}$ if* (i) *there exists a scalar field ζ such that* (13.3b) *holds, and* (ii) *for each $\hat{\boldsymbol{T}}_{\mathrm{A}} \in \mathcal{M}$ there exists a scalar field π such that* (13.2a) *holds. Let $\mathcal{M}$ have the further mild property that if $\hat{\boldsymbol{T}}_{\mathrm{A}} \in \mathcal{M}$, then there is a real number $\alpha \neq 1$ such that $\alpha\hat{\boldsymbol{T}}_{\mathrm{A}} \in \mathcal{M}$. If $\boldsymbol{p}$ is universal for this $\mathcal{M}$, then* (i) *and* (ii) *hold.*

Proof. If (i) and (ii) hold, then we add (13.2a) and (13.3b) to get (13.1a) (for arbitrary $\hat{\boldsymbol{T}}_{\mathrm{A}}$) with $p = \pi - \zeta$. Conversely, if $\boldsymbol{p}$ is universal for the restricted $\mathcal{M}$, then by definition there is a scalar pressure field β such that

$$\nabla \cdot \left[-\beta \boldsymbol{p}_{\boldsymbol{z}}^{-1} + \alpha \hat{\boldsymbol{T}}_{\mathrm{A}}(\boldsymbol{p}_{\boldsymbol{z}}, p, \cdot)^{*}\right] = \rho\, \boldsymbol{p}_{tt}. \tag{13.6}$$

From (13.1a) and (13.6) we immediately find that (13.2a) and (13.3b) hold (for arbitrary $\hat{\boldsymbol{T}}_{\mathrm{A}}$) with

$$\pi = \frac{p-\beta}{1-\alpha}, \quad \zeta = \frac{\alpha p - \beta}{1-\alpha}. \quad \square \tag{13.7}$$

If we apply this result to homogeneous, isotropic, incompressible bodies, then we get universal motions corresponding to each of the definitive cases (9.7)–(9.12) by replacing each of the constants $\alpha, \ldots$ by functions of t and by adding to $\tilde{\boldsymbol{p}}$ a rigid motion.

Let us show how we can use this theory to get an alternative derivation of the equations of motion of Sec. 12. As a special case corresponding to (9.10), we use (12.1) and (12.3):

$$\tilde{\boldsymbol{p}}(\mathbf{x}, t) = f(s,t)\boldsymbol{j}_1(\phi) + z\boldsymbol{j}_3, \quad f(s,t) \equiv \sqrt{s^2 + r(t)^2 - s_0^2}. \tag{13.8}$$

Then (13.3b) reduces to

$$\frac{1}{\rho}\frac{\tilde{\zeta}_s}{s} = \frac{(r r_{tt} + r_t^2) f^2 - r^2 r_t^2}{f^4}, \quad \tilde{\zeta}_\phi = 0 = \tilde{\zeta}_z \tag{13.9}$$

where $\tilde{\zeta}(\mathbf{x}, t) \equiv \zeta(\boldsymbol{z}, t)$. We integrate (13.9) to get an expression for $\tilde{\zeta}$ depending on an arbitrary function of t. We use (13.3a) to obtain the total stress

$$\hat{\boldsymbol{T}}(\boldsymbol{F}, p) = \zeta \boldsymbol{F}^{-*} + \left[-\pi \boldsymbol{F}^{-*} + \hat{\boldsymbol{T}}_{\mathrm{A}}(\boldsymbol{F})\right]. \tag{13.10}$$

The bracketed term in (13.10) is just the total stress for the equilibrium problem, which is a special case of that described in Ex. 4.16. It can be found just as (4.6). System (13.9), (13.10) is equivalent to (12.9) once the arbitrary function of t in ζ is identified with prescribed boundary pressures.

13.11. Exercise. Derive (13.9) and prove this equivalence.

13.12. Exercise. Find equations of motion corresponding to the deformations (9.7)–(9.12). (For simplicity, do not append rigid motions to $\tilde{\boldsymbol{p}}$. These computations may be done either directly as in Sec. 12 or by use of the potential ζ as in this section.)

14. Standing Shear Waves in an Incompressible Layer

We now treat a dynamical semi-inverse problem. We take our body to be the layer consisting of all material points $\boldsymbol{z} = x\boldsymbol{i} + y\boldsymbol{j} + s\boldsymbol{k}$ for which $0 < s < 1$. We study shearing motions of the form

$$\boldsymbol{p}(\boldsymbol{z}, t) = [x + u(s,t)]\boldsymbol{i} + [y + v(s,t)]\boldsymbol{j} + s\boldsymbol{k}. \tag{14.1}$$

Then $\boldsymbol{F}$, $\boldsymbol{F}^{-*}$, and $\boldsymbol{C}$ have the following matrices with respect to $\{\boldsymbol{i}, \boldsymbol{j}, \boldsymbol{k}\}$:

$$\begin{gathered}[\boldsymbol{F}] = \begin{pmatrix} 1 & 0 & u_s \\ 0 & 1 & v_s \\ 0 & 0 & 1 \end{pmatrix}, \quad [\boldsymbol{F}^{-*}] = \begin{pmatrix} 1 & 0 & 0 \\ 0 & 1 & 0 \\ -u_s & -v_s & 1 \end{pmatrix}, \\ [\boldsymbol{C}] = \begin{pmatrix} 1 & 0 & u_s \\ 0 & 1 & v_s \\ u_s & v_s & 1 + (u_s)^2 + (v_s)^2 \end{pmatrix}. \end{gathered} \tag{14.2}$$

The principal invariants of $\boldsymbol{C}$ are $\mathrm{I}_{\boldsymbol{C}} = 3 + (u_s)^2 + (v_s)^2 = \mathrm{II}_{\boldsymbol{C}}$, $\mathrm{III}_{\boldsymbol{C}} = 1$.

We assume that the layer is an incompressible isotropic elastic body whose material properties can depend on the material point only through the coordinate s. We assume that the invariants ψ_0 and ψ_1, introduced in (1.9), are independent of p. If the layer is subject to zero body force, then the equations of motion (1.2) and (1.9) reduce to

$$-p_x + [\mu((u_s)^2 + (v_s)^2, s)u_s]_s = \rho(s)u_{tt}, \tag{14.3a}$$

$$-p_y + [\mu((u_s)^2 + (v_s)^2, s)v_s]_s = \rho(s)v_{tt}, \tag{14.3b}$$

$$p_x u_s + p_y v_s + \{-p + \psi_0 + \psi_1[1 + (u_s)^2 + (v_s)^2]\}_s = 0 \tag{14.4}$$

where the arguments of ψ_0 and ψ_1 are $\big(3+(u_s)^2+(v_s)^2, 3+(u_s)^2+(v_s)^2, s\big)$ and where

$$\mu(\chi, s) \equiv \psi_0(3+\chi, 3+\chi, s) + \psi_1(3+\chi, 3+\chi, s)[2+\chi]. \tag{14.5}$$

The term in braces in (14.4) is $\boldsymbol{k} \cdot \boldsymbol{T} \cdot \boldsymbol{k}$.

14.6. Exercise. Use the Strong Ellipticity Condition to show that the matrix

$$\begin{pmatrix} \mu + 2(u_s)^2\mu_\chi & 2u_s v_s \mu_\chi \\ 2u_s v_s \mu_\chi & \mu + 2(v_s)^2\mu_\chi \end{pmatrix} \quad \text{is positive-definite} \tag{14.7}$$

and thereby deduce that

$$\frac{\partial}{\partial \xi}[\xi\mu(\xi^2, s)] > 0, \quad \mu(\xi^2, s) > 0 \quad \forall \xi. \tag{14.8}$$

14.9. Exercise. Use (14.3) and (14.4) to prove that p must have the form

$$p(x, y, s, t) = A(t)x + B(t)y + \bar{p}(s, t). \tag{14.10}$$

Now let us prescribe the normal tractions $\boldsymbol{i}_3 \cdot \boldsymbol{T} \cdot \boldsymbol{i}^3$ on the top and bottom faces $s = 0, 1$ to be independent of x and y. Then (14.10) implies that $A = 0 = B$. It then follows from (14.4) that these normal tractions must be the same on each face. We may use (14.4) to determine p once u and v are found. Since $A = 0 = B$, Eq. (14.3) reduces to

$$[\mu((u_s)^2 + (v_s)^2, s)u_s]_s = \rho(s)u_{tt}, \tag{14.11a}$$

$$[\mu((u_s)^2 + (v_s)^2, s)v_s]_s = \rho(s)v_{tt}, \tag{14.11b}$$

which is a quasilinear hyperbolic system by virtue of (14.7).

We restrict out attention to special sets of boundary conditions. On each face $s = 0$ and $s = 1$, we require that either

$$u = 0 = v, \tag{14.12}$$

so that the face is fixed, or

$$u_s = 0 = v_s, \tag{14.13}$$

so that the shear component of the traction is zero.

We seek standing-wave solutions of (14.11) of the form

$$u(s, t) = f(s)\cos\omega t, \quad v(s, t) = f(s)\sin\omega t, \tag{14.14}$$

so that f satisfies the quasilinear eigenvalue problem

$$\frac{d}{ds}[\mu((f'(s)^2, s)f'(s)] + \omega^2\rho(s)f(s) = 0 \tag{14.15}$$

subject to the boundary conditions that for $s = 0$ and $s = 1$

$$\text{either } f = 0 \text{ or } f' = 0. \tag{14.16}$$

Let us append the natural growth conditions to (14.8). Let φ be the inverse of $\eta \mapsto \eta\mu(\eta^2, s)$. We can therefore replace (14.15) and (14.16) with the equivalent semilinear problem

$$f' = \varphi(g, s), \quad g' = -\omega^2\rho(s)f, \tag{14.17}$$

$$f = 0 \quad \text{or} \quad g = 0 \quad \text{for} \quad s = 0, 1. \tag{14.18}$$

14.19. Exercise. Use the methods of Chaps. V and VI to prove:

14.20. Theorem. *The eigenvalues* $\omega^2 = \lambda_k^0$ *of the linearization of* (14.17) *and* (14.18) *are countably infinite, are simple, and can be ordered thus:* $0 \le \lambda_0^0 < \lambda_1^0 < \lambda_2^0 < \cdots$ *with* $\lambda_k^0 \to \infty$ *as* $k \to \infty$. $\lambda_0^0 = 0$ *if and only if* $g(0) = 0 = g(1)$.

Bifurcating from the trivial branch at $(\omega^2, (f, g)) = (\lambda_k^0, (0,0))$ *is a solution branch* $\mathcal{S}(\lambda_k^0)$, *which does not touch* $\mathcal{S}(\lambda_l^0)$ *for* $k \neq l$, *which lies in* $[0,\infty) \times C^0[0,1] \times C^0[0,1]$, *and which is unbounded there. On each such branch,* $(\omega^2, (f,g))$ *is a classical solution of* (14.17) *and* (14.18), *and except at the bifurcation point* f *has exactly* k *zeros on* $(0,1)$, *each of which is simple. If* $(\omega^2, (f,g)) \in \mathcal{S}(\lambda_k^0)$, *then so is* $(\omega^2, (-f,-g))$.

If there is a number M *such that* $\mu(\eta^2, s) \le M$ *for all* η *and* s, *then there is a positive number* σ *depending only on the boundary conditions such that*

$$\min \rho\omega^2 \le \sigma M k^2 \quad \text{on } \mathcal{S}(\lambda_k^0). \tag{14.21}$$

If there is a number m *such that* $\mu(\eta^2, s) \ge m$ *for all* η *and* s, *then there is the same positive number* σ *depending only on the boundary conditions such that*

$$\max \rho\omega^2 \ge \sigma m k^2 \quad \text{on } \mathcal{S}(\lambda_k^0). \tag{14.22}$$

If $\mu(\eta^2, s) \to \infty$ *as* $\eta \to \infty$, *then* ω^2 *is unbounded on each bifurcating branch. If* $\mu(\eta^2, s) \to 0$ *as* $\eta \to \infty$, *then* $\omega^2 \to 0$ *as* $\max|f| + \max|g| \to \infty$ *on each bifurcating branch.*

The treatment of this section is based on that of Antman & Guo (1984) (©Martinus Nijhoff Publishers, Dordrecht, reprinted by permission of Kluwer Academic Publishers).

15. Linear Elasticity

To obtain the equations of linear elasticity, we study a family of initial-boundary-value problems of nonlinear elasticity depending on a small parameter ε. When $\varepsilon = 0$, the problem has a solution corresponding to a natural reference configuration. The equations of linear elasticity are formally obtained as the first perturbation of the initial-boundary-value problem with respect to ε. The process is the same as that of Sec. II.8 and is akin to that of Sec. 11, but here the expansion parameter is supplied with the problem.

Let ε be a small number. We accordingly consider (1.1)–(1.4) subject to the following restrictions: The body force intensity has the form

$$\boldsymbol{f}(\boldsymbol{z}, t) = \varepsilon \boldsymbol{f}^{(1)}(\boldsymbol{z}, t) \tag{15.1}$$

where $\boldsymbol{f}^{(1)}$ is independent of ε. For simplicity of exposition, we assume that $\partial\mathcal{B} = \mathcal{S}_0 \cup \mathcal{S}_3$ with $\mathcal{S}_0 \cap \mathcal{S}_3 = \emptyset$ (see (XII.8.1)) and with $\mathcal{S}_0$ and $\mathcal{S}_1$ independent of t. The boundary conditions are required to have the form

$$\boldsymbol{p}(\boldsymbol{z}, t) = \boldsymbol{z} + \varepsilon \bar{\boldsymbol{p}}^{(1)}(\boldsymbol{z}, t) \quad \text{for } \boldsymbol{z} \in \mathcal{S}_0, \tag{15.2a}$$

$$\boldsymbol{T}(\boldsymbol{z}, t) \cdot \boldsymbol{\nu}(\boldsymbol{z}) = \varepsilon \bar{\boldsymbol{\tau}}^{(1)}(\boldsymbol{z}, t) \quad \text{for } \boldsymbol{z} \in \mathcal{S}_3 \tag{15.2b}$$

where $\bar{\boldsymbol{p}}^{(1)}$ and $\bar{\boldsymbol{\tau}}^{(1)}$ are independent of ε. The initial conditions are required to have the form

$$\boldsymbol{p}(\boldsymbol{z},0) = \boldsymbol{z} + \varepsilon\mathring{\boldsymbol{u}}^{(1)}(\boldsymbol{z}), \quad \boldsymbol{p}_t(\boldsymbol{z},0) = \varepsilon\mathring{\boldsymbol{v}}^{(1)}(\boldsymbol{z}) \tag{15.3}$$

where $\mathring{\boldsymbol{u}}^{(1)}$ and $\mathring{\boldsymbol{v}}^{(1)}$ are independent of ε. (We could replace the right-hand sides of (15.1)–(15.3) with any continuously differentiable functions of ε that vanish at $\varepsilon = 0$.) The natural state is required to be stress-free:

$$\hat{\boldsymbol{S}}(\boldsymbol{I},\boldsymbol{z}) = \boldsymbol{O}. \tag{15.4}$$

We seek solutions of the initial-boundary-value problem (1.1)–(1.4) subject to (15.1)–(15.4) in the form

$$\boldsymbol{p}(\boldsymbol{z},t,\varepsilon) = \boldsymbol{z} + \varepsilon\boldsymbol{p}^{(1)}(\boldsymbol{z},t) + \frac{\varepsilon^2}{2!}\boldsymbol{p}^{(2)} + \cdots. \tag{15.5}$$

The number of terms in this finite Taylor expansion is one less than the number of derivatives of $\hat{\boldsymbol{S}}$. We expand all the other variables of the initial-boundary-value problem in a similar form.

We now differentiate each equation of the initial-boundary-value problem with respect to ε and then set $\varepsilon = 0$. In view of (15.4), we obtain

$$\boldsymbol{C}^{(1)} \equiv 2\boldsymbol{E}^{(1)} \equiv \boldsymbol{p}_{\boldsymbol{z}}^{(1)} + \left(\boldsymbol{p}_{\boldsymbol{z}}^{(1)}\right)^*, \tag{15.6}$$

$$\nabla \cdot \boldsymbol{S}^{(1)} + \boldsymbol{f}^{(1)} = \rho\boldsymbol{p}_{tt}^{(1)}, \tag{15.7}$$

$$\boldsymbol{S}^{(1)}(\boldsymbol{z},t) = \boldsymbol{S}_{\boldsymbol{C}}(\boldsymbol{I},\boldsymbol{z}) : \boldsymbol{C}^{(1)}(\boldsymbol{z},t) \quad \text{or} \quad S^{(1)ij} = H^{ijkl}E_{kl}^{(1)}, \tag{15.8}$$

$$\boldsymbol{p}^{(1)}(\boldsymbol{z},t) = \bar{\boldsymbol{p}}^{(1)}(\boldsymbol{z},t) \quad \text{for } \boldsymbol{z} \in \mathcal{S}_0, \tag{15.9a}$$

$$\boldsymbol{S}^{(1)}(\boldsymbol{z},t)\cdot\boldsymbol{\nu}(\boldsymbol{z}) = \bar{\boldsymbol{\tau}}^{(1)}(\boldsymbol{z},t) \quad \text{for } \boldsymbol{z} \in \mathcal{S}_3, \tag{15.9b}$$

$$\boldsymbol{p}^{(1)}(\boldsymbol{z},0) = \mathring{\boldsymbol{u}}^{(1)}(\boldsymbol{z}), \quad \boldsymbol{p}_t^{(1)}(\boldsymbol{z},0) = \mathring{\boldsymbol{v}}^{(1)}(\boldsymbol{z}). \tag{15.10}$$

Equations (15.6)–(15.10) constitute the initial-boundary-value problem for linear elasticity.

15.11. Exercise. Show that

$$H^{ijkl} = H^{jikl} = H^{ijlk} \tag{15.12}$$

so that there can be at most 32 independent components in the 81 components $\{H^{ijkl}\}$. Show that if the material is hyperelastic (see (1.10a)), then

$$H^{ijkl} = H^{klij}, \tag{15.13}$$

in which case there can be at most 21 independent components among $\{H^{ijkl}\}$. Show that if the material is isotropic, then there are scalar-valued functions λ and μ of $\boldsymbol{z}$, called the *Lamé coefficients*, such that (15.8) reduces to

$$\boldsymbol{S}^{(1)} = \lambda \operatorname{tr} \boldsymbol{E}^{(1)} + 2\mu\boldsymbol{E}^{(1)}. \tag{15.14}$$

It is customary in linear elasticity to require that (15.13) hold and that $H^{ijkl}E_{ij}^{(1)}E_{kl}^{(1)}$ be a positive-definite quadratic form. This second requirement is a local version of the convexity of the stored-energy for nonlinear hyperelasticity. It is consequently more restrictive than the strong Legendre-Hadamard condition. It is physically reasonable because it only applies to the derivative of the constitutive function $\hat{\boldsymbol{S}}$ at the natural state.

For a discussion of the lack of frame-indifference in linear elasticity, see Fosdick & Serrin (1979) and Casey & Naghdi (1985).

The theory of linear elasticity is in a very satisfactory state of completion. There is an extensive literature. The definitive exposition of the theory is that of Gurtin (1972). Standard comprehensive texts include the works of Love (1927), Lur'e (1970), Sokolnikoff (1956), and Timoshenko & Goodier (1951). A representative sampling of other works, showing the breadth of analytical work on the subject and having extensive bibliographies, would include Achenbach (1973), Carlson (1972), Fichera (1972a,b), Knops & Payne (1971), Kupradze (1963), Kupradze et al. (1968), Muskhelishvili (1966), Parton & Perlis (1981), and Villaggio (1977).

16. Commentary. Other Problems

Rivlin (1947–1949) showed that it was possible to analyze correctly set problems of nonlinear elasticity and to deduce physically illuminating information from the solutions. Parts of this chapter maintain this tradition by bringing analytic tools more powerful than those at Rivlin's disposal to bear on concrete problems. Compendia of specific solutions are given by Atkin & Fox (1980), Eringen & Suhubi (1974), Green & Adkins (1970), Green & Zerna (1968), Lur'e (1980), Ogden (1984), Truesdell & Noll (1965), and Wang & Truesdell (1973). The book of Truesdell & Noll may be consulted for historical commentary. These books give expositions of the basic theory. For a more mathematical treatment of the theory, see Ciarlet (1988), Gurtin (1981b), Hanyga (1985), Marsden & Hughes (1983), and Valent (1988).

There are several important research themes not mentioned in this chapter:

(i) Singularities. Solutions of both equilibrium and dynamic problems typically exhibit singular behavior where the boundary is not smooth. The physical (and numerical) importance of determining the nature of singularities, which depends crucially on the constitutive response of the body as our development in Sec. 5 suggests, is a compelling reason for suffering the difficulties of nonlinear material response. This problem is largely open. See Knowles & Sternberg (1973, 1974, 1975) for work in elasticity, and see Grisvard (1985, 1992) for a discussion of mathematical implications.

(ii) Contact Problems. There is a large body of work on contact problems for linear elasticity, varying from the concrete, as in Galin (1980) and Gladwell (1980), to the analytic, as in Fichera (1972b). The method of variational inequalities, presented by Duvaut & Lions (1972) and Kinderlehrer & Stampacchia (1980), is most easily applied to problems like those for simple models of membranes in which the graph of the solution is confined between two graphs representing obstacles above and below. In nonlinear elasticity, the body is described parametrically, so these methods do not directly apply.

(iii) Homogenization. There is an extensive literature, both practical and mathematical, for describing the limiting response of a linearly elastic material that is a mixture of two or more species in a periodic or random array as the dimension of a cell size goes to zero. Very little work has been carried out for nonlinear elastic materials, a notable exception being that of Geymonat, Müller, & Triantafyllides (1993), which can be consulted for references.

(iv) Phase Changes, Crystals. Much of the mathematical theory of phase changes and crystals is embodied in the assumption that the restricted convexity of the Legendre-Hadamard Condition is suspended on a bounded set. This theory is developed in the references given in the subsection on existence and regularity at the end of Sec. 2. The underlying mathematical formulation of the physical theory of phase changes in a way that is close to standard formulations from metallurgy is now being intensively cultivated. See Gurtin (1993a,b) and the references cited therein. For studies of crystals, see the references in Bhattacharya (1992) and Zanzotto (1992).

(v) St. Venant's Principle. In linear elastostatics, St. Venant's Principle roughly says that the difference in effect of two equipollent traction systems confined to a small part of a boundary becomes negligible away from the boundary. There is a long history of giving this statement a precise mathematical formulation that can be proved. The problem for nonlinear elasticity is much subtler, as the development of Sec. IX.4 suggests. See Horgan (1989) and Horgan & Knowles (1983) for discussion and references.

CHAPTER XIV

General Theories of Rods and Shells

1. Introduction. Curvilinear Coordinates

In Chap. IV we defined a *theory of rods* to be the characterization of the motion of slender solid bodies by a finite number of equations in which there is but one independent spatial variable, which we denote by s. There are several kinds of rod theories, reflecting different ways to construct them. Perhaps the most elegant are the intrinsic(ally one-dimensional) theories (Cosserat Theories), the simplest example of which is that presented in Chap. IV. In intrinsic theories, the configuration of a rod is defined as a geometric entity, equations of motion are laid down, and constitutive equations relating mechanical variables to geometrical variables are prescribed. But, as we saw in Chap. VIII, there are parts of the theory that are best developed under the inspiration of the three-dimensional theory. In the special Cosserat theory of Chap. VIII, the classical equations of motion, namely, the balances of linear and angular momentum, suffice to produce a complete theory. They are inadequate for more refined intrinsic theories. In our treatment of refined intrinsic theories in Sec. 7, we discuss the construction of the requisite additional equations of motion.

The first five sections of this chapter are devoted to the derivation of a hierarchy of rod theories, called *induced* theories, by a generalization of projection methods used in the numerical solution of partial differential equations. We pick up the thread begun in Secs. VIII.3 and VIII.4 (which should be reread). We now have what was lacking in Chap. VIII: full three-dimensional theories of stress and constitutive equations. We may regard the induced theories either as approximations of the three-dimensional theory or as constrained versions of it. The latter interpretation has the virtue that the governing equations are exact consequences of the three-dimensional theory obtained by the imposition of constitutive restrictions in the form of constraints and by the use of constitutive equations for appropriate stress resultants. A goal of our study of induced theories is to show that they possess a detailed mathematical structure consisting in the form of the equations, the form of the requirements ensuring the preservation of orientation, and the form of constitutive restrictions and to show that this structure is independent of the specific constraints used to generate the theories. The induced theories we obtain can be identified with corresponding intrinsic theories introduced in Sec. VIII.17 and in Sec. 7 below.

Other kinds of rod theories may be constructed by the asymptotic expansion (possibly formal) of the variables in one or more small thickness parameters. Corresponding to different ways in which the data may depend on the parameters are different theories of rods (and strings). In Sec. 8 we describe a rigorous asymptotic analysis of the equilibrium of a nonlinear elastic cylindrical rod. This analysis gives a distinguished position to the special Cosserat theory of Chap. VIII. An alternative asymptotic approach is treated in the context of shell theory in Sec. 14.

For geometrical reasons, the derivation of induced rod theories is actually more difficult than the derivation of induced shell theories. The reverse is true for intrinsic theories. (Of course, the equations for shell theories are more difficult to analyze than those for rod theories.) We derive the induced rod theories first.

In response to practical considerations, much of our presentation employs curvilinear coordinates. We devote the rest of this section to adapting the equations of Chap. XII to such coordinate systems. We adhere to the convention that diagonal pairs of Latin indices are summed from 1 to 3 and diagonal pairs of Greek indices are summed from 1 to 2.

We study bodies $\mathcal{B}$ that are closures of domains. We suppose that there is a continuously differentiable invertible mapping $\mathcal{B} \ni \boldsymbol{z} \mapsto \tilde{\mathbf{x}}(\boldsymbol{z}) \in \mathbb{R}^3$ such that the Jacobian

$$(1.1) \qquad \det \frac{\partial \tilde{\mathbf{x}}}{\partial \boldsymbol{z}}(\boldsymbol{z}) > 0 \quad \forall\, \boldsymbol{z} \in \mathcal{B}.$$

The function $\tilde{\mathbf{x}}$ assigns a triple of curvilinear coordinates $\mathbf{x} \equiv (x^1, x^2, x^3)$ to each $\boldsymbol{z}$ in $\mathcal{B}$. We represent the nth power of x^1, say, by $(x^1)^n$. We denote the inverse of $\tilde{\mathbf{x}}$ by $\tilde{\boldsymbol{z}}$. The Inverse Function Theorem implies that $\tilde{\boldsymbol{z}}$ is continuously differentiable and that

$$(1.2) \qquad \jmath(\mathbf{x}) \equiv \det \frac{\partial \tilde{\boldsymbol{z}}}{\partial \mathbf{x}}(\mathbf{x}) > 0 \quad \forall\, \mathbf{x} \in \tilde{\mathbf{x}}(\mathcal{B}).$$

We adopt the convention that

$$(1.3) \qquad \frac{\partial \mathsf{y}}{\partial x^k} \equiv \mathsf{y}_{,k}$$

for any function y and we adopt the following standard abbreviations (see Sec. XI.3):

$$(1.4) \qquad \boldsymbol{g}_k(\mathbf{x}) \equiv \tilde{\boldsymbol{z}}_{,k}(\mathbf{x}), \quad \boldsymbol{g}^k(\mathbf{x}) \equiv \frac{\partial \tilde{x}^k}{\partial \boldsymbol{z}}(\tilde{\boldsymbol{z}}(\mathbf{x})), \quad \jmath = (\boldsymbol{g}_1 \times \boldsymbol{g}_2) \cdot \boldsymbol{g}_3.$$

Thus $\boldsymbol{g}^k \cdot \boldsymbol{g}_l = \delta^k_l$, so that the bases $\{\boldsymbol{g}_k\}$ and $\{\boldsymbol{g}^k\}$ are dual to each other. (Indeed, $\jmath\, \boldsymbol{g}^1 = \boldsymbol{g}_2 \times \boldsymbol{g}_3$, etc.) We record the identities (XI.3.12) and (XI.3.14):

$$(1.5\text{a,b}) \qquad \nabla \equiv \boldsymbol{g}^k \frac{\partial}{\partial x^k}, \qquad \boldsymbol{g}^k \cdot \boldsymbol{g}_{l,k} = \frac{1}{\jmath} \jmath_{,l}.$$

We define

$$\tilde{\boldsymbol{p}}(\mathbf{x},t) \equiv \boldsymbol{p}(\tilde{\boldsymbol{z}}(\mathbf{x}),t), \quad \tilde{\rho}(\mathbf{x}) = \rho(\tilde{\boldsymbol{z}}(\mathbf{x})). \tag{1.6a,b}$$

Then

$$\boldsymbol{F} \equiv \frac{\partial \boldsymbol{p}}{\partial \boldsymbol{z}} = \frac{\partial \tilde{\boldsymbol{p}}}{\partial \mathbf{x}} \cdot \frac{\partial \tilde{\mathbf{x}}}{\partial \boldsymbol{z}} = \tilde{\boldsymbol{p}}_{,k}\boldsymbol{g}^k, \quad \boldsymbol{C} \equiv \boldsymbol{F}^* \cdot \boldsymbol{F} = (\tilde{\boldsymbol{p}}_{,k} \cdot \tilde{\boldsymbol{p}}_{,l})\, \boldsymbol{g}^k \boldsymbol{g}^l. \tag{1.6c,d}$$

We define the *Piola-Kirchhoff stress vectors*

$$\begin{aligned}\boldsymbol{\tau}^k(\mathbf{x},t) &\equiv \boldsymbol{T}(\tilde{\boldsymbol{z}}(\mathbf{x}),t) \cdot \boldsymbol{g}^k(\mathbf{x}) \equiv \boldsymbol{F} \cdot \boldsymbol{S}(\tilde{\boldsymbol{z}}(\mathbf{x}),t) \cdot \boldsymbol{g}^k(\mathbf{x}) \\ &= \left[\boldsymbol{g}^k(\mathbf{x}) \cdot \boldsymbol{S}(\tilde{\boldsymbol{z}}(\mathbf{x}),t) \cdot \boldsymbol{g}^l(\mathbf{x})\right] \tilde{\boldsymbol{p}}_{,l}(\mathbf{x},t),\end{aligned} \tag{1.7a}$$

from which it follows that

$$\boldsymbol{T} = \boldsymbol{\tau}^k \boldsymbol{g}_k. \tag{1.7b}$$

To interpret $\boldsymbol{\tau}^k$, we study the traction across the material surface defined by $x^3 = \text{const}$. The unit normal to this surface is

$$\frac{\tilde{\boldsymbol{z}}_{,1}(\mathbf{x}) \times \tilde{\boldsymbol{z}}_{,2}(\mathbf{x})}{|\tilde{\boldsymbol{z}}_{,1}(\mathbf{x}) \times \tilde{\boldsymbol{z}}_{,1}(\mathbf{x})|} = \frac{\boldsymbol{g}_1(\mathbf{x}) \times \boldsymbol{g}_2(\mathbf{x})}{|\boldsymbol{g}_1(\mathbf{x}) \times \boldsymbol{g}_2(\mathbf{x})|} = \frac{\boldsymbol{g}^3(\mathbf{x})}{|\boldsymbol{g}^3(\mathbf{x})|}. \tag{1.8}$$

Thus the traction across this surface is

$$\boldsymbol{T} \cdot \frac{\boldsymbol{g}^3}{|\boldsymbol{g}^3|} = \frac{\boldsymbol{\tau}^3}{|\boldsymbol{g}^3|}. \tag{1.9}$$

The presence of the $|\boldsymbol{g}^3|$ in the denominator reflects the dependence of $\boldsymbol{\tau}^k$ on the coordinate system used.

Let $\boldsymbol{z} \mapsto \boldsymbol{\eta}(\boldsymbol{z})$ be an arbitrary differentiable function and let $\overset{\vartriangle}{\boldsymbol{p}}(\mathbf{x}) \equiv \boldsymbol{\eta}(\tilde{\boldsymbol{z}}(\mathbf{x}))$. Then we have the following identities (for the *virtual stress power*):

$$\boldsymbol{T} : \frac{\partial \boldsymbol{\eta}}{\partial \boldsymbol{z}} = \boldsymbol{T} : \left[\frac{\partial \overset{\vartriangle}{\boldsymbol{p}}}{\partial \mathbf{x}} \cdot \frac{\partial \tilde{\mathbf{x}}}{\partial \boldsymbol{z}}\right] = \boldsymbol{T} : [\overset{\vartriangle}{\boldsymbol{p}}_{,k}\boldsymbol{g}^k] = [\boldsymbol{T} \cdot \boldsymbol{g}^k] \cdot \overset{\vartriangle}{\boldsymbol{p}}_{,k} = \boldsymbol{\tau}^k \cdot \overset{\vartriangle}{\boldsymbol{p}}_{,k}. \tag{1.10}$$

This identity will prove most useful in our treatment of the Principle of Virtual Power.

Let us examine the form of the linear and angular momentum balance (XII.7.21) and (XII.7.23) in terms of the $\boldsymbol{\tau}^k$. From (1.6) and (1.7) we obtain

$$\nabla \cdot \boldsymbol{T}^* = \boldsymbol{g}^l \cdot [\boldsymbol{g}_k \boldsymbol{\tau}^k]_{,l} = \frac{1}{j}[j\boldsymbol{\tau}^k]_{,k}, \tag{1.11a}$$

$$\nabla \cdot (\boldsymbol{p} \times \boldsymbol{T})^* = \frac{1}{j}[j\tilde{\boldsymbol{p}} \times \boldsymbol{\tau}^k]_{,k}. \tag{1.11b}$$

The substitution of (1.11a) into (XII.7.21) yields an alternative version of the equations of motion. The substitution of (1.11b) into (XII.7.23) yields the following alternative to the symmetry condition (XII.7.27):

$$\tilde{\boldsymbol{p}}_{,k} \times \boldsymbol{\tau}^k = \mathbf{0}. \tag{1.12}$$

The Principle of Virtual Power, boundary conditions. We study the motion of $\mathcal{B}$ subject to boundary conditions of the form (XII.8.11), (XII.8.12): We assume that there are given functions

$$\tilde{\mathbf{x}}(\partial\mathcal{B}) \times [0,\infty) \times \mathbb{R}^3 \ni (\mathbf{x},t,\mathsf{q}) \mapsto (\bar{\boldsymbol{p}}(\mathbf{x},t,\mathsf{q}), \bar{\boldsymbol{\tau}}(\mathbf{x},t,\mathsf{q})) \in \mathbb{E}^3 \times \mathbb{E}^3 \tag{1.13}$$

such that a complementary combination of components of position and traction are specified in terms of an unknown function

$$\tilde{\mathbf{x}}(\partial\mathcal{B}) \times [0,\infty) \ni (\mathbf{x},t) \mapsto \mathsf{q}(\mathbf{x},t) \in \mathbb{R}^3 \tag{1.14}$$

via

$$\tilde{\boldsymbol{p}}(\mathbf{x},t) = \bar{\boldsymbol{p}}(\mathbf{x},t,\mathsf{q}(\mathbf{x},t)), \tag{1.15a}$$

$$[\nu_k(\mathbf{x})\boldsymbol{\tau}^k(\mathbf{x},t) - \bar{\boldsymbol{\tau}}(\mathbf{x},t,\mathsf{q}(\mathbf{x},t))] \cdot \frac{\partial \bar{\boldsymbol{p}}}{\partial \mathsf{q}}(\mathbf{x},t,\mathsf{q}(\mathbf{x},t)) = 0 \tag{1.15b}$$

for $(\mathbf{x},t) \in \tilde{\mathbf{x}}(\partial\mathcal{B}) \times [0,\infty)$. Here $\nu_k \equiv \boldsymbol{\nu} \cdot \boldsymbol{g}_k$. These are the classical forms of the boundary conditions. We assume that $\tilde{\boldsymbol{p}}$ satisfies initial conditions of the form

$$\tilde{\boldsymbol{p}}(\mathbf{x},0) = \boldsymbol{p}_0(\mathbf{x}), \quad \tilde{\boldsymbol{p}}_t(\mathbf{x},0) = \boldsymbol{p}_1(\mathbf{x}). \tag{1.16a,b}$$

We have made slight and obvious changes in the notation of Chap. XII used for prescribed initial and boundary data. We specify the weak form of the equations of motion in terms of the Principle of Virtual Power (XII.9.5):

$$\begin{aligned} &\int_0^\infty \int_{\tilde{\mathbf{x}}(\mathcal{B})} \Big[\boldsymbol{\tau}^k(\mathbf{x},t) \cdot \overset{\vartriangle}{\boldsymbol{p}}_{,k}(\mathbf{x},t) - \boldsymbol{f}(\tilde{\boldsymbol{z}}(\mathbf{x}),t) \cdot \overset{\vartriangle}{\boldsymbol{p}}(\mathbf{x},t)\Big]\, dv(\mathbf{x})\, dt \\ &\quad - \int_0^\infty \int_{\tilde{\mathbf{x}}(\partial\mathcal{B})} \bar{\boldsymbol{\tau}}(\tilde{\boldsymbol{z}}(\mathbf{x}),t,\mathsf{q}(\tilde{\boldsymbol{z}}(\mathbf{x}),t)) \cdot \overset{\vartriangle}{\boldsymbol{p}}(\mathbf{x},t)\, da(\mathbf{x})\, dt \\ &- \int_{\tilde{\mathbf{x}}(\mathcal{B})} \tilde{\rho}(\mathbf{x}) \left[\int_0^\infty \tilde{\boldsymbol{p}}_t(\mathbf{x},t) \cdot \overset{\vartriangle}{\boldsymbol{p}}_t(\mathbf{x},t)\, dt + \boldsymbol{p}_1(\tilde{\boldsymbol{z}}(\mathbf{x})) \cdot \overset{\vartriangle}{\boldsymbol{p}}(\mathbf{x},0)\right] dv(\mathbf{x}) = 0 \end{aligned} \tag{1.17}$$

for all (nice enough) $\overset{\vartriangle}{\boldsymbol{p}}$ with compact support in $\tilde{\mathbf{x}}(\mathcal{B}) \times [0,\infty)$ that satisfy the boundary conditions

$$\boldsymbol{\eta}(\boldsymbol{z},t) = \frac{\partial \bar{\boldsymbol{p}}}{\partial \mathsf{q}}(\boldsymbol{z},t,\mathsf{q}(\boldsymbol{z},t)) \cdot \mathsf{a}(\boldsymbol{z},t) \quad \text{for } (\boldsymbol{z},t) \in \partial\mathcal{B} \times [0,\infty), \tag{1.18}$$

for all (nice enough) a, and that satisfy the initial conditions (1.16a) on the position. The function $\overset{\vartriangle}{\boldsymbol{p}}$ is a typical tangent vector to the constraint manifold defined by the boundary conditions (1.15) and the initial condition(1.16b). Note that the remaining boundary and initial conditions are incorporated into (1.17).

We assume that the material response of an unconstrained material is specified by the constitutive equation

$$\boldsymbol{\tau}^k(\mathbf{x},t) = \hat{\boldsymbol{\tau}}^k(\tilde{\boldsymbol{p}}^t_{,1}(\mathbf{x},\cdot),.,.,\mathbf{x}). \tag{1.19}$$

For constrained materials, we adopt constitutive equations of the form (XII.12.49)–(XII.12.51). In consonance with (1.7), we set

$$\boldsymbol{\tau}^k_{\mathrm{L}}(\mathbf{x},t) \equiv \boldsymbol{T}_{\mathrm{L}}(\tilde{\boldsymbol{z}}(\mathbf{x}),t) \cdot \boldsymbol{g}^k(\mathbf{x}), \quad \boldsymbol{\tau}^k_{\mathrm{A}}(\mathbf{x},t) \equiv \boldsymbol{T}_{\mathrm{A}}(\tilde{\boldsymbol{z}}(\mathbf{x}),t) \cdot \boldsymbol{g}^k(\mathbf{x}). \tag{1.20}$$

2. Rod Theories

Geometry of the reference configuration. We now set $x^3 \equiv s$. We may assume that the coordinate s is the arc-length parameter of a distinguished material curve, called the *base curve*, and that it ranges over an interval $[s_1, s_2]$ as $\mathbf{x}$ ranges over $\tilde{\mathbf{x}}(\mathcal{B})$, i.e., $[s_1, s_2] = \{s \colon \mathbf{x} \in \tilde{\mathbf{x}}(\mathcal{B})\}$. Alternatively, we may take s to be the arc-length parameter of a coordinate curve, often the curve $x^1 = 0 = x^2$. The *material section* $\mathcal{B}(a)$ is the set of all material points of $\mathcal{B}$ whose coordinates have the form (x^1, x^2, a), i.e.,

$$\mathcal{B}(a) \equiv \{\boldsymbol{z} \in \mathcal{B} \colon \tilde{\mathbf{x}}(\boldsymbol{z}) = (x^1, x^2, a)\}. \tag{2.1}$$

We set

$$\mathcal{A}(s) \equiv \tilde{\mathbf{x}}(\mathcal{B}(s)). \tag{2.2}$$

We assume that (1.1) holds on $\mathcal{B}$. We further assume that $\mathcal{B}(s)$ is bounded for each s in $[s_1, s_2]$. For any subinterval $\mathcal{I}$ of $[s_1, s_2]$, we define

$$\mathcal{B}(\mathcal{I}) \equiv \bigcup_{s \in \mathcal{I}} \mathcal{B}(s) \equiv \{\boldsymbol{z} \in \mathcal{B} \colon \tilde{\mathbf{x}}(\boldsymbol{z}) = (x^1, x^2, a), \;\; a \in \mathcal{I}\}. \tag{2.3}$$

We say that $\mathcal{B}$ is *rod-like* iff these three assumptions hold. We do not attempt to define a *rod* itself, which would be a member of a class of slender, solid, rod-like bodies. Nevertheless, we call $\mathcal{B}$ a *rod.*

The *lateral surface* of $\mathcal{B}$ is

$$\mathcal{L} \equiv \{\boldsymbol{z} \in \partial\mathcal{B} : s_1 < s < s_2\}. \tag{2.4}$$

The *ends* of $\mathcal{B}$ are $\mathcal{B}(s_1)$ and $\mathcal{B}(s_2)$. $\tilde{\boldsymbol{p}}(\cdot, \cdot, s, t)$ is the *configuration* of the section $\mathcal{B}(s)$ at time t.

Constraints. We generate rod theories by approximating the unknown $\tilde{\boldsymbol{p}}$ by an expression involving a finite number of unknown functions of s and t. We assume that there is a (thrice continuously differentiable) function $\mathbb{R}^N \times \tilde{\mathbf{x}}(\mathcal{B}) \times \mathbb{R} \ni (\mathsf{u}, \mathbf{w}, \tau) \mapsto \boldsymbol{\pi}(\mathsf{u}, \mathbf{w}, \tau)$ such that

$$\tilde{\boldsymbol{p}}(\mathbf{x}, t) = \boldsymbol{\pi}(\mathsf{u}(s, t), \mathbf{x}, t). \tag{2.5}$$

The simplest example of such a representation is given in Sec. VIII.4:

$$\boldsymbol{\pi}(\mathsf{u}, \mathbf{x}, t) = \boldsymbol{r}(s, t) + x^1 \boldsymbol{d}_1(s, t) + x^2 \boldsymbol{d}_2(s, t) \tag{2.6}$$

where we identify u with (the components of) $\boldsymbol{r}, \boldsymbol{d}_1, \boldsymbol{d}_2$. We require that (2.5) satisfy the position boundary conditions (1.15a) on $\mathcal{L}$ identically in u and q:

$$\boldsymbol{\pi}(\mathsf{u}, \mathbf{x}, t) = \bar{\boldsymbol{p}}(\mathbf{x}, t, \mathsf{q}) \quad \text{for } (\mathbf{x}, t) \in \tilde{\mathbf{x}}(\mathcal{L}) \times [0, \infty), \tag{2.7}$$

More generally, we can take the domain of $\boldsymbol{\pi}$ to be $\mathbb{M}^N \times \tilde{\mathbf{x}}(\mathcal{B}) \times \mathbb{R}$ where $\mathbb{M}^N$ is an N-dimensional manifold. If $\boldsymbol{r}, \boldsymbol{d}_1, \boldsymbol{d}_2$ of (2.6) are unconstrained, then the corresponding $\mathbf{u}$ lies in $\mathbb{R}^9$. If $\boldsymbol{d}_1, \boldsymbol{d}_2$ are constrained to be orthonormal, then $\mathbf{u}$ lies in the six-dimensional manifold $\mathbb{E}^3 \times \mathrm{SO}(3)$. There are but few occasions when it is useful to allow $\boldsymbol{\pi}$ to depend on τ.

We regard (2.5) as imposing an infinite number of holonomic constraints on $\mathcal{A}(s) \ni (x^1, x^2) \mapsto \tilde{\boldsymbol{p}}(\mathbf{x}, t)$ that reduce the number of its degrees of freedom to N. In the terminology of rigid-body mechanics, $\mathbf{u}$ is the generalized position field for the constrained system. To ensure that there is no redundancy in the components of $\mathbf{u}$, we may impose the *independency condition*:

$$\text{(2.8)} \qquad \text{If} \quad [\partial \boldsymbol{\pi}(\mathbf{u}, \mathbf{x}, t)/\partial \mathbf{u}] \cdot \mathbf{v} = \mathbf{0} \quad \forall\, (x^1, x^2) \in \mathcal{B}(s), \quad \text{then } \mathbf{v} = \mathbf{0}.$$

To avoid notational difficulties with total partial derivatives, we have introduced the variables $\mathbf{w}$ and τ in the specification of $\boldsymbol{\pi}$ preceding (2.5). Thus without ambiguity we may write

$$\text{(2.9)} \qquad \frac{\partial \boldsymbol{\pi}(\mathbf{u}(s,t), \mathbf{x}, t)}{\partial s} = \frac{\partial \boldsymbol{\pi}}{\partial \mathbf{u}}(\mathbf{u}(s,t), \mathbf{x}, t) \cdot \mathbf{u}_s(s,t) + \frac{\partial \boldsymbol{\pi}}{\partial w^3}(\mathbf{u}(s,t), \mathbf{x}, t).$$

The total time derivative is defined analogously.

Preservation of orientation. Let us set

$$\text{(2.10)} \quad \delta(\mathbf{u}, \mathbf{v}, \mathbf{w}, \tau) \equiv \frac{\{[\partial \boldsymbol{\pi}/\partial w^1] \times [\partial \boldsymbol{\pi}/\partial w^2]\} \cdot \{[\partial \boldsymbol{\pi}/\partial \mathbf{u}] \cdot \mathbf{v} + \partial \boldsymbol{\pi}/\partial w^3\}}{j(\mathbf{w})}$$

where the arguments of the derivatives of $\boldsymbol{\pi}$ are $\mathbf{u}, \mathbf{w}, \tau$. Note that δ is affine in $\mathbf{v}$. In consonance with (2.5), we replace the requirement that the deformation preserve orientation with its constrained form:

$$\text{(2.11)} \qquad \delta(\mathbf{u}(s,t), \mathbf{u}_s(s,t), \mathbf{x}, t) > 0 \quad \forall \mathbf{x} \in \tilde{\mathbf{x}}(\mathcal{B}).$$

Let $\mathbb{T}^N$ denote the N-dimensional (tangent-bundle) space of the derivatives $\mathbf{u}_s$ of members of $\mathbb{M}^N$. By following the proof of Theorem VIII.6.2, we obtain the following result of Antman (1976b):

2.12. Theorem. *There is a subset $\mathcal{V}(s,t)$ of the space $\mathbb{M}^N \times \mathbb{T}^N$ with the property that its section $\{(\mathbf{u}, \mathbf{v}) \in \mathcal{V}(s,t) : \mathbf{u} = \mathbf{a}\}$ is a convex subset of $\mathbb{T}^N$ for each $\mathbf{a}, s, t$ such that* (2.11) *holds if and only if*

$$\text{(2.13)} \qquad (\mathbf{u}(s,t), \mathbf{u}_s(s,t)) \in \mathcal{V}(s,t) \quad \forall\, s, t.$$

Thus (2.13) plays the same role for rod theories as the requirement $\det \boldsymbol{F} > 0$ plays for the three-dimensional theory.

The constraint (2.5) does not allow arbitrary position boundary conditions on $\mathbf{u}$ to be prescribed at the ends $s = s_1, s_2$. We therefore either restrict the position boundary conditions at the ends or else approximate arbitrary position boundary conditions at the ends by those consistent with

(2.5). We assume that the constraint (2.5) generates position boundary conditions in the following parametric form analogous to (1.15a):

$$\mathbf{u}(s_\alpha, t) = \mathbf{u}_\alpha(t, \mathbf{v}_\alpha(t)), \quad \alpha = 1, 2, \tag{2.14}$$

where the $\mathbf{v}_\alpha$ are unknowns and $\partial \mathbf{u}_\alpha / \partial \mathbf{v}_\alpha$, $\alpha = 1, 2$, has rank $\leq N$ (cf. Sec. VIII.12). Alternatively, if the body is toroidal, i.e., if the ends consist of the same material points, then we impose periodicity conditions

$$\mathbf{u}(s + s_2 - s_1, t) = \mathbf{u}(s, t). \tag{2.15}$$

Other possibilities can be handled likewise. We finally assume that initial conditions have a form compatible with (2.5).

Multipliers, stress resultants, and the equations of motion. We now turn to the heart of our analysis. We may choose (the test function or virtual velocity) $\overset{\vartriangle}{\boldsymbol{p}}$ to be tangent to the constraint manifold defined by (2.5):

$$\overset{\vartriangle}{\boldsymbol{p}}(\mathbf{x}, t) = \frac{\partial \boldsymbol{\pi}}{\partial \mathbf{u}}(\mathbf{u}(s, t), \mathbf{x}, t) \cdot \overset{\vartriangle}{\mathbf{u}}(s, t), \tag{2.16}$$

with $\overset{\vartriangle}{\mathbf{u}}$ smooth enough. Just as the Principle of Virtual Power (XII.9.5) yields (1.17), condition (XII.12.50b) yields

$$\int_{\tilde{\mathbf{x}}(\mathcal{B})} \boldsymbol{\tau}^k(\mathbf{x}, t) \cdot \overset{\vartriangle}{\boldsymbol{p}}_{,k}(\mathbf{x}, t)\, dv(\mathbf{x}) = \int_{s_1}^{s_2} \left(\mathbf{m}_{\mathrm{L}} \cdot \overset{\vartriangle}{\mathbf{u}}_s + \mathbf{n}_{\mathrm{L}} \cdot \overset{\vartriangle}{\mathbf{u}} \right) ds = 0 \tag{2.17a}$$

for all smooth $\overset{\vartriangle}{\mathbf{u}}$, i.e.,

$$\partial_s \mathbf{m}_{\mathrm{L}} - \mathbf{n}_{\mathrm{L}} = \mathbf{0} \tag{2.17b}$$

in the sense of distributions, where

$$\mathbf{m}_{\mathrm{L}}(s, t) \equiv \int_{\mathcal{A}(s)} \boldsymbol{\tau}_{\mathrm{L}}^3(\mathbf{x}, t) \cdot \frac{\partial \boldsymbol{\pi}}{\partial \mathbf{u}}(\mathbf{u}(s, t), \mathbf{x}, t)\, j(\mathbf{x})\, dx^1\, dx^2, \tag{2.18}$$

$$\mathbf{n}_{\mathrm{L}}(s, t) \equiv \int_{\mathcal{A}(s)} \boldsymbol{\tau}_{\mathrm{L}}^k(\mathbf{x}, t) \cdot \frac{\partial}{\partial x^k} \left[\frac{\partial \boldsymbol{\pi}}{\partial \mathbf{u}}(\mathbf{u}(s, t), \mathbf{x}, t) \right] j(\mathbf{x})\, dx^1\, dx^2. \tag{2.19}$$

We now further require that $\overset{\vartriangle}{\mathbf{u}}$ satisfy the end conditions

$$\overset{\vartriangle}{\mathbf{u}}(s_\alpha, t) = \frac{\partial \mathbf{u}_\alpha}{\partial \mathbf{v}_\alpha}(t, \mathbf{v}_\alpha(t)) \cdot \mathbf{w}_\alpha(t), \quad \mathbf{w}_\alpha(t) \in \mathbb{R}^N, \quad \alpha = 1, 2, \tag{2.20}$$

when (2.14) holds, and satisfy

$$\overset{\vartriangle}{\mathbf{u}}(s + s_2 - s_1, t) = \overset{\vartriangle}{\mathbf{u}}(s, t) \tag{2.21}$$

when (2.15) holds. We now substitute (2.16) into the Principle of Virtual Power (1.17) and use (2.17) to obtain

$$(2.22)\quad \int_0^\infty \int_{s_1}^{s_2} [\mathbf{m}_A \cdot \overset{\Delta}{\mathbf{u}}_s + \mathbf{n}_A \cdot \overset{\Delta}{\mathbf{u}} - \mathbf{f} \cdot \overset{\Delta}{\mathbf{u}} - \mathbf{b} \cdot \overset{\Delta}{\mathbf{u}}_t - \mathbf{c} \cdot \overset{\Delta}{\mathbf{u}}]\, ds\, dt \\ - \int_0^\infty [\bar{\mathbf{m}}_2(t) \cdot \overset{\Delta}{\mathbf{u}}(s_2, t) - \bar{\mathbf{m}}_1(t) \cdot \overset{\Delta}{\mathbf{u}}(s_1, t)]\, dt - \int_{s_1}^{s_2} \bar{\mathbf{b}}(s) \cdot \overset{\Delta}{\mathbf{u}}(s, 0)\, ds = 0$$

for all sufficiently smooth $\overset{\Delta}{\mathbf{u}}$ satisfying (2.20) or (2.21) where

$$(2.23)\quad \mathbf{m}_A(s,t) \equiv \int_{\mathcal{A}(s)} \boldsymbol{\tau}_A^3(\mathbf{x}, t) \cdot \frac{\partial \boldsymbol{\pi}}{\partial \mathbf{u}}(\mathbf{u}(s,t), \mathbf{x}, t)\, j(\mathbf{x})\, dx^1\, dx^2,$$

$$(2.24)\quad \mathbf{n}_A(s,t) \equiv \int_{\mathcal{A}(s)} \boldsymbol{\tau}_A^k(\mathbf{x}, t) \cdot \frac{\partial}{\partial x^k}\left[\frac{\partial \boldsymbol{\pi}}{\partial \mathbf{u}}(\mathbf{u}(s,t), \mathbf{x}, t)\right] j(\mathbf{x})\, dx^1\, dx^2,$$

$$(2.25)\quad \mathbf{f}(s,t) \equiv \int_{\mathcal{A}(s)} \boldsymbol{f}(\tilde{\boldsymbol{z}}(\mathbf{x}), t) \cdot \frac{\partial \boldsymbol{\pi}}{\partial \mathbf{u}}(\mathbf{u}(s,t), \mathbf{x}, t)\, j(\mathbf{x})\, dx^1\, dx^2 \\ + \int_{\partial\mathcal{A}(s)} \bar{\boldsymbol{\tau}}(\mathbf{x}, t, \mathbf{q}(\mathbf{x}, t)) \cdot \frac{\partial \boldsymbol{\pi}}{\partial \mathbf{u}}(\mathbf{u}(s,t), \mathbf{x}, t) \frac{\nu^1\, dx^2 - \nu^2\, dx^1}{1 - \nu^3 \nu_3},$$

$$(2.26)\quad \mathbf{b}(s,t) \equiv \int_{\mathcal{A}(s)} \tilde{\rho}(\mathbf{x}) \tilde{\boldsymbol{p}}_t(\mathbf{x}, t) \cdot \frac{\partial \boldsymbol{\pi}}{\partial \mathbf{u}}(\mathbf{u}(s,t), \mathbf{x}, t)\, j(\mathbf{x})\, dx^1\, dx^2,$$

$$(2.27)\quad \mathbf{c}(s,t) \equiv \int_{\mathcal{A}(s)} \tilde{\rho}(\mathbf{x}) \tilde{\boldsymbol{p}}_t(\mathbf{x}, t) \cdot \frac{\partial}{\partial t}\left[\frac{\partial \boldsymbol{\pi}}{\partial \mathbf{u}}(\mathbf{u}(s,t), \mathbf{x}, t)\right] j(\mathbf{x})\, dx^1\, dx^2,$$

$$(2.28)\quad \bar{\mathbf{m}}_\alpha(t) \equiv \int_{\mathcal{A}(s)} \bar{\boldsymbol{\tau}}(\mathbf{x}, t, \mathbf{q}(\mathbf{x}, t)) \cdot \frac{\partial \boldsymbol{\pi}}{\partial \mathbf{u}}(\mathbf{u}(s,t), \mathbf{x}, t)\, j(\mathbf{x})\, dx^1\, dx^2 \bigg|_{s = s_\alpha},$$

$$(2.29)\quad \bar{\mathbf{b}}(s) \equiv \int_{\mathcal{A}(s)} \tilde{\rho}(\mathbf{x})\, \boldsymbol{p}_1(\mathbf{x}) \cdot \frac{\partial \boldsymbol{\pi}}{\partial \mathbf{u}}(\mathbf{u}(s,0), \mathbf{x}, 0)\, j(\mathbf{x})\, dx^1\, dx^2.$$

In (2.25) the components ν^k of the unit outer normal vector $\boldsymbol{\nu}$ to $\mathcal{L}$ at $\tilde{\boldsymbol{z}}(\mathbf{x})$ are defined by $\boldsymbol{\nu} = \nu^k \boldsymbol{g}_k$.

$\mathbf{m} = \mathbf{m}_L + \mathbf{m}_A$ and $\mathbf{n} = \mathbf{n}_L + \mathbf{n}_A$ are the *stress resultants*, i.e., weighted averages of the stress across a section. $\mathbf{f}$ is the applied load and contains contributions from both the body force and the tractions on the lateral surface. The requirements (2.7) and (2.14) ensure that the integrals for $\mathbf{f}$ and $\bar{\mathbf{m}}_\alpha$ are well defined, i.e., that they contain contributions only from those components of $\bar{\boldsymbol{\tau}}$ that are prescribed.

If the integrands appearing in the weak equation (2.22) are continuous, then (2.22) yields the classical versions of the equations of motion

$$(2.30)\qquad \partial_s \mathbf{m}_A - \mathbf{n}_A + \mathbf{f} = \mathbf{b}_t - \mathbf{c},$$

which are subject either to the position boundary conditions (2.14) and the natural boundary conditions

$$(2.31)\qquad \mathbf{m}_A(s_\alpha, t) \cdot \frac{\partial \mathbf{u}_\alpha}{\partial \mathbf{v}_\alpha} = \bar{\mathbf{m}}_\alpha \cdot \frac{\partial \mathbf{u}_\alpha}{\partial \mathbf{v}_\alpha} \quad \text{(no sum)},$$

or to (2.15) and the requirement that $\mathbf{m}_A$ and $\mathbf{n}_A$ have period $s_2 - s_1$ in s (or to appropriate combinations of these conditions). It is important to note that *the weak equation* (2.22) *is an exact consequence of the Principle of Virtual Power*: The choice of $\overset{\vartriangle}{\boldsymbol{p}}$ was inspired by the constraint (2.5), but the constraint itself was never introduced into the derivation of (2.22). We obtain equations for a rod theory by replacing $\tilde{\boldsymbol{p}}$ wherever it appears in the expressions for $\mathbf{b}$, $\mathbf{c}$, $\mathbf{f}$, and $\bar{\mathbf{m}}$ by its constrained form (2.5). (We could allow $\mathbf{f}$ to depend on s and t also through a dependence on $\mathbf{u}$ because we could allow $\boldsymbol{f}$ to depend on $\mathbf{x}$ and t through a dependence on $\tilde{\boldsymbol{p}}$.) In particular,

$$\mathbf{b} \equiv \int_{\mathcal{A}(s)} \rho \left[\frac{\partial \boldsymbol{\pi}}{\partial \mathbf{u}} \cdot \mathbf{u}_t + \frac{\partial \boldsymbol{\pi}}{\partial \tau} \right] \cdot \frac{\partial \boldsymbol{\pi}}{\partial \mathbf{u}}\, j \, dx^1 \, dx^2, \tag{2.32}$$

$$\mathbf{c} \equiv \int_{\mathcal{A}(s)} \rho \left[\frac{\partial \boldsymbol{\pi}}{\partial \mathbf{u}} \cdot \mathbf{u}_t + \frac{\partial \boldsymbol{\pi}}{\partial \tau} \right] \cdot \frac{\partial}{\partial t} \frac{\partial \boldsymbol{\pi}}{\partial \mathbf{u}} j \, dx^1 \, dx^2 \tag{2.33}$$

give these momenta as specific functions of $\mathbf{u}$ and $\mathbf{u}_t$.

Constitutive relations. Let us now replace the $\boldsymbol{\tau}_A^k$ in (2.23) and (2.24) with their constitutive equations of the form (1.19) corresponding to (XII.12.50) and (XII.12.51) in which $\boldsymbol{F}$ is replaced with its constrained form based on (2.5). We then get constitutive functions for rods of the form

$$\begin{aligned} \mathbf{m}_A(s,t) &= \hat{\mathbf{m}}_A \left(\mathbf{u}^t(s,\cdot), \mathbf{u}_s^t(s,\cdot), \tilde{\boldsymbol{\Lambda}}^t((\cdot,\cdot,s),\cdot), s, t \right), \\ \mathbf{n}_A(s,t) &= \hat{\mathbf{n}}_A \left(\mathbf{u}^t(s,\cdot), \mathbf{u}_s^t(s,\cdot), \tilde{\boldsymbol{\Lambda}}^t((\cdot,\cdot,s),\cdot), s, t \right) \end{aligned} \tag{2.34}$$

where

$$\tilde{\boldsymbol{\Lambda}}(\mathbf{x},t) \equiv \boldsymbol{\Lambda}(\boldsymbol{z},t). \tag{2.35}$$

The domain of the constitutive functions $\hat{\mathbf{m}}_A(\cdot,\cdot,\tilde{\boldsymbol{\Lambda}}^t,s,t)$ and $\hat{\mathbf{n}}_A(\cdot,\cdot,\tilde{\boldsymbol{\Lambda}}^t,s,t)$ consists of those histories $\left(\mathbf{u}^t(s,\cdot),\mathbf{u}_s^t(s,\cdot)\right)$ for which $\left(\mathbf{u}(s,t),\mathbf{u}_s(s,t)\right) \in \mathcal{V}(s,t)$ for all t. We have thus formulated a hierarchy of initial-boundary-value problems for rods, characterized by differential equations for functions depending only on s and t. Note that if we use a frame-indifferent version of (1.19), then the resulting form of (2.34) is also frame-indifferent.

2.36. Remark. In view of (2.17), there is no loss of generality in replacing $\mathbf{m}_A$ and $\mathbf{n}_A$ in (2.23), (2.24), (2.30), (2.31), and(2.34) with $\mathbf{m}$ and $\mathbf{n}$. We shall frequently do so.

2.37. Exercise. Let N be a positive integer. Let $\mathcal{I}$ be the collection of ordered pairs (α,β) of integers with $\alpha \geq 0$, $\beta \geq 0$, $\alpha+\beta \leq N$. We identify $\mathbf{u} = \{\boldsymbol{d}_{(\alpha,\beta)} : (\alpha,\beta) \in \mathcal{I}\}$ and set

$$\boldsymbol{\pi}(\mathbf{u}(s,t),\mathbf{x}) = \sum_{(\alpha,\beta)\in\mathcal{I}} (x^1)^\alpha (x^2)^\beta \boldsymbol{d}_{(\alpha,\beta)}(s,t). \tag{2.38}$$

Evaluate (2.23)–(2.29) when (2.38) is used. Interpret the resultants (2.23) and (2.24) for small values of α and β.

Note that the m and n used in this section are not the same as the $\boldsymbol{m}$ and $\boldsymbol{n}$ or the m and n used in Chap. VIII.

Our governing equations (2.30) could have been obtained by substituting (1.11a) into (XI.7.21), dotting this equation with (2.16), and integrating the resulting equation by parts subject to appropriate boundary conditions. This approach employing the classical equations of motion, which is tantamount to the naive derivation of the Principle of Virtual Power, sacrifices the generality inherent in this principle.

The boundary-value problem for the reactions. Let us suppose that $\partial\bar{\mathsf{p}}/\partial\mathsf{q}$ of (1.18) has rank 3 for $\mathbf{x} \in \tilde{\mathbf{x}}(\mathcal{L})$. Thus we are prescribing pure traction boundary conditions on $\mathcal{L}$. It follows that (2.5) is consistent with (1.18).

In addition to satisfying (2.17)–(2.19), the reactive stress vectors $\boldsymbol{\tau}_{\mathrm{L}}$ satisfy the underdetermined problem

$$(j\boldsymbol{\tau}^k_{\mathrm{L}})_{,k} = -(j\boldsymbol{\tau}^k_{\mathrm{A}})_{,k} - j\boldsymbol{f} + \tilde{\rho}j\partial_{tt}\boldsymbol{\pi}, \tag{2.39a}$$

$$\boldsymbol{\tau}^k_{\mathrm{L}}\nu_k = -\boldsymbol{\tau}^k_{\mathrm{A}}\nu_k + \bar{\boldsymbol{\tau}} \quad \text{for } \mathbf{x} \in \tilde{\mathbf{x}}(\mathcal{L}) \tag{2.39b}$$

in the sense of distributions. Note that the terms on the right-hand sides of (2.39) are typically known from the solution of the primary problem. Representations for the reactive stresses can be obtained in terms of the Betti-Gwyther-Finzi-Maxwell-Morera potentials, which allow for the arbitrariness inherent in (2.39) (see Truesdell & Toupin (1960, Chap. D.IV)). These representations simplify greatly for two-dimensional problems.

It is important to note that (2.39b) (and its generalization (1.15b)) ensure that the traction boundary conditions on $\mathcal{L}$ hold exactly. Thus the reactive traction compensates exactly for any deficiencies of the active traction, which is determined by the constitutive equations. This observation, first made by Podio-Guidugli (1989), puts to rest traditional objections (see Novozhilov (1948) and Green, Laws, & Naghdi (1967, 1968), e.g.) to the approach described in this section on the grounds that the active stress $\boldsymbol{T}_{\mathrm{A}}$, confounded with the total stress $\boldsymbol{T}$, when given by its constitutive equations typically fails to satisfy (1.15b). Analogous remarks apply to the shell theories introduced in Sec. 9.

In shell theories, rod theories, and in discrete models, the Lagrange multipliers serve as measures of the discrepancy between the solution of the constrained problem and the original unconstrained problem. Numerical methods that exploit the Lagrange multipliers and the underlying duality are termed 'mixed methods' (see Brezzi & Fortin (1991) and works cited therein).

3. Rods with Two Directors

We now specialize the results of the last section to the case in which u is the triple $\{\boldsymbol{r}, \boldsymbol{d}_1, \boldsymbol{d}_2\}$ of vectors. We make a very special choice of $\boldsymbol{\pi}$ of (2.5), leading to a very special class of rod theories. These rod theories are worthy of study not merely because they might furnish an accurate description of the deformed shape of a rod, but rather because they indicate how constitutive functions might be constructed. We discuss the issue of approximation in Sec. 16.

We introduce the *directors*

$$\boldsymbol{d}_1 = \delta_1\boldsymbol{a}_1, \quad \boldsymbol{d}_2 = \delta_2\boldsymbol{a}_2 \tag{3.1}$$

where $\boldsymbol{a}_1$ and $\boldsymbol{a}_2$ are independent unit vectors not necessarily orthogonal. We define

$$\boldsymbol{a}^3 \equiv \boldsymbol{a}_3 \equiv \frac{\boldsymbol{a}_1 \times \boldsymbol{a}_2}{|\boldsymbol{a}_1 \times \boldsymbol{a}_2|}. \tag{3.2}$$

We take $\boldsymbol{\pi}$ to have one of the increasingly specialized forms (cf. Sec. VIII.4)

(3.3a) $$\boldsymbol{\pi}(\mathsf{u}, \mathbf{w}, \tau) = \boldsymbol{r} + \boldsymbol{\kappa}(\boldsymbol{d}_1, \boldsymbol{d}_2, \mathbf{w}),$$

(3.3b) $$\boldsymbol{\pi}(\mathsf{u}, \mathbf{w}, \tau) = \boldsymbol{r} + \kappa^\alpha(\delta_1, \delta_2, \mathbf{w})\boldsymbol{a}_\alpha,$$

(3.3c) $$\boldsymbol{\pi}(\mathsf{u}, \mathbf{w}, \tau) = \boldsymbol{r} + \varphi^\alpha(\mathbf{w})\boldsymbol{d}_\alpha,$$

where $\boldsymbol{\kappa}$, the κ^α, and the φ^α are given functions. To put (3.3b) into form (3.3a), we just use $\delta_\alpha = |\boldsymbol{d}_\alpha|$, $\boldsymbol{a}_1 = \boldsymbol{d}_1/\delta_1$, etc. By constraining the $\boldsymbol{d}_\alpha$ to be orthonormal, we recover representations (VIII.4.1), associated with the special Cosserat theory of rods. In addition to the kinds of deformations permitted for such rods, Eqs. (3.3) allow thickness changes associated with changes in δ_1 and δ_2 and allow an internal shear associated with a change in the angle between $\boldsymbol{d}_1$ and $\boldsymbol{d}_2$.

We assume that in each case the reference configuration is defined by

(3.4) $$\tilde{\boldsymbol{z}}(\mathbf{x}) = \mathring{\boldsymbol{r}}(s) + \varphi^\alpha(\mathbf{x})\mathring{\boldsymbol{a}}_\alpha(s).$$

We denote values of variables in the reference configuration by the superscript $\circ$. We assume that $\{\mathring{\boldsymbol{a}}_1, \mathring{\boldsymbol{a}}_2, \mathring{\boldsymbol{a}}_3 \equiv \mathring{\boldsymbol{r}}'\}$ is orthonormal and therefore equals its dual basis $\{\mathring{\boldsymbol{a}}^1, \mathring{\boldsymbol{a}}^2, \mathring{\boldsymbol{a}}^3\}$. We take $\mathring{\delta}_1 = 1 = \mathring{\delta}_2$.

We readily compute that

(3.5a)
$$\boldsymbol{g}_1 = \tilde{\boldsymbol{z}}_{,1} = \varphi^\alpha_{,1}\mathring{\boldsymbol{a}}_\alpha, \quad \boldsymbol{g}_2 = \tilde{\boldsymbol{z}}_{,2} = \varphi^\alpha_{,2}\mathring{\boldsymbol{a}}_\alpha, \quad \boldsymbol{g}_3 = \tilde{\boldsymbol{z}}_{,3} = \mathring{\boldsymbol{a}}_3 + \varphi^\alpha_{,3}\mathring{\boldsymbol{a}}_\alpha + \varphi^\alpha\mathring{\boldsymbol{a}}'_\alpha$$

and either use the identity $\boldsymbol{I} = (\frac{\partial \tilde{\boldsymbol{z}}}{\partial x^k})(\frac{\partial \tilde{x}^k}{\partial \boldsymbol{z}}) = \boldsymbol{g}_k\boldsymbol{g}^k$ or else use the formulas $j\boldsymbol{g}^1 = \boldsymbol{g}_2 \times \boldsymbol{g}_3$, etc., to obtain

$$\boldsymbol{g}^1 = \frac{1}{j}\det\begin{pmatrix} (1+\varphi^\alpha\mathring{\boldsymbol{a}}^3\cdot\mathring{\boldsymbol{a}}'_\alpha)\mathring{\boldsymbol{a}}^1 - (\varphi^1_{,3} + \varphi^2\mathring{\boldsymbol{a}}^1\cdot\mathring{\boldsymbol{a}}'_2)\mathring{\boldsymbol{a}}^3 & \varphi^1_{,2} \\ (1+\varphi^\alpha\mathring{\boldsymbol{a}}^3\cdot\mathring{\boldsymbol{a}}'_\alpha)\mathring{\boldsymbol{a}}^2 - (\varphi^2_{,3} + \varphi^1\mathring{\boldsymbol{a}}^2\cdot\mathring{\boldsymbol{a}}'_1)\mathring{\boldsymbol{a}}^3 & \varphi^2_{,2} \end{pmatrix},$$

(3.5b) $$\boldsymbol{g}^2 = \frac{1}{j}\det\begin{pmatrix} \varphi^1_{,1} & (1+\varphi^\alpha\mathring{\boldsymbol{a}}^3\cdot\mathring{\boldsymbol{a}}'_\alpha)\mathring{\boldsymbol{a}}^1 - (\varphi^1_{,3} + \varphi^2\mathring{\boldsymbol{a}}^1\cdot\mathring{\boldsymbol{a}}'_2)\mathring{\boldsymbol{a}}^3 \\ \varphi^2_{,1} & (1+\varphi^\alpha\mathring{\boldsymbol{a}}^3\cdot\mathring{\boldsymbol{a}}'_\alpha)\mathring{\boldsymbol{a}}^2 - (\varphi^2_{,3} + \varphi^1\mathring{\boldsymbol{a}}^2\cdot\mathring{\boldsymbol{a}}'_1)\mathring{\boldsymbol{a}}^3 \end{pmatrix},$$

$$\boldsymbol{g}^3 = \frac{1}{j}\det\left(\varphi^\alpha_{,\beta}\right)\mathring{\boldsymbol{a}}^3 = \frac{\mathring{\boldsymbol{a}}^3}{1+\varphi^\alpha\mathring{\boldsymbol{a}}^3\cdot\mathring{\boldsymbol{a}}'_\alpha}$$

where

(3.5c) $$j = (\boldsymbol{g}_1 \times \boldsymbol{g}_2)\cdot\boldsymbol{g}_3 = \det\left(\varphi^\beta_{,\gamma}\right)\left[1+\varphi^\alpha\mathring{\boldsymbol{a}}^3\cdot\mathring{\boldsymbol{a}}'_\alpha\right].$$

Thus $\boldsymbol{F} = \frac{\partial \tilde{\boldsymbol{p}}}{\partial x^k}\frac{\partial \tilde{x}^k}{\partial \boldsymbol{z}} = \tilde{\boldsymbol{p}}_{,k}\boldsymbol{g}^k$ has the respective forms

(3.6a) $$\boldsymbol{F} = \frac{\partial \boldsymbol{\kappa}}{\partial x^\alpha}\boldsymbol{g}^\alpha + \left[\boldsymbol{r}_s + \frac{\partial \boldsymbol{\kappa}}{\partial \boldsymbol{d}_\alpha}\cdot(\partial_s\boldsymbol{d}_\alpha) + \frac{\partial \boldsymbol{\kappa}}{\partial w^3}\right]\boldsymbol{g}^3,$$

(3.6b) $$\boldsymbol{F} = \frac{\partial \kappa^\alpha}{\partial x^\gamma}\boldsymbol{a}_\alpha\boldsymbol{g}^\gamma + \left[\boldsymbol{r}_s + \left(\frac{\partial \kappa^\alpha}{\partial \delta_\gamma}\partial_s\delta_\gamma + \frac{\partial \kappa^\alpha}{\partial w^3}\right)\boldsymbol{a}_\alpha + \kappa^\alpha\partial_s\boldsymbol{a}_\alpha\right]\boldsymbol{g}^3,$$

(3.6c) $$\boldsymbol{F} = \varphi^\alpha_{,\gamma}\boldsymbol{d}_\alpha\boldsymbol{g}^\gamma + \left[\boldsymbol{r}_s + \varphi^\alpha_{,3}\boldsymbol{d}_\alpha + \varphi^\alpha\boldsymbol{d}_{\alpha,3}\right]\boldsymbol{g}^3.$$

Note that the form of $\boldsymbol{F}$ is very much dependent on the curvature and torsion of the reference configuration, which is manifested by the presence of the derivatives of the basis $\{\mathring{\boldsymbol{a}}^k\}$.

3.7. Exercise. Find explicit forms of $\mathcal{V}$, introduced in Theorem 2.12, corresponding to each version of (3.3).

Corresponding to the constraint (3.3a), the virtual displacement $\overset{\vartriangle}{\boldsymbol{p}}$ is taken to have the form

$$\overset{\vartriangle}{\boldsymbol{p}} = \overset{\vartriangle}{\boldsymbol{r}} + \frac{\partial \boldsymbol{\kappa}}{\partial \boldsymbol{d}_\alpha} \cdot \overset{\vartriangle}{\boldsymbol{d}}_\alpha \tag{3.8}$$

where $\overset{\vartriangle}{\boldsymbol{r}}$ and $\overset{\vartriangle}{\boldsymbol{d}}_\alpha$ are arbitrary vector-valued functions of s and t. We can either specialize the results of Sec. 2 to (3.8) or, equivalently, substitute (3.8) into (1.17). In particular, for $\overset{\vartriangle}{\boldsymbol{p}} = \overset{\vartriangle}{\boldsymbol{r}}$ we obtain the weak form of (VIII.2.15):

$$\boldsymbol{n}_s + \boldsymbol{f}^0 = \rho A \boldsymbol{r}_{tt} + \boldsymbol{q}_{tt} \tag{3.9}$$

where

$$\begin{aligned} \boldsymbol{n} &\equiv \int_{\mathcal{A}} \boldsymbol{\tau}^3_{\mathrm{A}} j \, dx^1 \, dx^2 = \int_{\mathcal{A}} \boldsymbol{T}_{\mathrm{A}} \cdot \boldsymbol{g}^3 j \, dx^1 \, dx^2 \\ &= \int_{\mathcal{A}} \boldsymbol{T}_{\mathrm{A}} \cdot \frac{\boldsymbol{g}^3}{|\boldsymbol{g}^3|} |\boldsymbol{g}_1 \times \boldsymbol{g}_2| \, dx^1 \, dx^2 \\ &= \int_{\mathcal{A}} \boldsymbol{T}_{\mathrm{A}} \cdot \mathring{\boldsymbol{a}}^3 \det\left(\varphi^\alpha_\beta\right) \, dx^1 \, dx^2, \end{aligned} \tag{3.10}$$

$$\boldsymbol{f}^0 \equiv \int_{\mathcal{A}} \boldsymbol{f} \, j \, dx^1 \, dx^2 + \int_{\partial \mathcal{A}} \bar{\boldsymbol{\tau}} \frac{\nu^1 \, dx^2 - \nu^2 \, dx^1}{1 - \nu^3 \nu_3}, \tag{3.11}$$

$$\boldsymbol{q} \equiv \int_{\mathcal{A}} \tilde{\rho} \, (\tilde{\boldsymbol{p}} - \boldsymbol{r}) \, j \, dx^1 \, dx^2, \tag{3.12}$$

$$\rho A \equiv \int_{\mathcal{A}} \tilde{\rho} \, j \, dx^1 \, dx^2. \tag{3.13}$$

In keeping with Remark 2.36, we drop the subscript A from $\boldsymbol{n}$ and from $\boldsymbol{m}^\alpha, \boldsymbol{n}^\alpha, \boldsymbol{m}$ defined below. Note that $|\boldsymbol{g}_1(\mathbf{x}) \times \boldsymbol{g}_2(\mathbf{x})| \, dx^1 \, dx^2$ is the differential area at $\mathbf{x}$ of the section s. Thus the third integral of (3.10) implies that $\boldsymbol{n}(s,t)$ is the resultant contact force across the section s at time t, in consonance with its definition in Chap. VIII. The methods of Ex. VIII.4.8 can be used to show that when $\tilde{\boldsymbol{p}} - \boldsymbol{r}$ is replaced by $\boldsymbol{\kappa}$, it can often be chosen to make $\boldsymbol{q} = \boldsymbol{0}$.

Likewise, for $\overset{\vartriangle}{\boldsymbol{p}} = [\partial \boldsymbol{\kappa} / \partial \boldsymbol{d}_\alpha] \cdot \overset{\vartriangle}{\boldsymbol{d}}_\alpha$ we obtain

$$\partial_s \boldsymbol{m}^\alpha - \boldsymbol{n}^\alpha + \boldsymbol{f}^\alpha = \partial_t \boldsymbol{b}^\alpha - \boldsymbol{c}^\alpha \tag{3.14}$$

where

$$(3.15)\qquad \boldsymbol{m}^\alpha \equiv \int_{\mathcal{A}} \underset{\text{A}}{\boldsymbol{\tau}}^3 \cdot \frac{\partial \boldsymbol{\kappa}}{\partial \boldsymbol{d}_\alpha}\, j\, dx^1\, dx^2,$$

$$(3.16)\qquad \boldsymbol{n}^\alpha \equiv \int_{\mathcal{A}} \underset{\text{A}}{\boldsymbol{\tau}}^k \cdot \frac{\partial}{\partial x^k}\left[\frac{\partial \boldsymbol{\kappa}}{\partial \boldsymbol{d}_\alpha}\right] j\, dx^1\, dx^2,$$

$$(3.17)\qquad \boldsymbol{f}^\alpha \equiv \int_{\mathcal{A}} \boldsymbol{f}\frac{\partial \boldsymbol{\kappa}}{\partial \boldsymbol{d}_\alpha} j\, dx^1\, dx^2 + \int_{\partial\mathcal{A}} \bar{\boldsymbol{\tau}} \cdot \frac{\partial \boldsymbol{\kappa}}{\partial \boldsymbol{d}_\alpha}\frac{\nu^1\, dx^2 - \nu^2\, dx^1}{1 - \nu^3\nu_3},$$

$$(3.18)\qquad \boldsymbol{b}^\alpha \equiv \int_{\mathcal{A}} \tilde{\rho}\tilde{\boldsymbol{p}}_t \cdot \frac{\partial \boldsymbol{\kappa}}{\partial \boldsymbol{d}_\alpha}\, j\, dx^1\, dx^2,$$

$$(3.19)\qquad \boldsymbol{c}^\alpha \equiv \int_{\mathcal{A}} \tilde{\rho}\tilde{\boldsymbol{p}}_t \cdot \frac{\partial}{\partial t}\left[\frac{\partial \boldsymbol{\kappa}}{\partial \boldsymbol{d}_\alpha}\right] j\, dx^1\, dx^2.$$

Equations (3.9) and (3.14) are three vectorial equations of motion appropriate for the variables appearing in (3.3). Equation (3.9) is of course standard, but Eqs. (3.14) are not. We wish to determine how (3.14) is related to the classical angular momentum balance (VIII.2.17). For this purpose, we derive (VIII.2.17) as an exact consequence of (1.17) by choosing

$$(3.20)\qquad \overset{\vartriangle}{\boldsymbol{p}}(\mathbf{x}, t) = [\tilde{\boldsymbol{p}}(\mathbf{x}, t) - \boldsymbol{r}(s, t)] \times \boldsymbol{\omega}(s, t)$$

where $\boldsymbol{\omega}$ is arbitrary. By using (1.12), we then obtain (VIII.2.17):

$$(3.21)\qquad \boldsymbol{m}_s + \boldsymbol{r}_s \times \boldsymbol{n} + \boldsymbol{l} = \boldsymbol{q} \times \boldsymbol{r}_{tt} + \boldsymbol{h}_t$$

where

$$(3.22)\qquad \boldsymbol{m} \equiv \int_{\mathcal{A}} (\tilde{\boldsymbol{p}} - \boldsymbol{r}) \times \underset{\text{A}}{\boldsymbol{\tau}}^3 j\, dx^1\, dx^2$$
$$= \int_{\mathcal{A}} (\tilde{\boldsymbol{p}} - \boldsymbol{r}) \times \boldsymbol{T}_{\text{A}} \cdot \overset{\circ}{\boldsymbol{a}}{}^3 \det\left(\varphi^\alpha_\beta\right)\, dx^1\, dx^2,$$

$$(3.23)\qquad \boldsymbol{l} \equiv \int_{\mathcal{A}} (\tilde{\boldsymbol{p}} - \boldsymbol{r}) \times \boldsymbol{f}\, j\, dx^1\, dx^2 + \int_{\partial\mathcal{A}} (\tilde{\boldsymbol{p}} - \boldsymbol{r}) \times \bar{\boldsymbol{\tau}}\frac{\nu^1\, dx^2 - \nu^2\, dx^1}{1 - \nu^3\nu_3},$$

$$(3.24)\qquad \boldsymbol{h} \equiv \int_{\mathcal{A}} \tilde{\rho}\,(\tilde{\boldsymbol{p}} - \boldsymbol{r}) \times (\tilde{\boldsymbol{p}} - \boldsymbol{r})_t\, j\, dx^1\, dx^2$$

(cf. (VIII.3.17)). Equation (3.22) says that $\boldsymbol{m}(s, t)$ is the resultant contact torque about $\boldsymbol{r}(s, t)$ acting across the section s at time t.

Note that (3.9), (3.14), and (3.21) are exact consequences of the three-dimensional equations of motion. We get equations for rods by replacing $\tilde{\boldsymbol{p}}$ with (3.3).

3.25. Exercise. Carry out all the steps in (3.9)–(3.24).

Our derivation of (3.9), (3.14), and(3.21) furnishes an alternative to that of Green (1959) (which was generalized by Antman & Warner (1966)). A motivation for the choice of $\overset{\vartriangle}{\boldsymbol{p}}$ leading to (3.20) is furnished by (3.3c), which produces a version of (3.8), one of whose terms is $\sum \varphi^\alpha \delta_\alpha \overset{\vartriangle}{\boldsymbol{a}}_\alpha$. Since the $\boldsymbol{a}_\alpha$ are unit vectors, it follows that $\boldsymbol{a}_1 \cdot \overset{\vartriangle}{\boldsymbol{a}}_1 = 0 = \boldsymbol{a}_2 \cdot \overset{\vartriangle}{\boldsymbol{a}}_2$, so that each $\overset{\vartriangle}{\boldsymbol{a}}_\alpha$ has the form $\boldsymbol{a}_\alpha \times \boldsymbol{\omega}$ where $\boldsymbol{\omega}$ is arbitrary. Thus if $\overset{\vartriangle}{\boldsymbol{r}} = \boldsymbol{0}$ and $\overset{\vartriangle}{\delta}_\alpha = 0$, then $\overset{\vartriangle}{\boldsymbol{p}} = (\sum \varphi^\alpha \delta_\alpha \boldsymbol{a}_\alpha) \times \boldsymbol{\omega} = (\boldsymbol{p} - \boldsymbol{r}) \times \boldsymbol{\omega}$.

If we substitute (3.3a) into (3.22)–(3.24), we discern no obvious connection between the resulting approximations and the integrals of (3.15)–(3.19). (While this fact is somewhat disturbing, it is not fatal for analysis because (3.9) and (3.14) when supplemented with constitutive equations lead to well-set problems.) If, however, we substitute (3.3c) into (3.22), we immediately obtain

$$\boldsymbol{m} = \boldsymbol{d}_\alpha \times \boldsymbol{m}^\alpha. \tag{3.26}$$

Thus (3.21) is a consequence of (3.14) when (3.3c) holds. The use of (3.3b) leads to a more complicated version of (3.26).

When we use (3.21), we confront the question of determining which components of (3.14) should complement (3.21). The answer can be found by the observation that $\boldsymbol{m}$ is primarily responsible for changes in $\boldsymbol{a}_3$, so that the remaining components of $\boldsymbol{m}^1$ and $\boldsymbol{m}^2$ should be responsible for changes in δ_1, δ_2, and the angle between $\boldsymbol{d}_1$ and $\boldsymbol{d}_2$. We get mechanical variables corresponding to δ_1 and δ_2 by substituting (3.1) into (3.3a) and taking

$$\overset{\vartriangle}{\boldsymbol{p}} = \sum \frac{\partial \boldsymbol{\kappa}}{\partial \boldsymbol{d}_\alpha} \cdot \boldsymbol{a}_\alpha \overset{\vartriangle}{\delta}_\alpha \tag{3.27}$$

where the $\overset{\vartriangle}{\delta}_\alpha$ are arbitrary.

3.28. Exercise. Construct the components of $\boldsymbol{m}^\alpha$ and $\boldsymbol{n}^\alpha$ corresponding to $\partial_s \delta_\alpha$ and δ_α by this process, and give their physical interpretations. Show that the resulting equations can be obtained by dotting the two equations of (3.14) respectively with the $\boldsymbol{a}_\alpha$.

When (3.1) is substituted into (3.3a), care must be taken in computing $\overset{\vartriangle}{\boldsymbol{p}}$ by (3.8). The variations $\overset{\vartriangle}{\boldsymbol{a}}_\alpha$ are *not* arbitrary because they must satisfy the constraints that $\overset{\vartriangle}{\boldsymbol{a}}_1 \cdot \boldsymbol{a}_1 = 0$, etc. Accordingly, Lagrange multipliers must be introduced into the Principle of Virtual Power. These may be removed from the resulting classical equations by taking suitable cross products. Such cross products are implicit in (3.20). It is not so easy to get an equation for the angle between $\boldsymbol{d}_1$ and $\boldsymbol{d}_2$. One strategy is to represent $\boldsymbol{a}_3$ in terms of spherical coordinates with respect to a fixed basis depending on s, e.g., $\{\mathring{\boldsymbol{a}}_k\}$. Then $\boldsymbol{a}_1$ and $\boldsymbol{a}_2$ can be located with respect to the nodal line formed by the intersection of the planes normal to $\boldsymbol{a}_3$ and $\mathring{\boldsymbol{a}}_3$ in the manner of Sec. VIII.12. Such a representation (with its attendant polar singularity) would produce an explicit formula for the angle between $\boldsymbol{d}_1$ and $\boldsymbol{d}_2$, which could then be varied in a corresponding representation for $\overset{\vartriangle}{\boldsymbol{p}}$. (A related strategy was proposed in a slightly different context by Simo, Marsden, & Krishnaprasad (1988).) Whenever we can isolate equations for variables corresponding to changes in δ_1, δ_2, and the angle between $\boldsymbol{d}_1$ and $\boldsymbol{d}_2$, we can take our governing equations to be these equations together with (3.9) and (3.21).

Let us substitute (3.3c) into (1.12), multiply it by j, and integrate the resulting expression over $\mathcal{A}(s)$. Using (3.10), (3.15), and (3.16), we get

$$\boldsymbol{r}' \times \boldsymbol{n} + \partial_s \boldsymbol{d}_\alpha \times \boldsymbol{m}^\alpha + \boldsymbol{d}_\alpha \times \boldsymbol{n}^\alpha = \boldsymbol{0}. \tag{3.29}$$

This consequence of the symmetry condition (1.12) can be regarded as a constitutive restriction, just like (1.12). It is automatically satisfied whenever constitutive functions for $\boldsymbol{T}$ satisfy (XII.7.27). Likewise, the consequences of (1.12) for the theory of Sec. 2 and for the theory of Sec. 3 under the assumptions (3.3a,b) can be completely incorporated into its constitutive equations, even though no simple relation like (3.29) results. (It is possible to obtain other one-dimensional consequences of (1.12) by multiplying it by various functions before integrating it over $\mathcal{A}(s)$.) The consequences of isotropy for the three-dimensional theory are immediately carried over to rod theories. (See the treatment in Secs. 5 and 14.)

4. Elastic Rods

For elastic rods, the constitutive equations (2.34) reduce to

$$\begin{aligned} \mathsf{m}_{\mathrm{A}}(s,t) &= \hat{\mathsf{m}}_{\mathrm{A}}(\mathsf{u}(s,t), \mathsf{u}_s(s,t), \tilde{\boldsymbol{\Lambda}}((\cdot,\cdot,s),t), s, t), \\ \mathsf{n}_{\mathrm{A}}(s,t) &= \hat{\mathsf{n}}_{\mathrm{A}}(\mathsf{u}(s,t), \mathsf{u}_s(s,t), \tilde{\boldsymbol{\Lambda}}((\cdot,\cdot,s),t), s, t) \end{aligned} \tag{4.1}$$

where

$$\begin{aligned} &\hat{\mathsf{m}}_{\mathrm{A}}(\mathsf{u},\mathsf{v},\tilde{\boldsymbol{\Lambda}}((\cdot,\cdot,s),t),s,t) \\ &\qquad = \int_{\mathcal{A}(s)} \left[\hat{\boldsymbol{T}}_{\mathrm{A}} \cdot \boldsymbol{g}^3(\mathbf{x})\right] \cdot \frac{\partial \boldsymbol{\pi}}{\partial \mathsf{u}} \, j(\mathbf{x}) \, dx^1 \, dx^2, \end{aligned} \tag{4.2}$$

$$\begin{aligned} &\hat{\mathsf{n}}_{\mathrm{A}}(\mathsf{u},\mathsf{v},\tilde{\boldsymbol{\Lambda}}((\cdot,\cdot,s),t),s,t) \\ &\qquad = \int_{\mathcal{A}(s)} \left[\hat{\boldsymbol{T}}_{\mathrm{A}} \cdot \boldsymbol{g}^l(\mathbf{x})\right] \cdot \left[\frac{\partial^2 \boldsymbol{\pi}}{\partial \mathsf{u} \partial \mathsf{u}} \cdot \mathsf{v} + \frac{\partial}{\partial w^l} \frac{\partial \boldsymbol{\pi}}{\partial \mathsf{u}}\right] j(\mathbf{x}) \, dx^1 \, dx^2 \end{aligned} \tag{4.3}$$

where the arguments of the derivatives of $\boldsymbol{\pi}$ are $\mathsf{u}, \mathbf{x}, t$ and where the arguments of $\hat{\boldsymbol{T}}_{\mathrm{A}}$ are $\check{\boldsymbol{F}}(\mathsf{u},\mathsf{v},\mathbf{x},t), \tilde{\boldsymbol{\Lambda}}(\mathbf{x},t), \tilde{\boldsymbol{z}}(\mathbf{x})$ with

$$\check{\boldsymbol{F}}(\mathsf{u},\mathsf{v},\mathbf{x},t) \equiv \left[\frac{\partial \boldsymbol{\pi}}{\partial \mathsf{u}}(\mathsf{u},\mathbf{x},t) \cdot \mathsf{v}\delta_k^3 + \frac{\partial \boldsymbol{\pi}}{\partial w^k}(\mathsf{u},\mathbf{x},t)\right] \boldsymbol{g}^k(\mathbf{x}). \tag{4.4}$$

We define

$$\hat{\mathsf{m}}_{\mathrm{L}}(\mathsf{u},\mathsf{v},\tilde{\boldsymbol{\Lambda}}((\cdot,\cdot,s),t),s,t) = \int_{\mathcal{A}(s)} \left[\hat{\boldsymbol{T}}_{\mathrm{L}} \cdot \boldsymbol{g}^3(\mathbf{x})\right] \cdot \frac{\partial \boldsymbol{\pi}}{\partial \mathsf{u}} \, j(\mathbf{x}) \, dx^1 \, dx^2 \tag{4.5}$$

with the same arguments as for (4.2) and (4.3), and define $\hat{\mathsf{n}}_{\mathrm{L}}$ in analogy with (4.3) and (4.5).

We now determine the implications for elastic rods of the Strong Ellipticity Condition (XIII.2.28), (XIII.2.29) for constrained elastic media, which requires that

$$\begin{aligned} [\hat{\boldsymbol{T}}_{\mathrm{L}}(\boldsymbol{G}+\alpha\boldsymbol{ab},\boldsymbol{\Lambda},\boldsymbol{z}) + \hat{\boldsymbol{T}}_{\mathrm{A}}(\boldsymbol{G}+\alpha\boldsymbol{ab},\boldsymbol{\Lambda},\boldsymbol{z}) \\ - \hat{\boldsymbol{T}}_{\mathrm{L}}(\boldsymbol{G},\boldsymbol{\Lambda},\boldsymbol{z}) - \hat{\boldsymbol{T}}_{\mathrm{A}}(\boldsymbol{G},\boldsymbol{\Lambda},\boldsymbol{z})] : \boldsymbol{ab} > 0 \end{aligned} \tag{4.6}$$

$$\begin{aligned}(4.7)\qquad &\forall \boldsymbol{G},\ \forall \boldsymbol{ab} \neq \boldsymbol{O},\ \forall \alpha \in [0,1] \text{ such that}\\ &\boldsymbol{G} + \alpha \boldsymbol{ab} \text{ satisfies the constraints.}\end{aligned}$$

It follows from Theorem 2.12 that if $\det \check{\boldsymbol{F}}(\mathbf{u}, \mathbf{v}, \mathbf{x}, t) > 0$ and $\det \check{\boldsymbol{F}}(\mathbf{u}, \mathbf{v} + \mathbf{w}, \mathbf{x}, t) > 0$, then $\det \check{\boldsymbol{F}}(\mathbf{u}, \mathbf{v} + \alpha\mathbf{w}, \mathbf{x}, t) > 0$ for all $\alpha \in [0,1]$. Since

$$\mathbf{a} \cdot (\boldsymbol{T} \cdot \boldsymbol{g}^3) \cdot \frac{\partial \boldsymbol{\pi}}{\partial \mathbf{u}} = \left\{ \left[\left(\frac{\partial \boldsymbol{\pi}}{\partial \mathbf{u}} \right) \cdot \mathbf{a} \right] \boldsymbol{g}^3 \right\} : \boldsymbol{T}, \tag{4.8}$$

$$\check{\boldsymbol{F}}(\mathbf{u}, \mathbf{v} + \alpha\mathbf{w}, \mathbf{x}, t) = \check{\boldsymbol{F}}(\mathbf{u}, \mathbf{v}, \mathbf{x}, t) + \alpha \left[\frac{\partial \boldsymbol{\pi}}{\partial \mathbf{u}}(\mathbf{u}, \mathbf{x}, t) \cdot \mathbf{w} \right] \boldsymbol{g}^3, \tag{4.9}$$

we deduce from (4.2)–(4.7) that

(4.10)

$$\begin{aligned}
&[\hat{\mathbf{m}}_{\mathrm{A}}(\mathbf{u}, \mathbf{v} + \alpha\mathbf{w}, \tilde{\boldsymbol{\Lambda}}, s, t) - \hat{\mathbf{m}}_{\mathrm{A}}(\mathbf{u}, \mathbf{v}, \tilde{\boldsymbol{\Lambda}}, s, t)] \cdot \mathbf{w}\\
&+[\hat{\mathbf{m}}_{\mathrm{L}}(\mathbf{u}, \mathbf{v} + \alpha\mathbf{w}, \tilde{\boldsymbol{\Lambda}}, s, t) - \hat{\mathbf{m}}_{\mathrm{L}}(\mathbf{u}, \mathbf{v}, \tilde{\boldsymbol{\Lambda}}, s, t)] \cdot \mathbf{w}\\
&= \int_{\mathcal{A}(s)} \Big[\hat{\boldsymbol{T}}_{\mathrm{A}} \left(\check{\boldsymbol{F}}(\mathbf{u}, \mathbf{v} + \alpha\mathbf{w}, \mathbf{x}), \tilde{\boldsymbol{\Lambda}}, \tilde{\boldsymbol{z}}(\mathbf{x}) \right) - \hat{\boldsymbol{T}}_{\mathrm{A}} \left(\check{\boldsymbol{F}}(\mathbf{u}, \mathbf{v}, \mathbf{x}), \tilde{\boldsymbol{\Lambda}}, \tilde{\boldsymbol{z}}(\mathbf{x}) \right)\\
&\qquad + \hat{\boldsymbol{T}}_{\mathrm{L}} \left(\check{\boldsymbol{F}}(\mathbf{u}, \mathbf{v} + \alpha\mathbf{w}, \mathbf{x}), \tilde{\boldsymbol{\Lambda}}, \tilde{\boldsymbol{z}}(\mathbf{x}) \right) - \hat{\boldsymbol{T}}_{\mathrm{L}} \left(\check{\boldsymbol{F}}(\mathbf{u}, \mathbf{v}, \mathbf{x}), \tilde{\boldsymbol{\Lambda}}, \tilde{\boldsymbol{z}}(\mathbf{x}) \right) \Big]\\
&\qquad : \left[\frac{\partial \boldsymbol{\pi}}{\partial \mathbf{u}}(\mathbf{u}, \mathbf{x}, t) \cdot \mathbf{w} \right] \boldsymbol{g}^3(\mathbf{x})\, j(\mathbf{x})\, dx^1\, dx^2\\
&> 0 \quad \text{for } \alpha\mathbf{w} \neq \mathbf{0}.
\end{aligned}$$

Now we use the Global Constraint Principle to simplify (4.10). Equation (2.17a) implies that

(4.11)

$$\begin{aligned}
\int_{s_1}^{s_2} \Big\{ &\left[\hat{\mathbf{m}}_{\mathrm{L}}(\mathbf{u} + \alpha\mathring{\mathbf{u}}, \mathbf{u}_s + \alpha\mathring{\mathbf{u}}_s, \tilde{\boldsymbol{\Lambda}}, s, t) - \hat{\mathbf{m}}_{\mathrm{L}}(\mathbf{u}, \mathbf{u}_s, \tilde{\boldsymbol{\Lambda}}, s, t) \right] \cdot \mathring{\mathbf{u}}_s\\
&+ \left[\hat{\mathbf{n}}_{\mathrm{L}}(\mathbf{u} + \alpha\mathring{\mathbf{u}}, \mathbf{u}_s + \alpha\mathring{\mathbf{u}}_s, \tilde{\boldsymbol{\Lambda}}, s, t) - \hat{\mathbf{n}}_{\mathrm{L}}(\mathbf{u}, \mathbf{u}_s, \tilde{\boldsymbol{\Lambda}}, s, t) \right] \cdot \mathring{\mathbf{u}} \Big\}\, ds = 0
\end{aligned}$$

for all $\mathring{\mathbf{u}}$. Let $s_1 < \xi < \xi + \varepsilon < \eta < s_2$ and let $\mathbf{a}$ be an arbitrary fixed element of $\mathbb{R}^N$. Let

$$\omega(s, \varepsilon, \xi, \eta) = \begin{cases} 0 & \text{for } s_1 \leq s \leq \xi,\\ s - \xi & \text{for } \xi \leq s \leq \xi + \varepsilon,\\ \frac{\varepsilon(\eta - s)}{\eta - \sigma}(\eta - s - \varepsilon) & \text{for } \xi + \varepsilon \leq s \leq \eta,\\ 0 & \text{for } \eta \leq s \leq s_2. \end{cases} \tag{4.12}$$

We substitute $\mathring{\mathbf{u}} = \omega(s, \varepsilon, \xi, \eta)\mathbf{a}$ into (4.11), divide the resulting equation by ε, and then let $\varepsilon \searrow 0$ to obtain

$$\begin{aligned}
&\Big[\hat{\mathbf{m}}_{\mathrm{L}}(\mathbf{u}(\xi, t), \mathbf{u}_s(\xi, t) + \alpha\mathbf{a}, \tilde{\boldsymbol{\Lambda}}((\cdot, \cdot, \xi), t), \xi, t)\\
&\qquad - \hat{\mathbf{m}}_{\mathrm{L}}(\mathbf{u}(\xi, t), \mathbf{u}_s(\xi, t), \tilde{\boldsymbol{\Lambda}}((\cdot, \cdot, \xi), t), \xi, t) \Big] \cdot \mathbf{a} = 0,
\end{aligned} \tag{4.13}$$

which must hold for all $\xi, t, \mathsf{a}, \mathsf{u}$. Since we can vary u and u_s independently at ξ, we deduce from (4.13) that

$$\left[\hat{\mathsf{m}}_{\mathrm{L}}(\mathsf{u}, \mathsf{v} + \alpha\mathsf{a}, \tilde{\boldsymbol{\Lambda}}, s, t) - \hat{\mathsf{m}}_{\mathrm{L}}(\mathsf{u}, \mathsf{v}, \tilde{\boldsymbol{\Lambda}}, s, t)\right] \cdot \mathsf{a} = 0, \tag{4.14}$$

for all $s, \mathsf{u}, \mathsf{v}, \mathsf{a}$. It then follows from (4.14) that (4.10) reduces to

$$[\hat{\mathsf{m}}_{\mathrm{A}}(\mathsf{u}, \mathsf{v} + \alpha\mathsf{w}, \tilde{\boldsymbol{\Lambda}}, s, t) - \hat{\mathsf{m}}_{\mathrm{A}}(\mathsf{u}, \mathsf{v}, \tilde{\boldsymbol{\Lambda}}, s, t)] \cdot \mathsf{w} > 0 \tag{4.15}$$

for all $\mathsf{u}, \mathsf{v}, \mathsf{w}$ with $\alpha\mathsf{w} \neq \mathbf{0}$, i.e., $\hat{\mathsf{m}}_{\mathrm{A}}(\mathsf{u}, \cdot, \tilde{\boldsymbol{\Lambda}}((\cdot, \cdot, s), t), s, t)$ is strictly monotone. If we divide (4.15) by α and then let $\alpha \searrow 0$, we obtain an inequality whose strict form is

$$\mathsf{a} \cdot \frac{\partial \hat{\mathsf{m}}_{\mathrm{A}}}{\partial \mathsf{v}} \cdot \mathsf{a} > 0 \quad \forall\, \mathsf{a} \neq \mathbf{0}. \tag{4.16}$$

Thus $\mathsf{v} \mapsto \hat{\mathsf{m}}(\mathsf{u}, \mathsf{v}, s, t)$ is strictly monotone.

4.17. Exercise. Prove that the Strong Ellipticity Condition yields the Monotonicity Condition (VIII.8.1) when $\boldsymbol{n}$ and $\boldsymbol{m}$ are defined by (3.10) and (3.22) and when (3.3a) holds with the $\{\boldsymbol{d}_\alpha\}$ orthonormal.

4.18. Exercise. Let the three-dimensional material be hyperelastic so that (XIII.1.10) holds. Impose the constraint (2.5) where $\boldsymbol{\pi}$ does not depend explicitly on t. Define the stored energy function for the rod by

$$\hat{w}(\mathsf{u}, \mathsf{v}, s) \equiv \int_{\mathcal{A}(s)} W\big(\check{\boldsymbol{F}}(\mathsf{u}, \mathsf{v}, \mathsf{x})^* \cdot \check{\boldsymbol{F}}(\mathsf{u}, \mathsf{v}, \mathsf{x}), \check{z}(\mathsf{x})\big) \jmath(\mathsf{x})\, dx^1\, dx^2 \tag{4.19}$$

where $\check{\boldsymbol{F}}$ is defined in (4.4). Prove that

$$\hat{\mathsf{n}}(\mathsf{u}, \mathsf{v}, s) = \frac{\partial \hat{w}}{\partial \mathsf{u}}, \quad \hat{\mathsf{m}}(\mathsf{u}, \mathsf{v}, s) = \frac{\partial \hat{w}}{\partial \mathsf{v}}. \tag{4.20}$$

State a rod-theoretic version of Hamilton's Principle that delivers the equations of motion.

The constitutive equations (4.2) and (4.3) make explicit the role of cross-sectional geometry and can therefore be used to introduce weightings like those of (VIII.10.3), especially in the important and most common case that $\boldsymbol{\pi}$ is affine in u. As a useful rule of thumb, these weightings can be taken to be those that would hold for the linearization of (4.2) and (4.3) about the reference configuration.

If the thickness of the rod tapers down to zero at an end, then that end may not be able to sustain certain kinds of boundary conditions, e.g., the end might not be able to be welded to a rigid wall. Mathematically speaking, in the notation of Chap. VIII and Sec. 3, a solution of a boundary-value problem is sought in which the director $\boldsymbol{d}_3$ is taken to lie in a a function space (typically a Sobolev space) chosen to reflect the growth of an energy, which depends on the tapering. In a natural space, it might not be possible to define suitable boundary values (traces) of $\boldsymbol{d}_3$. This issue can be treated by the methods described in Sec. XIII.5. It leads to some surprising conclusions (see Antman (1976b, Fig. 2.1)), which indicate deficiencies of rod theories. Instances in which the thickness of a rod drops to zero at places other than its ends arise in problems of optimal design against buckling; see Cox & Overton (1992) and the references cited therein.

There are a host of rod theories constructed by approximation from the three-dimensional theory, the relative advantages of each often being the object of a tiresome discussion. The results of these last two sections show that if the rod theory is developed systematically from any approximation of the form (2.5), then it inherits a distinctive mathematical structure from the three-dimensional theory, which is independent of the particular assumptions underlying the theory. This structure resides in the properties of $\mathcal{V}$, specified in Theorem 2.12, in the form of the equations of motion, and in such constitutive restrictions as the monotonicity condition (4.15).

Condition (4.15) supports a full existence and regularity theory for hyperelastic rods (see Antman (1976b), Ball (1981a), and Seidman & Wolfe (1988)). This theory can no doubt be extended to Cauchy elastic rods by the methods of Antman (1983a).

Consider a specific boundary-value problem for the partial differential equations of equilibrium of a hyperelastic three-dimensional rod-like body and the corresponding ordinary differential equations for a hierarchy of rod theories parametrized by the number of degrees of freedom N. If the total energy functional for the three-dimensional theory is weakly lower semicontinuous on a suitable Sobolev space (this is the crucial assumption supporting the proof of the existence of a minimizer of the energy), then the minimizers of the constrained problems for rods, which are classical solutions of the corresponding boundary-value problems, converge weakly to the minimizer of the energy for the three-dimensional problem (see Antman (1976a)). These facts give a precise position to the hierarchy of rod theories.

They do not offer the ideal resolution of the validity of rod theories as approximations to unconstrained three-dimensional bodies. For this purpose, sharp error estimates are sought. Some reasonable estimates have been obtained for various linear theories of elastic rods and plates by methods that exploit the convexity of the stored-energy function for such theories. Such techniques are not available for nonlinearly elastic rods. (Cf. Morgenstern & Szabo (1961, Sec. 9), Koiter (1970), Mielke (1990), and Ciarlet (1990).) The models for which estimates have been found are very similar to, but not necessarily identical to, those one would obtain by adopting the simplest constraints in Secs. 7 and 8 for the position field; the simplest models for which there are estimates typically differ from these constrained models by employing slightly different moduli for the stress resultants.

For thin bodies with unloaded lateral surfaces, these estimates typically show that the constrained position field accurately approximates the position field for an unconstrained body and that the resultants are accurate, but offer no similar statement about the stress field computed from the constrained position field (i.e., about the active stress). The role of the reactive stress is not considered in such analyses. These estimates are thus compatible with our development, in which the role of the reactive stress is central.

5. Planar Problems for Elastic Rods, Problems with Symmetries

We now specialize the constitutive equations for elastic rods with two directors to those for planar motions. This theory and its specializations deliver many of the formulations for the problems studied in Chaps. IV and VI, and in the next section. We adhere as closely as possible to the notational schemes introduced in Chap. IV, in Sec. VIII.16, and in Sec. 3.

Let $\{\boldsymbol{i}, \boldsymbol{j}, \boldsymbol{k}\}$ be a fixed right-handed orthonormal basis. We take span $\{\boldsymbol{i}, \boldsymbol{j}\}$ to be the plane of motion by setting

$$\boldsymbol{a}_3 \equiv \boldsymbol{a} \equiv \cos\theta\boldsymbol{i} + \sin\theta\boldsymbol{j}, \quad \boldsymbol{a}_1 \equiv \boldsymbol{b} \equiv -\sin\theta\boldsymbol{i} + \cos\theta\boldsymbol{j}, \quad \boldsymbol{a}_2 \equiv \boldsymbol{k}, \tag{5.1}$$

$$\boldsymbol{r}_s = \nu\boldsymbol{a} + \eta\boldsymbol{b}, \quad \boldsymbol{d}_1 = \delta_1\boldsymbol{b}, \quad \boldsymbol{d}_2 = \delta_2\boldsymbol{k}. \tag{5.2}$$

We define

$$\mu \equiv \theta_s, \quad \omega_\alpha \equiv \partial_s\delta_\alpha. \tag{5.3}$$

The strain variables for our problem are $(\nu, \eta, \mu, \omega_1, \omega_2, \delta_1, \delta_2) \equiv \mathbf{q}$. We denote values of variables in the reference configuration by superposed circles. The specialization of (3.6b) is

$$\boldsymbol{F} = \left[\frac{\partial \kappa^1}{\partial x^\alpha}\boldsymbol{b} + \frac{\partial \kappa^2}{\partial x^\alpha}\boldsymbol{k}\right]\boldsymbol{g}^\alpha + \left[(\nu - \kappa^1\mu)\boldsymbol{a} + \zeta^1\boldsymbol{b} + \zeta^2\boldsymbol{k}\right]\boldsymbol{g}^3, \tag{5.4}$$

from which we find

$$\begin{aligned} \boldsymbol{C} = &\left[\frac{\partial \kappa^1}{\partial x^\alpha}\frac{\partial \kappa^1}{\partial x^\beta} + \frac{\partial \kappa^2}{\partial x^\alpha}\frac{\partial \kappa^2}{\partial x^\beta}\right]\boldsymbol{g}^\alpha\boldsymbol{g}^\beta \\ &+ \left[\frac{\partial \kappa^1}{\partial x^\alpha}z^1 + \frac{\partial \kappa^2}{\partial x^\alpha}\zeta^2\right](\boldsymbol{g}^\alpha\boldsymbol{g}^3 + \boldsymbol{g}^3\boldsymbol{g}^\alpha) \\ &+ \left[(\nu - \kappa^1\mu)^2 + (\zeta^1)^2 + (\zeta^2)^2\right]\boldsymbol{g}^3\boldsymbol{g}^3 \end{aligned} \tag{5.5}$$

where

$$\begin{aligned} \zeta^1 &\equiv \eta + \frac{\partial \kappa^1}{\partial \delta_\beta}\omega_\beta + \frac{\partial \kappa^1}{\partial w^3}, \\ \zeta^2 &\equiv \frac{\partial \kappa^2}{\partial \delta_\beta}\omega_\beta + \frac{\partial \kappa^2}{\partial w^3}, \\ \boldsymbol{g}^1 &= \frac{\varphi^2_{,2}\mathring{\boldsymbol{b}} - \varphi^1_{,2}\boldsymbol{k}}{\det(\varphi^\alpha_{,\beta})} + \frac{\varphi^2_{,3}\varphi^1_{,2} - \varphi^1_{,3}\varphi^2_{,2}}{\det(\varphi^\alpha_{,\beta})\left(1 - \varphi^1\mathring{\mu}\right)}\mathring{\boldsymbol{a}}, \\ \boldsymbol{g}^2 &= \frac{-\varphi^2_{,1}\mathring{\boldsymbol{b}} + \varphi^1_{,1}\boldsymbol{k}}{\det(\varphi^\alpha_{,\beta})} - \frac{\varphi^2_{,3}\varphi^1_{,1} - \varphi^1_{,3}\varphi^2_{,1}}{\det(\varphi^\alpha_{,\beta})\left(1 - \varphi^1\mathring{\mu}\right)}\mathring{\boldsymbol{a}}, \\ \boldsymbol{g}^3 &= \frac{\mathring{\boldsymbol{a}}}{\left(1 - \varphi^1\mathring{\mu}\right)}, \\ j &= \det(\varphi^\alpha_{,\beta})\left(1 - \varphi^1\mathring{\mu}\right). \end{aligned} \tag{5.6}$$

Note that $\mathring{\mu} \equiv \mathring{\theta}'$ is the curvature of $\mathring{\boldsymbol{r}}$.

We require that the following *symmetry conditions* hold: $\mathcal{A}(s)$ is symmetric about the x^1-axis for each s, $\hat{\rho}$ is even in x^2, $\kappa^1 \equiv \boldsymbol{\kappa}\cdot\boldsymbol{a}^1 \equiv \boldsymbol{\kappa}\cdot\boldsymbol{b}$ is even in x^2, φ^1 is even in x^2, $\kappa^2 \equiv \boldsymbol{\kappa}\cdot\boldsymbol{a}^2 \equiv \boldsymbol{\kappa}\cdot\boldsymbol{k}$ is odd in x^2, and φ^2 is odd in x^2. Thus j is even in x^2.

Following the proposal in Sec. 3, we take as our governing equations of motion (3.9), (3.21), and two equations corresponding to the variables δ_1 and δ_2, obtained in Ex. 3.28. Dropping the subscript A in accord with Remark 2.36, we set

$$\begin{gathered} \boldsymbol{n} \equiv N\boldsymbol{a} + H\boldsymbol{b}, \quad \boldsymbol{m} \equiv M\boldsymbol{k}, \\ \Omega^\alpha \equiv \boldsymbol{m}^\alpha \cdot \boldsymbol{a}_\alpha \ \text{(no sum)}, \quad \Delta^\alpha \equiv \boldsymbol{n}^\alpha \cdot \boldsymbol{a}_\alpha + \boldsymbol{m}^\alpha \cdot \boldsymbol{a}_{\alpha,3} \ \text{(no sum)}, \\ l^\alpha \equiv \boldsymbol{f}^\alpha \cdot \boldsymbol{a}_\alpha \ \text{(no sum)}, \quad \beta^\alpha \equiv \boldsymbol{b}^\alpha \cdot \boldsymbol{a}_\alpha \ \text{(no sum)}, \end{gathered} \tag{5.7}$$

$$\gamma^\alpha \equiv \boldsymbol{c}^\alpha \cdot \boldsymbol{a}_\alpha + \boldsymbol{b}^\alpha \cdot \partial_t \boldsymbol{a}_\alpha \text{ (no sum).} \tag{5.8}$$

Our mechanical variables, corresponding to the elements of $\mathbf{q}$, are

$$\mathbf{Q} \equiv (N, H, M, \Omega_1, \Omega_2, \Delta_1, \Delta_2).$$

Then (3.9), (3.21), and the $\boldsymbol{a}_\alpha$-components of (3.14) reduce to

$$\partial_s(N\boldsymbol{a} + H\boldsymbol{b}) + \boldsymbol{f}^0 = \rho A \boldsymbol{r}_{tt} + \boldsymbol{q}_{tt}, \tag{5.9}$$

$$\partial_s M + \boldsymbol{k} \cdot (\boldsymbol{r}_s \times \boldsymbol{n}) + \boldsymbol{k} \cdot \boldsymbol{l} = \boldsymbol{k} \cdot (\boldsymbol{q} \times \boldsymbol{r}_{tt} + \boldsymbol{h}_t), \tag{5.10}$$

$$\partial_s \Omega^\alpha - \Delta^\alpha + l^\alpha = \partial_t \beta^\alpha - \gamma^\alpha. \tag{5.11}$$

5.12. Exercise. Prove that

$$\partial_t \beta^\alpha - \gamma^\alpha = \int_{\mathcal{A}} \tilde{\rho}\, [\boldsymbol{r}_{tt} + \boldsymbol{\kappa}_{tt}] \cdot \frac{\partial \boldsymbol{\kappa}}{\partial \delta_\alpha} j \, dx^1 \, dx^2. \tag{5.13}$$

5.14. Exercise. Let (3.3b) hold. Use the symmetry conditions on $\mathcal{A}$, $\tilde{\rho}$, and $\boldsymbol{\kappa}$ to show that

$$\begin{aligned}
\boldsymbol{q} &= \left[\int_{\mathcal{A}} \tilde{\rho}\, \kappa^1 j \, dx^1 \, dx^2\right] \boldsymbol{b}, \\
\boldsymbol{k} \cdot \boldsymbol{h}_t &= \left[\int_{\mathcal{A}} \tilde{\rho}\, (\kappa^1)^2 j \, dx^1 \, dx^2\right] \theta_{tt} + 2 \left[\int_{\mathcal{A}} \tilde{\rho} \kappa^1 \frac{\partial \kappa^1}{\partial \delta_\alpha} j \, dx^1 \, dx^2\right] (\partial_t \delta_\alpha) \theta_t, \\
\partial_t \beta^\alpha - \gamma^\alpha &= \left[\int_{\mathcal{A}} \tilde{\rho} \frac{\partial \kappa^1}{\partial \delta_\alpha} j \, dx^1 \, dx^2\right] \boldsymbol{r}_{tt} \cdot \boldsymbol{b} \\
&\quad + \left[\int_{\mathcal{A}} \tilde{\rho} \left(\frac{\partial \kappa^1}{\partial \delta_\mu} \frac{\partial \kappa^1}{\partial \delta_\alpha} + \frac{\partial \kappa^2}{\partial \delta_\mu} \frac{\partial \kappa^2}{\partial \delta_\alpha}\right) j \, dx^1 \, dx^2\right] \partial_{tt} \delta_\mu \\
&\quad + \left[\int_{\mathcal{A}} \tilde{\rho} \left(\frac{\partial^2 \kappa^1}{\partial \delta_\mu \partial \delta_\nu} \frac{\partial \kappa^1}{\partial \delta_\alpha} + \frac{\partial^2 \kappa^2}{\partial \delta_\mu \partial \delta_\nu} \frac{\partial \kappa^2}{\partial \delta_\alpha}\right) j \, dx^1 \, dx^2\right] (\partial_t \delta_\mu)(\partial_t \delta_\nu) \\
&\quad - \left[\int_{\mathcal{A}} \tilde{\rho} \kappa^1 \frac{\partial \kappa^1}{\partial \delta_\alpha} j \, dx^1 \, dx^2\right] (\theta_t)^2.
\end{aligned} \tag{5.15a}$$

Specialize these expressions when (3.3c) is used. In particular, show that if φ^1 can be chosen so that $\int_{\mathcal{A}} \rho \varphi^1 j \, dx^1 \, dx^2 = 0$ (cf. Ex. VIII.4.8), then (5.15a) reduces to

$$\begin{aligned}
\boldsymbol{q} &= \boldsymbol{0}, \\
\boldsymbol{k} \cdot \boldsymbol{h}_t &= \left[\int_{\mathcal{A}} \tilde{\rho}\, (\varphi^1)^2 \det(\varphi^\alpha_{,\beta}) \left(1 - \varphi^1 \mathring{\mu}\right) \, dx^1 \, dx^2\right] ({\delta_1}^2 \theta_t)_t, \\
\partial_t \beta^1 - \gamma^1 &= \left[\int_{\mathcal{A}} \tilde{\rho}\, (\varphi^1)^2 \det(\varphi^\alpha_{,\beta}) \left(1 - \varphi^1 \mathring{\mu}\right) \, dx^1 \, dx^2\right] \left(\partial_{tt} \delta_1 - \delta_1 {\theta_t}^2\right), \\
\partial_t \beta^2 - \gamma^2 &= \left[\int_{\mathcal{A}} \tilde{\rho}\, (\varphi^2)^2 \det(\varphi^\alpha_{,\beta}) \left(1 - \varphi^1 \mathring{\mu}\right) \, dx^1 \, dx^2\right] \partial_{tt} \delta_2.
\end{aligned} \tag{5.15b}$$

Using (3.10), (3.15), (3.16), (3.22), (5.4), and (5.7), we obtain constitutive representations for $(\mathsf{q}, s) \mapsto \hat{\mathsf{Q}}(\mathsf{q}, s)$ for elastic rods:

$$(5.16)\quad \hat{N} = \boldsymbol{a} \cdot \int_{\mathcal{A}} \boldsymbol{F} \cdot \hat{\boldsymbol{S}}_{\mathrm{A}} \cdot \mathring{\boldsymbol{a}} \det(\varphi^{\alpha}_{,\beta})\, dx^1\, dx^2$$

$$= \int_{\mathcal{A}} \frac{\nu - \kappa^1 \mu}{1 - \varphi^1 \mathring{\mu}} \mathring{\boldsymbol{a}} \cdot \hat{\boldsymbol{S}}_{\mathrm{A}} \cdot \mathring{\boldsymbol{a}} \det(\varphi^{\alpha}_{,\beta})\, dx^1\, dx^2,$$

$$(5.17)\quad \hat{H} = \int_{\mathcal{A}} \left[\zeta^1 \frac{\mathring{\boldsymbol{a}}}{1 - \varphi^1 \mathring{\mu}} + \frac{\partial \kappa^1}{\partial x^\alpha} \boldsymbol{g}^\alpha \right] \cdot \hat{\boldsymbol{S}}_{\mathrm{A}} \cdot \mathring{\boldsymbol{a}} \det(\varphi^{\alpha}_{,\beta})\, dx^1\, dx^2,$$

$$(5.18)\quad \hat{M} = -\int_{\mathcal{A}} \kappa^1 \frac{\nu - \kappa^1 \mu}{1 - \varphi^1 \mathring{\mu}} \mathring{\boldsymbol{a}} \cdot \hat{\boldsymbol{S}}_{\mathrm{A}} \cdot \mathring{\boldsymbol{a}} \det(\varphi^{\alpha}_{,\beta})\, dx^1\, dx^2,$$

$$(5.19)\quad \hat{\Omega}^\sigma = \int_{\mathcal{A}} \left\{ \frac{\partial \kappa^1}{\partial \delta_\sigma} \left[\zeta^1 \frac{\mathring{\boldsymbol{a}}}{1 - \varphi^1 \mathring{\mu}} + \frac{\partial \kappa^1}{\partial x^\alpha} \boldsymbol{g}^\alpha \right] + \frac{\partial \kappa^2}{\partial \delta_\sigma} \left[\zeta^2 \frac{\mathring{\boldsymbol{a}}}{1 - \varphi^1 \mathring{\mu}} + \frac{\partial \kappa^2}{\partial x^\alpha} \boldsymbol{g}^\alpha \right] \right\} \cdot \hat{\boldsymbol{S}}_{\mathrm{A}} \cdot \mathring{\boldsymbol{a}} \det(\varphi^{\alpha}_{,\beta})\, dx^1\, dx^2,$$

$$(5.20)\quad \hat{\Delta}^\sigma = \int_{\mathcal{A}} \frac{\partial}{\partial x^k} \left(\frac{\partial \kappa^1}{\partial \delta_\sigma} \boldsymbol{b} + \frac{\partial \kappa^2}{\partial \delta_\sigma} \boldsymbol{k} \right) \cdot \boldsymbol{F} \cdot \hat{\boldsymbol{S}}_{\mathrm{A}} \cdot \boldsymbol{g}^k\, j\, dx^1\, dx^2$$

where the argument $\boldsymbol{C}$ of $\hat{\boldsymbol{S}}_{\mathrm{A}}$ is given by (5.5).

Our objective, motivated by the problems of Chaps. IV, VI, and IX, is to determine conditions stating exactly when these constitutive functions vanish. In particular, we wish to study these functions when $\mathring{\boldsymbol{r}}$ is a circle and when it is a straight line. The problems we study in the former case are typically uniform. We accordingly introduce a number of simplifications.

5.21. Exercise. Decompose $\boldsymbol{C}$ with respect to the basis $\{\mathring{\boldsymbol{b}}, \boldsymbol{k}, \mathring{\boldsymbol{a}}\}$ and use the representation (XIII.1.7) to prove:

5.22. Proposition. *Let the symmetry conditions hold, let κ^1 and κ^2 be independent of x^3, and let $\hat{\boldsymbol{S}}_{\mathrm{A}}$ be an isotropic function of $\boldsymbol{C}$. Then*

$$(5.23a)\qquad \hat{H}(\mathsf{q}, s) = 0 = \hat{\Omega}^\alpha(\mathsf{q}, s)$$

if

$$(5.23b)\qquad \eta = 0 = \omega_\alpha.$$

5.24. Exercise. Prove:

5.25. Proposition. *Let (3.3c) hold, let $\mathcal{A}(s)$ be symmetric about both the x^1- and x^2-axes, let φ^1 and φ^2 be independent of x^3, let φ^1 be odd in*

x^1 and even in x^2, let φ^2 be even in x^1 and odd in x^2, and let $\hat{\boldsymbol{S}}_{\text{A}}$ be an isotropic function of $\boldsymbol{C}$. Then

$$\hat{M}(\mathbf{q}, s) = 0 \tag{5.26a}$$

if (5.23b) *holds and if*

$$\delta_1 \mu = \nu \mathring{\mu}. \tag{5.26b}$$

Note that μ/ν, which appears in (5.26b), is the curvature of $\boldsymbol{r}$ at any point at which $\eta = 0 = \eta_s$. The Monotonicity Condition (4.16) implies that $\hat{M}$ is a strictly increasing function of μ (but does not imply that $\hat{M}$ is a strictly increasing function of the actual curvature). In general, as Proposition 5.25 shows, $\hat{M}$ does not vanish when $\mu = \mathring{\mu}$. If $\mathring{\mu}$, which is the curvature of $\mathring{\boldsymbol{r}}$, is constant, so that the $\mathring{\boldsymbol{r}}$ is a circle, then $\mu - \mathring{\mu}$ measures the departure from circularity. Thus we must expect that a pure inflational deformation taking $\mathring{\boldsymbol{r}}$ to a circle of different radius must be maintained by a nonzero bending couple $\hat{M}$. If we were to follow traditional practice and treat expressions like $\varphi^1 \mathring{\mu}$ and $\varphi^1 \delta_1 \mu/\nu$ as negligible with respect to 1 in (5.18) (i.e., if we were to treat the ratio of thickness to radius of curvature as small in both the reference and deformed configurations), then we would obtain a constitutive function $\hat{M}$ that vanishes when $\mu = \mathring{\mu}$. Such functions lead to more elegant results for problems for rings. Let us note that the requirements of symmetry in Proposition 5.25, leading to the simplifications embodied in (5.26), prevent concomitant simplifications in inertia terms (cf. Ex. VIII.4.8).

For straight rods, the formulas (5.16)–(5.20) are greatly simplified. As a consequence, we obtain much sharper constitutive restrictions, such as

5.27. Proposition. *Let* (3.3b) *hold, let $\mathring{\mu} = 0$, let $\mathcal{A}(s)$ be symmetric about both the x^1- and x^2-axes, let κ^1 be odd in x^1 and even in x^2, let κ^2 be even in x^1 and odd in x^2, and let $\hat{\boldsymbol{S}}_{\text{A}}$ be an isotropic function of $\boldsymbol{C}$. Then under the transformation*

$$(\eta, \mu) \mapsto -(\eta, \mu) \tag{5.28}$$

the induced constitutive functions $\hat{H}$ and $\hat{M}$ change sign, while the other components of $\hat{\mathbf{Q}}$ remain unchanged.

Let us note that the symmetries in this result are physically reasonable. What is not evident is whether there should be more symmetry than this. The method for proving this result is that to be used for Theorem 11.30.

5.29. Exercise. Show that the double symmetry of $\mathcal{A}$ in Proposition 5.27 is essential by specializing (5.18) when (3.3c) holds with $\varphi^\alpha = x^\alpha$, with $\mathring{\mu} = 0$, with the centroid of $\mathcal{A}(s)$ at $x^1 = 0 = x^2$, and with the $\boldsymbol{d}_\alpha$ constrained to be orthonormal. Suppose that $\mathcal{A}$ is not symmetric about the x^2-axis. Show that in general $\hat{H}$ is odd in η, $\hat{M}$ and $\hat{N}$ are even in η, $\hat{M}(\nu, \eta, 0, s) = 0$, but that $\hat{M}(\nu, \eta, \cdot, s)$ is not odd. (For the last step, it suffices to show that $\hat{M}_{\mu\mu}(\nu, 0, 0, s) \neq 0$.) Illustrate the computations when $\mathcal{A}$ is an isosceles triangle.

5.30. Exercise. Carry out a perturbation analysis of the the problem treated in Sec. VI.5 under the sole symmetry assumption that $\hat{H}$ is odd in η. For a uniform rod, show that $\lambda_{(1)}$ (cf. Sec. VI.6) typically vanishes, despite the lack of symmetry, so that bifurcation is not (strictly) transcritical. To get a transcritical bifurcation for a uniform rod, a more refined theory may be required (see Buzano, Geymonat, & Poston (1985) and Antman & Marlow (1992)). Show that for a nonuniform rod, transcritical bifurcation is possible. (By applying the Alternative Theorem, show that $\lambda_{(1)}$ depends on integrals of products of powers of the eigenfunctions with second derivatives of the constitutive functions evaluated at the trivial solution. The eigenfunctions are determined just by the first derivatives of the constitutive functions evaluated at the trivial solution.) Note that in optimal designs (see Cox & Overton (1992)), the rod is not uniform. Therefore, transcritical bifurcations are possible. Their presence may deprive the optimal design (based on maximizing the minimum buckling load) of any physical meaning.

We can get explicit constitutive functions for $\hat{\mathbf{Q}}$ by simply choosing an explicit form for $\hat{S}_{\mathrm{A}}$. In particular, for an isotropic hyperelastic material, we can take the stored-energy function to be a sum of the invariants of $\boldsymbol{C}$ raised to various powers (see Antman & Calderer (1985a), Antman & Negrón-Marrero (1987), and Ogden (1984)). The constitutive assumptions supporting the treatment in Sec. VI.6, e.g., are inspired by such constructions.

5.31. Exercise. Let $W(\boldsymbol{C}) = V(\boldsymbol{C}) - V(\boldsymbol{I}) - [\partial V(\boldsymbol{I})/\partial \boldsymbol{C}] : (\boldsymbol{C} - \boldsymbol{I})$ with

$$V(\boldsymbol{C}) = A(\mathrm{III}_{\boldsymbol{C}})^{-1} + B\mathrm{I}_{\boldsymbol{C}} + \tfrac{1}{2}[(\mathrm{I}_{\boldsymbol{C}})^2 - 2\mathrm{II}_{\boldsymbol{C}}] + \tfrac{1}{2}D(\mathrm{III}_{\boldsymbol{C}})^2$$

where A, B, C, D are positive constants. Find the two leading terms for the $\bar{N}$ and the leading terms for $\bar{H}$ and $\bar{M}$ in the asymptotic representations (IV.4.17).

5.32. Exercise. Let (3.3b) hold. Let $\boldsymbol{f}^0 = \mathbf{0}$, $\boldsymbol{l} = \mathbf{0}$, $l^\alpha = 0$. Let $\boldsymbol{e}_1(s) \equiv \cos s\boldsymbol{i} + \sin s\boldsymbol{j}$. A *free radial motion* of a ring is a solution (5.1)–(5.3), (5.9)–(5.11) for which $\boldsymbol{r}(s,t) = r(t)\boldsymbol{e}_1(s)$, $\boldsymbol{d}_1(s,t) = \delta_1(t)\boldsymbol{e}_1(s)$, $\boldsymbol{d}_2(s,t) = \delta_2(t)\boldsymbol{k}$, $\theta(s,t) = s$. Find the ordinary differential equations governing this motion for an elastic ring of reference radius 1. Specialize the results to the case that (3.3c) holds.

5.33. Problem. Formulate the equations for travelling waves in a ring for which $\delta_1 = 1 = \delta_2$. Analyze their qualitative behavior (see Sec. IX.3). How do these results differ from those for straight rods? Suppose that the ring is subjected to a hydrostatic pressure of a form discussed in Sec. III.5. Analyze the qualitative behavior of the governing equations.

For simplicity, let us now limit our attention to the special Cosserat theory, which is obtained in Sec. 3 by taking $\boldsymbol{a}_1$ and $\boldsymbol{a}_2$ orthonormal and by taking $\delta_1 = 1 = \delta_2$. The methods developed above can be applied to any class of rod problems. We illustrate this principle in the following two important exercises.

5.34. Exercise. Suppose that the reference configuration, the material response, and the representation (3.3c) for an elastic rod are symmetric with respect to the $\{\boldsymbol{i}, \boldsymbol{j}\}$-plane. In the reference configuration, let

$$\mathring{\boldsymbol{r}}_s = \mathring{\boldsymbol{d}}_3, \quad \mathring{\boldsymbol{d}}_1 = -\sin\mathring{\theta}\boldsymbol{i} + \cos\mathring{\theta}\boldsymbol{j}, \quad \mathring{\boldsymbol{d}}_2 = \boldsymbol{k}, \quad \mathring{\boldsymbol{d}}_3 = \cos\mathring{\theta}\boldsymbol{i} + \sin\mathring{\theta}\boldsymbol{j}. \tag{5.35}$$

As in Chap. VIII, set

$$\begin{gathered} \boldsymbol{r}_s = v_k\boldsymbol{d}_k, \quad \partial_s\boldsymbol{d}_k = \boldsymbol{u}\times\boldsymbol{d}_k, \quad \boldsymbol{u} = u_k\boldsymbol{d}_k, \\ \boldsymbol{n} = n_k\boldsymbol{d}_k, \quad \boldsymbol{m} = m_k\boldsymbol{d}_k \end{gathered} \tag{5.36}$$

where $\boldsymbol{n}$ and $\boldsymbol{m}$ are defined in (3.10) and (3.22). Prove that if (3.3c) holds, then the constitutive functions $\hat{m}_k, \hat{n}_k$, which depend on the u_l, v_l and s, have the following symmetries:

(i) $\hat{n}_1$ is odd in v_1; the other constitutive functions are even in v_1.
(ii) $\hat{m}_1$ is odd in u_1; the other constitutive functions are even in u_1.
(iii) $\hat{n}_2$ and $\hat{m}_3$ change sign when v_2 and u_3 change sign; the other constitutive functions are unchanged when v_2 and u_3 change sign.

If the constitutive equations are invertible, then these symmetries hold if and only if there are scalar-valued functions $\sigma_1, \sigma_{32}, \sigma_3, \tau_1, \tau_2, \tau_{23}$ such that

$$(5.37) \qquad \begin{aligned} u_1 &= \sigma_1(\iota, s)m_1, \quad & u_3 &= \sigma_{32}(\iota, s)n_2 + \sigma_3(\iota, s)m_3, \\ v_1 &= \tau_1(\iota, s)n_1, \quad & v_2 &= \tau_2(\iota, s)n_2 + \tau_{23}(\iota, s)m_3 \end{aligned}$$

where $\iota \equiv (n_1^2, n_3, m_1^2, m_2, n_2^2 + m_3^2, n_2 m_3)$. Show that for a hyperelastic rod

$$(5.38) \qquad \tau_2 = \sigma_3, \quad \tau_{23} = \sigma_{32}.$$

(These are the equations used in Sec. IX.5.) Prove that if furthermore $\mathcal{A}$ is also symmetric about the x^2-axis, then $\hat{m}^2$ is odd in u_2, and the other constitutive functions are even in u_2. Note that the specialization of these results gives (IV.2.9).

5.39. Exercise. Let (3.3c) hold and let the reference configuration of the rod be straight and axisymmetric. Let φ^1 be odd in x^1 and even in x^2, and let φ^1 be even in x^1 and odd in x^2. Let $\hat{\boldsymbol{S}}$ be an isotropic function of $\boldsymbol{C}$. Prove that the resulting constitutive functions are transversely isotropic in the sense of Sec. VIII.9.

5.40. Exercise. The form of a rod theory depends upon the choice of $\mathring{\boldsymbol{r}}$ within the three-dimensional body. Determine the formulas relating the rod theories for two different choices of $\mathring{\boldsymbol{r}}$ when (3.3c) holds. (These formulas can be very useful when one choice of $\mathring{\boldsymbol{r}}$ gives the constitutive equations a lot of symmetry, whereas a different choice is particularly convenient for a special class of problems, e.g., contact problems, in which $\mathring{\boldsymbol{r}}$ might be chosen to be a material curve on the boundary of the body $\mathcal{B}$ that can come into unilateral contact with some prescribed surface.)

6. Necking

We study a special class of planar equilibrium problems for an initially straight rod having two directors. We seek solutions of the equilibrium equations of Sec. 5 having the form

$$(6.1) \qquad \boldsymbol{r}' = \nu \boldsymbol{i}, \quad \boldsymbol{d}_1 = \delta \boldsymbol{j}, \quad \boldsymbol{d}_2 = \delta \boldsymbol{k}.$$

These restrictions correspond to deformations in which the rod can locally change its length and its thickness, but cannot bend, twist, or shear.

6.2. Exercise. Show that if $\hat{\boldsymbol{S}}_{\text{A}}$ and (3.3) are isotropic (or, more generally, transversely isotropic) and if $\boldsymbol{f}^0 = \boldsymbol{0}$, $\boldsymbol{f}^\alpha = \boldsymbol{0}$, $\boldsymbol{l} = \boldsymbol{0}$, then the nontrivial equilibrium equations of (5.9)–(5.11), (5.16)–(5.20) reduce to

$$(6.3) \qquad N' = 0,$$

$$(6.4) \qquad \Omega' - \Delta = 0,$$

$$(6.5) \qquad \begin{aligned} N(s) &\equiv \boldsymbol{n}(s) \cdot \boldsymbol{i} = \hat{N}(\nu(s), \delta'(s), \delta(s), s), \\ \Omega(s) &\equiv \boldsymbol{m}^1(s) \cdot \boldsymbol{j} \equiv \boldsymbol{m}^2(s) \cdot \boldsymbol{k} = \hat{\Omega}(\nu(s), \delta'(s), \delta(s), s), \\ \Delta(s) &\equiv \boldsymbol{n}^1(s) \cdot \boldsymbol{j} \equiv \boldsymbol{n}^2(s) \cdot \boldsymbol{k} = \hat{\Delta}(\nu(s), \delta'(s), \delta(s), s). \end{aligned}$$

A combination of the methods used to prove Propositions 5.22 and 5.27 implies that

$$(6.6) \qquad \hat{\Omega}(\nu, \cdot, \delta, s) \text{ is odd, } \hat{N}(\nu, \cdot, \delta, s) \text{ and } \hat{\Delta}(\nu, \cdot, \delta, s) \text{ are even.}$$

The Strong Ellipticity Condition implies that

$$\text{(6.7a,b)} \qquad \frac{\partial(\hat{N},\hat{\Omega})}{\partial(\nu,\omega)} \text{ is positive-definite}, \quad \frac{\partial\hat{\Delta}}{\partial\delta} > 0.$$

But the Strong Ellipticity Condition does not imply that

$$\text{(6.7c)} \qquad \frac{\partial(\hat{N},\hat{\Delta})}{\partial(\nu,\delta)} \text{ is positive-definite}.$$

We take $[s_1, s_2] = [0, 1]$. We assume that the deformed length l of the rod is prescribed:

$$\text{(6.8)} \qquad \boldsymbol{r}(0) = \boldsymbol{0}, \quad \boldsymbol{r}(1) = l\boldsymbol{i}.$$

We could weld the ends to rigid walls perpendicular to the $\boldsymbol{i}$-axis:

$$\text{(6.9)} \qquad \delta(0) = 1 = \delta(1)$$

or we could assume that there is no restraint preventing the rod from changing its thickness at the ends:

$$\text{(6.10a)} \qquad \Omega(0) = 0 = \Omega(1).$$

Conditions (6.6) and (6.7) ensure that (6.10a) is equivalent to

$$\text{(6.10b)} \qquad \delta'(0) = 0 = \delta'(1).$$

Let us strengthen (6.7a) by assuming that

$$\text{(6.11)} \qquad \hat{N}(\nu,\omega,\delta,s)\frac{\nu-1}{|\nu-1|} + \hat{\Omega}(\nu,\omega,\delta,s)\frac{\omega}{|\omega|} \to \infty \quad \text{as } \nu^2+\omega^2 \to \infty \text{ or as } \nu \to 0$$

for each fixed δ and s. It then follows from Theorem XIX.2.30 that

$$(\nu,\omega) \mapsto (\hat{N}(\nu,\omega,\delta,s), \hat{\Omega}(\nu,\omega,\delta,s))$$

has the inverse

$$(N,\Omega) \mapsto (\nu^\sharp(N,\Omega,\delta,s), \omega^\sharp(N,\Omega,\delta,s)).$$

We define

$$\text{(6.12)} \qquad \Delta^\sharp(N,\Omega,\delta,s) \equiv \hat{\Delta}(\nu^\sharp(N,\Omega,\delta,s), \omega^\sharp(N,\Omega,\delta,s), \delta, s).$$

Condition (6.6) implies that $\omega^\sharp$ is odd in Ω and that $\nu^\sharp$ and $\Delta^\sharp$ are even in Ω.

Thus (6.3)–(6.5) is equivalent to

$$N = \text{const.}, \tag{6.13a}$$

$$\delta' = \omega^\sharp(N, \Omega, \delta, s), \tag{6.13b}$$

$$\Omega' = \Delta^\sharp(N, \Omega, \delta, s). \tag{6.13c}$$

Then (6.8) implies that

$$l = \int_0^1 \nu^\sharp(N, \Omega(s), \delta(s), s)\, ds. \tag{6.14}$$

Let us seek a trivial solution of (6.13) and (6.14) in which $\Omega = 0$. Then (6.13b) implies that $\delta = $ const. and (6.13c) reduces to $\Delta^\sharp(N, 0, \delta, s) = 0$, which is generally incompatible with the constancy of N and δ unless $\Delta^\sharp$ is independent of s. Let us accordingly assume that the rod is uniform, so that the constitutive functions are independent of s. In this case, there is a trivial solution of (6.13) and (6.14) if and only if there are numbers N and δ such that

$$\nu^\sharp(N, 0, \delta) = l, \quad \Delta^\sharp(N, 0, \delta) = 0. \tag{6.15}$$

This system is equivalent to

$$N = \hat{N}(l, 0, \delta), \quad 0 = \hat{\Delta}(l, 0, \delta). \tag{6.16a,b}$$

Let us strengthen (6.7b) by requiring that

$$\hat{\Delta}(\nu, 0, \delta) \to \left\{ \begin{matrix} \infty \\ -\infty \end{matrix} \right\} \quad \text{as} \quad \delta \to \left\{ \begin{matrix} \infty \\ 0 \end{matrix} \right\} \tag{6.17}$$

for each fixed ν. Then (6.7b) and (6.17) imply that (6.16b) has a unique solution $\delta_0(l)$, from which we obtain $N = N^0(l) \equiv \hat{N}(l, 0, \delta_0(l))$. In general, $\delta_0(l) \neq 1$ so that (6.9) cannot be expected to hold. Thus trivial solutions are associated with the boundary condition (6.10). We adopt the convention that if $(\nu, \omega, \delta) \mapsto \hat{R}(\nu, \omega, \delta)$ is any constitutive function, then $R^0(l) \equiv \hat{R}(l, 0, \delta_0(l))$. Using (6.16), we then find that

$$\frac{dN^0}{dl} = \frac{N^0_\nu \Delta^0_\delta - N^0_\delta \Delta^0_\nu}{\Delta^0_\delta}. \tag{6.18}$$

Since we do not adopt (6.7c), this derivative need not be positive.

Let us first study (6.13), (6.14), and (6.10) by the perturbation method. We seek nontrivial solution branches in the form

$$\begin{aligned} l(\varepsilon) &= l_0 + \varepsilon l_1 + \frac{\varepsilon^2}{2!} l_2 + \cdots, \\ \Omega(s, \varepsilon) &= \varepsilon \Omega_1(s) + \frac{\varepsilon^2}{2!} \Omega_2(s) + \cdots, \quad \text{etc.} \end{aligned} \tag{6.19}$$

In light of the symmetry condition (6.6), we find that the first-order terms satisfy

(6.20) $$\delta_1' = \omega_\Omega^\sharp(N^0(l_0), 0, \delta_0(l_0))\Omega_1,$$

(6.21) $$\Omega_1' = \Delta_N^\sharp(N^0(l_0), 0, \delta_0(l_0))N_1 + \Delta_\delta^\sharp(N^0(l_0), 0, \delta_0(l_0))\delta_1,$$

(6.22) $$l_1 = \nu_N^\sharp(N^0(l_0), 0, \delta_0(l_0))N_1 + \nu_\delta^\sharp(N^0(l_0), 0, \delta_0(l_0)) \int_0^1 \delta_1(s)\, ds,$$

(6.23) $$\Omega_1(0) = 0 = \Omega_1(1)$$

where N_1 is a constant. From (6.20) and (6.21) we obtain

(6.24) $$\Omega_1'' + q(l_0)\Omega_1 = 0$$

where

(6.25) $$\begin{aligned} q(l) &\equiv -\Delta_\delta^\sharp(N^0(l_0), 0, \delta_0(l_0))\omega_\Omega^\sharp(N^0(l_0), 0, \delta_0(l_0)) \\ &= -\frac{N_\nu^0 \Delta_\delta^0 - N_\delta^0 \Delta_\nu^0}{N_\nu^0 \Omega_\omega^0} = -\frac{\Delta_\delta^0}{N_\nu^0 \Omega_\omega^0}\frac{dN^0}{dl}. \end{aligned}$$

Note that Ω is a variable ideally suited for measuring the deviation from triviality. The boundary-value problem (6.23), (6.24) has a nontrivial solution if and only if there is a positive integer n such that

(6.26) $$q(l_0) = n^2\pi^2.$$

To interpret the significance of (6.26), we sketch a typical graph of N^0 in Fig. 6.27 drawn with a solid line. Bifurcation theory says that a nontrivial branch bifurcates from the trivial branch at any value of l_0 satisfying (6.26) if either l_0 is a simple solution or if the material is hyperelastic. In view of (6.18), bifurcation thus cannot occur until dN^0/dl descends to a negative threshold. The first bifurcating branch is sketched in Fig. 6.27 with a dashed line. Corresponding to it is a necked state, i.e., a configuration with variable thickness. The graph consisting of that for the trivial solution up to the first critical value of l and that for the bifurcating branch beyond that value is presumably observed in experiment.

6.28. Exercise. Carry out a perturbation analysis of (6.13), (6.14), and (6.10) that accounts for deviations from the solutions of the linear problem (6.20)–(6.23).

6.29. Exercise. Discuss alterations necessary to treat problems in which N rather than l is prescribed.

6.30. Exercise. Carry out a global bifurcation analysis of (6.13), (6.14), and (6.10), showing that nodal properties of Ω inherited from the linearization at a simple eigenvalue l are preserved globally.

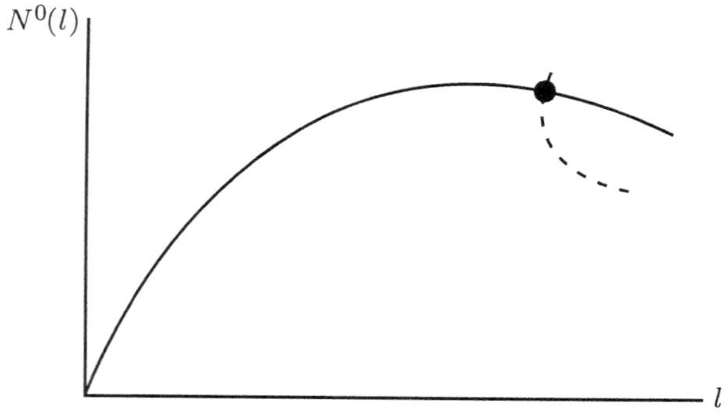

Fig. 6.27. Force-elongation curves for an elastic rod under tension. The dashed curve corresponds to a bifurcating branch corresponding to a necked configuration.

6.31. Exercise. Describe how the phase portrait of (6.13b,c) behaves for different ranges of l and relate its behavior to that of (6.18). Use the phase portrait to describe what δ must look like for any possible solution of (6.13b,c). In particular, describe the qualitative behavior of solutions satisfying (6.9).

6.32. Problem. Analyze deformations in which the rod can twist as well as change its length and thickness.

The material in this section is based upon that of Antman (1973a) and Antman & Carbone (1977, ©Noordhoff International Publishing, Leyden). Excerpts of this last work are reprinted by permission of Kluwer Academic Publishers.

7. Intrinsic Theories of Rods

In Chap. VIII we studied the special intrinsic theory of rods in which a configuration is defined by a position vector function, locating the base curve, and by an orthonormal pair of vector functions, the directors, which characterize the orientation of a cross section. The corresponding equations of motion are just the balances of linear and angular momentum. The main difficulties we faced in constructing this theory were to account for the preservation of orientation, to obtain suitable expressions for the linear and angular momenta, and to lay down suitable constitutive assumptions. We motivated our treatment of these issues by interpreting the deformation as that for a constrained three-dimensional body. A considerable refinement and generalization of this approach, including a treatment of constitutive relations not possible in Chap. VIII, is carried out in Secs. 1–6 of the present chapter.

We now address the question of constructing intrinsic theories of rods with a level of generality approaching that for induced theories presented in Secs. 1–6. (Cf. Sec. VIII.17.) In particular, to what extent can we produce an intrinsic analog of the two-director theory described in Sec. 3?

We begin our study of intrinsic theories by defining the configuration of a rod by a set of vector-valued functions

$$(s,t) \mapsto \boldsymbol{r}(s,t) \equiv \boldsymbol{d}_0(s,t), \boldsymbol{d}_1(s,t), \dots, \boldsymbol{d}_K(s,t). \tag{7.1}$$

This set of functions may be interpreted as a special choice of u introduced in Sec. 2. We could specialize (7.1) by requiring that under the rigid motion $\boldsymbol{z} \mapsto \boldsymbol{c}(t) + \boldsymbol{Q}(t) \cdot \boldsymbol{z}$

the vectors of (7.1) transform according to

$$\text{(7.2)} \qquad \boldsymbol{r}(s,t) \mapsto \boldsymbol{c}(t) + \boldsymbol{Q}(t)\cdot\boldsymbol{r}(s,t), \quad \boldsymbol{d}_k(s,t) \mapsto \boldsymbol{Q}(t)\cdot\boldsymbol{d}_k(s,t), \quad k = 1,\dots,K.$$

This condition is suggested by (3.3c). By examining (2.5) and (3.3), we see that (7.2) restricts the interpretations of (7.1).

Corresponding to (7.1) we can adopt as the weak form of our governing equations of motion the Principle of Virtual Power:

(7.3)

$$\int_0^\infty \int_{s_1}^{s_2} \sum_{k=0}^{K} \left(\boldsymbol{m}^k \cdot \partial_s \overset{\vartriangle}{\boldsymbol{d}}_k + \boldsymbol{n}^k \cdot \overset{\vartriangle}{\boldsymbol{d}}_k - \boldsymbol{f}^k \cdot \overset{\vartriangle}{\boldsymbol{d}}_k - \boldsymbol{b}^k \cdot \partial_t \overset{\vartriangle}{\boldsymbol{d}}_k - \boldsymbol{c}^k \cdot \overset{\vartriangle}{\boldsymbol{d}}_k \right) ds\, dt$$
$$- \int_0^\infty \sum_{k=0}^{K} \left[\bar{\boldsymbol{m}}_2^k(t) \cdot \overset{\vartriangle}{\boldsymbol{d}}_k(s_2,t) - \bar{\boldsymbol{m}}_1^k(t) \cdot \overset{\vartriangle}{\boldsymbol{d}}_k(s_1,t) \right] dt - \int_{s_1}^{s_2} \sum_{k=0}^{K} \bar{\boldsymbol{b}}^k(t) \cdot \overset{\vartriangle}{\boldsymbol{d}}_k(s,0)\, ds = 0$$

for sufficiently smooth $\overset{\vartriangle}{\boldsymbol{d}}_k$ satisfying appropriate boundary conditions and initial conditions (cf. (2.22)). If (7.2) holds, then we are motivated by (2.5), (2.26), (2.27), and (3.3) to choose

$$\text{(7.4)} \qquad \boldsymbol{b}^k = \sum_{l=0}^{K} J^{kl}(s) \partial_t \boldsymbol{d}_l, \quad \boldsymbol{c}^k = \boldsymbol{0}$$

where the J^{kl} are prescribed. In this case, we require that the quadratic form

$$\text{(7.5)} \qquad \tfrac{1}{2} \sum_{k,l=0}^{K} J^{kl} \boldsymbol{v}_k \cdot \boldsymbol{v}_l$$

corresponding to the kinetic energy be positive-definite.

We can define constitutive equations for an elastic material by requiring the $\boldsymbol{m}^k$ and $\boldsymbol{n}^k$ to depend on the

$$\text{(7.6)} \qquad \boldsymbol{d}_0, \dots, \boldsymbol{d}_K, \partial_s \boldsymbol{d}_0, \dots, \partial_s \boldsymbol{d}_K.$$

If (7.2) holds, we can treat frame-indifference by the methods of Sec. VIII.7.

We can deduce the governing equations for a hyperelastic rod by using Hamilton's principle, corresponding to (7.3). In particular, if (7.2) holds, we can take the kinetic energy to be (7.5) with $\boldsymbol{v}_k$ replaced with $\partial_t \boldsymbol{d}_k$ and the stored energy to depend upon the scalar invariants of (7.6).

Thus the use of (7.2) enables us to produce intrinsic theories of rods with a simple and natural mathematical structure. These theories may be regarded as abstractions of those presented in Secs. 1–6. In this case, the development in Secs. 1–6 may be used to motivate the criteria for the preservation of orientation, the expressions for the linear and angular momenta, and constitutive assumptions.

8. Mielke's Treatment of St. Venant's Principle

In this section we describe how the rod theory of Chap. VIII can be given a precise mathematical justification as furnishing an approximation to certain equilibrium problems of the three-dimensional theory. We emphasize the structure of the governing equations, sketching but briefly how the underlying analytic tools can be used to yield the basic Theorem 8.45.

We study the behavior of equilibrium solutions of a nonlinearly elastic body having as its natural reference configuration the cylindrical segment

$$\text{(8.1)} \qquad \mathcal{B} = \{\boldsymbol{z} = z^\alpha \boldsymbol{i}_\alpha + s\boldsymbol{k} \colon (z^1, z^2) \in \mathcal{A},\ -l < s < l\}$$

where $\mathcal{A}$ is a given domain in $\mathbb{R}^2$ having centroid at $(0,0)$ (so that $\int_{\mathcal{A}} z^\alpha \, dz^1 \, dz^2 = 0$) and having z^1 and z^2 as principal axes of inertia (so that $\int_{\mathcal{A}} z^1 z^2 \, dz^1 \, dz^2 = 0$). We assume that the material properties of the body do not vary with s. We assume that the body force $\boldsymbol{f} = \boldsymbol{0}$ and that the traction on the lateral surface $\mathcal{L}$ is $\boldsymbol{0}$. *St. Venant's problem* is to find all equilibrium configurations of this body when the resultant force and moment are prescribed at each end. Thus we confront the problem of finding all solutions $\boldsymbol{p}$ of

$$\nabla \cdot \boldsymbol{T}^* = \boldsymbol{0} \quad \text{in} \quad \mathcal{B}, \tag{8.2}$$

$$\boldsymbol{T} \cdot \boldsymbol{\nu} = \boldsymbol{0} \quad \text{on} \quad \mathcal{L}, \tag{8.3}$$

$$\boldsymbol{T}(\boldsymbol{z}) = \hat{\boldsymbol{T}}\big(\boldsymbol{F}(\boldsymbol{z}), z^1, z^2\big) = \boldsymbol{F}(\boldsymbol{z}) \cdot \tilde{\boldsymbol{S}}\big(\boldsymbol{E}(\boldsymbol{z}), z^1, z^2\big), \tag{8.4}$$

$$\int_{\mathcal{A}} \boldsymbol{T}(\pm l, z^1, z^2) \cdot \boldsymbol{k} \, dz^1 \, dz^2 \quad \text{is prescribed,} \tag{8.5a}$$

$$\int_{\mathcal{A}} \boldsymbol{p}(\pm l, z^1, z^2) \times [\boldsymbol{T}(\pm l, z^1, z^2) \cdot \boldsymbol{k}] \, dz^1 \, dz^2 \quad \text{is prescribed} \tag{8.5b}$$

where $2\boldsymbol{E} \equiv \boldsymbol{C} - \boldsymbol{I}$ and where $\tilde{\boldsymbol{S}}\big(\boldsymbol{O}, z^1, z^2\big) = \boldsymbol{O}$ because the reference configuration is natural. We assume that $\tilde{\boldsymbol{S}}(\cdot, z^1, z^2) \in C^2$.

We shall rewrite (8.2)–(8.5) as a perturbation of the equilibrium equations for the special Cosserat theory of rods. We accordingly decompose the position field in a form that gives a distinguished role to the variable s:

$$\boldsymbol{p}(\boldsymbol{z}) = \boldsymbol{r}(s) + z^\alpha \boldsymbol{d}_\alpha(s) + w^k(\boldsymbol{z}) \boldsymbol{d}_k(s) \tag{8.6}$$

where $\{\boldsymbol{d}_k(s)\}$ is an orthonormal right-handed basis. We require that $\boldsymbol{r}(s)$ be the mean position of the section s:

$$\boldsymbol{r}(s) = \frac{1}{a(\mathcal{A})} \int_{\mathcal{A}} \boldsymbol{p}(z^1, z^2, s) \, dz^1 \, dz^2 \tag{8.7}$$

where $a(\mathcal{A})$ is the area of $\mathcal{A}$. We likewise require that the pair $\boldsymbol{d}_1, \boldsymbol{d}_2$ give a mean configuration of the deformed section. To express this restriction in a concrete mathematical form, we first observe that the cross product $z^\alpha \boldsymbol{d}_\alpha(s) \times [\boldsymbol{p}(\boldsymbol{z}) - \boldsymbol{r}(s)]$ is a measure of the discrepancy between the two vectors in the product. We wish to choose the pair $\boldsymbol{d}_1, \boldsymbol{d}_2$ so that the integral of this cross product over $\mathcal{A}$ vanishes. It then follows from (8.6) that

$$\int_{\mathcal{A}} z^\alpha \boldsymbol{d}_\alpha(s) \times w^k(\boldsymbol{z}) \boldsymbol{d}_k(s) \, dz^1 \, dz^2 = \boldsymbol{0}. \tag{8.8}$$

Thus (8.6)–(8.8) imply the six conditions

$$\int_{\mathcal{A}} w^k \, dz^1 \, dz^2 = 0, \quad \int_{\mathcal{A}} z^\alpha w^3 \, dz^1 \, dz^2 = 0, \quad \int_{\mathcal{A}} (z^1 w^2 - z^2 w^1) \, dz^1 \, dz^2 = 0, \tag{8.9}$$

which make (8.6) well-defined provided we can find $\boldsymbol{d}_1$ and $\boldsymbol{d}_2$.

Our aim is to express the $\boldsymbol{d}_\alpha$ and the w^k in terms of $\boldsymbol{p}$ so that we can identify conditions under which the last summand in (8.6) is small. We first indicate the kinds of conditions that would enable us to express $\boldsymbol{d}_1$ and $\boldsymbol{d}_2$ in terms of $\boldsymbol{p}$. Equations (8.6)–(8.8) and the properties of z^α imply that

$$\boldsymbol{d}_\alpha \times \boldsymbol{q}^\alpha = \boldsymbol{0}, \quad \boldsymbol{d}_3 \cdot \boldsymbol{q}^\alpha = 0, \quad \boldsymbol{q}^\alpha \equiv \int_{\mathcal{A}} z^\alpha \boldsymbol{p} \, dz^1 \, dz^2. \tag{8.10a,b,c}$$

If $\boldsymbol{q}^1$ and $\boldsymbol{q}^2$ are independent, i.e., if $\boldsymbol{q}^1 \times \boldsymbol{q}^2 \neq \boldsymbol{0}$, then $\boldsymbol{d}_1$ and $\boldsymbol{d}_2$ lie in the plane spanned by $\boldsymbol{q}^1$ and $\boldsymbol{q}^2$, and (8.10a) locates $\boldsymbol{d}_1$ and $\boldsymbol{d}_2$ in this plane to within an angle π. To show that this independence is plausible, suppose that $\boldsymbol{p}(\boldsymbol{z}) - \boldsymbol{r}(s) - \boldsymbol{p}_{,\alpha}(0,0,s)z^\alpha$ is 'small'. Then $\boldsymbol{q}^1 \times \boldsymbol{q}^2$ approximately equals

$$[\boldsymbol{p}_{,1}(0,0,s) \times \boldsymbol{p}_{,2}(0,0,s)] \left[\int_{\mathcal{A}} (z^1)^2\, dz^1\, dz^2\right] \left[\int_{\mathcal{A}} (z^2)^2\, dz^1\, dz^2\right],$$

which does not vanish by virtue of (XII.1.1).

Since $\{\boldsymbol{d}_k\}$ is orthonormal, we can write

$$\text{(8.11a)} \qquad \boldsymbol{d}_k'(s) = \boldsymbol{u}(s) \times \boldsymbol{d}_k(s), \quad \boldsymbol{u} = u^l \boldsymbol{d}_l.$$

Let us set

$$\text{(8.11b)} \qquad \boldsymbol{p}_{,3}(\boldsymbol{z}) = \boldsymbol{d}_3(s) + y^k(\boldsymbol{z})\boldsymbol{d}_k(s).$$

Equating (8.11b) to the corresponding derivative of (8.6) we obtain

$$\text{(8.12)} \qquad \boldsymbol{r}' + z^\alpha \boldsymbol{u} \times \boldsymbol{d}_\alpha + w^k \boldsymbol{u} \times \boldsymbol{d}_k + w^k_{,3}\boldsymbol{d}_k = \boldsymbol{d}_3 + y^k \boldsymbol{d}_k.$$

Integrating (8.12) over $\mathcal{A}$ and using (8.9) and the properties of the z^α, we obtain

$$\text{(8.13)} \qquad \boldsymbol{r}' = \boldsymbol{d}_3 + \left[\frac{1}{a(\mathcal{A})} \int_{\mathcal{A}} y^k(z^1, z^2, s)\, dz^1\, dz^2\right] \boldsymbol{d}_k.$$

Now we operate on (8.12) with $\int_{\mathcal{A}} dz^1\, dz^2 z^\beta \boldsymbol{d}_\beta \times$ and use (8.9) to obtain

$$\text{(8.14)} \qquad \left[\int_{\mathcal{A}} z^\alpha (z^\beta + w^\beta) dz^1\, dz^2\right] (\boldsymbol{u}\delta_{\alpha\beta} - u_\alpha \boldsymbol{d}_\beta)$$
$$= \left[\int_{\mathcal{A}} \varepsilon_{\alpha kl} z^\alpha (y^k - \delta^k_\beta w^\beta_{,3})\, dz^1\, dz^2\right] \boldsymbol{d}^l = \left[\int_{\mathcal{A}} \varepsilon_{\alpha kl} z^\alpha y^k\, dz^1\, dz^2\right] \boldsymbol{d}^l.$$

By successively dotting (8.14) with the $\boldsymbol{d}_m$, we can easily solve (8.14) for the u^k provided that the the integrals $\int_{\mathcal{A}} z^\alpha w^\beta\, dz^1\, dz^2$ are sufficiently small. We denote this solution by

$$\text{(8.15)} \qquad \mathsf{u} = \mathsf{u}[\mathsf{w}, \mathsf{y}],$$

where $\mathsf{w} \equiv (w^1, w^2, w^3)$, etc. It is easy to see that $\mathsf{u}[\mathsf{w}, \mathsf{y}] = O(|\mathsf{w}| + |\mathsf{y}|)$ as $|\mathsf{w}| + |\mathsf{y}| \to 0$. Thus we can replace (8.11a) with

$$\text{(8.16)} \qquad \boldsymbol{d}_k' = u^l[\mathsf{w}, \mathsf{y}]\boldsymbol{d}_l \times \boldsymbol{d}_k.$$

If w and y are given, then we can solve the ordinary differential equations (8.13) and (8.16) globally (because (8.16) admits the integrals that the $\boldsymbol{d}_k \cdot \boldsymbol{d}_l$ are constants).

We now obtain equations for w and y. From (8.12) and (8.13) we immediately obtain

$$\text{(8.17)} \quad w^k_{,3}(\boldsymbol{z}) = y^k(\boldsymbol{z}) - \left[\frac{1}{a(\mathcal{A})} \int_{\mathcal{A}} y^k(\boldsymbol{z})\, dz^1\, dz^2\right] - \varepsilon_{kml}\{w^l(\boldsymbol{z}) + \delta^l_\alpha(s) z^\alpha\} u^m[\mathsf{w}, \mathsf{y}](s).$$

Now (8.6) and (8.11) imply that

$$\text{(8.18)} \qquad \boldsymbol{F} = \boldsymbol{p}_{,l}\boldsymbol{i}^l = (\boldsymbol{d}_\alpha + w^l_{,\alpha}\boldsymbol{d}_l)\boldsymbol{i}^\alpha + (\boldsymbol{d}_3 + y^l \boldsymbol{d}_l)\boldsymbol{k},$$

$$\text{(8.19)} \qquad 2\boldsymbol{E} = (w_{\alpha,\beta} + w_{\beta,\alpha} + w^k_{,\alpha} w_{k,\beta})\boldsymbol{i}^\alpha \boldsymbol{i}^\beta$$
$$+ (y_\alpha + w^3_{,\alpha} + w^k_{,\alpha} y_k)(\boldsymbol{i}^\alpha \boldsymbol{k} + \boldsymbol{k}\, \boldsymbol{i}^\alpha) + (2y^3 + y^k y_k)\boldsymbol{k}\boldsymbol{k}.$$

Thus

$$\text{(8.20)} \qquad \boldsymbol{F}_{,3} = y^l_{,3}\boldsymbol{d}_l \boldsymbol{k} + \cdots$$

where the ellipsis stands for terms independent of the $y^l_{,3}$. By using (8.16) and (8.17), we see that the ellipsis can be made to depend upon $\boldsymbol{d}_k$, w, y, $\mathsf{w}_{,\alpha}$, $\mathsf{y}_{,\alpha}$.

Let us write (8.2) and (8.4) in the form

$$\boldsymbol{k}\frac{\partial}{\partial s}\cdot\hat{\boldsymbol{T}}^* \equiv \frac{\partial(\hat{\boldsymbol{T}}\cdot\boldsymbol{k})}{\partial \boldsymbol{F}} : \boldsymbol{F}_{,3} = -\boldsymbol{i}^\alpha\frac{\partial}{\partial z^\alpha}\cdot\hat{\boldsymbol{T}}^*, \tag{8.21}$$

which (8.20) converts to

$$\frac{\partial(\hat{\boldsymbol{T}}\cdot\boldsymbol{k})}{\partial(\boldsymbol{F}\cdot\boldsymbol{k})}\cdot\left(y^l_{,3}\boldsymbol{d}_l\right) = \cdots \tag{8.22}$$

where the ellipsis has the same meaning as in (8.20). The Strong Ellipticity Condition implies that $\partial(\hat{\boldsymbol{T}}\cdot\boldsymbol{k})/\partial(\boldsymbol{F}\cdot\boldsymbol{k})$ is positive-definite. Therefore (8.22) can be written in the form

$$y^l_{,3}(\boldsymbol{z}) = \cdots. \tag{8.23}$$

Our governing equations for $\boldsymbol{r}$, $\boldsymbol{d}_k$, w, and y are (8.13), (8.16), (8.17), and (8.23). It is convenient to rewrite (8.17) and (8.23) in the form

$$\frac{d}{ds}w^l(\cdot,\cdot,s) = \cdots, \quad \frac{d}{ds}y^l(\cdot,\cdot,s) = \cdots, \tag{8.24}$$

which are ordinary differential equations for the functions $s \mapsto w^l(\cdot,\cdot,s), y^l(\cdot,\cdot,s)$, which assume values in infinite-dimensional spaces. Let us observe that (8.2)–(8.4) and therefore (8.23) admit the integrals

$$\boldsymbol{n} = \boldsymbol{n}(\pm l) \quad \text{(const.)}, \qquad \boldsymbol{m} + \boldsymbol{r}\times\boldsymbol{n} = \boldsymbol{m}(\pm l) + \boldsymbol{r}(\pm l)\times\boldsymbol{n}(\pm l) \quad \text{(const.)} \tag{8.25}$$

where

$$\begin{aligned} \boldsymbol{n} &= \tilde{n}^k(\mathsf{w},\mathsf{y})\boldsymbol{d}_k \equiv \int_{\mathcal{A}} \hat{\boldsymbol{T}}(\boldsymbol{F}, z^1, z^2)\cdot\boldsymbol{k}\,dz^1\,dz^2, \\ \boldsymbol{m} &= \tilde{m}^k(\mathsf{w},\mathsf{y})\boldsymbol{d}_k \equiv \int_{\mathcal{A}} (\boldsymbol{p}-\boldsymbol{r})\times[\hat{\boldsymbol{T}}(\boldsymbol{F}, z^1, z^2)\cdot\boldsymbol{k}]\,dz^1\,dz^2. \end{aligned} \tag{8.26}$$

(Cf. (3.9), (3.21), and (8.5).)

Our objective is to cast our governing 'ordinary differential equations' into a form that exposes the underlying role of rod theory and that is amenable to mathematical analysis. The basic mathematical tool we use is the Center-Manifold Theorem, a version of the Implicit-Function Theorem. Let us pause to examine how this theorem can be applied to a system of ordinary differential equations.

Consider a system of ordinary differential equations in the form

$$\begin{aligned} \mathsf{x}' &= \mathsf{A}\cdot\mathsf{x} + \mathsf{a}(\mathsf{x},\mathsf{y},\mathsf{z}), \\ \mathsf{y}' &= \mathsf{B}\cdot\mathsf{y} + \mathsf{b}(\mathsf{x},\mathsf{y},\mathsf{z}), \\ \mathsf{z}' &= \mathsf{C}\cdot\mathsf{z} + \mathsf{c}(\mathsf{x},\mathsf{y},\mathsf{z}) \end{aligned} \tag{8.27a,b,c}$$

where $\mathsf{x}\in\mathbb{R}^a$, $\mathsf{y}\in\mathbb{R}^b$, $\mathsf{z}\in\mathbb{R}^c$, where A, B, C are constant square matrices of dimensions $a\times a$, $b\times b$, $c\times c$, respectively, and where a, b, c are twice continuously differentiable functions from $\mathbb{R}^{a+b+c}$ to $\mathbb{R}^a$, $\mathbb{R}^b$, $\mathbb{R}^c$, respectively, that together with their first derivatives vanish at $(\mathbf{0},\mathbf{0},\mathbf{0})$. An *invariant manifold* $\mathcal{M}$ for (8.27) is a set with the property that if $(\mathsf{x}(0),\mathsf{y}(0),\mathsf{z}(0))\in\mathcal{M}$, then $(\mathsf{x}(s),\mathsf{y}(s),\mathsf{z}(s))\in\mathcal{M}$ for all s for which the solution exists. Suppose that all the eigenvalues of A have negative real parts, that all the eigenvalues of B have positive real parts, and that all the eigenvalues of C have zero real parts. Then the behavior of x and y is largely determined from this spectral information, so that the behavior of the system (8.27) devolves on the behavior of z. This intuition is made precise by the

8.28. Center-Manifold Theorem. *System* (8.27) *has twice continuously differentiable invariant manifolds* $\mathcal{A}$, $\mathcal{B}$, *and* $\mathcal{C}$, *called the stable, unstable, and center manifolds, that are respectively tangent to the generalized null spaces of* A, B, *and* C *at* $(\mathbf{0},\mathbf{0},\mathbf{0})$.

It follows from the Center-Manifold Theorem that for $|\mathsf{z}|$ sufficiently small, the center manifold can be described by $\mathsf{x} = \tilde{\mathsf{x}}(\mathsf{z})$, $\mathsf{y} = \tilde{\mathsf{y}}(\mathsf{z})$ where $\tilde{\mathsf{x}}$ and $\tilde{\mathsf{y}}$ are twice continuously differentiable and satisfy $\tilde{\mathsf{x}}(\mathbf{0}) = \mathbf{0} = \tilde{\mathsf{y}}(\mathbf{0})$, $(\partial\tilde{\mathsf{x}}/\partial\mathsf{z})(\mathbf{0}) = \mathbf{0} = (\partial\tilde{\mathsf{y}}/\partial\mathsf{z})(\mathbf{0})$. The flow on the center manifold is governed by the c-dimensional system

$$\mathsf{z}' = \mathsf{C}\cdot\mathsf{z} + \mathsf{c}(\tilde{\mathsf{x}}(\mathsf{z}),\tilde{\mathsf{y}}(\mathsf{z}),\mathsf{z}). \tag{8.29}$$

The next theorem says that the stability of solutions of (8.27) corresponds to that of (8.29).

8.30. Theorem. *Let* $b = 0$. *If the zero solution of* (8.29) *is stable or asymptotically stable or unstable as* $s \to \infty$, *then the zero solution of* (8.27) *is respectively stable or asymptotically stable or unstable. Moreover, if the zero solution of* (8.29) *is stable and if* (x,z) *is a solution of* (8.27) *with* $(\mathsf{x}(0),\mathsf{z}(0))$ *sufficiently small, then there is a solution* $\tilde{\mathsf{z}}$ *of* (8.29) *and there is a positive number* γ *such that*

$$\mathsf{x}(s) = \tilde{\mathsf{x}}\left(\tilde{\mathsf{z}}(s)\right) + O\left(e^{-\gamma s}\right), \quad \mathsf{z}(s) = \tilde{\mathsf{z}}(s) + O\left(e^{-\gamma s}\right) \tag{8.31}$$

as $s \to \infty$.

The simple example $x' = -x + x^2 + xz - z^3$, $z' = az^3 + xz^2$ of Carr (1981) illustrates the basic ideas of these theorems. For full discussions of center-manifold theory, see Carr (1981), Guckenheimer & Holmes (1983), and Marsden & McCracken (1976).

With appropriate technical adjustments, which we do not spell out, the theory can be extended to a class of partial differential equations. We now put our equations in a form to which we can apply such a theory.

Let us set

$$\tilde{S}(\boldsymbol{E}, z^1, z^2) = \mathcal{H}(z^1, z^2) : \boldsymbol{E} + O\left(|\boldsymbol{E}|^2\right) \quad \text{as } \boldsymbol{E} \to \boldsymbol{O}. \tag{8.32}$$

$\mathcal{H}$ is the fourth-order tensor of elasticities (see Sec. XIII.15). We assume it to be positive-definite: There is a number $\eta > 0$ such that

$$\boldsymbol{E} : \mathcal{H} : \boldsymbol{E} \geq \eta|\boldsymbol{E}|^2. \tag{8.33}$$

We now write (8.24) and (8.3) in the form

$$\frac{d}{ds}(\mathsf{w},\mathsf{y}) - \mathsf{L}\cdot(\mathsf{w},\mathsf{y}) = \phi(\mathsf{w},\mathsf{y}) = O\left(|\mathsf{w}|^2 + |\mathsf{y}|^2 + \mathsf{w}_{,\alpha}\cdot\mathsf{w}_{,\alpha} + \mathsf{y}_{,\alpha}\cdot\mathsf{y}_{,\alpha}\right), \tag{8.34}$$

$$\mathsf{B}\cdot(\mathsf{w},\mathsf{y}) = \psi(\mathsf{w},\mathsf{y}) = O\left(|\mathsf{w}|^2 + |\mathsf{y}|^2 + \mathsf{w}_{,\alpha}\cdot\mathsf{w}_{,\alpha} + \mathsf{y}_{,\alpha}\cdot\mathsf{y}_{,\alpha}\right) \tag{8.35}$$

on $\mathcal{L}$ where L and B are linear (partial differential) operators depending only on $\mathcal{H}$. We wish to replace (8.35) with a linear constraint so that we can readily employ center-manifold theory.

An analysis of a problem of linear elasticity yields

8.36. Proposition. L *has a discrete spectrum with* 0 *as the only eigenvalue on the imaginary axis.*

Thus there is a real number δ such that the linear problem

$$(\mathsf{L} + \delta\mathsf{I})\cdot(\mathsf{w},\mathsf{y}) = (\mathbf{0},\mathbf{0}), \qquad \mathsf{B}\cdot(\mathsf{w},\mathsf{y}) = \omega \tag{8.37}$$

has a unique solution denoted by

$$(\mathsf{w},\mathsf{y}) = \mathsf{G}\cdot\omega. \tag{8.38}$$

Let us now define

$$(\tilde{\mathsf{w}}, \tilde{\mathsf{y}}) \equiv (\mathsf{w}, \mathsf{y}) - \mathsf{G} \cdot \psi(\mathsf{w}, \mathsf{y}). \tag{8.39}$$

It follows from (8.38) that if (w, y) is a solution of (8.34), (8.35), then $(\tilde{\mathsf{w}}, \tilde{\mathsf{y}})$ satisfies the linear boundary conditions

$$\mathsf{B} \cdot (\tilde{\mathsf{w}}, \tilde{\mathsf{y}}) = \mathsf{0}. \tag{8.40}$$

Since (8.39) is invertible, we can replace (8.34) with an equation of the form

$$\frac{d}{ds}(\tilde{\mathsf{w}}, \tilde{\mathsf{y}}) - \tilde{\mathsf{L}} \cdot (\tilde{\mathsf{w}}, \tilde{\mathsf{y}}) = \tilde{\phi}(\tilde{\mathsf{w}}, \tilde{\mathsf{y}}). \tag{8.41}$$

Since (8.40) is linear, we can treat it as restricting the domain of definition $\mathcal{E}$ of the operator $\tilde{\mathsf{L}}$ while preserving its linear structure. Thus (8.41) is an ordinary differential equation for functions taking values in $\mathcal{E}$. For a suitable choice of $\mathcal{E}$, made precise below, we can employ the Center-Manifold Theorem.

It can be shown that the generalized null space of $\tilde{\mathsf{L}}$ is six-dimensional and is spanned by vectors associated with solutions of the St. Venant problem of linear elasticity. We can therefore decompose (8.41) into a form like (8.27) with the equation corresponding to (8.27c) obtained by projecting (8.41) onto the generalized null space of $\tilde{\mathsf{L}}$. We denote this projection of $(\tilde{\mathsf{w}}, \tilde{\mathsf{y}})$ by $(\bar{\mathsf{w}}, \bar{\mathsf{y}})$ and denote the equation corresponding to (8.27c) by

$$\frac{d}{ds}(\bar{\mathsf{w}}, \bar{\mathsf{y}}) - \bar{\mathsf{L}}(\bar{\mathsf{w}}, \bar{\mathsf{y}}) = \bar{\phi}(\bar{\mathsf{w}}, \bar{\mathsf{y}}, \tilde{\mathsf{w}} - \bar{\mathsf{w}}, \tilde{\mathsf{y}} - \bar{\mathsf{y}}). \tag{8.42}$$

This is a sixth-order ordinary differential equation for $(\bar{\mathsf{w}}, \bar{\mathsf{y}})$. The restriction of (8.42) to the center manifold, which corresponds to (8.29), is obtained by setting $\tilde{\mathsf{w}} = \bar{\mathsf{w}}$, $\tilde{\mathsf{y}} = \bar{\mathsf{y}}$ in (8.42).

Using (8.13) and (8.16), we identify the u and v of Chap. VIII with $(\bar{\mathsf{w}}, \bar{\mathsf{y}})$ by

$$\boldsymbol{r}' = v^k \boldsymbol{d}_k = \boldsymbol{d}_3 + \bar{y}^k \boldsymbol{d}_k, \quad u^k = u^k[\bar{\mathsf{w}}, \bar{\mathsf{y}}]. \tag{8.43}$$

Now for $\boldsymbol{E}$ small enough, Eqs. (8.32) and (8.39) can be used to show that $(\bar{\mathsf{w}}, \bar{\mathsf{y}}) \mapsto (\bar{\mathsf{n}}(\bar{\mathsf{w}}, \bar{\mathsf{y}}), \bar{\mathsf{m}}(\bar{\mathsf{w}}, \bar{\mathsf{y}}))$, defined by (8.26), is invertible. Thus (8.42) is equivalent to the ordinary differential equations for $\bar{\mathsf{n}}$ and $\bar{\mathsf{m}}$ obtained from (8.25). Setting $\bar{\mathsf{n}}(\bar{\mathsf{w}}, \bar{\mathsf{y}}) = \hat{\mathsf{n}}(\mathsf{u}, \mathsf{v})$, etc., we see (8.42) is equivalent to the classical equilibrium problem for the special Cosserat theory of nonlinearly elastic rods, namely,

$$\boldsymbol{n}' = \boldsymbol{0}, \quad \boldsymbol{m}' + \boldsymbol{r}' \times \boldsymbol{n} = \boldsymbol{0}, \quad \boldsymbol{n} = \hat{n}^k(\mathsf{u}, \mathsf{v})\boldsymbol{d}_k, \quad \boldsymbol{m} = \hat{m}^k(\mathsf{u}, \mathsf{v})\boldsymbol{d}_k. \tag{8.44}$$

We can obtain an estimate like (8.31) by restricting the size of solutions. A precise statement of the results informally presented in this section is

8.45. Theorem. *Let* (8.1) *and* (8.33) *hold. Then there is a positive number* γ^+, *depending on* $\mathcal{A}$ *and* $\tilde{\boldsymbol{S}}$, *with the property that for each* $\gamma \in (0, \gamma^+)$ *and for each* $\beta > 2$ *there are positive numbers* C *and* δ^+ *such that the following assertion holds: For each* $l > 0$ *and for each solution* $\boldsymbol{p}$ *of* (8.2)–(8.5) *satisfying*

$$\| [(\partial \boldsymbol{p}/\partial \boldsymbol{z})(\cdot, s)]^* \cdot [(\partial \boldsymbol{p}/\partial \boldsymbol{z})(\cdot, s)], W^1_\beta(\mathcal{A}) \| \le \delta \tag{8.46}$$

with $\delta \le \delta^+$, *there is a (St. Venant) solution* $\bar{\boldsymbol{p}}$, *generated by the solution of the center-manifold equation* (8.42) *or, equivalently, by* (8.44), *satisfying* (8.25) *such that*

$$\|\boldsymbol{p} - \bar{\boldsymbol{p}}, W^2_\beta(\mathcal{A} \times (-\lambda, \lambda))\| \le C\delta e^{-\gamma(l-\lambda)} \quad \forall \lambda \in (0, l). \tag{8.47}$$

The presentation in this section is based upon the work of Mielke (1988, 1990). While the development here gives a rigorous position to the theory of Chap. VIII, it is restricted by the smallness of the strain tensor, which justifies the use of (8.33), a condition far more stringent than the Strong Ellipticity Condition.

Very pretty work on the three-dimensional characterization of solutions of the equilibrium equations for rods was carried out by Ericksen (1977a, 1983) and Muncaster (1979, 1983); See Sec. XIII.8. Their work furnished one source of inspiration for Mielke.

9. Shell Theories

There is a duality in the derivation of rod and shell theories from the three-dimensional theory that is effected by interchanging the roles of (x^1, x^2) with x^3. Accordingly, we merely sketch the main steps in constructing shell theories, emphasizing those places where the theory differs in its particulars from rod theories. Let us remark that a plate is just a special kind of shell, so that we make no mathematical distinction for the former.

Geometry of the reference configuration. Let $\mathcal{B}$ and $\mathbf{x}$ have the properties described in the paragraph containing (1.1) and (1.2). But now we set $(x^1, x^2) \equiv (s^1, s^2) \equiv \mathbf{s}$. We assume that

$$\mathcal{M} \equiv \{\mathbf{s} : \tilde{\boldsymbol{z}}(\mathbf{x}) \in \mathcal{B}\} \tag{9.1}$$

is the closure of a domain in $\mathbb{R}^2$. We further assume that $\tilde{x}^3$ is bounded on $\mathcal{B}$. $\mathcal{B}$ is said to be *shell-like* iff these two assumptions hold. The *edge* of the shell is $\{\tilde{\boldsymbol{z}}(\mathbf{x}) \in \partial\mathcal{B} : \mathbf{s} \in \partial\mathcal{M}\}$. For simplicity, we assume that $\partial\mathcal{B}$ is the union of the edge and of the two surfaces

$$\mathcal{L}_1 \equiv \{\tilde{\boldsymbol{z}}(\mathbf{x}) \in \partial\mathcal{B} : x^3 = h_1(\mathbf{s})\}, \quad \mathcal{L}_2 \equiv \{\tilde{\boldsymbol{z}}(\mathbf{x}) \in \partial\mathcal{B} : x^3 = h_2(\mathbf{s})\}, \tag{9.2}$$

with $h_1(\mathbf{s}) < h_2(\mathbf{s})$ for $\mathbf{s} \in \mathcal{M}$. $\mathcal{L}_1$ and $\mathcal{L}_2$ are called the *faces* of the shell. We introduce a material *base surface* $\tilde{\boldsymbol{z}}(\mathcal{M}_\sharp)$ of the form

$$\mathcal{M}_\sharp = \{\mathbf{x} : x^3 = h(\mathbf{s}),\ \mathbf{s} \in \mathcal{M}\} \tag{9.3}$$

where $h \in C^2(\mathcal{M})$.

9.4. Exercise. Prove that the differential area of the base surface is

$$da(\mathbf{s}) = \Gamma(\mathbf{s})\, ds^1\, ds^2 \tag{9.5}$$

with

$$\begin{aligned} \Gamma(\mathbf{s}) &= j(\mathbf{s}, h(\mathbf{s}))|\boldsymbol{g}^3(\mathbf{s}, h(\mathbf{s})) - \boldsymbol{g}^\beta(\mathbf{s}, h(\mathbf{s}))h_{,\beta}(\mathbf{s})| \\ &= \frac{j(\mathbf{s}, h(\mathbf{s}))}{\alpha_3(\mathbf{s})} = j(\mathbf{s}, h(\mathbf{s}))[\alpha^3(\mathbf{s}) - \alpha^\beta(\mathbf{s})h_{,\beta}(\mathbf{s})] \end{aligned} \tag{9.6a}$$

where $\boldsymbol{\alpha}(\mathbf{s}) = \alpha_k(\mathbf{s})\boldsymbol{g}^k(\mathbf{s}, h(\mathbf{s})) = \alpha^k(\mathbf{s})\boldsymbol{g}_k(\mathbf{s}, h(\mathbf{s}))$ is the unit normal to the base surface at $\tilde{\boldsymbol{z}}(\mathbf{s}, h(\mathbf{s}))$. Consequently, the differential volume of $\mathcal{B}$ is $[j(\mathbf{x})/\Gamma(\mathbf{s})]da(\mathbf{s})\, dx^3$. In the special case in which $\mathcal{M}_\sharp$ is a coordinate surface and is given by $x^3 = h_0$ (const.), we find that

$$\Gamma(\mathbf{s}) = j(\mathbf{s}, h_0)|\boldsymbol{g}^3(\mathbf{s}, h_0)| = |\boldsymbol{g}_1(\mathbf{s}, h_0) \times \boldsymbol{g}_2(\mathbf{s}, h_0)|. \tag{9.6b}$$

Let σ be an arc-length parameter for the boundary $\tilde{\boldsymbol{z}}(\partial\mathcal{M}_\sharp)$ of the base surface. This curve is defined parametrically by

$$\sigma \mapsto \tilde{\boldsymbol{z}}\big(\bar{\mathbf{s}}(\sigma), h(\bar{\mathbf{s}}(\sigma))\big). \tag{9.7}$$

The definition of arc length implies that

$$\left| [\boldsymbol{g}_\alpha(\bar{\mathbf{s}}(\sigma), h(\bar{\mathbf{s}}(\sigma))) + \boldsymbol{g}_3(\bar{\mathbf{s}}(\sigma), h(\bar{\mathbf{s}}(\sigma)))h_{,\alpha}(\mathbf{s})] \frac{d\bar{\mathbf{s}}^\alpha}{d\sigma}(\sigma) \right| = 1 \tag{9.8}$$

wherever the left-hand side is defined. The edge, described parametrically by

$$(\sigma, x^3) \mapsto \tilde{\boldsymbol{z}}(\bar{\mathbf{s}}(\sigma), x^3), \tag{9.9}$$

has differential area

$$\begin{aligned} &\left| \boldsymbol{g}_\alpha(\bar{\mathbf{s}}(\sigma), x^3) \frac{d\bar{\mathbf{s}}^\alpha}{d\sigma}(\sigma) \times \boldsymbol{g}_3(\bar{\mathbf{s}}(\sigma), x^3) \right| \, d\sigma \, dx^3 \\ &= j(\bar{\mathbf{s}}(\sigma), x^3) \left| -\boldsymbol{g}^2(\bar{\mathbf{s}}(\sigma), x^3) \frac{d\bar{\mathbf{s}}^1}{d\sigma}(\sigma) + \boldsymbol{g}^1(\bar{\mathbf{s}}(\sigma), x^3) \frac{d\bar{\mathbf{s}}^2}{d\sigma}(\sigma) \right| \, d\sigma \, dx^3. \end{aligned} \tag{9.10}$$

With a slight abuse of notation, we denote the differential arc length of $\tilde{\boldsymbol{z}}(\partial\mathcal{M}_\sharp)$ at $(\mathbf{s}, h(\mathbf{s}))$ by $d\sigma(\mathbf{s})$.

Constraints. We generate an induced theory of shells by constraining $\boldsymbol{p}$ by the obvious analog of (2.5), obtained by replacing s with $\mathbf{s}$:

$$\tilde{\boldsymbol{p}}(\mathbf{x}, t) = \boldsymbol{\pi}(\mathsf{u}(\mathbf{s}, t), \mathbf{x}, t). \tag{9.11}$$

The simplest example of such a representation is

$$\boldsymbol{\pi}(\mathsf{u}, \mathbf{x}, t) = \boldsymbol{r}(\mathbf{s}, t) + x^3 \boldsymbol{d}(\mathbf{s}, t) \tag{9.12}$$

where we identify u with $(\boldsymbol{r}, \boldsymbol{d})$. We assume that an independency condition like (2.8) holds. We can readily compute the strains from (9.11) and (9.12).

We adopt the analog of (2.7) for the faces. On the edge, we assume that (9.11) generates position boundary conditions in the form

$$\mathsf{u}(\mathbf{s}, t) = \bar{\mathsf{u}}(\mathbf{s}, t, \mathsf{v}(\mathbf{s}, t)), \qquad \mathbf{s} \in \partial\mathcal{M}, \tag{9.13}$$

or periodicity conditions of the form

$$\mathsf{u}(s^1 + \xi^1, s^2, t) = \mathsf{u}(s^1, s^2, t), \quad \mathsf{u}(s^1, s^2 + \xi^2, t) = \mathsf{u}(s^1, s^2, t), \tag{9.14}$$

or combinations of the components of (9.13) and (9.14).

We treat the preservation of orientation as in Sec. 2. The structure of $\mathcal{V}(\mathbf{s}, t)$ is not so specific as that of Theorem 2.12.

Stress resultants and the equations of motion. By the same procedure as that used in Sec. 2, we find that the Lagrange multipliers satisfy

$$\Gamma^{-1}(\Gamma \mathsf{m}^\alpha_{\mathrm{L}})_{,\alpha} - \mathsf{n}_{\mathrm{L}} = \mathbf{0} \tag{9.15}$$

in the sense of distributions, where

$$\Gamma(\mathbf{s})\mathbf{m}^{\alpha}_{\mathrm{L}}(\mathbf{s},t) \equiv \int_{h_1(\mathbf{s})}^{h_2(\mathbf{s})} \boldsymbol{\tau}^{\alpha}_{\mathrm{L}}(\mathbf{x},t)\cdot\frac{\partial\boldsymbol{\pi}}{\partial\mathbf{u}}(\mathbf{u}(\mathbf{s},t),\mathbf{x},t)\,j(\mathbf{x})\,dx^3, \tag{9.16}$$

$$\Gamma(\mathbf{s})\mathbf{n}_{\mathrm{L}}(\mathbf{s},t) \equiv \int_{h_1(\mathbf{s})}^{h_2(\mathbf{s})} \boldsymbol{\tau}^{k}_{\mathrm{L}}(\mathbf{x},t)\cdot\frac{\partial}{\partial x^k}\left[\frac{\partial\boldsymbol{\pi}}{\partial\mathbf{u}}(\mathbf{u}(\mathbf{s},t),\mathbf{x},t)\right]j(\mathbf{x})\,dx^3. \tag{9.17}$$

From (1.17) and (XII.12.50b) we obtain

$$\begin{aligned}&\int_0^\infty\int_{\mathcal{M}}[\mathbf{m}^{\alpha}_{\mathrm{A}}\cdot\overset{\vartriangle}{\mathbf{u}}_{,\alpha}+\mathbf{n}_{\mathrm{A}}\cdot\overset{\vartriangle}{\mathbf{u}}-\mathbf{f}\cdot\overset{\vartriangle}{\mathbf{u}}-\mathbf{b}\cdot\overset{\vartriangle}{\mathbf{u}}_t-\mathbf{c}\cdot\overset{\vartriangle}{\mathbf{u}}]\,da(\mathbf{s})\,dt\\ &-\int_0^\infty\int_{\partial\mathcal{M}}\Gamma\bar{\mathbf{m}}\cdot\overset{\vartriangle}{\mathbf{u}}\,d\sigma\,dt-\int_{\mathcal{M}}\bar{\mathbf{b}}(\mathbf{s})\cdot\overset{\vartriangle}{\mathbf{u}}(\mathbf{s},0)\,da(\mathbf{s})=0\end{aligned} \tag{9.18}$$

for all sufficiently smooth $\overset{\vartriangle}{\mathbf{u}}$ satisfying

$$\overset{\vartriangle}{\mathbf{u}}(s_\alpha,t)=\frac{\partial\bar{\mathbf{u}}}{\partial\mathbf{v}}(\mathbf{s},t,\mathbf{v}(\mathbf{s},t))\cdot\mathbf{w}(\mathbf{s},t),\quad \mathbf{w}(\mathbf{s},t)\in\mathbb{R}^N,\quad \mathbf{s}\in\partial\mathcal{M}, \tag{9.19}$$

when (9.13) holds and satisfying appropriate periodicity conditions when (9.14) holds. Here

(9.20)
$$\Gamma(\mathbf{s})\mathbf{m}^{\alpha}_{\mathrm{A}}(\mathbf{s},t) \equiv \int_{h_1(\mathbf{s})}^{h_2(\mathbf{s})} \boldsymbol{\tau}^{\alpha}_{\mathrm{A}}(\mathbf{x},t)\cdot\frac{\partial\boldsymbol{\pi}}{\partial\mathbf{u}}(\mathbf{u}(\mathbf{s},t),\mathbf{x},t)\,j(\mathbf{x})\,dx^3,$$

(9.21)
$$\Gamma(\mathbf{s})\mathbf{n}_{\mathrm{A}}(s,t) \equiv \int_{h_1(\mathbf{s})}^{h_2(\mathbf{s})} \boldsymbol{\tau}^{k}_{\mathrm{A}}(\mathbf{x},t)\cdot\frac{\partial}{\partial x^k}\left[\frac{\partial\boldsymbol{\pi}}{\partial\mathbf{u}}(\mathbf{u}(\mathbf{s},t),\mathbf{x},t)\right]j(\mathbf{x})\,dx^3,$$

(9.22)
$$\begin{aligned}\Gamma(\mathbf{s})\mathbf{f}(\mathbf{s},t) \equiv{}& \int_{h_1(\mathbf{s})}^{h_2(\mathbf{s})} \boldsymbol{f}(\mathbf{x},t)\cdot\frac{\partial\boldsymbol{\pi}}{\partial\mathbf{u}}(\mathbf{u}(\mathbf{s},t),\mathbf{x},t)\,j(\mathbf{x})\,dx^3\\ &+\Big[\bar{\boldsymbol{\tau}}(\tilde{\boldsymbol{z}}(\mathbf{x}),t,\mathbf{q}(\tilde{\boldsymbol{z}}(\mathbf{x}),t))\cdot\frac{\partial\boldsymbol{\pi}}{\partial\mathbf{u}}(\mathbf{u}(\mathbf{s},t),\mathbf{x},t)\\ &\qquad j(\mathbf{x})(\nu^3(\mathbf{x})-\nu^{\alpha}(\mathbf{x})h_{,\alpha}(\mathbf{s}))\Big]\Big|_{x^3=h_1(\mathbf{s})}^{x^3=h_2(\mathbf{s})},\end{aligned}$$

(9.23)
$$\Gamma(\mathbf{s})\mathbf{b}(\mathbf{s},t) \equiv \int_{h_1(\mathbf{s})}^{h_2(\mathbf{s})} \tilde{\rho}(\mathbf{x})\,\tilde{\boldsymbol{p}}_t(\mathbf{x},t)\cdot\frac{\partial\boldsymbol{\pi}}{\partial\mathbf{u}}(\mathbf{u}(\mathbf{s},t),\mathbf{x},t)\,j(\mathbf{x})\,dx^3,$$

(9.24)
$$\Gamma(\mathbf{s})\mathbf{c}(\mathbf{s},t) \equiv \int_{h_1(\mathbf{s})}^{h_2(\mathbf{s})} \tilde{\rho}(\mathbf{x})\,\tilde{\boldsymbol{p}}_t(\mathbf{x},t)\cdot\frac{\partial}{\partial t}\left[\frac{\partial\boldsymbol{\pi}}{\partial\mathbf{u}}(\mathbf{u}(\mathbf{s},t),\mathbf{x},t)\right]j(\mathbf{x})\,dx^3,$$

(9.25)

$$\Gamma(\mathbf{s})\bar{\mathsf{m}}(\mathbf{s},t) \equiv \int_{h_1(\mathbf{s})}^{h_2(\mathbf{s})} \bar{\boldsymbol{\tau}}(\tilde{\boldsymbol{z}}(\mathbf{x}),t,\mathsf{q}(\tilde{\boldsymbol{z}}(\mathbf{x}),t)) \cdot \frac{\partial \boldsymbol{\pi}}{\partial \mathsf{u}}(\mathsf{u}(\mathbf{s},t),\mathbf{x},t)\, j(\mathbf{x}) \left| -\boldsymbol{g}^2(\bar{\mathbf{s}}(\sigma),x^3)\frac{d\bar{s}^1}{d\sigma}(\sigma) + \boldsymbol{g}^1(\bar{\mathbf{s}}(\sigma),x^3)\frac{d\bar{s}^2}{d\sigma}(\sigma) \right| dx^3,$$

(9.26)

$$\Gamma(\mathbf{s})\bar{\mathsf{b}}(\mathbf{s}) \equiv \int_{h_1(\mathbf{s})}^{h_2(\mathbf{s})} \tilde{\rho}(\mathbf{x})\, \boldsymbol{p}_1(\mathbf{x}) \cdot \frac{\partial \boldsymbol{\pi}}{\partial \mathsf{u}}(\mathsf{u}(\mathbf{s},0),\mathbf{x},0)\, j(\mathbf{x})\, dx^1\, dx^2.$$

In (9.22), h stands for h_1 or h_2. In (9.25), σ is an arc-length parameter for the boundary $\tilde{\boldsymbol{z}}(\partial\mathcal{M}_\sharp)$ of the base surface. This curve is defined parametrically by $\sigma \mapsto \tilde{\boldsymbol{z}}(\bar{\mathbf{s}}(\sigma), h(\bar{\mathbf{s}}(\sigma)))$. The edge is described parametrically by $(\sigma, x^3) \mapsto \tilde{\boldsymbol{z}}(\bar{\mathbf{s}}(\sigma), x^3)$.

The classical form of (9.18) is

$$\Gamma^{-1}(\Gamma \mathsf{m}^\alpha_{\mathrm{A}})_{,\alpha} - \mathsf{n}_{\mathrm{A}} + \mathsf{f} = \mathsf{b}_t - \mathsf{c}. \tag{9.27}$$

It is appropriate to introduce further weightings like those of (VIII.10.3) to account for thickness variations. Let us emphasize that (9.18)–(9.27) are exact consequences of the Principle of Virtual Power. In the sequel we often drop the subscript A from the stress resultants (see Remark 2.36). Just as in the discussion centered on (2.39), the reactive stress $\boldsymbol{\tau}_{\mathrm{L}}$ ensures that traction boundary conditions on the faces are satisfied exactly.

9.28. Exercise. Carry out the analog of Sec. 4 for shells. In particular, determine the shell-theoretic consequences of the Strong Ellipticity Condition. (The resulting inequality has the features of both (4.16) and the Strong Ellipticity Condition for the three-dimensional theory.)

10. Shells with One Director, Special Cosserat Theory, Kirchhoff Theory

In this section we specialize the treatment of Sec. 9 to shells that can bend, twist, stretch, shear, and change their thickness. We then treat shells that are constrained not to change their thickness, the theory for which corresponds to the special Cosserat theory. We finally introduce the Kirchhoff constraints, which prevent certain kinds of shearing. There are some subtleties in these constrained theories, which merit our attention.

We identify u with the pair $\{\boldsymbol{r}, \boldsymbol{d}\}$ of vectors and we take $\boldsymbol{d} = \delta \boldsymbol{a}_3$ where $\boldsymbol{a}_3$ is a unit vector. $\boldsymbol{d}$ is the *director*. We generate the theory with a constraint of the form (9.11) by taking $\boldsymbol{\pi}$ to have one of the increasingly specialized forms

$$\boldsymbol{\pi}(\mathsf{u},\mathbf{w},\tau) = \boldsymbol{r} + \boldsymbol{\kappa}(\boldsymbol{d},\mathbf{w}), \tag{10.1a}$$

$$\boldsymbol{\pi}(\mathsf{u},\mathbf{w},\tau) = \boldsymbol{r} + \kappa(\delta,\mathbf{w})\boldsymbol{a}_3, \tag{10.1b}$$

$$\boldsymbol{\pi}(\mathsf{u},\mathbf{w},\tau) = \boldsymbol{r} + \varphi(\mathbf{w})\boldsymbol{d}, \tag{10.1c}$$

where $\boldsymbol{\kappa}$, κ, and φ are given functions. We assume that in each case the reference configuration is defined by

$$\tilde{\boldsymbol{z}}(\mathbf{x}) = \overset{\circ}{\boldsymbol{r}}(\mathbf{s}) + \varphi(\mathbf{x})\overset{\circ}{\boldsymbol{a}}_3(\mathbf{s}). \tag{10.2}$$

We assume that $\{\overset{\circ}{\boldsymbol{a}}_1 \equiv \overset{\circ}{\boldsymbol{r}}_{,1},\ \overset{\circ}{\boldsymbol{a}}_2 \equiv \overset{\circ}{\boldsymbol{r}}_{,2},\ \overset{\circ}{\boldsymbol{a}}_3\}$ is orthonormal and therefore equals its dual basis $\{\overset{\circ}{\boldsymbol{a}}{}^1, \overset{\circ}{\boldsymbol{a}}{}^2, \overset{\circ}{\boldsymbol{a}}{}^3\}$. We take $\overset{\circ}{\delta} = 1$.

In accord with (10.1a), we may take

$$\overset{\vartriangle}{\boldsymbol{p}} = \overset{\vartriangle}{\boldsymbol{r}} + \frac{\partial \boldsymbol{\kappa}}{\partial \boldsymbol{d}} \cdot \overset{\vartriangle}{\boldsymbol{d}}, \tag{10.3a}$$

which reduces to

$$\overset{\vartriangle}{\boldsymbol{p}} = \overset{\vartriangle}{\boldsymbol{r}} - \kappa \overset{\vartriangle}{\boldsymbol{a}}_3 + \kappa_\delta \overset{\vartriangle}{\delta} \boldsymbol{a}_3 \tag{10.3b}$$

when (10.1b) is used. We first choose $\overset{\vartriangle}{\boldsymbol{p}} = \overset{\vartriangle}{\boldsymbol{r}}$, in which case (1.17) reduces to the weak form of the balance of linear momentum:

$$\Gamma^{-1}(\Gamma \boldsymbol{n}^\alpha)_{,\alpha} + \boldsymbol{f}^0 = 2\rho h \boldsymbol{r}_{tt} + \boldsymbol{q}_{tt} \tag{10.4}$$

where

$$\Gamma \boldsymbol{n}^\alpha \equiv \int_{h_1}^{h_2} \boldsymbol{\tau}_{\mathrm{A}}^\alpha j \, dx^3, \tag{10.5}$$

$$\Gamma \boldsymbol{f}^0 \equiv \int_{h_1}^{h_2} \boldsymbol{f} \, j \, dx^3 + \ \left[\bar{\boldsymbol{\tau}}(j\alpha^3 - \alpha^\alpha h_{,\alpha})\right]\Big|_{h_1}^{h_2}, \tag{10.6}$$

$$\Gamma \boldsymbol{q} \equiv \int_{h_1}^{h_2} \tilde{\rho} \boldsymbol{\kappa} \, j \, dx^3, \tag{10.7}$$

$$2\Gamma \rho h \equiv \int_{h_1}^{h_2} \tilde{\rho} \, j \, dx^3. \tag{10.8}$$

Imitating Sec. 9 directly, we now take $\overset{\vartriangle}{\boldsymbol{p}} = [\partial \boldsymbol{\kappa}/\partial \boldsymbol{d}] \cdot \overset{\vartriangle}{\boldsymbol{d}}$, and reduce (1.17) to the weak form of

$$\Gamma^{-1}(\Gamma \boldsymbol{\mu}^\alpha)_{,\alpha} - \boldsymbol{\xi} + \boldsymbol{f}^1 = \boldsymbol{b}_t^1 - \boldsymbol{c}^1 \tag{10.9}$$

where

$$\Gamma \boldsymbol{\mu}^\alpha \equiv \int_{h_1}^{h_2} \boldsymbol{\tau}_{\mathrm{A}}^\alpha \cdot \frac{\partial \boldsymbol{\kappa}}{\partial \boldsymbol{d}} j \, dx^3, \tag{10.10a}$$

$$\Gamma \boldsymbol{\xi} \equiv \int_{h_1}^{h_2} \boldsymbol{\tau}_{\mathrm{A}}^k \cdot \frac{\partial}{\partial x^k} \left[\frac{\partial \boldsymbol{\kappa}}{\partial \boldsymbol{d}}\right] j \, dx^3, \tag{10.10b}$$

$$\boldsymbol{b}_t^1 - \boldsymbol{c}^1 \equiv \int_{h_1}^{h_2} \tilde{\rho} \tilde{\boldsymbol{p}}_{tt} \cdot \frac{\partial \boldsymbol{\kappa}}{\partial \boldsymbol{d}} j \, dx^3, \quad \text{etc.} \tag{10.10c}$$

Note that (10.9) and (10.10) are consequences of the arbitrariness of $\overset{\vartriangle}{\boldsymbol{d}}$. Equations (10.4) and (10.9), together with appropriate constitutive equations, form the basis of a mathematically satisfactory system of six equations for the six unknown components of $\boldsymbol{r}$ and $\boldsymbol{d}$. We, however, wish to relate the variables $\boldsymbol{\mu}^\alpha$ and $\boldsymbol{\xi}$ of this system to couple and force resultants. For this purpose, we must curtail the generality inherent in (10.1a).

To illuminate our results, we decompose $\overset{\vartriangle}{\boldsymbol{d}} = \overset{\vartriangle}{\delta}\boldsymbol{a}_3 + \delta\overset{\vartriangle}{\boldsymbol{a}}_3$. Since $\boldsymbol{a}_3$ is a unit vector, $\overset{\vartriangle}{\boldsymbol{a}}_3$ is any vector perpendicular to $\boldsymbol{a}_3$ and therefore has the form $\overset{\vartriangle}{\boldsymbol{a}}_3 = \boldsymbol{\omega} \times \boldsymbol{a}_3$ where $\boldsymbol{\omega}$ is an arbitrary function of s and t.

If we take the dot product of (10.9) with $\boldsymbol{a}_3$, we obtain

$$\Gamma^{-1}(\Gamma\Omega^\alpha)_{,\alpha} - \Delta + f^1 = \beta_t - \gamma \tag{10.11}$$

where

(10.12)
$$\Gamma\Omega^\alpha \equiv \int_{h_1}^{h_2} \boldsymbol{\tau}^\alpha_{\mathrm{A}} \cdot \frac{\partial \boldsymbol{\kappa}}{\partial \boldsymbol{d}} \cdot \boldsymbol{a}_3\, j\, dx^3 = \Gamma\boldsymbol{\mu}^\alpha \cdot \boldsymbol{a}_3,$$

(10.13)
$$\Gamma\Delta \equiv \int_{h_1}^{h_2} \boldsymbol{\tau}^k_{\mathrm{A}} \cdot \frac{\partial}{\partial x^k}\left[\frac{\partial \boldsymbol{\kappa}}{\partial \boldsymbol{d}} \cdot \boldsymbol{a}_3\right] j\, dx^3 = \Gamma(\boldsymbol{\xi}\cdot\boldsymbol{a}_3 + \boldsymbol{\mu}^\alpha \cdot \boldsymbol{a}_{3,\alpha}),$$

(10.14)
$$\Gamma f^1 \equiv \int_{h_1}^{h_2} \boldsymbol{f} \cdot \frac{\partial \boldsymbol{\kappa}}{\partial \boldsymbol{d}} \cdot \boldsymbol{a}_3\, j\, dx^3 + \left[\bar{\boldsymbol{\tau}} \cdot \frac{\partial \boldsymbol{\kappa}}{\partial \boldsymbol{d}} \cdot \boldsymbol{a}_3\, j(\alpha^3 - \alpha^\alpha h_{,\alpha})\right]\Bigg|_{h_1}^{h_2} = \Gamma\boldsymbol{f}^1 \cdot \boldsymbol{a}_3,$$

(10.15)
$$\Gamma\beta \equiv \int_{h_1}^{h_2} \tilde{\rho}\tilde{\boldsymbol{p}}_t \cdot \frac{\partial \boldsymbol{\kappa}}{\partial \boldsymbol{d}} \cdot \boldsymbol{a}_3\, j\, dx^3,$$

(10.16)
$$\Gamma\gamma \equiv \int_{h_1}^{h_2} \tilde{\rho}\tilde{\boldsymbol{p}}_t \cdot \frac{\partial}{\partial t}\left[\frac{\partial \boldsymbol{\kappa}}{\partial \boldsymbol{d}} \cdot \boldsymbol{a}_3\right] j\, dx^3.$$

Alternatively, we obtain the weak form of (10.11) from (1.17) by choosing $\overset{\vartriangle}{\boldsymbol{p}} = [\partial\boldsymbol{\kappa}/\partial\boldsymbol{d}] \cdot \boldsymbol{a}_3\overset{\vartriangle}{\delta}$.

To get equations complementing these, we merely take the cross product of (10.9) with $\boldsymbol{d}$ to obtain

$$\Gamma^{-1}(\Gamma\boldsymbol{d} \times \boldsymbol{\mu}^\alpha)_{,\alpha} - \boldsymbol{d}_{,\alpha} \times \boldsymbol{\mu}^\alpha - \boldsymbol{d} \times \boldsymbol{\xi} + \boldsymbol{d} \times \boldsymbol{f}^1 = \boldsymbol{d} \times \boldsymbol{b}^1_t - \boldsymbol{d} \times \boldsymbol{c}^1. \tag{10.17}$$

Alternatively, we obtain the weak form of (10.17) from (1.17) by choosing

$$\overset{\vartriangle}{\boldsymbol{p}} = [\partial\boldsymbol{\kappa}/\partial\boldsymbol{d}] \cdot (\boldsymbol{\omega} \times \delta\boldsymbol{a}_3) \tag{10.18a}$$

where $\boldsymbol{\omega}$ is an arbitrary function of s and t.

Now for the choice

$$\stackrel{\Delta}{\boldsymbol{p}} = (\tilde{\boldsymbol{p}} - \boldsymbol{r}) \times \boldsymbol{\omega} \tag{10.18b}$$

where $\boldsymbol{\omega}$ is an arbitrary function of $\mathbf{s}$ and t, we obtain from (1.17) and (1.12) the weak form of the balance of angular momentum

$$\Gamma^{-1}(\Gamma \boldsymbol{m}^\alpha)_{,\alpha} + \boldsymbol{r}_{,\alpha} \times \boldsymbol{n}^\alpha + \boldsymbol{l} = \boldsymbol{q} \times \boldsymbol{r}_{tt} + \boldsymbol{h}_t \tag{10.19}$$

where

$$\Gamma \boldsymbol{m}^\alpha \equiv \int_{h_1}^{h_2} (\tilde{\boldsymbol{p}} - \boldsymbol{r}) \times \boldsymbol{\tau}_{\mathrm{A}}^\alpha \, j \, dx^3, \tag{10.20}$$

$$\Gamma \boldsymbol{l} \equiv \int_{h_1}^{h_2} (\tilde{\boldsymbol{p}} - \boldsymbol{r}) \times \boldsymbol{f} \, j \, dx^3 + [(\tilde{\boldsymbol{p}} - \boldsymbol{r}) \times \bar{\boldsymbol{\tau}} j(\alpha^3 - \alpha^\alpha h_{,\alpha})]\Big|_{h_1}^{h_2}, \tag{10.21}$$

$$\Gamma \boldsymbol{h} \equiv \int_{h_1}^{h_2} \tilde{\rho} \, (\tilde{\boldsymbol{p}} - \boldsymbol{r}) \times (\tilde{\boldsymbol{p}} - \boldsymbol{r})_t \, j \, dx^3. \tag{10.22}$$

The physical meaning of $\boldsymbol{m}^\alpha$ can be read off from (10.20).

Observe that if (10.1b) holds, then $\boldsymbol{p} - \boldsymbol{r}$ is constrained to equal $\kappa \boldsymbol{a}_3$, (10.3b) is valid, and (10.18a,b) have the same form, namely, a cross product of $\boldsymbol{d}$ with an arbitrary vector. In this case, (10.17) is equivalent to (10.19), and we adopt (10.4), (10.11), and (10.19) as our governing equations. Note that (10.20) shows that the $\boldsymbol{m}^\alpha$ have no components in the $\boldsymbol{a}_3$-direction. Observe as before that (10.4), (10.9), (10.11), (10.17), and (10.19) are exact consequences of the Principle of Virtual Power. We get shell-theoretic versions by replacing $\tilde{\boldsymbol{p}}$ with (10.1).

The generality of (10.1a) is adopted to exhibit the extent to which the mathematical structure of the governing equations is independent of the particular constraint. If (10.1b) does not hold, then I surmise that the replacement of (10.17) with (10.19) might lead to a problem that is mathematically unattractive.

10.23. Exercise. Derive the following analog of (3.29) when (10.1c) holds:

$$\boldsymbol{r}_{,\alpha} \times \boldsymbol{n}^\alpha + \boldsymbol{d}_{,\alpha} \times \boldsymbol{m}^\alpha + \boldsymbol{d} \times \int_{h_1}^{h_2} \varphi_{,k} \boldsymbol{\tau}^k j \, dx^3 = \mathbf{0}. \tag{10.23}$$

Now let us suppose that $\boldsymbol{d}$ is constrained to have unit length, so that $\delta = 1$. In this case, the configuration is determined by five unknown scalar functions determining $\boldsymbol{r}$ and the unit vector $\boldsymbol{d}$. If (10.1b) holds, then the classical equations of motion are the vector equations (10.4) and (10.19). Since (10.19) then agrees with (10.17), which has no component in the $\boldsymbol{d}$-direction, this system, when supplemented by appropriate constitutive equations, contains but five equations, and is therefore formally determinate. (If (10.1b) does not hold, then the governing equations are (10.4) and (10.17), as follows from (1.17) and (10.3a) with $\stackrel{\Delta}{\delta} = 0$.)

Let us further suppose that $\boldsymbol{d} = \boldsymbol{a}_3$ is perpendicular to the $\boldsymbol{r}_{,\alpha}$:

$$\boldsymbol{d} = \boldsymbol{a}_3 = \frac{\boldsymbol{r}_{,1} \times \boldsymbol{r}_{,2}}{|\boldsymbol{r}_{,1} \times \boldsymbol{r}_{,2}|}. \tag{10.24}$$

These are the *Kirchhoff assumptions.* They ensure that the configuration of the shell is determined solely by $\boldsymbol{r}$.

10.25. Exercise. Derive the governing equations under the Kirchhoff assumptions by using (10.24) directly, and alternatively by replacing (10.24) with the constraints

$$\boldsymbol{d} \cdot \boldsymbol{d} = 1, \quad \boldsymbol{d} \cdot \boldsymbol{r}_{,\alpha} = 0 \tag{10.26}$$

and employing the Multiplier Rule XII.12.54.

See Sec. 13 for further discussion of the Kirchhoff assumptions.

11. Axisymmetric Deformations of Axisymmetric Shells

We now study the axisymmetric deformation of axisymmetric shells with one director. We illustrate the ideas of the preceding section. In the next section we treat a global bifurcation problem. Let $\{\boldsymbol{i}, \boldsymbol{j}, \boldsymbol{k}\}$ be a fixed right-handed orthonormal basis for Euclidean 3-space. We set $\mathbf{s} \equiv (s, \phi)$, $\mathbf{x} \equiv (s, \phi, x^3)$, and

$$\boldsymbol{e}_1(\phi) = \cos\phi \boldsymbol{i} + \sin\phi \boldsymbol{j}, \quad \boldsymbol{e}_2(\phi) = -\sin\phi \boldsymbol{i} + \cos\phi \boldsymbol{j}, \quad \boldsymbol{e}_3 = \boldsymbol{k}. \tag{11.1}$$

We assume that the body has the form

$$\mathcal{B} \equiv \{\tilde{\boldsymbol{z}}(\mathbf{x}) \colon s \in [s_1, s_2],\ \phi \in [0, 2\pi),\ x^3 \in [h_1(s), h_2(s)]\} \tag{11.2}$$

where $\tilde{\boldsymbol{z}}$ has the form (10.2) with

$$\begin{gathered} \mathring{\boldsymbol{r}} = \mathring{r}(s)\boldsymbol{e}_1(\phi) + \mathring{z}(s)\boldsymbol{k}, \quad \varphi(\mathbf{x}) = \varphi(s, x^3), \\ \mathring{\boldsymbol{r}}_s(s,\phi) \equiv \mathring{\boldsymbol{a}}(s,\phi) \equiv \cos\mathring{\theta}(s)\boldsymbol{e}_1(\phi) + \sin\mathring{\theta}(s)\boldsymbol{k}, \\ \mathring{\boldsymbol{a}}_3(s,\phi) \equiv \mathring{\boldsymbol{b}}(s,\phi) \equiv -\sin\mathring{\theta}(s)\boldsymbol{e}_1(\phi) + \cos\mathring{\theta}(s)\boldsymbol{k}. \end{gathered} \tag{11.3}$$

We assume that $\mathcal{M}_\sharp$ corresponds to the coordinate surface $x^3 = h_0$ (see (9.3)) and that φ is independent of ϕ. Then

$$\begin{gathered} \Gamma(s) = \mathring{r}(s)[1 - \varphi(s, h_0)\mathring{\sigma}(s)][1 - \varphi(s, h_0)\mathring{\mu}(s)], \\ j = \mathring{r}(1 - \varphi\mathring{\sigma})(1 - \varphi\mathring{\mu})\varphi_{,3} \end{gathered} \tag{11.4}$$

where $\mathring{\sigma} \equiv (\sin\mathring{\theta})/\mathring{r}$ and $\mathring{\mu} \equiv \mathring{\theta}'$. Note that $\Gamma(s)$ reduces to $\mathring{r}(s)$ if φ vanishes for $x^3 = h_0$. We require that $\varphi_{,3} > 0$ everywhere, $1 - \varphi\mathring{\sigma} > 0$ everywhere, and $1 - \varphi\mathring{\mu} > 0$ everywhere.

We assume that the components with respect to the $\boldsymbol{e}_k$ of all the scalar-, vector-, and tensor-valued functions that serve as data for our problems are independent of ϕ, so that our problems are invariant under the group SO(2) of rotations about $\boldsymbol{k}$. We must impose further conditions so that our problems are invariant under the full group O(2) of rotations and reflections, in which case we term our problems *axisymmetric*. (Note that the (continuous) vector field of radially disposed unit vectors on a circle centered at the origin is invariant under O(2), whereas a (continuous) field of tangentially disposed unit vectors is invariant only under SO(2).) In this section we formulate the equations for the axisymmetric deformation of axisymmetric shells with one director.

We adopt (10.1) with $\boldsymbol{\kappa}(\boldsymbol{d}(\mathbf{s}), \mathbf{x}) \cdot \boldsymbol{e}_k(\phi)$ independent of ϕ, with $\boldsymbol{\kappa} \cdot \boldsymbol{e}_2 = 0$, and with

$$
\begin{aligned}
\boldsymbol{r}(s,\phi,t) &= r(s,t)\boldsymbol{e}_1(\phi) + z(s,t)\boldsymbol{k}, \\
\boldsymbol{d}(s,\phi,t) &= \delta(s,t)\boldsymbol{b}(s,\phi,t), \\
\boldsymbol{a}(s,\phi,t) &\equiv \boldsymbol{a}_1(s,\phi,t) \equiv \cos\theta(s,t)\boldsymbol{e}_1(\phi) + \sin\theta(s,t)\boldsymbol{k}, \\
\boldsymbol{a}_2(s,\phi,t) &\equiv \boldsymbol{e}_2(\phi), \\
\boldsymbol{b}(s,\phi,t) &\equiv \boldsymbol{a}_3(s,\phi,t) = -\sin\theta(s,t)\boldsymbol{e}_1(\phi) + \cos\theta(s,t)\boldsymbol{k}.
\end{aligned}
\tag{11.5}
$$

Thus such an axisymmetric configuration is determined by the four real-valued functions r, z, δ, and θ. We set

$$
\begin{gathered}
\boldsymbol{r}_s \equiv \nu\boldsymbol{a} + \eta\boldsymbol{b}, \quad \tau(s) \equiv r(s)/s, \quad \sigma(s) \equiv \sin\theta(s)/s, \\
\mu \equiv \theta', \quad \omega \equiv \delta'.
\end{gathered}
\tag{11.6}
$$

If we use (10.1b) with $\kappa_{,3} > 0$, then

$$
\boldsymbol{F} = \frac{[(\nu - \kappa\mu)\boldsymbol{a} + (\eta + \kappa_s + \kappa_\delta\omega - \kappa_{,3}\varphi_s/\varphi_{,3})\boldsymbol{b}]\mathring{\boldsymbol{a}}}{1 - \varphi\mathring{\mu}} + \frac{\tau - \kappa\sigma}{1 - \varphi\mathring{\sigma}}\boldsymbol{e}_2\boldsymbol{e}_2 + \frac{\kappa_{,3}}{\varphi_{,3}}\boldsymbol{b}\mathring{\boldsymbol{b}}.
\tag{11.7}
$$

Here $\kappa_s = \partial\kappa/\partial w_1$. Then

$$
\boldsymbol{C} = C_{kl}\mathring{\boldsymbol{a}}^k\mathring{\boldsymbol{a}}^l, \quad C_{kl} \equiv \mathring{\boldsymbol{a}}_k \cdot \boldsymbol{C} \cdot \mathring{\boldsymbol{a}}_l
\tag{11.8}
$$

where

$$
\begin{gathered}
C_{11} \equiv \frac{(\nu - \kappa\mu)^2 + (\eta + \kappa_s + \kappa_\delta\omega - \varphi_s\kappa_{,3}/\varphi_{,3})^2}{(1 - \varphi\mathring{\mu})^2}, \\
C_{22} \equiv \frac{(\tau - \kappa\sigma)^2}{(1 - \varphi\mathring{\sigma})^2}, \quad C_{33} \equiv \left(\frac{\kappa_{,3}}{\varphi_{,3}}\right)^2, \\
C_{13} \equiv \frac{\eta + \kappa_s + \kappa_\delta\omega}{1 - \varphi\mathring{\sigma}}\frac{\kappa_{,3}}{\varphi_{,3}}, \quad C_{21} = 0 = C_{23}.
\end{gathered}
\tag{11.9}
$$

If (10.1b) holds, then the requirement that $\boldsymbol{p}$ preserve orientation, i.e., that its Jacobian be positive everywhere, is equivalent to the inequalities

$$\begin{gathered} \tau > \max_{x^3} \kappa(\delta, s, x^3)\sigma, \quad \nu > \max_{x^3} \kappa(\delta, s, x^3)\mu, \\ \min_{x^3} \kappa_{,3}(\delta, s, x^3) > 0. \end{gathered} \tag{11.10}$$

If $h_1 = -h,\ h_2 = h$, and if κ is odd in x^3, then (11.10) reduces to

$$\begin{gathered} \tau > \kappa(\delta, s, h(s))|\sigma|, \quad \nu > \kappa(\delta, s, h(s))|\mu|, \\ \min_{x^3} \kappa_{,3}(\delta, s, x^3) > 0. \end{gathered} \tag{11.11}$$

The set of $\mathsf{q} \equiv (\tau, \nu, \eta, \sigma, \mu, \delta, \omega)$ satisfying (11.10) is denoted $\mathcal{V}(s)$. For fixed δ, this set is convex.

In consonance with our assumption of axisymmetry, we require that

$$\begin{gathered} \boldsymbol{e}_2 \cdot \boldsymbol{T} \cdot \mathring{\boldsymbol{a}} = 0, \quad \boldsymbol{e}_2 \times \boldsymbol{T} \cdot \boldsymbol{e}_2 = \boldsymbol{0}, \quad \boldsymbol{e}_2 \cdot \boldsymbol{T} \cdot \mathring{\boldsymbol{b}} = 0, \\ \boldsymbol{f}^0 \cdot \boldsymbol{e}_2 = 0, \quad \boldsymbol{e}_2 \times \boldsymbol{l} = \boldsymbol{0} \end{gathered} \tag{11.12}$$

with the components of $\boldsymbol{f}^0$ and $\boldsymbol{l}$ independent of f. We accordingly set

$$\begin{gathered} \boldsymbol{n}^1 \equiv N\boldsymbol{a} + H\boldsymbol{b}, \quad \Gamma\boldsymbol{n}^2 \equiv T\boldsymbol{e}_2, \quad \boldsymbol{m}^1 \equiv -M\boldsymbol{e}_2, \\ \Gamma\boldsymbol{m}^2 \equiv \Sigma\boldsymbol{a}, \quad \Omega^1 \equiv \Omega, \quad \Omega^2 = 0. \end{gathered} \tag{11.13}$$

Then (10.5), (10.20), (10.12), (11.4), and (11.12) yield

$$\Gamma(N\boldsymbol{a} + H\boldsymbol{b}) \equiv \mathring{r} \int_{h_1}^{h_2} \boldsymbol{T}_{\mathrm{A}} \cdot \mathring{\boldsymbol{a}}(1 - \varphi\mathring{\sigma})\varphi_{,3}\, dx^3, \tag{11.14a}$$

$$T \equiv \int_{h_1}^{h_2} \boldsymbol{e}_2 \cdot \boldsymbol{T}_{\mathrm{A}} \cdot \boldsymbol{e}_2(1 - \varphi\mathring{\mu})\varphi_{,3}\, dx^3, \tag{11.14b}$$

$$\Gamma M \equiv -\mathring{r}\boldsymbol{e}_2 \cdot \int_{h_1}^{h_2} (\tilde{\boldsymbol{p}} - \boldsymbol{r}) \times \boldsymbol{T}_{\mathrm{A}} \cdot \mathring{\boldsymbol{a}}(1 - \varphi\mathring{\sigma})\varphi_{,3}\, dx^3, \tag{11.14c}$$

$$\Sigma \equiv \boldsymbol{a} \cdot \int_{h_1}^{h_2} (\boldsymbol{p} - \boldsymbol{r}) \times \boldsymbol{T}_{\mathrm{A}} \cdot \boldsymbol{e}_2(1 - \varphi\mathring{\mu})\varphi_{,3}\, dx^3, \tag{11.14d}$$

$$\Gamma\Omega \equiv \mathring{r}\boldsymbol{b} \cdot \int_{h_1}^{h_2} \left(\boldsymbol{T}_{\mathrm{A}} \cdot \mathring{\boldsymbol{a}}\right) \cdot \frac{\partial \boldsymbol{\kappa}}{\partial \boldsymbol{d}}(1 - \varphi\mathring{\sigma})\varphi_{,3}\, dx^3, \tag{11.14e}$$

etc. These variables satisfy the equations of motion

$$\frac{\partial}{\partial s}[\Gamma(N\boldsymbol{a} + H\boldsymbol{b})] - T\boldsymbol{e}_1 + \Gamma\boldsymbol{f}^0 = 2\Gamma\rho h\boldsymbol{r}_{tt} + \Gamma\boldsymbol{q}_{tt}, \tag{11.15}$$

$$\begin{aligned} &\frac{\partial}{\partial s}(\Gamma M) - \Sigma\cos\theta + \Gamma(\nu H - \eta N) - \Gamma\boldsymbol{l} \cdot \boldsymbol{e}_2 \\ &\qquad = -\Gamma(\boldsymbol{q} \times \boldsymbol{r}_{tt} + \boldsymbol{h}_t) \cdot \boldsymbol{e}_2, \end{aligned} \tag{11.16}$$

$$\frac{\partial}{\partial s}(\Gamma\Omega) - \Gamma\Delta + \Gamma f^1 = \Gamma\beta_t - \Gamma\gamma, \tag{11.17}$$

which are specializations of (10.4), (10.19), and (10.11).

Constitutive equations for elastic shells. A Cosserat shell, constrained to undergo axisymmetric deformations, is *nonlinearly elastic* if there are functions $\hat{T}$, $\hat{N}$, $\hat{H}$, $\hat{\Sigma}$, $\hat{M}$, $\hat{\Delta}$, $\hat{\Omega}$ defined on $\{(\mathbf{q}, s) \colon \mathbf{q} \in \mathcal{V}(s), s \in [s_1, s_2]\}$ such that

$$T(s,t) = \hat{T}(\mathbf{q}(s,t), s), \qquad \text{etc.} \tag{11.18}$$

We assume that these constitutive functions are continuously differentiable. To carry out analyses of these equations, we require that the constitutive functions (11.18) satisfy certain monotonicity, growth, and symmetry properties. The monotonicity conditions can be chosen so that the substitution of (11.18) into the equations of motion yields a quasilinear system of hyperbolic partial differential equations. We follow the approach of Sec. 4 to show that these monotonicity conditions are shell-theoretic consequences of the Strong Ellipticity Condition. We adopt natural compatible growth conditions. It is easy to impose symmetry conditions that ensure that we can carry out our global qualitative analysis of the bifurcation problem to be treated in Sec. 12. But it is not evident whether such conditions, designed merely to meet mathematical exigencies, are physically natural or artificial. To resolve this issue, we again resort to our three-dimensional interpretation of the Cosserat theory.

Let us assume that (10.1b) holds. We obtain constitutive equations for our constrained three-dimensional theory by substituting (11.7)–(11.9) into the constitutive equation (XIII.1.3b) for elastic materials and then substituting the resulting expression into (11.14), (11.18), and (2.1). If we set $S^{ij} \equiv \mathring{\boldsymbol{a}}^i \cdot \hat{\mathbf{S}} \cdot \mathring{\boldsymbol{a}}^j$ and assume in consonance with (11.12) that these components are independent of ϕ, then the versions of (11.18) so obtained, called the *induced constitutive equations*, have the form

$$\begin{aligned}
\hat{N} &= \frac{\mathring{r}}{\Gamma} \int_{h_1}^{h_2} \frac{\nu - \kappa\mu}{1 - \varphi\mathring{\mu}} \hat{S}^{11}_{\mathrm{A}} (1 - \varphi\mathring{\sigma}) \varphi_{,3} \, dx^3, \\
\hat{M} &= -\frac{\mathring{r}}{\Gamma} \int_{h_1}^{h_2} \kappa \frac{\nu - \kappa\mu}{1 - \varphi\mathring{\mu}} \hat{S}^{11}_{\mathrm{A}} (1 - \varphi\mathring{\sigma}) \varphi_{,3} \, dx^3, \\
\hat{H} &= \frac{\mathring{r}}{\Gamma} \int_{h_1}^{h_2} \boldsymbol{\beta} \cdot \hat{\boldsymbol{S}}_{\mathrm{A}} \cdot \mathring{\boldsymbol{a}} \, (1 - \varphi\mathring{\sigma}) \, dx^3, \\
\hat{\Omega} &= \frac{\mathring{r}}{\Gamma} \int_{h_1}^{h_2} \kappa_\delta \boldsymbol{\beta} \cdot \hat{\boldsymbol{S}}_{\mathrm{A}} \cdot \mathring{\boldsymbol{a}} (1 - \varphi\mathring{\sigma}) \, dx^3, \\
\hat{T} &= \int_{h_1}^{h_2} \frac{\tau - \kappa\sigma}{1 - \varphi\mathring{\sigma}} \hat{S}^{22}_{\mathrm{A}} (1 - \varphi\mathring{\mu}) \varphi_{,3} \, dx^3, \\
\hat{\Sigma} &= -\int_{h_1}^{h_2} \kappa \frac{\tau - \kappa\sigma}{1 - \varphi\mathring{\sigma}} \hat{S}^{22}_{\mathrm{A}} (1 - \varphi\mathring{\mu}) \varphi_{,3} \, dx^3,
\end{aligned} \tag{11.19}$$

$$\hat{\Delta} = \frac{\overset{\circ}{r}}{\Gamma} \int_{h_1}^{h_2} \boldsymbol{\beta} \cdot \hat{\boldsymbol{S}}_{\mathrm{A}} \cdot \left\{ \partial_s(\kappa_\delta)\overset{\circ}{\boldsymbol{a}} + \frac{(\kappa_\delta)_{,3}}{\varphi_{,3}}[(1-\varphi\overset{\circ}{\mu})\overset{\circ}{\boldsymbol{b}} - \varphi_s \overset{\circ}{\boldsymbol{a}}] \right\} (1-\varphi\overset{\circ}{\sigma})\, dx^3$$
$$- \frac{\overset{\circ}{r}}{\Gamma} \int_{h_1}^{h_2} \kappa_\delta \left[\mu \frac{\nu - \kappa\mu}{1-\varphi\overset{\circ}{\mu}} \hat{S}^{11}_{\mathrm{A}}(1-\varphi\overset{\circ}{\sigma}) + \sigma \frac{\tau - \kappa\sigma}{1-\varphi\overset{\circ}{\sigma}} \hat{S}^{22}_{\mathrm{A}}(1-\varphi\overset{\circ}{\mu}) \right] \varphi_{,3}\, dx^3,$$

where
$$\boldsymbol{\beta} \equiv \frac{\varphi_{,3}(\eta + \kappa_s + \kappa_\delta\omega) - \kappa_{,3}\varphi_s}{1-\varphi\overset{\circ}{\mu}} \overset{\circ}{\boldsymbol{a}} + \kappa_{,3}\overset{\circ}{\boldsymbol{b}},$$

and where the $\hat{S}^{ij}_{\mathrm{A}}$ depend on (11.9), $\boldsymbol{\Lambda}$, and (s, x^3).

11.20. Exercise. Prove:

11.21. Proposition. *If $\hat{\boldsymbol{T}}$ satisfies the Strong Ellipticity Condition (XIII.2.28), (XIII.2.29), then the induced constitutive functions $\hat{T}$, etc., satisfy the following monotonicity conditions: The matrices*

(11.22a,b,c) $$\frac{\partial(\hat{N}, \hat{H}, \hat{M}, \hat{\Omega})}{\partial(\nu, \eta, \mu, \omega)}, \quad \frac{\partial(\hat{T}, \hat{\Sigma})}{\partial(\tau, \sigma)}, \quad \frac{\partial \hat{\Delta}}{\partial \delta} \quad \textit{are positive-definite.}$$

We adopt (11.22) as a fundamental constitutive restriction. It ensures that an increase in the bending strain μ be accompanied by an increase in the bending couple M, etc. Rather than supplement (11.22) with a compatible system of growth conditions (requiring that infinite resultants accompany extreme strains; see Negrón-Marrero & Antman (1990), e.g.), we content ourselves with prescribing a major consequence of a suitable set of such conditions:

11.23. Hypothesis. *For any given numbers N, H, M, Ω, the equations*

(11.24) $$\hat{N}(\mathsf{q}, s) = N, \ \hat{H}(\mathsf{q}, s) = H, \ \hat{M}(\mathsf{q}, s) = M, \ \hat{\Omega}(\mathsf{q}, s) = \Omega$$

have a unique solution for (ν, η, μ, ω) as a function of $(\tau, N, H, \sigma, M, \delta, \Omega, s)$.

We denote this solution by

(11.25a) $$\nu = \nu^\sharp(\tau, N, H, \sigma, M, \delta, \Omega, s), \ldots$$

where the ellipsis stands for analogous expressions for $\eta^\sharp$, $\mu^\sharp$, $\omega^\sharp$. Let us set

(11.25b) $$T^\sharp(\tau, N, H, \sigma, M, \delta, \Omega, s) \equiv \hat{T}(\tau, \nu^\sharp, \eta^\sharp, \sigma, \mu^\sharp, \delta, \omega^\sharp, s), \ldots$$

where the ellipsis stands for analogous definitions of $\Sigma^\sharp$, $\Delta^\sharp$, and where the arguments of $\nu^\sharp$, $\eta^\sharp$, $\mu^\sharp$, $\omega^\sharp$ in (11.25b) are those shown in (11.25a). The functions $\nu^\sharp$, etc., of (11.25) are necessarily continuously differentiable. The constitutive equations (11.25) are equivalent to (11.18).

Now we turn to the symmetry conditions, which will be used in the next section and which consequently constitute our immediate motivation for our development of the Cosserat theory from the three-dimensional theory. We assume that the constitutive function $\hat{\boldsymbol{S}}_{\mathrm{A}}$ satisfies the symmetry conditions:

(11.26a) $$\hat{S}^{21}_{\mathrm{A}} = 0 = \hat{S}^{23}_{\mathrm{A}} \quad \text{if } C_{21} = 0 = C_{23},$$

(11.26b)
$$\hat{S}^{11}_{\mathrm{A}},\ \hat{S}^{22}_{\mathrm{A}},\ \hat{S}^{33}_{\mathrm{A}} \text{ are even functions of } C_{13} \text{ if } C_{21} = 0 \text{ and } C_{23} = 0,$$

(11.26c) $\hat{S}^{13}_{\mathrm{A}}$ is an odd function of C_{13} if $C_{21} = 0$ and $C_{23} = 0$.

An isotropic material meets these conditions. Note that (11.3), (11.7), (11.9), and (11.26) ensure (11.12).

We adopt the *symmetry conditions*: $\tilde{\rho}$ is even in x^3, κ is odd in x^3, $h_1 = -h$, and $h_2 = h$.

The analog of Proposition 5.22 is

11.27. Proposition. *Let* (10.1b) *hold, let the symmetry conditions hold, let* κ *be independent of* s, *and let* (11.26) *hold. Then*

(11.28a) $$\hat{H}(\mathbf{q}, s) = 0 = \hat{\Omega}(\mathbf{q}, s)$$

if

(11.28b) $$\eta = 0 = \omega.$$

The proof of this proposition is straightforward. The form of M prevents us from obtaining a result analogous to Proposition 5.25.

The simplifications of our constitutive equations for plates, which have the defining property that

(11.29) $$\overset{\circ}{\sigma} = 0 = \overset{\circ}{\mu},$$

enable us to get an analog of Proposition 5.27. The central result of our analysis is

11.30. Theorem. *Let* (11.29) *hold and let the hypotheses of Proposition* 11.27 *hold. Then under the transformation*

(11.31) $$(\eta, \sigma, \mu) \mapsto -(\eta, \sigma, \mu),$$

the induced constitutive functions $\hat{H}$, $\hat{\Sigma}$, $\hat{M}$ *change sign while the other induced constitutive functions remain unchanged.*

Proof. We replace the arguments η, σ, μ of $\hat{M}$ in (11.18) with their negatives. We then make the change of variable $x^3 \mapsto -x^3$ in the resulting integral. Since $\kappa(\delta, s, \cdot)$ is odd, $\kappa(\delta, s, x^3)$ is thereby replaced with $-\kappa(\delta, s, x^3)$, etc. The arguments now occupying the C_{11}-, C_{22}-, C_{33}-slots in S^{11} are unchanged, while the argument now occupying the C_{31}-slot is the negative of that originally occupying this slot. We use (11.26b) to restore the original sign. We thereby find that $\hat{M}$ changes sign under (11.31). The other symmetries are proved likewise. □

11.32. Exercise. Carry out the proofs of the other symmetries.

The induced constitutive equations enjoy many other symmetries, which can be determined by similar methods.

The symmetry condition of Theorem 11.30 is much weaker than the analogous conditions for plates for which δ is constrained to equal 1 (cf. Sec. X.2). In such theories, the shear response is uncoupled from the flexural response. Equation (11.28a) is exploited below in a way that renders Theorem 11.30 as useful as the stronger results for the more restricted theory.

11.33. Exercise. Carry out the analog of Ex. VI.7.13. for the problem of Sec. X.2.

11.34. Exercise. Carry out the analog of Ex. VI.7.13. for the problem of Sec. X.5.

12. Global Buckled States of a Cosserat Plate

The boundary-value problem. We now treat an axisymmetric equilibrium problem for a plate which we take to be annular in order to avoid bothering with the polar singularity. We assume that (11.22), Hypothesis 11.23, and the conclusion of Theorem 11.30 hold. We take $\Gamma(s) = s$ and assume that $\boldsymbol{f}^0 = \boldsymbol{0}$, $\boldsymbol{l} \cdot \boldsymbol{e}_2 = 0$, $f^1 = 0$. We assume that the edge $s = 1$ of the plate is constrained to be parallel to $\boldsymbol{k}$, so that

$$\theta(1) = 0. \tag{12.1}$$

A normal pressure of intensity λ units of force per unit reference length is applied to the edge $s = 1$, so that

$$N(1) = -\lambda, \quad H(1) = 0. \tag{12.2}$$

We finally assume that either the thickness of the plate at this edge is fixed, so that there is a prescribed number δ_1 such that

$$\delta(1) = \delta_1 \tag{12.3a}$$

or else there is no restraint preventing such a change, so that

$$\Omega(1) = 0. \tag{12.3b}$$

The inner edge is subjected to similar boundary conditions:

$$r(a) = a, \quad \theta(a) = 0, \quad H(a) = 0, \quad \Omega(a) = 0. \tag{12.4}$$

We can readily handle variants of these boundary conditions. We must solve equilibrium versions of (11.4), (11.15)–(11.18), (12.1)–(12.4), which constitute our boundary value problem.

Alternative formulation of the governing equations. We recast our boundary-value problem in a form that makes the role of symmetry explicit and that promotes the ensuing analysis. We integrate (11.15) subject to the boundary condition (12.2) and use (11.28) to obtain

$$-N(s) = \Lambda^\sharp(s)\cos\theta(s), \tag{12.5a}$$

$$H(s) = \Lambda^\sharp(s)\sin\theta(s) \tag{12.5b}$$

where

$$s\Lambda^\sharp(s) \equiv \lambda + \int_s^1 T(\tau(\xi),\dots)\,d\xi. \tag{12.5c}$$

We next deduce from (11.6), (11.16), (11.17), (11.25), and (12.5) that

$$\theta' = \mu^\sharp(\tau, -\Lambda^\sharp\cos\theta, \Lambda^\sharp\sin\theta, s^{-1}\sin\theta, M, \delta, \Omega, s), \tag{12.6}$$

$$(sM)' = \Sigma^\sharp\cos\theta - s\Lambda^\sharp(\nu^\sharp\sin\theta + \eta^\sharp\cos\theta), \tag{12.7}$$

$$\delta' = \delta^\sharp(\tau, -\Lambda^\sharp\cos\theta, \Lambda^\sharp\sin\theta, s^{-1}\sin\theta, M, \delta, \Omega, s), \tag{12.8}$$

$$(s\Omega)' = s\Delta^\sharp \tag{12.9}$$

where the arguments of $\Sigma^\sharp$, $\Delta^\sharp$, $\nu^\sharp$, $\eta^\sharp$ in (12.7) and (12.9) are the same as those of $\mu^\sharp$ in (12.6).

Hypothesis 11.23 and the consequences of Theorem 11.30 imply that under the transformation

$$(H,\sigma,M) \mapsto -(H,\sigma,M) \tag{12.10}$$

the constitutive functions $\eta^\sharp$, $\Sigma^\sharp$, $\mu^\sharp$ of (11.25) change sign while the other constitutive functions of (11.25) remain unchanged. It then follows from the Mean Value Theorem that $\eta^\sharp$, $\Sigma^\sharp$, $\mu^\sharp$ can be represented as

$$\eta^\sharp = \overline{\eta_H}H + \overline{\eta_\sigma}\sigma + \overline{\eta_M}M, \quad \text{etc.,} \tag{12.11a}$$

where

(12.11b)

$$\overline{\eta_H}(\tau, N, H, \sigma, M, \delta, \Omega, s) \equiv \int_0^1 \eta_H(\tau, N, \alpha H, \alpha\sigma, \alpha M, \delta, \Omega, s)\,d\alpha, \quad \text{etc.}$$

Then equations (12.6), (12.7) assume the form

$$\theta' = A\sin\theta + \overline{\mu_M}M, \tag{12.12}$$

$$(sM)' = C\sin\theta + DM. \tag{12.13}$$

where

$$A \equiv (\overline{\mu_H}\Lambda^\sharp + \overline{\mu_\sigma}s^{-1}),$$
$$C \equiv [\overline{\Sigma_H}\Lambda^\sharp + \overline{\Sigma_\sigma}s^{-1} - s\Lambda^\sharp(\overline{\eta_H}\Lambda^\sharp + \overline{\eta_\sigma}s^{-1})]\cos\theta - s\Lambda^\sharp\nu^\sharp,$$
$$D \equiv [\overline{\Sigma_M} - s\Lambda^\sharp\overline{\eta_M}]\cos\theta$$

Our boundary-value problem is thus equivalent to the system (12.5), (12.8), (12.9), (12.12), (12.13), (12.1)–(12.4) for

$$\mathbf{u} \equiv (r, N, H, \theta, M, \delta, \Omega). \tag{12.14}$$

Unbuckled states. If we regard the $\Lambda^\sharp$, δ, Ω appearing in (12.12), (12.13) as given well-behaved functions, then this second-order system admits the solution $\theta = 0 = M$, which describes an *unbuckled state.*

Unbuckled states are typically not unique. For λ positive there can be such states with a necking instability (see Antman (1973a), Antman & Carbone (1977), and Negrón-Marrero (1989)). Nonuniqueness can be manifested in far more prosaic circumstances, as we now show. It is easy to see that unbuckled states are governed by the system

$$\hat{N} = -\lambda - \int_s^1 \hat{T}\, d\xi, \tag{12.15a}$$

$$(s\hat{\Omega})' - s\hat{\Delta} = 0 \tag{12.15b}$$

where the arguments of $\hat{N}$, $\hat{\Omega}$, and $\hat{\Delta}$ are $\big(r(s)/s, r'(s), 0, 0, 0, \delta(s), \delta'(s), s\big)$ and those of $\hat{T}$ are the same with s replaced by ξ. Let us now make several simplifying assumptions. We suppose that these constitutive functions are independent of s, as would happen if we construct our constitutive equations as in Sec. 3 by using three-dimensional constitutive equations for a homogeneous material and by taking the plate thickness h to be constant. Next we suppose that $\hat{\Omega} = 0$ when $\eta = \sigma = \mu = \omega = 0$. This property is a consequence of (11.29) for induced constitutive equations. Finally, we assume that $\hat{T}(\tau, \nu, 0, 0, 0, \delta, 0) = \hat{N}(\nu, \tau, 0, 0, 0, \delta, 0)$. A transversely isotropic plate would have this property.

Suppose that boundary condition (12.3b) holds. Then we can seek unbuckled states with $r' = \nu_0$ (const.), $\delta = \delta_0$ (const.). These numbers satisfy

$$\hat{N}(\nu_0, \nu_0, 0, 0, 0, \delta_0, 0) = -\lambda, \quad \hat{\Delta}(\nu_0, \nu_0, 0, 0, 0, \delta_0, 0) = 0. \tag{12.16}$$

It is easy to impose natural growth conditions that ensure that (12.16) has a solution (by a degree-theoretic argument), but our monotonicity conditions are insufficient to ensure that such a solution is unique. (The positive-definiteness of the matrix of partial derivatives of $\hat{T}$, $\hat{N}$, $\hat{\Delta}$ with respect to τ, ν, δ would suffice to ensure the uniqueness. This positive-definiteness is not a consequence of the Strong Ellipticity Condition, although it would follow from the Coleman-Noll Condition XIII.2.21.)

Were we to impose conditions less restrictive than those that led to (12.16), then the study of the mere existence of unbuckled states could lead to challenging exercises in analysis. For example, an inkling of the complexity that can occur for plates not satisfying the transverse isotropy condition can be found in Secs. X.3 and XIII.7. The unbuckled state need not depend continuously on the load parameter λ. The corresponding analysis for nonuniform plates for our problem would be far more formidable. The requisite tools for treating these singular problems can be modelled on those used by Negrón-Marrero (1985).

Nodal properties of buckled plates. Let C^0 denote the space of continuous functions on $[a, 0]$. We use analogous expressions for other spaces. We denote the Cartesian product of m copies of C^0 by $[C^0]^m$.

Let us suppose that all the arguments of the overlined functions in (12.12), (12.13) are continuous functions of s. (These arguments are the same as those of $\mu^\sharp$ in (12.6).) Thus (12.12) implies that $\theta \in C^1$ if and only if $(\theta, M) \in [C^0]^2$. Moreover, θ has a simple zero at s^* if and only if $\theta(s^*) = 0$, $M(s^*) \neq 0$; and θ has a double zero at s^* if and only if $\theta(s^*) = 0$, $M(s^*) = 0$. Let

$$\mathcal{Z}_k \equiv \{(\theta, M) \in [C^0]^2 \colon \theta(a) = 0 = \theta(1), \theta \text{ has exactly } k+1 \text{ zeros on } [a,1], \text{ each of which is simple}\}. \tag{12.17}$$

$\mathcal{Z}_k$ is an open set in $C^0 \times C^0$. If (θ, M) belongs to its boundary ∂Z_k, then θ has a double zero on $[a, 1]$.

Let $\mathcal{S}$ be a connected family of solution pairs containing a solution pair $(\tilde{\lambda}, \tilde{\mathbf{u}})$ with $(\tilde{\theta}, \tilde{M})$ in $\mathcal{Z}_k$. Since $\mathcal{Z}_k$ is open in $[C^0]^2$, there is a neighborhood $\mathcal{N}$ of $(\tilde{\lambda}, \tilde{\mathbf{u}})$ in the topology of $[C^0]^7 \times \mathbb{R}$ such that if $(\lambda, \mathbf{u}) \in \mathcal{N} \cap \mathcal{S}$, then $(\theta, M) \in \mathcal{Z}_k$. Let $\tilde{\mathcal{S}}$ be the largest connected component of $\mathcal{S}$ containing $(\tilde{\lambda}, \tilde{\mathbf{u}})$ and having the further property that there is no θ corresponding to one of its solution pairs having a double zero. From the preceding paragraph we know that (θ, M) in $\mathcal{Z}_k$ for each solution pair of $\tilde{\mathcal{S}}$.

Now if θ has a double zero, then the corresponding zero values for (θ, M) can be taken as initial data for (12.12), (12.13). The resulting initial-value problem has a unique solution, which must be the zero solution (by virtue of the special form of (12.12), (12.13) consequent on the constitutive symmetry under the transformation (11.31)). This conclusion holds no matter what continuous values the hidden variables in (12.12), (12.13) assume). Hence we have

12.18. Theorem. *If a connected set $\mathcal{S}$ of solution pairs contains no unbuckled state and if it has one solution pair with $(\tilde{\theta}, \tilde{M})$ in $\mathcal{Z}_k$, then each of its solution pairs (θ, M) is in $\mathcal{Z}_k$. Thus the number of simple zeros of θ is constant on such a set of solution pairs and characterizes it globally.*

To prove the existence of branches of buckled states we first need to show that we can write our boundary-value problem as a fixed-point problem involving a compact operator on $[C^0]^7$. We can easily do this by integrating (12.8), (12.9), (12.12), and (12.13) and using (11.25) to show that the arbitrary constants of integration can be adjusted to ensure that all the boundary conditions are satisfied. (For details of a much more difficult version of this process, see Negrón-Marrero & Antman (1990) and the works cited therein.) We can now invoke Corollary V.4.18 and Theorem V.4.19 to obtain

12.19. Theorem. *let $\bar{\lambda}$ be an eigenvalue of odd algebraic multiplicity of the linearization of the boundary-value problem about a branch of unbuckled states and let $\bar{\mathbf{u}}(\lambda)$ be the corresponding unbuckled state. From $(\bar{\lambda}, \bar{\mathbf{u}}(\bar{\lambda}))$ there bifurcates a branch of buckled states that is either unbounded in $[C^0]^7 \times \mathbb{R}$ or else returns to another such pair $(\bar{\kappa}, \bar{\mathbf{u}}(\bar{\kappa}))$. If the eigenvalue is simple, then the bifurcating branch inherits the nodal properties of the eigenfunction.*

13. Intrinsic Theory of Special Cosserat Shells

In this section we outline the direct formulation of the special Cosserat theory of shells as a description of a class of two-dimensional bodies. We proceed formally, leaving as informal exercises the treatment of refinements and the full statement of regularity restrictions corresponding to those of Chaps. VIII and XII. The *motion of a special Cosserat shell* is the motion of a material surface together with the motion of an independent unit vector field defined on the surface. We assume that the material surface is continuously differentiable. It is convenient, though somewhat inelegant, to introduce curvilinear coordinates $\mathbf{s} \equiv (s^1, s^2)$ for the material surface by identifying it with a continuously differentiable one-to-one mapping $\mathbf{s} \mapsto \mathring{\boldsymbol{r}}(\mathbf{s})$ of a region $\mathcal{M}$ of $\mathbb{R}^2$ into $\mathbb{E}^3$. $\mathcal{M}$ is assumed to be a domain together with none, some, or all of its boundary. We use the notation

$$\frac{\partial \boldsymbol{r}}{\partial s^\alpha} \equiv \boldsymbol{r}_{,\alpha}, \qquad \alpha = 1, 2. \tag{13.1}$$

The requirement that the coordinate system $\mathbf{s}$ be nonsingular (except possibly at isolated points and curves) is expressed by

$$\mathring{\boldsymbol{r}}_{,1} \times \mathring{\boldsymbol{r}}_{,2} \neq \mathbf{0} \quad \forall \mathbf{s} \in \operatorname{int} \mathcal{M}. \tag{13.2}$$

Geometry of deformation. Formally, a *motion of a special Cosserat shell* is a mapping

$$\operatorname{cl} \mathcal{M} \times \mathbb{R} \ni (\mathbf{s}, t) \mapsto (\boldsymbol{r}(\mathbf{s}, t), \boldsymbol{d}(\mathbf{s}, t)) \in \mathbb{E}^3 \times \mathbb{S}^2, \tag{13.3}$$

where $\mathbb{S}^2$ is the unit sphere in $\mathbb{E}^3$. The pair $(\boldsymbol{r}(\cdot, t), \boldsymbol{d}(\cdot, t))$ is the *configuration at time t* of the shell.

The *reference configuration* is $(\mathring{\boldsymbol{r}}, \mathring{\boldsymbol{d}})$ where $\mathring{\boldsymbol{d}}$ is taken to be the oriented unit normal field:

$$\mathring{\boldsymbol{d}} \equiv \frac{\mathring{\boldsymbol{r}}_{,1} \times \mathring{\boldsymbol{r}}_{,2}}{|\mathring{\boldsymbol{r}}_{,1} \times \mathring{\boldsymbol{r}}_{,2}|}. \tag{13.4}$$

We may interpret $\boldsymbol{r}(\cdot, t)$ as the material base surface at time t and $\boldsymbol{d}(\mathbf{s}, t)$ as characterizing the deformation at time t of the material fiber normal to the material base surface at $\mathring{\boldsymbol{r}}(\mathbf{s})$ in the reference configuration. Consequently we assume that $\boldsymbol{r} \mapsto \boldsymbol{c} + \boldsymbol{Q} \cdot \boldsymbol{r}$ and $\boldsymbol{d} \mapsto \boldsymbol{Q} \cdot \boldsymbol{d}$ under the rigid rotation $\boldsymbol{Q}$ and the translation $\boldsymbol{c}$. The interpretation of this assumption in the light of (10.1) shows that it corresponds to (10.1c).

To model the three-dimensional requirement that orientation be preserved, we may adopt a shell-theoretic analog of (2.11), which is directly motivated by three-dimensional considerations, or we may simply content ourselves with the following primitive condition, which reflects the interpretation given in the last paragraph:

$$\frac{(\boldsymbol{r}_{,1} \times \boldsymbol{r}_{,2}) \cdot \boldsymbol{d}}{(\mathring{\boldsymbol{r}}_{,1} \times \mathring{\boldsymbol{r}}_{,2}) \cdot \mathring{\boldsymbol{d}}} > 0 \quad \text{on } \operatorname{int} \mathcal{M}. \tag{13.5}$$

Thus $\{\boldsymbol{r}_{,1}, \boldsymbol{r}_{,2}, \boldsymbol{d}\}$ is a basis for $\mathbb{E}^3$. It is generally not an orthonormal basis, however. This fact is the source of technical difficulty.

The *strains* for (13.3) are any independent set of functions from the set of all scalar and triple scalar products formed from the five vector-valued functions $\boldsymbol{r}_{,1}, \boldsymbol{r}_{,2}, \boldsymbol{d}, \boldsymbol{d}_{,1}, \boldsymbol{d}_{,2}$ subject to the requirement that $\boldsymbol{d} \cdot \boldsymbol{d} =$

1. In view of the interpretation of $\boldsymbol{r}, \boldsymbol{d}$, these strains are unaffected by rigid motions and determine the configuration of the shell to within a rigid motion.

13.6. Exercise. Show that a suitable set of strains is

$$\boldsymbol{r}_{,\alpha} \cdot \boldsymbol{r}_{,\beta}, \quad \boldsymbol{r}_{,\alpha} \cdot \boldsymbol{d}, \quad \boldsymbol{r}_{,\alpha} \cdot \boldsymbol{d}_{,\beta}, \quad \boldsymbol{d} \cdot \boldsymbol{a}^3 \tag{13.7}$$

where $\boldsymbol{a}^3 \equiv (\boldsymbol{r}_{,1} \times \boldsymbol{r}_{,2})/|\boldsymbol{r}_{,1} \times \boldsymbol{r}_{,2}|$. Show that these quantities determine the configuration of the shell to within a rigid motion.

Mechanics. Let $\mathcal{A}$ be a subdomain of int $\mathcal{M}$ with a smooth boundary. Let $\boldsymbol{\gamma}$ be the unit normal field to $\partial\mathring{\boldsymbol{r}}(\mathcal{A})$ that is tangent to $\mathring{\boldsymbol{r}}(\mathcal{A})$ and points outward from it. The resultant contact force per unit length of $\partial\mathring{\boldsymbol{r}}(\mathcal{A})$ exerted at $\mathring{\boldsymbol{r}}(\mathrm{s})$ at time t by the material outside $\mathrm{cl}\,\mathring{\boldsymbol{r}}(\mathcal{A})$ on it is denoted $\boldsymbol{n}(\mathrm{s}, t, \boldsymbol{\gamma}(\mathrm{s}))$. Similarly, the resultant contact torque per unit length of $\partial\mathring{\boldsymbol{r}}(\mathcal{A})$ exerted at $\mathring{\boldsymbol{r}}(\mathrm{s})$ at time t by the material outside $\mathrm{cl}\,\mathring{\boldsymbol{r}}(\mathcal{A})$ on it is denoted $\boldsymbol{r}(\mathrm{s}, t) \times \boldsymbol{n}(\mathrm{s}, t, \boldsymbol{\gamma}(\mathrm{s})) + \boldsymbol{m}(\mathrm{s}, t, \boldsymbol{\gamma}(\mathrm{s}))$. Here we are following Cauchy's postulate of Sec. XII.7 in assuming that these contact loads depend on $\partial\mathring{\boldsymbol{r}}(\mathcal{A})$ only through the normal vector $\boldsymbol{\gamma}$. (We emphasize that $\boldsymbol{\gamma}$ has no intrinsic geometrical significance: It merely identifies the material plane across which the traction acts.) Let $\boldsymbol{f}(\mathrm{s}, t)$ denote the force and $\boldsymbol{l}(\mathrm{s}, t)$ the couple per unit area of $\mathring{\boldsymbol{r}}(\mathcal{A})$ exerted on this body at (s, t) by all other agencies. Thus the resultant force and torque on $\mathring{\boldsymbol{r}}(\mathcal{A})$ at time t are

$$\int_{\partial\mathcal{A}} \boldsymbol{n}(\mathrm{s}, t, \boldsymbol{\gamma}(\mathrm{s}))\, d\sigma(\mathrm{s}) + \int_{\mathcal{A}} \boldsymbol{f}(\mathrm{s}, t)\, da(\mathrm{s}), \tag{13.8}$$

$$\begin{aligned} &\int_{\partial\mathcal{A}} [\boldsymbol{r}(\mathrm{s}, t) \times \boldsymbol{n}(\mathrm{s}, t, \boldsymbol{\gamma}(\mathrm{s})) + \boldsymbol{m}(\mathrm{s}, t, \boldsymbol{\gamma}(\mathrm{s}))]\, d\sigma(\mathrm{s}) \\ &\qquad + \int_{\mathcal{A}} [\boldsymbol{r}(\mathrm{s}, t) \times \boldsymbol{f}(\mathrm{s}, t) + \boldsymbol{l}(\mathrm{s}, t)]\, da(\mathrm{s}), \end{aligned} \tag{13.9}$$

where $d\sigma(\mathrm{s})$ is the differential arc length of $\partial\mathring{\boldsymbol{r}}(\mathcal{A})$ at s, and where $da(\mathrm{s}) = |\mathring{\boldsymbol{r}}_{,1}(\mathrm{s}) \times \mathring{\boldsymbol{r}}_{,2}(\mathrm{s})|\, ds^1\, ds^2 \equiv \Gamma(\mathrm{s})\, ds^1\, ds^2$ is the differential area of $\mathring{\boldsymbol{r}}(\mathcal{A})$ at s.

We denote the linear and angular momenta of $\mathring{\boldsymbol{r}}(\mathcal{A})$ at s at time t by

$$\int_{\mathcal{A}} [2(\rho h)(\mathrm{s})\boldsymbol{r}_t(\mathrm{s}, t) + \boldsymbol{q}_t(\mathrm{s}, t)]\, da(\mathrm{s}), \tag{13.10}$$

$$\begin{aligned} &\int_{\mathcal{A}} [2(\rho h)(\mathrm{s})\boldsymbol{r}(\mathrm{s}, t) \times \boldsymbol{r}_t(\mathrm{s}, t) \\ &\qquad + \boldsymbol{r}(\mathrm{s}, t) \times \boldsymbol{q}_t(\mathrm{s}, t) + \boldsymbol{q}(\mathrm{s}, t) \times \boldsymbol{r}_t(\mathrm{s}, t) + \boldsymbol{h}(\mathrm{s}, t)]\, da(\mathrm{s}). \end{aligned} \tag{13.11}$$

In keeping with our interpretations of $\boldsymbol{r}$ and $\boldsymbol{d}$, we take $\boldsymbol{q}$ and $\boldsymbol{h}$ to have the general forms of (10.7) and (10.22) when (10.1c) holds. (As explained in Secs. VIII.2–VIII.4, the inertial terms typically require motivation from the three-dimensional theory.) In light of the results of Sec. VIII.4, we can often choose

$$q = \mathbf{0}, \quad h = \rho J d \times d_t \tag{13.12}$$

(cf. (VIII.2.10)–(VIII.2.12)). The integral forms of the *equations of motion* are obtained by equating (13.8) to the time derivative of (13.10) and equating (13.9) to the time derivative of (13.11). The impulse-momentum laws are then obtained by integrating these integral laws with respect to time.

We now imitate the proof of Cauchy's Stress Theorem (XII.7.14) to determine how $\boldsymbol{n}$ and $\boldsymbol{m}$ depend on $\boldsymbol{\gamma}$. The advantages of two-dimensionality are counterbalanced by the curvilinearity.

Let us choose $\mathcal{A}$ to be the region in int $\mathcal{M}$ bounded by the right triangle with vertices at (s^1, s^2), $(s^1 + \beta_2 l, s^2)$, $(s^1, s^2 + \beta_1 l)$. Note that the hypotenuse has length l and unit outer normal (β_1, β_2). The image of $\mathcal{A}$ under $\mathring{\boldsymbol{r}}$ is the region bounded by the curvilinear triangle with oriented sides

$$\begin{aligned}
&[0,1] \ni \alpha \mapsto \mathring{\boldsymbol{r}}(\mathbf{s} + \alpha(\beta_2 l, 0)),\\
&[0,1] \ni \alpha \mapsto \mathring{\boldsymbol{r}}(\mathbf{s} + \alpha(0, \beta_1 l) + (1-\alpha)(\beta_2 l, 0)),\\
&[0,1] \ni \alpha \mapsto \mathring{\boldsymbol{r}}(\mathbf{s} + (1-\alpha)(0, \beta_1 l)).
\end{aligned}$$

The unit outer normal to the image of the hypotenuse that is tangent to $\mathring{\boldsymbol{r}}(\mathcal{A})$ at $\mathring{\boldsymbol{r}}(\mathbf{s} + \boldsymbol{\zeta}(\alpha))$ where $\boldsymbol{\zeta}(\alpha) \equiv \alpha(0, \beta_1 l) + (1-\alpha)(\beta_2 l, 0)$ is

$$\boldsymbol{\gamma}(\mathbf{s}) = \frac{\frac{d}{d\alpha}[\mathring{\boldsymbol{r}}(\mathbf{s} + \boldsymbol{\zeta}(\alpha))] \times \mathring{\boldsymbol{d}}(\mathbf{s} + \boldsymbol{\zeta}(\alpha))}{|\frac{d}{d\alpha}[\mathring{\boldsymbol{r}}(\mathbf{s} + \boldsymbol{\zeta}(\alpha))] \times \mathring{\boldsymbol{d}}(\mathbf{s} + \boldsymbol{\zeta}(\alpha))|}. \tag{13.13}$$

Let us define $\{\bar{\boldsymbol{g}}^\beta\}$ to be the basis dual to $\{\mathring{\boldsymbol{r}}_{,\mu}\}$ on span $\{\mathring{\boldsymbol{r}}_{,1}, \mathring{\boldsymbol{r}}_{,2}\}$, and observe that $\Gamma \bar{\boldsymbol{g}}^1 = \mathring{\boldsymbol{r}}_{,2} \times \mathring{\boldsymbol{d}}$, $\Gamma \bar{\boldsymbol{g}}^2 = \mathring{\boldsymbol{d}} \times \mathring{\boldsymbol{r}}_{,1}$.

13.14. Exercise. Prove that (13.13) reduces to

$$\boldsymbol{\gamma} = \frac{\Gamma[\beta_1 \bar{\boldsymbol{g}}^1 + \beta_2 \bar{\boldsymbol{g}}^2]}{|\beta_2 \mathring{\boldsymbol{r}}_{,1} - \beta_1 \mathring{\boldsymbol{r}}_{,2}|} \equiv \gamma_\alpha \bar{\boldsymbol{g}}^\alpha, \tag{13.15}$$

and show by example that in general $\beta_\alpha \neq \gamma_\alpha$.

We denote by $-\tilde{\boldsymbol{n}}^2(\mathbf{s} + \alpha(\beta_2 l, 0), t)$ and $-\tilde{\boldsymbol{n}}^1(\mathbf{s} + (1-\alpha)(0, \beta_1 l), t)$ the forces per unit length acting across the material curves

$$\begin{aligned}
&[0,1] \ni \alpha \to \mathring{\boldsymbol{r}}(\mathbf{s} + \alpha(\beta_2 l, 0)),\\
&[0,1] \ni \alpha \to \mathring{\boldsymbol{r}}(\mathbf{s} + (1-\alpha)(0, \beta_1 l)),
\end{aligned}$$

respectively. For our special choice of $\mathcal{A}$ as a triangular region, we obtain

(13.16)

$$\begin{aligned}\int_{\partial\mathcal{A}} \boldsymbol{n}(\mathsf{s},t,\boldsymbol{\gamma}(\mathsf{s}))\,d\sigma(\mathsf{s}) \\ = -\int_0^1 \tilde{\boldsymbol{n}}^2(\mathsf{s}+\alpha(\beta_2 l,0),t),t)\left|\mathring{\boldsymbol{r}}_{,1}(\mathsf{s}+\alpha(\beta_2 l,0),t)\right|\beta_2 l\,d\alpha \\ - \int_0^1 \tilde{\boldsymbol{n}}^1(\mathsf{s}+(1-\alpha)(0,\beta_1 l),t)\left|\mathring{\boldsymbol{r}}_{,2}(\mathsf{s}+(1-\alpha)(0,\beta_1 l),t)\right|\beta_1 l\,d\alpha \\ + \int_0^1 \boldsymbol{n}(\mathsf{s},t,\boldsymbol{\gamma}(\mathsf{s}))\left|\beta_2\mathring{\boldsymbol{r}}_{,1}(\mathsf{s}+\boldsymbol{\zeta}(\alpha))-\beta_1\mathring{\boldsymbol{r}}_{,2}(\mathsf{s}+\boldsymbol{\zeta}(\alpha))\right| l\,d\alpha.\end{aligned}$$

We now substitute this expression into the first equation of motion, divide it by l, and let $l \to 0$ to obtain

$$|\beta_2\mathring{\boldsymbol{r}}_{,1}-\beta_1\mathring{\boldsymbol{r}}_{,2}|\boldsymbol{n}(\cdot,\cdot,\boldsymbol{\gamma}) = \tilde{\boldsymbol{n}}^1|\mathring{\boldsymbol{r}}_{,2}|\beta_1 + \tilde{\boldsymbol{n}}^2|\mathring{\boldsymbol{r}}_{,1}|\beta_2. \tag{13.17}$$

Using (13.15), we obtain from (13.17) a version of Cauchy's Stress Theorem:

$$\boldsymbol{n}(\cdot,\cdot,\boldsymbol{\gamma}) = \left(|\bar{\boldsymbol{g}}^1\cdot\bar{\boldsymbol{g}}^1|\tilde{\boldsymbol{n}}^1\mathring{\boldsymbol{r}}_1 + |\bar{\boldsymbol{g}}^2\cdot\bar{\boldsymbol{g}}^2|\tilde{\boldsymbol{n}}^2\mathring{\boldsymbol{r}}_2\right)\cdot\boldsymbol{\gamma}. \tag{13.18}$$

Thus $\boldsymbol{n}$ is linear in $\boldsymbol{\gamma}$. We now set

$$\boldsymbol{n}^1 = \sqrt{\bar{\boldsymbol{g}}^1\cdot\bar{\boldsymbol{g}}^1}\,\tilde{\boldsymbol{n}}^1, \quad \boldsymbol{n}^2 = \sqrt{\bar{\boldsymbol{g}}^2\cdot\bar{\boldsymbol{g}}^2}\,\tilde{\boldsymbol{n}}^2, \tag{13.19}$$

so that

$$\boldsymbol{n}(\cdot,\cdot,\boldsymbol{\gamma}) = \boldsymbol{n}^\alpha\gamma_\alpha = \frac{\Gamma\boldsymbol{n}^\alpha\beta_\alpha}{|\beta_2\mathring{\boldsymbol{r}}_{,1}-\beta_1\mathring{\boldsymbol{r}}_{,2}|}. \tag{13.20}$$

Now we consider an arbitrary $\mathcal{A}$ in int $\mathcal{M}$, which we assume is bounded by a sufficiently smooth curve $[0,\mu] \ni \lambda \mapsto \check{\mathsf{s}}(\lambda)$ where λ is the arc-length parameter. From (13.20) and the Divergence Theorem applied to $\mathcal{A}$, we obtain

$$\begin{aligned}\int_{\partial\mathcal{A}} \boldsymbol{n}(\mathsf{s},t,\boldsymbol{\gamma}(\mathsf{s}))\,d\sigma(\mathsf{s}) &= \int_0^\mu \frac{\Gamma\boldsymbol{n}^\alpha\beta_\alpha}{|\beta_2\mathring{\boldsymbol{r}}_{,1}-\beta_1\mathring{\boldsymbol{r}}_{,2}|}\left|\frac{d}{d\lambda}\mathring{\boldsymbol{r}}(\check{\mathsf{s}}(\lambda))\right| d\lambda \\ &= \int_0^\mu \Gamma\boldsymbol{n}^\alpha\beta_\alpha\,d\lambda \\ &= \int_{\mathcal{A}} (\Gamma\boldsymbol{n}^\alpha)_{,\alpha}\,ds^1\,ds^2 \\ &= \int_{\mathcal{A}} \Gamma^{-1}(\Gamma\boldsymbol{n}^\alpha)_{,\alpha}\,da.\end{aligned} \tag{13.21}$$

In deducing the third equality, we have used the identities $\beta_1 = d\check{s}^2/d\lambda$, $\beta_2 = -d\check{s}^1/d\lambda$.

To make the underlying geometrical ideas used in this derivation completely precise, we have pulled back the integrals to $\mathcal{A}$ in $\mathbb{R}^2$ where we applied the Divergence Theorem. (We could have used these same ideas to formulate a version of the Divergence Theorem on $\mathring{\boldsymbol{r}}(\mathcal{A})$.)

Let us substitute (13.21) into the integral form of the equations of motion, from which we obtain

$$\Gamma^{-1}(\Gamma \boldsymbol{n}^\alpha)_{,\alpha} + \boldsymbol{f} = 2\rho h \boldsymbol{r}_{tt} + \boldsymbol{q}_{tt}. \tag{13.22}$$

Similarly, we find the balance of moments

$$\Gamma^{-1}(\Gamma \boldsymbol{m}^\alpha)_{,\alpha} + \boldsymbol{r}_{,\alpha} \times \boldsymbol{n}^\alpha + \boldsymbol{l} = \boldsymbol{q} \times \boldsymbol{r}_{tt} + \boldsymbol{h}_t. \tag{13.23}$$

The material of the shell is *elastic* iff $\boldsymbol{n}^\alpha$ and $\boldsymbol{m}^\alpha$ are prescribed functions of $\boldsymbol{r}_{,\alpha}, \boldsymbol{d}, \boldsymbol{d}_{,\alpha}$, and $\mathbf{s}$. To express the form of the constitutive equations invariant under rigid motions, we introduce one of the bases $\{\boldsymbol{r}_{,1}, \boldsymbol{r}_{,2}, \boldsymbol{d}\}$ or $\{\boldsymbol{r}_{,1}, \boldsymbol{r}_{,2}, \boldsymbol{a}^3\}$ where $\boldsymbol{a}^3 \equiv (\boldsymbol{r}_{,1} \times \boldsymbol{r}_{,2})/|\boldsymbol{r}_{,1} \times \boldsymbol{r}_{,2}|$.

13.24. Exercise. (i) Show that if the components of $\boldsymbol{n}^\alpha$ and $\boldsymbol{m}^\alpha$ with respect to either of these bases depend only on (13.7) and $\mathbf{s}$, then the constitutive equations are invariant under rigid motions.

The definition and characterization of a transversely isotropic special Cosserat shell (with the transverse direction in the reference configuration taken to be $\mathring{\boldsymbol{d}}$) is more delicate than those for rods in Sec. VIII.9 and for three-dimensional bodies in Sec. XII.13, because the reference configuration of a shell may be curved as is that for a rod and because the number of independent spatial variables exceeds 1 as for a three-dimensional body. The underlying concepts are of course the same. It is intuitively clear that the response at a material point $\mathbf{s}$ cannot be transversely isotropic unless $\mathring{\boldsymbol{r}}$ is spherical (or planar) near $\mathbf{s}$ (or unless the constitutive equations are unrealistically degenerate; cf. the discussion at the end of Sec. VIII.9). We omit the details of the development, referring to Carroll & Naghdi (1972), Cohen & Wang (1989), Ericksen (1970, 1972), Murdoch & Cohen (1979), and Wang (1972, 1973). As we show in Sec. 14, we can finesse the difficulty (and the beauty) of this theory by choosing constitutive equations constructed as in Secs. 9 and 10 from the constitutive equations for isotropic three-dimensional materials.

13.25. Exercise. Imitate the treatment of Sec. 7 to formulate a Cosserat theory for shells with one unconstrained director.

The material of such a shell is *hyperelastic* iff there is a scalar-valued function $\hat{w}$ of $\boldsymbol{r}_{,\alpha}, \boldsymbol{d}, \boldsymbol{d}_{,\alpha}$, and $\mathbf{s}$ such that

$$\begin{aligned} \boldsymbol{n}^\alpha(\mathbf{s},t) &= \frac{\partial \hat{w}}{\partial \boldsymbol{r}_{,\alpha}}(\boldsymbol{r}_{,1}(\mathbf{s},t), \boldsymbol{r}_{,2}(\mathbf{s},t), \boldsymbol{d}(\mathbf{s},t), \boldsymbol{d}_{,1}(\mathbf{s},t), \boldsymbol{d}_{,2}(\mathbf{s},t), \mathbf{s}), \\ \boldsymbol{m}^\alpha(\mathbf{s},t) &= \boldsymbol{d} \times \frac{\partial \hat{w}}{\partial \boldsymbol{d}_{,\alpha}}(\boldsymbol{r}_{,1}(\mathbf{s},t), \boldsymbol{r}_{,2}(\mathbf{s},t), \boldsymbol{d}(\mathbf{s},t), \boldsymbol{d}_{,1}(\mathbf{s},t), \boldsymbol{d}_{,2}(\mathbf{s},t), \mathbf{s}). \end{aligned} \tag{13.26}$$

The frame-indifferent form of (13.26) is obtained when $\hat{w}$ is a function of appropriate invariants and $\mathbf{s}$.

13.27. Exercise. Prove that if $\hat{w}$ depends only on these invariants and on s, then it satisfies the identity

$$\text{(13.28)} \qquad \boldsymbol{r}_{,\alpha} \times \frac{\partial \hat{w}}{\partial \boldsymbol{r}_{,\alpha}} + \boldsymbol{d}_{,\alpha} \times \frac{\partial \hat{w}}{\partial \boldsymbol{d}_{,\alpha}} + \boldsymbol{d} \times \frac{\partial \hat{w}}{\partial \boldsymbol{d}} = \boldsymbol{0},$$

which corresponds to (10.23).

13.29. Exercise. Formulate Hamilton's Principle for hyperelastic shells satisfying the constraint $\boldsymbol{d} \cdot \boldsymbol{d} = 1$ and verify that the Euler-Lagrange equations for the Lagrangian functional are the governing equations (13.22) and (13.23).

Note that the entire treatment of this section is consistent with that of Sec. 10 provided that (10.1c) is employed.

The buckling problem for rectangular plates. We now specialize our preceding development to formulate a family of buckling problems for rectangular special Cosserat plates, which is typical of problems governed by partial differential equations that could be readily analyzed by perturbation methods. We take

$$\text{(13.30)} \qquad \mathcal{M} = \{\mathbf{s} : |s^1| \leq l^1, \ |s^2| \leq l^2\}$$

where l^1, l^2 are prescribed positive numbers. We assume that $\mathring{\boldsymbol{r}}$ lies in the $\{\boldsymbol{i}^1, \boldsymbol{i}^2\}$-plane. We study problems for which

$$\text{(13.31)} \qquad \boldsymbol{n}^1(\pm l^1, s^2) \cdot \boldsymbol{i}_1 = -\lambda^1, \quad \boldsymbol{n}^2(s^1, \pm l^2) \cdot \boldsymbol{i}_2 = -\lambda^2.$$

We allow λ^1, λ^2 to be either positive or negative. We have four bifurcation parameters $l^1, l^2, \lambda^1, \lambda^2$, which can be reduced by suitable scaling.

To express the constitutive equations, we adopt a suitably specialized version of (13.26). In particular, we adopt an analog of (4.19) for an isotropic W. Employing (10.1c) with $\varphi = x^3$, we find that

$$\text{(13.32)} \quad \boldsymbol{C} = (\boldsymbol{r}_{,\alpha} + x^3 \boldsymbol{d}_{,\alpha}) \cdot (\boldsymbol{r}_{,\beta} + x^3 \boldsymbol{d}_{,\beta}) \boldsymbol{i}^\alpha \boldsymbol{i}^\beta + (\boldsymbol{r}_{,\alpha} + x^3 \boldsymbol{d}_{,\alpha}) \cdot \boldsymbol{d} \, (\boldsymbol{i}^\alpha \boldsymbol{i}^3 + \boldsymbol{i}^3 \boldsymbol{i}^\alpha) + \boldsymbol{i}^3 \boldsymbol{i}^3.$$

Thus $\boldsymbol{C}$ is determined by $\boldsymbol{r}_{,\alpha} \cdot \boldsymbol{r}_{,\beta}$, $\boldsymbol{r}_{,\alpha} \cdot \boldsymbol{d}$, $\boldsymbol{r}_{,\alpha} \cdot \boldsymbol{d}_{,\beta} + \boldsymbol{r}_{,\beta} \cdot \boldsymbol{d}_{,\alpha}$, $\boldsymbol{d}_{,\alpha} \cdot \boldsymbol{d}_{,\beta}$, which correspond to (13.7). For an isotropic, hyperelastic three-dimensional material, the stored-energy function depends on the principal invariants of (13.32). The first invariant is quadratic in x^3 and the other two are quartics in x^3. The thirteen coefficients of the different powers of x^3 in these invariants are the two-dimensional invariants for the plate problem.

13.33 Exercise. Find these invariants. Next suppose that the plate is merely transversely isotropic. Use a representation theorem for such materials to find the corresponding plate-theoretic invariants and compare these with the invariants found from the plate-theoretic invariants corresponding to a three-dimensional isotropic material.

13.34. Problem. Formulate buckling problems for a plate with a stored-energy function depending on (13.33) by supplementing (13.31) with a variety of other suitable boundary conditions. Analyze these by using perturbation methods. Determine the eigensurface consisting of those $l^1, l^2, \lambda^1, \lambda^2$ for which the linearized problem has nontrivial solutions. Discuss the geometric and algebraic multiplicities of the eigenvalues. Discuss the role of shearability. Determine whether buckling is supercritical or subcritical.

The methods of Golubitsky, Stewart, & Schaeffer (1988) can be profitably applied to parts of this problem. See Schaeffer & Golubitsky (1979) and the discussion in Sec. XIII.11. An important open question is to adapt the methods of Healey & Kielhofer (1991) to problems like these governed by systems of partial differential equations.

13.35. Problem. Carry out the analogs of Problem 13.34 for cylindrical and spherical shells.

Kirchhoff shells. We specialize these formulations to Kirchhoff shells by assuming that $\boldsymbol{d}$ cannot be sheared with respect to the surface $\boldsymbol{r}$. Our main goal is to determine appropriate forms of boundary conditions from a suitable version of the Principle of Virtual Power. By methods just like those of Sec. XII.9, we can show the equivalence of impulse-momentum laws for (13.23), (13.24) to the Principle of Virtual Power in the form

(13.36)

$$\begin{aligned}
&\int_0^\infty \int_{\mathcal{M}} \Big[-\boldsymbol{n}^\alpha \cdot \overset{\vartriangle}{\boldsymbol{r}}_{,\alpha} - \boldsymbol{m}^\alpha \cdot \boldsymbol{\omega}_{,\alpha} + \boldsymbol{f} \cdot \overset{\vartriangle}{\boldsymbol{r}} + (\overset{\vartriangle}{\boldsymbol{r}}_{,\alpha} \times \boldsymbol{n}^\alpha + \boldsymbol{l}) \cdot \boldsymbol{\omega} \Big] \Gamma \, ds^1 \, ds^2 \, dt \\
&+ \int_0^\infty \int_{\partial\mathcal{M}} \Big[\boldsymbol{n}^\alpha \cdot \overset{\vartriangle}{\boldsymbol{r}} + \boldsymbol{m}^\alpha \cdot \boldsymbol{\omega} \Big] \beta_\alpha \Gamma \, d\sigma \, dt \\
&+ \int_0^\infty \int_{\mathcal{M}} \Big[(2\rho h \boldsymbol{r}_t + \boldsymbol{q}_t) \cdot \overset{\vartriangle}{\boldsymbol{r}}_t + (\boldsymbol{\omega} \times \boldsymbol{q})_t \cdot \boldsymbol{r}_t + \boldsymbol{h} \cdot \boldsymbol{\omega}_t \Big] \Gamma \, ds^1 \, ds^2 \, dt \\
&+ \int_{\mathcal{M}} \Big[(2\rho h \boldsymbol{r}_t + \boldsymbol{q}_t) \cdot \overset{\vartriangle}{\boldsymbol{r}} + (\boldsymbol{\omega} \times \boldsymbol{q}) \cdot \boldsymbol{r}_t + \boldsymbol{h} \cdot \boldsymbol{\omega} \Big] \Big|_{t=0} \Gamma \, ds^1 \, ds^2 = 0
\end{aligned}$$

for all $\overset{\vartriangle}{\boldsymbol{r}}, \boldsymbol{\omega}$. Here we interpret $\boldsymbol{\omega}$ as coming from $\overset{\vartriangle}{\boldsymbol{d}} = \boldsymbol{\omega} \times \boldsymbol{d}$, which is valid since $\boldsymbol{d}$ is assumed to be a unit vector. (We can, of course, derive (13.36) by purely formal methods.)

Now we impose the Kirchhoff constraint (10.24):

$$\boldsymbol{d} = \frac{\boldsymbol{r}_{,1} \times \boldsymbol{r}_{,2}}{|\boldsymbol{r}_{,1} \times \boldsymbol{r}_{,2}|}. \tag{13.37}$$

In this case, the $\boldsymbol{n}^\alpha \cdot \boldsymbol{d}$ are Lagrange multipliers.

13.38. Exercise. Deduce from (13.37) that

$$\overset{\vartriangle}{\boldsymbol{d}} = \boldsymbol{\omega} \times \boldsymbol{d}, \quad \boldsymbol{\omega} \equiv \boldsymbol{d} \times \frac{\overset{\vartriangle}{\boldsymbol{r}}_{,1} \times \boldsymbol{r}_{,2} + \boldsymbol{r}_{,1} \times \overset{\vartriangle}{\boldsymbol{r}}_{,2}}{|\boldsymbol{r}_{,1} \times \boldsymbol{r}_{,2}|}. \tag{13.39}$$

The substitution of this $\boldsymbol{\omega}$ into (13.36) yields an equation whose boundary term is

$$\begin{aligned}
\int_0^\infty \int_{\partial\mathcal{M}} \Big[(\boldsymbol{n}^\alpha \beta_\alpha) \cdot \overset{\vartriangle}{\boldsymbol{r}} &+ \frac{[(\boldsymbol{m}^\alpha \beta_\alpha) \cdot \boldsymbol{r}_{,2}]\boldsymbol{d}}{|\boldsymbol{r}_{,1} \times \boldsymbol{r}_{,2}|} \cdot \overset{\vartriangle}{\boldsymbol{r}}_{,1} \\
&- \frac{[(\boldsymbol{m}^\alpha \beta_\alpha) \cdot \boldsymbol{r}_{,1}]\boldsymbol{d}}{|\boldsymbol{r}_{,1} \times \boldsymbol{r}_{,2}|} \cdot \overset{\vartriangle}{\boldsymbol{r}}_{,2} \Big] \Gamma \, d\sigma \, dt.
\end{aligned} \tag{13.40}$$

We use (13.40) to determine suitable versions of boundary conditions analogous to (1.15). Let us study boundary conditions near a place where $\mathring{\boldsymbol{r}}(\partial\mathcal{M})$ is continuously differentiable. Without loss of generality, we take this part of the boundary to be defined by $s^1 = s_0^1$, a constant, and assume that $\mathcal{M}$ lies locally to the left of this boundary so that $\beta_1 = 1$, $\beta_2 = 0$.

We could prescribe the position $\boldsymbol{r}(s_0^1, \cdot, t)$ at time t of this part of the boundary. In this case, $\overset{\vartriangle}{\boldsymbol{r}}(s_0^1, \cdot, t) = \boldsymbol{0}$ and $\overset{\vartriangle}{\boldsymbol{r}}_{,2}(s_0^1, \cdot, t) = \boldsymbol{0}$, and the form of (13.40) says that we can either *clamp* this part of the boundary, i.e., prescribe $\boldsymbol{r}_{,1}(s_0^1, \cdot, t)$, in which case $\overset{\triangle}{\boldsymbol{r}}_{,1}(s_0^1, \cdot, t) = \boldsymbol{0}$, or else prescribe the complementary couple

$$(13.41) \qquad \frac{(\boldsymbol{m}^1 \cdot \boldsymbol{r}_{,2})\boldsymbol{d}}{|\boldsymbol{r}_{,1} \times \boldsymbol{r}_{,2}|},$$

which is the coefficient of $\overset{\triangle}{\boldsymbol{r}}_{,1}$ in the integrand of (13.40).

Now suppose that we do not prescribe $\boldsymbol{r}(s_0^1, \cdot, t)$. We take $\overset{\triangle}{\boldsymbol{r}}(s_0^1, \cdot, t)$ to have compact support and integrate the third term in the integrand of (13.40) by parts over this support. Then the arbitrariness of $\overset{\triangle}{\boldsymbol{r}}(s_0^1, \cdot, t)$ says that we should prescribe its coefficient in the resulting integrand of (13.40):

$$(13.42) \qquad \boldsymbol{n}^1 + \frac{\partial}{\partial s^2}\left\{\frac{(\boldsymbol{m}^1 \cdot \boldsymbol{r}_{,2})\boldsymbol{d}}{|\boldsymbol{r}_{,1} \times \boldsymbol{r}_{,1}|}\right\}.$$

We could further clamp this edge or prescribe (13.41).

Finally, we could prescribe some components of $\boldsymbol{r}(s_0^1, \cdot, t)$ and leave the remaining components free. The strategy for prescribing boundary conditions is the same.

13.43. Exercise. Suppose that $\overset{\circ}{\boldsymbol{r}}(\mathcal{M})$ lies in the $\{\boldsymbol{i}, \boldsymbol{j}\}$-plane. Suppose that those boundary values of $\boldsymbol{r}$ and $\boldsymbol{r}_{,\alpha}$ that are prescribed agree with their values in the reference configuration. Suppose that $\boldsymbol{f}$, $\boldsymbol{l}$, the prescribed boundary values of $\boldsymbol{n}^\alpha \beta_\alpha$ and $\boldsymbol{m}^\alpha \beta_\alpha$, and the prescribed initial values of $\boldsymbol{r} - \overset{\circ}{\boldsymbol{r}}$ are each proportional to a small parameter ε. Find the linear problem governing the leading term in a formal perturbation expansion of the initial-boundary-value problem for the motion of an isotropic Kirchhoff plate. In particular, find the equation governing the component of linearized displacement in the $\boldsymbol{k}$-direction. It differs from the classical equation in that it accounts for rotatory inertia.

Membranes. A degenerate version of the equations for a shell are those for a membrane (just as a degenerate version of the equations for a rod are those for a string). The membrane equations are obtained from the shell equations by introducing the constitutive assumptions that there is neither flexural stiffness nor flexural inertia, so that $\boldsymbol{m}^\alpha = \boldsymbol{0}$ and $\boldsymbol{h} = \boldsymbol{0}$; by taking $\boldsymbol{q} = \boldsymbol{0}$; and by taking the applied couple $\boldsymbol{l} = \boldsymbol{0}$. In this case, (13.23) reduces to

$$(13.44) \qquad \boldsymbol{r}_{,\alpha} \times \boldsymbol{n}^\alpha = \boldsymbol{0},$$

which together with (13.22) are the governing equations for a membrane. We regard (13.44) as a constitutive restriction on the $\boldsymbol{n}^\alpha$. To determine its consequences, let us decompose $\boldsymbol{n}^\alpha$ by

$$(13.45) \qquad \boldsymbol{n}^\alpha = n^{\alpha\beta}\boldsymbol{r}_{,\beta} + n^{\alpha 3}\boldsymbol{a}^3$$

where $\boldsymbol{a}^3 = \boldsymbol{d}$ of (13.37). The independence of the base vectors appearing on the right-hand side of (13.45) enables us to deduce from (13.44) that

$$(13.46\text{a,b}) \qquad n^{\alpha 3} = 0, \quad n^{12} = n^{21}.$$

Conditions (13.46a) and (13.45) imply that $\boldsymbol{n}^\alpha \in \operatorname{span}\{\boldsymbol{r}_{,1}, \boldsymbol{r}_{,2}\}$. Thus the contact force vectors are tangent to the surface $\boldsymbol{r}$. Condition (13.46a) also says that a membrane cannot sustain certain components of shear force. To interpret the symmetry condition (13.46b) ,let us use (1.7a) and (10.5) to obtain

$$\Gamma \boldsymbol{n}^\alpha = \int_{h_1}^{h_2} \boldsymbol{p}_{,k}\boldsymbol{g}^k \cdot \boldsymbol{S}_{\mathrm{A}} \cdot \boldsymbol{g}^\alpha \, j \, dx^3 \tag{13.47a}$$

or, equivalently,

$$n^{\alpha\beta} = \boldsymbol{n}^\alpha \cdot \bar{\boldsymbol{g}}^\beta = \frac{1}{\Gamma}\int_{h_1}^{h_2} \bar{\boldsymbol{g}}^\beta \cdot \boldsymbol{p}_{,k}\, \boldsymbol{g}^k \cdot \boldsymbol{S}_{\mathrm{A}} \cdot \boldsymbol{g}^\alpha \, j \, dx^3 \tag{13.47b}$$

for a three-dimensional shell-like body. We conceive of a membrane as being an extremely thin shell, in which case we could hope to approximate $\boldsymbol{p}$ and $\boldsymbol{g}^k$ in (13.47) with $\boldsymbol{r}$ and $\bar{\boldsymbol{g}}^k$. Doing so, we can make the identification

$$n^{\alpha\beta} \sim \frac{1}{\Gamma}\int_{h_1}^{h_2} \bar{\boldsymbol{g}}^\alpha \cdot \boldsymbol{S}_{\mathrm{A}} \cdot \bar{\boldsymbol{g}}^\beta \, j \, dx^3. \tag{13.48}$$

Thus condition (13.46b) is a reflection of the symmetry of $\boldsymbol{S}$ and of the thinness of the membrane.

The constitutive equations for membranes are like those for shells, except that now (13.7) is reduced to $\{\boldsymbol{r}_{,\alpha} \cdot \boldsymbol{r}_{,\beta}\}$. If the membrane is hyperelastic, then it has a stored-energy function $\hat{\omega}$ depending on these strains. If the membrane is also isotropic, then the stored-energy function $\hat{w}$ is a symmetric function of the eigenvalues of the matrix $(\boldsymbol{r}_{,\alpha} \cdot \boldsymbol{r}_{,\beta})$ with respect to the matrix $(\mathring{\boldsymbol{r}}_{,\alpha} \cdot \mathring{\boldsymbol{r}}_{,\beta})$. Consequently, $\hat{w}$ depends on the strains only through $\lambda_1 + \lambda_2$ and $\lambda_1\lambda_2$, where λ_1 and λ_2 are solutions of

$$\det(\boldsymbol{r}_{,\alpha} \cdot \boldsymbol{r}_{,\beta} - \lambda \mathring{\boldsymbol{r}}_{,\alpha} \cdot \mathring{\boldsymbol{r}}_{,\beta}) = 0. \tag{13.49}$$

Thus $\hat{w}$ depends on

$$\frac{(\boldsymbol{r}_{,1} \cdot \boldsymbol{r}_{,1})(\mathring{\boldsymbol{r}}_{,2} \cdot \mathring{\boldsymbol{r}}_{,2}) + (\boldsymbol{r}_{,2} \cdot \boldsymbol{r}_{,2})(\mathring{\boldsymbol{r}}_{,1} \cdot \mathring{\boldsymbol{r}}_{,1}) - 2(\boldsymbol{r}_{,1} \cdot \boldsymbol{r}_{,2})(\mathring{\boldsymbol{r}}_{,1} \cdot \mathring{\boldsymbol{r}}_{,2})}{|\mathring{\boldsymbol{r}}_{,1} \times \mathring{\boldsymbol{r}}_{,2}|^2}, \quad \frac{\boldsymbol{r}_{,1} \times \boldsymbol{r}_{,2}}{\mathring{\boldsymbol{r}}_{,1} \times \mathring{\boldsymbol{r}}_{,2}}, \tag{13.50a,b}$$

and $\mathbf{s}$. Note that (13.50b) is the local ratio of deformed to reference area. The equations of motion for a hyperelastic membrane are the Euler-Lagrange equations for the Lagrangian functional. In particular, the equilibrium of a hyperelastic isotropic membrane under zero body force is governed by the Euler-Lagrange equations for the total energy functional

$$\int_{\mathcal{M}} \hat{w} \, da(\mathbf{s}). \tag{13.51}$$

The minimal surface equation. The minimal surface spanning a given contour $\mathcal{C}$ is that surface through $\mathcal{C}$ having the least area. Its equations are the Euler-Lagrange equations for the area functional

$$
\begin{aligned}
\int_{\boldsymbol{r}(\mathcal{M})} da &= \int_{\mathcal{M}} |\boldsymbol{r}_{,1} \times \boldsymbol{r}_{,2}|\, ds^1\, ds^2 = \int_{\mathcal{M}} \left| \frac{\boldsymbol{r}_{,1} \times \boldsymbol{r}_{,2}}{\mathring{\boldsymbol{r}}_{,1} \times \mathring{\boldsymbol{r}}_{,2}} \right| |\mathring{\boldsymbol{r}}_{,1} \times \mathring{\boldsymbol{r}}_{,2}|\, ds^1\, ds^2 \\
&= \int_{\mathcal{M}} \left| \frac{\boldsymbol{r}_{,1} \times \boldsymbol{r}_{,2}}{\mathring{\boldsymbol{r}}_{,1} \times \mathring{\boldsymbol{r}}_{,2}} \right| da(\mathbf{s}).
\end{aligned}
\tag{13.52}
$$

Thus the minimal surface equations are a special case of the equations of equilibrium for hyperelastic isotropic membranes. The version of the minimal surface equation most studied is that in which $\boldsymbol{r}$ is the graph of a function. In this case, the equations have a spatial formulation, which removes it from the setting for nonlinear elasticity promoted in this book.

13.53. Exercise. Prove that the natural state for the membrane with total energy functional (13.52) is the degenerate state in which the area of the membrane is zero. (This situation is the analog of that discussed in Sec. II.7.) Prove the same result for membranes with stored-energy functions of the form

$$
w = \Omega\left(\left| \frac{\boldsymbol{r}_{,1} \times \boldsymbol{r}_{,2}}{\mathring{\boldsymbol{r}}_{,1} \times \mathring{\boldsymbol{r}}_{,2}} \right|, \mathbf{s} \right)
\tag{13.54}
$$

when $\Omega(\cdot, \mathbf{s})$ is strictly increasing.

If $\Omega(\cdot, \mathbf{s})$ is not strictly increasing, then there is at least one nondegenerate natural state. A simple example is $\Omega(\xi, \mathbf{s}) = \xi^{-1} + \xi$, in which area reductions are penalized, just as in most of this book. In the absence of bending stiffness, nonincreasing Ω's can cause instabilities associated with a loss of ellipticity. If $\Omega(\cdot, \mathbf{s})$ is not affine, then the reference quantity $|\mathring{\boldsymbol{r}}_{,1} \times \mathring{\boldsymbol{r}}_{,2}|$ must appear explicitly in the integrand of the total stored-energy functional (13.51). Thus, the minimal surface equation has special properties not shared by any other membrane equations.

Note that the second integral of (13.52) is quadratic in derivatives of $\boldsymbol{r}$, but it is not coercive in any useful sense. (The corresponding nonparametric form of this area functional, in which $\boldsymbol{r}$ is the graph of a real-valued function, has asymptotically linear growth, which is the source of analytic difficulties.) Useful coercivity could attend the introduction of a further dependence on (13.50a).

14. Asymptotic Methods. The von Kármán Equations

In this section we outline effective methods for constructing theories of plates corresponding to formal asymptotic approximations of the three-dimensional equations of equilibrium. The small parameter is the (half-) thickness h. In particular, we show how the von Kármán plate equations can be given a rational position within nonlinear elasticity.

Let $\{\boldsymbol{i}_k\}$ be a fixed right-handed orthonormal basis for $\mathbb{E}^3$. We set $\boldsymbol{i}_3 = \boldsymbol{k}$. We take all components of tensors with respect to this basis. We sum twice-repeated indices over their ranges, whether the indices are diagonally disposed or not. Let $\mathcal{M}$ be a domain in $\mathbb{R}^2$. We study *plates*, i.e., bodies of the form

$$
\mathcal{B}(h) \equiv \{\boldsymbol{z} = z^k \boldsymbol{i}_k : (z^1, z^2) \in \operatorname{cl} \mathcal{M},\ z^3 \in [-h, h]\}.
\tag{14.1}
$$

We can identify $\mathcal{M}$ with $\overset{\circ}{\boldsymbol{r}}(\mathcal{M})$ as a domain in $\operatorname{span}\{\boldsymbol{i}_1, \boldsymbol{i}_2\}$.

We shall express the resulting equations in terms of the displacement $\boldsymbol{u}$ defined by

$$\boldsymbol{p}(\boldsymbol{z}) = \boldsymbol{z} + \boldsymbol{u}(\boldsymbol{z}). \tag{14.2}$$

Thus

$$\boldsymbol{F} = \boldsymbol{I} + \boldsymbol{u}_{\boldsymbol{z}}. \tag{14.3}$$

We adopt the constitutive equation (XIII.1.7) for a homogeneous isotropic elastic body whose reference configuration is the natural undeformed state. Its linearization about this state is

$$\overset{\vartriangle}{\boldsymbol{S}} = \lambda (\operatorname{tr} \overset{\vartriangle}{\boldsymbol{E}})\boldsymbol{I} + 2\mu \overset{\vartriangle}{\boldsymbol{E}} \tag{14.4a}$$

where λ and μ, the *Lamé constants*, are assumed positive. ($\overset{\vartriangle}{\boldsymbol{E}}$ is the linearization of $\boldsymbol{E} = \frac{1}{2}(\boldsymbol{C} - \boldsymbol{I})$.) Then (14.4a) is equivalent to an expression of the form

$$\overset{\vartriangle}{\boldsymbol{E}} = \boldsymbol{A}(\overset{\vartriangle}{\boldsymbol{S}}) \equiv \frac{1+\nu}{E}\overset{\vartriangle}{\boldsymbol{S}} - \frac{\nu}{E}(\operatorname{tr} \overset{\vartriangle}{\boldsymbol{S}})\boldsymbol{I} \tag{14.4b}$$

with $E > 0$, $0 < \nu < \frac{1}{2}$. (E is the *elastic modulus*, or *Young's modulus*, and ν is *Poisson's ratio*.) We can now apply the invertible linear operator $\boldsymbol{A}$ to (XIII.1.7) to obtain the equivalent constitutive equation

(14.5a)

$$\boldsymbol{A}(\boldsymbol{S}) = \frac{1-2\nu}{E}\alpha_0 \boldsymbol{I} - \frac{\nu}{E}[\alpha_1 \operatorname{tr} \boldsymbol{C} + \alpha_2 \operatorname{tr}(\boldsymbol{C}^2)]\boldsymbol{I} + \frac{1+\nu}{E}(\alpha_1 \boldsymbol{C} + \alpha_2 \boldsymbol{C}^2)$$

where the α's depend upon the principal invariants of $\boldsymbol{C}$. For simplicity, we assume that the α's are infinitely differentiable. A weak form of (14.5a) is obtained by integrating the inner product of it with an arbitrary symmetric tensor-valued function $\tilde{\boldsymbol{H}}$:

$$\begin{aligned}\int_{\mathcal{B}(h)} \boldsymbol{A}(\boldsymbol{S}) : \tilde{\boldsymbol{H}}\, dv = {} & \frac{1-2\nu}{E}\int_{\mathcal{B}(h)} \alpha_0 \operatorname{tr} \tilde{\boldsymbol{H}}\, dv \\ & - \frac{\nu}{E}\int_{\mathcal{B}(h)} [\alpha_1 \operatorname{tr} \boldsymbol{C} + \alpha_2 \operatorname{tr}(\boldsymbol{C}^2)] \operatorname{tr} \tilde{\boldsymbol{H}}\, dv \\ & + \frac{1+\nu}{E}\int_{\mathcal{B}(h)} (\alpha_1 \boldsymbol{C} + \alpha_2 \boldsymbol{C}^2) : \tilde{\boldsymbol{H}}\, dv \end{aligned} \tag{14.5b}$$

for all symmetric $\tilde{\boldsymbol{H}}$. Note that $\tilde{\boldsymbol{H}}$ should be regarded as a 'virtual stress'. Here and throughout this section, we proceed formally by tacitly assuming that all integrals make sense. It is easy to impose sufficient conditions ensuring that they do.

Although it is possible to handle fairly general classes of boundary conditions, it is illuminating to restrict our attention to the special and slightly artificial conditions on the edge that

(14.6a,b) $$u^{\alpha}_{,3} = 0, \quad u^3 = 0 \quad \text{for } (z^1, z^2) \in \partial\mathcal{M},$$

(14.7) $$\int_{-h}^{h} \boldsymbol{k} \times \boldsymbol{T} \cdot \boldsymbol{\gamma}\, dz^3 = \boldsymbol{0} \quad \text{for } (z^1, z^2) \in \partial\mathcal{M}$$

where $\boldsymbol{\gamma}$ is the unit outer normal to $\partial\mathcal{M}$. The weak form of the equilibrium equations is the Principle of Virtual Power (which incorporates (14.7)):

(14.8) $$\int_{\mathcal{B}(h)} [\boldsymbol{S} : \tilde{\boldsymbol{\eta}}_{\boldsymbol{z}} + \boldsymbol{u}_{\boldsymbol{z}} \cdot \boldsymbol{S} : \tilde{\boldsymbol{\eta}}_{\boldsymbol{z}} - \boldsymbol{f} \cdot \tilde{\boldsymbol{\eta}}]\, dv - \int_{\mathcal{L}^+ \cup \mathcal{L}^-} \bar{\boldsymbol{\tau}} \cdot \tilde{\boldsymbol{\eta}}\, da = 0$$

for all $\tilde{\boldsymbol{\eta}}$ satisfying boundary conditions (14.6a,b).

Below, we prescribe the traction on the faces.

We now transform our problem from one on a domain parametrized by the thickness h to a set of equations parametrized by h on a fixed domain by the simple device of setting

(14.9) $$z^1 = y^1, \quad z^2 = y^2, \quad z^3 = hy^3,$$

so that $\mathbf{y} \equiv (y^1, y^2, y^3)$ ranges over

$$\operatorname{cl}\mathcal{M} \times [-1,1] \equiv \{\mathbf{y} : (y^1, y^2) \in \operatorname{cl} M,\ y^3 \in [-1,1]\}.$$

We also use the notation $\mathbf{s} \equiv (s^1, s^2) \equiv (y^1, y^2)$.

For our asymptotic approach to yield equations of von Kármán type, it is essential that the magnitudes of the applied loads be scaled appropriately in the thickness. A special case of such scalings, sufficient for our goal of merely illustrating the technique, is to take

(14.10) $$\boldsymbol{f}(\boldsymbol{z}) = h^3 f(\mathbf{y})\boldsymbol{k},$$

(14.11) $$\bar{\boldsymbol{\tau}}(\mathbf{x}) = \pm h^4 \tau^{\pm}(\mathbf{s})\boldsymbol{k} \quad \text{for } y^3 = \pm 1.$$

We now scale the stress, the virtual stress, displacement, and virtual displacement fields by defining

(14.12)
$$S^{\alpha\beta}(\boldsymbol{z}) \equiv h^2 \Sigma^{\alpha\beta}(\mathbf{y}, h), \quad S^{\alpha 3}(\boldsymbol{z}) \equiv h^3 \Sigma^{\alpha 3}(\mathbf{y}, h), \quad S^{33}(\boldsymbol{z}) \equiv h^4 \Sigma^{33}(\mathbf{y}, h),$$
$$\tilde{H}^{\alpha\beta}(\boldsymbol{z}) \equiv h^2 H^{\alpha\beta}(\mathbf{y}, h), \quad \tilde{H}^{\alpha 3}(\boldsymbol{z}) \equiv h^3 H^{\alpha 3}(\mathbf{y}, h), \quad \tilde{H}^{33}(\boldsymbol{z}) \equiv h^4 H^{33}(\mathbf{y}, h),$$

(14.13)
$$u_\alpha(\boldsymbol{z}) = h^2 v_\alpha(\mathbf{y}, h), \quad u_3(\boldsymbol{z}) = h v_3(\mathbf{y}, h),$$
$$\tilde{\eta}_\alpha(\boldsymbol{z}) = h^2 \eta_\alpha(\mathbf{y}, h), \quad \tilde{\eta}_3(\boldsymbol{z}) = h \eta_3(\mathbf{y}, h).$$

We set

$$\boldsymbol{\Sigma} \equiv \Sigma^{jk} \boldsymbol{i}_j \boldsymbol{i}_k, \quad \boldsymbol{H} \equiv H^{jk} \boldsymbol{i}_j \boldsymbol{i}_k, \quad \boldsymbol{v} \equiv v_k \boldsymbol{i}^k, \quad \boldsymbol{\eta} \equiv \eta_k \boldsymbol{i}^k.$$

If the boundary-value problem (14.5)–(14.8), (14.10), (14.11) has a solution $(\boldsymbol{S}, \boldsymbol{u})$, then $(\boldsymbol{\Sigma}, \boldsymbol{v})$ satisfies the following versions of (14.5b) and (14.8):

$$A_0(\boldsymbol{\Sigma}, \boldsymbol{H}) + h^2 A_2(\boldsymbol{\Sigma}, \boldsymbol{H}) + h^4 A_4(\boldsymbol{\Sigma}, \boldsymbol{H}) = N(\boldsymbol{v}, \boldsymbol{H}, h) \quad \forall\, \mathbf{y} \mapsto \boldsymbol{H}(\mathbf{y}) \in \mathrm{Sym}, \tag{14.14}$$

$$B(\boldsymbol{\Sigma}, \boldsymbol{\eta}) + 2C_0(\boldsymbol{\Sigma}, \boldsymbol{v}, \boldsymbol{\eta}) + 2h^2 C_2(\boldsymbol{\Sigma}, \boldsymbol{v}, \boldsymbol{\eta}) = F(\boldsymbol{\eta}) \quad \forall\, \mathbf{y} \mapsto \boldsymbol{\eta}(\mathbf{y}) \quad \text{satisfying (14.6)} \tag{14.15}$$

where

(14.16)
$$A_0(\boldsymbol{\Sigma}, \boldsymbol{H}) \equiv \int_{\mathcal{M}\times[-1,1]} \left[\frac{1+\nu}{E} \Sigma^{\alpha\beta} H_{\alpha\beta} - \frac{\nu}{E} \Sigma^{\alpha\alpha} H_{\beta\beta} \right] dv(\mathbf{y}),$$

(14.17)
$$A_2(\boldsymbol{\Sigma}, \boldsymbol{H}) \equiv \int_{\mathcal{M}\times[-1,1]} \left[2\frac{1+\nu}{E} \Sigma^{\alpha 3} H_{\alpha 3} - \frac{\nu}{E} (\Sigma^{33} H_{\beta\beta} + \Sigma^{\beta\beta} H_{33}) \right] dv(\mathbf{y}),$$

(14.18)
$$A_4(\boldsymbol{\Sigma}, \boldsymbol{H}) \equiv \frac{1}{E} \int_{\mathcal{M}\times[-1,1]} \Sigma^{33} H_{33}\, dv(\mathbf{y}),$$

(14.19)
$$\begin{aligned} N(\boldsymbol{v}, \boldsymbol{H}, h) \equiv \int_{\mathcal{M}\times[-1,1]} \bigg\{ & \frac{1-2\nu}{E} h^{-2} \alpha_0 (H_{\alpha\alpha} + h^2 H_{33}) \\ & - \frac{\nu}{E} h^{-2} (\alpha_1 \operatorname{tr} \boldsymbol{C} + \alpha_2 \operatorname{tr} \boldsymbol{C}^2)(H_{\alpha\alpha} + h^2 H_{33}) \\ & + \frac{1+\nu}{E} h^{-2} (\alpha_1 C_{\alpha\beta} + \alpha_2 C_{\alpha k} C_{k\beta}) H_{\alpha\beta} \\ & + 2\frac{1+\nu}{E} h^{-1} (\alpha_1 C_{\alpha 3} + \alpha_2 C_{\alpha k} C_{k3}) H_{\alpha 3} \\ & + \frac{1+\nu}{E} (\alpha_1 C_{33} + \alpha_2 C_{33}^2) H_{33} \bigg\}\, dv(\mathbf{y}), \end{aligned}$$

(14.20)
$$B(\boldsymbol{\Sigma}, \boldsymbol{\eta}) \equiv -\int_{\mathcal{M}\times[-1,1]} \boldsymbol{\Sigma} : \frac{\partial \boldsymbol{\eta}}{\partial \mathbf{y}}\, dv(\mathbf{y}),$$

(14.21)
$$C_0(\boldsymbol{\Sigma}, \boldsymbol{v}, \boldsymbol{\eta}) \equiv -\frac{1}{2} \int_{\mathcal{M}\times[-1,1]} \Sigma^{kl} \frac{\partial v_3}{\partial y^k} \frac{\partial \eta_3}{\partial y^l}\, dv(\mathbf{y}),$$

(14.22)
$$C_2(\boldsymbol{\Sigma}, \boldsymbol{v}, \boldsymbol{\eta}) \equiv -\frac{1}{2} \int_{\mathcal{M}\times[-1,1]} \Sigma^{kl} \frac{\partial v_\alpha}{\partial y^k} \frac{\partial \eta_\alpha}{\partial y^l}\, dv(\mathbf{y}),$$

(14.23)

$$F(\boldsymbol{\eta}) \equiv -\int_{\mathcal{M}\times[-1,1]} f\eta_3\, dv(\mathbf{y}) - \int_{\mathcal{M}} \tau^+(\mathbf{s})\eta_3(\mathbf{s},1)\, ds^1\, ds^2 + \int_{\mathcal{M}} \tau^-(\mathbf{s})\eta_3(\mathbf{s},-1)\, ds^1\, ds^2,$$

with

$$C_{\alpha\beta} = \delta_{\alpha\beta} + h^2(v_{\alpha,\beta} + v_{\beta,\alpha} + v_{3,\alpha}v_{3,\beta}) + h^4 v_{\mu,\alpha}v_{\mu,\beta}, \tag{14.24a}$$

$$C_{\alpha 3} = h(v_{\alpha,3} + v_{3,\alpha} + v_{3,\alpha}v_{3,3}) + h^3 v_{\mu,\alpha}v_{\mu,3}, \tag{14.24b}$$

$$C_{33} = 1 + v_{3,3}(2 + v_{3,3}) + h^2 v_{\mu,3}v_{\mu,3}, \tag{14.24c}$$

and where $\alpha_0, \alpha_1, \alpha_2$ depend upon the principal invariants of $\boldsymbol{C}$ of the form (14.24). Note that the scaling $dv(\mathbf{z}) = h\, dv(\mathbf{y})$ affects every integral except for the last two terms of (14.23), where the distinction in scales between (14.10) and (14.11) is manifested.

We seek solutions of (14.14) and (14.15) having formal expansions of the form

$$\begin{aligned} \boldsymbol{\Sigma}(\mathbf{y},h) &= \boldsymbol{\Sigma}_{(0)}(\mathbf{y}) + h\boldsymbol{\Sigma}_{(1)}(\mathbf{y}) + \tfrac{1}{2}h^2\boldsymbol{\Sigma}_{(2)}(\mathbf{y}) + \cdots, \\ \boldsymbol{v}(\mathbf{y},h) &= \boldsymbol{v}^{(0)}(\mathbf{y}) + h\boldsymbol{v}^{(1)}(\mathbf{y}) + \tfrac{1}{2}h^2\boldsymbol{v}^{(2)}(\mathbf{y}) + \cdots. \end{aligned} \tag{14.25}$$

We limit our attention to the leading term $(\Sigma_{(0)}, v^{(0)})$, which, as we shall show, satisfies the (nonlinear) von Kármán equations. Once the leading term is determined, subsequent corrections are obtained as solutions of linear equations with coefficients depending on the leading terms. We denote the (two-dimensional) Laplacian by Δ and consequently we denote the (two-dimensional) biharmonic operator by Δ^2. Our fundamental result is

14.26. Theorem. *Let the data in* (14.10) *and* (14.11) *be sufficiently smooth and let these equations have a solution of the form* (14.25) *with* $v_3^{(0)} \in C^4(\mathcal{M}\times[-1,1]) \cap C^0(\operatorname{cl}\mathcal{M}\times[-1,1])$. *Then* $\boldsymbol{v}^{(0)}$ *is of Kirchhoff-Love type, i.e., there is a function* $\mathcal{M} \ni \mathbf{s} \mapsto \boldsymbol{\omega}(\mathbf{s})$ *such that*

$$v_3^{(0)}(\mathbf{y}) = \omega_3(\mathbf{s}), \quad v_\alpha^{(0)}(\mathbf{y}) = \omega_\alpha(\mathbf{s}) - y^3\omega_{3,\alpha}(\mathbf{s}). \tag{14.27a,b}$$

Moreover, the stresses are given by

$$\frac{1+\nu}{E}\Sigma_{(0)}^{\alpha\beta} - \frac{\nu}{E}\Sigma_{(0)}^{\mu\mu}\delta^{\alpha\beta} = \tfrac{1}{2}\left[v_{\alpha,\beta}^{(0)} + v_{\beta,\alpha}^{(0)} + v_{3,\alpha}^{(0)}v_{3,\beta}^{(0)}\right], \tag{14.28}$$

$$\Sigma_{(0)}^{\alpha 3} = -\frac{E}{2(1-\nu^2)}[1-(y^3)^2]\Delta\omega_{3,\alpha}, \tag{14.29}$$

$$(14.30)\quad 2\Sigma^{33}_{(0)} = (y^3+1)\tau^+ + (y^3-1)\tau^- + (1+y^3)\int_{-1}^{1} f(\mathbf{s},\eta)\,d\eta - 2\int_{-1}^{y^3} f(\mathbf{s},\eta)\,d\eta + \frac{E[1-(y^3)^2]}{1-\nu^2}\left[\frac{y^3}{3}\Delta^2\omega_3 - (1-\nu)\omega_{\alpha,\beta}\omega_{\alpha,\beta} - \nu(\Delta\omega_3)^2\right]$$

where $\boldsymbol{\omega}$ satisfies the two-dimensional problem

$$(14.31)\quad \frac{2E}{3(1-\nu^2)}\Delta^2\omega_3 - 2\Omega^{\alpha\beta}\omega_{3,\alpha\beta} = \tau^+ + \tau^- + \int_{-1}^{1} f(\mathbf{s},\eta)\,d\eta \quad \text{in } \mathcal{M},$$

$$(14.32)\quad \Omega^{\alpha\beta}_{,\beta} = 0 \quad \text{in } \mathcal{M},$$

$$(14.33)\quad \omega_3 = 0, \quad \omega_{3,\alpha}\gamma_\alpha = 0 \quad \text{on } \partial\mathcal{M},$$

$$(14.34)\quad \Omega^{\alpha\beta}\gamma_\beta = 0 \quad \text{on } \partial\mathcal{M},$$

with

$$(14.35)\quad \frac{2(1-\nu^2)}{E}\Omega^{\alpha\beta} = (1-\nu)(\omega_{\alpha,\beta}+\omega_{\beta,\alpha}) + 2\nu\omega_{\mu,\mu}\delta_{\alpha\beta} + (1-\nu)\omega_{3,\alpha}\omega_{3,\beta} + \nu\omega_{3,\mu}\omega_{3,\mu}\delta_{\alpha\beta}.$$

We sketch the proof below. We now show that (14.31) and (14.32) can be converted to von Kármán's equations. Let us assume for simplicity that $\mathcal{M}$ is simply-connected and that $\Omega^{\alpha\beta}$ is continuously differentiable. We apply Theorem XII.3.5 to each equation of (14.32) to deduce that there are twice continuously differentiable functions ψ_1 and ψ_2 such that

$$(14.36)\quad \Omega^{11} = \psi_{1,2}, \quad \Omega^{12} = -\psi_{1,1}, \quad \Omega^{21} = \psi_{2,2}, \quad \Omega^{22} = -\psi_{2,2}.$$

Since $\Omega^{12} = \Omega^{21}$, it follows that $\psi_{\alpha,\alpha} = 0$. We now apply Theorem XII.3.5 to this equation to conclude that there is a thrice continuously differentiable function Φ such that $\psi_1 = \Phi_{,2}$, $\psi_2 = -\Phi_{,2}$. Thus

$$(14.37)\quad \Omega^{\alpha\beta} = \varepsilon_{\alpha\gamma}\varepsilon_{\beta\delta}\Phi_{,\gamma\delta},$$

which ensures that (14.32) is identically satisfied.

14.38. Exercise. Show that (14.31), (14.35), and (14.37) imply that ω_3 and Φ satisfy the semilinear *von Kármán equations*:

$$(14.39)\quad \frac{2E}{3(1-\nu^2)}\Delta^2\omega_3 - 2[\Phi,\omega_3] = \tau^+ + \tau^- + \int_{-1}^{1} f(\mathbf{s},\eta)\,d\eta \quad \text{in } \mathcal{M},$$

$$(14.40)\quad \Delta^2\Phi = -\tfrac{1}{2}[\omega_3,\omega_3]$$

where the *Monge-Ampère form* $[\cdot,\cdot]$ is defined by

$$(14.41)\quad [f,g] \equiv f_{,11}g_{,22} + f_{,22}g_{,11} - 2f_{,12}g_{,12}.$$

(In fact, (14.31), (14.32), and (14.37) are equivalent to (14.39) and (14.40).)

Sketch of the Proof of Theorem 14.26. We introduce the abbreviations

$$\begin{aligned}\xi &\equiv v^{(0)}_{3,3}(1+\tfrac{1}{2}v^{(0)}_{3,3}), \quad \chi \equiv v^{(1)}_{3,3}(1+v^{(0)}_{3,3}) \\ \kappa_{\alpha\beta} &\equiv v^{(0)}_{\alpha,\beta}+v^{(0)}_{\beta,\alpha}+v^{(0)}_{3,\alpha}v^{(0)}_{3,\beta}, \quad \lambda_\alpha \equiv v^{(0)}_{\alpha,3}+v^{(0)}_{3,\alpha}+v^{(0)}_{3,\alpha}v^{(0)}_{3,3}.\end{aligned} \tag{14.42}$$

Then (14.24) and (14.25) imply that

$$\begin{aligned}\mathrm{I}_C &= 3+2\xi+2h\chi+o(h), \\ \mathrm{II}_C &= 3+4\xi+4h\chi+o(h), \\ \mathrm{III}_C &= 1+2\xi+4h\chi(1+v^{(0)}_{3,3})+o(h), \\ \boldsymbol{C}:\boldsymbol{C} &= 3+4\xi+4\xi^2+4h\chi(1+2\xi)+o(h), \\ C_{\alpha\beta}H_{\alpha\beta} &= H_{\alpha\alpha}+h^2\kappa_{\alpha\beta}H_{\alpha\beta}+o(h^2), \\ C_{\alpha 3}H_{\alpha 3} &= h\lambda_\alpha H_{\alpha 3}+o(h), \\ C_{33}H_{33} &= (1+2\xi)H_{33}+o(h), \\ C_{\alpha k}C_{k\beta}H_{\alpha\beta} &= H_{\alpha\alpha}+h^2(2\kappa_{\alpha\beta}+\lambda_\alpha\lambda_\beta)H_{\alpha\beta}+o(h^2), \\ C_{\alpha k}C_{k3}H_{\alpha 3} &= 2h(1+\xi)\lambda_\alpha H_{\alpha 3}+o(h), \\ C_{3k}C_{k3}H_{33} &= (1+2\xi)^2H_{33}+o(h).\end{aligned} \tag{14.43}$$

Proof of (14.27a). We take $\boldsymbol{H}=H_{33}\boldsymbol{kk}$ in (14.14) and then let $h\to 0$. It follows from (14.19) and (14.43) that

$$\int_{\mathcal{M}\times[-1,1]} \Theta(\xi)H_{33}\,dv = 0 \quad \forall\, H_{33}, \tag{14.44a}$$

so that

$$\Theta(\xi(\mathbf{y})) = 0 \quad \forall\, \mathbf{y}\in\mathcal{M}\times[-1,1] \tag{14.44b}$$

where

$$\begin{aligned}\Theta(\xi) &\equiv \frac{1-2\nu}{E}\sum_{i=0}^{2}\tilde{\alpha}_i(\xi)+\frac{2}{E}\xi[\tilde{\alpha}_1(\xi)+2\tilde{\alpha}_2(\xi)]+\frac{4}{E}\xi^2\tilde{\alpha}_2(\xi), \\ \tilde{\alpha}_i(\xi) &\equiv \alpha_i(3+2\xi,3+4\xi,1+2\xi).\end{aligned} \tag{14.45}$$

Since the reference configuration is natural, it follows from (XIII.1.7) that $\Theta(0)=0$. The definitions of E and ν implicit in (14.4) ensure that $\Theta'(0)=1$, so that 0 is an isolated zero. Therefore, the only continuous solution ξ of (14.44b) satisfying the boundary condition that $\xi=0$ for $\mathbf{s}\in\partial\mathcal{M}$ (which comes from (14.6b)) is $\xi=0$. It then follows from (14.42), (14.6b), and the smoothness of $v^{(0)}_3$ that (14.27a) must hold. □

Proof of (14.27b). We take $\boldsymbol{H}=H_{\alpha 3}\boldsymbol{i}_\alpha\boldsymbol{k}$ in (14.14), let $h\to 0$, and use (14.27a) to obtain

$$\int_{\mathcal{M}\times[-1,1]}\left[v^{(0)}_{\alpha,3}+v^{(0)}_{3,\alpha}\right]H_{\alpha 3}\,dv = 0 \quad \forall\, H_{\alpha 3}, \tag{14.46a}$$

which implies that

$$v^{(0)}_{\alpha,3}+v^{(0)}_{3,\alpha}=0. \tag{14.46b}$$

We immediately deduce (14.27b) from (14.46b). □

Using (14.27), we can refine (14.43):

$$\mathrm{I}_C = 3+2hv^{(1)}_{3,3}+h^2\left[2v^{(0)}_{\alpha,\alpha}+2v^{(0)}_{3,\alpha}v^{(0)}_{3,\alpha}++(v^{(1)}_{3,3})^2+2v^{(2)}_{3,3}\right]+o(h^2), \quad \text{etc.} \tag{14.47}$$

We again take $\boldsymbol{H} = H_{33}\boldsymbol{kk}$ in (14.14), use (14.47), divide the resulting expression by h, and then let $h \to 0$. We find that

$$v^{(1)}_{3,3} = 0. \tag{14.48}$$

Proof of (14.28). We take $\boldsymbol{H} = H_{\alpha\beta}\boldsymbol{i}^{\alpha}\boldsymbol{i}^{\beta}$ in (14.14), use (14.47) and (14.48), and then let $h \to 0$. We exploit properties of the invariants α_i to show that the resulting equation is the weak form of (14.28). □

The remaining statements of Theorem 14.26 are proved in a similar fashion (see Ciarlet (1980)).

This section is based upon the refinement of the work of Ciarlet (1980) by Davet (1986). These authors employ a more general and precise functional setting than that used here. Related expositions are given by Ciarlet & Rabier (1980) and by Ciarlet (1990).

This work was the first that gave the von Kármán equations a rational position within the general theory of nonlinear elasticity. All previous derivations of these equations, beginning with von Kármán's (1910), employed a variety of ad hoc assumptions about the negligibility of certain terms. A virtue of Ciarlet's approach is that these equations characterize the leading term of a formal asymptotic expansion of the solution for the full three-dimensional equations. The von Kármán equations have the important mathematical virtue that they are semilinear. Consequently their analysis is much simpler than that of the quasilinear equations typical of other problems of nonlinear elasticity. (But the linearizations of the quasilinear equations for Cosserat shells, which are needed for bifurcation analyses, are scarcely any more complicated that those for the von Kármán equations.)

It is important to note that the formal validity of the von Kármán equations is predicated on very special scalings of the loads in the thickness variables. It is very easy for mathematical analyses of these equations to be conducted for parameters outside the range of physical validity (cf. Sec. X.5).

Recent work of Fox, Raoult, & Simo (1993) suggest that asymptotic methods using a broader repertoire of scalings yield asymptotic models having the same form as those discussed in Secs. 10 and 13.

15. Commentary. Historical Notes

In the first part of this book, we treated strings, rods, membranes, and shells as certain kinds of one- and two-dimensional deformable bodies, whose deformations are governed by equations directly representing geometrical laws, fundamental mechanical principles, and properly invariant constitutive equations. Despite their generality, these models proved to be quite tractable; we were able to solve a wide variety of problems and to deduce useful physical information from the solutions. We note that simplified versions of these models are used to describe most phenomena of technological importance in structural mechanics and physiology.

In the present chapter we derived rod and shell theories from the three-dimensional theory by using constraint methods and asymptotic methods with the aims of deepening our understanding of the theories derived earlier by direct methods and of showing how these theories can be refined. (We postponed a treatment of the intrinsic (special Cosserat) theory of shells to this section only because the derivation of (13.18) is more difficult than the analogous derivation of Cauchy's Theorem in Sec. XII.7.) The constraint methods deliver a hierarchy of properly invariant models of increasing precision and complexity whose mathematical structure is independent of the details of the actual system of constraints. Thus the voluminous literature disputing the relative merits of scarcely distinguishable rod and shell theories is of little consequence.

Another purpose of rod and shell theories constructed by either constraint methods or asymptotic methods is to model accurately the behavior of the body as described by the full three-dimensional theory. The traditional justification of this purpose has been that the full three-dimensional theory, even when linear, typically is deemed intractable in the sense that illuminating closed-form solutions are not available. There is a growing literature, begun by Morgenstern (1959), giving rigorous estimates of the errors between solutions of the three-dimensional equilibrium equations of linear elasticity and of those for the rod or shell theory. (Most available results are for linear plate theory. Morgenstern and many of his successors used dual variational principles, which rely on the convexity of the stored-energy function, and are accordingly not readily adapted for nonlinear elasticity. These errors are typically bounded by some constant (not computed) times the thickness of the body to a certain positive power. See Ciarlet (1990, 1994) and Le Dret (1991) for a treatment and extensive references for such bounds. Different sorts of estimates were developed by John (1965, 1971) for nonlinear problems. Based on a careful study of the underlying geometry and applicable to nonlinear problems, they express the validity of various shell models in terms of a priori bounds on the strain.

Bounds like those of Morgenstern have to be interpreted with care: The norm used for the error is critical, pointwise errors being both desirable and difficult. In particular, for rod and shell theories constructed by an asymptotic expansion in a small thickness parameter, important boundary-layer discrepancies occur. The nature of the boundary layer depends crucially on the rod or shell model. The issue of boundary layers is intimately associated with the St. Venant Principle; see the brief discussion in Sec. XIII.16. Virtually nothing is known about the behavior of boundary layers in corners. The situation can be complicated by the presence of other small parameters. (See, e.g., Sanchez-Palencia (1990) and Sanchez-Palencia & Vassiliev (1992) for treatments of asymptotic problems for shells when the curvature is small.) For a discussion of the many mathematical subtleties of constructing reliable plate theories, see Babuška & Li (1992). For a definitive mathematical treatment of asymptotic questions encompassing those arising in beam and plate theory, see Maz'ja, Nazarov, & Plamenevskii (1991).

Of course, the asymptotic limit of the problem for a body as one or two dimensions become small is a very singular kind of problem, and we should expect a host of difficulties. To control these singularities, at least to obtain classical theories such as von Kármán's, and to obtain error bounds it is necessary to suppose that load parameters and other data are appropriately scaled in the thickness parameter. Karwowski (1990) and Fox, Raoult & Simo (1993) have made a broader study of scalings of other parameters in terms of the thickness, and have thereby obtained asymptotic theories much closer to those obtained by the direct methods and by the method of constraints.

Antman (1976a) proved that the absolute minimizer of the potential-energy functional for a hyperelastic body under conservative loading is the weak limit in a suitable Sobolev space of a subsequence of (weak) solutions of the equilibrium equations for a hierarchy of rod or shell models with increasing degrees of freedom, constructed by the method of constraints. (This work has the flavor of that showing that a problem can be effectively treated by numerical methods.) On the other hand, Simmonds (1992) has emphasized that adding more degrees of freedom to a shell model can actually cause it to approximate the solution of the three-dimensional problem worse than the original simpler model. The explanation of this apparent paradox is merely that the result of Antman applies to bodies of a fixed shape and thickness, and treats a limit as the number of degrees of freedom goes to infinity, whereas Simmonds is balancing two limit processes, that treated by Antman and the asymptotic process as the thickness goes to zero. Moreover, it is well known that taking more terms in an asymptotic series need not improve its accuracy.

In summary, much remains to be done to determine the errors between solutions of rod or shell equations and the solutions of the three-dimensional equations. Little is known about nonlinearly elastic materials, about other materials, about dynamical problems linear or not, about behavior near corners, about laminated materials, and about the scaling of other parameters. Standard demonstrations of the utility of a given rod or shell theory for effectively approximating a limited number of problems should not be embraced with blind enthusiasm: In linear theories, rod and shell theories can be constructed to deliver exact solutions of the three-dimensional problem by suitably endowing the rod or shell theory with more parameters than the three-dimensional problem.

Much recent work on rod and shell theory has been motivated by developments in numerical analysis and computational techniques. The progress in the numerical analysis of problems of solid mechanics suggests that a day will come when rod and shell theories will lose their distinctive identities within computational mechanics and be subsumed under a general theory for the numerical treatment of three-dimensional problems, endowed with useful error estimates. Such a theory would automatically take advantage of thinness where it is appropriate to do so and would adaptively use refined computational methods to account for boundary layers and singularities at edges, corners, and cracks. Of course, for appropriate systems of loads, such computational schemes might well be equivalent to classical rod and shell theories in regions remote from boundaries. But such an observation would be relegated to a historical curiosity.

Would such a consummation signify the death of rod shell theory? I think not, for the simple reason that computational mechanics is not coextensive with mechanics. There are worthy roles for rod and shell theories besides that of furnishing a collection of approximations rich enough to be both technologically useful and capable of provoking endless controversy. One such role is to provide an illuminating collection of problems amenable to global analysis in which the effects of general material response can be studied. The analysis of such problems, to which much of this book is devoted, exactly complements numerical analysis, which is incapable of treating a whole family of materials at one time. (This and the preceding paragraph are adapted from Antman (1989) by permission of the American Society of Mechanical Engineers).

An account of the development of intrinsic theories of rods is given in Sec. VIII.18. The method of constraints can be traced back to Leibniz, who introduced the notion of averaging the stress over a cross section. Discussion of the further development of rod theories by Jas. Bernoulli, Euler, Cauchy, Poisson, Kirchhoff, Clebsch, Love, and others is given by Dill (1992), Love (1927), Szabó (1979), Timoshenko (1953), Truesdell (1959, 1960, 1983), and Truesdell & Toupin (1960, Secs. 63A, 214). The method of constraints for rods was developed for linear problems by Medick (1966), Mindlin & Herrmann (1952), Mindlin & McNiven (1960), and Volterra (1956, 1961) among others. The extension of these methods to nonlinear problems was carried out by Antman (1972), Antman & Warner (1966), Green, Laws, & Naghdi (1968), Green & Naghdi (1970), and Green, Naghdi, & Wenner (1974). The general setting used in this chapter was developed by Antman (1976a) and further refined by Antman & Marlow (1991). Adjustments to this approach making the equations readily amenable to numerical analysis were made by Simo (1985), Simo, Marsden, & Krishnaprasad (1988), and Simo & L. Vu-Quoc (1986a,b, 1988, 1991), the last paper accounting for cross-sectional warping.

Among the formulations of the asymptotic theory of rods, besides that of Mielke presented in Sec. 8, are those of Cimitière, Geymonat, Le Dret, Raoult, & Tutek (1988), Karwowski (1990), and Rigolot (1976). See Ciarlet (1990) for a comprehensive survey.

For accounts of the foundation of plate and shell theories by Germain, Lagrange, Poisson, Navier, Kirchhoff, Love, and others, see Love (1927), Naghdi (1972), Szabó (1979), Timoshenko (1953), Truesdell (1983, 1991b), and Truesdell & Toupin (1960, Sec. 212). A modern version of the method used by Kirchhoff (1850) to determine appropriate boundary conditions for the Kirchhoff shell is given in the paragraphs following (13.34).

The theory of what are here called special Cosserat shells was formulated by the Cosserats (1908, 1909). Synge & Chien (1941) gave a direct formulation of what we call the Kirchhoff theory. Ericksen & Truesdell (1958) developed a full analysis of strain and formulated the equations for the equilibrium of forces and torques (see the discussion in Sec. VIII.18). Green, Naghdi, & Wainwright (1965) formulated the full theory with one extensible director and Cohen & DeSilva (1966a,b) did the same for a theory with three directors. (Theories with three directors have the virtue neither of simplicity nor of generality.) See Naghdi (1972) for a thermodynamical treatment of the 1-director theory. A modern exposition of the Cosserat theory is given by Cohen & Wang (1989).

It should be noted that the special Cosserat theories of rods and shells are particularly attractive because the classical balances of force and torque deliver the right number of equations of motion for the kinematic variables admitted into the theory. Moreover, these theories furnish natural, geometrically exact generalizations of the classical equations. As our work in this chapter shows, the more complicated kinematical structure of even the 1-director theory leads to new kinds of governing equations involving stress resultants that are not so easily interpreted.

Versions of the constraint method for nonlinear problems for shells were obtained by Green, Laws, & Naghdi (1968) and Green & Naghdi (1970). The approach shown in this chapter is that of Antman (1976a) and Antman & Marlow (1991). Refinements of this method oriented toward numerical and stability analysis were made by Fox & Simo (1992), Simo & Fox (1989), Simo, Fox, & Rifai (1989, 1990), Simo & Kennedy (1991), Simo, Marsden, & Krishnaprasad (1988), and Simo, Rifai, & Fox (1990, 1992).

For accounts of the manifold approximate theories of shells (detailing the extensive contributions of E. Reissner) together with studies of concrete problems, written in a style not too dissimilar from that of this book, see Dikmen (1982), Libai & Simmonds (1988), and Reissner (1985). The main source for the mathematical analysis of nonlinear problems within this class of theories is Vorovich (1989). A comprehensive bibliography of texts and survey articles on shells is given by Noor (1989). An interesting collection of concrete problems for geometrically exact theories of shells can be obtained by adapting those for traditional theories treated by Timoshenko & Gere (1961) and Timoshenko & Woinowsky-Krieger (1959).

CHAPTER XV

Nonlinear Plasticity

1. Introduction

In this chapter we discuss a general class of materials with memory, the plastic materials, which are useful in describing the behavior of metals. Our purpose is to present the basic theory, in which some concepts of Chap. XII are further developed and illustrated, and in which the theory of elasticity plays a central role, in as simple a context as is compatible with the underlying physics. The exposition is simpler than that of most treatments because we consistently use internal variables in the material formulation, which obviates the need for a complicated treatment of frame-indifference, as is necessary in the spatial treatment of theories involving stress rates. To illustrate the nature of such theories, we treat a model with a lot of physical structure in Secs. 2 and 3. In Sec. 4 we show how to formulate a natural numerical approach for the solution of a particular dynamical problem. In Sec. 5 we give a general formulation of antiplane problems, whose degeneracies illuminate subtle difficulties with the concept of permanent plastic deformation. We conclude this chapter with a brief discussion of discrete models, which are used to motivate various theories.

The various theories of small-strain plasticity (see Geiringer (1973), Hill (1950), Kachanov (1969), and Prager & Hodge (1951), among others), which had reached a reasonable level of completion by the 1950's, can present a bewildering vista to the novice. Because they are restricted to quasistatic processes, many treatments of small-strain plasticity are fraught with differentials used as infinitesimals to preclude the appearance of partial derivatives with respect to time (in a manner reminiscent of traditional treatments of thermostatics). Readers disappointed with the mixture of basic principle, ad hoc approximation, and primitive analysis in such expositions might conclude that the subject is unworthy of serious mathematical effort. This conclusion is false: Small-strain plasticity can be precisely formulated, poses worthy challenges to the analyst (see Temam (1983) and T. W. Ting (1973)), and is of practical utility.

Nevertheless, such theories are too special to guide the development of useful generalizations. (Serious problems arise with geometric descriptions of large strain and rotation and can lead to consequent difficulties with the invariance of material properties under rigid motions.) In response to these questions, an effort was begun in the 1950's to develop a correct *dynamical* theory for large plastic deformation, in which the roles of geometry, mechanical principle, and properly invariant constitutive equations are clear (and in which time is not regarded as unmentionable). At first, the goal was simply to construct something clean, not necessarily something tractable.

Truesdell (1955a,b, 1956) introduced the theory of *hypoelasticity* in which the present mechanical state of a material depends on the past history of deformation but is independent of the rate at which the deformation is effected. A criterion of yielding (a concept to be explained below) is not introduced a priori, but can arise in a natural way for certain problems. A central difficulty in formulating the theory is to account for the requirement that material properties are invariant under rigid motions. This theory was accompanied by a number of illuminating solved problems. Green & Naghdi (1965) produced a theory for large plastic deformations inspired on one hand by the treatment of invariance in hypoelasticity and on the other hand by the concept of yielding in small-strain plasticity. This theory was refined in subsequent years (see Casey & Naghdi (1984), Naghdi (1990), Simo & Ortiz (1985), and works cited therein), but for a long time these theories were not applied to problems, either concrete or theoretical. Pipkin & Rivlin (1965), D. R. Owen (1968, 1970), and Šilhavý (1977) developed the alternative theory of *materials with elastic range* to account for the phenomena of plasticity. This theory has grown quite abstract and has not yet attracted analysis. (But the recent work of Lucchesi & Podio-Guidugli (1988, 1990) and Lucchesi, Owen, & Podio-Guidugli (1992) is an effort to make this theory useful and accessible.)

It was not until the 1980's that variants of the theory of Green & Naghdi were seriously considered for numerical study. The availability of powerful computational methods makes it now feasible to attack well-formulated problems for large deformations (see Simo (1994)), as our treatment in Sec. 4 suggests.

The exposition of Sections 1–4, together with Figs. 4.7 and 4.20, is adapted from Antman & Szymczak (1989) by kind permission of the American Mathematical Society.

2. Constitutive Equations

Recall that if there is a function $\hat{\boldsymbol{S}}$ such that the second Piola-Kirchhoff stress tensor is specified constitutively by

$$\boldsymbol{S}(\boldsymbol{z},t) = \hat{\boldsymbol{S}}\big(\boldsymbol{C}(\boldsymbol{z},t),\boldsymbol{z}\big), \tag{2.1}$$

then the material is called *elastic*. Otherwise, it is called *inelastic* or possibly *viscoelastic*, the latter term often being loosely defined. All such materials are special cases of (*simple*) *materials with memory*, which have constitutive equations delivering the second Piola-Kirchhoff stress as a functional of the past history of the strain:

$$\boldsymbol{S}(\boldsymbol{z},t) = \hat{\boldsymbol{S}}\big(\boldsymbol{C}^t(\boldsymbol{z},\cdot),\boldsymbol{z}\big). \tag{2.2}$$

Clearly, (2.1) is a degenerate special case of (2.2). Both of these equations have forms compatible with the requirement that the material properties be invariant under rigid motions. We are interested in materials with memory that describe the behavior of metals (because this is a basic goal of plasticity theory).

From the mathematical viewpoint, there is a very easy way to construct a family of materials with memory: We take $\boldsymbol{S}$ to have the form

$$\boldsymbol{S}(\boldsymbol{z},t) = \hat{\boldsymbol{S}}\big(\boldsymbol{C}(\boldsymbol{z},t),\boldsymbol{\Pi}(\boldsymbol{z},t),\boldsymbol{C}_t(\boldsymbol{z},t),\boldsymbol{z}\big), \tag{2.3}$$

where $\boldsymbol{\Pi}$, called an *internal variable*, is defined by the ordinary differential equation

$$\boldsymbol{\Pi}_t = \boldsymbol{\Phi}(\boldsymbol{C},\boldsymbol{\Pi},\boldsymbol{C}_t,\boldsymbol{z}). \tag{2.4}$$

The formal solution of (2.4) for $\boldsymbol{\Pi}(\boldsymbol{z},\cdot)$ delivers it as a tensor-valued functional of the past history of $\boldsymbol{C}(\boldsymbol{z},\cdot)$ and of an initial datum. The substitution of this solution into (2.3) converts it into a special form of (2.2). We interpret $\boldsymbol{\Pi}$ below. The presence of $\boldsymbol{C}_t$ in (2.3) can account for internal friction. A virtue of the system (2.3), (2.4) is that material properties are characterized by just two tensor-valued functions $\hat{\boldsymbol{S}}$ and $\boldsymbol{\Phi}$ that depend on just a finite number of arguments. In contrast, the function $\hat{\boldsymbol{S}}$ of (2.2) depends on a variable $\boldsymbol{C}^t$ belonging to an infinite-dimensional function space. In principle, a system of the form (2.3), (2.4) can be easily adjusted to describe a rich variety of real plastic materials. (In practice, there is no consensus as to what are accurate constitutive equations for materials undergoing large, fast deformations.) To describe sensible classes of constitutive functions for (2.3), (2.4), we must first discuss the basic physical processes of plastic deformation that are to be modelled.

Let us study the deformation of a virgin paper clip. If we deform it slightly, we can observe that the amount of stress needed to effect such a deformation depends on the amount of deformation from its natural reference configuration. This property is typical of elastic materials. Such materials have a degenerate memory that only recalls their natural states. If we release the paper clip from a small deformation, it rapidly returns to its natural state, internal friction damping out the motion. (Elastic materials do not have a mechanism for internal friction, but it is easy to incorporate one into a more general model.) If we now subject the paper clip to a large deformation and release it, it does not return to its original natural state but to a state of *permanent plastic deformation*. In small deformations from this state of permanent plastic deformation, the clip behaves elastically, but typically with a different function relating the deformation to the stress. Thus the large deformation has changed the nature of the elastic response to small deformations by changing the configuration (of permanent plastic deformation) in which the stresses are zero. Thus in modelling these phenomena, we must show how elastic constitutive equations change with the change of the state of permanent plastic deformation and we must account for the evolution of this state. Further experiments on the paper clip might show that the behavior is roughly independent of the rate at which the deformations are performed, but depends primarily on the order in which they are performed. Thus we can incorporate this feature in our model by requiring the constitutive equations to be *rate-independent*.

We now translate these qualitative experimental results into a specific mathematical model. We begin with a conceptually simple model, relegating refinements to the next section.

It is convenient, but neither mathematically essential nor always physically desirable, to interpret $\boldsymbol{\Pi}$ as the *plastic strain* defining the state of permanent plastic deformation. Specifically, we study an 'infinitesimal neighborhood' of material point $\boldsymbol{z}$ that has undergone a strain $\boldsymbol{C}(\boldsymbol{z},t)$. We imagine that the material around this neighborhood is excised, so that the traction exerted on the boundary of this neighborhood by the material in contact with it is zero. If the material is plastic, the material at $\boldsymbol{z}$ unloads elastically to a strain $\boldsymbol{\Pi}(\boldsymbol{z},t)$. This gedankenexperiment of excision describes a process by which to localize the considerations discussed in the deformation of the paper clip.

We now introduce the *yield function*

$$(2.5) \qquad (\boldsymbol{C},\boldsymbol{\Pi},\boldsymbol{z}) \mapsto \gamma(\boldsymbol{C},\boldsymbol{\Pi},\boldsymbol{z}),$$

which will tell us when the material behaves elastically and when the deformation is large enough to shift the value of $\boldsymbol{\Pi}$. The set

$$(2.6) \qquad \mathcal{E}(\boldsymbol{\Pi},\boldsymbol{z}) \equiv \{\boldsymbol{C} : \gamma(\boldsymbol{C},\boldsymbol{\Pi},\boldsymbol{z}) < 0\},$$

called the *elastic region*, is assumed to be homeomorphic to an open ball (in the six-dimensional space of symmetric tensors) for each fixed $\boldsymbol{\Pi},\boldsymbol{z}$. The deformation tensor $\boldsymbol{C}$ is required to satisfy $\gamma(\boldsymbol{C},\boldsymbol{\Pi},\boldsymbol{z}) \leq 0$. This requirement does *not* mean that there is an a priori bound imposed on $\boldsymbol{C}$: Rather, it means that if $\boldsymbol{C}$ becomes large, then $\boldsymbol{\Pi}$ must change accordingly. (We would expect it also to grow.) For given $\boldsymbol{\Pi}$, $\boldsymbol{z}$, the *yield surface* (*in strain space*) is $\partial\mathcal{E}(\boldsymbol{\Pi},\boldsymbol{z})$. It is a five-dimensional surface in the six-dimensional space of symmetric tensors. Below we give conditions ensuring that the evolution of $\boldsymbol{C}$ and $\boldsymbol{\Pi}$ is always consistent with the inequality $\gamma(\boldsymbol{C},\boldsymbol{\Pi},\boldsymbol{z}) \leq 0$.

We say that:

If $\gamma\big(\boldsymbol{C}(\boldsymbol{z}),\boldsymbol{\Pi}(\boldsymbol{z}),\boldsymbol{z}\big) < 0$, then $\boldsymbol{z}$ is behaving *elastically.*

If $\gamma\big(\boldsymbol{C}(\boldsymbol{z}),\boldsymbol{\Pi}(\boldsymbol{z}),\boldsymbol{z}\big) = 0$ and $\gamma_{\boldsymbol{C}}\big(\boldsymbol{C}(\boldsymbol{z},t),\boldsymbol{\Pi}(\boldsymbol{z},t),\boldsymbol{z}\big) : \boldsymbol{C}_t(\boldsymbol{z},t) < 0$, then $\boldsymbol{z}$ is undergoing *plastic unloading.*

If $\gamma\big(\boldsymbol{C}(\boldsymbol{z}),\boldsymbol{\Pi}(\boldsymbol{z}),\boldsymbol{z}\big) = 0$ and $\gamma_{\boldsymbol{C}}\big(\boldsymbol{C}(\boldsymbol{z},t),\boldsymbol{\Pi}(\boldsymbol{z},t),\boldsymbol{z}\big) : \boldsymbol{C}_t(\boldsymbol{z},t) = 0$, then $\boldsymbol{z}$ is undergoing *plastic neutral loading.*

If $\gamma\big(\boldsymbol{C}(\boldsymbol{z}),\boldsymbol{\Pi}(\boldsymbol{z}),\boldsymbol{z}\big) = 0$ and $\gamma_{\boldsymbol{C}}\big(\boldsymbol{C}(\boldsymbol{z},t),\boldsymbol{\Pi}(\boldsymbol{z},t),\boldsymbol{z}\big) : \boldsymbol{C}_t(\boldsymbol{z},t) > 0$, then $\boldsymbol{z}$ is undergoing *plastic loading.*

We now use this yield function to specialize (2.4) in a useful way:

$$(2.7) \qquad \boldsymbol{\Pi}_t = \begin{cases} \boldsymbol{\Omega}(\boldsymbol{C},\boldsymbol{\Pi},\boldsymbol{C}_t,\boldsymbol{z}) & \text{during plastic loading,} \\ \boldsymbol{O} & \text{otherwise.} \end{cases}$$

Note that if there is no plastic loading at $\boldsymbol{z}$, then $\boldsymbol{\Pi}(\boldsymbol{z},\cdot)$ is a constant. In this case, the constitutive equation (2.3) at $\boldsymbol{z}$ is that for a viscoelastic material of differential type; if the argument $\boldsymbol{C}_t$ of $\hat{\boldsymbol{S}}$ were absent, then this constitutive equation is that for an elastic material. The use of (2.7), which describes the evolution of $\boldsymbol{\Pi}$, affords a very simple mechanism by which to obtain different elastic or viscoelastic constitutive equations, parametrized by $\boldsymbol{\Pi}$, to describe the elastic or viscoelastic response for small deformations about different states of permanent plastic deformation.

An attribute of elastoplastic materials not enjoyed by 'viscoplastic' and 'viscoelastic' materials is that the response is rate-independent. This means that the response of elastoplastic materials depends on the past history of deformation only through the order in which the deformations occur and not through the rate. For example, if a component of $\tau \mapsto \boldsymbol{C}^t(\boldsymbol{z},\tau) \equiv \boldsymbol{C}(\boldsymbol{z}, t-\tau)$ were given by $\tau \mapsto \sin(t-\tau)$, then its effect on the stress at time t would be the same as that produced by $\tau \mapsto \sin(t-\tau^2)$, $\tau \mapsto \sin(t-e^\tau-1)$, etc. Formally, the response (2.2) is *rate-independent* at $\boldsymbol{z}$ iff

$$\hat{\boldsymbol{S}}(\boldsymbol{C}(\boldsymbol{z}, t-\cdot), \boldsymbol{z}) = \hat{\boldsymbol{S}}(\boldsymbol{C}(\boldsymbol{z}, t-\varphi(\cdot)), \boldsymbol{z}) \tag{2.8}$$

for all φ on $[0, \infty]$ that strictly increase from 0 to ∞. Let us apply this condition to (2.3) and (2.7). We first observe the identity

$$\frac{\partial}{\partial \tau}\boldsymbol{C}(\boldsymbol{z}, t-\varphi(\tau)) = -\varphi'(\tau)\boldsymbol{C}_t(\boldsymbol{z}, t-\varphi(\tau)) \tag{2.9}$$

when φ is differentiable. Now the past history of $\boldsymbol{C}$ affects $\hat{\boldsymbol{S}}$ of (2.3) directly through the argument $\boldsymbol{C}_t$ and indirectly through the dependence of the solution $\boldsymbol{\Pi}$ of the ordinary differential equation (2.7) on the past history of $\boldsymbol{C}$. An arbitrary change of the past history of $\boldsymbol{C}$ in a rate-independent way has no effect on the argument $\boldsymbol{C}$ of $\hat{\boldsymbol{S}}$ and $\boldsymbol{\Omega}$. To see its effect on $\boldsymbol{C}_t$, we first note that for $\varphi(\tau) = \tau$, the right-hand side of (2.9) at $\tau = 0$ reduces to $-\boldsymbol{C}_t(\boldsymbol{z}, t)$, while for an arbitrary φ this right-hand side at $\tau = 0$ reduces to $-\varphi'(0)\boldsymbol{C}_t(\boldsymbol{z}, t)$. Thus an arbitrary φ replaces the argument $\boldsymbol{C}_t$ with $\alpha\boldsymbol{C}_t$ where $\alpha = \varphi'(0)$ can be any positive number. The effect of such a change in the past history on $\boldsymbol{\Pi}$ is the same, so that (2.7) is replaced with $\alpha\boldsymbol{\Pi}_t = \boldsymbol{\Omega}(\boldsymbol{C}, \boldsymbol{\Pi}, \alpha\boldsymbol{C}_t, \boldsymbol{z})$. Thus if (2.3) and (2.7) are rate-independent, then

$$\hat{\boldsymbol{S}}\big(\boldsymbol{C}, \boldsymbol{\Pi}, \alpha\boldsymbol{C}_t, \boldsymbol{z}\big) = \hat{\boldsymbol{S}}\big(\boldsymbol{C}, \boldsymbol{\Pi}, \boldsymbol{C}_t, \boldsymbol{z}\big), \tag{2.10a}$$

$$\boldsymbol{\Omega}\big(\boldsymbol{C}, \boldsymbol{\Pi}, \alpha\boldsymbol{C}_t, \boldsymbol{z}\big) = \alpha\boldsymbol{\Omega}\big(\boldsymbol{C}, \boldsymbol{\Pi}, \boldsymbol{C}_t, \boldsymbol{z}\big) \tag{2.10b}$$

for all $\alpha > 0$, i.e., $\hat{\boldsymbol{S}}$ is positively homogeneous of degree 0 and $\boldsymbol{\Omega}$ is positively homogeneous of degree 1 in $\boldsymbol{C}_t$. Let us assume that $\hat{\boldsymbol{S}}$ is continuous in $\boldsymbol{C}_t$ (at least for $\boldsymbol{C}_t$ near $\boldsymbol{O}$). Then we can let $\alpha \searrow 0$ in (2.10a) and thereby deduce that $\hat{\boldsymbol{S}}\big(\boldsymbol{C}, \boldsymbol{\Pi}, \boldsymbol{C}_t, \boldsymbol{z}\big) = \hat{\boldsymbol{S}}\big(\boldsymbol{C}, \boldsymbol{\Pi}, \boldsymbol{O}, \boldsymbol{z}\big)$, i.e., that $\hat{\boldsymbol{S}}$ is independent of $\boldsymbol{C}_t$. Similarly, we assume that $\boldsymbol{\Omega}$ is continuously differentiable in $\boldsymbol{C}_t$ (for $\boldsymbol{C}_t$ near $\boldsymbol{O}$). Then (2.10b) implies that $\partial\boldsymbol{\Omega}/\partial\boldsymbol{C}_t$ is positively homogeneous of degree 0 in $\boldsymbol{C}_t$ and is therefore independent of $\boldsymbol{C}_t$, i.e., that $\boldsymbol{\Omega}$ is linear in $\boldsymbol{C}_t$, so that there is a fourth-order tensor $\mathcal{A}$ such that

$$\boldsymbol{\Pi}_t = \mathcal{A}(\boldsymbol{C}, \boldsymbol{\Pi}, \boldsymbol{z}) : \boldsymbol{C}_t \quad \text{during plastic loading.} \tag{2.11}$$

(We briefly discuss rate-dependent materials in Secs. 3 and 6.) We now obtain a useful specialization of the form of (2.11).

2.12. Proposition. *If* (2.7) *has a continuous right-hand side, then* $\mathcal{A}$ *is such that* (2.11) *has the form*

$$\boldsymbol{\Pi}_t = \boldsymbol{M}(\boldsymbol{C}, \boldsymbol{\Pi}, \boldsymbol{z})[\gamma_{\boldsymbol{C}}(\boldsymbol{C}, \boldsymbol{\Pi}, \boldsymbol{z}) : \boldsymbol{C}_t] \quad \textit{during plastic loading.} \tag{2.13}$$

Proof. The continuity of the right-hand side of (2.7) implies that

$$\mathcal{A}(\boldsymbol{C}, \boldsymbol{\Pi}, \boldsymbol{z}) : \boldsymbol{C}_t = \boldsymbol{O} \tag{2.14}$$

for all $\boldsymbol{C}, \boldsymbol{\Pi}, \boldsymbol{C}_t$ such that

$$\gamma\big(\boldsymbol{C}, \boldsymbol{\Pi}, \boldsymbol{z}\big) = 0, \quad \gamma_{\boldsymbol{C}}\big(\boldsymbol{C}, \boldsymbol{\Pi}, \boldsymbol{z}\big) : \boldsymbol{C}_t(\boldsymbol{z}, t) = 0. \tag{2.15a,b}$$

Now we decompose $\mathcal{A}$ thus:

$$\mathcal{A} = \boldsymbol{M}\gamma_{\boldsymbol{C}} + \mathcal{B} \tag{2.16}$$

where $\boldsymbol{M}$ is a second-order tensor and $\mathcal{B}$ is a fourth-order tensor orthogonal to all tensors of the form $\boldsymbol{N}\gamma_{\boldsymbol{C}}$. To express this property mathematically, we denote the inner product on the space of fourth-order tensors by '::'. (Our computations are exactly the same as those for second-order tensors acting on vectors, except that we systematically replace vectors by second-order tensors, second-order tensors by fourth-order tensors, '·' by ':', and ':' by '::'). Thus the orthogonality condition is

$$0 = \mathcal{B} :: (\boldsymbol{N}\gamma_{\boldsymbol{C}}) = \boldsymbol{N} : \mathcal{B} : \gamma_{\boldsymbol{C}} \quad \forall\, \boldsymbol{N}. \tag{2.17}$$

The arbitrariness of $\boldsymbol{N}$ implies that

$$\mathcal{B} : \gamma_{\boldsymbol{C}} = \boldsymbol{O}. \tag{2.18}$$

Now we substitute (2.16) into (2.14) and use (2.15b) to obtain

$$(\boldsymbol{M}\gamma_{\boldsymbol{C}} + \mathcal{B}) : \boldsymbol{C}_t = \mathcal{B} : \boldsymbol{C}_t = \boldsymbol{O} \tag{2.19}$$

for all $\boldsymbol{C}_t$ satisfying (2.15b). Thus (2.18) and (2.19) say that $\mathcal{B}$ annihilates $\gamma_{\boldsymbol{C}}$ and every second-order tensor orthogonal to $\gamma_{\boldsymbol{C}}$, i.e., $\mathcal{B}$ annihilates every second-order tensor. Thus $\mathcal{B} = \mathcal{O}$, the zero fourth-order tensor. (This proof could have been replaced with an application of the general Multiplier Rule XVII.2.24.) □

During plastic loading on any time interval, γ must vanish identically, so that

$$0 = \gamma_{\boldsymbol{C}} : \boldsymbol{C}_t + \gamma_{\boldsymbol{\Pi}} : \boldsymbol{\Pi}_t. \tag{2.20}$$

By substituting (2.13) into (2.20), we obtain the *compatibility condition*

$$1 + \gamma_{\boldsymbol{\Pi}} : \boldsymbol{M} = 0 \quad \text{when } \gamma(\boldsymbol{C}, \boldsymbol{\Pi}, \boldsymbol{z}) = 0, \tag{2.21}$$

which restricts the form of $\boldsymbol{M}$. Thus the yield function influences the form of (2.13).

We give a concrete example of such constitutive equations in Sec. 4 and we describe some specific generalized constitutive equations in Sec. 5. We discuss generalizations and refinements of the present theory in the next section.

In any computational process, it is possible that the differential equations governing the evolution of the arguments $\boldsymbol{C}$ and $\boldsymbol{\Pi}$ of γ could cause these arguments to assume values outside the domain of γ, i.e., values for which $\gamma > 0$. We can easily prohibit this possibility by the following refinement: We simply extend γ and $\boldsymbol{M}$ in any convenient way to the complement of $\{(\boldsymbol{C}, \boldsymbol{\Pi}, \boldsymbol{z}) : \gamma(\boldsymbol{C}, \boldsymbol{\Pi}, \boldsymbol{z}) \leq 0\}$ and replace (2.21) with

$$(2.22) \qquad 1 + \gamma_{\boldsymbol{\Pi}} : \boldsymbol{M} = -(\gamma_{\boldsymbol{C}} : \boldsymbol{C}_t)\,\omega\gamma \quad \text{when } \gamma(\boldsymbol{C}, \boldsymbol{\Pi}, \boldsymbol{z}) > 0$$

where ω is any positive-valued function of $(\boldsymbol{C}, \boldsymbol{\Pi}, \boldsymbol{z})$ on this complement. Of course, (2.22) implies (2.21). We assume that (2.13) holds on this complement. Then

$$(2.23) \qquad \frac{\partial}{\partial t}\gamma(\boldsymbol{C}, \boldsymbol{\Pi}, \boldsymbol{z}) = (1 + \gamma_{\boldsymbol{\Pi}} : \boldsymbol{M})\,\gamma_{\boldsymbol{C}} : \boldsymbol{C}_t = -(\gamma_{\boldsymbol{C}} : \boldsymbol{C}_t)^2\,\omega\gamma \leq 0$$

when $\gamma(\boldsymbol{C}, \boldsymbol{\Pi}, \boldsymbol{z}) \geq 0$. Thus if γ initially does not exceed 0, then γ can never exceed 0.

These considerations suggest that we can readily generalize our model in the following way: We retain a yield function γ as before and define the elastic region as before, but we no longer require that $\boldsymbol{C}$ and $\boldsymbol{\Pi}$ be such that $\gamma(\boldsymbol{C}, \boldsymbol{\Pi}, \boldsymbol{z}) \leq 0$. Instead, we merely replace (2.7) with

$$(2.24) \qquad \boldsymbol{\Pi}_t = \begin{cases} \boldsymbol{O} & \text{if } \gamma(\boldsymbol{C}, \boldsymbol{\Pi}, \boldsymbol{z}) < 0, \\ \boldsymbol{\Omega}(\boldsymbol{C}, \boldsymbol{\Pi}, \boldsymbol{C}_t, \boldsymbol{z}) & \text{otherwise.} \end{cases}$$

If we assume that the material response is rate-independent and if $\boldsymbol{\Omega}$ is continuously differentiable in $\boldsymbol{C}_t$, then, as before, we find that $\boldsymbol{\Omega}$ is linear in $\boldsymbol{C}_t$. We do not, however, deduce a representation like (2.13).

3. Refinements and Generalizations

In this section we examine several simple ways to extend the theory of Sec. 2 to make it accommodate a wider range of phenomena observed in real materials. It must be pointed out that there has been an extensive body of experiment on the plastic response of materials. Nevertheless, even for commonly used kinds of aluminum and steel, the experimental results typically give data only for very special deformations, usually very small. Thus analytical work, to be of potential physical value, should allow a wide range of constitutive behavior.

An important generalization of the theory of Sec. 2 is obtained by letting the material be rate-dependent. Such dependence is observed when high-speed effects are present. A striking manifestation of rate-dependence can be seen in the variety of buckling patterns found in numerous studies of the dynamic axial compression of thin metal cylinders. A survey of the effects of rate-dependent behavior, in the context of wave propagation (of relevance to our discussion in the next section), is given by Clifton (1974, 1983). In adopting such a generalization, we would be forced to sacrifice the specific constitutive restrictions associated with (2.11) and (2.13). There is a corresponding loss of specificity in the thermodynamical restrictions on rate-dependent response, which we discuss below.

We can readily generalize our results of Sec. 2 by adding other internal variables to $\boldsymbol{\Pi}$. One such variable is the *hardening parameter* k. It accounts for changes in the yield function due to working (i.e., to a history of deformation). Another variable is the *back stress* $\boldsymbol{K}$, which is a symmetric tensor. It accounts for a loss of symmetry of the yield surface due to working. We construct a theory involving these additional internal variables by the simple device of replacing $\boldsymbol{\Pi}$ wherever it appears in Sec. 2 with the triple $\mathsf{V} \equiv (\boldsymbol{\Pi}, k, \boldsymbol{K})$ and by making obvious adjustments in the inner products appearing there. These adjustments make sense because V lies in a thirteen-dimensional inner-product space. When the constitutive equations are rate-independent, we obtain the obvious generalizations of (2.11) and (2.13). Clearly, we could incorporate further internal variables, such as damage parameters, by the same process.

Were we to follow the lead of small-strain plasticity, we would incorporate the parameters k and $\boldsymbol{K}$ into our more general theory by replacing $\gamma(\boldsymbol{C}, \boldsymbol{\Pi}, \boldsymbol{z})$ and $\hat{\boldsymbol{S}}(\boldsymbol{C}, \boldsymbol{\Pi}, \boldsymbol{z})$ with $\gamma^\sharp(\boldsymbol{C}, \boldsymbol{\Pi}, \boldsymbol{z}) - k$ and $\boldsymbol{S}^\sharp(\boldsymbol{C}, \boldsymbol{\Pi}, \boldsymbol{z}) + \boldsymbol{K}$.

We now formulate a full thermodynamical theory of rate-independent plasticity with internal variables (cf. Coleman & Gurtin (1967)). We recall the developments of Sec. XII.14. Let $\theta(\boldsymbol{z}, t)$, $\boldsymbol{g}(\boldsymbol{z}, t) \equiv \theta_{\boldsymbol{z}}(\boldsymbol{z}, t)$, $\eta(\boldsymbol{z}, t)$, and $\psi(\boldsymbol{z}, t)$ denote the absolute temperature, temperature gradient, entropy density, and free-energy density at $(\boldsymbol{z}, t)$. The Clausius-Duhem version (XII.14.23) of the entropy inequality is

$$-\psi_t - \eta\theta_t + \tfrac{1}{2}\boldsymbol{S} : \boldsymbol{C}_t + \theta^{-1}\boldsymbol{q} \cdot \boldsymbol{g} \geq 0. \tag{3.1}$$

Let us set

$$\mathsf{U} \equiv (\boldsymbol{C}, \theta, \boldsymbol{g}). \tag{3.2}$$

U lies in a ten-dimensional space with inner product $\circ$. Our constitutive equations for rate-independent thermoplasticity are

$$\begin{aligned} \boldsymbol{S}(\boldsymbol{z}, t) &= \hat{\boldsymbol{S}}\big(\mathsf{U}(\boldsymbol{z}, t), \mathsf{V}(\boldsymbol{z}, t), \boldsymbol{z}\big), \\ \boldsymbol{q}(\boldsymbol{z}, t) &= \hat{\boldsymbol{q}}\,\big(\mathsf{U}(\boldsymbol{z}, t), \mathsf{V}(\boldsymbol{z}, t), \boldsymbol{z}\big), \\ \psi(\boldsymbol{z}, t) &= \hat{\psi}\,\big(\mathsf{U}(\boldsymbol{z}, t), \mathsf{V}(\boldsymbol{z}, t), \boldsymbol{z}\big), \\ \eta(\boldsymbol{z}, t) &= \hat{\eta}\,\big(\mathsf{U}(\boldsymbol{z}, t), \mathsf{V}(\boldsymbol{z}, t), \boldsymbol{z}\big), \end{aligned} \tag{3.3}$$

where V is an internal variable lying in a finite-dimensional space with inner product $\bullet$. The evolution of V is governed by an equation like (2.13). In analogy with the development that follows (2.13), we say that *thermoplastic loading* occurs iff

$$\frac{\partial\gamma}{\partial\mathsf{U}} \circ \mathsf{U}_t > 0 \quad \text{when } \gamma = 0 \tag{3.4}$$

and we assume that this evolution is governed by

(3.5)
$$\mathsf{V}_t = \begin{cases} \mathsf{M}(\mathsf{U},\mathsf{V},\boldsymbol{z})\,[\partial\gamma(\mathsf{U},\mathsf{V},\boldsymbol{z})/\partial\mathsf{U}]\circ\mathsf{U}_t & \text{during thermoplastic loading,} \\ \mathsf{O} & \text{otherwise.} \end{cases}$$

Here M takes values in the space of V.

Note that we could include the stress $\boldsymbol{S}$ in V, provided we reduce the first equation of (3.5) to an identity. That is, we can characterize a material of the rate type as a material of the internal-variable type (see Sec. XII.10). Many theories of plasticity have constitutive equations of the rate type, most of which employ spatial formulations. The requirement of frame-indifference makes these equations complicated in form.

We now follow the approach of Sec. XII.14, which requires that for continuous processes the constitutive equations must be such that the entropy inequality (3.1) is identically satisfied for all histories of U. We deduce the restrictions this requirement imposes on our materials. These restrictions ensure that the material response meets some minimal level of dissipativity. Let us substitute (3.3) and (3.5) into (3.1) to obtain

(3.6)
$$\begin{aligned} 0 \le {} & \left[-\hat{\psi}_C - \left(\frac{\partial\hat{\psi}}{\partial\mathsf{V}}\bullet\mathsf{M}\right)\gamma_C + \tfrac{1}{2}\hat{\boldsymbol{S}}\right] : \boldsymbol{C}_t + \left[-\hat{\psi}_\theta - \left(\frac{\partial\hat{\psi}}{\partial\mathsf{V}}\bullet\mathsf{M}\right)\gamma_\theta - \hat{\eta}\right]\theta_t \\ & + \left[-\hat{\psi}_{\boldsymbol{g}} - \left(\frac{\partial\hat{\psi}}{\partial\mathsf{V}}\bullet\mathsf{M}\right)\gamma_{\boldsymbol{g}}\right]\cdot\boldsymbol{g}_t + \theta^{-1}\hat{\boldsymbol{q}}\cdot\boldsymbol{g} \\ \equiv {} & \mathsf{H}\circ\mathsf{U}_t + \frac{\hat{\boldsymbol{q}}\cdot\boldsymbol{g}}{\theta} \\ & \forall\ \mathsf{U}(\boldsymbol{z},\cdot) \quad \text{satisfying (3.4) during thermoplastic loading.} \end{aligned}$$

The only place that $\mathsf{U}_t \equiv (\boldsymbol{C}_t, \theta_t, \boldsymbol{g}_t)$ appears in (3.6) is where it is visible. Let us decompose U_t thus:

(3.7)
$$\mathsf{U}_t = \alpha\frac{\partial\gamma}{\partial\mathsf{U}} + \mathsf{W} \quad \text{where } \mathsf{W}\circ\frac{\partial\gamma}{\partial\mathsf{U}} = 0.$$

The only condition imposed on U_t by (3.4) is that $\alpha > 0$; W is an arbitrary vector orthogonal to $\partial\gamma/\partial\mathsf{U}$. Thus (3.6) reduces to

(3.8)
$$0 \le \alpha\mathsf{H}\circ\frac{\partial\gamma}{\partial\mathsf{U}} + \mathsf{H}\circ\mathsf{W} + \theta^{-1}\hat{\boldsymbol{q}}\cdot\boldsymbol{g}$$

for all $\alpha > 0$ and for all W orthogonal to $\partial\gamma/\partial\mathsf{U}$. By making α sufficiently small and by setting $\mathsf{W} = \mathsf{O}$, we deduce from (3.8) the Fourier heat conduction inequality

(3.9a)
$$\hat{\boldsymbol{q}}\cdot\boldsymbol{g} \ge 0.$$

It then follows from (3.8) that

(3.9b,c)
$$\mathsf{H}\circ\mathsf{W} = 0 \quad \forall\mathsf{W} \quad \text{satisfying} \quad \mathsf{W}\circ\frac{\partial\gamma}{\partial\mathsf{U}} = 0, \qquad \mathsf{H}\circ\frac{\partial\gamma}{\partial\mathsf{U}} \ge 0.$$

Using a simple version of the argument given in the proof of Proposition 2.12, we deduce from (3.9b) that there is a scalar field λ such that

$$\mathsf{H} = \lambda \frac{\partial \gamma}{\partial \mathsf{U}}. \tag{3.9d}$$

(Condition (3.9c) implies that $\lambda \geq 0$.) In summary, we obtain from (3.9d) that

(3.10a)

$$\begin{pmatrix} \hat{\boldsymbol{S}} \\ -\hat{\eta} \\ \mathbf{0} \end{pmatrix} = \begin{pmatrix} \hat{\psi}_{\boldsymbol{C}} \\ \hat{\psi}_{\theta} \\ \hat{\psi}_{\boldsymbol{g}} \end{pmatrix} + \begin{cases} [\lambda + (\partial\hat{\psi}/\partial\mathsf{V}) \bullet \mathsf{M}]\partial\gamma/\partial\mathsf{U} & \text{during thermoplastic loading,} \\ \begin{pmatrix} \boldsymbol{O} \\ 0 \\ \mathbf{0} \end{pmatrix} & \text{otherwise,} \end{cases}$$

If we now require that the right-hand side of (3.10a) be continuous, then we choose λ so that the coefficient of $\partial\gamma/\partial\mathsf{U}$ vanishes. In this case, (3.10a) reduces to

$$\begin{aligned} \hat{\boldsymbol{S}}(\mathsf{U}, \mathsf{V}, \boldsymbol{z}) &= \hat{\psi}_{\boldsymbol{C}}(\boldsymbol{C}, \theta, \mathsf{V}, \boldsymbol{z}), \\ -\hat{\eta}(\mathsf{U}, \mathsf{V}, \boldsymbol{z}) &= \hat{\psi}_{\theta}(\boldsymbol{C}, \theta, \mathsf{V}, \boldsymbol{z}). \end{aligned} \tag{3.10b}$$

Note that $\hat{\psi}$ is independent of $\boldsymbol{g}$. Conditions (3.9) and (3.10) also imply (3.6).

This approach to deducing constitutive restrictions from thermodynamical principles works in general. The richer the material response, the less specific the consequences of the entropy inequality. The primary references for this and the preceding section are Antman & Szymczak (1989), Casey & Naghdi (1984), Green & Naghdi (1965), and Simo & Ortiz (1985).

4. Example: Longitudinal Motion of a Bar

We now give an example of elastoplastic constitutive equations describing the one-dimensional theory of the longitudinal motion of a bar. In this theory we ignore thickness changes that would accompany the longitudinal motion of a three-dimensional bar. Consequently, in our model, the vector $\boldsymbol{z}$ of Sec. 2 is replaced with the scalar s, the vector $\boldsymbol{p}(\boldsymbol{z})$ with the scalar $s + u(s)$, the scalar $\rho(\boldsymbol{z})$ with the scalar $(\rho A)(s)$, and the tensors $\boldsymbol{C}^{1/2} - \boldsymbol{I}$, $\boldsymbol{\Pi}$, $\boldsymbol{S}$, $\boldsymbol{T}$ with the scalars $E \equiv u_s$, Π, S, T. (The variable T is the same as N in Chaps. II–IV.) The elastic region $\mathcal{E}(\boldsymbol{\Pi}, \boldsymbol{z})$ in E-space and the yield surface $\partial\mathcal{E}(\boldsymbol{\Pi}, \boldsymbol{z})$ respectively reduce to an interval $\big(E^-(\Pi, s), E^+(\Pi, s)\big)$ and its two end points. This degeneracy of the yield surface must be borne in mind when the constitutive model we now present is used to illuminate the formalism of Sec. 2.

We now describe a one-dimensional constitutive model, which offers a concrete illustration of the concepts developed in Sec. 2. For simplicity, we assume that the material is homogeneous so that the constitutive functions do not depend explicitly on s.

It is convenient on physical and mathematical grounds to formulate our equations for the longitudinal first Piola-Kirchhoff stress T rather than for the longitudinal second Piola-Kirchhoff stress S. (They are related by $T = (1+E)S$.) Let Λ be a given number in $(-1, 0)$. Our analog of the rate-independent version of (2.3) is the constitutive equation

$$T(s,t) = \hat{T}\big(E(s,t), \Pi(s,t)\big) \tag{4.1}$$

where $\hat{T}$ is defined on $(-1,\infty) \times [\Lambda, \infty)$. $\hat{T}$ is required to satisfy

$$\hat{T}_E(E,\Pi) > 0, \quad \hat{T}(\Pi,\Pi) = 0, \tag{4.2a,b}$$

$$\hat{T}_E(E,\Pi) \to -\infty \quad \text{as } E \to -1, \ \Pi \to \Lambda. \tag{4.2c}$$

Equation (4.2b) says that Π represents the permanent plastic strain. We shall express the yield function in terms of two strictly increasing functions $(-1,\infty) \ni E \mapsto T^-(E)$, $[\Lambda,\infty) \ni E \mapsto T^+(E)$ with

$$T^-(E) < 0 < T^+(E), \quad T^-(E) \to -\infty \text{ as } E \to -1. \tag{4.3a,b,c}$$

We assume that

(4.4) for each $\Pi > \Lambda$, the curves $T = \hat{T}(E,\Pi)$ and $T = T^-(E)$ intersect at exactly one point, where $E = E^-(\Pi)$,

(4.5) the curves $T = \hat{T}(E,\Lambda)$ and $T = T^-(E)$ do not intersect for $E > -1$,

(4.6) for each $\Pi \geq \Lambda$, the curves $T = \hat{T}(E,\Pi)$ and $T = T^+(E)$ intersect at exactly one point, where $E = E^+(\Pi)$.

See Fig. 4.7. (It suffices for T^+ to be defined as a positive function only on $[E^+(\Lambda),\infty)$.)

We now produce specific choices for the functions $\hat{T}$, T^-, T^+ that meet the conditions we have imposed on them. Let $A^{\pm}$, $B^{\pm}$, C, D, b, c, d be positive constants. We take

$$T^-(E) = -A^- - B^-(E+1)^{-b}, \quad T^+(E) = A^+ - B^+(E+2)^{-b}, \tag{4.8a,b}$$

$$\hat{T}(E,\Pi) = -C[E+1-(\Pi-\Lambda)]^{-c} + D[E+1-(\Pi-\Lambda)]^{d}. \tag{4.9}$$

Note that (4.2a,c), (4.3a,c), and the requirement that $T^{\pm}$ be increasing are met by (4.8) and (4.9). To ensure that (4.6) holds, we take $c < b$. Unfortunately, this condition makes it more difficult to satisfy (4.4) and (4.5). Nevertheless, it is not hard to show that the available parameters can be adjusted to meet not only all the requirements we have imposed so far, but also to accommodate the prescription of the elastic modulus $\hat{T}_E(0,0)$, the yield strains $E^{\pm}(0)$, and the yield stresses $\hat{T}(E^{\pm}(0),0)$. These numbers constitute the most readily accessible experimental data for solids.

Let us observe that the elastic region has the form

$$(E^-(\Pi), E^+(\Pi)) = \{E : T^-(E) < \hat{T}(E,\Pi) < T^+(E)\}. \tag{4.10}$$

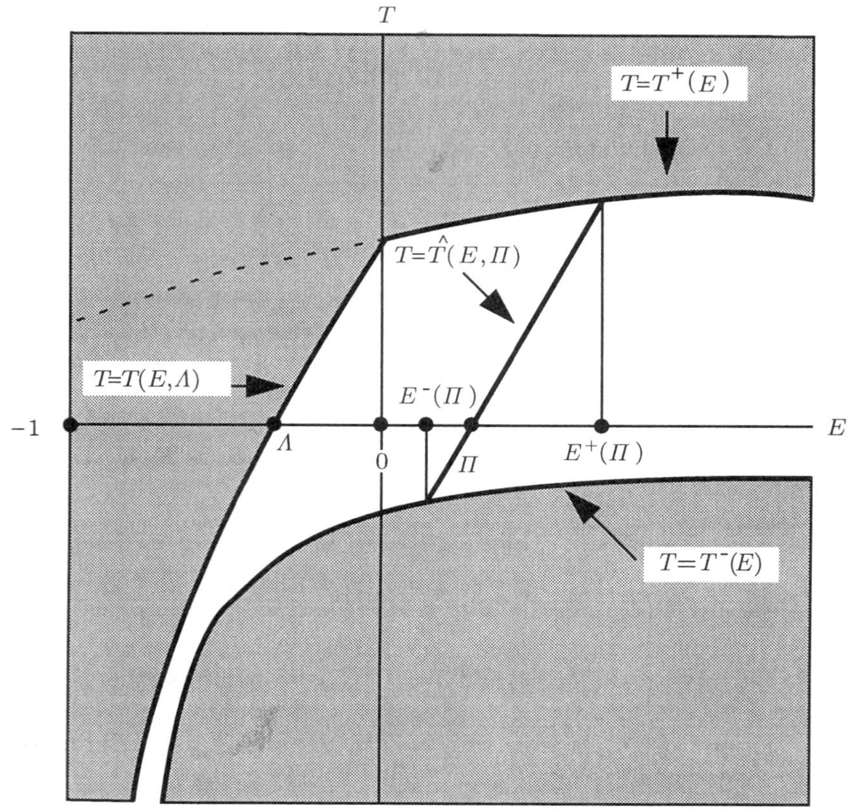

Fig. 4.7. Stress-strain curves. Note that there is a most compressive elastic response defined by the curve $T = \hat{T}(E, \Lambda)$. The elastic region is the interval $(E^-(\Pi), E^+(\Pi))$. We can define $E^-(\Lambda) = -\infty$. The shaded region consists of inadmissible pairs (E, T).

We may accordingly define yield functions

$$(4.11) \qquad \gamma^-(E, \Pi) \equiv T^-(E) - \hat{T}(E, \Pi), \quad \gamma^+(E, \Pi) \equiv \hat{T}(E, \Pi) - T^+(E).$$

(To make contact with the presentation of Sec. 2, we could define the yield function by $\gamma(E, \Pi) \equiv \gamma^-(E, \Pi)\gamma^+(E, \Pi)$, but this function is not particularly convenient for dealing with the disconnectedness of the yield surface.) By imitating the development leading to (2.13), we obtain

$$(4.12) \qquad \Pi_t = M^\pm(E, \Pi) g_E^\pm(E, \Pi) E_t \quad \text{if } E = E^\pm(\Pi), \quad g_E^\pm(E, \Pi) E_t > 0,$$

which describes the evolution of Π during plastic loading on each of the yield surfaces. Of course, $\Pi_t = 0$ otherwise. Moreover, the constitutive functions $M^\pm$ are related to the yield functions $g^{\pm}$ by the analog of (2.21):

$$0 = 1 + M^{\pm}(E, \Pi) g^{\pm}_{\Pi}(E, \Pi) \quad \text{if } E = E^{\pm}(\Pi), \quad g^{\pm}_{E}(E, \Pi) E_t > 0. \tag{4.13}$$

Thus (4.12) reduces to

$$\Pi_t = \begin{cases} -\frac{g^{\pm}_{E}(E,\Pi)}{g^{\pm}_{\Pi}(E,\Pi)} E_t & \text{if } E = E^{\pm}(\Pi), \quad g^{\pm}_{E}(E, \Pi) E_t > 0, \\ 0 & \text{otherwise.} \end{cases} \tag{4.14}$$

Then the governing equations consist of the equation of motion

$$\hat{T}(u_s, \Pi)_s = \rho A u_{tt} \tag{4.15}$$

and (4.14). It is convenient to set $v(s,t) = u_t(s,t)$ and write (4.15) as the system

$$E_t = v_s, \tag{4.16a}$$

$$(\rho A v)_t = \hat{T}(E, \Pi)_s. \tag{4.16b}$$

Thus the governing equations consist of (4.16) and (4.14). In our analysis we choose *not* to assign the ordinary differential equation (4.14) a status equivalent to (4.16) because we seek methods of broad applicability, capable of handling significant generalizations of (4.14), such as

$$\Pi_t = \hat{\Pi}(E^t(s, \cdot), s). \tag{4.17}$$

This policy is physically natural because (4.15) or (4.16) represents a fundamental balance law of physics, while (4.14) or (4.17) represents a constitutive equation that varies with each problem.

For both analytical work and numerical analysis, it is natural to replace (4.16b) with the Impulse-Momentum Law (II.3.1):

$$\left. \int_{s^-}^{s^+} \rho A v \, ds \right|_{t^-}^{t^+} = \left. \int_{t^-}^{t^+} \hat{T}(E, \Pi) \, dt \right|_{s^-}^{s^+}, \tag{4.18}$$

which is to hold for (almost) all intervals (t^-, t^+), (s^-, s^+). We complement (4.18) with a corresponding version of (4.16a):

$$\left. \int_{s^-}^{s^+} E \, ds \right|_{t^-}^{t^+} = \left. \int_{t^-}^{t^+} v \, dt \right|_{s^-}^{s^+}. \tag{4.19}$$

A basic philosophy for the numerical solution of initial-value problems for (4.14), (4.18), and (4.19) is first to freeze Π and treat (4.18), (4.19) as a hyperbolic system for a short time interval. Then update Π by an ordinary differential equation solver for (4.14) and repeat the process. In solving the hyperbolic equations, we could use the Godunov scheme, which is a natural discretization of the physical laws implicit in (4.18), (4.19) and is a very effective numerical method.

Initial-value problems for special versions of these systems have been treated numerically by Antman & Szymczak (1989), Trangenstein & Colella (1991), Trangenstein & Pember (1991, 1992). Some of the results of Antman & Szymczak for large deformations are illustrated in Fig. 4.20. They show the surprising result that the solution for an artificial elastic material corresponding to (4.1) without memory effects is much closer to that for (4.1) than is the solution for an elastic-perfectly plastic material of the sort used for small-strain plasticity.

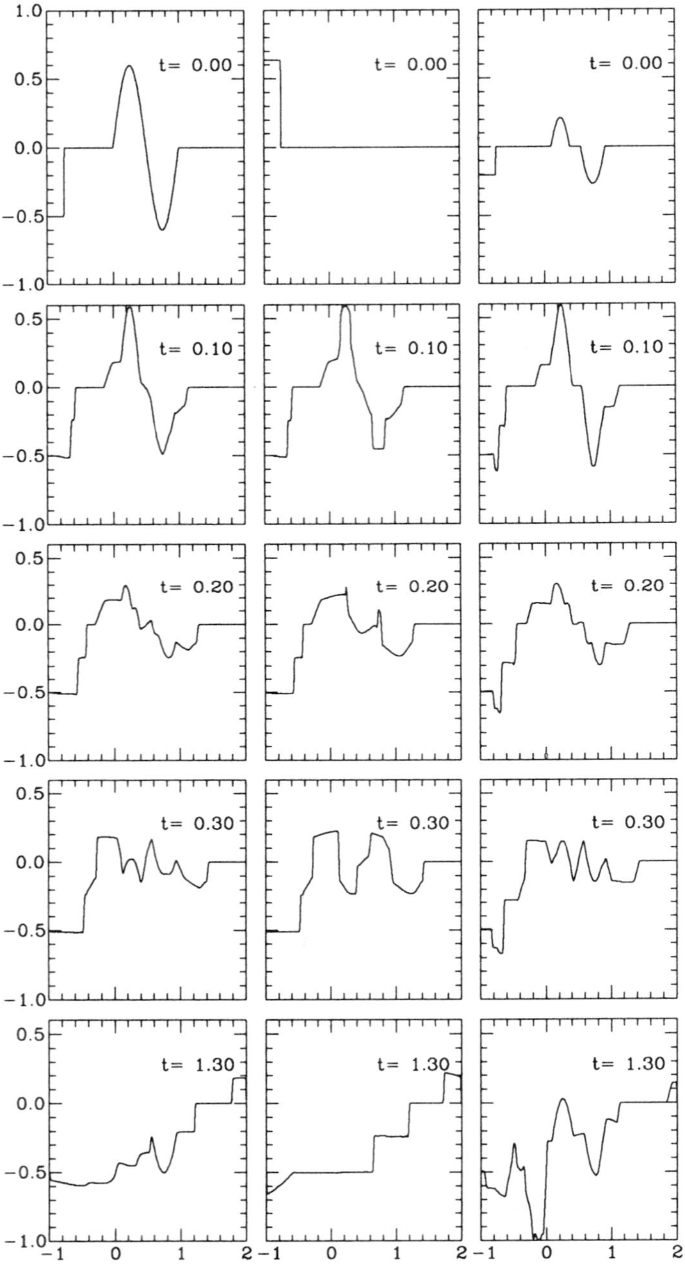

Fig. 4.20. Strain histories coming from the solution of an initial-value problem. The first column is for a special case of the nonlinear elastoplastic material (4.8), (4.9). The second column is for the corresponding artificial purely elastic material in which the constitutive equation is $T = T^-(E)$ for $E \leq E^-(0)$, $T = \hat{T}(E,0)$ for $E^-(0) \leq E \leq E^+(0)$, $T = T^+(E)$ for $E^+(0) \leq E$. The third column is for the corresponding linearly elastic-perfectly plastic material in which $\hat{T}(E,\Pi) = (cC+dD)(E-\Pi)$, $T^\pm(E) = \pm A^\pm$. Note that the last row of graphs is at a much later time.

There is a subtle analytic question masked in our formulation: Our system admits shocks. The Impulse-Momentum Laws (4.18), (4.19) or the equivalent Principle of Virtual Power show how to interpret these discontinuities for the system (4.16). But there is no natural weak form for (4.14), in which E_t could be like a Dirac delta where a shock occurs. Plohr & Sharp (1992) have handled such problems in the special case when (4.14) can be incorporated in a Principle of Virtual Power. More generally, Renardy (1989) has shown how the introduction of suitable small dissipative mechanisms obviates the difficulty.

5. Antiplane Shearing Motions

Let $\{\boldsymbol{i}_1, \boldsymbol{i}_2, \boldsymbol{k}\}$ be a fixed right-handed, orthonormal basis for Euclidean 3-space. Since our basis is Cartesian, we use only subscripts for our vectors and tensors, with Greek indices ranging over 1,2 and with repeated Greek indices summed over this range. Let $\boldsymbol{z} = x_\nu \boldsymbol{i}_\nu + z\boldsymbol{k}$. We study antiplane motions, which have the form

$$\boldsymbol{p}(x_1, x_2, z, t) = x^\nu \boldsymbol{i}_\nu + [z + w(x_1, x_2, t)]\boldsymbol{k}. \tag{5.1}$$

We set $\dot{w}_{,\nu} = \partial_t w_{,\nu}$. From (5.1) we readily compute

$$\boldsymbol{F} = \boldsymbol{I} + w_{,\nu}\boldsymbol{k}\,\boldsymbol{i}_\nu, \tag{5.2}$$

$$\boldsymbol{C} = \boldsymbol{I} + w_{,\mu}w_{,\nu}\boldsymbol{i}_\mu\boldsymbol{i}_\nu + w_{,\nu}(\boldsymbol{k}\,\boldsymbol{i}_\nu + \boldsymbol{i}_\nu\boldsymbol{k}), \tag{5.3}$$

$$\boldsymbol{H} \equiv \boldsymbol{I} - \boldsymbol{C}^{-1} = w_{,\nu}(\boldsymbol{k}\,\boldsymbol{i}_\nu + \boldsymbol{i}_\nu\boldsymbol{k}) - w_{,\nu}w_{,\nu}\boldsymbol{k}\boldsymbol{k}, \tag{5.4}$$

$$\dot{\boldsymbol{H}} = \dot{w}_{,\nu}\,(\boldsymbol{k}\boldsymbol{i}_\nu + \boldsymbol{i}_\nu\boldsymbol{k}) - 2\,w_{,\nu}\,\dot{w}_{,\nu}\,\boldsymbol{k}\boldsymbol{k}. \tag{5.5}$$

Note that $\det \boldsymbol{F} = 1$. We introduce a deformation tensor $\boldsymbol{M}$ and a strain tensor $\boldsymbol{P}$ of permanent plastic deformation having the same forms as $\boldsymbol{F}$ and $\boldsymbol{H}$:

$$\boldsymbol{M} = \boldsymbol{I} + m_\nu\boldsymbol{k}\,\boldsymbol{i}_\nu, \tag{5.6a}$$

$$\boldsymbol{P} \equiv \boldsymbol{I} - (\boldsymbol{M}^* \cdot \boldsymbol{M})^{-1} = m_\nu(\boldsymbol{k}\,\boldsymbol{i}_\nu + \boldsymbol{i}_\nu\boldsymbol{k}) - m_\nu m_\nu\boldsymbol{k}\boldsymbol{k}, \tag{5.6b}$$

where the m_ν are assumed to be independent of z.

In general, $\boldsymbol{F}$ cannot be determined from $\boldsymbol{C}$. Under our assumption of antiplanarity, however, these two tensors are equivalent. The same remarks apply to $\boldsymbol{M}$ and $\boldsymbol{P}$.

We assume that the material is an incompressible, isotropic, nonlinearly viscoplastic material. The incompressibility implies that there is a Lagrange multiplier p such that the second Piola-Kirchhoff stress tensor $\boldsymbol{S}$ has the form

$$\boldsymbol{S} = -p\boldsymbol{C}^{-1} + \hat{\boldsymbol{S}}_{\mathrm{A}} \tag{5.7}$$

where the extra stress $\hat{\boldsymbol{S}}_{\mathrm{A}}$ is a prescribed constitutive function of volume-preserving deformation histories. We specifically assume that $\hat{\boldsymbol{S}}_{\mathrm{A}}$ is an isotropic function of $\boldsymbol{H}$, $\dot{\boldsymbol{H}}$, $\boldsymbol{P}$ that does not depend explicitly on the coordinates x^ν, z. We shall informally refer to p as the *pressure*, without requiring that it represent the mean normal Cauchy stress (see (XII.12.33)). It is not prescribed by a constitutive equation. We shall let it absorb a variety of other terms having the same form. Thus its meaning will change in our exposition. Nevertheless, these changes affect neither the mathematics nor the mechanics. We could use Theorem XII.13.32 to represent $\hat{\boldsymbol{S}}_{\text{A}}$ as a sum of products of scalar functions of the joint invariants of $\boldsymbol{H}$, $\dot{\boldsymbol{H}}$, $\boldsymbol{P}$ with symmetric parts of certain products of these tensors. The set of joint invariants is

$$
\begin{aligned}
\mathcal{I} \equiv \{I_1 \equiv w_{,\nu}\, w_{,\nu}\,, I_2 \equiv 2w_{,\nu}\, m_\nu, I_3 \equiv m_\nu m_\nu,\\
I_4 \equiv 2w_{,\nu}\, \dot{w}_{,\nu}\,, I_5 \equiv 2m_\nu \dot{w}_{,\nu}\,, I_6 \equiv \dot{w}_{,\nu}\, \dot{w}_{,\nu}\,\}.
\end{aligned}
\tag{5.8}
$$

We shall restrict the form of our constitutive functions to be consistent with antiplane motions after we examine the equations of motion.

Let $\boldsymbol{T} \equiv \boldsymbol{F} \cdot \boldsymbol{S}$ denote the first Piola-Kirchhoff stress tensor. We similarly define $\hat{\boldsymbol{T}}_{\text{A}} \equiv \boldsymbol{F} \cdot \hat{\boldsymbol{S}}_{\text{A}}$. Without loss of generality, let us assume that the density is unity. Under the assumption of antiplanarity and under the constitutive assumptions we have made, the general equations of motion

$$
\left(\boldsymbol{i}_\nu \frac{\partial}{\partial x^\nu} + \boldsymbol{k} \frac{\partial}{\partial x_3} \right) \cdot \boldsymbol{T}^* = \ddot{w} \boldsymbol{k}
\tag{5.9}
$$

reduce to

$$
-p_{,\nu} + p_{,3} w_{,\nu} + T_{\text{A}\nu\mu,\mu} = 0,
\tag{5.10}
$$

$$
-p_{,3} + T_{\text{A}3\mu,\mu} = \ddot{w}.
\tag{5.11}
$$

Equation (5.11) implies that $p_{,33} = 0$, so that p has the form

$$
p(x_1, x_2, z, t) = h(x_1, x_2, t) z + g(x_1, x_2, t).
$$

We assume that the pressure (either p itself or some mean normal stress, which depends on z only through p) is bounded as $z \to \pm\infty$. Thus $h = 0$ and $p_{,3} = 0$. Hence (5.11) reduces to

$$
T_{\text{A}3\mu,\mu} = \ddot{w},
\tag{5.12}
$$

which is our fundamental equation of motion for w (and m_ν). Once these variables are found, (5.10) can be integrated to produce p.

In order that the mixed partial derivatives of p as given by (5.10) be equal, the stresses $T_{\text{A}\mu\nu}$ must satisfy the *compatibility conditions*:

$$
T_{\text{A}1\nu,\nu 2} = T_{\text{A}2\nu,\nu 1}.
\tag{5.13}
$$

We interpret (5.13) as a constitutive restriction on $\hat{\boldsymbol{S}}_{\text{A}}$. For hyperelastic media, Knowles (1976) has determined necessary and sufficient conditions on material response for (5.13) to hold (see Theorem XIII.10.9). For our class of viscoplastic materials we content ourselves with a sufficient condition for (5.13) to hold, namely, that $\hat{\boldsymbol{S}}_{\mathrm{A}}$ has the form $\hat{\boldsymbol{S}}_{\mathrm{A}}(\boldsymbol{H}, \dot{\boldsymbol{H}}, \boldsymbol{P}) = \alpha_0(\mathcal{I})\boldsymbol{H} + \alpha_1(\mathcal{I})\dot{\boldsymbol{H}} + \alpha_2(\mathcal{I})\boldsymbol{P}$. Since we can absorb the $\boldsymbol{C}^{-1}$ from $\boldsymbol{H}$ into the pressure term, this choice of constitutive function is equivalent to that in which $\boldsymbol{H}$ is replaced with $\boldsymbol{I}$:

$$\hat{\boldsymbol{S}}_{\mathrm{A}}(\boldsymbol{H}, \dot{\boldsymbol{H}}, \boldsymbol{P}) = \alpha_0(\mathcal{I})\boldsymbol{I} + \alpha_1(\mathcal{I})\dot{\boldsymbol{H}} + \alpha_2(\mathcal{I})\boldsymbol{P}. \tag{5.14}$$

Here $\alpha_0, \alpha_1, \alpha_2$ are given constitutive functions. A straightforward calculation shows that (5.14) satisfies (5.13) and reduces (5.12) to

$$\ddot{w} = (\alpha_0 w_{,1} + \alpha_1 \dot{w}_{,1} + \alpha_2 m_1)_{,1} + (\alpha_0 w_{,2} + \alpha_1 \dot{w}_{,2} + \alpha_2 m_2)_{,2}. \tag{5.15}$$

(Since we shall limit our attention to (5.14), which clearly represents an isotropic function, we actually have no need for the remarks on representation theorems preceding (5.8).)

Let the reference configuration occupied by the material be the cylindrical domain with cross section Ω in (x_1, x_2)-space. Let the ν_μ be the components of the outer unit normal to $\partial\Omega$ (which we assume exists almost everywhere). We may prescribe the motion w or the traction $T_{3\mu}\nu_\mu$ at each point of the boundary $\partial\Omega$ of Ω. We may also prescribe initial values for w and $\dot{w}$.

Yielding. We now choose our measures of deformation and permanent plastic deformation to be $\boldsymbol{F}$ and $\boldsymbol{M}$. For the antiplane motions we study, these tensors are equivalent to $\boldsymbol{C}$ and $\boldsymbol{P}$. In fact, $\boldsymbol{F}$ is equivalent to the two-dimensional gradient $\nabla w = (w_{,1}, w_{,2})$ and $\boldsymbol{M}$ is equivalent to $\boldsymbol{m} \equiv (m_1, m_2)$. We formulate the constitutive equations for yielding in terms of ∇w and $\boldsymbol{m}$. At the end of this section, we discuss the difficulties that arise when we formulate these equations in terms of the tensors $\boldsymbol{H}$ and $\boldsymbol{P}$.

We assume that the yield function γ is an isotropic scalar-valued function of ∇w and $\boldsymbol{m}$, and is independent of $\nabla\dot{w}$. Thus, by Cauchy's Representation Theorem VIII.7.8, γ is a function of the invariants

$$\mathcal{J} \equiv \{I_1, I_2, I_3\}. \tag{5.16}$$

We set

$$\gamma_K \equiv \frac{\partial \gamma}{\partial I_K} \tag{5.17}$$

and use an analogous notation for the derivatives of other invariants. The constitutive equation (2.13) thus has the form

$$\dot{\boldsymbol{m}} = \boldsymbol{n}(\nabla w, \boldsymbol{m})[I_4\gamma_1(\mathcal{J}) + I_5\gamma_2(\mathcal{J})] \tag{5.18}$$

where $\boldsymbol{n}$ is assumed to be an isotropic vector-valued function of its arguments. Standard representation theorems imply that there are scalar-valued functions η_1 and η_2 of $\mathcal{J}$ such that $\boldsymbol{n}$ has the form

$$\boldsymbol{n}(\nabla w, \boldsymbol{m}) = \eta_1(\mathcal{J})\nabla w + \eta_2(\mathcal{J})\boldsymbol{m}. \tag{5.19}$$

In terms of the yield function γ depending on $\mathcal{J}$, plastic loading is defined by

$$\gamma(\mathcal{J}) = 0, \quad I_4\gamma_1(\mathcal{J}) + I_5\gamma_2(\mathcal{J}) > 0. \tag{5.20a,b}$$

It follows from (5.20b) that during plastic loading, $\nabla\dot{w}$ cannot vanish anywhere, and the invariants I_1 and I_3 cannot vanish simultaneously. During plastic loading on any time interval, γ is identically 0, whence

$$\gamma_1\dot{I}_1 + \gamma_2\dot{I}_2 + \gamma_3\dot{I}_3 = 0, \tag{5.21}$$

which corresponds to (2.20). Replacing $\dot{\boldsymbol{m}}$ wherever it appears in (5.21) with its representation by (5.18) and (5.19), we use (5.20b) to reduce (5.21) to

$$1 + (2I_1\eta_1 + I_2\eta_2)\gamma_2 + (I_2\eta_1 + 2I_3\eta_2)\gamma_3 = 0, \tag{5.22}$$

which corresponds to (2.21). Condition (5.22) relates the constitutive function (5.19) to the yield function γ.

The essential governing equations for antiplane motion are (5.15), (5.18), and (5.19), with (5.19) restricted by (5.22).

Let us now reexamine this problem when we choose the strain to be $\boldsymbol{H}$ and the internal variable to be $\boldsymbol{P}$, which at first sight is at least as attractive as the choice made above. We assume that γ is an isotropic scalar-valued function of $\boldsymbol{H}$ and $\boldsymbol{P}$ and is independent of $\dot{\boldsymbol{H}}$. Then (not surprisingly by the Representation Theorem XII.13.32) γ again depends on (5.16), and (5.20b) characterizes plastic loading. Under these conditions, (2.13) yields

$$\dot{\boldsymbol{P}} = \boldsymbol{K}(\boldsymbol{H}, \boldsymbol{P})[I_4\gamma_1(\mathcal{J}) + I_5\gamma_2(\mathcal{J})] \tag{5.23}$$

during plastic loading, where $\boldsymbol{K}$ is assumed to be an isotropic tensor-valued function. We can invoke Theorem XII.13.32 to show that $\boldsymbol{K}$ has a representation of the form

(5.24)
$$\begin{aligned}\boldsymbol{K}(\boldsymbol{H}, \boldsymbol{P}) = \xi_0(\mathcal{J})\boldsymbol{I} &+ \xi_1(\mathcal{J})\boldsymbol{H} + \xi_2(\mathcal{J})\boldsymbol{P} + \xi_3(\mathcal{J})\boldsymbol{H}\cdot\boldsymbol{H} + \xi_4(\mathcal{J})\boldsymbol{P}\cdot\boldsymbol{P} \\ &+ \xi_5(\mathcal{J})(\boldsymbol{H}\cdot\boldsymbol{P} + \boldsymbol{P}\cdot\boldsymbol{H}) + \xi_6(\mathcal{J})(\boldsymbol{H}\cdot\boldsymbol{H}\cdot\boldsymbol{P} + \boldsymbol{P}\cdot\boldsymbol{H}\cdot\boldsymbol{H}) \\ &+ \xi_7(\mathcal{J})(\boldsymbol{H}\cdot\boldsymbol{P}\cdot\boldsymbol{P} + \boldsymbol{P}\cdot\boldsymbol{P}\cdot\boldsymbol{H}).\end{aligned}$$

Suppose that we now wish to interpret $\boldsymbol{P}$ as the permanent plastic deformation and accordingly require that $\boldsymbol{P}$ have the same form, namely (5.6b), as $\boldsymbol{H}$ has in (5.4). It then follows from (5.23) that the components $K_{\alpha\beta}$ of $\boldsymbol{K}$ must equal 0. Let us assume that the ξ_k are continuous. By substituting (5.4) and (5.6) into (5.24), we find that $K_{\alpha\beta} = 0$ if and only if

$$\xi_0 = 0, \quad \xi_3 = 0, \quad \xi_4 = 0, \tag{5.25a}$$

$$2\xi_5 - 2\xi_6 I_1 - 2\xi_7 I_3 = 0. \tag{5.25b}$$

Substituting (5.4), (5.6b), (5.24), and (5.25) into (5.23), we first obtain (5.18) and (5.19) from the (3,1)- and (3,2)-slots of (5.23) with

$$\eta_1 = \xi_1 + \tfrac{1}{2}\xi_6 I_2 + \xi_7 I_3, \tag{5.26a}$$

$$\eta_2 = \xi_2 + \xi_6 I_1 + \tfrac{1}{2}\xi_7 I_2. \tag{5.26b}$$

Equating the terms in the (3,3)-slot, we obtain

$$\begin{aligned}2m_\nu \dot{m}_\nu = \{&\xi_1 I_1 + \xi_2 I_3 - \xi_5(I_2 + 2I_1 I_3) + \xi_6[I_1 I_2 + 2I_3(I_1 + {I_1}^2)] \\ &+ \xi_7[I_3 I_2 + 2I_1(I_3 + {I_3}^2)]\}[I_4 g_1(\mathcal{J}) + I_5 g_2(\mathcal{J})].\end{aligned} \tag{5.27}$$

By substituting (5.26) into (5.27), we obtain yet another restriction on the ξ_k in addition to (5.25), which is a consequence of the antiplanarity:

$$\begin{aligned}&\xi_1(I_2 - I_1) + \xi_2 I_3 + \xi_5 I_2(1 - I_3) \\ &+ \xi_6 I_2[\tfrac{1}{2}I_2 + I_1(I_3 - 1)] + \xi_7 I_3(I_2 + I_2 I_3 - 2I_1 I_3) = 0.\end{aligned} \tag{5.28}$$

If $I_1 = 0$, then $I_2 = 0$ and (5.22), (5.26b), and (5.28) reduce to

$$1 + 2I_3\eta_2\gamma_3 = 0, \quad \eta_2 = \xi_2, \quad I_3\xi_2 = 0, \tag{5.29a,b,c}$$

which are patently inconsistent when γ is well behaved. The source of the difficulty is (5.29c), which comes from the requirement that $\boldsymbol{P}$ have the same form as $\boldsymbol{H}$ in the (3,3)-slot. By sacrificing this condition, we get rid of the inconsistency. We can then adopt a variety of strategies for choosing $\boldsymbol{P}$ and its evolution equation (5.23), but none seems simpler than our use of $\boldsymbol{m}$ as above.

This example indicates a subtle difficulty in the interpretation of $\boldsymbol{\Pi}$ as the permanent plastic deformation. (Only the degeneracy of antiplane motions allows us to use the variable $\boldsymbol{m}$ to escape the inconsistency of (5.29).) When we adopt a set of constitutive equations, say (2.3) and (2.7), involving an internal variable $\boldsymbol{\Pi}$, the entire physical meaning of $\boldsymbol{\Pi}$ inheres in these equations. (In the viewpoint advanced at the beginning of Sec. 2, the internal variable is just a convenient mathematical device for introducing memory.) When we attempt to give an internal variable a physical interpretation, say, as a permanent plastic deformation, we risk the imposition of further restrictions on the constitutive equations that can lead to inconsistencies. This danger is especially pronounced when the motion is restricted as it is here by the assumption of antiplanarity, and as it would be in the construction of rod and shell theories by the imposition of constraints as in Chap. XIV.

Constitutive functions for one-dimensional shearing. We specialize the foregoing development to one-dimensional problems by setting $w_{,2} = 0 = m_2$. In this conceptually simple setting, we discuss concrete ways of representing the constitutive functions. In the next subsection, we use isotropy to construct the natural two-dimensional constitutive equations from the concrete one-dimensional equations.

Under these assumptions, we obtain from (5.2)–(5.6) and (5.14) that

$$T_{31} = \hat{T}(w_{,1}, \dot{w}_{,1}, m_1) \equiv w_{,1} S_{A11} + S_{A31} = \alpha_0 w_{,1} + \alpha_1 \dot{w}_{,1} + \alpha_2 m_1 \tag{5.30}$$

where $\alpha_0, \alpha_1, \alpha_2$ depend on the specializations of (5.8). This dependence endows (5.30) with a certain amount of oddness.

To specify the yield process, we introduce an auxiliary function Γ, which is a positive-valued and increasing function of $w_{,1}$. We define the odd and even parts of Γ by

$$\mu(w_{,1}{}^2) w_{,1} \equiv \frac{\Gamma(w_{,1}) - \Gamma(-w_{,1})}{2}, \quad \sqrt{\nu(w_{,1}{}^2)} \equiv \frac{\Gamma(w_{,1}) + \Gamma(-w_{,1})}{2}. \tag{5.31}$$

Note that μ must be positive-valued and that the even part is positive. These functions are illustrated in Fig. 5.32.

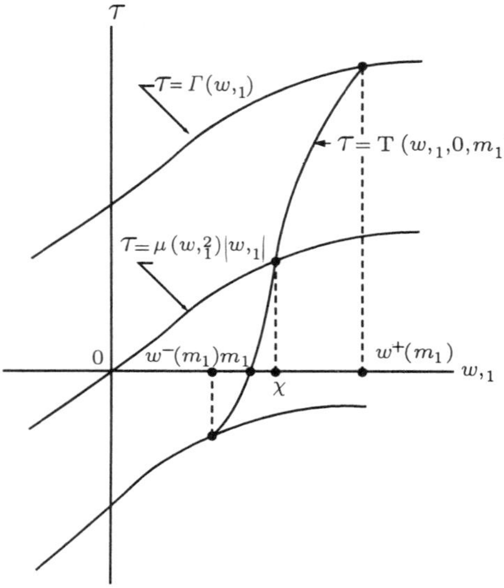

Fig. 5.32. Stress-strain curves for one-dimensional elastoplastic shear.

We assume, as in Sec. 4, that the yielding process is unaffected by the strain rates. We accordingly define the *elastic region* in strain space for a given m_1 to consist of all $w_{,1}$ satisfying

$$|\hat{T}(w_{,1}, 0, m_1) - \mu(w_{,1}{}^2) w_{,1}| < \sqrt{\nu(w_{,1}{}^2)}. \tag{5.33}$$

We define the yield function γ by

$$\begin{aligned} \gamma(w_{,1}{}^2, w_{,1} m_1, m_1^2) &\equiv [\hat{T}(w_{,1}, 0, m_1) - \mu(w_{,1}{}^2) w_{,1}]^2 - \nu(w_{,1}{}^2) \\ &= [(\alpha_0 - \mu) w_{,1} + \alpha_2 m_1]^2 - \nu(w_{,1}{}^2) \end{aligned} \tag{5.34}$$

where the arguments of α_0 have the restricted one-dimensional form.

We now specialize the form of our constitutive equations. We assume that $\hat{T}(\cdot, \dot{w}_{,1}, m_1)$ has a positive derivative for any fixed values of the remaining arguments. We further assume that for each fixed m_1 the equations

$$\hat{T}(\omega, 0, m_1) = \pm\Gamma(\pm\omega) \tag{5.35}$$

have unique solutions, denoted $\omega^{\pm}(m_1)$. Then the elastic region is the interval $(\omega^{-}(m_1), \omega^{+}(m_1))$. We assume that it is the domain of definition of the function $\hat{T}(\cdot, \dot{w}_{,1}, m_1)$. (Thus this domain does not depend on $\dot{w}_{,1}$.)

The symmetries of our constitutive functions make it attractive to replace m_1, which measures permanent plastic deformation, by another parameter χ, defined to be the unique solution of

$$\hat{T}(\chi, 0, m_1) = \mu(\chi^2)\chi \tag{5.36}$$

under the assumption that this equation has a unique solution for every m_1. In this case, we can prescribe $\hat{T}(w_{,1}, 0, m_1)$ implicitly to be the solution T of the form

$$w_{,1} - \chi = h(T - \mu(\chi^2)\chi), \quad m_1 - \chi = h(-\mu(\chi^2)\chi). \tag{5.37a,b}$$

In consonance with our previous assumptions, we require that $h' > 0$, that (5.37b) have a unique solution for each m_1, and that for each χ the curve (5.37a) intersects the curves $T = \pm\Gamma(\pm w_{,1})$ uniquely. Thus the equations

$$\omega - \chi = h(\pm\Gamma(\pm\omega) - \mu(\chi^2)\chi) \tag{5.38}$$

have unique solutions $\omega^{\pm}(m_1)$.

Constitutive functions for two-dimensional shearing. Each of the constitutive functions just introduced involves scalar-valued functions of one-dimensional specializations of the invariants (5.8). We generate corresponding constitutive equations for two-dimensional shearing by replacing each of these specializations with the full invariants. This procedure enables us to deduce two-dimensional equations from one-dimensional equations. In particular, we expand the square in (5.34) and use it to define

$$\gamma(\mathcal{J}) = [\alpha_0(\mathcal{J}) - \mu(I_1)]^2 I_1 + 2[\alpha_0(\mathcal{J}) - \mu(I_1)]\alpha_2(\mathcal{J}) I_2 + \alpha_2(\mathcal{J})^2 I_3 - \nu(I_1). \tag{5.39}$$

Similarly, we extend (5.30) by taking it to have the form (5.14), which we rewrite as

$$T_{3\nu} = \alpha_0(\mathcal{I}) w_{,\nu} + \alpha_1(\mathcal{I}) \dot{w}_{,\nu} + \alpha_2(\mathcal{I}) m_\nu. \tag{5.40}$$

Axisymmetric shearing problems. Let r and θ be polar coordinates for the (x_1, x_2)-plane, and let Ω be a disk. We study antiplane motions for which w and the permanent plastic deformation are independent of θ. We accordingly assume that (5.6) has the form generated by

$$m_1 = m\cos\theta, \quad m_2 = m\sin\theta \tag{5.41}$$

where m depends only on r. We denote differentiation with respect to r by a subscript r.

Under these conditions, (5.15) reduces to

$$r\ddot{w} = [r(\alpha_0 w_r + \alpha_1 \dot{w}_r + \alpha_2 m)]_r \tag{5.42}$$

where α_0, α_1, and α_2 depend on

$$\mathcal{I} \equiv \{I_1 \equiv {w_r}^2,\ I_2 \equiv 2w_r m,\ I_3 \equiv m^2, I_4 \equiv 2w_r\dot{w}_r,\ I_5 \equiv 2m\dot{w}_r,\ I_6 \equiv {\dot{w}_r}^2\}. \tag{5.43}$$

This equation gives us the general form of the yield function γ. Equation (5.18) reduces to

$$\dot{m} = [\eta_1(\mathcal{J}) w_r + \eta_2(\mathcal{J}) m][I_4 g_1(\mathcal{J}) + I_5 g_2(\mathcal{J})]. \tag{5.44}$$

The specific constitutive functions for one-dimensional problems are specialized by using (5.43) and replacing $w_{,1}$ and m_1 with w_r and m. These same constitutive equations can be constructed by making these replacements in the equations for two-dimensional problems. The shear stresses $T_{3\nu}$ are replaced with a single component corresponding to the directions of z and r.

6. Discrete Models

It is traditional, sometimes illuminating, and sometimes misleading to motivate constitutive assumptions for inelastic media from the equations for the discrete mechanics of mass points joined by massless springs and massless dissipative devices. Here we describe the simplest model that motivates certain continuum models of viscoplasticity of the form described above. It is illustrated in Fig. 6.1.

Since only the tip mass M has mass, the equations of motion for any other part of the system in Fig. 6.1 show that the tension N is independent of position. We assume that the tension in the spring, i.e., the force exerted by the piston on the spring, is given by the constitutive equation $N(t) = \hat{N}(y(t))$ and that the tension in the dashpot, i.e., the force exerted by the cylinder on the piston, is given by the constitutive equation $N(t) = G(\dot{\pi}(t))$ (or by a generalization of this described below). Here and below the superposed dot stands for the time derivative. We assume that $\hat{N}$ is continuously differentiable with $\hat{N}'$ everywhere positive and that $\hat{N}(0) = 0$, as in Chap. II. The equation of motion, the constitutive equations, and the relation between x, y, and π are

$$(6.2) \qquad m\ddot{x} = -N, \quad N = \hat{N}(y) = G(\dot{\pi}), \quad x - y = \pi.$$

This system reduces to

$$(6.3a) \qquad m\ddot{x} + \hat{N}(x - \pi) = 0,$$

$$(6.3b) \qquad G(\dot{\pi}) = \hat{N}(x - \pi).$$

We choose to regard π as an internal variable. In this case, (6.3b) is analogous to (2.4), and (6.3a) is analogous to the equation of motion into which (2.3) is substituted.

Of course, we can recast (6.2) into alternative forms, such as the following second-order equation for $\dot{\pi}$:

$$(6.4) \qquad m\left[\frac{d^2}{dt^2}\hat{N}^{-1}(G(\dot{\pi})) + \ddot{\pi}\right] + G(\dot{\pi}) = 0.$$

If $\hat{N}$ and G are each continuously differentiable and strictly increasing, then we can convert (6.2) to an equation of Lienard type:

$$(6.5) \qquad m\left[\ddot{y} + \frac{\hat{N}'(y)\dot{y}}{G'(G^{-1}(\hat{N}(y)))}\right] + \hat{N}(y) = 0.$$

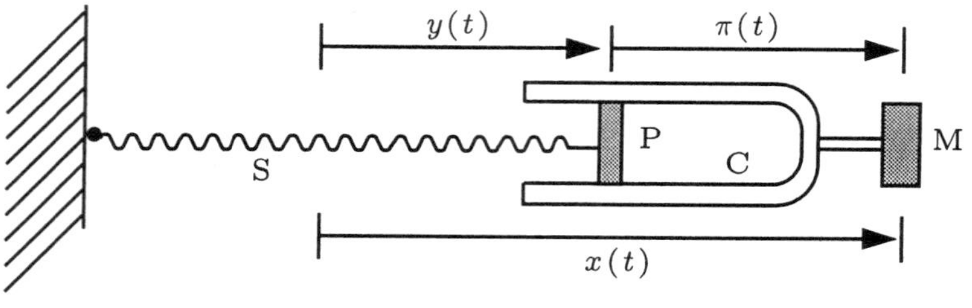

Fig. 6.1. Typical configuration of a mass-dashpot-spring system. One end of a massless spring S is attached to a fixed point. The other end is attached to a moving massless piston P. The origin of coordinates is located at the position of the piston when the spring occupies its natural reference configuration. The position of the piston at time t relative to the origin is $y(t)$. The piston slides in a massless cylinder C. The piston and cylinder constitute a dashpot, which can offer viscous *and* frictional resistance to the motion. A mass point M with mass m is rigidly attached to the cylinder. Its position relative to the origin at time t is $x(t)$ and its position relative to that of the piston is $\pi(t) = x(t) - y(t)$.

If G is a continuous strictly increasing function, then (6.3) is equivalent to a system of ordinary differential equations in standard form, which could be analyzed by standard methods. In this case, we could say that the behavior of the spring-dashpot system is viscoelastic. We are here concerned with an alternative situation in which the dashpot is a source of Coulomb friction. Then G is a certain kind of multivalued function (see Fig. 6.6). Precisely, G is related to $\dot{\pi}$ by the requirement that $(\dot{\pi}, G)$ lie on a graph $\mathcal{G}$ of a form shown in Fig. 6.6. In this case, we generalize (6.3b) by requiring that

$$G = \hat{N}(x - \pi), \quad (\dot{\pi}, G) \in \mathcal{G}. \tag{6.7}$$

We shall consider only cases (a) and (b) of Fig. 6.6.

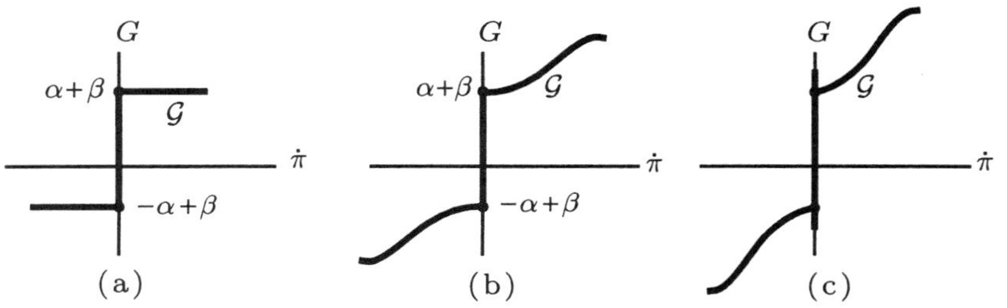

Fig. 6.6. (a) Constitutive graph for Coulomb friction. (b) A generalization of (a). (c) Graph for slip-stick friction.

Let us suppose that the graph $\mathcal{G}$ is vertical only on the G-axis, which it intersects on an interval of the form $[-\alpha + \beta, \alpha + \beta]$ with $\alpha > |\beta|$. It then follows from (6.7) that if $-\alpha + \beta < \hat{N}(x(t) - \pi(t)) < \alpha + \beta$ or, equivalently, if $\gamma(x(t), \pi(t)) < 0$, where

$$\gamma(x, \pi) \equiv [\hat{N}(x - \pi) - \beta]^2 - \alpha^2, \tag{6.8}$$

then $\dot{\pi}(t) = 0$. In this case, we can say that $x(t)$ is in the *elastic region*. Otherwise, we say that the system is behaving *plastically*.

Let us first analyze the situation in Fig. 6.6a. By our preceding remarks, (i) $\dot{\pi}(t) = 0$ if $\gamma(x(t), \pi(t)) < 0$, and (ii) $\dot{\pi}(t)$ has the same sign as $\hat{N}(x(t) - \pi(t))$, which in turn has the same sign as $x(t) - \pi(t)$ and as $\hat{N}(x(t) - \pi(t)) - \beta$, if $\gamma(x(t), \pi(t)) = 0$. Thus

$$\dot{\pi} = |\dot{\pi}| \operatorname{sign}(x - \pi) = |\dot{\pi}| \operatorname{sign} \hat{N}(x - \pi) = |\dot{\pi}| \operatorname{sign}[\hat{N}(x - \pi) - \beta], \quad |\dot{\pi}|\gamma = 0. \tag{6.9a,b}$$

Let us study the case in which $|\dot{\pi}(t)| > 0$ for all t in some time interval. Here

$$0 = \frac{d}{dt}\gamma(x(t), \pi(t)) = 2[\hat{N}(x(t) - \pi(t)) - \beta]\hat{N}'(x(t) - \pi(t))(\dot{x}(t) - \dot{\pi}(t)) \tag{6.10}$$

on this time interval. Thus we conclude that

$$\dot{\pi} = \begin{cases} 0 & \text{if } \gamma(x, \pi) < 0, \\ \dot{x} & \text{if } \gamma(x, \pi) = 0. \end{cases} \tag{6.11}$$

This constitutive equation for the evolution of π, which is analogous to (2.7), (2.11), is rate-independent. Let us also note that if $\frac{d}{dt}\gamma(x(t), \pi(t)) < 0$ for all t on some open time interval, then $\gamma(x(t), \pi(t)) \neq 0$ here, so that it follows from (6.9b) that $|\dot{\pi}|(t) = 0$ here. Thus we are led to require that

$$|\dot{\pi}(t)| \frac{d}{dt}\gamma(x(t), \pi(t)) = 0 \quad \forall t. \tag{6.12}$$

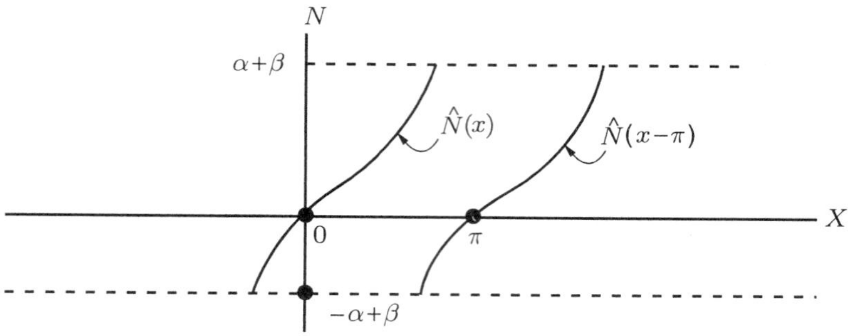

Fig. 6.13. Tension-elongation graphs for the the spring-dashpot system (6.3) corresponding to Fig. 6.6a.

In Fig. 6.13 we sketch the elasto-plastic response of the spring-dashpot system corresponding to Fig. 6.6a. It is analogous to Fig. 4.7. Note that when $\gamma < 0$, the piston is locked in the cylinder because the tension is not large enough to overcome the Coulomb friction and thereby permit relative motion.

We now turn to the dissipative mechanism corresponding to Fig. 6.6b. We assume that G is strictly increasing and continuous on $\mathbb{R} \setminus \{0\}$. In place of (6.11) and its predecessors, we obtain from (6.7) that

$$\dot{\pi} = \begin{cases} 0 & \text{if} \quad \gamma(x, \pi) < 0, \\ G^{-1}(\hat{N}(x - \pi)) & \text{if} \quad \gamma(x, \pi) = 0. \end{cases} \tag{6.14}$$

This constitutive equation is not rate-independent. In this case, we can solve (6.14) for π as a functional $\hat{\pi}$ of the past history of x. Then the constitutive equation for the spring-dashpot system (6.3) has the form $N(t) = \hat{H}(x(t), \hat{\pi}[x^t](t))$. For this material, N is not confined to the interval $[-\alpha + \beta, \alpha + \beta]$. When N does lie in this interval, the constitutive response is captured by Fig. 6.13. Outside this interval, the complicated behavior is dictated by (6.14). For the model of Fig. 6.6a, the corresponding complicated behavior occurs when $N = \pm\alpha + \beta$.

The constitutive response captured by Fig. 6.6b corresponds to the sum of a Coulomb graph Fig. 6.6a and a graph of a genuine increasing, continuous function. The latter describes purely viscous effects. We could make this superposition very explicit in Fig. 6.1 by replacing the single dashpot with a pair of dashpots in series, one exerting a purely viscous resistance and the other pure Coulomb friction. The motion of the piston in the former is impeded by a viscous fluid contained in the cylinder, as in a shock absorber. In the other dashpot, the piston is regarded as rubbing against the cylinder, without lubrication. (For the study of the corresponding continua, as in Fig. 4.20, such viscous dissipation plays an important role in the analysis of shocks.)

Alternatively, we could replace the single dashpot in Fig. 6.1 by the same pair of dashpots in parallel. The effect of such a combination of dashpots is to ensure that viscous dissipation comes into play only when π varies, i.e., only when N is outside $[-\alpha + \beta, \alpha + \beta]$. In other words, dissipation only occurs when the system is *behaving plastically.*

We note, however, that the response for Fig. 6.6b does not correspond to that of Fig. 4.7, which has hardening, which is rate-independent, and which has a yield surface within which the strain is confined. We now show how to obtain a discrete model for such a theory.

We revert back to the Coulomb law of Fig. 6.6a, but now we allow the parameters α and β to depend on a time-dependent internal variable ζ. We must supplement the theory with a rule describing the evolution of ζ. We might regard ζ as a temperature. The Coulomb friction in the dashpot could increase with increased temperature because the moving parts might have a tighter fit at higher temperature. In turn, the increase of temperature, properly described by a heat conduction equation (expressing the balance of energy), could be attributed to an accumulation of plastic working and therefore depend on $\dot{\pi}$. Thus we are led to a constitutive equation for ζ giving $\dot{\zeta}$ as a function of $x, \pi, \zeta, \dot{x}, \dot{\pi}$. We limit our attention to a rate-independent constitutive equation of the form

$$\dot{\zeta} = \phi(x, \pi, \zeta)|\dot{x}| + \psi(x, \pi, \zeta)|\dot{\pi}|. \tag{6.15}$$

In particular, we could take $\phi = 0$ and $\psi = 1$. We note that if $\phi = 0$, then (6.15) operates only during plastic loading. We replace (6.8) with

$$\gamma(x, \pi, \zeta) \equiv [\hat{N}(x - \pi) - \beta(\zeta)]^2 - \alpha(\zeta)^2. \tag{6.16}$$

Following the preceding development, we are again led to (6.9) and (6.12), but now, in place of (6.10), we find that

$$\begin{aligned} 0 &= \frac{d}{dt}\gamma(x(t), \pi(t), \zeta(t)) \\ &= 2\Big[\hat{N}(x(t) - \pi(t)) - \beta(\zeta(t))\Big]\Big[\hat{N}'(x(t) - \pi(t))\,(\dot{x}(t) - \dot{\pi}(t)) - \beta'(\zeta(t))\dot{\zeta}(t)\Big] \\ &\quad - 2\alpha(\zeta(t))\alpha'(\zeta(t))\dot{\zeta}(t) \end{aligned} \tag{6.17}$$

on any time-interval on which $|\dot{\pi}|$ is everywhere positive. Substituting (6.9a), (6.15), and (6.16) into (6.17), we obtain (in place of (6.11)) that during plastic loading

$$\begin{aligned} &\dot{\pi}\Big[\hat{N}'(x - \pi) + A(\zeta, x - \pi)\psi(x, \pi, \zeta)\Big] \\ &\qquad = \dot{x}\hat{N}'(x - \pi) - |\dot{x}|A(\zeta, x - \pi)\phi(x, \pi, \zeta)\,\mathrm{sign}\,(x - \pi) \end{aligned} \tag{6.18a}$$

where

$$A(\zeta, x - \pi) \equiv \alpha'(\zeta) + \beta'(\zeta)\,\mathrm{sign}\,(x - \pi). \tag{6.18b}$$

The constitutive equation (6.18a) is rate-independent. For $\phi = 0, \psi = 1, \beta = 0$, it reduces to

$$\dot{\pi}\Big[\hat{N}'(x - \pi) + \alpha'(\zeta)\Big] = \dot{x}\hat{N}'(x - \pi). \tag{6.19}$$

From (6.18) we obtain

$$\begin{aligned} \dot{N} &= \hat{N}'(x - \pi)(\dot{x} - \dot{\pi}) \\ &= \frac{\dot{x}A(\zeta, x - \pi)\hat{N}'(x - \pi)}{\hat{N}'(x - \pi) + A(\zeta, x - \pi)\psi(x, \pi, \zeta)}\,[\psi(x, \pi, \zeta) + \phi(x, \pi, \zeta)\,\mathrm{sign}\,\dot{x}\,\mathrm{sign}\,(x - \pi)] \end{aligned} \tag{6.20}$$

when $\gamma(x, \pi, \zeta) = 0$. (We note that a great many theories of plasticity and viscoplasticity start with a constitutive equation, analogous to (6.20), for the stress rate. See the comments following (3.5).)

If we now make the reasonable assumption that

$$\big[|\alpha'(\zeta)| + |\beta'(\zeta)|\big]\,|\phi(x, \pi, \zeta)|, \quad \big[|\alpha'(\zeta)| + |\beta'(\zeta)|\big]\,|\psi(x, \pi, \zeta)| \le \hat{N}'(x - \pi), \tag{6.21}$$

then (6.9a) and (6.18) imply that

$$\mathrm{sign}\,\dot{x} = \mathrm{sign}\,\dot{\pi} = \mathrm{sign}\,\hat{N}(x - \pi) = \mathrm{sign}\,(x - \pi). \tag{6.22}$$

In this case, the bracketed term in (6.20) reduces to $\psi + \phi$.

Let us suppose that $\gamma = 0$ on a time interval. Then (6.22) ensures that $\dot{x}$ has fixed sign on this interval. We can accordingly convert (6.15) and (6.18) to a pair of ordinary differential equations of the form

$$\frac{d\zeta}{dx} = f(x, \pi, \zeta), \quad \frac{d\pi}{dx} = g(x, \pi, \zeta). \tag{6.23}$$

We can readily impose restrictions on the constitutive functions entering in this theory to ensure that the initial-value problem for (6.23) subject to $\zeta(0) = 0 = \pi(0)$ has a unique global solution $x \mapsto \zeta^\star(x), \pi^\star(x)$. In this case, we find that the yield 'surface' (cf. (6.8)) is given explicitly by

$$N = \pm\alpha(\zeta^\star(x)) + \beta(\zeta^\star(x)). \tag{6.24}$$

This equation corresponds to those for T^+ and T^- of Sec. 4.

We could continue to generalize our discrete models. But we see that the process of introducing an internal variable like ζ for discrete models is in fact more delicate than the corresponding process described above for continuous models, which have a good chance of directly describing observed phenomena. Of course, a more complicated model is necessary to describe shearing motions.

The development of this section was inspired by the treatment of discrete problems by Simo & Hughes (1991), from which it differs in numerous particulars. The ordinary differential equations of discrete plasticity can offer very serious challenges to analysis; see Buhite & Owen (1979) and D. R. Owen (1987).

CHAPTER XVI

Dynamical Problems

In this chapter we treat a collection of elementary but illuminating dynamical problems for elastic and viscoelastic bodies. This material merely serves as an entrée to some parts of the rich and fascinating modern research on the quasilinear hyperbolic and parabolic systems applicable to elasticity. We develop the theory only in the context of concrete problems, so that we can concentrate on the effects of constitutive hypotheses.

We shall encounter expressions like $\frac{\partial}{\partial s}\hat{N}\big(u(s,t),s\big) \equiv \hat{N}\big(u(s,t),s\big)_s$. To distinguish notationally this 'total partial derivative' from the partial derivative of $\hat{N}$ with respect to its second argument, we present the functions $\hat{N}$ and u formally as

$$(y,x) \mapsto \hat{N}(y,x), \qquad (s,t) \mapsto u(s,t).$$

By convention we henceforth use y and x as names for the first and second arguments of $\hat{N}$ and use s and t as names for the first and second arguments of u. We can then write

$$\frac{\partial}{\partial s}\hat{N}\big(u(s,t),s\big) = \hat{N}_y\big(u(s,t),s\big)u_s(s,t) + \hat{N}_x\big(u(s,t),s\big).$$

1. The One-Dimensional Quasilinear Wave Equation

The *one-dimensional quasilinear wave equation*

$$(1.1) \qquad (\rho A)(s)w_{tt} = \hat{N}(w_s, s)_s$$

describes (i) the longitudinal motion of a straight elastic rod, (ii) the pure shearing of an incompressible elastic layer, and (iii) the one-dimensional motion of a gas. In each case, $(\rho A)(s)$ is the mass per unit length at s in the reference configuration. In case (i), $w(s,t)$ is the position of material point s at time t, and $\hat{N}(y,s)$, the tension at s corresponding to the stretch y, has the form shown in Fig. 1.2a. (See Sec. IV.1.) In case (ii), $w(s,t)$ is the displacement transverse to the s-axis, and $\hat{N}(y,s)$, the shear stress at s corresponding to the shear strain y, has the form shown in Fig. 1.2b. (See Sec. XIII.14.) In case (iii), w has the same meaning as in case (i), although the fluid-dynamical interpretation of w_s is the specific volume, and $\hat{N}(y,s)$, the negative of the pressure at s, corresponding to the specific volume y,

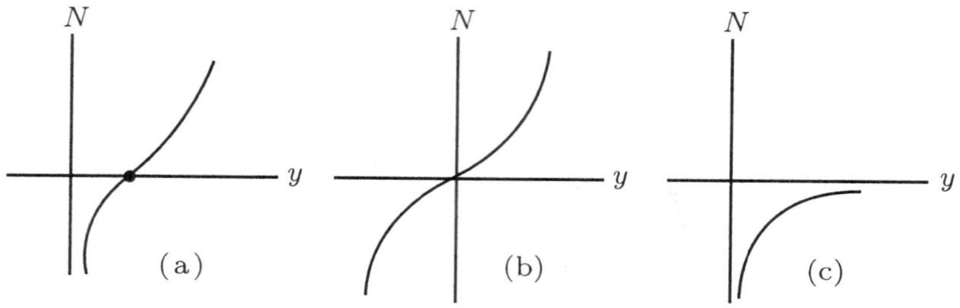

Fig. 1.2. (a) Tension-stretch law for the longitudinal motion of an elastic rod. (b) Shear stress-shear strain law for the shearing motion of an incompressible elastic body. (c) Negative pressure-specific volume law for a gas.

has the form shown in Fig. 1.2c. (We shall not pursue this case.) In each instance, it is reasonable to impose the monotonicity condition

$$\hat{N}_y(y,x) > 0 \quad \forall x, y. \tag{1.3}$$

Though each case satisfies this condition, their mathematical differences are profound. In particular, the mathematical theory is very well developed if $\hat{N}_{yy}(y,s)$ is either everywhere positive or everywhere negative. This assumption conforms to most constitutive laws for gases (case (iii)), but it is never satisfied for the odd functions for shear response in case (ii), and there is no apparent physical reason why it should hold in case (i). We regard (1.1) as an abbreviation for a suitable generalization in the form of a principle of virtual power or as an impulse-momentum law.

It is convenient to write (1.1) as a system of two first-order equations by setting

$$u = w_s, \quad v = w_t \tag{1.4}$$

so that (1.1) is equivalent to the matrix equation

$$\begin{aligned} \begin{pmatrix} u \\ (\rho A)(s)v \end{pmatrix}_t \equiv \mathsf{J}(s)\cdot\begin{pmatrix} u \\ v \end{pmatrix}_t &= \begin{pmatrix} v \\ \hat{N}(u,s) \end{pmatrix}_s \\ &\equiv \mathsf{A}(u,s)\cdot\begin{pmatrix} u \\ v \end{pmatrix}_s + \begin{pmatrix} 0 \\ \hat{N}_x(u,s) \end{pmatrix} \end{aligned} \tag{1.5a}$$

where

$$\mathsf{J}(s) \equiv \begin{pmatrix} 1 & 0 \\ 0 & (\rho A)(s) \end{pmatrix}, \quad \mathsf{A}(u,s) \equiv \begin{pmatrix} 0 & 1 \\ \hat{N}_y(u,s) & 0 \end{pmatrix}. \tag{1.5b}$$

The eigenvalues of $\mathbf{A}$ relative to $\mathbf{J}$, i.e., the eigenvalues of $\mathbf{J}^{-1}\cdot\mathbf{A}$, are

$$\lambda_1(u,s) = -\lambda(u,s), \quad \lambda_2(u,s) = \lambda(u,s), \quad \lambda(u,s) \equiv \sqrt{\frac{\hat{N}_y(u,s)}{(\rho A)(s)}}. \tag{1.6}$$

For any classical solution (u,v) of (1.5), the *characteristics* are the curves $t \mapsto s_\alpha(t)$ in the (s,t)-plane defined by

$$\frac{ds_\alpha}{dt}(t) = \lambda_\alpha(u(s_\alpha,t),s_\alpha), \quad \alpha = 1,2. \tag{1.7}$$

We define the *Riemann invariants* r_α for (1.5) by

$$r_\alpha(s,t) \equiv \tilde{r}_\alpha(u(s,t),v(s,t)) \tag{1.8a}$$

where

$$\tilde{r}_\alpha(u,v) \equiv v - \int_{u_0}^{u} \lambda_\alpha(y,s)\,dy = v - (-1)^\alpha \int_{u_0}^{u} \lambda(y,s)\,dy. \tag{1.8b}$$

For case (i) we choose $u_0 = 1$, and for case (ii) we choose $u_0 = 0$. The Riemann invariants satisfy the partial differential equations

(1.9)
$$\partial_t r_\alpha + \lambda_\alpha(u,s)\partial_s r_\alpha = \tfrac{\hat{N}_x(u,s)}{(\rho A)(s)} - (-1)^\alpha \lambda(u,s)\int_{u_0}^{u} \lambda_x(y,s)\,dy \equiv f_\alpha(u,s).$$

In particular, on the characteristics (1.7), these equations reduce to the ordinary differential equations

$$\tfrac{d}{dt} r_\alpha(s_\alpha(t),t) = f_\alpha(u(s_\alpha(t),t),s_\alpha(t)). \tag{1.10}$$

Note that $f_\alpha = 0$ for uniform bodies, in which ρA is a constant and $\hat{N}$ is independent of s, so that the Riemann invariants are constants on the corresponding characteristics.

1.11. Exercise. Prove that $(\partial_u r_\alpha, \partial_v r_\alpha)$ is orthogonal to the eigenvector of $\mathbf{J}^{-1}\cdot\mathbf{A}$ corresponding to λ_α.

As usual, we assume that the range of $\hat{N}(\cdot,s)$ is $\mathbb{R}$. Then we can solve (1.8) uniquely for $v = \frac{1}{2}(r_1+r_2)$ and $u = \tilde{u}\left(\frac{1}{2}(r_1-r_2)\right)$, where $\tilde{u}$ is a strictly increasing function. It is these representations that we substitute into the right-hand sides of (1.7), (1.9), and (1.10) to get equations for the s_α and the r_α alone. We set $\lambda_\alpha(\tilde{u}(\frac{1}{2}(r_1-r_2)),s) \equiv \tilde{\lambda}_\alpha(r_1,r_2,s)$. (We keep the arguments of $\tilde{\lambda}_\alpha$ distinct to allow the subsequent treatment of blowup to be unfettered by the special form of (1.6).)

We study the constitutive restriction that $\hat{N}(\cdot,x)$ be strictly concave:

$$\hat{N}_{yy}(y,x) < 0 \Leftrightarrow \lambda_y(y,x) < 0 \quad \forall\, y,x. \tag{1.12}$$

This restriction is not inconsistent with case (i), is totally inconsistent with case (ii) (for odd, nonlinear functions $\hat{N}(\cdot, x)$), and is satisfied for all standard constitutive equations of gas dynamics in case (iii). We study it merely because it is analytically easier to handle than more general conditions and because it illuminates the source of difficulties for the more general conditions. We now show that if (1.12) holds, then no matter how smooth the data are, solutions of initial-value problems for (1.1) have derivatives that become infinite in finite time, except in special circumstances. For this purpose, we assume that the material is uniform, so that $f_\alpha = 0$, and that $\hat{N}$ is twice continuously differentiable.

From (1.9) we obtain

$$\partial_{ts} r_2 + \tilde{\lambda}_2 \partial_{ss} r_2 + \frac{\partial \tilde{\lambda}_2}{\partial r_1} \partial_s r_1 \partial_s r_2 + \frac{\partial \tilde{\lambda}_2}{\partial r_2} (\partial_s r_2)^2 = 0. \tag{1.13}$$

Let $p \equiv \partial_s r_2$. Then (1.13) implies that

$$\begin{aligned} &\tfrac{d}{dt} p(s_2(t), t) + \left[\frac{\partial \tilde{\lambda}_2}{\partial r_1} (r_1(s_2(t), t), r_2(s_2(t), t)) \right] \partial_s r_1(s_2(t), t)\, p(s_2(t), t) \\ &\quad + \left[\frac{\partial \tilde{\lambda}_2}{\partial r_2} (r_1(s_2(t), t), r_2(s_2(t), t)) \right] p(s_2(t), t)^2 = 0 \end{aligned} \tag{1.14}$$

where s_2 is any 2-characteristic. We write the first equation of (1.9) as

$$\partial_t r_1 + \tilde{\lambda}_2 \partial_s r_1 + (\tilde{\lambda}_1 - \tilde{\lambda}_2) \partial_s r_1 = 0.$$

Since $\tilde{\lambda}_2 - \tilde{\lambda}_1$ is everywhere positive, we can use this expression to replace $\partial_s r_1$ in (1.14). Now let $(r_1, r_2) \mapsto h(r_1, r_2)$ be any function satisfying

$$\frac{\partial h}{\partial r_1} = \frac{(\partial \tilde{\lambda}_2 / \partial r_1)}{\tilde{\lambda}_2 - \tilde{\lambda}_1}. \tag{1.15}$$

It follows from (1.10) that $\frac{d}{dt} h\,(r_1(s_2(t), t), r_2(s_2(t), t)) = \frac{\partial h}{\partial r_1}(s_2(t), t) \frac{dr_1}{dt}(s_2(t), t)$. Thus (1.14) reduces to

$$\begin{aligned} &\tfrac{d}{dt} p(s_2(t), t) + \left[\tfrac{d}{dt} h\,(r_1(s_2(t), t), r_2(s_2(t), t)) \right] p(s_2(t), t) \\ &\quad + \left[\frac{\partial \tilde{\lambda}_2}{\partial r_2} (r_1(s_2(t), t), r_2(s_2(t), t)) \right] p(s_2(t), t)^2 = 0. \end{aligned} \tag{1.16}$$

1.17. Exercise. Let $q(t) = \exp\left[h\,(r_1(s_2(t), t), r_2(s_2(t), t))\right] p(s_2(t), t)$. Prove that

$$q(t) = \frac{q(0)}{1 + q(0) \int_0^t e^{-h} (\partial \tilde{\lambda}_2 / \partial r_2)\, d\tau}. \tag{1.18}$$

Here the arguments of the integrand are those of (1.16) with t replaced with τ.

q and therefore p remain bounded as long as the denominator in (1.18) does not vanish. We now specialize our attention to case (i) when (1.12) holds. Suppose that the initial values for u lie in a compact subset of $(0, \infty)$ and that the initial values for v are bounded. Then the initial values for r_1 and r_2 are bounded. It follows from (1.10) that r_1 and r_2 are everywhere bounded.

1.19. Exercise. Prove that

$$\frac{\partial \tilde{\lambda}_2}{\partial r_2} = -\frac{\hat{N}_{yy}}{2\hat{N}_y}. \tag{1.20}$$

In view of (1.12) and the bounds on r_1 and r_2, we conclude that there is a positive number K such that $e^{-h}\partial\tilde{\lambda}_2/\partial r_2 \geq K$ everywhere, so that $\int_0^t e^{-h}(\partial\tilde{\lambda}_2/\partial r_2)\,d\tau \geq Kt$ everywhere. From (1.18) we therefore obtain

1.21. Theorem. *Let $\hat{N}$ for case* (i) *be twice continuously differentiable and satisfy* (1.12). *Let the initial values for u lie in a compact subset of* $(0, \infty)$ *and let the initial values for v be bounded. If there is an s_0 such that $\partial_s r_2(s_0, 0) < 0$, then $\partial_s r_2$ becomes unbounded after a finite time.*

1.22. Exercise. Formulate an analogous statement for r_1 and formulate conditions ensuring that the solution is continuously differentiable for all time.

This development is based on that of Lax (1964; 1973, Sec. 6), who terms equations *genuinely nonlinear* if they satisfy suitable generalizations of (1.12). For the more complicated treatment of problems that include case (ii), see MacCamy & Mizel (1967).

Let us now find a condition ensuring that u has a positive lower bound in case (i). We study the boundary-value problem consisting of (1.5) and

$$\begin{aligned} &u(s,0) = \bar{u}(s), \quad v(s,0) = \bar{v}(s) \quad \text{for } s \in [0,1], \\ &\text{either} \quad v(0,t) = v^0 \quad \text{or} \quad \hat{N}(u(0,t),0) = n^0, \\ &\text{either} \quad v(1,t) = v^1 \quad \text{or} \quad \hat{N}(u(1,t),1) = n^1, \end{aligned} \tag{1.23}$$

where $\bar{u}$ and $\bar{v}$ are prescribed continuous functions with $\bar{u}$ everywhere positive on $[0,1]$, and v^0, n^0, v^1, n^1 are prescribed real numbers. Now we could use results like those obtained by Greenberg (1973), Nishida (1968), Nishida & Smoller (1973) and others who show how Riemann invariants are controlled in a shock. It is more convenient, however, to exploit an idea of Venttsel' (1981). We extend our boundary-value problem to all space by periodicity, thus obtaining an initial-value problem with periodic initial data. Now we add $\varepsilon\partial_{ss}r_\alpha$ where $\varepsilon > 0$ to the right-hand sides of (1.9) to make each of its equations parabolic. We apply the Maximum Principle (for weakly coupled parabolic systems; see Protter & Weinberger (1967)) to the resulting equations to get bounds independent of ε on the Riemann invariants in terms of the initial data. In particular,

$$\int_1^{u(s,t)} \sqrt{\frac{\hat{N}_y(y,s)}{\rho A(s)}}\,dy = \tfrac{1}{2}[r_2(s,t) - r_1(s,t)] \geq \tfrac{1}{2}[r_2(s,0) - r_1(s,0)] > -\infty. \tag{1.24}$$

If $\hat{N}_y(\cdot, s)$ is not integrable on $(0,1)$, then it follows from (1.24) that $u(s,t) \geq \min \bar{u}$ for all (s,t).

2. The Riemann Problem. Uniqueness and Admissibility of Weak Solutions

Since (1.1) is regarded as giving a reasonable description of a class of physical processes that are not expected to terminate at a finite time, we interpret the breakdown of classical solutions at a finite time as indicating that a physically reasonable solution must be a suitably defined weak solution. As we showed in (II.5.6), if (1.5) has a weak solution (u, v) that is classical to the left and right of a continuously differentiable curve $t \mapsto \sigma(t)$

in (s,t)-space with u and v having limits from the left and right on this curve, then (u,v) must satisfy the Rankine-Hugoniot jump conditions

$$\sigma'(t)[\![u]\!] = -[\![v]\!], \quad \sigma'[\![v]\!] = -[\![\hat{N}(u,\sigma(t))/(\rho A)(\sigma(t))]\!] \tag{2.1a,b}$$

across this curve. If the limits (u_l, v_l) from the left are given, then the elimination of σ' from (2.1) shows that the limits (u_r, v_r) from the right for which there is a σ' satisfying (2.1) lie on the curves

$$\sqrt{\rho A}(v - v_l) = \pm\sqrt{[\hat{N}(u) - \hat{N}(u_l)](u - u_l)} \tag{2.2}$$

in the (u,v)-plane. Note that the monotonicity of $\hat{N}$ ensures that the right-hand side of (2.2) is real-valued.

We shall see that (2.1) is not sufficient to define a unique solution for initial-value problems. To get a physically natural unique solution we shall exclude 'unsuitable' parts of (2.2). Now if the states to the left and the right of a discontinuity are to be determined by other data given at an earlier time, then characteristics from the left and the right should impinge on the shock curve $t \mapsto \sigma(t)$ with increasing time. (Otherwise, the characteristics could not be used for the local construction of solutions because their initial values on the shock curve would not be known.) For our problem these conditions reduce to the requirement that σ' satisfy one of the following *Lax Entropy Conditions*:

$$-\lambda(u_r,\sigma) \equiv \lambda_1(u_r,\sigma) < \sigma' < \lambda_1(u_l,\sigma) \equiv -\lambda(u_l,\sigma), \tag{2.3a}$$

$$\lambda(u_r,\sigma) \equiv \lambda_2(u_r,\sigma) < \sigma' < \lambda_2(u_l,\sigma) \equiv \lambda(u_l,\sigma). \tag{2.3b}$$

A solution for which (2.3a) holds is said to have a *back shock* or a 1-*shock* at σ and a solution for which (2.3b) holds is said to have a *front shock* or a 2-*shock* at σ.

2.4. Exercise. Deduce (2.1) directly from (1.5).

Throughout the rest of this section we consider a uniform rod (for which $\hat{N}_x = 0$ and ρA is constant) satisfying (1.12) and the further technical requirement that

$$\liminf_{y\to\infty} \hat{N}_y(y) > 0, \tag{2.5}$$

which is reasonable for elastic rods.

The basic building blocks of (weak) solutions of (1.5) are solutions of *Riemann problems*, which are initial-value problems with initial data constant to the left and right of the origin:

$$\big(u(s,0), v(s,0)\big) = \begin{cases} (u_l, v_l) & \text{for } s < 0, \\ (u_r, v_r) & \text{for } s > 0 \end{cases} \tag{2.6}$$

where u_l, v_l, u_r, v_r are constants. (Both practical and theoretical methods for treating initial-value problems for (1.5) begin by replacing the initial data with approximate piecewise constant data, which locally lead to a collection of Riemann problems.) Since the initial data are not smooth, a solution of the Riemann problem for $t > 0$ may fail to be classical ab initio. We now show how to construct solutions of Riemann problems.

By using (1.10), we easily see that

$$(2.7)\qquad \big(u(s,t), v(s,t)\big) = \begin{cases} (u_l, v_l) & \text{for } s \le \lambda_1(u_l)t, \\ (u_r, v_r) & \text{for } s \ge \lambda_2(u_r)t \end{cases}$$

is a solution of the Riemann problem on the indicated region of (s,t)-space. But it is not immediately obvious how to use the characteristics to construct a solution in the rest of the half space $t > 0$. For this purpose, we first seek shock solutions of (1.5), (2.7) of the form

$$(2.8)\qquad \big(u(s,t), v(s,t)\big) = \begin{cases} (u_l, v_l) & \text{for } s \le \sigma' t, \\ (u_r, v_r) & \text{for } s \ge \sigma' t \end{cases}$$

where σ' is a constant. We suppose that (u_l, v_l) is given. We now determine all the states (u_r, v_r) and shock speeds σ' for which (2.8) represents a back shock, i.e., a weak solution of (1.15) satisfying (2.3a). From (2.3a) and (1.6) we obtain $\hat{N}_y(u_r) > \hat{N}_y(u_l)$. The concavity of $\hat{N}$ ensured by (1.12) then implies that $u_r < u_l$. Since σ' is negative by (2.3a), it follows from (2.1a) that $v_r < v_l$. Thus we conclude from (2.2) that the set of all states (u_r, v_r) for which (2.8) represents a back shock must lie on the curve

$$(2.9a)\quad \mathcal{S}_1(u_l, u_r): \quad \sqrt{\rho A}(v - v_l) = -\sqrt{[\hat{N}(u) - \hat{N}(u_l)](u - u_l)}, \quad u < u_l,$$

in (u, v)-space, which is illustrated in Fig. 2.10. For any (u_r, v_r) on this curve, we find the corresponding σ' from (2.1). Likewise, we find that the set of all states (u_r, v_r) for which (2.8) represents a front shock must lie on the curve

$$(2.9b)\quad \mathcal{S}_2(u_l, u_r): \quad \sqrt{\rho A}(v - v_l) = -\sqrt{[\hat{N}(u) - \hat{N}(u_l)](u - u_l)}, \quad u > u_l,$$

in (u, v)-space, which is also illustrated in Fig. 2.10.

We have just constructed special solutions for special Riemann problems. For example, if (u_l, v_l) is given and if (u_r, v_r) lies on the curve (2.9b), then (2.8) with σ' given by (2.1) satisfies the Riemann problem. This solution is illustrated in Fig. 2.11.

2.12. Exercise. Show that (1.12) implies that the curve (2.9a) intersects every straight line through (u_l, v_l) at most once.

A *rarefaction wave* is a solution of (1.5) of the form

$$(2.13)\qquad (u(s,t), v(s,t)) = (U(s/t), V(s/t)).$$

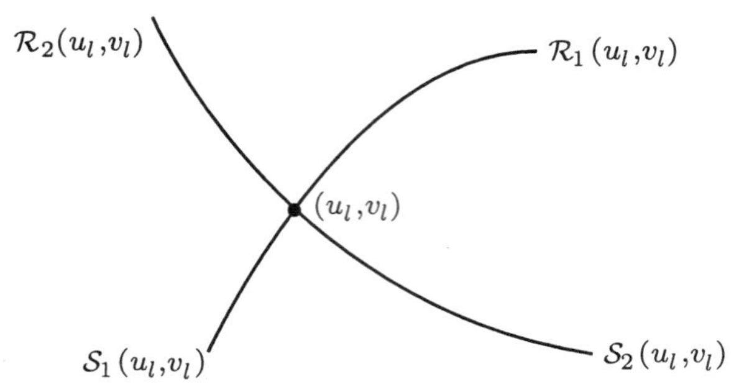

Fig. 2.10. Graphs of the shock and rarefaction curves $\mathcal{S}_1(u_l, u_r)$, $\mathcal{S}_2(u_l, u_r)$, $\mathcal{R}_1(u_l, u_r)$, $\mathcal{R}_2(u_l, u_r)$.

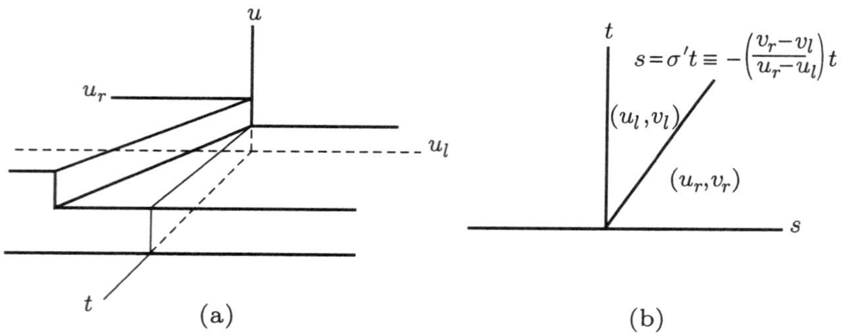

Fig. 2.11. (a) Graph of the 2-shock solution (2.8) of the Riemann problem (1.5), (2.6). (b) Level curves of (a). This is the traditional mode of displaying the shock in(a).

Let $\xi = s/t$. The substitution of this expression into (1.5) yields the system of ordinary differential equations

$$\begin{pmatrix} \xi & 1 \\ \hat{N}_y(U) & \rho A \xi \end{pmatrix} \begin{pmatrix} U' \\ V' \end{pmatrix} = \begin{pmatrix} 0 \\ 0 \end{pmatrix}. \tag{2.14}$$

If (2.14) has nontrivial solutions, which we seek, then $-\xi$ must be an eigenvalue of $\mathbf{A}$ relative to $\mathbf{J}$, so that $-\xi = \pm\lambda(U)$, and (U', V') must be a corresponding eigenvector. Let us first take

$$\xi = -\lambda(U(\xi)), \tag{2.15}$$

so that the corresponding eigenvector (U', V') satisfies

$$V'(\xi) = \lambda(U(\xi))U'(\xi). \tag{2.16}$$

Now (1.12) implies that (2.15) can be uniquely solved for $U(\xi)$ and that U' is everywhere positive. Thus $\xi \mapsto \lambda_1(U(\xi)) \equiv -\lambda(U(\xi))$ is increasing, in consonance with (2.15). We take $\lambda_1(u_l) < \xi < \lambda_1(u_r)$. Since we can solve (2.15) for $U(\xi)$, it follows that (2.14) is equivalent to an equation of the form $d\tilde{V}/dU = -\lambda(U)$, so that $\tilde{V}(U) - v_l = \int_{u_l}^{U} \lambda(y)\,dy$ gives the rarefaction wave $(U, \tilde{V}(U))$ to the right of the ray $s = \lambda_1(u_l)t \equiv -\lambda(u_1)t$. In particular, any solution value (u_r, v_r) of a rarefaction wave lying on the ray $s = \lambda_1(u_r)t$ must lie on the 1-*rarefaction curve*

$$\mathcal{R}_1(u_l, v_l): \quad v - v_l = \int_{u_l}^{u} \lambda(y)\,dy \tag{2.17a}$$

in (u, v)-space, which is illustrated in Fig. 2.10. Likewise, for the choice $\xi = \lambda U(\xi)$, we take $\lambda_2(u_l) < \xi < \lambda_2(u_r)$. Any solution value (u_r, v_r) lying on the ray $s = \lambda_2(u_r)t$ must lie on the 2-*rarefaction curve*

$$\mathcal{R}_2(u_l, v_l): \quad v - v_l = -\int_{u_l}^{u} \lambda(y)\,dy \tag{2.17b}$$

in (u, v)-space, which is illustrated in Fig. 2.10.

2.18. Exercise. Show that the curves $\mathcal{S}_1(u_l, v_l)$ and $\mathcal{R}_1(u_l, v_l)$ osculate at (u_l, v_l). (The same is true of the curves indexed by 2.)

We have just constructed solutions for another class of special Riemann problems. For example, if (u_l, v_l) is given and if (u_r, v_r) lies on the curve (2.17b), then

$$(u(s,t), v(s,t)) = \begin{cases} (u_l, v_l) & \text{for } s \le \lambda_2(u_l)t, \\ (U(\tfrac{s}{t}), V(\tfrac{s}{t})) & \text{for } \lambda_2(u_l)t \le s \le \lambda_2(u_r)t, \\ (u_r, v_r) & \text{for } s \ge \lambda_2(u_r)t \end{cases} \tag{2.19}$$

with $\big(U(\tfrac{s}{t}), V(\tfrac{s}{t})\big)$ lying on $\mathcal{R}_2(u_l, v_l)$ satisfies the Riemann problem. It constitutes a weak solution of (1.5), (2.6) because it is a classical solution on each of the open triangles defining each case in (2.19) and because it is continuous, so that it automatically satisfies the Rankine-Hugoniot conditions (2.1). This solution is illustrated in Fig. 2.20.

Now we show how to construct a solution to the Riemann problem for arbitrary data.

2.21. Exercise. Prove

2.22. Proposition. *Let* (1.12) *and* (2.5) *hold. Let* (u_l, v_l) *be fixed. To each* (u_r, v_r) *in the* (u, v)*-plane there corresponds a unique point* $(u_i, v_i) \in \mathcal{S}_1(u_l, v_l) \cup \mathcal{R}_1(u_l, v_l)$ *such that* $(u_r, v_r) \in \mathcal{S}_2(u_i, v_i) \cup \mathcal{R}_2(u_i, v_i)$.

The recipe for constructing solutions is straightforward: Given the Riemann data (2.6), construct (u_i, v_i) as in Proposition 2.21. The solution

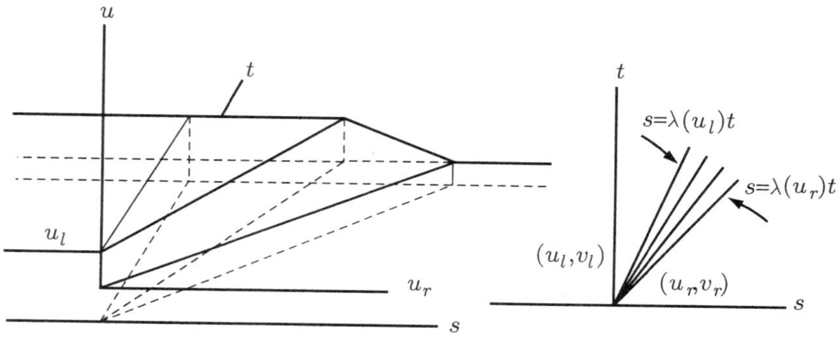

Fig. 2.20. (a) Graph of the solution of the Riemann problem (1.5), (2.6) when (u_r, v_r) lies on $\mathcal{R}_2(u_l, v_l)$. (b) Level curves of Fig. 2.20a. This is the traditional mode of displaying critical information about the rarefaction wave of (a).

corresponds to the trajectory going from (u_l,v_l) to (u_i,v_i) along $\mathcal{S}_1(u_l,v_l) \cup \mathcal{R}_1(u_l,v_l)$ and then going from (u_i,v_i) to (u_r,v_r) along $\mathcal{S}_2(u_i,v_i) \cup \mathcal{R}_2(u_i,v_i)$. For example, if (u_r, v_r) is in the region bounded by $\mathcal{S}_1(u_l, v_l)$ and $\mathcal{R}_2(u_l, v_l)$, as in Fig. 2.23a, then the first segment of this trajectory corresponds to a back shock with $\sigma' = -(v_i - v_l)/(u_i - u_l)$ (as a consequence of (2.1)), the point (u_i, v_i) corresponds to a constant state over the region $\sigma' t < s < \lambda_2(u_l)t$, and the third segment corresponds to the 2-rarefaction wave over the region $\lambda_2(u_l)t < s < \lambda_2(u_r)t$. The level curves of this solution are shown in Fig. 2.23b. It can be shown that all such solutions are unique in a properly defined sense.

2.24. Exercise. Verify that this construction generates weak solutions of the Riemann problem.

2.25. Exercise. Suppose that the Lax entropy conditions are not imposed. Prove that there are Riemann problems not having unique solutions.

Except for the use of the Riemann invariants, many of the techniques and results just described in these two sections carry over to quasilinear hyperbolic systems of any order, e.g., those that describe the motion of an elastic rod in space (see Chap. VIII). We encounter difficulties, however, if we suspend (1.12), in which case the efficacy of (2.3) may be reduced. Since we have no reason to expect that physical processes ostensibly well described by equations such as (1.5) admit multiple states, since the Rankine-Hugoniot conditions fail to ensure uniqueness of weak solutions,

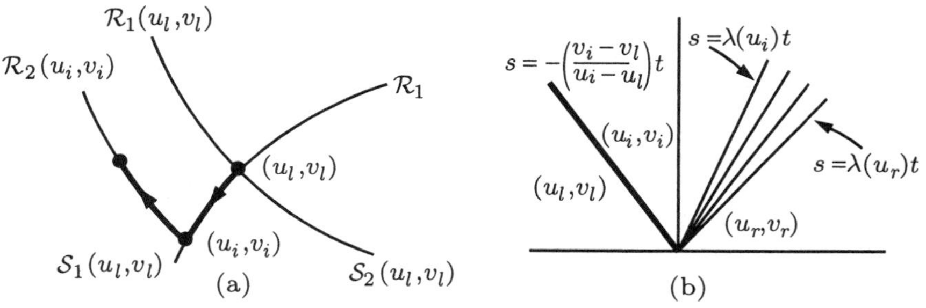

Fig. 2.23. (a) Shock and rarefaction curves for the solution of the Riemann problem when (u_r, v_r) is in the region bounded by $\mathcal{S}_1(u_l, v_l)$ and $\mathcal{R}_2(u_l, v_l)$. (b) Level curves of the solution defined by (a).

as is illustrated by Ex. 2.25, and since (2.3) may be inadequate for this purpose, we should like to refine our description of the underlying physics to find more effective replacements for (2.3) that ensure uniqueness. Such replacements, called *admissibility* or *entropy* conditions, are of three kinds:

(i) Purely analytic restrictions that ensure that initial-value problems for (genuinely nonlinear) systems have unique weak solutions. This approach as developed and refined by Oleĭnik (1957), Lax (1957), and Liu (1975) has not been created in a vacuum, but has been designed to be consistent with gas dynamics.

(ii) Thermodynamic restrictions, e.g., the use of jump conditions coming from the integral form of the entropy inequality (XII.14.18). It should be noted that in a Newtonian (linearly viscous) fluid with a Fourier heat conduction law, the Clausius-Duhem inequality is exactly equivalent to the standard mathematical and physical assumptions imposed on the dissipative terms. On the other hand, the consequences of the Clausius-Duhem inequality for nonlinear viscoelasticity are weaker than the physically natural and standard assumption that the dependence of stress on strain rate gives the equations a parabolic character.

(iii) Restrictions obtained from the analysis of singular perturbations of systems obtained from the hyperbolic system by appending 'strong' dissipative terms depending on a small parameter. The dissipative term traditionally regarded as desirable from the viewpoint of mechanics has been in the form of the viscous dissipation arising for Newtonian fluids; a somewhat stronger mechanism has been introduced for mathematical studies and is implicit in numerical studies. Except for very special cases, there is no full mathematical justification of the asymptotics underlying the singular perturbation.

These remarks indicate that the theory of admissibility conditions is well understood for equations of gas dynamics, although important analytic questions remain open. Here the various approaches (i)–(iii) are essentially equivalent and are consistent with the underlying fluid mechanics. For solid mechanics, however, the situation is far less clear: There is no extant version of (i) that can regarded as having an obvious and valid physical meaning. The admissibility condition coming from the Clausius-Duhem inequality, as described in (ii), has proved to be inadequate, for many applications. Accordingly, it has been proposed to supplement it with other conditions, such as Dafermos's (1973) entropy-rate admissibility condition. Some admissibility conditions of the form (iii) allow an unwelcome nonuniqueness, as we show in Sec. 5. Most studies of dissipative mechanisms for solids have adopted dissipative forces linear in the strain rate, just as for gas dynamics.

In view of these remarks, we are led to study those admissibility conditions for solids that are obtained as jump conditions that remain in the formal nonuniform limit as the strength of a physically natural family of dissipative mechanisms is reduced to zero. (We illustrate the basic concepts in Sec. 4.) The actual admissibility conditions therefore depend on the dissipative mechanism. Although the full mathematical justification for these limit processes is still lacking, we are able to analyze them in the important special case of travelling waves. In the next section we describe a family of viscous dissipative mechanisms for (1.5). In Sec. 4 we study admissibility conditions obtained by method (iii).

Quasilinear hyperbolic equations are presently undergoing an intensive development, parts of which employ technical methods of modern analysis, not yet described in research monographs. There is no definitive text on modern developments. An accessible modern introduction to the subject is given by Smoller (1983). Other introductions, strongly influenced by gas dynamics, are Courant & Friedrichs (1948), Jeffrey (1976), Lax (1973), Rozhdestvenskiĭ & Yanenko (1968), von Mises (1958), and Whitham (1974). That of Hanyga (1985) is applicable to nonlinear elasticity. For an illuminating discussion of admissibility conditions, see Dafermos (1984). For a definitive discussion of singular surfaces, see Wang & Truesdell (1973), Truesdell (1961), Truesdell & Toupin (1960, Chap. C), and the classical text of Hadamard (1903). A detailed treatment of the nonlinear equations for the motion of an elastic string is given by Shearer (1985a,b, 1986).

3. Dissipative Mechanisms for Longitudinal Motion. Preclusion of Total Compression

In this section we study an initial-boundary-value problem for an analog of (1.1) that governs the longitudinal motion of a viscoelastic rod:

$$(\rho A)(s)w_{tt} = N_s \tag{3.1}$$

with

$$N(s,t) = \hat{N}(w_s(s,t), w_{st}(s,t), s). \tag{3.2a}$$

Here $(0,\infty) \times \mathbb{R} \times [0,1] \ni (y,z,x) \mapsto \hat{N}(y,z,x) \equiv \phi_y(y,x) + \hat{g}(y,z,x) \in \mathbb{R}$ is a continuously differentiable constitutive function with

$$\phi_y(x,y) \equiv \hat{N}(y,0,x), \quad \hat{g}(y,0,x) = 0, \quad \hat{N}(1,0,x) = \phi_y(1,x) = 0. \tag{3.2b}$$

ϕ is the stored-energy function for the elastic response.

We shall seek solutions of the governing equations that never suffer a total compression, i.e., solutions for which the stretch $u = w_s$ is everywhere positive. Under suitable restrictions on the constitutive functions and on the data, we shall show that if w_s is initially positive, then it is always positive.

The analysis of (3.1), (3.2) was initiated by Greenberg, MacCamy, & Mizel (1967), who studied functions $\hat{N}$ affine in their last arguments:

$$\hat{N}(y, z, x) = f(y) + mz. \tag{3.3}$$

Here m is a positive constant. Major advances in the theory for this equation were made by Andrews (1980), Andrews & Ball (1982), and Pego (1987). Kanel' (1969) and MacCamy (1970) independently developed the theory for problems in which there is a positive number m such that

$$\hat{N}(y, z, x) = f(y) + g(y)z \quad \text{with } g \geq m. \tag{3.4}$$

Andrews (1979) (see Andrews & Ball (1982)) developed a very simple device for proving that w_s is everywhere positive for solutions of (3.1) when (3.4) holds.

The material formulation of the one-dimensional Navier-Stokes equations for compressible fluids has the form (3.4) with $g(y) = \mu(y)/y$, where μ is the viscosity, allowed to depend on the specific volume y, which is the reciprocal of the actual density. Now the three-dimensional Navier-Stokes equations for fluids can be generalized by adopting a constitutive equation for non-Newtonian fluids in place of those that are affine in the velocity gradient. It happens that the naive replacement of the affine dependence on the velocity gradient with a nonlinear dependence yields a theory with properties that do not reflect those of a single known real fluid. Consequently, generalizations of (3.4) in which $\hat{N}$ is allowed to be nonlinear in y, which are the object of our study, are apparently irrelevant for the dynamics of viscous gases. On the other hand, the form of (3.4) is not ideally suited for solids. For example, on physical grounds one might wish to postulate that $\hat{N}_y$ be positive everywhere, or at least that $\hat{N}_y$ be positive in certain large regions. But if g is not constant, then there are regions of (y, z)-space where $\hat{N}_y(y, z, x)$ is not positive. In particular, if $g(y) = 1/y$ or, more generally, if g is a strictly decreasing function of y, then for any $y > 0$, there is a positive z such that $\hat{N}_y(y, z, x) < 0$, a result that is counterintuitive.

Dafermos (1969) studied the fully nonlinear problem (3.1), (3.2) for functions $(y, z) \mapsto \hat{N}(y, z)$ having the property that there are positive numbers m and M such that

$$\hat{N}_z \geq m, \quad (\hat{N}_y)^2 \leq M\hat{N}_z. \tag{3.5}$$

This assumption (3.5) is perfectly suitable for the description of shearing motions of solids, but, as is demonstrated by Antman & Malek-Madani (1987), it is incompatible with physically natural restrictions on the longitudinal motions of solids.

While equations modelled on (3.4) are eminently reasonable for problems describing compressible Navier-Stokes fluids, there seems to be no physical reason to restrict the description of solids to (3.4). Thus it is desirable to generalize (3.4), but in doing so we cannot use (3.5) for longitudinal motions. Since the analyses of the papers just described depend crucially on the specific forms of (3.3)–(3.5), we must construct effective alternatives to the procedures of these papers.

It is again convenient to use (1.4) to recast (3.1), (3.2) as a system:

$$u_t = v_s, \tag{3.6a}$$

$$\rho A v_t = \hat{N}(u, v_s, s)_s. \tag{3.6b}$$

We assume that the rod has reference length 1 and that $s \in [0,1]$. We assume that the end $s = 0$ of the rod is fixed and that a small body of mass μ is attached to the end $s = 1$. Thus

$$(3.7\text{a,b}) \qquad v(0,t) = 0, \quad \mu v_t(0,t) = -\hat{N}(u(1,t), v_s(1,t), 1).$$

We impose the initial conditions

$$(3.8) \qquad u(s,0) = \bar{u}(s), \quad v(s,0) = \bar{v}(s) \quad \forall s \in [0,1]$$

where $\bar{u}$ and $\bar{v}$ are prescribed functions with $\bar{v}(0) = 0$. The classical form of our initial-boundary-value problem is (3.6)–(3.8).

We assume that an increase in the contact force must accompany an increase in the rate of stretch $v_s = w_{st}$ by requiring that there be a positive number m such that

$$(3.9) \qquad \hat{N}_z \equiv \hat{g}_z \geq m.$$

This condition, generalizing (3.4) and (3.5), ensures that the mechanical process described by (3.6) is dissipative in a uniform way and that for a fixed function u, (3.6b) is parabolic. We impose a further restriction, inspired partly by (3.4) and the work of Andrews (1979) and Antman (1988b), that says that viscous effects become more pronounced as the stretch w_s becomes small:

3.10. Hypothesis. *There are numbers $y_* \in (0,1)$, $M \geq 0$, and $A \geq 0$, and there is a continuously differentiable function ψ on $(0, y_*)$ with*

$$(3.11\text{a}) \qquad \psi(y) \to \infty \quad \text{as } y \to 0, \quad \psi \geq 0$$

such that

$$(3.11\text{b}) \qquad \hat{N}(y,z,x) \leq -\psi'(y)z + M\psi(y) + A$$

$$(3.11\text{c}) \qquad \text{for } y \leq y_* \text{ and for } z \in \mathbb{R}.$$

It is clear that this hypothesis prohibits a superlinear dependence of $\hat{N}$ on z for large z. This restriction can be removed at the expense of a more complicated analysis (see Antman & Seidman (1995)). Note that there is no loss of generality in taking $A = 0$ when $M > 0$, because we can always take ψ to have a positive lower bound. Since $\phi_y(y) < 0$ for $y < y_*$, condition (3.11b) is much milder than that obtained by replacing $\hat{N}$ with $\hat{g}$. (Kanel' (1969), MacCamy (1970), and Andrews (1979) each take $\hat{g}(y,z) = -\psi'(y)z$ for all y.)

Since we shall impose strong conditions on the viscous response, we can get by with restrictions on the elastic response, characterized by ϕ, weaker than usual. All we require is that $y \mapsto \phi(y,x)$ have a minimum 0 at $y = 1$.

We now obtain the energy equation. We define the *kinetic energy* K, the *strain energy* Φ, and the *dissipative stress work* W by

$$
\begin{aligned}
K(t) &\equiv \tfrac{1}{2}\int_0^1 (\rho A)(s)v(s,t)^2\,ds + \tfrac{1}{2}\mu v(1,t)^2, \\
\Phi(t) &\equiv \int_0^1 \phi\big(s,u(s,t)\big)\,ds, \qquad (3.12)\\
W(t) &\equiv \int_0^t\int_0^1 \hat{g}\big(s,u(s,\tau),v_s(s,\tau)\big)v_s(s,\tau)\,ds\,d\tau.
\end{aligned}
$$

Let us integrate the product of (3.6b) and v by parts over $[0,1]\times[0,t)$ and use (3.2) to obtain the *energy equation*

$$K(t)+\Phi(t)+W(t)=K(0)+\Phi(0). \tag{3.13}$$

Note that each term on the left-hand side of (3.13) is nonnegative.

If we assume that

$$\phi(y,x)\to\infty \quad \text{as } y\to 0, \tag{3.14}$$

then for each t, (3.13) implies that $u(\cdot,t)$ can vanish only on a set of measure 0. We shall use Hypothesis 3.10 to obtain a stronger result. (For *static* one-dimensional problems, (3.14) itself supports a delicate analysis showing that u can vanish nowhere; see Chap. VII.)

We now show that u has a positive lower bound. The existence theory of Antman & Seidman (1995) shows that if the data are sufficiently regular, then so is the solution as long as it exists. We accordingly assume that u is continuous. We suppose that $\min_s \bar{u}(s)>0$. Then without loss of generality we may choose the number y_* introduced in Hypothesis 3.10 so that $y_*\le 1$ and

$$y_*\le \min_s \bar{u}(s). \tag{3.15}$$

Of course, to prove that $u(s,t)$ is positive for all s, t, it suffices to show that $u(s,t)$ is positive only for those (s,t) for which $u(s,t)<y_*$. Thus suppose that there is a point (ξ,ω) such that $u(\xi,\omega)<y_*$. If $\xi=1$, then the continuity of u enables us to show that there is a $\xi<1$ with the same property. Since u is continuous, there is a $\vartheta\in(0,\omega)$ such that $u(\xi,\vartheta)=y_*$ and $u(\xi,t)<y_*$ for $\vartheta<t\le\omega$. We integrate (3.6b) over $[\xi,1]\times[\vartheta,t]$ and use (3.6a) and (3.11b) to obtain

$$
\begin{aligned}
\int_\xi^1 (\rho A)(s)v(s,\tau)\,ds\Big|_{\tau=\vartheta}^{\tau=t} &= \int_\vartheta^t \hat{N}\big(u(s,\tau),v_s(s,\tau),s\big)\,d\tau\Big|_{s=\xi}^{s=1}\\
&\le -\mu\int_\vartheta^t v_t(1,\tau)\,d\tau - \int_\vartheta^t \psi'(u(\xi,\tau))u_t(\xi,\tau)\,d\tau\\
&\quad + M\int_\vartheta^t \psi(u(\xi,\tau))\,d\tau + A(t-\vartheta) \qquad (3.16)\\
&\le \mu v(1,\vartheta)-\mu v(1,t)+\psi\big(u(\xi,\vartheta)\big)-\psi\big(u(\xi,t)\big)\\
&\quad + M\int_\vartheta^t \psi(u(\xi,\tau))\,d\tau + A(t-\vartheta)
\end{aligned}
$$

for all $t \in [\vartheta, \omega]$. From the Cauchy-Bunyakovskiĭ-Schwarz inequality and the energy equation (3.13), we find that

(3.17)
$$\left|\int_\xi^1 (\rho A)(s)v(s,t)\,ds\right| \le \sqrt{\int_0^\xi (\rho A)(s)\,ds}\sqrt{\int_0^\xi (\rho A)(s)v(s,t)^2\,ds} \le \text{const.}$$

for all $t \in [\vartheta, \omega]$. The energy equation also implies that $|v(1,t)| \le \text{const.}$ Thus (3.16) and (3.17) yield

$$\psi(u(\xi,t)) \le \psi(y_*) + M\int_\vartheta^t \psi\big(u(\xi,\tau)\big)\,d\tau + A(t-\vartheta) + \text{const.} \tag{3.18}$$

for all $t \in [\vartheta, \omega]$. It follows from the Gronwall inequality (see Ex. 3.20 below) that there is a positive number $C(\omega)$ such that

$$\psi\big(u(\xi,\omega)\big) \le C(\omega). \tag{3.19a}$$

Thus (3.11a) implies that there exists a positive-valued continuous function c such that

$$u(\xi,\omega) \ge c(\omega). \tag{3.19b}$$

(The continuity of u ensures that (3.18) and (3.19a,b) hold for $\xi = 1$.) Thus we obtain the pointwise bound

$$u(s,t) \ge c(t) \quad \forall\ (s,t). \tag{3.19c}$$

3.20. Exercise. Suppose that f, g, h are continuous and nowhere negative on $[0,\infty)$ and that y is a nonnegative-valued function satisfying

$$y(t) \le f(t) + g(t)\int_0^t h(\tau)y(\tau)\,d\tau \quad \text{for } t \ge 0. \tag{3.21a}$$

Let $Y(t) \equiv \int_0^t h(\tau)y(\tau)\,d\tau$ so that

$$Y'(t) \le h(t)[f(t) + g(t)Y(t)] \tag{3.21b}$$

or, equivalently,

$$\frac{d}{dt}\left[\exp\left(-\int_0^t hg\,d\tau\right)Y\right] \le \exp\left(-\int_0^t hg\,d\tau\right)hf. \tag{3.21c}$$

Obtain an estimate for Y and then for y. This is the *Gronwall inequality.*

3.22. Exercise. Let a given function $(s,t) \mapsto f(s,t)$ be added to the right-hand side of (3.6b). Suppose that f is square-integrable on every rectangle of the form $[0,1]\times[0,T]$. Use the inequality $2|ab| \le \varepsilon^{-1}a^2 + \varepsilon b^2$ for $\varepsilon > 0$ to deduce (3.19c).

3.23. Exercise. Let (3.7) be replaced with

$$v(0,t) = v_0(t), \quad \hat{N}(u(1,t), v_s(1,t), 1) = n_1(t). \tag{3.24}$$

Impose mild restrictions on the given functions v_0 and n_1, and deduce (3.19c). Repeat this process when (3.7) is replaced with

$$\hat{N}(u(0,t), v_s(0,t), 0) = n_0(t), \quad \hat{N}(u(1,t), v_s(1,t), 1) = n_1(t). \tag{3.25}$$

The problem with time-dependent position boundary conditions

$$v(0,t) = v_0(t), \quad v(1,t) = v_1(t) \tag{3.26}$$

is much more difficult: In the energy equation (3.13) we now get $\int_0^t n(s,\tau)v(s,\tau)d\tau\Big|_{s=0}^{s=1}$ on the right-hand side. Note that the constitutive equations indicate that this term has a contribution from ϕ_y and that the left-hand side of the energy equation contains an integral of ϕ. Since for any realistic material $\phi_y(y,x)$ is much more badly behaved than $\phi(y,x)$ itself for y small, the resulting form of the energy equation merely indicates that something small is dominated by something large, a fact both obvious and useless. We only get useful energy estimates when ϕ_y and $\hat{g}$ meet further conditions, which are described by Antman & Seidman (1995). This is not surprising: If the material strongly resists having its length changed, then changing its length against a large resistance might cause the generation of so much work that the right-hand side of the energy equation cannot be effectively controlled.

The following simple example is illuminating. Suppose that $f = 0$, $\hat{N}_x = 0$, and ρA is constant. Then

$$u(s,t) = 1 - t, \quad v(s,t) = -s \tag{3.27}$$

defines a solution of (3.6) satisfying the initial conditions $u(s,0) = 1$, $v(s,0) = -s$, and boundary conditions $v(0,t) = 0$, $v(1,t) = -1$. This solution corresponds to a total compression at time $t = 1$. Note that, as a function of t, the second boundary condition is ostensibly innocuous. Thus we cannot expect to get a useful energy inequality without further restrictions. For this example, we could replace the second boundary condition with the 'equivalent' condition that $n(1,t) = \hat{N}(1-t, -1, 1)$, which may be expected to become infinite as $t \nearrow 1$. Clearly, this boundary condition is not innocuous.

The results of this section are adapted from the work of Antman & Seidman (1995), by kind permission of Academic Press. They show how to weaken Hypothesis 3.10 and they furnish a treatment of the very difficult (3.26). The availability of a positive lower bound for u enables the existence and regularity of solutions to be deduced, e.g., as a simple consequence of a modification of the theory of Dafermos (1969), whose hypotheses would otherwise fail to be physically reasonable.

4. Shock Structure. Admissibility and Travelling Waves

We now give a formal treatment of admissibility conditions obtained by taking the asymptotic limit of the governing equations as the dissipative terms go to zero. This is the approach (iii) described near the end of Sec. 2. To be specific, we limit our attention to system (3.6) when $\hat{N}$ has the form

$$\hat{N}(y,z,x) = \check{N}(y, \varepsilon z, x, \varepsilon) \tag{4.1}$$

where ε is a small positive parameter. We assume that $\check{N}$ is a well-behaved function of its last argument. Thus we are suspending (3.9). When we insert (4.1) into (3.6), we get

$$u_t = v_s, \quad \rho A v_t = \check{N}(u, \varepsilon v_s, x, \varepsilon)_s. \tag{4.2a,b}$$

A system, like (4.2), in which a small parameter multiplies the highest-order derivatives is said to be *singularly perturbed*. We have encountered such a problem in Sec. VI.9, in which $1/\lambda$ plays the role of the small parameter. For small values of the parameter, the solutions, like those of Sec. VI.9, may possess *transition layers*, i.e., small regions in which the solution undergoes a large change. We expect to encounter transition layers in our study of (4.2), because we know that the hyperbolic equation obtained from (4.2) by setting $\varepsilon = 0$ has solutions with shocks, i.e., with discontinuities, at which the solutions certainly undergo large changes in small regions.

To motivate our treatment of (4.2), we examine a completely elementary, but illuminating, boundary-value problem for a linear ordinary differential equation:

$$(4.3a,b,c) \qquad \varepsilon \frac{d^2u}{ds^2} + \frac{du}{ds} = 0 \quad \text{on } (0,1), \quad u(0) = 0, \quad u(1) = 1$$

where ε is a small positive number. This problem has a transition layer at the boundary $s = 0$, i.e., a *boundary layer*. If we formally set $\varepsilon = 0$ in (4.3), then we find that $u = \text{const.}$, a function that cannot satisfy both boundary conditions of (4.3). On the other hand, we can immediately obtain an explicit solution of (4.3):

$$(4.4) \qquad u(s, \varepsilon) = \frac{1 - e^{-s/\varepsilon}}{1 - e^{-1/\varepsilon}},$$

which has the form shown in Fig. 4.5 and which clearly exhibits boundary-layer behavior near $s = 0$.

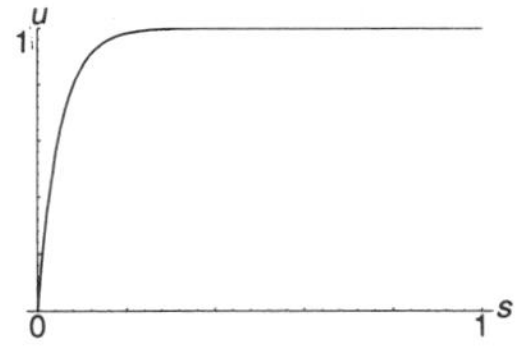

Fig. 4.5. Graph of (4.4) for ε =0.05.

In general, however, we cannot find explicit solutions to singular perturbation problems. Effective ways to approximate solutions of generalizations of (4.3) exploit the observation that (4.3) and (4.4) are each simplified by the introduction of a scaled length variable $\xi = s/\varepsilon$. We now illustrate one such general method by applying it to (4.3).

We set $\tilde{u}(\xi, \varepsilon) \equiv u(\varepsilon\xi, \varepsilon)$. Then $\tilde{u}(\cdot, \varepsilon)$ satisfies

$$(4.6a,b,c) \qquad \frac{d^2\tilde{u}}{d\xi^2} + \frac{d\tilde{u}}{d\xi} = 0 \quad \text{on } (0, 1/\varepsilon), \quad \tilde{u}(0) = 0, \quad \tilde{u}(1/\varepsilon) = 1.$$

We seek an asymptotic representation for the solution of (4.3) valid away from $s = 0$ in the form

$$u(s, \varepsilon) = \sum_{k=0}^{K} u_k(s)\varepsilon^k + o(\varepsilon^K) \tag{4.7}$$

for any nonnegative integer K, and we seek an asymptotic representation solution of (4.3) valid (in the boundary layer) near $s = 0$ in the form

$$\tilde{u}(\xi, \varepsilon) = \sum_{k=0}^{K} \tilde{u}_k(\xi)\varepsilon^k + o(\varepsilon^K) \tag{4.8}$$

for any nonnegative integer K. The coefficient functions u_k and $\tilde{u}_k$ are assumed smooth.

Let us substitute (4.7) into (4.3a). We find that the leading term of (4.7) that satisfies (4.3c) is given by $u_0(s) = 1$. Let us substitute (4.8) into (4.6a). We find that the leading term of (4.8) is given by $\tilde{u}_0(\xi) = A + Be^{-\xi}$, where A and B are constants. The boundary condition (4.6b) forces $A + B = 0$. We assume that the representations (4.7) and (4.8) are valid in some common interval $\mathcal{I}(\varepsilon)$ of the s-axis. Taking $K = 0$ in (4.7) and (4.8), we thus find that

$$u_0(\varepsilon\xi) = \tilde{u}_0(\xi) + o(1) \tag{4.9}$$

when $\varepsilon\xi$ is in this interval. We finally assume that there are points $s^\sharp(\varepsilon) \in \mathcal{I}(\varepsilon)$ such that $s^\sharp(\varepsilon) \to 0$ and $\xi = s^\sharp(\varepsilon)/\varepsilon \to \infty$ as $\varepsilon \to 0$ (as would happen, e.g., if $s^\sharp(\varepsilon) = \sqrt{\varepsilon}$ and $\mathcal{I} = (\frac{1}{2}\sqrt{\varepsilon}, 2\sqrt{\varepsilon})$). Then (4.9) yields the *matching condition*

$$u_0(0) = \tilde{u}_0(\infty), \tag{4.10}$$

from which we deduce that $\tilde{u}_0(\xi) = 1 - e^{-\xi}$. Thus u_0 and $\tilde{u}_0$ give excellent approximations to (4.4) in their respective regions of validity. Higher-order terms, which are trivial for (4.3), are readily constructed for more challenging problems. For ordinary differential equations, this method and others like it can be justified (see O'Malley (1974), Smith (1985), Vasil'eva & Butuzov (1973), and Wasow (1965), among others).

Let us now study (4.2) by a similar approach. We expect that solutions of (4.2) are well approximated by solutions of the hyperbolic system obtained by setting $\varepsilon = 0$ wherever v_s is bounded. Since the hyperbolic system can have shocks in which v_s becomes infinite, we expect that (4.2) has a transition layer near a shock path $s = \sigma(t)$ of the reduced hyperbolic system. We are accordingly led to introduce the scaled independent variable $\xi = [s - \sigma(t)]/\varepsilon$. For any function such as $(s, t) \mapsto u(s, t)$, we define $\tilde{u}(\xi, t) \equiv u(\sigma(t) + \varepsilon\xi, t)$. Then (4.2) is equivalent to

$$\begin{gathered} \varepsilon\tilde{u}_t - \sigma'(t)\tilde{u}_\xi = \tilde{v}_\xi, \\ \rho A(\sigma(t) + \varepsilon\xi)[\varepsilon\tilde{v}_t - \sigma'(t)\tilde{u}_\xi] = \check{N}(\tilde{u}, \tilde{v}_\xi, \sigma(t) + \varepsilon\xi, \varepsilon)_\xi. \end{gathered} \tag{4.11}$$

Now we construct representations for u and v like (4.7) and (4.8), obtaining one ostensibly valid away from $s = \sigma(t)$ and one valid near it. The leading term of the latter is governed by the equations obtained from (4.11) by formally setting $\varepsilon = 0$:

$$\tag{4.12} \begin{aligned} -\sigma'(t)\tilde{u}_\xi &= \tilde{v}_\xi, \\ -\rho A(\sigma(t))\sigma'(t)\tilde{u}_\xi &= \check{N}(\tilde{u}, \tilde{v}_\xi, \sigma(t), 0)_\xi. \end{aligned}$$

We likewise obtain the matching conditions

$$\tag{4.13} \tilde{u}(\pm\infty, t) = \lim_{s \to \sigma(t)\pm} u_0(s,t), \quad \tilde{v}(\pm\infty, t) = \lim_{s \to \sigma(t)\pm} v_0(s,t)$$

where (u_0, v_0) is the leading term of the expansion away from the shock, i.e., (u_0, v_0) satisfies (4.2) with $\varepsilon = 0$.

Note that (4.12) is a system of autonomous ordinary differential equations parametrized by t. Now *travelling-wave solutions* of a spatially autonomous version of (4.2), which have the form $u(s,t) = \tilde{u}(s - ct)$, $v(s,t) = \tilde{v}(s - ct)$, satisfy (4.12) with the shock speed $\sigma'(t)$ replaced by the speed c:

$$\tag{4.14} -c\tilde{u}' = \tilde{v}', \quad -\rho A c \tilde{u}' = \check{N}(\tilde{u}, \tilde{v}_\xi, 0)',$$

the prime denoting the derivative with respect to ξ. Thus travelling-wave solutions of (4.2) describe in the stretched variable ξ the large changes that are described by shocks in the less refined theory governed by the reduced hyperbolic equations. The travelling-wave solutions (infelicitously called *viscous shocks*) are therefore said to determine the *shock structure.*

We identify the states (u_l, v_l) and (u_r, v_r) to the left and right of a shock with the right-hand sides of (4.13). In view of the foregoing considerations, we can say that the shock determined by these states is *admissible* according to the travelling-wave criterion if there is a solution of the travelling-wave equations that respectively approaches (u_l, v_l) and (u_r, v_r) as $\xi \to -\infty$ and $\xi \to \infty$. Thus this criterion is reduced to a geometrical problem of studying whether trajectories of solutions of an autonomous ordinary differential equation can join two prescribed points in the phase space.

It is important to note that the solutions of (4.13) depend on the limiting behavior of $\check{N}$. Indeed, if $\check{N}$ were replaced with another function in which the parameter ε entered in a different manner, then (4.13) could be different. (We examine an aspect of this issue in the next section. There are many different ways to introduce ε, with as much prima facie validity as (4.1).) *Thus, the admissibility of shocks in a hyperbolic system according to the travelling-wave criterion is not a property of the hyperbolic system itself: It depends critically on the way the hyperbolic system is embedded into a system with dissipation, i.e., it depends on a description of the underlying physics more refined than that delivered by the hyperbolic system alone.*

Note that (4.14) admits the integrals

$$\tag{4.15} -c\tilde{u} = \tilde{v} + a, \quad -\rho A c \tilde{u} = \check{N}(\tilde{u}, \tilde{v}_\xi, 0) + b$$

where a and b are constants of integration. Thus (4.15) is equivalent to an autonomous first-order ordinary differential equation parametrized by a, b, c. The qualitative properties of such equations are easy to determine. (Note that the physically reasonable constitutive assumption that the derivative of $\check{N}$ with respect to its second argument have a positive lower bound ensures that this ordinary differential equation can be put into standard form.)

4.16. Problem. Use the travelling-wave criterion to study the admissibility of shocks for (4.2).

In this section we have treated the asymptotics underlying the travelling-wave criterion in a purely formal way. Important steps toward a comprehensive, rigorous theory were made by Foy (1964), DiPerna (1983a,b), Goodman & Xin (1992), and others.

5. Travelling Shear Waves in Viscoelastic Media

Formulation of the governing equations. In light of the discussion of Sec. 4, we examine travelling waves for an analog of the problem treated in Sec. XIII.14. Let an incompressible, isotropic, homogeneous, viscoelastic material of differential type of complexity 1 fill the entire space, a typical material point of which has the form $x\boldsymbol{i} + y\boldsymbol{j} + s\boldsymbol{k}$. We study shearing motions of the form

$$\boldsymbol{p}(x, y, s, t) = [x + u(s,t)]\boldsymbol{i} + [y + v(s,t)]\boldsymbol{j} + s\boldsymbol{k}. \tag{5.1}$$

Relative to the basis $\{\boldsymbol{i}, \boldsymbol{j}, \boldsymbol{k}\}$, the Green deformation tensor $\boldsymbol{C}$, material strain tensor $\boldsymbol{E}$, and their derivatives have matrices given by

$$\begin{aligned} 2[\boldsymbol{E}] \equiv [\boldsymbol{C} - \boldsymbol{I}] &= \begin{pmatrix} 0 & 0 & u_s \\ 0 & 0 & v_s \\ u_s & v_s & u_s^2 + v_s^2 \end{pmatrix}, \\ 2[\boldsymbol{E}_t] \equiv [\boldsymbol{C}_t] &= \begin{pmatrix} 0 & 0 & u_{st} \\ 0 & 0 & v_{st} \\ u_{st} & v_{st} & 2(u_s u_{st} + v_s v_{st}) \end{pmatrix} \end{aligned} \tag{5.2}$$

(cf. (XIII.14.2)). Note that $\det \boldsymbol{C} = 1$ so that (5.1) describes volume-preserving motion.

For this material, the first Piola-Kirchhoff stress $\boldsymbol{T}$ is given by a constitutive equation of the form

$$\boldsymbol{T} = -p\boldsymbol{F}^{-*} + \boldsymbol{F} \cdot \hat{\boldsymbol{S}}_{\mathrm{A}}(\boldsymbol{E}, \boldsymbol{E}_t) \equiv -p\boldsymbol{F}^{-*} + \hat{\boldsymbol{T}}_{\mathrm{A}}(\boldsymbol{E}, \boldsymbol{E}_t) \tag{5.3}$$

where $\hat{\boldsymbol{S}}_{\mathrm{A}}$, the extra stress, is an isotropic tensor function of its two arguments. $\hat{\boldsymbol{S}}_{\mathrm{A}}$ could also depend on p (see Sec. XII.12), but such dependence will not affect our analysis. The General Representation Theorem XII.13.32 shows that $\hat{\boldsymbol{S}}_{\mathrm{A}}$ has the form

$$\begin{aligned} \hat{\boldsymbol{S}}_{\mathrm{A}}(\boldsymbol{E}, \boldsymbol{E}_t) = \psi_1 \boldsymbol{E} &+ \psi_2 \boldsymbol{E}_t + \psi_3 \boldsymbol{E}^2 + \psi_4 \boldsymbol{E}_t^2 \\ &+ \psi_5[\boldsymbol{E} \cdot \boldsymbol{E}_t + \boldsymbol{E}_t \cdot \boldsymbol{E}] + \psi_6[\boldsymbol{E}^2 \cdot \boldsymbol{E}_t + \boldsymbol{E}_t \cdot \boldsymbol{E}^2] \\ &+ \psi_7[\boldsymbol{E} \cdot \boldsymbol{E}_t^2 + \boldsymbol{E}_t^2 \cdot \boldsymbol{E}] + \psi_8[\boldsymbol{E}^2 \cdot \boldsymbol{E}_t^2 + \boldsymbol{E}_t^2 \cdot \boldsymbol{E}^2] \end{aligned} \tag{5.4}$$

where $\psi_1, \ldots, \psi_8$ depend on the invariants

$$\begin{gathered} \operatorname{tr} \boldsymbol{E},\ \operatorname{tr} \boldsymbol{E}^2,\ \operatorname{tr} \boldsymbol{E}^3,\ \operatorname{tr} \boldsymbol{E}_t,\ \operatorname{tr} \boldsymbol{E}_t^2,\ \operatorname{tr} \boldsymbol{E}_t^3, \\ \operatorname{tr}(\boldsymbol{E} \cdot \boldsymbol{E}_t),\ \operatorname{tr}(\boldsymbol{E} \cdot \boldsymbol{E}_t^2),\ \operatorname{tr}(\boldsymbol{E}_t \cdot \boldsymbol{E}^2),\ \operatorname{tr}(\boldsymbol{E}^2 \cdot \boldsymbol{E}_t^2). \end{gathered} \tag{5.5}$$

When $\boldsymbol{E}$ and $\boldsymbol{E}_t$ are given by (5.2), a straightforward computation shows that all entries of (5.5) depend upon $\boldsymbol{\eta} = (\eta_0, \eta_1, \eta_2)$ where

$$\eta_0 = u_s^2 + v_s^2, \quad \eta_1 = 2(u_s u_{st} + v_s v_{st}), \quad \eta_2 = u_{st}^2 + v_{st}^2. \tag{5.6}$$

Moreover, a further computation shows that (5.4) yields constitutive equations of the form

$$\begin{aligned} \boldsymbol{i} \cdot \hat{\boldsymbol{T}}_{\mathrm{A}} \cdot \boldsymbol{k} &= \mu(\boldsymbol{\eta}) u_s + \nu(\boldsymbol{\eta}) u_{st}, \\ \boldsymbol{j} \cdot \hat{\boldsymbol{T}}_{\mathrm{A}} \cdot \boldsymbol{k} &= \mu(\boldsymbol{\eta}) v_s + \nu(\boldsymbol{\eta}) v_{st}, \\ \boldsymbol{k} \cdot \hat{\boldsymbol{T}}_{\mathrm{A}} \cdot \boldsymbol{k} &= \zeta(\boldsymbol{\eta}). \end{aligned} \tag{5.7}$$

We assume that the scalar functions μ, ν, ζ are continuously differentiable.

If the medium is subject to zero body force, then (5.3) and (5.7) imply that the equations of motion are

$$-p_x + [\mu u_s + \nu u_{st}]_s = \rho u_{tt}, \tag{5.8}$$

$$-p_y + [\mu v_s + \nu v_{st}]_s = \rho v_{tt}, \tag{5.9}$$

$$[-p + \zeta]_s = 0. \tag{5.10}$$

Since u and v are independent of x and y, Eqs. (5.8) and (5.9) imply that p is affine in x and y, while (5.10) implies that p_s is independent of x and y. Thus p must have the form

$$p(x, y, s, t) = m(t)x + n(t)y + q(s, t). \tag{5.11}$$

If we assume that the pressure field (at infinity) is independent of x and y, then $m = 0 = n$ and (5.8) and (5.9) reduce to the autonomous system

$$[\mu(\boldsymbol{\eta}) u_s + \nu(\boldsymbol{\eta}) u_{st}]_s = \rho u_{tt}, \quad [\mu(\boldsymbol{\eta}) v_s + \nu(\boldsymbol{\eta}) v_{st}]_s = \rho v_{tt}. \tag{5.12}$$

After (5.12) is solved, q can be found from (5.10) to within a function of t, which is determined by conditions at infinity.

More generally, we could set

$$U = u_s, \quad V = v_s. \tag{5.13}$$

Then (5.8), (5.9), and (5.11) yield, in place of (5.12), the related autonomous system

$$[\mu U + \nu U_t]_{ss} = \rho U_{tt}, \quad [\mu V + \nu V_t]_{ss} = \rho V_{tt}. \tag{5.14}$$

In this case, m, n and the various functions of t that arise from integration are determined by conditions at infinity. Note that (5.14) can be valid in the presence of certain body forces. To avoid minor technical difficulties, we limit our attention to (5.12).

We assume that (XIII.14.8) holds:

$$\partial_R[R\mu(R^2,0,0)] > 0, \quad \mu(R^2,0,0) > 0 \quad \forall R. \tag{5.15a,b}$$

We also require that (5.12) be parabolic in the sense that the matrix

$$\begin{pmatrix} \partial_{\dot U}(\mu U + \nu \dot U) & \partial_{\dot V}(\mu U + \nu \dot U) \\ \partial_{\dot U}(\mu V + \nu \dot V) & \partial_{\dot V}(\mu V + \nu \dot V) \end{pmatrix} \quad \text{is positive-definite.} \tag{5.16}$$

Here the arguments of μ and ν are $(\eta_0, 2U\dot U + 2V\dot V, \dot U^2 + \dot V^2)$. This parabolicity condition, which has thermodynamic significance (see Sec. XII.14), ensures that the constitutive equations (5.7) describe a dissipative material.

If we let $\mu_{,0} = \partial\mu/\partial\eta_0$, $\mu_{,1} = \partial\mu/\partial\eta_1$, etc., then (5.16) is equivalent to

$$\nu + 2[\mu_{,1}U^2 + (\mu_{,2} + \nu_{,1})U\dot U + \nu_{,2}\dot U^2] > 0, \tag{5.17a}$$

$$\nu + 2[\mu_{,1}V^2 + (\mu_{,2} + \nu_{,1})V\dot V + \nu_{,2}\dot V^2] > 0, \tag{5.17b}$$

$$\begin{aligned} &\nu^2 + \nu[2\mu_{,1}\eta_0 + (\mu_{,2} + \nu_{,1})\eta_1 + 2\nu_{,2}\eta_2] \\ &\quad + 4(\mu_{,1}\nu_{,2} - \mu_{,2}\nu_{,1})(U\dot V - V\dot U)^2 > 0. \end{aligned} \tag{5.18}$$

From (5.17) we obtain

$$\nu + 2\mu_{,1}\eta_0 + (\mu_{,2} + \nu_{,1})\eta_1 + 2\nu_{,2}\eta_2 > 0. \tag{5.19}$$

We further impose the coercivity condition that

$$\dot U(\mu U + \nu \dot U) + \dot V(\mu V + \nu \dot V) \to \infty \quad \text{as } \dot U^2 + \dot V^2 \to \infty. \tag{5.20}$$

Travelling waves. We first study travelling-wave solutions of (5.12) for their intrinsic value as a class of special solutions of these equations. Then we examine their significance for the shock structure of the associated hyperbolic system. We seek travelling-wave solutions of (5.12) of the form

$$u(s,t) = \tilde u\left(\frac{s-ct}{\gamma}\right), \quad v(s,t) = \tilde v\left(\frac{s-ct}{\gamma}\right) \tag{5.21}$$

where c (without loss of generality) is a positive constant and γ is a positive constant to be assigned later. We denote derivatives of $\tilde u$ and $\tilde v$ with respect to their arguments by primes. Substituting (5.21) into (5.12), integrating the resulting ordinary differential equations, setting

$$\tilde u' \equiv \gamma \tilde U, \quad \tilde v' \equiv \gamma \tilde V, \tag{5.22}$$

and dropping the tildes, we obtain the system

$$c\nu U' = \gamma(\mu - c^2\rho)U - \gamma a, \quad c\nu V' = \gamma(\mu - c^2\rho)V \tag{5.23a,b}$$

where a is a constant of integration and where the arguments of ν and μ are

$$\text{(5.23c)}\quad \eta_0 = U^2+V^2, \quad \eta_1 = -2\frac{c}{\gamma}(UU'+VV'), \quad \eta_2 = \frac{c^2}{\gamma^2}[(U')^2+(V')^2].$$

Since our problem is invariant under rotations of the (U, V)-plane, we take a to be nonnegative and take the constant of integration for (5.23b) to be zero. The scaling (5.22) is quite natural and is consistent with (5.13). Note that derivatives of U and V are hidden as arguments of μ and ν in (5.23). But (5.16) and (5.20) support the Global Implicit Function Theorem XIX.2.30, which says that (5.23) is equivalent to a system in standard form in which U', V' are expressed as functions of U, V, c, a, γ:

$$\text{(5.24)}\qquad U' = f(U, V, c, a, \gamma), \quad V' = g(U, V, c, a, \gamma)V.$$

System (5.23) implies that g must be even in V.

We now study how the portrait of (5.23) in the (U, V) phase plane depends on a, c, and the constitutive functions μ and ν. We first examine the degenerate case in which $a = 0$. The trajectories of (5.23) lie on rays through the origin of the (U, V)-plane. The singular points of (5.23) consist of the origin together with those circles of radius A about the origin for which

$$\text{(5.25)}\qquad \mu(A^2, 0, 0) = c^2\rho.$$

There can be any number of such circles because the only restriction we have imposed on $\mu(\cdot, 0, 0)$ is (5.15).

To find the actual nature of these radial trajectories, we observe that (5.23) causes the last term in the left-hand side of (5.18) to vanish. Let $\mathcal{S}(c, \gamma)$ be the set of all $(U, V, U', V') \in \mathbb{R}^4$ satisfying (5.23). Then (5.18) implies that

$$\text{(5.26)}\qquad \nu[\nu + 2\mu_{,1}\eta_0 + (\mu_{,2} + \nu_{,1})\eta_{,1} + \nu_{,2}\eta_2] > 0$$

on $\mathcal{S}(c, \gamma)$. Thus ν cannot vanish on $\mathcal{S}(c, \gamma)$. Now (5.19) implies that $\nu(0, 0, 0) > 0$. Since $(0, 0, 0, 0) \in \mathcal{S}(c, \gamma)$ and since ν is continuous, we can conclude that ν is positive on the connected component of $\mathcal{S}(c, \gamma)$ containing $(0, 0, 0, 0)$. That (5.23) can be put into standard form implies that $\mathcal{S}(c, \gamma)$ is connected. Thus ν is positive on $\mathcal{S}(c, \gamma)$. From (5.23) (or better yet, from its version in polar coordinates), we then find that trajectories move radially outward where $\mu(\boldsymbol{\eta}) > c^2\rho$ and $U^2 + V^2 \neq 0$, and inward where $\mu(\boldsymbol{\eta}) < c^2\rho$ and $U^2 + V^2 \neq 0$ (as the phase $(s - ct)/\gamma$ increases). Moreover, (5.23) implies that $\mu(\boldsymbol{\eta}) - c^2\rho$ can vanish only on the singular circles defined by (5.25) and possibly at the origin. Thus we have a complete phase portrait of (5.23). A typical example is illustrated in Fig. 5.27.

Now we turn our attention to the general case in which $a > 0$. Then (5.23) implies that

$$\text{(5.28)}\qquad c\nu[UV' - VU'] = \gamma a V.$$

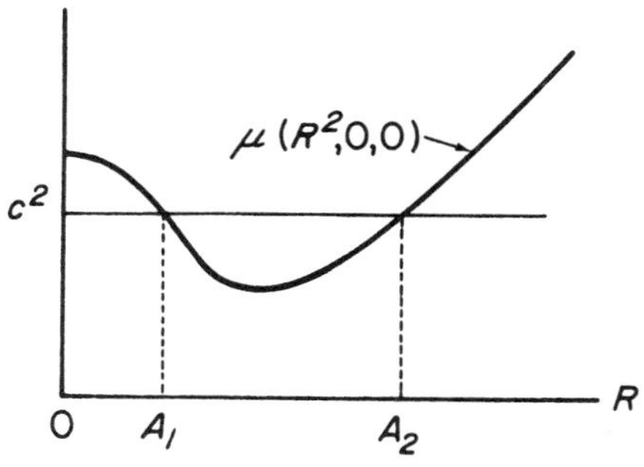

Fig. 5.27a. A typical graph of the constitutive function $\mu(\cdot, 0, 0)$ showing the two roots A_1 and A_2 of (5.30) when c^2 has the indicated value. Since μ is continuously differentiable, the slope of $R \mapsto \mu(R^2, 0, 0)$ is 0 at $R = 0$.

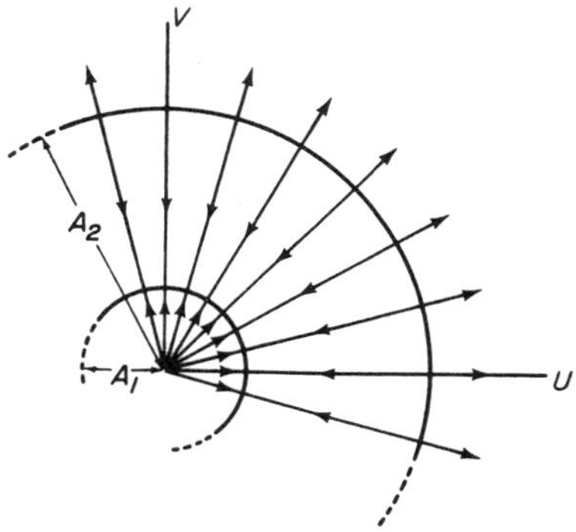

Fig. 5.27b. Phase portrait of (5.23) with $a = 0$ corresponding to Fig 5.27a. The two families of trajectories attracted toward the circle of radius A_1 are heteroclinic. Note that as the c^2 of Fig 5.27a is reduced to zero, the two singular circles approach each other, coalesce, and then disappear, whereupon the phase portrait is globally that of a node and contains no heteroclinic trajectories.

We introduce polar coordinates R, Θ by

$$(5.29) \qquad U = R\cos\Theta, \quad V = R\sin\Theta,$$

in terms of which (5.23) becomes

$$(5.30) \qquad c\nu R' = \gamma(\mu - c^2\rho)R - \gamma a\cos\Theta, \quad c\nu R\Theta' = \gamma a \sin\Theta$$

where $\boldsymbol{\eta} = (R^2, -2c\gamma^{-1}RR', c^2\gamma^{-2}[(R')^2 + (R\Theta')^2])$.

The singular points of (5.23) occur when $U' = 0 = V'$ or possibly where $\nu = 0$. In the former case, the positivity of a readily shows that the corresponding singular points are of the form $(U_*, 0)$ where U_* is a solution of

$$[\mu(U^2, 0, 0) - c^2\rho]U = a. \tag{5.31}$$

Suppose $\nu = 0$. Then (5.28) implies that $V = 0$, and (5.24) implies that $V' = 0$ when $V = 0$. On the other hand, (5.17b) implies that $(V')^2\nu_{,2} > 0$ when $\nu = 0$ (and $V = 0$), which contradicts the fact that $V' = 0$. We conclude that not only are there no singular points corresponding to the vanishing of ν, but that ν can vanish nowhere on solutions of (5.23). Note that the singular points are determined solely by the elastic response. Hypothesis (5.15b) says nothing about the number and disposition of solutions of (5.28). In Fig. 5.32 we illustrate the construction of roots $U_1, ..., U_5$ of (5.31) when $\mu(\cdot, 0, 0)$ has the form shown in Fig. 5.27a.

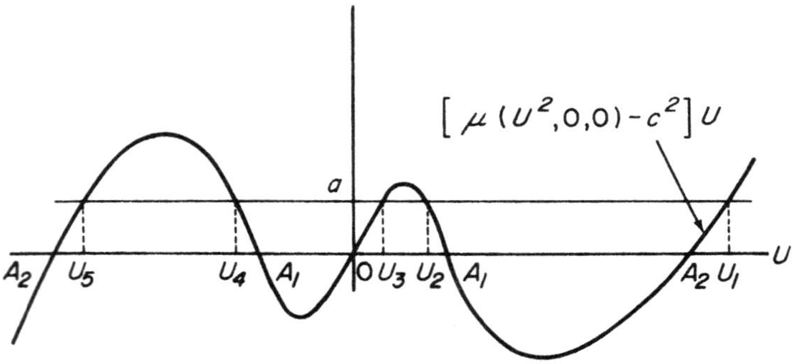

Fig. 5.32. If $\mu(\cdot, 0, 0)$ has the form shown in Fig 5.27a, then $U \mapsto [\mu(U^2, 0, 0) - c^2\rho]U$ has the form shown here and the intersection of its graph with a horizontal line with a small enough ordinate a determines the five roots $U_1, ..., U_5$ of (5.31).

It is important to note that the singular points are collinear. They accordingly correspond to states with constant and parallel (U, V)'s. Condition (5.15a) implies that c^2 and a can always be adjusted so that (5.31) has two prescribed roots.

To classify these singular points, we linearize (5.23) about them, obtaining the uncoupled system

$$c(\nu + 2\mu_{,1}U_*^2)\overset{\Delta}{U}{}' = \gamma(\mu - c^2\rho + 2\mu_{,0}U_*^2)\overset{\Delta}{U}, \tag{5.33a}$$

$$c\nu\overset{\Delta}{V}{}' = \gamma(\mu - c^2\rho)\overset{\Delta}{V} \tag{5.33b}$$

for the variations $\overset{\Delta}{U}, \overset{\Delta}{V}$. Here $\nu, \mu, \mu_{,0}, \mu_{,1}$ have arguments $U_*^2, 0, 0$. Note that (5.17) ensures that the coefficient of $\overset{\Delta}{U}{}'$ in (5.33a) is positive, while

(5.18) and (5.19) ensure that $\nu > 0$. The roots of the characteristic equation for (5.33) are

$$\frac{\gamma[\mu - c^2\rho + 2\mu_{,0}U_*^2]}{c(\nu + 2\mu_{,1}U_*^2)}, \quad \frac{\gamma(\mu - c^2\rho)}{c\nu}. \tag{5.34a,b}$$

Note that the numerator of (5.34a) is just γ times the derivative with respect to U of the left-hand side of (5.31) at roots of (5.31). It changes sign at simple roots. The denominator of (5.34a) has fixed sign, by our preceding remarks. From Fig. 5.32, which is typical, we find that $\mu - c^2\rho$ is positive at positive roots of (5.31) and negative at negative roots. Thus if (5.31) has only simple roots, then the singular points are either saddle points or nodes and their determination follows immediately from (5.31). That (5.31) is uncoupled means that the stable and unstable separatrices through each saddle point are parallel to the coordinate axes at the singular point and that if the two roots in (5.34) are not equal, then each node has axes parallel to the coordinate axes. Thus, we have

5.35. Proposition. *The location and the type of the singular points of* (5.23) *depend only on the elastic response* $\mu(\cdot, 0, 0)$.

When the situation of Fig. 5.32 holds, U_1 is an unstable node, U_2 a saddle point, U_3 an unstable node, U_4 a stable node, and U_5 a saddle point. See Fig. 5.44 below.

Asymptotics. Now we examine more carefully the asymptotics underlying the travelling-wave criterion for our problem. We first set $U_1 = u_s$, $U_2 = v_s$, $U_3 = u_t$, $U_4 = v_t$, and thus convert (5.12) to the system

$$\partial_t U_1 = \partial_s U_3, \quad \partial_t U_2 = \partial_s U_4, \tag{5.36a,b}$$

$$\rho \partial_t U_3 = \partial_s[\mu(\boldsymbol{\eta})U_1 + \nu(\boldsymbol{\eta})\partial_s U_3], \tag{5.36c}$$

$$\rho \partial_t U_4 = \partial_s[\mu(\boldsymbol{\eta})U_2 + \nu(\boldsymbol{\eta})\partial_s U_4] \tag{5.36d}$$

where $\eta_0 = U_1^2 + U_2^2$, $\eta_1 = 2(U_1\partial_s U_3 + U_2\partial_s U_4)$, $\eta_2 = (\partial_s U_3)^2 + (\partial_s U_4)^2$. The analog of (4.2) consists of (5.36a,b) and

$$\rho \partial_t U_3 = \partial_s[\mu(\eta_0, \varepsilon\eta_1, \varepsilon^2\eta_2)U_1 + \varepsilon\nu(\eta_0, \varepsilon\eta_1, \varepsilon^2\eta_2)\partial_s U_3], \quad \text{etc.} \tag{5.37}$$

With this introduction of the small parameter ε, the formal asymptotics of the shock layer leads to the admissibility conditions governed by the travelling-wave equations (5.23), just as in Sec. 4.

Let us now see how to treat the alternative parametrization

$$\mu(\boldsymbol{\eta}) = \bar{\mu}(\eta_0) + \varepsilon\hat{\mu}(\boldsymbol{\eta}), \quad \nu(\boldsymbol{\eta}) = \varepsilon\hat{\nu}(\boldsymbol{\eta}). \tag{5.38}$$

In this form, the parameter ε cannot in general be scaled out of (5.36) by our simple stretching transformation. To find the right asymptotics, we must know the details of the constitutive functions in (5.38). Suppose for large values of $|\eta_1|$ and $|\eta_2|$ that $\hat{\mu}$ and $\hat{\nu}$ have the form

$$\begin{aligned}\hat{\mu}(\boldsymbol{\eta}) &= M_1(\eta_0)|\eta_1|^k \operatorname{sign}(\eta_1)^\beta + M_2(\eta_0)|\eta_2|^{l/2} + \dots,\\ \hat{\nu}(\boldsymbol{\eta}) &= N_1(\eta_0)|\eta_1|^p + N_2(\eta_0)|\eta_2|^{q/2} + \dots\end{aligned} \tag{5.39}$$

where M_1, M_2, N_1, N_2 are prescribed functions, where $\beta = 0$ or 1, where $k, l, p, q > 0$, and where the ellipses denote lower-order terms.

Let us set

$$\begin{aligned}&\xi = \frac{s - \sigma(t)}{\gamma(\varepsilon)}, \quad \tilde{U}_1(\xi, t) = U_1(\sigma(t) + \gamma(\varepsilon)\xi, t), \quad \text{etc.},\\ &\tilde{\eta}_1 = 2(U_1 \partial_\xi \tilde{U}_3 + U_2 \partial_\xi \tilde{U}_4), \quad \eta_2 = (\partial_\xi \tilde{U}_3)^2 + (\partial_\xi \tilde{U}_4)^2\end{aligned} \tag{5.40}$$

where γ is to be chosen appropriately with $\gamma(0) = 0$. We substitute (5.39) and (5.40) into (5.36) to get expressions whose leading terms contain powers of ε and γ. We accordingly seek γ in the form ε^α where α is a positive number to be determined. In this case, (5.36c) has the form

(5.41)
$$\begin{aligned}-\rho\sigma'(t)\tilde{U}_3 = \partial_\xi \Big\{ &\Big[\bar{\mu} + \varepsilon^{1-\alpha k} M_1 |\tilde{\eta}_1|^k \operatorname{sign}(\tilde{\eta}_1)^\beta + \varepsilon^{1-\alpha l} M_2 |\tilde{\eta}_2|^{l/2} \Big] \tilde{U}_1\\ &+ \Big[\varepsilon^{1-\alpha(1+p)} N_1 |\tilde{\eta}_1|^p + \varepsilon^{1-\alpha(1+q)} N_2 |\tilde{\eta}_2|^{q/2} \Big] \tilde{\partial}_\xi U_3 \Big\} + \cdots\end{aligned}$$

where the ellipses denote terms of order lower than those exhibited here. We want to choose α so that the terms in (5.41) with ε to the lowest power vanish, i.e., so that the leading term of (5.41) is satisfied. To be specific, let us suppose that $q > p, k-1, l-1$; the treatment of all other cases is similar. Then the candidates for the smallest exponent in (5.41) are 0 and $1 - \alpha(1+q)$. If they are not equal, then the coefficients of the smaller must vanish, producing an inadmissible degeneracy. We consequently take $\alpha = 1/(1+q)$. A lengthy calculation shows that the resulting form of the leading-order equation of (5.41) is equivalent to a travelling-wave equation for a degenerate version of (5.23), namely,

$$U' = \frac{(\bar{\mu} - c^2\rho)U - a}{D}, \quad V' = \frac{(\bar{\mu} - c^2\rho)V}{D} \tag{5.42a,b}$$

with

$$D = cN_2^{1/(1+q)} \left\{ [(\bar{\mu} - c^2\rho)U - a]^2 + [(\bar{\mu} - c^2\rho)V]^2 \right\}^{q/2(q+1)}. \tag{5.42c}$$

Note that (5.42) is in the standard form (5.24). Since the denominators of (5.42a,b) are the same, we conclude that the only effect of the terms N_2 and q is merely a nonlinear rescaling of the parameters along the orbits of the phase portrait. These observations lead to

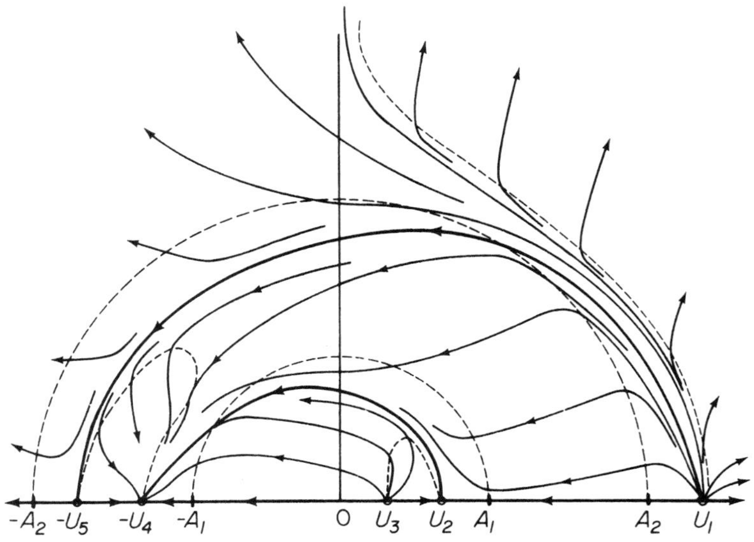

Fig. 5.44. Typical phase portrait for the reduced system (5.42) when $\bar{\mu}$ has the form of $\mu(\cdot, 0, 0)$ of Fig. 5.27a. The horizontal and vertical isoclines are shown as dashed curves. Note that the horizontal isoclines other than the U-axis are circles. Separatrices are shown as heavy curves. The portrait is symmetric about the U-axis. For this problem, the disposition of the horizontal and vertical isoclines completely determines the topological character of the phase portrait.

5.43. Theorem. *The qualitative properties of the phase portrait of* (5.42) *are completely determined by the elastic response* $\bar{\mu}$.

In Fig. 5.44, we exhibit the phase portrait of (5.42) when $\bar{\mu}$ has the form of $\mu(\cdot, 0, 0)$ shown in Fig. 5.27a. In this case, Fig. 5.32 is valid. It is a straightforward exercise to determine the qualitative behavior of U and V on any trajectory of Fig. 5.44. We discuss this behavior below.

A more careful analysis, which we do not undertake, leads to

5.45. Theorem. *On any compact subset of* $\{(U, V) : V \neq 0\}$, *the horizontal and vertical isoclines of the phase portrait of* (5.23), (5.38) *with* $\gamma = \varepsilon^\alpha$ *approach those of* (5.42) *uniformly as* $\varepsilon \to 0$. *If* $R \mapsto \bar{\mu}(R^2) - c^2\rho$ *has only simple zeros, then the horizontal isoclines of the phase portrait of* (5.23), (5.38) *with* $\gamma = \varepsilon^\alpha$ *approach those of* (5.42) *uniformly as* $\varepsilon \to 0$ *on any compact subset of the phase plane. If* $U \mapsto [\bar{\mu}(U^2) - c^2\rho]U - a$ *has only simple zeros, the same statement applies to the vertical isoclines.*

This is a theorem on the structural stability of the isoclines of the phase portrait of (5.23), (5.38) with $\gamma = \varepsilon^\alpha$. A similar theorem can be obtained for those orbits that connect nondegenerate equilibrium points.

It is important to observe that in general the phase portrait Fig. 5.44, which we have just shown to typify that for (5.23), (5.38) with $\gamma = \varepsilon^\alpha$, is *not* the same as that for (5.23), which corresponds to the parametrization (5.36a,b), (5.37). The singular points of Fig. 5.44 correspond to states to the left and to the right of a shock. The shock structure is determined by the orbit connecting such singular points. Figure 5.44 shows that there can be an infinity of such orbits. Thus the travelling-wave criterion can test the admissibility of shocks, but does not identify a unique shock structure. What may be required is a much deeper investigation of admissibility conditions generated by the evanescence of a whole array of dissipative mechanisms including viscosity, heat conduction, and strain-gradient effects (see Dafermos (1982) and Hagan & Slemrod (1983)).

In the standard numerical schemes for computing shocks are hidden numerical dissipative mechanisms, which typically add dissipation to 'compatibility' equations like (4.2a) and (5.36a,b). These dissipative mechanisms consequently are not of the form we have treated. Hagen & Slemrod (1983) identify some such mechanisms with special surface-tension (strain-gradient) effects. A considerable amount of the mathematical study of dissipative mechanisms has dealt with such mechanisms that act with equal effect on the compatibility equations and the equations of motion.

The technical parts of this section represent a refinement of Antman & Malek-Madani (1988). Some of the interpretations have been changed. Excerpts and Figs. 5.27a,b, 5.32, and 5.44 from this paper are reprinted with the kind permission of the *Quarterly of Aplied Mathematics.*

6. Blowup in Three-Dimensional Hyperelasticity

In this section we describe concavity methods that enable us to prove that classical solutions of certain initial-boundary-value problems of nonlinear elasticity cannot exist for all time. Our results suggest that a certain norm of a solution becomes infinite in finite time. Our results also apply to weak solutions, but for simplicity of exposition we restrict our attention to classical solutions.

We study the motion of a hyperelastic body $\mathcal{B}$ under zero body forces. We take constitutive functions in the form

$$\hat{\boldsymbol{T}}\big(\boldsymbol{p}_z(\boldsymbol{z},t),\boldsymbol{z}\big) = \frac{\partial \hat{W}}{\partial \boldsymbol{F}}\big(\boldsymbol{p}_z(\boldsymbol{z},t),\boldsymbol{z}\big). \tag{6.1}$$

We assume that $\hat{W} \geq 0$. We tacitly assume that $\hat{W}$ is in frame-indifferent form (see Sec. XIII.1). The equations of motion are

$$\rho(\boldsymbol{z})\boldsymbol{p}_{tt} = \nabla \cdot \hat{\boldsymbol{T}}\big(\boldsymbol{p}_z(\boldsymbol{z},t),\boldsymbol{z}\big). \tag{6.2}$$

Our initial conditions are

$$\boldsymbol{p}(\boldsymbol{z},0) = \boldsymbol{p}_0(\boldsymbol{z}), \quad \boldsymbol{p}_t(\boldsymbol{z},0) = \boldsymbol{p}_1(\boldsymbol{z}). \tag{6.3}$$

We assume that the boundary is traction-free:

$$\hat{\boldsymbol{T}}\big(\boldsymbol{p_z}(\boldsymbol{z},t)\big)\cdot\boldsymbol{\nu}(\boldsymbol{z}) = \boldsymbol{0} \quad \text{for } \boldsymbol{z}\in\partial\mathcal{B}. \tag{6.4}$$

Our initial-boundary-value problem is (6.1)–(6.4).

We shall be especially concerned with materials satisfying a constitutive restriction of the form

$$\boldsymbol{F} : \frac{\partial \hat{W}}{\partial \boldsymbol{F}}(\boldsymbol{F},\boldsymbol{z}) \le \alpha \hat{W}(\boldsymbol{F},\boldsymbol{z}) \tag{6.5}$$

where α is a positive number. We could readily weaken (6.5) by adding a positive constant to the right-hand side, since we can always redefine $\hat{W}$ to include an additive constant without affecting the underlying mechanics. (The effect of such a constant would be absorbed by the initial energy in the analysis that follows.) To see the significance of (6.5), we could take $\hat{W}$ to be a linear combination of powers of the invariants of $\boldsymbol{C}$ and then determine which exponents are compatible with the choice of α. A simpler approach is the following: Let $\boldsymbol{A}$ be an arbitrary constant tensor with $\det \boldsymbol{A} > 0$. We can characterize the behavior of $\hat{W}(\cdot,\boldsymbol{z})$ for large strains by stating how $(0,\infty)\ni\lambda\mapsto\hat{W}(\lambda\boldsymbol{A},\boldsymbol{z})$ behaves for large λ. In particular, the infimum β_* of the β's for which

$$\limsup_{\lambda\to\infty} \lambda^{-\beta}\hat{W}(\lambda\boldsymbol{A},\boldsymbol{z}) < \infty \tag{6.6}$$

characterizes how weak the material is under large extensions. (β and β_* may depend on $\boldsymbol{A}$.) The smaller the β_*, the weaker the material. In linear elasticity, $\hat{W}$ is quadratic in $\boldsymbol{F}$, so $\beta_* = 2$. A property like (6.5) represents an upper bound on the rate of growth of $\hat{W}$; it complements a restriction like (XIII.2.34). Suppose we strengthen (6.6) by requiring that

$$\frac{d}{d\lambda}\lambda^{-\beta}\hat{W}(\lambda\boldsymbol{A},\boldsymbol{z}) \le 0. \tag{6.7}$$

By carrying out the differentiation in (6.7), we readily find that (6.5) is equivalent to (6.7) with $\beta = \alpha$.

Now we use a blowup argument to prove that for appropriate initial conditions, a classical solution cannot exist for all time under the assumption that (6.5) holds with $\alpha > 2$. (This restriction on α is introduced merely to prevent technical difficulties.)

Let $\boldsymbol{p}$ be a classical solution of (6.1)–(6.4). It conserves energy:

$$E(t) \equiv \tfrac{1}{2}\int_{\mathcal{B}} \rho \boldsymbol{p}_t\cdot\boldsymbol{p}_t\,dv + \int_{\mathcal{B}} \hat{W}(\boldsymbol{p_z},\boldsymbol{z})\,dv = E(0). \tag{6.8}$$

We define

$$G(t) \equiv \int_{\mathcal{B}} \rho(\boldsymbol{z})\boldsymbol{p}(\boldsymbol{z},t)\cdot\boldsymbol{p}(\boldsymbol{z},t)\,dv(\boldsymbol{z}). \tag{6.9}$$

We seek a differential inequality for G that ensures that G blows up in finite time. We have not bothered to fix the translation of $\boldsymbol{p}$ by requiring that the mass center be fixed at the origin: $\int_{\mathcal{B}} \rho \boldsymbol{p}\, dv = \boldsymbol{0}$. But we know that $\int_{\mathcal{B}} \rho \boldsymbol{p}_{tt}\, dv = \boldsymbol{0}$, so that $\int_{\mathcal{B}} \rho \boldsymbol{p}\, dv$ can grow at most linearly in time. Therefore, a blowup of G really implies a blowup in position relative to the mass center.

By differentiating (6.9) and using the Cauchy-Bunyakovskiĭ-Schwarz inequality, we readily obtain

(6.10a,b) $$G'(t) = 2\int_{\mathcal{B}} \rho \boldsymbol{p}_t \cdot \boldsymbol{p}\, dv \leq 2\sqrt{\int_{\mathcal{B}} \rho \boldsymbol{p}_t \cdot \boldsymbol{p}_t\, dv}\sqrt{G(t)}.$$

We now use (6.2), integration by parts, inequality (6.5), and (6.8) to obtain

(6.11) $$\begin{aligned} G''(t) &= 2\int_{\mathcal{B}} \rho \boldsymbol{p}_{tt} \cdot \boldsymbol{p}\, dv + 2\int_{\mathcal{B}} \rho \boldsymbol{p}_t \cdot \boldsymbol{p}_t\, dv \\ &= -2\int_{\mathcal{B}} \boldsymbol{T} : \boldsymbol{p}_z\, dv + 2\int_{\mathcal{B}} \rho \boldsymbol{p}_t \cdot \boldsymbol{p}_t\, dv \\ &\geq -2\alpha \int_{\mathcal{B}} \hat{W}(\boldsymbol{p}_z, \boldsymbol{z})\, dv + 2\int_{\mathcal{B}} \rho \boldsymbol{p}_t \cdot \boldsymbol{p}_t\, dv \\ &= -2\alpha E(0) + (2+\alpha)\int_{\mathcal{B}} \rho \boldsymbol{p}_t \cdot \boldsymbol{p}_t\, dv. \end{aligned}$$

We multiply (6.11) by $G(t)$ and use (6.10b) to obtain

(6.12) $$GG'' - (\gamma+1)(G')^2 \geq -2\alpha E(0)G \quad \text{where } \gamma = \frac{\alpha-2}{4} > 0.$$

Let us assume that

(6.13) $$E(0) > 0, \quad G(0) > 0, \quad G'(0) > 0.$$

We suppose that a classical solution of our initial-boundary-value problem exists for $t \in [0,T)$. By continuity, there is an interval $[0,\tau) \subset [0,T)$ on which $G'(t) > 0$, so that $G(t) > G(0)$ here. It follows from (6.12) that

(6.14) $$\frac{d^2}{dt^2} G^{-\gamma} \leq 4\gamma(1+2\gamma)E(0)G^{-(\gamma+1)} \quad \text{on } [0,\tau).$$

Let us multiply (6.14) by $d[G(t)^{-\gamma}]/dt$, which is negative on $[0,\tau)$, and integrate the resulting inequality from 0 to t with $t < \tau$ to obtain an inequality that can be reduced to

(6.15) $$G'(t)^2 \geq G(t)H(G(t)) \quad \text{on } [0,\tau)$$

where

(6.16) $$H(G) \equiv 8E(0) + \left[\frac{G(t)}{G(0)}\right]^{2\gamma+1}\left[\frac{G'(0)^2}{G(0)} - 8E(0)\right].$$

Let us now further assume that the initial conditions are such that

$$G'(0)^2 > 8E(0)G(0), \tag{6.17}$$

which says that the initial velocity reinforces the initial displacement. Then H is a strictly increasing function on $[0, \infty)$. From (6.15) we thus obtain that $G'(t)^2 \geq G(t)H(G(0)) > 0$ on $[0, \tau)$. From the definition of τ and from the continuity of G' it follows that we can take $\tau = T$.

From (6.15) and (6.16), now valid on $[0, T)$, we obtain

$$T \leq \int_0^{G(T)} \frac{dG}{\sqrt{GH(G)}} \leq \int_0^1 \frac{dG}{\sqrt{8E(0)G}} + C \int_1^\infty G^{-\gamma-1/2}\, dG < K \tag{6.18}$$

where C and K are positive numbers, which are readily computed. Were a classical solution to exist for all time, we could let $T \to \infty$ in (6.18) to derive a contradiction. In summary, we have

6.19. Theorem. *Let* (6.5) *hold with* $\alpha > 2$. *Let the initial data satisfy* (6.13) *and* (6.17). *Then a classical solution of* (6.1)–(6.4) (*which is readily shown to be unique by an energy argument*) *cannot exist for all time.*

6.20. Exercise. Prove that (6.5) is frame-indifferent.

6.21. Exercise. Let $\boldsymbol{A}$ be a constant tensor and let $h : (0, \infty) \to (0, \infty)$ be a given function. For the material defined by

$$\hat{W}(\boldsymbol{F}) = \boldsymbol{A} : [\boldsymbol{F}^* \cdot \boldsymbol{F}] + h(\det \boldsymbol{F}), \tag{6.22}$$

find conditions on α and h that ensure that (6.5) holds.

6.23. Problem. Generalize Theorem 6.19 to handle boundary conditions of the form

$$\boldsymbol{p}(\boldsymbol{z}, t) = \boldsymbol{z} \quad \text{for } \boldsymbol{z} \in \mathcal{S}_0, \quad \hat{\boldsymbol{T}}(\boldsymbol{p}_{\boldsymbol{z}}(\boldsymbol{z}, t)) \cdot \boldsymbol{\nu}(\boldsymbol{z}) = \boldsymbol{0} \quad \text{for } \boldsymbol{z} \in \mathcal{S}_3 \tag{6.24}$$

where $\partial\mathcal{B} = \mathcal{S}_0 \cup \mathcal{S}_3$, $\mathcal{S}_0 \cap \mathcal{S}_3 = \emptyset$. Since we do not require the initial configuration to be a natural state, the prescription of zero boundary data for the displacement is essentially no more restrictive than the prescription of arbitrary (sufficiently smooth) time-independent data. (Cf. the treatment of Knops (1973), who uses an inequality that is similar to (6.5), but is not frame-indifferent.)

6.25. Problem. Carry out the analogous treatment of problem (6.1)–(6.4) when the assumption that (6.5) holds with $\alpha > 2$ is replaced with the assumption that (6.6) holds with $\beta > 2$. This problem is important because (6.7) would be a consequence of (6.6) if the unacceptable strict monotonicity condition (XIII.2.1), (XIII.2.2) were in force, but it does not seem to be a consequence of (6.6) if the Strong Ellipticity Condition holds.

We now study a related case of blowup in which each component of the boundary of body $\mathcal{B}$ is subjected to a hydrostatic pressure. We assume that there is a number $k \geq 0$ such that

$$\partial\mathcal{B} = \bigcup_{r=0}^{k} \Gamma_r \tag{6.26}$$

where the Γ_r are closed piecewise smooth surfaces not connected to each other. We assume that $\Gamma_1, \dots, \Gamma_k$ lie within Γ_0. (For example, $\mathcal{B}$ could be a spherical shell with Γ_1 the inner surface and Γ_0 the outer surface.) We replace (6.4) with the requirement that there is a set of constant pressures $\pi_0, \dots, \pi_k$, not necessarily positive, such that

$$(6.27) \qquad \boldsymbol{\Sigma}\big(\boldsymbol{p}(\boldsymbol{z},t),t\big)\cdot\boldsymbol{\xi}\big(\boldsymbol{p}(\boldsymbol{z},t),t\big) = -\pi_r\boldsymbol{\xi}\big(\boldsymbol{p}(\boldsymbol{z},t),t\big) \quad \text{for } \boldsymbol{z}\in\Gamma_r.$$

Here $\boldsymbol{\Sigma}(\boldsymbol{y},t)$ is the Cauchy stress tensor at $(\boldsymbol{y},t)$ and $\boldsymbol{\xi}\big(\boldsymbol{p}(\boldsymbol{z},t),t\big)$ is the outer unit normal vector to $\boldsymbol{p}(\Gamma_r,t)$ at $\boldsymbol{y}$. (See (XII.15.27)–(XII.15.34).) We define the function $\pi \in C^\infty(\mathcal{B})$ to be the solution of Laplace's equation $\Delta\pi = 0$ in $\mathcal{B}$ subject to the Dirichlet boundary conditions $\pi(\boldsymbol{z}) = \pi_r$ on Γ_r.

We assume that for each $t \in [0,T)$ the initial-boundary-value problem (6.1)–(6.3), (6.27) has a classical solution $\boldsymbol{p}$ with $\boldsymbol{p}(\cdot,t) : \operatorname{cl}\mathcal{B} \mapsto \boldsymbol{p}(\operatorname{cl}\mathcal{B},t)$ having an inverse $\boldsymbol{q}(\cdot,t) \in C^2(\mathcal{B})\cap C^1(\operatorname{cl}\mathcal{B})$. To extend the methods above to the present problem, we shall need a few identities for the pressure terms, which we now derive.

Let $\boldsymbol{z},t) \mapsto \boldsymbol{\eta}(z,t)$ be any function in $C^1(\mathcal{B})\cap C^0(\operatorname{cl}\mathcal{B})$. Using (6.27) and the results of Sec. XII.15 we can show that

$$\begin{aligned}(6.28) \qquad &\int_{\partial\mathcal{B}} \boldsymbol{\eta}(\boldsymbol{z},t)\cdot\boldsymbol{T}(\boldsymbol{z},t)\cdot\boldsymbol{\nu}(\boldsymbol{z})\,da(\boldsymbol{z})\\
&= \int_{\boldsymbol{p}(\partial\mathcal{B},t)} \boldsymbol{\eta}(\boldsymbol{q}(\boldsymbol{y},t),t)\cdot\boldsymbol{\Sigma}(\boldsymbol{y},t)\cdot\boldsymbol{\xi}(\boldsymbol{y})\,da(\boldsymbol{y})\\
&= -\int_{\boldsymbol{p}(\partial\mathcal{B},t)} \pi(\boldsymbol{q}(\boldsymbol{y},t))\,\boldsymbol{\eta}(\boldsymbol{q}(\boldsymbol{y},t),t)\cdot\boldsymbol{\xi}(\boldsymbol{y})\,da(\boldsymbol{y})\\
&= -\int_{\boldsymbol{p}(\mathcal{B},t)} \nabla_{\boldsymbol{y}}\cdot[\pi(\boldsymbol{q}(\boldsymbol{y},t))\,\boldsymbol{\eta}(\boldsymbol{q}(\boldsymbol{y},t),t)]\,dv(\boldsymbol{y})\\
&= -\int_{\mathcal{B}} \frac{\partial[\pi(\boldsymbol{z})\boldsymbol{\eta}(\boldsymbol{z},t)]}{\partial\boldsymbol{z}} : \boldsymbol{F}(\boldsymbol{z},t)^{-*}\det\boldsymbol{F}(\boldsymbol{z},t)\,dv(\boldsymbol{z}).\end{aligned}$$

The easiest way to prove the next result would be to extend $\boldsymbol{p}(\cdot,t)$ for each t continuously from $\operatorname{cl}\mathcal{B}$ to the region enclosed by Γ_0. (Such an extension could be effected by solving Dirichlet's problem for the Laplace equation for each component of $\boldsymbol{p}$ in each hole.) Suppose this extension is made. Let the region enclosed by Γ_r be denoted Ω_r. Let $\varepsilon_0 = 1$ and $\varepsilon_r = -1$ for $r = 1,\dots,k$. Taking due account of orientation, we have

$$\begin{aligned}(6.29) \qquad &\frac{d}{dt}\int_{\partial\boldsymbol{p}(\mathcal{B},t)} \pi(\boldsymbol{q}(\boldsymbol{y},t))\boldsymbol{p}(\boldsymbol{q}(\boldsymbol{y},t),t)\cdot\boldsymbol{\xi}(\boldsymbol{y})\,da(\boldsymbol{y})\\
&= \sum_{r=0}^{k}\pi_r\frac{d}{dt}\int_{\boldsymbol{p}(\Gamma_r,t)} \boldsymbol{y}\cdot\boldsymbol{\xi}(\boldsymbol{y})\,da(\boldsymbol{y})\end{aligned}$$

$$= \sum_{r=0}^{k} \varepsilon_r \pi_r \frac{d}{dt} \int_{\boldsymbol{p}(\Omega_r,t)} 3\, dv(\boldsymbol{y})$$

$$= 3 \sum_{r=0}^{k} \pi_r \int_{\boldsymbol{p}(\Gamma_r,t)} \boldsymbol{p}_t(\boldsymbol{q}(\boldsymbol{y},t),t) \cdot \boldsymbol{\xi}(\boldsymbol{y})\, da(\boldsymbol{y})$$

$$= 3 \int_{\boldsymbol{p}(\Gamma_r,t)} \pi(\boldsymbol{q}(\boldsymbol{y},t)) \boldsymbol{p}_t(\boldsymbol{q}(\boldsymbol{y},t),t) \cdot \boldsymbol{\xi}(\boldsymbol{y})\, da(\boldsymbol{y})$$

where the penultimate equality is a consequence of the Transport Theorem XII.15.23.

6.30. Exercise. Prove (6.29) directly without using the extensions of $\boldsymbol{p}$. (Hint: Use (XII.15.33) to pull the integral back to $\partial\mathcal{B}$, introduce Cartesian coordinates, and invoke Kelvin's (Stokes') Theorem to eliminate some terms; see Ball (1977a, Eq. (1.37)) and Sewell (1967).)

Taking the dot product of (6.2) with $\boldsymbol{p}_t$, integrating the resulting expression by parts, and using (6.28) and (6.29), we deduce the energy equality

$$\begin{aligned} E(t) \equiv \tfrac{1}{2} \int_{\mathcal{B}} \rho \boldsymbol{p}_t \cdot \boldsymbol{p}_t\, dv + \int_{\mathcal{B}} \hat{W}(\boldsymbol{p}_z, \boldsymbol{z})\, dv \\ + \tfrac{1}{3} \int_{\mathcal{B}} \frac{\partial(\pi \boldsymbol{p})}{\partial \boldsymbol{z}} : \boldsymbol{F}^{-*} \det \boldsymbol{F}\, dv(\boldsymbol{z}) = E(0). \end{aligned} \tag{6.31}$$

6.32. Exercise. Derive (6.31).

We are now ready to begin our analysis. We assume that (6.5) holds for $\alpha = 3$. In contrast to the technical requirement that $\alpha > 2$, which was used above, this restriction is substantive: It states that the material must be sufficiently weak. Defining G as in (6.9), we obtain (6.10) and (6.11) with $\alpha = 3$:

$$G''(t) \geq -6E(0) + 5 \int_{\mathcal{B}} \rho(\boldsymbol{z}) \boldsymbol{p}_t(\boldsymbol{z},t) \cdot \boldsymbol{p}_t(\boldsymbol{z},t)\, dv. \tag{6.33}$$

6.34. Exercise. Derive (6.33).

Let us now suppose that $E(0) \leq 0$, which can be effected by suitable choices of the pressures. From (6.33) we obtain as above that

$$g'' \leq 0 \quad \text{where } g(t) \equiv G^{-1/4}(t) \tag{6.35}$$

(in place of (6.14)), which says that $G^{-1/4}$ is concave and which implies that

$$g(t) \leq g(0) + g'(0)t \quad \text{or, equivalently,} \quad G(t) \geq \frac{1}{[g(0) + g'(0)t]^4}. \tag{6.36}$$

If the initial conditions are such that $G'(0) > 0$, then $g'(0) < 0$, and (6.36) implies that G blows up in finite time.

Now let us adopt the stronger restriction that $E(0) < 0$, while suspending the requirement that $G'(0) > 0$. From (6.33) we find that $G''(t) > -6E(0)$, so that $G'(t) > G'(0) - 6E(0)t$. Thus $G'(t_1) > 0$ if $t_1 > G'(0)/6E(0)$. We now reproduce the preceding argument by replacing $t = 0$ with $t = t_1$. Thus we have

6.37. Theorem. *Let* (6.5) *hold with* $\alpha = 3$. *Let the initial data satisfy either* $E(0) < 0$ *or* $E(0) \leq 0$ *and* $G'(0) > 0$. *Then a classical solution of* (6.1)–(6.3), (6.27) *cannot exist for all time.*

6.38. Problem. Use the methods leading to Theorem 6.19 to find a blowup result for (6.1)–(6.3), (6.27) when $E(0) > 0$.

6.39. Problem. Carry out the analogous treatment of problem (6.1), (6.2), (6.4), (6.27) when the assumption that (6.5) holds with $\alpha = 3$ is replaced with the assumption that (6.6) holds with $\beta = 3$.

The presentation in this section is based on those of Ball (1978), Knops (1973), and Knops, Levine, & Payne (1974). These articles should be consulted for further developments and for pertinent references. Also see Calderer (1983, 1986). As these references indicate, and as should be apparent from the use of integrals, most of these results apply to properly defined weak solutions of the governing equations.

We have given conditions under which a sort of energy of solutions blows up. However, we have not shown that the solution survives long enough for this energy to blow up. It is conceivable that it does not (see Ball (1978)). For ordinary differential equations, this difficulty cannot occur.

CHAPTER XVII

Appendix. Topics in Linear Analysis

1. Banach Spaces

A Banach space is a vector space with very attractive convergence properties. For our purposes, the most important Banach spaces are spaces of functions. Formally, a *Banach space* is a complete, normed, vector space. Let us now define each of these terms. By a *scalar* we mean a real or complex number.

A *vector space* (or equivalently a *linear space*) consists of a collection $\mathcal{X}$ of elements $\mathsf{x}, \mathsf{y}, \mathsf{z}, \dots$, called *vectors*, together with the operation of addition that associates with each pair x, y of elements of $\mathcal{X}$ its *sum* $\mathsf{x}+\mathsf{y}$ in $\mathcal{X}$, and the operation of scalar multiplication that associates with each scalar α and each $\mathsf{x} \in \mathcal{X}$ the *multiple* $\alpha\mathsf{x}$ in $\mathcal{X}$, such that

(V1) $\mathsf{x} + \mathsf{y} = \mathsf{y} + \mathsf{x} \quad \forall \mathsf{x}, \mathsf{y} \in \mathcal{X}$.

(V2) $\mathsf{x} + (\mathsf{y} + \mathsf{z}) = (\mathsf{x} + \mathsf{y}) + \mathsf{z} \quad \forall \mathsf{x}, \mathsf{y}, \mathsf{z} \in \mathcal{X}$.

(V3) $\exists \mathbf{0} \in \mathcal{X}$ such that $\mathsf{x} + \mathbf{0} = \mathsf{x} \quad \forall \mathsf{x} \in \mathcal{X}$.

(V4) To each vector $\mathsf{x} \in \mathcal{X}$ there corresponds a unique vector $(-\mathsf{x})$ such that $\mathsf{x}+(-\mathsf{x}) = \mathbf{0}$,

(V5) $\alpha(\beta\mathsf{x}) = (\alpha\beta)\mathsf{x}$ for every scalar α, β and for every vector x.

(V6) $1\mathsf{x} = \mathsf{x}$ for every vector x.

(V7) $\alpha(\mathsf{x} + \mathsf{y}) = \alpha\mathsf{x} + \alpha\mathsf{y}$ for every scalar α and for every vector x, y,

(V8) $(\alpha + \beta)\mathsf{x} = \alpha\mathsf{x} + \beta\mathsf{x}$ for every scalar α, β and for every vector x.

Since the operations of addition and scalar multiplication are invariably obvious, we refer to $\mathcal{X}$ itself as the vector space. A vector space (and correspondingly a Banach space) is called *real* or *complex* if the field of scalars is respectively the real numbers or the complex numbers. In elasticity, complex vector spaces arise primarily in the study of dynamical problems.

A vector space has *dimension* n where n is a nonnegative integer iff it has a set of n linearly independent vectors, but every set of $n + 1$ vectors is dependent. A vector space is *infinite-dimensional* iff for each positive integer k it has a set of k independent vectors.

A vector space is *normed* iff it is endowed with a function that associates with each x in $\mathcal{X}$ the real number $\|\mathsf{x}, \mathcal{X}\|$, usually abbreviated $\|\mathsf{x}\|$, called the *norm* of x, such that

(N1) If $\|\mathsf{x}\| = 0$, then $\mathsf{x} = \mathbf{0}$.

(N2) $\|\alpha\mathsf{x}\| = |\alpha|\, \|\mathsf{x}\| \quad \forall$ scalars α, $\forall \mathsf{x} \in \mathcal{X}$.

(N3) The *triangle inequality* holds: $\|\mathsf{x} + \mathsf{y}\| \le \|\mathsf{x}\| + \|\mathsf{y}\| \ \forall \mathsf{x}, \mathsf{y} \in \mathcal{X}$.

Note that the properties of a norm ensure that $\|\mathsf{x}\| \ge 0$ for all $\mathsf{x} \in \mathcal{X}$. The *distance* between x and y in a normed space $\mathcal{X}$ is $\|\mathsf{x} - \mathsf{y}\|$. We use this distance to define all the topological notions such as open set, closed set, etc., just as in Euclidean space.

A sequence $\{\mathsf{x}_k\}$ in a normed space $\mathcal{X}$ *converges* iff there exists an element x, called the *limit* of the sequence, in $\mathcal{X}$ with the property that for arbitrary $\varepsilon > 0$ there is a number $M(\varepsilon) > 0$ such that

$$\|\mathsf{x}_k - \mathsf{x}\| < \varepsilon \quad \text{when } k > M(\varepsilon). \tag{1.1}$$

A sequence $\{\mathsf{x}_k\}$ in a normed space $\mathcal{X}$ is a *Cauchy sequence* iff for arbitrary $\varepsilon > 0$ there is a number $N(\varepsilon) > 0$ such that

$$\|\mathsf{x}_k - \mathsf{x}_l\| < \varepsilon \quad \text{when } k, l > N(\varepsilon). \tag{1.2}$$

It is easy to see that every convergent sequence in a normed space is a Cauchy sequence. A normed space is *complete* iff every Cauchy sequence of elements in it converges in it. The importance of this concept is that in a normed space we can show that a sequence converges merely by showing that it is a Cauchy sequence: We do not have to have a candidate for a limit.

Examples.

i. Some finite-dimensional spaces. Let $p \geq 1$. Then $\mathbb{R}^n$, which consists of n-tuples $\mathsf{x} = (x_1, \dots, x_n)$ of real numbers, when endowed with any of the norms

$$\|\mathsf{x}, l_p\| \equiv \left[\sum_{k=1}^{n} |x_k|^p\right]^{1/p} \quad \text{or} \quad \|\mathsf{x}, l_\infty\| \equiv \max\{|x_k| : k = 1, \dots, n\}, \tag{1.3}$$

is a real Banach space. To show that the functions of (1.3) are norms, it is convenient to use the Hölder inequality:

$$\sum_{k=1}^{n} |x_k y_k| \leq \|\mathsf{x}, l_p\| \, \|\mathsf{y}, l_{p^*}\| \tag{1.4}$$

for $p \in (1, \infty)$ and for $\frac{1}{p} + \frac{1}{p^*} = 1$. (Several proofs of this inequality are given by Hardy, Littlewood, & Polya (1952), e.g.) The completeness of these spaces is an immediate consequence of the Cauchy Convergence Criterion for sequences of real numbers.

It is not hard to show that all norms on a finite-dimensional vector space are *equivalent*, i.e., if $\|\cdot\|_1$ and $\|\cdot\|_2$ are norms on such a space, then there are positive numbers c and C such that $c\|\mathsf{x}\|_1 \leq \|\mathsf{x}\|_2 \leq C\|\mathsf{x}\|_1$. Thus, if a sequence in a finite-dimensional vector space converges in one norm, then it converges in every other norm.

Recall from Chap. VII that a set $\mathcal{K}$ in a normed space is (*sequentially*) *compact* iff every sequence in $\mathcal{K}$ has a subsequence converging to a limit in $\mathcal{K}$. It follows from the Bolzano-Weierstrass Theorem that a set in a finite-dimensional normed space is compact if and only if it is closed and bounded. A theorem of F. Riesz says that a normed linear space is finite-dimensional if and only if its closed unit ball is compact.

ii. Spaces of continuous functions. Let Ω be a domain in $\mathbb{R}^n$ or $\mathbb{E}^n$. Let $[C^0(\operatorname{cl}\Omega)]^m$ denote the set of all uniformly continuous and bounded functions u from Ω to $\mathbb{R}^m$. It is a Banach space with norm

$$\|\mathsf{u}, [C^0(\operatorname{cl}\Omega)]^m\| \equiv \sup\{|\mathsf{u}(\mathbf{x})|, \mathbf{x} \in \operatorname{cl}\Omega\}. \tag{1.5}$$

Thus convergence in this space is uniform convergence. Completeness is equivalent to the proof of the Cauchy Convergence Criterion for continuous functions, which is found in all books on advanced calculus. (If Ω is bounded, then, because a continuous real-valued function on a compact subset of $\mathbb{R}^n$ or $\mathbb{E}^n$ is bounded and uniformly continuous, it follows that $[C^0(\operatorname{cl}\Omega)]^m$ is equivalent to the the set of continuous functions u on $\operatorname{cl}\Omega$, and the sup in (1.5) can be replaced with max.)

iii. Spaces of continuously differentiable functions. Let Ω be a domain in $\mathbb{R}^n$ or $\mathbb{E}^n$. Let $[C^1(\operatorname{cl}\Omega)]^m$ denote the set of all continuously differentiable functions u from Ω to $\mathbb{R}^m$ that together with their derivatives are bounded and uniformly continuous. It is a Banach space with norm

$$\|\mathsf{u}, [C^1(\operatorname{cl}\Omega)]^m\| \equiv \|\mathsf{u}, [C^0(\operatorname{cl}\Omega)]^m\| + \|\partial\mathsf{u}/\partial\mathbf{x}, [C^0(\operatorname{cl}\Omega)]^{mn}\|. \tag{1.6}$$

iv. Lebesgue spaces. Let $p \in [1, \infty)$ and let Ω be a domain (or more generally a measurable set) in $\mathbb{R}^n$ or in $\mathbb{E}^n$. Let $[L_p(\Omega)]^m$ denote the set of all (equivalence classes of almost everywhere equal) functions from Ω to $\mathbb{R}^m$ such that

$$\|\mathbf{u}, [L_p(\Omega)]^m\| \equiv \left\{ \int_\Omega |\mathbf{u}(\mathbf{x})|^p \, dv(\mathbf{x}) \right\}^{1/p} < \infty. \tag{1.7}$$

That this norm satisfies the triangle inequality follows from the Hölder inequality (I.8.2) for integrals, which is a consequence of (1.4).

v. The space of bounded functions. Let Ω be a a measurable set in $\mathbb{R}^n$ or in $\mathbb{E}^n$. The set $[L_\infty(\Omega)]^m$ of *essentially bounded* functions from Ω to $\mathbb{R}^m$ has the norm

$$\begin{aligned} &\|\mathbf{u}, [L_\infty(\Omega)]^m\| \\ &\qquad \equiv \inf\{\alpha \in \mathbb{R} : \text{ the set of } \mathbf{x}\text{'s for which } |\mathbf{u}(\mathbf{x})| > \alpha \text{ has measure } 0\}. \end{aligned} \tag{1.8}$$

vi. Sobolev spaces. Let Ω be a domain in $\mathbb{R}^n$ or in $\mathbb{E}^n$. To define Banach spaces of functions whose derivatives are in $L_p(\Omega)$, we first define a generalized derivative of a function that is merely locally integrable, i.e., integrable on compact subsets of Ω. Let $C_0^\infty(\Omega)$ denote that class of all scalar-valued functions that are infinitely differentiable on Ω and have compact support there (so that for each function in $C_0^\infty(\Omega)$ there is a compact subset of Ω outside of which the function vanishes). Let $\mathbf{u}$ be locally integrable on Ω. If there exists a locally integrable function $\mathbf{v}_k$ such that

$$\int_\Omega \mathbf{u}(\mathbf{x}) \frac{\partial \varphi(\mathbf{x})}{\partial x^k} \, dv(\mathbf{x}) = -\int_\Omega \mathbf{v}_k(\mathbf{x}) \varphi(\mathbf{x}) \, dv(\mathbf{x}) \tag{1.9}$$

for all $\varphi \in C_0^\infty(\Omega)$, then $\mathbf{v}_k$ is called the *distributional derivative* of $\mathbf{u}$ and is denoted $\partial \mathbf{u}/\partial x_k$. If $\mathbf{u}$ is differentiable on Ω, then its distributional derivative is readily shown to equal its classical derivative.

Let $p \geq 1$. The *Sobolev space* $[W_p^1(\Omega)]^m$ consists of (equivalence classes of almost everywhere equal) functions $\mathbf{u} \in [L_p(\Omega)]^m$ whose distributional derivatives $\partial \mathbf{u}/\partial \mathbf{x}$ belong to $[L_p(\Omega)]^{mn}$. The space $[W_p^1(\Omega)]^m$ is equipped with the norm

$$\|\mathbf{u}, [W_p^1(\Omega)]^m\| \equiv \{\|\mathbf{u}, [L_p(\Omega)]^m\|^p + \|\partial \mathbf{u}/\partial \mathbf{x}, [L_p(\Omega)]^{mn}\|^p\}^{1/p}. \tag{1.10}$$

There is an extensive literature devoted to the regularity and boundary behavior of functions in Sobolev spaces (see Adams (1975), Maz'ya (1985), and Nečas (1967), e.g.). The simplest such property of Sobolev spaces is the inequality (I.8.5).

A *real inner product* on a real vector space $\mathcal{X}$ is a real-valued function $\mathbf{x}, \mathbf{y} \mapsto \langle \mathbf{x}, \mathbf{y} \rangle$ such that

(I1) $\langle \mathbf{x}, \mathbf{y} \rangle = \langle \mathbf{y}, \mathbf{x} \rangle \quad \forall \mathbf{x}, \mathbf{y} \in \mathcal{X}$.

(I2) $\langle \alpha\mathbf{x} + \beta\mathbf{y}, \mathbf{z} \rangle = \alpha\langle \mathbf{x}, \mathbf{z} \rangle + \beta\langle \mathbf{y}, \mathbf{z} \rangle \quad \forall \alpha, \beta \in \mathbb{R},\ \forall \mathbf{x}, \mathbf{y}, \mathbf{z} \in \mathcal{X}$.

(I3) $\langle \mathbf{x}, \mathbf{x} \rangle \geq 0 \quad \forall \mathbf{x} \in \mathcal{X}$.

(I4) $\langle \mathbf{x}, \mathbf{x} \rangle = 0$ if and only if $\mathbf{x} = \mathbf{0}$.

A space with a real inner product is a real normed space with norm defined by $\|\mathbf{x}\| = \sqrt{\langle \mathbf{x}, \mathbf{x} \rangle}$. A complete inner-product space is called a *Hilbert space*. Examples of real Hilbert spaces are $\mathbb{R}^n$ endowed with the inner product $\langle \mathbf{x}, \mathbf{y} \rangle \equiv \sum_{k=1}^n x_k y_k$, which generates the norm (1.3) with $p = 2$, and the space $L_2(\Omega)$ endowed with the inner product $\langle u, v \rangle \equiv \int_\Omega u(\mathbf{x}) v(\mathbf{x}) \, dv(\mathbf{x})$, which generates the norm of (1.7) with $p = 2$. (We define the Euclidean space $\mathbb{E}^n$ to be abstract n-dimensional real inner-product space.)

A *linear manifold* $\mathcal{A}$ in a vector space is a set with the property that if $\mathbf{x}_1, \mathbf{x}_2 \in \mathcal{A}$, then $\alpha_1\mathbf{x}_1 + \alpha_2\mathbf{x}_2 \in \mathcal{A}$ for all scalars α_1, α_2. A linear manifold is itself a vector space, inheriting the operations of addition and scalar multiplication from its parent space, but it need not be a Banach space even if its parent space is a Banach space because it need not be complete. It is complete and therefore a Banach space if and only if it is closed.

2. Linear Operators and Linear Equations

Let $\mathcal{X}$ and $\mathcal{Y}$ be real Banach spaces, and let $\mathcal{D}$ be a linear manifold in $\mathcal{X}$. A mapping $\mathsf{f} : \mathcal{D} \to \mathcal{Y}$ is called a *linear operator* iff

$$\mathsf{f}(\alpha_1 \mathsf{x}_1 + \alpha_2 \mathsf{x}_2) = \alpha_1 \mathsf{f}(\mathsf{x}_1) + \alpha_2 \mathsf{f}(\mathsf{x}_2) \quad \forall\, \mathsf{x}_1, \mathsf{x}_2 \in \mathcal{D}, \quad \forall\, \alpha_1, \alpha_2 \in \mathbb{R}. \tag{2.1}$$

We denote abstract linear operators by uppercase sans-serif symbols $\mathsf{A}, \mathsf{B}, \dots$ and we denote the value of an operator A at x by $\mathsf{A} \cdot \mathsf{x}$. (We use the dot solely because it corresponds to the Gibbsian notation we use for linear operators, i.e., tensors, in Euclidean spaces.) The domain $\mathcal{D}$ of definition of a linear operator A is denoted $\mathcal{D}(\mathsf{A})$. We always assume that $\mathcal{D}(\mathsf{A})$ is dense in $\mathcal{X}$, i.e., the closure of $\mathcal{D}(\mathsf{A})$ in $\mathcal{X}$ is $\mathcal{X}$ itself. The *range* $\mathsf{A} \cdot \mathcal{D}(\mathsf{A})$, here denoted $\mathcal{R}(\mathsf{A})$, is defined by

$$\mathcal{R}(\mathsf{A}) \equiv \{\mathsf{y} \in \mathcal{Y} : \exists\, \mathsf{x} \in \mathcal{D}(\mathsf{A}) \quad \text{such that} \quad \mathsf{y} = \mathsf{A} \cdot \mathsf{x}\}. \tag{2.2}$$

The *null space* of A is defined to be

$$\mathcal{N}(\mathsf{A}) \equiv \{\mathsf{x} \in \mathcal{D}(\mathsf{A}) : \mathsf{A} \cdot \mathsf{x} = \mathbf{0}\}. \tag{2.3}$$

2.4. Example. Let $\mathcal{X} = L_2(0,1)$, $\mathcal{D} = C^2[0,1]$, $\mathcal{Y} = L_2[0,1]$. Then $\mathcal{D} \ni u \mapsto u'' \in \mathcal{Y}$ is a linear operator, with range $C^0[0,1]$ and with null space consisting of all affine functions u.

A fundamental objective of linear operator theory is to solve linear equations of the form

$$\mathsf{A} \cdot \mathsf{x} = \mathsf{f} \tag{2.5}$$

when A is a given linear operator and f is a given element of $\mathcal{Y}$. We now introduce further terminology, concepts, and theorems that promote this goal.

A linear operator A is *bounded* or, equivalently, *continuous* iff there is a real number α such that

$$\|\mathsf{A} \cdot \mathsf{x}, \mathcal{Y}\| \leq \alpha \|\mathsf{x}, \mathcal{X}\| \quad \forall\, \mathsf{x} \in \mathcal{D}(\mathsf{A}). \tag{2.6}$$

(Note that (2.6) immediately implies the continuity of A in the usual sense.) The $\mathcal{X}$ and $\mathcal{Y}$ could be dropped from the norms in (2.6), as we do below, because the meaning of the norms is clear from the context. The norm $\|\mathsf{A}\|$ of a bounded linear operator A is defined to be

$$\|\mathsf{A}\| \equiv \sup \left\{ \frac{\|\mathsf{A} \cdot \mathsf{x}\|}{\|\mathsf{x}\|} : \mathbf{0} \neq \mathsf{x} \in \mathcal{D}(\mathsf{A}) \right\} \equiv \sup \left\{ \|\mathsf{A} \cdot \mathsf{x}\| : \mathsf{x} \in \mathcal{D}(\mathsf{A}),\ \|\mathsf{x}\| = 1 \right\}. \tag{2.7}$$

(The proof of the second identity in (2.7) is elementary.) The linear operator of Ex. 2.4 is not bounded. A simple example of a bounded linear operator is the operator of integration on any reasonable space of integrable functions.

The set of all bounded linear operators $\mathsf{A} : \mathcal{D}(\mathsf{A}) \to \mathcal{Y}$ is denoted $\mathcal{L}(\mathcal{D}(\mathsf{A}), \mathcal{Y})$. That $\mathcal{D}(\mathsf{A})$ is normed and $\mathcal{Y}$ is a Banach space is easily shown to imply that $\mathcal{L}(\mathcal{D}(\mathsf{A}), \mathcal{Y})$, when equipped with the norm in (2.7), is itself a Banach space (with the obvious rules for addition and scalar multiplication). It can be shown that if A is a bounded linear operator, then it can be extended to the closure of $\mathcal{D}(\mathsf{A})$ in $\mathcal{X}$ without its norm being increased. This closure is itself a Banach space, which we might as well call $\mathcal{X}$. Thus we always regard the domain of a bounded linear operator as being a Banach space.

If $\mathcal{Y}$ is $\mathbb{R}$, then a member of $\mathcal{L}(\mathcal{D}(\mathsf{A}), \mathbb{R})$ is called a (real) *bounded linear functional* and $\mathcal{L}(\mathcal{D}(\mathsf{A}), \mathbb{R})$, which is denoted by $\mathcal{X}^*$, is called the *dual space* of $\mathcal{X}$. By our preceding remarks, $\mathcal{X}^*$ is a Banach space. For many useful Banach spaces there are explicit representations of the elements of $\mathcal{X}^*$. For example, if $\mathcal{H}$ is a real Hilbert space with inner product $\langle \cdot, \cdot \rangle$, then the Riesz Representation Theorem asserts that to each A in $\mathcal{H}^*$ there corresponds a unique vector x^* in $\mathcal{H}$ such that

$$\mathsf{A} \cdot \mathsf{x} = \langle \mathsf{x}, \mathsf{x}^* \rangle. \tag{2.8a}$$

Conversely, each vector x^* in $\mathcal{H}$ generates a bounded linear functional by (2.8a). Thus for a Hilbert space $\mathcal{H}$, the dual space $\mathcal{H}^*$ can always be identified with $\mathcal{H}$ itself. (For Sobolev spaces that are Hilbert spaces, this identification is not always convenient.)

We denote a typical element of $\mathcal{X}^*$ sometimes by x^*, but more often by $\mathcal{X} \ni \mathsf{x} \mapsto \langle \mathsf{x}, \mathsf{x}^* \rangle$, a notation that is motivated by (2.8a) and that generalizes it. In this notation, another representation theorem of Riesz says that if $p \in (1, \infty)$, then $L_p(\Omega)^*$ can be identified with $L_{p^*}(\Omega)$, where $p^* = p/(p-1)$, with

$$\langle u, u^* \rangle \equiv \int_\Omega u(\mathbf{x}) u^*(\mathbf{x}) \, dv(\mathbf{x}), \tag{2.8b}$$

the integral converging by the Hölder inequality.

Since $\mathcal{X}^*$ is a Banach space, we can define its dual space $\mathcal{X}^{**} \equiv (\mathcal{X}^*)^*$. Now each x in $\mathcal{X}$ defines a bounded linear functional on $\mathcal{X}^*$ by $\mathsf{x}^* \mapsto \langle \mathsf{x}, \mathsf{x}^* \rangle$. Thus we can identify $\mathcal{X}$ with a subset of $\mathcal{X}^{**}$. When $\mathcal{X}$ can be identified with $\mathcal{X}^{**}$, $\mathcal{X}$ is said to be *reflexive*, and there is a symmetry between $\mathcal{X}$ and $\mathcal{X}^*$. Hilbert spaces, the L_p spaces with $1 < p < \infty$, and the Sobolev spaces W_p^1 with $1 < p < \infty$ are reflexive.

The notions of dual space and reflexivity are essential for the definition of weak convergence. This topic is developed in detail in Sec. VII.3.

Adjoints. We now turn our attention to the study of the solvability of (2.5). Let us first consider as a concrete example the boundary-value problem

$$u'' + \pi^2 u = \sin \pi s \quad \text{on } (0,1), \quad u(0) = 0 = u(1). \tag{2.9a,b,c}$$

We may formulate this problem as an operator equation of the form (2.5) by taking $\mathcal{X} = C^2[0,1]$, $\mathcal{D} = \{x \in \mathcal{X} : x(0) = 0 = x(1)\}$, the linear operator A to be $u \mapsto u'' + \pi^2 u$, and f to be the function with values $\sin \pi s$. We can find the general solution of (2.9a) in terms of two arbitrary constants. Upon trying to use the boundary conditions (2.9b,c) to evaluate the constants, we find that (2.9) has no solution. This naive procedure fails for more complicated problems because it depends upon the availability of a specific representation for the solutions of the equation. A slicker demonstration of the same conclusion is obtained by assuming that (2.9) has a solution, then by multiplying (2.9a) by $\sin \pi s$, and by integrating the resulting equation by parts over $(0,1)$. The boundary conditions reduce this equation to the absurdity: $0 = \int_0^1 \sin^2 \pi s \, ds$.

We now abstract this procedure to make it applicable to a whole range of problems of the form (2.5). We seek a simple test on f to tell whether (2.5) is solvable. For this purpose, we need the concept of an adjoint operator.

Let A be a linear operator (not necessarily bounded) from $\mathcal{D}(\mathsf{A})$ to $\mathcal{Y}$. Let $\mathcal{D}(\mathsf{A}^*)$ consist of all the elements y^* in $\mathcal{Y}^*$ with the property that there exists an x^* in $\mathcal{X}^*$ such that

$$\langle \mathsf{A} \cdot \mathsf{x}, \mathsf{y}^* \rangle = \langle \mathsf{x}, \mathsf{x}^* \rangle \quad \forall \mathsf{x} \in \mathcal{D}(\mathsf{A}). \tag{2.10}$$

Here $\langle \mathsf{A} \cdot \mathsf{x}, \mathsf{y}^* \rangle$ is the value of the bounded linear functional y^* at $\mathsf{A} \cdot \mathsf{x}$ in $\mathcal{Y}$, and $\langle \mathsf{x}, \mathsf{x}^* \rangle$ is the value of the bounded linear functional x^* at x in $\mathcal{X}$. It is easy to show that $\mathcal{D}(\mathsf{A}^*)$ is a vector subspace of $\mathcal{Y}^*$ and that x^* is uniquely determined by y^*. We accordingly write $\mathsf{x}^* = \mathsf{A}^* \cdot \mathsf{y}^*$. It is easy to see that $\mathcal{Y}^* \supset \mathcal{D}(\mathsf{A}^*) \ni \mathsf{y}^* \mapsto \mathsf{A}^* \cdot \mathsf{y}^* \in \mathcal{X}^*$ is a linear operator, called the *adjoint* of A.

Let $\mathcal{M}$ be any subset of Banach space $\mathcal{X}$ and let $\mathcal{M}_*$ be any subset of Banach space $\mathcal{X}^*$. We define the *annihilators*

$$\begin{aligned} \mathcal{M}^\perp &\equiv \{\mathsf{x}^* \in \mathcal{X}^* : \langle \mathsf{x}, \mathsf{x}^* \rangle = 0 \ \forall \mathsf{x} \in \mathcal{M}\}, \\ {}^\perp\mathcal{M}_* &\equiv \{\mathsf{x} \in \mathcal{X} : \langle \mathsf{x}, \mathsf{x}^* \rangle = 0 \ \forall \mathsf{x}^* \in \mathcal{M}_*\}. \end{aligned} \tag{2.11}$$

Thus $\mathcal{M}^\perp$ consists of all bounded linear functionals on $\mathcal{X}$ that vanish on $\mathcal{M}$, and ${}^\perp\mathcal{M}_*$ consists of all the elements of $\mathcal{X}$ on which every element (bounded linear functional) of $\mathcal{M}_*$ vanishes. By applying the definition of closure and of span, we can easily prove

2.12. Proposition. *$\mathcal{M}^\perp$ and ${}^\perp\mathcal{M}_*$ are closed linear manifolds, i.e., Banach spaces, with*

$$\mathcal{M}^\perp = [\mathrm{cl}\,(\mathrm{span}\,\mathcal{M})]^\perp, \; {}^\perp\mathcal{M}_* = {}^\perp[\mathrm{cl}\,(\mathrm{span}\,\mathcal{M}_*)].$$

We begin our study of the solvability of (2.5) by making the trivial observations that if (2.5) has a solution, then f is in $\mathcal{R}(\mathsf{A})$, and that if (2.5) has a solution for every f in $\mathcal{Y}$, then $\mathcal{R}(\mathsf{A}) = \mathcal{Y}$. We accordingly seek to describe $\mathcal{R}(\mathsf{A})$. Our treatment of (2.9) suggests that $\mathcal{R}(\mathsf{A})$ is related to a suitable null space. Indeed,

2.13. Lemma. *If* (2.5) *has a solution, then* f *annihilates the null space of* A^*, *i.e.,*

$$\langle \mathsf{f}, \mathsf{y}^* \rangle = 0 \quad \forall \mathsf{y}^* \quad \textit{such that} \quad \mathsf{A}^* \cdot \mathsf{y}^* = \mathbf{0} \tag{2.14a}$$

or, equivalently,

$$\mathcal{R}(\mathsf{A}) \subset {}^\perp\mathcal{N}(\mathsf{A}^*). \tag{2.14b}$$

Proof. We are given that there exists an x such that $\mathsf{A} \cdot \mathsf{x} = \mathsf{f}$. For any $\mathsf{y}^* \in \mathcal{D}(\mathsf{A}^*)$, it therefore follows from (2.10) that

$$\langle \mathsf{f}, \mathsf{y}^* \rangle = \langle \mathsf{A} \cdot \mathsf{x}, \mathsf{y}^* \rangle = \langle \mathsf{x}, \mathsf{A}^* \cdot \mathsf{y}^* \rangle. \tag{2.15}$$

If we now take $\mathsf{y}^* \in \mathcal{N}(\mathsf{A}^*)$, then the rightmost term of (2.15) vanishes by definition of $\mathcal{N}(\mathsf{A}^*)$. □

For much of our work on perturbation methods, Lemma 2.13 suffices. We need stronger results to treat material constraints. In the process of developing these results, we obtain some powerful generalizations of Lemma 2.13.

Since the right-hand side of (2.14b) is closed by Proposition 2.12, we can replace the left-hand side of this containment by $\mathrm{cl}\,\mathcal{R}(\mathsf{A})$. We can actually do more. Our first step is

2.16. Theorem. $[\mathrm{cl}\,\mathcal{R}(\mathsf{A})]^\perp = \mathcal{R}(\mathsf{A})^\perp = \mathcal{N}(\mathsf{A}^*)$.

Proof. The first equality follows from Proposition 2.12. To prove the second, let $\mathsf{y}^* \in \mathcal{N}(\mathsf{A}^*)$, so that $\mathsf{A}^* \cdot \mathsf{y}^* = \mathbf{0}$. Thus $0 = \langle \mathsf{x}, \mathsf{A}^* \cdot \mathsf{y}^* \rangle = \langle \mathsf{A} \cdot \mathsf{x}, \mathsf{y}^* \rangle$ for all $\mathsf{x} \in \mathcal{D}(\mathsf{A})$. This says that $\mathsf{y}^* \in \mathcal{R}(\mathsf{A})^\perp$. Conversely, let $\mathsf{y}^* \in \mathcal{R}(\mathsf{A})^\perp$. Then by definition of range, $\langle \mathsf{A} \cdot \mathsf{x}, \mathsf{y}^* \rangle = 0 = \langle \mathsf{x}, \mathbf{0} \rangle$ for all $\mathsf{x} \in \mathcal{D}(\mathsf{A})$. The last equality shows that $\mathsf{y}^* \in \mathcal{D}(\mathsf{A}^*)$, so that (2.10) then yields $\mathsf{A}^* \cdot \mathsf{y}^* = \mathbf{0}$. □

Now by a direct application of our definitions, we can easily demonstrate that $\mathrm{cl}\,(\mathrm{span}\,\mathcal{M}) \subset {}^\perp(\mathcal{M}^\perp)$. A proof of the reverse containment, the details of which we do not pause to spell out, relies on the Hahn-Banach Theorem. Thus

2.17. Proposition. $\mathrm{cl}\,(\mathrm{span}\,\mathcal{M}) = {}^\perp(\mathcal{M}^\perp)$.

Applying this proposition to Theorem 2.16, we immediately obtain

2.18. Theorem. $\mathrm{cl}\,\mathcal{R}(\mathsf{A}) = {}^\perp[\mathcal{R}(\mathsf{A})^\perp] = {}^\perp\mathcal{N}(\mathsf{A}^*)$.

If $\mathcal{R}(\mathsf{A})$ is known to be closed, then Theorem 2.18 is clearly true with $\mathcal{R}(\mathsf{A})$ replacing $\mathrm{cl}\,\mathcal{R}(\mathsf{A})$. Thus

2.19. Corollary. *If* A *has closed range, then* (2.5) *has a solution if and only if* f *annihilates the null space of* A^*.

To exploit the powerful Corollary 2.19, we just have to demonstrate that A has closed range. It can be shown that if $\mathsf{A} = \mathsf{I} - \mathsf{K}$, where K is compact (see Sec. V.4), then A has closed range.

In general, these results are not symmetric in A and A^*: In place of a statement completely dual to Theorem 2.18, we have the easily proved

2.20. Theorem.

$$\mathrm{cl}\,\mathcal{R}(\mathbf{A}^*) \subset \mathcal{N}(\mathbf{A})^{\perp}. \tag{2.21}$$

If $\mathcal{X}$ is reflexive, then

$$\mathrm{cl}\,\mathcal{R}(\mathbf{A}^*) = \mathcal{N}(\mathbf{A})^{\perp}. \tag{2.22}$$

The proof of (2.22) is a direct consequence of Theorem 2.18. We can get (2.22) without using the reflexivity of $\mathcal{X}$ by restricting the class of operators $\mathbf{A}$. The following result, which we do not prove, is a special case of Banach's Closed Range Theorem:

2.23. Theorem. *Let $\mathbf{A} \in \mathcal{L}(\mathcal{X}, \mathcal{Y})$. Then the following statements are equivalent:*

(i) *$\mathcal{R}(\mathbf{A})$ is closed.*
(ii) *$\mathcal{R}(\mathbf{A}^*)$ is closed.*
(iii) *$\mathcal{R}(\mathbf{A}^*) = {}^{\perp}\mathcal{N}(\mathbf{A}^*)$.*
(iv) *$\mathcal{R}(\mathbf{A}^*) = \mathcal{N}(\mathbf{A})^{\perp}$.*

Lemma 2.13, Theorems 2.16, 2.18, 2.20, Corollary 2.19, and Theorem 2.23 are versions of the *(Fredholm) Alternative Theorem.*

2.24. Multiplier Rule. *Let $\mathbf{A}$ be a bounded linear operator from $\mathcal{X}$ onto $\mathcal{Y}$ (so that $\mathcal{R}(\mathbf{A}) = \mathcal{Y}$). Let $\mathbf{x}^* \in \mathcal{N}(\mathbf{A})^{\perp}$, i.e., let*

$$\langle \mathbf{x}, \mathbf{x}^* \rangle = 0 \quad \forall\, \mathbf{x} \in \mathcal{X} \quad \textit{such that} \quad \mathbf{A}\cdot\mathbf{x} = \mathbf{0}. \tag{2.25}$$

Then there exists a (Lagrange multiplier) $\mathbf{y}^* \in \mathcal{Y}^*$ such that

$$\langle \mathbf{x}, \mathbf{x}^* \rangle - \langle \mathbf{A}\cdot\mathbf{x}, \mathbf{y}^* \rangle = 0 \quad \forall\, \mathbf{x} \in \mathcal{X}. \tag{2.26}$$

Proof. Since $\mathcal{R}(\mathbf{A})$ is closed, we deduce from statement (iv) of Theorem 2.23 that there is a $\mathbf{y}^* \in \mathcal{R}(\mathbf{A}^*)$ such that $\mathbf{x}^* = \mathbf{A}^* \cdot \mathbf{y}^*$. Thus $\langle \mathbf{x}, \mathbf{x}^* \rangle = \langle \mathbf{x}, \mathbf{A}^* \cdot \mathbf{y}^* \rangle$. We use the definition of adjoint to obtain (2.26). □

This proof is but a special case of the proof of Theorem 9.31 (Lyusternik's Theorem) of Luenberger (1969).

2.27. Exercise. Use the Alternative Theorems to given necessary and sufficient conditions on the continuous function f so that the boundary-value problem

$$u'' + \pi^2 u = f \quad \text{on } (0,1), \quad u(0) = 0 = u(1) \tag{2.28}$$

has a unique solution. Perhaps the easiest way to do this is to use a Green function to convert (2.28) to an integral equation, which can be studied on $L_2(0,1)$ by the Alternative Theorems. In this approach, it is necessary to show that the integral equation is equivalent to the boundary-value problem (2.28).

Inverses. If $\mathbf{A} : \mathcal{X} \to \mathcal{Y}$ and $\mathbf{B} : \mathcal{Y} \to \mathcal{Z}$, then their *product* $\mathbf{A}\cdot\mathbf{B} : \mathcal{X} \to \mathcal{Z}$ is defined by $(\mathbf{A}\cdot\mathbf{B})\cdot\mathbf{x} \equiv \mathbf{A}\cdot(\mathbf{B}\cdot\mathbf{x})$. If $\mathbf{A} : \mathcal{X} \to \mathcal{X}$, then we write $\mathbf{A}^2 = \mathbf{A}\cdot\mathbf{A}$, etc.

If (2.5) has a unique solution for each $\mathbf{f} \in \mathcal{Y}$, then we represent the solution corresponding to a given $\mathbf{f}$ by $\mathbf{A}^{-1}\cdot\mathbf{f}$. Thus $\mathbf{A}\cdot\mathbf{A}^{-1}\cdot\mathbf{f} = \mathbf{f}$ for all $\mathbf{f} \in \mathcal{Y}$ or, equivalently, $\mathbf{A}\cdot\mathbf{A}^{-1} = \mathbf{I}$, where $\mathbf{I}$ is the identity operator from $\mathcal{Y}$ to itself. In other words, a one-to-one mapping $\mathbf{A}$ of $\mathcal{D}(\mathbf{A})$ onto $\mathcal{Y}$ has an inverse mapping $\mathbf{A} : \mathcal{Y} \to \mathcal{D}(\mathbf{A})$. It is readily seen that $\mathbf{A}^{-1}$ is a linear operator. The mapping $\mathbf{A}$ is said to be *invertible* if $\mathbf{A}$ has an inverse. When $\mathbf{A}$ is invertible, (2.5) is equivalent to $\mathbf{x} = \mathbf{A}^{-1}\cdot\mathbf{f}$. Thus (2.5) yields $\mathbf{A}^{-1}\cdot\mathbf{A}\cdot\mathbf{x} = \mathbf{A}^{-1}\cdot\mathbf{f} = \mathbf{x}$ for all $\mathbf{X} \in \mathcal{D}(\mathbf{A})$, so that $\mathbf{A}^{-1}\cdot\mathbf{A} = \mathbf{I}$, where $\mathbf{I}$ is the identity operator from $\mathcal{D}(\mathbf{A})$ to itself.

The following basic theorem, which we do not prove, is useful for some of the ensuing applications.

2.29. Banach's Open Mapping Theorem. *If* A *is a bounded linear operator from a Banach space* $\mathcal{X}$ *onto a Banach space* $\mathcal{Y}$, *then* A *maps open sets onto open sets. If, furthermore,* A *is a one-to-one (so that* A *is invertible with its range a Banach space), then* A^{-1} *is also a bounded linear operator.*

Projections. Let $\mathcal{A}$ and $\mathcal{B}$ be linear manifolds in a Banach space $\mathcal{X}$. Then $\mathcal{A}+\mathcal{B}$ denotes the set of all vectors in $\mathcal{X}$ of the form $\mathbf{a}+\mathbf{b}$ with $\mathbf{a}\in\mathcal{A}$ and $\mathbf{b}\in\mathcal{B}$. If, furthermore, $\mathcal{A}\cap\mathcal{B}=\{\mathbf{0}\}$, then we write $\mathcal{A}+\mathcal{B}$ as $\mathcal{A}\oplus\mathcal{B}$, which is called the *direct sum* of $\mathcal{A}$ and $\mathcal{B}$. It is easy to prove that $\mathcal{X}=\mathcal{A}\oplus\mathcal{B}$ if and only if each $\mathbf{x}\in\mathcal{X}$ has a unique representation of the form $\mathbf{x}=\mathbf{a}+\mathbf{b}$ with $\mathbf{a}\in\mathcal{A}$ and $\mathbf{b}\in\mathcal{B}$. When $\mathcal{X}$ admits the direct sum decomposition $\mathcal{X}=\mathcal{A}\oplus\mathcal{B}$, we say that $\mathcal{A}$ and $\mathcal{B}$ are *complementary linear manifolds* of $\mathcal{X}$ and that $\mathcal{A}$ is a *complement* of $\mathcal{B}$, and we denote $\operatorname{codim}\mathcal{A}\equiv\dim\mathcal{B}$. (It can be shown that $\operatorname{codim}\mathcal{A}$ is independent of the complementing space $\mathcal{B}$.)

Let P be the operator that associates with each such direct-sum decomposition $\mathbf{x}=\mathbf{a}+\mathbf{b}\in\mathcal{X}$ the vector $\mathbf{a}\in\mathcal{A}$. It is easy to see that P is a linear operator from $\mathcal{X}$ to itself and that $\mathsf{P}\cdot\mathsf{P}=\mathsf{P}$. P is called the *projection of* $\mathcal{X}$ *onto* $\mathcal{A}$ *(along* $\mathcal{B}$*)*. It follows that the projection of $\mathcal{X}$ onto the complementary linear manifold $\mathcal{B}$ along $\mathcal{A}$ is $\mathsf{I}-\mathsf{P}$. It can be shown that if $\mathcal{A}$ and $\mathcal{B}$ are Banach subspaces of $\mathcal{X}$ with $\mathcal{X}=\mathcal{A}\oplus\mathcal{B}$, then the projections of $\mathcal{X}$ onto $\mathcal{A}$ and $\mathcal{B}$ are bounded. This occurs if either $\mathcal{A}$ or $\mathcal{B}$ are finite-dimensional.

In general, any linear operator $\mathsf{P}:\mathcal{X}\to\mathcal{X}$ with $\mathsf{P}\cdot\mathsf{P}=\mathsf{P}$ is called a *projection*. Each such projection generates the decomposition $\mathcal{X}=\mathcal{R}(\mathsf{P})\oplus\mathcal{N}(\mathsf{P})$. To show this, we first observe that $\mathcal{X}=\mathcal{R}(\mathsf{P})+\mathcal{N}(\mathsf{P})$ because each $\mathbf{x}$ can be written in the form $\mathbf{x}-\mathsf{P}\cdot\mathbf{x}+(\mathsf{I}-\mathsf{P})\cdot\mathbf{x}$. Thus, if $\mathbf{x}\in\mathcal{R}(\mathsf{P})$, then $\mathbf{x}=\mathsf{P}\cdot\mathbf{x}$, so that if $\mathbf{x}\in\mathcal{R}(\mathsf{P})\cap\mathcal{N}(\mathsf{P})$, then $\mathbf{x}=\mathsf{P}\cdot\mathbf{x}=\mathbf{0}$, whence $\mathcal{R}(\mathsf{P})\cap\mathcal{N}(\mathsf{P})=\{\mathbf{0}\}$.

We state without proof some important properties about projections on Banach spaces:

2.30. Theorem. (i) *Let* P *be a bounded projection of Banach space* $\mathcal{X}$ *onto* $\mathcal{A}$. *Then* $\mathcal{A}$ *is a closed linear manifold of* $\mathcal{X}$, *i.e., it is a Banach subspace of* $\mathcal{X}$. (ii) *Let* $\mathcal{A}$ *be a Banach subspace of* $\mathcal{X}$. *Then there is a bounded projection of* $\mathcal{X}$ *onto* $\mathcal{A}$ *if and only if there is another Banach subspace* $\mathcal{B}$ *of* $\mathcal{X}$ *such that* $\mathcal{X}=\mathcal{A}\oplus\mathcal{B}$. *(In general, there need not be a bounded projection onto a Banach subspace* $\mathcal{A}$ *of* $\mathcal{X}$.) (iii) *If* $\mathcal{A}$ *is a finite-dimensional linear manifold of* $\mathcal{X}$, *or if it is a closed linear manifold of finite codimension, then there is a bounded projection of* $\mathcal{X}$ *onto* $\mathcal{A}$. (iv) *If* $\mathcal{X}$ *is a Hilbert space and if* $\mathcal{A}$ *is a closed linear manifold in it, then there is a bounded projection of* $\mathcal{X}$ *onto* $\mathcal{A}$.

Fredholm operators. Let A be a bounded linear operator from $\mathcal{X}$ to $\mathcal{Y}$. A is called a *Fredholm* operator iff $\dim\mathcal{N}(\mathsf{A})<\infty$ and $\operatorname{codim}\mathcal{R}(\mathsf{A})<\infty$. The difference

$$\operatorname{ind}\mathsf{A}\equiv\dim\mathcal{N}(\mathsf{A})-\operatorname{codim}\mathcal{R}(\mathsf{A})<\infty$$

is called the *Fredholm index* of A. We state the following properties of Fredholm operators without proof.

2.31. Theorem. *If* A *is a Fredholm operator, then*

(i) $\mathcal{R}(\mathsf{A})$ *is closed.*

(ii) A^* *is a Fredholm operator with index* $\operatorname{ind}\mathsf{A}^*=-\operatorname{ind}\mathsf{A}$.

(iii) *If* K *is a compact operator, then* $\mathsf{A}+\mathsf{K}$ *is a Fredholm operator with the same index as* A.

2.32. Exercise. Prove statements (i) and (ii).

2.33. Theorem. A *is a Fredholm operator of index* 0 *if and only if* A *is the sum of an invertible operator and a compact operator.*

2.34. Theorem. *If* A *is a Fredholm operator of index* 0 *and if* $\mathcal{N}(\mathsf{A}) = \{0\}$, *then the equation* $\mathsf{A} \cdot \mathsf{x} = \mathsf{f}$ *has a unique solution for each* $\mathsf{f} \in \mathcal{Y}$, *and* $\mathsf{A}^{-1} \in \mathcal{L}(\mathcal{Y}, \mathcal{X})$. *(This means that uniqueness implies existence, just as for this equation when* A *is a linear operator from* $\mathbb{R}^n$ *to itself.)*

From Theorem 2.30(iii) we immediately obtain that if A is a Fredholm operator, then there are bounded projections $\mathsf{P} \in \mathcal{L}(\mathcal{X}, \mathcal{X})$ and $\mathsf{Q} \in \mathcal{L}(\mathcal{Y}, \mathcal{Y})$ such that

$$\mathcal{R}(\mathsf{P}) \equiv \mathsf{P} \cdot \mathcal{X} = \mathcal{N}(\mathsf{A}), \quad \mathcal{R}(\mathsf{I} - \mathsf{Q}) = (\mathsf{I} - \mathsf{Q}) \cdot \mathcal{Y} = \mathcal{R}(\mathsf{A}). \tag{2.35}$$

As Theorems 2.31, 2.33, and 2.34 suggest, the Alternative Theorems have useful specializations for Fredholm operators, which furnished the motivation for the definition of these operators. We do not pursue this topic.

Most of the material in this section can be found in standard books on functional analysis. Among the more accessible references are Taylor & Lay (1980) and the following books, which are oriented toward differential equations and applications: Brezis (1983), Goldberg (1966), Hutson & Pym (1980), Kantorovich & Akilov (1977), Kreyszig (1978), and Luenberger (1969). For discussions of Fredholm operators, see Gohberg, Goldberg, & Kaashoek (1991) and Zeidler (1986).

CHAPTER XVIII

Appendix. Local Nonlinear Analysis

1. The Contraction Mapping Principle and the Implicit Function Theorem

In many of the problems treated in this text, we have detailed information about special solutions, especially those termed trivial. We can often determine solutions in a neighborhood of the special solutions or determine solutions of problems with nearby data by using methods relying on versions of the Implicit Function Theorem. In this chapter we develop these methods of local nonlinear analysis. Each of our results is a consequence of the Contraction Mapping Principle, which we now state and prove.

Let $\mathcal{F}$ be a closed subset of a Banach space $\mathcal{X}$ with norm $\|\cdot\|$. Let $\mathsf{f} : \mathcal{F} \to \mathcal{X}$. We seek solutions x in $\mathcal{F}$ of the nonlinear operator equation of the form

$$\mathsf{x} = \mathsf{f}(\mathsf{x}). \tag{1.1}$$

(A solution of (1.1) is called a *fixed point* of f.) The mapping f is called a (*strict*) *contraction* on $\mathcal{F}$ iff there is a number $\kappa \in [0, 1)$ such that

$$\|\mathsf{f}(\mathsf{x}) - \mathsf{f}(\mathsf{y})\| \leq \kappa \|\mathsf{x} - \mathsf{y}\| \quad \forall\, \mathsf{x}, \mathsf{y} \in \mathcal{F}. \tag{1.2}$$

The Contraction Mapping Principle ensures the existence and uniqueness of a solution to (1.1) when f is a contraction taking $\mathcal{F}$ into itself. Much of the rest of this chapter is devoted to putting concrete classes of equations into the form of (1.1), (1.2).

1.3. Contraction Mapping Principle (Banach-Cacciopoli Fixed-Point Theorem). *Let $\mathcal{F}$ be a closed subset of a Banach space $\mathcal{X}$ and let $\mathsf{f} : \mathcal{F} \to \mathcal{F}$ be a contraction. Then* (1.1) *has a unique solution $\bar{\mathsf{x}}$ in $\mathcal{F}$.*

Proof. Let x_0 be an arbitrary element of $\mathcal{F}$. We define the sequence $\{\mathsf{x}_k\}$ of elements in $\mathcal{F}$ by

$$\mathsf{x}_{k+1} = \mathsf{f}(\mathsf{x}_k), \; k = 0, 1, \ldots. \tag{1.4}$$

Then (1.2) implies that

$$\|\mathsf{x}_{k+1} - \mathsf{x}_k\| = \|\mathsf{f}(\mathsf{x}_k) - \mathsf{f}(\mathsf{x}_{k-1})\| \leq \kappa \|\mathsf{x}_k - \mathsf{x}_{k-1}\| \leq \kappa^k \|\mathsf{x}_1 - \mathsf{x}_0\|. \tag{1.5}$$

Thus for $k > l$,

$$\begin{aligned} \|\mathsf{x}_k - \mathsf{x}_l\| &\leq \|\mathsf{x}_k - \mathsf{x}_{k-1}\| + \|\mathsf{x}_{k-1} - \mathsf{x}_{k-2}\| + \cdots + \|\mathsf{x}_{l+1} - \mathsf{x}_l\| \\ &\leq \kappa^l (1 + \kappa + \cdots + \kappa^{k-l-1}) \|\mathsf{x}_1 - \mathsf{x}_0\| \\ &= \frac{\kappa^l (1 - \kappa^{k-l})}{1 - \kappa} \|\mathsf{x}_1 - \mathsf{x}_0\| \leq \frac{\kappa^l}{1 - \kappa} \|\mathsf{x}_1 - \mathsf{x}_0\|. \end{aligned} \tag{1.6}$$

Since $\kappa^l \to 0$ as $l \to \infty$ because $\kappa \in [0,1)$, it follows that $\{\mathsf{x}_k\}$ is a Cauchy sequence. Since $\mathcal{X}$ is a Banach space, this sequence converges to a limit $\bar{\mathsf{x}}$ in $\mathcal{X}$, and since $\mathcal{F}$ is closed, $\bar{\mathsf{x}} \in \mathcal{F}$. Since (1.2) implies that f is continuous, we now let $k \to \infty$ in (1.4) to show that $\bar{\mathsf{x}}$ is a solution of (1.1). To show that $\bar{\mathsf{x}}$ is unique, suppose that $\bar{\mathsf{x}}$ and $\bar{\mathsf{y}}$ are two solutions of (1.1). Then (1.1) and (1.2) imply that $\|\bar{\mathsf{x}} - \bar{\mathsf{y}}\| = \|\mathsf{f}(\bar{\mathsf{x}}) - \mathsf{f}(\bar{\mathsf{y}})\| \leq \kappa\|\bar{\mathsf{x}} - \bar{\mathsf{y}}\|$, so that $\|\bar{\mathsf{x}} - \bar{\mathsf{y}}\| = 0$. □

This proof is constructive. Indeed, if we let $k \to \infty$ in (1.6), then we obtain an error estimate:

$$\|\bar{\mathsf{x}} - \mathsf{x}_l\| \leq \frac{\kappa^l}{1-\kappa}\|\mathsf{f}(\mathsf{x}_0) - \mathsf{x}_0\|. \tag{1.7}$$

This theorem holds when $\mathcal{X}$ is a complete metric space.

1.8. Exercise. Define $\mathsf{f}^2(\mathsf{x}) \equiv \mathsf{f}(\mathsf{f}(\mathsf{x}))$, etc., where f is defined on a closed subset $\mathcal{F}$ of a Banach space $\mathcal{X}$. Suppose that there is a positive integer r such that f^r is a contraction on $\mathcal{F}$. Let $\bar{\mathsf{x}}$ be its fixed point. Prove that f (which need not be a contraction on $\mathcal{F}$) has the unique fixed point $\bar{\mathsf{x}}$.

Now we study the generalization

$$\mathsf{x} = \mathsf{f}(\mathsf{x}, \mathsf{y}) \tag{1.9}$$

of (1.1) obtained by introducing a parameter y into it.

1.10. Theorem. *Let $\mathcal{F}$ be a closed subset of a Banach space $\mathcal{X}$ and let $\mathcal{G}$ be a subset of a Banach space $\mathcal{Y}$. Let $\mathsf{f} : \mathcal{F} \times \mathcal{G} \to \mathcal{F}$ be a uniform contraction in the sense that there is a number $\kappa \in [0,1)$ independent of y such that*

$$\|\mathsf{f}(\mathsf{x}_1, \mathsf{y}) - \mathsf{f}(\mathsf{x}_2, \mathsf{y})\| \leq \kappa\|\mathsf{x} - \mathsf{y}\| \quad \forall\, \mathsf{x}_1, \mathsf{x}_2 \in \mathcal{F},\ \forall\, \mathsf{y} \in \mathcal{G}. \tag{1.11}$$

Let $\mathsf{f}(\mathsf{x}, \cdot)$ be continuous on $\mathcal{G}$ for each fixed $\mathsf{x} \in \mathcal{F}$. For each $\mathsf{y} \in \mathcal{G}$ let $\mathsf{h}(\mathsf{y})$ be the unique fixed point of $\mathsf{f}(\cdot, \mathsf{y})$. Then h is continuous from $\mathcal{G}$ to $\mathcal{F}$.

Proof. By definition of h,

$$\begin{aligned} \mathsf{h}(\mathsf{y}+\mathsf{z}) - \mathsf{h}(\mathsf{y}) &= \mathsf{f}(\mathsf{h}(\mathsf{y}+\mathsf{z}), \mathsf{y}+\mathsf{z}) - \mathsf{f}(\mathsf{h}(\mathsf{y}), \mathsf{y}) \\ &= \mathsf{f}(\mathsf{h}(\mathsf{y}+\mathsf{z}), \mathsf{y}+\mathsf{z}) - \mathsf{f}(\mathsf{h}(\mathsf{y}), \mathsf{y}+\mathsf{z}) \\ &\quad + \mathsf{f}(\mathsf{h}(\mathsf{y}), \mathsf{y}+\mathsf{z}) - \mathsf{f}(\mathsf{h}(\mathsf{y}), \mathsf{y}). \end{aligned} \tag{1.12}$$

Thus (1.11) implies that

$$\|\mathsf{h}(\mathsf{y}+\mathsf{z}) - \mathsf{h}(\mathsf{y})\| \leq \kappa\|\mathsf{h}(\mathsf{y}+\mathsf{z}) - \mathsf{h}(\mathsf{y})\| + \|\mathsf{f}(\mathsf{h}(\mathsf{y}), \mathsf{y}+\mathsf{z}) - \mathsf{f}(\mathsf{h}(\mathsf{y}), \mathsf{y})\|. \tag{1.13}$$

Since $\kappa \in [0,1)$, the continuity of $\mathsf{f}(\mathsf{x}, \cdot)$ implies that of h. □

We want to get results on the differentiability of h. Since the (Fréchet) derivative is a bounded linear operator, we can use the following result in reaching this goal.

1.14. Corollary. *Let $\mathcal{X}$ be a Banach space and let $\mathcal{G}$ be a subset of a Banach space $\mathcal{Y}$. For each y in $\mathcal{G}$, let $\mathsf{A}(\mathsf{y}) : \mathcal{X} \to \mathcal{X}$ be a bounded linear operator that is a uniform contraction in the sense that there is a number $\kappa \in [0,1)$ independent of y such that $\|\mathsf{A}(\mathsf{y})\| \leq \kappa$ for all y in $\mathcal{G}$. Let $\mathsf{A}(\cdot) \cdot \mathsf{x}$ be continuous on $\mathcal{G}$ for each fixed $\mathsf{x} \in \mathcal{X}$. Then the linear operator $\mathsf{I} - \mathsf{A}(\mathsf{y})$ has a bounded inverse, which depends continuously on y.*

1.15. Exercise. Prove this corollary (by showing that the equation $\mathsf{x} - \mathsf{A}(\mathsf{y}) \cdot \mathsf{x} = \mathsf{z}$ has a unique solution for each $\mathsf{z} \in \mathcal{X}$ and that this solution depends continuously on the parameters y, z).

Our major applications of the Contraction Mapping Principle are based on its following consequences. See Sec. IV.3 for the definition of Fréchet differentiability.

1.16. Theorem. *Let the hypotheses of Theorem 1.10 hold. Furthermore, let $\mathcal{F}$ and $\mathcal{G}$ be closures of the open sets* $\operatorname{int}\mathcal{F}$ *and* $\operatorname{int}\mathcal{G}$, *and let* f *be continuously Fréchet differentiable on* $\operatorname{int}\mathcal{F} \times \operatorname{int}\mathcal{G}$. *Then* h *is continuously Fréchet differentiable on* $\operatorname{int}\mathcal{G}$.

Proof. To prove this theorem, we need a candidate w for the differential of h in an arbitrary direction z. We shall get an equation for w by substituting $\mathsf{x} = \mathsf{h}(\mathsf{y})$ into (1.9) and formally differentiating this equation with respect to y. We shall then use Corollary 1.14 to prove that this equation has a solution depending continuously on y. Finally, from (1.9) we shall get a representation for the difference between $\mathsf{h}(\mathsf{y}+\mathsf{z}) - \mathsf{h}(\mathsf{y})$ and w and use it to prove the differentiability of h. We now carry out these steps.

Provisionally assuming that h is differentiable, we differentiate (1.9) with respect to y to obtain that w defined to equal $[\partial \mathsf{h}(\mathsf{y})/\partial \mathsf{y}] \cdot \mathsf{z}$, where z is an arbitrary element of $\mathcal{Y}$, would satisfy the linear equation

$$\mathsf{w} = \frac{\partial \mathsf{f}}{\partial \mathsf{x}}(\mathsf{h}(\mathsf{y}), \mathsf{y}) \cdot \mathsf{w} + \frac{\partial \mathsf{f}}{\partial \mathsf{y}}(\mathsf{h}(\mathsf{y}), \mathsf{y}) \cdot \mathsf{z}. \tag{1.17}$$

We identify the linear operator $\partial \mathsf{f}(\mathsf{h}(\mathsf{y}), \mathsf{y})/\partial \mathsf{x}$ with $\mathsf{A}(\mathsf{y})$ of Corollary 1.14, and readily verify that it satisfies the hypotheses of that corollary. Thus (1.17) has a unique solution, which we easily find to be linear in z. We denote this solution by $\mathsf{w} = \mathsf{B}(\mathsf{y}) \cdot \mathsf{z}$. This well-defined expression is our candidate for the differential of h in direction z. Note that Corollary 1.14 implies that $\mathsf{B}(\cdot)$ is continuous.

To show that $\mathsf{B}(\mathsf{y}) = \partial \mathsf{h}(\mathsf{y})/\partial \mathsf{y}$, we observe that (1.9), (1.17), and the definition of differentiability imply that

$$\begin{aligned}
\mathsf{u} &\equiv \mathsf{h}(\mathsf{y}+\mathsf{z}) - \mathsf{h}(\mathsf{y}) - \mathsf{B}(\mathsf{y}) \cdot \mathsf{z} \\
&= \mathsf{f}(\mathsf{h}(\mathsf{y}+\mathsf{z}), \mathsf{y}+\mathsf{z}) - \mathsf{f}(\mathsf{h}(\mathsf{y}), \mathsf{y}) - \frac{\partial \mathsf{f}}{\partial \mathsf{x}}(\mathsf{h}(\mathsf{y}), \mathsf{y}) \cdot \mathsf{B}(\mathsf{y}) \cdot \mathsf{z} - \frac{\partial \mathsf{f}}{\partial \mathsf{y}}(\mathsf{h}(\mathsf{y}), \mathsf{y}) \cdot \mathsf{z} \\
&= \mathsf{f}(\mathsf{h}(\mathsf{y}+\mathsf{z}), \mathsf{y}) + \frac{\partial \mathsf{f}}{\partial \mathsf{y}}(\mathsf{h}(\mathsf{y}+\mathsf{z}), \mathsf{y}) \cdot \mathsf{z} + o(\|\mathsf{z}\|) \\
&\quad - \mathsf{f}(\mathsf{h}(\mathsf{y}), \mathsf{y}) - \frac{\partial \mathsf{f}}{\partial \mathsf{x}}(\mathsf{h}(\mathsf{y}), \mathsf{y}) \cdot \mathsf{B}(\mathsf{y}) \cdot \mathsf{z} - \frac{\partial \mathsf{f}}{\partial \mathsf{y}}(\mathsf{h}(\mathsf{y}), \mathsf{y}) \cdot \mathsf{z} \\
&= \frac{\partial \mathsf{f}}{\partial \mathsf{x}}(\mathsf{h}(\mathsf{y}), \mathsf{y}) \cdot \mathsf{u} + \frac{\partial \mathsf{f}}{\partial \mathsf{y}}(\mathsf{h}(\mathsf{y}+\mathsf{z}), \mathsf{y}) \cdot \mathsf{z} - \frac{\partial \mathsf{f}}{\partial \mathsf{y}}(\mathsf{h}(\mathsf{y}), \mathsf{y}) \cdot \mathsf{z} \\
&\quad + o(\|\mathsf{h}(\mathsf{y}+\mathsf{z}) - \mathsf{h}(\mathsf{y})\|) + o(\|\mathsf{z}\|).
\end{aligned} \tag{1.18}$$

Since h and $\partial \mathsf{f}(\cdot, \mathsf{y})/\partial \mathsf{y}$ are continuous, we can write (1.18) in the form

$$\left[\mathsf{I} - \frac{\partial \mathsf{f}}{\partial \mathsf{x}}(\mathsf{h}(\mathsf{y}), \mathsf{y})\right] \cdot \mathsf{u} = o(\|\mathsf{z}\|). \tag{1.19}$$

We apply Corollary 1.14 to the operator on the left-hand side of (1.19) to obtain $\mathsf{u} = o(\|\mathsf{z}\|)$, which says that h is differentiable and that $\mathsf{B}(\cdot)$, which is continuous, is its derivative. □

It is much easier to prove

1.20. Theorem. *Let the hypotheses of Theorem 1.16 hold and let* f *be* k *times continuously differentiable on* $\operatorname{int}\mathcal{F} \times \operatorname{int}\mathcal{G}$. *Then* h *is itself* k *times continuously differentiable on* $\operatorname{int}\mathcal{G}$.

1.21. Exercise. Prove Theorem 1.20.

1.22. Exercise. Consider the following initial-value problem for a system of ordinary differential equations:

$$\mathsf{u}'(t) = \mathsf{g}(\mathsf{u}(t), t), \quad \mathsf{u}(0) = \mathbf{0} \tag{1.23}$$

where $\mathbf{u}$ takes values in $\mathbb{R}^n$, where $\mathbf{g}$ is continuous from $\mathbb{R}^n \times \mathbb{R}$ to $\mathbb{R}^n$, and where $\mathbf{g}$ satisfies the Lipschitz condition: There is a number K such that

$$|\mathbf{g}(\mathbf{u},t) - \mathbf{g}(\mathbf{v},t)| \leq K|\mathbf{u} - \mathbf{v}| \quad \forall \mathbf{u}, \mathbf{v} \in \mathbb{R}^n, \ \forall t. \tag{1.24}$$

Noting that (1.23) is equivalent to the integral equation

$$\mathbf{u}(t) = \int_0^t \mathbf{g}(\mathbf{u}(\tau), \tau)\, d\tau, \tag{1.25}$$

prove that (1.23) has a unique solution on a sufficiently small interval $[-r, r]$ by the following method: Let a and b be given positive numbers, let $M = \max\{|\mathbf{g}(\mathbf{u},t)| : |t| \leq a, |\mathbf{u}| \leq b\}$, and let $\mathcal{F}$ be the set of all continuous functions $\mathbf{v}$ on $[-r, r]$ such that $\mathbf{v}(0) = \mathbf{0}$, $\sup\{|\mathbf{v}(t)| : |t| \leq r\} \leq b$. Show that r can be chosen so small in terms of the parameters a, b, K, M that (1.25) has the form (1.1) where $\mathbf{f}$ is a contraction. (If (1.23) were to depend on parameters, including initial data, then Theorems 1.11, 1.16, and 1.20 would imply that the solution of the initial-value problem would depend on the parameters with the same smoothness with which the problem depends on the parameters.)

There are many fixed-point theorems related to the Contraction Mapping Principle. For discussions and references, see Istratescu (1981) and Zeidler (1986). We discuss an alternative class of fixed-point theorems in the next chapter. The much deeper Nash-Moser Theorem, which can be applied to otherwise intractable problems of dynamics and which we do not discuss, is treated in J. T. Schwartz (1969).

Our chief application of the Contraction Mapping Principle is to the local solution of equations of the form

$$\mathbf{g}(\mathbf{x}, \mathbf{y}) = \mathbf{0}. \tag{1.26}$$

1.27. Implicit Function Theorem (of Hildebrandt & Graves). *Let $\mathcal{X}, \mathcal{Y}, \mathcal{Z}$ be Banach spaces and let $\mathcal{E}$ be a neighborhood of $(\mathbf{0}, \mathbf{0})$ in $\mathcal{X} \times \mathcal{Y}$. Let $\mathcal{E} \ni (\mathbf{x}, \mathbf{y}) \mapsto \mathbf{g}(\mathbf{x}, \mathbf{y}) \in \mathcal{Z}$ be continuous, let $\mathbf{g}(\mathbf{0}, \mathbf{0}) = \mathbf{0}$, let the Fréchet derivative $\partial\mathbf{g}/\partial\mathbf{x}$ exist and be continuous on $\mathcal{E}$, and let $\partial\mathbf{g}(\mathbf{0},\mathbf{0})/\partial\mathbf{x}$ have a bounded inverse. Then there exists a neighborhood $\mathcal{G}$ of $\mathbf{0}$ in $\mathcal{Y}$ on which (1.26) has a unique solution for $\mathbf{x}$ given by a continuous operator $\mathbf{h} : \mathcal{G} \to \mathcal{X}$ with $\mathbf{h}(\mathbf{0}) = \mathbf{0}$ such that*

$$\mathbf{g}(\mathbf{h}(\mathbf{y}), \mathbf{y}) = \mathbf{0} \quad \forall \mathbf{y} \in \mathcal{G}. \tag{1.28}$$

If, furthermore, $\mathbf{g}$ is k times continuously differentiable on $\mathcal{E}$, $k \geq 1$, then so is $\mathbf{h}$.

Proof. We write (1.26) in the form

$$\mathbf{x} = -\mathbf{A}^{-1} \cdot [\mathbf{g}(\mathbf{x}, \mathbf{y}) - \mathbf{A} \cdot \mathbf{x}] \equiv \mathbf{f}(\mathbf{x}, \mathbf{y}), \quad \mathbf{A} \equiv \frac{\partial \mathbf{g}(\mathbf{0}, \mathbf{0})}{\partial \mathbf{x}}. \tag{1.29}$$

To show that $\mathbf{f}(\cdot, \mathbf{y})$ is a uniform contraction, we observe that the Mean Value Theorem implies that

$$\mathbf{g}(\mathbf{x}, \mathbf{y}) - \mathbf{g}(\mathbf{w}, \mathbf{y}) - \mathbf{A} \cdot (\mathbf{x} - \mathbf{w}) = \left\{ \int_0^1 \left[\frac{\partial \mathbf{g}(t\mathbf{x} + (1-t)\mathbf{w}, \mathbf{y})}{\partial \mathbf{x}} - \frac{\partial \mathbf{g}(\mathbf{0}, \mathbf{0})}{\partial \mathbf{x}} \right] dt \right\} \cdot (\mathbf{x} - \mathbf{w}). \tag{1.30}$$

Since the term in braces can be made arbitrarily small for $(\mathbf{x}, \mathbf{w}, \mathbf{y})$ sufficiently close to $(\mathbf{0}, \mathbf{0}, \mathbf{0})$, it follows that $\mathbf{f}$ is a contraction for small enough arguments. We complete the proof by invoking Theorems 1.16 and 1.20. □

Of course, a simple change of variables enables us to replace $(\mathbf{x}, \mathbf{y}) = (\mathbf{0}, \mathbf{0})$ in this theorem with any given point in $\mathcal{X} \times \mathcal{Y}$.

2. The Lyapunov-Schmidt Method. The Poincaré Shooting Method

In our study of bifurcation problems we encounter operators whose linearizations fail to satisfy the hypotheses of the Implicit Function Theorem. The Lyapunov-Schmidt Method, which we now describe, handles many such problems by projecting the equations onto two complementary spaces, with one projection amenable to the Implicit Function Theorem and the other consisting of equations in a finite-dimensional space. We then describe an alternative approach, the Poincaré Shooting Method, which is applicable only to boundary-value problems for ordinary differential equations.

The Lyapunov-Schmidt Method. We now study (1.26) under the hypotheses of Theorem 1.27, except that we now replace the requirement that $\mathsf{A} \equiv \partial \mathsf{g}(\mathsf{0},\mathsf{0})/\partial \mathsf{x}$ have a bounded inverse with the requirement that A be a Fredholm operator.

Since $\mathcal{N}(\mathsf{A})$ is a finite-dimensional subspace of $\mathcal{X}$ (by definition of a Fredholm operator), Theorem XVII.2.30(iii) says that there is a bounded projection P of $\mathcal{X}$ onto $\mathcal{N}(\mathsf{A})$, which induces the decomposition of $\mathsf{x} \in \mathcal{X}$ as the direct sum

$$\text{(2.1)} \qquad \mathsf{x} = \mathsf{u} + \mathsf{v}, \quad \mathsf{u} = \mathsf{P} \cdot \mathsf{x} \in \mathcal{N}(\mathsf{A}), \quad \mathsf{v} = (\mathsf{I} - \mathsf{P}) \cdot \mathsf{x}.$$

Since Theorem XVII.2.31(i) implies that $\mathcal{R}(\mathsf{A})$ is a closed subspace of $\mathcal{Z}$ with finite codimension, Theorem XVII.2.30(iii) says that there is a bounded projection $\mathsf{I} - \mathsf{Q}$ of $\mathcal{Y}$ onto $\mathcal{R}(\mathsf{A})$. Thus we can write (1.26) as

$$\text{(2.2a,b)} \qquad \mathsf{h}(\mathsf{u},\mathsf{v},\mathsf{y}) \equiv (\mathsf{I} - \mathsf{Q}) \cdot \mathsf{g}(\mathsf{u}+\mathsf{v},\mathsf{y}) = \mathsf{0}, \quad \mathsf{Q} \cdot \mathsf{g}(\mathsf{u}+\mathsf{v},\mathsf{y}) = \mathsf{0}.$$

From our hypotheses on g, it follows that h maps a neighborhood $\mathcal{O}$ of $(\mathsf{0},\mathsf{0},\mathsf{0})$ in $[\mathsf{P} \cdot \mathcal{X}] \times [(\mathsf{I} - \mathsf{P}) \cdot \mathcal{X}] \times \mathcal{Y}$ into $(\mathsf{I} - \mathsf{Q}) \cdot \mathcal{Z} = \mathcal{R}(\mathsf{A})$, that $\mathsf{h}(\mathsf{0},\mathsf{0},\mathsf{0}) = \mathsf{0}$, and that

$$\frac{\partial \mathsf{h}}{\partial \mathsf{v}}(\mathsf{0},\mathsf{0},\mathsf{0}) = (\mathsf{I} - \mathsf{Q})\frac{\partial \mathsf{g}}{\partial \mathsf{x}}(\mathsf{0},\mathsf{0}) = (\mathsf{I} - \mathsf{Q})\mathsf{A} = \mathsf{A}$$

is a bounded one-to-one mapping of $(\mathsf{I} - \mathsf{P}) \cdot \mathcal{X}$ onto $\mathcal{R}(\mathsf{A})$, this last observation following from the properties of P and Q. Now $\mathcal{R}(\mathsf{A})$ is a Banach space, by definition of a Fredholm mapping, and $(\mathsf{I} - \mathsf{P}) \cdot \mathcal{X}$ is a Banach space since P is bounded, by Theorem XVII.2.30(i). By the Open Mapping Theorem XVII.2.29, $\partial \mathsf{h}(\mathsf{0},\mathsf{0},\mathsf{0})/\partial \mathsf{v}$ has a bounded inverse. We can therefore use the Implicit Function Theorem 1.27 to solve (2.2a) for v as a unique function of u and y for (u,y) near $(\mathsf{0},\mathsf{0})$. We denote this solution by $\mathsf{v} = \hat{\mathsf{v}}(\mathsf{u},\mathsf{y})$. We now substitute this solution into (2.2b) to obtain

$$\text{(2.3)} \qquad \mathsf{Q} \cdot \mathsf{g}(\mathsf{u} + \hat{\mathsf{v}}(\mathsf{u},\mathsf{y}),\mathsf{y}) = \mathsf{0},$$

which is a system of d equations for the n-dimensional u where $d = \operatorname{codim} \mathcal{R}(\partial \mathsf{g}(\mathsf{0},\mathsf{0})/\partial \mathsf{x})$ and $n = \dim \mathcal{N}(\partial \mathsf{g}(\mathsf{0},\mathsf{0})/\partial \mathsf{x})$. The Implicit Function Theorem implies that $\hat{\mathsf{v}}$ is k times continuously differentiable near $(\mathsf{0},\mathsf{0})$ when g is k times continuously differentiable. Thus the operator in (2.3) has the same smoothness, and we can use its k-degree Taylor polynomial about $(\mathsf{0},\mathsf{0})$ in our study of (2.3). We can expect solutions of (2.3) when $n \geq d$, which occurs when $\partial \mathsf{g}(\mathsf{0},\mathsf{0})/\partial \mathsf{x}$ has a nonnegative Fredholm index. The most favorable case occurs when $\partial \mathsf{g}(\mathsf{0},\mathsf{0})/\partial \mathsf{x}$ has Fredholm index 0. In carrying out the detailed analysis of (2.3), it is convenient to decompose u with respect to a biorthogonal system, which is defined in terms of dual spaces.

Now in studying bifurcation problems for ordinary differential equations, it is convenient to use spaces of continuous functions or continuously differentiable functions, because they are appropriate for describing detailed nodal properties. The duality theory for such spaces is technical; one way to avoid the concomitant difficulties is to use devices suggested in Ex. XVII.2.27. It is even easier to use the Poincaré Shooting Method, which we soon describe.

For variants of the Lyapunov-Schmidt Method, see Chow & Hale (1982). For extensive applications, see Vaĭnberg & Trenogin (1969) and Zeidler (1986). For the application of this method to problems with group symmetry, see Golubitsky & Schaeffer (1985), Golubitsky, Stewart, & Schaeffer (1988), and Vanderbauwhede (1982).

The Poincaré Shooting Method. Let $\mathbf{u} = (u_1, \ldots, u_n)$. We study nth-order systems of ordinary differential equations of the form

$$\mathbf{u}' = \mathbf{f}(s, \mathbf{u}, \boldsymbol{\lambda}) \quad \text{for } 0 < s < 1 \tag{2.4}$$

subject to n side conditions (e.g., boundary conditions) of the form

$$\mathbf{b}[\mathbf{u}(\cdot), \boldsymbol{\lambda}] = \mathbf{0}. \tag{2.5}$$

Here $\boldsymbol{\lambda}$ designates an l-tuple of real parameters, and $\mathbf{f}$ is a given continuous function, defined for simplicity on all of $[0,1] \times \mathbb{R}^n \times \mathbb{R}^l$, with $\mathbf{f}(s, \cdot, \cdot)$ assumed to be k times continuously differentiable on $\mathbb{R}^n \times \mathbb{R}^l$ for each s, $k \geq 1$. $\mathbf{b}$ is a given k times continuously differentiable mapping from $C^0[0,1] \times \mathbb{R}^l$ to $\mathbb{R}^n$.

Our goal is to reduce the solution of (2.4), (2.5) to the solution of a finite number of equations in a finite number of unknowns. For this purpose, we exploit the fact that we can readily solve initial-value problems (see Ex. 1.22), at least locally. Consider (2.4) subject to the initial condition

$$\mathbf{u}(0) = \boldsymbol{\alpha}. \tag{2.6}$$

Suppose that (2.4), (2.6) has a solution $\hat{\mathbf{u}}(\cdot, \boldsymbol{\alpha}, \boldsymbol{\lambda})$ defined on the whole interval $[0,1]$. Its substitution into (2.5) yields the desideratum of a system of n equations for the n unknown components of $\boldsymbol{\alpha}$:

$$\mathbf{b}[\hat{\mathbf{u}}(\cdot, \boldsymbol{\alpha}, \boldsymbol{\lambda}), \boldsymbol{\lambda}] = \mathbf{0}. \tag{2.7}$$

For many classes of equations, one can show that all solutions of an initial-value problem exist on any bounded interval (e.g., by using energy estimates). When such results are not available, we can appeal to

2.8. Theorem. *Let the assumptions on $\mathbf{f}$ made above hold. Let the initial-value problem consisting of* (2.4) *with $\boldsymbol{\lambda} = \boldsymbol{\lambda}_0$ and $\mathbf{u}(0) = \boldsymbol{\alpha}_0$ have a (necessarily unique) solution $\hat{\mathbf{u}}(\cdot, \boldsymbol{\alpha}_0, \boldsymbol{\lambda}_0)$ defined on* $[0,1]$. *Then there are neighborhoods of $\boldsymbol{\alpha}_0$ in $\mathbb{R}^n$ and of $\boldsymbol{\lambda}_0$ in $\mathbb{R}^l$ such that if $\boldsymbol{\alpha}$ and $\boldsymbol{\lambda}$ are in these neighborhoods, then* (2.4), (2.6) *has a unique solution $\hat{\mathbf{u}}(\cdot, \boldsymbol{\alpha}, \boldsymbol{\lambda})$ defined on* $[0,1]$ *with $\hat{\mathbf{u}}$ continuous in s and k times continuously differentiable in $\boldsymbol{\alpha}$ and $\boldsymbol{\lambda}$.*

2.9. Exercise. Combine methods of Ex. 1.22 and Theorem 1.27 to prove this theorem.

By converting the initial-value problem (2.4), (2.6) to an integral equation like (1.25), we can easily prove that the derivatives up to order k of $\hat{\mathbf{u}}$ with respect to $\boldsymbol{\alpha}$ and $\boldsymbol{\lambda}$ at $(\boldsymbol{\alpha}_0, \boldsymbol{\lambda}_0)$ satisfy linear initial-value problems obtained by the formal differentiation of (2.4), (2.6) with respect to these parameters. For an example, see Sec. VI.6.

The finite-dimensional system (2.7) can often be simplified further by applying the Lyapunov-Schmidt Method to it. For a discussion of a variant of this approach, see J. B. Keller (1969). Such a reduction is not needed when (2.7) is to be analyzed by singularity theory.

If an ordinary differential equation is totally integrable, then its integrals can be used to reduce the solution of a boundary-value problem to the solution of a finite-dimensional problem without the local restrictions that support the Lyapunov-Schmidt Method and the Poincaré Shooting Method. The price for the global validity of this finite-dimensional problem is that the resulting finite-dimensional problem typically involves singular integrals. For an application of this approach, see Sec. VI.9.

Solution methods for the finite-dimensional problem. If the finite-dimensional problem reduces to a single equation for two unknowns, say in the form

$$f(x, y) = 0 \quad \text{where } f(0,0) = 0, \tag{2.10}$$

and if f is $(n-1)$ times continuously differentiable near $(0,0)$, then all small real solutions of (2.10) for y in terms of x for $x > 0$ can be found by a *Puiseux series*, which has the form

$$y = \sum_{k=1}^{n} A_k x^{\alpha_k} + o(x^{\alpha_n}) \tag{2.11}$$

where $0 < \alpha_1 < \cdots < \alpha_n$. An analogous representation holds for $x < 0$. The α_k's can be found systematically by means of the Newton Polygon. For an account of the basic theory, see Dieudonné (1949) and for numerous examples see Vaĭnberg & Trenogin (1969). Some of the underlying ideas are used in the treatment in Sec. IV.4, which is more complicated than that for (2.10) because the nonlinear operator in Sec. IV.4 need not have a polynomial approximation.

Detailed results comparable to those delivered by (2.11) for finite-dimensional systems more complicated than (2.10) are very hard to obtain. Very useful qualitative information can, however, be found by using the singularity theory of Golubitsky & Schaeffer (1985). Some of the fundamental concepts are discussed in Sec. VI.6.

CHAPTER XIX

Appendix. Degree Theory

1. Definition of the Brouwer Degree

Throughout this section, we take Ω to be a bounded open subset of $\mathbb{R}^n$ and take $\operatorname{cl}\Omega \ni \mathsf{x} \mapsto \mathsf{f}(\mathsf{x}) \in \mathbb{R}^n$ to be continuous. We wish to estimate the number of solutions x in $\operatorname{cl}\Omega$ of the equation

$$\mathsf{f}(\mathsf{x}) = \mathbf{0}. \tag{1.1}$$

We shall often be content with demonstrating that there is (at least) one solution. We denote the collection of solutions of (1.1) lying in $\operatorname{cl}\Omega$ by $\mathsf{f}^{-1}(\{\mathbf{0}\})$.

Our plan is first to consider a somewhat nicer collection of f's for which (1.1) has a finite number of 'simple' solutions. With each such solution, we associate the number ± 1. Let the sum of these numbers be d. Then clearly (1.1) has at least $|d|$ solutions. We shall show that there is a special choice of d, denoted $\deg[\mathsf{f}, \Omega]$, that (i) can be extended to continuous f's and can be used to estimate the number of solutions of (1.1), (ii) depends continuously on f and Ω, (iii) can be determined from the restriction of f to $\partial\Omega$, and (iv) is stable under a class of large perturbations of f. Since our aim is to expose the simplicity of the logical framework of the theory, we merely outline the main ideas of certain of the more technical aspects of the development.

For $\mathsf{f} \in C^1(\Omega)$, we set

$$J(\mathsf{x}) = \det \frac{\partial \mathsf{f}}{\partial \mathsf{x}}(\mathsf{x}). \tag{1.2}$$

Let $J^{-1}(\{0\})$ denote the set of all points x *in* Ω for which $J(\mathsf{x}) = 0$. To characterize the range of $J^{-1}(\{0\})$ under f, we use a special case of Sard's Theorem:

1.3. Lemma. *Let* $\mathsf{f} \in C^1(\Omega)$. *Then* $\mathsf{f}\left(J^{-1}(\{0\})\right)$ *has Lebesgue measure* 0, *so that* $\operatorname{int}\mathsf{f}\left(J^{-1}(\{0\})\right) = \emptyset$, *i.e., the image of the set of points at which the Jacobian of* f *vanishes is so sparse that it has no interior.*

Sketch of the proof. First let us suppose that f is affine, so that there is a constant linear transformation A and a constant vector b such that $\mathsf{f}(\mathsf{x}) = \mathsf{A}\cdot\mathsf{x}+\mathsf{b}$ and $J = \det \mathsf{A}$. If $\det \mathsf{A} \neq 0$, then $J^{-1}(\{0\}) = \emptyset$, so that $\mathsf{f}(J^{-1}(\{0\})) = \emptyset$. If $\det \mathsf{A} = 0$, then $J^{-1}(\{0\}) = \Omega$. In this case, f takes $J^{-1}(\{0\})$ into a plane in $\mathbb{R}^n$ of dimension $< n$, which has empty interior. Thus if f is affine, the conclusion of the lemma holds. If f is not affine, then its continuous differentiability allows us to approximate it locally by affine maps, whose effect on $J^{-1}(\{0\})$ we know. Careful estimates then show that the volume of $\mathsf{f}(J^{-1}(\{0\}))$ is bounded above by an arbitrarily small number. □

We are now ready to begin our definition of Brouwer degree. We always restrict our attention to continuous f's and Ω's for which (1.1) has no solution on $\partial\Omega$:

$$\mathbf{0} \notin \mathsf{f}(\partial\Omega). \tag{1.4}$$

We first consider the special case that

$$\mathsf{f} \in C^1(\Omega) \cap C^0(\operatorname{cl}\Omega), \tag{1.5}$$

$$J(\mathsf{x}) \neq 0 \text{ for } \mathsf{x} \in \mathsf{f}^{-1}(\{\mathbf{0}\}), \text{ i.e., } J^{-1}(\{0\}) \cap \mathsf{f}^{-1}(\{\mathbf{0}\}) = \emptyset. \tag{1.6}$$

(Condition (1.6) says that solutions of (1.1) are *simple*.) Under these conditions, we define the *Brouwer degree* of f to be

$$\deg[\mathsf{f}, \Omega] \equiv \sum \operatorname{sign} J(\mathsf{x}) \tag{1.7}$$

where the sum is taken over all $\mathsf{x} \in \mathsf{f}^{-1}(\{\mathbf{0}\})$. (This sum is taken to be zero if $\mathsf{f}^{-1}(\{\mathbf{0}\})$ is empty.) Let us show that this sum is well defined by showing that it is finite: The Inverse Function Theorem implies that $\mathsf{f}^{-1}(\{\mathbf{0}\})$ is discrete, i.e., consists solely of isolated points, because f must be one-to-one in a neighborhood of every x in $\mathsf{f}^{-1}(\{\mathbf{0}\})$. Since $\operatorname{cl}\Omega$ is compact, $\mathsf{f}^{-1}(\{\mathbf{0}\})$ must be finite, for if not, it would have an accumulation point in $\operatorname{cl}\Omega$, in violation of the discreteness. We illustrate the significance of the definition (1.7) in Fig. 1.8. Our aim is now to relax the restrictions (1.5) and (1.6).

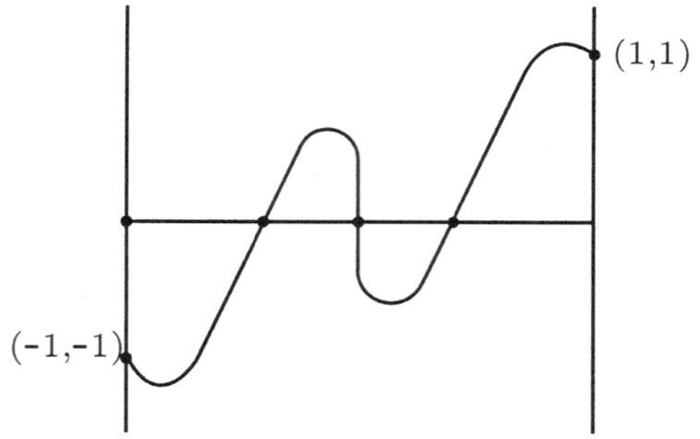

FIG. 1.8. Graph of a C^1 function $f : [-1, 1] \to \mathbb{R}$ required to satisfy the boundary conditions $f(\pm 1) = \pm 1$ and required to have simple zeros. It follows from (1.7) that $\deg(f, [-1, 1]) = 1$. For this function, Eq. (1.1) has exactly three solutions. For arbitrary C^1 perturbations of f satisfying the same boundary conditions, it is only certain that (1.1) has at least one solution, a conclusion that is ensured for any continuous f satisfying these boundary conditions by the Intermediate Value Theorem. Thus degree gives an estimate of the number of solutions that is stable under perturbations. This example illustrates that this degree is determined solely by the boundary conditions.

To handle the difficult condition (ii), we now obtain an integral representation for (1.7), to which $\mathsf{f}\left(J^{-1}(\{0\})\right)$ makes no contribution because it has zero measure, by Sard's Lemma: Let $\mathbb{B}(\mathbf{a}, \varepsilon)$ denote the open ball of radius ε about $\mathbf{a}$ in $\mathbb{R}^n$. Let $\{j_\varepsilon\}$ be a family of infinitely differentiable functions from $\mathbb{R}^n$ to $\mathbb{R}$ such that j_ε vanishes outside $\mathbb{B}(\mathbf{0}, \varepsilon)$ and such that

$$\int_{\mathbb{R}^n} j_\varepsilon(\mathsf{y})\, dv(\mathsf{y}) \equiv \int_{\mathbb{B}(\mathbf{0},\varepsilon)} j_\varepsilon(\mathsf{y})\, dv(\mathsf{y}) = 1.$$

(We can take $j_\varepsilon(\mathsf{y}) = \varepsilon^{-n} j_1(\mathsf{y}/\varepsilon)$ where $j_1(\mathsf{y}) = c\exp\left(-1/(1-|\mathsf{x}|^2)\right)$ for $|\mathsf{x}| < 1$, where $j_1(\mathsf{y}) = 0$ otherwise, and where the positive constant c is adjusted so that $\int_{\mathbb{R}^n} j_1(\mathsf{y})\, dv(\mathsf{y}) = 1$.) Then we have

1.9. Lemma (Heinz). *Let conditions* (1.4), (1.5), *and* (1.6) *hold. Then there is a number* $\bar{\varepsilon}(\mathsf{f}) > 0$ *such that*

$$\deg[\mathsf{f}, \Omega] = I_\varepsilon(\mathsf{f}, \Omega) \equiv \int_\Omega j_\varepsilon(\mathsf{f}(\mathsf{x}))J(\mathsf{x})\, dv(\mathsf{x}) \tag{1.10}$$

for $0 < \varepsilon < \bar{\varepsilon}(\mathsf{f})$.

Proof. Let $\{\mathsf{x}_1, \dots, \mathsf{x}_k\} = \mathsf{f}^{-1}(\{\mathbf{0}\})$. For small enough ε, the Inverse Function Theorem applied to the equation $\mathsf{f}(\mathsf{x}) = \mathsf{y}$ for $\mathsf{y} \in \mathbb{B}(\mathbf{0}, \varepsilon)$ implies that there are neighborhoods $\mathcal{A}_\varepsilon^1, \dots, \mathcal{A}_\varepsilon^k$ that are mapped homeomorphically by f onto $\mathbb{B}(\mathbf{0}, \varepsilon)$. Now

$$j_\varepsilon(\mathsf{f}(\mathsf{x})) = 0 \quad \text{if} \quad \mathsf{x} \notin \cup \mathcal{A}_\varepsilon^i \tag{1.11}$$

because $\mathsf{f}(\mathsf{x}) \notin \mathbb{B}(\mathbf{0}, \varepsilon)$ for such x. Therefore,

$$I_\varepsilon(\mathsf{f}, \Omega) = \sum_i \int_{\mathcal{A}_\varepsilon^i} j_\varepsilon(\mathsf{f}(\mathsf{x})) J(\mathsf{x})\, dv(\mathsf{x}). \tag{1.12}$$

Since $J(\mathsf{x}_i) \neq 0$ and since $\mathsf{f} \in C^1(\Omega)$, it follows that ε can be taken so small that J vanishes nowhere on each $\mathcal{A}_\varepsilon^i$. Thus sign $J(\mathsf{x})$ is defined and constant on $\mathcal{A}_\varepsilon^i$. We can now use the formula for change of variables to show that

$$\begin{aligned}\int_{\mathcal{A}_\varepsilon^i} j_\varepsilon(\mathsf{f}(\mathsf{x})) J(\mathsf{x})\, dv(\mathsf{x}) &= \operatorname{sign} J(\mathsf{x}_i) \int_{\mathcal{A}_\varepsilon^i} j_\varepsilon(\mathsf{f}(\mathsf{x})) |J(\mathsf{x})|\, dv(\mathsf{x}) \\ &= \operatorname{sign} J(\mathsf{x}_i) \int_{\mathbb{B}(\mathbf{0},\varepsilon)} j_\varepsilon(\mathsf{y})\, dv(\mathsf{y}) = \operatorname{sign} J(\mathsf{x}_i). \quad \square\end{aligned} \tag{1.13}$$

We now obtain some identities that enable us to define deg without using (1.6), but at the expense of initially replacing (1.5) with the more restrictive condition

$$\mathsf{f} \in C^2(\Omega) \cap C^0(\operatorname{cl}\Omega). \tag{1.14}$$

Under this condition, we first show that $\deg[\mathsf{f}, \Omega]$ is not affected by certain perturbations $\mathsf{f} - \mathsf{a}$ of f by constant vectors a. (We cannot use (1.10) for this purpose because $\bar{\varepsilon}(\mathsf{f} - \mathsf{a})$ could go to 0 as a approaches $\mathsf{f}\left(J^{-1}(\{0\})\right)$.)

1.15. Lemma. *Let* (1.4) *and* (1.14) *hold. Let* $\delta = \operatorname{dist}(0, \mathsf{f}(\partial\Omega))$. *Let* a_1 *and* a_2 *be two points of* $\mathbb{B}(\mathbf{0}, \delta)$ *with* $\mathsf{a}_1, \mathsf{a}_2 \notin \mathsf{f}\left(J^{-1}(\{0\})\right)$. *Then*

$$\deg[\mathsf{f} - \mathsf{a}_1, \Omega] = \deg[\mathsf{f} - \mathsf{a}_2, \Omega].$$

Proof. Lemma 1.9 implies that there is an $\varepsilon < \delta - \max\{|\mathsf{a}_1|, |\mathsf{a}_2|\}$ such that (1.10) holds with f replaced with $\mathsf{f} - \mathsf{a}_1$ and with $\mathsf{f} - \mathsf{a}_2$. To show that these integrals are equal, i.e., to show that

$$\int_\Omega \left[j_\varepsilon\left(\mathsf{f}(\mathsf{x}) - \mathsf{a}_2\right) - j_\varepsilon\left(\mathsf{f}(\mathsf{x}) - \mathsf{a}_1\right)\right] J(\mathsf{x})\, dv(\mathsf{x}) = 0, \tag{1.16}$$

we express the integrand of (1.16) as a divergence and then apply the Divergence Theorem. Toward this end, we first use the Mean Value Theorem with Integral Remainder (i.e., the Fundamental Theorem of Calculus) to write

$$\begin{aligned}j_\varepsilon(\mathsf{y} - \mathsf{a}_2) - j_\varepsilon(\mathsf{y} - \mathsf{a}_1) &= \int_0^1 \partial_t j_\varepsilon\left(\mathsf{y} - (1-t)\mathsf{a}_1 - t\mathsf{a}_2\right) dt \\ &= \operatorname{div} \mathsf{w}(\mathsf{y}) \equiv \sum_1^n \partial w_k(\mathsf{y}) / \partial y_k\end{aligned} \tag{1.17}$$

where

$$\mathsf{w}(\mathsf{y}) = (\mathsf{a}_1 - \mathsf{a}_2) \int_0^1 j_\varepsilon\left(\mathsf{y} - (1-t)\mathsf{a}_1 - t\mathsf{a}_2\right) dt.$$

Now let C_{ij} be the components of the cofactor matrix of $(\partial f_i / \partial x_j)$. We have the classical identity

$$\sum_j (\partial f_i / \partial x_j) C_{kj} = \delta_{ik} J \tag{1.18}$$

where δ_{ij} is the Kronecker delta. A careful treatment of the cofactor matrix (see many of the references cited at the end of Sec. 3) leads to the identity

$$\sum_i \frac{\partial C_{ij}}{\partial x_i} = 0. \tag{1.19}$$

We define

$$v_i(\mathsf{x}) \equiv \sum_j w_j(\mathsf{f}(\mathsf{x}))C_{ji}(\mathsf{x}) \tag{1.20}$$

for $\mathsf{x} \in \operatorname{cl}\Omega$ and define $v_i(\mathsf{x}) = 0$ for all other x. Using (1.17)–(1.19), we immediately find that $\operatorname{div}\mathsf{v}$ is the integrand of (1.16). Our choice of ε shows that w and therefore v have compact support on Ω. Since v is continuously differentiable, we use the Divergence Theorem to obtain (1.16). □

When (1.4) and (1.14) hold (but when (1.6) need not hold), we define

$$\deg[\mathsf{f}, \Omega] \equiv \deg[\mathsf{f} - \mathsf{a}, \Omega] \tag{1.21}$$

where $\mathsf{a} \notin \mathsf{f}\left(J^{-1}(\{0\})\right)$, $\mathsf{a} \in \mathbb{B}(\mathbf{0}, \delta))$ with $\delta = \operatorname{dist}(0, \mathsf{f}(\partial\Omega))$. This definition makes sense because Sard's Lemma 1.3 implies that such points a are dense in every neighborhood of $\mathbf{0}$ and because Lemma 1.15 implies that $\deg[\mathsf{f} - \mathsf{a}, \Omega]$ has the same value for all such a.

1.22. Exercise. Let (1.14) hold. Show that if $\mathsf{f}^{-1}(\{\mathbf{0}\}) = \emptyset$, then $\deg(\mathsf{f}, \Omega) = 0$.

The essential step in lifting the restriction (1.14) is the following approximation result:

1.23. Lemma. *Let* f *satisfy* (1.4) *and* (1.14) *and let* g *satisfy* (1.14). *Then there is a number* $\bar{\eta}(\mathsf{f}, \mathsf{g})$ *such that*

$$\deg(\mathsf{f} + t\mathsf{g}, \Omega) = \deg(\mathsf{f}, \Omega) \quad \text{for} \quad |t| < \bar{\eta}(\mathsf{f}, \mathsf{g}). \tag{1.24}$$

Proof. If $\mathsf{f}^{-1}(\{\mathbf{0}\}) = \emptyset$, then there cannot be any solutions of $\mathsf{f}(\mathsf{x}) + t\mathsf{g}(\mathsf{x}) = \mathbf{0}$ in Ω for $|t|$ sufficiently small, in which case the degrees in (1.24) are each equal to 0.

Now we establish (1.24) under the restrictive assumptions that

$$\mathsf{f}^{-1}(\{\mathbf{0}\}) = \{\mathsf{x}_1, \dots, \mathsf{x}_k\}, \quad J(\mathsf{x}_i) \neq 0. \tag{1.25}$$

Set $\mathsf{h}(t, \mathsf{x}) \equiv \mathsf{f}(\mathsf{x}) + t\mathsf{g}(\mathsf{x})$. Since $\mathsf{h}(0, \mathsf{x}_i) = 0$ and $\partial\mathsf{h}(0, \mathsf{x}_i)/\partial\mathsf{x} = J(\mathsf{x}_i) \neq 0$, the Implicit Function Theorem implies that there exist numbers $\varepsilon > 0, \eta > 0$ and continuous functions $\tilde{\mathsf{x}}_i : (-\eta, \eta) \to \mathbb{B}(\mathsf{x}_i, \varepsilon)$ such that $\tilde{\mathsf{x}}_i(0) = \mathsf{x}_i$ and $\tilde{\mathsf{x}}_i$ is the only solution of $\mathsf{h}(t, \mathsf{x}) = \mathbf{0}$ lying in $\mathbb{B}(\mathsf{x}_i, \varepsilon)$. We take ε so small that the $\operatorname{cl}\mathbb{B}(\mathsf{x}_i, \varepsilon)$ are disjoint and so that $\operatorname{sign} J(\mathsf{x}) = \operatorname{sign} J(\mathsf{x}_i)$ for $\mathsf{x} \in \operatorname{cl}\mathbb{B}(\mathsf{x}_i, \varepsilon)$.

Let $\mathcal{U} \equiv \cup_i \mathbb{B}(\mathsf{x}_i, \varepsilon)$. Since $\operatorname{cl}\Omega \setminus \mathcal{U}$ is a compact set on which the continuous function f vanishes nowhere, it follows that $|\mathsf{f}|$ has a positive minimum α on this set. For $|t| < \min\{\eta, \alpha/2\max_{\operatorname{cl}\Omega}|\mathsf{g}|\} \equiv \eta_1$, we find that $\mathsf{h}(t, \cdot)$ cannot vanish on $\operatorname{cl}\Omega \setminus \mathcal{U}$, so that the inverse image of $\mathbf{0}$ under $\mathsf{h}(t, \cdot)$ is $\{\tilde{\mathsf{x}}_1, \dots \tilde{\mathsf{x}}_k\}$.

Since $\det \partial\mathsf{h}/\partial\mathsf{x}$ is continuous, there is an $\eta_2 < \eta_1$ such that

$$\left|\det \frac{\partial\mathsf{h}}{\partial\mathsf{x}}(t, \mathsf{x}) - J(\mathsf{x})\right| < \min\{J(\mathsf{z}) : \mathsf{z} \in \operatorname{cl}\mathcal{U}\} \tag{1.26}$$

for $|t| < \eta_2$ and $\mathsf{x} \in \operatorname{cl}\mathcal{U}$. Were $\det \partial\mathsf{h}(t, \mathsf{x})/\partial\mathsf{x}$ and $J(\mathsf{x})$ to have opposite sign for $|t| < \eta_2$ and $\mathsf{x} \in \operatorname{cl}\mathcal{U}$, then (1.26) would be violated. Thus $\operatorname{sign}\det \partial\mathsf{h}(t, \tilde{\mathsf{x}}(t))/\partial\mathsf{x} = \operatorname{sign} J(\tilde{\mathsf{x}}(t)) = \operatorname{sign} J(\mathsf{x}_i)$, and (1.24) follows from the definition (1.7).

Now we drop conditions (1.25). By Sard's Lemma 1.3, we can choose an a in $\mathbb{B}(\mathbf{0}, \frac{1}{3}\delta)$ where $\delta = \operatorname{dist}(\mathbf{0}, \mathsf{f}(\partial\Omega))$ such that (1.25) holds with f replaced by $\mathsf{f} - \mathsf{a}$. Then (1.24) and (1.21) imply that there exists an $\eta_3 > 0$ such that

$$\deg(\mathsf{h}(t, \cdot) - \mathsf{a}, \Omega) = \deg(\mathsf{f} - \mathsf{a}, \Omega) = \deg(\mathsf{f}, \Omega) \quad \text{for} \quad |t| < \eta_3. \tag{1.27}$$

Now let $\eta_4 < \min\{\eta_3, \delta/3 \max_{\mathrm{cl}\,\Omega} |\mathbf{g}|\}$. Then

$$|\mathbf{h}(t,\mathbf{x}) - \mathbf{a}| \geq |\mathbf{f}(\mathbf{x})| - |\mathbf{a}| - |t|\,|\mathbf{g}(\mathbf{x})| > \delta - \tfrac{1}{3}\delta - \tfrac{1}{3}\delta,$$

so that $|\mathbf{a}| \leq \operatorname{dist}(0, \mathbf{h}(t, \partial\Omega))$. Thus we can use (1.21) to obtain $\deg(\mathbf{h}(t,\cdot),\Omega) = \deg(\mathbf{h}(t,\cdot) - \mathbf{a}, \Omega)$, which when combined with (1.27) yields (1.24). □

We now show that the degree is constant for all C^2 functions in a sufficiently small C^0-neighborhood of a function $\mathbf{f} \in C^0(\mathrm{cl}\,\Omega)$ for which $\mathbf{0} \notin \mathbf{f}(\partial\Omega)$. (It can be shown that a continuous function $\mathbf{f}$ on a compact set can be uniformly approximated by infinitely differentiable functions (cf. Adams (1975), e.g.), so that there certainly are C^2 functions in every neighborhood of $\mathbf{f}$. We shall tacitly use this observation in the sequel.) Let $\mathbf{g}_1, \mathbf{g}_2 \in C^2(\Omega) \cap C^0(\mathrm{cl}\,\Omega)$ with

$$\max_{\mathrm{cl}\,\Omega} |\mathbf{g}_1(\mathbf{x}) - \mathbf{f}(\mathbf{x})|,\ \max_{\mathrm{cl}\,\Omega} |\mathbf{g}_2(\mathbf{x}) - \mathbf{f}(\mathbf{x})| \leq \delta = \operatorname{dist}(0, \mathbf{f}(\partial\Omega)).$$

Let $\mathbf{h}(t,\mathbf{x}) \equiv (1-t)\mathbf{g}_1(\mathbf{x}) + t\mathbf{g}_2(\mathbf{x})$. Since $\mathbf{h}(t,\cdot) = \mathbf{h}(t_0,\cdot) + (t-t_0)(\mathbf{g}_2 - \mathbf{g}_1)$ for any fixed t_0, we deduce from Lemma 1.23 that $t \mapsto d(t) \equiv \deg\big(\mathbf{h}(t,\cdot),\Omega\big)$ is constant and a fortiori continuous on a neighborhood of the arbitrary point t_0. Thus d is continuous on the connected set $[0,1]$, so that $d([0,1])$ is connected, and d must be constant. Therefore $\deg(\mathbf{g}_1,\Omega) = \deg(\mathbf{g}_2,\Omega)$.

When $\mathbf{f} \in C^0(\mathrm{cl}\,\Omega)$ and (1.4) hold, we accordingly define

$$\deg(\mathbf{f},\Omega) \equiv \deg(\mathbf{g},\Omega) \tag{1.28}$$

where $\mathbf{g} \in C^2(\Omega) \cap C^0(\mathrm{cl}\,\Omega)$ is any function with $\max_{\mathrm{cl}\,\Omega} |\mathbf{g}(\mathbf{x}) - \mathbf{f}(\mathbf{x})| < \operatorname{dist}(\mathbf{0}, \mathbf{f}(\partial\Omega))$ and where $\deg(\mathbf{g},\Omega)$ is defined by (1.21).

2. Properties of the Brouwer Degree

In this section we present a collection of important properties of our most general definitions (1.28) of degree. They roughly fall into three groups: fundamental properties (which are enjoyed by definition (1.7) under the special restrictions (1.5) and (1.6)), properties that permit the effective evaluation of degree, and useful existence theorems for finite-dimensional problems. A statement given without proof follows immediately from the material preceding it. The basic technique for many of the proofs is to use definitions (1.21) and (1.28) to reduce the proof to the regular case in which (1.5) and (1.6) hold, so that the degree can be computed explicitly. This technique is illustrated in the proof of Theorem 2.2; analogous proofs of subsequent results are omitted.

Definition (1.7) immediately yields

2.1. Proposition. *Let* $\mathbf{I}$ *be the identity mapping on* Ω. *Then* $\deg[\mathbf{I},\Omega] = 1$ *if* $\mathbf{0} \in \Omega$ *and* $\deg[\mathbf{I},\Omega] = 0$ *if* $\mathbf{0} \notin \mathrm{cl}\,\Omega$.

The crucial property of degree is the following:

2.2. Theorem. *Let* $\mathbf{f} \in C^0(\mathrm{cl}\,\Omega)$ *and let* $\mathbf{0} \notin \mathbf{f}(\partial\Omega)$. *If* $\deg[\mathbf{f},\Omega] \neq 0$, *then* (1.1) *has a solution.*

Proof. Suppose not. By Sard's Lemma 1.3 (cf. (1.21)) we can choose a C^2 function $\mathbf{g}$ with simple zeros so that $\max_{\mathrm{cl}\,\Omega} |\mathbf{g}(\mathbf{x}) - \mathbf{f}(\mathbf{x})| < \operatorname{dist}(0, \mathbf{f}(\partial\Omega))$. Definition (1.28) implies that $\deg[\mathbf{g},\Omega] = \deg[\mathbf{f},\Omega] \neq \mathbf{0}$. On the other hand, the equation $\mathbf{g}(\mathbf{x}) = \mathbf{0}$ has no solutions, so that (1.7) implies that $\deg[\mathbf{g},\Omega] = 0$, a contradiction. □

The contrapositive of the second statement of Theorem 2.2 is: *If* (1.1) *does not have a solution, then* $\deg[\mathbf{f},\Omega] = 0$.

2.3. Exercise. Prove

2.4. Proposition (Continuity). deg *is a continuous mapping from* $C^0(\mathrm{cl}\,\Omega)$ *to the integers in the sense that if* $\mathbf{0} \notin \mathbf{f}(\partial\Omega)$ *and if* $\max_{\mathrm{cl}\,\Omega} |\mathbf{g}(\mathbf{x}) - \mathbf{f}(\mathbf{x})| < \mathrm{dist}\,(0, \mathbf{f}(\partial\Omega))$, *then* $\deg[\mathbf{f}, \Omega] = \deg[\mathbf{g}, \Omega]$.

2.5. Corollary. *The mapping* $\mathbf{a} \mapsto \deg[\mathbf{f} - \mathbf{a}, \Omega]$ *is constant on each connected component of* $\mathbb{R}^n \setminus \mathbf{f}(\partial\Omega)$.

2.6. Corollary (Homotopy Invariance). *Let*

$$[0,1] \ni t \mapsto \mathbf{h}(t, \cdot) \in C^0(\mathrm{cl}\,\Omega)$$

be continuous. Suppose that $\mathbf{h}(t, \mathbf{x}) \neq \mathbf{0}$ *for each* $\mathbf{x} \in \partial\Omega$ *and for each* $t \in [0,1]$. *Then*

$$\deg[\mathbf{h}(0, \cdot), \Omega] = \deg[\mathbf{h}(1, \cdot), \Omega].$$

Let $\mathbf{f} = \mathbf{h}(0, \cdot)$ and $\mathbf{g} = \mathbf{h}(1, \cdot)$. Then $\mathbf{h}$ is said to be a *homotopy* between $\mathbf{f}$ and $\mathbf{g}$, and $\mathbf{f}$ and $\mathbf{g}$ are said to be *homotopically equivalent.* Corollary 2.6 has the very useful consequence that in computing the degree we can replace the actual function $\mathbf{f}$ we are given with a much simpler function $\mathbf{g}$, provided that we can connect $\mathbf{g}$ to $\mathbf{f}$ with a homotopy meeting the hypotheses of this corollary.

Corollary 2.6 applies to homotopies $\mathbf{h}$ defined on cylindrical regions $[0,1] \times \Omega$ of $[0,1] \times \mathbb{R}^n$. To handle global bifurcation problems, Rabinowitz (1971a) extended this corollary to arbitrary regions $\mathcal{A}$ of $\mathbb{R} \times \mathbb{R}^n$:

2.7. Theorem (Generalized Homotopy Invariance). *For each* $t \in [a, b]$, *let* $\mathcal{A}(t)$ *be a bounded open set of* $\mathbb{R}^n$. *(It could be empty.) Define*

$$\mathcal{A} = \cup\{\{t\} \times \mathcal{A}(t) : t \in [a, b]\}$$

(so that $\{t\} \times \mathcal{A}(t)$ *is a section of* $\mathcal{A}$*). Let* $\mathcal{A} \setminus [\{a\} \times \mathcal{A}(a)] \setminus [\{b\} \times \mathcal{A}(b)]$ *be an open connected subset of* $\mathbb{R} \times \mathbb{R}^n$. *Let* $\mathbf{h} : \mathrm{cl}\,\mathcal{A} \to \mathbb{R}^n$ *be continuous and let*

$$\mathbf{0} \notin \mathbf{h}\,(\partial\mathcal{A} \setminus [\{a\} \times \mathcal{A}(a)] \setminus [\{b\} \times \mathcal{A}(b)])$$

(so that $\deg[\mathbf{h}(t, \cdot), \mathcal{A}(t)]$ *is defined for all* $t \in [a, b]$*). Then*

$$t \mapsto \deg[\mathbf{h}(t, \cdot), \mathcal{A}(t)] \quad \textit{is constant on} \quad [a, b]. \tag{2.8}$$

The proof of this theorem is given toward the end of this section.

We now list a collection of very useful properties of degree, which support further developments of the theory and which enable the degree to be easily computed. We omit the proofs when they have the character of that of Theorem 2.2.

2.9. Proposition (Additivity). *Let* Ω_1 *and* Ω_2 *be disjoint open sets with* $\Omega = \Omega_1 \cup \Omega_2$ *and with* $\mathbf{0} \notin \mathbf{f}(\partial\Omega_1) \cup \mathbf{f}(\partial\Omega_2)$. *Then*

$$\deg[\mathbf{f}, \Omega] = \deg[\mathbf{f}, \Omega_1] + \deg[\mathbf{f}, \Omega_2].$$

2.10. Proposition (Excision). *Let* $\mathcal{K}$ *be a closed subset of* $\mathrm{cl}\,\Omega$ *for which* $\mathbf{0} \notin \mathbf{f}(\mathcal{K})$. *Then*

$$\deg[\mathbf{f}, \Omega] = \deg[\mathbf{f}, \Omega \setminus \mathcal{K}].$$

2.11. Proposition. *If* $\deg[\mathbf{f}, \Omega] \neq 0$, *then* $\mathbf{f}(\Omega)$ *contains a nonempty open set in* $\mathbb{R}^n$ *containing* $\mathbf{0}$. *Consequently, if* $\mathbf{f}(\Omega)$ *is contained in a plane in* $\mathbb{R}^n$ *of dimension less than* n, *then* $\deg[\mathbf{f}, \Omega] = 0$.

Proof. By Theorem 2.2, we know that (1.1) has a solution, so that $\mathbf{f}^{-1}(\{\mathbf{0}\})$ is not empty. Let $\mathcal{C}$ denote the connected component of $\mathbb{R}^n \setminus \mathbf{f}(\partial\Omega)$ containing $\mathbf{0}$ consisting of all $\mathbf{a}$ for which $\deg[\mathbf{f} - \mathbf{a}, \Omega] = \deg[\mathbf{f}, \Omega] \neq 0$ (see Corollary 2.5). By Theorem 2.2, we also know that the equation $\mathbf{f}(\mathbf{x}) = \mathbf{a}$ has a solution for each such $\mathbf{a}$. Thus $\mathcal{C} \in \mathbf{f}(\Omega)$. Since $\mathcal{C}$ is open, it is the requisite nonempty set. The second statement of the proposition is then immediate. □

2.12. Proposition (Boundary Dependence). *If* f *and* g *agree on* $\partial\Omega$ *and if* $\mathbf{0} \notin \mathsf{f}(\partial\Omega)$, *then* $\deg[\mathsf{f}, \Omega] = \deg[\mathsf{g}, \Omega]$.

Proof. Set $\mathsf{h}(t, \mathsf{x}) = t\mathsf{f}(\mathsf{x}) + (1-t)\mathsf{g}(\mathsf{x})$. Since $\mathsf{h}(t, \mathsf{x}) = \mathsf{f}(\mathsf{x})$ for $\mathsf{x} \in \partial\Omega$, we can apply the homotopy invariance of degree of Corollary 2.6 to deduce the conclusion. □

2.13. Poincaré-Bohl Theorem. *If for each* $\mathsf{x} \in \partial\Omega$, *the vectors* $\mathsf{f}(\mathsf{x})$ *and* $\mathsf{g}(\mathsf{x})$ *do not point in opposite directions, then* $\deg[\mathsf{f}, \Omega] = \deg[\mathsf{g}, \Omega]$.

Proof. Set $\mathsf{h}(t, \mathsf{x}) = t\mathsf{f}(\mathsf{x}) + (1-t)\mathsf{g}(\mathsf{x})$. We assert that $\mathsf{h}(t, \mathsf{x}) \neq \mathbf{0}$ for $\mathsf{x} \in \partial\Omega$ and $t \in [0, 1]$, for if not, there would be an x and t such that $t\mathsf{f}(\mathsf{x}) = -(1-t)\mathsf{g}(\mathsf{x})$ and f and g would point in opposite directions. We can now apply the homotopy invariance of Corollary 2.6 to deduce the conclusion. □

2.14. Corollary (T. B. Benjamin). *If there is a fixed nonzero vector* $\mathsf{a} \in \mathbb{R}^n$ *such that* $\mathsf{f}(\mathsf{x}) \neq \lambda\mathsf{a}$ *for all* $\mathsf{x} \in \partial\Omega$ *and for all* $\lambda > 0$, *then* $\deg[\mathsf{f}, \Omega] = 0$.

Proof. f and a never point in opposite directions on $\partial\Omega$ and therefore by the Poincaré-Bohl Theorem have the same degree. But a is a constant function and therefore has degree 0. □

2.15. Borsuk Odd Mapping Theorem. *Let* Ω *be symmetric about* $\mathbf{0}$. *Let* $\mathsf{f} \in C^0(\operatorname{cl}\Omega)$. *Let* $\mathsf{f}(-\mathsf{x}) = -\mathsf{f}(\mathsf{x})$ *on* $\partial\Omega$ *with* f *not vanishing on* $\partial\Omega$. *Then* $\deg[\mathsf{f}, \Omega]$ *is an odd integer.*

2.16. Exercise. Prove this result for continuously differentiable f's with simple zeros. (The proof for the general case is not as simple as one might expect because all the approximation apparatus developed in Sec. 1 must be adapted to odd mappings. For details, see the references at the end of Sec. 3.)

We are now able to apply these results to the solvability of nonlinear finite-dimensional equations.

2.17. Brouwer Fixed-Point Theorem. *Let* $\operatorname{cl}\Omega$ *be homeomorphic to a compact convex set in* $\mathbb{R}^n$. *If* $\mathsf{f} : \operatorname{cl}\Omega \to \operatorname{cl}\Omega$ *is continuous, then it has a fixed point, i.e., a point* $\mathsf{x}_0 \in \operatorname{cl}\Omega$ *such that* $\mathsf{f}(\mathsf{x}_0) = \mathsf{x}_0$.

Proof. We first prove this theorem for the case that Ω is the unit ball $\mathbb{B}(\mathbf{0}, 1)$ centered at the origin. Now if there is an $\mathsf{x}_0 \in \partial\mathbb{B}(\mathbf{0}, 1)$ such that $\mathsf{f}(\mathsf{x}_0) = \mathsf{x}_0$, then we are done. We accordingly assume that there is no such x_0. We set

$$\mathsf{h}(t, \mathsf{x}) = \mathsf{x} - t\mathsf{f}(\mathsf{x}), \quad t \in [0, 1]. \tag{2.18}$$

We assert that $\mathsf{h}(t, \cdot)$ does not vanish on $\partial\mathbb{B}(\mathbf{0}, 1)$. If it were to vanish, then there would be an x with $|\mathsf{x}| = 1$ such that $\mathsf{x} = t\mathsf{f}(\mathsf{x})$. Clearly t can be neither 0 nor 1. Thus $\mathsf{f}(\mathsf{x}) = \mathsf{x}/t$ so that $|\mathsf{f}(\mathsf{x})| = t^{-1} > 1$, which is impossible because $\mathsf{f} : \mathbb{B}(\mathbf{0}, 1) \to \mathbb{B}(\mathbf{0}, 1)$. From the homotopy invariance of degree it follows that

$$\begin{aligned} 1 &= \deg[\mathsf{I}, \mathbb{B}(\mathbf{0}, 1)] = \deg[\mathsf{h}(0, \cdot), \mathbb{B}(\mathbf{0}, 1)] \\ &= \deg[\mathsf{h}(1, \cdot), \mathbb{B}(\mathbf{0}, 1)] = \deg[\mathsf{I} - \mathsf{f}, \mathbb{B}(\mathbf{0}, 1)], \end{aligned} \tag{2.19}$$

and the conclusion follows from Theorem 2.2.

Now let $\operatorname{cl}\Omega$ be homeomorphic to $\operatorname{cl}\mathbb{B}(0, 1)$, i.e., there exists a continuous function $\mathsf{g} : \operatorname{cl}\mathbb{B}(0, 1) \to \operatorname{cl}\Omega$ with a continuous inverse. Then $\mathsf{y} \mapsto \mathsf{g}^{-1}(\mathsf{f}(\mathsf{g}(\mathsf{y})))$ is a continuous mapping of $\operatorname{cl}\mathbb{B}(0, 1)$ into itself, and therefore has a fixed point. The invertibility of g immediately implies that f itself has a fixed point.

Next let $\operatorname{cl}\Omega$ be a closed convex set. We can assume that $\operatorname{cl}\Omega$ has a non-empty interior, for if not, it must lie in a plane of dimension $m < n$ in which it does have a non-empty interior, and we identify this plane with $\mathbb{R}^m$. By making a suitable translation, we may assume that $\mathbf{0}$ is in the interior of Ω. For any $\mathsf{x} \neq \mathbf{0}$ in Ω, the ray from $\mathbf{0}$ to x pierces $\partial\Omega$ at a unique point at the distance $r(\mathsf{x})$ from $\mathbf{0}$. Then $\mathsf{x} \mapsto \mathsf{x}/r(\mathsf{x})$ is a homeomorphism from $\text{cl}\,\Omega$ to $\mathbb{B}(0,1)$. It follows from the preceding paragraph that f has a fixed point on $\text{cl}\,\Omega$. The result for homeomorphic images of closed convex sets is proved just as in the last paragraph. □

The next few corollaries of the Poincaré-Bohl Theorem give concrete criteria for the existence of solutions to (1.1), which we employ throughout this book.

2.20. Proposition. *Let* $\mathsf{f} : \mathbb{R}^n \to \mathbb{R}^n$ *be continuous and satisfy the coercivity condition*

$$\mathsf{f}(\mathsf{x}) \cdot \frac{\mathsf{x}}{|\mathsf{x}|} \to \infty \quad \text{as} \quad |\mathsf{x}| \to \infty. \tag{2.21}$$

Then $\mathsf{f}(\mathbb{R}^n) = \mathbb{R}^n$, *i.e., the equation* $\mathsf{f}(\mathsf{x}) = \mathsf{a}$ *has a solution for each* $\mathsf{a} \in \mathbb{R}^n$.

Proof. Since

$$\mathsf{x} \cdot [\mathsf{f}(\mathsf{x}) - \mathsf{a}] \geq |\mathsf{x}| \left[\mathsf{f}(\mathsf{x}) \cdot \frac{\mathsf{x}}{|\mathsf{x}|} - |\mathsf{a}|\right], \tag{2.22}$$

Hypothesis (2.21) implies that the left-hand side of (2.22) is positive on the sphere $\partial\mathbb{B}(\mathbf{0}, R) \equiv \{\mathsf{x} \colon |\mathsf{x}| = R\}$ for sufficiently large R. Thus $\mathsf{f}(\mathsf{x}) - \mathsf{a}$ and x do not point in opposite directions on this sphere. The Poincaré-Bohl Theorem then implies that $\deg[\mathsf{f}-\mathsf{a}, \mathbb{B}(\mathbf{0}, R)] = \deg[\mathsf{I}, \mathbb{B}(\mathbf{0}, R)] = 1$, and the conclusion follows from Theorem 2.2. □

2.23. Exercise. Prove the following generalization of Proposition 2.20:

2.24. Proposition. *Let* $\Phi : \mathbb{R}^n \to \mathbb{R}$ *be continuously differentiable, let* $\Phi(\mathsf{x}) \to \infty$ *as* $|\mathsf{x}| \to \infty$, $\partial\Phi(\mathbf{0})/\partial\mathsf{x} = \mathbf{0}$, *and let* Φ *have a strictly monotone gradient:*

$$\left[\frac{\partial\Phi(\mathsf{x})}{\partial\mathsf{x}}(\mathsf{x}_1) - \frac{\partial\Phi(\mathsf{x})}{\partial\mathsf{x}}(\mathsf{x}_2)\right] \cdot (\mathsf{x}_1 - \mathsf{x}_2) > 0 \quad \text{for} \quad \mathsf{x}_1 \neq \mathsf{x}_2 \tag{2.25}$$

(so that Φ *is convex). Let* $\mathsf{f} : \mathbb{R}^n \to \mathbb{R}^n$ *be continuous and satisfy the coercivity condition*

$$\mathsf{f}(\mathsf{x}) \cdot \frac{\partial\Phi(\mathsf{x})/\partial\mathsf{x}}{|\partial\Phi(\mathsf{x})/\partial\mathsf{x}|} \to \infty \quad \text{as} \quad |\mathsf{x}| \to \infty. \tag{2.26}$$

Then the equation $\mathsf{f}(\mathsf{x}) = \mathsf{a}$ *has a solution for each* $\mathsf{a} \in \mathbb{R}^n$.

We can generalize Proposition 2.24 with the following results, used at several places in this book.

2.27. Exercise. Prove the following

2.28. Theorem. *Let* Ω *be an open convex subset of* $\mathbb{R}^n$ *and let* $\mathcal{Y}$ *be an arbitrary set. Let* $\mathsf{f} : \Omega \times \mathcal{Y} \to \mathbb{R}^n$ *with* $\mathsf{f}(\cdot, \mathsf{y})$ *continuous for each* $\mathsf{y} \in \mathcal{Y}$. *Let* $\Phi : \Omega \to \mathbb{R}$ *be continuously differentiable, let* $\Phi(\mathsf{x}) \to \infty$ *as* $|\mathsf{x}| \to \infty$ *or as* $\mathsf{x} \to \partial\Omega$, *let there be a point* $\mathsf{x}_0 \in \Omega$ *such that* $\partial\Phi(\mathbf{0})/\partial\mathsf{x}_0 = \mathbf{0}$, *and let* Φ *satisfy* (2.25). *Suppose that for each* $\mathsf{y} \in \mathcal{Y}$ *there is a number* $R(\mathsf{y})$ *in the range of* Φ *such that*

$$\mathsf{f}(\mathsf{x}, \mathsf{y}) \cdot \frac{\partial\Phi(\mathsf{x})}{\partial\mathsf{x}}(\mathsf{x}) \geq 0 \quad \forall\, \mathsf{x} \quad \text{such that} \quad \Phi(\mathsf{x}) = r \quad \forall\, r \geq R(\mathsf{y}). \tag{2.29}$$

Then for each $\mathsf{y} \in \mathcal{Y}$ *the equation* $\mathsf{f}(\mathsf{x}, \mathsf{y}) = \mathbf{0}$ *has a solution* $\mathsf{x} \in \Omega$.

We immediately obtain

2.30. Global Implicit Function Theorem. *Let the hypotheses of Theorem 2.28 hold and further let* $\mathsf{f}(\cdot, \mathsf{y})$ *be strictly monotone:*

$$[\mathsf{f}(\mathsf{x}_1) - \mathsf{f}(\mathsf{x}_2)] \cdot (\mathsf{x}_1 - \mathsf{x}_2) > 0 \quad \text{for} \quad \mathsf{x}_1 \neq \mathsf{x}_2. \tag{2.31}$$

Then for each $\mathsf{y} \in \mathcal{Y}$ *the equation* $\mathsf{f}(\mathsf{x}, \mathsf{y}) = \mathbf{0}$ *has a unique solution* $\mathsf{x} \in \Omega$.

By introducing further hypotheses on the smoothness of f in both of its arguments, we can assert that the solution has a corresponding level of smoothness in y, merely by invoking such results for the Local Implicit Function Theorem XVIII.1.27.

The following result is typical of those ensuring the existence of an eigenvalue for a nonlinear operator equation.

2.32. Theorem. *Let* $\mathsf{f} : \partial\mathbb{B}(\mathbf{0},1) \to \mathbb{R}^n$ *be continuous, and let* $\mathsf{f}(\mathsf{x}) \neq \mathbf{0}$ *for* $\mathsf{x} \in \partial\mathbb{B}(\mathbf{0},1)$. *If* n *is odd, then there is a direction normal to* $\partial\mathbb{B}(\mathbf{0},1)$ *at some point* $\mathsf{x}_0 \in \partial\mathbb{B}(\mathbf{0},1)$ *that is unchanged under* f, *i.e., there is a nonzero real number* λ *and an* x_0 *in* $\partial\mathbb{B}(\mathbf{0},1)$ *such that* $\mathsf{f}(\mathsf{x}_0) = \lambda\mathsf{x}_0$. *Equivalently, there is an* x_0 *in* $\partial\mathbb{B}(\mathbf{0},1)$ *at which the tangential component of* f *to* $\partial\mathbb{B}(\mathbf{0},1)$ *vanishes.*

Proof. Extend f to the ball $\mathbb{B}(\mathbf{0},1)$ by setting $\mathsf{f}(\alpha\mathsf{x}) = \alpha\mathsf{f}(\mathsf{x})$ for $\mathsf{x} \in \partial\mathbb{B}(\mathbf{0},1)$. $\deg[\mathsf{f},\Omega]$ is well defined (by its boundary behavior). Let us set

$$\mathsf{h}^+(t,\mathsf{x}) = t\mathsf{f}(\mathsf{x}) + (1-t)\mathsf{x}, \quad \mathsf{h}^-(t,\mathsf{x}) = t\mathsf{f}(\mathsf{x}) + (1-t)(-\mathsf{x}). \tag{2.33}$$

Now one of these homotopies must vanish for some $\mathsf{x}_0 \in \partial\mathbb{B}(\mathbf{0},1)$, for if not, homotopy invariance (Corollary 2.5) would imply that

$$\deg[\mathsf{f},\Omega] = \deg[\mathsf{I},\Omega] = 1 \quad \text{and} \quad \deg[\mathsf{f},\Omega] = \deg[-\mathsf{I},\Omega] = -1, \tag{2.34}$$

with the last equality of (2.34) a consequence of the oddness of n. Now neither of the homotopies of (2.33) can vanish on $\partial\mathbb{B}(\mathbf{0},1)$ for $t = 0$ because $|\mathsf{x}| = 1$, and neither can vanish here for $t = 1$ because f does not vanish on $\partial\mathbb{B}(\mathbf{0},1)$. Since one of the homotopies of (2.33) must vanish on $\partial\mathbb{B}(\mathbf{0},1)$, it thus follows that there is an x_0 in $\partial\mathbb{B}(\mathbf{0},1)$ and a t_0 in $(0,1)$ such that

$$\mathsf{f}(\mathsf{x}_0) = \pm\frac{(1-t_0)}{t_0}\mathsf{x}_0. \quad \square \tag{2.35}$$

An interesting related geometrical result, not of central importance here, is

2.36. Proposition. *A nowhere vanishing tangent field exists on* $\partial\mathbb{B}(\mathbf{0},1)$ *if and only if* n *is even.*

We do not pause to give the proof.

The index. Let $\mathsf{f} : \Omega \to \mathbb{R}^n$ be continuous and let $\mathbf{0} \notin \mathsf{f}(\partial\Omega)$. If x_0 is an isolated solution of the equation $\mathsf{f}(\mathsf{x}) = \mathbf{0}$, then by definition there is an open ball $\mathbb{B}(\mathsf{x}_0, r)$ about it in which there are no other solutions. By the excision property (Proposition 2.10), $\deg[\mathsf{f},\mathbb{B}(\mathsf{x}_0,\varepsilon)] = \deg[\mathsf{f},\mathbb{B}(\mathsf{x}_0,r)]$ for all $\varepsilon \in (0,r)$. This value of the degree is denoted $\operatorname{ind}[\mathsf{f},\mathsf{x}_0]$ and is called the *index* of f at x_0.

Let $\lambda \mapsto \mathsf{A}(\lambda)$ be a function that assigns an $n \times n$ matrix $\mathsf{A}(\lambda)$ to each λ in an interval. Let λ^0 be an eigenvalue of A. Then λ^0 has *algebraic multiplicity* m iff A is m-times differentiable at λ^0 and

$$\left.\frac{d^j}{d\lambda^j}\det\mathsf{A}(\lambda)\right|_{\lambda=\lambda^0} = 0 \quad \text{for} \quad j = 0,\dots,m-1, \qquad \left.\frac{d^m}{d\lambda^m}\det\mathsf{A}(\lambda)\right|_{\lambda=\lambda^0} \neq 0, \tag{2.37}$$

(See Sec. V.3 for definitions of eigenvalues.)

The following result is very useful for the study of bifurcation problems.

2.38. Proposition. *Let* $\mathsf{x} \mapsto \mathsf{f}(\lambda,\mathsf{x})$ *be continuously differentiable for all* $\lambda \in \mathbb{R}$ *and let* $\mathsf{f}(\lambda,\mathbf{0}) = \mathbf{0}$. *Suppose that* $\lambda \mapsto \partial\mathsf{f}(\lambda,\mathbf{0})/\partial\mathsf{x}$ *has isolated eigenvalues* λ_k, *each of which has a well-defined algebraic multiplicity* m_k. *Then* $\operatorname{ind}[\mathsf{f}(\lambda,\cdot),\mathbf{0}]$ *is defined for each* $\lambda \neq \lambda_k$ *and*

$$\operatorname{ind}[\mathsf{f}(\lambda_k+\varepsilon,\cdot),\mathbf{0}] = (-1)^{m_k}\operatorname{ind}[\mathsf{f}(\lambda_k-\varepsilon,\cdot),\mathbf{0}] \tag{2.39}$$

for ε *sufficiently small.*

2.40. Exercise. Prove Proposition 2.38.

Proof of Theorem 2.7. We reduce the proof to an application of the homotopy invariance of Corollary 2.6. Fix λ in $[a,b]$. Let

$$\mathcal{N}(\lambda) \equiv \{\mathsf{x} \in \mathcal{A}(\lambda) : \mathsf{h}(\lambda,\mathsf{x}) = \mathbf{0}\}.$$

By hypothesis, $\mathcal{N}(\lambda) \cap \{(t, \mathbf{x}) \in \partial A : t = \lambda\} = \emptyset$. Since $\mathcal{N}(\lambda)$ is the inverse image of the closed set $\{\mathbf{0}\}$ under the continuous mapping $\mathbf{h}(\lambda, \cdot)$, it is closed and therefore compact. Thus there is an open set $\mathcal{O}(\lambda)$ such that

$$\mathcal{N}(\lambda) \subset \mathcal{O}(\lambda) \subset \operatorname{cl} \mathcal{O}(\lambda) \subset \mathcal{A}(\lambda)$$

and there is an $\varepsilon > 0$ (depending on λ) such that

$$\{[\lambda - \varepsilon, \lambda + \varepsilon] \cap [a, b]\} \times \mathcal{O}(\lambda) \subset \mathcal{A}.$$

We assert that there is an ε so small that every solution pair $(\mu, \mathbf{y})$ of $\mathbf{h}(\mu, \mathbf{y}) = \mathbf{0}$ with $|\mu - \lambda| < \varepsilon$ lies in $\{[\lambda - \varepsilon, \lambda + \varepsilon] \cap [a, b]\} \times \mathcal{O}(\lambda)$. Were there no such ε, then there would be a sequence $\{\varepsilon_k, \lambda_k, \mathbf{x}_k\}$ with

$$\varepsilon_k \searrow 0, \quad \mathbf{h}(\lambda_k, \mathbf{x}_k) = \mathbf{0}, \quad |\lambda_k - \lambda| \leq \varepsilon_k,$$
$$(\lambda_k, \mathbf{x}_k) \notin \{[\lambda - \varepsilon_k, \lambda + \varepsilon_k] \cap [a, b]\} \times \mathcal{O}(\lambda).$$

By the Bolzano-Weierstrass Theorem, this sequence would have a subsequence, denoted the same way, with $\{(\lambda_k, \mathbf{x}_k)\}$ converging to a point of the form $(\lambda, \mathbf{z})$. Since $\mathbf{h}$ is continuous, it would follow that $\mathbf{h}(\lambda, \mathbf{z}) = \mathbf{0}$. Since $\mathbf{x}_k \notin \mathcal{O}(\lambda)$ for each K, it would follow from the openness of $\mathcal{O}(\lambda)$ that $\mathbf{y} \notin \mathcal{O}(\lambda)$, in contradiction of the fact that $\mathcal{N}(\lambda) \subset \mathcal{O}(\lambda)$.

For every $\mu \in [\lambda - \varepsilon, \lambda + \varepsilon] \cap [a, b]$, $\deg[\mathbf{h}(\mu, \cdot), \mathcal{O}(\mu)]$ is well-defined because $\mathbf{0} \notin \partial \mathcal{O}(\mu)$. From the homotopy invariance of degree on cylindrical regions (Corollary 2.6) it follows that $[\lambda - \varepsilon, \lambda + \varepsilon] \cap [a, b] \ni \mu \mapsto \deg[\mathbf{h}(\mu, \cdot), \mathcal{O}(\lambda)]$ is constant. The excision property (Proposition 2.10) implies that

$$\deg[\mathbf{h}(\mu, \cdot), \mathcal{O}(\lambda)] = \deg[\mathbf{h}(\mu, \cdot), \mathcal{A}(\mu)]$$

for all $\mu \in [\lambda - \varepsilon, \lambda + \varepsilon] \cap [a, b]$. Thus $\mu \mapsto \deg[\mathbf{h}(\mu, \cdot), \mathcal{A}(\mu)]$ is (locally) constant on $[\lambda - \varepsilon, \lambda + \varepsilon] \cap [a, b]$ and must therefore be constant on $[a, b]$. □

We recall that a *component* of a metric space is a maximal closed connected subset of it. We study components of solution pairs $(t, \mathbf{x})$ of equations of the form $\mathbf{h}(t, \mathbf{x}) = \mathbf{0}$.

2.41. Corollary. *Let the hypotheses of Theorem 2.7 hold. If*

$$d \equiv \deg[\mathbf{h}(t, \cdot), \mathcal{A}(t)] \neq 0$$

for some t, and consequently for all t in $[a, b]$, then there exists a component $\mathcal{C}$ of

$$\mathcal{S} \equiv \{(t, \mathbf{x}) \in \mathcal{A} \colon \mathbf{h}(t, \mathbf{x}) = \mathbf{0}\}$$

that meets both $\{a\} \times \mathcal{A}(a)$ and $\{b\} \times \mathcal{A}(b)$.

Proof. $\mathcal{S}$, being the inverse image of the closed set $\{\mathbf{0}\}$ under the continuous function $\mathbf{h}$, is closed, and being bounded, it is therefore compact. Now our assumption about the disposition of solutions of $\mathbf{h}(t, \mathbf{x}) = \mathbf{0}$ ensures that $\mathcal{S} \cap [\{t\} \times \mathcal{A}(t)] = \mathcal{S} \cap [\{t\} \times \operatorname{cl} \mathcal{A}(t)]$, so that this set is a closed subset of $\mathcal{S}$. The sets $\mathcal{K}(a) \equiv \mathcal{S} \cap [\{a\} \times \operatorname{cl} \mathcal{A}(a)]$ and $\mathcal{K}(b) \equiv \mathcal{S} \cap [\{b\} \times \operatorname{cl} \mathcal{A}(b)]$ are disjoint closed subsets of the metric space $\mathcal{S}$ and are nonempty by virtue of Theorem 2.2 and the assumption that $\deg[\mathbf{h}(t, \cdot), \mathcal{A}(t)] \neq \mathbf{0}$.

Now we invoke the following lemma from topology: *If $\mathcal{K}(a)$ and $\mathcal{K}(b)$ are disjoint closed subsets of a compact metric space $\mathcal{S}$, then either there exists a component of $\mathcal{S}$ that meets both $\mathcal{K}(a)$ and $\mathcal{K}(b)$, or else there are two disjoint compact sets $\mathcal{S}(a)$ and $\mathcal{S}(b)$ such that $\mathcal{S} = \mathcal{S}(a) \cup \mathcal{S}(b)$ with $\mathcal{K}(a) \subset \mathcal{S}(a)$, $\mathcal{K}(b) \subset \mathcal{S}(b)$.* We need only exclude the second alternative to establish this corollary. Let us suppose for the sake of contradiction that the second alternative holds. We set

$$2\varepsilon \equiv \operatorname{dist}(\mathcal{S}(a), \mathcal{S}(b)),$$
$$\mathcal{O} \equiv \{(t, \mathbf{x}) \in \mathcal{A} : \operatorname{dist}((t, \mathbf{x}), \mathcal{S}(a)) < \varepsilon\},$$
$$\mathcal{O}(t) \equiv \{\mathbf{x} : (t, \mathbf{x}) \in \mathcal{O}\}.$$

Clearly, ε is positive and $\mathcal{O} \cap \mathcal{S}(b) \neq \emptyset$. We define $\mathcal{O}(t)$ for all $t \in [a, b]$ by defining $\mathcal{O}(\lambda)$ to be the empty set wherever $\mathcal{O}$ does not intersect the plane $t = \lambda$. We identify $\mathcal{O}$ with $\mathcal{A}$ of Theorem 2.7, from which we conclude that

$$\deg[\mathbf{h}(t, \cdot), \mathcal{O}(t)] = \text{const.} \quad \forall\, t \in [a, b].$$

From the excision property (Proposition 2.10) it follows that

$$\deg[\mathbf{h}(a, \cdot), \mathcal{O}(a)] = \deg[\mathbf{h}(a, \cdot), \mathcal{A}(a)] = d.$$

But by our hypothesis,

$$\deg[\mathbf{h}(b, \cdot), \mathcal{O}(b)] = \deg[\mathbf{h}(b, \cdot), \emptyset)] = 0.$$

These last three identities are contradictory. □

3. Leray-Schauder Degree

Mathematical results valid for arbitrary finite-dimensional spaces do not automatically carry over to infinite-dimensional spaces. In particular, such consequences of the properties of the Brouwer degree as the Brouwer Fixed-Point Theorem need not hold without further restriction.

3.1. Example. Let l_2 be the space of real sequences $\mathbf{x} = (x_1, x_2, \dots)$ with $\|\mathbf{x}\|^2 = \sum x_k^2 < \infty$. Let $\mathbf{f}(\mathbf{x}) \equiv (\sqrt{1 - \|\mathbf{x}\|^2}, x_1, x_2, \dots)$. It is easy to see that $\mathbf{f} : l_2 \to l_2$ is continuous and takes the ball $\{\mathbf{x} : \|\mathbf{x}\| \leq 1\}$ not only into itself, but actually into its boundary, because $\|\mathbf{f}(\mathbf{x})\|^2 = 1 - \|\mathbf{x}\|^2 + \sum x_k^2 = 1$. Were $\mathbf{f}$ to have a fixed point $\bar{\mathbf{x}}$, then $\|\bar{\mathbf{x}}\| = 1$, and $\mathbf{f}(\bar{\mathbf{x}}) = (0, \bar{x}_1, \bar{x}_2, \dots) = (\bar{x}_1, \bar{x}_2, \dots)$. This equation implies that $\bar{\mathbf{x}} = \mathbf{0}$, in contradiction to the equality $\|\bar{\mathbf{x}}\| = 1$.

To extend the Brouwer degree to infinite-dimensional Banach spaces, we restrict the class of functions $\mathbf{f}$ to those of the form $\mathbf{I} - \mathbf{g}$ where $\mathbf{g}$ is compact and continuous. Concomitantly, the Schauder Fixed-Point Theorem, which is the extension of the Brouwer Fixed-Point Theorem, applies to $\mathbf{g}$'s that are compact and continuous. Recall that a mapping $\mathbf{k}$ from a Banach space $\mathcal{X}$ to a Banach space $\mathcal{Y}$ is *compact* iff $\mathbf{k}$ maps every bounded sequence in $\mathcal{X}$ into a sequence in $\mathcal{Y}$ having a convergent subsequence there. Mappings that are compact and continuous are 'nearly finite-dimensional' in the following sense.

3.2. Lemma (Schauder). *Let $\mathcal{X}$ and $\mathcal{Y}$ be Banach spaces and let $\mathbf{k} : \mathcal{X} \to \mathcal{Y}$ be compact and continuous. Then for every bounded $\mathcal{E} \subset \mathcal{X}$ and for every $\varepsilon > 0$ there is a finite-dimensional subspace $\mathcal{Y}_\varepsilon$ of $\mathcal{Y}$ and a continuous mapping $\mathbf{k}_\varepsilon$ from $\mathcal{X}$ to $\mathcal{Y}_n$ such that*

$$\|\mathbf{k}(\mathbf{x}) - \mathbf{k}_\varepsilon(\mathbf{x}), \mathcal{Y}\| \leq \varepsilon \quad \forall\, \mathbf{x} \in \mathcal{E}.$$

We omit the proof of this lemma (which can be found in standard books on functional analysis and in the references cited below). We now proceed to extend the definition of Brouwer degree to Banach spaces by exploiting the approximation property of Lemma 3.2. We denote the restriction of a function $\mathbf{f}$ to a subset $\mathcal{A}$ of its domain of definition by $\mathbf{f}|_{\mathcal{A}}$.

3.3. Lemma. *Let Ω be a bounded open subset of $\mathbb{R}^{n+m}$ with $m > 0$. Let $\operatorname{cl}\Omega \ni \mathbf{x} \mapsto \mathbf{g}(\mathbf{x}) \in \mathbb{R}^n$ be continuous. Let $\mathbf{0} \notin (\mathbf{I} - \mathbf{g})^{-1}(\partial\Omega)$. Then*

$$\deg[\mathbf{I} - \mathbf{g}, \Omega] = \deg[(\mathbf{I} - \mathbf{g})|_{\operatorname{cl}\Omega \cap \mathbb{R}^n}, \Omega \cap \mathbb{R}^n]. \tag{3.4}$$

Proof. A simple computation shows that the degree on the right-hand side is well-defined because there are no $\mathbf{x}$'s on $\partial\Omega \cap \mathbb{R}^n$ taken by $(\mathbf{I} - \mathbf{g})|_{\operatorname{cl}\Omega \cap \mathbb{R}^n}$ to $\mathbf{0}$. Let us first suppose that $\mathbf{g} \in C^1(\mathrm{cl}\,\Omega)$ and that $\det(\mathbf{I} - \partial\mathbf{g}/\partial\mathbf{x})(\mathbf{x}) \neq 0$ for every $\mathbf{x}$ satisfying $\mathbf{g}(\mathbf{x}) = \mathbf{x}$. By hypothesis,

$$\begin{aligned} \mathbf{g}(\mathbf{x}) &= (g_1(\mathbf{x}), \dots, g_n(\mathbf{x}), 0, \dots, 0), \\ (\mathbf{I} - \mathbf{g})(\mathbf{x}) &= (x_1 - g_1(\mathbf{x}), \dots, x_n - g_n(\mathbf{x}), x_{n+1}, \dots, x_{n+m}). \end{aligned} \tag{3.5}$$

We can readily compute $\mathbf{I} - \partial\mathbf{g}/\partial\mathbf{x}$ from (3.5), showing that its matrix can be partitioned into four submatrices with that in the upper left-hand corner being the $n \times n$ matrix with components $\delta_{ij} - \partial g_i/\partial x_j$, $i, j = 1, \dots, n$, with that in the lower right-hand corner being the $m \times m$ identity matrix, and with that in the lower left-hand corner being the $m \times n$ zero matrix. Consequently, the determinant of this matrix reduces to the determinant of the $n \times n$ matrix in the upper left-hand corner. This is just the Jacobian of $\mathbf{I} - \mathbf{g}$ restricted to $\Omega \cap \mathbb{R}^n$. We use definition (1.7) to establish (3.4) under our special assumptions. We now follow the approximation arguments leading to definitions 1.21 and 1.28 to extend this result to the general class of $\mathbf{g}$'s described by the hypotheses. □

Now consider the question of defining degree for a mapping $\mathbf{x} \mapsto \mathbf{f}(\mathbf{x})$ from the closure of a bounded open set Ω in a real n-dimensional normed space $\mathcal{F}$ to a real normed space $\tilde{\mathcal{F}}$ with the same dimension. We can identify any real normed n-dimensional space $\mathcal{F}$ with $\mathbb{R}$ by choosing a basis for $\mathcal{F}$ because the components of any vector in $\mathcal{F}$ form an n-tuple of real numbers. Is the resulting degree independent of the basis? To answer this question, we introduce bases $\{\mathbf{a}_k\}$, $\{\mathbf{b}_k\}$ for $\mathcal{F}$ and bases $\{\tilde{\mathbf{a}}_k\}$, $\{\tilde{\mathbf{b}}_k\}$ for $\tilde{\mathcal{F}}$. Using the summation convention, we let $\mathbf{b}_k = \mathbf{M} \cdot \mathbf{a}_k = M^j{}_k \mathbf{a}_j$ and $\tilde{\mathbf{b}}_k = \tilde{\mathbf{M}} \cdot \tilde{\mathbf{a}}_k = \tilde{M}^j{}_k \tilde{\mathbf{a}}_j$. We set

$$\begin{aligned} \mathbf{x} &= \xi^j \mathbf{a}_j = \eta^k \mathbf{b}_k \equiv M^j{}_k \eta^k \mathbf{a}_j, \\ \mathbf{f}(\mathbf{x}) &= \phi^j(\{\xi^p\})\tilde{\mathbf{a}}_j = \psi^k(\{\eta^q\})\tilde{\mathbf{b}}_j \equiv \tilde{M}^j{}_k \psi^k(\{\eta^q\})\tilde{\mathbf{a}}_j, \end{aligned}$$

so that

$$\tilde{M}^j{}_k \psi^k(\{\eta^q\}) = \phi^j(\{M^p{}_q \eta^q\}).$$

Thus the Chain Rule implies that

$$\det\left(\tilde{M}^j{}_k\right) \det\left(\frac{\partial \psi^k}{\partial \eta^j}(\{\eta^q\})\right) = \det\left(\frac{\partial \phi^j}{\partial \xi^p}\right) \det(M^p{}_q)$$

for continuously differentiable $\mathbf{f}$. Under the special conditions (1.5) and (1.6), this identity shows that the degree for $(\phi^1, \dots, \phi^n)$ equals that for $(\psi^1, \dots, \psi^n)$ on the corresponding images of Ω when $\det\left(\tilde{M}^j{}_k\right)$ has the same sign as $\det(M^p{}_q)$, i.e., when the orientation of $\{\mathbf{b}_k\}$ with respect to $\{\mathbf{a}_k\}$ is the same as that of $\{\tilde{\mathbf{b}}_k\}$ with respect to $\{\tilde{\mathbf{a}}_k\}$. Otherwise, these degrees have opposite sign. Our standard approximation procedures enable us to make these assertions in the general case. Thus, modulo orientation, degrees for mappings between normed spaces of the same finite dimension are independent of basis and accordingly well defined. For our purposes (see (3.6)), we have $\mathcal{F} = \tilde{\mathcal{F}}$, and we avoid any difficulty by consistently using the same basis for the domain space and the target space.

These considerations enable us to replace $\mathbb{R}^n$ and $\mathbb{R}^{n+m}$ in Lemma 3.3 with $\mathcal{F}$ and $\mathcal{F} \times \mathcal{E}$, respectively, where $\mathcal{F}$ and $\mathcal{E}$ are real normed spaces with dimensions n and m.

In the rest of this section, $\mathcal{X}$ denotes a real Banach space with norm $\|\cdot\|$, $\mathcal{G}$ denotes a bounded open subset of $\mathcal{X}$, and $\mathbf{g} : \mathrm{cl}\,\mathcal{G} \to \mathcal{X}$ is continuous. $\mathcal{X}$, $\mathcal{G}$, and $\mathbf{g}$ replace $\mathbb{R}^n$, Ω, and $\mathbf{I} - \mathbf{f}$ of Sec. 1.

We begin our process of defining $\deg[\mathbf{I} - \mathbf{g}, \mathcal{G}]$ by first studying a continuous mapping $\mathbf{g} : \mathcal{G} \to \mathcal{G} \cap \mathcal{F}$ where $\mathcal{F}$ is a finite-dimensional subspace of $\mathcal{X}$. For such $\mathbf{g}$'s, we define

$$\deg[\mathbf{I} - \mathbf{g}, \mathcal{G}] \equiv \deg[(\mathbf{I} - \mathbf{g})|_{\mathrm{cl}\,\mathcal{G} \cap \mathcal{F}}, \mathcal{G} \cap \mathcal{F}]. \tag{3.6}$$

To show that this definition makes sense, we must show that the right-hand side of (3.6) is independent of $\mathcal{F}$. Given $\mathbf{g}$, we let $\mathcal{F}_1$ and $\mathcal{F}_2$ be two finite-dimensional spaces so that the range of $\mathbf{g}$ is in both $\mathcal{F}_1$ and $\mathcal{F}_2$ and therefore in $\mathcal{F}_1 \cup \mathcal{F}_2$. Applying our generalized interpretation of Lemma 3.3, we obtain

$$\deg[(\mathbf{I}-\mathbf{g})|_{\mathrm{cl}\,\Omega\cap\mathcal{F}_\alpha}, \Omega\cap\mathcal{F}_\alpha] = \deg[(\mathbf{I}-\mathbf{g})|_{\mathrm{cl}\,\Omega\cap\mathcal{F}_1\cap F_2}, \Omega\cap\mathcal{F}_1\cap F_2], \quad \alpha = 1,2,$$

which gives the requisite independence.

We can now consider $\mathbf{g}: \mathcal{G} \to \mathcal{X}$ that are continuous and compact. For any such $\mathbf{g}$, we define the *Leray-Schauder degree* of $\mathbf{I}-\mathbf{g}$ to be

$$\deg[\mathbf{I}-\mathbf{g}, \mathcal{G}] \equiv \lim_{\varepsilon\to 0} \deg[\mathbf{I}-\mathbf{g}_\varepsilon, \mathcal{G}] \tag{3.7}$$

where the $\mathbf{g}_\varepsilon$ are a family of approximations to $\mathbf{g}$ in the sense of Lemma 3.2. We must show that the right-hand side of (3.7) is well defined:

3.8. Lemma. *Let $\mathcal{G}$ be a bounded open set in $\mathcal{X}$. Let $\mathbf{g}: \mathrm{cl}\,\mathcal{G} \to \mathcal{X}$ be compact and continuous. Let $\mathbf{0} \notin (\mathbf{I}-\mathbf{g})(\partial\mathcal{G})$. Then the Leray-Schauder degree is a well-defined integer, which is independent of the approximants $\mathbf{g}_\varepsilon$ appearing in* (3.7).

Proof. We need to show that the absence of fixed points of $\mathbf{g}$ on the boundary of $\mathcal{G}$ implies the absence of fixed points of $\mathbf{g}_\varepsilon$ on the boundary for ε sufficiently small. Toward this end, we first establish that

$$\exists\, \varepsilon > 0 \quad \text{such that} \quad \|\mathbf{x}-\mathbf{g}(\mathbf{x})\| > 2\varepsilon \quad \forall\, \mathbf{x} \in \partial\mathcal{G}. \tag{3.9}$$

Let us suppose that (3.9) does not hold. Then there is a sequence $\mathbf{x}_k \in \partial\mathcal{G}$, necessarily bounded, such that $\|\mathbf{x}_k - \mathbf{g}(\mathbf{x}_k)\| \to 0$. Since $\mathbf{g}$ is compact, there is a subsequence of $\{\mathbf{x}_k\}$, which we continue to denote by $\{\mathbf{x}_k\}$, such that $\mathbf{g}(\mathbf{x}_k)$ converges to a limit, which we denote by $\mathbf{y}$. Therefore, for this subsequence, we obtain from the triangle inequality that

$$\|\mathbf{x}_k - \mathbf{y}\| \le \|\mathbf{x}_k - \mathbf{g}(\mathbf{x}_k)\| + \|\mathbf{g}(\mathbf{x}_k) - \mathbf{y}\| \to 0.$$

Thus $\mathbf{x}_k \to \mathbf{y}$. Since $\mathbf{x}_k \in \partial\mathcal{G}$ and since $\partial\mathcal{G}$ is closed, it follows that $\mathbf{y} \in \partial\mathcal{G}$. Since $\mathbf{g}$ is continuous, it follows that $\mathbf{g}(\mathbf{x}_k) \to \mathbf{g}(\mathbf{y})$. Since limits are unique, we have $\mathbf{g}(\mathbf{y}) = \mathbf{y}$, which says that $\mathbf{g}$ has a fixed point on $\partial\mathcal{G}$, in contradiction to our hypothesis. Thus (3.9) holds.

By Lemma 3.2, there is a continuous mapping $\mathbf{g}_\varepsilon$ with range $\mathcal{X}_\varepsilon$ having finite dimension $k(\varepsilon)$ such that

$$\|\mathbf{g}_\varepsilon(\mathbf{x}) - \mathbf{g}(\mathbf{x})\| \le \varepsilon \quad \text{for} \quad \mathbf{x} \in \mathrm{cl}\,\mathcal{G}. \tag{3.10}$$

It follows from (3.9), (3.10), and the triangle inequality that

$$\|\mathbf{x} - \mathbf{g}_\varepsilon(\mathbf{x})\| \ge \|\mathbf{x} - \mathbf{g}(\mathbf{x})\| - \|\mathbf{g}(\mathbf{x}) - \mathbf{g}_\varepsilon(\mathbf{x})\| \ge \varepsilon > 0 \quad \forall\, \mathbf{x} \in \partial\mathcal{G}.$$

Thus $\mathbf{x} - \mathbf{g}_\varepsilon(\mathbf{x}) \neq \mathbf{0}$ for each $\mathbf{x} \in \partial(\mathcal{G}\cap\mathcal{X}_e)$, so that

$$\deg[\mathbf{I}-\mathbf{g}_\varepsilon, \mathcal{G}] \equiv \deg[(\mathbf{I}-\mathbf{g}_\varepsilon)|\mathrm{cl}\,\mathcal{G}\cap\mathcal{X}_e, \mathcal{G}\cap\mathcal{X}_e]$$

is well-defined.

Now for given ε, this degree is independent of the particular approximant $\mathbf{g}_\varepsilon$ introduced in (3.10). Indeed, suppose that $\tilde{\mathbf{g}}_\varepsilon$ with range $\tilde{\mathcal{X}}_\varepsilon$ also satisfies (3.10). We introduce the homotopy

$$\mathbf{h}(t,\mathbf{x}) \equiv t[\mathbf{x} - \tilde{\mathbf{g}}_\varepsilon(\mathbf{x})] + (1-t)[\mathbf{x} - \mathbf{g}_\varepsilon(\mathbf{x})].$$

Thus for $\mathbf{x} \in \partial(\mathcal{G}\cap\mathcal{X}_\varepsilon)$, (3.9) and (3.10) imply that

$$\begin{aligned}\|\mathbf{h}(t,\mathbf{x})\| &= \|[\mathbf{x}-\mathbf{g}(\mathbf{x})] - t[\mathbf{x}-\mathbf{g}(\mathbf{x})] - (1-t)[\mathbf{x}-\mathbf{g}(\mathbf{x})] + \mathbf{h}(t,\mathbf{x})\| \\ &\ge \|\mathbf{x}-\mathbf{g}(\mathbf{x})\| - t\|\mathbf{g}(\mathbf{x}) - \mathbf{g}_\varepsilon(\mathbf{x})\| - (1-t)\|\mathbf{g}(\mathbf{x}) - \tilde{\mathbf{g}}_\varepsilon(\mathbf{x})\| \\ &\ge 2\varepsilon - t\varepsilon - (1-t)\varepsilon = \varepsilon > 0.\end{aligned} \tag{3.11}$$

By the homotopy invariance (Corollary 2.6) and the boundary dependence (Proposition 2.12) of degree, we therefore conclude that $\deg[\mathbf{I}-\mathbf{g}_\varepsilon, \mathcal{G}] = \deg[\mathbf{I}-\tilde{\mathbf{g}}_\varepsilon, \mathcal{G}]$. We now observe that we can replace our $\tilde{\mathbf{g}}_\varepsilon$, required to satisfy (3.10), with $\mathbf{g}_\eta$ where η is any number in $(0, \varepsilon]$. Thus we conclude that $\deg[\mathbf{I}-\mathbf{g}_\varepsilon, \mathcal{G}]$ is actually independent of ε (for ε satisfying (3.9)). Thus the limit in (3.7) is trivially obtained. □

We now show that the analog of Theorem 2.2 holds:

3.12. Theorem. *Let the hypotheses of Lemma 3.8 hold. If* $\deg[\mathsf{I} - \mathsf{g}, \mathcal{G}] \neq 0$, *then* g *has a fixed point, i.e., there is an* x *that satisfies the equation* $\mathsf{g}(\mathsf{x}) = \mathsf{x}$.

Proof. Let g_ε be a 'sequence' of approximants to g in the sense of Lemma 3.2. By the last observation in the proof of Lemma 3.8, we know that

$$\deg[\mathsf{I} - \mathsf{g}_\varepsilon, \mathcal{G}] = \deg[\mathsf{I} - \mathsf{g}, \mathcal{G}] \neq 0 \tag{3.13}$$

for ε sufficiently small. Definition 3.6 and Theorem 2.2 imply that there is an $\mathsf{x}_\varepsilon \in \mathcal{G}$ satisfying

$$\mathsf{g}_\varepsilon(\mathsf{x}_\varepsilon) = \mathsf{x}_\varepsilon. \tag{3.14}$$

Since $\mathcal{G}$ is bounded, the compactness of g implies that $\mathsf{g}(\mathsf{x}_\varepsilon)$ has a subsequence, denoted the same way, that converges to a limit y in $\mathcal{X}$. Since

$$\|\mathsf{x}_\varepsilon - \mathsf{y}\| = \|\mathsf{g}_\varepsilon(\mathsf{x}_\varepsilon) - \mathsf{y}\| \leq \|\mathsf{g}_\varepsilon(\mathsf{x}_\varepsilon) - \mathsf{g}(\mathsf{x}_\varepsilon)\| + \|\mathsf{g}(\mathsf{x}_\varepsilon) - \mathsf{y}\| \tag{3.15}$$

for this subsequence, it follows from Lemma 3.2 and the definition of y that $\mathsf{x}_\varepsilon \to \mathsf{y}$. Since g is continuous, it follows that $\mathsf{y} = \lim \mathsf{g}(\mathsf{x}_\varepsilon) = \mathsf{g}(\mathsf{y})$. Thus g has a fixed point $\mathsf{y} \in \operatorname{cl} \mathcal{G}$. By hypothesis, $\mathsf{y} \notin \partial\mathcal{G}$, so that $\mathsf{y} \in \mathcal{G}$. □

Using the method of proof of Theorem 3.12, we can readily prove:

3.16. Theorem. *All the properties of the Brouwer degree described in Sec. 2 are valid for the Leray-Schauder degree under the standard hypotheses of Lemma 3.8 together with the further assumption that homotopies are compact and continuous on their domains in* (t, x)*-space.*

In particular, the analog of the Brouwer Fixed-Point Theorem is

3.17. Schauder Fixed-Point Theorem. *Let* $\operatorname{cl} \mathcal{G}$ *be homeomorphic to a bounded closed convex set in* $\mathcal{X}$. *If* $\mathsf{g} : \operatorname{cl} \mathcal{G} \to \operatorname{cl} \mathcal{G}$ *is compact and continuous, then it has a fixed point. Equivalently, if* $\operatorname{cl} \mathcal{G}$ *is homeomorphic to a compact convex set in* $\mathcal{X}$ *and if* $\mathsf{g} : \operatorname{cl} \mathcal{G} \to \operatorname{cl} \mathcal{G}$ *is continuous, then it has a fixed point.*

3.18. Exercise. Prove the first statement of this theorem when $\mathcal{G}$ is the open unit ball in $\mathcal{X}$. (Technical difficulties attend the proof of its generalization; see the references below.)

Appropriately framed analogs of Propositions 2.20, 2.24, 2.28, 2.30, and 2.38 are also valid. Many of the properties of degree can be extended to classes of mappings even larger than the compact perturbations of the identity.

The definition of the *Leray-Schauder index* is just like that of the (Brouwer) index given in Sec. 2.

Remarks on the bibliography. Our approach to degree theory, which is purely analytic, was pioneered by Nagumo (1951) and Heinz (1959). The presentation above follows the refinements of their treatment by Berger & Berger (1968), Deimling (1985), Lloyd (1978), Rabinowitz (1975), and J. T. Schwartz (1969). Among the many other sources giving an extensive development of degree theory by analytic means are Berger (1977), Browder (1976), Eisenack & Fenske (1978), Istratescu (1981), Mawhin (1979), Milnor (1969), Nirenberg (1974), Ortega & Rheinboldt (1970), and Zeidler (1986). Books presenting degree theory from a topological viewpoint include standard texts on algebraic topology, together with the works of Cronin (1964), Dugundji & Granas (1982), Krasnosel'skiĭ (1956), and Krasnosel'skiĭ & Zabreĭko (1975), and Rothe (1986), which have numerous applications to nonlinear equations. The pioneering paper of Leray & Schauder (1934) continues to merit careful study.

4. One-Parameter Global Bifurcation Theorem

In this section we use degree theory to prove the main parts of Theorem V.4.19, due to Rabinowitz (1971a), which we restate thus:

4.1. Theorem. *Let $\mathcal{X}$ be a real Banach space with norm $\|\cdot\|$ and let $\mathcal{D}$ be an open subset of $\mathbb{R} \times \mathcal{X}$. Let the intersection of $\mathcal{D}$ with $\mathbb{R} \times \{\mathbf{0}\}$ be an open interval $\mathcal{I}$. Let $\mathcal{D} \ni (\lambda, \mathsf{u}) \mapsto \mathsf{f}[\lambda, \mathsf{u}]$ have the form*

$$\mathsf{f}[\lambda, \mathsf{u}] = \mathsf{u} - \mathsf{L}(\lambda) \cdot \mathsf{u} - \mathsf{g}[\lambda, \mathsf{u}] \tag{4.2}$$

where $\mathsf{L}(\lambda) : \mathcal{X} \to \mathcal{X}$ is linear, $\mathsf{g}[\lambda, \mathsf{u}] = o(\|\mathsf{u}\|)$ as $\mathsf{u} \to \mathbf{0}$ uniformly for λ in any bounded set, and $(\lambda, \mathsf{u}) \mapsto \mathsf{L}(\lambda) \cdot \mathsf{u}, \mathsf{g}[\lambda, \mathsf{u}]$ are compact and continuous. Let $\alpha, \beta \in \mathcal{I} \setminus \mathcal{E}$, where $\mathcal{E}$ is the set of eigenvalues of $\lambda \mapsto \mathsf{I} - \mathsf{L}(\lambda)$. Let the Leray-Schauder indices of $\mathsf{I} - \mathsf{L}(\alpha)$ and $\mathsf{I} - \mathsf{L}(\beta)$ at $\mathbf{0}$ be different. (Note that if λ^0 is an (isolated) eigenvalue of $\lambda \mapsto \mathsf{I} - \mathsf{L}(\lambda)$ at which the index of $\mathsf{f}[\lambda, \cdot]$ at $\mathbf{0}$ changes, then there are numbers α, β with the properties just specified.) Then between α and β there is an eigenvalue λ^0 of $\lambda \mapsto \mathsf{I} - \mathsf{L}(\lambda)$ with the property that $(\lambda^0, \mathbf{0})$ belongs to a maximal connected subset $\mathcal{C}$ of the closure $\mathcal{S}$ of the set of nontrivial solution pairs. Moreover, $\mathcal{C}$ satisfies at least one of the following two alternatives:

(i) $\mathcal{C}$ does not lie in a closed and bounded subset of $\mathcal{D}$. (In particular, if $\mathcal{D} = \mathbb{R} \times \mathcal{X}$, then $\mathcal{C}$ is unbounded in $\mathbb{R} \times \mathcal{X}$.)

(ii) There is another eigenvalue μ of $\lambda \mapsto \mathsf{I} - \mathsf{L}(\lambda)$ on $\mathcal{I}$ such that $\mathcal{C}$ contains $(\mu, \mathbf{0})$.

In Theorem V.4.19, the set $\mathcal{C}$ is denoted $\mathcal{S}(\lambda^0)$. In this section, as in Sec. 2, we adhere to the convention that if $\mathcal{E}$ is a subset of $\mathbb{R} \times \mathcal{X}$, then $\mathcal{E}(t)$ is its section $\{\mathsf{x} : (t, \mathsf{x}) \in \mathcal{E}\}$. We accordingly change the notation to avoid ambiguity.

Proof. We first show that there must be a bifurcation point $(\lambda^0, \mathbf{0})$ on (α, β). It then follows from Theorem V.4.1 that λ^0 is in the spectrum of $\lambda \mapsto \mathsf{I} - \mathsf{L}(\lambda)$. The compactness of $\mathsf{L}(\lambda)$ ensures that the spectrum consists entirely of eigenvalues (see Riesz & Nagy (1955) or Stakgold (1979)).

Let us assume for contradiction that there is no bifurcation point on $[\alpha, \beta] \times \{\mathbf{0}\}$. From the definition of a bifurcation point in Sec. V.3, it follows that for each $\lambda \in [\alpha, \beta]$ there is a positive number $r(\lambda)$ such that if $\|\mathsf{u}\|, |\mu - \lambda| < r(\lambda)$, then (μ, u) is not a solution pair of

$$\mathsf{f}(\lambda, \mathsf{u}) = \mathbf{0}. \tag{4.3}$$

Now the segment $[\alpha, \beta] \times \{\mathbf{0}\}$ is covered by the open sets of the form $\{(\mu, \mathsf{u}) : \|\mathsf{u}\|, |\mu - \lambda| < r(\lambda)\}$, $\lambda \in [\alpha, \beta]$. Since this segment is essentially just a closed segment of the real line, it is compact. (To establish this formally, we need only use the Bolzano-Weierstrass Theorem to show that any sequence $(\lambda_k, \mathbf{0})$ has a convergent subsequence.) Therefore the open covering has a finite subcovering. Let r denote the minimum of the $r(\lambda)$ for this subcovering. Thus (4.3) has no solutions satisfying $\|\mathsf{u}\| = r$, $\lambda \in [\alpha, \beta]$. From the homotopy invariance (Corollary 2.6 and Theorem 3.16) it follows that $\deg[\mathsf{f}(\lambda, \cdot), \{\mathsf{u} : \|\mathsf{u}\| < r\}]$ is independent of λ, in contradiction to the hypothesis that the index changes.

Let us now suppose for contradiction that $\mathcal{C}$ satisfies neither of the properties (i) and (ii). Since f is a compact perturbation of the identity, we find that any bounded part of $\mathcal{S}$ is precompact, so that $\mathcal{C}$, which would be a closed and bounded subset of $\mathcal{S} \cup (\mathcal{I} \times \{\mathbf{0}\})$, would be compact. Therefore, by elementary properties of compact sets in metric spaces, its distance to $\mathcal{S} \setminus \mathcal{C}$ would be positive. Thus we could enclose $\mathcal{C}$ in an open set $\mathcal{U}$ whose closure contains no point of $\mathcal{S} \setminus \mathcal{C}$. If necessary, by shaving off small pieces of $\mathcal{U}$, we can ensure that the λ-axis does not intersect the boundary $\partial\mathcal{U}(\lambda)$ of any of its sections $\mathcal{U}(\lambda)$ (i.e., we can take the parts of $\partial\mathcal{U}$ near the λ-axis to be planes perpendicular to this axis). See Fig. 4.4.

Since we are assuming that $\mathcal{C}$ does not intersect the λ-axis anywhere but at $(\lambda^0, \mathbf{0})$, it follows that

$$r(\lambda) \equiv \operatorname{dist}(\mathcal{C}(\lambda), (\lambda, \mathbf{0}))$$

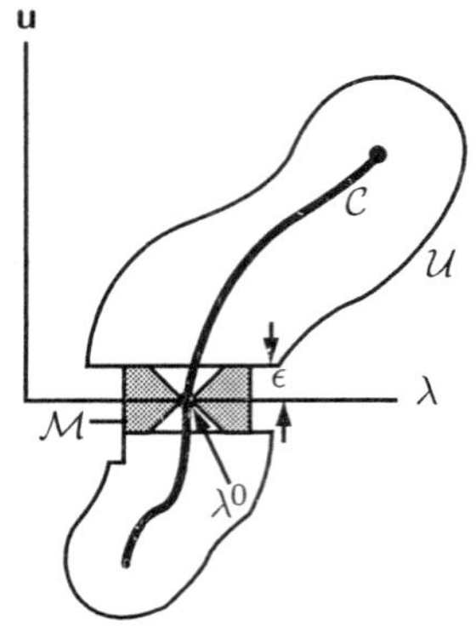

FIG. 4.4. Schematic diagram of the sets $\mathcal{C}$, $\mathcal{U}$, and $\mathcal{M}$ if neither properties (i) nor (ii) of Theorem 4.1 were to hold.

is positive for all $\lambda \neq \lambda^0$ at which $\mathcal{C}(\lambda)$ is not empty. Let ε be a small positive number. We set

$$\mathcal{M} \equiv \mathrm{cl}\,\{(\lambda, \mathbf{u}) \in \mathcal{U} : 2\|\mathbf{u}\| \leq r(\lambda), 2\varepsilon\}, \quad \mathcal{V} \equiv \mathcal{U} \setminus \mathcal{M}. \tag{4.5}$$

See Fig. 4.4. By the generalized homotopy invariance (Theorems 2.7 and 3.16), we find that the restrictions of $\lambda \mapsto \deg\,[\mathbf{f}(\lambda, \cdot), \mathcal{V}(\lambda)]$ to the subsets of its domain where $\lambda < \lambda^0$ and where $\lambda > \lambda^0$ are each constant. Since $\mathcal{V}(\lambda) = \emptyset$ for $|\lambda|$ large, we find that these constant values are each 0. The generalized homotopy invariance also implies that $\lambda \mapsto \deg\,[\mathbf{f}(\lambda, \cdot), \mathcal{U}(\lambda)]$ is a constant d. From the additivity property (Proposition 2.9 and Theorem 3.16) it follows that

$$d = \deg\,[\mathbf{f}(\lambda, \cdot), \mathcal{U}(\lambda)] = \mathrm{ind}\,[\mathbf{f}(\lambda, \cdot), \mathbf{0}] + \deg\,[\mathbf{f}(\lambda, \cdot), \mathcal{V}(\lambda)] = \mathrm{ind}\,[\mathbf{f}(\lambda, \cdot), \mathbf{0}] \tag{4.6}$$

wherever the index is defined. At such places it can be shown that

$$\mathrm{ind}\,[\mathbf{f}(\lambda, \cdot), \mathbf{0}] = \mathrm{ind}\,[\mathbf{I} - \mathbf{L}(\lambda, \mathbf{0})]. \tag{4.7}$$

Setting $\lambda = \alpha, \beta$ in (4.6) and (4.7), we find that $\mathrm{ind}\,[\mathbf{f}(\alpha, \cdot), \mathbf{0}] = d = \mathrm{ind}\,[\mathbf{f}(\beta, \cdot), \mathbf{0}]$ in contradiction to our hypotheses. □

4.8. Exercise. Prove (4.7).

References

The numbers in parentheses following a reference indicate the sections in which it is cited. The letter P refers to the Preface. Since many of the references describe work complementing that described in this book, the following listing is not representative of the emphasis placed here on the various topics.

R. Abraham & J.E. Marsden (1978), *Foundations of Mechanics,* 2nd edn., Benjamin/Cummings. (VIII.12)

R. Abraham & J. Robbin (1967), *Transversal Mappings and Flows*, Benjamin. (VI.9)

J. D. Achenbach (1973), *Wave Propagation in Elastic Solids*, North Holland. (XIII.15)

R. A. Adams (1975), *Sobolev Spaces*, Academic Press. (II.3, XVII.1, XIX.1)

S. A. Adeleke (1983), On the problem of eversion for incompressible elastic materials, *J. Elasticity* **13**, 63–69. (XIII.7)

J. E. Adkins (1954), Some generalizations of the shear problem for isotropic incompressible materials., *Proc. Camb. Phil. Soc.* **50**, 334–345. (XIII.10)

J. E. Adkins (1955), Finite deformation of materials exhibiting curvilinear aeolotropy, *Proc. Roy. Soc. Lond.* **A229**, 119–134. (XIII.7)

N. I. Akhiezer (1955), *Lectures on Variational Calculus* (in Russian), Nauka, English transl: *The Calculus of Variations*, Blaisdell, 1962. (VII.8)

J. C. Alexander (1981), A primer on connectivity, in *Fixed Point Theory*, E. Fadell & G. Fournier, eds., Springer Lect. Notes Math. **886**, 455–483. (V.3, V.4, V.8)

J. C. Alexander & S. S. Antman (1981), Global and local behavior of bifurcating multidimensional continua of solutions for multiparameter nonlinear eigenvalue problems, *Arch. Rational Mech. Anal.* **76**, 339–354. (III.3, V.4, XIII.6)

J. C. Alexander & S. S. Antman (1982), The ambiguous twist of Love, *Quart. Appl. Math.* **40**, 83–92. (VIII.12, VIII.18, XII.1)

J. C. Alexander & S. S. Antman (1983), Global behavior of solutions of nonlinear equations depending on infinite-dimensional parameters, *Indiana Univ. Math. J.* **32**, 39–62. (III.3, III.8, V.4, XIII.6)

J. C. Alexander, S. S. Antman & S.-T. Deng (1983), Nonlinear eigenvalue problems for the whirling of heavy elastic strings, II: New methods of global bifurcation theory, *Proc. Roy. Soc. Edin.* **93A**, 197–227. (VI.3)

J. C. Alexander & P. M. Fitzpatrick (1979), The homotopy of certain spaces of non-linear operators and its relation with global bifurcation of fixed points of parametrized condensing operators, *J. Funct. Anal.* **34**, 87–106. (V.4)

J. C. Alexander & P. M. Fitzpatrick (1980), Galerkin approximations in several parameter bifurcation problems, *Math. Proc. Camb. Phil. Soc.* **87**, 489–500. (V.4)

J. C. Alexander & P. M. Fitzpatrick (1981), Global bifurcation of solutions of equations involving several parameter multivalued condensing mappings, in *Fixed*

Point Theory, E. Fadell & G. Fournier, eds., Springer Lect. Notes Math. **886**, 1–20. (V.4)

J. C. Alexander & J. A. Yorke (1976), The implicit function theorem and global methods of cohomology, *J. Funct. Anal.* **21**, 330–339. (III.3, V.4, XIII.6)

J.-J. Alibert & B. Dacorogna (1992), An example of a quasiconvex function that is not polyconvex in two dimensions, *Arch. Rational Mech. Anal.* **117**, 155–166. (XIII.2)

G. Andrews (1979), On the Existence and Asymptotic Behavior of Solutions to a Damped Nonlinear Wave Equation, Heriot-Watt University, Dissertation. (XVI.3)

G. Andrews (1980), On the existence of solutions to the equation $u_{tt} = u_{xxt} + \sigma(u_x)_x$, *J. Diff. Eqs.* **35**, 200–231. (XVI.3)

G. Andrews & J. M. Ball (1982), Asymptotic behaviour and changes of phase in one-dimensional nonlinear viscoelasticity, *J. Diff. Eqs.* **44**, 306–341. (VIII.8, XVI.3)

S. S. Antman (1968), General solutions for extensible elasticae having nonlinear stress-strain laws, *Quart. Appl. Math.* **26**, 35–47. (IV.2, VIII.15, XII.12)

S. S. Antman (1970a), The shape of buckled nonlinearly elastic rings, *Z. angew. Math. Phys.* **21**, 422–438. (IV.3)

S. S. Antman (1970b), Existence of solutions of the equilibrium equations for nonlinearly elastic rings and arches, *Indiana Univ. Math. J.* **20**, 281–302. (VII.8)

S. S. Antman (1971), Existence and nonuniqueness of axisymmetric equilibrium states of nonlinearly elastic shells, *Arch. Rational Mech. Anal.* **40**, 329–371. (VII.8)

S. S. Antman (1972), The Theory of Rods, in *Handbuch der Physik* VIa/2, C. Truesdell, ed., Springer-Verlag, 641–703. (III.5, VII.8, VIII.4, VIII.18, XIV.15)

S. S. Antman (1973a), Nonuniqueness of equilibrium states for bars in tension, *J. Math. Anal. Appl.* **44**, 333–349. (XIV.6, XIV.12)

S. S. Antman (1973b), Monotonicity and invertibility conditions in one-dimensional nonlinear elasticity, in *Symposium on Nonlinear Elasticity*, R. W. Dickey, ed., Academic Press, 57–92. (IV.3)

S. S. Antman (1974a), Qualitative theory of the ordinary differential equations of nonlinear elasticity, in *Mechanics Today,* Vol. 1, 1972, S. Nemat-Nasser, ed., Pergamon Press, 58–101. (IV.2, IV.3)

S. S. Antman (1974b), Kirchhoff's problem for nonlinearly elastic rods, *Quart. Appl. Math.* **32**, 221–240. (VIII.8, VIII.18, IX.2)

S. S. Antman (1976a), Ordinary differential equations of one-dimensional nonlinear elasticity I: Foundations of the theories of nonlinearly elastic rods and shells, *Arch. Rational Mech. Anal.* **61**, 307–351. (VIII.5, XII.1, XIII.2, XIV.2, XIV.4, XIV.15)

S. S. Antman (1976b), Ordinary differential equations of one-dimensional nonlinear elasticity II: Existence and regularity theory for conservative problems, *Arch. Rational Mech. Anal.* **61**, 353–393. (VII.5, VII.8, VIII.12, XIV.2, XIV.4)

S. S. Antman (1977), Bifurcation problems for nonlinearly elastic structures, in *Symposium on Applications of Bifurcation Theory*, P. H. Rabinowitz, ed., Academic Press, 73–125. (V.2)

S. S. Antman (1978a), Buckled states of nonlinearly elastic plates, *Arch. Rational Mech. Anal.* **67**, 111–149. (X.2)

S. S. Antman (1978b), A family of semi-inverse problems of nonlinear elasticity, in *Contemporary Developments in Continuum Mechanics and Partial Differential*

Equations, G. M. de la Penha & L. A. Madeiros, eds., North Holland, 1–24. (XIII.6)

S. S. Antman (1979a), The eversion of thick spherical shells, *Arch. Rational Mech. Anal.* **70**, 113–123. (XIII.7)

S. S. Antman (1979b), Multiple equilibrium states for nonlinearly elastic strings, *SIAM J. Appl. Math.* **37**, 588–604. (III.3, III.8)

S. S. Antman (1980a), Nonlinear eigenvalue problems for the whirling of heavy elastic strings, *Proc. Roy. Soc. Edin.* **85A**, 59–85. (VI.2, VI.3)

S. S. Antman (1980b), The equations for the large vibrations of strings, *Amer. Math. Monthly* **87**, 359–370. (II.1)

S. S. Antman (1980c), Geometric aspects of global bifurcation in nonlinear elasticity, in *Geometric Methods in Physics*, G. Kaiser & J. E. Marsden, eds., Springer Lect. Notes Math. **775**, 1–29. (V.2)

S. S. Antman (1981), Global analysis of problems from nonlinear elastostatics, in *Applications of Nonlinear Analysis in the Physical Sciences*, H. Amann, N. Bazley & K. Kirchgässner, eds., Pitman, 245–270. (V.2, X.8)

S. S. Antman (1982), Material constraints in continuum mechanics, *Atti Accad. Naz. Lincei Rend. Cl. Sci. Fis. Mat. Natur. (8)* **70**, 256–264. (VIII.15, XII.12)

S. S. Antman (1983a), Regular and singular problems for large elastic deformations of tubes, wedges, and cylinders, *Arch. Rational Mech. Anal.* **83**, 1–52, Corrigenda, ibid. **95** (1986) 391–393. (VII.8, XIII.2, XIII.5, XIII.6, XIII.8, XIV.4)

S. S. Antman (1983b), The influence of elasticity on analysis: Modern developments, *Bull. Amer. Math. Soc.* (New Series) **9**, 267–291. (V.4)

S. S. Antman (1984), Large lateral buckling of nonlinearly elastic beams, *Arch. Rational Mech. Anal.* **84**, 293–305. (IX.6)

S. S. Antman (1988a), The paradoxical asymptotic status of massless springs, *SIAM J. Appl. Math.* **48**, 1319–1334. (III.7)

S. S. Antman (1988b), A zero-dimensional shock, *Quart. Appl. Math.* **46**, 569–581. (III.7, III.8, XVI.3)

S. S. Antman (1989), Nonlinear problems of geometrically exact shell theories, in *Analytic and Computational Models for Shells*, A. K. Noor, T. Belytschko & J. C. Simo, eds., Amer. Soc. Mech. Engrs., 109–131. (V.2, XIV.15)

S. S. Antman (1995), Asymptotic analysis of quasilinear parabolic equations describing viscoelastic bars, in preparation. (III.7)

S. S. Antman & H. Brezis (1978), The existence of orientation-preserving deformations in nonlinear elasticity, in *Nonlinear Analysis and Mechanics, Heriot-Watt Symposium,* Vol. II, R. J. Knops, ed., Pitman, 1–29. (VII.8)

S. S. Antman & M. C. Calderer (1985a), Asymptotic shapes of inflated noncircular elastic rings, *Math. Proc. Camb. Phil. Soc.* **97**, 357–379, Corrigendum, ibid. **101** (1987) 383. (IV.4, XIV.5)

S. S. Antman & M. C. Calderer (1985b), Asymptotic shapes of inflated spheroidal elastic shells, *Math. Proc. Camb. Phil. Soc.* **97**, 541–549, Corrigendum, ibid. **101** (1987) 383. (X.7)

S. S. Antman & E. R. Carbone (1977), Shear and necking instabilities in nonlinear elasticity, *J. Elasticity* **7**, 125–151. (VIII.8, XIV.6, XIV.12)

S. S. Antman & J. E. Dunn (1980), Qualitative behavior of buckled nonlinearly elastic arches, *J. Elasticity* **10**, 225–239. (IV.3, VI.12)

S. S. Antman & Guo Zhong-heng (1984), Large shearing oscillations of incompressible nonlinearly elastic bodies, *J. Elasticity* **14**, 249–262. (XIII.14)

S. S. Antman & K. B. Jordan (1975), Qualitative aspects of the spatial deformation of nonlinearly elastic rods, *Proc. Roy. Soc. Edin.* **73A**, 85–105. (IX.3)

S. S. Antman & C. S. Kenney (1981), Large buckled states of nonlinearly elastic rods under torsion, thrust, and gravity, *Arch. Rational Mech. Anal.* **76**, 289–338. (VIII.12, IX.5)

S. S. Antman & M. Lanza de Cristoforis (1995), Peculiar instabilities and apparent paradoxes due to the clamping of shearable rods. (IV.1)

S. S. Antman & T.-P. Liu (1979), Travelling waves in hyperelastic rods, *Quart. Appl. Math.* **36**, 377–399. (IX.4)

S. S. Antman & R. Malek-Madani (1987), Dissipative mechanics, in *Metastability and Incompletely Posed Problems*, S. S. Antman, J. L. Erickson, D. Kinderlehrer & I. Müller, eds., IMA Volumes in Mathematics and Its Applications **3**, Springer-Verlag, 1–16. (XVI.3)

S. S. Antman & R. Malek-Madani (1988), Travelling waves in nonlinearly viscoelastic media and shock structure in elastic media, *Quart. Appl. Math.* **46**, 77–93. (XVI.5)

S. S. Antman & R. S. Marlow (1991), Material constraints, Lagrange multipliers, and compatibility, *Arch. Rational Mech. Anal.* **116**, 257–299. (XII.12, XIV.15)

S. S. Antman & R. S. Marlow (1992), Transcritical buckling of columns, *Z. angew. Math. Phys.* **43**, 7–27. (VIII.9, XIV.5)

S. S. Antman & R. S. Marlow (1993), New phenomena in the buckling of arches described by refined theories, *Int. J. Solids Structures* **30**, 2213–2241. (VI.12)

S. S. Antman & A. Nachman (1980), Large buckled states of rotating rods, *Nonlin. Anal. T. M. A.* **4**, 303–327. (IV.5, VI.13, VII.8)

S. S. Antman & P. V. Negrón-Marrero (1987), The remarkable nature of radially symmetric equilibrium states of aeolotropic nonlinearly elastic bodies, *J. Elasticity* **18**, 131–164. (X.3, XIII.7)

S. S. Antman & J. E. Osborn (1979), The principle of virtual work and integral laws of motion, *Arch. Rational Mech. Anal.* **69**, 231–262. (II.4, XII.8, XII.9)

S. S. Antman & J. F. Pierce (1990), The intricate global structure of buckled states of compressible columns, *SIAM J. Appl. Math.* **50**, 395–419. (V.3, VI.9)

S. S. Antman & M. Reeken (1987), The drawing and whirling of strings: Singular global multiparameter bifurcation problems, *SIAM J. Math. Anal.* **18**, 337–365. (VI.4)

S. S. Antman & G. Rosenfeld (1978), Global behavior of buckled states of nonlinearly elastic rods, *SIAM Review* **20**, 513–566, Corrections and additions, ibid. **22** (1980), 186-187. (V.5, VI.9, VI.10)

S. S. Antman & T. I. Seidman (1995), Quasilinear hyperbolic-parabolic equations of nonlinear viscoelasticity, *J. Diff. Eqs.,* to appear. (II.3, II.11, XVI.3)

S. S. Antman & W. G. Szymczak (1989), Nonlinear elasto-plastic waves, *Contemporary Mathematics* **100**, 27–54. (XV.1, XV.3, XV.4)

S. S. Antman & W. G. Szymczak (1995), Large antiplane shearing motions of nonlinearly viscoplastic materials, in preparation. (XIII.10)

S. S. Antman & W. H. Warner (1966), Dynamical theory of hyperelastic rods, *Arch. Rational Mech. Anal.* **23**, 135–162. (XIV.3, XIV.15)

S. S. Antman & P. Wolfe (1983), Multiple equilibria of elastic strings under central forces: Highly singular nonlinear boundary value problems of the Bernoullis, *J. Diff. Eqs.* **47**, 180–213. (III.5, III.6, III.8)

V. I. Arnol'd (1974), *Mathematical Methods of Classical Mechanics* (in Russian), Nauka, English transl: Springer-Verlag, 1978. (VIII.12)

R. J. Atkin & N. Fox (1980), *An Introduction to the Theory of Elasticity*, Longman. (XIII.16)

C. D. Babcock (1983), Shell instability, *J. Appl. Mech.* **50**, 935–940. (V.2)

I. Babuška & L. Li (1992), The problem of plate modeling: Theoretical and computational results, *Comp. Meths. in Appl. Mech. Engg.* **100**, 249–273. (XIV.15)

C. Baiocchi & A. Capelo (1978), *Disequazioni variazionali e quasi variazionali. Applicazioni a problemi di frontiera libera*, Pitagora, English transl: *Variational and Quasivariational Problems*, Wiley, 1984. (VII.8)

A. K. Bajaj (1988), Bifurcations in flow-induced oscillations in tubes carrying a fluid, in *Dynamical Systems Approaches to Nonlinear Problems in Systems and Circuits*, F. M. A. Salam & M. L. Levi, eds., SIAM, 333–351. (VI.11)

A. K. Bajaj & P. R. Sethna (1984), Flow induced bifurcations to three-dimensional oscillatory motions in continuous tubes, *SIAM J. Appl. Math.* **44**, 270–286. (VI.11)

A. K. Bajaj, P. R. Sethna & T. S. Lundgren (1980), Hopf bifurcation in tubes carrying a fluid, *SIAM J. Appl. Math.* **39**, 213–230. (VI.11)

J. M. Ball (1977a), Convexity conditions and existence theorems in nonlinear elasticity, *Arch. Rational Mech. Anal.* **63**, 337–403. (XIII.2, XVI.6)

J. M. Ball (1977b), Constitutive inequalities and existence theorems in nonlinear elastostatics, in *Nonlinear Analysis and Mechanics: Heriot-Watt Symposium*, Vol. I, R. J. Knops, ed., Pitman Res. Notes Math. **17**, 187–241. (XIII.2, XIII.8)

J. M. Ball (1978), Finite-time blow-up in nonlinear problems, in *Nonlinear Evolution Equations*, M. G. Crandall, ed., Academic Press, 189–205. (XIII.12, XVI.6)

J. M. Ball (1980), Strict convexity, strong ellipticity, and regularity in the calculus of variations, *Math. Proc. Camb. Phil. Soc.* **87**, 501–513. (XIII.2)

J. M. Ball (1981a), Remarques sur l'existence et la régularité des solutions d'élastostatique nonlinéaire, in *Recent Contributions to Nonlinear Partial Differential Equations*, H. Berestycki & H. Brezis, eds., Pitman, 50–62. (VII.8, XIV.4)

J. M. Ball (1981b), Global invertibility of Sobolev functions and the interpenetration of matter, *Proc. Roy. Soc. Edin.* **88A**, 315–328. (XIII.2)

J. M. Ball (1982), Discontinuous equilibrium solutions and cavitation in nonlinear elasticity, *Phil. Trans. Roy. Soc. Lond.* **A306**, 557–611. (XIII.4, XIII.7)

J. M. Ball & R. D. James (1987), Fine phase mixtures as minimizers of energy, *Arch. Rational Mech. Anal.* **100**, 13–52. (VIII.8, XII.10, XIII.2)

J. M. Ball & R. D. James (1992), Proposed experimental tests of a theory of fine microstructure and the two-well problem, *Phil. Trans. Roy. Soc. Lond.* **A 338**, 389–450. (XIII.2)

J. M. Ball, R. J. Knops & J.E. Marsden (1978), Two examples in nonlinear elasticity, in *Journées d'analyse non linéaire*, Ph. Benilan & J. Robert, eds., Springer Lect. Notes Math. **665**, 41–48. (V.7)

J. M. Ball & F. Murat (1984), $W^{1,p}$-quasiconvexity and variational problems for multiple integrals, *J. Funct. Anal.* **58**, 225–253. (XIII.2)

J. M. Ball & D. G. Schaeffer (1983), Bifurcation and stability of homogeneous equilibrium configurations of an elastic body under dead-load tractions, *Math. Proc. Camb. Phil. Soc.* **94**, 315–339. (XIII.11)

L. Bauer, H. B. Keller & E. L. Reiss (1975), Multiple eigenvalues lead to secondary bifurcation, *SIAM Rev.* **17**, 101–122. (V.3)

L. Bauer, E. L. Reiss & H. B. Keller (1970), Axisymmetric buckling of hollow spheres and hemispheres, *Comm. Pure Appl. Math.* **23**, 529–568. (X.5)

P. Bauman, N. Owen & D. Phillips (1992), Maximum principles and a priori estimates for an incompressible material in nonlinear elasticity, *Comm. Partial Diff. Eqs.* **17**, 1185–1212. (XIII.2)

P. Bauman & D. Phillips (1994), Univalent minimizers of polyconvex functions in two dimensions, *Arch. Rational Mech. Anal.*, to appear. (XIII.2)

N. Bazley & B. Zwahlen (1968), Remarks on the bifurcation of solutions of a nonlinear eigenvalue problem, *Arch. Rational Mech. Anal.* **28**, 51–58. (VI.13)

M. F. Beatty & M. A. Hayes (1992), Deformations of an elastic, internally constrained material. Part 1: Homogeneous deformations, *J. Elasticity* **29**, 1–84. (XII.12)

M. Beck (1952), Die Knicklast des einseitig eingespannten tangential gedrückten Stabes, *Z. angew. Math. Phys.* **3**, 225–228. (V.2, V.7)

C. E. Beevers & R. E. Craine (1988), Asymptotic stability in nonlinear elastic materials with dissipation, *J. Elasticity* **19**, 101–110. (V.7)

C. E. Beevers & M. Šilhavý (1988), The asymptotic behavior of classical solutions to the mixed initial-boundary value problem in finite thermo-viscoelasticity, *Quart. Appl. Math.* **46**, 319–329. (V.7)

J. B. Bell, P. Colella & J. A. Trangenstein (1989), Higher-order Godunov methods for general systems of conservation laws, *J. Comp. Phys.* **82**, 362–397. (II.11)

J. F. Bell (1985), Contemporary perspectives in finite strain plasticity, *Int. J. Plasticity* **1**, 3–27. (XII.12)

M. S. Berger (1977), *Nonlinearity and Functional Analysis*, Academic Press. (V.8, VII.8, XIX.3)

M. S. Berger & M. S. Berger (1968), *Perspectives in Nonlinearity*, Benjamin. (XIX.3)

D. Bernoulli (1732), Methodus universalis determinandae curvaturae fili a potentiis quamcunque legem inter se, observantibus extensi, una cum solutione problematum quorundam novorum eo pertinentium, *Comm. Acad. Sci. Petrop.* **3** (1728), 62–69. (IV.1, VIII.16, VIII.18)

Jas. Bernoulli (1694), Curvatura laminae elasticae, *Acta. erud.*, in *Opera* **1**, 576-600. (IV.1, VIII.16, VIII.18)

Joh. Bernoulli (1729), Theoremata selecta pro conservatione virium vivarum demonstranda et experimentis confirmanda, *Comm. Acad. Petrop.* **2** (1727), 200–207, in *Opera Omnia* **3**, 124-130. (II.2)

S. Bharatha & M. Levinson (1978), Signorini's scheme for a general reference configuration in finite elastostatics, *Arch. Rational Mech. Anal.* **67**, 365–394. (XIII.11)

K. Bhattarcharya (1992), Self-accomodation in martensite, *Arch. Rational Mech. Anal.* **120**, 201–244. (XIII.16)

C. B. Biezeno & R. Grammel (1953), *Technische Dynamik*, 2nd edn., Springer-Verlag, English transl: *Engineering Dynamics*, Blackie, Van Nostrand, 1956. (V.2, VI.10)

M. A. Biot (1976), Hodograph method of nonlinear stress analysis of thick-walled cylinders and spheres, including porous materials, *Int. J. Solids Structures* **12**, 613–618. (X.3)

F. Bleich (1952), *Buckling Strength of Metal Structures*, McGraw-Hill. (V.2, VI.10)

G. A. Bliss (1946), *Lectures on the Calculus of Variations*, Univ. of Chicago Press. (VII.2, VII.8, VIII.15)

R. Böhme (1972), Die Lösung der Verzweigungsgleichungen für nichtlineare Eigenwertprobleme, *Math. Z.* **127**, 105–126. (V.3)

V. V. Bolotin (1956), *Dynamic Stability of Elastic Systems* (in Russian), Gostekhteorizdat, English transl: Holden-Day, 1964. (V.2)

V. V. Bolotin (1961), *Nonconservative Problems of the Theory of Elastic Stability* (in Russian), GIFML, English transl: Macmillan, 1963. (V.2, V.7)

M. Born (1906), *Untersuchungen über die Stabilität der elastischen Linie in Ebene und Raum, unter verschiedenen Grenzbedingungen*, Dieterich Univ. Buchdruckerei. (IV.6, IX.3, IX.4)

R. Bowen & C.-C. Wang (1976), *Introduction to Vectors and Tensors*, Plenum. (XI.2)

H. Brezis (1983), *Analyse fonctionelle*, Masson. (XVII.2)

F. Brezzi, J. Descloux, J. Rappaz & B. Zwahlen (1984), On the rotating beam: Some theoretical and numerical results, *Calcolo* **21**, 345–367. (VI.13)

F. Brezzi & M. Fortin (1991), *Mixed and Hybrid Finite Element Methods*, Springer-Verlag. (II.11, XIV.2)

P. W. Bridgman (1941), *The Nature of Thermodynamics*, Harvard Univ. Press. (XII.14)

F. E. Browder (1976), Nonlinear Operators and Nonlinear Equations of Evolution in Banach Spaces, in *Proc. Symp. Pure Math.* **18**, Part 2, Amer. Math. Soc. (XIX.3)

R. C. Browne (1979), Dynamic stability of one-dimensional visco-elastic bodies, *Arch. Rational Mech. Anal.* **68**, 231–262. (V.7)

D. O. Brush & B. O. Almroth (1975), *Buckling of Bars, Plates, and Shells*, McGraw-Hill. (V.2)

B. Budiansky (1974), Theory of buckling and post-buckling behavior of elastic structures, in *Advances in Applied Mechanics* **14**, C.-S. Yin, ed., Academic Press, 1–65. (V.2)

J. L. Buhite & D. R. Owen (1979), An ordinary differential equation from the theory of plasticity, *Arch. Rational Mech. Anal.* **71**, 357–383. (XV.6)

G. Burgermeister, H. Steup & H. Kretzschmar (1957, 1963), *Stabilitätstheorie,* Vols. I, II, Deutscher Verlag der Wissenschaften, Berlin. (V.2)

R. Burridge (1969), Spherically symmetric differential equations, the rotation group, and tensor spherical functions, *Proc. Camb. Phil. Soc.* **65**, 157–175. (XIII.11)

R. Burridge (1984), The group of motions in the plane and separation of variables in cylindrical coordinates, in *Mathematical Methods in Energy Research*, K. I. Gross, ed., SIAM, 22–39. (XIII.11)

G. R. Burton (1986), Multiple steady states for rotating rods, *Nonlin. Anal. T. M. A.* **10**, 1069–1076. (IV.5)

D. Bushnell (1981), Buckling of shells—Pitfalls for designers, *A.I.A.A. J.* **19**, 1183–1226. (V.2)

E. Buzano, G. Geymonat & T. Poston (1985), Post-buckling behavior of a non-linearly hyperelastic rod with cross-section invariant under the dihedral group, *Arch. Rational Mech. Anal.* **89**, 307–388. (VIII.9, XIV.5)

R. E. Caflisch & J. H. Maddocks (1984), Nonlinear dynamical theory of the elastica, *Proc. Roy. Soc. Edin.* **99A**, 1–23. (V.7, VIII.16)

M. C. Calderer (1983), The dynamic behavior of nonlinear elastic spherical shells, *J. Elasticity* **13**, 17–47. (XIII.12, XVI.6)

M. C. Calderer (1986), The dynamic behavior of viscoelastic spherical shells, *J. Diff. Eqs.* **63**, 289–305. (XIII.12, XVI.6)

A. J. Callegari, E. L. Reiss & H. B. Keller (1971), Membrane buckling: A study of solution multiplicity, *Comm. Pure Appl. Math.* **24**, 499–521. (X.3, X.8)

M. D. Cannon, D. D. Cullum & E. Polak (1970), *Theory of Optimal Control and Mathematical Programming*, McGraw-Hill. (VII.5)

G. Capriz & P. Podio-Guidugli (1974), On Signorini's perturbation method in finite elasticity, *Arch. Rational Mech. Anal.* **57**, 1–30. (XIII.11)

G. Capriz & P. Podio-Guidugli (1979), The role of Fredholm conditions in Signorini's perturbation method, *Arch. Rational Mech. Anal.* **70**, 261–288. (XIII.11)

G. Capriz & P. Podio-Guidugli (1982), A generalization of Signorini's perturbation method suggested by two problems of Grioli, *Rend. Sem. Mat. Padova* **68**, 149–162. (XIII.11)

D. E. Carlson (1972), Linear Thermoelasticity, in *Handbuch der Physik* VIa/2, C. Truesdell, ed., Springer-Verlag, 297–345. (XIII.15)

J. Carr (1981), *Applications of Centre Manifold Theory*, Springer-Verlag. (V.8, XIV.8)

J. Carr, M. E. Gurtin & M. Slemrod (1984), Structured phase transitions on a finite interval, *Arch. Rational Mech. Anal.* **86**, 317–351. (II.2, VIII.8, XII.10)

J. Carr & M. Z. M. Malhardeen (1979), Beck's problem, *SIAM J. Appl. Math.* **37**, 261–262. (VI.11)

G. F. Carrier (1945), On the nonlinear vibration problem of the elastic string, *Quart. Appl. Math.* **3**, 157–165. (II.8)

G. F. Carrier (1949), A note on the vibrating string, *Quart. Appl. Math.* **7**, 97–101. (II.8)

M. M. Carroll (1967), Controllable deformations of incompressible simple materials, *Int. J. Engg. Sci.* **5**, 515–525. (XIII.9)

M. M. Carroll (1988), Finite strain solutions in compressible isotropic elasticity, *J. Elasticity* **20**, 65–92. (XIII.2)

M. M. Carroll & P. M. Naghdi (1972), The influence of the reference geometry on the response of elastic shells, *Arch. Rational Mech. Anal.* **48**, 302–318. (XIV.13)

J. Casey & P. M. Naghdi (1984), Constitutive results for finitely deforming elastic-plastic materials, in *Constitutive Equations: Macro and Computational Aspects*, K. J. Willem, ed., Amer. Soc. Mech. Engrs., 53–71. (XV.1, XV.3)

J. Casey & P. M. Naghdi (1985), Physically nonlinear and related approximate theories of elasticity, and their invariance properties, *Arch. Rational Mech. Anal.* **88**, 59–82. (XIII.15)

A.-L. Cauchy (1850), Mémoire sur les systèmes isotropes de points matériels, *Mém. Acad. Sci. Paris* **22**, 615–654, in *Œuvres* (1) **2**, 351-386. (VIII.7)

L. Cesari (1983), *Optimization—Theory and Applications*, Springer-Verlag. (VII.8)

P. Chadwick (1976), *Continuum Mechanics*, Wiley. (XII.1)

C. Chen & W. von Wahl (1982), Die Rand-Anfangswertprobleme für quasilineare Wellengleichungen in Sobolewräumen niedriger Ordnung, *J. reine angew. Math.* **337**, 77–112. (XIII.2)

M.-S. Chen (1987), Hopf bifurcation in Beck's problem, *Nonlin. Anal. T. M. A.* **11**, 1061–1073. (VI.11)

P. J. Chen (1973), Growth and Decay of Waves in Solids, in *Handbuch der Physik* VIa/3, C. Truesdell, ed., 303–402. (XIII.2)

Y.-C. Chen (1991), On strong ellipticity and the Legendre-Hadamard condition, *Arch. Rational Mech. Anal.* **113**, 165–175. (XIII.2)

D. R. J. Chillingworth, J. E. Marsden & Y. H. Wan (1982), Symmetry and bifurcation in three-dimensional elasticity, I, *Arch. Rational Mech. Anal.* **80**, 295–331. (XIII.11)

D. R. J. Chillingworth, J. E. Marsden & Y. H. Wan (1983), Symmetry and bifurcation in three-dimensional elasticity, II, *Arch. Rational Mech. Anal.* **83**, 363–395. (XIII.11)

M. Chipot & D. Kinderlehrer (1988), Equilibrium configurations of crystals, *Arch. Rational Mech. Anal.* **103**, 237–277. (XIII.2)

S.-N. Chow & J. K. Hale (1982), *Methods of Bifurcation Theory*, Springer-Verlag. (IV.4, IV.8, V.4, V.8, XVIII.2)

P. G. Ciarlet (1978), *The Finite Element Method for Elliptic Problems*, North Holland. (II.11)

P. G. Ciarlet (1980), A justification of the von Kármán equations, *Arch. Rational Mech. Anal.* **73**, 349–389. (XIV.14)

P. G. Ciarlet (1988), *Mathematical Elasticity, Vol. I: Three-Dimensional Elasticity*, North Holland. (P, XIII.2, XIII.11, XIII.16)

P. G. Ciarlet (1990), *Plates and Junctions in Elastic Multi-Structures*, Springer-Verlag. (XIV.4, XIV.14, XIV.15)

P. G. Ciarlet (1994), *Mathematical Elasticity,* Vol. II, North-Holland, in preparation. (XIV.15)

P. G. Ciarlet & J.-L. Lions, eds. (1991), *Handbook of Numerical Analysis, Vol. II, Finite Element Methods*, North Holland. (II.11)

P. G. Ciarlet & J. Nečas (1987), Injectivity and self-contact in nonlinear elasticity, *Arch. Rational Mech. Anal.* **97**, 171–188. (XII.1)

P. G. Ciarlet & P. Rabier (1980), *Les Équations de von Kármán*, Springer Lect. Notes Math. **826**. (XIV.14)

A. Cimetière, G. Geymonat, H. Le Dret, A. Raoult & Z. Tutek (1988), Asymptotic theory and analysis for displacements and stress distribution in nonlinear elastic straight slender rods, *J. Elasticity* **19**, 111–161. (XIV.15)

A. Clebsch (1862), *Theorie der Elasticität fester Körper*, Teubner. (VIII.18)

P. Clément & J. Decloux (1984), Continuation and nodal properties of solutions of an eigenvalue problem for rotating rods, Dept. Math., Delft, Rept. 84-51. (VI.13)

P. Clément & J. Decloux (1991), A variational approach to a problem of rotating rods, *Arch. Rational Mech. Anal.* **114**, 1–13. (VI.13)

R. J. Clifton (1974), Plastic waves: Theory and experiment, in *Mechanics Today,* Vol. 1, 1972, S. Nemat-Nasser, ed., Pergamon Press, 102–167. (XV.3)

R. J. Clifton (1983), Dynamic plasticity, *J. Appl. Mech.* **50**, 941–952. (XV.3)

E. A. Coddington & N. Levinson, *Theory of Ordinary Differential Equations*, McGraw-Hill. (II.8, II.9, III.7, V.4, V.5, VI.2, VIII.5)

H. Cohen (1966), A nonlinear theory of elastic directed curves, *Int. J. Engg. Sci.* **4**, 511–524. (VIII.18)

H. Cohen & C. N. DeSilva (1966a), Nonlinear theory of elastic surfaces, *J. Math. Phys.* **7**, 246–253. (XIV.15)

H. Cohen & C. N. DeSilva (1966b), Theory of directed surfaces, *J. Math. Phys.* **7**, 960–966. (XIV.15)

H. Cohen & C.-C. Wang (1989), Direct models for rods and shells, *Arch. Rational Mech. Anal.* **108**, 35–81. (VIII.18, XIV.13, XIV.15)

B. D. Coleman (1964), Thermodynamics of materials with memory, *Arch. Rational Mech. Anal.* **17**, 1–46. (XII.14)

B. D. Coleman (1985), On the cold drawing of polymers, *Comp. Maths. with Appls.* **11**, 35–65. (VIII.8)

B. D. Coleman & M. E. Gurtin (1967), Thermodynamics with internal state variables, *J. Chem. Phys.* **47**, 597–613. (XV.3)

B. D. Coleman & V. J. Mizel (1964), Existence of caloric equations of state in thermodynamics, *J. Chem. Phys.* **40**, 1116–1125. (XII.14)

B. D. Coleman & W. Noll (1963), The thermodynamics of elastic materials with heat conduction and viscosity, *Arch. Rational Mech. Anal.* **13**, 167–178. (XII.14)

B. D. Coleman & D. R. Owen (1974), A mathematical foundation for thermodynamics, *Arch. Rational Mech. Anal.* **54**, 1–104. (XII.14)

J. F. Colombeau (1990), Multiplication of distributions, *Bull. Amer. Math. Soc.* **23**, 251–268. (II.4)

E. & F. Cosserat (1907), Sur la statique de la ligne déformable, *C. R. Acad. Sci. Paris* **145**, 1409–1412. (VIII.18)

E. & F. Cosserat (1908), Sur la théorie des corps minces, *C. R. Acad. Sci. Paris* **146**, 169–172. (XIV.15)

E. & F. Cosserat (1909), *Théorie des corps déformables*, Hermann. (VIII.18, XIV.15)

R. Courant & K. O. Friedrichs (1948), *Supersonic Flow and Shock Waves*, Interscience. (XVI.2)

R. Courant & D. Hilbert (1953), *Methods of Mathematical Physics,* Vol. I, Interscience. (V.5, VI.2)

R. Courant & D. Hilbert (1961), *Methods of Mathematical Physics,* Vol. II, Interscience. (VIII.8, XIII.2)

S. J. Cox & M. L. Overton (1992), On the optimal design of columns against buckling, *SIAM J. Math. Anal.* **23**, 287–325. (XIV.4, XIV.5)

M. G. Crandall & P. H. Rabinowitz (1970), Nonlinear Sturm-Liouville problems and topological degree, *J. Math. Mech.* **19**, 1083–1102. (V.4, V.5)

M. G. Crandall & P. H. Rabinowitz (1971), Bifurcation from simple eigenvalues, *J. Funct. Anal.* **8**, 321–340. (V.4)

J. Cronin (1964), *Fixed Points and Topological Degree in Nonlinear Analysis*, Amer. Math. Soc. (XIX.3)

P. K. Currie & M. Hayes (1981), On non-universal finite elastic deformations, in *Proceedings of the IUTAM Symposium on Finite Elasticity*, D. E. Carlson & R. T. Shield, eds., Martinus Nijhoff, 143–150. (XIII.9)

B. Dacorogna (1981), A relaxation theorem and its application to the equilibrium of gases, *Arch. Rational Mech. Anal.* **77**, 359–386. (XIII.2)

B. Dacorogna (1989), *Direct Methods in the Calculus of Variations*, Springer-Verlag. (P, VII.8, XIII.2)

C. M. Dafermos (1969), The mixed initial-boundary value problem for the equations of nonlinear one-dimensional viscoelasticity, *J. Diff. Eqs.* **6**, 71–86. (II.3, VIII.8, XVI.3)

C. M. Dafermos (1973), The entropy rate admissibility criterion for solutions of hyperbolic conservation laws, *J. Diff. Eqs.* **14**, 202–212. (XVI.2)

C. M. Dafermos (1982), Global smooth solutions to initial value problems for equations of one-dimensional nonlinear thermoviscoelasticity, *SIAM J. Math. Anal.* **13**, 397–408. (XVI.5)

C. M. Dafermos (1983), Hyperbolic systems of conservation laws, in *Systems of Nonlinear Partial Differential Equations*, J. M. Ball, ed., Reidel, 25–70. (XII.14)

C. M. Dafermos (1984), *Discontinuous thermokinetic processes,* Appendix 4B in Truesdell (1984), Springer-Verlag. (XII.14, XVI.2)

C. M. Dafermos (1985), Dissipation stabilization and the second law of thermodynamics, in *Thermodynamics and Constitutive Equations*, G. Grioli, ed., Springer Lect. Notes Phys. **228**, 44–88. (XII.14)

C. M. Dafermos & W. J. Hrusa (1985), Energy methods for quasilinear hyperbolic initial-boundary value problems. Applications to elastodynamics, *Arch. Rational Mech. Anal.* **87**, 267–292. (XIII.2)

J. Le Rond d'Alembert (1743), *Traité de dynamique*, David l'aîne, Paris. (II.2)

J. Le Rond d'Alembert (1749), Recherches sur la courbe que forme une corde tendue mise en vibration, *Hist. Acad. Sci. Berlin* **3** (1747), 214–249. (II.2)

E. N. Dancer (1973), Global structure of the solutions of non-linear real analytic eigenvalue problems, *Proc. Lond. Math. Soc. (3)* **27**, 747–765. (V.4)

E. N. Dancer (1974a), On the structure of solutions of non-linear eigenvalue problems, *Indiana Univ. Math. J.* **23**, 1069–1076. (V.4)

E. N. Dancer (1974b), A note on bifurcation from infinity, *Quart. J. Math. Oxford Ser.* **25**, 81–84. (VI.2)

J.-L. Davet (1986), Justification de modèles de plaques nonlinéaires pour des lois de comportment générales, *Mod. Math. Anal. Num.* **20**, 225–249. (XIV.14)

P. J. Davis (1989), Buckling and barrelling instabilities in finite elasticity, *J. Elasticity* **21**, 147–192. (XIII.11)

W. A. Day (1972), *The Thermodynamics of Simple Materials with Fading Memory*, Springer-Verlag. (XII.14)

K. Deimling (1985), *Nonlinear Functional Analysis*, Springer-Verlag. (V.8, VII.8, XIX.3)

C. N. DeSilva & A. B. Whitman (1969), A dynamical theory of elastic directed curves, *Z. angew. Math. Phys.* **20**, 200–212. (VIII.18)

C. N. DeSilva & A. B. Whitman (1971), A thermodynamic theory of directed curves, *J. Math. Phys.* **12**, 1603–1609. (VIII.2, VIII.18)

R. W. Dickey (1969), The nonlinear string under a vertical force, *SIAM J. Appl. Math.* **17**, 172–178. (III.1, III.3, III.8)

R. W. Dickey (1973), A quasilinear evolution equation and the method of Galerkin, *Proc. Amer. Math. Soc.* **37**, 149–156. (II.11)

R. W. Dickey (1983), The nonlinear circular membrane under a vertical force, *Quart. Appl. Math.* **41**, 331–338. (X.8)

R. W. Dickey & J. J. Roseman (1993), Equilibria of the circular elastica under a uniform central force field, *Quart. Appl. Math.* **51**, 201–216. (IV.3, VI.12)

J. Dieudonné (1949), Sur le polygone de Newton, *Archiv der Math.* **2**, 49–55. (IV.4, XIII.7, XVIII.2)

J. Dieudonné (1960), *Foundations of Modern Analysis*, Academic Press. (XI.2)

M. Dikmen (1982), *Theory of Thin Elastic Shells*, Pitman. (XIV.15)

E. H. Dill (1992), Kirchhoff's theory of rods, *Arch. Hist. Exact Sci.* **44**, 1–23. (VIII.18, XIV.15)

A. N. Dinnik (1935), *Stability of Elastic Systems* (in Russian), ONTI. (V.2)

R. DiPerna (1983a), Convergence of approximate solutions to conservation laws, *Arch. Rational Mech. Anal.* **82**, 27–70. (XVI.4)

R. DiPerna (1983b), Convergence of the viscosity method for isentropic gas dynamics, *Comm. Math. Phys.* **91**, 1–30. (XVI.4)

E. H. Dowell (1975), *Aeroelasticity of Plates and Shells*, Noordhoff. (V.2)

E. H. Dowell & M. Il'gamov (1988), *Studies in Nonlinear Aeroelasticity*, Springer-Verlag. (V.2)

T. C. Doyle & J. L. Ericksen (1956), Nonlinear elasticity, in *Advances in Applied Mechanics* **4**, H. L. Dryden & T. von Kármán, eds., Academic Press, 53–115. (XI.3)

J. Dugundji & A. Granas (1982), *Fixed Point Theory*, PWN-Polish Scientific Publishers. (XIX.3)

J. E. Dunn & J. B. Serrin (1985), On the thermodynamics of interstitial working, *Arch. Rational Mech. Anal.* **88**, 95–133. (XII.10)

G. Duvaut & J.-L. Lions (1972), *Les inéquations en mecanique et en physique*, Dunod, English transl: Springer-Verlag, 1976. (VII.8, XIII.16)

C. L. Dym (1974), *Stability Theory and its Applications to Structural Mechanics*, Noordhoff. (V.2)

D. G. Ebin & R. A. Saxton (1986), The initial-value problem for elastodynamics of incompressible bodies, *Arch. Rational Mech. Anal.* **94**, 15–38. (XII.15, XIII.2)

D. G. Ebin & S. R. Simanca (1992), Deformation of incompressible bodies with free boundary, *Arch. Rational Mech. Anal.* **119**, 61–97. (XIII.2)

G. Eisenack & C. Fenske (1978), *Fixpunkttheorie*, BI-Wissenschaftsverlag. (XIX.3)

L. P. Eisenhart (1926), *Riemannian Geometry*, Princeton Univ. Press. (XII.3)

J. G. Eisley (1963), Nonlinear deformation of elastic beams, rings, and strings, *Appl. Mech. Revs.* **16**, 677–680. (V.2)

I. Ekeland (1990), *Convexity Methods in Hamiltonian Mechanics*, Springer-Verlag. (II.10)

I. Ekeland & R. Temam (1972), *Analyse convexe et problèmes variationnels*, Dunod, English transl: North Holland, 1976. (VII.8)

H. Engler (1989), Global regular solutions for the dynamic antiplane shear problem in nonlinear viscoelasticity, *Math. Z.* **202**, 251–259. (XIII.10)

J. L. Ericksen (1953), On the propagation of waves in isotropic incompressible perfectly elastic materials, *J. Rational Mech. Anal.* **2**, 329–337. (XIII.2)

J. L. Ericksen (1954), Deformations possible in every isotropic incompressible perfectly elastic body, *Z. angew. Math. Phys.* **5**, 466–486. (XII.3, XIII.9)

J. L. Ericksen (1955a), Eversion of a perfectly elastic spherical shell, *Z. angew. Math. Mech.* **35**, 381–385. (XIII.7)

J. L. Ericksen (1955b), Deformations possible in every compressible, perfectly elastic material, *J. Math. & Phys.* **34**, 126–128. (XIII.9)

J. L. Ericksen (1960), Tensor Fields, in *Handbuch der Physik* III/1, C. Truesdell, ed., Springer-Verlag, 794–858. (XI.3)

J. L. Ericksen (1970), Uniformity in shells, *Arch. Rational Mech. Anal.* **37**, 73–84. (XIV.13)

J. L. Ericksen (1972), Symmetry transformations for thin elastic shells, *Arch. Rational Mech. Anal.* **47**, 1–14. (XIV.13)

J. L. Ericksen (1975), Equilibrium of bars, *J. Elasticity* **5**, 191–202. (II.2, VIII.8)

J. L. Ericksen (1977a), On the formulation of Saint-Venant's problem, in *Nonlinear Analysis and Mechanics, Heriot-Watt Symposium,* Vol. I, R. J. Knops, ed., Pitman, 158–186. (XIII.8, XIV.8)

J. L. Ericksen (1977b), Special topics in nonlinear elastostatics, in *Advances in Applied Mechanics* **17**, C.-S. Yih, ed., Academic Press. (II.2, VIII.8, XIII.8)

J. L. Ericksen (1980a), Some phase transitions in crystals, *Arch. Rational Mech. Anal.* **73**, 99–124. (VIII.8, XIII.2)

J. L. Ericksen (1980b), Periodic solutions for elastic prisms, *Quart. Appl. Math.* **37**, 443–446. (XIII.8)

J. L. Ericksen (1983), Ill-posed problems in thermoelasticity theory, in *Systems of Nonlinear Partial Differential Equations*, J. M. Ball, ed., Reidel, 71–93. (XIII.8, XIV.8)

J. L. Ericksen (1986), Constitutive theory for some constrained elastic crystals, *Int. J. Solids Structures* **22**, 951–964. (XII.12)

J. L. Ericksen (1991), *Introduction to the Thermodynamics of Solids*, Chapman and Hall. (XII.14)

J. L. Ericksen & R. S. Rivlin (1954), Large elastic deformations of homogeneous anisotropic materials, *J. Rational Mech. Anal.* **3**, 281–301. (XII.12)

J. L. Ericksen & C. Truesdell (1958), Exact theory of stress and strain in rods and shells, *Arch. Rational Mech. Anal.* **1**, 295–323. (VIII.18, XIV.15)

A. C. Eringen (1962), *Non-linear Theory of Continuous Media*, McGraw-Hill. (XI.3)

A. C. Eringen & E. S. Suhubi (1974), *Elastodynamics, Vol. I, Finite Motions*, Academic Press. (XIII.16)

L. Euler (1727), De oscillationibus annulorum elasticorum, in *Opera Omnia* II, Vol. 11, 378–382. (IV.1, VIII.16, VIII.18)

L. Euler (1732), Solutio problematis de invenienda curva quam format lamina utcunque elastica in singulis punctis a potentiis quibuscunque sollicitata, *Comm. Acad. Sci. Petrop.* **3** (1728), 70–84, in *Opera Omnia* II, Vol. 10, 1–16. (IV.1, VIII.16, VIII.18)

L. Euler (1744), *Additamentum I de curvis elasticis, methodus inveniendi lineas curvas maximi minimivi proprietate gaudentes*, Bousquent, Lausanne, in *Opera Omnia* I, Vol. 24, 231–297. (IV.2, V.2, VIII.18)

L. Euler (1751), De motu corporum flexibilum, *Comm. Acad. Sci. Petrop.* **14** (1744/1746), 182–196, in *Opera Omnia* I, Vol. 10, 165–176. (II.1, II.2)

L. Euler (1752), Découverte d'un nouveau principe de mécanique, *Hist. Acad. Sci. Berlin* **6** (1750), 185–217, in *Opera Omnia* II, Vol. 5. (II.1, II.2)

L. Euler (1771), Genuina principia doctrinae de statu aequilibrii et motu corporum tam perfecte flexibilium quam elasticorum, *Novi Comm. Acad. Sci. Petrop.* **15** (1770), 381–413 (1770), in *Opera Omnia* II, Vol. 11, 37–61. (II.1, II.2, VIII.18)

L. Euler (1774), De genuina methodo tam aequilibrium quam motum corporum flexibilium determinandi et utriusque egregio consensu, *Novi Comm. Acad. Sci. Petrop.* **20**, 286–303, in *Opera Omnia* II, Vol. 11, 180–193. (VIII.18)

L. Euler (1775), De motu turbinatorio chordarum musicarum ubi simul universa theoria tam aequilibrii quam motus corporum flexibilium simulque etiam elasticorum breviter explicatur, *Novi Comm. Acad. Sci. Petrop.* **19**, 340–370, in *Opera Omnia* II, Vol. 11, 158–179. (VIII.18)

L. Euler (1780a), Determinatio onerum, quae columnae gestare valent, *Acta Acad. Sci. Petrop.* **2**, 121–145, in *Opera Omnia* II, Vol. 17, 232–251. (V.2)

L. Euler (1780b), Examen insignis paradoxi in theoria columnarum occurentis, *Acta Acad. Sci. Petrop.* **2**, 146–162, in *Opera Omnia* II, Vol. 17, 252–265. (V.2)

L. Euler (1780c), De altitudine columnarum sub proprio pondere corruentium, *Acta Acad. Sci. Petrop.* **2**, 163–193, in *Opera Omnia* II, Vol. 17, 166–293. (V.2)

L. C. Evans & R. F. Gariepy (1991), *Measure Theory and Fine Properties of Functions*, CRC Press. (XI.2)

G. M. Ewing (1969), *Calculus of Variations with Applications*, Norton. (VII.8)

H. Federer (1969), *Geometric Measure Theory*, Springer-Verlag. (XI.2)

G. Fichera (1972a), Existence Theorems in Elasticity, in *Handbuch der Physik* VIa/2, C. Truesdell, ed., 347–389. (XIII.15)

G. Fichera (1972b), Boundary Value Problems of Elasticity with Unilateral Constraints, in *Handbuch der Physik* VIa/2, C. Truesdell, ed., Springer-Verlag, 391–424. (VII.8, XIII.15, XIII.16)

D. Fisher (1987), Conservative configuration dependent loads on bounded surfaces, *Z. angew. Math. Phys.* **38**, 883–892. (VI.12)

D. Fisher (1988), Configuration dependent pressure potentials, *J. Elasticity* **19**, 77–84. (VI.12)

D. Fisher (1989), Conservative loads on shells, *Z. angew. Math. Phys.* **40**, 39–50. (VI.12)

P. M. Fitzpatrick, I. Massabò & J. Pejsachowicz (1983), Global several-parameter bifurcation and continuation theorems: a unified approach via complementing maps, *Math. Ann.* **263**, 61–73. (III.3, V.4, XIII.6)

P. M. Fitzpatrick & J. Pejsachowicz (1991), Parity and generalized multiplicity, *Trans. Amer. Math. Soc.* **326**, 281–305. (V.4)

P. M. Fitzpatrick & J. Pejsachowicz (1993), Orientation and the Leray-Schauder Theory for Fully Nonlinear Elliptic Boundary Value Problems, *Mem. Amer. Math. Soc.* **101**. (V.4)

P. M. Fitzpatrick, P. Rabier & J. Pejsachowicz (1994), Orientability of Fredholm families and topological degree for orientable nonlinear Fredholm maps, *J. Funct. Anal.*, to appear. (V.4)

B. Fleishman (1959), Unpublished work, quoted by J. B. Keller (1959). (II.7)

I. Fonseca (1987), Variational methods for elastic crystals, *Arch. Rational Mech. Anal.* **97**, 189–220. (XIII.2)

I. Fonseca (1988), The lower quasiconvex envelope of the stored energy function for an elastic crystal, *J. Math. Pures Appl.* **67**, 175–195. (XIII.2)

I. Fonseca (1989a), Interfacial energy and the Maxwell rule, *Arch. Rational Mech. Anal.* **106**, 63–95. (XIII.2)

I. Fonseca (1989b), Phase transitions of elastic solid materials, *Arch. Rational Mech. Anal.* **107**, 195–223. (XIII.2)

R. L. Fosdick (1966), Remarks on compatibility, in *Modern Developments in the Mechanics of Continua*, S. Eskinazi, ed., Academic Press, 109–127. (XIII.9)

R. L. Fosdick (1971), Statically possible radially symmetric deformations in isotropic, incompressible elastic solids, *Z. angew. Math. Phys.* **22**, 590–607. (XIII.9)

R. L. Fosdick & R. D. James (1981), The elastica and the problem of pure bending for a non-convex stored energy function, *J. Elasticity* **11**, 165–186. (VII.8)

R. L. Fosdick & G. P. MacSithigh (1986), Minimization in incompressible nonlinear elasticity theory, *J. Elasticity* **16**, 267–301. (XIII.2)

R. L. Fosdick & K. W. Schuler (1969), On Ericksen's problem for plane deformations with uniform transverse stretch, *Int. J. Engg. Sci.* **7**, 217–233. (XIII.9)

R. L. Fosdick & J. B. Serrin (1979), On the impossibility of linear Cauchy and Piola-Kirchhoff constitutive theories for stress in solids, *J. Elasticity* **9**, 83–89. (XIII.15)

D. D. Fox, A. Raoult & J. C. Simo (1993), A justification of nonlinear properly invariant plate theories, *Arch. Rational Mech. Anal.* **124**, 157–199. (XIV.14. XIV.15)

D. D. Fox & J. C. Simo (1992), A nonlinear geometrically exact shell model incorporating independent (drill) rotations, *Comp. Meths. Appl. Mech. Engg.* **98**, 329–343. (XIV.15)

L. R. Foy (1964), Steady state solutions of hyperbolic systems of conservation laws with viscosity terms, *Comm. Pure Appl. Math.* **17**, 177–188. (XVI.4)

R. Frisch-Fay (1962), *Flexible Bars*, Butterworths. (IV.2, IV.6, V.1, IX.4)

D. Fu, K. R. Rajagopal & A. Z. Szeri (1990), Non-homogeneous deformations in a wedge of Mooney-Rivlin material, *Int. J. Non-Linear Mech.* **25**, 375–387. (XIII.8)

S. Fučik, J. Nečas, J. Souček & V. Souček (1973), *Spectral Analysis of Nonlinear Operators*, Springer Lect. Notes Math. **346**. (V.8)

P. Funk (1970), *Variationsrechnung und ihre Anwendung in Physik und Technik*, 2nd edn., Springer-Verlag. (IV.6, IX.4)

A. Gajewski & M. Życzkowski (1970), Optimal design of elastic columns subject to the general conservative behavior of loading, *Z. angew. Math. Phys.* **21**, 806–818. (VI.10)

L. A. Galin (1980), *Contact Problems of the Theories of Elasticity and Viscoelasticity*, Nauka. (XIII.16)

R. C. Gauss & S. S. Antman (1984), Large thermal buckling of nonuniform beams and plates, *Int. J. Solids Structures* **20**, 979–1000. (XIII.7)

H. Geiringer (1973), Ideal Plasticity, in *Handbuch der Physik* VIa/3, C. Truesdell, ed., Springer-Verlag, 403–534. (XV.1)

I. M. Gel'fand & S. V. Fomin (1961), *Variational Calculus* (in Russian), Fizmatgiz, English transl: *Calculus of Variations*, Prentice-Hall, 1963. (VII.8)

G. Geymonat, S. Müller & N. Triantafyllidis (1993), Homogenization of nonlinearly elastic materials, microscopic bifurcation, and macroscopic loss of rank-one convexity, *Arch. Rational Mech. Anal.* **122**, 231–290. (XIII.16)

M. Giaquinta (1983), *Multiple Integrals in the Calculus of Variations and Nonlinear Elliptic Systems*, Princeton Univ. Press. (XIII.2)

M. Giaquinta, G. Modica & J. Souček (1989), Cartesian currents, weak diffeomorphisms and existence theorems in nonlinear elasticity, *Arch. Rational Mech. Anal.* **105**, 97–159, Erratum and Addendum, ibid. **109** (1990), 385-392. (XIII.2, XIII.7)

J. W. Gibbs & E. B. Wilson (1901), *Vector Analysis*, Yale Univ. Press. (VIII.5, XI.2)

G. M. L. Gladwell (1980), *Contact Problems in the Classical Theory of Elasticity*, Sijthoff & Noordhoff. (XIII.16)

I. Gohberg, S. Goldberg & M. Kaashoek (1991), *Classes of Linear Operators*, Vol. I, Birkhäuser. (XVII.2)

S. Goldberg (1966), *Unbounded Linear Operators*, McGraw-Hill. (XVII.2)

H. Goldstein (1980), *Classical Mechanics*, 2nd edn., Addison-Wesley. (VIII.5)

M. Golubitsky & D. G. Schaeffer (1979), A theory for imperfect bifurcation via singularity theory, *Comm. Pure Appl. Math.* **32**, 21–98. (VI.6)

M. Golubitsky & D. G. Schaeffer (1985), *Singularities and Groups in Bifurcation Theory*, Vol. I, Springer-Verlag. (V.1, V.8, VI.6, VI.8, XVIII.2)

M. Golubitsky, I. Stewart & D. G. Schaeffer (1988), *Singularities and Groups in Bifurcation Theory*, Vol. II, Springer-Verlag. (V.8, VI.6, XIII.11, XIV.13, XVIII.2)

J. Goodman & Z. Xin (1992), Viscous limits for piecewise smooth solutions to systems of conservation laws, *Arch. Rational Mech. Anal.* **121**, 235–265. (XVI.4)

A. E. Green (1955), Finite elastic deformation of compressible isotropic bodies, *Proc. Roy. Soc. Lond.* **A227**, 271–278. (XIII.7)

A. E. Green (1959), The equilibrium of rods, *Arch. Rational Mech. Anal.* **3**, 417–421. (XIV.3)

A. E. Green & J. E. Adkins (1970), *Large Elastic Deformations*, 2nd edn., Oxford Univ. Press. (P, X.8, XII.13, XIII.7, XIII.11, XIII.16)

A. E. Green & N. Laws (1966), A general theory of rods, *Proc. Roy. Soc. Lond.* **A293**, 145–155. (VIII.18)

A. E. Green, N. Laws & P. M. Naghdi (1967), A linear theory of straight elastic rods, *Arch. Rational Mech. Anal.* **25**, 285–298. (XIV.2)

A. E. Green, N. Laws & P. M. Naghdi (1968), Rods, plates and shells, *Proc. Camb. Phil. Soc.* **64**, 895–913. (XIV.2, XIV.15)

A. E. Green & P. M. Naghdi (1965), A general theory of an elastic-plastic continuum, *Arch. Rational Mech. Anal.* **18**, 251–281. (XV.1, XV.3)

A. E. Green & P. M. Naghdi (1970), Non-isothermal theory of rods, plates, and shells, *Int. J. Solids Structures* **6**, 209–244. (XIV.15)

A. E. Green, P. M. Naghdi & L. J. A. Trapp (1970), Thermodynamics of a continuum with internal constraints, *Int. J. Engg. Sci.* **8**, 891–908. (XII.12)

A. E. Green, P. M. Naghdi & W. L. Wainwright (1965), A general theory of a Cosserat surface, *Arch. Rational Mech. Anal.* **20**, 287–308. (XIV.15)

A. E. Green, P. M. Naghdi & M. L. Wenner (1974), On the theory of rods, *Proc. Roy. Soc. Lond.* **A337**, 451–507. (XIV.15)

A. E. Green & R. T. Shield (1950), Finite elastic deformation of compressible isotropic bodies, *Proc. Roy. Soc. Lond.* **A202**, 407–419. (XIII.7)

A. E. Green & E. B. Spratt (1954), Second-order effects in the deformation of elastic bodies, *Proc. Roy. Soc. Lond.* **A224**, 347–361. (XIII.11)

A. E. Green & W. Zerna (1968), *Theoretical Elasticity,* 2nd edn., Oxford Univ. Press. (P, XI.3, XIII.3, XIII.7, XIII.11, XIII.16)

J. M. Greenberg (1973), Estimates for fully developed shock solutions to the equation $\frac{\partial u}{\partial t} - \frac{\partial v}{\partial x} = 0$ and $\frac{\partial v}{\partial t} - \frac{\partial \sigma(u)}{\partial x} = 0$, *Indiana Univ. Math. J.* **22**, 989–1003. (XVI.1)

J. M. Greenberg (1989), Continuum limits of discrete gases, *Arch. Rational Mech. Anal.* **105**, 367–376. (II.11)

J. M. Greenberg (1992), The shock generation problem for a discrete gas with short range repulsive forces, *Comm. Pure Appl. Math.* **45**, 1125–1139. (II.11)

J. M. Greenberg, R. C. MacCamy & V. J. Mizel (1968), On the existence, uniqueness, and stability of solutions of the equation $\sigma'(u_x)u_{xx} + \lambda u_{txt} = \rho_0 u_{tt}$, *J. Math. Mech.* **17**, 707–728. (II.3, XVI.3)

A. G. Greenhill (1883), On the strength of shafting when exposed both to torsion and to end thrust, *Inst. Mech. Engineers, Proc.,* 182–225. (V.2)

E. I. Grigolyuk & V. V. Kabanov (1978), *Stability of Shells* (in Russian), Nauka, Moscow. (V.2)

G. Grioli (1962), *Mathematical Theory of Elastic Equilibrium*, Springer-Verlag. (XIII.11)

G. Grioli (1983), Mathematical problems in elastic equilibrium with finite deformations, *Appl. Anal.* **15**, 171–186. (XIII.11)

P. Grisvard (1985), *Elliptic Problems in Nonsmooth Domains*, Pitman. (XIII.16)

P. Grisvard (1992), *Singularities in Boundary Value Problems*, Masson, Springer-Verlag. (XIII.16)

J. Guckenheimer & P. Holmes (1983), *Nonlinear Oscillations, Dynamical Systems, and Bifurcations of Vector Fields*, Springer-Verlag. (V.8, XIV.8)

Guo Zhong-heng & R. Solecki (1963), Free and forced finite-amplitude oscillations of an elastic thick-walled hollow sphere made of incompressible material, *Arch. Mech. Stos.* **15**, 427–433. (XIII.12)

M. E. Gurtin (1968), On the thermodynamics of materials with memory, *Arch. Rational Mech. Anal.* **28**, 40–50. (XII.14)

M. E. Gurtin (1972), The Linear Theory of Elasticity, in *Handbuch der Physik* VIa/2, C. Truesdell, ed., 1–295. (XIII.15)

M. E. Gurtin (1974), Modern continuum thermodynamics, in *Mechanics Today,* Vol. 1, 1972, S. Nemat-Nasser, ed., 168–213. (XII.7)

M. E. Gurtin (1981a), *An Introduction to Continuum Mechanics*, Academic Press. (XI.2, XII.1, XII.7, XII.13)

M. E. Gurtin (1981b), *Topics in Finite Elasticity*, SIAM. (XIII.16)

M. E. Gurtin (1993a), *Thermomechanics of Evolving Phase Boundaries in the Plane*, Oxford Univ. Press. (XIII.16)

M. E. Gurtin (1993b), The dynamics of solid-solid phase transitions. 1. Coherent interfaces, *Arch. Rational Mech. Anal.* **123**, 305–335. (XIII.16)

M. E. Gurtin & L. C. Martins (1976), Cauchy's theorem in classical physics, *Arch. Rational Mech. Anal.* **60**, 305–324. (XII.7)

M. E. Gurtin & P. Podio-Guidugli (1973), The thermodynamics of constrained materials, *Arch. Rational Mech. Anal.* **51**, 192–208. (XII.12)

M. E. Gurtin & S. J. Spector (1979), On stability and uniqueness in finite elasticity, *Arch. Rational Mech. Anal.* **70**, 153–165. (XIII.2)

M. E. Gurtin & R. Temam (1981), On the anti-plane shear problem in finite elasticity, *J. Elasticity* **2**, 197–206. (VII.8, VIII.8, XIII.10)

J. Hadamard (1903), *Leçons sur la propagation des ondes*, Chelsea reprint. (XVI.2)

R. Hagan & M. Slemrod (1983), The viscosity-capillarity admissibility criterion for shocks and phase transitions, *Arch. Rational Mech. Anal.* **83**, 333–361. (II.2, XII.10, XVI.5)

J. K. Hale (1969), *Ordinary Differential Equations*, Wiley-Interscience. (II.9, VI.2)

P. R. Halmos (1958), *Finite-Dimensional Vector Spaces*, Van Nostrand. (XI.2)

G. H. Halphen (1884), Sur une courbe élastique, *C. R. Acad. Sci. Paris* **98**, 422–425. (IV.3)

A. Hanyga (1985), *Mathematical Theory of Non-Linear Elasticity*, Ellis Horwood. (P, XIII.16, XVI.2)

G. H. Hardy, J. E. Littlewood & G. Pólya (1952), *Inequalities*, Cambridge Univ. Press. (XVII.1)

J. A. Haringx (1942), On the buckling and the lateral rigidity of helical compression springs, *Proc. Nederl. Akad. Wet.* **45**, 533–539, 650–654. (IV.6)

J. A. Haringx (1948-1949), On highly compressible helical springs and rubber rods, and their application for vibration-free mountings, *Phillips Res. Reports* **3**, 401–449; **4**, 49–80, 206–220, 261–290, 375–400, 407–448. (IV.6)

P. Hartman (1964), *Ordinary Differential Equations*, Wiley, New York. (VI.2)

F. Hartmann (1937), *Knickung, Kippung, Beulung*, F. Deutike, Leipzig. (V.2)

B. D. Hassard, N. D. Kazarinoff & Y.-H. Wan (1981), *Theory and Applications of Hopf Bifurcation*, Cambridge Univ. Press. (V.8)

M. Hayes (1969), Static implications of the strong-ellipticity condition, *Arch. Rational Mech. Anal.* **33**, 181–191. (XIII.2)

M. Hayes & R. S. Rivlin (1961), Propagation of a plane wave in an isotropic elastic material subject to pure homogeneous deformation, *Arch. Rational Mech. Anal.* **8**, 15–22. (XIII.2)

T. J. Healey (1990), Large rotating states of a conducting elastic wire in a magnetic field: Subtle symmetry and multiparameter bifurcation, *J. Elasticity* **24**, 211–227. (III.5)

T. J. Healey (1992), Large rotatory oscillations of transversely isotropic rods: Spatiotemporal symmetry breaking bifurcation, *SIAM J. Appl. Math.* **52**, 1120–1135. (V.2, VI.13)

T. J. Healey & H. Kielhofer (1991), Symmetry and nodal properties in the global bifurcation analysis of quasi-linear elliptic equations, *Arch. Rational Mech. Anal.* **113**, 299–311. (XIV.13)

T. J. Healey & J. N. Papadopoulos (1990), Steady axial motion of strings, *J. Appl. Mech.* **57**, 785–787. (III.7)

E. Heinz (1959), An elementary analytic theory of the degree of a mapping in n-dimensional space, *J. Math. Mech.* **8**, 231–247. (XIX.3)

D. Henry (1981), *Geometric Theory of Semilinear Parabolic Equations*, Springer Lect. Notes Math. **840**. (V.8)

G. Herrmann (1967a), Stability of equilibrium of elastic systems subjected to nonconservative forces, *Appl. Mech. Revs.* **20**, 103–108. (V.2, V.7)

G. Herrmann, ed. (1967b), *Proc. Int. Conf. Dynamic Stability of Structures*, Pergamon, 1965. (V.2)

W. Hess (1884), Über die Biegung und Drillung eines unendlich dünnen elastischen Stabes, dessen eines Ende von einem Kräftepaar angegriffen wird, *Math. Ann.* **23**, 181–212. (IX.3, IX.6)

W. Hess (1885), Über die Biegung und Drillung eines unendlich dünnen elastischen Stabes mit zwei gleichen Widerständen, auf dessen freies Ende eine Kraft und ein um die Hauptaxe ungleichen Widerstandes drehendes Kräftepaar einwirkt, *Math. Ann.* **25**, 1–38. (IX.3)

M. Hestenes (1966), *Calculus of Variations and Optimal Control*, Wiley. (VII.8)

R. Hill (1950), *The Mathematical Theory of Plasticity*, Oxford Univ. Press. (XV.1)

R. Hill (1957), On uniqueness and stability in the theory of finite elastic strain, *J. Mech. Phys. Solids* **5**, 229–241. (XIII.2)

I. Hlaváček, J. Haslinger, J. Nečas & J. Lovišek (1983), *Solutions of Variational Inequalities in Mechanics* (in Czech), SNTL, English transl: Springer-Verlag, 1988. (VII.8)

P. J. Holmes, ed. (1980), *New Approaches to Nonlinear Problems in Dynamics*, SIAM. (V.8)

C. O. Horgan (1989), Recent developments concerning Saint-Venant's principle: An update, *Appl. Mech. Revs.* **42**, 295–304. (XIII.16)

C. O. Horgan & R. Abeyaratne (1986), A bifurcation problem for a compressible nonlinearly elastic medium: growth of a micro-void, *J. Elasticity* **16**, 189–200. (XIII.7)

C. O. Horgan & J. K. Knowles (1983), Recent developments concerning Saint-Venant's principle, in *Advances in Applied Mechanics* **24**, J. W. Hutchinson, ed., Academic Press, 179–269. (XIII.16)

T. Y. Hou & P. D. Lax (1991), Dispersion approximation in fluid dynamics, *Comm. Pure Appl. Math.* **44**, 1–40. (II.11)

W. J. Hrusa & M. Renardy (1988), An existence theorem for the Dirichlet problem in the elastodynamics of incompressible materials, *Arch. Rational Mech. Anal.* **102**, 95–117, Corrections, ibid. **110** (1990), 373. (XIII.2)

T. J. R. Hughes (1987), *The Finite Element Method: Linear Static and Dynamic Finite Element Analysis*, Prentice-Hall. (II.11)

T. J. R. Hughes, T. Kato & J. E. Marsden (1977), Well-posed quasilinear second-order hyperbolic systems with applications to nonlinear elastodynamics and general relativity, *Arch. Rational Mech. Anal.* **63**, 273–294. (XIII.2)

W. Hurewicz & H. Wallman (1948), *Dimension Theory*, rev. edn., Princeton Univ. Press. (V.4)

K. Huseyin (1975), *Nonlinear Theory of Elastic Stability*, Sijthoff & Noordhoff. (V.2)

J. Hutchinson & W. T. Koiter (1970), Postbuckling theory, *Appl. Mech. Revs.* **23**, 1353–1366. (V.2, V.8)

V. Hutson & J. Pym (1980), *Applications of Functional Analysis and Operator Theory*, Academic Press. (XVII.2)

A. A. Ilyukhin (1979), *Spatial Problems of the Nonlinear Theory of Elastic Rods* (in Russian), Naukova Dumka. (IV.6, IX.3, IX.4)

E. L. Ince (1926), *Ordinary Differential Equations*, Longmans, Green. (II.8, V.5)

A. D. Ioffe & V. M. Tihomirov (1974), *Theory of Extremal Problems* (in Russian), Nauka, English transl: North Holland, 1979. (VII.8)

G. Iooss (1979), *Bifurcation of Maps and Applications*, North-Holland. (V.8)

G. Iooss & D. D. Joseph (1990), *Elementary Stability and Bifurcation Theory*, 2nd edn., Springer-Verlag. (II.8, V.6, V.8)

E. Isaacson (1965), The shape of a balloon, *Comm. Pure Appl. Math.* **18**, 163–166. (X.7)

V. I. Istratescu (1981), *Fixed Point Theory*, Reidel. (XVIII.1, XIX.3)

J. Ize (1976), Bifurcation Theory for Fredholm Operators, *Mem. Amer. Math. Soc.* **7**. (V.4, V.8)

J. Ize, I. Massabò, J. Pejsachowicz & A. Vignoli (1986), Nonlinear multiparamter equations: Structure and dimension of global branches of solutions to multiparameter nonlinear equations, *Trans. Amer. Math. Soc.* **291**, 383–435. (III.3, V.4, XIII.6)

S. B. Jackson (1944), Vertices of plane curves, *Bull. Amer. Math. Soc.* **50**, 564–578. (IV.3)

R. D. James (1979), Co-existent phases in the one-dimensional static theory of elastic bars, *Arch. Rational Mech. Anal.* **72**, 99–140. (II.2, VIII.8)

R. D. James (1980), The propagation of phase boundaries in elastic bars, *Arch. Rational Mech. Anal.* **73**, 125–158. (II.2, VIII.8)

R. D. James (1981), The equilibrium and post-buckling behavior of an elastic curve governed by a non-convex energy, *J. Elasticity* **11**, 239–269. (VII.8)

A. Jeffrey (1976), *Quasilinear Hyperbolic Systems and Waves*, Pitman. (XVI.2)

F. John (1965), Estimates for the derivatives of the stresses in a thin shell and interior shell equations, *Comm. Pure Appl. Math.* **18**, 235–267. (XIV.15)

F. John (1971), Refined interior equations for thin elastic shells, *Comm. Pure Appl. Math.* **24**, 583–615. (XIV.15)

C. Johnson (1987), *Numerical Solution of Partial Differential Equations by the Finite Element Method*, Cambridge Univ. Press. (II.11)

M. Kachanov (1969), *Fundamentals of the Theory of Plasticity* (in Russian), Nauka, English transl: Mir, 1974. (XV.1)

C. B. Kafadar (1972), On Ericksen's problem, *Arch. Rational Mech. Anal.* **47**, 15–27. (XIII.9)

Ya. I. Kanel' (1969), On a model system of equations of one-dimensional gas motion (in Russian), *Diff. Urav.* **4**, 721–734, English transl: *Diff. Eqs.* **4** (1969), 374–380. (II.3, XVI.2, XVI.3)

L. V. Kantorovich & G. P. Akilov (1977), *Functional Analysis* (in Russian), 2nd. edn., Nauka, English transl: Pergamon, 1982. (XVII.2)

A. J. Karwowski (1990), Asymptotic models for a long, elastic cylinder, *J. Elasticity* **24**, 229–287. (XIV.15)

W. L. Kath (1985), Slowly varying phase planes and boundary-layer theory, *Studies in Appl. Math.* **72**, 221–239. (VI.9)

T. Kato (1976), *Perturbation Theory for Linear Operators,* 2nd edn., Springer-Verlag. (V.4)

J. P. Keener (1979), Secondary bifurcation and multiple eigenvalues, *SIAM J. Appl. Math.* **37**, 330–349. (V.3)

H. B. Keller (1981), Two new bifurcation phenomena, in *Applications of Nonlinear Analysis in the Physical Sciences*, H. Amann, N. Bazley & K. Kirchgässner, eds., Pitman, 60–73. (V.3)

J. B. Keller (1959), Large amplitude motion of a string, *Amer. J. Phys.* **27**, 584–586. (II.7)

J. B. Keller (1968), *Perturbation Theory,* Lecture Notes, Dept. of Math., Michigan State Univ. (II.8, V.6)

J. B. Keller (1969), Bifurcation theory for ordinary differential equations, in *Bifurcation Theory and Nonlinear Eigenvalue Problems*, J. B. Keller & S. S. Antman, eds., Benjamin, 17–48. (XVIII.2)

J. B. Keller & L. Ting (1966), Periodic vibrations of systems governed by nonlinear partial differential equations, *Comm. Pure Appl. Math.* **19**, 371–420. (II.8)

O. D. Kellogg (1929), *Foundations of Potential Theory*, Springer-Verlag. (XI.2)

D. Kinderlehrer & P. Pedregal (1991), Characterization of Young measures generated by gradients, *Arch. Rational Mech. Anal.* **115**, 329–365. (XIII.2)

D. Kinderlehrer & G. Stampacchia (1980), *Introduction to Variational Inequalities and Their Applications*, Academic Press. (VII.8, XIII.16)

G. Kirchhoff (1850), Über das Gleichgewicht und die Bewegung einer elastischen Scheibe, *J. reine angew. Math.* **40**, 51–58, in *Gesammelte Abhandlungen*, 237–279. (XIV.15)

G. Kirchhoff (1859), Über das Gleichgewicht und die Bewegung eines unendlich dünnen elastischen Stabes, *J. reine angew. Math.* **56**, 285–316, in *Gesammelte Abhandlungen*, 285–316. (VIII.9, VIII.10, VIII.16, VIII.18, IX.2)

W. W. Klingbeil & R. T. Shield (1966), On a class of solutions in plane finite elasticity, *Z. angew. Math. Phys.* **17**, 489–511. (XIII.9)

R. J. Knops (1973), Logarithmic convexity and other techniques applied to problems in continuum mechanics, in *Symposium on Non-Well-Posed Problems and Logarithmic Convexity*, R. J. Knops, ed., Springer Lect. Notes Math. **316**, 31–54. (XVI.6)

R. J. Knops, H. A. Levine & L. E. Payne (1974), Non-existence, instability, and growth theorems for solutions of a class of abstract nonlinear equations with applications to nonlinear elastodynamics, *Arch. Rational Mech. Anal.* **55**, 52–72. (XVI.6)

R. J. Knops & L. E. Payne (1971), *Uniqueness Theorems in Linear Elasticity*, Springer-Verlag. (XIII.15)

R. J. Knops & E. W. Wilkes (1973), Theory of Elastic Stability, in *Handbuch der Physik* VIa/3, Springer-Verlag, 125–302. (V.2, V.7, V.8)

J. K. Knowles (1960), Large amplitude oscillations of a tube of incompressible elastic material, *Quart. Appl. Math.* **18**, 71–77. (XIII.12)

J. K. Knowles (1976), On finite anti-plane shear for incompressible elastic materials, *J. Austral. Math. Soc. Ser. B* **19**, 400–415. (XIII.10, XV.5)

J. K. Knowles (1977a), A note on anti-plane shear for compressible materials in finite elastostatics, *J. Austral. Math. Soc. Ser. B* **20**, 1–7. (XIII.10)

J. K. Knowles (1977b), The finite antiplane shear field near the tip of a crack for a class of incompressible elastic solids, *Int. J. Fracture* **13**, 611–639. (XIII.10)

J. K. Knowles & M. T. Jakub (1965), Finite dynamic deformations of an incompressible elastic medium containing a spherical cavity, *Arch. Rational Mech. Anal.* **18**, 376–387. (XIII.12)

J. K. Knowles & E. Sternberg (1973), An asymptotic finite-deformation analysis of the elastostatic field near the tip of a crack, *J. Elasticity* **3**, 67–107. (XIII.16)

J. K. Knowles & E. Sternberg (1974), Finite-deformation analysis of the elastostatic field near the tip of a crack: Reconsideration and higher-order results, *J. Elasticity* **4**, 201–233. (XIII.16)

J. K. Knowles & E. Sternberg (1975), On the singularity induced by certain mixed boundary conditions in linearized and nonlinear elastostatics, *Int. J. Solids Structures* **11**, 1173–1201. (XIII.16)

W. T. Koiter (1945), Over de stabiliteit van het elastisch evenwicht, Delft, Thesis, English transl: NASA Tech. Transl. **F10**, (1967). (V.8)

W. T. Koiter (1963), Elastic stability and post-buckling behavior, in *Nonlinear Problems*, R. E. Langer, ed., Univ. of Wisconsin Press, 257–275. (V.8)

W. T. Koiter (1970), On the foundations of the linear theory of thin elastic shells, *Proc. Kon. Ned. Akad. Wetensch.* **B73**, 169–195. (XIV.4)

C. F. Kollbrunner & M. Meister (1961), *Knicken, Biegedrillknicken, Kippen, Theorie und Berechnung von Knickstäben, Knickvorschriften,* 2nd edn., Springer-Verlag. (V.2)

I. I. Kolodner (1955), Heavy rotating string—a nonlinear eigenvalue problem, *Comm. Pure Appl. Math.* **8**, 395–408. (VI.2, VI.3)

Z. Kordas & M. Życzkowski (1963), On the loss of stability of a rod under a supertangential force, *Arch. Mech. Stos.* **15**, 7–31. (VI.11)

D. Kovári (1969), Räumliche Verzweigungsprobleme des dünnen elastichen Stabes mit endlichen Verformungen, *Ing. Arch.* **37**, 393–416. (IX.5)

M. A. Krasnosel'skiĭ (1956), *Topological Methods in the Theory of Nonlinear Integral Equations* (in Russian), Gostekhteorizdat, English transl: Pergamon Press, 1964. (V.4, V.8, VI.2, XIX.3)

M. A. Krasnosel'skiĭ (1962), *Positive Solutions of Operator Equations* (in Russian), Fizmatgiz, English transl: Noordhoff, 1964. (V.8)

M. A. Krasnoselskiĭ & P. P. Zabreĭko (1975), *Geometric Methods of Nonlinear Analysis* (in Russian), Nauka, English transl: Springer-Verlag, 1984. (V.8, XIX.3)

E. Kreyszig (1978), *Introductory Functional Analysis with Applications,* Wiley. (XVII.2)

V. D. Kupradze (1963), *Potential Methods in the Theory of Elasticity* (in Russian), Fizmatgiz, English transl: Israel Program for Scientific Translation, 1965. (XIII.15)

V. D. Kupradze, T. G. Gegelia, M. O. Basheleĭshvili & T. B. Burchuladze (1968), *Three-Dimensional Problems of the Mathematical Theory of Elasticity* (in Russian), Tbilisi University, English transl: Elsevier, North Holland, 1979. (XIII.15)

O. A. Ladyženskaja, V. A. Solonnikov & N. N. Ural'ceva (1967), *Linear and Quasilinear Equations of Parabolic Type* (in Russian), Nauka, English transl: Amer. Math. Soc., 1968. (II.11)

O. A. Ladyzhenskaya (1969), *The Mathematical Theory of Viscous Incompressible Flow,* 2nd edn., Gordon and Breach. (XII.3)

O. A. Ladyzhenskaya (1973), *The Boundary Value Problems of Mathematical Physics* (in Russian), Nauka, English transl: Springer-Verlag, 1985. (II.11)

J. L. Lagrange (1762), Application de la méthode exposée précédente à la solution de différens problèmes de dynamique, *Misc. Taur.* $\mathbf{2}_2$ (1760/1761), 196–298, *Œuvres* **1**, 365-468. (II.1, II.2)

J. L. Lagrange (1788), *Méchanique analytique,* Veuve Desaint, Paris. (II.2)

M. Lanza de Cristoforis & S. S. Antman (1991), The large deformation of elastic tubes in two-dimensional flows, *SIAM J. Math. Anal.* **22**, 1193–1221. (IV.1, V.4, VIII.4, XIII.6)

P. D. Lax (1957), Hyperbolic systems of conservation laws, II, *Comm. Pure Appl. Math.* **10**, 537–566. (XVI.2)

P. D. Lax (1964), Development of singularities of solutions of nonlinear hyperbolic partial differential equations, *J. Math. Phys.* **5**, 611–613. (II.8, XVI.1)

P. D. Lax (1973), *Hyperbolic Systems of Conservation Laws and the Mathematical Theory of Shock Waves,* SIAM. (XVI.1, XVI.2)

H. Le Dret (1985), Constitutive laws and existence questions in incompressible nonlinear elasticity, *J. Elasticity* **15**, 369–387. (XII.12)

H. Le Dret (1991), *Problèmes Variationnels dans les Multi-Domaines: Modélisation des Jonctions et Applications,* Masson. (XIV.15)

G. W. Leibniz (1684), Demonstrationes novae de resistentia solidorum, *Acta erud.*, Leibnizens Math. Schriften, **6**, 106-112. (VIII.18)

H. Leipholz (1968), *Stabilitätstheorie*, Teubner, English transl: Academic Press, 1970. (V.2, V.7)

H. H. E. Leipholz, ed. (1971), *IUTAM Symposium on Instability of Continuous Systems*, Springer-Verlag. (V.2)

H. H. E. Leipholz (1978a), *Stabilität elastischen Systeme*, Braun-Verlag. (V.2, V.7)

H. H. E. Leipholz (1978b), *Stability of Elastic Structures*, Springer-Verlag. (V.2)

J. Leray & J. Schauder (1934), Topologie et équations fonctionelles, *Ann. Sci. École Norm. Sup. (3)* **51**, 45–78. (XIX.3)

P. Le Tallec & J. T. Oden (1981), Existence and characterization of hydrostatic pressure in finite deformations of incompressible elastic bodies, *J. Elasticity* **11**, 341–357. (XII.12)

A. Lev (1978), Branching of solutions of equations in Banach space without multiplicity assumptions, *Proc. Lond. Math. Soc. (3)* **37**, 308–341. (V.4)

M. Lévy (1884), Memoire sur un nouveau cas intégrable du problème de l'élastique et l'une de ses applications, *J. de Math. Pures Appl. (3)* **10**, 5–42. (IV.3)

A. Libai & J. G. Simmonds (1988), *The Nonlinear Theory of Elastic Shells: One Space Dimension*, Academic Press. (XIV.15)

J.-L. Lions (1969), *Quelques méthodes de résolution des problèmes aux limites non linéaires*, Dunod, Gauthier-Villars. (II.11)

T.-P. Liu (1975), The Riemann problem for general systems of conservation laws, *J. Diff. Eqs.* **18**, 218–234. (XVI.2)

N. G. Lloyd (1978), *Degree Theory*, Cambridge Univ. Press. (XIX.3)

B. V. Loginov (1985), *Theory of the Bifurcation of Solutions of Nonlinear Equations under Conditions of Group Invariance*, FAN, Tashkent. (V.8)

A. E. H. Love (1893), *A Treatise on the Mathematical Theory of Elasticity,* 1st edn., Cambridge Univ. Press. (VIII.18)

A. E. H. Love (1927), *A Treatise on the Mathematical Theory of Elasticity,* 4th edn., Cambridge Univ. Press. (IV.2, IV.6, V.1, IX.2, IX.4, XIII.15, XIV.15)

M. Lucchesi, D. R. Owen & P. Podio-Guidugli (1992), Materials with elastic range: A theory with a view toward applications. Part III: Approximate constitutive relations, *Arch. Rational Mech. Anal.* **117**, 53–96. (XV.1)

M. Lucchesi & P. Podio-Guidugli (1988), Materials with elastic range: A theory with a view toward applications, I, *Arch. Rational Mech. Anal.* **102**, 23–43. (XV.1)

M. Lucchesi & P. Podio-Guidugli (1990), Materials with elastic range: A theory with a view toward applications, Part II, *Arch. Rational Mech. Anal.* **110**, 9–42. (XV.1)

D. G. Luenberger (1969), *Optimization by Vector Space Methods*, Wiley. (XVII.2)

A. I. Lur'e (1955), *Spatial Problems of Elasticity* (in Russian), Gostekhizdat, English transl: Interscience, 1964. (XIII.15)

A. I. Lur'e (1970), *Theory of Elasticity* (in Russian), Nauka. (XIII.15, XIII.16)

A. I. Lur'e (1980), *Nonlinear Theory of Elasticity* (in Russian), Nauka, English transl: North-Holland, 1990. (XIII.16)

L. A. Lyusternik (1937), Sur une classe d'équations différentielles non linéaires, *Mat. Sb.* **44**, 1145–1167. (V.3)

L. A. Lyusternik (1938), Quelques remarques supplémentaires sur les équations nonlinéaires du type de Sturm-Liouville, *Mat. Sb.* **46**, 227–232. (V.3)

R. C. MacCamy (1970), Existence, uniqueness and stability of $u_{tt} = \frac{\partial}{\partial x}[\sigma(u_x) + \lambda(u_x)u_{xt}]$, *Indiana Univ. Math. J.* **20**, 231–238. (II.3, XVI.3)

R. C. MacCamy & V. J. Mizel (1967), Existence and nonexistence in the large of solutions of quasilinear wave equations, *Arch. Rational Mech. Anal.* **25**, 299–320. (XVI.1)

J. H. Maddocks (1984), Stability of nonlinearly elastic rods, *Arch. Rational Mech. Anal.* **85**, 311–354. (VII.8, IX.5)

J. H. Maddocks (1987), Stability and folds, *Arch. Rational Mech. Anal.* **99**, 301–328. (VII.8)

J. H. Maddocks (1994), On second-order conditions in constrained variational problems, *J. Optim. Th. Appl.*, to appear. (VII.8)

R. Magnus (1976), A generalization of multiplicity and the problem of bifurcation, *Proc. Lond. Math. Soc.* **32**, 251–278. (V.4)

R. Magnus & T. Poston (1979), Infinite dimensions and the fold catastrophe, in *Structural Stability in Physics*, W. Güttinger & H. Eikemeier, eds., Springer-Verlag, 63–83. (II.2)

R. S. Marlow (1992), On the stress in an internally constrained elastic material, *J. Elasticity* **27**, 97–131. (XII.12)

A. W. Marris (1975), Universal deformations in incompressible isotropic elastic materials, *J. Elasticity* **5**, 111–128. (XIII.9)

A. W. Marris (1982), Two new theorems on Ericksen's problem, *Arch. Rational Mech. Anal.* **79**, 131–173. (XIII.9)

A. W. Marris & J. F. Shiau (1970), Universal deformations in isotropic incompressible hyperelastic materials when the deformation tensor has equal proper values, *Arch. Rational Mech. Anal.* **36**, 135–160. (XIII.9)

J. E. Marsden (1978), Qualitative methods in bifurcation theory, *Bull. Amer. Math. Soc.* **84**, 1125–1148. (V.8)

J. E. Marsden & T. J. R. Hughes (1983), *Mathematical Foundations of Elasticity*, Prentice-Hall. (P, XIII.11, XIII.16)

J. E. Marsden & M. McCracken (1976), *The Hopf Bifurcation and its Applications*, Springer-Verlag. (V.8, XIV.8)

J. Mawhin (1979), *Topological Degree Methods in Nonlinear Boundary Value Problems*, Conf. Bd. Math. Sci. **40**, Amer. Math. Sci. (V.8, XIX.3)

V. G. Maz'ja, S. A. Nazarov & B. A. Plamenevskii (1991), *Asymptotische Theorie elliptischer Randwertaufgaben in singulär gestörten Gebieten,* Vol. 2, Akademie Verlag. (XIV.15)

V. G. Maz'ya (1985), *Spaces of S. L. Sobolev* (in Russian), Leningrad Univ. Press, English transl. with coauthor T. O. Šapošnikova: *Sobolev Spaces*, Springer-Verlag, 1985. (XVII.1)

M. A. Medick (1966), One dimensional theories of wave propagation and vibrations in elastic bars of rectangular cross-section, *J. Appl. Mech.* **33**, 489–495. (XIV.15)

A. Mielke (1988), Saint-Venant's problem and semi-inverse solutions in nonlinear elasticity, *Arch. Rational Mech. Anal.* **102**, 205–229, Corrigendum, ibid. **110** (1990), 351-352. (XIV.8)

A. Mielke (1990), Normal hyperbolicity of center manifolds and Saint-Venant's principle, *Arch. Rational Mech. Anal.* **110**, 353–372. (XIV.4, XIV.8)

A. Mielke & P. Holmes (1988), Spatially complex equilibria of buckled rods, *Arch. Rational Mech. Anal.* **101**, 319–348. (IX.4)

M. Millman & J. B. Keller (1969), Perturbation theory of nonlinear boundary value problems, *J. Math. Phys.* **10**, 342–361. (II.8)

J. Milnor (1969), *Topology from the Differentiable Point of View*, Univ. of Virginia Press. (XIX.3)

G. M. Minchin (1887), *A Treatise on Statics,* Vol. 1, 1st edn., Longmans, Green. (III.8)

R. D. Mindlin & G. Herrmann (1952), A one-dimensional theory of compressional waves in an elastic rod, in *Proc. First U. S. Natl. Cong. Appl. Mech.*, Amer. Soc. Mech. Engrs., 187–191. (XIV.15)

R. D. Mindlin & H. D. McNiven (1960), Axially symmetric waves in elastic rods, *J. Appl. Mech.* **27**, 145–151. (XIV.15)

L. Modica (1987), The gradient theory of phase transitions and the minimal interface condition, *Arch. Rational Mech. Anal.* **98**, 123–142. (XIII.2)

F. C. Moon (1978), Problems in magneto-solid mechanics, in *Mechanics Today,* Vol. 4, S. Nemat-Nasser, ed., Pergamon. (V.2)

F. C. Moon (1984), *Magneto-Solid Mechanics*, Wiley. (V.2)

F. Morgan (1988), *Geometric Measure Theory: A Beginner's Guide*, Academic Press. (XI.2)

D. Morgenstern (1959), Herleitung der Plattentheorie aus der dreidimensionalle Elastizitätstheorie, *Arch. Rational Mech. Anal.* **4**, 145–152. (XIV.15)

D. Morgenstern & I. Szabó (1961), *Vorlesungen über theoretische Mechanik*, Springer-Verlag. (XIV.4)

C. B. Morrey, Jr. (1952), Quasiconvexity and the lower semicontinuity of multiple integrals, *Pac. J. Math.* **2**, 25–53. (XIII.2)

C. B. Morrey, Jr. (1966), *Multiple Integrals in the Calculus of Variations*, Springer-Verlag. (XIII.2)

S. Müller (1988), Weak continuity of determinants and nonlinear elasticity, *C. R. Acad. Sci. Paris, Sér. I* **307**, 501–506. (XIII.2)

S. Müller (1990), Higher integrability of determinants and weak convergence in L^1, *J. reine angew. Math.* **412**, 20–34. (XIII.2)

S. Müller & S. Spector (1995), An existence theory for elasticity that allows for cavitation, to appear. (XIII.7)

S. Müller, Q. Tang & B. S. Yau (1995), On a new class of elastic deformations not allowing for cavitation, to appear. (XIII.7)

R. G. Muncaster (1979), St. Venant's problem in nonlinear elasticity: A study of cross-sections, in *Nonlinear Analysis and Mechanics: Heriot-Watt Symposium,* Vol. IV, R. J. Knops, ed., Pitman, 17–75. (XIII.8, XIV.8)

R. G. Muncaster (1983), Invariant manifolds in mechanics, *Arch. Rational Mech. Anal.* **84**, 353–392. (XIV.8)

A. I. Murdoch & H. Cohen (1979), Symmetry considerations for material surfaces, *Arch. Rational Mech. Anal.* **72**, 61–98, Addendum, ibid. **76** (1981), 393-400. (XIV.13)

N. I. Muskhelishvili (1966), *Some Fundamental Problems of the Mathematical Theory of Elasticity,* 5th edn., Nauka, English translation of 4th edn: Noordhoff, 1963. (XIII.15)

A. Nadai (1950), *Theory of Flow and Fracture of Solids*, McGraw-Hill. (V.2)

P. M. Naghdi (1972), The Theory of Shells, in *Handbuch der Physik* VIa/2, C. Truesdell, ed., Springer-Verlag, 425–640. (XIV.15)

P. M. Naghdi (1990), A critical review of the state of finite plasticity, *Z. angew. Math. Phys.* **41**, 315–387. (XV.1)

M. Nagumo (1951), A theory of degree of mapping based on infinitesimal analysis, *Amer. J. Math.* **73**, 485–496. (XIX.3)

M. A. Naimark (1969), *Linear Differential Operators,* 2nd edn. (in Russian), Nauka, English transl. of 1st edn: Unger, 1967. (V.6)

I. P. Natanson (1961), *Theory of Functions of a Real Variable,* Vol. I, translated from Russian, Unger. (III.3)

J. Nečas (1967), *Les méthodes directes en théorie des équations elliptiques*, Masson. (II.3, V.6, XVII.1)

P. V. Negrón-Marrero (1985), Large Buckling of Circular Plates with Singularities due to anisotropy, Univ. of Maryland, Dissertation. (XIII.7, XIV.12)

P. V. Negrón-Marrero (1989), Necked states of nonlinearly elastic plates, *Proc. Roy. Soc. Edin.* **112A**, 277–291. (XIV.12)

P. V. Negrón-Marrero & S. S. Antman (1990), Singular global bifurcation problems for the buckling of anisotropic plates, *Proc. Roy. Soc. Lond.* **A427**, 95–137. (X.3, X.4, XIV.11, XIV.12)

S. Nemat-Nasser (1970), Thermoelastic stability under general loads, *Appl. Mech. Revs.* **23**, 615–624. (V.2)

H. K. Nickerson, D. C. Spencer & N. E. Steenrod (1959), *Advanced Calculus*, Van Nostrand. (XI.2)

E. L. Nikolai (1916), On the Problem of Elastic Lines of Double Curvature (in Russian), Petrograd, Dissertation, reprinted in E. L. Nikolai (1955). (IX.3)

E. L. Nikolai (1928), On the stability of the straight equilibrium form of a compressed and twisted rod (in Russian), *Izv. Leningr. Politekh. Inst.* **31**, 357–387, reprinted in E. L. Nikolai (1955). (V.7)

E. L. Nikolai (1929), On the question of the stability of a twisted rod (in Russian), *Vestnik. Prikl. Mat. i Mekh.* **1**, 388–406, reprinted in E. L. Nikolai (1955). (V.7)

E. L. Nikolai (1955), *Works on Mechanics* (in Russian), G.I.T.T.L. (IV.6, V.2, IX.4)

L. Nirenberg (1974), *Topics in Nonlinear Functional Analysis,* Lecture Notes, Courant Inst., New York Univ. (V.4, V.8, XIX.3)

T. Nishida (1968), Global solution for an initial boundary value problem of a quasilinear hyperbolic system, *Proc. Japan Acad. Sci.* **44**, 642–646. (XVI.1)

T. Nishida & J. Smoller (1973), Solutions in the large for some nonlinear hyperbolic conservation laws, *Comm. Pure Appl. Math.* **26**, 183–200. (XVI.1)

W. Noll (1958), A mathematical theory of the mechanical behavior of continuous media, *Arch. Rational Mech. Anal.* **2**, 197–226. (XII.11)

W. Noll (1959), The foundations of classical mechanics in the light of recent advances in continuum mechanics, in *The Axiomatic Method, with Special Reference to Geometry and Physics*, L. Henkin, P. Suppes & A. Tarski, eds., North-Holland, 266–281. (XII.7)

W. Noll (1966), The foundations of mechanics, in *Non-Linear Continuum Theories* (C.I.M.E. Conference), G. Grioli & C. Truesdell, eds., Cremonese, 159–200. (XII.1, XII.7, XII.12)

W. Noll (1987), *Finite-Dimensional Spaces: Algebra, Geometry, and Analysis,* Vol. I, Kluwer. (XI.2)

W. Noll (1993), The geometry of contact, separation, and reformation of continuous bodies, *Arch. Rational Mech. Anal.* **122**, 197–212. (XII.1)

W. Noll & E. G. Virga (1988), Fit regions and functions of bounded variation, *Arch. Rational Mech. Anal.* **102**, 1–21. (XII.1)

A. K. Noor (1989), List of books, monographs, conference proceedings and survey papers on shells, in *Analytic and Computational Models of Shells*, A. K. Noor, T. Belytschko & J. C. Simo, eds., Amer. Soc. Mech. Engrs., vii–xxxiii. (XIV.15)

V. V. Novozhilov (1948), *Foundations of the Nonlinear Theory of Elasticity* (in Russian), Gostekhteorizdat, English transl: Graylock Press, 1953. (XIV.2)

F. Odeh & I. Tadjbakhsh (1965), A nonlinear eigenvalue problem for rotating rods, *Arch. Rational Mech. Anal.* **20**, 81–94. (VI.13)

J. T. Oden & G. F. Carey (1981-1984), *Finite Elements,* Vols. I–V, Prentice Hall. (II.11)

R. W. Ogden (1984), *Non-Linear Elastic Deformations*, Ellis Horwood. (P, XIII.2, XIII.7, XIII.11, XIII.16, XIV.5)

O. A. Oleĭnik (1957), On the uniqueness of the generalized solution of the Cauchy problem for a nonlinear system of equations occurring in mechanics (in Russian), *Usp. Mat. Nauk* **12**, 169–176. (XVI.2)

W. E. Olmstead (1977), Extent of the left branching solution to certain bifurcation problems, *SIAM J. Math. Anal.* **8**, 392–401. (VI.9)

R. E. O'Malley (1974), *Introduction to Singular Perturbations*, Academic Press. (XVI.4)

R. E. O'Malley (1976), Phase-plane solutions to some singular perturbation problems, *J. Math. Anal. Appl.* **54**, 449–466. (VI.9)

J. Ortega & W. Rheinboldt (1970), *Iterative Solutions of Nonlinear Equations in Several Variables*, Academic Press. (XIX.3)

D. R. Owen (1968), Thermodynamics of materials with elastic range, *Arch. Rational Mech. Anal.* **31**, 91–112. (XV.1)

D. R. Owen (1970), A mechanical theory of materials with elastic range, *Arch. Rational Mech. Anal.* **37**, 85–110. (XV.1)

D. R. Owen (1984), *A First Course in the Mathematical Foundations of Thermodynamics*, Springer-Verlag. (XII.14)

D. R. Owen (1987), Weakly decaying energy separation and uniqueness of motions of an elastic-plastic oscillator with work-hardening, *Arch. Rational Mech. Anal.* **98**, 95–114. (XV.6)

N. C. Owen (1987), Existence and stability of necking deformations for nonlinearly elastic rods, *Arch. Rational Mech. Anal.* **98**, 357–383. (VIII.8)

P. D. Panagiotopoulos (1985), *Inequality Problems in Mechanics and Applications*, Birkhäuser. (VII.8)

Ya. G. Panovko & I. I. Gubanova (1979), *Stability and Vibrations of Elastic Systems* (in Russian), 3rd edn., Nauka, English translation of 1st edn: Consultants Bureau, 1965. (V.2, V.7)

S. V. Parter (1970), Nonlinear eigenvalue problems for some fourth order equations. I: Maximal solutions, II: Fixed-point methods, *SIAM J. Math. Anal.* **1**, 437–457, 458-478. (VI.13)

V. Z. Parton & P. I. Perlis (1981), *Mathematical Methods of the Theory of Elasticity* (in Russian), Nauka, English transl: Mir, 1984. (XIII.15)

R. L. Pego (1987), Phase transitions in one-dimensional nonlinear viscoelasticity: Admissibility and stability, *Arch. Rational Mech. Anal.* **97**, 353–394. (VIII.8, XVI.3)

H.-O. Peitgen, D. Saupe & K. Schmitt (1981), Nonlinear elliptic boundary value problems versus their finite difference approximations. Numerically irrelevant solutions, *J. reine angew. Math.* **322**, 74–117. (V.3)

K. A. Pericak-Spector & S. J. Spector (1988), Nonuniqueness for a hyperbolic system: Cavitation in nonlinear elastodynamics, *Arch. Rational Mech. Anal.* **101**, 293–317. (XIII.7)

A. Pflüger (1964), *Stabilitätsprobleme der Elastostatik,* 2nd edn., Springer-Verlag. (V.2)

J. F. Pierce (1989), *Singularity Theory, Rod Theory, and Symmetry-Breaking Loads*, Springer Lect. Notes Math. **1377**. (XIII.11)

J. F. Pierce & A. P. Whitman (1980), Topological properties of the manifolds of configurations for several simple deformable bodies, *Arch. Rational Mech. Anal.* **74**, 101–113. (XII.1)

G. H. Pimbley, Jr. (1962), A sublinear Sturm-Liouville problem, *J. Math. Mech.* **11**, 121–138. (VI.2)

G. H. Pimbley, Jr. (1963), A superlinear Sturm-Liouville problem, *Trans. Amer. Math. Soc.* **103**, 229–248. (VI.2)

G. H. Pimbley, Jr. (1969), *Eigenfunction Branches of Nonlinear Operators, and their Bifurcations*, Springer Lect. Notes Math. **104**. (V.3, V.8)

A. C. Pipkin & R. S. Rivlin (1965), Mechanics of rate-independent materials, *Z. angew. Math. Phys.* **16**, 313–326. (XV.1)

A. C. Pipkin & R. S. Rivlin (1966), Electrical, thermal, and magnetic constitutive equations for deformed isotropic materials, *Atti. Accad. Naz. Lincei Mem. Cl. Sci. Fis. Mat. Natur.* **8**, 1–29. (XII.13)

B. J. Plohr & D. Sharp (1992), A conservative formulation of plasticity, *Adv. Appl. Math.* **13**, 462–493. (XV.4)

P. Podio-Guidugli (1989), An exact derivation of the thin plate equation, *J. Elasticity* **22**, 121–133. (XIV.2)

P. Podio-Guidugli (1990), Constrained elasticity, *Atti Accad. Naz. Lincei Rend. Cl. Fis. Mat. Natur.* **(9) 1**, 341–350. (XII.12)

P. Podio-Guidugli & G. Vergara Cafarelli (1991), Extreme elastic deformations, *Arch. Rational Mech. Anal.* **115**, 311–328. (XIII.2)

P. Podio-Guidugli, G. Vergara Caffarelli & E. G. Virga (1986), Discontinuous energy minimizers in nonlinear elastostatics. An example of J. Ball revisited, *J. Elasticity* **16**, 75–96. (XIII.7)

P. Podio-Guidugli & M. Vianello (1989), Constraint manifolds for isotropic solids, *Arch. Rational Mech. Anal.* **104**, 105–121. (XII.12)

A. V. Pogorelov & V. I. Babenko (1992), Geometrical methods in the theory of stability of thin shells (Review), *Prikl. Mekh.* **28**, 3–21, English transl: *Int. Appl. Mech.* **28**, 1–17. (V.2)

E. P. Popov (1948), *Nonlinear Problems of the Statics of Rods* (in Russian), OGIZ. (IV.6, IX.4)

T. Poston & I. Stewart (1978), *Catastrophe Theory and its Applications*, Pitman. (V.8, VI.6)

M. Potier-Ferry (1981), The linearization principle for the stability of solutions of quasilinear parabolic equations, I, *Arch. Rational Mech. Anal.* **77**, 301–320. (V.7)

M. Potier-Ferry (1982), On the mathematical foundations of elastic stability theory, I, *Arch. Rational Mech. Anal.* **77**, 55–72. (V.7)

W. Prager & P. G. Hodge, Jr. (1951), *Theory of Perfectly Plastic Solids*, Wiley. (XV.1)

J. Prescott (1924), *Applied Elasticity*, Longmans, Green. (V.2)

G. Prodi, ed. (1974), *Eigenvalues of Nonlinear Problems, C.I.M.E. Conference*, Cremonese. (V.8)

M. H. Protter & H. F. Weinberger (1967), *Maximum Principles in Differential Equations*, Prentice-Hall. (IV.5, XVI.1)

P. H. Rabinowitz (1971a), Some global results for nonlinear eigenvalue problems, *J. Funct. Anal.* **7**, 487–513. (V.4, XIX.2, XIX.4)

P. H. Rabinowitz (1971b), A global theorem for nonlinear eigenvalue problems and applications, in *Contributions to Nonlinear Functional analysis*, E. H. Zarantonello, ed., Academic Press, 11–36. (V.4)

P. H. Rabinowitz (1973a), Some aspects of nonlinear eigenvalue problems, *Rocky Mountain J. Math.* **3**, 161–202. (V.3, V.4, V.8. VI.2)
P. H. Rabinowitz (1973b), On bifurcation from infinity, *J. Diff. Eqs.* **14**, 462–475. (VI.2)
P. H. Rabinowitz (1975), *Théorie du degré topologique et applications à des problèmes aux limites non linéaires,* Lecture Notes, Univ. Paris VI. (V.4, V.8, XIX.3)
P. H. Rabinowitz (1986), *Minimax Methods in Critical Point Theory with Applications to Differential Equations*, Conf. Bd. Math. Sci. **65**, Amer. Math. Soc. (VII.8)
P. H. Rabinowitz, A. Ambrosetti, I. Ekeland & E J. Zehnder, eds. (1987), *Periodic Solutions of Hamiltonian Systems,* NATO Advanced Research Workshop, Reidel. (II.10)
K. R. Rajagopal & M. M. Carroll (1992), Inhomogeneous deformations of nonlinearly elastic wedges, *Int. J. Solids Structures* **29**, 735–744. (XIII.8)
K. R. Rajagopal & A. S. Wineman (1984), On a class of deformations of materials with nonconvex stored energy function, *J. Struct. Mech.* **12**, 471–482. (XIII.9)
K. R. Rajagopal & A. S. Wineman (1985), New exact solutions in non-linear elasticity, *Int. J. Engg. Sci.* **23**, 217–234. (XIII.9)
M. Reeken (1977), The equation of motion of a chain, *Math. Z.* **155**, 219–237. (VI.3)
M. Reeken (1979), Classical solutions of the chain equation I, II, *Math. Z.* **165, 166**, 143–169, 67–82. (VI.3)
M. Reeken (1980), Rotating chain fixed at two points vertically above each other, *Rocky Mountain J. Math.* **10**, 409–427. (VI.3)
M. Reeken (1984a), Exotic equilibrium states of the elastic string, *Proc. Roy. Soc. Edin.* **96A**, 289–302. (II.6)
M. Reeken (1984b), The rotating string, *Math. Ann.* **268**, 59–84. (VI.3)
E. L. Reiss (1969), Column buckling—an elementary example of bifurcation, in *Bifurcation Theory and Nonlinear Eigenvalue Problems*, J. B Keller & S. Antman, eds., Benjamin, 1–16. (IV.2, V.1)
E. L. Reiss & B. J. Matkowsky (1971), Nonlinear dynamic buckling of a compressed elastic column, *Quart. Appl. Math.* **29**, 245–260. (V.2)
E. Reissner (1985), Reflections on the theory of elastic plates, *Appl. Mech. Revs.* **38**, 1453–1464. (XIV.15)
E. Reissner (1989), Lateral buckling of beams, *Computers and Structures* **33**, 1289–1306. (IX.6)
M. Renardy (1986), Some remarks on the Navier-Stokes equations with a pressure-dependent viscosity, *Comm. Partial Diff. Eqs.* **11**, 779–793. (XII.12)
M. Renardy (1989), On Rankine-Hugoniot conditions for Maxwell liquids, *J. Non-Newtonian Fluids* **32**, 69–77. (XV.4)
M. Renardy, W. S. Hrusa & J. Nohel (1987), *Mathematical Problems in Viscoelasticity,* Longman. (II.2, II.3)
W. C. Rheinboldt (1986), *Numerical Analysis of Parametrized Nonlinear Equations,* Wiley. (V.8)
O. Richmond & W. A. Spitzig (1980), Pressure dependence and dilatancy of plastic flow, in *Theoretical and Applied Mechanics, Proc. XV Intl. Cong.*, F. P. J. Rimrott & B. Tabarrok, eds., North Holland, 377–386. (XII.12)
H. Richter (1948), Das isotrope Elastizitätsgesetz, *Z. Angew. Math. Mech.* **28**, 205–209. (XII.13)
F. Riesz & B. Sz. Nagy (1955), *Functional Analysis,* 2nd edn., Unger. (VI.2, XIX.4)

A. Rigolot (1976), Sur une théorie asymptotique des poutres droites, Univ. Paris VI, Dissertation. (XIV.15)

R. S. Rivlin (1947), Torsion of a rubber cylinder, *J. Appl. Phys.* **18**, 444–449, 837. (XIII.16)

R. S. Rivlin (1948a), Large elastic deformations of isotropic materials, I. Fundamental concepts, *Phil. Trans. Roy. Soc. Lond.* **A 240**, 459–490. (XIII.16)

R. S. Rivlin (1948b), Large elastic deformations of isotropic materials, II. Some uniqueness theorems for pure, homogeneous deformations, *Phil. Trans. Roy. Soc. Lond.* **A 240**, 491–508. (XIII.11, XIII.16)

R. S. Rivlin (1948c), Large elastic deformations of isotropic materials, III. Some simple problems in cylindrical polar coordinates, *Phil. Trans. Roy. Soc. Lond.* **A 240**, 509–525. (XIII.16)

R. S. Rivlin (1948d), Large elastic deformations of isotropic materials, IV. Further developments of the general theory, *Phil. Trans. Roy. Soc. Lond.* **A 241**, 379–397. (XIII.16)

R. S. Rivlin (1949a), Large elastic deformations of isotropic materials, V. The problem of flexure, *Proc. Roy. Soc. Lond.* **A 195**, 463–473. (XIII.16)

R. S. Rivlin (1949b), Large elastic deformations of isotropic materials, VI. The results in the theory of torsion, shear, and flexure, *Phil. Trans. Roy. Soc. Lond.* **A 242**, 173–195. (XIII.16)

R. S. Rivlin (1949c), A note on torsion of an incompressible, highly-elastic cylinder, *Proc. Camb. Phil. Soc.* **45**, 485–587. (XIII.16)

R. S. Rivlin (1974), Stability of pure homogeneous deformation of an elastic cube under dead loading, *Quart. Appl. Math.* **32**, 265–271. (XIII.11)

R. S. Rivlin & J. L. Ericksen (1955), Stress-deformation relations for isotropic materials, *J. Rational Mech. Anal.* **4**, 323–425. (XII.13)

R. S. Rivlin & D. W. Saunders (1951), Large elastic deformations of isotropic materials, VII. Experiments on the deformation of rubber, *Phil. Trans. Roy. Soc. Lond.* **A 243**, 251–288. (XII.10, XIII.9)

R. S. Rivlin & K. N. Sawyers (1977), On the speed of propagation of waves in a deformed elastic material, *Z. angew. Math. Phys.* **28**, 1045–1057. (XIII.2)

R. S. Rivlin & K. N. Sawyers (1978), On the speed of propagation of waves in a deformed compressible elastic material, *Z. angew. Math. Phys.* **29**, 245–251. (XIII.2)

R. S. Rivlin & C. Topakoglu (1954), A theorem in the theory of finite elastic deformations, *J. Rational Mech. Anal.* **3**, 581–589. (XIII.11)

P. Rosakis (1990), Ellipticity and deformations with discontinuous gradients in finite elasticity, *Arch. Rational Mech. Anal.* **109**, 1–37. (XIII.2)

P. Rosakis & H. C. Simpson (1994), On the relation between polyconvexity and rank-one convexity in nonlinear elasticity, to appear. (XIII.2)

E. E. Rossinger (1987), *Generalized Solutions to Nonlinear PDE*, North Holland. (II.4)

R. Rostamian (1978), Internal constraints in boundary value problems of continuum mechanics, *Indiana Univ. Math. J.* **27**, 637–656. (XII.12)

R. Rostamian (1981), Internal constraints in linear elasticity, *J. Elasticity* **11**, 11–31. (XII.12)

E. H. Rothe (1986), *Introduction to Various Aspects of Degree Theory in Banach Spaces*, Amer. Math. Soc. (XIX.3)

E. J. Routh (1891), *A Treatise on Analytical Statics,* Vol. 1, 1st edn., Cambridge Univ. Press. (III.8)

E. J. Routh (1905), *The Advanced Part of a Treatise on the Dynamics of a System of Rigid Bodies,* 6th edn., Macmillan. (III.7)
B. L. Rozhdestvenskiĭ & N. N. Yanenko (1968), *Systems of Quasilinear Equations and their Applications to Gas Dynamics* (in Russian), Nauka, English transl: Amer. Math. Soc., 1983. (XVI.2)
B. Russell (1945), *A History of Western Philosophy,* Simon & Schuster. (V.7)
A. R. Rzhanitsyn (1955), *Stability of Equilibrium of Elastic Systems* (in Russian), G.I.T.T.L. (V.2)
L. Saalschütz (1880), *Der belastete Stab,* Teubner. (IV.6, IX.4)
A.-J.-C. B. de Saint Venant (1843), Mémoire sur le calcul de la résistance et de la flexion des pièces solides à simple ou à double courbure, en prenant simultanément en considération les divers efforts auxquels elles peuvent être soumises dans tous le! s sens, *C. R. Acad. Sci. Paris* **17**, 942–954, 1020–1031. (VIII.18)
A.-J.-C. B. de Saint Venant (1845), Note sur l'état d'équilibre d'une verge élastique à double courbure lorsque les déplacements éprouvés par ses points, par suite de l'action des forces qui la sollicitent, ne sont pas très petits, *C. R. Acad. Sci. Paris* **19**, 36–44, 181–187. (VIII.18)
F. M. A. Salam & M. L. Levi, eds. (1988), *Dynamical Systems Approaches to Nonlinear Problems in Systems and Circuits,* SIAM. (V.8)
E. Sanchez-Palencia (1990), Passage à la limite de l'élasticité tridimensionelle à la théorie asymptotique des coques minces, *C. R. Acad. Sci. Paris* **311,** Sér. II, 909–916. (XIV.15)
E. Sanchez-Palencia & D. G. Vassiliev (1992), Remarks on vibration of thin elastic shells and their numerical computation, *C. R. Acad. Sci. Paris* **314,** Sér. II, 445–452. (XIV.15)
D. H. Sattinger (1973), *Topics in Stability and Bifurcation Theory,* Springer Lect. Notes Math. **309**. (V.8)
D. H. Sattinger (1979), *Group Theoretic Methods in Bifurcation Theory,* Springer-Verlag. (V.8)
D. H. Sattinger (1983), *Branching in the Presence of Symmetry,* SIAM. (V.8)
K. N. Sawyers (1976), Stability of an elastic cube under dead loading: Two equal forces, *Int. J. Non-Lin. Mech.* **11**, 11–23. (XIII.11)
D. G. Schaeffer & M. Golubitsky (1979), Boundary conditions and mode jumping in the buckling of a rectangular plate, *Comm. Math. Phys.* **69**, 209–236. (XIV.13)
S. Schochet (1985), The incompressible limit in nonlinear elasticity, *Comm. Math. Phys.* **102**, 207–215. (VIII.15, XII.12)
J. T. Schwartz (1969), *Nonlinear Functional Analysis,* Gordon & Breach. (XVIII.1, XIX.3)
L. Schwartz (1966), *Théorie des Distributions,* rev. edn., Hermann. (XII.3)
N. Scott (1975), Acceleration waves in constrained elastic materials, *Arch. Rational Mech. Anal.* **58**, 57–75. (XIII.2)
N. H. Scott & M. Hayes (1985), A note on wave propagation in internally constrained hyperelastic materials, *Wave Motion* **7**, 601–605. (XIII.2)
T. I. Seidman & P. Wolfe (1988), Equilibrium states of an elastic conducting rod in a magnetic field, *Arch. Rational Mech. Anal.* **102**, 307–329. (VII.8, XIV.4)
C. B. Sensenig (1963), Instability of thick elastic solids, *Comm. Pure Appl. Math.* **17**, 451–491. (XIII.11)
J. B. Serrin (1959), Mathematical Principles of Classical Fluid Mechanics, in *Handbuch der Physik* VIII/1, C. Truesdell, ed., Springer-Verlag, 125–263. (XII.13)
J. B. Serrin (1986), An outline of thermodynamic structure, in *New Perspectives in Thermodynamics,* J. B. Serrin, ed., Springer-Verlag, 3–32. (XII.14)

M. J. Sewell (1967), On configuration-dependent loading, *Arch. Rational Mech. Anal.* **23**, 327–351. (XVI.6)

M. Shearer (1985a), Elementary wave solutions of the equations describing the motion of an elastic string, *SIAM J. Math. Anal.* **16**, 447–459. (XVI.2)

M. Shearer (1985b), The nonlinear interaction of smooth travelling waves in an elastic string, *Wave Motion* **7**, 169–175. (XVI.2)

M. Shearer (1986), The Riemann problem for the planar motion of an elastic string, *J. Diff. Eqs.* **61**, 149–163. (XVI.2)

K.-G. Shih & S. S. Antman (1986), Qualitative properties of large buckled states of spherical shells, *Arch. Rational Mech. Anal.* **93**, 357–384. (X.5)

F. Sidoroff (1978), Sur l'équation tensorielle $AX + XA = H$, *C. R. Acad. Sci. Paris* **A286**, 71–73. (XI.2, XII.12)

A. Signorini (1930), Sulle deformazione termoelastiche finite, in *Proc. 3rd Int. Cong. Appl. Mech.,* Vol. 2, 80–89. (XIII.11)

A. Signorini (1949), Trasformazioni termoelastiche finite. Memoria 2^a, *Ann. di Mat. Pura Appl.* **30**, 1–72. (XIII.11)

A. Signorini (1955), Trasformazioni termoelastiche finite, *Ann. di Mat. Pura Appl.* **39**, 147–201. (XIII.11)

M. Šilhavý (1977), On transformation laws for plastic deformations of materials with elastic range, *Arch. Rational Mech. Anal.* **63**, 169–182. (XV.1)

M. Šilhavý (1985), The existence of the flux vector and the divergence theorem for general Cauchy fluxes, *Arch. Rational Mech. Anal.* **90**, 195–212. (XII.7)

M. Šilhavý (1991), Cauchy's stress theorem and tensor fields with divergences in L^p, *Arch. Rational Mech. Anal.* **116**, 223–255. (XII.7, XII.9)

L. B. Sills & B. Budiansky (1978), Postbuckling ring analysis, *J. Appl. Mech.* **100**, 208–210. (VI.12)

J. G. Simmonds (1992), Asymptotic analysis of end effects in the axisymmetric deformation of elastic tubes weak in shear: Higher-order shell theories are inadequate and unnecessary, *Int. J. Solids Structures* **29**, 2441–2469. (XIV.15)

J. C. Simo (1985), A finite strain beam formulation. The three-dimensional dynamical problem. Part I, *Comp. Meths. Appl. Mech. Engg.* **49**, 55–70. (XIV.15)

J. C. Simo (1994), Topics on the Numerical Analysis and Simulation of Plasticity, in *Handbook of Numerical Analysis,* Vol. III, P. G. Ciarlet & J.-L. Lions, eds., Elsevier. (XV.1)

J. C. Simo & D. D. Fox (1989), On a stress resultant geometrically exact shell model. Part I: Formulation and optimal parametrization, *Comp. Meths. Appl. Mech. Engg.* **72**, 267–304. (XIV.15)

J. C. Simo, D. D. Fox & M. S. Rifai (1989), On a stress resultant geometrically exact shell model. Part II: The linear theory; computational aspects, *Comp. Meths. Appl. Mech. Engg.* **73**, 53–92. (XIV.15)

J. C. Simo, D. D. Fox & M. S. Rifai (1990), On a stress resultant geometrically exact shell model. Part III: Computational aspects of the nonlinear theory, *Comp. Meths. Appl. Mech. Engg.* **79**, 21–70. (XIV.15)

J. C. Simo & T. J. R. Hughes (1991), *Elasticity and Viscoplasticity, Computational Aspects,* Lecture Notes, Stanford Univ. (XV.6)

J. C. Simo & J. G. Kennedy (1991), On a stress resultant geometrically exact shell model. Part V: Nonlinear plasticity: Formulation and integration algorithms, *Comp. Meths. Appl. Mech. Engg.* **96**, 133–171. (XIV.15)

J. C. Simo, J. E. Marsden & P. S. Krishnaprasad (1988), The Hamiltonian structure of nonlinear elasticity: The material and convective representation of solids, rods, and plates, *Arch. Rational Mech. Anal.* **104**, 125–183. (XIV.3, XIV.15)

J. C. Simo & M. Ortiz (1985), A unified approach to finite deformation elastoplastic analysis based on the use of hyperelastic constitutive equations, *Comp. Meths. Appl. Mech. Engg.* **49**, 221–245. (XV.1, XV.3)

J. C. Simo, T. Posbergh & J. E. Marsden (1991), Stability of relative equilibria. Part II: Application to nonlinear elasticity, *Arch. Rational Mech. Anal.* **115**, 61–100. (II.10)

J. C. Simo, M. S. Rifai & D. D. Fox (1990), On a stress resultant geometrically exact shell model. Part IV: Variable thickness shells with through-the-thickness stretching, *Comp. Meths. Appl. Mech. Engg.* **81**, 91–126. (XIV.15)

J. C. Simo, M. S. Rifai & D. D. Fox (1992), On a stress resultant geometrically exact shell model. Part VI: Conserving algorithms for nonlinear dynamics, *Int. J. Num. Meths. Engg.* **34**, 117–164. (XIV.15)

J. C. Simo & L. Vu-Quoc (1986a), Three-dimensional finite strain rod model. Part I: Computational aspects, *Comp. Meths. Appl. Mech. Engg.* **58**, 79–116. (XIV.15)

J. C. Simo & L. Vu-Quoc (1986b), On the dynamics of flexible beams under large overall motions—The plane case: Part I, Part II, *J. Appl. Mech.* **53**, 849–863. (XIV.15)

J. C. Simo & L. Vu-Quoc (1988), On the dynamics in space of rods undergoing large motions—A geometrically exact approach, *Comp. Meths. Appl. Mech. Engg.* **66**, 125–161. (VIII.5, VIII.11, XIV.15)

J. C. Simo & L. Vu-Quoc (1991), A geometrically exact rod model incorporating shear and torsion-warping deformation, *Int. J. Solids Structures* **27**, 371–393. (XIV.15)

H. C. Simpson & S. J. Spector (1983), On copositive matrices and strong ellipticity for isotropic elastic materials, *Arch. Rational Mech. Anal.* **84**, 55–68. (XIII.2)

H. C. Simpson & S. J. Spector (1985), On barrelling instabilities in finite elasticity, *J. Elasticity* **14**, 103–125. (XIII.11)

H. C. Simpson & S. J. Spector (1987), On the positivity of the second variation in finite elasticity, *Arch. Rational Mech. Anal.* **98**, 1–30. (XIII.2)

M. Singh & A. C. Pipkin (1965), Note on Ericksen's problem, *Z. angew. Math. Phys.* **16**, 706–709. (XIII.9)

J. Sivaloganathan (1986), Uniqueness of regular and singular equilibria for spherically symmetric problems of nonlinear elasticity, *Arch. Rational Mech. Anal.* **96**, 97–136. (X.3, XIII.7)

D. R. Smith (1985), *Singular Perturbation Theory*, Cambridge Univ. Press. (XVI.4)

G. F. Smith (1971), On isotropic functions of symmetric tensors, skew-symmetric tensors and vectors, *Int. J. Engg. Sci.* **9**, 899–916. (XII.7)

J. Smoller (1983), *Shock Waves and Reaction-Diffusion Equations*, Springer-Verlag. (XVI.2)

I. S. Sokolnikoff (1956), *The Mathematical Theory of Elasticity*, McGraw-Hill. (XII.3, XIII.15)

S. J. Spector (1984), On the absence of bifurcation for elastic bars in uniaxial tension, *Arch. Rational Mech. Anal.* **85**, 171–199. (XIII.11)

A. J. M. Spencer (1964), Finite deformation of an almost incompressible solid, in *I.U.T.A.M. International Symposium on Second Order Effects in Elasticity, Plasticity, and Fluid Dynamics, Haifa, Israel, 1962*, M. Reiner & D. Abir, eds., Macmillan, 200–216. (VIII.15, XII.12)

I. Stakgold (1971), Branching of solutions of nonlinear equations, *SIAM Rev.* **13**, 289–332. (V.8)

I. Stakgold (1979), *Green's Functions and Boundary Value Problems*, Wiley-Interscience. (II.8, V.1, V.5, V.6, VI.2, X.2, XI.1, XIX.4)

J. Stern (1979), Der Gelenkstab bei grossen elastischen Verformungen, *Ing. Arch.* **48**, 173–184. (VI.10)

F. Stoppelli (1954), Un teorema di esistenza e di unicità relativo alle equazioni dell'elastostatica isoterma per deformazioni finite, *Ricerche Mat.* **3**, 247–267. (XIII.11)

F. Stoppelli (1955), Sulla svillupabilità in serie di potenze di un parametro delle soluzioni delle equazioni dell'elastostatica isoterma, *Ricerche Mat.* **4**, 58–73. (XIII.11)

F. Stoppelli (1957-1958), Sull'esistenza di soluzioni delle equazioni dell'elastostatica isoterma nel caso di sollecitazioni dotate di assi di equilibrio, I, II, III, *Ricerche Mat.* **6**, 241–287, **7**, 71–101, 138–152. (XIII.11)

M. Struwe (1990), *Variational Methods*, Springer-Verlag. (VII.8)

C. A. Stuart (1973a), Some bifurcation theory for k-set contractions, *Proc. Lond. Math. Soc.* **(3)27**, 531–550. (V.4)

C. A. Stuart (1973b), Solutions of large norm for nonlinear Sturm-Liouville problems, *Quart. J. Math. Oxford Ser.* **24**, 129–139. (VI.2)

C. A. Stuart (1975), Spectral theory of rotating chains, *Proc. Roy. Soc. Edin.* **73A**, 199–214. (VI.3)

C. A. Stuart (1976), Spectral theory of rotating chains, in *Applications of Methods of Functional Analysis to Problems in Mechanics*, P. Germain & B. Nayroles, eds., Springer Lect. Notes Math. **503**, 159–165. (VI.2, VI.3)

C. A. Stuart (1985), Radially symmetric cavitation for hyperelastic materials, *Anal. non linéaire* **2**, 33–66. (X.3, XIII.7)

V. Šverák (1991), Quasiconvex functions with subquadratic growth, *Proc. Roy. Soc. Lond.* **A433**, 723–725. (XIII.2)

V. Šverák (1992a), Rank-one convexity does not imply quasiconvexity, *Proc. Roy. Soc. Edin.* **120**, 185–189. (XIII.2)

V. Šverák (1992b), New examples of quasiconvex functions, *Arch. Rational Mech. Anal.* **119**, 293–300. (XIII.2)

J. L. Synge & W. Z. Chien (1941), The intrinsic theory of elastic plates and shells, in *Th. von Kármán Anniversary Album*, California Inst. of Technology, 103–120. (XIV.15)

B. Szabó & I. Babuška (1991), *Finite Element Analysis*, Wiley. (II.11)

I. Szabó (1979), *Geschichte der mechanischen Prinzipien*, Birkhäuser. (XIV.15)

A. J. Szeri (1990), On the everted states of spherical and cylindrical shells, *Quart. Appl. Math.* **47**, 49–58. (X.3, XIII.7)

I. Tadjbakhsh & F. Odeh (1967), Equilibrium states of elastic rings, *J. Math. Anal. Appl.* **18**, 59–74. (IV.3)

L. Tao, K. R. Rajagopal & A. S. Wineman (1990), On an inhomogeneous deformation of an isotropic compressible elastic material, *Arch. Mech.* **42**, 729–734. (XIII.10)

A. E. Taylor & D. C. Lay (1980), *Introduction to Functional Analysis,* 2nd edn., Wiley. (V.4, V.6, XVII.2)

B. Taylor (1714), De motu nervi tensi, *Phil. Trans. Lond.* **28** (1713), 26–32. (II.2)

G. I. Taylor (1963), On the shape of parachutes, in *The Scientific Papers of Sir Geoffrey Ingram Taylor,* Vol. III, G. K Batchelor, ed., Cambridge Univ. Press, 26–37. (X.3)

R. Temam (1977), *Navier-Stokes Equations*, North-Holland. (XII.3)

R. Temam (1983), *Problèmes mathématiques en plasticité*, Bordas, English transl: Gauthier-Villars, 1985. (XV.1)

J. M. T. Thompson & G. W. Hunt (1973), *A General Theory of Elastic Stability,* Wiley. (V.2)

J. M. T. Thompson & G. W. Hunt (1984), *Elastic Instability Phenomena*, Wiley. (V.2, V.8, VI.6)

W. Thomson (Lord Kelvin) & P. G. Tait (1867), *Treatise on Natural Philosophy,* Part I, 1st. edn., Cambridge Univ. Press. (VIII.18)

S. P. Timoshenko (1921), On the correction for shear of the differential equations for transverse vibrations of prismatic bars, *Phil. Mag.* **41**, 744–746. (VIII.10, VIII.18)

S. P. Timoshenko (1953), *History of the Strength of Materials*, McGraw-Hill. (XIV.15)

S. P. Timoshenko & J. M. Gere (1961), *Theory of Elastic Stability,* 2nd edn., McGraw-Hill. (V.2, V.7, XIV.15)

S. P. Timoshenko & J. N. Goodier (1951), *Theory of Elasticity*, McGraw-Hill. (XIII.15)

S. P. Timoshenko & S. Woinowsky-Krieger (1959), *Theory of Plates and Shells,* 2nd edn., McGraw-Hill. (XIV.15)

T. C. T. Ting (1985), Determination of $\mathbf{C}^{1/2}$, $\mathbf{C}^{-1/2}$ and more general isotropic tensor functions of **C**, *J. Elasticity* **15**, 319–323. (XII.12)

T. W. Ting (1973), Topics in the Mathematical Theory of Plasticity, in *Handbuch der Physik* VIa/3, C. Truesdell, ed., Springer-Verlag, 535–590. (XV.1)

T. W. Ting (1974), St. Venant's compatibility conditions, *Tensor, N. S.* **28**, 5–12. (XII.3)

I. Todhunter (1853), *A Treatise on Analytical Statics,* 1st edn., Macmillan. (III.8)

J. F. Toland (1973), Asymptotic linearity and nonlinear eigenvalue problems, *Quart. J. Math. Oxford Ser.* **24**, 241–250. (VI.2)

J. F. Toland (1976), Global bifurcations of k-set contractions without multiplicity assumptions, *Quart. J. Math. Oxford Ser.* **27**, 199–216. (V.4)

J. F. Toland (1977), Global bifurcation via Galerkin's method, *Nonlin. Anal. T. M. A.* **1**, 305–317. (V.4)

J. F. Toland (1979a), A duality principle for non-convex optimization and the calculus of variations, *Arch. Rational Mech. Anal.* **71**, 41–61. (VI.2, VI.3)

J. F. Toland (1979b), On the stability of rotating heavy chains, *J. Diff. Eqs.* **32**, 15–31. (VI.2, VI.3)

E. Tonti (1969), Variational formulation of nonlinear differential equations, *Acad. Roy. Belg. Bull. Cl. Sci.* **55**, 137–165, 262-278. (II.10)

J. A. Trangenstein & P. Colella (1991), A Godunov method for elastic-plastic deformations, *Comm. Pure Appl. Math.* **44**, 41–100. (XV.4)

J. A. Trangenstein & R. B. Pember (1991), The Riemann problem for the longitudinal motion of an elastic-plastic bar, *SIAM J. Sci. Stat. Comp.* **12**, 180–207. (XV.4)

J. A. Trangenstein & R. B. Pember (1992), Numerical algorithms for strong discontinuities in elastic-plastic solids, *J. Comp. Phys.* **103**, 63–89. (XV.4)

C. Truesdell (1952), The mechanical foundations of elasticity and fluid dynamics, *J. Rational Mech. Anal.* **1**, 125–300. (XIII.2)

C. Truesdell (1954), A new chapter in the theory of the elastica, in *Proc. 1st Midwestern Conf. Solid Mech.*, 52–55. (IV.6)

C. Truesdell (1955a), Hypo-elasticity, *J. Rational Mech. Anal.* **4**, 83–133, 1019–1020. (XV.1)

C. Truesdell (1955b), The simplest rate theory of pure elasticity, *Comm. Pure Appl. Math.* **8**, 123–132. (XV.1)

C. Truesdell (1956), Hypo-elastic shear, *J. Appl. Phys.* **27**, 441–447. (XV.1)

C. Truesdell (1959), The rational mechanics of materials—past, present, future, *Appl. Mech. Revs.* **12**, 75–80. (XIV.15)

C. Truesdell (1960), The Rational Mechanics of Flexible or Elastic Bodies 1638-1788, in *L. Euleri Opera Omnia* II, $\mathbf{11}_2$, Füssli, Zürich. (II.2, III.8, IV.1, IV.2, V.2, VIII.16, VIII.18, XIV.15)

C. Truesdell (1961), General and exact theory of waves in finite elastic strain, *Arch. Rational Mech. Anal.* **8**, 263–296. (XVI.2)

C. Truesdell (1962), Solutio generalis et accurata problematum quamplurimorum de motu corporum elasticorum incomprimibilium in deformationibus valde magnis, *Arch. Rational Mech. Anal.* **11**, 106–113, Addendum, ibid. **12**, 427-428. (XIII.13)

C. Truesdell (1966), Existence of longitudinal waves, *J. Acoust. Soc. Amer.* **40**, 729–730. (XIII.2)

C. Truesdell (1978), Some challenges offered to analysis by rational thermomechanics, in *Contemporary Developments in Continuum Mechanics and Partial Differential Equations*, G. M. de la Penha & L. A. Madeiros, eds., North-Holland, 495–603. (XIII.7)

C. Truesdell (1980), *The Tragicomical History of Thermodynamics 1822–1854*, Springer-Verlag. (XII.14)

C. Truesdell (1983), The influence of elasticity on analysis: the classical heritage, *Bull. Amer. Math. Soc.* **9**, 293–310. (XIV.15)

C. Truesdell (1984), *Rational Thermodynamics,* 2nd edn., Springer-Verlag. (XII.14)

C. Truesdell (1991a), *A First Course in Rational Continuum Mechanics,* Vol. 1, 2nd edn., Academic Press. (VIII.3, XII.1, XII.7, XII.12)

C. Truesdell (1991b), Sophie Germain: Fame earned by stubborn error, *Boll. Storia Scienze Mat.* **11**, 3–24. (XIV.15)

C. Truesdell & S. Bharatha (1977), *The Concepts and Logic of Classical Thermodynamics as a Theory of Heat Engines, Rigorously Constructed upon the Foundation Laid by S. Carnot and F. Reech*, Springer-Verlag. (XII.14)

C. Truesdell & W. Noll (1965), The Non-linear Field Theories of Mechanics, in *Handbuch der Physik* III/3, Springer-Verlag. (P, VIII.7, IX.2, XI.2, XII.1, XII.10, XII.12, XII.13, XIII.2, XIII.4, XIII.7, XIII.11, XIII.16)

C. Truesdell & R. A. Toupin (1960), The Classical Field Theories, in *Handbuch der Physik* III/1, Springer-Verlag. (XI.3, XII.4, XII.5, XII.7, XII.12, XIII.13, XIV.2, XIV.15, XVI.2)

C. Truesdell & R. A. Toupin (1963), Static grounds for inequalities in finite elastic strain, *Arch. Rational Mech. Anal.* **12**, 1–33. (XIII.2)

R. E. L. Turner (1970), Nonlinear eigenvalue problems with nonlocal operators, *Comm. Pure Appl. Math.* **23**, 963–972. (V.4)

R. E. L. Turner (1973), Superlinear Sturm-Liouville problems, *J. Diff. Eqs.* **13**, 157–171. (VI.2)

M. M. Vaĭnberg (1956), *Variational Methods for the Study of Nonlinear Operators* (in Russian), Gostekhteorizdat, English transl: Holden-Day, 1964. (II.10, VII.8)

M. M. Vaĭnberg & P. G. Aizengendler (1966), *The theory and methods of investigation of branch points of solutions* (in Russian), Matematicheskiĭ Analiz **12**, Nauka, English transl: Progress in Mathematics **2**, Plenum, 1968. (V.8)

M. M. Vaĭnberg & V. A. Trenogin (1969), *Theory of Solution Branching of Nonlinear Equations* (in Russian), Nauka, English transl: Noordhoff, 1974. (IV.4,V.4, V.8, XVIII.2)

T. Valent (1988), *Boundary Value Problems of Finite Elasticity*, Springer-Verlag. (P, XIII.11, XIII.16)

W. Van Buren (1968), On the Existence and Uniqueness of Solutions to Boundary Value Problems in Finite Elasticity, Carnegie-Mellon Univ., Thesis, Res. Rpt. 68-1D7-MEKMA-RI, Westinghouse Research Labs, Pittsburgh. (XIII.11)

A. Vanderbauwhede (1982), *Local Bifurcation and Symmetry*, Pitman. (V.8, XVIII.2)

A. B. Vasil'eva & V. F. Butuzov (1973), *Asymptotic Expansions of Solutions of Singularly Perturbed Equations*, Nauka. (XVI.4)

T. D. Venttsel' (1981), A one-sided maximum principle for spatial derivatives of the solution of the one-dimensional system of gas dynamics with "viscosity", *Vest. Mosk. Univ. Mat.* **36**, 41–44, English transl: *Moscow Univ. Math. Bull.* **36**, 50–53. (XVI.1)

M. Vianello (1990), On the active part of the stress for elastic materials with internal constraints, *J. Elasticity* **24**, 289–294. (XII.13)

P. Villaggio (1977), *Qualitative Methods in Elasticity*, Noordhoff. (XIII.15)

A. S. Vol'mir (1967), *Stability of Deformable Systems*, 2nd edn. (in Russian), Nauka. (V.2)

A. S. Vol'mir (1979), *Shells in the Flow of Fluids and Gases* (in Russian), Nauka. (V.2)

E. Volterra (1956), Equations of motion for curved and twisted elastic bars deduced by the "method of internal constraints", *Ing. Arch.* **24**, 392–400. (XIV.15)

E. Volterra (1961), Second approximation of the method of internal constraints and its applications, *Int. J. Mech. Sci.* **3**, 47–67. (XIV.15)

T. von Kármán (1910), Festigkeitsprobleme im Maschinenbau, in *Encyclopädie der Mathematischen Wissenschaften IV/4C*, 311–385. (XIV.14)

R. von Mises (1958), *Mathematical Theory of Compressible Fluid Flow*, Academic Press. (XVI.2)

J. von Neumann (1944), Proposal and analysis of a new numerical method in the treatment of hydrodynamic shock problems, in *Collected Works,* Vol. VI, Pergamon. (II.11)

I. I. Vorovich (1989), *Mathematical Problems of the Nonlinear Theory of Shallow Shells* (in Russian), Nauka. (X.1, XIV.15)

Y. H. Wan & J. E. Marsden (1984), Symmetry and bifurcation in three-dimensional elasticity, III, *Arch. Rational Mech. Anal.* **84**, 203–233. (XIII.11)

C.-C. Wang (1970a), Proof that all dynamic universal solutions are quasi-equilibrated, *Z. angew. Math. Mech.* **50**, 311–314. (XIII.13)

C.-C. Wang (1970b), A new representation theorem for isotropic functions: An answer to Professor G. F. Smith's criticism of my papers on representations for isotropic functions, *Arch. Rational Mech. Anal.* **36**, 166–223, Corrigendum, ibid. **43** (1971) 392-395. (XII.13)

C.-C. Wang (1972), Material uniformity and homogeneity in shells, *Arch. Rational Mech. Anal.* **47**, 343–368. (XIV.13)

C.-C. Wang (1973), On the response functions of isotropic elastic shells, *Arch. Rational Mech. Anal.* **50**, 81–98. (XIV.13)

C.-C. Wang & C. Truesdell (1973), *Introduction to Rational Elasticity*, Noordhoff. (P, XIII.2, XIII.9, XIII.11, XIII.16, XVI.2)

W. Wasow (1965), *Asymptotic Expansions for Ordinary Differential Equations*, Wiley-Interscience. (XVI.4)

H. F. Weinberger (1965), *A First Course in Partial Differential Equations*, Blaisdell. (II.1)

H. J. Weinitschke (1987), On finite displacements of circular elastic membranes, *Math. Meth. Appl. Sci.* **9**, 76–98. (X.8)

H. J. Weinitschke & H. Grabmüller (1992), Recent mathematical results in the nonlinear theory of flat and curved elastic membranes of revolution, *J. Engg. Math.* **26**, 159–194. (X.8)

Z. Wesołowski (1962), Stability in some cases of tension in the light of the theory of finite strain, *Arch. Mech. Stos.* **14**, 875–900. (XIII.11)

Z. Wesołowski (1963), The axially symmetric problem of stability loss of an elastic bar subject to tension, *Arch. Mech. Stos.* **15**, 383–395. (XIII.11)

G. B. Whitham (1974), *Linear and Nonlinear Waves*, Wiley-Interscience. (XVI.2)

A. B. Whitman & C. N. DeSilva (1974), An exact solution in a nonlinear theory of rods, *J. Elasticity* **4**, 265–280. (IX.2)

E. T. Whittaker (1937), *A Treatise on the Analytical Dynamics of Particles and Rigid Bodies,* 4th edn., Cambridge Univ. Press. (VIII.12)

G. T. Whyburn (1964), *Topological Analysis,* rev. edn., Princeton Univ. Press. (V.3)

S. Wiggins (1990), *Introduction to Applied Nonlinear Dynamical Systems and Chaos*, Springer-Verlag. (V.8)

J. T. Wissmann (1991), Material Response and Inverse Problems of Elasticity, Univ. of Maryland, Dissertation. (XIII.2, XIII.4)

P. Wolfe (1983), Equilibrium states of an elastic conductor in a magnetic field: a paradigm of bifurcation theory, *Trans. Amer. Math. Soc.* **278**, 377–387. (III.5)

P. Wolfe (1985), Rotating states of an elastic conductor, in *Physical Mathematics and Nonlinear Partial Differential Equations*, J. Lightbourne & S. Rankin, eds., Dekker. (III.5)

P. Wolfe (1990), Bifurcation theory of an elastic conducting wire subject to magnetic forces, *J. Elasticity* **23**, 201–217. (III.5)

J. H. Wolkowisky (1969), Nonlinear Sturm-Liouville problems, *Arch. Rational Mech. Anal.* **35**, 299–320. (V.3, VI.2)

C.-H. Wu (1972), Whirling of a string at large angular speeds—a nonlinear eigenvalue problem with moving boundary layers, *SIAM J. Appl. Math.* **22**, 1–13. (VI.3)

C.-H. Wu (1979), Large finite-strain membrane problems, *Quart. Appl. Math.* **36**, 347–360. (X.7)

G. Zanzotto (1992), On the material symmetry group of elastic crystals and the Born rule, *Arch. Rational Mech. Anal.* **121**, 1–36. (XIII.16)

L. Zee & E. Sternberg (1983), Ordinary and strong ellipticity in the equilibrium theory of incompressible hyperelastic solids, *Arch. Rational Mech. Anal.* **83**, 53–90. (XIII.2)

E. Zeidler (1985), *Nonlinear Functional Analysis and it Applications, Vol. III, Variational Methods and Optimization*, Springer-Verlag. (VIII.8)

E. Zeidler (1986), *Nonlinear Functional Analysis and it Applications, Vol. I, Fixed-Point Theorems*, Springer-Verlag. (V.8, XVII.2, XVIII.1, XVIII.2, XIX.3)

K. Zhang (1991), Energy minimizers in nonlinear elastostatics and the implicit function theorem, *Arch. Rational Mech. Anal.* **114**, 95–117. (XIII.2)

H. Ziegler (1977), *Principles of Structural Stability* 2nd edn., Birkhäuser. (V.2, V.7, IX.5)

W. P. Ziemer (1983), Cauchy flux and sets of finite perimeter, *Arch. Rational Mech. Anal.* **84**, 189–201. (XII.7)

W. P. Ziemer (1989), *Weakly Differentiable Functions*, Springer-Verlag. (XI.2)

M. Życzkowski & A. Gajewski (1971), Optimal structural design in non-conservative problems of elastic stability, in *IUTAM Symposium on Instability of Continuous Systems*, H. H. E. Leipholz, ed., Springer-Verlag. (VI.11)

Index

Applied Mathematical Sciences

(continued from page ii)

52. *Chipot:* Variational Inequalities and Flow in Porous Media.
53. *Majda:* Compressible Fluid Flow and System of Conservation Laws in Several Space Variables.
54. *Wasow:* Linear Turning Point Theory.
55. *Yosida:* Operational Calculus: A Theory of Hyperfunctions.
56. *Chang/Howes:* Nonlinear Singular Perturbation Phenomena: Theory and Applications.
57. *Reinhardt:* Analysis of Approximation Methods for Differential and Integral Equations.
58. *Dwoyer/Hussaini/Voigt (eds):* Theoretical Approaches to Turbulence.
59. *Sanders/Verhulst:* Averaging Methods in Nonlinear Dynamical Systems.
60. *Ghil/Childress:* Topics in Geophysical Dynamics: Atmospheric Dynamics, Dynamo Theory and Climate Dynamics.
61. *Sattinger/Weaver:* Lie Groups and Algebras with Applications to Physics, Geometry, and Mechanics.
62. *LaSalle:* The Stability and Control of Discrete Processes.
63. *Grasman:* Asymptotic Methods of Relaxation Oscillations and Applications.
64. *Hsu:* Cell-to-Cell Mapping: A Method of Global Analysis for Nonlinear Systems.
65. *Rand/Armbruster:* Perturbation Methods, Bifurcation Theory and Computer Algebra.
66. *Hlavácek/Haslinger/Necasl/Lovísek:* Solution of Variational Inequalities in Mechanics.
67. *Cercignani:* The Boltzmann Equation and Its Applications.
68. *Temam:* Infinite Dimensional Dynamical Systems in Mechanics and Physics.
69. *Golubitsky/Stewart/Schaeffer:* Singularities and Groups in Bifurcation Theory, Vol. II.
70. *Constantin/Foias/Nicolaenko/Temam:* Integral Manifolds and Inertial Manifolds for Dissipative Partial Differential Equations.
71. *Catlin:* Estimation, Control, and the Discrete Kalman Filter.
72. *Lochak/Meunier:* Multiphase Averaging for Classical Systems.
73. *Wiggins:* Global Bifurcations and Chaos.
74. *Mawhin/Willem:* Critical Point Theory and Hamiltonian Systems.
75. *Abraham/Marsden/Ratiu:* Manifolds, Tensor Analysis, and Applications, 2nd ed.
76. *Lagerstrom:* Matched Asymptotic Expansions: Ideas and Techniques.
77. *Aldous:* Probability Approximations via the Poisson Clumping Heuristic.
78. *Dacorogna:* Direct Methods in the Calculus of Variations.
79. *Hernández-Lerma:* Adaptive Markov Processes.
80. *Lawden:* Elliptic Functions and Applications.
81. *Bluman/Kumei:* Symmetries and Differential Equations.
82. *Kress:* Linear Integral Equations.
83. *Bebernes/Eberly:* Mathematical Problems from Combustion Theory.
84. *Joseph:* Fluid Dynamics of Viscoelastic Fluids.
85. *Yang:* Wave Packets and Their Bifurcations in Geophysical Fluid Dynamics.
86. *Dendrinos/Sonis:* Chaos and Socio-Spatial Dynamics.
87. *Weder:* Spectral and Scattering Theory for Wave Propagation in Perturbed Stratified Media.
88. *Bogaevski/Povzner:* Algebraic Methods in Nonlinear Perturbation Theory.
89. *O'Malley:* Singular Perturbation Methods for Ordinary Differential Equations.
90. *Meyer/Hall:* Introduction to Hamiltonian Dynamical Systems and the N-body Problem.
91. *Straughan:* The Energy Method, Stability, and Nonlinear Convection.
92. *Naber:* The Geometry of Minkowski Spacetime.
93. *Colton/Kress:* Inverse Acoustic and Electromagnetic Scattering Theory.
94. *Hoppensteadt:* Analysis and Simulation of Chaotic Systems.
95. *Hackbusch:* Iterative Solution of Large Sparse Systems of Equations.
96. *Marchioro/Pulvirenti:* Mathematical Theory of Incompressible Nonviscous Fluids.
97. *Lasota/Mackey:* Chaos, Fractals, and Noise: Stochastic Aspects of Dynamics, 2nd ed.
98. *de Boor/Höllig/Riemenschneider:* Box Splines.
99. *Hale/Lunel:* Introduction to Functional Differential Equations.
100. *Sirovich (ed):* Trends and Perspectives in Applied Mathematics.

(continued on next page)

101. *Nusse/Yorke:* Dynamics: Numerical Explorations.
102. *Chossat/Iooss:* The Couette-Taylor Problem.
103. *Chorin:* Vorticity and Turbulence.
104. *Farkas:* Periodic Motions.
105. *Wiggins:* Normally Hyperbolic Invariant Manifolds in Dynamical Systems.
106. *Cercignani/Illner/Pulvirenti:* The Mathematical Theory of Dilute Gases.
107. *Antman:* Nonlinear Problems of Elasticity.